Student Solutions Manual
for

SINGLE VARIABLE CALCULUS
EARLY TRANSCENDENTALS

SIXTH EDITION

DANIEL ANDERSON
University of Iowa

JEFFERY A. COLE
Anoka-Ramsey Community College

DANIEL DRUCKER
Wayne State University

BROOKS/COLE
CENGAGE Learning·

Australia • Brazil • Japan • Korea • Mexico • Singapore • Spain • United Kingdom • United States

ISBN-13: 978-0-495-01240-5
ISBN-10: 0-495-01240-8

Brooks/Cole
10 Davis Drive
Belmont, CA 94002-3098
USA

Cengage Learning is a leading provider of customized learning solutions with office locations around the globe, including Singapore, the United Kingdom, Australia, Mexico, Brazil, and Japan. Locate your local office at:
international.cengage.com/region

Cengage Learning products are represented in Canada by Nelson Education, Ltd.

For your course and learning solutions, visit
academic.cengage.com

Purchase any of our products at your local college store or at our preferred online store **www.ichapters.com**

Cover image: Amelie Fear, Folkmusicians violins

Printed in Canada
3 4 5 6 7 12 11 10 09

□ PREFACE

This *Student Solutions Manual* contains strategies for solving and solutions to selected exercises in the text *Single Variable Calculus, Early Transcendentals,* Sixth Edition, by James Stewart. It contains solutions to the odd-numbered exercises in each section, the review sections, the True-False Quizzes, and the Problem Solving sections, as well as solutions to all the exercises in the Concept Checks.

We use some non-standard notation in order to save space. If you see a symbol which you don't recognize, refer to the Table of Abbreviations and symbols on page iv.

This manual is a text supplement and should be read along with the text. You should read all exercise solutions in this manual because many concept explanations are given and then used in subsequent solutions. All concepts necessary to solve a particular problem are not reviewed for every exercise. If you are having difficulty with a previously covered concept, refer back to the section where it was covered for more complete help.

A significant number of today's students are involved in various outside activities, and find it difficult, if not impossible, to attend all class sessions; this manual should help meet the needs of these students. In addition, it is our hope that this manual's solutions will enhance the understanding of all readers of the material and provide insights to solving other exercises.

We use some non-standard notation in order to save space. If you see a symbol which you don't recognize, refer to the Table of Abbreviations and Symbols on page v.

We appreciate feedback concerning errors, solution correctness or style, and manual style. Any comments may be sent directly to jeff.cole@anokaramsey.edu, or in care of the publisher: Brooks/Cole, Cengage Learning, 10 Davis Drive, Belmont CA 94002-3098.

We would like to thank Brian Betsill, Stephanie Kuhns, and Kathi Townes, of TECHarts, for their production services; and Stacy Green, of Brooks/Cole, Cengage Learning, for her patience and support. All of these people have provided invaluable help in creating this manual.

Jeffery A. Cole
Anoka-Ramsey Community College

James Stewart
McMaster University

Daniel Drucker
Wayne State University

Daniel Anderson
University of Iowa

ABREVIATIONS AND SYMBOLS

CD	concave downward
CU	concave upward
D	the domain of f
FDT	First Derivative Test
HA	horizontal asymptote(s)
I	interval of convergence
IP	inflection point(s)
R	radius of convergence
VA	vertical asymptote(s)
$\stackrel{CAS}{=}$	indicates the use of a computer algebra system.
$\stackrel{H}{=}$	indicates the use of l'Hospital's Rule.
$\stackrel{j}{=}$	indicates the use of Formula j in the Table of Integrals in the back endpapers.
$\stackrel{s}{=}$	indicates the use of the substitution $\{u = \sin x, du = \cos x\, dx\}$.
$\stackrel{c}{=}$	indicates the use of the substitution $\{u = \cos x, du = -\sin x\, dx\}$.

CONTENTS

☐ **DIAGNOSTIC TESTS**   **1**

---

1 ☐   **FUNCTIONS AND MODELS**   **9**

1.1   Four Ways to Represent a Function   9

1.2   Mathematical Models: A Catalog of Essential Functions   13

1.3   New Functions from Old Functions   17

1.4   Graphing Calculators and Computers   23

1.5   Exponential Functions   26

1.6   Inverse Functions and Logarithms   28

     Review   33

Principles of Problem Solving   **39**

---

2 ☐   **LIMITS AND DERIVATIVES**   **41**

2.1   The Tangent and Velocity Problems   41

2.2   The Limit of a Function   43

2.3   Calculating Limits Using the Limit Laws   46

2.4   The Precise Definition of a Limit   51

2.5   Continuity   55

2.6   Limits at Infinity; Horizontal Asymptotes   60

2.7   Derivatives and Rates of Change   66

2.8   The Derivative as a Function   71

     Review   78

Problems Plus   **85**

---

3 ☐   **DIFFERENTIATION RULES**   **87**

3.1   Derivatives of Polynomials and Exponential Functions   87

3.2   The Product and Quotient Rules   92

3.3   Derivatives of Trigonometric Functions   96

3.4   The Chain Rule   99

3.5   Implicit Differentiation   105

3.6   Derivatives of Logarithmic Functions   110

3.7   Rates of Change in the Natural and Social Sciences   113

3.8   Exponential Growth and Decay   117

3.9   Related Rates   119

3.10   Linear Approximations and Differentials   123

3.11   Hyperbolic Functions   126

        Review   129

Problems Plus   139

4 □ APPLICATIONS OF DIFFERENTIATION   147

4.1   Maximum and Minimum Values   147

4.2   The Mean Value Theorem   152

4.3   How Derivatives Affect the Shape of a Graph   155

4.4   Indeterminate Forms and L'Hospital's Rule   167

4.5   Summary of Curve Sketching   172

4.6   Graphing with Calculus *and* Calculators   184

4.7   Optimization Problems   193

4.8   Newton's Method   204

4.9   Antiderivatives   209

        Review   214

Problems Plus   227

5 □ INTEGRALS   233

5.1   Areas and Distances   233

5.2   The Definite Integral   237

5.3   The Fundamental Theorem of Calculus   242

5.4   Indefinite Integrals and the Net Change Theorem   247

5.5   The Substitution Rule   250

        Review   255

Problems Plus   261

6 □ APPLICATIONS OF INTEGRATION   263

6.1   Areas Between Curves   263

6.2   Volumes   270

6.3   Volumes by Cylindrical Shells   278

6.4   Work   282

6.5   Average Value of a Function   284

        Review   286

Problems Plus   291

7   □   TECHNIQUES OF INTEGRATION   295

7.1   Integration by Parts   295

7.2   Trigonometric Integrals   300

7.3   Trigonometric Substitution   304

7.4   Integration of Rational Functions by Partial Fractions   310

7.5   Strategy for Integration   318

7.6   Integration Using Tables and Computer Algebra Systems   324

7.7   Approximate Integration   328

7.8   Improper Integrals   335

        Review   343

Problems Plus   351

8   □   FURTHER APPLICATIONS OF INTEGRATION   355

8.1   Arc Length   355

8.2   Area of a Surface of Revolution   359

8.3   Applications to Physics and Engineering   363

8.4   Applications to Economics and Biology   370

8.5   Probability   371

        Review   374

Problems Plus   377

9   □   DIFFERENTIAL EQUATIONS   381

9.1   Modeling with Differential Equations   381

9.2   Direction Fields and Euler's Method   382

9.3   Separable Equations   386

9.5   Models for Population Growth   392

9.6   Linear Equations   396

9.7   Predator-Prey Systems   399

        Review   401

Problems Plus   407

10   □   PARAMETRIC EQUATIONS AND POLAR COORDINATES   411

10.1   Curves Defined by Parametric Equations   411

10.2   Calculus with Parametric Curves   416

10.3   Polar Coordinates   422

10.4   Areas and Lengths in Polar Coordinates   431

10.5   Conic Sections   437

10.6   Conic Sections in Polar Coordinates   442
        Review   445

Problems Plus   453

11   □   INFINITE SEQUENCES AND SERIES   455

11.1   Sequences   455

11.2   Series   460

11.3   The Integral Test and Estimates of Sums   466

11.4   The Comparison Tests   470

11.5   Alternating Series   473

11.6   Absolute Convergence and the Ratio and Root Tests   475

11.7   Strategy for Testing Series   478

11.8   Power Series   480

11.9   Representations of Functions as Power Series   484

11.10   Taylor and Maclaurin Series   489

11.11   Applications of Taylor Polynomials   496
        Review   502

Problems Plus   509

□   APPENDIXES   515

A   Numbers, Inequalities, and Absolute Values   515

B   Coordinate Geometry and Lines   517

C   Graphs of Second-Degree Equations   520

D   Trigonometry   523

E   Sigma Notation   527

G   The Logarithm Defined as an Integral   529

H   Complex Numbers   529

☐ DIAGNOSTIC TESTS

Test A   Algebra

1. (a) $(-3)^4 = (-3)(-3)(-3)(-3) = 81$

   (b) $-3^4 = -(3)(3)(3)(3) = -81$

   (c) $3^{-4} = \dfrac{1}{3^4} = \dfrac{1}{81}$

   (d) $\dfrac{5^{23}}{5^{21}} = 5^{23-21} = 5^2 = 25$

   (e) $\left(\dfrac{2}{3}\right)^{-2} = \left(\dfrac{3}{2}\right)^2 = \dfrac{9}{4}$

   (f) $16^{-3/4} = \dfrac{1}{16^{3/4}} = \dfrac{1}{\left(\sqrt[4]{16}\right)^3} = \dfrac{1}{2^3} = \dfrac{1}{8}$

2. (a) Note that $\sqrt{200} = \sqrt{100 \cdot 2} = 10\sqrt{2}$ and $\sqrt{32} = \sqrt{16 \cdot 2} = 4\sqrt{2}$. Thus $\sqrt{200} - \sqrt{32} = 10\sqrt{2} - 4\sqrt{2} = 6\sqrt{2}$.

   (b) $(3a^3b^3)(4ab^2)^2 = 3a^3b^3 16a^2b^4 = 48a^5b^7$

   (c) $\left(\dfrac{3x^{3/2}y^3}{x^2y^{-1/2}}\right)^{-2} = \left(\dfrac{x^2y^{-1/2}}{3x^{3/2}y^3}\right)^2 = \dfrac{(x^2y^{-1/2})^2}{(3x^{3/2}y^3)^2} = \dfrac{x^4y^{-1}}{9x^3y^6} = \dfrac{x^4}{9x^3y^6y} = \dfrac{x}{9y^7}$

3. (a) $3(x+6) + 4(2x-5) = 3x + 18 + 8x - 20 = 11x - 2$

   (b) $(x+3)(4x-5) = 4x^2 - 5x + 12x - 15 = 4x^2 + 7x - 15$

   (c) $\left(\sqrt{a} + \sqrt{b}\right)\left(\sqrt{a} - \sqrt{b}\right) = \left(\sqrt{a}\right)^2 - \sqrt{a}\sqrt{b} + \sqrt{a}\sqrt{b} - \left(\sqrt{b}\right)^2 = a - b$

   Or: Use the formula for the difference of two squares to see that $\left(\sqrt{a} + \sqrt{b}\right)\left(\sqrt{a} - \sqrt{b}\right) = \left(\sqrt{a}\right)^2 - \left(\sqrt{b}\right)^2 = a - b$.

   (d) $(2x+3)^2 = (2x+3)(2x+3) = 4x^2 + 6x + 6x + 9 = 4x^2 + 12x + 9$.

   Note: A quicker way to expand this binomial is to use the formula $(a+b)^2 = a^2 + 2ab + b^2$ with $a = 2x$ and $b = 3$:
   $(2x+3)^2 = (2x)^2 + 2(2x)(3) + 3^2 = 4x^2 + 12x + 9$

   (e) See Reference Page 1 for the binomial formula $(a+b)^3 = a^3 + 3a^2b + 3ab^2 + b^3$. Using it, we get
   $(x+2)^3 = x^3 + 3x^2(2) + 3x(2^2) + 2^3 = x^3 + 6x^2 + 12x + 8$.

4. (a) Using the difference of two squares formula, $a^2 - b^2 = (a+b)(a-b)$, we have
   $4x^2 - 25 = (2x)^2 - 5^2 = (2x+5)(2x-5)$.

   (b) Factoring by trial and error, we get $2x^2 + 5x - 12 = (2x-3)(x+4)$.

   (c) Using factoring by grouping and the difference of two squares formula, we have
   $x^3 - 3x^2 - 4x + 12 = x^2(x-3) - 4(x-3) = (x^2-4)(x-3) = (x-2)(x+2)(x-3)$.

   (d) $x^4 + 27x = x(x^3 + 27) = x(x+3)(x^2 - 3x + 9)$

   This last expression was obtained using the sum of two cubes formula, $a^3 + b^3 = (a+b)(a^2 - ab + b^2)$ with $a = x$ and $b = 3$. [See Reference Page 1 in the textbook.]

   (e) The smallest exponent on x is $-\frac{1}{2}$, so we will factor out $x^{-1/2}$.
   $3x^{3/2} - 9x^{1/2} + 6x^{-1/2} = 3x^{-1/2}(x^2 - 3x + 2) = 3x^{-1/2}(x-1)(x-2)$

   (f) $x^3y - 4xy = xy(x^2 - 4) = xy(x-2)(x+2)$

5. (a) $\dfrac{x^2+3x+2}{x^2-x-2}=\dfrac{(x+1)(x+2)}{(x+1)(x-2)}=\dfrac{x+2}{x-2}$

(b) $\dfrac{2x^2-x-1}{x^2-9}\cdot\dfrac{x+3}{2x+1}=\dfrac{(2x+1)(x-1)}{(x-3)(x+3)}\cdot\dfrac{x+3}{2x+1}=\dfrac{x-1}{x-3}$

(c) $\dfrac{x^2}{x^2-4}-\dfrac{x+1}{x+2}=\dfrac{x^2}{(x-2)(x+2)}-\dfrac{x+1}{x+2}=\dfrac{x^2}{(x-2)(x+2)}-\dfrac{x+1}{x+2}\cdot\dfrac{x-2}{x-2}=\dfrac{x^2-(x+1)(x-2)}{(x-2)(x+2)}$

$=\dfrac{x^2-(x^2-x-2)}{(x+2)(x-2)}=\dfrac{x+2}{(x+2)(x-2)}=\dfrac{1}{x-2}$

(d) $\dfrac{\frac{y}{x}-\frac{x}{y}}{\frac{1}{y}-\frac{1}{x}}=\dfrac{\frac{y}{x}-\frac{x}{y}}{\frac{1}{y}-\frac{1}{x}}\cdot\dfrac{xy}{xy}=\dfrac{y^2-x^2}{x-y}=\dfrac{(y-x)(y+x)}{-(y-x)}=\dfrac{y+x}{-1}=-(x+y)$

6. (a) $\dfrac{\sqrt{10}}{\sqrt{5}-2}=\dfrac{\sqrt{10}}{\sqrt{5}-2}\cdot\dfrac{\sqrt{5}+2}{\sqrt{5}+2}=\dfrac{\sqrt{50}+2\sqrt{10}}{\left(\sqrt{5}\right)^2-2^2}=\dfrac{5\sqrt{2}+2\sqrt{10}}{5-4}=5\sqrt{2}+2\sqrt{10}$

(b) $\dfrac{\sqrt{4+h}-2}{h}=\dfrac{\sqrt{4+h}-2}{h}\cdot\dfrac{\sqrt{4+h}+2}{\sqrt{4+h}+2}=\dfrac{4+h-4}{h\left(\sqrt{4+h}+2\right)}=\dfrac{h}{h\left(\sqrt{4+h}+2\right)}=\dfrac{1}{\sqrt{4+h}+2}$

7. (a) $x^2+x+1=\left(x^2+x+\frac{1}{4}\right)+1-\frac{1}{4}=\left(x+\frac{1}{2}\right)^2+\frac{3}{4}$

(b) $2x^2-12x+11=2(x^2-6x)+11=2(x^2-6x+9-9)+11=2(x^2-6x+9)-18+11=2(x-3)^2-7$

8. (a) $x+5=14-\frac{1}{2}x\iff x+\frac{1}{2}x=14-5\iff\frac{3}{2}x=9\iff x=\frac{2}{3}\cdot9\iff x=6$

(b) $\dfrac{2x}{x+1}=\dfrac{2x-1}{x}\implies 2x^2=(2x-1)(x+1)\iff 2x^2=2x^2+x-1\iff x=1$

(c) $x^2-x-12=0\iff(x+3)(x-4)=0\iff x+3=0\text{ or }x-4=0\iff x=-3\text{ or }x=4$

(d) By the quadratic formula, $2x^2+4x+1=0\iff$

$$x=\frac{-4\pm\sqrt{4^2-4(2)(1)}}{2(2)}=\frac{-4\pm\sqrt{8}}{4}=\frac{-4\pm2\sqrt{2}}{4}=\frac{2\left(-2\pm\sqrt{2}\right)}{4}=\frac{-2\pm\sqrt{2}}{2}=-1\pm\frac{1}{2}\sqrt{2}.$$

(e) $x^4-3x^2+2=0\iff(x^2-1)(x^2-2)=0\iff x^2-1=0\text{ or }x^2-2=0\iff x^2=1\text{ or }x^2=2\iff$

$x=\pm1\text{ or }x=\pm\sqrt{2}$

(f) $3\,|x-4|=10\iff|x-4|=\frac{10}{3}\iff x-4=-\frac{10}{3}\text{ or }x-4=\frac{10}{3}\iff x=\frac{2}{3}\text{ or }x=\frac{22}{3}$

(g) Multiplying through $2x(4-x)^{-1/2}-3\sqrt{4-x}=0$ by $(4-x)^{1/2}$ gives $2x-3(4-x)=0\iff$

$2x-12+3x=0\iff5x-12=0\iff5x=12\iff x=\frac{12}{5}.$

9. (a) $-4<5-3x\le17\iff-9<-3x\le12\iff3>x\ge-4\text{ or }-4\le x<3.$

In interval notation, the answer is $[-4,3)$.

(b) $x^2<2x+8\iff x^2-2x-8<0\iff(x+2)(x-4)<0.$ Now, $(x+2)(x-4)$ will change sign at the critical

values $x=-2$ and $x=4$. Thus the possible intervals of solution are $(-\infty,-2),(-2,4)$, and $(4,\infty)$. By choosing a

single test value from each interval, we see that $(-2,4)$ is the only interval that satisfies the inequality.

(c) The inequality $x(x-1)(x+2) > 0$ has critical values of $-2, 0$, and 1. The corresponding possible intervals of solution are $(-\infty, -2)$, $(-2, 0)$, $(0, 1)$ and $(1, \infty)$. By choosing a single test value from each interval, we see that both intervals $(-2, 0)$ and $(1, \infty)$ satisfy the inequality. Thus, the solution is the union of these two intervals: $(-2, 0) \cup (1, \infty)$.

(d) $|x - 4| < 3$ $\Leftrightarrow$ $-3 < x - 4 < 3$ $\Leftrightarrow$ $1 < x < 7$. In interval notation, the answer is $(1, 7)$.

(e) $\dfrac{2x - 3}{x + 1} \le 1$ $\Leftrightarrow$ $\dfrac{2x - 3}{x + 1} - 1 \le 0$ $\Leftrightarrow$ $\dfrac{2x - 3}{x + 1} - \dfrac{x + 1}{x + 1} \le 0$ $\Leftrightarrow$ $\dfrac{2x - 3 - x - 1}{x + 1} \le 0$ $\Leftrightarrow$ $\dfrac{x - 4}{x + 1} \le 0$.

Now, the expression $\dfrac{x - 4}{x + 1}$ may change signs at the critical values $x = -1$ and $x = 4$, so the possible intervals of solution are $(-\infty, -1)$, $(-1, 4]$, and $[4, \infty)$. By choosing a single test value from each interval, we see that $(-1, 4]$ is the only interval that satisfies the inequality.

10. (a) False. In order for the statement to be true, it must hold for all real numbers, so, to show that the statement is false, pick $p = 1$ and $q = 2$ and observe that $(1 + 2)^2 \ne 1^2 + 2^2$. In general, $(p + q)^2 = p^2 + 2pq + q^2$.

(b) True as long as a and b are positive real numbers. To see this, think in terms of the laws of exponents:
$$\sqrt{ab} = (ab)^{1/2} = a^{1/2}b^{1/2} = \sqrt{a}\,\sqrt{b}.$$

(c) False. To see this, let $p = 1$ and $q = 2$, then $\sqrt{1^2 + 2^2} \ne 1 + 2$.

(d) False. To see this, let $T = 1$ and $C = 2$, then $\dfrac{1 + 1(2)}{2} \ne 1 + 1$.

(e) False. To see this, let $x = 2$ and $y = 3$, then $\dfrac{1}{2 - 3} \ne \dfrac{1}{2} - \dfrac{1}{3}$.

(f) True since $\dfrac{1/x}{a/x - b/x} \cdot \dfrac{x}{x} = \dfrac{1}{a - b}$, as long as $x \ne 0$ and $a - b \ne 0$.

Test B   Analytic Geometry

1. (a) Using the point $(2, -5)$ and $m = -3$ in the point-slope equation of a line, $y - y_1 = m(x - x_1)$, we get
$$y - (-5) = -3(x - 2) \quad \Rightarrow \quad y + 5 = -3x + 6 \quad \Rightarrow \quad y = -3x + 1.$$

(b) A line parallel to the x-axis must be horizontal and thus have a slope of 0. Since the line passes through the point $(2, -5)$, the y-coordinate of every point on the line is -5, so the equation is $y = -5$.

(c) A line parallel to the y-axis is vertical with undefined slope. So the x-coordinate of every point on the line is 2 and so the equation is $x = 2$.

(d) Note that $2x - 4y = 3$ $\Rightarrow$ $-4y = -2x + 3$ $\Rightarrow$ $y = \frac{1}{2}x - \frac{3}{4}$. Thus the slope of the given line is $m = \frac{1}{2}$. Hence, the slope of the line we're looking for is also $\frac{1}{2}$ (since the line we're looking for is required to be parallel to the given line). So the equation of the line is $y - (-5) = \frac{1}{2}(x - 2)$ $\Rightarrow$ $y + 5 = \frac{1}{2}x - 1$ $\Rightarrow$ $y = \frac{1}{2}x - 6$.

2. First we'll find the distance between the two given points in order to obtain the radius, r, of the circle:
$$r = \sqrt{[3 - (-1)]^2 + (-2 - 4)^2} = \sqrt{4^2 + (-6)^2} = \sqrt{52}.$$ Next use the standard equation of a circle, $(x - h)^2 + (y - k)^2 = r^2$, where (h, k) is the center, to get $(x + 1)^2 + (y - 4)^2 = 52$.

3. We must rewrite the equation in standard form in order to identify the center and radius. Note that

$x^2 + y^2 - 6x + 10y + 9 = 0 \Rightarrow x^2 - 6x + 9 + y^2 + 10y = 0$. For the left-hand side of the latter equation, we

factor the first three terms and complete the square on the last two terms as follows: $x^2 - 6x + 9 + y^2 + 10y = 0 \Rightarrow$

$(x - 3)^2 + y^2 + 10y + 25 = 25 \Rightarrow (x - 3)^2 + (y + 5)^2 = 25$. Thus, the center of the circle is $(3, -5)$ and the radius is 5.

4. (a) $A(-7, 4)$ and $B(5, -12) \Rightarrow m_{AB} = \dfrac{-12 - 4}{5 - (-7)} = \dfrac{-16}{12} = -\dfrac{4}{3}$

(b) $y - 4 = -\frac{4}{3}[x - (-7)] \Rightarrow y - 4 = -\frac{4}{3}x - \frac{28}{3} \Rightarrow 3y - 12 = -4x - 28 \Rightarrow 4x + 3y + 16 = 0$. Putting $y = 0$,

we get $4x + 16 = 0$, so the x-intercept is -4, and substituting 0 for x results in a y-intercept of $-\frac{16}{3}$.

(c) The midpoint is obtained by averaging the corresponding coordinates of both points: $\left(\frac{-7+5}{2}, \frac{4+(-12)}{2}\right) = (-1, -4)$.

(d) $d = \sqrt{[5 - (-7)]^2 + (-12 - 4)^2} = \sqrt{12^2 + (-16)^2} = \sqrt{144 + 256} = \sqrt{400} = 20$

(e) The perpendicular bisector is the line that intersects the line segment $\overline{AB}$ at a right angle through its midpoint. Thus the

perpendicular bisector passes through $(-1, -4)$ and has slope $\frac{3}{4}$ [the slope is obtained by taking the negative reciprocal of

the answer from part (a)]. So the perpendicular bisector is given by $y + 4 = \frac{3}{4}[x - (-1)]$ or $3x - 4y = 13$.

(f) The center of the required circle is the midpoint of $\overline{AB}$, and the radius is half the length of $\overline{AB}$, which is 10. Thus, the

equation is $(x + 1)^2 + (y + 4)^2 = 100$.

5. (a) Graph the corresponding horizontal lines (given by the equations $y = -1$ and

$y = 3$) as solid lines. The inequality $y \geq -1$ describes the points (x, y) that lie

on or *above* the line $y = -1$. The inequality $y \leq 3$ describes the points (x, y)

that lie on or *below* the line $y = 3$. So the pair of inequalities $-1 \leq y \leq 3$

describes the points that lie on or *between* the lines $y = -1$ and $y = 3$.

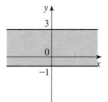

(b) Note that the given inequalities can be written as $-4 < x < 4$ and $-2 < y < 2$,

respectively. So the region lies between the vertical lines $x = -4$ and $x = 4$ and

between the horizontal lines $y = -2$ and $y = 2$. As shown in the graph, the

region common to both graphs is a rectangle (minus its edges) centered at the

origin.

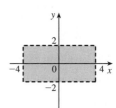

(c) We first graph $y = 1 - \frac{1}{2}x$ as a dotted line. Since $y < 1 - \frac{1}{2}x$, the points in the

region lie *below* this line.

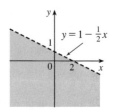

(d) We first graph the parabola $y = x^2 - 1$ using a solid curve. Since $y \geq x^2 - 1$, the points in the region lie on or *above* the parabola.

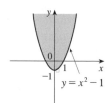

(e) We graph the circle $x^2 + y^2 = 4$ using a dotted curve. Since $\sqrt{x^2 + y^2} < 2$, the region consists of points whose distance from the origin is less than 2, that is, the points that lie *inside* the circle.

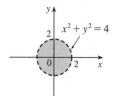

(f) The equation $9x^2 + 16y^2 = 144$ is an ellipse centered at $(0, 0)$. We put it in standard form by dividing by 144 and get $\dfrac{x^2}{16} + \dfrac{y^2}{9} = 1$. The x-intercepts are located at a distance of $\sqrt{16} = 4$ from the center while the y-intercepts are a distance of $\sqrt{9} = 3$ from the center (see the graph).

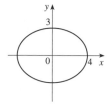

Test C   Functions

1. (a) Locate -1 on the x-axis and then go down to the point on the graph with an x-coordinate of -1. The corresponding y-coordinate is the value of the function at $x = -1$, which is -2. So, $f(-1) = -2$.

(b) Using the same technique as in part (a), we get $f(2) \approx 2.8$.

(c) Locate 2 on the y-axis and then go left and right to find all points on the graph with a y-coordinate of 2. The corresponding x-coordinates are the x-values we are searching for. So $x = -3$ and $x = 1$.

(d) Using the same technique as in part (c), we get $x \approx -2.5$ and $x \approx 0.3$.

(e) The domain is all the x-values for which the graph exists, and the range is all the y-values for which the graph exists. Thus, the domain is $[-3, 3]$, and the range is $[-2, 3]$.

2. Note that $f(2 + h) = (2 + h)^3$ and $f(2) = 2^3 = 8$. So the difference quotient becomes

$$\frac{f(2 + h) - f(2)}{h} = \frac{(2 + h)^3 - 8}{h} = \frac{8 + 12h + 6h^2 + h^3 - 8}{h} = \frac{12h + 6h^2 + h^3}{h} = \frac{h(12 + 6h + h^2)}{h} = 12 + 6h + h^2.$$

3. (a) Set the denominator equal to 0 and solve to find restrictions on the domain: $x^2 + x - 2 = 0$ $\Rightarrow$ $(x - 1)(x + 2) = 0$ $\Rightarrow$ $x = 1$ or $x = -2$. Thus, the domain is all real numbers except 1 or -2 or, in interval notation, $(-\infty, -2) \cup (-2, 1) \cup (1, \infty)$.

(b) Note that the denominator is always greater than or equal to 1, and the numerator is defined for all real numbers. Thus, the domain is $(-\infty, \infty)$.

(c) Note that the function h is the sum of two root functions. So h is defined on the intersection of the domains of these two root functions. The domain of a square root function is found by setting its radicand greater than or equal to 0. Now,

$4 - x \geq 0 \quad \Rightarrow \quad x \leq 4$ and $x^2 - 1 \geq 0 \quad \Rightarrow \quad (x-1)(x+1) \geq 0 \quad \Rightarrow \quad x \leq -1$ or $x \geq 1$. Thus, the domain of h is $(-\infty, -1] \cup [1, 4]$.

4. (a) Reflect the graph of f about the x-axis.

(b) Stretch the graph of f vertically by a factor of 2, then shift 1 unit downward.

(c) Shift the graph of f right 3 units, then up 2 units.

5. (a) Make a table and then connect the points with a smooth curve:

x	-2	-1	0	1	2
y	-8	-1	0	1	8

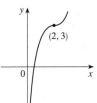

(b) Shift the graph from part (a) left 1 unit.

(c) Shift the graph from part (a) right 2 units and up 3 units.

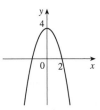

(d) First plot $y = x^2$. Next, to get the graph of $f(x) = 4 - x^2$, reflect f about the x-axis and then shift it upward 4 units.

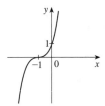

(e) Make a table and then connect the points with a smooth curve:

x	0	1	4	9
y	0	1	2	3

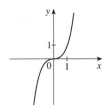

(f) Stretch the graph from part (e) vertically by a factor of two.

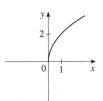

(g) First plot $y = 2^x$. Next, get the graph of $y = -2^x$ by reflecting the graph of $y = 2^x$ across the x-axis.

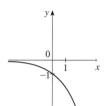

(h) Note that $y = 1 + x^{-1} = 1 + 1/x$. So first plot $y = 1/x$ and then shift it upward 1 unit.

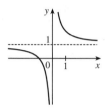

6. (a) $f(-2) = 1 - (-2)^2 = -3$ and $f(1) = 2(1) + 1 = 3$

(b) For $x \le 0$ plot $f(x) = 1 - x^2$ and, on the same plane, for $x > 0$ plot the graph of $f(x) = 2x + 1$.

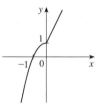

7. (a) $(f \circ g)(x) = f(g(x)) = f(2x - 3) = (2x - 3)^2 + 2(2x - 3) - 1 = 4x^2 - 12x + 9 + 4x - 6 - 1 = 4x^2 - 8x + 2$

(b) $(g \circ f)(x) = g(f(x)) = g(x^2 + 2x - 1) = 2(x^2 + 2x - 1) - 3 = 2x^2 + 4x - 2 - 3 = 2x^2 + 4x - 5$

(c) $(g \circ g \circ g)(x) = g(g(g(x))) = g(g(2x - 3)) = g(2(2x - 3) - 3) = g(4x - 9) = 2(4x - 9) - 3$

$\qquad = 8x - 18 - 3 = 8x - 21$

Test D   Trigonometry

1. (a) $300° = 300° \left(\dfrac{\pi}{180°} \right) = \dfrac{300\pi}{180} = \dfrac{5\pi}{3}$ (b) $-18° = -18° \left(\dfrac{\pi}{180°} \right) = -\dfrac{18\pi}{180} = -\dfrac{\pi}{10}$

2. (a) $\dfrac{5\pi}{6} = \dfrac{5\pi}{6} \left(\dfrac{180°}{\pi} \right) = 150°$ (b) $2 = 2 \left(\dfrac{180°}{\pi} \right) = \dfrac{360°}{\pi} \approx 114.6°$

3. We will use the arc length formula, $s = r\theta$, where s is arc length, r is the radius of the circle, and θ is the measure of the central angle in radians. First, note that $30° = 30° \left(\dfrac{\pi}{180°} \right) = \dfrac{\pi}{6}$. So $s = (12)\left(\dfrac{\pi}{6} \right) = 2\pi$ cm.

4. (a) $\tan(\pi/3) = \sqrt{3}$ [You can read the value from a right triangle with sides 1, 2, and $\sqrt{3}$.]

(b) Note that $7\pi/6$ can be thought of as an angle in the third quadrant with reference angle $\pi/6$. Thus, $\sin(7\pi/6) = -\frac{1}{2}$, since the sine function is negative in the third quadrant.

(c) Note that $5\pi/3$ can be thought of as an angle in the fourth quadrant with reference angle $\pi/3$. Thus,

$\sec(5\pi/3) = \dfrac{1}{\cos(5\pi/3)} = \dfrac{1}{1/2} = 2$, since the cosine function is positive in the fourth quadrant.

5. $\sin \theta = a/24 \ \Rightarrow \ a = 24 \sin \theta$ and $\cos \theta = b/24 \ \Rightarrow \ b = 24 \cos \theta$

6. $\sin x = \frac{1}{3}$ and $\sin^2 x + \cos^2 x = 1$ $\Rightarrow$ $\cos x = \sqrt{1 - \frac{1}{9}} = \dfrac{2\sqrt{2}}{3}$. Also, $\cos y = \frac{4}{5}$ $\Rightarrow$ $\sin y = \sqrt{1 - \frac{16}{25}} = \frac{3}{5}$.

So, using the sum identity for the sine, we have

$$\sin(x + y) = \sin x \, \cos y + \cos x \, \sin y = \frac{1}{3} \cdot \frac{4}{5} + \frac{2\sqrt{2}}{3} \cdot \frac{3}{5} = \frac{4 + 6\sqrt{2}}{15} = \frac{1}{15}\left(4 + 6\sqrt{2}\right)$$

7. (a) $\tan \theta \, \sin \theta + \cos \theta = \dfrac{\sin \theta}{\cos \theta} \sin \theta + \cos \theta = \dfrac{\sin^2 \theta}{\cos \theta} + \dfrac{\cos^2 \theta}{\cos \theta} = \dfrac{1}{\cos \theta} = \sec \theta$

(b) $\dfrac{2 \tan x}{1 + \tan^2 x} = \dfrac{2 \sin x / (\cos x)}{\sec^2 x} = 2 \dfrac{\sin x}{\cos x} \cos^2 x = 2 \sin x \, \cos x = \sin 2x$

8. $\sin 2x = \sin x$ $\Leftrightarrow$ $2 \sin x \, \cos x = \sin x$ $\Leftrightarrow$ $2 \sin x \, \cos x - \sin x = 0$ $\Leftrightarrow$ $\sin x \, (2 \cos x - 1) = 0$ $\Leftrightarrow$

$\sin x = 0$ or $\cos x = \frac{1}{2}$ $\Rightarrow$ $x = 0, \frac{\pi}{3}, \pi, \frac{5\pi}{3}, 2\pi$.

9. We first graph $y = \sin 2x$ (by compressing the graph of $\sin x$ by a factor of 2) and then shift it upward 1 unit.

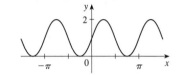

1 □ FUNCTIONS AND MODELS

1.1 Four Ways to Represent a Function

In exercises requiring estimations or approximations, your answers may vary slightly from the answers given here.

1. (a) The point $(-1, -2)$ is on the graph of f, so $f(-1) = -2$.

   (b) When $x = 2$, y is about 2.8, so $f(2) \approx 2.8$.

   (c) $f(x) = 2$ is equivalent to $y = 2$. When $y = 2$, we have $x = -3$ and $x = 1$.

   (d) Reasonable estimates for x when $y = 0$ are $x = -2.5$ and $x = 0.3$.

   (e) The domain of f consists of all x-values on the graph of f. For this function, the domain is $-3 \le x \le 3$, or $[-3, 3]$.
   The range of f consists of all y-values on the graph of f. For this function, the range is $-2 \le y \le 3$, or $[-2, 3]$.

   (f) As x increases from -1 to 3, y increases from -2 to 3. Thus, f is increasing on the interval $[-1, 3]$.

3. From Figure 1 in the text, the lowest point occurs at about $(t, a) = (12, -85)$. The highest point occurs at about $(17, 115)$. Thus, the range of the vertical ground acceleration is $-85 \le a \le 115$. Written in interval notation, we get $[-85, 115]$.

5. No, the curve is not the graph of a function because a vertical line intersects the curve more than once. Hence, the curve fails the Vertical Line Test.

7. Yes, the curve is the graph of a function because it passes the Vertical Line Test. The domain is $[-3, 2]$ and the range is $[-3, -2) \cup [-1, 3]$.

9. The person's weight increased to about 160 pounds at age 20 and stayed fairly steady for 10 years. The person's weight dropped to about 120 pounds for the next 5 years, then increased rapidly to about 170 pounds. The next 30 years saw a gradual increase to 190 pounds. Possible reasons for the drop in weight at 30 years of age: diet, exercise, health problems.

11. The water will cool down almost to freezing as the ice melts. Then, when the ice has melted, the water will slowly warm up to room temperature.

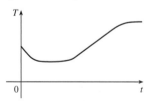

13. Of course, this graph depends strongly on the geographical location!

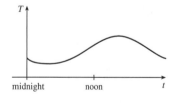

15. As the price increases, the amount sold decreases.

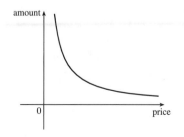

17.

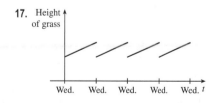

9

19. (a)

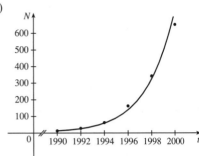

(b) From the graph, we estimate the number of cell-phone subscribers worldwide to be about 92 million in 1995 and 485 million in 1999.

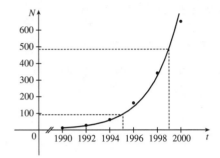

21. $f(x) = 3x^2 - x + 2$.

$f(2) = 3(2)^2 - 2 + 2 = 12 - 2 + 2 = 12$.

$f(-2) = 3(-2)^2 - (-2) + 2 = 12 + 2 + 2 = 16$.

$f(a) = 3a^2 - a + 2$.

$f(-a) = 3(-a)^2 - (-a) + 2 = 3a^2 + a + 2$.

$f(a+1) = 3(a+1)^2 - (a+1) + 2 = 3(a^2 + 2a + 1) - a - 1 + 2 = 3a^2 + 6a + 3 - a + 1 = 3a^2 + 5a + 4$.

$2f(a) = 2 \cdot f(a) = 2(3a^2 - a + 2) = 6a^2 - 2a + 4$.

$f(2a) = 3(2a)^2 - (2a) + 2 = 3(4a^2) - 2a + 2 = 12a^2 - 2a + 2$.

$f(a^2) = 3(a^2)^2 - (a^2) + 2 = 3(a^4) - a^2 + 2 = 3a^4 - a^2 + 2$.

$[f(a)]^2 = [3a^2 - a + 2]^2 = (3a^2 - a + 2)(3a^2 - a + 2)$
$= 9a^4 - 3a^3 + 6a^2 - 3a^3 + a^2 - 2a + 6a^2 - 2a + 4 = 9a^4 - 6a^3 + 13a^2 - 4a + 4$.

$f(a+h) = 3(a+h)^2 - (a+h) + 2 = 3(a^2 + 2ah + h^2) - a - h + 2 = 3a^2 + 6ah + 3h^2 - a - h + 2$.

23. $f(x) = 4 + 3x - x^2$, so $f(3+h) = 4 + 3(3+h) - (3+h)^2 = 4 + 9 + 3h - (9 + 6h + h^2) = 4 - 3h - h^2$,

and $\dfrac{f(3+h) - f(3)}{h} = \dfrac{(4 - 3h - h^2) - 4}{h} = \dfrac{h(-3 - h)}{h} = -3 - h$.

25. $\dfrac{f(x) - f(a)}{x - a} = \dfrac{\dfrac{1}{x} - \dfrac{1}{a}}{x - a} = \dfrac{\dfrac{a - x}{xa}}{x - a} = \dfrac{a - x}{xa(x - a)} = \dfrac{-1(x - a)}{xa(x - a)} = -\dfrac{1}{ax}$

27. $f(x) = x/(3x - 1)$ is defined for all x except when $0 = 3x - 1 \Leftrightarrow x = \frac{1}{3}$, so the domain is $\{x \in \mathbb{R} \mid x \neq \frac{1}{3}\} = (-\infty, \frac{1}{3}) \cup (\frac{1}{3}, \infty)$.

29. $f(t) = \sqrt{t} + \sqrt[3]{t}$ is defined when $t \geq 0$. These values of t give real number results for $\sqrt{t}$, whereas any value of t gives a real number result for $\sqrt[3]{t}$. The domain is $[0, \infty)$.

31. $h(x) = 1/\sqrt[4]{x^2 - 5x}$ is defined when $x^2 - 5x > 0 \Leftrightarrow x(x - 5) > 0$. Note that $x^2 - 5x \neq 0$ since that would result in division by zero. The expression $x(x - 5)$ is positive if $x < 0$ or $x > 5$. (See Appendix A for methods for solving inequalities.) Thus, the domain is $(-\infty, 0) \cup (5, \infty)$.

33. $f(x) = 5$ is defined for all real numbers, so the domain is $\mathbb{R}$, or $(-\infty, \infty)$.

The graph of f is a horizontal line with y-intercept 5.

35. $f(t) = t^2 - 6t$ is defined for all real numbers, so the domain is $\mathbb{R}$, or $(-\infty, \infty)$. The graph of f is a parabola opening upward since the coefficient of t^2 is positive. To find the t-intercepts, let $y = 0$ and solve for t.

$0 = t^2 - 6t = t(t - 6) \quad \Rightarrow \quad t = 0$ and $t = 6$. The t-coordinate of the vertex is halfway between the t-intercepts, that is, at $t = 3$. Since $f(3) = 3^2 - 6 \cdot 3 = -9$, the vertex is $(3, -9)$.

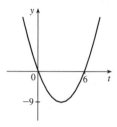

37. $g(x) = \sqrt{x - 5}$ is defined when $x - 5 \geq 0$ or $x \geq 5$, so the domain is $[5, \infty)$.

Since $y = \sqrt{x - 5} \quad \Rightarrow \quad y^2 = x - 5 \quad \Rightarrow \quad x = y^2 + 5$, we see that g is the top half of a parabola.

39. $G(x) = \dfrac{3x + |x|}{x}$. Since $|x| = \begin{cases} x & \text{if } x \geq 0 \\ -x & \text{if } x < 0 \end{cases}$, we have

$$G(x) = \begin{cases} \dfrac{3x + x}{x} & \text{if } x > 0 \\ \dfrac{3x - x}{x} & \text{if } x < 0 \end{cases} = \begin{cases} \dfrac{4x}{x} & \text{if } x > 0 \\ \dfrac{2x}{x} & \text{if } x < 0 \end{cases} = \begin{cases} 4 & \text{if } x > 0 \\ 2 & \text{if } x < 0 \end{cases}$$

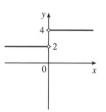

Note that G is not defined for $x = 0$. The domain is $(-\infty, 0) \cup (0, \infty)$.

41. $f(x) = \begin{cases} x + 2 & \text{if } x < 0 \\ 1 - x & \text{if } x \geq 0 \end{cases}$

The domain is $\mathbb{R}$.

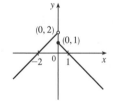

43. $f(x) = \begin{cases} x + 2 & \text{if } x \leq -1 \\ x^2 & \text{if } x > -1 \end{cases}$

Note that for $x = -1$, both $x + 2$ and x^2 are equal to 1. The domain is $\mathbb{R}$.

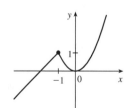

45. Recall that the slope m of a line between the two points (x_1, y_1) and (x_2, y_2) is $m = \dfrac{y_2 - y_1}{x_2 - x_1}$ and an equation of the line

connecting those two points is $y - y_1 = m(x - x_1)$. The slope of this line segment is $\dfrac{7 - (-3)}{5 - 1} = \dfrac{5}{2}$, so an equation is

$y - (-3) = \frac{5}{2}(x - 1)$. The function is $f(x) = \frac{5}{2}x - \frac{11}{2}$, $1 \leq x \leq 5$.

47. We need to solve the given equation for y. $x + (y-1)^2 = 0 \iff (y-1)^2 = -x \iff y - 1 = \pm\sqrt{-x} \iff$ $y = 1 \pm \sqrt{-x}$. The expression with the positive radical represents the top half of the parabola, and the one with the negative radical represents the bottom half. Hence, we want $f(x) = 1 - \sqrt{-x}$. Note that the domain is $x \le 0$.

49. For $0 \le x \le 3$, the graph is the line with slope -1 and y-intercept 3, that is, $y = -x + 3$. For $3 < x \le 5$, the graph is the line with slope 2 passing through $(3, 0)$; that is, $y - 0 = 2(x - 3)$, or $y = 2x - 6$. So the function is

$$f(x) = \begin{cases} -x + 3 & \text{if } 0 \le x \le 3 \\ 2x - 6 & \text{if } 3 < x \le 5 \end{cases}$$

51. Let the length and width of the rectangle be L and W. Then the perimeter is $2L + 2W = 20$ and the area is $A = LW$. Solving the first equation for W in terms of L gives $W = \dfrac{20 - 2L}{2} = 10 - L$. Thus, $A(L) = L(10 - L) = 10L - L^2$. Since lengths are positive, the domain of A is $0 < L < 10$. If we further restrict L to be larger than W, then $5 < L < 10$ would be the domain.

53. Let the length of a side of the equilateral triangle be x. Then by the Pythagorean Theorem, the height y of the triangle satisfies $y^2 + \left(\frac{1}{2}x\right)^2 = x^2$, so that $y^2 = x^2 - \frac{1}{4}x^2 = \frac{3}{4}x^2$ and $y = \frac{\sqrt{3}}{2}x$. Using the formula for the area A of a triangle, $A = \frac{1}{2}(\text{base})(\text{height})$, we obtain $A(x) = \frac{1}{2}(x)\left(\frac{\sqrt{3}}{2}x\right) = \frac{\sqrt{3}}{4}x^2$, with domain $x > 0$.

55. Let each side of the base of the box have length x, and let the height of the box be h. Since the volume is 2, we know that $2 = hx^2$, so that $h = 2/x^2$, and the surface area is $S = x^2 + 4xh$. Thus, $S(x) = x^2 + 4x(2/x^2) = x^2 + (8/x)$, with domain $x > 0$.

57. The height of the box is x and the length and width are $L = 20 - 2x$, $W = 12 - 2x$. Then $V = LWx$ and so

$$V(x) = (20 - 2x)(12 - 2x)(x) = 4(10 - x)(6 - x)(x) = 4x(60 - 16x + x^2) = 4x^3 - 64x^2 + 240x.$$

The sides L, W, and x must be positive. Thus, $L > 0 \iff 20 - 2x > 0 \iff x < 10$; $W > 0 \iff 12 - 2x > 0 \iff x < 6$; and $x > 0$. Combining these restrictions gives us the domain $0 < x < 6$.

59. (a)

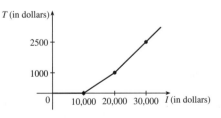

(b) On $14,000, tax is assessed on $4000, and $10\%(\$4000) = \$400$.

On $26,000, tax is assessed on $16,000, and

$10\%(\$10,000) + 15\%(\$6000) = \$1000 + \$900 = \$1900$.

(c) As in part (b), there is $1000 tax assessed on $20,000 of income, so the graph of T is a line segment from $(10,000, 0)$ to $(20,000, 1000)$. The tax on $30,000 is $2500, so the graph of T for $x > 20,000$ is the ray with initial point $(20,000, 1000)$ that passes through $(30,000, 2500)$.

61. f is an odd function because its graph is symmetric about the origin. g is an even function because its graph is symmetric with respect to the y-axis.

63. (a) Because an even function is symmetric with respect to the y-axis, and the point $(5, 3)$ is on the graph of this even function, the point $(-5, 3)$ must also be on its graph.

(b) Because an odd function is symmetric with respect to the origin, and the point $(5, 3)$ is on the graph of this odd function, the point $(-5, -3)$ must also be on its graph.

65. $f(x) = \dfrac{x}{x^2 + 1}$.

$f(-x) = \dfrac{-x}{(-x)^2 + 1} = \dfrac{-x}{x^2 + 1} = -\dfrac{x}{x^2 + 1} = -f(x)$.

So f is an odd function.

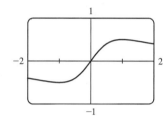

67. $f(x) = \dfrac{x}{x + 1}$, so $f(-x) = \dfrac{-x}{-x + 1} = \dfrac{x}{x - 1}$.

Since this is neither $f(x)$ nor $-f(x)$, the function f is neither even nor odd.

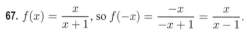
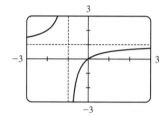

69. $f(x) = 1 + 3x^2 - x^4$.

$f(-x) = 1 + 3(-x)^2 - (-x)^4 = 1 + 3x^2 - x^4 = f(x)$.

So f is an even function.

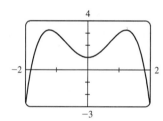

1.2 Mathematical Models: A Catalog of Essential Functions

1. (a) $f(x) = \sqrt[5]{x}$ is a root function with $n = 5$.

(b) $g(x) = \sqrt{1 - x^2}$ is an algebraic function because it is a root of a polynomial.

(c) $h(x) = x^9 + x^4$ is a polynomial of degree 9.

(d) $r(x) = \dfrac{x^2 + 1}{x^3 + x}$ is a rational function because it is a ratio of polynomials.

(e) $s(x) = \tan 2x$ is a trigonometric function.

(f) $t(x) = \log_{10} x$ is a logarithmic function.

3. We notice from the figure that g and h are even functions (symmetric with respect to the y-axis) and that f is an odd function (symmetric with respect to the origin). So (b) $\left[y = x^5\right]$ must be f. Since g is flatter than h near the origin, we must have (c) $\left[y = x^8\right]$ matched with g and (a) $\left[y = x^2\right]$ matched with h.

5. (a) An equation for the family of linear functions with slope 2

is $y = f(x) = 2x + b$, where b is the y-intercept.

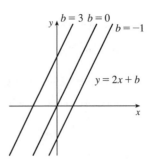

(b) $f(2) = 1$ means that the point $(2, 1)$ is on the graph of f. We can use the

point-slope form of a line to obtain an equation for the family of linear

functions through the point $(2, 1)$. $y - 1 = m(x - 2)$, which is equivalent

to $y = mx + (1 - 2m)$ in slope-intercept form.

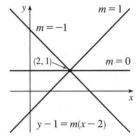

(c) To belong to both families, an equation must have slope $m = 2$, so the equation in part (b), $y = mx + (1 - 2m)$,

becomes $y = 2x - 3$. It is the *only* function that belongs to both families.

7. All members of the family of linear functions $f(x) = c - x$ have graphs

that are lines with slope -1. The y-intercept is c.

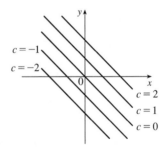

9. Since $f(-1) = f(0) = f(2) = 0$, f has zeros of -1, 0, and 2, so an equation for f is $f(x) = a[x - (-1)](x - 0)(x - 2)$,

or $f(x) = ax(x + 1)(x - 2)$. Because $f(1) = 6$, we'll substitute 1 for x and 6 for $f(x)$.

$6 = a(1)(2)(-1)$ $\Rightarrow$ $-2a = 6$ $\Rightarrow$ $a = -3$, so an equation for f is $f(x) = -3x(x + 1)(x - 2)$.

11. (a) $D = 200$, so $c = 0.0417D(a + 1) = 0.0417(200)(a + 1) = 8.34a + 8.34$. The slope is 8.34, which represents the

change in mg of the dosage for a child for each change of 1 year in age.

(b) For a newborn, $a = 0$, so $c = 8.34$ mg.

13. (a)

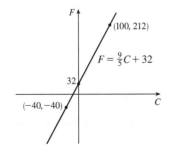

(b) The slope of $\frac{9}{5}$ means that F increases $\frac{9}{5}$ degrees for each increase

of $1°$C. (Equivalently, F increases by 9 when C increases by 5

and F decreases by 9 when C decreases by 5.) The F-intercept of

32 is the Fahrenheit temperature corresponding to a Celsius

temperature of 0.

15. (a) Using N in place of x and T in place of y, we find the slope to be $\dfrac{T_2 - T_1}{N_2 - N_1} = \dfrac{80 - 70}{173 - 113} = \dfrac{10}{60} = \dfrac{1}{6}$. So a linear

equation is $T - 80 = \frac{1}{6}(N - 173) \quad \Leftrightarrow \quad T - 80 = \frac{1}{6}N - \frac{173}{6} \quad \Leftrightarrow \quad T = \frac{1}{6}N + \frac{307}{6} \; \left[\frac{307}{6} = 51.1\overline{6}\right]$.

(b) The slope of $\frac{1}{6}$ means that the temperature in Fahrenheit degrees increases one-sixth as rapidly as the number of cricket
chirps per minute. Said differently, each increase of 6 cricket chirps per minute corresponds to an increase of $1°F$.

(c) When $N = 150$, the temperature is given approximately by $T = \frac{1}{6}(150) + \frac{307}{6} = 76.1\overline{6}\,°F \approx 76\,°F$.

17. (a) We are given $\dfrac{\text{change in pressure}}{10 \text{ feet change in depth}} = \dfrac{4.34}{10} = 0.434$. Using P for pressure and d for depth with the point

$(d, P) = (0, 15)$, we have the slope-intercept form of the line, $P = 0.434d + 15$.

(b) When $P = 100$, then $100 = 0.434d + 15 \quad \Leftrightarrow \quad 0.434d = 85 \quad \Leftrightarrow \quad d = \frac{85}{0.434} \approx 195.85$ feet. Thus, the pressure is

100 lb/in^2 at a depth of approximately 196 feet.

19. (a) The data appear to be periodic and a sine or cosine function would make the best model. A model of the form

$f(x) = a\cos(bx) + c$ seems appropriate.

(b) The data appear to be decreasing in a linear fashion. A model of the form $f(x) = mx + b$ seems appropriate.

Some values are given to many decimal places. These are the results given by several computer algebra systems — rounding is left to the reader.

21. (a)

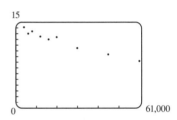

A linear model does seem appropriate.

(b) Using the points $(4000, 14.1)$ and $(60{,}000, 8.2)$, we obtain

$$y - 14.1 = \frac{8.2 - 14.1}{60{,}000 - 4000}(x - 4000) \text{ or, equivalently,}$$

$$y \approx -0.000105357x + 14.521429.$$

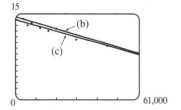

(c) Using a computing device, we obtain the least squares regression line $y = -0.0000997855x + 13.950764$.

The following commands and screens illustrate how to find the least squares regression line on a TI-83 Plus.

Enter the data into list one (L1) and list two (L2). Press $\boxed{\text{STAT}}\,\boxed{1}$ to enter the editor.

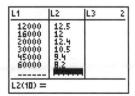

Find the regession line and store it in Y_1. Press $\boxed{\text{2nd}}\,\boxed{\text{QUIT}}\,\boxed{\text{STAT}}\,\boxed{\blacktriangleright}\,\boxed{4}\,\boxed{\text{VARS}}\,\boxed{\blacktriangleright}\,\boxed{1}\,\boxed{1}\,\boxed{\text{ENTER}}$.

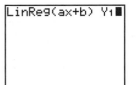 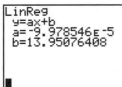 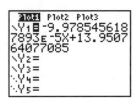

Note from the last figure that the regression line has been stored in Y_1 and that Plot1 has been turned on (Plot1 is highlighted). You can turn on Plot1 from the Y= menu by placing the cursor on Plot1 and pressing ENTER or by pressing 2nd STAT PLOT 1 ENTER .

Now press ZOOM 9 to produce a graph of the data and the regression line. Note that choice 9 of the ZOOM menu automatically selects a window that displays all of the data.

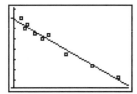

(d) When $x = 25{,}000$, $y \approx 11.456$; or about 11.5 per 100 population.

(e) When $x = 80{,}000$, $y \approx 5.968$; or about a 6% chance.

(f) When $x = 200{,}000$, y is negative, so the model does not apply.

23. (a)

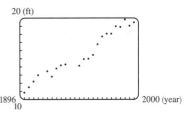

A linear model does seem appropriate.

(b)

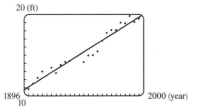

Using a computing device, we obtain the least squares regression line $y = 0.089119747x - 158.2403249$, where x is the year and y is the height in feet.

(c) When $x = 2000$, the model gives $y \approx 20.00$ ft. Note that the actual winning height for the 2000 Olympics is *less than* the winning height for 1996—so much for that prediction.

(d) When $x = 2100$, $y \approx 28.91$ ft. This would be an increase of 9.49 ft from 1996 to 2100. Even though there was an increase of 8.59 ft from 1900 to 1996, it is unlikely that a similar increase will occur over the next 100 years.

25.

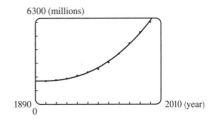

Using a computing device, we obtain the cubic function $y = ax^3 + bx^2 + cx + d$ with $a = 0.0012937$, $b = -7.06142$, $c = 12{,}823$, and $d = -7{,}743{,}770$. When $x = 1925$, $y \approx 1914$ (million).

1.3   New Functions from Old Functions

1. (a) If the graph of f is shifted 3 units upward, its equation becomes $y = f(x) + 3$.

(b) If the graph of f is shifted 3 units downward, its equation becomes $y = f(x) - 3$.

(c) If the graph of f is shifted 3 units to the right, its equation becomes $y = f(x - 3)$.

(d) If the graph of f is shifted 3 units to the left, its equation becomes $y = f(x + 3)$.

(e) If the graph of f is reflected about the x-axis, its equation becomes $y = -f(x)$.

(f) If the graph of f is reflected about the y-axis, its equation becomes $y = f(-x)$.

(g) If the graph of f is stretched vertically by a factor of 3, its equation becomes $y = 3f(x)$.

(h) If the graph of f is shrunk vertically by a factor of 3, its equation becomes $y = \frac{1}{3}f(x)$.

3. (a) (graph 3) The graph of f is shifted 4 units to the right and has equation $y = f(x - 4)$.

(b) (graph 1) The graph of f is shifted 3 units upward and has equation $y = f(x) + 3$.

(c) (graph 4) The graph of f is shrunk vertically by a factor of 3 and has equation $y = \frac{1}{3}f(x)$.

(d) (graph 5) The graph of f is shifted 4 units to the left and reflected about the x-axis. Its equation is $y = -f(x + 4)$.

(e) (graph 2) The graph of f is shifted 6 units to the left and stretched vertically by a factor of 2. Its equation is
$y = 2f(x + 6)$.

5. (a) To graph $y = f(2x)$ we shrink the graph of f horizontally by a factor of 2.

(b) To graph $y = f\left(\frac{1}{2}x\right)$ we stretch the graph of f horizontally by a factor of 2.

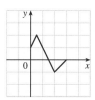

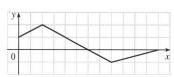

The point $(4, -1)$ on the graph of f corresponds to the point $\left(\frac{1}{2} \cdot 4, -1\right) = (2, -1)$.

The point $(4, -1)$ on the graph of f corresponds to the point $(2 \cdot 4, -1) = (8, -1)$.

(c) To graph $y = f(-x)$ we reflect the graph of f about the y-axis.

(d) To graph $y = -f(-x)$ we reflect the graph of f about the y-axis, then about the x-axis.

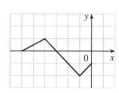

The point $(4, -1)$ on the graph of f corresponds to the point $(-1 \cdot 4, -1) = (-4, -1)$.

The point $(4, -1)$ on the graph of f corresponds to the point $(-1 \cdot 4, -1 \cdot -1) = (-4, 1)$.

7. The graph of $y = f(x) = \sqrt{3x - x^2}$ has been shifted 4 units to the left, reflected about the x-axis, and shifted downward 1 unit. Thus, a function describing the graph is

$$y = \underbrace{-1 \cdot}_{\substack{\text{reflect} \\ \text{about } x\text{-axis}}} \quad \underbrace{f\ (x+4)}_{\substack{\text{shift} \\ \text{4 units left}}} \quad \underbrace{-\ 1}_{\substack{\text{shift} \\ \text{1 unit left}}}$$

This function can be written as

$$y = -f(x+4) - 1 = -\sqrt{3(x+4) - (x+4)^2} - 1 = -\sqrt{3x + 12 - (x^2 + 8x + 16)} - 1 = -\sqrt{-x^2 - 5x - 4} - 1$$

9. $y = -x^3$: Start with the graph of $y = x^3$ and reflect about the x-axis. Note: Reflecting about the y-axis gives the same result since substituting $-x$ for x gives us $y = (-x)^3 = -x^3$.

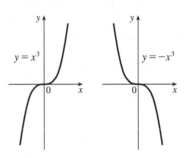

11. $y = (x+1)^2$: Start with the graph of $y = x^2$ and shift 1 unit to the left.

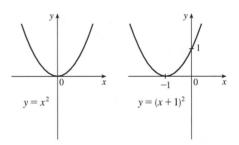

13. $y = 1 + 2\cos x$: Start with the graph of $y = \cos x$, stretch vertically by a factor of 2, and then shift 1 unit upward.

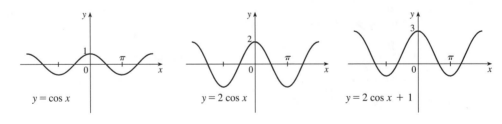

15. $y = \sin(x/2)$: Start with the graph of $y = \sin x$ and stretch horizontally by a factor of 2.

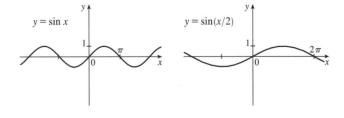

17. $y = \sqrt{x+3}$: Start with the graph of $y = \sqrt{x}$ and shift 3 units to the left.

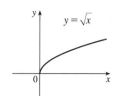

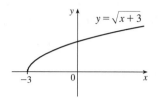

19. $y = \frac{1}{2}(x^2 + 8x) = \frac{1}{2}(x^2 + 8x + 16) - 8 = \frac{1}{2}(x+4)^2 - 8$: Start with the graph of $y = x^2$, compress vertically by a

factor of 2, shift 4 units to the left, and then shift 8 units downward.

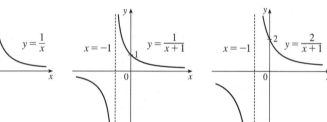

$y = x^2$ $y = \frac{1}{2}x^2$ $y = \frac{1}{2}(x+4)^2$ $y = \frac{1}{2}(x+4)^2 - 8$

21. $y = 2/(x+1)$: Start with the graph of $y = 1/x$, shift 1 unit to the left, and then stretch vertically by a factor of 2.

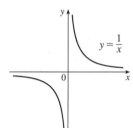

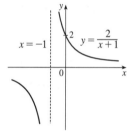

23. $y = |\sin x|$: Start with the graph of $y = \sin x$ and reflect all the parts of the graph below the x-axis about the x-axis.

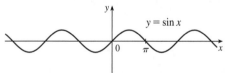

25. This is just like the solution to Example 4 except the amplitude of the curve (the 30°N curve in Figure 9 on June 21) is

$14 - 12 = 2$. So the function is $L(t) = 12 + 2\sin\left[\frac{2\pi}{365}(t - 80)\right]$. March 31 is the 90th day of the year, so the model gives

$L(90) \approx 12.34$ h. The daylight time (5:51 AM to 6:18 PM) is 12 hours and 27 minutes, or 12.45 h. The model value differs

from the actual value by $\frac{12.45 - 12.34}{12.45} \approx 0.009$, less than 1%.

27. (a) To obtain $y = f(|x|)$, the portion of the graph of $y = f(x)$ to the right of the y-axis is reflected about the y-axis.

(b) $y = \sin|x|$

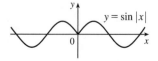

(c) $y = \sqrt{|x|}$

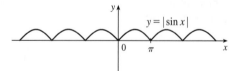

29. $f(x) = x^3 + 2x^2$; $g(x) = 3x^2 - 1$. $D = \mathbb{R}$ for both f and g.

$(f + g)(x) = (x^3 + 2x^2) + (3x^2 - 1) = x^3 + 5x^2 - 1$, $D = \mathbb{R}$.

$(f - g)(x) = (x^3 + 2x^2) - (3x^2 - 1) = x^3 - x^2 + 1$, $D = \mathbb{R}$.

$(fg)(x) = (x^3 + 2x^2)(3x^2 - 1) = 3x^5 + 6x^4 - x^3 - 2x^2$, $D = \mathbb{R}$.

$\left(\dfrac{f}{g}\right)(x) = \dfrac{x^3 + 2x^2}{3x^2 - 1}$, $D = \left\{x \mid x \neq \pm\dfrac{1}{\sqrt{3}}\right\}$ since $3x^2 - 1 \neq 0$.

31. $f(x) = x^2 - 1$, $D = \mathbb{R}$; $g(x) = 2x + 1$, $D = \mathbb{R}$.

(a) $(f \circ g)(x) = f(g(x)) = f(2x + 1) = (2x + 1)^2 - 1 = (4x^2 + 4x + 1) - 1 = 4x^2 + 4x$, $D = \mathbb{R}$.

(b) $(g \circ f)(x) = g(f(x)) = g(x^2 - 1) = 2(x^2 - 1) + 1 = (2x^2 - 2) + 1 = 2x^2 - 1$, $D = \mathbb{R}$.

(c) $(f \circ f)(x) = f(f(x)) = f(x^2 - 1) = (x^2 - 1)^2 - 1 = (x^4 - 2x^2 + 1) - 1 = x^4 - 2x^2$, $D = \mathbb{R}$.

(d) $(g \circ g)(x) = g(g(x)) = g(2x + 1) = 2(2x + 1) + 1 = (4x + 2) + 1 = 4x + 3$, $D = \mathbb{R}$.

33. $f(x) = 1 - 3x$; $g(x) = \cos x$. $D = \mathbb{R}$ for both f and g, and hence for their composites.

(a) $(f \circ g)(x) = f(g(x)) = f(\cos x) = 1 - 3\cos x$.

(b) $(g \circ f)(x) = g(f(x)) = g(1 - 3x) = \cos(1 - 3x)$.

(c) $(f \circ f)(x) = f(f(x)) = f(1 - 3x) = 1 - 3(1 - 3x) = 1 - 3 + 9x = 9x - 2$.

(d) $(g \circ g)(x) = g(g(x)) = g(\cos x) = \cos(\cos x)$ [Note that this is *not* $\cos x \cdot \cos x$.]

35. $f(x) = x + \dfrac{1}{x}$, $D = \{x \mid x \neq 0\}$; $g(x) = \dfrac{x + 1}{x + 2}$, $D = \{x \mid x \neq -2\}$

(a) $(f \circ g)(x) = f(g(x)) = f\left(\dfrac{x + 1}{x + 2}\right) = \dfrac{x + 1}{x + 2} + \dfrac{1}{\dfrac{x + 1}{x + 2}} = \dfrac{x + 1}{x + 2} + \dfrac{x + 2}{x + 1}$

$= \dfrac{(x + 1)(x + 1) + (x + 2)(x + 2)}{(x + 2)(x + 1)} = \dfrac{(x^2 + 2x + 1) + (x^2 + 4x + 4)}{(x + 2)(x + 1)} = \dfrac{2x^2 + 6x + 5}{(x + 2)(x + 1)}$

Since $g(x)$ is not defined for $x = -2$ and $f(g(x))$ is not defined for $x = -2$ and $x = -1$, the domain of $(f \circ g)(x)$ is $D = \{x \mid x \neq -2, -1\}$.

(b) $(g \circ f)(x) = g(f(x)) = g\left(x + \dfrac{1}{x}\right) = \dfrac{\left(x + \dfrac{1}{x}\right) + 1}{\left(x + \dfrac{1}{x}\right) + 2} = \dfrac{\dfrac{x^2 + 1 + x}{x}}{\dfrac{x^2 + 1 + 2x}{x}} = \dfrac{x^2 + x + 1}{x^2 + 2x + 1} = \dfrac{x^2 + x + 1}{(x + 1)^2}$

Since $f(x)$ is not defined for $x = 0$ and $g(f(x))$ is not defined for $x = -1$,

the domain of $(g \circ f)(x)$ is $D = \{x \mid x \neq -1, 0\}$.

(c) $(f \circ f)(x) = f(f(x)) = f\left(x + \dfrac{1}{x}\right) = \left(x + \dfrac{1}{x}\right) + \dfrac{1}{x + \dfrac{1}{x}} = x + \dfrac{1}{x} + \dfrac{1}{\dfrac{x^2 + 1}{x}} = x + \dfrac{1}{x} + \dfrac{x}{x^2 + 1}$

$= \dfrac{x(x)(x^2 + 1) + 1(x^2 + 1) + x(x)}{x(x^2 + 1)} = \dfrac{x^4 + x^2 + x^2 + 1 + x^2}{x(x^2 + 1)}$

$= \dfrac{x^4 + 3x^2 + 1}{x(x^2 + 1)}$, $D = \{x \mid x \neq 0\}$

(d) $(g \circ g)(x) = g(g(x)) = g\left(\dfrac{x+1}{x+2}\right) = \dfrac{\dfrac{x+1}{x+2}+1}{\dfrac{x+1}{x+2}+2} = \dfrac{\dfrac{x+1+1(x+2)}{x+2}}{\dfrac{x+1+2(x+2)}{x+2}} = \dfrac{x+1+x+2}{x+1+2x+4} = \dfrac{2x+3}{3x+5}$

Since $g(x)$ is not defined for $x = -2$ and $g(g(x))$ is not defined for $x = -\frac{5}{3}$,

the domain of $(g \circ g)(x)$ is $D = \left\{x \mid x \neq -2, -\frac{5}{3}\right\}$.

37. $(f \circ g \circ h)(x) = f(g(h(x))) = f(g(x-1)) = f(2(x-1)) = 2(x-1) + 1 = 2x - 1$

39. $(f \circ g \circ h)(x) = f(g(h(x))) = f(g(x^3 + 2)) = f[(x^3 + 2)^2]$

$\qquad = f(x^6 + 4x^3 + 4) = \sqrt{(x^6 + 4x^3 + 4) - 3} = \sqrt{x^6 + 4x^3 + 1}$

41. Let $g(x) = x^2 + 1$ and $f(x) = x^{10}$. Then $(f \circ g)(x) = f(g(x)) = f(x^2 + 1) = (x^2 + 1)^{10} = F(x)$.

43. Let $g(x) = \sqrt[3]{x}$ and $f(x) = \dfrac{x}{1+x}$. Then $(f \circ g)(x) = f(g(x)) = f(\sqrt[3]{x}) = \dfrac{\sqrt[3]{x}}{1 + \sqrt[3]{x}} = F(x)$.

45. Let $g(t) = \cos t$ and $f(t) = \sqrt{t}$. Then $(f \circ g)(t) = f(g(t)) = f(\cos t) = \sqrt{\cos t} = u(t)$.

47. Let $h(x) = x^2$, $g(x) = 3^x$, and $f(x) = 1 - x$. Then

$\qquad (f \circ g \circ h)(x) = f(g(h(x))) = f(g(x^2)) = f\left(3^{x^2}\right) = 1 - 3^{x^2} = H(x)$.

49. Let $h(x) = \sqrt{x}$, $g(x) = \sec x$, and $f(x) = x^4$. Then

$\qquad (f \circ g \circ h)(x) = f(g(h(x))) = f(g(\sqrt{x})) = f(\sec \sqrt{x}) = (\sec \sqrt{x})^4 = \sec^4(\sqrt{x}) = H(x)$.

51. (a) $g(2) = 5$, because the point $(2, 5)$ is on the graph of g. Thus, $f(g(2)) = f(5) = 4$, because the point $(5, 4)$ is on the graph of f.

(b) $g(f(0)) = g(0) = 3$

(c) $(f \circ g)(0) = f(g(0)) = f(3) = 0$

(d) $(g \circ f)(6) = g(f(6)) = g(6)$. This value is not defined, because there is no point on the graph of g that has x-coordinate 6.

(e) $(g \circ g)(-2) = g(g(-2)) = g(1) = 4$

(f) $(f \circ f)(4) = f(f(4)) = f(2) = -2$

53. (a) Using the relationship *distance = rate · time* with the radius r as the distance, we have $r(t) = 60t$.

(b) $A = \pi r^2 \quad \Rightarrow \quad (A \circ r)(t) = A(r(t)) = \pi(60t)^2 = 3600\pi t^2$. This formula gives us the extent of the rippled area (in cm^2) at any time t.

55. (a) From the figure, we have a right triangle with legs 6 and d, and hypotenuse s.

By the Pythagorean Theorem, $d^2 + 6^2 = s^2 \quad \Rightarrow \quad s = f(d) = \sqrt{d^2 + 36}$.

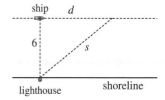

(b) Using $d = rt$, we get $d = (30 \text{ km/hr})(t \text{ hr}) = 30t$ (in km). Thus,

$d = g(t) = 30t$.

(c) $(f \circ g)(t) = f(g(t)) = f(30t) = \sqrt{(30t)^2 + 36} = \sqrt{900t^2 + 36}$. This function represents the distance between the lighthouse and the ship as a function of the time elapsed since noon.

57. (a)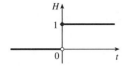

$$H(t) = \begin{cases} 0 & \text{if } t < 0 \\ 1 & \text{if } t \geq 0 \end{cases}$$

(b)

$$V(t) = \begin{cases} 0 & \text{if } t < 0 \\ 120 & \text{if } t \geq 0 \end{cases} \quad \text{so } V(t) = 120H(t).$$

(c)

Starting with the formula in part (b), we replace 120 with 240 to reflect the different voltage. Also, because we are starting 5 units to the right of $t = 0$, we replace t with $t - 5$. Thus, the formula is $V(t) = 240H(t - 5)$.

59. If $f(x) = m_1 x + b_1$ and $g(x) = m_2 x + b_2$, then

$$(f \circ g)(x) = f(g(x)) = f(m_2 x + b_2) = m_1(m_2 x + b_2) + b_1 = m_1 m_2 x + m_1 b_2 + b_1.$$

So $f \circ g$ is a linear function with slope $m_1 m_2$.

61. (a) By examining the variable terms in g and h, we deduce that we must square g to get the terms $4x^2$ and $4x$ in h. If we let

$f(x) = x^2 + c$, then $(f \circ g)(x) = f(g(x)) = f(2x + 1) = (2x + 1)^2 + c = 4x^2 + 4x + (1 + c)$. Since

$h(x) = 4x^2 + 4x + 7$, we must have $1 + c = 7$. So $c = 6$ and $f(x) = x^2 + 6$.

(b) We need a function g so that $f(g(x)) = 3(g(x)) + 5 = h(x)$. But

$h(x) = 3x^2 + 3x + 2 = 3(x^2 + x) + 2 = 3(x^2 + x - 1) + 5$, so we see that $g(x) = x^2 + x - 1$.

63. (a) If f and g are even functions, then $f(-x) = f(x)$ and $g(-x) = g(x)$.

(i) $(f + g)(-x) = f(-x) + g(-x) = f(x) + g(x) = (f + g)(x)$, so $f + g$ is an *even* function.
(ii) $(fg)(-x) = f(-x) \cdot g(-x) = f(x) \cdot g(x) = (fg)(x)$, so fg is an *even* function.

(b) If f and g are odd functions, then $f(-x) = -f(x)$ and $g(-x) = -g(x)$.

(i) $(f + g)(-x) = f(-x) + g(-x) = -f(x) + [-g(x)] = -[f(x) + g(x)] = -(f + g)(x)$,
so $f + g$ is an *odd* function.
(ii) $(fg)(-x) = f(-x) \cdot g(-x) = -f(x) \cdot [-g(x)] = f(x) \cdot g(x) = (fg)(x)$, so fg is an *even* function.

65. We need to examine $h(-x)$.

$$h(-x) = (f \circ g)(-x) = f(g(-x)) = f(g(x)) \quad \text{[because } g \text{ is even]} \quad = h(x)$$

Because $h(-x) = h(x)$, h is an even function.

1.4 Graphing Calculators and Computers

1. $f(x) = \sqrt{x^3 - 5x^2}$

(a) $[-5, 5]$ by $[-5, 5]$ (b) $[0, 10]$ by $[0, 2]$ (c) $[0, 10]$ by $[0, 10]$

(There is no graph shown.)

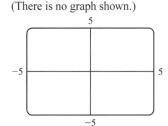

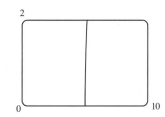

The most appropriate graph is produced in viewing rectangle (c).

3. Since the graph of $f(x) = 5 + 20x - x^2$ is a parabola opening downward, an appropriate viewing rectangle should include the maximum point.

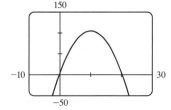

5. $f(x) = \sqrt[4]{81 - x^4}$ is defined when $81 - x^4 \geq 0 \iff x^4 \leq 81 \iff |x| \leq 3$, so the domain of f is $[-3, 3]$. Also $0 \leq \sqrt[4]{81 - x^4} \leq \sqrt[4]{81} = 3$, so the range is $[0, 3]$.

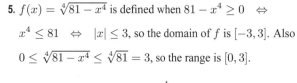

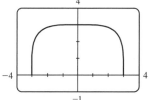

7. The graph of $f(x) = x^3 - 225x$ is symmetric with respect to the origin. Since $f(x) = x^3 - 225x = x(x^2 - 225) = x(x + 15)(x - 15)$, there are x-intercepts at 0, -15, and 15. $f(20) = 3500$.

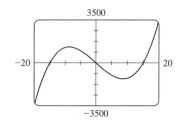

9. The period of $g(x) = \sin(1000x)$ is $\frac{2\pi}{1000} \approx 0.0063$ and its range is $[-1, 1]$. Since $f(x) = \sin^2(1000x)$ is the square of g, its range is $[0, 1]$ and a viewing rectangle of $[-0.01, 0.01]$ by $[0, 1.1]$ seems appropriate.

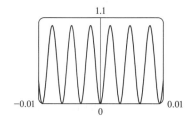

11. The domain of $y = \sqrt{x}$ is $x \geq 0$, so the domain of $f(x) = \sin\sqrt{x}$ is $[0, \infty)$ and the range is $[-1, 1]$. With a little trial-and-error experimentation, we find that an Xmax of 100 illustrates the general shape of f, so an appropriate viewing rectangle is $[0, 100]$ by $[-1.5, 1.5]$.

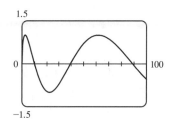

13. The first term, $10 \sin x$, has period 2π and range $[-10, 10]$. It will be the dominant term in any "large" graph of

$y = 10 \sin x + \sin 100x$, as shown in the first figure. The second term, $\sin 100x$, has period $\frac{2\pi}{100} = \frac{\pi}{50}$ and range $[-1, 1]$.

It causes the bumps in the first figure and will be the dominant term in any "small" graph, as shown in the view near the

origin in the second figure.

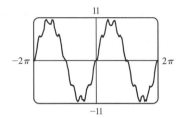

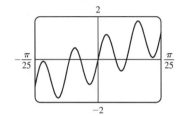

15. We must solve the given equation for y to obtain equations for the upper and
lower halves of the ellipse.

$$4x^2 + 2y^2 = 1 \quad \Leftrightarrow \quad 2y^2 = 1 - 4x^2 \quad \Leftrightarrow \quad y^2 = \frac{1 - 4x^2}{2} \quad \Leftrightarrow$$

$$y = \pm\sqrt{\frac{1 - 4x^2}{2}}$$

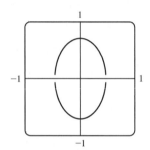

17. From the graph of $y = 3x^2 - 6x + 1$
and $y = 0.23x - 2.25$ in the viewing
rectangle $[-1, 3]$ by $[-2.5, 1.5]$, it is
difficult to see if the graphs intersect.
If we zoom in on the fourth quadrant,
we see the graphs do not intersect.

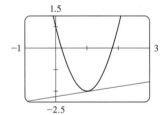

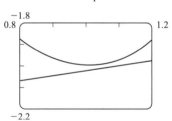

19. From the graph of $f(x) = x^3 - 9x^2 - 4$, we see that there is one solution
of the equation $f(x) = 0$ and it is slightly larger than 9. By zooming in or
using a root or zero feature, we obtain $x \approx 9.05$.

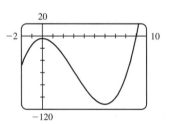

21. We see that the graphs of $f(x) = x^2$ and $g(x) = \sin x$ intersect twice. One
solution is $x = 0$. The other solution of $f = g$ is the x-coordinate of the
point of intersection in the first quadrant. Using an intersect feature or
zooming in, we find this value to be approximately 0.88. Alternatively, we
could find that value by finding the positive zero of $h(x) = x^2 - \sin x$.

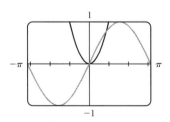

Note: After producing the graph on a TI-83 Plus, we can find the approximate value 0.88 by using the following keystrokes:

2nd CALC 5 ENTER ENTER 1 ENTER . The "1" is just a guess for 0.88.

23. $g(x) = x^3/10$ is larger than $f(x) = 10x^2$ whenever $x > 100$.

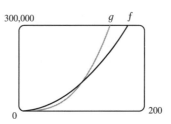

25.

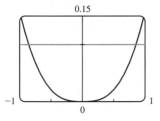

We see from the graphs of $y = |\sin x - x|$ and $y = 0.1$ that there are two solutions to the equation $|\sin x - x| = 0.1$: $x \approx -0.85$ and $x \approx 0.85$. The condition $|\sin x - x| < 0.1$ holds for any x lying between these two values, that is, $-0.85 < x < 0.85$.

27. (a) The root functions $y = \sqrt{x}$, $y = \sqrt[4]{x}$ and $y = \sqrt[6]{x}$

(b) The root functions $y = x$, $y = \sqrt[3]{x}$ and $y = \sqrt[5]{x}$

(c) The root functions $y = \sqrt{x}$, $y = \sqrt[3]{x}$, $y = \sqrt[4]{x}$ and $y = \sqrt[5]{x}$

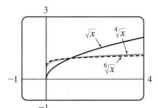

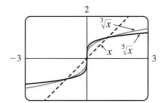

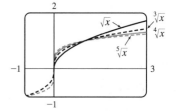

(d) ● For any n, the nth root of 0 is 0 and the nth root of 1 is 1; that is, all nth root functions pass through the points $(0,0)$ and $(1,1)$.

● For odd n, the domain of the nth root function is $\mathbb{R}$, while for even n, it is $\{x \in \mathbb{R} \mid x \geq 0\}$.

● Graphs of even root functions look similar to that of $\sqrt{x}$, while those of odd root functions resemble that of $\sqrt[3]{x}$.

● As n increases, the graph of $\sqrt[n]{x}$ becomes steeper near 0 and flatter for $x > 1$.

29. $f(x) = x^4 + cx^2 + x$. If $c < -1.5$, there are three humps: two minimum points and a maximum point. These humps get flatter as c increases, until at $c = -1.5$ two of the humps disappear and there is only one minimum point. This single hump then moves to the right and approaches the origin as c increases.

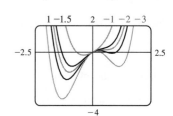

31. $y = x^n 2^{-x}$. As n increases, the maximum of the function moves further from the origin, and gets larger. Note, however, that regardless of n, the function approaches 0 as $x \to \infty$.

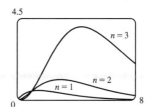

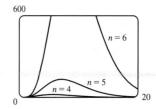

33. $y^2 = cx^3 + x^2$. If $c < 0$, the loop is to the right of the origin, and if c is positive, it is to the left. In both cases, the closer c is to 0, the larger the loop is. (In the limiting case, $c = 0$, the loop is "infinite," that is, it doesn't close.) Also, the larger $|c|$ is, the steeper the slope is on the loopless side of the origin.

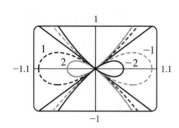

35. The graphing window is 95 pixels wide and we want to start with $x = 0$ and end with $x = 2\pi$. Since there are 94 "gaps" between pixels, the distance between pixels is $\frac{2\pi - 0}{94}$. Thus, the x-values that the calculator actually plots are $x = 0 + \frac{2\pi}{94} \cdot n$, where $n = 0, 1, 2, \ldots, 93, 94$. For $y = \sin 2x$, the actual points plotted by the calculator are $\left(\frac{2\pi}{94} \cdot n, \sin\left(2 \cdot \frac{2\pi}{94} \cdot n\right)\right)$ for $n = 0, 1, \ldots, 94$. For $y = \sin 96x$, the points plotted are $\left(\frac{2\pi}{94} \cdot n, \sin\left(96 \cdot \frac{2\pi}{94} \cdot n\right)\right)$ for $n = 0, 1, \ldots, 94$. But

$$\sin\left(96 \cdot \tfrac{2\pi}{94} \cdot n\right) = \sin\left(94 \cdot \tfrac{2\pi}{94} \cdot n + 2 \cdot \tfrac{2\pi}{94} \cdot n\right) = \sin\left(2\pi n + 2 \cdot \tfrac{2\pi}{94} \cdot n\right)$$

$$= \sin\left(2 \cdot \tfrac{2\pi}{94} \cdot n\right) \quad \text{[by periodicity of sine]}, \quad n = 0, 1, \ldots, 94$$

So the y-values, and hence the points, plotted for $y = \sin 96x$ are identical to those plotted for $y = \sin 2x$. *Note:* Try graphing $y = \sin 94x$. Can you see why all the y-values are zero?

1.5 Exponential Functions

1. (a) $f(x) = a^x$, $a > 0$ (b) $\mathbb{R}$ (c) $(0, \infty)$ (d) See Figures 4(c), 4(b), and 4(a), respectively.

3. All of these graphs approach 0 as $x \to -\infty$, all of them pass through the point $(0, 1)$, and all of them are increasing and approach ∞ as $x \to \infty$. The larger the base, the faster the function increases for $x > 0$, and the faster it approaches 0 as $x \to -\infty$.

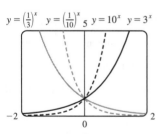

Note: The notation "$x \to \infty$" can be thought of as "x becomes large" at this point. More details on this notation are given in Chapter 2.

5. The functions with bases greater than 1 (3^x and 10^x) are increasing, while those with bases less than 1 $\left[\left(\frac{1}{3}\right)^x \text{ and } \left(\frac{1}{10}\right)^x\right]$ are decreasing. The graph of $\left(\frac{1}{3}\right)^x$ is the reflection of that of 3^x about the y-axis, and the graph of $\left(\frac{1}{10}\right)^x$ is the reflection of that of 10^x about the y-axis. The graph of 10^x increases more quickly than that of 3^x for $x > 0$, and approaches 0 faster as $x \to -\infty$.

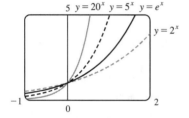

7. We start with the graph of $y = 4^x$ (Figure 3) and then shift 3 units downward. This shift doesn't affect the domain, but the range of $y = 4^x - 3$ is $(-3, \infty)$. There is a horizontal asymptote of $y = -3$.

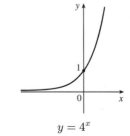

$y = 4^x$

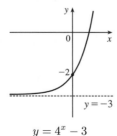

$y = 4^x - 3$

9. We start with the graph of $y = 2^x$ (Figure 3), reflect it about the y-axis, and then about the x-axis (or just rotate $180°$ to handle both reflections) to obtain the graph of $y = -2^{-x}$. In each graph, $y = 0$ is the horizontal asymptote.

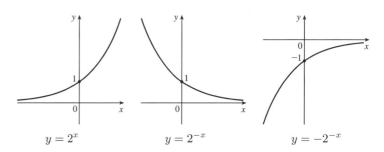

$$y = 2^x \qquad\qquad y = 2^{-x} \qquad\qquad y = -2^{-x}$$

11. We start with the graph of $y = e^x$ (Figure 13) and reflect about the y-axis to get the graph of $y = e^{-x}$. Then we compress the graph vertically by a factor of 2 to obtain the graph of $y = \frac{1}{2}e^{-x}$ and then reflect about the x-axis to get the graph of $y = -\frac{1}{2}e^{-x}$. Finally, we shift the graph upward one unit to get the graph of $y = 1 - \frac{1}{2}e^{-x}$.

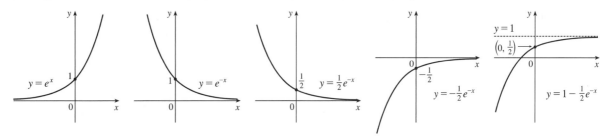

13. (a) To find the equation of the graph that results from shifting the graph of $y = e^x$ 2 units downward, we subtract 2 from the original function to get $y = e^x - 2$.

(b) To find the equation of the graph that results from shifting the graph of $y = e^x$ 2 units to the right, we replace x with $x - 2$ in the original function to get $y = e^{(x-2)}$.

(c) To find the equation of the graph that results from reflecting the graph of $y = e^x$ about the x-axis, we multiply the original function by -1 to get $y = -e^x$.

(d) To find the equation of the graph that results from reflecting the graph of $y = e^x$ about the y-axis, we replace x with $-x$ in the original function to get $y = e^{-x}$.

(e) To find the equation of the graph that results from reflecting the graph of $y = e^x$ about the x-axis and then about the y-axis, we first multiply the original function by -1 (to get $y = -e^x$) and then replace x with $-x$ in this equation to get $y = -e^{-x}$.

15. (a) The denominator $1 + e^x$ is never equal to zero because $e^x > 0$, so the domain of $f(x) = 1/(1 + e^x)$ is $\mathbb{R}$.

(b) $1 - e^x = 0 \iff e^x = 1 \iff x = 0$, so the domain of $f(x) = 1/(1 - e^x)$ is $(-\infty, 0) \cup (0, \infty)$.

17. Use $y = Ca^x$ with the points $(1, 6)$ and $(3, 24)$. $\quad 6 = Ca^1 \quad \left[C = \frac{6}{a}\right] \quad$ and $24 = Ca^3 \implies 24 = \left(\dfrac{6}{a}\right)a^3 \implies$

$4 = a^2 \implies a = 2 \quad$ [since $a > 0$] $\quad$ and $C = \frac{6}{2} = 3$. The function is $f(x) = 3 \cdot 2^x$.

19. If $f(x) = 5^x$, then $\dfrac{f(x+h) - f(x)}{h} = \dfrac{5^{x+h} - 5^x}{h} = \dfrac{5^x 5^h - 5^x}{h} = \dfrac{5^x\left(5^h - 1\right)}{h} = 5^x\left(\dfrac{5^h - 1}{h}\right).$

21. 2 ft = 24 in, $f(24) = 24^2$ in = 576 in = 48 ft. $g(24) = 2^{24}$ in = $2^{24}/(12 \cdot 5280)$ mi $\approx$ 265 mi

23. The graph of g finally surpasses that of f at $x \approx 35.8$.

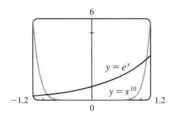 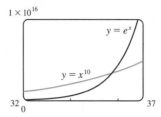

25. (a) Fifteen hours represents 5 doubling periods (one doubling period is three hours). $100 \cdot 2^5 = 3200$

(b) In t hours, there will be $t/3$ doubling periods. The initial population is 100,

so the population y at time t is $y = 100 \cdot 2^{t/3}$.

(c) $t = 20 \Rightarrow y = 100 \cdot 2^{20/3} \approx 10{,}159$

(d) We graph $y_1 = 100 \cdot 2^{x/3}$ and $y_2 = 50{,}000$. The two curves intersect at
$x \approx 26.9$, so the population reaches 50,000 in about 26.9 hours.

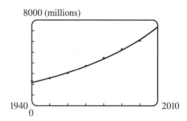

27. An exponential model is $y = ab^t$, where $a = 3.154832569 \times 10^{-12}$
and $b = 1.017764706$. This model gives $y(1993) \approx 5498$ million and
$y(2010) \approx 7417$ million.

29.

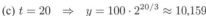

From the graph, it appears that f is an odd function (f is undefined for $x = 0$).
To prove this, we must show that $f(-x) = -f(x)$.

$$f(-x) = \frac{1 - e^{1/(-x)}}{1 + e^{1/(-x)}} = \frac{1 - e^{(-1/x)}}{1 + e^{(-1/x)}} = \frac{1 - \dfrac{1}{e^{1/x}}}{1 + \dfrac{1}{e^{1/x}}} \cdot \frac{e^{1/x}}{e^{1/x}} = \frac{e^{1/x} - 1}{e^{1/x} + 1}$$

$$= -\frac{1 - e^{1/x}}{1 + e^{1/x}} = -f(x)$$

so f is an odd function.

1.6 Inverse Functions and Logarithms

1. (a) See Definition 1.

(b) It must pass the Horizontal Line Test.

3. f is not one-to-one because $2 \neq 6$, but $f(2) = 2.0 = f(6)$.

5. No horizontal line intersects the graph of f more than once. Thus, by the Horizontal Line Test, f is one-to-one.

7. The horizontal line $y = 0$ (the x-axis) intersects the graph of f in more than one point. Thus, by the Horizontal Line Test, f is not one-to-one.

9. The graph of $f(x) = x^2 - 2x$ is a parabola with axis of symmetry $x = -\dfrac{b}{2a} = -\dfrac{-2}{2(1)} = 1$. Pick any x-values equidistant from 1 to find two equal function values. For example, $f(0) = 0$ and $f(2) = 0$, so f is not one-to-one.

11. $g(x) = 1/x$. $x_1 \neq x_2 \ \Rightarrow \ 1/x_1 \neq 1/x_2 \ \Rightarrow \ g(x_1) \neq g(x_2)$, so g is one-to-one.

Geometric solution: The graph of g is the hyperbola shown in Figure 14 in Section 1.2. It passes the Horizontal Line Test, so g is one-to-one.

13. A football will attain every height h up to its maximum height twice: once on the way up, and again on the way down. Thus, even if t_1 does not equal t_2, $f(t_1)$ may equal $f(t_2)$, so f is not 1-1.

15. Since $f(2) = 9$ and f is 1-1, we know that $f^{-1}(9) = 2$. Remember, if the point $(2, 9)$ is on the graph of f, then the point $(9, 2)$ is on the graph of f^{-1}.

17. First, we must determine x such that $g(x) = 4$. By inspection, we see that if $x = 0$, then $g(x) = 4$. Since g is 1-1 (g is an increasing function), it has an inverse, and $g^{-1}(4) = 0$.

19. We solve $C = \frac{5}{9}(F - 32)$ for F: $\frac{9}{5}C = F - 32 \ \Rightarrow \ F = \frac{9}{5}C + 32$. This gives us a formula for the inverse function, that is, the Fahrenheit temperature F as a function of the Celsius temperature C. $F \geq -459.67 \ \Rightarrow \ \frac{9}{5}C + 32 \geq -459.67 \ \Rightarrow$
$\frac{9}{5}C \geq -491.67 \ \Rightarrow \ C \geq -273.15$, the domain of the inverse function.

21. $f(x) = \sqrt{10 - 3x} \ \Rightarrow \ y = \sqrt{10 - 3x} \ \ (y \geq 0) \ \Rightarrow \ y^2 = 10 - 3x \ \Rightarrow \ 3x = 10 - y^2 \ \Rightarrow \ x = -\frac{1}{3}y^2 + \frac{10}{3}$.
Interchange x and y: $y = -\frac{1}{3}x^2 + \frac{10}{3}$. So $f^{-1}(x) = -\frac{1}{3}x^2 + \frac{10}{3}$. Note that the domain of f^{-1} is $x \geq 0$.

23. $y = f(x) = e^{x^3} \ \Rightarrow \ \ln y = x^3 \ \Rightarrow \ x = \sqrt[3]{\ln y}$. Interchange x and y: $y = \sqrt[3]{\ln x}$. So $f^{-1}(x) = \sqrt[3]{\ln x}$.

25. $y = f(x) = \ln(x + 3) \ \Rightarrow \ x + 3 = e^y \ \Rightarrow \ x = e^y - 3$. Interchange x and y: $y = e^x - 3$. So $f^{-1}(x) = e^x - 3$.

27. $y = f(x) = x^4 + 1 \ \Rightarrow \ y - 1 = x^4 \ \Rightarrow \ x = \sqrt[4]{y - 1}$ (not $\pm$ since $x \geq 0$). Interchange x and y: $y = \sqrt[4]{x - 1}$. So $f^{-1}(x) = \sqrt[4]{x - 1}$. The graph of $y = \sqrt[4]{x - 1}$ is just the graph of $y = \sqrt[4]{x}$ shifted right one unit. From the graph, we see that f and f^{-1} are reflections about the line $y = x$.

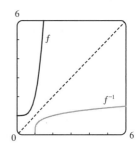

29. Reflect the graph of f about the line $y = x$. The points $(-1, -2)$, $(1, -1)$, $(2, 2)$, and $(3, 3)$ on f are reflected to $(-2, -1)$, $(-1, 1)$, $(2, 2)$, and $(3, 3)$ on f^{-1}.

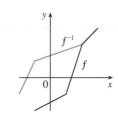

31. (a) It is defined as the inverse of the exponential function with base a, that is, $\log_a x = y \iff a^y = x$.

(b) $(0, \infty)$ (c) $\mathbb{R}$ (d) See Figure 11.

33. (a) $\log_5 125 = 3$ since $5^3 = 125$.   (b) $\log_3 \frac{1}{27} = -3$ since $3^{-3} = \frac{1}{3^3} = \frac{1}{27}$.

35. (a) $\log_2 6 - \log_2 15 + \log_2 20 = \log_2\left(\frac{6}{15}\right) + \log_2 20$ [by Law 2]

$$= \log_2\left(\frac{6}{15} \cdot 20\right) \qquad \text{[by Law 1]}$$

$$= \log_2 8, \text{ and } \log_2 8 = 3 \text{ since } 2^3 = 8.$$

(b) $\log_3 100 - \log_3 18 - \log_3 50 = \log_3\left(\frac{100}{18}\right) - \log_3 50 = \log_3\left(\frac{100}{18 \cdot 50}\right)$

$$= \log_3\left(\tfrac{1}{9}\right), \text{ and } \log_3\left(\tfrac{1}{9}\right) = -2 \text{ since } 3^{-2} = \tfrac{1}{9}.$$

37. $\ln 5 + 5\ln 3 = \ln 5 + \ln 3^5$ [by Law 3]

$$= \ln(5 \cdot 3^5) \qquad \text{[by Law 1]}$$

$$= \ln 1215$$

39. $\ln(1 + x^2) + \frac{1}{2}\ln x - \ln\sin x = \ln(1 + x^2) + \ln x^{1/2} - \ln\sin x = \ln[(1 + x^2)\sqrt{x}] - \ln\sin x = \ln\dfrac{(1 + x^2)\sqrt{x}}{\sin x}$

41. To graph these functions, we use $\log_{1.5} x = \dfrac{\ln x}{\ln 1.5}$ and $\log_{50} x = \dfrac{\ln x}{\ln 50}$.

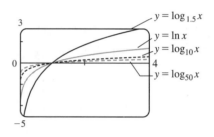

These graphs all approach $-\infty$ as $x \to 0^+$, and they all pass through the point $(1, 0)$. Also, they are all increasing, and all approach ∞ as $x \to \infty$. The functions with larger bases increase extremely slowly, and the ones with smaller bases do so somewhat more quickly. The functions with large bases approach the y-axis more closely as $x \to 0^+$.

43. 3 ft $= 36$ in, so we need x such that $\log_2 x = 36 \iff x = 2^{36} = 68{,}719{,}476{,}736$. In miles, this is

$$68{,}719{,}476{,}736 \text{ in} \cdot \frac{1 \text{ ft}}{12 \text{ in}} \cdot \frac{1 \text{ mi}}{5280 \text{ ft}} \approx 1{,}084{,}587.7 \text{ mi.}$$

45. (a) Shift the graph of $y = \log_{10} x$ five units to the left to obtain the graph of $y = \log_{10}(x + 5)$. Note the vertical asymptote of $x = -5$.

(b) Reflect the graph of $y = \ln x$ about the x-axis to obtain the graph of $y = -\ln x$.

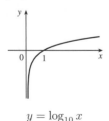

$y = \log_{10} x$

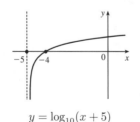

$y = \log_{10}(x + 5)$

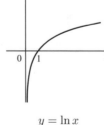

$y = \ln x$

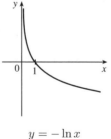

$y = -\ln x$

47. (a) $2\ln x = 1 \Rightarrow \ln x = \frac{1}{2} \Rightarrow x = e^{1/2} = \sqrt{e}$

(b) $e^{-x} = 5 \Rightarrow -x = \ln 5 \Rightarrow x = -\ln 5$

49. (a) $2^{x-5} = 3 \iff \log_2 3 = x - 5 \iff x = 5 + \log_2 3$.

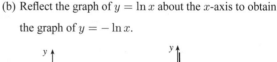

Or: $2^{x-5} = 3 \iff \ln(2^{x-5}) = \ln 3 \iff (x-5)\ln 2 = \ln 3 \iff x - 5 = \dfrac{\ln 3}{\ln 2} \iff x = 5 + \dfrac{\ln 3}{\ln 2}$

(b) $\ln x + \ln(x-1) = \ln(x(x-1)) = 1 \Leftrightarrow x(x-1) = e^1 \Leftrightarrow x^2 - x - e = 0$. The quadratic formula (with $a = 1$,

$b = -1$, and $c = -e$) gives $x = \frac{1}{2}\left(1 \pm \sqrt{1+4e}\right)$, but we reject the negative root since the natural logarithm is not

defined for $x < 0$. So $x = \frac{1}{2}\left(1 + \sqrt{1+4e}\right)$.

51. (a) $e^x < 10 \Rightarrow \ln e^x < \ln 10 \Rightarrow x < \ln 10 \Rightarrow x \in (-\infty, \ln 10)$

(b) $\ln x > -1 \Rightarrow e^{\ln x} > e^{-1} \Rightarrow x > e^{-1} \Rightarrow x \in (1/e, \infty)$

53. (a) For $f(x) = \sqrt{3 - e^{2x}}$, we must have $3 - e^{2x} \ge 0 \Rightarrow e^{2x} \le 3 \Rightarrow 2x \le \ln 3 \Rightarrow x \le \frac{1}{2}\ln 3$. Thus, the domain

of f is $(-\infty, \frac{1}{2}\ln 3]$.

(b) $y = f(x) = \sqrt{3 - e^{2x}}$ [note that $y \ge 0$] $\Rightarrow y^2 = 3 - e^{2x} \Rightarrow e^{2x} = 3 - y^2 \Rightarrow 2x = \ln(3 - y^2) \Rightarrow$

$x = \frac{1}{2}\ln(3 - y^2)$. Interchange x and y: $y = \frac{1}{2}\ln(3 - x^2)$. So $f^{-1}(x) = \frac{1}{2}\ln(3 - x^2)$. For the domain of f^{-1}, we must

have $3 - x^2 > 0 \Rightarrow x^2 < 3 \Rightarrow |x| < \sqrt{3} \Rightarrow -\sqrt{3} < x < \sqrt{3} \Rightarrow 0 \le x < \sqrt{3}$ since $x \ge 0$. Note that the

domain of f^{-1}, $[0, \sqrt{3})$, equals the range of f.

55. We see that the graph of $y = f(x) = \sqrt{x^3 + x^2 + x + 1}$ is increasing, so f is 1-1.

Enter $x = \sqrt{y^3 + y^2 + y + 1}$ and use your CAS to solve the equation for y.

Using Derive, we get two (irrelevant) solutions involving imaginary expressions,

as well as one which can be simplified to the following:

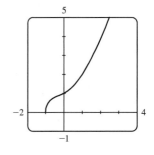

$$y = f^{-1}(x) = -\frac{\sqrt[3]{4}}{6}\left(\sqrt[3]{D - 27x^2 + 20} - \sqrt[3]{D + 27x^2 - 20} + \sqrt[3]{2}\right)$$

where $D = 3\sqrt{3}\sqrt{27x^4 - 40x^2 + 16}$.

Maple and Mathematica each give two complex expressions and one real expression, and the real expression is equivalent

to that given by Derive. For example, Maple's expression simplifies to $\dfrac{1}{6}\dfrac{M^{2/3} - 8 - 2M^{1/3}}{2M^{1/3}}$, where

$M = 108x^2 + 12\sqrt{48 - 120x^2 + 81x^4} - 80$.

57. (a) $n = 100 \cdot 2^{t/3} \Rightarrow \dfrac{n}{100} = 2^{t/3} \Rightarrow \log_2\left(\dfrac{n}{100}\right) = \dfrac{t}{3} \Rightarrow t = 3\log_2\left(\dfrac{n}{100}\right)$. Using formula (10), we can write

this as $t = f^{-1}(n) = 3 \cdot \dfrac{\ln(n/100)}{\ln 2}$. This function tells us how long it will take to obtain n bacteria (given the number n).

(b) $n = 50,000 \Rightarrow t = f^{-1}(50,000) = 3 \cdot \dfrac{\ln\left(\frac{50,000}{100}\right)}{\ln 2} = 3\left(\dfrac{\ln 500}{\ln 2}\right) \approx 26.9$ hours

59. (a) $\sin^{-1}\left(\frac{\sqrt{3}}{2}\right) = \frac{\pi}{3}$ since $\sin\frac{\pi}{3} = \frac{\sqrt{3}}{2}$ and $\frac{\pi}{3}$ is in $\left[-\frac{\pi}{2}, \frac{\pi}{2}\right]$.

(b) $\cos^{-1}(-1) = \pi$ since $\cos\pi = -1$ and π is in $[0, \pi]$.

61. (a) $\arctan 1 = \frac{\pi}{4}$ since $\tan\frac{\pi}{4} = 1$ and $\frac{\pi}{4}$ is in $\left(-\frac{\pi}{2}, \frac{\pi}{2}\right)$.

(b) $\sin^{-1}\frac{1}{\sqrt{2}} = \frac{\pi}{4}$ since $\sin\frac{\pi}{4} = \frac{1}{\sqrt{2}}$ and $\frac{\pi}{4}$ is in $\left[-\frac{\pi}{2}, \frac{\pi}{2}\right]$.

63. (a) In general, $\tan(\arctan x) = x$ for any real number x. Thus, $\tan(\arctan 10) = 10$.

(b) $\sin^{-1}\left(\sin \frac{7\pi}{3}\right) = \sin^{-1}\left(\sin \frac{\pi}{3}\right) = \sin^{-1}\frac{\sqrt{3}}{2} = \frac{\pi}{3}$ since $\sin \frac{\pi}{3} = \frac{\sqrt{3}}{2}$ and $\frac{\pi}{3}$ is in $\left[-\frac{\pi}{2}, \frac{\pi}{2}\right]$.

[Recall that $\frac{7\pi}{3} = \frac{\pi}{3} + 2\pi$ and the sine function is periodic with period 2π.]

65. Let $y = \sin^{-1} x$. Then $-\frac{\pi}{2} \le y \le \frac{\pi}{2} \quad \Rightarrow \quad \cos y \ge 0$, so $\cos(\sin^{-1} x) = \cos y = \sqrt{1 - \sin^2 y} = \sqrt{1 - x^2}$.

67. Let $y = \tan^{-1} x$. Then $\tan y = x$, so from the triangle we see that

$$\sin(\tan^{-1} x) = \sin y = \frac{x}{\sqrt{1 + x^2}}.$$

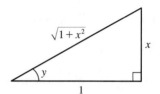

69.

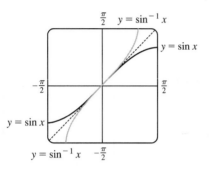

The graph of $\sin^{-1} x$ is the reflection of the graph of $\sin x$ about the line $y = x$.

71. $g(x) = \sin^{-1}(3x + 1)$.

Domain $(g) = \{x \mid -1 \le 3x + 1 \le 1\} = \{x \mid -2 \le 3x \le 0\} = \{x \mid -\frac{2}{3} \le x \le 0\} = \left[-\frac{2}{3}, 0\right]$.

Range $(g) = \{y \mid -\frac{\pi}{2} \le y \le \frac{\pi}{2}\} = \left[-\frac{\pi}{2}, \frac{\pi}{2}\right]$.

73. (a) If the point (x, y) is on the graph of $y = f(x)$, then the point $(x - c, y)$ is that point shifted c units to the left. Since f is

1-1, the point (y, x) is on the graph of $y = f^{-1}(x)$ and the point corresponding to $(x - c, y)$ on the graph of f is

$(y, x - c)$ on the graph of f^{-1}. Thus, the curve's reflection is shifted *down* the same number of units as the curve itself is

shifted to the left. So an expression for the inverse function is $g^{-1}(x) = f^{-1}(x) - c$.

(b) If we compress (or stretch) a curve horizontally, the curve's reflection in the line $y = x$ is compressed (or stretched)

vertically by the same factor. Using this geometric principle, we see that the inverse of $h(x) = f(cx)$ can be expressed as

$h^{-1}(x) = (1/c) f^{-1}(x)$.

1 Review

CONCEPT CHECK

1. (a) A **function** f is a rule that assigns to each element x in a set A exactly one element, called $f(x)$, in a set B. The set A is called the **domain** of the function. The **range** of f is the set of all possible values of $f(x)$ as x varies throughout the domain.

   (b) If f is a function with domain A, then its **graph** is the set of ordered pairs $\{(x, f(x)) \mid x \in A\}$.

   (c) Use the Vertical Line Test on page 16.

2. The four ways to represent a function are: verbally, numerically, visually, and algebraically. An example of each is given below.

   Verbally: An assignment of students to chairs in a classroom (a description in words)

   Numerically: A tax table that assigns an amount of tax to an income (a table of values)

   Visually: A graphical history of the Dow Jones average (a graph)

   Algebraically: A relationship between distance, rate, and time: $d = rt$ (an explicit formula)

3. (a) An **even function** f satisfies $f(-x) = f(x)$ for every number x in its domain. It is symmetric with respect to the y-axis.

   (b) An **odd function** g satisfies $g(-x) = -g(x)$ for every number x in its domain. It is symmetric with respect to the origin.

4. A function f is called **increasing** on an interval I if $f(x_1) < f(x_2)$ whenever $x_1 < x_2$ in I.

5. A **mathematical model** is a mathematical description (often by means of a function or an equation) of a real-world phenomenon.

6. (a) Linear function: $f(x) = 2x + 1$, $f(x) = ax + b$

   (b) Power function: $f(x) = x^2$, $f(x) = x^a$

   (c) Exponential function: $f(x) = 2^x$, $f(x) = a^x$

   (d) Quadratic function: $f(x) = x^2 + x + 1$, $f(x) = ax^2 + bx + c$

   (e) Polynomial of degree 5: $f(x) = x^5 + 2$

   (f) Rational function: $f(x) = \dfrac{x}{x+2}$, $f(x) = \dfrac{P(x)}{Q(x)}$ where $P(x)$ and $Q(x)$ are polynomials

7.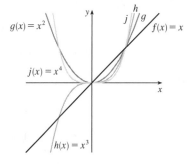

8. (a)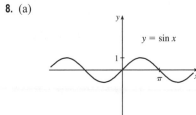

   (b)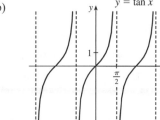

(c)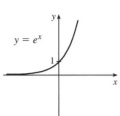

(d)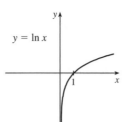

(e)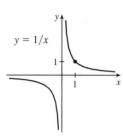

(f)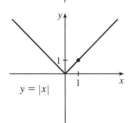

(g)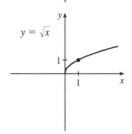

(h)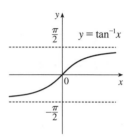

9. (a) The domain of $f + g$ is the intersection of the domain of f and the domain of g; that is, $A \cap B$.

(b) The domain of fg is also $A \cap B$.

(c) The domain of f/g must exclude values of x that make g equal to 0; that is, $\{x \in A \cap B \mid g(x) \neq 0\}$.

10. Given two functions f and g, the **composite** function $f \circ g$ is defined by $(f \circ g)(x) = f(g(x))$. The domain of $f \circ g$ is the set of all x in the domain of g such that $g(x)$ is in the domain of f.

11. (a) If the graph of f is shifted 2 units upward, its equation becomes $y = f(x) + 2$.

(b) If the graph of f is shifted 2 units downward, its equation becomes $y = f(x) - 2$.

(c) If the graph of f is shifted 2 units to the right, its equation becomes $y = f(x - 2)$.

(d) If the graph of f is shifted 2 units to the left, its equation becomes $y = f(x + 2)$.

(e) If the graph of f is reflected about the x-axis, its equation becomes $y = -f(x)$.

(f) If the graph of f is reflected about the y-axis, its equation becomes $y = f(-x)$.

(g) If the graph of f is stretched vertically by a factor of 2, its equation becomes $y = 2f(x)$.

(h) If the graph of f is shrunk vertically by a factor of 2, its equation becomes $y = \frac{1}{2}f(x)$.

(i) If the graph of f is stretched horizontally by a factor of 2, its equation becomes $y = f\left(\frac{1}{2}x\right)$.

(j) If the graph of f is shrunk horizontally by a factor of 2, its equation becomes $y = f(2x)$.

12. (a) A function f is called a *one-to-one function* if it never takes on the same value twice; that is, if $f(x_1) \neq f(x_2)$ whenever $x_1 \neq x_2$. (Or, f is 1-1 if each output corresponds to only one input.)

Use the Horizontal Line Test: A function is one-to-one if and only if no horizontal line intersects its graph more than once.

(b) If f is a one-to-one function with domain A and range B, then its *inverse function* f^{-1} has domain B and range A and is defined by

$$f^{-1}(y) = x \quad \Leftrightarrow \quad f(x) = y$$

for any y in B. The graph of f^{-1} is obtained by reflecting the graph of f about the line $y = x$.

13. (a) The inverse sine function $f(x) = \sin^{-1} x$ is defined as follows:

$$\sin^{-1} x = y \quad \Leftrightarrow \quad \sin y = x \qquad \text{and} \qquad -\frac{\pi}{2} \le y \le \frac{\pi}{2}$$

Its domain is $-1 \le x \le 1$ and its range is $-\frac{\pi}{2} \le y \le \frac{\pi}{2}$.

(b) The inverse cosine function $f(x) = \cos^{-1} x$ is defined as follows:

$$\cos^{-1} x = y \quad \Leftrightarrow \quad \cos y = x \qquad \text{and} \qquad 0 \le y \le \pi$$

Its domain is $-1 \le x \le 1$ and its range is $0 \le y \le \pi$.

(c) The inverse tangent function $f(x) = \tan^{-1} x$ is defined as follows:

$$\tan^{-1} x = y \quad \Leftrightarrow \quad \tan y = x \qquad \text{and} \qquad -\frac{\pi}{2} < y < \frac{\pi}{2}$$

Its domain is $\mathbb{R}$ and its range is $-\frac{\pi}{2} < y < \frac{\pi}{2}$.

TRUE-FALSE QUIZ

1. False. Let $f(x) = x^2$, $s = -1$, and $t = 1$. Then $f(s+t) = (-1+1)^2 = 0^2 = 0$, but
$f(s) + f(t) = (-1)^2 + 1^2 = 2 \ne 0 = f(s+t)$.

3. False. Let $f(x) = x^2$. Then $f(3x) = (3x)^2 = 9x^2$ and $3f(x) = 3x^2$. So $f(3x) \ne 3f(x)$.

5. True. See the Vertical Line Test.

7. False. Let $f(x) = x^3$. Then f is one-to-one and $f^{-1}(x) = \sqrt[3]{x}$. But $1/f(x) = 1/x^3$, which is not equal to $f^{-1}(x)$.

9. True. The function $\ln x$ is an increasing function on $(0, \infty)$.

11. False. Let $x = e^2$ and $a = e$. Then $\dfrac{\ln x}{\ln a} = \dfrac{\ln e^2}{\ln e} = \dfrac{2 \ln e}{\ln e} = 2$ and $\ln \dfrac{x}{a} = \ln \dfrac{e^2}{e} = \ln e = 1$, so in general the statement
is false. What *is* true, however, is that $\ln \dfrac{x}{a} = \ln x - \ln a$.

13. False. For example, $\tan^{-1} 20$ is defined; $\sin^{-1} 20$ and $\cos^{-1} 20$ are not.

EXERCISES

1. (a) When $x = 2$, $y \approx 2.7$. Thus, $f(2) \approx 2.7$.

(b) $f(x) = 3 \quad \Rightarrow \quad x \approx 2.3, 5.6$

(c) The domain of f is $-6 \le x \le 6$, or $[-6, 6]$.

(d) The range of f is $-4 \le y \le 4$, or $[-4, 4]$.

(e) f is increasing on $[-4, 4]$, that is, on $-4 \le x \le 4$.

(f) f is not one-to-one since it fails the Horizontal Line Test.

(g) f is odd since its graph is symmetric about the origin.

3. $f(x) = x^2 - 2x + 3$, so $f(a+h) = (a+h)^2 - 2(a+h) + 3 = a^2 + 2ah + h^2 - 2a - 2h + 3$, and

$$\frac{f(a+h) - f(a)}{h} = \frac{(a^2 + 2ah + h^2 - 2a - 2h + 3) - (a^2 - 2a + 3)}{h} = \frac{h(2a + h - 2)}{h} = 2a + h - 2.$$

5. $f(x) = 2/(3x-1)$.   Domain: $3x - 1 \neq 0 \;\Rightarrow\; 3x \neq 1 \;\Rightarrow\; x \neq \frac{1}{3}.$ $D = \left(-\infty, \frac{1}{3}\right) \cup \left(\frac{1}{3}, \infty\right)$

Range: all reals except 0 ($y = 0$ is the horizontal asymptote for f.) $R = (-\infty, 0) \cup (0, \infty)$

7. $h(x) = \ln(x+6)$.   Domain: $x + 6 > 0 \;\Rightarrow\; x > -6.$ $D = (-6, \infty)$

Range: $x + 6 > 0$, so $\ln(x+6)$ takes on all real numbers and, hence, the range is $\mathbb{R}$.

$R = (-\infty, \infty)$

9. (a) To obtain the graph of $y = f(x) + 8$, we shift the graph of $y = f(x)$ up 8 units.

(b) To obtain the graph of $y = f(x+8)$, we shift the graph of $y = f(x)$ left 8 units.

(c) To obtain the graph of $y = 1 + 2f(x)$, we stretch the graph of $y = f(x)$ vertically by a factor of 2, and then shift the resulting graph 1 unit upward.

(d) To obtain the graph of $y = f(x-2) - 2$, we shift the graph of $y = f(x)$ right 2 units (for the "-2" inside the parentheses), and then shift the resulting graph 2 units downward.

(e) To obtain the graph of $y = -f(x)$, we reflect the graph of $y = f(x)$ about the x-axis.

(f) To obtain the graph of $y = f^{-1}(x)$, we reflect the graph of $y = f(x)$ about the line $y = x$ (assuming f is one-to-one).

11. $y = -\sin 2x$: Start with the graph of $y = \sin x$, compress horizontally by a factor of 2, and reflect about the x-axis.

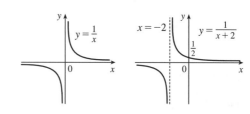

13. $y = \frac{1}{2}(1 + e^x)$:

Start with the graph of $y = e^x$, shift 1 unit upward, and compress vertically by a factor of 2.

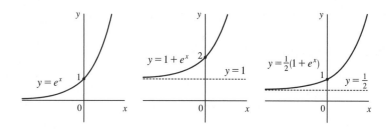

15. $f(x) = \dfrac{1}{x+2}$:

Start with the graph of $f(x) = 1/x$ and shift 2 units to the left.

17. (a) The terms of f are a mixture of odd and even powers of x, so f is neither even nor odd.

(b) The terms of f are all odd powers of x, so f is odd.

(c) $f(-x) = e^{-(-x)^2} = e^{-x^2} = f(x)$, so f is even.

(d) $f(-x) = 1 + \sin(-x) = 1 - \sin x$. Now $f(-x) \neq f(x)$ and $f(-x) \neq -f(x)$, so f is neither even nor odd.

19. $f(x) = \ln x, \quad D = (0, \infty); \quad g(x) = x^2 - 9, \quad D = \mathbb{R}$.

(a) $(f \circ g)(x) = f(g(x)) = f(x^2 - 9) = \ln(x^2 - 9)$.

　　Domain: $x^2 - 9 > 0 \;\Rightarrow\; x^2 > 9 \;\Rightarrow\; |x| > 3 \;\Rightarrow\; x \in (-\infty, -3) \cup (3, \infty)$

(b) $(g \circ f)(x) = g(f(x)) = g(\ln x) = (\ln x)^2 - 9.$　Domain: $x > 0$, or $(0, \infty)$

(c) $(f \circ f)(x) = f(f(x)) = f(\ln x) = \ln(\ln x).$　Domain: $\ln x > 0 \;\Rightarrow\; x > e^0 = 1$, or $(1, \infty)$

(d) $(g \circ g)(x) = g(g(x)) = g(x^2 - 9) = (x^2 - 9)^2 - 9.$　Domain: $x \in \mathbb{R}$, or $(-\infty, \infty)$

21.

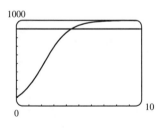

Many models appear to be plausible. Your choice depends on whether you think medical advances will keep increasing life expectancy, or if there is bound to be a natural leveling-off of life expectancy. A linear model, $y = 0.2493x - 423.4818$, gives us an estimate of 77.6 years for the year 2010.

23. We need to know the value of x such that $f(x) = 2x + \ln x = 2$. Since $x = 1$ gives us $y = 2$, $f^{-1}(2) = 1$.

25. (a) $e^{2\ln 3} = \left(e^{\ln 3}\right)^2 = 3^2 = 9$

(b) $\log_{10} 25 + \log_{10} 4 = \log_{10}(25 \cdot 4) = \log_{10} 100 = \log_{10} 10^2 = 2$

(c) $\tan\left(\arcsin \frac{1}{2}\right) = \tan \frac{\pi}{6} = \frac{1}{\sqrt{3}}$

(d) Let $\theta = \cos^{-1} \frac{4}{5}$, so $\cos \theta = \frac{4}{5}$. Then $\sin\left(\cos^{-1} \frac{4}{5}\right) = \sin \theta = \sqrt{1 - \cos^2 \theta} = \sqrt{1 - \left(\frac{4}{5}\right)^2} = \sqrt{\frac{9}{25}} = \frac{3}{5}$.

27. (a)

The population would reach 900 in about 4.4 years.

(b) $P = \dfrac{100{,}000}{100 + 900e^{-t}} \;\Rightarrow\; 100P + 900Pe^{-t} = 100{,}000 \;\Rightarrow\; 900Pe^{-t} = 100{,}000 - 100P \;\Rightarrow$

$e^{-t} = \dfrac{100{,}000 - 100P}{900P} \;\Rightarrow\; -t = \ln\left(\dfrac{1000 - P}{9P}\right) \;\Rightarrow\; t = -\ln\left(\dfrac{1000 - P}{9P}\right)$, or $\ln\left(\dfrac{9P}{1000 - P}\right)$; this is the time required for the population to reach a given number P.

(c) $P = 900 \;\Rightarrow\; t = \ln\left(\dfrac{9 \cdot 900}{1000 - 900}\right) = \ln 81 \approx 4.4$ years, as in part (a).

PRINCIPLES OF PROBLEM SOLVING

1.

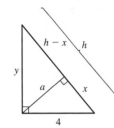

By using the area formula for a triangle, $\frac{1}{2}$ (base) (height), in two ways, we see that

$\frac{1}{2}(4)(y) = \frac{1}{2}(h)(a)$, so $a = \dfrac{4y}{h}$. Since $4^2 + y^2 = h^2$, $y = \sqrt{h^2 - 16}$, and

$$a = \frac{4\sqrt{h^2 - 16}}{h}.$$

3. $|2x - 1| = \begin{cases} 2x - 1 & \text{if } x \geq \frac{1}{2} \\ 1 - 2x & \text{if } x < \frac{1}{2} \end{cases}$ and $|x + 5| = \begin{cases} x + 5 & \text{if } x \geq -5 \\ -x - 5 & \text{if } x < -5 \end{cases}$

Therefore, we consider the three cases $x < -5$, $-5 \leq x < \frac{1}{2}$, and $x \geq \frac{1}{2}$.

If $x < -5$, we must have $1 - 2x - (-x - 5) = 3 \iff x = 3$, which is false, since we are considering $x < -5$.

If $-5 \leq x < \frac{1}{2}$, we must have $1 - 2x - (x + 5) = 3 \iff x = -\frac{7}{3}$.

If $x \geq \frac{1}{2}$, we must have $2x - 1 - (x + 5) = 3 \iff x = 9$.

So the two solutions of the equation are $x = -\frac{7}{3}$ and $x = 9$.

5. $f(x) = |x^2 - 4|x| + 3|$. If $x \geq 0$, then $f(x) = |x^2 - 4x + 3| = |(x - 1)(x - 3)|$.

    Case (i): If $0 < x \leq 1$, then $f(x) = x^2 - 4x + 3$.

    Case (ii): If $1 < x \leq 3$, then $f(x) = -(x^2 - 4x + 3) = -x^2 + 4x - 3$.

    Case (iii): If $x > 3$, then $f(x) = x^2 - 4x + 3$.

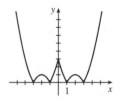

This enables us to sketch the graph for $x \geq 0$. Then we use the fact that f is an even function to reflect this part of the graph about the y-axis to obtain the entire graph. Or, we could consider also the cases $x < -3$, $-3 \leq x < -1$, and $-1 \leq x < 0$.

7. Remember that $|a| = a$ if $a \geq 0$ and that $|a| = -a$ if $a < 0$. Thus,

$$x + |x| = \begin{cases} 2x & \text{if } x \geq 0 \\ 0 & \text{if } x < 0 \end{cases} \quad \text{and} \quad y + |y| = \begin{cases} 2y & \text{if } y \geq 0 \\ 0 & \text{if } y < 0 \end{cases}$$

We will consider the equation $x + |x| = y + |y|$ in four cases.

(1) $x \geq 0, y \geq 0$	(2) $x \geq 0, y < 0$	(3) $x < 0, y \geq 0$	(4) $x < 0, y < 0$
$2x = 2y$	$2x = 0$	$0 = 2y$	$0 = 0$
$x = y$	$x = 0$	$0 = y$	

Case 1 gives us the line $y = x$ with nonnegative x and y.

Case 2 gives us the portion of the y-axis with y negative.

Case 3 gives us the portion of the x-axis with x negative.

Case 4 gives us the entire third quadrant.

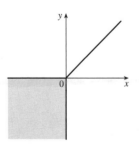

9. $|x| + |y| \le 1$. The boundary of the region has equation $|x| + |y| = 1$. In quadrants I, II, III, and IV, this becomes the lines $x + y = 1$, $-x + y = 1$, $-x - y = 1$, and $x - y = 1$ respectively.

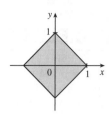

11. $(\log_2 3)(\log_3 4)(\log_4 5) \cdots (\log_{31} 32) = \left(\dfrac{\ln 3}{\ln 2}\right)\left(\dfrac{\ln 4}{\ln 3}\right)\left(\dfrac{\ln 5}{\ln 4}\right) \cdots \left(\dfrac{\ln 32}{\ln 31}\right) = \dfrac{\ln 32}{\ln 2} = \dfrac{\ln 2^5}{\ln 2} = \dfrac{5\ln 2}{\ln 2} = 5$

13. $\ln(x^2 - 2x - 2) \le 0 \;\Rightarrow\; x^2 - 2x - 2 \le e^0 = 1 \;\Rightarrow\; x^2 - 2x - 3 \le 0 \;\Rightarrow\; (x-3)(x+1) \le 0 \;\Rightarrow\; x \in [-1, 3]$. Since the argument must be positive, $x^2 - 2x - 2 > 0 \;\Rightarrow\; \left[x - \left(1 - \sqrt{3}\right)\right]\left[x - \left(1 + \sqrt{3}\right)\right] > 0 \;\Rightarrow\; x \in \left(-\infty, 1 - \sqrt{3}\right) \cup \left(1 + \sqrt{3}, \infty\right)$. The intersection of these intervals is $\left[-1, 1 - \sqrt{3}\right) \cup \left(1 + \sqrt{3}, 3\right]$.

15. Let d be the distance traveled on each half of the trip. Let t_1 and t_2 be the times taken for the first and second halves of the trip. For the first half of the trip we have $t_1 = d/30$ and for the second half we have $t_2 = d/60$. Thus, the average speed for the

entire trip is $\dfrac{\text{total distance}}{\text{total time}} = \dfrac{2d}{t_1 + t_2} = \dfrac{2d}{\dfrac{d}{30} + \dfrac{d}{60}} \cdot \dfrac{60}{60} = \dfrac{120d}{2d + d} = \dfrac{120d}{3d} = 40$. The average speed for the entire trip

is 40 mi/h.

17. Let S_n be the statement that $7^n - 1$ is divisible by 6.

- S_1 is true because $7^1 - 1 = 6$ is divisible by 6.
- Assume S_k is true, that is, $7^k - 1$ is divisible by 6. In other words, $7^k - 1 = 6m$ for some positive integer m. Then $7^{k+1} - 1 = 7^k \cdot 7 - 1 = (6m + 1) \cdot 7 - 1 = 42m + 6 = 6(7m + 1)$, which is divisible by 6, so S_{k+1} is true.
- Therefore, by mathematical induction, $7^n - 1$ is divisible by 6 for every positive integer n.

19. $f_0(x) = x^2$ and $f_{n+1}(x) = f_0(f_n(x))$ for $n = 0, 1, 2, \ldots$.

$f_1(x) = f_0(f_0(x)) = f_0\left(x^2\right) = \left(x^2\right)^2 = x^4$, $f_2(x) = f_0(f_1(x)) = f_0(x^4) = (x^4)^2 = x^8$,

$f_3(x) = f_0(f_2(x)) = f_0(x^8) = (x^8)^2 = x^{16}, \ldots$. Thus, a general formula is $f_n(x) = x^{2^{n+1}}$.

2 □ LIMITS AND DERIVATIVES

2.1 The Tangent and Velocity Problems

1. (a) Using $P(15, 250)$, we construct the following table:

t	Q	slope $= m_{PQ}$
5	$(5, 694)$	$\frac{694-250}{5-15} = -\frac{444}{10} = -44.4$
10	$(10, 444)$	$\frac{444-250}{10-15} = -\frac{194}{5} = -38.8$
20	$(20, 111)$	$\frac{111-250}{20-15} = -\frac{139}{5} = -27.8$
25	$(25, 28)$	$\frac{28-250}{25-15} = -\frac{222}{10} = -22.2$
30	$(30, 0)$	$\frac{0-250}{30-15} = -\frac{250}{15} = -16.\overline{6}$

(b) Using the values of t that correspond to the points closest to P ($t = 10$ and $t = 20$), we have

$$\frac{-38.8 + (-27.8)}{2} = -33.3$$

(c) From the graph, we can estimate the slope of the tangent line at P to be $\frac{-300}{9} = -33.\overline{3}$.

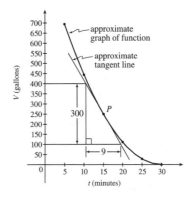

3. (a)

	x	Q	m_{PQ}
(i)	0.5	$(0.5, 0.333333)$	0.333333
(ii)	0.9	$(0.9, 0.473684)$	0.263158
(iii)	0.99	$(0.99, 0.497487)$	0.251256
(iv)	0.999	$(0.999, 0.499750)$	0.250125
(v)	1.5	$(1.5, 0.6)$	0.2
(vi)	1.1	$(1.1, 0.523810)$	0.238095
(vii)	1.01	$(1.01, 0.502488)$	0.248756
(viii)	1.001	$(1.001, 0.500250)$	0.249875

(b) The slope appears to be $\frac{1}{4}$.

(c) $y - \frac{1}{2} = \frac{1}{4}(x - 1)$ or $y = \frac{1}{4}x + \frac{1}{4}$.

5. (a) $y = y(t) = 40t - 16t^2$. At $t = 2$, $y = 40(2) - 16(2)^2 = 16$. The average velocity between times 2 and $2 + h$ is

$$v_{\text{ave}} = \frac{y(2 + h) - y(2)}{(2 + h) - 2} = \frac{[40(2 + h) - 16(2 + h)^2] - 16}{h} = \frac{-24h - 16h^2}{h} = -24 - 16h, \text{ if } h \neq 0.$$

(i) $[2, 2.5]$: $h = 0.5$, $v_{\text{ave}} = -32$ ft/s   (ii) $[2, 2.1]$: $h = 0.1$, $v_{\text{ave}} = -25.6$ ft/s

(iii) $[2, 2.05]$: $h = 0.05$, $v_{\text{ave}} = -24.8$ ft/s   (iv) $[2, 2.01]$: $h = 0.01$, $v_{\text{ave}} = -24.16$ ft/s

(b) The instantaneous velocity when $t = 2$ (h approaches 0) is -24 ft/s.

7. (a) (i) On the interval $[1, 3]$, $v_{\text{ave}} = \dfrac{s(3) - s(1)}{3 - 1} = \dfrac{10.7 - 1.4}{2} = \dfrac{9.3}{2} = 4.65$ m/s.

(ii) On the interval $[2, 3]$, $v_{\text{ave}} = \dfrac{s(3) - s(2)}{3 - 2} = \dfrac{10.7 - 5.1}{1} = 5.6$ m/s.

(iii) On the interval $[3, 5]$, $v_{\text{ave}} = \dfrac{s(5) - s(3)}{5 - 3} = \dfrac{25.8 - 10.7}{2} = \dfrac{15.1}{2} = 7.55$ m/s.

(iv) On the interval $[3, 4]$, $v_{\text{ave}} = \dfrac{s(4) - s(3)}{4 - 3} = \dfrac{17.7 - 10.7}{1} = 7$ m/s.

(b)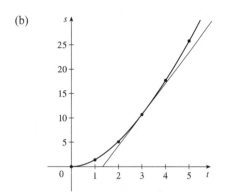

Using the points $(2, 4)$ and $(5, 23)$ from the approximate tangent line, the instantaneous velocity at $t = 3$ is about $\dfrac{23 - 4}{5 - 2} \approx 6.3$ m/s.

9. (a) For the curve $y = \sin(10\pi/x)$ and the point $P(1, 0)$:

x	Q	m_{PQ}
2	$(2, 0)$	0
1.5	$(1.5, 0.8660)$	1.7321
1.4	$(1.4, -0.4339)$	-1.0847
1.3	$(1.3, -0.8230)$	-2.7433
1.2	$(1.2, 0.8660)$	4.3301
1.1	$(1.1, -0.2817)$	-2.8173

x	Q	m_{PQ}
0.5	$(0.5, 0)$	0
0.6	$(0.6, 0.8660)$	-2.1651
0.7	$(0.7, 0.7818)$	-2.6061
0.8	$(0.8, 1)$	-5
0.9	$(0.9, -0.3420)$	3.4202

As x approaches 1, the slopes do not appear to be approaching any particular value.

(b)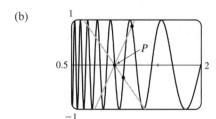

We see that problems with estimation are caused by the frequent oscillations of the graph. The tangent is so steep at P that we need to take x-values much closer to 1 in order to get accurate estimates of its slope.

(c) If we choose $x = 1.001$, then the point Q is $(1.001, -0.0314)$ and $m_{PQ} \approx -31.3794$. If $x = 0.999$, then Q is $(0.999, 0.0314)$ and $m_{PQ} = -31.4422$. The average of these slopes is -31.4108. So we estimate that the slope of the tangent line at P is about -31.4.

2.2   The Limit of a Function

1. As x approaches 2, $f(x)$ approaches 5. [Or, the values of $f(x)$ can be made as close to 5 as we like by taking x sufficiently close to 2 (but $x \neq 2$).] Yes, the graph could have a hole at $(2, 5)$ and be defined such that $f(2) = 3$.

3. (a) $\lim\limits_{x \to -3} f(x) = \infty$ means that the values of $f(x)$ can be made arbitrarily large (as large as we please) by taking x sufficiently close to -3 (but not equal to -3).

 (b) $\lim\limits_{x \to 4^+} f(x) = -\infty$ means that the values of $f(x)$ can be made arbitrarily large negative by taking x sufficiently close to 4 through values larger than 4.

5. (a) $f(x)$ approaches 2 as x approaches 1 from the left, so $\lim\limits_{x \to 1^-} f(x) = 2$.

 (b) $f(x)$ approaches 3 as x approaches 1 from the right, so $\lim\limits_{x \to 1^+} f(x) = 3$.

 (c) $\lim\limits_{x \to 1} f(x)$ does not exist because the limits in part (a) and part (b) are not equal.

 (d) $f(x)$ approaches 4 as x approaches 5 from the left and from the right, so $\lim\limits_{x \to 5} f(x) = 4$.

 (e) $f(5)$ is not defined, so it doesn't exist.

7. (a) $\lim\limits_{t \to 0^-} g(t) = -1$ 　　　　　　　　　　 (b) $\lim\limits_{t \to 0^+} g(t) = -2$

 (c) $\lim\limits_{t \to 0} g(t)$ does not exist because the limits in part (a) and part (b) are not equal.

 (d) $\lim\limits_{t \to 2^-} g(t) = 2$ 　　　　　　　　　　 (e) $\lim\limits_{t \to 2^+} g(t) = 0$

 (f) $\lim\limits_{t \to 2} g(t)$ does not exist because the limits in part (d) and part (e) are not equal.

 (g) $g(2) = 1$ 　　　　　　　　　　　　　　 (h) $\lim\limits_{t \to 4} g(t) = 3$

9. (a) $\lim\limits_{x \to -7} f(x) = -\infty$ 　　　　 (b) $\lim\limits_{x \to -3} f(x) = \infty$ 　　　　 (c) $\lim\limits_{x \to 0} f(x) = \infty$

 (d) $\lim\limits_{x \to 6^-} f(x) = -\infty$ 　　　　 (e) $\lim\limits_{x \to 6^+} f(x) = \infty$

 (f) The equations of the vertical asymptotes are $x = -7$, $x = -3$, $x = 0$, and $x = 6$.

11. (a) $\lim\limits_{x \to 0^-} f(x) = 1$

 (b) $\lim\limits_{x \to 0^+} f(x) = 0$

 (c) $\lim\limits_{x \to 0} f(x)$ does not exist because the limits in part (a) and part (b) are not equal.

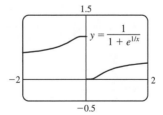

13. $\lim\limits_{x \to 1^-} f(x) = 2$, 　$\lim\limits_{x \to 1^+} f(x) = -2$, 　$f(1) = 2$

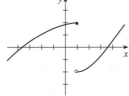

15. $\lim\limits_{x \to 3^+} f(x) = 4$, 　$\lim\limits_{x \to 3^-} f(x) = 2$, $\lim\limits_{x \to -2} f(x) = 2$,

 $f(3) = 3$, 　$f(-2) = 1$

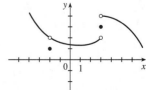

17. For $f(x) = \dfrac{x^2 - 2x}{x^2 - x - 2}$:

x	$f(x)$
2.5	0.714286
2.1	0.677419
2.05	0.672131
2.01	0.667774
2.005	0.667221
2.001	0.666778

x	$f(x)$
1.9	0.655172
1.95	0.661017
1.99	0.665552
1.995	0.666110
1.999	0.666556

It appears that $\lim\limits_{x \to 2} \dfrac{x^2 - 2x}{x^2 - x - 2} = 0.\bar{6} = \frac{2}{3}$.

19. For $f(x) = \dfrac{e^x - 1 - x}{x^2}$:

x	$f(x)$
1	0.718282
0.5	0.594885
0.1	0.517092
0.05	0.508439
0.01	0.501671

x	$f(x)$
-1	0.367879
-0.5	0.426123
-0.1	0.483742
-0.05	0.491770
-0.01	0.498337

It appears that $\lim\limits_{x \to 0} \dfrac{e^x - 1 - x}{x^2} = 0.5 = \frac{1}{2}$.

21. For $f(x) = \dfrac{\sqrt{x+4} - 2}{x}$:

x	$f(x)$
1	0.236068
0.5	0.242641
0.1	0.248457
0.05	0.249224
0.01	0.249844

x	$f(x)$
-1	0.267949
-0.5	0.258343
-0.1	0.251582
-0.05	0.250786
-0.01	0.250156

It appears that $\lim\limits_{x \to 0} \dfrac{\sqrt{x+4} - 2}{x} = 0.25 = \frac{1}{4}$.

23. For $f(x) = \dfrac{x^6 - 1}{x^{10} - 1}$:

x	$f(x)$
0.5	0.985337
0.9	0.719397
0.95	0.660186
0.99	0.612018
0.999	0.601200

x	$f(x)$
1.5	0.183369
1.1	0.484119
1.05	0.540783
1.01	0.588022
1.001	0.598800

It appears that $\lim\limits_{x \to 1} \dfrac{x^6 - 1}{x^{10} - 1} = 0.6 = \frac{3}{5}$.

25. $\lim\limits_{x \to -3^+} \dfrac{x + 2}{x + 3} = -\infty$ since the numerator is negative and the denominator approaches 0 from the positive side as $x \to -3^+$.

27. $\lim\limits_{x \to 1} \dfrac{2 - x}{(x - 1)^2} = \infty$ since the numerator is positive and the denominator approaches 0 through positive values as $x \to 1$.

29. Let $t = x^2 - 9$. Then as $x \to 3^+$, $t \to 0^+$, and $\lim\limits_{x \to 3^+} \ln(x^2 - 9) = \lim\limits_{t \to 0^+} \ln t = -\infty$ by (3).

31. $\lim\limits_{x \to 2\pi^-} x \csc x = \lim\limits_{x \to 2\pi^-} \dfrac{x}{\sin x} = -\infty$ since the numerator is positive and the denominator approaches 0 through negative values as $x \to 2\pi^-$.

33. (a) $f(x) = \dfrac{1}{x^3 - 1}$.

From these calculations, it seems that

$$\lim\limits_{x \to 1^-} f(x) = -\infty \quad \text{and} \quad \lim\limits_{x \to 1^+} f(x) = \infty.$$

x	$f(x)$
0.5	-1.14
0.9	-3.69
0.99	-33.7
0.999	-333.7
0.9999	-3333.7
0.99999	$-33{,}333.7$

x	$f(x)$
1.5	0.42
1.1	3.02
1.01	33.0
1.001	333.0
1.0001	3333.0
1.00001	33{,}333.3

(b) If x is slightly smaller than 1, then $x^3 - 1$ will be a negative number close to 0, and the reciprocal of $x^3 - 1$, that is, $f(x)$, will be a negative number with large absolute value. So $\lim\limits_{x \to 1^-} f(x) = -\infty$.

If x is slightly larger than 1, then $x^3 - 1$ will be a small positive number, and its reciprocal, $f(x)$, will be a large positive number. So $\lim\limits_{x \to 1^+} f(x) = \infty$.

(c) It appears from the graph of f that

$$\lim\limits_{x \to 1^-} f(x) = -\infty \text{ and } \lim\limits_{x \to 1^+} f(x) = \infty.$$

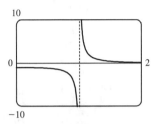

35. (a) Let $h(x) = (1+x)^{1/x}$.

(b)

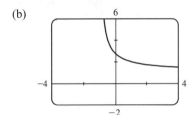

x	$h(x)$
-0.001	2.71964
-0.0001	2.71842
-0.00001	2.71830
-0.000001	2.71828
0.000001	2.71828
0.00001	2.71827
0.0001	2.71815
0.001	2.71692

It appears that $\lim\limits_{x \to 0} (1+x)^{1/x} \approx 2.71828$, which is approximately e.

In Section 3.6 we will see that the value of the limit is exactly e.

37. For $f(x) = x^2 - (2^x/1000)$:
(a)

x	$f(x)$
1	0.998000
0.8	0.638259
0.6	0.358484
0.4	0.158680
0.2	0.038851
0.1	0.008928
0.05	0.001465

It appears that $\lim\limits_{x \to 0} f(x) = 0$.

(b)

x	$f(x)$
0.04	0.000572
0.02	-0.000614
0.01	-0.000907
0.005	-0.000978
0.003	-0.000993
0.001	-0.001000

It appears that $\lim\limits_{x \to 0} f(x) = -0.001$.

39. No matter how many times we zoom in toward the origin, the graphs of $f(x) = \sin(\pi/x)$ appear to consist of almost-vertical lines. This indicates more and more frequent oscillations as $x \to 0$.

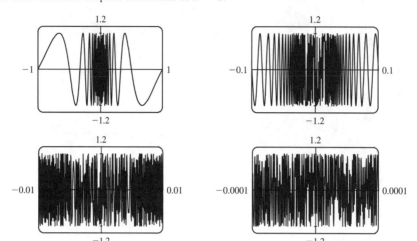

41.

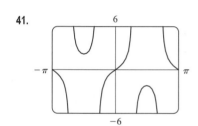

There appear to be vertical asymptotes of the curve $y = \tan(2\sin x)$ at $x \approx \pm 0.90$ and $x \approx \pm 2.24$. To find the exact equations of these asymptotes, we note that the graph of the tangent function has vertical asymptotes at $x = \frac{\pi}{2} + \pi n$. Thus, we must have $2\sin x = \frac{\pi}{2} + \pi n$, or equivalently, $\sin x = \frac{\pi}{4} + \frac{\pi}{2} n$. Since $-1 \leq \sin x \leq 1$, we must have $\sin x = \pm \frac{\pi}{4}$ and so $x = \pm \sin^{-1} \frac{\pi}{4}$ (corresponding to $x \approx \pm 0.90$). Just as $150°$ is the reference angle for $30°$, $\pi - \sin^{-1} \frac{\pi}{4}$ is the reference angle for $\sin^{-1} \frac{\pi}{4}$. So $x = \pm \left(\pi - \sin^{-1} \frac{\pi}{4} \right)$ are also equations of vertical asymptotes (corresponding to $x \approx \pm 2.24$).

2.3 Calculating Limits Using the Limit Laws

1. (a) $\lim\limits_{x \to 2} [f(x) + 5g(x)] = \lim\limits_{x \to 2} f(x) + \lim\limits_{x \to 2} [5g(x)]$ [Limit Law 1]

$\qquad\qquad\qquad\qquad\quad = \lim\limits_{x \to 2} f(x) + 5 \lim\limits_{x \to 2} g(x)$ [Limit Law 3]

$\qquad\qquad\qquad\qquad\quad = 4 + 5(-2) = -6$

(b) $\lim\limits_{x \to 2} [g(x)]^3 = \left[\lim\limits_{x \to 2} g(x) \right]^3$ [Limit Law 6]

$\qquad\qquad\quad = (-2)^3 = -8$

(c) $\lim\limits_{x \to 2} \sqrt{f(x)} = \sqrt{\lim\limits_{x \to 2} f(x)}$ [Limit Law 11]

$\qquad\qquad\quad = \sqrt{4} = 2$

(d) $\displaystyle\lim_{x\to2}\frac{3f(x)}{g(x)}=\frac{\displaystyle\lim_{x\to2}[3f(x)]}{\displaystyle\lim_{x\to2}g(x)}$ [Limit Law 5]

$\displaystyle=\frac{3\displaystyle\lim_{x\to2}f(x)}{\displaystyle\lim_{x\to2}g(x)}$ [Limit Law 3]

$\displaystyle=\frac{3(4)}{-2}=-6$

(e) Because the limit of the denominator is 0, we can't use Limit Law 5. The given limit, $\displaystyle\lim_{x\to2}\frac{g(x)}{h(x)}$, does not exist because the denominator approaches 0 while the numerator approaches a nonzero number.

(f) $\displaystyle\lim_{x\to2}\frac{g(x)\,h(x)}{f(x)}=\frac{\displaystyle\lim_{x\to2}[g(x)\,h(x)]}{\displaystyle\lim_{x\to2}f(x)}$ [Limit Law 5]

$\displaystyle=\frac{\displaystyle\lim_{x\to2}g(x)\cdot\lim_{x\to2}h(x)}{\displaystyle\lim_{x\to2}f(x)}$ [Limit Law 4]

$\displaystyle=\frac{-2\cdot0}{4}=0$

3. $\displaystyle\lim_{x\to-2}(3x^4+2x^2-x+1)=\lim_{x\to-2}3x^4+\lim_{x\to-2}2x^2-\lim_{x\to-2}x+\lim_{x\to-2}1$ [Limit Laws 1 and 2]

$\displaystyle=3\lim_{x\to-2}x^4+2\lim_{x\to-2}x^2-\lim_{x\to-2}x+\lim_{x\to-2}1$ [3]

$=3(-2)^4+2(-2)^2-(-2)+(1)$ [9, 8, and 7]

$=48+8+2+1=59$

5. $\displaystyle\lim_{x\to8}(1+\sqrt[3]{x})(2-6x^2+x^3)=\lim_{x\to8}(1+\sqrt[3]{x})\cdot\lim_{x\to8}(2-6x^2+x^3)$ [Limit Law 4]

$\displaystyle=\left(\lim_{x\to8}1+\lim_{x\to8}\sqrt[3]{x}\right)\cdot\left(\lim_{x\to8}2-6\lim_{x\to8}x^2+\lim_{x\to8}x^3\right)$ [1, 2, and 3]

$=(1+\sqrt[3]{8})\cdot(2-6\cdot8^2+8^3)$ [7, 10, 9]

$=(3)(130)=390$

7. $\displaystyle\lim_{x\to1}\left(\frac{1+3x}{1+4x^2+3x^4}\right)^3=\left(\lim_{x\to1}\frac{1+3x}{1+4x^2+3x^4}\right)^3$ [6]

$\displaystyle=\left[\frac{\displaystyle\lim_{x\to1}(1+3x)}{\displaystyle\lim_{x\to1}(1+4x^2+3x^4)}\right]^3$ [5]

$\displaystyle=\left[\frac{\displaystyle\lim_{x\to1}1+3\lim_{x\to1}x}{\displaystyle\lim_{x\to1}1+4\lim_{x\to1}x^2+3\lim_{x\to1}x^4}\right]^3$ [2, 1, and 3]

$\displaystyle=\left[\frac{1+3(1)}{1+4(1)^2+3(1)^4}\right]^3=\left[\frac{4}{8}\right]^3=\left(\frac{1}{2}\right)^3=\frac{1}{8}$ [7, 8, and 9]

9. $\displaystyle\lim_{x\to4^-}\sqrt{16-x^2}=\sqrt{\lim_{x\to4^-}(16-x^2)}$ [11]

$\displaystyle=\sqrt{\lim_{x\to4^-}16-\lim_{x\to4^-}x^2}$ [2]

$=\sqrt{16-(4)^2}=0$ [7 and 9]

11. $\displaystyle\lim_{x \to 2} \frac{x^2 + x - 6}{x - 2} = \lim_{x \to 2} \frac{(x + 3)(x - 2)}{x - 2} = \lim_{x \to 2} (x + 3) = 2 + 3 = 5$

13. $\displaystyle\lim_{x \to 2} \frac{x^2 - x + 6}{x - 2}$ does not exist since $x - 2 \to 0$ but $x^2 - x + 6 \to 8$ as $x \to 2$.

15. $\displaystyle\lim_{t \to -3} \frac{t^2 - 9}{2t^2 + 7t + 3} = \lim_{t \to -3} \frac{(t + 3)(t - 3)}{(2t + 1)(t + 3)} = \lim_{t \to -3} \frac{t - 3}{2t + 1} = \frac{-3 - 3}{2(-3) + 1} = \frac{-6}{-5} = \frac{6}{5}$

17. $\displaystyle\lim_{h \to 0} \frac{(4 + h)^2 - 16}{h} = \lim_{h \to 0} \frac{(16 + 8h + h^2) - 16}{h} = \lim_{h \to 0} \frac{8h + h^2}{h} = \lim_{h \to 0} \frac{h(8 + h)}{h} = \lim_{h \to 0} (8 + h) = 8 + 0 = 8$

19. By the formula for the sum of cubes, we have

$$\lim_{x \to -2} \frac{x + 2}{x^3 + 8} = \lim_{x \to -2} \frac{x + 2}{(x + 2)(x^2 - 2x + 4)} = \lim_{x \to -2} \frac{1}{x^2 - 2x + 4} = \frac{1}{4 + 4 + 4} = \frac{1}{12}.$$

21. $\displaystyle\lim_{t \to 9} \frac{9 - t}{3 - \sqrt{t}} = \lim_{t \to 9} \frac{\left(3 + \sqrt{t}\right)\left(3 - \sqrt{t}\right)}{3 - \sqrt{t}} = \lim_{t \to 9} \left(3 + \sqrt{t}\right) = 3 + \sqrt{9} = 6$

23. $\displaystyle\lim_{x \to 7} \frac{\sqrt{x + 2} - 3}{x - 7} = \lim_{x \to 7} \frac{\sqrt{x + 2} - 3}{x - 7} \cdot \frac{\sqrt{x + 2} + 3}{\sqrt{x + 2} + 3} = \lim_{x \to 7} \frac{(x + 2) - 9}{(x - 7)\left(\sqrt{x + 2} + 3\right)}$

$$= \lim_{x \to 7} \frac{x - 7}{(x - 7)\left(\sqrt{x + 2} + 3\right)} = \lim_{x \to 7} \frac{1}{\sqrt{x + 2} + 3} = \frac{1}{\sqrt{9} + 3} = \frac{1}{6}$$

25. $\displaystyle\lim_{x \to -4} \frac{\frac{1}{4} + \frac{1}{x}}{4 + x} = \lim_{x \to -4} \frac{\frac{x + 4}{4x}}{4 + x} = \lim_{x \to -4} \frac{x + 4}{4x(4 + x)} = \lim_{x \to -4} \frac{1}{4x} = \frac{1}{4(-4)} = -\frac{1}{16}$

27. $\displaystyle\lim_{x \to 16} \frac{4 - \sqrt{x}}{16x - x^2} = \lim_{x \to 16} \frac{(4 - \sqrt{x})(4 + \sqrt{x})}{(16x - x^2)(4 + \sqrt{x})} = \lim_{x \to 16} \frac{16 - x}{x(16 - x)(4 + \sqrt{x})}$

$$= \lim_{x \to 16} \frac{1}{x(4 + \sqrt{x})} = \frac{1}{16\left(4 + \sqrt{16}\right)} = \frac{1}{16(8)} = \frac{1}{128}$$

29. $\displaystyle\lim_{t \to 0} \left(\frac{1}{t\sqrt{1 + t}} - \frac{1}{t}\right) = \lim_{t \to 0} \frac{1 - \sqrt{1 + t}}{t\sqrt{1 + t}} = \lim_{t \to 0} \frac{\left(1 - \sqrt{1 + t}\right)\left(1 + \sqrt{1 + t}\right)}{t\sqrt{t + 1}\left(1 + \sqrt{1 + t}\right)} = \lim_{t \to 0} \frac{-t}{t\sqrt{1 + t}\left(1 + \sqrt{1 + t}\right)}$

$$= \lim_{t \to 0} \frac{-1}{\sqrt{1 + t}\left(1 + \sqrt{1 + t}\right)} = \frac{-1}{\sqrt{1 + 0}\left(1 + \sqrt{1 + 0}\right)} = -\frac{1}{2}$$

31. (a)

$$\lim_{x \to 0} \frac{x}{\sqrt{1 + 3x} - 1} \approx \frac{2}{3}$$

(b)

x	$f(x)$
-0.001	0.6661663
-0.0001	0.6666167
-0.00001	0.6666617
-0.000001	0.6666662
0.000001	0.6666672
0.00001	0.6666717
0.0001	0.6667167
0.001	0.6671663

The limit appears to be $\dfrac{2}{3}$.

(c) $\displaystyle\lim_{x\to0}\left(\frac{x}{\sqrt{1+3x}-1}\cdot\frac{\sqrt{1+3x}+1}{\sqrt{1+3x}+1}\right)=\lim_{x\to0}\frac{x\left(\sqrt{1+3x}+1\right)}{(1+3x)-1}=\lim_{x\to0}\frac{x\left(\sqrt{1+3x}+1\right)}{3x}$

$$=\frac{1}{3}\lim_{x\to0}\left(\sqrt{1+3x}+1\right) \qquad\qquad\text{[Limit Law 3]}$$

$$=\frac{1}{3}\left[\sqrt{\lim_{x\to0}(1+3x)}+\lim_{x\to0}1\right] \qquad\qquad\text{[1 and 11]}$$

$$=\frac{1}{3}\left(\sqrt{\lim_{x\to0}1+3\lim_{x\to0}x}+1\right) \qquad\qquad\text{[1, 3, and 7]}$$

$$=\frac{1}{3}\left(\sqrt{1+3\cdot0}+1\right) \qquad\qquad\text{[7 and 8]}$$

$$=\frac{1}{3}(1+1)=\frac{2}{3}$$

33. Let $f(x)=-x^2$, $g(x)=x^2\cos20\pi x$ and $h(x)=x^2$. Then

$-1\le\cos20\pi x\le1 \quad\Rightarrow\quad -x^2\le x^2\cos20\pi x\le x^2 \quad\Rightarrow\quad f(x)\le g(x)\le h(x)$.

So since $\displaystyle\lim_{x\to0}f(x)=\lim_{x\to0}h(x)=0$, by the Squeeze Theorem we have

$\displaystyle\lim_{x\to0}g(x)=0$.

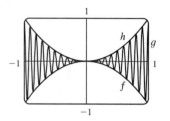

35. We have $\displaystyle\lim_{x\to4}(4x-9)=4(4)-9=7$ and $\displaystyle\lim_{x\to4}\left(x^2-4x+7\right)=4^2-4(4)+7=7$. Since $4x-9\le f(x)\le x^2-4x+7$

for $x\ge0$, $\displaystyle\lim_{x\to4}f(x)=7$ by the Squeeze Theorem.

37. $-1\le\cos(2/x)\le1 \quad\Rightarrow\quad -x^4\le x^4\cos(2/x)\le x^4$. Since $\displaystyle\lim_{x\to0}\left(-x^4\right)=0$ and $\displaystyle\lim_{x\to0}x^4=0$, we have

$\displaystyle\lim_{x\to0}\left[x^4\cos(2/x)\right]=0$ by the Squeeze Theorem.

39. $|x-3|=\begin{cases}x-3 & \text{if } x-3\ge0 \\ -(x-3) & \text{if } x-3<0\end{cases}=\begin{cases}x-3 & \text{if } x\ge3 \\ 3-x & \text{if } x<3\end{cases}$

Thus, $\displaystyle\lim_{x\to3^+}(2x+|x-3|)=\lim_{x\to3^+}(2x+x-3)=\lim_{x\to3^+}(3x-3)=3(3)-3=6$ and

$\displaystyle\lim_{x\to3^-}(2x+|x-3|)=\lim_{x\to3^-}(2x+3-x)=\lim_{x\to3^-}(x+3)=3+3=6$. Since the left and right limits are equal,

$\displaystyle\lim_{x\to3}(2x+|x-3|)=6$.

41. $\left|2x^3-x^2\right|=\left|x^2(2x-1)\right|=\left|x^2\right|\cdot|2x-1|=x^2|2x-1|$

$|2x-1|=\begin{cases}2x-1 & \text{if } 2x-1\ge0 \\ -(2x-1) & \text{if } 2x-1<0\end{cases}=\begin{cases}2x-1 & \text{if } x\ge0.5 \\ -(2x-1) & \text{if } x<0.5\end{cases}$

So $\left|2x^3-x^2\right|=x^2[-(2x-1)]$ for $x<0.5$.

Thus, $\displaystyle\lim_{x\to0.5^-}\frac{2x-1}{\left|2x^3-x^2\right|}=\lim_{x\to0.5^-}\frac{2x-1}{x^2[-(2x-1)]}=\lim_{x\to0.5^-}\frac{-1}{x^2}=\frac{-1}{(0.5)^2}=\frac{-1}{0.25}=-4$.

43. Since $|x| = -x$ for $x < 0$, we have $\lim\limits_{x \to 0^-} \left(\dfrac{1}{x} - \dfrac{1}{|x|} \right) = \lim\limits_{x \to 0^-} \left(\dfrac{1}{x} - \dfrac{1}{-x} \right) = \lim\limits_{x \to 0^-} \dfrac{2}{x}$, which does not exist since the

denominator approaches 0 and the numerator does not.

45. (a)

(b) (i) Since $\operatorname{sgn} x = 1$ for $x > 0$, $\lim\limits_{x \to 0^+} \operatorname{sgn} x = \lim\limits_{x \to 0^+} 1 = 1$.

(ii) Since $\operatorname{sgn} x = -1$ for $x < 0$, $\lim\limits_{x \to 0^-} \operatorname{sgn} x = \lim\limits_{x \to 0^-} -1 = -1$.

(iii) Since $\lim\limits_{x \to 0^-} \operatorname{sgn} x \neq \lim\limits_{x \to 0^+} \operatorname{sgn} x$, $\lim\limits_{x \to 0} \operatorname{sgn} x$ does not exist.

(iv) Since $|\operatorname{sgn} x| = 1$ for $x \neq 0$, $\lim\limits_{x \to 0} |\operatorname{sgn} x| = \lim\limits_{x \to 0} 1 = 1$.

47. (a) (i) $\lim\limits_{x \to 1^+} F(x) = \lim\limits_{x \to 1^+} \dfrac{x^2 - 1}{|x - 1|} = \lim\limits_{x \to 1^+} \dfrac{x^2 - 1}{x - 1} = \lim\limits_{x \to 1^+} (x + 1) = 2$

(c)

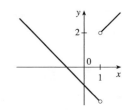

(ii) $\lim\limits_{x \to 1^-} F(x) = \lim\limits_{x \to 1^-} \dfrac{x^2 - 1}{|x - 1|} = \lim\limits_{x \to 1^-} \dfrac{x^2 - 1}{-(x - 1)} = \lim\limits_{x \to 1^-} -(x + 1) = -2$

(b) No, $\lim\limits_{x \to 1} F(x)$ does not exist since $\lim\limits_{x \to 1^+} F(x) \neq \lim\limits_{x \to 1^-} F(x)$.

49. (a) (i) $[\![x]\!] = -2$ for $-2 \leq x < -1$, so $\lim\limits_{x \to -2^+} [\![x]\!] = \lim\limits_{x \to -2^+} (-2) = -2$

(ii) $[\![x]\!] = -3$ for $-3 \leq x < -2$, so $\lim\limits_{x \to -2^-} [\![x]\!] = \lim\limits_{x \to -2^-} (-3) = -3$.

The right and left limits are different, so $\lim\limits_{x \to -2} [\![x]\!]$ does not exist.

(iii) $[\![x]\!] = -3$ for $-3 \leq x < -2$, so $\lim\limits_{x \to -2.4} [\![x]\!] = \lim\limits_{x \to -2.4} (-3) = -3$.

(b) (i) $[\![x]\!] = n - 1$ for $n - 1 \leq x < n$, so $\lim\limits_{x \to n^-} [\![x]\!] = \lim\limits_{x \to n^-} (n - 1) = n - 1$.

(ii) $[\![x]\!] = n$ for $n \leq x < n + 1$, so $\lim\limits_{x \to n^+} [\![x]\!] = \lim\limits_{x \to n^+} n = n$.

(c) $\lim\limits_{x \to a} [\![x]\!]$ exists $\Leftrightarrow$ a is not an integer.

51. The graph of $f(x) = [\![x]\!] + [\![-x]\!]$ is the same as the graph of $g(x) = -1$ with holes at each integer, since $f(a) = 0$ for any

integer a. Thus, $\lim\limits_{x \to 2^-} f(x) = -1$ and $\lim\limits_{x \to 2^+} f(x) = -1$, so $\lim\limits_{x \to 2} f(x) = -1$. However,

$f(2) = [\![2]\!] + [\![-2]\!] = 2 + (-2) = 0$, so $\lim\limits_{x \to 2} f(x) \neq f(2)$.

53. Since $p(x)$ is a polynomial, $p(x) = a_0 + a_1 x + a_2 x^2 + \cdots + a_n x^n$. Thus, by the Limit Laws,

$$\lim\limits_{x \to a} p(x) = \lim\limits_{x \to a} \left(a_0 + a_1 x + a_2 x^2 + \cdots + a_n x^n \right) = a_0 + a_1 \lim\limits_{x \to a} x + a_2 \lim\limits_{x \to a} x^2 + \cdots + a_n \lim\limits_{x \to a} x^n$$

$$= a_0 + a_1 a + a_2 a^2 + \cdots + a_n a^n = p(a)$$

Thus, for any polynomial p, $\lim\limits_{x \to a} p(x) = p(a)$.

55. $\lim\limits_{x\to1}[f(x)-8] = \lim\limits_{x\to1}\left[\dfrac{f(x)-8}{x-1}\cdot(x-1)\right] = \lim\limits_{x\to1}\dfrac{f(x)-8}{x-1}\cdot\lim\limits_{x\to1}(x-1) = 10\cdot0 = 0.$

Thus, $\lim\limits_{x\to1}f(x) = \lim\limits_{x\to1}\{[f(x)-8]+8\} = \lim\limits_{x\to1}[f(x)-8] + \lim\limits_{x\to1}8 = 0+8 = 8.$

Note: The value of $\lim\limits_{x\to1}\dfrac{f(x)-8}{x-1}$ does not affect the answer since it's multiplied by 0. What's important is that $\lim\limits_{x\to1}\dfrac{f(x)-8}{x-1}$

exists.

57. Observe that $0 \le f(x) \le x^2$ for all x, and $\lim\limits_{x\to0}0 = 0 = \lim\limits_{x\to0}x^2$. So, by the Squeeze Theorem, $\lim\limits_{x\to0}f(x) = 0.$

59. Let $f(x) = H(x)$ and $g(x) = 1 - H(x)$, where H is the Heaviside function defined in Exercise 1.3.57.

Thus, either f or g is 0 for any value of x. Then $\lim\limits_{x\to0}f(x)$ and $\lim\limits_{x\to0}g(x)$ do not exist, but $\lim\limits_{x\to0}[f(x)g(x)] = \lim\limits_{x\to0}0 = 0.$

61. Since the denominator approaches 0 as $x \to -2$, the limit will exist only if the numerator also approaches

0 as $x \to -2$. In order for this to happen, we need $\lim\limits_{x\to-2}\left(3x^2 + ax + a + 3\right) = 0 \quad\Leftrightarrow$

$3(-2)^2 + a(-2) + a + 3 = 0 \quad\Leftrightarrow\quad 12 - 2a + a + 3 = 0 \quad\Leftrightarrow\quad a = 15.$ With $a = 15$, the limit becomes

$\lim\limits_{x\to-2}\dfrac{3x^2 + 15x + 18}{x^2 + x - 2} = \lim\limits_{x\to-2}\dfrac{3(x+2)(x+3)}{(x-1)(x+2)} = \lim\limits_{x\to-2}\dfrac{3(x+3)}{x-1} = \dfrac{3(-2+3)}{-2-1} = \dfrac{3}{-3} = -1.$

2.4   The Precise Definition of a Limit

1. On the left side of $x = 2$, we need $|x - 2| < \left|\frac{10}{7} - 2\right| = \frac{4}{7}$. On the right side, we need $|x - 2| < \left|\frac{10}{3} - 2\right| = \frac{4}{3}$. For both of

these conditions to be satisfied at once, we need the more restrictive of the two to hold, that is, $|x - 2| < \frac{4}{7}$. So we can choose

$\delta = \frac{4}{7}$, or any smaller positive number.

3. The leftmost question mark is the solution of $\sqrt{x} = 1.6$ and the rightmost, $\sqrt{x} = 2.4$. So the values are $1.6^2 = 2.56$ and

$2.4^2 = 5.76$. On the left side, we need $|x - 4| < |2.56 - 4| = 1.44$. On the right side, we need $|x - 4| < |5.76 - 4| = 1.76$.

To satisfy both conditions, we need the more restrictive condition to hold—namely, $|x - 4| < 1.44$. Thus, we can choose

$\delta = 1.44$, or any smaller positive number.

5.

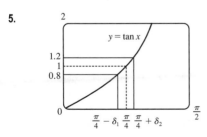

From the graph, we find that $\tan x = 0.8$ when $x \approx 0.675$, so

$\frac{\pi}{4} - \delta_1 \approx 0.675 \quad\Rightarrow\quad \delta_1 \approx \frac{\pi}{4} - 0.675 \approx 0.1106.$ Also, $\tan x = 1.2$

when $x \approx 0.876$, so $\frac{\pi}{4} + \delta_2 \approx 0.876 \quad\Rightarrow\quad \delta_2 = 0.876 - \frac{\pi}{4} \approx 0.0906.$

Thus, we choose $\delta = 0.0906$ (or any smaller positive number) since this is

the smaller of δ_1 and δ_2.

7. For $\varepsilon = 1$, the definition of a limit requires that we find δ such that $\left|(4 + x - 3x^3) - 2\right| < 1$ $\Leftrightarrow$ $1 < 4 + x - 3x^3 < 3$

whenever $0 < |x - 1| < \delta$. If we plot the graphs of $y = 1$, $y = 4 + x - 3x^3$ and $y = 3$ on the same screen, we see that we

need $0.86 \le x \le 1.11$. So since $|1 - 0.86| = 0.14$ and $|1 - 1.11| = 0.11$, we choose $\delta = 0.11$ (or any smaller positive

number). For $\varepsilon = 0.1$, we must find δ such that $\left|(4 + x - 3x^3) - 2\right| < 0.1$ $\Leftrightarrow$ $1.9 < 4 + x - 3x^3 < 2.1$ whenever

$0 < |x - 1| < \delta$. From the graph, we see that we need $0.988 \le x \le 1.012$. So since $|1 - 0.988| = 0.012$ and

$|1 - 1.012| = 0.012$, we choose $\delta = 0.012$ (or any smaller positive number) for the inequality to hold.

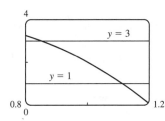
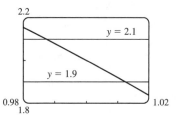

9. (a)

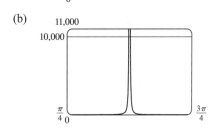

From the graph, we find that $y = \tan^2 x = 1000$ when $x \approx 1.539$ and

$x \approx 1.602$ for x near $\frac{\pi}{2}$. Thus, we get $\delta \approx 1.602 - \frac{\pi}{2} \approx 0.031$ for

$M = 1000$.

(b)

From the graph, we find that $y = \tan^2 x = 10{,}000$ when $x \approx 1.561$ and

$x \approx 1.581$ for x near $\frac{\pi}{2}$. Thus, we get $\delta \approx 1.581 - \frac{\pi}{2} \approx 0.010$ for

$M = 10{,}000$.

11. (a) $A = \pi r^2$ and $A = 1000 \text{ cm}^2$ $\Rightarrow$ $\pi r^2 = 1000$ $\Rightarrow$ $r^2 = \frac{1000}{\pi}$ $\Rightarrow$ $r = \sqrt{\frac{1000}{\pi}}$ $(r > 0)$ $\approx 17.8412 \text{ cm}$.

(b) $|A - 1000| \le 5$ $\Rightarrow$ $-5 \le \pi r^2 - 1000 \le 5$ $\Rightarrow$ $1000 - 5 \le \pi r^2 \le 1000 + 5$ $\Rightarrow$

$\sqrt{\frac{995}{\pi}} \le r \le \sqrt{\frac{1005}{\pi}}$ $\Rightarrow$ $17.7966 \le r \le 17.8858$. $\sqrt{\frac{1000}{\pi}} - \sqrt{\frac{995}{\pi}} \approx 0.04466$ and $\sqrt{\frac{1005}{\pi}} - \sqrt{\frac{1000}{\pi}} \approx 0.04455$. So

if the machinist gets the radius within 0.0445 cm of 17.8412, the area will be within 5 cm^2 of 1000.

(c) x is the radius, $f(x)$ is the area, a is the target radius given in part (a), L is the target area (1000), ε is the tolerance in the

area (5), and δ is the tolerance in the radius given in part (b).

13. (a) $|4x - 8| = 4|x - 2| < 0.1$ $\Leftrightarrow$ $|x - 2| < \dfrac{0.1}{4}$, so $\delta = \dfrac{0.1}{4} = 0.025$.

(b) $|4x - 8| = 4|x - 2| < 0.01$ $\Leftrightarrow$ $|x - 2| < \dfrac{0.01}{4}$, so $\delta = \dfrac{0.01}{4} = 0.0025$.

15. Given $\varepsilon > 0$, we need $\delta > 0$ such that if $0 < |x - 1| < \delta$, then

$|(2x + 3) - 5| < \varepsilon$. But $|(2x + 3) - 5| < \varepsilon$ ⇔

$|2x - 2| < \varepsilon$ ⇔ $2|x - 1| < \varepsilon$ ⇔ $|x - 1| < \varepsilon/2$.

So if we choose $\delta = \varepsilon/2$, then $0 < |x - 1| < \delta$ ⇒

$|(2x + 3) - 5| < \varepsilon$. Thus, $\lim\limits_{x \to 1} (2x + 3) = 5$ by the definition of a limit.

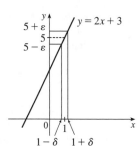

17. Given $\varepsilon > 0$, we need $\delta > 0$ such that if $0 < |x - (-3)| < \delta$, then

$|(1 - 4x) - 13| < \varepsilon$. But $|(1 - 4x) - 13| < \varepsilon$ ⇔

$|-4x - 12| < \varepsilon$ ⇔ $|-4||x + 3| < \varepsilon$ ⇔ $|x - (-3)| < \varepsilon/4$.

So if we choose $\delta = \varepsilon/4$, then $0 < |x - (-3)| < \delta$ ⇒

$|(1 - 4x) - 13| < \varepsilon$. Thus, $\lim\limits_{x \to -3} (1 - 4x) = 13$ by the definition of

a limit.

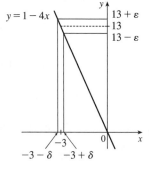

19. Given $\varepsilon > 0$, we need $\delta > 0$ such that if $0 < |x - 3| < \delta$, then $\left| \dfrac{x}{5} - \dfrac{3}{5} \right| < \varepsilon$ ⇔ $\frac{1}{5}|x - 3| < \varepsilon$ ⇔ $|x - 3| < 5\varepsilon$.

So choose $\delta = 5\varepsilon$. Then $0 < |x - 3| < \delta$ ⇒ $|x - 3| < 5\varepsilon$ ⇒ $\dfrac{|x - 3|}{5} < \varepsilon$ ⇒ $\left| \dfrac{x}{5} - \dfrac{3}{5} \right| < \varepsilon$. By the definition

of a limit, $\lim\limits_{x \to 3} \dfrac{x}{5} = \dfrac{3}{5}$.

21. Given $\varepsilon > 0$, we need $\delta > 0$ such that if $0 < |x - 2| < \delta$, then $\left| \dfrac{x^2 + x - 6}{x - 2} - 5 \right| < \varepsilon$ ⇔

$\left| \dfrac{(x + 3)(x - 2)}{x - 2} - 5 \right| < \varepsilon$ ⇔ $|x + 3 - 5| < \varepsilon$ $[x \neq 2]$ ⇔ $|x - 2| < \varepsilon$. So choose $\delta = \varepsilon$.

Then $0 < |x - 2| < \delta$ ⇒ $|x - 2| < \varepsilon$ ⇒ $|x + 3 - 5| < \varepsilon$ ⇒ $\left| \dfrac{(x + 3)(x - 2)}{x - 2} - 5 \right| < \varepsilon$ $[x \neq 2]$ ⇒

$\left| \dfrac{x^2 + x - 6}{x - 2} - 5 \right| < \varepsilon$. By the definition of a limit, $\lim\limits_{x \to 2} \dfrac{x^2 + x - 6}{x - 2} = 5$.

23. Given $\varepsilon > 0$, we need $\delta > 0$ such that if $0 < |x - a| < \delta$, then $|x - a| < \varepsilon$. So $\delta = \varepsilon$ will work.

25. Given $\varepsilon > 0$, we need $\delta > 0$ such that if $0 < |x - 0| < \delta$, then $|x^2 - 0| < \varepsilon$ ⇔ $x^2 < \varepsilon$ ⇔ $|x| < \sqrt{\varepsilon}$. Take $\delta = \sqrt{\varepsilon}$.

Then $0 < |x - 0| < \delta$ ⇒ $|x^2 - 0| < \varepsilon$. Thus, $\lim\limits_{x \to 0} x^2 = 0$ by the definition of a limit.

27. Given $\varepsilon > 0$, we need $\delta > 0$ such that if $0 < |x - 0| < \delta$, then $||x| - 0| < \varepsilon$. But $||x|| = |x|$. So this is true if we pick $\delta = \varepsilon$.

Thus, $\lim\limits_{x \to 0} |x| = 0$ by the definition of a limit.

29. Given $\varepsilon > 0$, we need $\delta > 0$ such that if $0 < |x - 2| < \delta$, then $\left|(x^2 - 4x + 5) - 1\right| < \varepsilon$ $\Leftrightarrow$ $\left|x^2 - 4x + 4\right| < \varepsilon$ $\Leftrightarrow$

$\left|(x - 2)^2\right| < \varepsilon$. So take $\delta = \sqrt{\varepsilon}$. Then $0 < |x - 2| < \delta$ $\Leftrightarrow$ $|x - 2| < \sqrt{\varepsilon}$ $\Leftrightarrow$ $\left|(x - 2)^2\right| < \varepsilon$. Thus,

$\lim_{x \to 2} (x^2 - 4x + 5) = 1$ by the definition of a limit.

31. Given $\varepsilon > 0$, we need $\delta > 0$ such that if $0 < |x - (-2)| < \delta$, then $\left|(x^2 - 1) - 3\right| < \varepsilon$ or upon simplifying we need

$\left|x^2 - 4\right| < \varepsilon$ whenever $0 < |x + 2| < \delta$. Notice that if $|x + 2| < 1$, then $-1 < x + 2 < 1$ $\Rightarrow$ $-5 < x - 2 < -3$ $\Rightarrow$

$|x - 2| < 5$. So take $\delta = \min\{\varepsilon/5, 1\}$. Then $0 < |x + 2| < \delta$ $\Rightarrow$ $|x - 2| < 5$ and $|x + 2| < \varepsilon/5$, so

$\left|(x^2 - 1) - 3\right| = |(x + 2)(x - 2)| = |x + 2|\,|x - 2| < (\varepsilon/5)(5) = \varepsilon$. Thus, by the definition of a limit, $\lim_{x \to -2} (x^2 - 1) = 3$.

33. Given $\varepsilon > 0$, we let $\delta = \min\left\{2, \frac{\varepsilon}{8}\right\}$. If $0 < |x - 3| < \delta$, then $|x - 3| < 2$ $\Rightarrow$ $-2 < x - 3 < 2$ $\Rightarrow$

$4 < x + 3 < 8$ $\Rightarrow$ $|x + 3| < 8$. Also $|x - 3| < \frac{\varepsilon}{8}$, so $\left|x^2 - 9\right| = |x + 3|\,|x - 3| < 8 \cdot \frac{\varepsilon}{8} = \varepsilon$. Thus, $\lim_{x \to 3} x^2 = 9$.

35. (a) The points of intersection in the graph are $(x_1, 2.6)$ and $(x_2, 3.4)$

with $x_1 \approx 0.891$ and $x_2 \approx 1.093$. Thus, we can take δ to be the

smaller of $1 - x_1$ and $x_2 - 1$. So $\delta = x_2 - 1 \approx 0.093$.

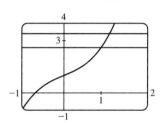

(b) Solving $x^3 + x + 1 = 3 + \varepsilon$ gives us two nonreal complex roots and one real root, which is

$$x(\varepsilon) = \dfrac{\left(216 + 108\varepsilon + 12\sqrt{336 + 324\varepsilon + 81\varepsilon^2}\,\right)^{2/3} - 12}{6\left(216 + 108\varepsilon + 12\sqrt{336 + 324\varepsilon + 81\varepsilon^2}\,\right)^{1/3}}. \text{ Thus, } \delta = x(\varepsilon) - 1.$$

(c) If $\varepsilon = 0.4$, then $x(\varepsilon) \approx 1.093\,272\,342$ and $\delta = x(\varepsilon) - 1 \approx 0.093$, which agrees with our answer in part (a).

37. *1. Guessing a value for δ*   Given $\varepsilon > 0$, we must find $\delta > 0$ such that $|\sqrt{x} - \sqrt{a}| < \varepsilon$ whenever $0 < |x - a| < \delta$. But

$|\sqrt{x} - \sqrt{a}| = \dfrac{|x - a|}{\sqrt{x} + \sqrt{a}} < \varepsilon$ (from the hint). Now if we can find a positive constant C such that $\sqrt{x} + \sqrt{a} > C$ then

$\dfrac{|x - a|}{\sqrt{x} + \sqrt{a}} < \dfrac{|x - a|}{C} < \varepsilon$, and we take $|x - a| < C\varepsilon$. We can find this number by restricting x to lie in some interval

centered at a. If $|x - a| < \frac{1}{2}a$, then $-\frac{1}{2}a < x - a < \frac{1}{2}a$ $\Rightarrow$ $\frac{1}{2}a < x < \frac{3}{2}a$ $\Rightarrow$ $\sqrt{x} + \sqrt{a} > \sqrt{\frac{1}{2}a} + \sqrt{a}$, and so

$C = \sqrt{\frac{1}{2}a} + \sqrt{a}$ is a suitable choice for the constant. So $|x - a| < \left(\sqrt{\frac{1}{2}a} + \sqrt{a}\right)\varepsilon$. This suggests that we let

$\delta = \min\left\{\frac{1}{2}a, \left(\sqrt{\frac{1}{2}a} + \sqrt{a}\right)\varepsilon\right\}$.

2. Showing that δ works   Given $\varepsilon > 0$, we let $\delta = \min\left\{\frac{1}{2}a, \left(\sqrt{\frac{1}{2}a} + \sqrt{a}\right)\varepsilon\right\}$. If $0 < |x - a| < \delta$, then

$|x - a| < \frac{1}{2}a$ $\Rightarrow$ $\sqrt{x} + \sqrt{a} > \sqrt{\frac{1}{2}a} + \sqrt{a}$ (as in part 1). Also $|x - a| < \left(\sqrt{\frac{1}{2}a} + \sqrt{a}\right)\varepsilon$, so

$|\sqrt{x} - \sqrt{a}| = \dfrac{|x - a|}{\sqrt{x} + \sqrt{a}} < \dfrac{\left(\sqrt{a/2} + \sqrt{a}\right)\varepsilon}{\left(\sqrt{a/2} + \sqrt{a}\right)} = \varepsilon$. Therefore, $\lim_{x \to a} \sqrt{x} = \sqrt{a}$ by the definition of a limit.

39. Suppose that $\lim_{x \to 0} f(x) = L$. Given $\varepsilon = \frac{1}{2}$, there exists $\delta > 0$ such that $0 < |x| < \delta \;\Rightarrow\; |f(x) - L| < \frac{1}{2}$. Take any rational

number r with $0 < |r| < \delta$. Then $f(r) = 0$, so $|0 - L| < \frac{1}{2}$, so $L \le |L| < \frac{1}{2}$. Now take any irrational number s with

$0 < |s| < \delta$. Then $f(s) = 1$, so $|1 - L| < \frac{1}{2}$. Hence, $1 - L < \frac{1}{2}$, so $L > \frac{1}{2}$. This contradicts $L < \frac{1}{2}$, so $\lim_{x \to 0} f(x)$ does not

exist.

41. $\dfrac{1}{(x+3)^4} > 10{,}000 \;\Leftrightarrow\; (x+3)^4 < \dfrac{1}{10{,}000} \;\Leftrightarrow\; |x+3| < \dfrac{1}{\sqrt[4]{10{,}000}} \;\Leftrightarrow\; |x - (-3)| < \dfrac{1}{10}$

43. Given $M < 0$ we need $\delta > 0$ so that $\ln x < M$ whenever $0 < x < \delta$; that is, $x = e^{\ln x} < e^M$ whenever $0 < x < \delta$. This

suggests that we take $\delta = e^M$. If $0 < x < e^M$, then $\ln x < \ln e^M = M$. By the definition of a limit, $\lim_{x \to 0^+} \ln x = -\infty$.

2.5   Continuity

1. From Definition 1, $\lim_{x \to 4} f(x) = f(4)$.

3. (a) The following are the numbers at which f is discontinuous and the type of discontinuity at that number: -4 (removable),

-2 (jump), 2 (jump), 4 (infinite).

(b) f is continuous from the left at -2 since $\lim\limits_{x \to -2^-} f(x) = f(-2)$. f is continuous from the right at 2 and 4 since

$\lim\limits_{x \to 2^+} f(x) = f(2)$ and $\lim\limits_{x \to 4^+} f(x) = f(4)$. It is continuous from neither side at -4 since $f(-4)$ is undefined.

5. The graph of $y = f(x)$ must have a discontinuity at $x = 3$ and must show that $\lim\limits_{x \to 3^-} f(x) = f(3)$.

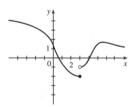

7. (a)

(b) There are discontinuities at times $t = 1, 2, 3,$ and 4. A person

parking in the lot would want to keep in mind that the charge will

jump at the beginning of each hour.

9. Since f and g are continuous functions,

$$\lim_{x \to 3} [2f(x) - g(x)] = 2 \lim_{x \to 3} f(x) - \lim_{x \to 3} g(x) \qquad \text{[by Limit Laws 2 and 3]}$$

$$= 2f(3) - g(3) \qquad \text{[by continuity of } f \text{ and } g \text{ at } x = 3\text{]}$$

$$= 2 \cdot 5 - g(3) = 10 - g(3)$$

Since it is given that $\lim\limits_{x \to 3} [2f(x) - g(x)] = 4$, we have $10 - g(3) = 4$, so $g(3) = 6$.

11. $\lim\limits_{x \to -1} f(x) = \lim\limits_{x \to -1} \left(x + 2x^3\right)^4 = \left(\lim\limits_{x \to -1} x + 2 \lim\limits_{x \to -1} x^3\right)^4 = \left[-1 + 2(-1)^3\right]^4 = (-3)^4 = 81 = f(-1)$.

By the definition of continuity, f is continuous at $a = -1$.

13. For $a > 2$, we have

$$\lim\limits_{x \to a} f(x) = \lim\limits_{x \to a} \frac{2x + 3}{x - 2} = \frac{\lim\limits_{x \to a} (2x + 3)}{\lim\limits_{x \to a} (x - 2)} \qquad \text{[Limit Law 5]}$$

$$= \frac{2 \lim\limits_{x \to a} x + \lim\limits_{x \to a} 3}{\lim\limits_{x \to a} x - \lim\limits_{x \to a} 2} \qquad \text{[1, 2, and 3]}$$

$$= \frac{2a + 3}{a - 2} \qquad \text{[7 and 8]}$$

$$= f(a)$$

Thus, f is continuous at $x = a$ for every a in $(2, \infty)$; that is, f is continuous on $(2, \infty)$.

15. $f(x) = \ln|x - 2|$ is discontinuous at 2 since $f(2) = \ln 0$ is not defined.

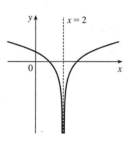

17. $f(x) = \begin{cases} e^x & \text{if } x < 0 \\ x^2 & \text{if } x \geq 0 \end{cases}$

The left-hand limit of f at $a = 0$ is $\lim\limits_{x \to 0^-} f(x) = \lim\limits_{x \to 0^-} e^x = 1$. The

right-hand limit of f at $a = 0$ is $\lim\limits_{x \to 0^+} f(x) = \lim\limits_{x \to 0^+} x^2 = 0$. Since these

limits are not equal, $\lim\limits_{x \to 0} f(x)$ does not exist and f is discontinuous at 0.

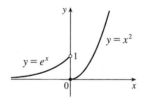

19. $f(x) = \begin{cases} \cos x & \text{if } x < 0 \\ 0 & \text{if } x = 0 \\ 1 - x^2 & \text{if } x > 0 \end{cases}$

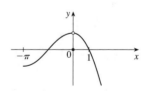

$\lim\limits_{x \to 0} f(x) = 1$, but $f(0) = 0 \neq 1$, so f is discontinuous at 0.

21. $F(x) = \dfrac{x}{x^2 + 5x + 6}$ is a rational function. So by Theorem 5 (or Theorem 7), F is continuous at every number in its domain,

$\left\{x \mid x^2 + 5x + 6 \neq 0\right\} = \left\{x \mid (x + 3)(x + 2) \neq 0\right\} = \left\{x \mid x \neq -3, \, -2\right\}$ or $(-\infty, -3) \cup (-3, -2) \cup (-2, \infty)$.

23. By Theorem 5, the polynomials x^2 and $2x - 1$ are continuous on $(-\infty, \infty)$. By Theorem 7, the root function $\sqrt{x}$ is

continuous on $[0, \infty)$. By Theorem 9, the composite function $\sqrt{2x - 1}$ is continuous on its domain, $[\frac{1}{2}, \infty)$.

By part 1 of Theorem 4, the sum $R(x) = x^2 + \sqrt{2x - 1}$ is continuous on $[\frac{1}{2}, \infty)$.

25. By Theorem 7, the exponential function e^{-5t} and the trigonometric function $\cos 2\pi t$ are continuous on $(-\infty, \infty)$.

By part 4 of Theorem 4, $L(t) = e^{-5t} \cos 2\pi t$ is continuous on $(-\infty, \infty)$.

27. By Theorem 5, the polynomial $t^4 - 1$ is continuous on $(-\infty, \infty)$. By Theorem 7, $\ln x$ is continuous on its domain, $(0, \infty)$.

By Theorem 9, $\ln(t^4 - 1)$ is continuous on its domain, which is

$$\{t \mid t^4 - 1 > 0\} = \{t \mid t^4 > 1\} = \{t \mid |t| > 1\} = (-\infty, -1) \cup (1, \infty)$$

29. The function $y = \dfrac{1}{1 + e^{1/x}}$ is discontinuous at $x = 0$ because the

left- and right-hand limits at $x = 0$ are different.

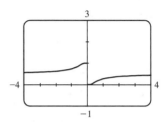

31. Because we are dealing with root functions, $5 + \sqrt{x}$ is continuous on $[0, \infty)$, $\sqrt{x + 5}$ is continuous on $[-5, \infty)$, so the

quotient $f(x) = \dfrac{5 + \sqrt{x}}{\sqrt{5 + x}}$ is continuous on $[0, \infty)$. Since f is continuous at $x = 4$, $\lim\limits_{x \to 4} f(x) = f(4) = \frac{7}{3}$.

33. Because $x^2 - x$ is continuous on $\mathbb{R}$, the composite function $f(x) = e^{x^2 - x}$ is continuous on $\mathbb{R}$, so

$$\lim\limits_{x \to 1} f(x) = f(1) = e^{1 - 1} = e^0 = 1.$$

35. $f(x) = \begin{cases} x^2 & \text{if } x < 1 \\ \sqrt{x} & \text{if } x \geq 1 \end{cases}$

By Theorem 5, since $f(x)$ equals the polynomial x^2 on $(-\infty, 1)$, f is continuous on $(-\infty, 1)$. By Theorem 7, since $f(x)$

equals the root function $\sqrt{x}$ on $(1, \infty)$, f is continuous on $(1, \infty)$. At $x = 1$, $\lim\limits_{x \to 1^-} f(x) = \lim\limits_{x \to 1^-} x^2 = 1$ and

$\lim\limits_{x \to 1^+} f(x) = \lim\limits_{x \to 1^+} \sqrt{x} = 1$. Thus, $\lim\limits_{x \to 1} f(x)$ exists and equals 1. Also, $f(1) = \sqrt{1} = 1$. Thus, f is continuous at $x = 1$.

We conclude that f is continuous on $(-\infty, \infty)$.

37. $f(x) = \begin{cases} 1 + x^2 & \text{if } x \leq 0 \\ 2 - x & \text{if } 0 < x \leq 2 \\ (x - 2)^2 & \text{if } x > 2 \end{cases}$

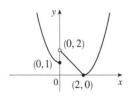

f is continuous on $(-\infty, 0)$, $(0, 2)$, and $(2, \infty)$ since it is a polynomial on

each of these intervals. Now $\lim\limits_{x \to 0^-} f(x) = \lim\limits_{x \to 0^-} (1 + x^2) = 1$ and $\lim\limits_{x \to 0^+} f(x) = \lim\limits_{x \to 0^+} (2 - x) = 2$, so f is

discontinuous at 0. Since $f(0) = 1$, f is continuous from the left at 0. Also, $\lim\limits_{x \to 2^-} f(x) = \lim\limits_{x \to 2^-} (2 - x) = 0$,

$\lim\limits_{x \to 2^+} f(x) = \lim\limits_{x \to 2^+} (x - 2)^2 = 0$, and $f(2) = 0$, so f is continuous at 2. The only number at which f is discontinuous is 0.

39. $f(x) = \begin{cases} x+2 & \text{if } x < 0 \\ e^x & \text{if } 0 \le x \le 1 \\ 2-x & \text{if } x > 1 \end{cases}$

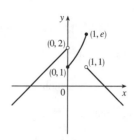

f is continuous on $(-\infty, 0)$ and $(1, \infty)$ since on each of these intervals

it is a polynomial; it is continuous on $(0, 1)$ since it is an exponential.

Now $\lim\limits_{x \to 0^-} f(x) = \lim\limits_{x \to 0^-} (x+2) = 2$ and $\lim\limits_{x \to 0^+} f(x) = \lim\limits_{x \to 0^+} e^x = 1$, so f is discontinuous at 0. Since $f(0) = 1$, f is

continuous from the right at 0. Also $\lim\limits_{x \to 1^-} f(x) = \lim\limits_{x \to 1^-} e^x = e$ and $\lim\limits_{x \to 1^+} f(x) = \lim\limits_{x \to 1^+} (2-x) = 1$, so f is discontinuous

at 1. Since $f(1) = e$, f is continuous from the left at 1.

41. $f(x) = \begin{cases} cx^2 + 2x & \text{if } x < 2 \\ x^3 - cx & \text{if } x \ge 2 \end{cases}$

f is continuous on $(-\infty, 2)$ and $(2, \infty)$. Now $\lim\limits_{x \to 2^-} f(x) = \lim\limits_{x \to 2^-} (cx^2 + 2x) = 4c + 4$ and

$\lim\limits_{x \to 2^+} f(x) = \lim\limits_{x \to 2^+} (x^3 - cx) = 8 - 2c$. So f is continuous $\Leftrightarrow$ $4c + 4 = 8 - 2c$ $\Leftrightarrow$ $6c = 4$ $\Leftrightarrow$ $c = \frac{2}{3}$. Thus, for f

to be continuous on $(-\infty, \infty)$, $c = \frac{2}{3}$.

43. (a) $f(x) = \dfrac{x^4 - 1}{x - 1} = \dfrac{(x^2 + 1)(x^2 - 1)}{x - 1} = \dfrac{(x^2 + 1)(x + 1)(x - 1)}{x - 1} = (x^2 + 1)(x + 1)$ [or $x^3 + x^2 + x + 1$]

for $x \ne 1$. The discontinuity is removable and $g(x) = x^3 + x^2 + x + 1$ agrees with f for $x \ne 1$ and is continuous on $\mathbb{R}$.

(b) $f(x) = \dfrac{x^3 - x^2 - 2x}{x - 2} = \dfrac{x(x^2 - x - 2)}{x - 2} = \dfrac{x(x - 2)(x + 1)}{x - 2} = x(x + 1)$ [or $x^2 + x$]   for $x \ne 2$. The discontinuity

is removable and $g(x) = x^2 + x$ agrees with f for $x \ne 2$ and is continuous on $\mathbb{R}$.

(c) $\lim\limits_{x \to \pi^-} f(x) = \lim\limits_{x \to \pi^-} [\![\sin x]\!] = \lim\limits_{x \to \pi^-} 0 = 0$ and $\lim\limits_{x \to \pi^+} f(x) = \lim\limits_{x \to \pi^+} [\![\sin x]\!] = \lim\limits_{x \to \pi^+} (-1) = -1$, so $\lim\limits_{x \to \pi} f(x)$ does not

exist. The discontinuity at $x = \pi$ is a jump discontinuity.

45. $f(x) = x^2 + 10 \sin x$ is continuous on the interval $[31, 32]$, $f(31) \approx 957$, and $f(32) \approx 1030$. Since $957 < 1000 < 1030$,

there is a number c in $(31, 32)$ such that $f(c) = 1000$ by the Intermediate Value Theorem. *Note:* There is also a number c in

$(-32, -31)$ such that $f(c) = 1000$.

47. $f(x) = x^4 + x - 3$ is continuous on the interval $[1, 2]$, $f(1) = -1$, and $f(2) = 15$. Since $-1 < 0 < 15$, there is a number c

in $(1, 2)$ such that $f(c) = 0$ by the Intermediate Value Theorem. Thus, there is a root of the equation $x^4 + x - 3 = 0$ in the

interval $(1, 2)$.

49. $f(x) = \cos x - x$ is continuous on the interval $[0, 1]$, $f(0) = 1$, and $f(1) = \cos 1 - 1 \approx -0.46$. Since $-0.46 < 0 < 1$, there

is a number c in $(0, 1)$ such that $f(c) = 0$ by the Intermediate Value Theorem. Thus, there is a root of the equation

$\cos x - x = 0$, or $\cos x = x$, in the interval $(0, 1)$.

51. (a) $f(x) = \cos x - x^3$ is continuous on the interval $[0, 1]$, $f(0) = 1 > 0$, and $f(1) = \cos 1 - 1 \approx -0.46 < 0$. Since

$1 > 0 > -0.46$, there is a number c in $(0, 1)$ such that $f(c) = 0$ by the Intermediate Value Theorem. Thus, there is a root

of the equation $\cos x - x^3 = 0$, or $\cos x = x^3$, in the interval $(0, 1)$.

(b) $f(0.86) \approx 0.016 > 0$ and $f(0.87) \approx -0.014 < 0$, so there is a root between 0.86 and 0.87, that is, in the interval

$(0.86, 0.87)$.

53. (a) Let $f(x) = 100e^{-x/100} - 0.01x^2$. Then $f(0) = 100 > 0$ and

$f(100) = 100e^{-1} - 100 \approx -63.2 < 0$. So by the Intermediate

Value Theorem, there is a number c in $(0, 100)$ such that $f(c) = 0$.

This implies that $100e^{-c/100} = 0.01c^2$.

(b) Using the intersect feature of the graphing device, we find that the

root of the equation is $x = 70.347$, correct to three decimal places.

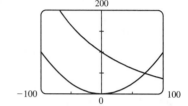

55. ($\Rightarrow$) If f is continuous at a, then by Theorem 8 with $g(h) = a + h$, we have

$$\lim_{h \to 0} f(a + h) = f\left(\lim_{h \to 0} (a + h)\right) = f(a).$$

($\Leftarrow$) Let $\varepsilon > 0$. Since $\lim_{h \to 0} f(a + h) = f(a)$, there exists $\delta > 0$ such that $0 < |h| < \delta \quad \Rightarrow$

$|f(a + h) - f(a)| < \varepsilon$. So if $0 < |x - a| < \delta$, then $|f(x) - f(a)| = |f(a + (x - a)) - f(a)| < \varepsilon$.

Thus, $\lim_{x \to a} f(x) = f(a)$ and so f is continuous at a.

57. As in the previous exercise, we must show that $\lim_{h \to 0} \cos(a + h) = \cos a$ to prove that the cosine function is continuous.

$$\lim_{h \to 0} \cos(a + h) = \lim_{h \to 0} (\cos a \cos h - \sin a \sin h) = \lim_{h \to 0} (\cos a \cos h) - \lim_{h \to 0} (\sin a \sin h)$$

$$= \left(\lim_{h \to 0} \cos a\right)\left(\lim_{h \to 0} \cos h\right) - \left(\lim_{h \to 0} \sin a\right)\left(\lim_{h \to 0} \sin h\right) = (\cos a)(1) - (\sin a)(0) = \cos a$$

59. $f(x) = \begin{cases} 0 & \text{if } x \text{ is rational} \\ 1 & \text{if } x \text{ is irrational} \end{cases}$ is continuous nowhere. For, given any number a and any $\delta > 0$, the interval $(a - \delta, a + \delta)$

contains both infinitely many rational and infinitely many irrational numbers. Since $f(a) = 0$ or 1, there are infinitely many

numbers x with $0 < |x - a| < \delta$ and $|f(x) - f(a)| = 1$. Thus, $\lim_{x \to a} f(x) \neq f(a)$. [In fact, $\lim_{x \to a} f(x)$ does not even exist.]

61. If there is such a number, it satisfies the equation $x^3 + 1 = x \iff x^3 - x + 1 = 0$. Let the left-hand side of this equation be

called $f(x)$. Now $f(-2) = -5 < 0$, and $f(-1) = 1 > 0$. Note also that $f(x)$ is a polynomial, and thus continuous. So by the

Intermediate Value Theorem, there is a number c between -2 and -1 such that $f(c) = 0$, so that $c = c^3 + 1$.

63. $f(x) = x^4 \sin(1/x)$ is continuous on $(-\infty, 0) \cup (0, \infty)$ since it is the product of a polynomial and a composite of a

trigonometric function and a rational function. Now since $-1 \leq \sin(1/x) \leq 1$, we have $-x^4 \leq x^4 \sin(1/x) \leq x^4$. Because

$\lim_{x \to 0}(-x^4) = 0$ and $\lim_{x \to 0} x^4 = 0$, the Squeeze Theorem gives us $\lim_{x \to 0}(x^4 \sin(1/x)) = 0$, which equals $f(0)$. Thus, f is

continuous at 0 and, hence, on $(-\infty, \infty)$.

65. Define $u(t)$ to be the monk's distance from the monastery, as a function of time, on the first day, and define $d(t)$ to be his

distance from the monastery, as a function of time, on the second day. Let D be the distance from the monastery to the top of

the mountain. From the given information we know that $u(0) = 0$, $u(12) = D$, $d(0) = D$ and $d(12) = 0$. Now consider the

function $u - d$, which is clearly continuous. We calculate that $(u - d)(0) = -D$ and $(u - d)(12) = D$. So by the

Intermediate Value Theorem, there must be some time t_0 between 0 and 12 such that $(u - d)(t_0) = 0 \quad \Leftrightarrow \quad u(t_0) = d(t_0)$.

So at time t_0 after 7:00 AM, the monk will be at the same place on both days.

2.6   Limits at Infinity; Horizontal Asymptotes

1. (a) As x becomes large, the values of $f(x)$ approach 5.

(b) As x becomes large negative, the values of $f(x)$ approach 3.

3. (a) $\lim_{x \to 2} f(x) = \infty$ (b) $\lim_{x \to -1^-} f(x) = \infty$ (c) $\lim_{x \to -1^+} f(x) = -\infty$

(d) $\lim_{x \to \infty} f(x) = 1$ (e) $\lim_{x \to -\infty} f(x) = 2$ (f) Vertical: $x = -1$, $x = 2$; Horizontal: $y = 1$, $y = 2$

5. $f(0) = 0$, $f(1) = 1$,

$\lim_{x \to \infty} f(x) = 0$,

f is odd

7. $\lim_{x \to 2} f(x) = -\infty$, $\lim_{x \to \infty} f(x) = \infty$,

$\lim_{x \to -\infty} f(x) = 0$, $\lim_{x \to 0^+} f(x) = \infty$,

$\lim_{x \to 0^-} f(x) = -\infty$

9. $f(0) = 3$, $\lim_{x \to 0^-} f(x) = 4$,

$\lim_{x \to 0^+} f(x) = 2$,

$\lim_{x \to -\infty} f(x) = -\infty$, $\lim_{x \to 4^-} f(x) = -\infty$,

$\lim_{x \to 4^+} f(x) = \infty$, $\lim_{x \to \infty} f(x) = 3$

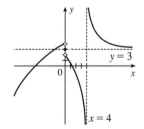

11. If $f(x) = x^2/2^x$, then a calculator gives $f(0) = 0$, $f(1) = 0.5$, $f(2) = 1$, $f(3) = 1.125$, $f(4) = 1$, $f(5) = 0.78125$,

$f(6) = 0.5625$, $f(7) = 0.3828125$, $f(8) = 0.25$, $f(9) = 0.158203125$, $f(10) = 0.09765625$, $f(20) \approx 0.00038147$,

$f(50) \approx 2.2204 \times 10^{-12}$, $f(100) \approx 7.8886 \times 10^{-27}$.

It appears that $\lim_{x \to \infty} \left(x^2/2^x\right) = 0$.

13. $\displaystyle\lim_{x\to\infty}\frac{3x^2-x+4}{2x^2+5x-8}=\lim_{x\to\infty}\frac{(3x^2-x+4)/x^2}{(2x^2+5x-8)/x^2}$

[divide both the numerator and denominator by x^2

(the highest power of x thatappears in the denominator)]

$$=\frac{\displaystyle\lim_{x\to\infty}(3-1/x+4/x^2)}{\displaystyle\lim_{x\to\infty}(2+5/x-8/x^2)}$$

[Limit Law 5]

$$=\frac{\displaystyle\lim_{x\to\infty}3-\lim_{x\to\infty}(1/x)+\lim_{x\to\infty}(4/x^2)}{\displaystyle\lim_{x\to\infty}2+\lim_{x\to\infty}(5/x)-\lim_{x\to\infty}(8/x^2)}$$

[Limit Laws 1 and 2]

$$=\frac{3-\displaystyle\lim_{x\to\infty}(1/x)+4\lim_{x\to\infty}(1/x^2)}{2+5\displaystyle\lim_{x\to\infty}(1/x)-8\lim_{x\to\infty}(1/x^2)}$$

[Limit Laws 7 and 3]

$$=\frac{3-0+4(0)}{2+5(0)-8(0)}$$

[Theorem 5 of Section 2.5]

$$=\frac{3}{2}$$

15. $\displaystyle\lim_{x\to\infty}\frac{1}{2x+3}=\lim_{x\to\infty}\frac{1/x}{(2x+3)/x}=\frac{\displaystyle\lim_{x\to\infty}(1/x)}{\displaystyle\lim_{x\to\infty}(2+3/x)}=\frac{\displaystyle\lim_{x\to\infty}(1/x)}{\displaystyle\lim_{x\to\infty}2+3\lim_{x\to\infty}(1/x)}=\frac{0}{2+3(0)}=\frac{0}{2}=0$

17. $\displaystyle\lim_{x\to-\infty}\frac{1-x-x^2}{2x^2-7}=\lim_{x\to-\infty}\frac{(1-x-x^2)/x^2}{(2x^2-7)/x^2}=\frac{\displaystyle\lim_{x\to-\infty}(1/x^2-1/x-1)}{\displaystyle\lim_{x\to-\infty}(2-7/x^2)}$

$$=\frac{\displaystyle\lim_{x\to-\infty}(1/x^2)-\lim_{x\to-\infty}(1/x)-\lim_{x\to-\infty}1}{\displaystyle\lim_{x\to-\infty}2-7\lim_{x\to-\infty}(1/x^2)}=\frac{0-0-1}{2-7(0)}=-\frac{1}{2}$$

19. Divide both the numerator and denominator by x^3 (the highest power of x that occurs in the denominator).

$$\lim_{x\to\infty}\frac{x^3+5x}{2x^3-x^2+4}=\lim_{x\to\infty}\frac{\dfrac{x^3+5x}{x^3}}{\dfrac{2x^3-x^2+4}{x^3}}=\lim_{x\to\infty}\frac{1+\dfrac{5}{x^2}}{2-\dfrac{1}{x}+\dfrac{4}{x^3}}=\frac{\displaystyle\lim_{x\to\infty}\left(1+\dfrac{5}{x^2}\right)}{\displaystyle\lim_{x\to\infty}\left(2-\dfrac{1}{x}+\dfrac{4}{x^3}\right)}$$

$$=\frac{\displaystyle\lim_{x\to\infty}1+5\lim_{x\to\infty}\dfrac{1}{x^2}}{\displaystyle\lim_{x\to\infty}2-\lim_{x\to\infty}\dfrac{1}{x}+4\lim_{x\to\infty}\dfrac{1}{x^3}}=\frac{1+5(0)}{2-0+4(0)}=\frac{1}{2}$$

21. First, multiply the factors in the denominator. Then divide both the numerator and denominator by u^4.

$$\lim_{u\to\infty}\frac{4u^4+5}{(u^2-2)(2u^2-1)}=\lim_{u\to\infty}\frac{4u^4+5}{2u^4-5u^2+2}=\lim_{u\to\infty}\frac{\dfrac{4u^4+5}{u^4}}{\dfrac{2u^4-5u^2+2}{u^4}}=\lim_{u\to\infty}\frac{4+\dfrac{5}{u^4}}{2-\dfrac{5}{u^2}+\dfrac{2}{u^4}}$$

$$=\frac{\displaystyle\lim_{u\to\infty}\left(4+\dfrac{5}{u^4}\right)}{\displaystyle\lim_{u\to\infty}\left(2-\dfrac{5}{u^2}+\dfrac{2}{u^4}\right)}=\frac{\displaystyle\lim_{u\to\infty}4+5\lim_{u\to\infty}\dfrac{1}{u^4}}{\displaystyle\lim_{u\to\infty}2-5\lim_{u\to\infty}\dfrac{1}{u^2}+2\lim_{u\to\infty}\dfrac{1}{u^4}}=\frac{4+5(0)}{2-5(0)+2(0)}=\frac{4}{2}=2$$

23. $\lim\limits_{x\to\infty} \dfrac{\sqrt{9x^6-x}}{x^3+1} = \lim\limits_{x\to\infty} \dfrac{\sqrt{9x^6-x}\,/x^3}{(x^3+1)/x^3} = \dfrac{\lim\limits_{x\to\infty}\sqrt{(9x^6-x)/x^6}}{\lim\limits_{x\to\infty}(1+1/x^3)}$ [since $x^3 = \sqrt{x^6}$ for $x>0$]

$= \dfrac{\lim\limits_{x\to\infty}\sqrt{9-1/x^5}}{\lim\limits_{x\to\infty}1+\lim\limits_{x\to\infty}(1/x^3)} = \dfrac{\sqrt{\lim\limits_{x\to\infty}9-\lim\limits_{x\to\infty}(1/x^5)}}{1+0}$

$= \sqrt{9-0} = 3$

25. $\lim\limits_{x\to\infty}\left(\sqrt{9x^2+x}-3x\right) = \lim\limits_{x\to\infty}\dfrac{\left(\sqrt{9x^2+x}-3x\right)\left(\sqrt{9x^2+x}+3x\right)}{\sqrt{9x^2+x}+3x} = \lim\limits_{x\to\infty}\dfrac{\left(\sqrt{9x^2+x}\right)^2-(3x)^2}{\sqrt{9x^2+x}+3x}$

$= \lim\limits_{x\to\infty}\dfrac{(9x^2+x)-9x^2}{\sqrt{9x^2+x}+3x} = \lim\limits_{x\to\infty}\dfrac{x}{\sqrt{9x^2+x}+3x}\cdot\dfrac{1/x}{1/x}$

$= \lim\limits_{x\to\infty}\dfrac{x/x}{\sqrt{9x^2/x^2+x/x^2}+3x/x} = \lim\limits_{x\to\infty}\dfrac{1}{\sqrt{9+1/x}+3} = \dfrac{1}{\sqrt{9}+3} = \dfrac{1}{3+3} = \dfrac{1}{6}$

27. $\lim\limits_{x\to\infty}\left(\sqrt{x^2+ax}-\sqrt{x^2+bx}\right) = \lim\limits_{x\to\infty}\dfrac{\left(\sqrt{x^2+ax}-\sqrt{x^2+bx}\right)\left(\sqrt{x^2+ax}+\sqrt{x^2+bx}\right)}{\sqrt{x^2+ax}+\sqrt{x^2+bx}}$

$= \lim\limits_{x\to\infty}\dfrac{(x^2+ax)-(x^2+bx)}{\sqrt{x^2+ax}+\sqrt{x^2+bx}} = \lim\limits_{x\to\infty}\dfrac{[(a-b)x]/x}{\left(\sqrt{x^2+ax}+\sqrt{x^2+bx}\right)/\sqrt{x^2}}$

$= \lim\limits_{x\to\infty}\dfrac{a-b}{\sqrt{1+a/x}+\sqrt{1+b/x}} = \dfrac{a-b}{\sqrt{1+0}+\sqrt{1+0}} = \dfrac{a-b}{2}$

29. $\lim\limits_{x\to\infty}\dfrac{x+x^3+x^5}{1-x^2+x^4} = \lim\limits_{x\to\infty}\dfrac{(x+x^3+x^5)/x^4}{(1-x^2+x^4)/x^4}$ [divide by the highest power of x in the denominator]

$= \lim\limits_{x\to\infty}\dfrac{1/x^3+1/x+x}{1/x^4-1/x^2+1} = \infty$

because $(1/x^3+1/x+x)\to\infty$ and $(1/x^4-1/x^2+1)\to1$ as $x\to\infty$.

31. $\lim\limits_{x\to-\infty}(x^4+x^5) = \lim\limits_{x\to-\infty}x^5(\tfrac{1}{x}+1)$ [factor out the largest power of x] $= -\infty$ because $x^5\to-\infty$ and $1/x+1\to1$

as $x\to-\infty$.

Or: $\lim\limits_{x\to-\infty}(x^4+x^5) = \lim\limits_{x\to-\infty}x^4(1+x) = -\infty$.

33. $\lim\limits_{x\to\infty}\dfrac{1-e^x}{1+2e^x} = \lim\limits_{x\to\infty}\dfrac{(1-e^x)/e^x}{(1+2e^x)/e^x} = \lim\limits_{x\to\infty}\dfrac{1/e^x-1}{1/e^x+2} = \dfrac{0-1}{0+2} = -\dfrac{1}{2}$

35. Since $-1\le\cos x\le1$ and $e^{-2x}>0$, we have $-e^{-2x}\le e^{-2x}\cos x\le e^{-2x}$. We know that $\lim\limits_{x\to\infty}(-e^{-2x})=0$ and

$\lim\limits_{x\to\infty}(e^{-2x})=0$, so by the Squeeze Theorem, $\lim\limits_{x\to\infty}(e^{-2x}\cos x)=0$.

37. (a)

From the graph of $f(x)=\sqrt{x^2+x+1}+x$, we estimate the value of $\lim\limits_{x\to-\infty}f(x)$ to be -0.5.

(b)

x	$f(x)$
$-10{,}000$	-0.4999625
$-100{,}000$	-0.4999962
$-1{,}000{,}000$	-0.4999996

From the table, we estimate the limit to be -0.5.

(c) $\displaystyle\lim_{x\to-\infty}\left(\sqrt{x^2+x+1}+x\right) = \lim_{x\to-\infty}\left(\sqrt{x^2+x+1}+x\right)\left[\frac{\sqrt{x^2+x+1}-x}{\sqrt{x^2+x+1}-x}\right] = \lim_{x\to-\infty}\frac{(x^2+x+1)-x^2}{\sqrt{x^2+x+1}-x}$

$\displaystyle = \lim_{x\to-\infty}\frac{(x+1)(1/x)}{\left(\sqrt{x^2+x+1}-x\right)(1/x)} = \lim_{x\to-\infty}\frac{1+(1/x)}{-\sqrt{1+(1/x)+(1/x^2)}-1}$

$\displaystyle = \frac{1+0}{-\sqrt{1+0+0}-1} = -\frac{1}{2}$

Note that for $x < 0$, we have $\sqrt{x^2} = |x| = -x$, so when we divide the radical by x, with $x < 0$, we get

$\displaystyle\frac{1}{x}\sqrt{x^2+x+1} = -\frac{1}{\sqrt{x^2}}\sqrt{x^2+x+1} = -\sqrt{1+(1/x)+(1/x^2)}.$

39. $\displaystyle\lim_{x\to\infty}\frac{2x+1}{x-2} = \lim_{x\to\infty}\frac{\dfrac{2x+1}{x}}{\dfrac{x-2}{x}} = \lim_{x\to\infty}\frac{2+\dfrac{1}{x}}{1-\dfrac{2}{x}} = \frac{\displaystyle\lim_{x\to\infty}\left(2+\frac{1}{x}\right)}{\displaystyle\lim_{x\to\infty}\left(1-\frac{2}{x}\right)} = \frac{\displaystyle\lim_{x\to\infty}2+\lim_{x\to\infty}\frac{1}{x}}{\displaystyle\lim_{x\to\infty}1-\lim_{x\to\infty}\frac{2}{x}}$

$\displaystyle = \frac{2+0}{1-0} = 2,$ so $y = 2$ is a horizontal asymptote.

The denominator $x - 2$ is zero when $x = 2$ and the numerator is not zero, so we

investigate $y = f(x) = \dfrac{2x+1}{x-2}$ as x approaches 2. $\displaystyle\lim_{x\to2^-}f(x) = -\infty$ because as

$x \to 2^-$ the numerator is positive and the denominator approaches 0 through

negative values. Similarly, $\displaystyle\lim_{x\to2^+}f(x) = \infty$. Thus, $x = 2$ is a vertical asymptote.

The graph confirms our work.

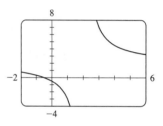

41. $\displaystyle\lim_{x\to\infty}\frac{2x^2+x-1}{x^2+x-2} = \lim_{x\to\infty}\frac{\dfrac{2x^2+x-1}{x^2}}{\dfrac{x^2+x-2}{x^2}} = \lim_{x\to\infty}\frac{2+\dfrac{1}{x}-\dfrac{1}{x^2}}{1+\dfrac{1}{x}-\dfrac{2}{x^2}} = \frac{\displaystyle\lim_{x\to\infty}\left(2+\frac{1}{x}-\frac{1}{x^2}\right)}{\displaystyle\lim_{x\to\infty}\left(1+\frac{1}{x}-\frac{2}{x^2}\right)}$

$\displaystyle = \frac{\displaystyle\lim_{x\to\infty}2+\lim_{x\to\infty}\frac{1}{x}-\lim_{x\to\infty}\frac{1}{x^2}}{\displaystyle\lim_{x\to\infty}1+\lim_{x\to\infty}\frac{1}{x}-2\lim_{x\to\infty}\frac{1}{x^2}} = \frac{2+0-0}{1+0-2(0)} = 2,$ so $y = 2$ is a horizontal asymptote.

$y = f(x) = \dfrac{2x^2+x-1}{x^2+x-2} = \dfrac{(2x-1)(x+1)}{(x+2)(x-1)},$ so $\displaystyle\lim_{x\to-2^-}f(x) = \infty,$

$\displaystyle\lim_{x\to-2^+}f(x) = -\infty,\ \lim_{x\to1^-}f(x) = -\infty,$ and $\displaystyle\lim_{x\to1^+}f(x) = \infty.$ Thus, $x = -2$

and $x = 1$ are vertical asymptotes. The graph confirms our work.

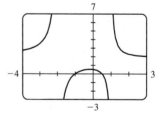

43. $y = f(x) = \dfrac{x^3-x}{x^2-6x+5} = \dfrac{x(x^2-1)}{(x-1)(x-5)} = \dfrac{x(x+1)(x-1)}{(x-1)(x-5)} = \dfrac{x(x+1)}{x-5} = g(x)$ for $x \neq 1.$

The graph of g is the same as the graph of f with the exception of a hole in the

graph of f at $x = 1$. By long division, $g(x) = \dfrac{x^2+x}{x-5} = x+6+\dfrac{30}{x-5}.$

As $x \to \pm\infty,\ g(x) \to \pm\infty,$ so there is no horizontal asymptote. The denominator

of g is zero when $x = 5$. $\displaystyle\lim_{x\to5^-}g(x) = -\infty$ and $\displaystyle\lim_{x\to5^+}g(x) = \infty,$ so $x = 5$ is a

vertical asymptote. The graph confirms our work.

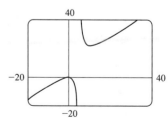

45. From the graph, it appears $y = 1$ is a horizontal asymptote.

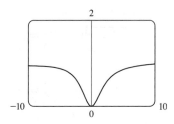

$$\lim_{x \to \infty} \frac{3x^3 + 500x^2}{x^3 + 500x^2 + 100x + 2000} = \lim_{x \to \infty} \frac{\dfrac{3x^3 + 500x^2}{x^3}}{\dfrac{x^3 + 500x^2 + 100x + 2000}{x^3}} = \lim_{x \to \infty} \frac{3 + (500/x)}{1 + (500/x) + (100/x^2) + (2000/x^3)}$$

$$= \frac{3 + 0}{1 + 0 + 0 + 0} = 3, \quad \text{so } y = 3 \text{ is a horizontal asymptote.}$$

The discrepancy can be explained by the choice of the viewing window. Try
$[-100{,}000,\ 100{,}000]$ by $[-1, 4]$ to get a graph that lends credibility to our
calculation that $y = 3$ is a horizontal asymptote.

47. Let's look for a rational function.

    (1) $\displaystyle\lim_{x \to \pm\infty} f(x) = 0 \;\Rightarrow\;$ degree of numerator $<$ degree of denominator

    (2) $\displaystyle\lim_{x \to 0} f(x) = -\infty \;\Rightarrow\;$ there is a factor of x^2 in the denominator (not just x, since that would produce a sign
        change at $x = 0$), and the function is negative near $x = 0$.

    (3) $\displaystyle\lim_{x \to 3^-} f(x) = \infty$ and $\displaystyle\lim_{x \to 3^+} f(x) = -\infty \;\Rightarrow\;$ vertical asymptote at $x = 3$; there is a factor of $(x - 3)$ in the
        denominator.

    (4) $f(2) = 0 \;\Rightarrow\;$ 2 is an x-intercept; there is at least one factor of $(x - 2)$ in the numerator.

Combining all of this information and putting in a negative sign to give us the desired left- and right-hand limits gives us
$f(x) = \dfrac{2 - x}{x^2(x - 3)}$ as one possibility.

49. $y = f(x) = x^4 - x^6 = x^4(1 - x^2) = x^4(1 + x)(1 - x)$. The y-intercept is

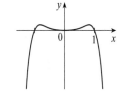

    $f(0) = 0$. The x-intercepts are 0, -1, and 1 [found by solving $f(x) = 0$ for x].

    Since $x^4 > 0$ for $x \neq 0$, f doesn't change sign at $x = 0$. The function does change

    sign at $x = -1$ and $x = 1$. As $x \to \pm\infty$, $f(x) = x^4(1 - x^2)$ approaches $-\infty$

    because $x^4 \to \infty$ and $(1 - x^2) \to -\infty$.

51. $y = f(x) = (3 - x)(1 + x)^2(1 - x)^4$. The y-intercept is $f(0) = 3(1)^2(1)^4 = 3$.

    The x-intercepts are 3, -1, and 1. There is a sign change at 3, but not at -1 and 1.

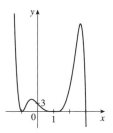

    When x is large positive, $3 - x$ is negative and the other factors are positive, so

    $\displaystyle\lim_{x \to \infty} f(x) = -\infty$. When x is large negative, $3 - x$ is positive, so

    $\displaystyle\lim_{x \to -\infty} f(x) = \infty$.

53. (a) Since $-1 \leq \sin x \leq 1$ for all x, $-\dfrac{1}{x} \leq \dfrac{\sin x}{x} \leq \dfrac{1}{x}$ for $x > 0$. As $x \to \infty$, $-1/x \to 0$ and $1/x \to 0$, so by the Squeeze

    Theorem, $(\sin x)/x \to 0$. Thus, $\displaystyle\lim_{x \to \infty} \frac{\sin x}{x} = 0$.

(b) From part (a), the horizontal asymptote is $y = 0$. The function
$y = (\sin x)/x$ crosses the horizontal asymptote whenever $\sin x = 0$;
that is, at $x = \pi n$ for every integer n. Thus, the graph crosses the
asymptote *an infinite number of times.*

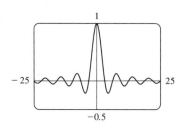

55. Divide the numerator and the denominator by the highest power of x in $Q(x)$.

(a) If $\deg P < \deg Q$, then the numerator $\to 0$ but the denominator doesn't. So $\lim\limits_{x\to\infty} [P(x)/Q(x)] = 0$.

(b) If $\deg P > \deg Q$, then the numerator $\to \pm\infty$ but the denominator doesn't, so $\lim\limits_{x\to\infty} [P(x)/Q(x)] = \pm\infty$

(depending on the ratio of the leading coefficients of P and Q).

57. $\lim\limits_{x\to\infty} \dfrac{5\sqrt{x}}{\sqrt{x-1}} \cdot \dfrac{1/\sqrt{x}}{1/\sqrt{x}} = \lim\limits_{x\to\infty} \dfrac{5}{\sqrt{1-(1/x)}} = \dfrac{5}{\sqrt{1-0}} = 5$ and

$\lim\limits_{x\to\infty} \dfrac{10e^x - 21}{2e^x} \cdot \dfrac{1/e^x}{1/e^x} = \lim\limits_{x\to\infty} \dfrac{10 - (21/e^x)}{2} = \dfrac{10-0}{2} = 5$. Since $\dfrac{10e^x - 21}{2e^x} < f(x) < \dfrac{5\sqrt{x}}{\sqrt{x-1}}$,

we have $\lim\limits_{x\to\infty} f(x) = 5$ by the Squeeze Theorem.

59. (a) $\lim\limits_{t\to\infty} v(t) = \lim\limits_{t\to\infty} v^*\left(1 - e^{-gt/v^*}\right) = v^*(1-0) = v^*$

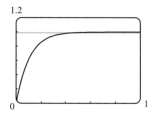

(b) We graph $v(t) = 1 - e^{-9.8t}$ and $v(t) = 0.99v^*$, or in this case,
$v(t) = 0.99$. Using an intersect feature or zooming in on the point of
intersection, we find that $t \approx 0.47$ s.

61. Let $g(x) = \dfrac{3x^2 + 1}{2x^2 + x + 1}$ and $f(x) = |g(x) - 1.5|$. Note that

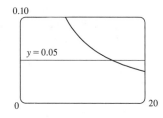

$\lim\limits_{x\to\infty} g(x) = \frac{3}{2}$ and $\lim\limits_{x\to\infty} f(x) = 0$. We are interested in finding the

x-value at which $f(x) < 0.05$. From the graph, we find that $x \approx 14.804$,
so we choose $N = 15$ (or any larger number).

63. For $\varepsilon = 0.5$, we need to find N such that $\left| \dfrac{\sqrt{4x^2 + 1}}{x+1} - (-2) \right| < 0.5$ $\Leftrightarrow$

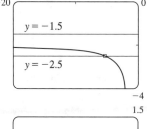

$-2.5 < \dfrac{\sqrt{4x^2 + 1}}{x+1} < -1.5$ whenever $x \le N$. We graph the three parts of this

inequality on the same screen, and see that the inequality holds for $x \le -6$. So we
choose $N = -6$ (or any smaller number).

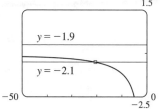

For $\varepsilon = 0.1$, we need $-2.1 < \dfrac{\sqrt{4x^2 + 1}}{x+1} < -1.9$ whenever $x \le N$. From the

graph, it seems that this inequality holds for $x \le -22$. So we choose $N = -22$

(or any smaller number).

65. (a) $1/x^2 < 0.0001 \iff x^2 > 1/0.0001 = 10\,000 \iff x > 100 \quad (x > 0)$

(b) If $\varepsilon > 0$ is given, then $1/x^2 < \varepsilon \iff x^2 > 1/\varepsilon \iff x > 1/\sqrt{\varepsilon}$. Let $N = 1/\sqrt{\varepsilon}$.

Then $x > N \implies x > \dfrac{1}{\sqrt{\varepsilon}} \implies \left| \dfrac{1}{x^2} - 0 \right| = \dfrac{1}{x^2} < \varepsilon$, so $\displaystyle\lim_{x \to \infty} \dfrac{1}{x^2} = 0$.

67. For $x < 0$, $|1/x - 0| = -1/x$. If $\varepsilon > 0$ is given, then $-1/x < \varepsilon \iff x < -1/\varepsilon$.

Take $N = -1/\varepsilon$. Then $x < N \implies x < -1/\varepsilon \implies |(1/x) - 0| = -1/x < \varepsilon$, so $\displaystyle\lim_{x \to -\infty} (1/x) = 0$.

69. Given $M > 0$, we need $N > 0$ such that $x > N \implies e^x > M$. Now $e^x > M \iff x > \ln M$, so take

$N = \max(1, \ln M)$. (This ensures that $N > 0$.) Then $x > N = \max(1, \ln M) \implies e^x > \max(e, M) \geq M$,

so $\displaystyle\lim_{x \to \infty} e^x = \infty$.

71. Suppose that $\displaystyle\lim_{x \to \infty} f(x) = L$. Then for every $\varepsilon > 0$ there is a corresponding positive number N such that $|f(x) - L| < \varepsilon$

whenever $x > N$. If $t = 1/x$, then $x > N \iff 0 < 1/x < 1/N \iff 0 < t < 1/N$. Thus, for every $\varepsilon > 0$ there is a

corresponding $\delta > 0$ (namely $1/N$) such that $|f(1/t) - L| < \varepsilon$ whenever $0 < t < \delta$. This proves that

$\displaystyle\lim_{t \to 0^+} f(1/t) = L = \lim_{x \to \infty} f(x)$.

Now suppose that $\displaystyle\lim_{x \to -\infty} f(x) = L$. Then for every $\varepsilon > 0$ there is a corresponding negative number N such that

$|f(x) - L| < \varepsilon$ whenever $x < N$. If $t = 1/x$, then $x < N \iff 1/N < 1/x < 0 \iff 1/N < t < 0$. Thus, for every

$\varepsilon > 0$ there is a corresponding $\delta > 0$ (namely $-1/N$) such that $|f(1/t) - L| < \varepsilon$ whenever $-\delta < t < 0$. This proves that

$\displaystyle\lim_{t \to 0^-} f(1/t) = L = \lim_{x \to -\infty} f(x)$.

2.7 Derivatives and Rates of Change

1. (a) This is just the slope of the line through two points: $m_{PQ} = \dfrac{\Delta y}{\Delta x} = \dfrac{f(x) - f(3)}{x - 3}$.

(b) This is the limit of the slope of the secant line PQ as Q approaches P: $m = \displaystyle\lim_{x \to 3} \dfrac{f(x) - f(3)}{x - 3}$.

3. (a) (i) Using Definition 1 with $f(x) = 4x - x^2$ and $P(1, 3)$,

$$m = \lim_{x \to a} \frac{f(x) - f(a)}{x - a} = \lim_{x \to 1} \frac{(4x - x^2) - 3}{x - 1} = \lim_{x \to 1} \frac{-(x^2 - 4x + 3)}{x - 1} = \lim_{x \to 1} \frac{-(x - 1)(x - 3)}{x - 1}$$

$$= \lim_{x \to 1} (3 - x) = 3 - 1 = 2$$

(ii) Using Equation 2 with $f(x) = 4x - x^2$ and $P(1, 3)$,

$$m = \lim_{h \to 0} \frac{f(a + h) - f(a)}{h} = \lim_{h \to 0} \frac{f(1 + h) - f(1)}{h} = \lim_{h \to 0} \frac{\left[4(1 + h) - (1 + h)^2\right] - 3}{h}$$

$$= \lim_{h \to 0} \frac{4 + 4h - 1 - 2h - h^2 - 3}{h} = \lim_{h \to 0} \frac{-h^2 + 2h}{h} = \lim_{h \to 0} \frac{h(-h + 2)}{h} = \lim_{h \to 0} (-h + 2) = 2$$

(b) An equation of the tangent line is $y - f(a) = f'(a)(x - a) \implies y - f(1) = f'(1)(x - 1) \implies y - 3 = 2(x - 1)$,

or $y = 2x + 1$.

(c)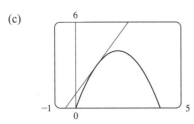

The graph of $y = 2x + 1$ is tangent to the graph of $y = 4x - x^2$ at the point $(1, 3)$. Now zoom in toward the point $(1, 3)$ until the parabola and the tangent line are indistiguishable.

5. Using (1) with $f(x) = \dfrac{x-1}{x-2}$ and $P(3, 2)$,

$$m = \lim_{x \to a} \frac{f(x) - f(a)}{x - a} = \lim_{x \to 3} \frac{\dfrac{x-1}{x-2} - 2}{x - 3} = \lim_{x \to 3} \frac{\dfrac{x-1-2(x-2)}{x-2}}{x - 3}$$

$$= \lim_{x \to 3} \frac{3 - x}{(x-2)(x-3)} = \lim_{x \to 3} \frac{-1}{x-2} = \frac{-1}{1} = -1$$

Tangent line: $y - 2 = -1(x - 3) \iff y - 2 = -x + 3 \iff y = -x + 5$

7. Using (1), $m = \lim\limits_{x \to 1} \dfrac{\sqrt{x} - \sqrt{1}}{x - 1} = \lim\limits_{x \to 1} \dfrac{(\sqrt{x} - 1)(\sqrt{x} + 1)}{(x-1)(\sqrt{x} + 1)} = \lim\limits_{x \to 1} \dfrac{x-1}{(x-1)(\sqrt{x}+1)} = \lim\limits_{x \to 1} \dfrac{1}{\sqrt{x}+1} = \dfrac{1}{2}$.

Tangent line: $y - 1 = \frac{1}{2}(x - 1) \iff y = \frac{1}{2}x + \frac{1}{2}$

9. (a) Using (2) with $y = f(x) = 3 + 4x^2 - 2x^3$,

$$m = \lim_{h \to 0} \frac{f(a+h) - f(a)}{h} = \lim_{h \to 0} \frac{3 + 4(a+h)^2 - 2(a+h)^3 - (3 + 4a^2 - 2a^3)}{h}$$

$$= \lim_{h \to 0} \frac{3 + 4(a^2 + 2ah + h^2) - 2(a^3 + 3a^2h + 3ah^2 + h^3) - 3 - 4a^2 + 2a^3}{h}$$

$$= \lim_{h \to 0} \frac{3 + 4a^2 + 8ah + 4h^2 - 2a^3 - 6a^2h - 6ah^2 - 2h^3 - 3 - 4a^2 + 2a^3}{h}$$

$$= \lim_{h \to 0} \frac{8ah + 4h^2 - 6a^2h - 6ah^2 - 2h^3}{h} = \lim_{h \to 0} \frac{h(8a + 4h - 6a^2 - 6ah - 2h^2)}{h}$$

$$= \lim_{h \to 0} (8a + 4h - 6a^2 - 6ah - 2h^2) = 8a - 6a^2$$

(b) At $(1, 5)$: $m = 8(1) - 6(1)^2 = 2$, so an equation of the tangent line is $y - 5 = 2(x - 1) \iff y = 2x + 3$.

At $(2, 3)$: $m = 8(2) - 6(2)^2 = -8$, so an equation of the tangent line is $y - 3 = -8(x - 2) \iff y = -8x + 19$.

(c)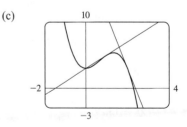

11. (a) The particle is moving to the right when s is increasing; that is, on the intervals $(0, 1)$ and $(4, 6)$. The particle is moving to the left when s is decreasing; that is, on the interval $(2, 3)$. The particle is standing still when s is constant; that is, on the intervals $(1, 2)$ and $(3, 4)$.

(b) The velocity of the particle is equal to the slope of the tangent line of the

graph. Note that there is no slope at the corner points on the graph. On the

interval $(0, 1)$, the slope is $\dfrac{3 - 0}{1 - 0} = 3$. On the interval $(2, 3)$, the slope is

$\dfrac{1 - 3}{3 - 2} = -2$. On the interval $(4, 6)$, the slope is $\dfrac{3 - 1}{6 - 4} = 1$.

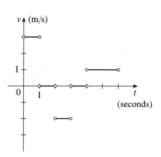

13. Let $s(t) = 40t - 16t^2$.

$$v(2) = \lim_{t \to 2} \frac{s(t) - s(2)}{t - 2} = \lim_{t \to 2} \frac{(40t - 16t^2) - 16}{t - 2} = \lim_{t \to 2} \frac{-16t^2 + 40t - 16}{t - 2} = \lim_{t \to 2} \frac{-8(2t^2 - 5t + 2)}{t - 2}$$

$$= \lim_{t \to 2} \frac{-8(t - 2)(2t - 1)}{t - 2} = -8 \lim_{t \to 2}(2t - 1) = -8(3) = -24$$

Thus, the instantaneous velocity when $t = 2$ is -24 ft/s.

15. $v(a) = \lim\limits_{h \to 0} \dfrac{s(a + h) - s(a)}{h} = \lim\limits_{h \to 0} \dfrac{\dfrac{1}{(a + h)^2} - \dfrac{1}{a^2}}{h} = \lim\limits_{h \to 0} \dfrac{\dfrac{a^2 - (a + h)^2}{a^2(a + h)^2}}{h} = \lim\limits_{h \to 0} \dfrac{a^2 - (a^2 + 2ah + h^2)}{ha^2(a + h)^2}$

$= \lim\limits_{h \to 0} \dfrac{-(2ah + h^2)}{ha^2(a + h)^2} = \lim\limits_{h \to 0} \dfrac{-h(2a + h)}{ha^2(a + h)^2} = \lim\limits_{h \to 0} \dfrac{-(2a + h)}{a^2(a + h)^2} = \dfrac{-2a}{a^2 \cdot a^2} = \dfrac{-2}{a^3}$ m/s

So $v(1) = \dfrac{-2}{1^3} = -2$ m/s, $v(2) = \dfrac{-2}{2^3} = -\dfrac{1}{4}$ m/s, and $v(3) = \dfrac{-2}{3^3} = -\dfrac{2}{27}$ m/s.

17. $g'(0)$ is the only negative value. The slope at $x = 4$ is smaller than the slope at $x = 2$ and both are smaller than the slope at

$x = -2$. Thus, $g'(0) < 0 < g'(4) < g'(2) < g'(-2)$.

19. We begin by drawing a curve through the origin with a

slope of 3 to satisfy $f(0) = 0$ and $f'(0) = 3$. Since

$f'(1) = 0$, we will round off our figure so that there is

a horizontal tangent directly over $x = 1$. Last, we

make sure that the curve has a slope of -1 as we pass

over $x = 2$. Two of the many possibilities are shown.

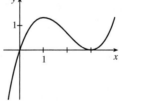 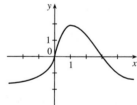

21. Using Definition 2 with $f(x) = 3x^2 - 5x$ and the point $(2, 2)$, we have

$$f'(2) = \lim_{h \to 0} \frac{f(2 + h) - f(2)}{h} = \lim_{h \to 0} \frac{[3(2 + h)^2 - 5(2 + h)] - 2}{h} = \lim_{h \to 0} \frac{(12 + 12h + 3h^2 - 10 - 5h) - 2}{h}$$

$$= \lim_{h \to 0} \frac{3h^2 + 7h}{h} = \lim_{h \to 0}(3h + 7) = 7$$

So an equation of the tangent line at $(2, 2)$ is $y - 2 = 7(x - 2)$ or $y = 7x - 12$.

23. (a) Using Definition 2 with $F(x) = 5x/(1 + x^2)$ and the point $(2, 2)$, we have

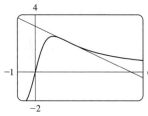

$$F'(2) = \lim_{h \to 0} \frac{F(2 + h) - F(2)}{h} = \lim_{h \to 0} \frac{\dfrac{5(2 + h)}{1 + (2 + h)^2} - 2}{h}$$

$$= \lim_{h \to 0} \frac{\dfrac{5h + 10}{h^2 + 4h + 5} - 2}{h} = \lim_{h \to 0} \frac{\dfrac{5h + 10 - 2(h^2 + 4h + 5)}{h^2 + 4h + 5}}{h}$$

$$= \lim_{h \to 0} \frac{-2h^2 - 3h}{h(h^2 + 4h + 5)} = \lim_{h \to 0} \frac{h(-2h - 3)}{h(h^2 + 4h + 5)} = \lim_{h \to 0} \frac{-2h - 3}{h^2 + 4h + 5} = \frac{-3}{5}$$

So an equation of the tangent line at $(2, 2)$ is $y - 2 = -\frac{3}{5}(x - 2)$ or $y = -\frac{3}{5}x + \frac{16}{5}$.

25. Use Definition 2 with $f(x) = 3 - 2x + 4x^2$.

$$f'(a) = \lim_{h \to 0} \frac{f(a + h) - f(a)}{h} = \lim_{h \to 0} \frac{[3 - 2(a + h) + 4(a + h)^2] - (3 - 2a + 4a^2)}{h}$$

$$= \lim_{h \to 0} \frac{(3 - 2a - 2h + 4a^2 + 8ah + 4h^2) - (3 - 2a + 4a^2)}{h}$$

$$= \lim_{h \to 0} \frac{-2h + 8ah + 4h^2}{h} = \lim_{h \to 0} \frac{h(-2 + 8a + 4h)}{h} = \lim_{h \to 0} (-2 + 8a + 4h) = -2 + 8a$$

27. Use Definition 2 with $f(t) = (2t + 1)/(t + 3)$.

$$f'(a) = \lim_{h \to 0} \frac{f(a + h) - f(a)}{h} = \lim_{h \to 0} \frac{\dfrac{2(a + h) + 1}{(a + h) + 3} - \dfrac{2a + 1}{a + 3}}{h} = \lim_{h \to 0} \frac{(2a + 2h + 1)(a + 3) - (2a + 1)(a + h + 3)}{h(a + h + 3)(a + 3)}$$

$$= \lim_{h \to 0} \frac{(2a^2 + 6a + 2ah + 6h + a + 3) - (2a^2 + 2ah + 6a + a + h + 3)}{h(a + h + 3)(a + 3)}$$

$$= \lim_{h \to 0} \frac{5h}{h(a + h + 3)(a + 3)} = \lim_{h \to 0} \frac{5}{(a + h + 3)(a + 3)} = \frac{5}{(a + 3)^2}$$

29. Use Definition 2 with $f(x) = 1/\sqrt{x + 2}$.

$$f'(a) = \lim_{h \to 0} \frac{f(a + h) - f(a)}{h} = \lim_{h \to 0} \frac{\dfrac{1}{\sqrt{(a + h) + 2}} - \dfrac{1}{\sqrt{a + 2}}}{h} = \lim_{h \to 0} \frac{\dfrac{\sqrt{a + 2} - \sqrt{a + h + 2}}{\sqrt{a + h + 2}\sqrt{a + 2}}}{h}$$

$$= \lim_{h \to 0} \left[\frac{\sqrt{a + 2} - \sqrt{a + h + 2}}{h\sqrt{a + h + 2}\sqrt{a + 2}} \cdot \frac{\sqrt{a + 2} + \sqrt{a + h + 2}}{\sqrt{a + 2} + \sqrt{a + h + 2}} \right] = \lim_{h \to 0} \frac{(a + 2) - (a + h + 2)}{h\sqrt{a + h + 2}\sqrt{a + 2}\left(\sqrt{a + 2} + \sqrt{a + h + 2}\right)}$$

$$= \lim_{h \to 0} \frac{-h}{h\sqrt{a + h + 2}\sqrt{a + 2}\left(\sqrt{a + 2} + \sqrt{a + h + 2}\right)} = \lim_{h \to 0} \frac{-1}{\sqrt{a + h + 2}\sqrt{a + 2}\left(\sqrt{a + 2} + \sqrt{a + h + 2}\right)}$$

$$= \frac{-1}{\left(\sqrt{a + 2}\right)^2 \left(2\sqrt{a + 2}\right)} = -\frac{1}{2(a + 2)^{3/2}}$$

Note that the answers to Exercises 31 – 36 are not unique.

31. By Definition 2, $\lim_{h \to 0} \dfrac{(1 + h)^{10} - 1}{h} = f'(1)$, where $f(x) = x^{10}$ and $a = 1$.

Or: By Definition 2, $\lim_{h \to 0} \dfrac{(1 + h)^{10} - 1}{h} = f'(0)$, where $f(x) = (1 + x)^{10}$ and $a = 0$.

33. By Equation 3, $\lim\limits_{x\to 5}\dfrac{2^x-32}{x-5}=f'(5)$, where $f(x)=2^x$ and $a=5$.

35. By Definition 2, $\lim\limits_{h\to 0}\dfrac{\cos(\pi+h)+1}{h}=f'(\pi)$, where $f(x)=\cos x$ and $a=\pi$.

Or: By Definition 2, $\lim\limits_{h\to 0}\dfrac{\cos(\pi+h)+1}{h}=f'(0)$, where $f(x)=\cos(\pi+x)$ and $a=0$.

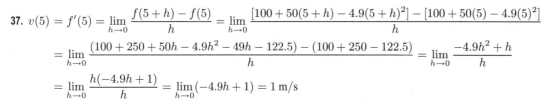

37. $v(5)=f'(5)=\lim\limits_{h\to 0}\dfrac{f(5+h)-f(5)}{h}=\lim\limits_{h\to 0}\dfrac{[100+50(5+h)-4.9(5+h)^2]-[100+50(5)-4.9(5)^2]}{h}$

$=\lim\limits_{h\to 0}\dfrac{(100+250+50h-4.9h^2-49h-122.5)-(100+250-122.5)}{h}=\lim\limits_{h\to 0}\dfrac{-4.9h^2+h}{h}$

$=\lim\limits_{h\to 0}\dfrac{h(-4.9h+1)}{h}=\lim\limits_{h\to 0}(-4.9h+1)=1\text{ m/s}$

The speed when $t=5$ is $|1|=1\text{ m/s}$.

39. The sketch shows the graph for a room temperature of $72°$ and a refrigerator temperature of $38°$. The initial rate of change is greater in magnitude than the rate of change after an hour.

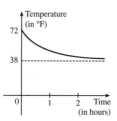

41. (a) (i) $[2000,2002]$: $\dfrac{P(2002)-P(2000)}{2002-2000}=\dfrac{77-55}{2}=\dfrac{22}{2}=11$ percent/year

(ii) $[2000,2001]$: $\dfrac{P(2001)-P(2000)}{2001-2000}=\dfrac{68-55}{1}=13$ percent/year

(iii) $[1999,2000]$: $\dfrac{P(2000)-P(1999)}{2000-1999}=\dfrac{55-39}{1}=16$ percent/year

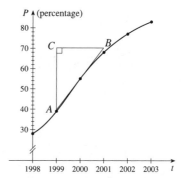

(b) Using the values from (ii) and (iii), we have $\dfrac{13+16}{2}=14.5$ percent/year.

(c) Estimating A as $(1999,40)$ and B as $(2001,70)$, the slope at 2000 is

$\dfrac{70-40}{2001-1999}=\dfrac{30}{2}=15$ percent/year.

43. (a) (i) $\dfrac{\Delta C}{\Delta x}=\dfrac{C(105)-C(100)}{105-100}=\dfrac{6601.25-6500}{5}=\$20.25/\text{unit}.$

(ii) $\dfrac{\Delta C}{\Delta x}=\dfrac{C(101)-C(100)}{101-100}=\dfrac{6520.05-6500}{1}=\$20.05/\text{unit}.$

(b) $\dfrac{C(100+h)-C(100)}{h}=\dfrac{[5000+10(100+h)+0.05(100+h)^2]-6500}{h}=\dfrac{20h+0.05h^2}{h}$

$=20+0.05h,\ h\neq 0$

So the instantaneous rate of change is $\lim\limits_{h\to 0}\dfrac{C(100+h)-C(100)}{h}=\lim\limits_{h\to 0}(20+0.05h)=\$20/\text{unit}.$

45. (a) $f'(x)$ is the rate of change of the production cost with respect to the number of ounces of gold produced. Its units are dollars per ounce.

(b) After 800 ounces of gold have been produced, the rate at which the production cost is increasing is $17/ounce. So the cost of producing the 800th (or 801st) ounce is about $17.

(c) In the short term, the values of $f'(x)$ will decrease because more efficient use is made of start-up costs as x increases. But eventually $f'(x)$ might increase due to large-scale operations.

47. $T'(10)$ is the rate at which the temperature is changing at 10:00 AM. To estimate the value of $T'(10)$, we will average the difference quotients obtained using the times $t = 8$ and $t = 12$. Let $A = \dfrac{T(8) - T(10)}{8 - 10} = \dfrac{72 - 81}{-2} = 4.5$ and

$B = \dfrac{T(12) - T(10)}{12 - 10} = \dfrac{88 - 81}{2} = 3.5$. Then $T'(10) = \lim\limits_{t \to 10} \dfrac{T(t) - T(10)}{t - 10} \approx \dfrac{A + B}{2} = \dfrac{4.5 + 3.5}{2} = 4°\text{F/h}$.

49. (a) $S'(T)$ is the rate at which the oxygen solubility changes with respect to the water temperature. Its units are $(\text{mg/L})/°\text{C}$.

(b) For $T = 16°\text{C}$, it appears that the tangent line to the curve goes through the points $(0, 14)$ and $(32, 6)$. So

$S'(16) \approx \dfrac{6 - 14}{32 - 0} = -\dfrac{8}{32} = -0.25 \ (\text{mg/L})/°\text{C}$. This means that as the temperature increases past $16°\text{C}$, the oxygen solubility is decreasing at a rate of $0.25 \ (\text{mg/L})/°\text{C}$.

51. Since $f(x) = x \sin(1/x)$ when $x \neq 0$ and $f(0) = 0$, we have

$f'(0) = \lim\limits_{h \to 0} \dfrac{f(0 + h) - f(0)}{h} = \lim\limits_{h \to 0} \dfrac{h \sin(1/h) - 0}{h} = \lim\limits_{h \to 0} \sin(1/h)$. This limit does not exist since $\sin(1/h)$ takes the values -1 and 1 on any interval containing 0. (Compare with Example 4 in Section 2.2.)

2.8 The Derivative as a Function

1. It appears that f is an odd function, so f' will be an even function—that is, $f'(-a) = f'(a)$.

(a) $f'(-3) \approx 1.5$ (b) $f'(-2) \approx 1$

(c) $f'(-1) \approx 0$ (d) $f'(0) \approx -4$

(e) $f'(1) \approx 0$ (f) $f'(2) \approx 1$

(g) $f'(3) \approx 1.5$

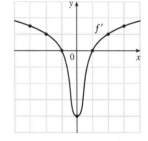

3. (a)$' =$ II, since from left to right, the slopes of the tangents to graph (a) start out negative, become 0, then positive, then 0, then negative again. The actual function values in graph II follow the same pattern.

(b)$' =$ IV, since from left to right, the slopes of the tangents to graph (b) start out at a fixed positive quantity, then suddenly become negative, then positive again. The discontinuities in graph IV indicate sudden changes in the slopes of the tangents.

(c)$' =$ I, since the slopes of the tangents to graph (c) are negative for $x < 0$ and positive for $x > 0$, as are the function values of graph I.

(d)$' =$ III, since from left to right, the slopes of the tangents to graph (d) are positive, then 0, then negative, then 0, then positive, then 0, then negative again, and the function values in graph III follow the same pattern.

Hints for Exercises 4–11: First plot x-intercepts on the graph of f' for any horizontal tangents on the graph of f. Look for any corners on the graph of f—there will be a discontinuity on the graph of f'. On any interval where f has a tangent with positive (or negative) slope, the graph of f' will be positive (or negative). If the graph of the function is linear, the graph of f' will be a horizontal line.

5.

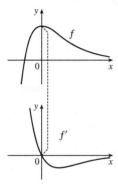

7.

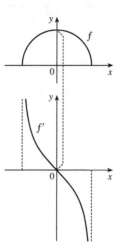

9.

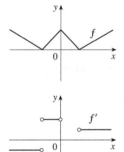

11.

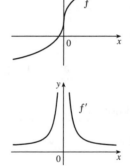

13. It appears that there are horizontal tangents on the graph of M for $t = 1963$ and $t = 1971$. Thus, there are zeros for those values of t on the graph of M'. The derivative is negative for the years 1963 to 1971.

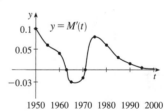

15.

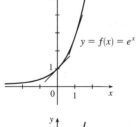

The slope at 0 appears to be 1 and the slope at 1 appears to be 2.7. As x decreases, the slope gets closer to 0. Since the graphs are so similar, we might guess that $f'(x) = e^x$.

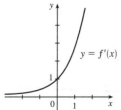

17. (a) By zooming in, we estimate that $f'(0) = 0$, $f'\left(\frac{1}{2}\right) = 1$, $f'(1) = 2$,

and $f'(2) = 4$.

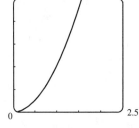

(b) By symmetry, $f'(-x) = -f'(x)$. So $f'\left(-\frac{1}{2}\right) = -1$, $f'(-1) = -2$,

and $f'(-2) = -4$.

(c) It appears that $f'(x)$ is twice the value of x, so we guess that $f'(x) = 2x$.

(d) $f'(x) = \lim\limits_{h \to 0} \dfrac{f(x+h) - f(x)}{h} = \lim\limits_{h \to 0} \dfrac{(x+h)^2 - x^2}{h}$

$= \lim\limits_{h \to 0} \dfrac{(x^2 + 2hx + h^2) - x^2}{h} = \lim\limits_{h \to 0} \dfrac{2hx + h^2}{h} = \lim\limits_{h \to 0} \dfrac{h(2x + h)}{h} = \lim\limits_{h \to 0} (2x + h) = 2x$

19. $f'(x) = \lim\limits_{h \to 0} \dfrac{f(x+h) - f(x)}{h} = \lim\limits_{h \to 0} \dfrac{\left[\frac{1}{2}(x+h) - \frac{1}{3}\right] - \left(\frac{1}{2}x - \frac{1}{3}\right)}{h} = \lim\limits_{h \to 0} \dfrac{\frac{1}{2}x + \frac{1}{2}h - \frac{1}{3} - \frac{1}{2}x + \frac{1}{3}}{h}$

$= \lim\limits_{h \to 0} \dfrac{\frac{1}{2}h}{h} = \lim\limits_{h \to 0} \dfrac{1}{2} = \dfrac{1}{2}$

Domain of f = domain of f' = $\mathbb{R}$.

21. $f'(t) = \lim\limits_{h \to 0} \dfrac{f(t+h) - f(t)}{h} = \lim\limits_{h \to 0} \dfrac{\left[5(t+h) - 9(t+h)^2\right] - (5t - 9t^2)}{h}$

$= \lim\limits_{h \to 0} \dfrac{5t + 5h - 9(t^2 + 2th + h^2) - 5t + 9t^2}{h} = \lim\limits_{h \to 0} \dfrac{5t + 5h - 9t^2 - 18th - 9h^2 - 5t + 9t^2}{h}$

$= \lim\limits_{h \to 0} \dfrac{5h - 18th - 9h^2}{h} = \lim\limits_{h \to 0} \dfrac{h(5 - 18t - 9h)}{h} = \lim\limits_{h \to 0} (5 - 18t - 9h) = 5 - 18t$

Domain of f = domain of f' = $\mathbb{R}$.

23. $f'(x) = \lim\limits_{h \to 0} \dfrac{f(x+h) - f(x)}{h} = \lim\limits_{h \to 0} \dfrac{\left[(x+h)^3 - 3(x+h) + 5\right] - (x^3 - 3x + 5)}{h}$

$= \lim\limits_{h \to 0} \dfrac{(x^3 + 3x^2h + 3xh^2 + h^3 - 3x - 3h + 5) - (x^3 - 3x + 5)}{h} = \lim\limits_{h \to 0} \dfrac{3x^2h + 3xh^2 + h^3 - 3h}{h}$

$= \lim\limits_{h \to 0} \dfrac{h(3x^2 + 3xh + h^2 - 3)}{h} = \lim\limits_{h \to 0} (3x^2 + 3xh + h^2 - 3) = 3x^2 - 3$

Domain of f = domain of f' = $\mathbb{R}$.

25. $g'(x) = \lim\limits_{h \to 0} \dfrac{g(x+h) - g(x)}{h} = \lim\limits_{h \to 0} \dfrac{\sqrt{1 + 2(x+h)} - \sqrt{1 + 2x}}{h} \left[\dfrac{\sqrt{1 + 2(x+h)} + \sqrt{1 + 2x}}{\sqrt{1 + 2(x+h)} + \sqrt{1 + 2x}}\right]$

$= \lim\limits_{h \to 0} \dfrac{(1 + 2x + 2h) - (1 + 2x)}{h\left[\sqrt{1 + 2(x+h)} + \sqrt{1 + 2x}\right]} = \lim\limits_{h \to 0} \dfrac{2}{\sqrt{1 + 2x + 2h} + \sqrt{1 + 2x}} = \dfrac{2}{2\sqrt{1 + 2x}} = \dfrac{1}{\sqrt{1 + 2x}}$

Domain of g = $\left[-\frac{1}{2}, \infty\right)$, domain of g' = $\left(-\frac{1}{2}, \infty\right)$.

27. $G'(t) = \lim\limits_{h \to 0} \dfrac{G(t+h) - G(t)}{h} = \lim\limits_{h \to 0} \dfrac{\dfrac{4(t+h)}{(t+h)+1} - \dfrac{4t}{t+1}}{h} = \lim\limits_{h \to 0} \dfrac{\dfrac{4(t+h)(t+1) - 4t(t+h+1)}{(t+h+1)(t+1)}}{h}$

$= \lim\limits_{h \to 0} \dfrac{\left(4t^2 + 4ht + 4t + 4h\right) - \left(4t^2 + 4ht + 4t\right)}{h(t+h+1)(t+1)} = \lim\limits_{h \to 0} \dfrac{4h}{h(t+h+1)(t+1)}$

$= \lim\limits_{h \to 0} \dfrac{4}{(t+h+1)(t+1)} = \dfrac{4}{(t+1)^2}$

Domain of G = domain of $G' = (-\infty, -1) \cup (-1, \infty)$.

29. $f'(x) = \lim\limits_{h \to 0} \dfrac{f(x+h) - f(x)}{h} = \lim\limits_{h \to 0} \dfrac{(x+h)^4 - x^4}{h} = \lim\limits_{h \to 0} \dfrac{\left(x^4 + 4x^3 h + 6x^2 h^2 + 4xh^3 + h^4\right) - x^4}{h}$

$= \lim\limits_{h \to 0} \dfrac{4x^3 h + 6x^2 h^2 + 4xh^3 + h^4}{h} = \lim\limits_{h \to 0} \left(4x^3 + 6x^2 h + 4xh^2 + h^3\right) = 4x^3$

Domain of f = domain of $f' = \mathbb{R}$.

31. (a) $f'(x) = \lim\limits_{h \to 0} \dfrac{f(x+h) - f(x)}{h} = \lim\limits_{h \to 0} \dfrac{\left[(x+h)^4 + 2(x+h)\right] - (x^4 + 2x)}{h}$

$= \lim\limits_{h \to 0} \dfrac{x^4 + 4x^3 h + 6x^2 h^2 + 4xh^3 + h^4 + 2x + 2h - x^4 - 2x}{h}$

$= \lim\limits_{h \to 0} \dfrac{4x^3 h + 6x^2 h^2 + 4xh^3 + h^4 + 2h}{h} = \lim\limits_{h \to 0} \dfrac{h(4x^3 + 6x^2 h + 4xh^2 + h^3 + 2)}{h}$

$= \lim\limits_{h \to 0} \left(4x^3 + 6x^2 h + 4xh^2 + h^3 + 2\right) = 4x^3 + 2$

(b) Notice that $f'(x) = 0$ when f has a horizontal tangent, $f'(x)$ is

positive when the tangents have positive slope, and $f'(x)$ is

negative when the tangents have negative slope.

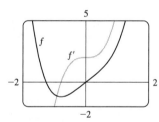

33. (a) $U'(t)$ is the rate at which the unemployment rate is changing with respect to time. Its units are percent per year.

(b) To find $U'(t)$, we use $\lim\limits_{h \to 0} \dfrac{U(t+h) - U(t)}{h} \approx \dfrac{U(t+h) - U(t)}{h}$ for small values of h.

For 1993: $U'(1993) \approx \dfrac{U(1994) - U(1993)}{1994 - 1993} = \dfrac{6.1 - 6.9}{1} = -0.80$

For 1994: We estimate $U'(1994)$ by using $h = -1$ and $h = 1$, and then average the two results to obtain a final estimate.

$h = -1 \quad \Rightarrow \quad U'(1994) \approx \dfrac{U(1993) - U(1994)}{1993 - 1994} = \dfrac{6.9 - 6.1}{-1} = -0.80;$

$h = 1 \quad \Rightarrow \quad U'(1994) \approx \dfrac{U(1995) - U(1994)}{1995 - 1994} = \dfrac{5.6 - 6.1}{1} = -0.50.$

So we estimate that $U'(1994) \approx \frac{1}{2}[(-0.80) + (-0.50)] = -0.65.$

t	1993	1994	1995	1996	1997	1998	1999	2000	2001	2002
$U'(t)$	-0.80	-0.65	-0.35	-0.35	-0.45	-0.35	-0.25	0.25	0.90	1.10

35. f is not differentiable at $x = -4$, because the graph has a corner there, and at $x = 0$, because there is a discontinuity there.

37. f is not differentiable at $x = -1$, because the graph has a vertical tangent there, and at $x = 4$, because the graph has a corner there.

39. As we zoom in toward $(-1, 0)$, the curve appears more and more like a

straight line, so $f(x) = x + \sqrt{|x|}$ is differentiable at $x = -1$. But no

matter how much we zoom in toward the origin, the curve doesn't straighten

out—we can't eliminate the sharp point (a cusp). So f is not differentiable

at $x = 0$.

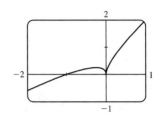

41. $a = f$, $b = f'$, $c = f''$. We can see this because where a has a horizontal tangent, $b = 0$, and where b has a horizontal tangent,

$c = 0$. We can immediately see that c can be neither f nor f', since at the points where c has a horizontal tangent, neither a

nor b is equal to 0.

43. We can immediately see that a is the graph of the acceleration function, since at the points where a has a horizontal tangent,

neither c nor b is equal to 0. Next, we note that $a = 0$ at the point where b has a horizontal tangent, so b must be the graph of

the velocity function, and hence, $b' = a$. We conclude that c is the graph of the position function.

45. $f'(x) = \lim\limits_{h \to 0} \dfrac{f(x+h) - f(x)}{h} = \lim\limits_{h \to 0} \dfrac{[1 + 4(x+h) - (x+h)^2] - (1 + 4x - x^2)}{h}$

$= \lim\limits_{h \to 0} \dfrac{(1 + 4x + 4h - x^2 - 2xh - h^2) - (1 + 4x - x^2)}{h} = \lim\limits_{h \to 0} \dfrac{4h - 2xh - h^2}{h} = \lim\limits_{h \to 0} (4 - 2x - h) = 4 - 2x$

$f''(x) = \lim\limits_{h \to 0} \dfrac{f'(x+h) - f'(x)}{h} = \lim\limits_{h \to 0} \dfrac{[4 - 2(x+h)] - (4 - 2x)}{h} = \lim\limits_{h \to 0} \dfrac{-2h}{h} = \lim\limits_{h \to 0} (-2) = -2$

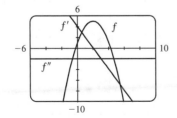

We see from the graph that our answers are reasonable because the graph of

f' is that of a linear function and the graph of f'' is that of a constant

function.

47. $f'(x) = \lim\limits_{h \to 0} \dfrac{f(x+h) - f(x)}{h} = \lim\limits_{h \to 0} \dfrac{[2(x+h)^2 - (x+h)^3] - (2x^2 - x^3)}{h}$

$= \lim\limits_{h \to 0} \dfrac{h(4x + 2h - 3x^2 - 3xh - h^2)}{h} = \lim\limits_{h \to 0} (4x + 2h - 3x^2 - 3xh - h^2) = 4x - 3x^2$

$f''(x) = \lim\limits_{h \to 0} \dfrac{f'(x+h) - f'(x)}{h} = \lim\limits_{h \to 0} \dfrac{[4(x+h) - 3(x+h)^2] - (4x - 3x^2)}{h} = \lim\limits_{h \to 0} \dfrac{h(4 - 6x - 3h)}{h}$

$= \lim\limits_{h \to 0} (4 - 6x - 3h) = 4 - 6x$

$f'''(x) = \lim\limits_{h \to 0} \dfrac{f''(x+h) - f''(x)}{h} = \lim\limits_{h \to 0} \dfrac{[4 - 6(x+h)] - (4 - 6x)}{h} = \lim\limits_{h \to 0} \dfrac{-6h}{h} = \lim\limits_{h \to 0} (-6) = -6$

$f^{(4)}(x) = \lim\limits_{h \to 0} \dfrac{f'''(x+h) - f'''(x)}{h} = \lim\limits_{h \to 0} \dfrac{-6 - (-6)}{h} = \lim\limits_{h \to 0} \dfrac{0}{h} = \lim\limits_{h \to 0} (0) = 0$

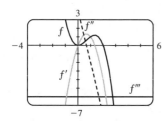

The graphs are consistent with the geometric interpretations of the derivatives because f' has zeros where f has a local minimum and a local maximum, f'' has a zero where f' has a local maximum, and f''' is a constant function equal to the slope of f''.

49. (a) Note that we have factored $x - a$ as the difference of two cubes in the third step.

$f'(a) = \lim\limits_{x \to a} \dfrac{f(x) - f(a)}{x - a} = \lim\limits_{x \to a} \dfrac{x^{1/3} - a^{1/3}}{x - a} = \lim\limits_{x \to a} \dfrac{x^{1/3} - a^{1/3}}{(x^{1/3} - a^{1/3})(x^{2/3} + x^{1/3}a^{1/3} + a^{2/3})}$

$= \lim\limits_{x \to a} \dfrac{1}{x^{2/3} + x^{1/3}a^{1/3} + a^{2/3}} = \dfrac{1}{3a^{2/3}}$ or $\tfrac{1}{3}a^{-2/3}$

(b) $f'(0) = \lim\limits_{h \to 0} \dfrac{f(0+h) - f(0)}{h} = \lim\limits_{h \to 0} \dfrac{\sqrt[3]{h} - 0}{h} = \lim\limits_{h \to 0} \dfrac{1}{h^{2/3}}$. This function increases without bound, so the limit does not

exist, and therefore $f'(0)$ does not exist.

(c) $\lim\limits_{x \to 0} |f'(x)| = \lim\limits_{x \to 0} \dfrac{1}{3x^{2/3}} = \infty$ and f is continuous at $x = 0$ (root function), so f has a vertical tangent at $x = 0$.

51. $f(x) = |x - 6| = \begin{cases} x - 6 & \text{if } x - 6 \ge 6 \\ -(x - 6) & \text{if } x - 6 < 0 \end{cases} = \begin{cases} x - 6 & \text{if } x \ge 6 \\ 6 - x & \text{if } x < 6 \end{cases}$

So the right-hand limit is $\lim\limits_{x \to 6^+} \dfrac{f(x) - f(6)}{x - 6} = \lim\limits_{x \to 6^+} \dfrac{|x - 6| - 0}{x - 6} = \lim\limits_{x \to 6^+} \dfrac{x - 6}{x - 6} = \lim\limits_{x \to 6^+} 1 = 1$, and the left-hand limit

is $\lim\limits_{x \to 6^-} \dfrac{f(x) - f(6)}{x - 6} = \lim\limits_{x \to 6^-} \dfrac{|x - 6| - 0}{x - 6} = \lim\limits_{x \to 6^-} \dfrac{6 - x}{x - 6} = \lim\limits_{x \to 6^-} (-1) = -1$. Since these limits are not equal,

$f'(6) = \lim\limits_{x \to 6} \dfrac{f(x) - f(6)}{x - 6}$ does not exist and f is not differentiable at 6.

However, a formula for f' is $f'(x) = \begin{cases} 1 & \text{if } x > 6 \\ -1 & \text{if } x < 6 \end{cases}$

Another way of writing the formula is $f'(x) = \dfrac{x - 6}{|x - 6|}$.

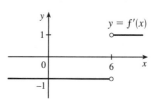

53. (a) $f(x) = x\,|x| = \begin{cases} x^2 & \text{if } x \geq 0 \\ -x^2 & \text{if } x < 0 \end{cases}$

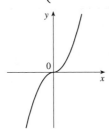

(b) Since $f(x) = x^2$ for $x \geq 0$, we have $f'(x) = 2x$ for $x > 0$.

[See Exercise 2.8.17(d).] Similarly, since $f(x) = -x^2$ for $x < 0$, we have $f'(x) = -2x$ for $x < 0$. At $x = 0$, we have

$$f'(0) = \lim_{x \to 0} \frac{f(x) - f(0)}{x - 0} = \lim_{x \to 0} \frac{x\,|x|}{x} = \lim_{x \to 0} |x| = 0.$$

So f is differentiable at 0. Thus, f is differentiable for all x.

(c) From part (b), we have $f'(x) = \begin{cases} 2x & \text{if } x \geq 0 \\ -2x & \text{if } x < 0 \end{cases} = 2\,|x|.$

55. (a) If f is even, then

$$f'(-x) = \lim_{h \to 0} \frac{f(-x + h) - f(-x)}{h} = \lim_{h \to 0} \frac{f[-(x - h)] - f(-x)}{h}$$

$$= \lim_{h \to 0} \frac{f(x - h) - f(x)}{h} = -\lim_{h \to 0} \frac{f(x - h) - f(x)}{-h} \qquad [\text{let } \Delta x = -h]$$

$$= -\lim_{\Delta x \to 0} \frac{f(x + \Delta x) - f(x)}{\Delta x} = -f'(x)$$

Therefore, f' is odd.

(b) If f is odd, then

$$f'(-x) = \lim_{h \to 0} \frac{f(-x + h) - f(-x)}{h} = \lim_{h \to 0} \frac{f[-(x - h)] - f(-x)}{h}$$

$$= \lim_{h \to 0} \frac{-f(x - h) + f(x)}{h} = \lim_{h \to 0} \frac{f(x - h) - f(x)}{-h} \qquad [\text{let } \Delta x = -h]$$

$$= \lim_{\Delta x \to 0} \frac{f(x + \Delta x) - f(x)}{\Delta x} = f'(x)$$

Therefore, f' is even.

57.

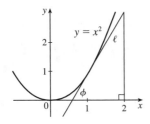

In the right triangle in the diagram, let Δy be the side opposite angle ϕ and Δx the side adjacent angle ϕ. Then the slope of the tangent line ℓ is $m = \Delta y/\Delta x = \tan \phi$. Note that $0 < \phi < \frac{\pi}{2}$. We know (see Exercise 17) that the derivative of $f(x) = x^2$ is $f'(x) = 2x$. So the slope of the tangent to the curve at the point $(1, 1)$ is 2. Thus, ϕ is the angle between 0 and $\frac{\pi}{2}$ whose tangent is 2; that is, $\phi = \tan^{-1} 2 \approx 63°$.

2 Review

1. (a) $\lim_{x \to a} f(x) = L$: See Definition 2.2.1 and Figures 1 and 2 in Section 2.2.

(b) $\lim_{x \to a^+} f(x) = L$: See the paragraph after Definition 2.2.2 and Figure 9(b) in Section 2.2.

(c) $\lim_{x \to a^-} f(x) = L$: See Definition 2.2.2 and Figure 9(a) in Section 2.2.

(d) $\lim_{x \to a} f(x) = \infty$: See Definition 2.2.4 and Figure 12 in Section 2.2.

(e) $\lim_{x \to \infty} f(x) = L$: See Definition 2.6.1 and Figure 2 in Section 2.6.

2. In general, the limit of a function fails to exist when the function does not approach a fixed number. For each of the following functions, the limit fails to exist at $x = 2$.

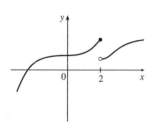
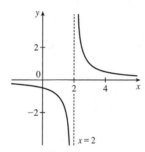
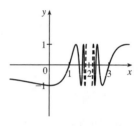

| The left- and right-hand limits are not equal. | There is an infinite discontinuity. | There are an infinite number of oscillations. |

3. (a)–(g) See the statements of Limit Laws 1–6 and 11 in Section 2.3.

4. See Theorem 3 in Section 2.3.

5. (a) See Definition 2.2.6 and Figures 12–14 in Section 2.2.

(b) See Definition 2.6.3 and Figures 3 and 4 in Section 2.6.

6. (a) $y = x^4$: No asymptote

(b) $y = \sin x$: No asymptote

(c) $y = \tan x$: Vertical asymptotes $x = \frac{\pi}{2} + \pi n$, n an integer

(d) $y = \tan^{-1} x$: Horizontal asymptotes $y = \pm \frac{\pi}{2}$

(e) $y = e^x$: Horizontal asymptote $y = 0$

$$\left(\lim_{x \to -\infty} e^x = 0 \right)$$

(f) $y = \ln x$: Vertical asymptote $x = 0$

$$\left(\lim_{x \to 0^+} \ln x = -\infty \right)$$

(g) $y = 1/x$: Vertical asymptote $x = 0$, horizontal asymptote $y = 0$

(h) $y = \sqrt{x}$: No asymptote

7. (a) A function f is continuous at a number a if $f(x)$ approaches $f(a)$ as x approaches a; that is, $\lim_{x \to a} f(x) = f(a)$.

(b) A function f is continuous on the interval $(-\infty, \infty)$ if f is continuous at every real number a. The graph of such a function has no breaks and every vertical line crosses it.

8. See Theorem 2.5.10.

9. See Definition 2.7.1.

10. See the paragraph containing Formula 3 in Section 2.7.

11. (a) The average rate of change of y with respect to x over the interval $[x_1, x_2]$ is $\dfrac{f(x_2) - f(x_1)}{x_2 - x_1}$.

(b) The instantaneous rate of change of y with respect to x at $x = x_1$ is $\lim\limits_{x_2 \to x_1} \dfrac{f(x_2) - f(x_1)}{x_2 - x_1}$.

12. See Definition 2.7.2. The pages following the definition discuss interpretations of $f'(a)$ as the slope of a tangent line to the graph of f at $x = a$ and as an instantaneous rate of change of $f(x)$ with respect to x when $x = a$.

13. See the paragraphs before and after Example 6 in Section 2.8.

14. (a) A function f is differentiable at a number a if its derivative f' exists at $x = a$; that is, if $f'(a)$ exists.

(b) See Theorem 2.8.4. This theorem also tells us that if f is *not* continuous at a, then f is *not* differentiable at a.

(c)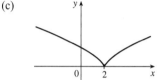

15. See the discussion and Figure 7 on page 159.

TRUE-FALSE QUIZ

1. False. Limit Law 2 applies only if the individual limits exist (these don't).

3. True. Limit Law 5 applies.

5. False. Consider $\lim\limits_{x \to 5} \dfrac{x(x-5)}{x-5}$ or $\lim\limits_{x \to 5} \dfrac{\sin(x-5)}{x-5}$. The first limit exists and is equal to 5. By Example 3 in Section 2.2, we know that the latter limit exists (and it is equal to 1).

7. True. A polynomial is continuous everywhere, so $\lim\limits_{x \to b} p(x)$ exists and is equal to $p(b)$.

9. True. See Figure 8 in Section 2.6.

11. False. Consider $f(x) = \begin{cases} 1/(x-1) & \text{if } x \neq 1 \\ 2 & \text{if } x = 1 \end{cases}$

13. True. Use Theorem 2.5.8 with $a = 2$, $b = 5$, and $g(x) = 4x^2 - 11$. Note that $f(4) = 3$ is not needed.

15. True, by the definition of a limit with $\varepsilon = 1$.

17. False. See the note after Theorem 4 in Section 2.8.

19. False. $\dfrac{d^2y}{dx^2}$ is the second derivative while $\left(\dfrac{dy}{dx}\right)^2$ is the first derivative squared. For example, if $y = x$, then $\dfrac{d^2y}{dx^2} = 0$, but $\left(\dfrac{dy}{dx}\right)^2 = 1$.

EXERCISES

1. (a) (i) $\lim\limits_{x \to 2^+} f(x) = 3$ (ii) $\lim\limits_{x \to -3^+} f(x) = 0$

(iii) $\lim\limits_{x \to -3} f(x)$ does not exist since the left and right limits are not equal. (The left limit is -2.)

(iv) $\lim\limits_{x \to 4} f(x) = 2$

(v) $\lim\limits_{x \to 0} f(x) = \infty$ (vi) $\lim\limits_{x \to 2^-} f(x) = -\infty$

(vii) $\lim\limits_{x \to \infty} f(x) = 4$ (viii) $\lim\limits_{x \to -\infty} f(x) = -1$

(b) The equations of the horizontal asymptotes are $y = -1$ and $y = 4$.

(c) The equations of the vertical asymptotes are $x = 0$ and $x = 2$.

(d) f is discontinuous at $x = -3, 0, 2$, and 4. The discontinuities are jump, infinite, infinite, and removable, respectively.

3. Since the exponential function is continuous, $\lim\limits_{x \to 1} e^{x^3 - x} = e^{1-1} = e^0 = 1$.

5. $\lim\limits_{x \to -3} \dfrac{x^2 - 9}{x^2 + 2x - 3} = \lim\limits_{x \to -3} \dfrac{(x+3)(x-3)}{(x+3)(x-1)} = \lim\limits_{x \to -3} \dfrac{x-3}{x-1} = \dfrac{-3-3}{-3-1} = \dfrac{-6}{-4} = \dfrac{3}{2}$

7. $\lim\limits_{h \to 0} \dfrac{(h-1)^3 + 1}{h} = \lim\limits_{h \to 0} \dfrac{(h^3 - 3h^2 + 3h - 1) + 1}{h} = \lim\limits_{h \to 0} \dfrac{h^3 - 3h^2 + 3h}{h} = \lim\limits_{h \to 0} (h^2 - 3h + 3) = 3$

Another solution: Factor the numerator as a sum of two cubes and then simplify.

$\lim\limits_{h \to 0} \dfrac{(h-1)^3 + 1}{h} = \lim\limits_{h \to 0} \dfrac{(h-1)^3 + 1^3}{h} = \lim\limits_{h \to 0} \dfrac{[(h-1) + 1]\left[(h-1)^2 - 1(h-1) + 1^2\right]}{h}$

$= \lim\limits_{h \to 0} \left[(h-1)^2 - h + 2\right] = 1 - 0 + 2 = 3$

9. $\lim\limits_{r \to 9} \dfrac{\sqrt{r}}{(r-9)^4} = \infty$ since $(r-9)^4 \to 0$ as $r \to 9$ and $\dfrac{\sqrt{r}}{(r-9)^4} > 0$ for $r \neq 9$.

11. $\lim\limits_{u \to 1} \dfrac{u^4 - 1}{u^3 + 5u^2 - 6u} = \lim\limits_{u \to 1} \dfrac{(u^2 + 1)(u^2 - 1)}{u(u^2 + 5u - 6)} = \lim\limits_{u \to 1} \dfrac{(u^2 + 1)(u+1)(u-1)}{u(u+6)(u-1)} = \lim\limits_{u \to 1} \dfrac{(u^2 + 1)(u+1)}{u(u+6)} = \dfrac{2(2)}{1(7)} = \dfrac{4}{7}$

13. Since x is positive, $\sqrt{x^2} = |x| = x$. Thus,

$$\lim_{x \to \infty} \frac{\sqrt{x^2 - 9}}{2x - 6} = \lim_{x \to \infty} \frac{\sqrt{x^2 - 9}/\sqrt{x^2}}{(2x - 6)/x} = \lim_{x \to \infty} \frac{\sqrt{1 - 9/x^2}}{2 - 6/x} = \frac{\sqrt{1 - 0}}{2 - 0} = \frac{1}{2}$$

15. Let $t = \sin x$. Then as $x \to \pi^-$, $\sin x \to 0^+$, so $t \to 0^+$. Thus, $\lim\limits_{x \to \pi^-} \ln(\sin x) = \lim\limits_{t \to 0^+} \ln t = -\infty$.

17. $\lim\limits_{x \to \infty} \left(\sqrt{x^2 + 4x + 1} - x\right) = \lim\limits_{x \to \infty} \left[\dfrac{\sqrt{x^2 + 4x + 1} - x}{1} \cdot \dfrac{\sqrt{x^2 + 4x + 1} + x}{\sqrt{x^2 + 4x + 1} + x}\right] = \lim\limits_{x \to \infty} \dfrac{(x^2 + 4x + 1) - x^2}{\sqrt{x^2 + 4x + 1} + x}$

$= \lim\limits_{x \to \infty} \dfrac{(4x + 1)/x}{(\sqrt{x^2 + 4x + 1} + x)/x} \qquad \left[\text{divide by } x = \sqrt{x^2} \text{ for } x > 0\right]$

$= \lim\limits_{x \to \infty} \dfrac{4 + 1/x}{\sqrt{1 + 4/x + 1/x^2} + 1} = \dfrac{4 + 0}{\sqrt{1 + 0 + 0} + 1} = \dfrac{4}{2} = 2$

19. Let $t = 1/x$. Then as $x \to 0^+$, $t \to \infty$, and $\lim\limits_{x \to 0^+} \tan^{-1}(1/x) = \lim\limits_{t \to \infty} \tan^{-1} t = \dfrac{\pi}{2}$.

21. From the graph of $y = (\cos^2 x)/x^2$, it appears that $y = 0$ is the horizontal

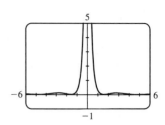

asymptote and $x = 0$ is the vertical asymptote. Now $0 \le (\cos x)^2 \le 1 \implies$

$\dfrac{0}{x^2} \le \dfrac{\cos^2 x}{x^2} \le \dfrac{1}{x^2} \implies 0 \le \dfrac{\cos^2 x}{x^2} \le \dfrac{1}{x^2}$. But $\lim\limits_{x \to \pm\infty} 0 = 0$ and

$\lim\limits_{x \to \pm\infty} \dfrac{1}{x^2} = 0$, so by the Squeeze Theorem, $\lim\limits_{x \to \pm\infty} \dfrac{\cos^2 x}{x^2} = 0$.

Thus, $y = 0$ is the horizontal asymptote. $\lim\limits_{x \to 0} \dfrac{\cos^2 x}{x^2} = \infty$ because $\cos^2 x \to 1$ and $x^2 \to 0$ as $x \to 0$, so $x = 0$ is the

vertical asymptote.

23. Since $2x - 1 \le f(x) \le x^2$ for $0 < x < 3$ and $\lim\limits_{x \to 1} (2x - 1) = 1 = \lim\limits_{x \to 1} x^2$, we have $\lim\limits_{x \to 1} f(x) = 1$ by the Squeeze Theorem.

25. Given $\varepsilon > 0$, we need $\delta > 0$ such that if $0 < |x - 2| < \delta$, then $|(14 - 5x) - 4| < \varepsilon$. But $|(14 - 5x) - 4| < \varepsilon \iff$

$|-5x + 10| < \varepsilon \iff |-5|\,|x - 2| < \varepsilon \iff |x - 2| < \varepsilon/5$. So if we choose $\delta = \varepsilon/5$, then $0 < |x - 2| < \delta \implies$

$|(14 - 5x) - 4| < \varepsilon$. Thus, $\lim\limits_{x \to 2} (14 - 5x) = 4$ by the definition of a limit.

27. Given $\varepsilon > 0$, we need $\delta > 0$ so that if $0 < |x - 2| < \delta$, then $\left|x^2 - 3x - (-2)\right| < \varepsilon$. First, note that if $|x - 2| < 1$, then

$-1 < x - 2 < 1$, so $0 < x - 1 < 2 \implies |x - 1| < 2$. Now let $\delta = \min\{\varepsilon/2, 1\}$. Then $0 < |x - 2| < \delta \implies$

$\left|x^2 - 3x - (-2)\right| = |(x - 2)(x - 1)| = |x - 2|\,|x - 1| < (\varepsilon/2)(2) = \varepsilon$.

Thus, $\lim\limits_{x \to 2} \left(x^2 - 3x\right) = -2$ by the definition of a limit.

29. (a) $f(x) = \sqrt{-x}$ if $x < 0$, $f(x) = 3 - x$ if $0 \le x < 3$, $f(x) = (x - 3)^2$ if $x > 3$.

    (i) $\lim\limits_{x \to 0^+} f(x) = \lim\limits_{x \to 0^+} (3 - x) = 3$ (ii) $\lim\limits_{x \to 0^-} f(x) = \lim\limits_{x \to 0^-} \sqrt{-x} = 0$

    (iii) Because of (i) and (ii), $\lim\limits_{x \to 0} f(x)$ does not exist.   (iv) $\lim\limits_{x \to 3^-} f(x) = \lim\limits_{x \to 3^-} (3 - x) = 0$

    (v) $\lim\limits_{x \to 3^+} f(x) = \lim\limits_{x \to 3^+} (x - 3)^2 = 0$ (vi) Because of (iv) and (v), $\lim\limits_{x \to 3} f(x) = 0$.

  (b) f is discontinuous at 0 since $\lim\limits_{x \to 0} f(x)$ does not exist.   (c)

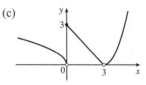

    f is discontinuous at 3 since $f(3)$ does not exist.

31. $\sin x$ is continuous on $\mathbb{R}$ by Theorem 7 in Section 2.5. Since e^x is continuous on $\mathbb{R}$, $e^{\sin x}$ is continuous on $\mathbb{R}$ by Theorem 9 in

Section 2.5. Lastly, x is continuous on $\mathbb{R}$ since it's a polynomial and the product $xe^{\sin x}$ is continuous on its domain $\mathbb{R}$ by

Theorem 4 in Section 2.5.

33. $f(x) = 2x^3 + x^2 + 2$ is a polynomial, so it is continuous on $[-2, -1]$ and $f(-2) = -10 < 0 < 1 = f(-1)$. So by the

Intermediate Value Theorem there is a number c in $(-2, -1)$ such that $f(c) = 0$, that is, the equation $2x^3 + x^2 + 2 = 0$ has a

root in $(-2, -1)$.

35. (a) The slope of the tangent line at $(2, 1)$ is

$$\lim_{x \to 2} \frac{f(x) - f(2)}{x - 2} = \lim_{x \to 2} \frac{9 - 2x^2 - 1}{x - 2} = \lim_{x \to 2} \frac{8 - 2x^2}{x - 2} = \lim_{x \to 2} \frac{-2(x^2 - 4)}{x - 2} = \lim_{x \to 2} \frac{-2(x - 2)(x + 2)}{x - 2}$$

$$= \lim_{x \to 2} \left[-2(x + 2) \right] = -2 \cdot 4 = -8$$

(b) An equation of this tangent line is $y - 1 = -8(x - 2)$ or $y = -8x + 17$.

37. (a) $s = s(t) = 1 + 2t + t^2/4$. The average velocity over the time interval $[1, 1 + h]$ is

$$v_{\text{ave}} = \frac{s(1 + h) - s(1)}{(1 + h) - 1} = \frac{1 + 2(1 + h) + (1 + h)^2/4 - 13/4}{h} = \frac{10h + h^2}{4h} = \frac{10 + h}{4}$$

So for the following intervals the average velocities are:

(i) $[1, 3]$: $h = 2$, $v_{\text{ave}} = (10 + 2)/4 = 3$ m/s   (ii) $[1, 2]$: $h = 1$, $v_{\text{ave}} = (10 + 1)/4 = 2.75$ m/s

(iii) $[1, 1.5]$: $h = 0.5$, $v_{\text{ave}} = (10 + 0.5)/4 = 2.625$ m/s   (iv) $[1, 1.1]$: $h = 0.1$, $v_{\text{ave}} = (10 + 0.1)/4 = 2.525$ m/s

(b) When $t = 1$, the instantaneous velocity is $\lim\limits_{h \to 0} \dfrac{s(1 + h) - s(1)}{h} = \lim\limits_{h \to 0} \dfrac{10 + h}{4} = \dfrac{10}{4} = 2.5$ m/s.

39. (a) $f'(2) = \lim\limits_{x \to 2} \dfrac{f(x) - f(2)}{x - 2} = \lim\limits_{x \to 2} \dfrac{x^3 - 2x - 4}{x - 2}$ (c)

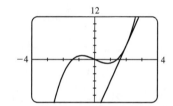

$$= \lim_{x \to 2} \frac{(x - 2)(x^2 + 2x + 2)}{x - 2}$$

$$= \lim_{x \to 2} \left(x^2 + 2x + 2 \right) = 10$$

(b) $y - 4 = 10(x - 2)$ or $y = 10x - 16$

41. (a) $f'(r)$ is the rate at which the total cost changes with respect to the interest rate. Its units are dollars/(percent per year).

(b) The total cost of paying off the loan is increasing by $1200/(percent per year) as the interest rate reaches 10%. So if the interest rate goes up from 10% to 11%, the cost goes up approximately $1200.

(c) As r increases, C increases. So $f'(r)$ will always be positive.

43.

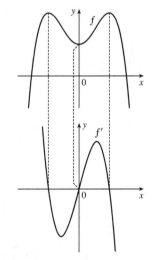

45. (a) $f'(x) = \lim\limits_{h \to 0} \dfrac{f(x+h) - f(x)}{h} = \lim\limits_{h \to 0} \dfrac{\sqrt{3 - 5(x+h)} - \sqrt{3 - 5x}}{h} \cdot \dfrac{\sqrt{3 - 5(x+h)} + \sqrt{3 - 5x}}{\sqrt{3 - 5(x+h)} + \sqrt{3 - 5x}}$

$= \lim\limits_{h \to 0} \dfrac{[3 - 5(x+h)] - (3 - 5x)}{h \left(\sqrt{3 - 5(x+h)} + \sqrt{3 - 5x} \right)} = \lim\limits_{h \to 0} \dfrac{-5}{\sqrt{3 - 5(x+h)} + \sqrt{3 - 5x}} = \dfrac{-5}{2\sqrt{3 - 5x}}$

(b) Domain of f: (the radicand must be nonnegative) $3 - 5x \ge 0 \;\;\Rightarrow$

$5x \le 3 \;\;\Rightarrow\;\; x \in \left(-\infty, \frac{3}{5} \right]$

Domain of f': exclude $\frac{3}{5}$ because it makes the denominator zero;

$x \in \left(-\infty, \frac{3}{5} \right)$

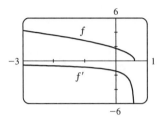

(c) Our answer to part (a) is reasonable because $f'(x)$ is always negative and f

is always decreasing.

47. f is not differentiable: at $x = -4$ because f is not continuous, at $x = -1$ because f has a corner, at $x = 2$ because f is not

continuous, and at $x = 5$ because f has a vertical tangent.

49. $C'(1990)$ is the rate at which the total value of US currency in circulation is changing in billions of dollars per year. To

estimate the value of $C'(1990)$, we will average the difference quotients obtained using the times $t = 1985$ and $t = 1995$.

Let $A = \dfrac{C(1985) - C(1990)}{1985 - 1990} = \dfrac{187.3 - 271.9}{-5} = \dfrac{-84.6}{-5} = 16.92$ and

$B = \dfrac{C(1995) - C(1990)}{1995 - 1990} = \dfrac{409.3 - 271.9}{5} = \dfrac{137.4}{5} = 27.48.$ Then

$C'(1990) = \lim\limits_{t \to 1990} \dfrac{C(t) - C(1990)}{t - 1990} \approx \dfrac{A + B}{2} = \dfrac{16.92 + 27.48}{2} = \dfrac{44.4}{2} = 22.2$ billion dollars/year.

51. $|f(x)| \le g(x) \;\;\Leftrightarrow\;\; -g(x) \le f(x) \le g(x)$ and $\lim\limits_{x \to a} g(x) = 0 = \lim\limits_{x \to a} -g(x).$

Thus, by the Squeeze Theorem, $\lim\limits_{x \to a} f(x) = 0.$

□ PROBLEMS PLUS

1. Let $t = \sqrt[6]{x}$, so $x = t^6$. Then $t \to 1$ as $x \to 1$, so

$$\lim_{x \to 1} \frac{\sqrt[3]{x} - 1}{\sqrt{x} - 1} = \lim_{t \to 1} \frac{t^2 - 1}{t^3 - 1} = \lim_{t \to 1} \frac{(t-1)(t+1)}{(t-1)(t^2+t+1)} = \lim_{t \to 1} \frac{t+1}{t^2+t+1} = \frac{1+1}{1^2+1+1} = \frac{2}{3}.$$

Another method: Multiply both the numerator and the denominator by $(\sqrt{x} + 1)\left(\sqrt[3]{x^2} + \sqrt[3]{x} + 1\right)$.

3. For $-\frac{1}{2} < x < \frac{1}{2}$, we have $2x - 1 < 0$ and $2x + 1 > 0$, so $|2x - 1| = -(2x - 1)$ and $|2x + 1| = 2x + 1$.

Therefore, $\lim_{x \to 0} \frac{|2x - 1| - |2x + 1|}{x} = \lim_{x \to 0} \frac{-(2x-1) - (2x+1)}{x} = \lim_{x \to 0} \frac{-4x}{x} = \lim_{x \to 0} (-4) = -4.$

5. Since $[\![x]\!] \le x < [\![x]\!] + 1$, we have $\dfrac{[\![x]\!]}{[\![x]\!]} \le \dfrac{x}{[\![x]\!]} < \dfrac{[\![x]\!] + 1}{[\![x]\!]}$ $\Rightarrow$ $1 \le \dfrac{x}{[\![x]\!]} < 1 + \dfrac{1}{[\![x]\!]}$ for $x \ge 1$. As $x \to \infty$, $[\![x]\!] \to \infty$,

so $\dfrac{1}{[\![x]\!]} \to 0$ and $1 + \dfrac{1}{[\![x]\!]} \to 1$. Thus, $\lim_{x \to \infty} \dfrac{x}{[\![x]\!]} = 1$ by the Squeeze Theorem.

7. f is continuous on $(-\infty, a)$ and (a, ∞). To make f continuous on $\mathbb{R}$, we must have continuity at a. Thus,

$$\lim_{x \to a^+} f(x) = \lim_{x \to a^-} f(x) \quad \Rightarrow \quad \lim_{x \to a^+} x^2 = \lim_{x \to a^-} (x + 1) \quad \Rightarrow \quad a^2 = a + 1 \quad \Rightarrow \quad a^2 - a - 1 = 0 \quad \Rightarrow$$

[by the quadratic formula] $\quad a = \left(1 \pm \sqrt{5}\right)/2 \approx 1.618$ or -0.618.

9. $\lim_{x \to a} f(x) = \lim_{x \to a} \left(\frac{1}{2}\left[f(x) + g(x)\right] + \frac{1}{2}\left[f(x) - g(x)\right]\right) = \frac{1}{2} \lim_{x \to a} \left[f(x) + g(x)\right] + \frac{1}{2} \lim_{x \to a} \left[f(x) - g(x)\right]$

$\qquad = \frac{1}{2} \cdot 2 + \frac{1}{2} \cdot 1 = \frac{3}{2},$

and $\lim_{x \to a} g(x) = \lim_{x \to a} \left(\left[f(x) + g(x)\right] - f(x)\right) = \lim_{x \to a} \left[f(x) + g(x)\right] - \lim_{x \to a} f(x) = 2 - \frac{3}{2} = \frac{1}{2}.$

So $\lim_{x \to a} \left[f(x)g(x)\right] = \left[\lim_{x \to a} f(x)\right] \left[\lim_{x \to a} g(x)\right] = \frac{3}{2} \cdot \frac{1}{2} = \frac{3}{4}.$

Another solution: Since $\lim_{x \to a} \left[f(x) + g(x)\right]$ and $\lim_{x \to a} \left[f(x) - g(x)\right]$ exist, we must have

$\lim_{x \to a} \left[f(x) + g(x)\right]^2 = \left(\lim_{x \to a} \left[f(x) + g(x)\right]\right)^2$ and $\lim_{x \to a} \left[f(x) - g(x)\right]^2 = \left(\lim_{x \to a} \left[f(x) - g(x)\right]\right)^2$, so

$\lim_{x \to a} \left[f(x)\,g(x)\right] = \lim_{x \to a} \frac{1}{4}\left(\left[f(x) + g(x)\right]^2 - \left[f(x) - g(x)\right]^2\right) \qquad$ [because all of the f^2 and g^2 cancel]

$\qquad = \frac{1}{4}\left(\lim_{x \to a} \left[f(x) + g(x)\right]^2 - \lim_{x \to a} \left[f(x) - g(x)\right]^2\right) = \frac{1}{4}\left(2^2 - 1^2\right) = \frac{3}{4}.$

11. (a) Consider $G(x) = T(x + 180°) - T(x)$. Fix any number a. If $G(a) = 0$, we are done: Temperature at $a =$ Temperature

at $a + 180°$. If $G(a) > 0$, then $G(a + 180°) = T(a + 360°) - T(a + 180°) = T(a) - T(a + 180°) = -G(a) < 0$.

Also, G is continuous since temperature varies continuously. So, by the Intermediate Value Theorem, G has a zero on the

interval $[a, a + 180°]$. If $G(a) < 0$, then a similar argument applies.

(b) Yes. The same argument applies.

(c) The same argument applies for quantities that vary continuously, such as barometric pressure. But one could argue that altitude above sea level is sometimes discontinuous, so the result might not always hold for that quantity.

13. (a) Put $x = 0$ and $y = 0$ in the equation: $f(0 + 0) = f(0) + f(0) + 0^2 \cdot 0 + 0 \cdot 0^2 \Rightarrow f(0) = 2f(0)$.

Subtracting $f(0)$ from each side of this equation gives $f(0) = 0$.

(b) $f'(0) = \lim\limits_{h \to 0} \dfrac{f(0 + h) - f(0)}{h} = \lim\limits_{h \to 0} \dfrac{\left[f(0) + f(h) + 0^2 h + 0h^2\right] - f(0)}{h} = \lim\limits_{h \to 0} \dfrac{f(h)}{h} = \lim\limits_{x \to 0} \dfrac{f(x)}{x} = 1$

(c) $f'(x) = \lim\limits_{h \to 0} \dfrac{f(x + h) - f(x)}{h} = \lim\limits_{h \to 0} \dfrac{\left[f(x) + f(h) + x^2 h + xh^2\right] - f(x)}{h} = \lim\limits_{h \to 0} \dfrac{f(h) + x^2 h + xh^2}{h}$

$= \lim\limits_{h \to 0} \left[\dfrac{f(h)}{h} + x^2 + xh\right] = 1 + x^2$

3 □ DIFFERENTIATION RULES

3.1 Derivatives of Polynomials and Exponential Functions

1. (a) e is the number such that $\lim\limits_{h \to 0} \dfrac{e^h - 1}{h} = 1$.

(b)

x	$\dfrac{2.7^x - 1}{x}$
-0.001	0.9928
-0.0001	0.9932
0.001	0.9937
0.0001	0.9933

x	$\dfrac{2.8^x - 1}{x}$
-0.001	1.0291
-0.0001	1.0296
0.001	1.0301
0.0001	1.0297

From the tables (to two decimal places),

$\lim\limits_{h \to 0} \dfrac{2.7^h - 1}{h} = 0.99$ and $\lim\limits_{h \to 0} \dfrac{2.8^h - 1}{h} = 1.03$.

Since $0.99 < 1 < 1.03$, $2.7 < e < 2.8$.

3. $f(x) = 186.5$ is a constant function, so its derivative is 0, that is, $f'(x) = 0$.

5. $f(t) = 2 - \frac{2}{3}t \quad \Rightarrow \quad f'(t) = 0 - \frac{2}{3} = -\frac{2}{3}$

7. $f(x) = x^3 - 4x + 6 \quad \Rightarrow \quad f'(x) = 3x^2 - 4(1) + 0 = 3x^2 - 4$

9. $f(t) = \frac{1}{4}(t^4 + 8) \quad \Rightarrow \quad f'(t) = \frac{1}{4}(t^4 + 8)' = \frac{1}{4}(4t^{4-1} + 0) = t^3$

11. $y = x^{-2/5} \quad \Rightarrow \quad y' = -\frac{2}{5}x^{(-2/5)-1} = -\frac{2}{5}x^{-7/5} = -\dfrac{2}{5x^{7/5}}$

13. $V(r) = \frac{4}{3}\pi r^3 \quad \Rightarrow \quad V'(r) = \frac{4}{3}\pi(3r^2) = 4\pi r^2$

15. $A(s) = -\dfrac{12}{s^5} = -12s^{-5} \quad \Rightarrow \quad A'(s) = -12(-5s^{-6}) = 60s^{-6} \quad$ or $\quad 60/s^6$

17. $G(x) = \sqrt{x} - 2e^x = x^{1/2} - 2e^x \quad \Rightarrow \quad G'(x) = \frac{1}{2}x^{-1/2} - 2e^x = \dfrac{1}{2\sqrt{x}} - 2e^x$

19. $F(x) = \left(\frac{1}{2}x\right)^5 = \left(\frac{1}{2}\right)^5 x^5 = \frac{1}{32}x^5 \quad \Rightarrow \quad F'(x) = \frac{1}{32}(5x^4) = \frac{5}{32}x^4$

21. $y = ax^2 + bx + c \quad \Rightarrow \quad y' = 2ax + b$

23. $y = \dfrac{x^2 + 4x + 3}{\sqrt{x}} = x^{3/2} + 4x^{1/2} + 3x^{-1/2} \quad \Rightarrow$

$y' = \frac{3}{2}x^{1/2} + 4\left(\frac{1}{2}\right)x^{-1/2} + 3\left(-\frac{1}{2}\right)x^{-3/2} = \frac{3}{2}\sqrt{x} + \dfrac{2}{\sqrt{x}} - \dfrac{3}{2x\sqrt{x}} \quad \left[\text{note that } x^{3/2} = x^{2/2} \cdot x^{1/2} = x\sqrt{x}\right]$

The last expression can be written as $\dfrac{3x^2}{2x\sqrt{x}} + \dfrac{4x}{2x\sqrt{x}} - \dfrac{3}{2x\sqrt{x}} = \dfrac{3x^2 + 4x - 3}{2x\sqrt{x}}$.

25. $y = 4\pi^2 \quad \Rightarrow \quad y' = 0$ since $4\pi^2$ is a constant.

27. We first expand using the Binomial Theorem (see Reference Page 1).

$$H(x) = (x + x^{-1})^3 = x^3 + 3x^2 x^{-1} + 3x(x^{-1})^2 + (x^{-1})^3 = x^3 + 3x + 3x^{-1} + x^{-3} \Rightarrow$$

$$H'(x) = 3x^2 + 3 + 3(-1x^{-2}) + (-3x^{-4}) = 3x^2 + 3 - 3x^{-2} - 3x^{-4}$$

29. $u = \sqrt[5]{t} + 4\sqrt{t^5} = t^{1/5} + 4t^{5/2} \Rightarrow u' = \frac{1}{5}t^{-4/5} + 4\left(\frac{5}{2}t^{3/2}\right) = \frac{1}{5}t^{-4/5} + 10t^{3/2}$ or $1/\left(5\sqrt[5]{t^4}\right) + 10\sqrt{t^3}$

31. $z = \dfrac{A}{y^{10}} + Be^y = Ay^{-10} + Be^y \Rightarrow z' = -10Ay^{-11} + Be^y = -\dfrac{10A}{y^{11}} + Be^y$

33. $y = \sqrt[4]{x} = x^{1/4} \Rightarrow y' = \frac{1}{4}x^{-3/4} = \dfrac{1}{4\sqrt[4]{x^3}}$. At $(1,1)$, $y' = \frac{1}{4}$ and an equation of the tangent line is

$y - 1 = \frac{1}{4}(x-1)$ or $y = \frac{1}{4}x + \frac{3}{4}$.

35. $y = x^4 + 2e^x \Rightarrow y' = 4x^3 + 2e^x$. At $(0,2)$, $y' = 2$ and an equation of the tangent line is $y - 2 = 2(x - 0)$

or $y = 2x + 2$. The slope of the normal line is $-\frac{1}{2}$ (the negative reciprocal of 2) and an equation of the normal line is

$y - 2 = -\frac{1}{2}(x - 0)$ or $y = -\frac{1}{2}x + 2$.

37. $y = 3x^2 - x^3 \Rightarrow y' = 6x - 3x^2$.

At $(1, 2)$, $y' = 6 - 3 = 3$, so an equation of the tangent line is

$y - 2 = 3(x - 1)$ or $y = 3x - 1$.

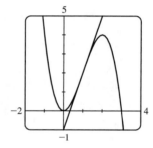

39. $f(x) = e^x - 5x \Rightarrow f'(x) = e^x - 5$.

Notice that $f'(x) = 0$ when f has a horizontal tangent, f' is positive
when f is increasing, and f' is negative when f is decreasing.

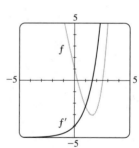

41. $f(x) = 3x^{15} - 5x^3 + 3 \Rightarrow f'(x) = 45x^{14} - 15x^2$.

Notice that $f'(x) = 0$ when f has a horizontal tangent, f' is positive
when f is increasing, and f' is negative when f is decreasing.

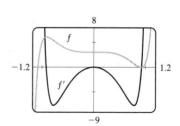

43. (a)

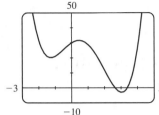

(b) From the graph in part (a), it appears that f' is zero at $x_1 \approx -1.25$, $x_2 \approx 0.5$, and $x_3 \approx 3$. The slopes are negative (so f' is negative) on $(-\infty, x_1)$ and (x_2, x_3). The slopes are positive (so f' is positive) on (x_1, x_2) and (x_3, ∞).

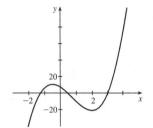

(c) $f(x) = x^4 - 3x^3 - 6x^2 + 7x + 30 \quad\Rightarrow$

$f'(x) = 4x^3 - 9x^2 - 12x + 7$

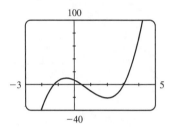

45. $f(x) = x^4 - 3x^3 + 16x \quad\Rightarrow\quad f'(x) = 4x^3 - 9x^2 + 16 \quad\Rightarrow\quad f''(x) = 12x^2 - 18x$

47. $f(x) = 2x - 5x^{3/4} \quad\Rightarrow\quad f'(x) = 2 - \frac{15}{4}x^{-1/4} \quad\Rightarrow\quad f''(x) = \frac{15}{16}x^{-5/4}$

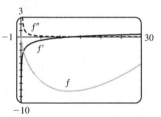

Note that f' is negative when f is decreasing and positive when f is increasing. f'' is always positive since f' is always increasing.

49. (a) $s = t^3 - 3t \quad\Rightarrow\quad v(t) = s'(t) = 3t^2 - 3 \quad\Rightarrow\quad a(t) = v'(t) = 6t$

(b) $a(2) = 6(2) = 12$ m/s^2

(c) $v(t) = 3t^2 - 3 = 0$ when $t^2 = 1$, that is, $t = 1$ and $a(1) = 6$ m/s^2.

51. The curve $y = 2x^3 + 3x^2 - 12x + 1$ has a horizontal tangent when $y' = 6x^2 + 6x - 12 = 0 \quad\Leftrightarrow\quad 6(x^2 + x - 2) = 0 \quad\Leftrightarrow\quad 6(x + 2)(x - 1) = 0 \quad\Leftrightarrow\quad x = -2$ or $x = 1$. The points on the curve are $(-2, 21)$ and $(1, -6)$.

53. $y = 6x^3 + 5x - 3 \quad\Rightarrow\quad m = y' = 18x^2 + 5$, but $x^2 \geq 0$ for all x, so $m \geq 5$ for all x.

55. The slope of the line $12x - y = 1$ (or $y = 12x - 1$) is 12, so the slope of both lines tangent to the curve is 12.

$y = 1 + x^3 \quad\Rightarrow\quad y' = 3x^2$. Thus, $3x^2 = 12 \quad\Rightarrow\quad x^2 = 4 \quad\Rightarrow\quad x = \pm 2$, which are the x-coordinates at which the tangent lines have slope 12. The points on the curve are $(2, 9)$ and $(-2, -7)$, so the tangent line equations are $y - 9 = 12(x - 2)$ or $y = 12x - 15$ and $y + 7 = 12(x + 2)$ or $y = 12x + 17$.

57. The slope of $y = x^2 - 5x + 4$ is given by $m = y' = 2x - 5$. The slope of $x - 3y = 5 \Leftrightarrow y = \frac{1}{3}x - \frac{5}{3}$ is $\frac{1}{3}$,

so the desired normal line must have slope $\frac{1}{3}$, and hence, the tangent line to the parabola must have slope -3. This occurs if

$2x - 5 = -3 \Rightarrow 2x = 2 \Rightarrow x = 1$. When $x = 1$, $y = 1^2 - 5(1) + 4 = 0$, and an equation of the normal line is

$y - 0 = \frac{1}{3}(x - 1)$ or $y = \frac{1}{3}x - \frac{1}{3}$.

59.

Let (a, a^2) be a point on the parabola at which the tangent line passes through the point $(0, -4)$. The tangent line has slope $2a$ and equation $y - (-4) = 2a(x - 0) \Leftrightarrow y = 2ax - 4$. Since (a, a^2) also lies on the line, $a^2 = 2a(a) - 4$, or $a^2 = 4$. So $a = \pm 2$ and the points are $(2, 4)$ and $(-2, 4)$.

61. $f'(x) = \lim\limits_{h \to 0} \dfrac{f(x+h) - f(x)}{h} = \lim\limits_{h \to 0} \dfrac{\frac{1}{x+h} - \frac{1}{x}}{h} = \lim\limits_{h \to 0} \dfrac{x - (x+h)}{hx(x+h)} = \lim\limits_{h \to 0} \dfrac{-h}{hx(x+h)} = \lim\limits_{h \to 0} \dfrac{-1}{x(x+h)} = -\dfrac{1}{x^2}$

63. Let $P(x) = ax^2 + bx + c$. Then $P'(x) = 2ax + b$ and $P''(x) = 2a$. $P''(2) = 2 \Rightarrow 2a = 2 \Rightarrow a = 1$.

$P'(2) = 3 \Rightarrow 2(1)(2) + b = 3 \Rightarrow 4 + b = 3 \Rightarrow b = -1$.

$P(2) = 5 \Rightarrow 1(2)^2 + (-1)(2) + c = 5 \Rightarrow 2 + c = 5 \Rightarrow c = 3$. So $P(x) = x^2 - x + 3$.

65. $y = f(x) = ax^3 + bx^2 + cx + d \Rightarrow f'(x) = 3ax^2 + 2bx + c$. The point $(-2, 6)$ is on f, so $f(-2) = 6 \Rightarrow$

$-8a + 4b - 2c + d = 6$ **(1)**. The point $(2, 0)$ is on f, so $f(2) = 0 \Rightarrow 8a + 4b + 2c + d = 0$ **(2)**. Since there are

horizontal tangents at $(-2, 6)$ and $(2, 0)$, $f'(\pm 2) = 0$. $f'(-2) = 0 \Rightarrow 12a - 4b + c = 0$ **(3)** and $f'(2) = 0 \Rightarrow$

$12a + 4b + c = 0$ **(4)**. Subtracting equation **(3)** from **(4)** gives $8b = 0 \Rightarrow b = 0$. Adding **(1)** and **(2)** gives $8b + 2d = 6$,

so $d = 3$ since $b = 0$. From **(3)** we have $c = -12a$, so **(2)** becomes $8a + 4(0) + 2(-12a) + 3 = 0 \Rightarrow 3 = 16a \Rightarrow$

$a = \frac{3}{16}$. Now $c = -12a = -12\left(\frac{3}{16}\right) = -\frac{9}{4}$ and the desired cubic function is $y = \frac{3}{16}x^3 - \frac{9}{4}x + 3$.

67. $f(x) = 2 - x$ if $x \le 1$ and $f(x) = x^2 - 2x + 2$ if $x > 1$. Now we compute the right- and left-hand derivatives defined in

Exercise 2.8.54:

$f'_-(1) = \lim\limits_{h \to 0^-} \dfrac{f(1+h) - f(1)}{h} = \lim\limits_{h \to 0^-} \dfrac{2 - (1+h) - 1}{h} = \lim\limits_{h \to 0^-} \dfrac{-h}{h} = \lim\limits_{h \to 0^-} -1 = -1$ and

$f'_+(1) = \lim\limits_{h \to 0^+} \dfrac{f(1+h) - f(1)}{h} = \lim\limits_{h \to 0^+} \dfrac{(1+h)^2 - 2(1+h) + 2 - 1}{h} = \lim\limits_{h \to 0^+} \dfrac{h^2}{h} = \lim\limits_{h \to 0^+} h = 0$.

Thus, $f'(1)$ does not exist since $f'_-(1) \ne f'_+(1)$, so f

is not differentiable at 1. But $f'(x) = -1$ for $x < 1$

and $f'(x) = 2x - 2$ if $x > 1$.

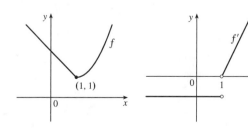

69. (a) Note that $x^2 - 9 < 0$ for $x^2 < 9 \iff |x| < 3 \iff -3 < x < 3$. So

$$f(x) = \begin{cases} x^2 - 9 & \text{if } x \le -3 \\ -x^2 + 9 & \text{if } -3 < x < 3 \\ x^2 - 9 & \text{if } x \ge 3 \end{cases} \implies f'(x) = \begin{cases} 2x & \text{if } x < -3 \\ -2x & \text{if } -3 < x < 3 \\ 2x & \text{if } x > 3 \end{cases} = \begin{cases} 2x & \text{if } |x| > 3 \\ -2x & \text{if } |x| < 3 \end{cases}$$

To show that $f'(3)$ does not exist we investigate $\lim\limits_{h \to 0} \dfrac{f(3+h) - f(3)}{h}$ by computing the left- and right-hand derivatives defined in Exercise 2.8.54.

$$f'_-(3) = \lim_{h \to 0^-} \frac{f(3+h) - f(3)}{h} = \lim_{h \to 0^-} \frac{[-(3+h)^2 + 9] - 0}{h} = \lim_{h \to 0^-} (-6 - h) = -6 \quad \text{and}$$

$$f'_+(3) = \lim_{h \to 0^+} \frac{f(3+h) - f(3)}{h} = \lim_{h \to 0^+} \frac{[(3+h)^2 - 9] - 0}{h} = \lim_{h \to 0^+} \frac{6h + h^2}{h} = \lim_{h \to 0^+} (6 + h) = 6.$$

Since the left and right limits are different,

$\lim\limits_{h \to 0} \dfrac{f(3+h) - f(3)}{h}$ does not exist, that is, $f'(3)$

does not exist. Similarly, $f'(-3)$ does not exist.

Therefore, f is not differentiable at 3 or at -3.

(b)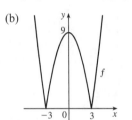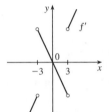

71. Substituting $x = 1$ and $y = 1$ into $y = ax^2 + bx$ gives us $a + b = 1$ **(1)**. The slope of the tangent line $y = 3x - 2$ is 3 and the slope of the tangent to the parabola at (x, y) is $y' = 2ax + b$. At $x = 1$, $y' = 3 \implies 3 = 2a + b$ **(2)**. Subtracting **(1)** from **(2)** gives us $2 = a$ and it follows that $b = -1$. The parabola has equation $y = 2x^2 - x$.

73. $y = f(x) = ax^2 \implies f'(x) = 2ax$. So the slope of the tangent to the parabola at $x = 2$ is $m = 2a(2) = 4a$. The slope of the given line, $2x + y = b \iff y = -2x + b$, is seen to be -2, so we must have $4a = -2 \iff a = -\frac{1}{2}$. So when $x = 2$, the point in question has y-coordinate $-\frac{1}{2} \cdot 2^2 = -2$. Now we simply require that the given line, whose equation is $2x + y = b$, pass through the point $(2, -2)$: $2(2) + (-2) = b \iff b = 2$. So we must have $a = -\frac{1}{2}$ and $b = 2$.

75. f is clearly differentiable for $x < 2$ and for $x > 2$. For $x < 2$, $f'(x) = 2x$, so $f'_-(2) = 4$. For $x > 2$, $f'(x) = m$, so $f'_+(2) = m$. For f to be differentiable at $x = 2$, we need $4 = f'_-(2) = f'_+(2) = m$. So $f(x) = 4x + b$. We must also have continuity at $x = 2$, so $4 = f(2) = \lim\limits_{x \to 2^+} f(x) = \lim\limits_{x \to 2^+} (4x + b) = 8 + b$. Hence, $b = -4$.

77. *Solution 1:*   Let $f(x) = x^{1000}$. Then, by the definition of a derivative, $f'(1) = \lim\limits_{x \to 1} \dfrac{f(x) - f(1)}{x - 1} = \lim\limits_{x \to 1} \dfrac{x^{1000} - 1}{x - 1}$.

But this is just the limit we want to find, and we know (from the Power Rule) that $f'(x) = 1000x^{999}$, so

$f'(1) = 1000(1)^{999} = 1000$. So $\lim\limits_{x \to 1} \dfrac{x^{1000} - 1}{x - 1} = 1000$.

Solution 2:   Note that $(x^{1000} - 1) = (x - 1)(x^{999} + x^{998} + x^{997} + \cdots + x^2 + x + 1)$. So

$$\lim_{x \to 1} \frac{x^{1000} - 1}{x - 1} = \lim_{x \to 1} \frac{(x - 1)(x^{999} + x^{998} + x^{997} + \cdots + x^2 + x + 1)}{x - 1} = \lim_{x \to 1} (x^{999} + x^{998} + x^{997} + \cdots + x^2 + x + 1)$$

$$= \underbrace{1 + 1 + 1 + \cdots + 1 + 1 + 1}_{1000 \text{ ones}} = 1000, \text{ as above.}$$

79. $y = x^2 \implies y' = 2x$, so the slope of a tangent line at the point (a, a^2) is $y' = 2a$ and the slope of a normal line is $-1/(2a)$,

for $a \neq 0$. The slope of the normal line through the points (a, a^2) and $(0, c)$ is $\dfrac{a^2 - c}{a - 0}$, so $\dfrac{a^2 - c}{a} = -\dfrac{1}{2a} \implies$

$a^2 - c = -\frac{1}{2} \implies a^2 = c - \frac{1}{2}$. The last equation has two solutions if $c > \frac{1}{2}$, one solution if $c = \frac{1}{2}$, and no solution if

$c < \frac{1}{2}$. Since the y-axis is normal to $y = x^2$ regardless of the value of c (this is the case for $a = 0$), we have three normal lines

if $c > \frac{1}{2}$ and one normal line if $c \leq \frac{1}{2}$.

3.2 The Product and Quotient Rules

1. Product Rule: $y = (x^2 + 1)(x^3 + 1) \implies$

$$y' = (x^2 + 1)(3x^2) + (x^3 + 1)(2x) = 3x^4 + 3x^2 + 2x^4 + 2x = 5x^4 + 3x^2 + 2x.$$

Multiplying first: $y = (x^2 + 1)(x^3 + 1) = x^5 + x^3 + x^2 + 1 \implies y' = 5x^4 + 3x^2 + 2x$ (equivalent).

3. By the Product Rule, $f(x) = (x^3 + 2x)e^x \implies$

$$f'(x) = (x^3 + 2x)(e^x)' + e^x(x^3 + 2x)' = (x^3 + 2x)e^x + e^x(3x^2 + 2)$$
$$= e^x[(x^3 + 2x) + (3x^2 + 2)] = e^x(x^3 + 3x^2 + 2x + 2)$$

5. By the Quotient Rule, $y = \dfrac{e^x}{x^2} \implies y' = \dfrac{x^2 \dfrac{d}{dx}(e^x) - e^x \dfrac{d}{dx}(x^2)}{(x^2)^2} = \dfrac{x^2(e^x) - e^x(2x)}{x^4} = \dfrac{xe^x(x - 2)}{x^4} = \dfrac{e^x(x - 2)}{x^3}$.

The notations $\overset{PR}{\implies}$ and $\overset{QR}{\implies}$ indicate the use of the Product and Quotient Rules, respectively.

7. $g(x) = \dfrac{3x - 1}{2x + 1} \overset{QR}{\implies} g'(x) = \dfrac{(2x + 1)(3) - (3x - 1)(2)}{(2x + 1)^2} = \dfrac{6x + 3 - 6x + 2}{(2x + 1)^2} = \dfrac{5}{(2x + 1)^2}$

9. $V(x) = (2x^3 + 3)(x^4 - 2x) \overset{PR}{\implies}$

$V'(x) = (2x^3 + 3)(4x^3 - 2) + (x^4 - 2x)(6x^2) = (8x^6 + 8x^3 - 6) + (6x^6 - 12x^3) = 14x^6 - 4x^3 - 6$

11. $F(y) = \left(\dfrac{1}{y^2} - \dfrac{3}{y^4} \right)(y + 5y^3) = (y^{-2} - 3y^{-4})(y + 5y^3) \overset{PR}{\implies}$

$F'(y) = (y^{-2} - 3y^{-4})(1 + 15y^2) + (y + 5y^3)(-2y^{-3} + 12y^{-5})$

$\quad = (y^{-2} + 15 - 3y^{-4} - 45y^{-2}) + (-2y^{-2} + 12y^{-4} - 10 + 60y^{-2})$

$\quad = 5 + 14y^{-2} + 9y^{-4}$ or $5 + 14/y^2 + 9/y^4$

13. $y = \dfrac{x^3}{1 - x^2} \overset{QR}{\implies} y' = \dfrac{(1 - x^2)(3x^2) - x^3(-2x)}{(1 - x^2)^2} = \dfrac{x^2(3 - 3x^2 + 2x^2)}{(1 - x^2)^2} = \dfrac{x^2(3 - x^2)}{(1 - x^2)^2}$

15. $y = \dfrac{t^2 + 2}{t^4 - 3t^2 + 1}$ $\overset{QR}{\Rightarrow}$

$$y' = \frac{(t^4 - 3t^2 + 1)(2t) - (t^2 + 2)(4t^3 - 6t)}{(t^4 - 3t^2 + 1)^2} = \frac{2t[(t^4 - 3t^2 + 1) - (t^2 + 2)(2t^2 - 3)]}{(t^4 - 3t^2 + 1)^2}$$

$$= \frac{2t(t^4 - 3t^2 + 1 - 2t^4 - 4t^2 + 3t^2 + 6)}{(t^4 - 3t^2 + 1)^2} = \frac{2t(-t^4 - 4t^2 + 7)}{(t^4 - 3t^2 + 1)^2}$$

17. $y = (r^2 - 2r)e^r$ $\overset{PR}{\Rightarrow}$ $y' = (r^2 - 2r)(e^r) + e^r(2r - 2) = e^r(r^2 - 2r + 2r - 2) = e^r(r^2 - 2)$

19. $y = \dfrac{v^3 - 2v\sqrt{v}}{v} = v^2 - 2\sqrt{v} = v^2 - 2v^{1/2}$ $\Rightarrow$ $y' = 2v - 2\left(\frac{1}{2}\right)v^{-1/2} = 2v - v^{-1/2}.$

We can change the form of the answer as follows: $2v - v^{-1/2} = 2v - \dfrac{1}{\sqrt{v}} = \dfrac{2v\sqrt{v} - 1}{\sqrt{v}} = \dfrac{2v^{3/2} - 1}{\sqrt{v}}$

21. $f(t) = \dfrac{2t}{2 + \sqrt{t}}$ $\overset{QR}{\Rightarrow}$ $f'(t) = \dfrac{(2 + t^{1/2})(2) - 2t\left(\frac{1}{2}t^{-1/2}\right)}{(2 + \sqrt{t})^2} = \dfrac{4 + 2t^{1/2} - t^{1/2}}{(2 + \sqrt{t})^2} = \dfrac{4 + t^{1/2}}{(2 + \sqrt{t})^2}$ or $\dfrac{4 + \sqrt{t}}{(2 + \sqrt{t})^2}$

23. $f(x) = \dfrac{A}{B + Ce^x}$ $\overset{QR}{\Rightarrow}$ $f'(x) = \dfrac{(B + Ce^x) \cdot 0 - A(Ce^x)}{(B + Ce^x)^2} = -\dfrac{ACe^x}{(B + Ce^x)^2}$

25. $f(x) = \dfrac{x}{x + c/x}$ $\Rightarrow$ $f'(x) = \dfrac{(x + c/x)(1) - x(1 - c/x^2)}{\left(x + \dfrac{c}{x}\right)^2} = \dfrac{x + c/x - x + c/x}{\left(\dfrac{x^2 + c}{x}\right)^2} = \dfrac{2c/x}{\dfrac{(x^2 + c)^2}{x^2}} \cdot \dfrac{x^2}{x^2} = \dfrac{2cx}{(x^2 + c)^2}$

27. $f(x) = x^4 e^x$ $\Rightarrow$ $f'(x) = x^4 e^x + e^x \cdot 4x^3 = (x^4 + 4x^3)e^x$ [or $x^3 e^x(x + 4)$] $\Rightarrow$

$f''(x) = (x^4 + 4x^3)e^x + e^x(4x^3 + 12x^2) = (x^4 + 4x^3 + 4x^3 + 12x^2)e^x$

$= (x^4 + 8x^3 + 12x^2)e^x$ [or $x^2 e^x(x + 2)(x + 6)$]

29. $f(x) = \dfrac{x^2}{1 + 2x}$ $\Rightarrow$ $f'(x) = \dfrac{(1 + 2x)(2x) - x^2(2)}{(1 + 2x)^2} = \dfrac{2x + 4x^2 - 2x^2}{(1 + 2x)^2} = \dfrac{2x^2 + 2x}{(1 + 2x)^2}$ $\Rightarrow$

$f''(x) = \dfrac{(1 + 2x)^2(4x + 2) - (2x^2 + 2x)(1 + 4x + 4x^2)'}{[(1 + 2x)^2]^2} = \dfrac{2(1 + 2x)^2(2x + 1) - 2x(x + 1)(4 + 8x)}{(1 + 2x)^4}$

$= \dfrac{2(1 + 2x)[(1 + 2x)^2 - 4x(x + 1)]}{(1 + 2x)^4} = \dfrac{2(1 + 4x + 4x^2 - 4x^2 - 4x)}{(1 + 2x)^3} = \dfrac{2}{(1 + 2x)^3}$

31. $y = \dfrac{2x}{x + 1}$ $\Rightarrow$ $y' = \dfrac{(x + 1)(2) - (2x)(1)}{(x + 1)^2} = \dfrac{2}{(x + 1)^2}.$

At $(1, 1)$, $y' = \frac{1}{2}$, and an equation of the tangent line is $y - 1 = \frac{1}{2}(x - 1)$, or $y = \frac{1}{2}x + \frac{1}{2}.$

33. $y = 2xe^x$ $\Rightarrow$ $y' = 2(x \cdot e^x + e^x \cdot 1) = 2e^x(x + 1).$

At $(0, 0)$, $y' = 2e^0(0 + 1) = 2 \cdot 1 \cdot 1 = 2$, and an equation of the tangent line is $y - 0 = 2(x - 0)$, or $y = 2x$. The slope of

the normal line is $-\frac{1}{2}$, so an equation of the normal line is $y - 0 = -\frac{1}{2}(x - 0)$, or $y = -\frac{1}{2}x.$

35. (a) $y = f(x) = \dfrac{1}{1+x^2}$ $\Rightarrow$

$f'(x) = \dfrac{(1+x^2)(0) - 1(2x)}{(1+x^2)^2} = \dfrac{-2x}{(1+x^2)^2}$. So the slope of the

tangent line at the point $\left(-1, \frac{1}{2}\right)$ is $f'(-1) = \dfrac{2}{2^2} = \frac{1}{2}$ and its

equation is $y - \frac{1}{2} = \frac{1}{2}(x+1)$ or $y = \frac{1}{2}x + 1$.

(b)

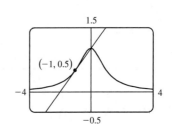

37. (a) $f(x) = \dfrac{e^x}{x^3}$ $\Rightarrow$ $f'(x) = \dfrac{x^3(e^x) - e^x(3x^2)}{(x^3)^2} = \dfrac{x^2 e^x(x-3)}{x^6} = \dfrac{e^x(x-3)}{x^4}$

(b)

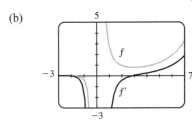

$f' = 0$ when f has a horizontal tangent line, f' is negative when f is decreasing, and f' is positive when f is increasing.

39. (a) $f(x) = (x-1)e^x$ $\Rightarrow$ $f'(x) = (x-1)e^x + e^x(1) = e^x(x-1+1) = xe^x$.

$f''(x) = x(e^x) + e^x(1) = e^x(x+1)$

(b)

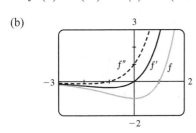

$f' = 0$ when f has a horizontal tangent and $f'' = 0$ when f' has a horizontal tangent. f' is negative when f is decreasing and positive when f is increasing. f'' is negative when f' is decreasing and positive when f' is increasing. f'' is negative when f is concave down and positive when f is concave up.

41. $f(x) = \dfrac{x^2}{1+x}$ $\Rightarrow$ $f'(x) = \dfrac{(1+x)(2x) - x^2(1)}{(1+x)^2} = \dfrac{2x + 2x^2 - x^2}{(1+x)^2} = \dfrac{x^2 + 2x}{x^2 + 2x + 1}$ $\Rightarrow$

$f''(x) = \dfrac{(x^2+2x+1)(2x+2) - (x^2+2x)(2x+2)}{(x^2+2x+1)^2} = \dfrac{(2x+2)(x^2+2x+1 - x^2 - 2x)}{[(x+1)^2]^2}$

$= \dfrac{2(x+1)(1)}{(x+1)^4} = \dfrac{2}{(x+1)^3}$,

so $f''(1) = \dfrac{2}{(1+1)^3} = \dfrac{2}{8} = \dfrac{1}{4}$.

43. We are given that $f(5) = 1$, $f'(5) = 6$, $g(5) = -3$, and $g'(5) = 2$.

(a) $(fg)'(5) = f(5)g'(5) + g(5)f'(5) = (1)(2) + (-3)(6) = 2 - 18 = -16$

(b) $\left(\dfrac{f}{g}\right)'(5) = \dfrac{g(5)f'(5) - f(5)g'(5)}{[g(5)]^2} = \dfrac{(-3)(6) - (1)(2)}{(-3)^2} = -\dfrac{20}{9}$

(c) $\left(\dfrac{g}{f}\right)'(5) = \dfrac{f(5)g'(5) - g(5)f'(5)}{[f(5)]^2} = \dfrac{(1)(2) - (-3)(6)}{(1)^2} = 20$

45. $f(x) = e^x g(x)$ $\Rightarrow$ $f'(x) = e^x g'(x) + g(x)e^x = e^x[g'(x) + g(x)]$. $f'(0) = e^0[g'(0) + g(0)] = 1(5+2) = 7$

47. (a) From the graphs of f and g, we obtain the following values: $f(1) = 2$ since the point $(1, 2)$ is on the graph of f;

$g(1) = 1$ since the point $(1, 1)$ is on the graph of g; $f'(1) = 2$ since the slope of the line segment between $(0, 0)$ and $(2, 4)$

is $\dfrac{4 - 0}{2 - 0} = 2$; $g'(1) = -1$ since the slope of the line segment between $(-2, 4)$ and $(2, 0)$ is $\dfrac{0 - 4}{2 - (-2)} = -1$.

Now $u(x) = f(x)g(x)$, so $u'(1) = f(1)g'(1) + g(1)f'(1) = 2 \cdot (-1) + 1 \cdot 2 = 0$.

(b) $v(x) = f(x)/g(x)$, so $v'(5) = \dfrac{g(5)f'(5) - f(5)g'(5)}{[g(5)]^2} = \dfrac{2\left(-\frac{1}{3}\right) - 3 \cdot \frac{2}{3}}{2^2} = \dfrac{-\frac{8}{3}}{4} = -\dfrac{2}{3}$.

49. (a) $y = xg(x) \quad \Rightarrow \quad y' = xg'(x) + g(x) \cdot 1 = xg'(x) + g(x)$

(b) $y = \dfrac{x}{g(x)} \quad \Rightarrow \quad y' = \dfrac{g(x) \cdot 1 - xg'(x)}{[g(x)]^2} = \dfrac{g(x) - xg'(x)}{[g(x)]^2}$

(c) $y = \dfrac{g(x)}{x} \quad \Rightarrow \quad y' = \dfrac{xg'(x) - g(x) \cdot 1}{(x)^2} = \dfrac{xg'(x) - g(x)}{x^2}$

51. If $y = f(x) = \dfrac{x}{x + 1}$, then $f'(x) = \dfrac{(x + 1)(1) - x(1)}{(x + 1)^2} = \dfrac{1}{(x + 1)^2}$. When $x = a$, the equation of the tangent line is

$y - \dfrac{a}{a + 1} = \dfrac{1}{(a + 1)^2}(x - a)$. This line passes through $(1, 2)$ when $2 - \dfrac{a}{a + 1} = \dfrac{1}{(a + 1)^2}(1 - a) \quad \Leftrightarrow$

$2(a + 1)^2 - a(a + 1) = 1 - a \quad \Leftrightarrow \quad 2a^2 + 4a + 2 - a^2 - a - 1 + a = 0 \quad \Leftrightarrow \quad a^2 + 4a + 1 = 0$.

The quadratic formula gives the roots of this equation as $a = \dfrac{-4 \pm \sqrt{4^2 - 4(1)(1)}}{2(1)} = \dfrac{-4 \pm \sqrt{12}}{2} = -2 \pm \sqrt{3}$,

so there are two such tangent lines. Since

$$f\left(-2 \pm \sqrt{3}\right) = \dfrac{-2 \pm \sqrt{3}}{-2 \pm \sqrt{3} + 1} = \dfrac{-2 \pm \sqrt{3}}{-1 \pm \sqrt{3}} \cdot \dfrac{-1 \mp \sqrt{3}}{-1 \mp \sqrt{3}}$$

$$= \dfrac{2 \pm 2\sqrt{3} \mp \sqrt{3} - 3}{1 - 3} = \dfrac{-1 \pm \sqrt{3}}{-2} = \dfrac{1 \mp \sqrt{3}}{2},$$

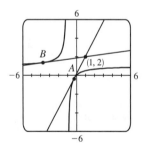

the lines touch the curve at $A\left(-2 + \sqrt{3}, \frac{1 - \sqrt{3}}{2}\right) \approx (-0.27, -0.37)$

and $B\left(-2 - \sqrt{3}, \frac{1 + \sqrt{3}}{2}\right) \approx (-3.73, 1.37)$.

53. If $P(t)$ denotes the population at time t and $A(t)$ the average annual income, then $T(t) = P(t)A(t)$ is the total personal

income. The rate at which $T(t)$ is rising is given by $T'(t) = P(t)A'(t) + A(t)P'(t) \quad \Rightarrow$

$$T'(1999) = P(1999)A'(1999) + A(1999)P'(1999) = (961{,}400)(\$1400/\text{yr}) + (\$30{,}593)(9200/\text{yr})$$
$$= \$1{,}345{,}960{,}000/\text{yr} + \$281{,}455{,}600/\text{yr} = \$1{,}627{,}415{,}600/\text{yr}$$

So the total personal income was rising by about $\$1.627$ billion per year in 1999.

The term $P(t)A'(t) \approx \$1.346$ billion represents the portion of the rate of change of total income due to the existing

population's increasing income. The term $A(t)P'(t) \approx \$281$ million represents the portion of the rate of change of total

income due to increasing population.

We will sometimes use the form $f'g + fg'$ rather than the form $fg' + gf'$ for the Product Rule.

55. (a) $(fgh)' = [(fg)h]' = (fg)'h + (fg)h' = (f'g + fg')h + (fg)h' = f'gh + fg'h + fgh'$

(b) Putting $f = g = h$ in part (a), we have $\dfrac{d}{dx}[f(x)]^3 = (fff)' = f'ff + ff'f + fff' = 3fff' = 3[f(x)]^2 f'(x)$.

(c) $\dfrac{d}{dx}(e^{3x}) = \dfrac{d}{dx}(e^x)^3 = 3(e^x)^2 e^x = 3e^{2x}e^x = 3e^{3x}$

57. For $f(x) = x^2 e^x$, $f'(x) = x^2 e^x + e^x(2x) = e^x(x^2 + 2x)$. Similarly, we have

$$f''(x) = e^x(x^2 + 4x + 2)$$
$$f'''(x) = e^x(x^2 + 6x + 6)$$
$$f^{(4)}(x) = e^x(x^2 + 8x + 12)$$
$$f^{(5)}(x) = e^x(x^2 + 10x + 20)$$

It appears that the coefficient of x in the quadratic term increases by 2 with each differentiation. The pattern for the constant terms seems to be $0 = 1 \cdot 0$, $2 = 2 \cdot 1$, $6 = 3 \cdot 2$, $12 = 4 \cdot 3$, $20 = 5 \cdot 4$. So a reasonable guess is that $f^{(n)}(x) = e^x[x^2 + 2nx + n(n-1)]$.

Proof: Let S_n be the statement that $f^{(n)}(x) = e^x[x^2 + 2nx + n(n-1)]$.

1. S_1 is true because $f'(x) = e^x(x^2 + 2x)$.

2. Assume that S_k is true; that is, $f^{(k)}(x) = e^x[x^2 + 2kx + k(k-1)]$. Then

$$f^{(k+1)}(x) = \frac{d}{dx}\left[f^{(k)}(x)\right] = e^x(2x + 2k) + [x^2 + 2kx + k(k-1)]e^x$$
$$= e^x[x^2 + (2k+2)x + (k^2 + k)] = e^x[x^2 + 2(k+1)x + (k+1)k]$$

This shows that S_{k+1} is true.

3. Therefore, by mathematical induction, S_n is true for all n; that is, $f^{(n)}(x) = e^x[x^2 + 2nx + n(n-1)]$ for every positive integer n.

3.3 Derivatives of Trigonometric Functions

1. $f(x) = 3x^2 - 2\cos x \Rightarrow f'(x) = 6x - 2(-\sin x) = 6x + 2\sin x$

3. $f(x) = \sin x + \frac{1}{2}\cot x \Rightarrow f'(x) = \cos x - \frac{1}{2}\csc^2 x$

5. $g(t) = t^3 \cos t \Rightarrow g'(t) = t^3(-\sin t) + (\cos t) \cdot 3t^2 = 3t^2 \cos t - t^3 \sin t$ or $t^2(3\cos t - t\sin t)$

7. $h(\theta) = \csc\theta + e^\theta \cot\theta \Rightarrow h'(\theta) = -\csc\theta \cot\theta + e^\theta(-\csc^2\theta) + (\cot\theta)e^\theta = -\csc\theta \cot\theta + e^\theta(\cot\theta - \csc^2\theta)$

9. $y = \dfrac{x}{2 - \tan x} \Rightarrow y' = \dfrac{(2 - \tan x)(1) - x(-\sec^2 x)}{(2 - \tan x)^2} = \dfrac{2 - \tan x + x\sec^2 x}{(2 - \tan x)^2}$

11. $f(\theta) = \dfrac{\sec\theta}{1 + \sec\theta} \Rightarrow$

$$f'(\theta) = \frac{(1 + \sec\theta)(\sec\theta \tan\theta) - (\sec\theta)(\sec\theta \tan\theta)}{(1 + \sec\theta)^2} = \frac{(\sec\theta \tan\theta)\,[(1 + \sec\theta) - \sec\theta]}{(1 + \sec\theta)^2} = \frac{\sec\theta \tan\theta}{(1 + \sec\theta)^2}$$

13. $y = \dfrac{\sin x}{x^2} \;\Rightarrow\; y' = \dfrac{x^2 \cos x - (\sin x)(2x)}{(x^2)^2} = \dfrac{x(x \cos x - 2 \sin x)}{x^4} = \dfrac{x \cos x - 2 \sin x}{x^3}$

15. Using Exercise 3.2.55(a), $f(x) = xe^x \csc x \;\Rightarrow$

$$f'(x) = (x)'e^x \csc x + x(e^x)' \csc x + xe^x(\csc x)' = 1e^x \csc x + xe^x \csc x + xe^x(-\cot x \csc x)$$

$$= e^x \csc x \,(1 + x - x \cot x)$$

17. $\dfrac{d}{dx}(\csc x) = \dfrac{d}{dx}\left(\dfrac{1}{\sin x}\right) = \dfrac{(\sin x)(0) - 1(\cos x)}{\sin^2 x} = \dfrac{-\cos x}{\sin^2 x} = -\dfrac{1}{\sin x} \cdot \dfrac{\cos x}{\sin x} = -\csc x \cot x$

19. $\dfrac{d}{dx}(\cot x) = \dfrac{d}{dx}\left(\dfrac{\cos x}{\sin x}\right) = \dfrac{(\sin x)(-\sin x) - (\cos x)(\cos x)}{\sin^2 x} = -\dfrac{\sin^2 x + \cos^2 x}{\sin^2 x} = -\dfrac{1}{\sin^2 x} = -\csc^2 x$

21. $y = \sec x \;\Rightarrow\; y' = \sec x \tan x$, so $y'\left(\frac{\pi}{3}\right) = \sec \frac{\pi}{3} \tan \frac{\pi}{3} = 2\sqrt{3}$.   An equation of the tangent line to the curve $y = \sec x$

at the point $\left(\frac{\pi}{3}, 2\right)$ is $y - 2 = 2\sqrt{3}\left(x - \frac{\pi}{3}\right)$ or $y = 2\sqrt{3}\,x + 2 - \frac{2}{3}\sqrt{3}\,\pi$.

23. $y = x + \cos x \;\Rightarrow\; y' = 1 - \sin x$. At $(0, 1)$, $y' = 1$, and an equation of the tangent line is $y - 1 = 1(x - 0)$, or $y = x + 1$.

25. (a) $y = 2x \sin x \;\Rightarrow\; y' = 2(x \cos x + \sin x \cdot 1)$.   At $\left(\frac{\pi}{2}, \pi\right)$,   (b)

$y' = 2\left(\frac{\pi}{2}\cos\frac{\pi}{2} + \sin\frac{\pi}{2}\right) = 2(0 + 1) = 2$, and an equation of the

tangent line is $y - \pi = 2\left(x - \frac{\pi}{2}\right)$, or $y = 2x$.

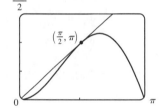

27. (a) $f(x) = \sec x - x \;\Rightarrow\; f'(x) = \sec x \tan x - 1$

(b)

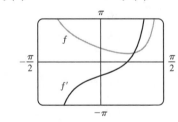

Note that $f' = 0$ where f has a minimum. Also note that f' is negative
when f is decreasing and f' is positive when f is increasing.

29. $H(\theta) = \theta \sin \theta \;\Rightarrow\; H'(\theta) = \theta\,(\cos \theta) + (\sin \theta) \cdot 1 = \theta \cos \theta + \sin \theta \;\Rightarrow$

$H''(\theta) = \theta\,(-\sin \theta) + (\cos \theta) \cdot 1 + \cos \theta = -\theta \sin \theta + 2 \cos \theta$

31. (a) $f(x) = \dfrac{\tan x - 1}{\sec x} \;\Rightarrow$

$$f'(x) = \dfrac{\sec x(\sec^2 x) - (\tan x - 1)(\sec x \tan x)}{(\sec x)^2} = \dfrac{\sec x(\sec^2 x - \tan^2 x + \tan x)}{\sec^2 x} = \dfrac{1 + \tan x}{\sec x}$$

(b) $f(x) = \dfrac{\tan x - 1}{\sec x} = \dfrac{\dfrac{\sin x}{\cos x} - 1}{\dfrac{1}{\cos x}} = \dfrac{\dfrac{\sin x - \cos x}{\cos x}}{\dfrac{1}{\cos x}} = \sin x - \cos x \;\Rightarrow\; f'(x) = \cos x - (-\sin x) = \cos x + \sin x$

(c) From part (a), $f'(x) = \dfrac{1 + \tan x}{\sec x} = \dfrac{1}{\sec x} + \dfrac{\tan x}{\sec x} = \cos x + \sin x$, which is the expression for $f'(x)$ in part (b).

33. $f(x) = x + 2\sin x$ has a horizontal tangent when $f'(x) = 0$ $\Leftrightarrow$ $1 + 2\cos x = 0$ $\Leftrightarrow$ $\cos x = -\frac{1}{2}$ $\Leftrightarrow$

$x = \frac{2\pi}{3} + 2\pi n$ or $\frac{4\pi}{3} + 2\pi n$, where n is an integer. Note that $\frac{4\pi}{3}$ and $\frac{2\pi}{3}$ are $\pm\frac{\pi}{3}$ units from π. This allows us to write the

solutions in the more compact equivalent form $(2n + 1)\pi \pm \frac{\pi}{3}$, n an integer.

35. (a) $x(t) = 8\sin t$ $\Rightarrow$ $v(t) = x'(t) = 8\cos t$ $\Rightarrow$ $a(t) = x''(t) = -8\sin t$

(b) The mass at time $t = \frac{2\pi}{3}$ has position $x\left(\frac{2\pi}{3}\right) = 8\sin\frac{2\pi}{3} = 8\left(\frac{\sqrt{3}}{2}\right) = 4\sqrt{3}$, velocity $v\left(\frac{2\pi}{3}\right) = 8\cos\frac{2\pi}{3} = 8\left(-\frac{1}{2}\right) = -4$,

and acceleration $a\left(\frac{2\pi}{3}\right) = -8\sin\frac{2\pi}{3} = -8\left(\frac{\sqrt{3}}{2}\right) = -4\sqrt{3}$. Since $v\left(\frac{2\pi}{3}\right) < 0$, the particle is moving to the left.

37.

From the diagram we can see that $\sin\theta = x/10$ $\Leftrightarrow$ $x = 10\sin\theta$. We want to find the rate

of change of x with respect to θ, that is, $dx/d\theta$. Taking the derivative of $x = 10\sin\theta$, we get

$dx/d\theta = 10(\cos\theta)$. So when $\theta = \frac{\pi}{3}$, $\frac{dx}{d\theta} = 10\cos\frac{\pi}{3} = 10\left(\frac{1}{2}\right) = 5$ ft/rad.

39. $\lim\limits_{x\to 0}\dfrac{\sin 3x}{x} = \lim\limits_{x\to 0}\dfrac{3\sin 3x}{3x}$ [multiply numerator and denominator by 3]

$= 3\lim\limits_{3x\to 0}\dfrac{\sin 3x}{3x}$ [as $x\to 0$, $3x\to 0$]

$= 3\lim\limits_{\theta\to 0}\dfrac{\sin\theta}{\theta}$ [let $\theta = 3x$]

$= 3(1)$ [Equation 2]

$= 3$

41. $\lim\limits_{t\to 0}\dfrac{\tan 6t}{\sin 2t} = \lim\limits_{t\to 0}\left(\dfrac{\sin 6t}{t}\cdot\dfrac{1}{\cos 6t}\cdot\dfrac{t}{\sin 2t}\right) = \lim\limits_{t\to 0}\dfrac{6\sin 6t}{6t}\cdot\lim\limits_{t\to 0}\dfrac{1}{\cos 6t}\cdot\lim\limits_{t\to 0}\dfrac{2t}{2\sin 2t}$

$= 6\lim\limits_{t\to 0}\dfrac{\sin 6t}{6t}\cdot\lim\limits_{t\to 0}\dfrac{1}{\cos 6t}\cdot\dfrac{1}{2}\lim\limits_{t\to 0}\dfrac{2t}{\sin 2t} = 6(1)\cdot\dfrac{1}{1}\cdot\dfrac{1}{2}(1) = 3$

43. $\lim\limits_{\theta\to 0}\dfrac{\sin(\cos\theta)}{\sec\theta} = \dfrac{\sin\left(\lim\limits_{\theta\to 0}\cos\theta\right)}{\lim\limits_{\theta\to 0}\sec\theta} = \dfrac{\sin 1}{1} = \sin 1$

45. Divide numerator and denominator by θ.   ($\sin\theta$ also works.)

$\lim\limits_{\theta\to 0}\dfrac{\sin\theta}{\theta + \tan\theta} = \lim\limits_{\theta\to 0}\dfrac{\dfrac{\sin\theta}{\theta}}{1 + \dfrac{\sin\theta}{\theta}\cdot\dfrac{1}{\cos\theta}} = \dfrac{\lim\limits_{\theta\to 0}\dfrac{\sin\theta}{\theta}}{1 + \lim\limits_{\theta\to 0}\dfrac{\sin\theta}{\theta}\lim\limits_{\theta\to 0}\dfrac{1}{\cos\theta}} = \dfrac{1}{1 + 1\cdot 1} = \dfrac{1}{2}$

47. $\lim\limits_{x\to\pi/4}\dfrac{1 - \tan x}{\sin x - \cos x} = \lim\limits_{x\to\pi/4}\dfrac{\left(1 - \dfrac{\sin x}{\cos x}\right)\cdot\cos x}{(\sin x - \cos x)\cdot\cos x} = \lim\limits_{x\to\pi/4}\dfrac{\cos x - \sin x}{(\sin x - \cos x)\cos x} = \lim\limits_{x\to\pi/4}\dfrac{-1}{\cos x} = \dfrac{-1}{1/\sqrt{2}} = -\sqrt{2}$

49. (a) $\dfrac{d}{dx}\tan x = \dfrac{d}{dx}\dfrac{\sin x}{\cos x}$ $\Rightarrow$ $\sec^2 x = \dfrac{\cos x\cos x - \sin x\,(-\sin x)}{\cos^2 x} = \dfrac{\cos^2 x + \sin^2 x}{\cos^2 x}$.   So $\sec^2 x = \dfrac{1}{\cos^2 x}$.

(b) $\dfrac{d}{dx}\sec x = \dfrac{d}{dx}\dfrac{1}{\cos x}$ $\Rightarrow$ $\sec x\tan x = \dfrac{(\cos x)(0) - 1(-\sin x)}{\cos^2 x}$.   So $\sec x\tan x = \dfrac{\sin x}{\cos^2 x}$.

(c) $\dfrac{d}{dx}\left(\sin x+\cos x\right)=\dfrac{d}{dx}\dfrac{1+\cot x}{\csc x}\quad\Rightarrow$

$$\cos x-\sin x=\frac{\csc x\left(-\csc^2 x\right)-(1+\cot x)(-\csc x\,\cot x)}{\csc^2 x}=\frac{\csc x\left[-\csc^2 x+(1+\cot x)\,\cot x\right]}{\csc^2 x}$$

$$=\frac{-\csc^2 x+\cot^2 x+\cot x}{\csc x}=\frac{-1+\cot x}{\csc x}$$

So $\cos x-\sin x=\dfrac{\cot x-1}{\csc x}$.

51. By the definition of radian measure, $s=r\theta$, where r is the radius of the circle. By drawing the bisector of the angle θ, we can

see that $\sin\dfrac{\theta}{2}=\dfrac{d/2}{r}\quad\Rightarrow\quad d=2r\sin\dfrac{\theta}{2}$. So $\displaystyle\lim_{\theta\to 0+}\frac{s}{d}=\lim_{\theta\to 0+}\frac{r\theta}{2r\sin(\theta/2)}=\lim_{\theta\to 0+}\frac{2\cdot(\theta/2)}{2\sin(\theta/2)}=\lim_{\theta\to 0}\frac{\theta/2}{\sin(\theta/2)}=1.$

[This is just the reciprocal of the limit $\displaystyle\lim_{x\to 0}\frac{\sin x}{x}=1$ combined with the fact that as $\theta\to 0$, $\frac{\theta}{2}\to 0$ also.]

3.4 The Chain Rule

1. Let $u=g(x)=4x$ and $y=f(u)=\sin u$. Then $\dfrac{dy}{dx}=\dfrac{dy}{du}\dfrac{du}{dx}=(\cos u)(4)=4\cos 4x$.

3. Let $u=g(x)=1-x^2$ and $y=f(u)=u^{10}$. Then $\dfrac{dy}{dx}=\dfrac{dy}{du}\dfrac{du}{dx}=(10u^9)(-2x)=-20x(1-x^2)^9$.

5. Let $u=g(x)=\sqrt{x}$ and $y=f(u)=e^u$. Then $\dfrac{dy}{dx}=\dfrac{dy}{du}\dfrac{du}{dx}=(e^u)\left(\tfrac12 x^{-1/2}\right)=e^{\sqrt{x}}\cdot\dfrac{1}{2\sqrt{x}}=\dfrac{e^{\sqrt{x}}}{2\sqrt{x}}.$

7. $F(x)=(x^4+3x^2-2)^5\quad\Rightarrow\quad F'(x)=5(x^4+3x^2-2)^4\cdot\dfrac{d}{dx}\left(x^4+3x^2-2\right)=5(x^4+3x^2-2)^4(4x^3+6x)$

$\left[\text{or }10x(x^4+3x^2-2)^4(2x^2+3)\right]$

9. $F(x)=\sqrt[4]{1+2x+x^3}=(1+2x+x^3)^{1/4}\quad\Rightarrow$

$$F'(x)=\tfrac14(1+2x+x^3)^{-3/4}\cdot\frac{d}{dx}\left(1+2x+x^3\right)=\frac{1}{4(1+2x+x^3)^{3/4}}\cdot(2+3x^2)=\frac{2+3x^2}{4(1+2x+x^3)^{3/4}}$$

$$=\frac{2+3x^2}{4\sqrt[4]{(1+2x+x^3)^3}}$$

11. $g(t)=\dfrac{1}{(t^4+1)^3}=(t^4+1)^{-3}\quad\Rightarrow\quad g'(t)=-3(t^4+1)^{-4}(4t^3)=-12t^3(t^4+1)^{-4}=\dfrac{-12t^3}{(t^4+1)^4}$

13. $y=\cos(a^3+x^3)\quad\Rightarrow\quad y'=-\sin(a^3+x^3)\cdot 3x^2\quad[a^3\text{ is just a constant}]\quad=-3x^2\sin(a^3+x^3)$

15. $y=xe^{-kx}\quad\Rightarrow\quad y'=x\left[e^{-kx}(-k)\right]+e^{-kx}\cdot 1=e^{-kx}(-kx+1)\quad\left[\text{or }(1-kx)e^{-kx}\right]$

17. $g(x)=(1+4x)^5(3+x-x^2)^8\quad\Rightarrow$

$g'(x)=(1+4x)^5\cdot 8(3+x-x^2)^7(1-2x)+(3+x-x^2)^8\cdot 5(1+4x)^4\cdot 4$

$\qquad=4(1+4x)^4(3+x-x^2)^7\left[2(1+4x)(1-2x)+5(3+x-x^2)\right]$

$\qquad=4(1+4x)^4(3+x-x^2)^7\left[(2+4x-16x^2)+(15+5x-5x^2)\right]=4(1+4x)^4(3+x-x^2)^7(17+9x-21x^2)$

19. $y = (2x - 5)^4(8x^2 - 5)^{-3}$ $\Rightarrow$

$\quad y' = 4(2x - 5)^3(2)(8x^2 - 5)^{-3} + (2x - 5)^4(-3)(8x^2 - 5)^{-4}(16x)$

$\quad\quad = 8(2x - 5)^3(8x^2 - 5)^{-3} - 48x(2x - 5)^4(8x^2 - 5)^{-4}$

[This simplifies to $8(2x - 5)^3(8x^2 - 5)^{-4}(-4x^2 + 30x - 5)$.]

21. $y = \left(\dfrac{x^2 + 1}{x^2 - 1}\right)^3$ $\Rightarrow$

$\quad y' = 3\left(\dfrac{x^2 + 1}{x^2 - 1}\right)^2 \cdot \dfrac{d}{dx}\left(\dfrac{x^2 + 1}{x^2 - 1}\right) = 3\left(\dfrac{x^2 + 1}{x^2 - 1}\right)^2 \cdot \dfrac{(x^2 - 1)(2x) - (x^2 + 1)(2x)}{(x^2 - 1)^2}$

$\quad\quad = 3\left(\dfrac{x^2 + 1}{x^2 - 1}\right)^2 \cdot \dfrac{2x[x^2 - 1 - (x^2 + 1)]}{(x^2 - 1)^2} = 3\left(\dfrac{x^2 + 1}{x^2 - 1}\right)^2 \cdot \dfrac{2x(-2)}{(x^2 - 1)^2} = \dfrac{-12x(x^2 + 1)^2}{(x^2 - 1)^4}$

23. $y = e^{x\cos x}$ $\Rightarrow$ $y' = e^{x\cos x} \cdot \dfrac{d}{dx}(x\cos x) = e^{x\cos x}[x(-\sin x) + (\cos x) \cdot 1] = e^{x\cos x}(\cos x - x\sin x)$

25. $F(z) = \sqrt{\dfrac{z - 1}{z + 1}} = \left(\dfrac{z - 1}{z + 1}\right)^{1/2}$ $\Rightarrow$

$\quad F'(z) = \dfrac{1}{2}\left(\dfrac{z - 1}{z + 1}\right)^{-1/2} \cdot \dfrac{d}{dz}\left(\dfrac{z - 1}{z + 1}\right) = \dfrac{1}{2}\left(\dfrac{z + 1}{z - 1}\right)^{1/2} \cdot \dfrac{(z + 1)(1) - (z - 1)(1)}{(z + 1)^2}$

$\quad\quad = \dfrac{1}{2}\dfrac{(z + 1)^{1/2}}{(z - 1)^{1/2}} \cdot \dfrac{z + 1 - z + 1}{(z + 1)^2} = \dfrac{1}{2}\dfrac{(z + 1)^{1/2}}{(z - 1)^{1/2}} \cdot \dfrac{2}{(z + 1)^2} = \dfrac{1}{(z - 1)^{1/2}(z + 1)^{3/2}}$

27. $y = \dfrac{r}{\sqrt{r^2 + 1}}$ $\Rightarrow$

$\quad y' = \dfrac{\sqrt{r^2 + 1}\,(1) - r \cdot \frac{1}{2}(r^2 + 1)^{-1/2}(2r)}{\left(\sqrt{r^2 + 1}\right)^2} = \dfrac{\sqrt{r^2 + 1} - \dfrac{r^2}{\sqrt{r^2 + 1}}}{\left(\sqrt{r^2 + 1}\right)^2} = \dfrac{\dfrac{\sqrt{r^2 + 1}\sqrt{r^2 + 1} - r^2}{\sqrt{r^2 + 1}}}{\left(\sqrt{r^2 + 1}\right)^2}$

$\quad\quad = \dfrac{(r^2 + 1) - r^2}{\left(\sqrt{r^2 + 1}\right)^3} = \dfrac{1}{(r^2 + 1)^{3/2}}$ or $(r^2 + 1)^{-3/2}$

Another solution: Write y as a product and make use of the Product Rule. $y = r(r^2 + 1)^{-1/2}$ $\Rightarrow$

$y' = r \cdot -\frac{1}{2}(r^2 + 1)^{-3/2}(2r) + (r^2 + 1)^{-1/2} \cdot 1 = (r^2 + 1)^{-3/2}[-r^2 + (r^2 + 1)^1] = (r^2 + 1)^{-3/2}(1) = (r^2 + 1)^{-3/2}$.

The step that students usually have trouble with is factoring out $(r^2 + 1)^{-3/2}$. But this is no different than factoring out x^2 from $x^2 + x^5$; that is, we are just factoring out a factor with the *smallest* exponent that appears on it. In this case, $-\frac{3}{2}$ is smaller than $-\frac{1}{2}$.

29. $y = \sin(\tan 2x)$ $\Rightarrow$ $y' = \cos(\tan 2x) \cdot \dfrac{d}{dx}(\tan 2x) = \cos(\tan 2x) \cdot \sec^2(2x) \cdot \dfrac{d}{dx}(2x) = 2\cos(\tan 2x)\sec^2(2x)$

31. Using Formula 5 and the Chain Rule, $y = 2^{\sin \pi x}$ $\Rightarrow$

$\quad y' = 2^{\sin \pi x}(\ln 2) \cdot \dfrac{d}{dx}(\sin \pi x) = 2^{\sin \pi x}(\ln 2) \cdot \cos \pi x \cdot \pi = 2^{\sin \pi x}(\pi \ln 2)\cos \pi x$

33. $y = \sec^2 x + \tan^2 x = (\sec x)^2 + (\tan x)^2 \Rightarrow$

$y' = 2(\sec x)(\sec x \tan x) + 2(\tan x)(\sec^2 x) = 2\sec^2 x \tan x + 2\sec^2 x \tan x = 4\sec^2 x \tan x$

35. $y = \cos\left(\dfrac{1 - e^{2x}}{1 + e^{2x}}\right) \Rightarrow$

$y' = -\sin\left(\dfrac{1 - e^{2x}}{1 + e^{2x}}\right) \cdot \dfrac{d}{dx}\left(\dfrac{1 - e^{2x}}{1 + e^{2x}}\right) = -\sin\left(\dfrac{1 - e^{2x}}{1 + e^{2x}}\right) \cdot \dfrac{(1 + e^{2x})(-2e^{2x}) - (1 - e^{2x})(2e^{2x})}{(1 + e^{2x})^2}$

$= -\sin\left(\dfrac{1 - e^{2x}}{1 + e^{2x}}\right) \cdot \dfrac{-2e^{2x}\left[(1 + e^{2x}) + (1 - e^{2x})\right]}{(1 + e^{2x})^2} = -\sin\left(\dfrac{1 - e^{2x}}{1 + e^{2x}}\right) \cdot \dfrac{-2e^{2x}(2)}{(1 + e^{2x})^2} = \dfrac{4e^{2x}}{(1 + e^{2x})^2} \cdot \sin\left(\dfrac{1 - e^{2x}}{1 + e^{2x}}\right)$

37. $y = \cot^2(\sin\theta) = [\cot(\sin\theta)]^2 \Rightarrow$

$y' = 2[\cot(\sin\theta)] \cdot \dfrac{d}{d\theta}[\cot(\sin\theta)] = 2\cot(\sin\theta) \cdot [-\csc^2(\sin\theta) \cdot \cos\theta] = -2\cos\theta \, \cot(\sin\theta) \, \csc^2(\sin\theta)$

39. $f(t) = \tan(e^t) + e^{\tan t} \Rightarrow f'(t) = \sec^2(e^t) \cdot \dfrac{d}{dt}(e^t) + e^{\tan t} \cdot \dfrac{d}{dt}(\tan t) = \sec^2(e^t) \cdot e^t + e^{\tan t} \cdot \sec^2 t$

41. $f(t) = \sin^2\left(e^{\sin^2 t}\right) = \left[\sin\left(e^{\sin^2 t}\right)\right]^2 \Rightarrow$

$f'(t) = 2\left[\sin\left(e^{\sin^2 t}\right)\right] \cdot \dfrac{d}{dt}\sin\left(e^{\sin^2 t}\right) = 2\sin\left(e^{\sin^2 t}\right) \cdot \cos\left(e^{\sin^2 t}\right) \cdot \dfrac{d}{dt}e^{\sin^2 t}$

$= 2\sin\left(e^{\sin^2 t}\right)\cos\left(e^{\sin^2 t}\right) \cdot e^{\sin^2 t} \cdot \dfrac{d}{dt}\sin^2 t = 2\sin\left(e^{\sin^2 t}\right)\cos\left(e^{\sin^2 t}\right)e^{\sin^2 t} \cdot 2\sin t \cos t$

$= 4\sin\left(e^{\sin^2 t}\right)\cos\left(e^{\sin^2 t}\right)e^{\sin^2 t}\sin t \cos t$

43. $g(x) = (2ra^{rx} + n)^p \Rightarrow$

$g'(x) = p(2ra^{rx} + n)^{p-1} \cdot \dfrac{d}{dx}(2ra^{rx} + n) = p(2ra^{rx} + n)^{p-1} \cdot 2ra^{rx}(\ln a) \cdot r = 2r^2 p(\ln a)(2ra^{rx} + n)^{p-1}a^{rx}$

45. $y = \cos\sqrt{\sin(\tan\pi x)} = \cos(\sin(\tan\pi x))^{1/2} \Rightarrow$

$y' = -\sin(\sin(\tan\pi x))^{1/2} \cdot \dfrac{d}{dx}(\sin(\tan\pi x))^{1/2} = -\sin(\sin(\tan\pi x))^{1/2} \cdot \tfrac{1}{2}(\sin(\tan\pi x))^{-1/2} \cdot \dfrac{d}{dx}(\sin(\tan\pi x))$

$= \dfrac{-\sin\sqrt{\sin(\tan\pi x)}}{2\sqrt{\sin(\tan\pi x)}} \cdot \cos(\tan\pi x) \cdot \dfrac{d}{dx}\tan\pi x = \dfrac{-\sin\sqrt{\sin(\tan\pi x)}}{2\sqrt{\sin(\tan\pi x)}} \cdot \cos(\tan\pi x) \cdot \sec^2(\pi x) \cdot \pi$

$= \dfrac{-\pi\cos(\tan\pi x)\sec^2(\pi x)\sin\sqrt{\sin(\tan\pi x)}}{2\sqrt{\sin(\tan\pi x)}}$

47. $h(x) = \sqrt{x^2 + 1} \Rightarrow h'(x) = \dfrac{1}{2}(x^2 + 1)^{-1/2}(2x) = \dfrac{x}{\sqrt{x^2 + 1}} \Rightarrow$

$h''(x) = \dfrac{\sqrt{x^2 + 1} \cdot 1 - x\left[\frac{1}{2}(x^2 + 1)^{-1/2}(2x)\right]}{\left(\sqrt{x^2 + 1}\right)^2} = \dfrac{(x^2 + 1)^{-1/2}\left[(x^2 + 1) - x^2\right]}{(x^2 + 1)^1} = \dfrac{1}{(x^2 + 1)^{3/2}}$

49. $y = e^{\alpha x} \sin \beta x \;\Rightarrow\; y' = e^{\alpha x} \cdot \beta \cos \beta x + \sin \beta x \cdot \alpha e^{\alpha x} = e^{\alpha x}(\beta \cos \beta x + \alpha \sin \beta x) \;\Rightarrow$

$y'' = e^{\alpha x}(-\beta^2 \sin \beta x + \alpha\beta \cos \beta x) + (\beta \cos \beta x + \alpha \sin \beta x) \cdot \alpha e^{\alpha x}$

$\quad = e^{\alpha x}(-\beta^2 \sin \beta x + \alpha\beta \cos \beta x + \alpha\beta \cos \beta x + \alpha^2 \sin \beta x) = e^{\alpha x}(\alpha^2 \sin \beta x - \beta^2 \sin \beta x + 2\alpha\beta \cos \beta x)$

$\quad = e^{\alpha x}\left[(\alpha^2 - \beta^2)\sin \beta x + 2\alpha\beta \cos \beta x\right]$

51. $y = (1 + 2x)^{10} \;\Rightarrow\; y' = 10(1 + 2x)^9 \cdot 2 = 20(1 + 2x)^9.$

At $(0, 1)$, $y' = 20(1 + 0)^9 = 20$, and an equation of the tangent line is $y - 1 = 20(x - 0)$, or $y = 20x + 1$.

53. $y = \sin(\sin x) \;\Rightarrow\; y' = \cos(\sin x) \cdot \cos x.$ At $(\pi, 0)$, $y' = \cos(\sin \pi) \cdot \cos \pi = \cos(0) \cdot (-1) = 1(-1) = -1$, and an equation of the tangent line is $y - 0 = -1(x - \pi)$, or $y = -x + \pi$.

55. (a) $y = \dfrac{2}{1 + e^{-x}} \;\Rightarrow\; y' = \dfrac{(1 + e^{-x})(0) - 2(-e^{-x})}{(1 + e^{-x})^2} = \dfrac{2e^{-x}}{(1 + e^{-x})^2}.$ (b)

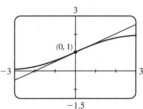

At $(0, 1)$, $y' = \dfrac{2e^0}{(1 + e^0)^2} = \dfrac{2(1)}{(1 + 1)^2} = \dfrac{2}{2^2} = \dfrac{1}{2}.$ So an equation of the

tangent line is $y - 1 = \frac{1}{2}(x - 0)$ or $y = \frac{1}{2}x + 1$.

57. (a) $f(x) = x\sqrt{2 - x^2} = x(2 - x^2)^{1/2} \;\Rightarrow$

$f'(x) = x \cdot \frac{1}{2}(2 - x^2)^{-1/2}(-2x) + (2 - x^2)^{1/2} \cdot 1 = (2 - x^2)^{-1/2}\left[-x^2 + (2 - x^2)\right] = \dfrac{2 - 2x^2}{\sqrt{2 - x^2}}$

(b)

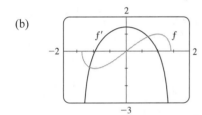

$f' = 0$ when f has a horizontal tangent line, f' is negative when f is decreasing, and f' is positive when f is increasing.

59. For the tangent line to be horizontal, $f'(x) = 0$. $f(x) = 2 \sin x + \sin^2 x \;\Rightarrow\; f'(x) = 2 \cos x + 2 \sin x \cos x = 0 \;\Leftrightarrow$

$2 \cos x(1 + \sin x) = 0 \;\Leftrightarrow\; \cos x = 0$ or $\sin x = -1$, so $x = \frac{\pi}{2} + 2n\pi$ or $\frac{3\pi}{2} + 2n\pi$, where n is any integer. Now

$f\left(\frac{\pi}{2}\right) = 3$ and $f\left(\frac{3\pi}{2}\right) = -1$, so the points on the curve with a horizontal tangent are $\left(\frac{\pi}{2} + 2n\pi, 3\right)$ and $\left(\frac{3\pi}{2} + 2n\pi, -1\right)$,

where n is any integer.

61. $F(x) = f(g(x)) \;\Rightarrow\; F'(x) = f'(g(x)) \cdot g'(x)$, so $F'(5) = f'(g(5)) \cdot g'(5) = f'(-2) \cdot 6 = 4 \cdot 6 = 24$

63. (a) $h(x) = f(g(x)) \;\Rightarrow\; h'(x) = f'(g(x)) \cdot g'(x)$, so $h'(1) = f'(g(1)) \cdot g'(1) = f'(2) \cdot 6 = 5 \cdot 6 = 30.$

(b) $H(x) = g(f(x)) \;\Rightarrow\; H'(x) = g'(f(x)) \cdot f'(x)$, so $H'(1) = g'(f(1)) \cdot f'(1) = g'(3) \cdot 4 = 9 \cdot 4 = 36.$

65. (a) $u(x) = f(g(x)) \;\Rightarrow\; u'(x) = f'(g(x))g'(x).$ So $u'(1) = f'(g(1))g'(1) = f'(3)g'(1).$ To find $f'(3)$, note that f is

linear from $(2, 4)$ to $(6, 3)$, so its slope is $\dfrac{3 - 4}{6 - 2} = -\dfrac{1}{4}.$ To find $g'(1)$, note that g is linear from $(0, 6)$ to $(2, 0)$, so its slope

is $\dfrac{0 - 6}{2 - 0} = -3.$ Thus, $f'(3)g'(1) = \left(-\frac{1}{4}\right)(-3) = \frac{3}{4}.$

(b) $v(x) = g(f(x)) \;\Rightarrow\; v'(x) = g'(f(x))f'(x).$ So $v'(1) = g'(f(1))f'(1) = g'(2)f'(1)$, which does not exist since

$g'(2)$ does not exist.

(c) $w(x) = g(g(x)) \quad \Rightarrow \quad w'(x) = g'(g(x))g'(x)$.   So $w'(1) = g'(g(1))g'(1) = g'(3)g'(1)$. To find $g'(3)$, note that g is

linear from $(2, 0)$ to $(5, 2)$, so its slope is $\dfrac{2-0}{5-2} = \dfrac{2}{3}$. Thus, $g'(3)g'(1) = \left(\frac{2}{3}\right)(-3) = -2$.

67. (a) $F(x) = f(e^x) \quad \Rightarrow \quad F'(x) = f'(e^x)\dfrac{d}{dx}(e^x) = f'(e^x)e^x$

(b) $G(x) = e^{f(x)} \quad \Rightarrow \quad G'(x) = e^{f(x)}\dfrac{d}{dx}f(x) = e^{f(x)}f'(x)$

69. $r(x) = f(g(h(x))) \quad \Rightarrow \quad r'(x) = f'(g(h(x))) \cdot g'(h(x)) \cdot h'(x)$, so

$r'(1) = f'(g(h(1))) \cdot g'(h(1)) \cdot h'(1) = f'(g(2)) \cdot g'(2) \cdot 4 = f'(3) \cdot 5 \cdot 4 = 6 \cdot 5 \cdot 4 = 120$

71. $F(x) = f(3f(4f(x))) \quad \Rightarrow$

$$F'(x) = f'(3f(4f(x))) \cdot \frac{d}{dx}(3f(4f(x))) = f'(3f(4f(x))) \cdot 3f'(4f(x)) \cdot \frac{d}{dx}(4f(x))$$

$$= f'(3f(4f(x))) \cdot 3f'(4f(x)) \cdot 4f'(x), \quad \text{so}$$

$$F'(0) = f'(3f(4f(0))) \cdot 3f'(4f(0)) \cdot 4f'(0) = f'(3f(4 \cdot 0)) \cdot 3f'(4 \cdot 0) \cdot 4 \cdot 2 = f'(3 \cdot 0) \cdot 3 \cdot 2 \cdot 4 \cdot 2 = 2 \cdot 3 \cdot 2 \cdot 4 \cdot 2 = 96.$$

73. $y = Ae^{-x} + Bxe^{-x} \quad \Rightarrow$

$y' = A(-e^{-x}) + B[x(-e^{-x}) + e^{-x} \cdot 1] = -Ae^{-x} + Be^{-x} - Bxe^{-x} = (B - A)e^{-x} - Bxe^{-x} \quad \Rightarrow$

$y'' = (B - A)(-e^{-x}) - B[x(-e^{-x}) + e^{-x} \cdot 1] = (A - B)e^{-x} - Be^{-x} + Bxe^{-x} = (A - 2B)e^{-x} + Bxe^{-x}$,

so $y'' + 2y' + y = (A - 2B)e^{-x} + Bxe^{-x} + 2[(B - A)e^{-x} - Bxe^{-x}] + Ae^{-x} + Bxe^{-x}$

$$= [(A - 2B) + 2(B - A) + A]e^{-x} + [B - 2B + B]xe^{-x} = 0.$$

75. The use of $D, D^2, \ldots, D^n$ is just a derivative notation (see text page 157). In general, $Df(2x) = 2f'(2x)$,

$D^2 f(2x) = 4f''(2x), \ldots, D^n f(2x) = 2^n f^{(n)}(2x)$. Since $f(x) = \cos x$ and $50 = 4(12) + 2$, we have

$f^{(50)}(x) = f^{(2)}(x) = -\cos x$, so $D^{50}\cos 2x = -2^{50}\cos 2x$.

77. $s(t) = 10 + \frac{1}{4}\sin(10\pi t) \quad \Rightarrow \quad$ the velocity after t seconds is $v(t) = s'(t) = \frac{1}{4}\cos(10\pi t)(10\pi) = \frac{5\pi}{2}\cos(10\pi t)$ cm/s.

79. (a) $B(t) = 4.0 + 0.35\sin\dfrac{2\pi t}{5.4} \quad \Rightarrow \quad \dfrac{dB}{dt} = \left(0.35\cos\dfrac{2\pi t}{5.4}\right)\left(\dfrac{2\pi}{5.4}\right) = \dfrac{0.7\pi}{5.4}\cos\dfrac{2\pi t}{5.4} = \dfrac{7\pi}{54}\cos\dfrac{2\pi t}{5.4}$

(b) At $t = 1$, $\dfrac{dB}{dt} = \dfrac{7\pi}{54}\cos\dfrac{2\pi}{5.4} \approx 0.16$.

81. $s(t) = 2e^{-1.5t}\sin 2\pi t \quad \Rightarrow$

$v(t) = s'(t) = 2[e^{-1.5t}(\cos 2\pi t)(2\pi) + (\sin 2\pi t)e^{-1.5t}(-1.5)] = 2e^{-1.5t}(2\pi\cos 2\pi t - 1.5\sin 2\pi t)$

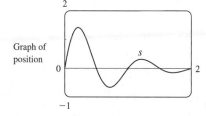

Graph of
position

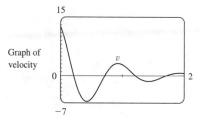

Graph of
velocity

83. By the Chain Rule, $a(t) = \dfrac{dv}{dt} = \dfrac{dv}{ds}\dfrac{ds}{dt} = \dfrac{dv}{ds}\,v(t) = v(t)\dfrac{dv}{ds}$. The derivative dv/dt is the rate of change of the velocity with respect to time (in other words, the acceleration) whereas the derivative dv/ds is the rate of change of the velocity with respect to the displacement.

85. (a) Using a calculator or CAS, we obtain the model $Q = ab^t$ with $a \approx 100.0124369$ and $b \approx 0.000045145933$.

(b) Use $Q'(t) = ab^t \ln b$ (from Formula 5) with the values of a and b from part (a) to get $Q'(0.04) \approx -670.63 \ \mu A$.

The result of Example 2 in Section 2.1 was $-670 \ \mu A$.

87. (a) Derive gives $g'(t) = \dfrac{45(t-2)^8}{(2t+1)^{10}}$ without simplifying. With either Maple or Mathematica, we first get

$g'(t) = 9\dfrac{(t-2)^8}{(2t+1)^9} - 18\dfrac{(t-2)^9}{(2t+1)^{10}}$, and the simplification command results in the expression given by Derive.

(b) Derive gives $y' = 2(x^3 - x + 1)^3(2x+1)^4(17x^3 + 6x^2 - 9x + 3)$ without simplifying. With either Maple or Mathematica, we first get $y' = 10(2x+1)^4(x^3 - x + 1)^4 + 4(2x+1)^5(x^3 - x + 1)^3(3x^2 - 1)$. If we use Mathematica's Factor or Simplify, or Maple's factor, we get the above expression, but Maple's simplify gives the polynomial expansion instead. For locating horizontal tangents, the factored form is the most helpful.

89. (a) If f is even, then $f(x) = f(-x)$. Using the Chain Rule to differentiate this equation, we get

$f'(x) = f'(-x)\dfrac{d}{dx}(-x) = -f'(-x)$. Thus, $f'(-x) = -f'(x)$, so f' is odd.

(b) If f is odd, then $f(x) = -f(-x)$. Differentiating this equation, we get $f'(x) = -f'(-x)(-1) = f'(-x)$, so f' is even.

91. (a) $\dfrac{d}{dx}(\sin^n x \cos nx) = n\sin^{n-1} x \cos x \cos nx + \sin^n x(-n\sin nx)$ [Product Rule]

$= n\sin^{n-1} x(\cos nx \cos x - \sin nx \sin x)$ [factor out $n\sin^{n-1} x$]

$= n\sin^{n-1} x \cos(nx + x)$ [Addition Formula for cosine]

$= n\sin^{n-1} x \cos[(n+1)x]$ [factor out x]

(b) $\dfrac{d}{dx}(\cos^n x \cos nx) = n\cos^{n-1} x(-\sin x)\cos nx + \cos^n x(-n\sin nx)$ [Product Rule]

$= -n\cos^{n-1} x(\cos nx \sin x + \sin nx \cos x)$ [factor out $-n\cos^{n-1} x$]

$= -n\cos^{n-1} x \sin(nx + x)$ [Addition Formula for sine]

$= -n\cos^{n-1} x \sin[(n+1)x]$ [factor out x]

93. Since $\theta° = \left(\frac{\pi}{180}\right)\theta$ rad, we have $\dfrac{d}{d\theta}(\sin\theta°) = \dfrac{d}{d\theta}\left(\sin\frac{\pi}{180}\theta\right) = \frac{\pi}{180}\cos\frac{\pi}{180}\theta = \frac{\pi}{180}\cos\theta°$.

95. The Chain Rule says that $\dfrac{dy}{dx} = \dfrac{dy}{du}\dfrac{du}{dx}$, so

$$\dfrac{d^2y}{dx^2} = \dfrac{d}{dx}\left(\dfrac{dy}{dx}\right) = \dfrac{d}{dx}\left(\dfrac{dy}{du}\dfrac{du}{dx}\right) = \left[\dfrac{d}{dx}\left(\dfrac{dy}{du}\right)\right]\dfrac{du}{dx} + \dfrac{dy}{du}\dfrac{d}{dx}\left(\dfrac{du}{dx}\right) \quad \text{[Product Rule]}$$

$$= \left[\dfrac{d}{du}\left(\dfrac{dy}{du}\right)\dfrac{du}{dx}\right]\dfrac{du}{dx} + \dfrac{dy}{du}\dfrac{d^2u}{dx^2} = \dfrac{d^2y}{du^2}\left(\dfrac{du}{dx}\right)^2 + \dfrac{dy}{du}\dfrac{d^2u}{dx^2}$$

3.5 Implicit Differentiation

1. (a) $\dfrac{d}{dx}\left(xy + 2x + 3x^2\right) = \dfrac{d}{dx}\,(4) \ \Rightarrow\ (x \cdot y' + y \cdot 1) + 2 + 6x = 0 \ \Rightarrow\ xy' = -y - 2 - 6x \ \Rightarrow$

$y' = \dfrac{-y - 2 - 6x}{x}$ or $y' = -6 - \dfrac{y+2}{x}$.

(b) $xy + 2x + 3x^2 = 4 \ \Rightarrow\ xy = 4 - 2x - 3x^2 \ \Rightarrow\ y = \dfrac{4 - 2x - 3x^2}{x} = \dfrac{4}{x} - 2 - 3x$, so $y' = -\dfrac{4}{x^2} - 3$.

(c) From part (a), $y' = \dfrac{-y - 2 - 6x}{x} = \dfrac{-(4/x - 2 - 3x) - 2 - 6x}{x} = \dfrac{-4/x - 3x}{x} = -\dfrac{4}{x^2} - 3$.

3. (a) $\dfrac{d}{dx}\left(\dfrac{1}{x} + \dfrac{1}{y}\right) = \dfrac{d}{dx}\,(1) \ \Rightarrow\ -\dfrac{1}{x^2} - \dfrac{1}{y^2}\,y' = 0 \ \Rightarrow\ -\dfrac{1}{y^2}\,y' = \dfrac{1}{x^2} \ \Rightarrow\ y' = -\dfrac{y^2}{x^2}$

(b) $\dfrac{1}{x} + \dfrac{1}{y} = 1 \ \Rightarrow\ \dfrac{1}{y} = 1 - \dfrac{1}{x} = \dfrac{x-1}{x} \ \Rightarrow\ y = \dfrac{x}{x-1}$, so $y' = \dfrac{(x-1)(1) - (x)(1)}{(x-1)^2} = \dfrac{-1}{(x-1)^2}$.

(c) $y' = -\dfrac{y^2}{x^2} = -\dfrac{[x/(x-1)]^2}{x^2} = -\dfrac{x^2}{x^2(x-1)^2} = -\dfrac{1}{(x-1)^2}$

5. $\dfrac{d}{dx}\left(x^3 + y^3\right) = \dfrac{d}{dx}\,(1) \ \Rightarrow\ 3x^2 + 3y^2 \cdot y' = 0 \ \Rightarrow\ 3y^2\,y' = -3x^2 \ \Rightarrow\ y' = -\dfrac{x^2}{y^2}$

7. $\dfrac{d}{dx}\left(x^2 + xy - y^2\right) = \dfrac{d}{dx}\,(4) \ \Rightarrow\ 2x + x \cdot y' + y \cdot 1 - 2y\,y' = 0 \ \Rightarrow$

$xy' - 2y\,y' = -2x - y \ \Rightarrow\ (x - 2y)\,y' = -2x - y \ \Rightarrow\ y' = \dfrac{-2x - y}{x - 2y} = \dfrac{2x + y}{2y - x}$

9. $\dfrac{d}{dx}\left[x^4(x+y)\right] = \dfrac{d}{dx}\left[y^2(3x - y)\right] \ \Rightarrow\ x^4(1 + y') + (x+y) \cdot 4x^3 = y^2(3 - y') + (3x - y) \cdot 2y\,y' \ \Rightarrow$

$x^4 + x^4\,y' + 4x^4 + 4x^3 y = 3y^2 - y^2\,y' + 6xy\,y' - 2y^2\,y' \ \Rightarrow\ x^4\,y' + 3y^2\,y' - 6xy\,y' = 3y^2 - 5x^4 - 4x^3 y \ \Rightarrow$

$(x^4 + 3y^2 - 6xy)\,y' = 3y^2 - 5x^4 - 4x^3 y \ \Rightarrow\ y' = \dfrac{3y^2 - 5x^4 - 4x^3 y}{x^4 + 3y^2 - 6xy}$

11. $\dfrac{d}{dx}\left(x^2 y^2 + x \sin y\right) = \dfrac{d}{dx}\,(4) \ \Rightarrow\ x^2 \cdot 2y\,y' + y^2 \cdot 2x + x \cos y \cdot y' + \sin y \cdot 1 = 0 \ \Rightarrow$

$2x^2 y\,y' + x \cos y \cdot y' = -2xy^2 - \sin y \ \Rightarrow\ (2x^2 y + x \cos y)y' = -2xy^2 - \sin y \ \Rightarrow\ y' = \dfrac{-2xy^2 - \sin y}{2x^2 y + x \cos y}$

13. $\dfrac{d}{dx}\left(4 \cos x \sin y\right) = \dfrac{d}{dx}\,(1) \ \Rightarrow\ 4\left[\cos x \cdot \cos y \cdot y' + \sin y \cdot (-\sin x)\right] = 0 \ \Rightarrow$

$y'(4 \cos x \cos y) = 4 \sin x \sin y \ \Rightarrow\ y' = \dfrac{4 \sin x \sin y}{4 \cos x \cos y} = \tan x \tan y$

15. $\dfrac{d}{dx}(e^{x/y}) = \dfrac{d}{dx}(x - y) \ \Rightarrow\ e^{x/y} \cdot \dfrac{d}{dx}\left(\dfrac{x}{y}\right) = 1 - y' \ \Rightarrow$

$e^{x/y} \cdot \dfrac{y \cdot 1 - x \cdot y'}{y^2} = 1 - y' \ \Rightarrow\ e^{x/y} \cdot \dfrac{1}{y} - \dfrac{xe^{x/y}}{y^2} \cdot y' = 1 - y' \ \Rightarrow\ y' - \dfrac{xe^{x/y}}{y^2} \cdot y' = 1 - \dfrac{e^{x/y}}{y} \ \Rightarrow$

$y'\left(1 - \dfrac{xe^{x/y}}{y^2}\right) = \dfrac{y - e^{x/y}}{y} \ \Rightarrow\ y' = \dfrac{\dfrac{y - e^{x/y}}{y}}{\dfrac{y^2 - xe^{x/y}}{y^2}} = \dfrac{y(y - e^{x/y})}{y^2 - xe^{x/y}}$

17. $\sqrt{xy} = 1 + x^2 y$ $\Rightarrow$ $\frac{1}{2}(xy)^{-1/2}(xy' + y \cdot 1) = 0 + x^2 y' + y \cdot 2x$ $\Rightarrow$ $\frac{x}{2\sqrt{xy}}y' + \frac{y}{2\sqrt{xy}} = x^2 y' + 2xy$ $\Rightarrow$

$y'\left(\frac{x}{2\sqrt{xy}} - x^2\right) = 2xy - \frac{y}{2\sqrt{xy}}$ $\Rightarrow$ $y'\left(\frac{x - 2x^2\sqrt{xy}}{2\sqrt{xy}}\right) = \frac{4xy\sqrt{xy} - y}{2\sqrt{xy}}$ $\Rightarrow$ $y' = \frac{4xy\sqrt{xy} - y}{x - 2x^2\sqrt{xy}}$

19. $\frac{d}{dx}(e^y \cos x) = \frac{d}{dx}[1 + \sin(xy)]$ $\Rightarrow$ $e^y(-\sin x) + \cos x \cdot e^y \cdot y' = \cos(xy) \cdot (xy' + y \cdot 1)$ $\Rightarrow$

$-e^y \sin x + e^y \cos x \cdot y' = x \cos(xy) \cdot y' + y \cos(xy)$ $\Rightarrow$ $e^y \cos x \cdot y' - x \cos(xy) \cdot y' = e^y \sin x + y \cos(xy)$ $\Rightarrow$

$[e^y \cos x - x \cos(xy)] y' = e^y \sin x + y \cos(xy)$ $\Rightarrow$ $y' = \frac{e^y \sin x + y \cos(xy)}{e^y \cos x - x \cos(xy)}$

21. $\frac{d}{dx}\{f(x) + x^2[f(x)]^3\} = \frac{d}{dx}(10)$ $\Rightarrow$ $f'(x) + x^2 \cdot 3[f(x)]^2 \cdot f'(x) + [f(x)]^3 \cdot 2x = 0$. If $x = 1$, we have

$f'(1) + 1^2 \cdot 3[f(1)]^2 \cdot f'(1) + [f(1)]^3 \cdot 2(1) = 0$ $\Rightarrow$ $f'(1) + 1 \cdot 3 \cdot 2^2 \cdot f'(1) + 2^3 \cdot 2 = 0$ $\Rightarrow$

$f'(1) + 12f'(1) = -16$ $\Rightarrow$ $13f'(1) = -16$ $\Rightarrow$ $f'(1) = -\frac{16}{13}$.

23. $\frac{d}{dy}(x^4 y^2 - x^3 y + 2xy^3) = \frac{d}{dy}(0)$ $\Rightarrow$ $x^4 \cdot 2y + y^2 \cdot 4x^3 x' - (x^3 \cdot 1 + y \cdot 3x^2 x') + 2(x \cdot 3y^2 + y^3 \cdot x') = 0$ $\Rightarrow$

$4x^3 y^2 x' - 3x^2 y x' + 2y^3 x' = -2x^4 y + x^3 - 6xy^2$ $\Rightarrow$ $(4x^3 y^2 - 3x^2 y + 2y^3) x' = -2x^4 y + x^3 - 6xy^2$ $\Rightarrow$

$x' = \frac{dx}{dy} = \frac{-2x^4 y + x^3 - 6xy^2}{4x^3 y^2 - 3x^2 y + 2y^3}$

25. $x^2 + xy + y^2 = 3$ $\Rightarrow$ $2x + xy' + y \cdot 1 + 2yy' = 0$ $\Rightarrow$ $xy' + 2yy' = -2x - y$ $\Rightarrow$ $y'(x + 2y) = -2x - y$ $\Rightarrow$

$y' = \frac{-2x - y}{x + 2y}$. When $x = 1$ and $y = 1$, we have $y' = \frac{-2 - 1}{1 + 2 \cdot 1} = \frac{-3}{3} = -1$, so an equation of the tangent line is

$y - 1 = -1(x - 1)$ or $y = -x + 2$.

27. $x^2 + y^2 = (2x^2 + 2y^2 - x)^2$ $\Rightarrow$ $2x + 2yy' = 2(2x^2 + 2y^2 - x)(4x + 4yy' - 1)$. When $x = 0$ and $y = \frac{1}{2}$, we have

$0 + y' = 2(\frac{1}{2})(2y' - 1)$ $\Rightarrow$ $y' = 2y' - 1$ $\Rightarrow$ $y' = 1$, so an equation of the tangent line is $y - \frac{1}{2} = 1(x - 0)$

or $y = x + \frac{1}{2}$.

29. $2(x^2 + y^2)^2 = 25(x^2 - y^2)$ $\Rightarrow$ $4(x^2 + y^2)(2x + 2yy') = 25(2x - 2yy')$ $\Rightarrow$

$4(x + yy')(x^2 + y^2) = 25(x - yy')$ $\Rightarrow$ $4yy'(x^2 + y^2) + 25yy' = 25x - 4x(x^2 + y^2)$ $\Rightarrow$

$y' = \frac{25x - 4x(x^2 + y^2)}{25y + 4y(x^2 + y^2)}$. When $x = 3$ and $y = 1$, we have $y' = \frac{75 - 120}{25 + 40} = -\frac{45}{65} = -\frac{9}{13}$,

so an equation of the tangent line is $y - 1 = -\frac{9}{13}(x - 3)$ or $y = -\frac{9}{13}x + \frac{40}{13}$.

31. (a) $y^2 = 5x^4 - x^2$ $\Rightarrow$ $2yy' = 5(4x^3) - 2x$ $\Rightarrow$ $y' = \frac{10x^3 - x}{y}$. (b)

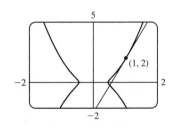

So at the point $(1, 2)$ we have $y' = \frac{10(1)^3 - 1}{2} = \frac{9}{2}$, and an equation

of the tangent line is $y - 2 = \frac{9}{2}(x - 1)$ or $y = \frac{9}{2}x - \frac{5}{2}$.

33. $9x^2 + y^2 = 9 \;\Rightarrow\; 18x + 2y\,y' = 0 \;\Rightarrow\; 2y\,y' = -18x \;\Rightarrow\; y' = -9x/y \;\Rightarrow$

$y'' = -9\left(\dfrac{y \cdot 1 - x \cdot y'}{y^2}\right) = -9\left(\dfrac{y - x(-9x/y)}{y^2}\right) = -9 \cdot \dfrac{y^2 + 9x^2}{y^3} = -9 \cdot \dfrac{9}{y^3}$ [since x and y must satisfy the original

equation, $9x^2 + y^2 = 9$]. Thus, $y'' = -81/y^3$.

35. $x^3 + y^3 = 1 \;\Rightarrow\; 3x^2 + 3y^2\,y' = 0 \;\Rightarrow\; y' = -\dfrac{x^2}{y^2} \;\Rightarrow$

$y'' = -\dfrac{y^2(2x) - x^2 \cdot 2y\,y'}{(y^2)^2} = -\dfrac{2xy^2 - 2x^2 y(-x^2/y^2)}{y^4} = -\dfrac{2xy^4 + 2x^4 y}{y^6} = -\dfrac{2xy(y^3 + x^3)}{y^6} = -\dfrac{2x}{y^5},$

since x and y must satisfy the original equation, $x^3 + y^3 = 1$.

37. (a) There are eight points with horizontal tangents: four at $x \approx 1.57735$ and

four at $x \approx 0.42265$.

(b) $y' = \dfrac{3x^2 - 6x + 2}{2(2y^3 - 3y^2 - y + 1)} \;\Rightarrow\; y' = -1$ at $(0, 1)$ and $y' = \frac{1}{3}$ at $(0, 2)$.

Equations of the tangent lines are $y = -x + 1$ and $y = \frac{1}{3}x + 2$.

(c) $y' = 0 \;\Rightarrow\; 3x^2 - 6x + 2 = 0 \;\Rightarrow\; x = 1 \pm \frac{1}{3}\sqrt{3}$

(d) By multiplying the right side of the equation by $x - 3$, we obtain the first
graph. By modifying the equation in other ways, we can generate the other
graphs.

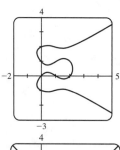

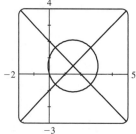

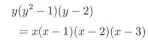

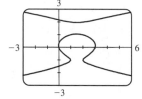

$$y(y^2 - 1)(y - 2)$$
$$= x(x - 1)(x - 2)(x - 3)$$

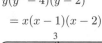

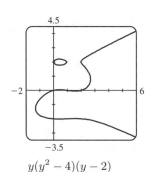

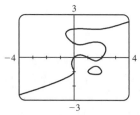

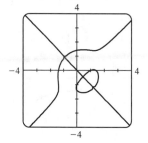

$$y(y^2 - 4)(y - 2)$$
$$= x(x - 1)(x - 2)$$

$$y(y + 1)(y^2 - 1)(y - 2)$$
$$= x(x - 1)(x - 2)$$

$$(y + 1)(y^2 - 1)(y - 2)$$
$$= (x - 1)(x - 2)$$

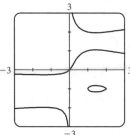

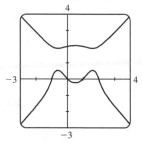

$$x(y + 1)(y^2 - 1)(y - 2)$$
$$= y(x - 1)(x - 2)$$

$$y(y^2 + 1)(y - 2)$$
$$= x(x^2 - 1)(x - 2)$$

$$y(y + 1)(y^2 - 2)$$
$$= x(x - 1)(x^2 - 2)$$

39. From Exercise 29, a tangent to the lemniscate will be horizontal if $y' = 0$ $\Rightarrow$ $25x - 4x(x^2 + y^2) = 0$ $\Rightarrow$

$x[25 - 4(x^2 + y^2)] = 0$ $\Rightarrow$ $x^2 + y^2 = \frac{25}{4}$ **(1)**. (Note that when x is 0, y is also 0, and there is no horizontal tangent

at the origin.) Substituting $\frac{25}{4}$ for $x^2 + y^2$ in the equation of the lemniscate, $2(x^2 + y^2)^2 = 25(x^2 - y^2)$, we get

$x^2 - y^2 = \frac{25}{8}$ **(2)**. Solving **(1)** and **(2)**, we have $x^2 = \frac{75}{16}$ and $y^2 = \frac{25}{16}$, so the four points are $\left(\pm\frac{5\sqrt{3}}{4}, \pm\frac{5}{4}\right)$.

41. $\dfrac{x^2}{a^2} - \dfrac{y^2}{b^2} = 1$ $\Rightarrow$ $\dfrac{2x}{a^2} - \dfrac{2yy'}{b^2} = 0$ $\Rightarrow$ $y' = \dfrac{b^2x}{a^2y}$ $\Rightarrow$ an equation of the tangent line at (x_0, y_0) is

$y - y_0 = \dfrac{b^2x_0}{a^2y_0}(x - x_0)$. Multiplying both sides by $\dfrac{y_0}{b^2}$ gives $\dfrac{y_0 y}{b^2} - \dfrac{y_0^2}{b^2} = \dfrac{x_0 x}{a^2} - \dfrac{x_0^2}{a^2}$. Since (x_0, y_0) lies on the hyperbola,

we have $\dfrac{x_0 x}{a^2} - \dfrac{y_0 y}{b^2} = \dfrac{x_0^2}{a^2} - \dfrac{y_0^2}{b^2} = 1$.

43. If the circle has radius r, its equation is $x^2 + y^2 = r^2$ $\Rightarrow$ $2x + 2yy' = 0$ $\Rightarrow$ $y' = -\dfrac{x}{y}$, so the slope of the tangent line

at $P(x_0, y_0)$ is $-\dfrac{x_0}{y_0}$. The negative reciprocal of that slope is $\dfrac{-1}{-x_0/y_0} = \dfrac{y_0}{x_0}$, which is the slope of OP, so the tangent line at

P is perpendicular to the radius OP.

45. $y = \tan^{-1}\sqrt{x}$ $\Rightarrow$ $y' = \dfrac{1}{1 + \left(\sqrt{x}\right)^2} \cdot \dfrac{d}{dx}\left(\sqrt{x}\right) = \dfrac{1}{1 + x}\left(\tfrac{1}{2}x^{-1/2}\right) = \dfrac{1}{2\sqrt{x}\,(1 + x)}$

47. $y = \sin^{-1}(2x + 1)$ $\Rightarrow$

$y' = \dfrac{1}{\sqrt{1 - (2x + 1)^2}} \cdot \dfrac{d}{dx}(2x + 1) = \dfrac{1}{\sqrt{1 - (4x^2 + 4x + 1)}} \cdot 2 = \dfrac{2}{\sqrt{-4x^2 - 4x}} = \dfrac{1}{\sqrt{-x^2 - x}}$

49. $G(x) = \sqrt{1 - x^2}\arccos x$ $\Rightarrow$ $G'(x) = \sqrt{1 - x^2} \cdot \dfrac{-1}{\sqrt{1 - x^2}} + \arccos x \cdot \tfrac{1}{2}(1 - x^2)^{-1/2}(-2x) = -1 - \dfrac{x\arccos x}{\sqrt{1 - x^2}}$

51. $h(t) = \cot^{-1}(t) + \cot^{-1}(1/t)$ $\Rightarrow$

$h'(t) = -\dfrac{1}{1 + t^2} - \dfrac{1}{1 + (1/t)^2} \cdot \dfrac{d}{dt}\dfrac{1}{t} = -\dfrac{1}{1 + t^2} - \dfrac{t^2}{t^2 + 1}\cdot\left(-\dfrac{1}{t^2}\right) = -\dfrac{1}{1 + t^2} + \dfrac{1}{t^2 + 1} = 0.$

Note that this makes sense because $h(t) = \dfrac{\pi}{2}$ for $t > 0$ and $h(t) = \dfrac{3\pi}{2}$ for $t < 0$.

53. $y = \cos^{-1}(e^{2x})$ $\Rightarrow$ $y' = -\dfrac{1}{\sqrt{1 - (e^{2x})^2}} \cdot \dfrac{d}{dx}(e^{2x}) = -\dfrac{2e^{2x}}{\sqrt{1 - e^{4x}}}$

55. $f(x) = \sqrt{1 - x^2}\arcsin x$ $\Rightarrow$ $f'(x) = \sqrt{1 - x^2} \cdot \dfrac{1}{\sqrt{1 - x^2}} + \arcsin x \cdot \tfrac{1}{2}\left(1 - x^2\right)^{-1/2}(-2x) = 1 - \dfrac{x\arcsin x}{\sqrt{1 - x^2}}$

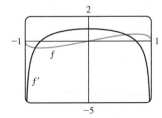

Note that $f' = 0$ where the graph of f has a horizontal tangent. Also note that f' is negative when f is decreasing and f' is positive when f is increasing.

57. Let $y = \cos^{-1} x$. Then $\cos y = x$ and $0 \le y \le \pi$ $\Rightarrow$ $-\sin y \dfrac{dy}{dx} = 1$ $\Rightarrow$

$$\frac{dy}{dx} = -\frac{1}{\sin y} = -\frac{1}{\sqrt{1 - \cos^2 y}} = -\frac{1}{\sqrt{1 - x^2}}. \quad \text{[Note that } \sin y \ge 0 \text{ for } 0 \le y \le \pi.\text{]}$$

59. $x^2 + y^2 = r^2$ is a circle with center O and $ax + by = 0$ is a line through O [assume a

and b are not both zero]. $x^2 + y^2 = r^2$ $\Rightarrow$ $2x + 2yy' = 0$ $\Rightarrow$ $y' = -x/y$, so the

slope of the tangent line at $P_0\,(x_0, y_0)$ is $-x_0/y_0$. The slope of the line OP_0 is y_0/x_0,

which is the negative reciprocal of $-x_0/y_0$. Hence, the curves are orthogonal, and the

families of curves are orthogonal trajectories of each other.

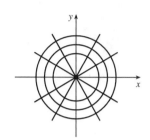

61. $y = cx^2$ $\Rightarrow$ $y' = 2cx$ and $x^2 + 2y^2 = k$ [assume $k > 0$] $\Rightarrow$ $2x + 4yy' = 0$ $\Rightarrow$

$$2yy' = -x \quad \Rightarrow \quad y' = -\frac{x}{2(y)} = -\frac{x}{2(cx^2)} = -\frac{1}{2cx}, \text{ so the curves are orthogonal if}$$

$c \ne 0$. If $c = 0$, then the horizontal line $y = cx^2 = 0$ intersects $x^2 + 2y^2 = k$ orthogonally

at $\left(\pm\sqrt{k}, 0\right)$, since the ellipse $x^2 + 2y^2 = k$ has vertical tangents at those two points.

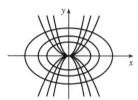

63. To find the points at which the ellipse $x^2 - xy + y^2 = 3$ crosses the x-axis, let $y = 0$ and solve for x.

$y = 0$ $\Rightarrow$ $x^2 - x(0) + 0^2 = 3$ $\Leftrightarrow$ $x = \pm\sqrt{3}$. So the graph of the ellipse crosses the x-axis at the points $\left(\pm\sqrt{3}, 0\right)$.

Using implicit differentiation to find y', we get $2x - xy' - y + 2yy' = 0$ $\Rightarrow$ $y'(2y - x) = y - 2x$ $\Leftrightarrow$ $y' = \dfrac{y - 2x}{2y - x}$.

So y' at $\left(\sqrt{3}, 0\right)$ is $\dfrac{0 - 2\sqrt{3}}{2(0) - \sqrt{3}} = 2$ and y' at $\left(-\sqrt{3}, 0\right)$ is $\dfrac{0 + 2\sqrt{3}}{2(0) + \sqrt{3}} = 2$. Thus, the tangent lines at these points are parallel.

65. $x^2 y^2 + xy = 2$ $\Rightarrow$ $x^2 \cdot 2yy' + y^2 \cdot 2x + x \cdot y' + y \cdot 1 = 0$ $\Leftrightarrow$ $y'(2x^2 y + x) = -2xy^2 - y$ $\Leftrightarrow$

$y' = -\dfrac{2xy^2 + y}{2x^2 y + x}$. So $-\dfrac{2xy^2 + y}{2x^2 y + x} = -1$ $\Leftrightarrow$ $2xy^2 + y = 2x^2 y + x$ $\Leftrightarrow$ $y(2xy + 1) = x(2xy + 1)$ $\Leftrightarrow$

$y(2xy + 1) - x(2xy + 1) = 0$ $\Leftrightarrow$ $(2xy + 1)(y - x) = 0$ $\Leftrightarrow$ $xy = -\frac{1}{2}$ or $y = x$. But $xy = -\frac{1}{2}$ $\Rightarrow$

$x^2 y^2 + xy = \frac{1}{4} - \frac{1}{2} \ne 2$, so we must have $x = y$. Then $x^2 y^2 + xy = 2$ $\Rightarrow$ $x^4 + x^2 = 2$ $\Leftrightarrow$ $x^4 + x^2 - 2 = 0$ $\Leftrightarrow$

$(x^2 + 2)(x^2 - 1) = 0$. So $x^2 = -2$, which is impossible, or $x^2 = 1$ $\Leftrightarrow$ $x = \pm 1$. Since $x = y$, the points on the curve

where the tangent line has a slope of -1 are $(-1, -1)$ and $(1, 1)$.

67. (a) If $y = f^{-1}(x)$, then $f(y) = x$. Differentiating implicitly with respect to x and remembering that y is a function of x,

$$\text{we get } f'(y)\frac{dy}{dx} = 1, \text{ so } \frac{dy}{dx} = \frac{1}{f'(y)} \quad \Rightarrow \quad \left(f^{-1}\right)'(x) = \frac{1}{f'(f^{-1}(x))}.$$

(b) $f(4) = 5$ $\Rightarrow$ $f^{-1}(5) = 4$. By part (a), $\left(f^{-1}\right)'(5) = \dfrac{1}{f'(f^{-1}(5))} = \dfrac{1}{f'(4)} = 1/\left(\frac{2}{3}\right) = \frac{3}{2}$.

69. $x^2 + 4y^2 = 5 \;\Rightarrow\; 2x + 4(2yy') = 0 \;\Rightarrow\; y' = -\dfrac{x}{4y}$. Now let h be the height of the lamp, and let (a, b) be the point of

tangency of the line passing through the points $(3, h)$ and $(-5, 0)$. This line has slope $(h - 0)/[3 - (-5)] = \frac{1}{8}h$. But the

slope of the tangent line through the point (a, b) can be expressed as $y' = -\dfrac{a}{4b}$, or as $\dfrac{b - 0}{a - (-5)} = \dfrac{b}{a + 5}$ [since the line

passes through $(-5, 0)$ and (a, b)], so $-\dfrac{a}{4b} = \dfrac{b}{a + 5} \;\Leftrightarrow\; 4b^2 = -a^2 - 5a \;\Leftrightarrow\; a^2 + 4b^2 = -5a$. But $a^2 + 4b^2 = 5$

[since (a, b) is on the ellipse], so $5 = -5a \;\Leftrightarrow\; a = -1$. Then $4b^2 = -a^2 - 5a = -1 - 5(-1) = 4 \;\Rightarrow\; b = 1$, since the

point is on the top half of the ellipse. So $\dfrac{h}{8} = \dfrac{b}{a + 5} = \dfrac{1}{-1 + 5} = \dfrac{1}{4} \;\Rightarrow\; h = 2$. So the lamp is located 2 units above the

x-axis.

3.6   Derivatives of Logarithmic Functions

1. The differentiation formula for logarithmic functions, $\dfrac{d}{dx}(\log_a x) = \dfrac{1}{x \ln a}$, is simplest when $a = e$ because $\ln e = 1$.

3. $f(x) = \sin(\ln x) \;\Rightarrow\; f'(x) = \cos(\ln x) \cdot \dfrac{d}{dx} \ln x = \cos(\ln x) \cdot \dfrac{1}{x} = \dfrac{\cos(\ln x)}{x}$

5. $f(x) = \log_2(1 - 3x) \;\Rightarrow\; f'(x) = \dfrac{1}{(1 - 3x) \ln 2} \dfrac{d}{dx}(1 - 3x) = \dfrac{-3}{(1 - 3x) \ln 2}$ or $\dfrac{3}{(3x - 1) \ln 2}$

7. $f(x) = \sqrt[5]{\ln x} = (\ln x)^{1/5} \;\Rightarrow\; f'(x) = \frac{1}{5}(\ln x)^{-4/5} \dfrac{d}{dx}(\ln x) = \dfrac{1}{5(\ln x)^{4/5}} \cdot \dfrac{1}{x} = \dfrac{1}{5x \sqrt[5]{(\ln x)^4}}$

9. $f(x) = \sin x \ln(5x) \;\Rightarrow\; f'(x) = \sin x \cdot \dfrac{1}{5x} \cdot \dfrac{d}{dx}(5x) + \ln(5x) \cdot \cos x = \dfrac{\sin x \cdot 5}{5x} + \cos x \ln(5x) = \dfrac{\sin x}{x} + \cos x \ln(5x)$

11. $F(t) = \ln \dfrac{(2t + 1)^3}{(3t - 1)^4} = \ln(2t + 1)^3 - \ln(3t - 1)^4 = 3\ln(2t + 1) - 4\ln(3t - 1) \;\Rightarrow$

$F'(t) = 3 \cdot \dfrac{1}{2t + 1} \cdot 2 - 4 \cdot \dfrac{1}{3t - 1} \cdot 3 = \dfrac{6}{2t + 1} - \dfrac{12}{3t - 1}$, or combined, $\dfrac{-6(t + 3)}{(2t + 1)(3t - 1)}$.

13. $g(x) = \ln\left(x\sqrt{x^2 - 1}\right) = \ln x + \ln(x^2 - 1)^{1/2} = \ln x + \frac{1}{2}\ln(x^2 - 1) \;\Rightarrow$

$g'(x) = \dfrac{1}{x} + \dfrac{1}{2} \cdot \dfrac{1}{x^2 - 1} \cdot 2x = \dfrac{1}{x} + \dfrac{x}{x^2 - 1} = \dfrac{x^2 - 1 + x \cdot x}{x(x^2 - 1)} = \dfrac{2x^2 - 1}{x(x^2 - 1)}$

15. $f(u) = \dfrac{\ln u}{1 + \ln(2u)} \;\Rightarrow$

$f'(u) = \dfrac{[1 + \ln(2u)] \cdot \frac{1}{u} - \ln u \cdot \frac{1}{2u} \cdot 2}{[1 + \ln(2u)]^2} = \dfrac{\frac{1}{u}[1 + \ln(2u) - \ln u]}{[1 + \ln(2u)]^2} = \dfrac{1 + (\ln 2 + \ln u) - \ln u}{u[1 + \ln(2u)]^2} = \dfrac{1 + \ln 2}{u[1 + \ln(2u)]^2}$

17. $y = \ln|2 - x - 5x^2| \;\Rightarrow\; y' = \dfrac{1}{2 - x - 5x^2} \cdot (-1 - 10x) = \dfrac{-10x - 1}{2 - x - 5x^2}$ or $\dfrac{10x + 1}{5x^2 + x - 2}$

19. $y = \ln(e^{-x} + xe^{-x}) = \ln(e^{-x}(1 + x)) = \ln(e^{-x}) + \ln(1 + x) = -x + \ln(1 + x) \;\Rightarrow$

$y' = -1 + \dfrac{1}{1 + x} = \dfrac{-1 - x + 1}{1 + x} = -\dfrac{x}{1 + x}$

21. $y = 2x \log_{10} \sqrt{x} = 2x \log_{10} x^{1/2} = 2x \cdot \frac{1}{2} \log_{10} x = x \log_{10} x \quad \Rightarrow \quad y' = x \cdot \frac{1}{x \ln 10} + \log_{10} x \cdot 1 = \frac{1}{\ln 10} + \log_{10} x$

Note: $\frac{1}{\ln 10} = \frac{\ln e}{\ln 10} = \log_{10} e$, so the answer could be written as $\frac{1}{\ln 10} + \log_{10} x = \log_{10} e + \log_{10} x = \log_{10} ex$.

23. $y = x^2 \ln(2x) \quad \Rightarrow \quad y' = x^2 \cdot \frac{1}{2x} \cdot 2 + \ln(2x) \cdot (2x) = x + 2x \ln(2x) \quad \Rightarrow$

$y'' = 1 + 2x \cdot \frac{1}{2x} \cdot 2 + \ln(2x) \cdot 2 = 1 + 2 + 2 \ln(2x) = 3 + 2 \ln(2x)$

25. $y = \ln\left(x + \sqrt{1 + x^2}\right) \quad \Rightarrow$

$y' = \frac{1}{x + \sqrt{1 + x^2}} \frac{d}{dx}\left(x + \sqrt{1 + x^2}\right) = \frac{1}{x + \sqrt{1 + x^2}}\left[1 + \frac{1}{2}(1 + x^2)^{-1/2}(2x)\right]$

$= \frac{1}{x + \sqrt{1 + x^2}}\left(1 + \frac{x}{\sqrt{1 + x^2}}\right) = \frac{1}{x + \sqrt{1 + x^2}} \cdot \frac{\sqrt{1 + x^2} + x}{\sqrt{1 + x^2}} = \frac{1}{\sqrt{1 + x^2}} \quad \Rightarrow$

$y'' = -\frac{1}{2}(1 + x^2)^{-3/2}(2x) = \frac{-x}{(1 + x^2)^{3/2}}$

27. $f(x) = \frac{x}{1 - \ln(x - 1)} \quad \Rightarrow$

$f'(x) = \frac{[1 - \ln(x-1)] \cdot 1 - x \cdot \dfrac{-1}{x-1}}{[1 - \ln(x-1)]^2} = \frac{\dfrac{(x-1)[1 - \ln(x-1)] + x}{x-1}}{[1 - \ln(x-1)]^2} = \frac{x - 1 - (x-1)\ln(x-1) + x}{(x-1)[1 - \ln(x-1)]^2}$

$= \frac{2x - 1 - (x-1)\ln(x-1)}{(x-1)[1 - \ln(x-1)]^2}$

$\text{Dom}(f) = \{x \mid x - 1 > 0 \quad \text{and} \quad 1 - \ln(x-1) \neq 0\} = \{x \mid x > 1 \quad \text{and} \quad \ln(x-1) \neq 1\}$

$= \{x \mid x > 1 \quad \text{and} \quad x - 1 \neq e^1\} = \{x \mid x > 1 \quad \text{and} \quad x \neq 1 + e\} = (1, 1+e) \cup (1+e, \infty)$

29. $f(x) = \ln(x^2 - 2x) \quad \Rightarrow \quad f'(x) = \frac{1}{x^2 - 2x}(2x - 2) = \frac{2(x-1)}{x(x-2)}.$

$\text{Dom}(f) = \{x \mid x(x-2) > 0\} = (-\infty, 0) \cup (2, \infty).$

31. $f(x) = \frac{\ln x}{x^2} \quad \Rightarrow \quad f'(x) = \frac{x^2(1/x) - (\ln x)(2x)}{(x^2)^2} = \frac{x - 2x \ln x}{x^4} = \frac{x(1 - 2\ln x)}{x^4} = \frac{1 - 2\ln x}{x^3},$

so $f'(1) = \frac{1 - 2\ln 1}{1^3} = \frac{1 - 2 \cdot 0}{1} = 1.$

33. $y = \ln\left(xe^{x^2}\right) = \ln x + \ln e^{x^2} = \ln x + x^2 \quad \Rightarrow \quad y' = \frac{1}{x} + 2x.$ At $(1, 1)$, the slope of the tangent line is

$y'(1) = 1 + 2 = 3$, and an equation of the tangent line is $y - 1 = 3(x - 1)$, or $y = 3x - 2.$

35. $f(x) = \sin x + \ln x \quad \Rightarrow \quad f'(x) = \cos x + 1/x.$

This is reasonable, because the graph shows that f increases when f' is

positive, and $f'(x) = 0$ when f has a horizontal tangent.

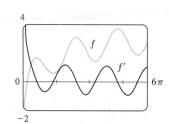

37. $y = (2x + 1)^5(x^4 - 3)^6$ $\Rightarrow$ $\ln y = \ln((2x + 1)^5(x^4 - 3)^6)$ $\Rightarrow$ $\ln y = 5\ln(2x + 1) + 6\ln(x^4 - 3)$ $\Rightarrow$

$\dfrac{1}{y}\,y' = 5 \cdot \dfrac{1}{2x + 1} \cdot 2 + 6 \cdot \dfrac{1}{x^4 - 3} \cdot 4x^3$ $\Rightarrow$

$y' = y\left(\dfrac{10}{2x + 1} + \dfrac{24x^3}{x^4 - 3}\right) = (2x + 1)^5(x^4 - 3)^6\left(\dfrac{10}{2x + 1} + \dfrac{24x^3}{x^4 - 3}\right).$

[The answer could be simplified to $y' = 2(2x + 1)^4(x^4 - 3)^5(29x^4 + 12x^3 - 15)$, but this is unnecessary.]

39. $y = \dfrac{\sin^2 x \tan^4 x}{(x^2 + 1)^2}$ $\Rightarrow$ $\ln y = \ln(\sin^2 x \tan^4 x) - \ln(x^2 + 1)^2$ $\Rightarrow$

$\ln y = \ln(\sin x)^2 + \ln(\tan x)^4 - \ln(x^2 + 1)^2$ $\Rightarrow$ $\ln y = 2\ln|\sin x| + 4\ln|\tan x| - 2\ln(x^2 + 1)$ $\Rightarrow$

$\dfrac{1}{y}\,y' = 2 \cdot \dfrac{1}{\sin x} \cdot \cos x + 4 \cdot \dfrac{1}{\tan x} \cdot \sec^2 x - 2 \cdot \dfrac{1}{x^2 + 1} \cdot 2x$ $\Rightarrow$ $y' = \dfrac{\sin^2 x \tan^4 x}{(x^2 + 1)^2}\left(2\cot x + \dfrac{4\sec^2 x}{\tan x} - \dfrac{4x}{x^2 + 1}\right)$

41. $y = x^x$ $\Rightarrow$ $\ln y = \ln x^x$ $\Rightarrow$ $\ln y = x\ln x$ $\Rightarrow$ $y'/y = x(1/x) + (\ln x) \cdot 1$ $\Rightarrow$ $y' = y(1 + \ln x)$ $\Rightarrow$

$y' = x^x(1 + \ln x)$

43. $y = x^{\sin x}$ $\Rightarrow$ $\ln y = \ln x^{\sin x}$ $\Rightarrow$ $\ln y = \sin x \ln x$ $\Rightarrow$ $\dfrac{y'}{y} = (\sin x) \cdot \dfrac{1}{x} + (\ln x)(\cos x)$ $\Rightarrow$

$y' = y\left(\dfrac{\sin x}{x} + \ln x \cos x\right)$ $\Rightarrow$ $y' = x^{\sin x}\left(\dfrac{\sin x}{x} + \ln x \, \cos x\right)$

45. $y = (\cos x)^x$ $\Rightarrow$ $\ln y = \ln(\cos x)^x$ $\Rightarrow$ $\ln y = x\ln\cos x$ $\Rightarrow$ $\dfrac{1}{y}\,y' = x \cdot \dfrac{1}{\cos x} \cdot (-\sin x) + \ln\cos x \cdot 1$ $\Rightarrow$

$y' = y\left(\ln\cos x - \dfrac{x\sin x}{\cos x}\right)$ $\Rightarrow$ $y' = (\cos x)^x(\ln\cos x - x\tan x)$

47. $y = (\tan x)^{1/x}$ $\Rightarrow$ $\ln y = \ln(\tan x)^{1/x}$ $\Rightarrow$ $\ln y = \dfrac{1}{x}\ln\tan x$ $\Rightarrow$

$\dfrac{1}{y}\,y' = \dfrac{1}{x} \cdot \dfrac{1}{\tan x} \cdot \sec^2 x + \ln\tan x \cdot \left(-\dfrac{1}{x^2}\right)$ $\Rightarrow$ $y' = y\left(\dfrac{\sec^2 x}{x\tan x} - \dfrac{\ln\tan x}{x^2}\right)$ $\Rightarrow$

$y' = (\tan x)^{1/x}\left(\dfrac{\sec^2 x}{x\tan x} - \dfrac{\ln\tan x}{x^2}\right)$ or $y' = (\tan x)^{1/x} \cdot \dfrac{1}{x}\left(\csc x \sec x - \dfrac{\ln\tan x}{x}\right)$

49. $y = \ln(x^2 + y^2)$ $\Rightarrow$ $y' = \dfrac{1}{x^2 + y^2}\dfrac{d}{dx}(x^2 + y^2)$ $\Rightarrow$ $y' = \dfrac{2x + 2yy'}{x^2 + y^2}$ $\Rightarrow$ $x^2 y' + y^2 y' = 2x + 2yy'$ $\Rightarrow$

$x^2 y' + y^2 y' - 2yy' = 2x$ $\Rightarrow$ $(x^2 + y^2 - 2y)y' = 2x$ $\Rightarrow$ $y' = \dfrac{2x}{x^2 + y^2 - 2y}$

51. $f(x) = \ln(x - 1)$ $\Rightarrow$ $f'(x) = \dfrac{1}{(x - 1)} = (x - 1)^{-1}$ $\Rightarrow$ $f''(x) = -(x - 1)^{-2}$ $\Rightarrow$ $f'''(x) = 2(x - 1)^{-3}$ $\Rightarrow$

$f^{(4)}(x) = -2 \cdot 3(x - 1)^{-4}$ $\Rightarrow$ $\cdots$ $\Rightarrow$ $f^{(n)}(x) = (-1)^{n-1} \cdot 2 \cdot 3 \cdot 4 \cdot \cdots \cdot (n - 1)(x - 1)^{-n} = (-1)^{n-1}\dfrac{(n - 1)!}{(x - 1)^n}$

53. If $f(x) = \ln(1 + x)$, then $f'(x) = \dfrac{1}{1 + x}$, so $f'(0) = 1$.

Thus, $\lim\limits_{x \to 0}\dfrac{\ln(1 + x)}{x} = \lim\limits_{x \to 0}\dfrac{f(x)}{x} = \lim\limits_{x \to 0}\dfrac{f(x) - f(0)}{x - 0} = f'(0) = 1$.

3.7 Rates of Change in the Natural and Social Sciences

1. (a) $s = f(t) = t^3 - 12t^2 + 36t \quad \Rightarrow \quad v(t) = f'(t) = 3t^2 - 24t + 36$

(b) $v(3) = 27 - 72 + 36 = -9$ ft/s

(c) The particle is at rest when $v(t) = 0$. $\quad 3t^2 - 24t + 36 = 0 \quad \Leftrightarrow \quad 3(t-2)(t-6) = 0 \quad \Leftrightarrow \quad t = 2$ s or 6 s.

(d) The particle is moving in the positive direction when $v(t) > 0$. $3(t-2)(t-6) > 0 \quad \Leftrightarrow \quad 0 \le t < 2$ or $t > 6$.

(e) Since the particle is moving in the positive direction and in the negative direction, we need to calculate the distance traveled in the intervals $[0, 2]$, $[2, 6]$, and $[6, 8]$ separately.

$|f(2) - f(0)| = |32 - 0| = 32.$

$|f(6) - f(2)| = |0 - 32| = 32.$

$|f(8) - f(6)| = |32 - 0| = 32.$

The total distance is $32 + 32 + 32 = 96$ ft.

(f)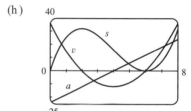

(g) $v(t) = 3t^2 - 24t + 36 \quad \Rightarrow$

$a(t) = v'(t) = 6t - 24.$

$a(3) = 6(3) - 24 = -6$ (ft/s)/s or ft/s^2.

(h)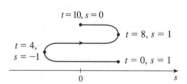

(i) The particle is speeding up when v and a have the same sign. This occurs when $2 < t < 4$ [v and a are both negative] and when $t > 6$ [v and a are both positive]. It is slowing down when v and a have opposite signs; that is, when $0 \le t < 2$ and when $4 < t < 6$.

3. (a) $s = f(t) = \cos(\pi t/4) \quad \Rightarrow \quad v(t) = f'(t) = -\sin(\pi t/4) \cdot (\pi/4)$

(b) $v(3) = -\frac{\pi}{4} \sin \frac{3\pi}{4} = -\frac{\pi}{4} \cdot \frac{\sqrt{2}}{2} = -\frac{\pi\sqrt{2}}{8}$ ft/s $[\approx -0.56]$

(c) The particle is at rest when $v(t) = 0$. $\quad -\frac{\pi}{4} \sin \frac{\pi t}{4} = 0 \quad \Rightarrow \quad \sin \frac{\pi t}{4} = 0 \quad \Rightarrow \quad \frac{\pi t}{4} = \pi n \quad \Rightarrow \quad t = 0, 4, 8$ s.

(d) The particle is moving in the positive direction when $v(t) > 0$. $\quad -\frac{\pi}{4} \sin \frac{\pi t}{4} > 0 \quad \Rightarrow \quad \sin \frac{\pi t}{4} < 0 \quad \Rightarrow \quad 4 < t < 8.$

(e) From part (c), $v(t) = 0$ for $t = 0, 4, 8$. As in Exercise 1, we'll find the distance traveled in the intervals $[0, 4]$ and $[4, 8]$.

$|f(4) - f(0)| = |-1 - 1| = 2$

$|f(8) - f(4)| = |1 - (-1)| = 2.$

The total distance is $2 + 2 = 4$ ft.

(f)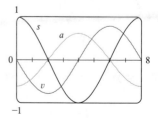

(g) $v(t) = -\dfrac{\pi}{4} \sin \dfrac{\pi t}{4} \quad \Rightarrow$

$a(t) = v'(t) = -\dfrac{\pi}{4} \cos \dfrac{\pi t}{4} \cdot \dfrac{\pi}{4} = -\dfrac{\pi^2}{16} \cos \dfrac{\pi t}{4}.$

$a(3) = -\dfrac{\pi^2}{16} \cos \dfrac{3\pi}{4} = -\dfrac{\pi^2}{16}\left(-\dfrac{\sqrt{2}}{2}\right) = \dfrac{\pi^2\sqrt{2}}{32}$ (ft/s)/s or ft/s^2.

(h)

(i) The particle is speeding up when v and a have the same sign. This occurs when $0 < t < 2$ or $8 < t < 10$ [v and a are both negative] and when $4 < t < 6$ [v and a are both positive]. It is slowing down when v and a have opposite signs; that is, when $2 < t < 4$ and when $6 < t < 8$.

5. (a) From the figure, the velocity v is positive on the interval $(0, 2)$ and negative on the interval $(2, 3)$. The acceleration a is positive (negative) when the slope of the tangent line is positive (negative), so the acceleration is positive on the interval $(0, 1)$, and negative on the interval $(1, 3)$. The particle is speeding up when v and a have the same sign, that is, on the interval $(0, 1)$ when $v > 0$ and $a > 0$, and on the interval $(2, 3)$ when $v < 0$ and $a < 0$. The particle is slowing down when v and a have opposite signs, that is, on the interval $(1, 2)$ when $v > 0$ and $a < 0$.

(b) $v > 0$ on $(0, 3)$ and $v < 0$ on $(3, 4)$. $a > 0$ on $(1, 2)$ and $a < 0$ on $(0, 1)$ and $(2, 4)$. The particle is speeding up on $(1, 2)$ [$v > 0$, $a > 0$] and on $(3, 4)$ [$v < 0$, $a < 0$]. The particle is slowing down on $(0, 1)$ and $(2, 3)$ [$v > 0$, $a < 0$].

7. (a) $s(t) = t^3 - 4.5t^2 - 7t$ $\Rightarrow$ $v(t) = s'(t) = 3t^2 - 9t - 7 = 5$ $\Leftrightarrow$ $3t^2 - 9t - 12 = 0$ $\Leftrightarrow$ $3(t - 4)(t + 1) = 0$ $\Leftrightarrow$ $t = 4$ or -1. Since $t \geq 0$, the particle reaches a velocity of 5 m/s at $t = 4$ s.

(b) $a(t) = v'(t) = 6t - 9 = 0$ $\Leftrightarrow$ $t = 1.5$. The acceleration changes from negative to positive, so the velocity changes from decreasing to increasing. Thus, at $t = 1.5$ s, the velocity has its minimum value.

9. (a) $h = 10t - 0.83t^2$ $\Rightarrow$ $v(t) = \dfrac{dh}{dt} = 10 - 1.66t$, so $v(3) = 10 - 1.66(3) = 5.02$ m/s.

(b) $h = 25$ $\Rightarrow$ $10t - 0.83t^2 = 25$ $\Rightarrow$ $0.83t^2 - 10t + 25 = 0$ $\Rightarrow$ $t = \frac{10 \pm \sqrt{17}}{1.66} \approx 3.54$ or 8.51.

The value $t_1 = \frac{10 - \sqrt{17}}{1.66}$ corresponds to the time it takes for the stone to rise 25 m and $t_2 = \frac{10 + \sqrt{17}}{1.66}$ corresponds to the time when the stone is 25 m high on the way down. Thus, $v(t_1) = 10 - 1.66\left(\frac{10 - \sqrt{17}}{1.66}\right) = \sqrt{17} \approx 4.12$ m/s.

11. (a) $A(x) = x^2$ $\Rightarrow$ $A'(x) = 2x$. $A'(15) = 30$ mm^2/mm is the rate at which the area is increasing with respect to the side length as x reaches 15 mm.

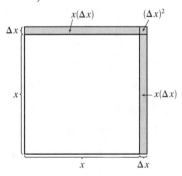

(b) The perimeter is $P(x) = 4x$, so $A'(x) = 2x = \frac{1}{2}(4x) = \frac{1}{2}P(x)$. The figure suggests that if Δx is small, then the change in the area of the square is approximately half of its perimeter (2 of the 4 sides) times Δx. From the figure, $\Delta A = 2x\,(\Delta x) + (\Delta x)^2$. If Δx is small, then $\Delta A \approx 2x\,(\Delta x)$ and so $\Delta A / \Delta x \approx 2x$.

13. (a) Using $A(r) = \pi r^2$, we find that the average rate of change is:

(i) $\dfrac{A(3) - A(2)}{3 - 2} = \dfrac{9\pi - 4\pi}{1} = 5\pi$

(ii) $\dfrac{A(2.5) - A(2)}{2.5 - 2} = \dfrac{6.25\pi - 4\pi}{0.5} = 4.5\pi$

(iii) $\dfrac{A(2.1) - A(2)}{2.1 - 2} = \dfrac{4.41\pi - 4\pi}{0.1} = 4.1\pi$

(b) $A(r) = \pi r^2$ $\Rightarrow$ $A'(r) = 2\pi r$, so $A'(2) = 4\pi$.

(c) The circumference is $C(r) = 2\pi r = A'(r)$. The figure suggests that if Δr is small,

then the change in the area of the circle (a ring around the outside) is approximately equal

to its circumference times Δr. Straightening out this ring gives us a shape that is approx-

imately rectangular with length $2\pi r$ and width Δr, so $\Delta A \approx 2\pi r(\Delta r)$. Algebraically,

$\Delta A = A(r + \Delta r) - A(r) = \pi(r + \Delta r)^2 - \pi r^2 = 2\pi r(\Delta r) + \pi(\Delta r)^2$.

So we see that if Δr is small, then $\Delta A \approx 2\pi r(\Delta r)$ and therefore, $\Delta A/\Delta r \approx 2\pi r$.

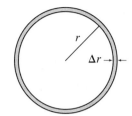

15. $S(r) = 4\pi r^2 \Rightarrow S'(r) = 8\pi r \Rightarrow$

(a) $S'(1) = 8\pi$ ft²/ft   (b) $S'(2) = 16\pi$ ft²/ft   (c) $S'(3) = 24\pi$ ft²/ft

As the radius increases, the surface area grows at an increasing rate. In fact, the rate of change is linear with respect to the

radius.

17. The mass is $f(x) = 3x^2$, so the linear density at x is $\rho(x) = f'(x) = 6x$.

(a) $\rho(1) = 6$ kg/m   (b) $\rho(2) = 12$ kg/m   (c) $\rho(3) = 18$ kg/m

Since ρ is an increasing function, the density will be the highest at the right end of the rod and lowest at the left end.

19. The quantity of charge is $Q(t) = t^3 - 2t^2 + 6t + 2$, so the current is $Q'(t) = 3t^2 - 4t + 6$.

(a) $Q'(0.5) = 3(0.5)^2 - 4(0.5) + 6 = 4.75$ A   (b) $Q'(1) = 3(1)^2 - 4(1) + 6 = 5$ A

The current is lowest when Q' has a minimum. $Q''(t) = 6t - 4 < 0$ when $t < \frac{2}{3}$. So the current decreases when $t < \frac{2}{3}$ and

increases when $t > \frac{2}{3}$. Thus, the current is lowest at $t = \frac{2}{3}$ s.

21. (a) To find the rate of change of volume with respect to pressure, we first solve for V in terms of P.

$$PV = C \Rightarrow V = \frac{C}{P} \Rightarrow \frac{dV}{dP} = -\frac{C}{P^2}.$$

(b) From the formula for dV/dP in part (a), we see that as P increases, the absolute value of dV/dP decreases.

Thus, the volume is decreasing more rapidly at the beginning.

(c) $\beta = -\frac{1}{V}\frac{dV}{dP} = -\frac{1}{V}\left(-\frac{C}{P^2}\right) = \frac{C}{(PV)P} = \frac{C}{CP} = \frac{1}{P}$

23. In Example 6, the population function was $n = 2^t n_0$. Since we are tripling instead of doubling and the initial population is

400, the population function is $n(t) = 400 \cdot 3^t$. The rate of growth is $n'(t) = 400 \cdot 3^t \cdot \ln 3$, so the rate of growth after

2.5 hours is $n'(2.5) = 400 \cdot 3^{2.5} \cdot \ln 3 \approx 6850$ bacteria/hour.

25. (a) **1920:** $m_1 = \dfrac{1860 - 1750}{1920 - 1910} = \dfrac{110}{10} = 11, m_2 = \dfrac{2070 - 1860}{1930 - 1920} = \dfrac{210}{10} = 21$,

$(m_1 + m_2)/2 = (11 + 21)/2 = 16$ million/year

1980: $m_1 = \dfrac{4450 - 3710}{1980 - 1970} = \dfrac{740}{10} = 74, m_2 = \dfrac{5280 - 4450}{1990 - 1980} = \dfrac{830}{10} = 83$,

$(m_1 + m_2)/2 = (74 + 83)/2 = 78.5$ million/year

(b) $P(t) = at^3 + bt^2 + ct + d$ (in millions of people), where $a \approx 0.0012937063$, $b \approx -7.061421911$, $c \approx 12{,}822.97902$,

and $d \approx -7{,}743{,}770.396$.

(c) $P(t) = at^3 + bt^2 + ct + d \Rightarrow P'(t) = 3at^2 + 2bt + c$ (in millions of people per year)

(d) $P'(1920) = 3(0.0012937063)(1920)^2 + 2(-7.061421911)(1920) + 12{,}822.97902$

≈ 14.48 million/year [smaller than the answer in part (a), but close to it]

$P'(1980) \approx 75.29$ million/year (smaller, but close)

(e) $P'(1985) \approx 81.62$ million/year, so the rate of growth in 1985 was about 81.62 million/year.

27. (a) Using $v = \dfrac{P}{4\eta l}(R^2 - r^2)$ with $R = 0.01$, $l = 3$, $P = 3000$, and $\eta = 0.027$, we have v as a function of r:

$v(r) = \dfrac{3000}{4(0.027)3}(0.01^2 - r^2)$. $v(0) = 0.\overline{925}$ cm/s, $v(0.005) = 0.69\overline{4}$ cm/s, $v(0.01) = 0$.

(b) $v(r) = \dfrac{P}{4\eta l}(R^2 - r^2) \;\Rightarrow\; v'(r) = \dfrac{P}{4\eta l}(-2r) = -\dfrac{Pr}{2\eta l}$. When $l = 3$, $P = 3000$, and $\eta = 0.027$, we have

$v'(r) = -\dfrac{3000r}{2(0.027)3}$. $v'(0) = 0$, $v'(0.005) = -92.\overline{592}$ (cm/s)/cm, and $v'(0.01) = -185.\overline{185}$ (cm/s)/cm.

(c) The velocity is greatest where $r = 0$ (at the center) and the velocity is changing most where $r = R = 0.01$ cm

(at the edge).

29. (a) $C(x) = 1200 + 12x - 0.1x^2 + 0.0005x^3 \;\Rightarrow\; C'(x) = 12 - 0.2x + 0.0015x^2$ \$/yard, which is the marginal cost

function.

(b) $C'(200) = 12 - 0.2(200) + 0.0015(200)^2 = \$32$/yard, and this is the rate at which costs are increasing with respect to

the production level when $x = 200$. $C'(200)$ predicts the cost of producing the 201st yard.

(c) The cost of manufacturing the 201st yard of fabric is $C(201) - C(200) = 3632.2005 - 3600 \approx \$32.20$, which is

approximately $C'(200)$.

31. (a) $A(x) = \dfrac{p(x)}{x} \;\Rightarrow\; A'(x) = \dfrac{xp'(x) - p(x) \cdot 1}{x^2} = \dfrac{xp'(x) - p(x)}{x^2}$.

$A'(x) > 0 \;\Rightarrow\; A(x)$ is increasing; that is, the average productivity increases as the size of the workforce increases.

(b) $p'(x)$ is greater than the average productivity $\;\Rightarrow\; p'(x) > A(x) \;\Rightarrow\; p'(x) > \dfrac{p(x)}{x} \;\Rightarrow\; xp'(x) > p(x) \;\Rightarrow$

$xp'(x) - p(x) > 0 \;\Rightarrow\; \dfrac{xp'(x) - p(x)}{x^2} > 0 \;\Rightarrow\; A'(x) > 0$.

33. $PV = nRT \;\Rightarrow\; T = \dfrac{PV}{nR} = \dfrac{PV}{(10)(0.0821)} = \dfrac{1}{0.821}(PV)$. Using the Product Rule, we have

$\dfrac{dT}{dt} = \dfrac{1}{0.821}[P(t)V'(t) + V(t)P'(t)] = \dfrac{1}{0.821}[(8)(-0.15) + (10)(0.10)] \approx -0.2436$ K/min.

35. (a) If the populations are stable, then the growth rates are neither positive nor negative; that is, $\dfrac{dC}{dt} = 0$ and $\dfrac{dW}{dt} = 0$.

(b) "The caribou go extinct" means that the population is zero, or mathematically, $C = 0$.

(c) We have the equations $\dfrac{dC}{dt} = aC - bCW$ and $\dfrac{dW}{dt} = -cW + dCW$. Let $dC/dt = dW/dt = 0$, $a = 0.05$, $b = 0.001$,

$c = 0.05$, and $d = 0.0001$ to obtain $0.05C - 0.001CW = 0$ **(1)** and $-0.05W + 0.0001CW = 0$ **(2)**. Adding 10 times

(2) to **(1)** eliminates the CW-terms and gives us $0.05C - 0.5W = 0 \;\Rightarrow\; C = 10W$. Substituting $C = 10W$ into **(1)**

results in $0.05(10W) - 0.001(10W)W = 0$ $\Leftrightarrow$ $0.5W - 0.01W^2 = 0$ $\Leftrightarrow$ $50W - W^2 = 0$ $\Leftrightarrow$

$W(50 - W) = 0$ $\Leftrightarrow$ $W = 0$ or 50. Since $C = 10W$, $C = 0$ or 500. Thus, the population pairs (C, W) that lead to

stable populations are $(0, 0)$ and $(500, 50)$. So it is possible for the two species to live in harmony.

3.8   Exponential Growth and Decay

1. The relative growth rate is $\dfrac{1}{P}\dfrac{dP}{dt} = 0.7944$, so $\dfrac{dP}{dt} = 0.7944P$ and, by Theorem 2, $P(t) = P(0)e^{0.7944t} = 2e^{0.7944t}$.

   Thus, $P(6) = 2e^{0.7944(6)} \approx 234.99$ or about 235 members.

3. (a) By Theorem 2, $P(t) = P(0)e^{kt} = 100e^{kt}$. Now $P(1) = 100e^{k(1)} = 420$ $\Rightarrow$ $e^k = \frac{420}{100}$ $\Rightarrow$ $k = \ln 4.2$.

   So $P(t) = 100e^{(\ln 4.2)t} = 100(4.2)^t$.

   (b) $P(3) = 100(4.2)^3 = 7408.8 \approx 7409$ bacteria

   (c) $dP/dt = kP$ $\Rightarrow$ $P'(3) = k \cdot P(3) = (\ln 4.2)(100(4.2)^3)$ [from part (a)] $\approx 10{,}632$ bacteria/hour

   (d) $P(t) = 100(4.2)^t = 10{,}000$ $\Rightarrow$ $(4.2)^t = 100$ $\Rightarrow$ $t = (\ln 100)/(\ln 4.2) \approx 3.2$ hours

5. (a) Let the population (in millions) in the year t be $P(t)$. Since the initial time is the year 1750, we substitute $t - 1750$ for t in

   Theorem 2, so the exponential model gives $P(t) = P(1750)e^{k(t-1750)}$. Then $P(1800) = 980 = 790e^{k(1800-1750)}$ $\Rightarrow$

   $\frac{980}{790} = e^{k(50)}$ $\Rightarrow$ $\ln \frac{980}{790} = 50k$ $\Rightarrow$ $k = \frac{1}{50}\ln \frac{980}{790} \approx 0.0043104$. So with this model, we have

   $P(1900) = 790e^{k(1900-1750)} \approx 1508$ million, and $P(1950) = 790e^{k(1950-1750)} \approx 1871$ million. Both of these

   estimates are much too low.

   (b) In this case, the exponential model gives $P(t) = P(1850)e^{k(t-1850)}$ $\Rightarrow$ $P(1900) = 1650 = 1260e^{k(1900-1850)}$ $\Rightarrow$

   $\ln \frac{1650}{1260} = k(50)$ $\Rightarrow$ $k = \frac{1}{50}\ln \frac{1650}{1260} \approx 0.005393$. So with this model, we estimate

   $P(1950) = 1260e^{k(1950-1850)} \approx 2161$ million. This is still too low, but closer than the estimate of $P(1950)$ in part (a).

   (c) The exponential model gives $P(t) = P(1900)e^{k(t-1900)}$ $\Rightarrow$ $P(1950) = 2560 = 1650e^{k(1950-1900)}$ $\Rightarrow$

   $\ln \frac{2560}{1650} = k(50)$ $\Rightarrow$ $k = \frac{1}{50}\ln \frac{2560}{1650} \approx 0.008785$. With this model, we estimate

   $P(2000) = 1650e^{k(2000-1900)} \approx 3972$ million. This is much too low. The discrepancy is explained by the fact that the

   world birth rate (average yearly number of births per person) is about the same as always, whereas the mortality rate

   (especially the infant mortality rate) is much lower, owing mostly to advances in medical science and to the wars in the first

   part of the twentieth century. The exponential model assumes, among other things, that the birth and mortality rates will

   remain constant.

7. (a) If $y = [N_2O_5]$ then by Theorem 2, $\dfrac{dy}{dt} = -0.0005y$ $\Rightarrow$ $y(t) = y(0)e^{-0.0005t} = Ce^{-0.0005t}$.

   (b) $y(t) = Ce^{-0.0005t} = 0.9C$ $\Rightarrow$ $e^{-0.0005t} = 0.9$ $\Rightarrow$ $-0.0005t = \ln 0.9$ $\Rightarrow$ $t = -2000 \ln 0.9 \approx 211$ s

9. (a) If $y(t)$ is the mass (in mg) remaining after t years, then $y(t) = y(0)e^{kt} = 100e^{kt}$.

$$y(30) = 100e^{30k} = \tfrac{1}{2}(100) \quad \Rightarrow \quad e^{30k} = \tfrac{1}{2} \quad \Rightarrow \quad k = -(\ln 2)/30 \quad \Rightarrow \quad y(t) = 100e^{-(\ln 2)t/30} = 100 \cdot 2^{-t/30}$$

(b) $y(100) = 100 \cdot 2^{-100/30} \approx 9.92$ mg

(c) $100e^{-(\ln 2)t/30} = 1 \quad \Rightarrow \quad -(\ln 2)t/30 = \ln \tfrac{1}{100} \quad \Rightarrow \quad t = -30 \frac{\ln 0.01}{\ln 2} \approx 199.3$ years

11. Let $y(t)$ be the level of radioactivity. Thus, $y(t) = y(0)e^{-kt}$ and k is determined by using the half-life:

$$y(5730) = \tfrac{1}{2}y(0) \quad \Rightarrow \quad y(0)e^{-k(5730)} = \tfrac{1}{2}y(0) \quad \Rightarrow \quad e^{-5730k} = \tfrac{1}{2} \quad \Rightarrow \quad -5730k = \ln \tfrac{1}{2} \quad \Rightarrow \quad k = -\frac{\ln \tfrac{1}{2}}{5730} = \frac{\ln 2}{5730}.$$

If 74% of the ^{14}C remains, then we know that $y(t) = 0.74y(0) \quad \Rightarrow \quad 0.74 = e^{-t(\ln 2)/5730} \quad \Rightarrow \quad \ln 0.74 = -\frac{t \ln 2}{5730} \quad \Rightarrow$

$$t = -\frac{5730(\ln 0.74)}{\ln 2} \approx 2489 \approx 2500 \text{ years.}$$

13. (a) Using Newton's Law of Cooling, $\dfrac{dT}{dt} = k(T - T_s)$, we have $\dfrac{dT}{dt} = k(T - 75)$. Now let $y = T - 75$, so

$y(0) = T(0) - 75 = 185 - 75 = 110$, so y is a solution of the initial-value problem $dy/dt = ky$ with $y(0) = 110$ and by

Theorem 2 we have $y(t) = y(0)e^{kt} = 110e^{kt}$.

$$y(30) = 110e^{30k} = 150 - 75 \quad \Rightarrow \quad e^{30k} = \tfrac{75}{110} = \tfrac{15}{22} \quad \Rightarrow \quad k = \tfrac{1}{30} \ln \tfrac{15}{22}, \text{ so } y(t) = 110e^{\frac{1}{30}t \ln\left(\frac{15}{22}\right)} \text{ and}$$

$$y(45) = 110e^{\frac{45}{30} \ln\left(\frac{15}{22}\right)} \approx 62°\text{F}. \text{ Thus, } T(45) \approx 62 + 75 = 137°\text{F}.$$

(b) $T(t) = 100 \quad \Rightarrow \quad y(t) = 25.$ $\quad y(t) = 110e^{\frac{1}{30}t \ln\left(\frac{15}{22}\right)} = 25 \quad \Rightarrow \quad e^{\frac{1}{30}t \ln\left(\frac{15}{22}\right)} = \tfrac{25}{110} \quad \Rightarrow \quad \tfrac{1}{30}t \ln \tfrac{15}{22} = \ln \tfrac{25}{110} \quad \Rightarrow$

$$t = \frac{30 \ln \tfrac{25}{110}}{\ln \tfrac{15}{22}} \approx 116 \text{ min.}$$

15. $\dfrac{dT}{dt} = k(T - 20)$. Letting $y = T - 20$, we get $\dfrac{dy}{dt} = ky$, so $y(t) = y(0)e^{kt}$. $\quad y(0) = T(0) - 20 = 5 - 20 = -15$, so

$y(25) = y(0)e^{25k} = -15e^{25k}$, and $y(25) = T(25) - 20 = 10 - 20 = -10$, so $-15e^{25k} = -10 \quad \Rightarrow \quad e^{25k} = \tfrac{2}{3}$. Thus,

$25k = \ln\left(\tfrac{2}{3}\right)$ and $k = \tfrac{1}{25} \ln\left(\tfrac{2}{3}\right)$, so $y(t) = y(0)e^{kt} = -15e^{(1/25) \ln(2/3)t}$. More simply, $e^{25k} = \tfrac{2}{3} \quad \Rightarrow \quad e^k = \left(\tfrac{2}{3}\right)^{1/25} \quad \Rightarrow$

$e^{kt} = \left(\tfrac{2}{3}\right)^{t/25} \quad \Rightarrow \quad y(t) = -15 \cdot \left(\tfrac{2}{3}\right)^{t/25}.$

(a) $T(50) = 20 + y(50) = 20 - 15 \cdot \left(\tfrac{2}{3}\right)^{50/25} = 20 - 15 \cdot \left(\tfrac{2}{3}\right)^2 = 20 - \tfrac{20}{3} = 13.\overline{3}°\text{C}$

(b) $15 = T(t) = 20 + y(t) = 20 - 15 \cdot \left(\tfrac{2}{3}\right)^{t/25} \quad \Rightarrow \quad 15 \cdot \left(\tfrac{2}{3}\right)^{t/25} = 5 \quad \Rightarrow \quad \left(\tfrac{2}{3}\right)^{t/25} = \tfrac{1}{3} \quad \Rightarrow$

$\quad (t/25) \ln\left(\tfrac{2}{3}\right) = \ln\left(\tfrac{1}{3}\right) \quad \Rightarrow \quad t = 25 \ln\left(\tfrac{1}{3}\right)/\ln\left(\tfrac{2}{3}\right) \approx 67.74 \text{ min.}$

17. (a) Let $P(h)$ be the pressure at altitude h. Then $dP/dh = kP \quad \Rightarrow \quad P(h) = P(0)e^{kh} = 101.3e^{kh}$.

$$P(1000) = 101.3e^{1000k} = 87.14 \quad \Rightarrow \quad 1000k = \ln\left(\tfrac{87.14}{101.3}\right) \quad \Rightarrow \quad k = \tfrac{1}{1000} \ln\left(\tfrac{87.14}{101.3}\right) \quad \Rightarrow$$

$$P(h) = 101.3 \, e^{\frac{1}{1000} h \ln\left(\frac{87.14}{101.3}\right)}, \text{ so } P(3000) = 101.3e^{3 \ln\left(\frac{87.14}{101.3}\right)} \approx 64.5 \text{ kPa.}$$

(b) $P(6187) = 101.3 \, e^{\frac{6187}{1000} \ln\left(\frac{87.14}{101.3}\right)} \approx 39.9 \text{ kPa}$

19. (a) Using $A = A_0\left(1 + \dfrac{r}{n}\right)^{nt}$ with $A_0 = 3000$, $r = 0.05$, and $t = 5$, we have:

(i) Annually: $n = 1$; $A = 3000\left(1 + \frac{0.05}{1}\right)^{1 \cdot 5} = \$3828.84$

(ii) Semiannually: $n = 2$; $A = 3000\left(1 + \frac{0.05}{2}\right)^{2 \cdot 5} = \$3840.25$

(iii) Monthly: $n = 12$; $A = 3000\left(1 + \frac{0.05}{12}\right)^{12 \cdot 5} = \$3850.08$

(iv) Weekly: $n = 52$; $A = 3000\left(1 + \frac{0.05}{52}\right)^{52 \cdot 5} = \$3851.61$

(v) Daily: $n = 365$; $A = 3000\left(1 + \frac{0.05}{365}\right)^{365 \cdot 5} = \$3852.01$

(vi) Continuously: $A = 3000e^{(0.05)5} = \$3852.08$

(b) $dA/dt = 0.05A$ and $A(0) = 3000$.

3.9   Related Rates

1. $V = x^3 \quad \Rightarrow \quad \dfrac{dV}{dt} = \dfrac{dV}{dx}\dfrac{dx}{dt} = 3x^2\dfrac{dx}{dt}$

3. Let s denote the side of a square. The square's area A is given by $A = s^2$. Differentiating with respect to t gives us

$\dfrac{dA}{dt} = 2s\dfrac{ds}{dt}$. When $A = 16$, $s = 4$. Substitution 4 for s and 6 for $\dfrac{ds}{dt}$ gives us $\dfrac{dA}{dt} = 2(4)(6) = 48$ cm^2/s.

5. $V = \pi r^2 h = \pi(5)^2 h = 25\pi h \quad \Rightarrow \quad \dfrac{dV}{dt} = 25\pi\dfrac{dh}{dt} \quad \Rightarrow \quad 3 = 25\pi\dfrac{dh}{dt} \quad \Rightarrow \quad \dfrac{dh}{dt} = \dfrac{3}{25\pi}$ m/min.

7. $y = x^3 + 2x \quad \Rightarrow \quad \dfrac{dy}{dt} = \dfrac{dy}{dx}\dfrac{dx}{dt} = (3x^2 + 2)(5) = 5(3x^2 + 2)$. When $x = 2$, $\dfrac{dy}{dt} = 5(14) = 70$.

9. $z^2 = x^2 + y^2 \quad \Rightarrow \quad 2z\dfrac{dz}{dt} = 2x\dfrac{dx}{dt} + 2y\dfrac{dy}{dt} \quad \Rightarrow \quad \dfrac{dz}{dt} = \dfrac{1}{z}\left(x\dfrac{dx}{dt} + y\dfrac{dy}{dt}\right)$. When $x = 5$ and $y = 12$,

$z^2 = 5^2 + 12^2 \quad \Rightarrow \quad z^2 = 169 \quad \Rightarrow \quad z = \pm 13$. For $\dfrac{dx}{dt} = 2$ and $\dfrac{dy}{dt} = 3$, $\dfrac{dz}{dt} = \dfrac{1}{\pm 13}(5 \cdot 2 + 12 \cdot 3) = \pm\dfrac{46}{13}$.

11. (a) Given: a plane flying horizontally at an altitude of 1 mi and a speed of 500 mi/h passes directly over a radar station.
If we let t be time (in hours) and x be the horizontal distance traveled by the plane (in mi), then we are given
that $dx/dt = 500$ mi/h.

(b) Unknown: the rate at which the distance from the plane to the station is increasing   (c)
when it is 2 mi from the station. If we let y be the distance from the plane to the station,
then we want to find dy/dt when $y = 2$ mi.

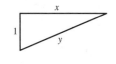

(d) By the Pythagorean Theorem, $y^2 = x^2 + 1 \quad \Rightarrow \quad 2y\,(dy/dt) = 2x\,(dx/dt)$.

(e) $\dfrac{dy}{dt} = \dfrac{x}{y}\dfrac{dx}{dt} = \dfrac{x}{y}(500)$. Since $y^2 = x^2 + 1$, when $y = 2$, $x = \sqrt{3}$, so $\dfrac{dy}{dt} = \dfrac{\sqrt{3}}{2}(500) = 250\sqrt{3} \approx 433$ mi/h.

13. (a) Given: a man 6 ft tall walks away from a street light mounted on a 15-ft-tall pole at a rate of 5 ft/s. If we let t be time (in s)
and x be the distance from the pole to the man (in ft), then we are given that $dx/dt = 5$ ft/s.

(b) Unknown: the rate at which the tip of his shadow is moving when he is 40 ft

from the pole. If we let y be the distance from the man to the tip of his

shadow (in ft), then we want to find $\frac{d}{dt}(x+y)$ when $x = 40$ ft.

(c)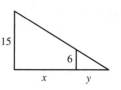

(d) By similar triangles, $\frac{15}{6} = \frac{x+y}{y}$ $\Rightarrow$ $15y = 6x + 6y$ $\Rightarrow$ $9y = 6x$ $\Rightarrow$ $y = \frac{2}{3}x.$

(e) The tip of the shadow moves at a rate of $\frac{d}{dt}(x+y) = \frac{d}{dt}\left(x + \frac{2}{3}x\right) = \frac{5}{3}\frac{dx}{dt} = \frac{5}{3}(5) = \frac{25}{3}$ ft/s.

15.

We are given that $\frac{dx}{dt} = 60$ mi/h and $\frac{dy}{dt} = 25$ mi/h. $z^2 = x^2 + y^2$ $\Rightarrow$

$2z\frac{dz}{dt} = 2x\frac{dx}{dt} + 2y\frac{dy}{dt}$ $\Rightarrow$ $z\frac{dz}{dt} = x\frac{dx}{dt} + y\frac{dy}{dt}$ $\Rightarrow$ $\frac{dz}{dt} = \frac{1}{z}\left(x\frac{dx}{dt} + y\frac{dy}{dt}\right).$

After 2 hours, $x = 2\,(60) = 120$ and $y = 2\,(25) = 50$ $\Rightarrow$ $z = \sqrt{120^2 + 50^2} = 130,$

so $\frac{dz}{dt} = \frac{1}{z}\left(x\frac{dx}{dt} + y\frac{dy}{dt}\right) = \frac{120(60) + 50(25)}{130} = 65$ mi/h.

17.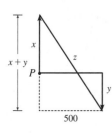

We are given that $\frac{dx}{dt} = 4$ ft/s and $\frac{dy}{dt} = 5$ ft/s. $z^2 = (x+y)^2 + 500^2$ $\Rightarrow$

$2z\frac{dz}{dt} = 2(x+y)\left(\frac{dx}{dt} + \frac{dy}{dt}\right).$ 15 minutes after the woman starts, we have

$x = (4\text{ ft/s})(20\text{ min})(60\text{ s/min}) = 4800$ ft and $y = 5 \cdot 15 \cdot 60 = 4500$ $\Rightarrow$

$z = \sqrt{(4800 + 4500)^2 + 500^2} = \sqrt{86{,}740{,}000},$ so

$\frac{dz}{dt} = \frac{x+y}{z}\left(\frac{dx}{dt} + \frac{dy}{dt}\right) = \frac{4800 + 4500}{\sqrt{86{,}740{,}000}}(4+5) = \frac{837}{\sqrt{8674}} \approx 8.99$ ft/s.

19. $A = \frac{1}{2}bh$, where b is the base and h is the altitude. We are given that $\frac{dh}{dt} = 1$ cm/min and $\frac{dA}{dt} = 2$ cm²/min. Using the

Product Rule, we have $\frac{dA}{dt} = \frac{1}{2}\left(b\frac{dh}{dt} + h\frac{db}{dt}\right).$ When $h = 10$ and $A = 100$, we have $100 = \frac{1}{2}b(10)$ $\Rightarrow$ $\frac{1}{2}b = 10$ $\Rightarrow$

$b = 20$, so $2 = \frac{1}{2}\left(20 \cdot 1 + 10\frac{db}{dt}\right)$ $\Rightarrow$ $4 = 20 + 10\frac{db}{dt}$ $\Rightarrow$ $\frac{db}{dt} = \frac{4 - 20}{10} = -1.6$ cm/min.

21.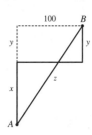

We are given that $\frac{dx}{dt} = 35$ km/h and $\frac{dy}{dt} = 25$ km/h. $z^2 = (x+y)^2 + 100^2$ $\Rightarrow$

$2z\frac{dz}{dt} = 2(x+y)\left(\frac{dx}{dt} + \frac{dy}{dt}\right).$ At 4:00 PM, $x = 4(35) = 140$ and $y = 4(25) = 100$ $\Rightarrow$

$z = \sqrt{(140 + 100)^2 + 100^2} = \sqrt{67{,}600} = 260,$ so

$\frac{dz}{dt} = \frac{x+y}{z}\left(\frac{dx}{dt} + \frac{dy}{dt}\right) = \frac{140 + 100}{260}(35 + 25) = \frac{720}{13} \approx 55.4$ km/h.

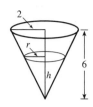

23. If C = the rate at which water is pumped in, then $\dfrac{dV}{dt} = C - 10{,}000$, where

$V = \frac{1}{3}\pi r^2 h$ is the volume at time t. By similar triangles, $\dfrac{r}{2} = \dfrac{h}{6}$ $\Rightarrow$ $r = \dfrac{1}{3}h$ $\Rightarrow$

$V = \frac{1}{3}\pi\left(\frac{1}{3}h\right)^2 h = \frac{\pi}{27}h^3$ $\Rightarrow$ $\dfrac{dV}{dt} = \dfrac{\pi}{9}h^2 \dfrac{dh}{dt}$. When $h = 200$ cm,

$\dfrac{dh}{dt} = 20$ cm/min, so $C - 10{,}000 = \dfrac{\pi}{9}(200)^2(20)$ $\Rightarrow$ $C = 10{,}000 + \dfrac{800{,}000}{9}\pi \approx 289{,}253$ cm^3/min.

25. 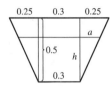 The figure is labeled in meters. The area A of a trapezoid is

$\frac{1}{2}(\text{base}_1 + \text{base}_2)(\text{height})$, and the volume V of the 10-meter-long trough is $10A$.

Thus, the volume of the trapezoid with height h is $V = (10)\frac{1}{2}[0.3 + (0.3 + 2a)]h$.

By similar triangles, $\dfrac{a}{h} = \dfrac{0.25}{0.5} = \dfrac{1}{2}$, so $2a = h$ $\Rightarrow$ $V = 5(0.6 + h)h = 3h + 5h^2$.

Now $\dfrac{dV}{dt} = \dfrac{dV}{dh}\dfrac{dh}{dt}$ $\Rightarrow$ $0.2 = (3 + 10h)\dfrac{dh}{dt}$ $\Rightarrow$ $\dfrac{dh}{dt} = \dfrac{0.2}{3 + 10h}$. When $h = 0.3$,

$\dfrac{dh}{dt} = \dfrac{0.2}{3 + 10(0.3)} = \dfrac{0.2}{6}$ m/min $= \dfrac{1}{30}$ m/min or $\dfrac{10}{3}$ cm/min.

27. We are given that $\dfrac{dV}{dt} = 30$ ft^3/min. $V = \frac{1}{3}\pi r^2 h = \frac{1}{3}\pi\left(\dfrac{h}{2}\right)^2 h = \dfrac{\pi h^3}{12}$ $\Rightarrow$

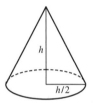

$\dfrac{dV}{dt} = \dfrac{dV}{dh}\dfrac{dh}{dt}$ $\Rightarrow$ $30 = \dfrac{\pi h^2}{4}\dfrac{dh}{dt}$ $\Rightarrow$ $\dfrac{dh}{dt} = \dfrac{120}{\pi h^2}$.

When $h = 10$ ft, $\dfrac{dh}{dt} = \dfrac{120}{10^2\pi} = \dfrac{6}{5\pi} \approx 0.38$ ft/min.

29. $A = \frac{1}{2}bh$, but $b = 5$ m and $\sin\theta = \dfrac{h}{4}$ $\Rightarrow$ $h = 4\sin\theta$, so $A = \frac{1}{2}(5)(4\sin\theta) = 10\sin\theta$.

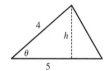

We are given $\dfrac{d\theta}{dt} = 0.06$ rad/s, so $\dfrac{dA}{dt} = \dfrac{dA}{d\theta}\dfrac{d\theta}{dt} = (10\cos\theta)(0.06) = 0.6\cos\theta$.

When $\theta = \dfrac{\pi}{3}$, $\dfrac{dA}{dt} = 0.6\left(\cos\dfrac{\pi}{3}\right) = (0.6)\left(\frac{1}{2}\right) = 0.3$ m^2/s.

31. Differentiating both sides of $PV = C$ with respect to t and using the Product Rule gives us $P\dfrac{dV}{dt} + V\dfrac{dP}{dt} = 0$ $\Rightarrow$

$\dfrac{dV}{dt} = -\dfrac{V}{P}\dfrac{dP}{dt}$. When $V = 600$, $P = 150$ and $\dfrac{dP}{dt} = 20$, so we have $\dfrac{dV}{dt} = -\dfrac{600}{150}(20) = -80$. Thus, the volume is

decreasing at a rate of 80 cm^3/min.

33. With $R_1 = 80$ and $R_2 = 100$, $\dfrac{1}{R} = \dfrac{1}{R_1} + \dfrac{1}{R_2} = \dfrac{1}{80} + \dfrac{1}{100} = \dfrac{180}{8000} = \dfrac{9}{400}$, so $R = \dfrac{400}{9}$. Differentiating $\dfrac{1}{R} = \dfrac{1}{R_1} + \dfrac{1}{R_2}$

with respect to t, we have $-\dfrac{1}{R^2}\dfrac{dR}{dt} = -\dfrac{1}{R_1^2}\dfrac{dR_1}{dt} - \dfrac{1}{R_2^2}\dfrac{dR_2}{dt}$ $\Rightarrow$ $\dfrac{dR}{dt} = R^2\left(\dfrac{1}{R_1^2}\dfrac{dR_1}{dt} + \dfrac{1}{R_2^2}\dfrac{dR_2}{dt}\right)$. When $R_1 = 80$ and

$R_2 = 100$, $\dfrac{dR}{dt} = \dfrac{400^2}{9^2}\left[\dfrac{1}{80^2}(0.3) + \dfrac{1}{100^2}(0.2)\right] = \dfrac{107}{810} \approx 0.132$ Ω/s.

35. We are given $d\theta/dt = 2°/\text{min} = \frac{\pi}{90}$ rad/min. By the Law of Cosines,

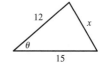

$$x^2 = 12^2 + 15^2 - 2(12)(15)\cos\theta = 369 - 360\cos\theta \quad \Rightarrow$$

$$2x\frac{dx}{dt} = 360\sin\theta\frac{d\theta}{dt} \quad \Rightarrow \quad \frac{dx}{dt} = \frac{180\sin\theta}{x}\frac{d\theta}{dt}. \text{ When } \theta = 60°,$$

$$x = \sqrt{369 - 360\cos 60°} = \sqrt{189} = 3\sqrt{21}, \text{ so } \frac{dx}{dt} = \frac{180\sin 60°}{3\sqrt{21}}\frac{\pi}{90} = \frac{\pi\sqrt{3}}{3\sqrt{21}} = \frac{\sqrt{7}\,\pi}{21} \approx 0.396 \text{ m/min}.$$

37. (a) By the Pythagorean Theorem, $4000^2 + y^2 = \ell^2$. Differentiating with respect to t,

we obtain $2y\dfrac{dy}{dt} = 2\ell\dfrac{d\ell}{dt}$. We know that $\dfrac{dy}{dt} = 600$ ft/s, so when $y = 3000$ ft,

$$\ell = \sqrt{4000^2 + 3000^2} = \sqrt{25,000,000} = 5000 \text{ ft}$$

and $\dfrac{d\ell}{dt} = \dfrac{y}{\ell}\dfrac{dy}{dt} = \dfrac{3000}{5000}(600) = \dfrac{1800}{5} = 360$ ft/s.

(b) Here $\tan\theta = \dfrac{y}{4000} \quad \Rightarrow \quad \dfrac{d}{dt}(\tan\theta) = \dfrac{d}{dt}\left(\dfrac{y}{4000}\right) \quad \Rightarrow \quad \sec^2\theta\dfrac{d\theta}{dt} = \dfrac{1}{4000}\dfrac{dy}{dt} \quad \Rightarrow \quad \dfrac{d\theta}{dt} = \dfrac{\cos^2\theta}{4000}\dfrac{dy}{dt}.$ When

$y = 3000$ ft, $\dfrac{dy}{dt} = 600$ ft/s, $\ell = 5000$ and $\cos\theta = \dfrac{4000}{\ell} = \dfrac{4000}{5000} = \dfrac{4}{5}$, so $\dfrac{d\theta}{dt} = \dfrac{(4/5)^2}{4000}(600) = 0.096$ rad/s.

39. $\cot\theta = \dfrac{x}{5} \quad \Rightarrow \quad -\csc^2\theta\dfrac{d\theta}{dt} = \dfrac{1}{5}\dfrac{dx}{dt} \quad \Rightarrow \quad -\left(\csc\dfrac{\pi}{3}\right)^2\left(-\dfrac{\pi}{6}\right) = \dfrac{1}{5}\dfrac{dx}{dt} \quad \Rightarrow$

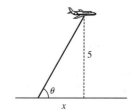

$$\frac{dx}{dt} = \frac{5\pi}{6}\left(\frac{2}{\sqrt{3}}\right)^2 = \frac{10}{9}\pi \text{ km/min } [\approx 130 \text{ mi/h}]$$

41. We are given that $\dfrac{dx}{dt} = 300$ km/h. By the Law of Cosines,

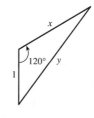

$$y^2 = x^2 + 1^2 - 2(1)(x)\cos 120° = x^2 + 1 - 2x\left(-\tfrac{1}{2}\right) = x^2 + x + 1, \text{ so}$$

$$2y\frac{dy}{dt} = 2x\frac{dx}{dt} + \frac{dx}{dt} \quad \Rightarrow \quad \frac{dy}{dt} = \frac{2x+1}{2y}\frac{dx}{dt}. \text{ After 1 minute, } x = \frac{300}{60} = 5 \text{ km } \Rightarrow$$

$$y = \sqrt{5^2 + 5 + 1} = \sqrt{31} \text{ km } \quad \Rightarrow \quad \frac{dy}{dt} = \frac{2(5)+1}{2\sqrt{31}}(300) = \frac{1650}{\sqrt{31}} \approx 296 \text{ km/h}.$$

43. Let the distance between the runner and the friend be ℓ. Then by the Law of Cosines,

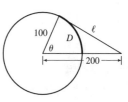

$$\ell^2 = 200^2 + 100^2 - 2 \cdot 200 \cdot 100 \cdot \cos\theta = 50,000 - 40,000\cos\theta \ (\star). \text{ Differentiating}$$

implicitly with respect to t, we obtain $2\ell\dfrac{d\ell}{dt} = -40,000(-\sin\theta)\dfrac{d\theta}{dt}$. Now if D is the

distance run when the angle is θ radians, then by the formula for the length of an arc

on a circle, $s = r\theta$, we have $D = 100\theta$, so $\theta = \dfrac{1}{100}D \quad \Rightarrow \quad \dfrac{d\theta}{dt} = \dfrac{1}{100}\dfrac{dD}{dt} = \dfrac{7}{100}$. To substitute into the expression for

$\dfrac{d\ell}{dt}$, we must know $\sin\theta$ at the time when $\ell = 200$, which we find from $(\star)$: $200^2 = 50,000 - 40,000\cos\theta \quad \Leftrightarrow$

$\cos\theta = \tfrac{1}{4} \quad \Rightarrow \quad \sin\theta = \sqrt{1 - \left(\tfrac{1}{4}\right)^2} = \dfrac{\sqrt{15}}{4}$. Substituting, we get $2(200)\dfrac{d\ell}{dt} = 40,000\dfrac{\sqrt{15}}{4}\left(\dfrac{7}{100}\right) \quad \Rightarrow$

$d\ell/dt = \dfrac{7\sqrt{15}}{4} \approx 6.78$ m/s. Whether the distance between them is increasing or decreasing depends on the direction in which

the runner is running.

3.10   Linear Approximations and Differentials

1. $f(x) = x^4 + 3x^2$ $\Rightarrow$ $f'(x) = 4x^3 + 6x$, so $f(-1) = 4$ and $f'(-1) = -10$.

Thus, $L(x) = f(-1) + f'(-1)(x - (-1)) = 4 + (-10)(x + 1) = -10x - 6$.

3. $f(x) = \cos x$ $\Rightarrow$ $f'(x) = -\sin x$, so $f\left(\frac{\pi}{2}\right) = 0$ and $f'\left(\frac{\pi}{2}\right) = -1$.

Thus, $L(x) = f\left(\frac{\pi}{2}\right) + f'\left(\frac{\pi}{2}\right)\left(x - \frac{\pi}{2}\right) = 0 - 1\left(x - \frac{\pi}{2}\right) = -x + \frac{\pi}{2}$.

5. $f(x) = \sqrt{1 - x}$ $\Rightarrow$ $f'(x) = \dfrac{-1}{2\sqrt{1 - x}}$, so $f(0) = 1$ and $f'(0) = -\frac{1}{2}$.

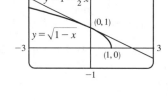

Therefore,

$$\sqrt{1 - x} = f(x) \approx f(0) + f'(0)(x - 0) = 1 + \left(-\tfrac{1}{2}\right)(x - 0) = 1 - \tfrac{1}{2}x.$$

So $\sqrt{0.9} = \sqrt{1 - 0.1} \approx 1 - \tfrac{1}{2}(0.1) = 0.95$

and $\sqrt{0.99} = \sqrt{1 - 0.01} \approx 1 - \tfrac{1}{2}(0.01) = 0.995$.

7. $f(x) = \sqrt[3]{1 - x} = (1 - x)^{1/3}$ $\Rightarrow$ $f'(x) = -\tfrac{1}{3}(1 - x)^{-2/3}$, so $f(0) = 1$

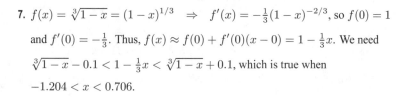

and $f'(0) = -\tfrac{1}{3}$. Thus, $f(x) \approx f(0) + f'(0)(x - 0) = 1 - \tfrac{1}{3}x$. We need

$\sqrt[3]{1 - x} - 0.1 < 1 - \tfrac{1}{3}x < \sqrt[3]{1 - x} + 0.1$, which is true when

$-1.204 < x < 0.706$.

9. $f(x) = \dfrac{1}{(1 + 2x)^4} = (1 + 2x)^{-4}$ $\Rightarrow$

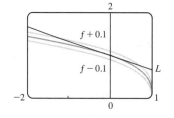

$f'(x) = -4(1 + 2x)^{-5}(2) = \dfrac{-8}{(1 + 2x)^5}$, so $f(0) = 1$ and $f'(0) = -8$.

Thus, $f(x) \approx f(0) + f'(0)(x - 0) = 1 + (-8)(x - 0) = 1 - 8x$.

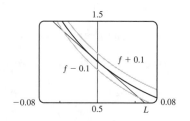

We need $\dfrac{1}{(1 + 2x)^4} - 0.1 < 1 - 8x < \dfrac{1}{(1 + 2x)^4} + 0.1$, which is true

when $-0.045 < x < 0.055$.

11. (a) The differential dy is defined in terms of dx by the equation $dy = f'(x)\, dx$. For $y = f(x) = x^2 \sin 2x$,

$f'(x) = x^2 \cos 2x \cdot 2 + \sin 2x \cdot 2x = 2x(x \cos 2x + \sin 2x)$, so $dy = 2x(x \cos 2x + \sin 2x)\, dx$.

(b) For $y = f(t) = \ln\sqrt{1 + t^2} = \tfrac{1}{2}\ln(1 + t^2)$, $f'(t) = \dfrac{1}{2} \cdot \dfrac{1}{1 + t^2} \cdot 2t = \dfrac{t}{1 + t^2}$, so $dy = \dfrac{t}{1 + t^2}\, dt$.

13. (a) For $y = f(u) = \dfrac{u + 1}{u - 1}$, $f'(u) = \dfrac{(u - 1)(1) - (u + 1)(1)}{(u - 1)^2} = \dfrac{-2}{(u - 1)^2}$, so $dy = \dfrac{-2}{(u - 1)^2}\, du$.

(b) For $y = f(r) = (1 + r^3)^{-2}$, $f'(r) = -2(1 + r^3)^{-3}(3r^2) = \dfrac{-6r^2}{(1 + r^3)^3}$, so $dy = \dfrac{-6r^2}{(1 + r^3)^3}\, dr$.

15. (a) $y = e^{x/10}$ $\Rightarrow$ $dy = e^{x/10} \cdot \tfrac{1}{10}\, dx = \tfrac{1}{10}e^{x/10} dx$

(b) $x = 0$ and $dx = 0.1$ $\Rightarrow$ $dy = \tfrac{1}{10}e^{0/10}(0.1) = 0.01$.

17. (a) $y = \tan x \quad \Rightarrow \quad dy = \sec^2 x \, dx$

(b) When $x = \pi/4$ and $dx = -0.1$, $dy = [\sec(\pi/4)]^2(-0.1) = \left(\sqrt{2}\right)^2(-0.1) = -0.2$.

19. $y = f(x) = 2x - x^2$, $x = 2$, $\Delta x = -0.4 \quad \Rightarrow$

$\Delta y = f(1.6) - f(2) = 0.64 - 0 = 0.64$

$dy = (2 - 2x)\, dx = (2 - 4)(-0.4) = 0.8$

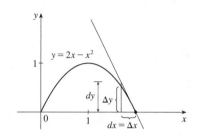

21. $y = f(x) = 2/x$, $x = 4$, $\Delta x = 1 \quad \Rightarrow$

$\Delta y = f(5) - f(4) = \frac{2}{5} - \frac{2}{4} = -0.1$

$dy = -\dfrac{2}{x^2}\, dx = -\dfrac{2}{4^2}(1) = -0.125$

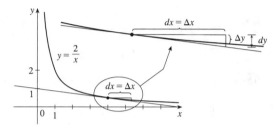

23. To estimate $(2.001)^5$, we'll find the linearization of $f(x) = x^5$ at $a = 2$. Since $f'(x) = 5x^4$, $f(2) = 32$, and $f'(2) = 80$, we have $L(x) = 32 + 80(x - 2) = 80x - 128$. Thus, $x^5 \approx 80x - 128$ when x is near 2 , so

$(2.001)^5 \approx 80(2.001) - 128 = 160.08 - 128 = 32.08$.

25. To estimate $(8.06)^{2/3}$, we'll find the linearization of $f(x) = x^{2/3}$ at $a = 8$. Since $f'(x) = \frac{2}{3}x^{-1/3} = 2/\left(3\sqrt[3]{x}\right)$,

$f(8) = 4$, and $f'(8) = \frac{1}{3}$, we have $L(x) = 4 + \frac{1}{3}(x - 8) = \frac{1}{3}x + \frac{4}{3}$. Thus, $x^{2/3} \approx \frac{1}{3}x + \frac{4}{3}$ when x is near 8, so

$(8.06)^{2/3} \approx \frac{1}{3}(8.06) + \frac{4}{3} = \frac{12.06}{3} = 4.02$.

27. $y = f(x) = \tan x \quad \Rightarrow \quad dy = \sec^2 x \, dx$. When $x = 45°$ and $dx = -1°$,

$dy = \sec^2 45°(-\pi/180) = \left(\sqrt{2}\right)^2(-\pi/180) = -\pi/90$, so $\tan 44° = f(44°) \approx f(45°) + dy = 1 - \pi/90 \approx 0.965$.

29. $y = f(x) = \sec x \quad \Rightarrow \quad f'(x) = \sec x \tan x$, so $f(0) = 1$ and $f'(0) = 1 \cdot 0 = 0$. The linear approximation of f at 0 is

$f(0) + f'(0)(x - 0) = 1 + 0(x) = 1$. Since 0.08 is close to 0, approximating $\sec 0.08$ with 1 is reasonable.

31. $y = f(x) = \ln x \quad \Rightarrow \quad f'(x) = 1/x$, so $f(1) = 0$ and $f'(1) = 1$. The linear approximation of f at 1 is

$f(1) + f'(1)(x - 1) = 0 + 1(x - 1) = x - 1$. Now $f(1.05) = \ln 1.05 \approx 1.05 - 1 = 0.05$, so the approximation

is reasonable.

33. (a) If x is the edge length, then $V = x^3 \quad \Rightarrow \quad dV = 3x^2\, dx$. When $x = 30$ and $dx = 0.1$, $dV = 3(30)^2(0.1) = 270$, so the

maximum possible error in computing the volume of the cube is about 270 cm^3. The relative error is calculated by dividing

the change in V, ΔV, by V. We approximate ΔV with dV.

$$\text{Relative error} = \frac{\Delta V}{V} \approx \frac{dV}{V} = \frac{3x^2\, dx}{x^3} = 3\frac{dx}{x} = 3\left(\frac{0.1}{30}\right) = 0.01.$$

Percentage error = relative error $\times 100\% = 0.01 \times 100\% = 1\%$.

(b) $S = 6x^2 \Rightarrow dS = 12x\,dx$. When $x = 30$ and $dx = 0.1$, $dS = 12(30)(0.1) = 36$, so the maximum possible error in computing the surface area of the cube is about 36 cm².

Relative error $= \dfrac{\Delta S}{S} \approx \dfrac{dS}{S} = \dfrac{12x\,dx}{6x^2} = 2\dfrac{dx}{x} = 2\left(\dfrac{0.1}{30}\right) = 0.00\overline{6}$.

Percentage error = relative error $\times 100\% = 0.00\overline{6} \times 100\% = 0.\overline{6}\%$.

35. (a) For a sphere of radius r, the circumference is $C = 2\pi r$ and the surface area is $S = 4\pi r^2$, so

$$r = \frac{C}{2\pi} \Rightarrow S = 4\pi\left(\frac{C}{2\pi}\right)^2 = \frac{C^2}{\pi} \Rightarrow dS = \frac{2}{\pi}C\,dC. \text{ When } C = 84 \text{ and } dC = 0.5, dS = \frac{2}{\pi}(84)(0.5) = \frac{84}{\pi},$$

so the maximum error is about $\dfrac{84}{\pi} \approx 27$ cm². Relative error $\approx \dfrac{dS}{S} = \dfrac{84/\pi}{84^2/\pi} = \dfrac{1}{84} \approx 0.012$

(b) $V = \dfrac{4}{3}\pi r^3 = \dfrac{4}{3}\pi\left(\dfrac{C}{2\pi}\right)^3 = \dfrac{C^3}{6\pi^2} \Rightarrow dV = \dfrac{1}{2\pi^2}C^2\,dC$. When $C = 84$ and $dC = 0.5$,

$dV = \dfrac{1}{2\pi^2}(84)^2(0.5) = \dfrac{1764}{\pi^2}$, so the maximum error is about $\dfrac{1764}{\pi^2} \approx 179$ cm³.

The relative error is approximately $\dfrac{dV}{V} = \dfrac{1764/\pi^2}{(84)^3/(6\pi^2)} = \dfrac{1}{56} \approx 0.018$.

37. (a) $V = \pi r^2 h \Rightarrow \Delta V \approx dV = 2\pi rh\,dr = 2\pi rh\,\Delta r$

(b) The error is

$$\Delta V - dV = [\pi(r + \Delta r)^2 h - \pi r^2 h] - 2\pi rh\,\Delta r = \pi r^2 h + 2\pi rh\,\Delta r + \pi(\Delta r)^2 h - \pi r^2 h - 2\pi rh\,\Delta r = \pi(\Delta r)^2 h.$$

39. $V = RI \Rightarrow I = \dfrac{V}{R} \Rightarrow dI = -\dfrac{V}{R^2}dR$. The relative error in calculating I is $\dfrac{\Delta I}{I} \approx \dfrac{dI}{I} = \dfrac{-(V/R^2)\,dR}{V/R} = -\dfrac{dR}{R}$.

Hence, the relative error in calculating I is approximately the same (in magnitude) as the relative error in R.

41. (a) $dc = \dfrac{dc}{dx}dx = 0\,dx = 0$ (b) $d(cu) = \dfrac{d}{dx}(cu)\,dx = c\dfrac{du}{dx}dx = c\,du$

(c) $d(u+v) = \dfrac{d}{dx}(u+v)\,dx = \left(\dfrac{du}{dx} + \dfrac{dv}{dx}\right)dx = \dfrac{du}{dx}dx + \dfrac{dv}{dx}dx = du + dv$

(d) $d(uv) = \dfrac{d}{dx}(uv)\,dx = \left(u\dfrac{dv}{dx} + v\dfrac{du}{dx}\right)dx = u\dfrac{dv}{dx}dx + v\dfrac{du}{dx}dx = u\,dv + v\,du$

(e) $d\left(\dfrac{u}{v}\right) = \dfrac{d}{dx}\left(\dfrac{u}{v}\right)dx = \dfrac{v\dfrac{du}{dx} - u\dfrac{dv}{dx}}{v^2}dx = \dfrac{v\dfrac{du}{dx}dx - u\dfrac{dv}{dx}dx}{v^2} = \dfrac{v\,du - u\,dv}{v^2}$

(f) $d(x^n) = \dfrac{d}{dx}(x^n)\,dx = nx^{n-1}\,dx$

43. (a) The graph shows that $f'(1) = 2$, so $L(x) = f(1) + f'(1)(x - 1) = 5 + 2(x - 1) = 2x + 3$.

$f(0.9) \approx L(0.9) = 4.8$ and $f(1.1) \approx L(1.1) = 5.2$.

(b) From the graph, we see that $f'(x)$ is positive and decreasing. This means that the slopes of the tangent lines are positive, but the tangents are becoming less steep. So the tangent lines lie *above* the curve. Thus, the estimates in part (a) are too large.

3.11 Hyperbolic Functions

1. (a) $\sinh 0 = \frac{1}{2}(e^0 - e^0) = 0$ (b) $\cosh 0 = \frac{1}{2}(e^0 + e^0) = \frac{1}{2}(1 + 1) = 1$

3. (a) $\sinh(\ln 2) = \dfrac{e^{\ln 2} - e^{-\ln 2}}{2} = \dfrac{e^{\ln 2} - (e^{\ln 2})^{-1}}{2} = \dfrac{2 - 2^{-1}}{2} = \dfrac{2 - \frac{1}{2}}{2} = \dfrac{3}{4}$

     (b) $\sinh 2 = \frac{1}{2}(e^2 - e^{-2}) \approx 3.62686$

5. (a) $\operatorname{sech} 0 = \dfrac{1}{\cosh 0} = \dfrac{1}{1} = 1$ (b) $\cosh^{-1} 1 = 0$ because $\cosh 0 = 1$.

7. $\sinh(-x) = \frac{1}{2}[e^{-x} - e^{-(-x)}] = \frac{1}{2}(e^{-x} - e^x) = -\frac{1}{2}(e^{-x} - e^x) = -\sinh x$

9. $\cosh x + \sinh x = \frac{1}{2}(e^x + e^{-x}) + \frac{1}{2}(e^x - e^{-x}) = \frac{1}{2}(2e^x) = e^x$

11. $\sinh x \cosh y + \cosh x \sinh y = \left[\frac{1}{2}(e^x - e^{-x})\right]\left[\frac{1}{2}(e^y + e^{-y})\right] + \left[\frac{1}{2}(e^x + e^{-x})\right]\left[\frac{1}{2}(e^y - e^{-y})\right]$

$$= \tfrac{1}{4}[(e^{x+y} + e^{x-y} - e^{-x+y} - e^{-x-y}) + (e^{x+y} - e^{x-y} + e^{-x+y} - e^{-x-y})]$$

$$= \tfrac{1}{4}(2e^{x+y} - 2e^{-x-y}) = \tfrac{1}{2}[e^{x+y} - e^{-(x+y)}] = \sinh(x + y)$$

13. Divide both sides of the identity $\cosh^2 x - \sinh^2 x = 1$ by $\sinh^2 x$:

$$\frac{\cosh^2 x}{\sinh^2 x} - \frac{\sinh^2 x}{\sinh^2 x} = \frac{1}{\sinh^2 x} \quad \Leftrightarrow \quad \coth^2 x - 1 = \operatorname{csch}^2 x.$$

15. Putting $y = x$ in the result from Exercise 11, we have

$$\sinh 2x = \sinh(x + x) = \sinh x \cosh x + \cosh x \sinh x = 2\sinh x \cosh x.$$

17. $\tanh(\ln x) = \dfrac{\sinh(\ln x)}{\cosh(\ln x)} = \dfrac{(e^{\ln x} - e^{-\ln x})/2}{(e^{\ln x} + e^{-\ln x})/2} = \dfrac{x - (e^{\ln x})^{-1}}{x + (e^{\ln x})^{-1}} = \dfrac{x - x^{-1}}{x + x^{-1}} = \dfrac{x - 1/x}{x + 1/x} = \dfrac{(x^2 - 1)/x}{(x^2 + 1)/x} = \dfrac{x^2 - 1}{x^2 + 1}$

19. By Exercise 9, $(\cosh x + \sinh x)^n = (e^x)^n = e^{nx} = \cosh nx + \sinh nx$.

21. $\operatorname{sech} x = \dfrac{1}{\cosh x} \quad \Rightarrow \quad \operatorname{sech} x = \dfrac{1}{5/3} = \dfrac{3}{5}.$

     $\cosh^2 x - \sinh^2 x = 1 \quad \Rightarrow \quad \sinh^2 x = \cosh^2 x - 1 = \left(\frac{5}{3}\right)^2 - 1 = \frac{16}{9} \quad \Rightarrow \quad \sinh x = \frac{4}{3}$ [because $x > 0$].

     $\operatorname{csch} x = \dfrac{1}{\sinh x} \quad \Rightarrow \quad \operatorname{csch} x = \dfrac{1}{4/3} = \dfrac{3}{4}.$

     $\tanh x = \dfrac{\sinh x}{\cosh x} \quad \Rightarrow \quad \tanh x = \dfrac{4/3}{5/3} = \dfrac{4}{5}.$

     $\coth x = \dfrac{1}{\tanh x} \quad \Rightarrow \quad \coth x = \dfrac{1}{4/5} = \dfrac{5}{4}.$

23. (a) $\displaystyle\lim_{x \to \infty} \tanh x = \lim_{x \to \infty} \frac{e^x - e^{-x}}{e^x + e^{-x}} \cdot \frac{e^{-x}}{e^{-x}} = \lim_{x \to \infty} \frac{1 - e^{-2x}}{1 + e^{-2x}} = \frac{1 - 0}{1 + 0} = 1$

     (b) $\displaystyle\lim_{x \to -\infty} \tanh x = \lim_{x \to -\infty} \frac{e^x - e^{-x}}{e^x + e^{-x}} \cdot \frac{e^x}{e^x} = \lim_{x \to -\infty} \frac{e^{2x} - 1}{e^{2x} + 1} = \frac{0 - 1}{0 + 1} = -1$

(c) $\displaystyle\lim_{x\to\infty} \sinh x = \lim_{x\to\infty} \frac{e^x - e^{-x}}{2} = \infty$

(d) $\displaystyle\lim_{x\to-\infty} \sinh x = \lim_{x\to-\infty} \frac{e^x - e^{-x}}{2} = -\infty$

(e) $\displaystyle\lim_{x\to\infty} \operatorname{sech} x = \lim_{x\to\infty} \frac{2}{e^x + e^{-x}} = 0$

(f) $\displaystyle\lim_{x\to\infty} \coth x = \lim_{x\to\infty} \frac{e^x + e^{-x}}{e^x - e^{-x}} \cdot \frac{e^{-x}}{e^{-x}} = \lim_{x\to\infty} \frac{1 + e^{-2x}}{1 - e^{-2x}} = \frac{1+0}{1-0} = 1$ [*Or:* Use part (a)]

(g) $\displaystyle\lim_{x\to 0^+} \coth x = \lim_{x\to 0^+} \frac{\cosh x}{\sinh x} = \infty$, since $\sinh x \to 0$ through positive values and $\cosh x \to 1$.

(h) $\displaystyle\lim_{x\to 0^-} \coth x = \lim_{x\to 0^-} \frac{\cosh x}{\sinh x} = -\infty$, since $\sinh x \to 0$ through negative values and $\cosh x \to 1$.

(i) $\displaystyle\lim_{x\to-\infty} \operatorname{csch} x = \lim_{x\to-\infty} \frac{2}{e^x - e^{-x}} = 0$

25. Let $y = \sinh^{-1} x$. Then $\sinh y = x$ and, by Example 1(a), $\cosh^2 y - \sinh^2 y = 1 \Rightarrow$ [with $\cosh y > 0$]

$\cosh y = \sqrt{1 + \sinh^2 y} = \sqrt{1 + x^2}$. So by Exercise 9, $e^y = \sinh y + \cosh y = x + \sqrt{1+x^2} \Rightarrow y = \ln\left(x + \sqrt{1+x^2}\right)$.

27. (a) Let $y = \tanh^{-1} x$. Then $x = \tanh y = \dfrac{\sinh y}{\cosh y} = \dfrac{(e^y - e^{-y})/2}{(e^y + e^{-y})/2} \cdot \dfrac{e^y}{e^y} = \dfrac{e^{2y} - 1}{e^{2y} + 1} \Rightarrow xe^{2y} + x = e^{2y} - 1 \Rightarrow$

$1 + x = e^{2y} - xe^{2y} \Rightarrow 1 + x = e^{2y}(1 - x) \Rightarrow e^{2y} = \dfrac{1+x}{1-x} \Rightarrow 2y = \ln\left(\dfrac{1+x}{1-x}\right) \Rightarrow y = \tfrac{1}{2}\ln\left(\dfrac{1+x}{1-x}\right)$.

(b) Let $y = \tanh^{-1} x$. Then $x = \tanh y$, so from Exercise 18 we have

$e^{2y} = \dfrac{1 + \tanh y}{1 - \tanh y} = \dfrac{1+x}{1-x} \Rightarrow 2y = \ln\left(\dfrac{1+x}{1-x}\right) \Rightarrow y = \tfrac{1}{2}\ln\left(\dfrac{1+x}{1-x}\right)$.

29. (a) Let $y = \cosh^{-1} x$. Then $\cosh y = x$ and $y \ge 0 \Rightarrow \sinh y \dfrac{dy}{dx} = 1 \Rightarrow$

$\dfrac{dy}{dx} = \dfrac{1}{\sinh y} = \dfrac{1}{\sqrt{\cosh^2 y - 1}} = \dfrac{1}{\sqrt{x^2 - 1}}$ [since $\sinh y \ge 0$ for $y \ge 0$].   *Or:* Use Formula 4.

(b) Let $y = \tanh^{-1} x$. Then $\tanh y = x \Rightarrow \operatorname{sech}^2 y \dfrac{dy}{dx} = 1 \Rightarrow \dfrac{dy}{dx} = \dfrac{1}{\operatorname{sech}^2 y} = \dfrac{1}{1 - \tanh^2 y} = \dfrac{1}{1 - x^2}$.

Or: Use Formula 5.

(c) Let $y = \operatorname{csch}^{-1} x$. Then $\operatorname{csch} y = x \Rightarrow -\operatorname{csch} y \coth y \dfrac{dy}{dx} = 1 \Rightarrow \dfrac{dy}{dx} = -\dfrac{1}{\operatorname{csch} y \coth y}$. By Exercise 13,

$\coth y = \pm\sqrt{\operatorname{csch}^2 y + 1} = \pm\sqrt{x^2 + 1}$. If $x > 0$, then $\coth y > 0$, so $\coth y = \sqrt{x^2 + 1}$. If $x < 0$, then $\coth y < 0$,

so $\coth y = -\sqrt{x^2 + 1}$. In either case we have $\dfrac{dy}{dx} = -\dfrac{1}{\operatorname{csch} y \coth y} = -\dfrac{1}{|x|\sqrt{x^2 + 1}}$.

(d) Let $y = \operatorname{sech}^{-1} x$. Then $\operatorname{sech} y = x \Rightarrow -\operatorname{sech} y \tanh y \dfrac{dy}{dx} = 1 \Rightarrow$

$\dfrac{dy}{dx} = -\dfrac{1}{\operatorname{sech} y \tanh y} = -\dfrac{1}{\operatorname{sech} y \sqrt{1 - \operatorname{sech}^2 y}} = -\dfrac{1}{x\sqrt{1 - x^2}}$.   [Note that $y > 0$ and so $\tanh y > 0$.]

(e) Let $y = \coth^{-1} x$. Then $\coth y = x \Rightarrow -\operatorname{csch}^2 y \dfrac{dy}{dx} = 1 \Rightarrow \dfrac{dy}{dx} = -\dfrac{1}{\operatorname{csch}^2 y} = \dfrac{1}{1 - \coth^2 y} = \dfrac{1}{1 - x^2}$

by Exercise 13.

31. $f(x) = x \sinh x - \cosh x \Rightarrow f'(x) = x (\sinh x)' + \sinh x \cdot 1 - \sinh x = x \cosh x$

33. $h(x) = \ln(\cosh x) \Rightarrow h'(x) = \dfrac{1}{\cosh x}(\cosh x)' = \dfrac{\sinh x}{\cosh x} = \tanh x$

35. $y = e^{\cosh 3x} \Rightarrow y' = e^{\cosh 3x} \cdot \sinh 3x \cdot 3 = 3e^{\cosh 3x} \sinh 3x$

37. $f(t) = \operatorname{sech}^2(e^t) = [\operatorname{sech}(e^t)]^2 \Rightarrow$

$f'(t) = 2[\operatorname{sech}(e^t)] [\operatorname{sech}(e^t)]' = 2\operatorname{sech}(e^t) \left[-\operatorname{sech}(e^t) \tanh(e^t) \cdot e^t\right] = -2e^t \operatorname{sech}^2(e^t) \tanh(e^t)$

39. $y = \arctan(\tanh x) \Rightarrow y' = \dfrac{1}{1 + (\tanh x)^2} (\tanh x)' = \dfrac{\operatorname{sech}^2 x}{1 + \tanh^2 x}$

41. $G(x) = \dfrac{1 - \cosh x}{1 + \cosh x} \Rightarrow$

$G'(x) = \dfrac{(1 + \cosh x)(-\sinh x) - (1 - \cosh x)(\sinh x)}{(1 + \cosh x)^2} = \dfrac{-\sinh x - \sinh x \cosh x - \sinh x + \sinh x \cosh x}{(1 + \cosh x)^2}$

$= \dfrac{-2\sinh x}{(1 + \cosh x)^2}$

43. $y = \tanh^{-1}\sqrt{x} \Rightarrow y' = \dfrac{1}{1 - \left(\sqrt{x}\right)^2} \cdot \dfrac{1}{2}x^{-1/2} = \dfrac{1}{2\sqrt{x}\,(1 - x)}$

45. $y = x \sinh^{-1}(x/3) - \sqrt{9 + x^2} \Rightarrow$

$y' = \sinh^{-1}\left(\dfrac{x}{3}\right) + x\dfrac{1/3}{\sqrt{1 + (x/3)^2}} - \dfrac{2x}{2\sqrt{9 + x^2}} = \sinh^{-1}\left(\dfrac{x}{3}\right) + \dfrac{x}{\sqrt{9 + x^2}} - \dfrac{x}{\sqrt{9 + x^2}} = \sinh^{-1}\left(\dfrac{x}{3}\right)$

47. $y = \coth^{-1}\sqrt{x^2 + 1} \Rightarrow y' = \dfrac{1}{1 - (x^2 + 1)}\dfrac{2x}{2\sqrt{x^2 + 1}} = -\dfrac{1}{x\sqrt{x^2 + 1}}$

49. As the depth d of the water gets large, the fraction $\dfrac{2\pi d}{L}$ gets large, and from Figure 3 or Exercise 23(a), $\tanh\left(\dfrac{2\pi d}{L}\right)$

approaches 1. Thus, $v = \sqrt{\dfrac{gL}{2\pi} \tanh\left(\dfrac{2\pi d}{L}\right)} \approx \sqrt{\dfrac{gL}{2\pi}(1)} = \sqrt{\dfrac{gL}{2\pi}}$.

51. (a) $y = 20\cosh(x/20) - 15 \Rightarrow y' = 20\sinh(x/20) \cdot \frac{1}{20} = \sinh(x/20)$. Since the right pole is positioned at $x = 7$,

we have $y'(7) = \sinh\frac{7}{20} \approx 0.3572$.

(b) If α is the angle between the tangent line and the x-axis, then $\tan\alpha = $ slope of the line $= \sinh\frac{7}{20}$, so

$\alpha = \tan^{-1}\left(\sinh\frac{7}{20}\right) \approx 0.343 \text{ rad} \approx 19.66°$. Thus, the angle between the line and the pole is $\theta = 90° - \alpha \approx 70.34°$.

53. (a) $y = A\sinh mx + B\cosh mx \Rightarrow y' = mA\cosh mx + mB\sinh mx \Rightarrow$

$y'' = m^2 A\sinh mx + m^2 B\cosh mx = m^2(A\sinh mx + B\cosh mx) = m^2 y$

(b) From part (a), a solution of $y'' = 9y$ is $y(x) = A \sinh 3x + B \cosh 3x$. So $-4 = y(0) = A \sinh 0 + B \cosh 0 = B$, so

$B = -4$. Now $y'(x) = 3A \cosh 3x - 12 \sinh 3x \quad \Rightarrow \quad 6 = y'(0) = 3A \quad \Rightarrow \quad A = 2$, so $y = 2 \sinh 3x - 4 \cosh 3x$.

55. The tangent to $y = \cosh x$ has slope 1 when $y' = \sinh x = 1 \quad \Rightarrow \quad x = \sinh^{-1} 1 = \ln(1 + \sqrt{2})$, by Equation 3.

Since $\sinh x = 1$ and $y = \cosh x = \sqrt{1 + \sinh^2 x}$, we have $\cosh x = \sqrt{2}$. The point is $(\ln(1 + \sqrt{2}), \sqrt{2})$.

57. If $ae^x + be^{-x} = \alpha \cosh(x + \beta) \quad$ [or $\alpha \sinh(x + \beta)$], $\quad$ then

$ae^x + be^{-x} = \frac{\alpha}{2}(e^{x+\beta} \pm e^{-x-\beta}) = \frac{\alpha}{2}(e^x e^\beta \pm e^{-x} e^{-\beta}) = (\frac{\alpha}{2} e^\beta) e^x \pm (\frac{\alpha}{2} e^{-\beta}) e^{-x}$. Comparing coefficients of e^x

and e^{-x}, we have $a = \frac{\alpha}{2} e^\beta$ **(1)** and $b = \pm \frac{\alpha}{2} e^{-\beta}$ **(2)**. We need to find α and β. Dividing equation **(1)** by equation **(2)**

gives us $\frac{a}{b} = \pm e^{2\beta} \quad \Rightarrow \quad (\star) \quad 2\beta = \ln(\pm \frac{a}{b}) \quad \Rightarrow \quad \beta = \frac{1}{2} \ln(\pm \frac{a}{b})$. Solving equations **(1)** and **(2)** for e^β gives us

$e^\beta = \dfrac{2a}{\alpha}$ and $e^\beta = \pm \dfrac{\alpha}{2b}$, so $\dfrac{2a}{\alpha} = \pm \dfrac{\alpha}{2b} \quad \Rightarrow \quad \alpha^2 = \pm 4ab \quad \Rightarrow \quad \alpha = 2\sqrt{\pm ab}$.

$(\star)$ If $\frac{a}{b} > 0$, we use the $+$ sign and obtain a cosh function, whereas if $\frac{a}{b} < 0$, we use the $-$ sign and obtain a sinh

function.

In summary, if a and b have the same sign, we have $ae^x + be^{-x} = 2\sqrt{ab} \cosh(x + \frac{1}{2} \ln \frac{a}{b})$, whereas, if a and b have the

opposite sign, then $ae^x + be^{-x} = 2\sqrt{-ab} \sinh(x + \frac{1}{2} \ln(-\frac{a}{b}))$.

3 Review

CONCEPT CHECK

1. (a) The Power Rule: If n is any real number, then $\dfrac{d}{dx}(x^n) = nx^{n-1}$. The derivative of a variable base raised to a constant

power is the power times the base raised to the power minus one.

(b) The Constant Multiple Rule: If c is a constant and f is a differentiable function, then $\dfrac{d}{dx}[cf(x)] = c\dfrac{d}{dx} f(x)$.

The derivative of a constant times a function is the constant times the derivative of the function.

(c) The Sum Rule: If f and g are both differentiable, then $\dfrac{d}{dx}[f(x) + g(x)] = \dfrac{d}{dx} f(x) + \dfrac{d}{dx} g(x)$. The derivative of a sum

of functions is the sum of the derivatives.

(d) The Difference Rule: If f and g are both differentiable, then $\dfrac{d}{dx}[f(x) - g(x)] = \dfrac{d}{dx} f(x) - \dfrac{d}{dx} g(x)$. The derivative of a

difference of functions is the difference of the derivatives.

(e) The Product Rule: If f and g are both differentiable, then $\dfrac{d}{dx}[f(x) g(x)] = f(x)\dfrac{d}{dx} g(x) + g(x)\dfrac{d}{dx} f(x)$. The

derivative of a product of two functions is the first function times the derivative of the second function plus the second

function times the derivative of the first function.

(f) The Quotient Rule: If f and g are both differentiable, then $\dfrac{d}{dx}\left[\dfrac{f(x)}{g(x)}\right] = \dfrac{g(x)\,\dfrac{d}{dx}\,f(x) - f(x)\,\dfrac{d}{dx}\,g(x)}{[g(x)]^2}$.

The derivative of a quotient of functions is the denominator times the derivative of the numerator minus the numerator times the derivative of the denominator, all divided by the square of the denominator.

(g) The Chain Rule: If f and g are both differentiable and $F = f \circ g$ is the composite function defined by $F(x) = f(g(x))$, then F is differentiable and F' is given by the product $F'(x) = f'(g(x))\,g'(x)$. The derivative of a composite function is the derivative of the outer function evaluated at the inner function times the derivative of the inner function.

2. (a) $y = x^n \;\Rightarrow\; y' = nx^{n-1}$

    (b) $y = e^x \;\Rightarrow\; y' = e^x$

    (c) $y = a^x \;\Rightarrow\; y' = a^x \ln a$

    (d) $y = \ln x \;\Rightarrow\; y' = 1/x$

    (e) $y = \log_a x \;\Rightarrow\; y' = 1/(x \ln a)$

    (f) $y = \sin x \;\Rightarrow\; y' = \cos x$

    (g) $y = \cos x \;\Rightarrow\; y' = -\sin x$

    (h) $y = \tan x \;\Rightarrow\; y' = \sec^2 x$

    (i) $y = \csc x \;\Rightarrow\; y' = -\csc x \cot x$

    (j) $y = \sec x \;\Rightarrow\; y' = \sec x \tan x$

    (k) $y = \cot x \;\Rightarrow\; y' = -\csc^2 x$

    (l) $y = \sin^{-1} x \;\Rightarrow\; y' = 1/\sqrt{1-x^2}$

    (m) $y = \cos^{-1} x \;\Rightarrow\; y' = -1/\sqrt{1-x^2}$

    (n) $y = \tan^{-1} x \;\Rightarrow\; y' = 1/(1+x^2)$

    (o) $y = \sinh x \;\Rightarrow\; y' = \cosh x$

    (p) $y = \cosh x \;\Rightarrow\; y' = \sinh x$

    (q) $y = \tanh x \;\Rightarrow\; y' = \operatorname{sech}^2 x$

    (r) $y = \sinh^{-1} x \;\Rightarrow\; y' = 1/\sqrt{1+x^2}$

    (s) $y = \cosh^{-1} x \;\Rightarrow\; y' = 1/\sqrt{x^2-1}$

    (t) $y = \tanh^{-1} x \;\Rightarrow\; y' = 1/(1-x^2)$

3. (a) e is the number such that $\displaystyle\lim_{h \to 0} \frac{e^h - 1}{h} = 1$.

    (b) $e = \displaystyle\lim_{x \to 0} (1+x)^{1/x}$

    (c) The differentiation formula for $y = a^x \quad [y' = a^x \ln a]$ is simplest when $a = e$ because $\ln e = 1$.

    (d) The differentiation formula for $y = \log_a x \quad [y' = 1/(x \ln a)]$ is simplest when $a = e$ because $\ln e = 1$.

4. (a) Implicit differentiation consists of differentiating both sides of an equation involving x and y with respect to x, and then solving the resulting equation for y'.

    (b) Logarithmic differentiation consists of taking natural logarithms of both sides of an equation $y = f(x)$, simplifying, differentiating implicitly with respect to x, and then solving the resulting equation for y'.

5. (a) The linearization L of f at $x = a$ is $L(x) = f(a) + f'(a)(x - a)$.

    (b) If $y = f(x)$, then the differential dy is given by $dy = f'(x)\,dx$.

    (c) See Figure 5 in Section 3.10.

TRUE-FALSE QUIZ

1. True. This is the Sum Rule.

3. True. This is the Chain Rule.

5. False. $\dfrac{d}{dx} f\left(\sqrt{x}\right) = \dfrac{f'\left(\sqrt{x}\right)}{2\sqrt{x}}$ by the Chain Rule.

7. False. $\dfrac{d}{dx} 10^x = 10^x \ln 10$

9. True. $\dfrac{d}{dx}\left(\tan^2 x\right) = 2\tan x \sec^2 x$, and $\dfrac{d}{dx}\left(\sec^2 x\right) = 2\sec x\,(\sec x \tan x) = 2\tan x \sec^2 x$.

Or: $\dfrac{d}{dx}\left(\sec^2 x\right) = \dfrac{d}{dx}\left(1 + \tan^2 x\right) = \dfrac{d}{dx}\left(\tan^2 x\right)$.

11. True. $g(x) = x^5 \;\Rightarrow\; g'(x) = 5x^4 \;\Rightarrow\; g'(2) = 5(2)^4 = 80$, and by the definition of the derivative,

$$\lim_{x \to 2} \frac{g(x) - g(2)}{x - 2} = g'(2) = 80.$$

EXERCISES

1. $y = (x^4 - 3x^2 + 5)^3 \;\Rightarrow\;$

$y' = 3(x^4 - 3x^2 + 5)^2\,\dfrac{d}{dx}\,(x^4 - 3x^2 + 5) = 3(x^4 - 3x^2 + 5)^2(4x^3 - 6x) = 6x(x^4 - 3x^2 + 5)^2(2x^2 - 3)$

3. $y = \sqrt{x} + \dfrac{1}{\sqrt[3]{x^4}} = x^{1/2} + x^{-4/3} \;\Rightarrow\; y' = \tfrac{1}{2}x^{-1/2} - \tfrac{4}{3}x^{-7/3} = \dfrac{1}{2\sqrt{x}} - \dfrac{4}{3\sqrt[3]{x^7}}$

5. $y = 2x\sqrt{x^2 + 1} \;\Rightarrow\;$

$y' = 2x \cdot \tfrac{1}{2}(x^2 + 1)^{-1/2}(2x) + \sqrt{x^2 + 1}\,(2) = \dfrac{2x^2}{\sqrt{x^2 + 1}} + 2\sqrt{x^2 + 1} = \dfrac{2x^2 + 2(x^2 + 1)}{\sqrt{x^2 + 1}} = \dfrac{2(2x^2 + 1)}{\sqrt{x^2 + 1}}$

7. $y = e^{\sin 2\theta} \;\Rightarrow\; y' = e^{\sin 2\theta}\,\dfrac{d}{d\theta}\,(\sin 2\theta) = e^{\sin 2\theta}(\cos 2\theta)(2) = 2\cos 2\theta\, e^{\sin 2\theta}$

9. $y = \dfrac{t}{1 - t^2} \;\Rightarrow\; y' = \dfrac{(1 - t^2)(1) - t(-2t)}{(1 - t^2)^2} = \dfrac{1 - t^2 + 2t^2}{(1 - t^2)^2} = \dfrac{t^2 + 1}{(1 - t^2)^2}$

11. $y = \sqrt{x}\cos\sqrt{x} \;\Rightarrow\;$

$y' = \sqrt{x}\left(\cos\sqrt{x}\right)' + \cos\sqrt{x}\,\left(\sqrt{x}\right)' = \sqrt{x}\left[-\sin\sqrt{x}\left(\tfrac{1}{2}x^{-1/2}\right)\right] + \cos\sqrt{x}\left(\tfrac{1}{2}x^{-1/2}\right)$

$= \tfrac{1}{2}x^{-1/2}\left(-\sqrt{x}\sin\sqrt{x} + \cos\sqrt{x}\right) = \dfrac{\cos\sqrt{x} - \sqrt{x}\sin\sqrt{x}}{2\sqrt{x}}$

13. $y = \dfrac{e^{1/x}}{x^2} \;\Rightarrow\; y' = \dfrac{x^2(e^{1/x})' - e^{1/x}\left(x^2\right)'}{(x^2)^2} = \dfrac{x^2(e^{1/x})(-1/x^2) - e^{1/x}(2x)}{x^4} = \dfrac{-e^{1/x}(1 + 2x)}{x^4}$

15. $\dfrac{d}{dx}\left(xy^4 + x^2 y\right) = \dfrac{d}{dx}\left(x + 3y\right) \quad \Rightarrow \quad x \cdot 4y^3 y' + y^4 \cdot 1 + x^2 \cdot y' + y \cdot 2x = 1 + 3y' \quad \Rightarrow$

$y'\left(4xy^3 + x^2 - 3\right) = 1 - y^4 - 2xy \quad \Rightarrow \quad y' = \dfrac{1 - y^4 - 2xy}{4xy^3 + x^2 - 3}$

17. $y = \dfrac{\sec 2\theta}{1 + \tan 2\theta} \quad \Rightarrow$

$y' = \dfrac{(1 + \tan 2\theta)(\sec 2\theta \, \tan 2\theta \cdot 2) - (\sec 2\theta)(\sec^2 2\theta \cdot 2)}{(1 + \tan 2\theta)^2} = \dfrac{2\sec 2\theta\,[(1 + \tan 2\theta)\tan 2\theta - \sec^2 2\theta]}{(1 + \tan 2\theta)^2}$

$= \dfrac{2\sec 2\theta\,(\tan 2\theta + \tan^2 2\theta - \sec^2 2\theta)}{(1 + \tan 2\theta)^2} = \dfrac{2\sec 2\theta\,(\tan 2\theta - 1)}{(1 + \tan 2\theta)^2} \qquad \left[1 + \tan^2 x = \sec^2 x\right]$

19. $y = e^{cx}(c\sin x - \cos x) \quad \Rightarrow$

$y' = e^{cx}(c\cos x + \sin x) + ce^{cx}(c\sin x - \cos x) = e^{cx}(c^2 \sin x - c\cos x + c\cos x + \sin x)$

$= e^{cx}(c^2 \sin x + \sin x) = e^{cx}\sin x\,(c^2 + 1)$

21. $y = 3^{x\ln x} \quad \Rightarrow \quad y' = 3^{x\ln x} \cdot \ln 3 \cdot \dfrac{d}{dx}(x\ln x) = 3^{x\ln x} \cdot \ln 3\left(x \cdot \dfrac{1}{x} + \ln x \cdot 1\right) = 3^{x\ln x} \cdot \ln 3(1 + \ln x)$

23. $y = \left(1 - x^{-1}\right)^{-1} \quad \Rightarrow$

$y' = -1\left(1 - x^{-1}\right)^{-2}\left[-(-1x^{-2})\right] = -(1 - 1/x)^{-2}x^{-2} = -((x-1)/x)^{-2}x^{-2} = -(x-1)^{-2}$

25. $\sin(xy) = x^2 - y \quad \Rightarrow \quad \cos(xy)(xy' + y \cdot 1) = 2x - y' \quad \Rightarrow \quad x\cos(xy)y' + y' = 2x - y\cos(xy) \quad \Rightarrow$

$y'[x\cos(xy) + 1] = 2x - y\cos(xy) \quad \Rightarrow \quad y' = \dfrac{2x - y\cos(xy)}{x\cos(xy) + 1}$

27. $y = \log_5(1 + 2x) \quad \Rightarrow \quad y' = \dfrac{1}{(1 + 2x)\ln 5}\dfrac{d}{dx}(1 + 2x) = \dfrac{2}{(1 + 2x)\ln 5}$

29. $y = \ln\sin x - \tfrac{1}{2}\sin^2 x \quad \Rightarrow \quad y' = \dfrac{1}{\sin x} \cdot \cos x - \tfrac{1}{2} \cdot 2\sin x \cdot \cos x = \cot x - \sin x \, \cos x$

31. $y = x\tan^{-1}(4x) \quad \Rightarrow \quad y' = x \cdot \dfrac{1}{1 + (4x)^2} \cdot 4 + \tan^{-1}(4x) \cdot 1 = \dfrac{4x}{1 + 16x^2} + \tan^{-1}(4x)$

33. $y = \ln|\sec 5x + \tan 5x| \quad \Rightarrow$

$y' = \dfrac{1}{\sec 5x + \tan 5x}(\sec 5x \, \tan 5x \cdot 5 + \sec^2 5x \cdot 5) = \dfrac{5\sec 5x\,(\tan 5x + \sec 5x)}{\sec 5x + \tan 5x} = 5\sec 5x$

35. $y = \cot(3x^2 + 5) \quad \Rightarrow \quad y' = -\csc^2(3x^2 + 5)(6x) = -6x\csc^2(3x^2 + 5)$

37. $y = \sin\left(\tan\sqrt{1 + x^3}\right) \quad \Rightarrow \quad y' = \cos\left(\tan\sqrt{1 + x^3}\right)\left(\sec^2\sqrt{1 + x^3}\right)\left[3x^2 / \left(2\sqrt{1 + x^3}\right)\right]$

39. $y = \tan^2(\sin\theta) = [\tan(\sin\theta)]^2 \quad \Rightarrow \quad y' = 2[\tan(\sin\theta)] \cdot \sec^2(\sin\theta) \cdot \cos\theta$

41. $y = \dfrac{\sqrt{x+1}\,(2-x)^5}{(x+3)^7}$ $\Rightarrow$ $\ln y = \frac{1}{2}\ln(x+1) + 5\ln(2-x) - 7\ln(x+3)$ $\Rightarrow$ $\dfrac{y'}{y} = \dfrac{1}{2(x+1)} + \dfrac{-5}{2-x} - \dfrac{7}{x+3}$ $\Rightarrow$

$y' = \dfrac{\sqrt{x+1}\,(2-x)^5}{(x+3)^7}\left[\dfrac{1}{2(x+1)} - \dfrac{5}{2-x} - \dfrac{7}{x+3}\right]$ or $y' = \dfrac{(2-x)^4(3x^2-55x-52)}{2\sqrt{x+1}\,(x+3)^8}$.

43. $y = x\sinh(x^2)$ $\Rightarrow$ $y' = x\cosh(x^2)\cdot 2x + \sinh(x^2)\cdot 1 = 2x^2\cosh(x^2) + \sinh(x^2)$

45. $y = \ln(\cosh 3x)$ $\Rightarrow$ $y' = (1/\cosh 3x)(\sinh 3x)(3) = 3\tanh 3x$

47. $y = \cosh^{-1}(\sinh x)$ $\Rightarrow$ $y' = \dfrac{1}{\sqrt{(\sinh x)^2 - 1}}\cdot \cosh x = \dfrac{\cosh x}{\sqrt{\sinh^2 x - 1}}$

49. $y = \cos\left(e^{\sqrt{\tan 3x}}\right)$ $\Rightarrow$

$y' = -\sin\left(e^{\sqrt{\tan 3x}}\right)\cdot \left(e^{\sqrt{\tan 3x}}\right)' = -\sin\left(e^{\sqrt{\tan 3x}}\right)e^{\sqrt{\tan 3x}}\cdot \frac{1}{2}(\tan 3x)^{-1/2}\cdot \sec^2(3x)\cdot 3$

$= \dfrac{-3\sin\left(e^{\sqrt{\tan 3x}}\right)e^{\sqrt{\tan 3x}}\sec^2(3x)}{2\sqrt{\tan 3x}}$

51. $f(t) = \sqrt{4t+1}$ $\Rightarrow$ $f'(t) = \frac{1}{2}(4t+1)^{-1/2}\cdot 4 = 2(4t+1)^{-1/2}$ $\Rightarrow$

$f''(t) = 2(-\frac{1}{2})(4t+1)^{-3/2}\cdot 4 = -4/(4t+1)^{3/2}$, so $f''(2) = -4/9^{3/2} = -\frac{4}{27}$.

53. $x^6 + y^6 = 1$ $\Rightarrow$ $6x^5 + 6y^5y' = 0$ $\Rightarrow$ $y' = -x^5/y^5$ $\Rightarrow$

$y'' = -\dfrac{y^5(5x^4) - x^5(5y^4y')}{(y^5)^2} = -\dfrac{5x^4y^4\left[y - x(-x^5/y^5)\right]}{y^{10}} = -\dfrac{5x^4\left[(y^6+x^6)/y^5\right]}{y^6} = -\dfrac{5x^4}{y^{11}}$

55. We first show it is true for $n=1$: $f(x) = xe^x$ $\Rightarrow$ $f'(x) = xe^x + e^x = (x+1)e^x$. We now assume it is true

for $n=k$: $f^{(k)}(x) = (x+k)e^x$. With this assumption, we must show it is true for $n = k+1$:

$f^{(k+1)}(x) = \dfrac{d}{dx}\left[f^{(k)}(x)\right] = \dfrac{d}{dx}\left[(x+k)e^x\right] = (x+k)e^x + e^x = [(x+k)+1]e^x = [x+(k+1)]e^x$.

Therefore, $f^{(n)}(x) = (x+n)e^x$ by mathematical induction.

57. $y = 4\sin^2 x$ $\Rightarrow$ $y' = 4\cdot 2\sin x\cos x$. At $\left(\frac{\pi}{6},1\right)$, $y' = 8\cdot\frac{1}{2}\cdot\frac{\sqrt3}{2} = 2\sqrt3$, so an equation of the tangent line

is $y - 1 = 2\sqrt3\left(x - \frac{\pi}{6}\right)$, or $y = 2\sqrt3\,x + 1 - \pi\sqrt3/3$.

59. $y = \sqrt{1+4\sin x}$ $\Rightarrow$ $y' = \frac{1}{2}(1+4\sin x)^{-1/2}\cdot 4\cos x = \dfrac{2\cos x}{\sqrt{1+4\sin x}}$.

At $(0,1)$, $y' = \dfrac{2}{\sqrt1} = 2$, so an equation of the tangent line is $y - 1 = 2(x-0)$, or $y = 2x + 1$.

61. $y = (2+x)e^{-x}$ $\Rightarrow$ $y' = (2+x)(-e^{-x}) + e^{-x}\cdot 1 = e^{-x}[-(2+x)+1] = e^{-x}(-x-1)$.

At $(0,2)$, $y' = 1(-1) = -1$, so an equation of the tangent line is $y - 2 = -1(x-0)$, or $y = -x + 2$.

The slope of the normal line is 1, so an equation of the normal line is $y - 2 = 1(x-0)$, or $y = x + 2$.

63. (a) $f(x) = x\sqrt{5-x}$ ⇒

$$f'(x) = x\left[\frac{1}{2}(5-x)^{-1/2}(-1)\right] + \sqrt{5-x} = \frac{-x}{2\sqrt{5-x}} + \sqrt{5-x} \cdot \frac{2\sqrt{5-x}}{2\sqrt{5-x}} = \frac{-x}{2\sqrt{5-x}} + \frac{2(5-x)}{2\sqrt{5-x}}$$

$$= \frac{-x + 10 - 2x}{2\sqrt{5-x}} = \frac{10 - 3x}{2\sqrt{5-x}}$$

(b) At $(1, 2)$: $f'(1) = \frac{7}{4}$.

So an equation of the tangent line is $y - 2 = \frac{7}{4}(x-1)$ or $y = \frac{7}{4}x + \frac{1}{4}$.

At $(4, 4)$: $f'(4) = -\frac{2}{2} = -1$.

So an equation of the tangent line is $y - 4 = -1(x-4)$ or $y = -x + 8$.

(c)

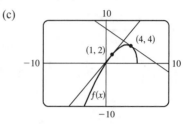

(d)

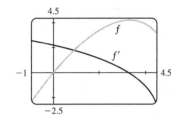

The graphs look reasonable, since f' is positive where f has tangents with positive slope, and f' is negative where f has tangents with negative slope.

65. $y = \sin x + \cos x$ ⇒ $y' = \cos x - \sin x = 0$ ⇔ $\cos x = \sin x$ and $0 \le x \le 2\pi$ ⇔ $x = \frac{\pi}{4}$ or $\frac{5\pi}{4}$, so the points are $\left(\frac{\pi}{4}, \sqrt{2}\right)$ and $\left(\frac{5\pi}{4}, -\sqrt{2}\right)$.

67. $f(x) = (x-a)(x-b)(x-c)$ ⇒ $f'(x) = (x-b)(x-c) + (x-a)(x-c) + (x-a)(x-b)$.

So $\dfrac{f'(x)}{f(x)} = \dfrac{(x-b)(x-c) + (x-a)(x-c) + (x-a)(x-b)}{(x-a)(x-b)(x-c)} = \dfrac{1}{x-a} + \dfrac{1}{x-b} + \dfrac{1}{x-c}$.

Or: $f(x) = (x-a)(x-b)(x-c)$ ⇒ $\ln|f(x)| = \ln|x-a| + \ln|x-b| + \ln|x-c|$ ⇒

$$\frac{f'(x)}{f(x)} = \frac{1}{x-a} + \frac{1}{x-b} + \frac{1}{x-c}$$

69. (a) $h(x) = f(x)\,g(x)$ ⇒ $h'(x) = f(x)\,g'(x) + g(x)\,f'(x)$ ⇒

$$h'(2) = f(2)\,g'(2) + g(2)\,f'(2) = (3)(4) + (5)(-2) = 12 - 10 = 2$$

(b) $F(x) = f(g(x))$ ⇒ $F'(x) = f'(g(x))\,g'(x)$ ⇒ $F'(2) = f'(g(2))\,g'(2) = f'(5)(4) = 11 \cdot 4 = 44$

71. $f(x) = x^2 g(x)$ ⇒ $f'(x) = x^2 g'(x) + g(x)(2x) = x[xg'(x) + 2g(x)]$

73. $f(x) = [g(x)]^2$ ⇒ $f'(x) = 2[g(x)] \cdot g'(x) = 2g(x)\,g'(x)$

75. $f(x) = g(e^x)$ ⇒ $f'(x) = g'(e^x)\,e^x$

77. $f(x) = \ln|g(x)|$ ⇒ $f'(x) = \dfrac{1}{g(x)}g'(x) = \dfrac{g'(x)}{g(x)}$

79. $h(x) = \dfrac{f(x)\,g(x)}{f(x) + g(x)} \quad \Rightarrow$

$$h'(x) = \frac{[f(x) + g(x)]\,[f(x)\,g'(x) + g(x)\,f'(x)] - f(x)\,g(x)\,[f'(x) + g'(x)]}{[f(x) + g(x)]^2}$$

$$= \frac{[f(x)]^2\,g'(x) + f(x)\,g(x)\,f'(x) + f(x)\,g(x)\,g'(x) + [g(x)]^2\,f'(x) - f(x)\,g(x)\,f'(x) - f(x)\,g(x)\,g'(x)}{[f(x) + g(x)]^2}$$

$$= \frac{f'(x)\,[g(x)]^2 + g'(x)\,[f(x)]^2}{[f(x) + g(x)]^2}$$

81. Using the Chain Rule repeatedly, $h(x) = f(g(\sin 4x)) \quad \Rightarrow$

$$h'(x) = f'(g(\sin 4x)) \cdot \frac{d}{dx}\,(g(\sin 4x)) = f'(g(\sin 4x)) \cdot g'(\sin 4x) \cdot \frac{d}{dx}\,(\sin 4x) = f'(g(\sin 4x))g'(\sin 4x)(\cos 4x)(4).$$

83. $y = [\ln(x + 4)]^2 \quad \Rightarrow \quad y' = 2[\ln(x+4)]^1 \cdot \dfrac{1}{x+4} \cdot 1 = 2\,\dfrac{\ln(x+4)}{x+4}$ and $y' = 0 \quad \Leftrightarrow \quad \ln(x+4) = 0 \quad \Leftrightarrow$

$x + 4 = e^0 \quad \Rightarrow \quad x + 4 = 1 \quad \Leftrightarrow \quad x = -3$, so the tangent is horizontal at the point $(-3, 0)$.

85. $y = f(x) = ax^2 + bx + c \quad \Rightarrow \quad f'(x) = 2ax + b$. We know that $f'(-1) = 6$ and $f'(5) = -2$, so $-2a + b = 6$ and $10a + b = -2$. Subtracting the first equation from the second gives $12a = -8 \quad \Rightarrow \quad a = -\frac{2}{3}$. Substituting $-\frac{2}{3}$ for a in the first equation gives $b = \frac{14}{3}$. Now $f(1) = 4 \quad \Rightarrow \quad 4 = a + b + c$, so $c = 4 + \frac{2}{3} - \frac{14}{3} = 0$ and hence, $f(x) = -\frac{2}{3}x^2 + \frac{14}{3}x$.

87. $s(t) = Ae^{-ct}\cos(\omega t + \delta) \quad \Rightarrow$

$v(t) = s'(t) = A\{e^{-ct}\,[-\omega\sin(\omega t + \delta)] + \cos(\omega t + \delta)(-ce^{-ct})\} = -Ae^{-ct}\,[\omega\sin(\omega t + \delta) + c\cos(\omega t + \delta)] \quad \Rightarrow$

$a(t) = v'(t) = -A\{e^{-ct}[\omega^2\cos(\omega t + \delta) - c\omega\sin(\omega t + \delta)] + [\omega\sin(\omega t + \delta) + c\cos(\omega t + \delta)](-ce^{-ct})\}$

$\qquad = -Ae^{-ct}[\omega^2\cos(\omega t + \delta) - c\omega\sin(\omega t + \delta) - c\omega\sin(\omega t + \delta) - c^2\cos(\omega t + \delta)]$

$\qquad = -Ae^{-ct}[(\omega^2 - c^2)\cos(\omega t + \delta) - 2c\omega\sin(\omega t + \delta)] = Ae^{-ct}[(c^2 - \omega^2)\cos(\omega t + \delta) + 2c\omega\sin(\omega t + \delta)]$

89. (a) $y = t^3 - 12t + 3 \quad \Rightarrow \quad v(t) = y' = 3t^2 - 12 \quad \Rightarrow \quad a(t) = v'(t) = 6t$

(b) $v(t) = 3(t^2 - 4) > 0$ when $t > 2$, so it moves upward when $t > 2$ and downward when $0 \le t < 2$.

(c) Distance upward $= y(3) - y(2) = -6 - (-13) = 7$,

Distance downward $= y(0) - y(2) = 3 - (-13) = 16$. Total distance $= 7 + 16 = 23$.

(d)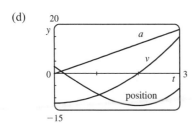

(e) The particle is speeding up when v and a have the same sign, that is, when $t > 2$. The particle is slowing down when v and a have opposite signs; that is, when $0 < t < 2$.

91. The linear density ρ is the rate of change of mass m with respect to length x.

$m = x\left(1 + \sqrt{x}\right) = x + x^{3/2} \quad \Rightarrow \quad \rho = dm/dx = 1 + \frac{3}{2}\sqrt{x}$, so the linear density when $x = 4$ is $1 + \frac{3}{2}\sqrt{4} = 4$ kg/m.

93. (a) $y(t) = y(0)e^{kt} = 200e^{kt}$ ⇒ $y(0.5) = 200e^{0.5k} = 360$ ⇒ $e^{0.5k} = 1.8$ ⇒ $0.5k = \ln 1.8$ ⇒

$k = 2\ln 1.8 = \ln(1.8)^2 = \ln 3.24$ ⇒ $y(t) = 200e^{(\ln 3.24)t} = 200(3.24)^t$

(b) $y(4) = 200(3.24)^4 \approx 22{,}040$ bacteria

(c) $y'(t) = 200(3.24)^t \cdot \ln 3.24$, so $y'(4) = 200(3.24)^4 \cdot \ln 3.24 \approx 25{,}910$ bacteria per hour

(d) $200(3.24)^t = 10{,}000$ ⇒ $(3.24)^t = 50$ ⇒ $t\ln 3.24 = \ln 50$ ⇒ $t = \ln 50/\ln 3.24 \approx 3.33$ hours

95. (a) $C'(t) = -kC(t)$ ⇒ $C(t) = C(0)e^{-kt}$ by Theorem 9.4.2. But $C(0) = C_0$, so $C(t) = C_0 e^{-kt}$.

(b) $C(30) = \frac{1}{2}C_0$ since the concentration is reduced by half. Thus, $\frac{1}{2}C_0 = C_0 e^{-30k}$ ⇒ $\ln\frac{1}{2} = -30k$ ⇒

$k = -\frac{1}{30}\ln\frac{1}{2} = \frac{1}{30}\ln 2$. Since 10% of the original concentration remains if 90% is eliminated, we want the value of t

such that $C(t) = \frac{1}{10}C_0$. Therefore, $\frac{1}{10}C_0 = C_0 e^{-t(\ln 2)/30}$ ⇒ $\ln 0.1 = -t(\ln 2)/30$ ⇒ $t = -\frac{30}{\ln 2}\ln 0.1 \approx 100$ h.

97. If $x =$ edge length, then $V = x^3$ ⇒ $dV/dt = 3x^2\, dx/dt = 10$ ⇒ $dx/dt = 10/(3x^2)$ and $S = 6x^2$ ⇒

$dS/dt = (12x)\, dx/dt = 12x[10/(3x^2)] = 40/x$. When $x = 30$, $dS/dt = \frac{40}{30} = \frac{4}{3}$ cm^2/min.

99. Given $dh/dt = 5$ and $dx/dt = 15$, find dz/dt. $z^2 = x^2 + h^2$ ⇒

$2z\dfrac{dz}{dt} = 2x\dfrac{dx}{dt} + 2h\dfrac{dh}{dt}$ ⇒ $\dfrac{dz}{dt} = \dfrac{1}{z}(15x + 5h)$. When $t = 3$,

$h = 45 + 3(5) = 60$ and $x = 15(3) = 45$ ⇒ $z = \sqrt{45^2 + 60^2} = 75$,

so $\dfrac{dz}{dt} = \dfrac{1}{75}[15(45) + 5(60)] = 13$ ft/s.

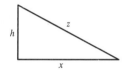

101. We are given $d\theta/dt = -0.25$ rad/h. $\tan\theta = 400/x$ ⇒

$x = 400\cot\theta$ ⇒ $\dfrac{dx}{dt} = -400\csc^2\theta\,\dfrac{d\theta}{dt}$. When $\theta = \frac{\pi}{6}$,

$\dfrac{dx}{dt} = -400(2)^2(-0.25) = 400$ ft/h.

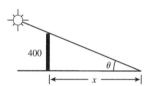

103. (a) $f(x) = \sqrt[3]{1 + 3x} = (1 + 3x)^{1/3}$ ⇒ $f'(x) = (1 + 3x)^{-2/3}$, so the linearization of f at $a = 0$ is

$L(x) = f(0) + f'(0)(x - 0) = 1^{1/3} + 1^{-2/3}x = 1 + x$. Thus, $\sqrt[3]{1 + 3x} \approx 1 + x$ ⇒

$\sqrt[3]{1.03} = \sqrt[3]{1 + 3(0.01)} \approx 1 + (0.01) = 1.01$.

(b) The linear approximation is $\sqrt[3]{1 + 3x} \approx 1 + x$, so for the required accuracy

we want $\sqrt[3]{1 + 3x} - 0.1 < 1 + x < \sqrt[3]{1 + 3x} + 0.1$. From the graph,

it appears that this is true when $-0.23 < x < 0.40$.

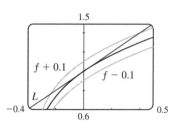

105. $A = x^2 + \frac{1}{2}\pi\left(\frac{1}{2}x\right)^2 = \left(1 + \frac{\pi}{8}\right)x^2 \;\Rightarrow\; dA = \left(2 + \frac{\pi}{4}\right)x\,dx$. When $x = 60$

and $dx = 0.1$, $dA = \left(2 + \frac{\pi}{4}\right)60(0.1) = 12 + \frac{3\pi}{2}$, so the maximum error is

approximately $12 + \frac{3\pi}{2} \approx 16.7$ cm^2.

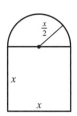

107. $\displaystyle\lim_{h \to 0} \frac{\sqrt[4]{16 + h} - 2}{h} = \left[\frac{d}{dx}\,\sqrt[4]{x}\right]_{x=16} = \frac{1}{4}x^{-3/4}\bigg|_{x=16} = \frac{1}{4\left(\sqrt[4]{16}\right)^3} = \frac{1}{32}$

109. $\displaystyle\lim_{x \to 0} \frac{\sqrt{1 + \tan x} - \sqrt{1 + \sin x}}{x^3} = \lim_{x \to 0} \frac{\left(\sqrt{1 + \tan x} - \sqrt{1 + \sin x}\right)\left(\sqrt{1 + \tan x} + \sqrt{1 + \sin x}\right)}{x^3\left(\sqrt{1 + \tan x} + \sqrt{1 + \sin x}\right)}$

$\displaystyle = \lim_{x \to 0} \frac{(1 + \tan x) - (1 + \sin x)}{x^3\left(\sqrt{1 + \tan x} + \sqrt{1 + \sin x}\right)} = \lim_{x \to 0} \frac{\sin x\,(1/\cos x - 1)}{x^3\left(\sqrt{1 + \tan x} + \sqrt{1 + \sin x}\right)} \cdot \frac{\cos x}{\cos x}$

$\displaystyle = \lim_{x \to 0} \frac{\sin x\,(1 - \cos x)}{x^3\left(\sqrt{1 + \tan x} + \sqrt{1 + \sin x}\right)\cos x} \cdot \frac{1 + \cos x}{1 + \cos x}$

$\displaystyle = \lim_{x \to 0} \frac{\sin x \cdot \sin^2 x}{x^3\left(\sqrt{1 + \tan x} + \sqrt{1 + \sin x}\right)\cos x\,(1 + \cos x)}$

$\displaystyle = \left(\lim_{x \to 0} \frac{\sin x}{x}\right)^3 \lim_{x \to 0} \frac{1}{\left(\sqrt{1 + \tan x} + \sqrt{1 + \sin x}\right)\cos x\,(1 + \cos x)}$

$\displaystyle = 1^3 \cdot \frac{1}{\left(\sqrt{1} + \sqrt{1}\right) \cdot 1 \cdot (1 + 1)} = \frac{1}{4}$

111. $\dfrac{d}{dx}\,[f(2x)] = x^2 \;\Rightarrow\; f'(2x) \cdot 2 = x^2 \;\Rightarrow\; f'(2x) = \frac{1}{2}x^2$. Let $t = 2x$. Then $f'(t) = \frac{1}{2}\left(\frac{1}{2}t\right)^2 = \frac{1}{8}t^2$, so $f'(x) = \frac{1}{8}x^2$.

1. Let a be the x-coordinate of Q. Since the derivative of $y = 1 - x^2$ is $y' = -2x$, the slope at Q is $-2a$. But since the triangle

is equilateral, $\overline{AO}/\overline{OC} = \sqrt{3}/1$, so the slope at Q is $-\sqrt{3}$. Therefore, we must have that $-2a = -\sqrt{3} \ \Rightarrow \ a = \frac{\sqrt{3}}{2}$.

Thus, the point Q has coordinates $\left(\frac{\sqrt{3}}{2}, 1 - \left(\frac{\sqrt{3}}{2}\right)^2\right) = \left(\frac{\sqrt{3}}{2}, \frac{1}{4}\right)$ and by symmetry, P has coordinates $\left(-\frac{\sqrt{3}}{2}, \frac{1}{4}\right)$.

3.

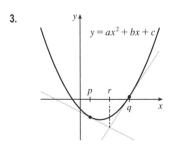

We must show that r (in the figure) is halfway between p and q, that is,

$r = (p + q)/2$. For the parabola $y = ax^2 + bx + c$, the slope of the tangent line is

given by $y' = 2ax + b$. An equation of the tangent line at $x = p$ is

$y - (ap^2 + bp + c) = (2ap + b)(x - p)$. Solving for y gives us

$$y = (2ap + b)x - 2ap^2 - bp + (ap^2 + bp + c)$$

or $\qquad y = (2ap + b)x + c - ap^2 \quad$ **(1)**

Similarly, an equation of the tangent line at $x = q$ is

$$y = (2aq + b)x + c - aq^2 \quad \textbf{(2)}$$

We can eliminate y and solve for x by subtracting equation **(1)** from equation **(2)**.

$$[(2aq + b) - (2ap + b)]x - aq^2 + ap^2 = 0$$

$$(2aq - 2ap)x = aq^2 - ap^2$$

$$2a(q - p)x = a(q^2 - p^2)$$

$$x = \frac{a(q + p)(q - p)}{2a(q - p)} = \frac{p + q}{2}$$

Thus, the x-coordinate of the point of intersection of the two tangent lines, namely r, is $(p + q)/2$.

5. Let $y = \tan^{-1} x$. Then $\tan y = x$, so from the triangle we see that

$\sin(\tan^{-1} x) = \sin y = \dfrac{x}{\sqrt{1 + x^2}}$. Using this fact we have that

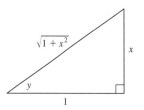

$$\sin(\tan^{-1}(\sinh x)) = \frac{\sinh x}{\sqrt{1 + \sinh^2 x}} = \frac{\sinh x}{\cosh x} = \tanh x.$$

Hence, $\sin^{-1}(\tanh x) = \sin^{-1}(\sin(\tan^{-1}(\sinh x))) = \tan^{-1}(\sinh x)$.

7. We use mathematical induction. Let S_n be the statement that $\dfrac{d^n}{dx^n}(\sin^4 x + \cos^4 x) = 4^{n-1}\cos(4x + n\pi/2)$.

S_1 is true because

$$\frac{d}{dx}(\sin^4 x + \cos^4 x) = 4\sin^3 x \, \cos x - 4\cos^3 x \, \sin x = 4\sin x \, \cos x \left(\sin^2 x - \cos^2 x\right) x$$

$$= -4\sin x \, \cos x \, \cos 2x = -2\sin 2x \, \cos 2 = -\sin 4x = \sin(-4x)$$

$$= \cos\left(\tfrac{\pi}{2} - (-4x)\right) = \cos\left(\tfrac{\pi}{2} + 4x\right) = 4^{n-1}\cos\left(4x + n\tfrac{\pi}{2}\right) \text{ when } n = 1$$

[continued]

Now assume S_k is true, that is, $\dfrac{d^k}{dx^k}\left(\sin^4 x + \cos^4 x\right) = 4^{k-1}\cos\left(4x + k\frac{\pi}{2}\right)$. Then

$$\frac{d^{k+1}}{dx^{k+1}}\left(\sin^4 x + \cos^4 x\right) = \frac{d}{dx}\left[\frac{d^k}{dx^k}(\sin^4 x + \cos^4 x)\right] = \frac{d}{dx}\left[4^{k-1}\cos\left(4x + k\frac{\pi}{2}\right)\right]$$

$$= -4^{k-1}\sin\left(4x + k\frac{\pi}{2}\right)\cdot\frac{d}{dx}\left(4x + k\frac{\pi}{2}\right) = -4^k\sin\left(4x + k\frac{\pi}{2}\right)$$

$$= 4^k\sin\left(-4x - k\frac{\pi}{2}\right) = 4^k\cos\left(\frac{\pi}{2} - \left(-4x - k\frac{\pi}{2}\right)\right) = 4^k\cos\left(4x + (k+1)\frac{\pi}{2}\right)$$

which shows that S_{k+1} is true.

Therefore, $\dfrac{d^n}{dx^n}\left(\sin^4 x + \cos^4 x\right) = 4^{n-1}\cos\left(4x + n\frac{\pi}{2}\right)$ for every positive integer n, by mathematical induction.

Another proof: First write

$$\sin^4 x + \cos^4 x = (\sin^2 x + \cos^2 x)^2 - 2\sin^2 x\cos^2 x = 1 - \tfrac{1}{2}\sin^2 2x = 1 - \tfrac{1}{4}(1 - \cos 4x) = \tfrac{3}{4} + \tfrac{1}{4}\cos 4x$$

Then we have $\dfrac{d^n}{dx^n}\left(\sin^4 x + \cos^4 x\right) = \dfrac{d^n}{dx^n}\left(\dfrac{3}{4} + \dfrac{1}{4}\cos 4x\right) = \dfrac{1}{4}\cdot 4^n\cos\left(4x + n\dfrac{\pi}{2}\right) = 4^{n-1}\cos\left(4x + n\dfrac{\pi}{2}\right)$.

9. We must find a value x_0 such that the normal lines to the parabola $y = x^2$ at $x = \pm x_0$ intersect at a point one unit from the

points $\left(\pm x_0, x_0^2\right)$. The normals to $y = x^2$ at $x = \pm x_0$ have slopes $-\dfrac{1}{\pm 2x_0}$ and pass through $\left(\pm x_0, x_0^2\right)$ respectively, so the

normals have the equations $y - x_0^2 = -\dfrac{1}{2x_0}(x - x_0)$ and $y - x_0^2 = \dfrac{1}{2x_0}(x + x_0)$. The common y-intercept is $x_0^2 + \dfrac{1}{2}$.

We want to find the value of x_0 for which the distance from $\left(0, x_0^2 + \frac{1}{2}\right)$ to $\left(x_0, x_0^2\right)$ equals 1. The square of the distance is

$(x_0 - 0)^2 + \left[x_0^2 - \left(x_0^2 + \frac{1}{2}\right)\right]^2 = x_0^2 + \frac{1}{4} = 1 \iff x_0 = \pm\frac{\sqrt{3}}{2}$. For these values of x_0, the y-intercept is $x_0^2 + \frac{1}{2} = \frac{5}{4}$, so

the center of the circle is at $\left(0, \frac{5}{4}\right)$.

Another solution: Let the center of the circle be $(0, a)$. Then the equation of the circle is $x^2 + (y - a)^2 = 1$.

Solving with the equation of the parabola, $y = x^2$, we get $x^2 + (x^2 - a)^2 = 1 \iff x^2 + x^4 - 2ax^2 + a^2 = 1 \iff$

$x^4 + (1 - 2a)x^2 + a^2 - 1 = 0$. The parabola and the circle will be tangent to each other when this quadratic equation in x^2

has equal roots; that is, when the discriminant is 0. Thus, $(1 - 2a)^2 - 4(a^2 - 1) = 0 \iff$

$1 - 4a + 4a^2 - 4a^2 + 4 = 0 \iff 4a = 5$, so $a = \frac{5}{4}$. The center of the circle is $\left(0, \frac{5}{4}\right)$.

11. We can assume without loss of generality that $\theta = 0$ at time $t = 0$, so that $\theta = 12\pi t$ rad. [The angular velocity of the wheel

is 360 rpm $= 360\cdot(2\pi\text{ rad})/(60\text{ s}) = 12\pi$ rad/s.] Then the position of A as a function of time is

$A = (40\cos\theta, 40\sin\theta) = (40\cos 12\pi t, 40\sin 12\pi t)$, so $\sin\alpha = \dfrac{y}{1.2\text{ m}} = \dfrac{40\sin\theta}{120} = \dfrac{\sin\theta}{3} = \dfrac{1}{3}\sin 12\pi t$.

(a) Differentiating the expression for $\sin\alpha$, we get $\cos\alpha\cdot\dfrac{d\alpha}{dt} = \dfrac{1}{3}\cdot 12\pi\cdot\cos 12\pi t = 4\pi\cos\theta$. When $\theta = \dfrac{\pi}{3}$, we have

$\sin\alpha = \dfrac{1}{3}\sin\theta = \dfrac{\sqrt{3}}{6}$, so $\cos\alpha = \sqrt{1 - \left(\dfrac{\sqrt{3}}{6}\right)^2} = \sqrt{\dfrac{11}{12}}$ and $\dfrac{d\alpha}{dt} = \dfrac{4\pi\cos\frac{\pi}{3}}{\cos\alpha} = \dfrac{2\pi}{\sqrt{11/12}} = \dfrac{4\pi\sqrt{3}}{\sqrt{11}} \approx 6.56$ rad/s.

(b) By the Law of Cosines, $|AP|^2 = |OA|^2 + |OP|^2 - 2\,|OA|\,|OP|\cos\theta$ $\Rightarrow$

$120^2 = 40^2 + |OP|^2 - 2\cdot 40\,|OP|\cos\theta$ $\Rightarrow$ $|OP|^2 - (80\cos\theta)\,|OP| - 12{,}800 = 0$ $\Rightarrow$

$|OP| = \frac{1}{2}\left(80\cos\theta \pm \sqrt{6400\cos^2\theta + 51{,}200}\,\right) = 40\cos\theta \pm 40\sqrt{\cos^2\theta + 8} = 40\left(\cos\theta + \sqrt{8 + \cos^2\theta}\,\right)$ cm

[since $|OP| > 0$].   As a check, note that $|OP| = 160$ cm when $\theta = 0$ and $|OP| = 80\sqrt{2}$ cm when $\theta = \frac{\pi}{2}$.

(c) By part (b), the x-coordinate of P is given by $x = 40\left(\cos\theta + \sqrt{8 + \cos^2\theta}\,\right)$, so

$$\frac{dx}{dt} = \frac{dx}{d\theta}\frac{d\theta}{dt} = 40\left(-\sin\theta - \frac{2\cos\theta\sin\theta}{2\sqrt{8 + \cos^2\theta}}\right)\cdot 12\pi = -480\pi\sin\theta\left(1 + \frac{\cos\theta}{\sqrt{8 + \cos^2\theta}}\right) \text{ cm/s.}$$

In particular, $dx/dt = 0$ cm/s when $\theta = 0$ and $dx/dt = -480\pi$ cm/s when $\theta = \frac{\pi}{2}$.

13. Consider the statement that $\dfrac{d^n}{dx^n}(e^{ax}\sin bx) = r^n e^{ax}\sin(bx + n\theta)$. For $n = 1$,

$\dfrac{d}{dx}(e^{ax}\sin bx) = ae^{ax}\sin bx + be^{ax}\cos bx$, and

$re^{ax}\sin(bx + \theta) = re^{ax}[\sin bx\cos\theta + \cos bx\sin\theta] = re^{ax}\left(\dfrac{a}{r}\sin bx + \dfrac{b}{r}\cos bx\right) = ae^{ax}\sin bx + be^{ax}\cos bx$

since $\tan\theta = \dfrac{b}{a}$ $\Rightarrow$ $\sin\theta = \dfrac{b}{r}$ and $\cos\theta = \dfrac{a}{r}$. So the statement is true for $n = 1$.

Assume it is true for $n = k$. Then

$$\frac{d^{k+1}}{dx^{k+1}}(e^{ax}\sin bx) = \frac{d}{dx}\left[r^k e^{ax}\sin(bx + k\theta)\right] = r^k ae^{ax}\sin(bx + k\theta) + r^k e^{ax}b\cos(bx + k\theta)$$

$$= r^k e^{ax}[a\sin(bx + k\theta) + b\cos(bx + k\theta)]$$

But

$\sin[bx + (k+1)\theta] = \sin[(bx + k\theta) + \theta] = \sin(bx + k\theta)\cos\theta + \sin\theta\cos(bx + k\theta) = \frac{a}{r}\sin(bx + k\theta) + \frac{b}{r}\cos(bx + k\theta)$.

Hence, $a\sin(bx + k\theta) + b\cos(bx + k\theta) = r\sin[bx + (k+1)\theta]$. So

$\dfrac{d^{k+1}}{dx^{k+1}}(e^{ax}\sin bx) = r^k e^{ax}[a\sin(bx + k\theta) + b\cos(bx + k\theta)] = r^k e^{ax}[r\sin(bx + (k+1)\theta)] = r^{k+1}e^{ax}[\sin(bx + (k+1)\theta)]$.

Therefore, the statement is true for all n by mathematical induction.

15. It seems from the figure that as P approaches the point $(0, 2)$ from the right, $x_T \to \infty$ and $y_T \to 2^+$. As P approaches the

point $(3, 0)$ from the left, it appears that $x_T \to 3^+$ and $y_T \to \infty$. So we guess that $x_T \in (3, \infty)$ and $y_T \in (2, \infty)$. It is

more difficult to estimate the range of values for x_N and y_N. We might perhaps guess that $x_N \in (0, 3)$,

and $y_N \in (-\infty, 0)$ or $(-2, 0)$.

In order to actually solve the problem, we implicitly differentiate the equation of the ellipse to find the equation of the

tangent line: $\dfrac{x^2}{9} + \dfrac{y^2}{4} = 1$ $\Rightarrow$ $\dfrac{2x}{9} + \dfrac{2y}{4}y' = 0$, so $y' = -\dfrac{4}{9}\dfrac{x}{y}$. So at the point (x_0, y_0) on the ellipse, an equation of the tangent line is $y - y_0 = -\dfrac{4}{9}\dfrac{x_0}{y_0}(x - x_0)$ or $4x_0x + 9y_0y = 4x_0^2 + 9y_0^2$. This can be written as $\dfrac{x_0x}{9} + \dfrac{y_0y}{4} = \dfrac{x_0^2}{9} + \dfrac{y_0^2}{4} = 1$,

because (x_0, y_0) lies on the ellipse. So an equation of the tangent line is $\dfrac{x_0x}{9} + \dfrac{y_0y}{4} = 1$.

Therefore, the x-intercept x_T for the tangent line is given by $\dfrac{x_0x_T}{9} = 1 \Leftrightarrow x_T = \dfrac{9}{x_0}$, and the y-intercept y_T is given

by $\dfrac{y_0y_T}{4} = 1 \Leftrightarrow y_T = \dfrac{4}{y_0}$.

So as x_0 takes on all values in $(0, 3)$, x_T takes on all values in $(3, \infty)$, and as y_0 takes on all values in $(0, 2)$, y_T takes on

all values in $(2, \infty)$. At the point (x_0, y_0) on the ellipse, the slope of the normal line is $-\dfrac{1}{y'(x_0, y_0)} = \dfrac{9}{4}\dfrac{y_0}{x_0}$, and its

equation is $y - y_0 = \dfrac{9}{4}\dfrac{y_0}{x_0}(x - x_0)$. So the x-intercept x_N for the normal line is given by $0 - y_0 = \dfrac{9}{4}\dfrac{y_0}{x_0}(x_N - x_0) \Rightarrow$

$x_N = -\dfrac{4x_0}{9} + x_0 = \dfrac{5x_0}{9}$, and the y-intercept y_N is given by $y_N - y_0 = \dfrac{9}{4}\dfrac{y_0}{x_0}(0 - x_0) \Rightarrow y_N = -\dfrac{9y_0}{4} + y_0 = -\dfrac{5y_0}{4}$.

So as x_0 takes on all values in $(0, 3)$, x_N takes on all values in $\left(0, \frac{5}{3}\right)$, and as y_0 takes on all values in $(0, 2)$, y_N takes on

all values in $\left(-\frac{5}{2}, 0\right)$.

17. (a) If the two lines L_1 and L_2 have slopes m_1 and m_2 and angles of

inclination ϕ_1 and ϕ_2, then $m_1 = \tan \phi_1$ and $m_2 = \tan \phi_2$. The triangle

in the figure shows that $\phi_1 + \alpha + (180° - \phi_2) = 180°$ and so

$\alpha = \phi_2 - \phi_1$. Therefore, using the identity for $\tan(x - y)$, we have

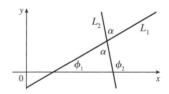

$\tan \alpha = \tan(\phi_2 - \phi_1) = \dfrac{\tan \phi_2 - \tan \phi_1}{1 + \tan \phi_2 \tan \phi_1}$ and so $\tan \alpha = \dfrac{m_2 - m_1}{1 + m_1m_2}$.

(b) (i) The parabolas intersect when $x^2 = (x - 2)^2 \Rightarrow x = 1$. If $y = x^2$, then $y' = 2x$, so the slope of the tangent

to $y = x^2$ at $(1, 1)$ is $m_1 = 2(1) = 2$. If $y = (x - 2)^2$, then $y' = 2(x - 2)$, so the slope of the tangent to

$y = (x - 2)^2$ at $(1, 1)$ is $m_2 = 2(1 - 2) = -2$. Therefore, $\tan \alpha = \dfrac{m_2 - m_1}{1 + m_1m_2} = \dfrac{-2 - 2}{1 + 2(-2)} = \dfrac{4}{3}$ and

so $\alpha = \tan^{-1}\left(\frac{4}{3}\right) \approx 53°$ [or $127°$].

(ii) $x^2 - y^2 = 3$ and $x^2 - 4x + y^2 + 3 = 0$ intersect when $x^2 - 4x + (x^2 - 3) + 3 = 0 \Leftrightarrow 2x(x - 2) = 0 \Rightarrow$

$x = 0$ or 2, but 0 is extraneous. If $x = 2$, then $y = \pm 1$. If $x^2 - y^2 = 3$ then $2x - 2yy' = 0 \Rightarrow y' = x/y$ and

$x^2 - 4x + y^2 + 3 = 0 \Rightarrow 2x - 4 + 2yy' = 0 \Rightarrow y' = \dfrac{2 - x}{y}$. At $(2, 1)$ the slopes are $m_1 = 2$ and

$m_2 = 0$, so $\tan \alpha = \dfrac{0 - 2}{1 + 2 \cdot 0} = -2 \Rightarrow \alpha \approx 117°$. At $(2, -1)$ the slopes are $m_1 = -2$ and $m_2 = 0$,

so $\tan \alpha = \dfrac{0 - (-2)}{1 + (-2)(0)} = 2 \Rightarrow \alpha \approx 63°$ [or $117°$].

19. Since $\angle ROQ = \angle OQP = \theta$, the triangle QOR is isosceles, so

$|QR| = |RO| = x$. By the Law of Cosines, $x^2 = x^2 + r^2 - 2rx\cos\theta$. Hence,

$2rx\cos\theta = r^2$, so $x = \dfrac{r^2}{2r\cos\theta} = \dfrac{r}{2\cos\theta}$. Note that as $y \to 0^+$, $\theta \to 0^+$ (since

$\sin\theta = y/r$), and hence $x \to \dfrac{r}{2\cos 0} = \dfrac{r}{2}$. Thus, as P is taken closer and closer

to the x-axis, the point R approaches the midpoint of the radius AO.

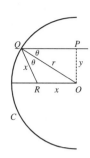

21. $\displaystyle\lim_{x\to 0} \frac{\sin(a+2x) - 2\sin(a+x) + \sin a}{x^2}$

$\displaystyle = \lim_{x\to 0} \frac{\sin a\,\cos 2x + \cos a\,\sin 2x - 2\sin a\,\cos x - 2\cos a\,\sin x + \sin a}{x^2}$

$\displaystyle = \lim_{x\to 0} \frac{\sin a\,(\cos 2x - 2\cos x + 1) + \cos a\,(\sin 2x - 2\sin x)}{x^2}$

$\displaystyle = \lim_{x\to 0} \frac{\sin a\,(2\cos^2 x - 1 - 2\cos x + 1) + \cos a\,(2\sin x\,\cos x - 2\sin x)}{x^2}$

$\displaystyle = \lim_{x\to 0} \frac{\sin a\,(2\cos x)(\cos x - 1) + \cos a\,(2\sin x)(\cos x - 1)}{x^2}$

$\displaystyle = \lim_{x\to 0} \frac{2(\cos x - 1)[\sin a\,\cos x + \cos a\,\sin x](\cos x + 1)}{x^2(\cos x + 1)}$

$\displaystyle = \lim_{x\to 0} \frac{-2\sin^2 x\,[\sin(a+x)]}{x^2(\cos x + 1)} = -2\lim_{x\to 0}\left(\frac{\sin x}{x}\right)^2 \cdot \frac{\sin(a+x)}{\cos x + 1} = -2(1)^2\frac{\sin(a+0)}{\cos 0 + 1} = -\sin a$

23.
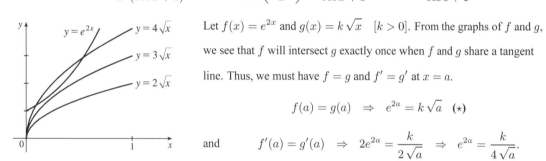
Let $f(x) = e^{2x}$ and $g(x) = k\sqrt{x}$ $[k > 0]$. From the graphs of f and g, we see that f will intersect g exactly once when f and g share a tangent line. Thus, we must have $f = g$ and $f' = g'$ at $x = a$.

$$f(a) = g(a) \;\Rightarrow\; e^{2a} = k\sqrt{a} \quad (\star)$$

and

$$f'(a) = g'(a) \;\Rightarrow\; 2e^{2a} = \frac{k}{2\sqrt{a}} \;\Rightarrow\; e^{2a} = \frac{k}{4\sqrt{a}}.$$

So we must have $k\sqrt{a} = \dfrac{k}{4\sqrt{a}} \;\Rightarrow\; \left(\sqrt{a}\right)^2 = \dfrac{k}{4k} \;\Rightarrow\; a = \tfrac{1}{4}$. From $(\star)$, $e^{2(1/4)} = k\sqrt{1/4} \;\Rightarrow$

$k = 2e^{1/2} = 2\sqrt{e} \approx 3.297$.

25. $y = \dfrac{x}{\sqrt{a^2 - 1}} - \dfrac{2}{\sqrt{a^2 - 1}}\arctan\dfrac{\sin x}{a + \sqrt{a^2 - 1} + \cos x}$. Let $k = a + \sqrt{a^2 - 1}$. Then

$$y' = \frac{1}{\sqrt{a^2-1}} - \frac{2}{\sqrt{a^2-1}} \cdot \frac{1}{1 + \sin^2 x/(k+\cos x)^2} \cdot \frac{\cos x(k + \cos x) + \sin^2 x}{(k+\cos x)^2}$$

$$= \frac{1}{\sqrt{a^2-1}} - \frac{2}{\sqrt{a^2-1}} \cdot \frac{k\cos x + \cos^2 x + \sin^2 x}{(k+\cos x)^2 + \sin^2 x} = \frac{1}{\sqrt{a^2-1}} - \frac{2}{\sqrt{a^2-1}} \cdot \frac{k\cos x + 1}{k^2 + 2k\cos x + 1}$$

$$= \frac{k^2 + 2k\cos x + 1 - 2k\cos x - 2}{\sqrt{a^2-1}\,(k^2 + 2k\cos x + 1)} = \frac{k^2 - 1}{\sqrt{a^2-1}\,(k^2 + 2k\cos x + 1)}$$

But $k^2 = 2a^2 + 2a\sqrt{a^2 - 1} - 1 = 2a\big(a + \sqrt{a^2 - 1}\big) - 1 = 2ak - 1$, so $k^2 + 1 = 2ak$, and $k^2 - 1 = 2(ak - 1)$.

[continued]

So $y' = \dfrac{2(ak-1)}{\sqrt{a^2-1}\,(2ak+2k\cos x)} = \dfrac{ak-1}{\sqrt{a^2-1}k\,(a+\cos x)}$. But $ak - 1 = a^2 + a\sqrt{a^2-1} - 1 = k\sqrt{a^2-1}$,

so $y' = 1/(a+\cos x)$.

27. $y = x^4 - 2x^2 - x \;\Rightarrow\; y' = 4x^3 - 4x - 1$. The equation of the tangent line at $x = a$ is

$y - (a^4 - 2a^2 - a) = (4a^3 - 4a - 1)(x - a)$ or $y = (4a^3 - 4a - 1)x + (-3a^4 + 2a^2)$ and similarly for $x = b$. So if at

$x = a$ and $x = b$ we have the same tangent line, then $4a^3 - 4a - 1 = 4b^3 - 4b - 1$ and $-3a^4 + 2a^2 = -3b^4 + 2b^2$. The first

equation gives $a^3 - b^3 = a - b \;\Rightarrow\; (a-b)(a^2 + ab + b^2) = (a-b)$. Assuming $a \neq b$, we have $1 = a^2 + ab + b^2$.

The second equation gives $3(a^4 - b^4) = 2(a^2 - b^2) \;\Rightarrow\; 3(a^2 - b^2)(a^2 + b^2) = 2(a^2 - b^2)$ which is true if $a = -b$.

Substituting into $1 = a^2 + ab + b^2$ gives $1 = a^2 - a^2 + a^2 \;\Rightarrow\; a = \pm 1$ so that $a = 1$ and $b = -1$ or vice versa. Thus,

the points $(1, -2)$ and $(-1, 0)$ have a common tangent line.

As long as there are only two such points, we are done. So we show that these are in fact the only two such points.

Suppose that $a^2 - b^2 \neq 0$. Then $3(a^2 - b^2)(a^2 + b^2) = 2(a^2 - b^2)$ gives $3(a^2 + b^2) = 2$ or $a^2 + b^2 = \frac{2}{3}$.

Thus, $ab = (a^2 + ab + b^2) - (a^2 + b^2) = 1 - \dfrac{2}{3} = \dfrac{1}{3}$, so $b = \dfrac{1}{3a}$. Hence, $a^2 + \dfrac{1}{9a^2} = \dfrac{2}{3}$, so $9a^4 + 1 = 6a^2 \;\Rightarrow\;$

$0 = 9a^4 - 6a^2 + 1 = (3a^2 - 1)^2$. So $3a^2 - 1 = 0 \;\Rightarrow\; a^2 = \dfrac{1}{3} \;\Rightarrow\; b^2 = \dfrac{1}{9a^2} = \dfrac{1}{3} = a^2$, contradicting our assumption

that $a^2 \neq b^2$.

29. Because of the periodic nature of the lattice points, it suffices to consider the points in the 5×2 grid shown. We can see that

the minimum value of r occurs when there is a line with slope $\frac{2}{5}$ which touches the circle centered at $(3, 1)$ and the circles

centered at $(0, 0)$ and $(5, 2)$.

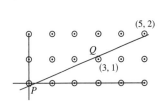

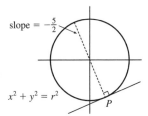

To find P, the point at which the line is tangent to the circle at $(0, 0)$, we simultaneously solve $x^2 + y^2 = r^2$ and

$y = -\frac{5}{2}x \;\Rightarrow\; x^2 + \frac{25}{4}x^2 = r^2 \;\Rightarrow\; x^2 = \frac{4}{29}r^2 \;\Rightarrow\; x = \frac{2}{\sqrt{29}}r, \; y = -\frac{5}{\sqrt{29}}r$. To find Q, we either use symmetry or

solve $(x-3)^2 + (y-1)^2 = r^2$ and $y - 1 = -\frac{5}{2}(x-3)$. As above, we get $x = 3 - \frac{2}{\sqrt{29}}r, \; y = 1 + \frac{5}{\sqrt{29}}r$. Now the slope of

the line PQ is $\frac{2}{5}$, so $m_{PQ} = \dfrac{1 + \frac{5}{\sqrt{29}}r - \left(-\frac{5}{\sqrt{29}}r\right)}{3 - \frac{2}{\sqrt{29}}r - \frac{2}{\sqrt{29}}r} = \dfrac{1 + \frac{10}{\sqrt{29}}r}{3 - \frac{4}{\sqrt{29}}r} = \dfrac{\sqrt{29} + 10r}{3\sqrt{29} - 4r} = \dfrac{2}{5} \;\Rightarrow\;$

$5\sqrt{29} + 50r = 6\sqrt{29} - 8r \;\Leftrightarrow\; 58r = \sqrt{29} \;\Leftrightarrow\; r = \frac{\sqrt{29}}{58}$. So the minimum value of r for which any line with slope $\frac{2}{5}$

intersects circles with radius r centered at the lattice points on the plane is $r = \frac{\sqrt{29}}{58} \approx 0.093$.

31.

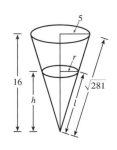

By similar triangles, $\dfrac{r}{5} = \dfrac{h}{16} \;\Rightarrow\; r = \dfrac{5h}{16}$. The volume of the cone is

$$V = \tfrac{1}{3}\pi r^2 h = \tfrac{1}{3}\pi \left(\dfrac{5h}{16}\right)^2 h = \dfrac{25\pi}{768}h^3,\ \text{so}\ \dfrac{dV}{dt} = \dfrac{25\pi}{256}h^2\dfrac{dh}{dt}.$$ Now the rate of

change of the volume is also equal to the difference of what is being added

$(2\ \text{cm}^3/\text{min})$ and what is oozing out ($k\pi rl$, where πrl is the area of the cone and k

is a proportionality constant). Thus, $\dfrac{dV}{dt} = 2 - k\pi rl.$

Equating the two expressions for $\dfrac{dV}{dt}$ and substituting $h = 10$, $\dfrac{dh}{dt} = -0.3$, $r = \dfrac{5(10)}{16} = \dfrac{25}{8}$, and $\dfrac{l}{\sqrt{281}} = \dfrac{10}{16}\;\Leftrightarrow$

$l = \dfrac{5}{8}\sqrt{281}$, we get $\dfrac{25\pi}{256}(10)^2(-0.3) = 2 - k\pi\dfrac{25}{8}\cdot\dfrac{5}{8}\sqrt{281}\;\Leftrightarrow\;\dfrac{125k\pi\sqrt{281}}{64} = 2 + \dfrac{750\pi}{256}.$ Solving for k gives us

$k = \dfrac{256 + 375\pi}{250\pi\sqrt{281}}$. To maintain a certain height, the rate of oozing, $k\pi rl$, must equal the rate of the liquid being poured in;

that is, $\dfrac{dV}{dt} = 0$. Thus, the rate at which we should pour the liquid into the container is

$$k\pi rl = \dfrac{256 + 375\pi}{250\pi\sqrt{281}}\cdot\pi\cdot\dfrac{25}{8}\cdot\dfrac{5\sqrt{281}}{8} = \dfrac{256 + 375\pi}{128} \approx 11.204\ \text{cm}^3/\text{min}$$

4 □ APPLICATIONS OF DIFFERENTIATION

4.1 Maximum and Minimum Values

1. A function f has an **absolute minimum** at $x = c$ if $f(c)$ is the smallest function value on the entire domain of f, whereas f has a **local minimum** at c if $f(c)$ is the smallest function value when x is near c.

3. Absolute maximum at s, absolute minimum at r, local maximum at c, local minima at b and r, neither a maximum nor a minimum at a and d.

5. Absolute maximum value is $f(4) = 5$; there is no absolute minimum value; local maximum values are $f(4) = 5$ and $f(6) = 4$; local minimum values are $f(2) = 2$ and $f(1) = f(5) = 3$.

7. Absolute minimum at 2, absolute maximum at 3, local minimum at 4

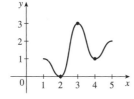

9. Absolute maximum at 5, absolute minimum at 2, local maximum at 3, local minima at 2 and 4

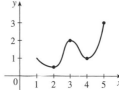

11. (a)

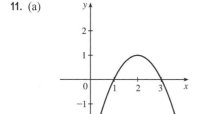

(b)

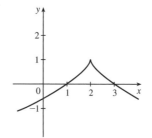

(c)

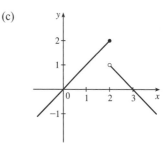

13. (a) *Note:* By the Extreme Value Theorem, f must *not* be continuous; because if it were, it would attain an absolute minimum.

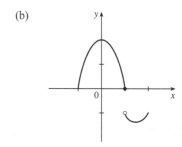

(b)

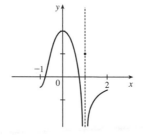

15. $f(x) = 8 - 3x$, $x \geq 1$. Absolute maximum $f(1) = 5$; no local maximum. No absolute or local minimum.

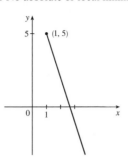

17. $f(x) = x^2$, $0 < x < 2$. No absolute or local maximum or minimum value.

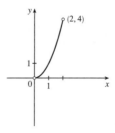

19. $f(x) = x^2$, $0 \leq x < 2$. Absolute minimum $f(0) = 0$; no local minimum. No absolute or local maximum.

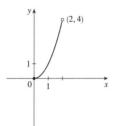

21. $f(x) = x^2$, $-3 \leq x \leq 2$. Absolute maximum $f(-3) = 9$. No local maximum. Absolute and local minimum $f(0) = 0$.

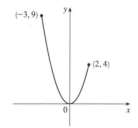

23. $f(x) = \ln x$, $0 < x \leq 2$. Absolute maximum $f(2) = \ln 2 \approx 0.69$; no local maximum. No absolute or local minimum.

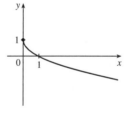

25. $f(x) = 1 - \sqrt{x}$. Absolute maximum $f(0) = 1$; no local maximum. No absolute or local minimum.

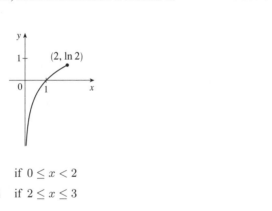

27. $f(x) = \begin{cases} 1 - x & \text{if } 0 \leq x < 2 \\ 2x - 4 & \text{if } 2 \leq x \leq 3 \end{cases}$

Absolute maximum $f(3) = 2$; no local maximum. No absolute or local minimum.

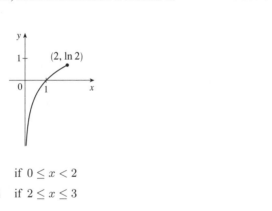

29. $f(x) = 5x^2 + 4x \ \Rightarrow \ f'(x) = 10x + 4. \ \ f'(x) = 0 \ \Rightarrow \ x = -\frac{2}{5}$, so $-\frac{2}{5}$ is the only critical number.

31. $f(x) = x^3 + 3x^2 - 24x \ \Rightarrow \ f'(x) = 3x^2 + 6x - 24 = 3(x^2 + 2x - 8)$.

$f'(x) = 0 \ \Rightarrow \ 3(x+4)(x-2) = 0 \ \Rightarrow \ x = -4, 2$. These are the only critical numbers.

33. $s(t) = 3t^4 + 4t^3 - 6t^2 \ \Rightarrow \ s'(t) = 12t^3 + 12t^2 - 12t. \ \ s'(t) = 0 \ \Rightarrow \ 12t(t^2 + t - 1) \ \Rightarrow$

$t = 0$ or $t^2 + t - 1 = 0$. Using the quadratic formula to solve the latter equation gives us

$t = \dfrac{-1 \pm \sqrt{1^2 - 4(1)(-1)}}{2(1)} = \dfrac{-1 \pm \sqrt{5}}{2} \approx 0.618, -1.618$. The three critical numbers are $0, \dfrac{-1 \pm \sqrt{5}}{2}$.

35. $g(y) = \dfrac{y - 1}{y^2 - y + 1} \ \Rightarrow$

$g'(y) = \dfrac{(y^2 - y + 1)(1) - (y - 1)(2y - 1)}{(y^2 - y + 1)^2} = \dfrac{y^2 - y + 1 - (2y^2 - 3y + 1)}{(y^2 - y + 1)^2} = \dfrac{-y^2 + 2y}{(y^2 - y + 1)^2} = \dfrac{y(2 - y)}{(y^2 - y + 1)^2}$.

$g'(y) = 0 \ \Rightarrow \ y = 0, 2$. The expression $y^2 - y + 1$ is never equal to 0, so $g'(y)$ exists for all real numbers.

The critical numbers are 0 and 2.

37. $h(t) = t^{3/4} - 2t^{1/4} \ \Rightarrow \ h'(t) = \frac{3}{4}t^{-1/4} - \frac{2}{4}t^{-3/4} = \frac{1}{4}t^{-3/4}(3t^{1/2} - 2) = \dfrac{3\sqrt{t} - 2}{4\sqrt[4]{t^3}}$.

$h'(t) = 0 \ \Rightarrow \ 3\sqrt{t} = 2 \ \Rightarrow \ \sqrt{t} = \frac{2}{3} \ \Rightarrow \ t = \frac{4}{9}$. $h'(t)$ does not exist at $t = 0$, so the critical numbers are 0 and $\frac{4}{9}$.

39. $F(x) = x^{4/5}(x - 4)^2 \ \Rightarrow$

$F'(x) = x^{4/5} \cdot 2(x - 4) + (x - 4)^2 \cdot \frac{4}{5}x^{-1/5} = \frac{1}{5}x^{-1/5}(x - 4)[5 \cdot x \cdot 2 + (x - 4) \cdot 4]$

$= \dfrac{(x - 4)(14x - 16)}{5x^{1/5}} = \dfrac{2(x - 4)(7x - 8)}{5x^{1/5}}$

$F'(x) = 0 \ \Rightarrow \ x = 4, \frac{8}{7}$. $F'(0)$ does not exist. Thus, the three critical numbers are $0, \frac{8}{7}$, and 4.

41. $f(\theta) = 2\cos\theta + \sin^2\theta \ \Rightarrow \ f'(\theta) = -2\sin\theta + 2\sin\theta\cos\theta. \ \ f'(\theta) = 0 \ \Rightarrow \ 2\sin\theta(\cos\theta - 1) = 0 \ \Rightarrow \ \sin\theta = 0$

or $\cos\theta = 1 \ \Rightarrow \ \theta = n\pi$ [n an integer] or $\theta = 2n\pi$. The solutions $\theta = n\pi$ include the solutions $\theta = 2n\pi$, so the critical

numbers are $\theta = n\pi$.

43. $f(x) = x^2e^{-3x} \ \Rightarrow \ f'(x) = x^2(-3e^{-3x}) + e^{-3x}(2x) = xe^{-3x}(-3x + 2). \ f'(x) = 0 \ \Rightarrow \ x = 0, \frac{2}{3}$

[e^{-3x} is never equal to 0]. $f'(x)$ always exists, so the critical numbers are 0 and $\frac{2}{3}$.

45. The graph of $f'(x) = 5e^{-0.1|x|}\sin x - 1$ has 10 zeros and exists

everywhere, so f has 10 critical numbers.

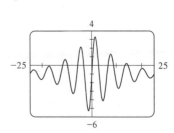

47. $f(x) = 3x^2 - 12x + 5$, $[0, 3]$. $f'(x) = 6x - 12 = 0$ $\Leftrightarrow$ $x = 2$. Applying the Closed Interval Method, we find that

$f(0) = 5$, $f(2) = -7$, and $f(3) = -4$. So $f(0) = 5$ is the absolute maximum value and $f(2) = -7$ is the absolute minimum

value.

49. $f(x) = 2x^3 - 3x^2 - 12x + 1$, $[-2, 3]$. $f'(x) = 6x^2 - 6x - 12 = 6(x^2 - x - 2) = 6(x - 2)(x + 1) = 0$ $\Leftrightarrow$

$x = 2, -1$. $f(-2) = -3$, $f(-1) = 8$, $f(2) = -19$, and $f(3) = -8$. So $f(-1) = 8$ is the absolute maximum value and

$f(2) = -19$ is the absolute minimum value.

51. $f(x) = x^4 - 2x^2 + 3$, $[-2, 3]$. $f'(x) = 4x^3 - 4x = 4x(x^2 - 1) = 4x(x + 1)(x - 1) = 0$ $\Leftrightarrow$ $x = -1, 0, 1$.

$f(-2) = 11$, $f(-1) = 2$, $f(0) = 3$, $f(1) = 2$, $f(3) = 66$. So $f(3) = 66$ is the absolute maximum value and $f(\pm 1) = 2$ is

the absolute minimum value.

53. $f(x) = \dfrac{x}{x^2 + 1}$, $[0, 2]$. $f'(x) = \dfrac{(x^2 + 1) - x(2x)}{(x^2 + 1)^2} = \dfrac{1 - x^2}{(x^2 + 1)^2} = 0$ $\Leftrightarrow$ $x = \pm 1$, but -1 is not in $[0, 2]$. $f(0) = 0$,

$f(1) = \frac{1}{2}$, $f(2) = \frac{2}{5}$. So $f(1) = \frac{1}{2}$ is the absolute maximum value and $f(0) = 0$ is the absolute minimum value.

55. $f(t) = t\sqrt{4 - t^2}$, $[-1, 2]$.

$f'(t) = t \cdot \frac{1}{2}(4 - t^2)^{-1/2}(-2t) + (4 - t^2)^{1/2} \cdot 1 = \dfrac{-t^2}{\sqrt{4 - t^2}} + \sqrt{4 - t^2} = \dfrac{-t^2 + (4 - t^2)}{\sqrt{4 - t^2}} = \dfrac{4 - 2t^2}{\sqrt{4 - t^2}}$.

$f'(t) = 0$ $\Rightarrow$ $4 - 2t^2 = 0$ $\Rightarrow$ $t^2 = 2$ $\Rightarrow$ $t = \pm\sqrt{2}$, but $t = -\sqrt{2}$ is not in the given interval, $[-1, 2]$.

$f'(t)$ does not exist if $4 - t^2 = 0$ $\Rightarrow$ $t = \pm 2$, but -2 is not in the given interval. $f(-1) = -\sqrt{3}$, $f(\sqrt{2}) = 2$, and

$f(2) = 0$. So $f(\sqrt{2}) = 2$ is the absolute maximum value and $f(-1) = -\sqrt{3}$ is the absolute minimum value.

57. $f(t) = 2\cos t + \sin 2t$, $[0, \pi/2]$.

$f'(t) = -2\sin t + \cos 2t \cdot 2 = -2\sin t + 2(1 - 2\sin^2 t) = -2(2\sin^2 t + \sin t - 1) = -2(2\sin t - 1)(\sin t + 1)$.

$f'(t) = 0$ $\Rightarrow$ $\sin t = \frac{1}{2}$ or $\sin t = -1$ $\Rightarrow$ $t = \frac{\pi}{6}$. $f(0) = 2$, $f(\frac{\pi}{6}) = \sqrt{3} + \frac{1}{2}\sqrt{3} = \frac{3}{2}\sqrt{3} \approx 2.60$, and $f(\frac{\pi}{2}) = 0$.

So $f(\frac{\pi}{6}) = \frac{3}{2}\sqrt{3}$ is the absolute maximum value and $f(\frac{\pi}{2}) = 0$ is the absolute minimum value.

59. $f(x) = xe^{-x^2/8}$, $[-1, 4]$. $f'(x) = x \cdot e^{-x^2/8} \cdot (-\frac{x}{4}) + e^{-x^2/8} \cdot 1 = e^{-x^2/8}(-\frac{x^2}{4} + 1)$. Since $e^{-x^2/8}$ is never 0,

$f'(x) = 0$ $\Rightarrow$ $-x^2/4 + 1 = 0$ $\Rightarrow$ $1 = x^2/4$ $\Rightarrow$ $x^2 = 4$ $\Rightarrow$ $x = \pm 2$, but -2 is not in the given interval, $[-1, 4]$.

$f(-1) = -e^{-1/8} \approx -0.88$, $f(2) = 2e^{-1/2} \approx 1.21$, and $f(4) = 4e^{-2} \approx 0.54$. So $f(2) = 2e^{-1/2}$ is the absolute maximum

value and $f(-1) = -e^{-1/8}$ is the absolute minimum value.

61. $f(x) = \ln(x^2 + x + 1)$, $[-1, 1]$. $f'(x) = \dfrac{1}{x^2 + x + 1} \cdot (2x + 1) = 0$ $\Leftrightarrow$ $x = -\frac{1}{2}$. Since $x^2 + x + 1 > 0$ for all x, the

domain of f and f' is $\mathbb{R}$. $f(-1) = \ln 1 = 0$, $f(-\frac{1}{2}) = \ln \frac{3}{4} \approx -0.29$, and $f(1) = \ln 3 \approx 1.10$. So $f(1) = \ln 3 \approx 1.10$ is

the absolute maximum value and $f(-\frac{1}{2}) = \ln \frac{3}{4} \approx -0.29$ is the absolute minimum value.

63. $f(x) = x^a(1-x)^b,\ \ 0 \le x \le 1, a > 0, b > 0.$

$f'(x) = x^a \cdot b(1-x)^{b-1}(-1) + (1-x)^b \cdot ax^{a-1} = x^{a-1}(1-x)^{b-1}[x \cdot b(-1) + (1-x) \cdot a]$

$\quad = x^{a-1}(1-x)^{b-1}(a - ax - bx)$

At the endpoints, we have $f(0) = f(1) = 0$ [the minimum value of f]. In the interval $(0,1)$, $f'(x) = 0\ \Leftrightarrow\ x = \dfrac{a}{a+b}.$

$f\left(\dfrac{a}{a+b}\right) = \left(\dfrac{a}{a+b}\right)^a\left(1 - \dfrac{a}{a+b}\right)^b = \dfrac{a^a}{(a+b)^a}\left(\dfrac{a+b-a}{a+b}\right)^b = \dfrac{a^a}{(a+b)^a} \cdot \dfrac{b^b}{(a+b)^b} = \dfrac{a^a b^b}{(a+b)^{a+b}}.$

So $f\left(\dfrac{a}{a+b}\right) = \dfrac{a^a b^b}{(a+b)^{a+b}}$ is the absolute maximum value.

65. (a)

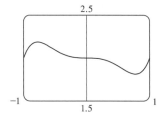

From the graph, it appears that the absolute maximum value is about

$f(-0.77) = 2.19$, and the absolute minimum value is about

$f(0.77) = 1.81.$

(b) $f(x) = x^5 - x^3 + 2\ \Rightarrow\ f'(x) = 5x^4 - 3x^2 = x^2(5x^2 - 3).$ So $f'(x) = 0\ \Rightarrow\ x = 0, \pm\sqrt{\tfrac{3}{5}}.$

$f\left(-\sqrt{\tfrac{3}{5}}\right) = \left(-\sqrt{\tfrac{3}{5}}\right)^5 - \left(-\sqrt{\tfrac{3}{5}}\right)^3 + 2 = -\left(\tfrac{3}{5}\right)^2\sqrt{\tfrac{3}{5}} + \tfrac{3}{5}\sqrt{\tfrac{3}{5}} + 2 = \left(\tfrac{3}{5} - \tfrac{9}{25}\right)\sqrt{\tfrac{3}{5}} + 2 = \tfrac{6}{25}\sqrt{\tfrac{3}{5}} + 2$ (maximum)

and similarly, $f\left(\sqrt{\tfrac{3}{5}}\right) = -\tfrac{6}{25}\sqrt{\tfrac{3}{5}} + 2$ (minimum).

67. (a)

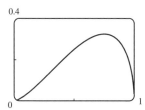

From the graph, it appears that the absolute maximum value is about

$f(0.75) = 0.32$, and the absolute minimum value is $f(0) = f(1) = 0;$

that is, at both endpoints.

(b) $f(x) = x\sqrt{x - x^2}\ \Rightarrow\ f'(x) = x \cdot \dfrac{1 - 2x}{2\sqrt{x - x^2}} + \sqrt{x - x^2} = \dfrac{(x - 2x^2) + (2x - 2x^2)}{2\sqrt{x - x^2}} = \dfrac{3x - 4x^2}{2\sqrt{x - x^2}}.$

So $f'(x) = 0\ \Rightarrow\ 3x - 4x^2 = 0\ \Rightarrow\ x(3 - 4x) = 0\ \Rightarrow\ x = 0$ or $\tfrac{3}{4}.$

$f(0) = f(1) = 0$ (minimum), and $f\left(\tfrac{3}{4}\right) = \tfrac{3}{4}\sqrt{\tfrac{3}{4} - \left(\tfrac{3}{4}\right)^2} = \tfrac{3}{4}\sqrt{\tfrac{3}{16}} = \tfrac{3\sqrt{3}}{16}$ (maximum).

69. The density is defined as $\rho = \dfrac{\text{mass}}{\text{volume}} = \dfrac{1000}{V(T)}$ (in g/cm^3). But a critical point of ρ will also be a critical point of V

$\left[\text{since } \dfrac{d\rho}{dT} = -1000V^{-2}\dfrac{dV}{dT} \text{ and } V \text{ is never } 0\right]$, and V is easier to differentiate than ρ.

$V(T) = 999.87 - 0.06426T + 0.0085043T^2 - 0.0000679T^3\ \Rightarrow\ V'(T) = -0.06426 + 0.0170086T - 0.0002037T^2.$

Setting this equal to 0 and using the quadratic formula to find T, we get

$T = \dfrac{-0.0170086 \pm \sqrt{0.0170086^2 - 4 \cdot 0.0002037 \cdot 0.06426}}{2(-0.0002037)} \approx 3.9665°\text{C}$ or $79.5318°\text{C}.$ Since we are only interested in the region $0°C \leq T \leq 30°C$, we check the density ρ at the endpoints and at $3.9665°C$: $\rho(0) \approx \dfrac{1000}{999.87} \approx 1.00013$;

$\rho(30) \approx \dfrac{1000}{1003.7628} \approx 0.99625$; $\rho(3.9665) \approx \dfrac{1000}{999.7447} \approx 1.000255$. So water has its maximum density at

about $3.9665°C$.

71. Let $a = -0.000\,032\,37$, $b = 0.000\,903\,7$, $c = -0.008\,956$, $d = 0.03629$, $e = -0.04458$, and $f = 0.4074$.

Then $S(t) = at^5 + bt^4 + ct^3 + dt^2 + et + f$ and $S'(t) = 5at^4 + 4bt^3 + 3ct^2 + 2dt + e$.

We now apply the Closed Interval Method to the continuous function S on the interval $0 \leq t \leq 10$. Since S' exists for all t,

the only critical numbers of S occur when $S'(t) = 0$. We use a rootfinder on a CAS (or a graphing device) to find that

$S'(t) = 0$ when $t_1 \approx 0.855$, $t_2 \approx 4.618$, $t_3 \approx 7.292$, and $t_4 \approx 9.570$. The values of S at these critical numbers are

$S(t_1) \approx 0.39$, $S(t_2) \approx 0.43645$, $S(t_3) \approx 0.427$, and $S(t_4) \approx 0.43641$. The values of S at the endpoints of the interval are

$S(0) \approx 0.41$ and $S(10) \approx 0.435$. Comparing the six numbers, we see that sugar was most expensive at $t_2 \approx 4.618$

(corresponding roughly to March 1998) and cheapest at $t_1 \approx 0.855$ (June 1994).

73. (a) $v(r) = k(r_0 - r)r^2 = kr_0r^2 - kr^3 \Rightarrow v'(r) = 2kr_0r - 3kr^2$. $v'(r) = 0 \Rightarrow kr(2r_0 - 3r) = 0 \Rightarrow$

$r = 0$ or $\frac{2}{3}r_0$ (but 0 is not in the interval). Evaluating v at $\frac{1}{2}r_0$, $\frac{2}{3}r_0$, and r_0, we get $v\left(\frac{1}{2}r_0\right) = \frac{1}{8}kr_0^3$, $v\left(\frac{2}{3}r_0\right) = \frac{4}{27}kr_0^3$,

and $v(r_0) = 0$. Since $\frac{4}{27} > \frac{1}{8}$, v attains its maximum value at $r = \frac{2}{3}r_0$. This supports the statement in the text.

(b) From part (a), the maximum value of v is $\frac{4}{27}kr_0^3$.

(c)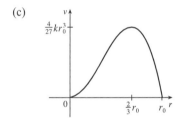

75. $f(x) = x^{101} + x^{51} + x + 1 \Rightarrow f'(x) = 101x^{100} + 51x^{50} + 1 \geq 1$ for all x, so $f'(x) = 0$ has no solution. Thus, $f(x)$

has no critical number, so $f(x)$ can have no local maximum or minimum.

77. If f has a local minimum at c, then $g(x) = -f(x)$ has a local maximum at c, so $g'(c) = 0$ by the case of Fermat's Theorem

proved in the text. Thus, $f'(c) = -g'(c) = 0$.

4.2 The Mean Value Theorem

1. $f(x) = 5 - 12x + 3x^2$, $[1, 3]$. Since f is a polynomial, it is continuous and differentiable on $\mathbb{R}$, so it is continuous on $[1, 3]$

and differentiable on $(1, 3)$. Also $f(1) = -4 = f(3)$. $f'(c) = 0 \Leftrightarrow -12 + 6c = 0 \Leftrightarrow c = 2$, which is in the open

interval $(1, 3)$, so $c = 2$ satisfies the conclusion of Rolle's Theorem.

3. $f(x) = \sqrt{x} - \frac{1}{3}x$, $[0, 9]$. f, being the difference of a root function and a polynomial, is continuous and differentiable

on $[0, \infty)$, so it is continuous on $[0, 9]$ and differentiable on $(0, 9)$. Also, $f(0) = 0 = f(9)$. $f'(c) = 0 \iff$

$\dfrac{1}{2\sqrt{c}} - \dfrac{1}{3} = 0 \iff 2\sqrt{c} = 3 \iff \sqrt{c} = \dfrac{3}{2} \implies c = \dfrac{9}{4}$, which is in the open interval $(0, 9)$, so $c = \dfrac{9}{4}$ satisfies the

conclusion of Rolle's Theorem.

5. $f(x) = 1 - x^{2/3}$. $f(-1) = 1 - (-1)^{2/3} = 1 - 1 = 0 = f(1)$. $f'(x) = -\frac{2}{3}x^{-1/3}$, so $f'(c) = 0$ has no solution. This

does not contradict Rolle's Theorem, since $f'(0)$ does not exist, and so f is not differentiable on $(-1, 1)$.

7. $\dfrac{f(8) - f(0)}{8 - 0} = \dfrac{6 - 4}{8} = \dfrac{1}{4}$. The values of c which satisfy $f'(c) = \frac{1}{4}$ seem to be about $c = 0.8, 3.2, 4.4$, and 6.1.

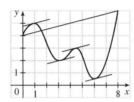

9. (a), (b) The equation of the secant line is

$y - 5 = \dfrac{8.5 - 5}{8 - 1}(x - 1) \iff y = \frac{1}{2}x + \frac{9}{2}$.

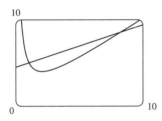

(c) $f(x) = x + 4/x \implies f'(x) = 1 - 4/x^2$.

So $f'(c) = \frac{1}{2} \implies c^2 = 8 \implies c = 2\sqrt{2}$, and

$f(c) = 2\sqrt{2} + \dfrac{4}{2\sqrt{2}} = 3\sqrt{2}$. Thus, an equation of the

tangent line is $y - 3\sqrt{2} = \frac{1}{2}(x - 2\sqrt{2}) \iff$

$y = \frac{1}{2}x + 2\sqrt{2}$.

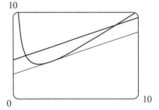

11. $f(x) = 3x^2 + 2x + 5$, $[-1, 1]$. f is continuous on $[-1, 1]$ and differentiable on $(-1, 1)$ since polynomials are continuous

and differentiable on $\mathbb{R}$. $f'(c) = \dfrac{f(b) - f(a)}{b - a} \iff 6c + 2 = \dfrac{f(1) - f(-1)}{1 - (-1)} = \dfrac{10 - 6}{2} = 2 \iff 6c = 0 \iff$

$c = 0$, which is in $(-1, 1)$.

13. $f(x) = e^{-2x}$, $[0, 3]$. f is continuous and differentiable on $\mathbb{R}$, so it is continuous on $[0, 3]$ and differentiable on $(0, 3)$.

$f'(c) = \dfrac{f(b) - f(a)}{b - a} \iff -2e^{-2c} = \dfrac{e^{-6} - e^0}{3 - 0} \iff e^{-2c} = \dfrac{1 - e^{-6}}{6} \iff -2c = \ln\left(\dfrac{1 - e^{-6}}{6}\right) \iff$

$c = -\dfrac{1}{2}\ln\left(\dfrac{1 - e^{-6}}{6}\right) \approx 0.897$, which is in $(0, 3)$.

15. $f(x) = (x-3)^{-2}$ $\Rightarrow$ $f'(x) = -2(x-3)^{-3}$. $f(4) - f(1) = f'(c)(4-1)$ $\Rightarrow$ $\dfrac{1}{1^2} - \dfrac{1}{(-2)^2} = \dfrac{-2}{(c-3)^3} \cdot 3$ $\Rightarrow$

$\dfrac{3}{4} = \dfrac{-6}{(c-3)^3}$ $\Rightarrow$ $(c-3)^3 = -8$ $\Rightarrow$ $c - 3 = -2$ $\Rightarrow$ $c = 1$, which is not in the open interval $(1,4)$. This does not

contradict the Mean Value Theorem since f is not continuous at $x = 3$.

17. Let $f(x) = 1 + 2x + x^3 + 4x^5$. Then $f(-1) = -6 < 0$ and $f(0) = 1 > 0$. Since f is a polynomial, it is continuous, so the

Intermediate Value Theorem says that there is a number c between -1 and 0 such that $f(c) = 0$. Thus, the given equation has

a real root. Suppose the equation has distinct real roots a and b with $a < b$. Then $f(a) = f(b) = 0$. Since f is a polynomial, it

is differentiable on (a,b) and continuous on $[a,b]$. By Rolle's Theorem, there is a number r in (a,b) such that $f'(r) = 0$. But

$f'(x) = 2 + 3x^2 + 20x^4 \geq 2$ for all x, so $f'(x)$ can never be 0. This contradiction shows that the equation can't have two

distinct real roots. Hence, it has exactly one real root.

19. Let $f(x) = x^3 - 15x + c$ for x in $[-2,2]$. If f has two real roots a and b in $[-2,2]$, with $a < b$, then $f(a) = f(b) = 0$. Since

the polynomial f is continuous on $[a,b]$ and differentiable on (a,b), Rolle's Theorem implies that there is a number r in (a,b)

such that $f'(r) = 0$. Now $f'(r) = 3r^2 - 15$. Since r is in (a,b), which is contained in $[-2,2]$, we have $|r| < 2$, so $r^2 < 4$.

It follows that $3r^2 - 15 < 3 \cdot 4 - 15 = -3 < 0$. This contradicts $f'(r) = 0$, so the given equation can't have two real roots

in $[-2,2]$. Hence, it has at most one real root in $[-2,2]$.

21. (a) Suppose that a cubic polynomial $P(x)$ has roots $a_1 < a_2 < a_3 < a_4$, so $P(a_1) = P(a_2) = P(a_3) = P(a_4)$.

By Rolle's Theorem there are numbers c_1, c_2, c_3 with $a_1 < c_1 < a_2$, $a_2 < c_2 < a_3$ and $a_3 < c_3 < a_4$ and

$P'(c_1) = P'(c_2) = P'(c_3) = 0$. Thus, the second-degree polynomial $P'(x)$ has three distinct real roots, which is

impossible.

(b) We prove by induction that a polynomial of degree n has at most n real roots. This is certainly true for $n = 1$. Suppose

that the result is true for all polynomials of degree n and let $P(x)$ be a polynomial of degree $n + 1$. Suppose that $P(x)$ has

more than $n + 1$ real roots, say $a_1 < a_2 < a_3 < \cdots < a_{n+1} < a_{n+2}$. Then $P(a_1) = P(a_2) = \cdots = P(a_{n+2}) = 0$.

By Rolle's Theorem there are real numbers $c_1, \ldots, c_{n+1}$ with $a_1 < c_1 < a_2, \ldots, a_{n+1} < c_{n+1} < a_{n+2}$ and

$P'(c_1) = \cdots = P'(c_{n+1}) = 0$. Thus, the nth degree polynomial $P'(x)$ has at least $n + 1$ roots. This contradiction shows

that $P(x)$ has at most $n + 1$ real roots.

23. By the Mean Value Theorem, $f(4) - f(1) = f'(c)(4 - 1)$ for some $c \in (1,4)$. But for every $c \in (1,4)$ we have

$f'(c) \geq 2$. Putting $f'(c) \geq 2$ into the above equation and substituting $f(1) = 10$, we get

$f(4) = f(1) + f'(c)(4 - 1) = 10 + 3f'(c) \geq 10 + 3 \cdot 2 = 16$. So the smallest possible value of $f(4)$ is 16.

25. Suppose that such a function f exists. By the Mean Value Theorem there is a number $0 < c < 2$ with

$f'(c) = \dfrac{f(2) - f(0)}{2 - 0} = \dfrac{5}{2}$. But this is impossible since $f'(x) \leq 2 < \frac{5}{2}$ for all x, so no such function can exist.

27. We use Exercise 26 with $f(x) = \sqrt{1+x}$, $g(x) = 1 + \frac{1}{2}x$, and $a = 0$. Notice that $f(0) = 1 = g(0)$ and

$$f'(x) = \frac{1}{2\sqrt{1+x}} < \frac{1}{2} = g'(x) \text{ for } x > 0. \text{ So by Exercise 26, } f(b) < g(b) \quad \Rightarrow \quad \sqrt{1+b} < 1 + \frac{1}{2}b \text{ for } b > 0.$$

Another method: Apply the Mean Value Theorem directly to either $f(x) = 1 + \frac{1}{2}x - \sqrt{1+x}$ or $g(x) = \sqrt{1+x}$ on $[0, b]$.

29. Let $f(x) = \sin x$ and let $b < a$. Then $f(x)$ is continuous on $[b, a]$ and differentiable on (b, a). By the Mean Value Theorem,

there is a number $c \in (b, a)$ with $\sin a - \sin b = f(a) - f(b) = f'(c)(a - b) = (\cos c)(a - b)$. Thus,

$|\sin a - \sin b| \le |\cos c|\,|b - a| \le |a - b|$. If $a < b$, then $|\sin a - \sin b| = |\sin b - \sin a| \le |b - a| = |a - b|$. If $a = b$, both

sides of the inequality are 0.

31. For $x > 0$, $f(x) = g(x)$, so $f'(x) = g'(x)$. For $x < 0$, $f'(x) = (1/x)' = -1/x^2$ and $g'(x) = (1 + 1/x)' = -1/x^2$, so

again $f'(x) = g'(x)$. However, the domain of $g(x)$ is not an interval [it is $(-\infty, 0) \cup (0, \infty)$] so we cannot conclude that

$f - g$ is constant (in fact it is not).

33. Let $f(x) = \arcsin\left(\dfrac{x-1}{x+1}\right) - 2\arctan\sqrt{x} + \dfrac{\pi}{2}$. Note that the domain of f is $[0, \infty)$. Thus,

$$f'(x) = \frac{1}{\sqrt{1 - \left(\dfrac{x-1}{x+1}\right)^2}} \cdot \frac{(x+1) - (x-1)}{(x+1)^2} - \frac{2}{1+x} \cdot \frac{1}{2\sqrt{x}} = \frac{1}{\sqrt{x}\,(x+1)} - \frac{1}{\sqrt{x}\,(x+1)} = 0.$$

Then $f(x) = C$ on $(0, \infty)$ by Theorem 5. By continuity of f, $f(x) = C$ on $[0, \infty)$. To find C, we let $x = 0 \Rightarrow$

$\arcsin(-1) - 2\arctan(0) + \frac{\pi}{2} = C \Rightarrow -\frac{\pi}{2} - 0 + \frac{\pi}{2} = 0 = C$. Thus, $f(x) = 0 \Rightarrow$

$\arcsin\left(\dfrac{x-1}{x+1}\right) = 2\arctan\sqrt{x} - \dfrac{\pi}{2}$.

35. Let $g(t)$ and $h(t)$ be the position functions of the two runners and let $f(t) = g(t) - h(t)$. By hypothesis,

$f(0) = g(0) - h(0) = 0$ and $f(b) = g(b) - h(b) = 0$, where b is the finishing time. Then by the Mean Value Theorem,

there is a time c, with $0 < c < b$, such that $f'(c) = \dfrac{f(b) - f(0)}{b - 0}$. But $f(b) = f(0) = 0$, so $f'(c) = 0$. Since

$f'(c) = g'(c) - h'(c) = 0$, we have $g'(c) = h'(c)$. So at time c, both runners have the same speed $g'(c) = h'(c)$.

4.3 How Derivatives Affect the Shape of a Graph

1. (a) f is increasing on $(1, 3)$ and $(4, 6)$.   (b) f is decreasing on $(0, 1)$ and $(3, 4)$.

(c) f is concave upward on $(0, 2)$.   (d) f is concave downward on $(2, 4)$ and $(4, 6)$.

(e) The point of inflection is $(2, 3)$.

3. (a) Use the Increasing/Decreasing (I/D) Test.   (b) Use the Concavity Test.

(c) At any value of x where the concavity changes, we have an inflection point at $(x, f(x))$.

5. (a) Since $f'(x) > 0$ on $(1, 5)$, f is increasing on this interval. Since $f'(x) < 0$ on $(0, 1)$ and $(5, 6)$, f is decreasing on these intervals.

(b) Since $f'(x) = 0$ at $x = 1$ and f' changes from negative to positive there, f changes from decreasing to increasing and has a local minimum at $x = 1$. Since $f'(x) = 0$ at $x = 5$ and f' changes from positive to negative there, f changes from increasing to decreasing and has a local maximum at $x = 5$.

7. There is an inflection point at $x = 1$ because $f''(x)$ changes from negative to positive there, and so the graph of f changes from concave downward to concave upward. There is an inflection point at $x = 7$ because $f''(x)$ changes from positive to negative there, and so the graph of f changes from concave upward to concave downward.

9. (a) $f(x) = 2x^3 + 3x^2 - 36x \Rightarrow f'(x) = 6x^2 + 6x - 36 = 6(x^2 + x - 6) = 6(x + 3)(x - 2)$.

We don't need to include the "6" in the chart to determine the sign of $f'(x)$.

Interval	$x + 3$	$x - 2$	$f'(x)$	f
$x < -3$	−	−	+	increasing on $(-\infty, -3)$
$-3 < x < 2$	+	−	−	decreasing on $(-3, 2)$
$x > 2$	+	+	+	increasing on $(2, \infty)$

(b) f changes from increasing to decreasing at $x = -3$ and from decreasing to increasing at $x = 2$. Thus, $f(-3) = 81$ is a local maximum value and $f(2) = -44$ is a local minimum value.

(c) $f'(x) = 6x^2 + 6x - 36 \Rightarrow f''(x) = 12x + 6$. $f''(x) = 0$ at $x = -\frac{1}{2}$, $f''(x) > 0 \Leftrightarrow x > -\frac{1}{2}$, and $f''(x) < 0 \Leftrightarrow x < -\frac{1}{2}$. Thus, f is concave upward on $\left(-\frac{1}{2}, \infty\right)$ and concave downward on $\left(-\infty, -\frac{1}{2}\right)$. There is an inflection point at $\left(-\frac{1}{2}, f\left(-\frac{1}{2}\right)\right) = \left(-\frac{1}{2}, \frac{37}{2}\right)$.

11. (a) $f(x) = x^4 - 2x^2 + 3 \Rightarrow f'(x) = 4x^3 - 4x = 4x(x^2 - 1) = 4x(x + 1)(x - 1)$.

Interval	$x + 1$	x	$x - 1$	$f'(x)$	f
$x < -1$	−	−	−	−	decreasing on $(-\infty, -1)$
$-1 < x < 0$	+	−	−	+	increasing on $(-1, 0)$
$0 < x < 1$	+	+	−	−	decreasing on $(0, 1)$
$x > 1$	+	+	+	+	increasing on $(1, \infty)$

(b) f changes from increasing to decreasing at $x = 0$ and from decreasing to increasing at $x = -1$ and $x = 1$. Thus, $f(0) = 3$ is a local maximum value and $f(\pm 1) = 2$ are local minimum values.

(c) $f''(x) = 12x^2 - 4 = 12\left(x^2 - \frac{1}{3}\right) = 12\left(x + 1/\sqrt{3}\right)\left(x - 1/\sqrt{3}\right)$. $f''(x) > 0 \Leftrightarrow x < -1/\sqrt{3}$ or $x > 1/\sqrt{3}$ and $f''(x) < 0 \Leftrightarrow -1/\sqrt{3} < x < 1/\sqrt{3}$. Thus, f is concave upward on $\left(-\infty, -\sqrt{3}/3\right)$ and $\left(\sqrt{3}/3, \infty\right)$ and concave downward on $\left(-\sqrt{3}/3, \sqrt{3}/3\right)$. There are inflection points at $\left(\pm\sqrt{3}/3, \frac{22}{9}\right)$.

13. (a) $f(x) = \sin x + \cos x$, $0 \le x \le 2\pi$. $f'(x) = \cos x - \sin x = 0$ $\Rightarrow$ $\cos x = \sin x$ $\Rightarrow$ $1 = \dfrac{\sin x}{\cos x}$ $\Rightarrow$

$\tan x = 1$ $\Rightarrow$ $x = \frac{\pi}{4}$ or $\frac{5\pi}{4}$. Thus, $f'(x) > 0$ $\Leftrightarrow$ $\cos x - \sin x > 0$ $\Leftrightarrow$ $\cos x > \sin x$ $\Leftrightarrow$ $0 < x < \frac{\pi}{4}$ or

$\frac{5\pi}{4} < x < 2\pi$ and $f'(x) < 0$ $\Leftrightarrow$ $\cos x < \sin x$ $\Leftrightarrow$ $\frac{\pi}{4} < x < \frac{5\pi}{4}$. So f is increasing on $\left(0, \frac{\pi}{4}\right)$ and $\left(\frac{5\pi}{4}, 2\pi\right)$ and f

is decreasing on $\left(\frac{\pi}{4}, \frac{5\pi}{4}\right)$.

(b) f changes from increasing to decreasing at $x = \frac{\pi}{4}$ and from decreasing to increasing at $x = \frac{5\pi}{4}$. Thus, $f\left(\frac{\pi}{4}\right) = \sqrt{2}$ is a

local maximum value and $f\left(\frac{5\pi}{4}\right) = -\sqrt{2}$ is a local minimum value.

(c) $f''(x) = -\sin x - \cos x = 0$ $\Rightarrow$ $-\sin x = \cos x$ $\Rightarrow$ $\tan x = -1$ $\Rightarrow$ $x = \frac{3\pi}{4}$ or $\frac{7\pi}{4}$. Divide the interval

$(0, 2\pi)$ into subintervals with these numbers as endpoints and complete a second derivative chart.

Interval	$f''(x) = -\sin x - \cos x$	Concavity
$\left(0, \frac{3\pi}{4}\right)$	$f''\left(\frac{\pi}{2}\right) = -1 < 0$	downward
$\left(\frac{3\pi}{4}, \frac{7\pi}{4}\right)$	$f''(\pi) = 1 > 0$	upward
$\left(\frac{7\pi}{4}, 2\pi\right)$	$f''\left(\frac{11\pi}{6}\right) = \frac{1}{2} - \frac{1}{2}\sqrt{3} < 0$	downward

There are inflection points at $\left(\frac{3\pi}{4}, 0\right)$ and $\left(\frac{7\pi}{4}, 0\right)$.

15. (a) $f(x) = e^{2x} + e^{-x}$ $\Rightarrow$ $f'(x) = 2e^{2x} - e^{-x}$. $f'(x) > 0$ $\Leftrightarrow$ $2e^{2x} > e^{-x}$ $\Leftrightarrow$ $e^{3x} > \frac{1}{2}$ $\Leftrightarrow$ $3x > \ln \frac{1}{2}$ $\Leftrightarrow$

$x > \frac{1}{3}(\ln 1 - \ln 2)$ $\Leftrightarrow$ $x > -\frac{1}{3}\ln 2$ $[\approx -0.23]$ and $f'(x) < 0$ if $x < -\frac{1}{3}\ln 2$. So f is increasing on $\left(-\frac{1}{3}\ln 2, \infty\right)$

and f is decreasing on $\left(-\infty, -\frac{1}{3}\ln 2\right)$.

(b) f changes from decreasing to increasing at $x = -\frac{1}{3}\ln 2$. Thus,

$$f\left(-\tfrac{1}{3}\ln 2\right) = f\left(\ln \sqrt[3]{1/2}\right) = e^{2\ln \sqrt[3]{1/2}} + e^{-\ln \sqrt[3]{1/2}} = e^{\ln \sqrt[3]{1/4}} + e^{\ln \sqrt[3]{2}} = \sqrt[3]{1/4} + \sqrt[3]{2} = 2^{-2/3} + 2^{1/3} \ [\approx 1.89]$$

is a local minimum value.

(c) $f''(x) = 4e^{2x} + e^{-x} > 0$ [the sum of two positive terms].   Thus, f is concave upward on $(-\infty, \infty)$ and there is no

point of inflection.

17. (a) $y = f(x) = \dfrac{\ln x}{\sqrt{x}}$. (Note that f is only defined for $x > 0$.)

$$f'(x) = \frac{\sqrt{x}\,(1/x) - \ln x\left(\frac{1}{2}x^{-1/2}\right)}{x} = \frac{\dfrac{1}{\sqrt{x}} - \dfrac{\ln x}{2\sqrt{x}}}{x} \cdot \frac{2\sqrt{x}}{2\sqrt{x}} = \frac{2 - \ln x}{2x^{3/2}} > 0 \Leftrightarrow 2 - \ln x > 0 \Leftrightarrow$$

$\ln x < 2$ $\Leftrightarrow$ $x < e^2$. Therefore f is increasing on $\left(0, e^2\right)$ and decreasing on $\left(e^2, \infty\right)$.

(b) f changes from increasing to decreasing at $x = e^2$, so $f(e^2) = \dfrac{\ln e^2}{\sqrt{e^2}} = \dfrac{2}{e}$ is a local maximum value.

(c) $f''(x) = \dfrac{2x^{3/2}(-1/x) - (2 - \ln x)(3x^{1/2})}{\left(2x^{3/2}\right)^2} = \dfrac{-2x^{1/2} + 3x^{1/2}(\ln x - 2)}{4x^3} = \dfrac{x^{1/2}(-2 + 3\ln x - 6)}{4x^3} = \dfrac{3\ln x - 8}{4x^{5/2}}$

$f''(x) = 0$ $\Leftrightarrow$ $\ln x = \frac{8}{3}$ $\Leftrightarrow$ $x = e^{8/3}$. $f''(x) > 0$ $\Leftrightarrow$ $x > e^{8/3}$, so f is concave upward on $\left(e^{8/3}, \infty\right)$ and

concave downward on $\left(0, e^{8/3}\right)$. There is an inflection point at $\left(e^{8/3}, \frac{8}{3}e^{-4/3}\right) \approx (14.39, 0.70)$.

19. $f(x) = x^5 - 5x + 3 \Rightarrow f'(x) = 5x^4 - 5 = 5(x^2 + 1)(x + 1)(x - 1)$.

First Derivative Test: $f'(x) < 0 \Rightarrow -1 < x < 1$ and $f'(x) > 0 \Rightarrow x > 1$ or $x < -1$. Since f' changes from positive to negative at $x = -1$, $f(-1) = 7$ is a local maximum value; and since f' changes from negative to positive at $x = 1$, $f(1) = -1$ is a local minimum value.

Second Derivative Test: $f''(x) = 20x^3$. $f'(x) = 0 \Leftrightarrow x = \pm 1$. $f''(-1) = -20 < 0 \Rightarrow f(-1) = 7$ is a local maximum value. $f''(1) = 20 > 0 \Rightarrow f(1) = -1$ is a local minimum value.

Preference: For this function, the two tests are equally easy.

21. $f(x) = x + \sqrt{1 - x} \Rightarrow f'(x) = 1 + \frac{1}{2}(1 - x)^{-1/2}(-1) = 1 - \dfrac{1}{2\sqrt{1 - x}}$. Note that f is defined for $1 - x \geq 0$; that is, for $x \leq 1$. $f'(x) = 0 \Rightarrow 2\sqrt{1 - x} = 1 \Rightarrow \sqrt{1 - x} = \frac{1}{2} \Rightarrow 1 - x = \frac{1}{4} \Rightarrow x = \frac{3}{4}$. f' does not exist at $x = 1$, but we can't have a local maximum or minimum at an endpoint.

First Derivative Test: $f'(x) > 0 \Rightarrow x < \frac{3}{4}$ and $f'(x) < 0 \Rightarrow \frac{3}{4} < x < 1$. Since f' changes from positive to negative at $x = \frac{3}{4}$, $f\left(\frac{3}{4}\right) = \frac{5}{4}$ is a local maximum value.

Second Derivative Test: $f''(x) = -\frac{1}{2}\left(-\frac{1}{2}\right)(1 - x)^{-3/2}(-1) = -\dfrac{1}{4\left(\sqrt{1 - x}\right)^3}$.

$f''\left(\frac{3}{4}\right) = -2 < 0 \Rightarrow f\left(\frac{3}{4}\right) = \frac{5}{4}$ is a local maximum value.

Preference: The First Derivative Test may be slightly easier to apply in this case.

23. (a) By the Second Derivative Test, if $f'(2) = 0$ and $f''(2) = -5 < 0$, f has a local maximum at $x = 2$.

(b) If $f'(6) = 0$, we know that f has a horizontal tangent at $x = 6$. Knowing that $f''(6) = 0$ does not provide any additional information since the Second Derivative Test fails. For example, the first and second derivatives of $y = (x - 6)^4$, $y = -(x - 6)^4$, and $y = (x - 6)^3$ all equal zero for $x = 6$, but the first has a local minimum at $x = 6$, the second has a local maximum at $x = 6$, and the third has an inflection point at $x = 6$.

25. $f'(0) = f'(2) = f'(4) = 0 \Rightarrow$ horizontal tangents at $x = 0, 2, 4$.

$f'(x) > 0$ if $x < 0$ or $2 < x < 4 \Rightarrow f$ is increasing on $(-\infty, 0)$ and $(2, 4)$.

$f'(x) < 0$ if $0 < x < 2$ or $x > 4 \Rightarrow f$ is decreasing on $(0, 2)$ and $(4, \infty)$.

$f''(x) > 0$ if $1 < x < 3 \Rightarrow f$ is concave upward on $(1, 3)$.

$f''(x) < 0$ if $x < 1$ or $x > 3 \Rightarrow f$ is concave downward on $(-\infty, 1)$ and $(3, \infty)$.

There are inflection points when $x = 1$ and 3.

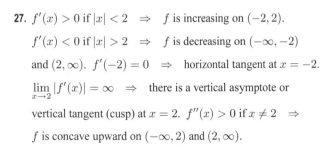

27. $f'(x) > 0$ if $|x| < 2 \Rightarrow f$ is increasing on $(-2, 2)$.

$f'(x) < 0$ if $|x| > 2 \Rightarrow f$ is decreasing on $(-\infty, -2)$ and $(2, \infty)$. $f'(-2) = 0 \Rightarrow$ horizontal tangent at $x = -2$.

$\lim\limits_{x \to 2} |f'(x)| = \infty \Rightarrow$ there is a vertical asymptote or vertical tangent (cusp) at $x = 2$. $f''(x) > 0$ if $x \neq 2 \Rightarrow$ f is concave upward on $(-\infty, 2)$ and $(2, \infty)$.

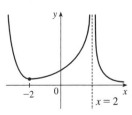

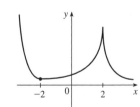

29. The function must be always decreasing (since the first derivative is always negative) and concave downward (since the second derivative is always negative).

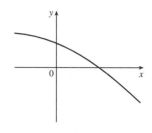

31. (a) f is increasing where f' is positive, that is, on $(0, 2)$, $(4, 6)$, and $(8, \infty)$; and decreasing where f' is negative, that is, on $(2, 4)$ and $(6, 8)$.

(b) f has local maxima where f' changes from positive to negative, at $x = 2$ and at $x = 6$, and local minima where f' changes from negative to positive, at $x = 4$ and at $x = 8$.

(c) f is concave upward (CU) where f' is increasing, that is, on $(3, 6)$ and $(6, \infty)$, and concave downward (CD) where f' is decreasing, that is, on $(0, 3)$.

(e)

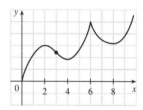

(d) There is a point of inflection where f changes from being CD to being CU, that is, at $x = 3$.

33. (a) $f(x) = 2x^3 - 3x^2 - 12x$ $\Rightarrow$ $f'(x) = 6x^2 - 6x - 12 = 6(x^2 - x - 2) = 6(x - 2)(x + 1)$.

$f'(x) > 0$ $\Leftrightarrow$ $x < -1$ or $x > 2$ and $f'(x) < 0$ $\Leftrightarrow$ $-1 < x < 2$. So f is increasing on $(-\infty, -1)$ and $(2, \infty)$, and f is decreasing on $(-1, 2)$.

(b) Since f changes from increasing to decreasing at $x = -1$, $f(-1) = 7$ is a local maximum value. Since f changes from decreasing to increasing at $x = 2$, $f(2) = -20$ is a local minimum value.

(d)

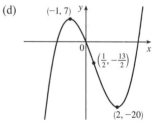

(c) $f''(x) = 6(2x - 1)$ $\Rightarrow$ $f''(x) > 0$ on $\left(\frac{1}{2}, \infty\right)$ and $f''(x) < 0$ on $\left(-\infty, \frac{1}{2}\right)$.

So f is concave upward on $\left(\frac{1}{2}, \infty\right)$ and concave downward on $\left(-\infty, \frac{1}{2}\right)$. There is a change in concavity at $x = \frac{1}{2}$, and we have an inflection point at $\left(\frac{1}{2}, -\frac{13}{2}\right)$.

35. (a) $f(x) = 2 + 2x^2 - x^4$ $\Rightarrow$ $f'(x) = 4x - 4x^3 = 4x(1 - x^2) = 4x(1 + x)(1 - x)$. $f'(x) > 0$ $\Leftrightarrow$ $x < -1$ or $0 < x < 1$ and $f'(x) < 0$ $\Leftrightarrow$ $-1 < x < 0$ or $x > 1$. So f is increasing on $(-\infty, -1)$ and $(0, 1)$ and f is decreasing on $(-1, 0)$ and $(1, \infty)$.

(b) f changes from increasing to decreasing at $x = -1$ and $x = 1$, so $f(-1) = 3$ and $f(1) = 3$ are local maximum values. f changes from decreasing to increasing at $x = 0$, so $f(0) = 2$ is a local minimum value.

(c) $f''(x) = 4 - 12x^2 = 4(1 - 3x^2)$. $f''(x) = 0 \Leftrightarrow 1 - 3x^2 = 0 \Leftrightarrow$

(d)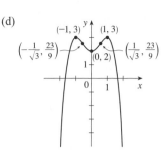

$x^2 = \frac{1}{3} \Leftrightarrow x = \pm 1/\sqrt{3}$. $f''(x) > 0$ on $\left(-1/\sqrt{3}, 1/\sqrt{3}\right)$ and $f''(x) < 0$

on $\left(-\infty, -1/\sqrt{3}\right)$ and $\left(1/\sqrt{3}, \infty\right)$. So f is concave upward on

$\left(-1/\sqrt{3}, 1/\sqrt{3}\right)$ and f is concave downward on $\left(-\infty, -1/\sqrt{3}\right)$ and

$\left(1/\sqrt{3}, \infty\right)$. $f\left(\pm 1/\sqrt{3}\right) = 2 + \frac{2}{3} - \frac{1}{9} = \frac{23}{9}$. There are points of inflection

at $\left(\pm 1/\sqrt{3}, \frac{23}{9}\right)$.

37. (a) $h(x) = (x+1)^5 - 5x - 2 \Rightarrow h'(x) = 5(x+1)^4 - 5$. $h'(x) = 0 \Leftrightarrow 5(x+1)^4 = 5 \Leftrightarrow (x+1)^4 = 1 \Rightarrow$

$(x+1)^2 = 1 \Rightarrow x + 1 = 1$ or $x + 1 = -1 \Rightarrow x = 0$ or $x = -2$. $h'(x) > 0 \Leftrightarrow x < -2$ or $x > 0$ and

$h'(x) < 0 \Leftrightarrow -2 < x < 0$. So h is increasing on $(-\infty, -2)$ and $(0, \infty)$ and h is decreasing on $(-2, 0)$.

(b) $h(-2) = 7$ is a local maximum value and $h(0) = -1$ is a local minimum value.

(d)

(c) $h''(x) = 20(x+1)^3 = 0 \Leftrightarrow x = -1$. $h''(x) > 0 \Leftrightarrow x > -1$ and

$h''(x) < 0 \Leftrightarrow x < -1$, so h is CU on $(-1, \infty)$ and h is CD on $(-\infty, -1)$.

There is a point of inflection at $(-1, h(-1)) = (-1, 3)$.

39. (a) $A(x) = x\sqrt{x+3} \Rightarrow A'(x) = x \cdot \frac{1}{2}(x+3)^{-1/2} + \sqrt{x+3} \cdot 1 = \dfrac{x}{2\sqrt{x+3}} + \sqrt{x+3} = \dfrac{x + 2(x+3)}{2\sqrt{x+3}} = \dfrac{3x+6}{2\sqrt{x+3}}$.

The domain of A is $[-3, \infty)$. $A'(x) > 0$ for $x > -2$ and $A'(x) < 0$ for $-3 < x < -2$, so A is increasing on $(-2, \infty)$

and decreasing on $(-3, -2)$.

(b) $A(-2) = -2$ is a local minimum value.

(d)

(c) $A''(x) = \dfrac{2\sqrt{x+3} \cdot 3 - (3x+6) \cdot \dfrac{1}{\sqrt{x+3}}}{\left(2\sqrt{x+3}\right)^2}$

$= \dfrac{6(x+3) - (3x+6)}{4(x+3)^{3/2}} = \dfrac{3x+12}{4(x+3)^{3/2}} = \dfrac{3(x+4)}{4(x+3)^{3/2}}$

$A''(x) > 0$ for all $x > -3$, so A is concave upward on $(-3, \infty)$. There is no inflection point.

41. (a) $C(x) = x^{1/3}(x+4) = x^{4/3} + 4x^{1/3} \Rightarrow C'(x) = \frac{4}{3}x^{1/3} + \frac{4}{3}x^{-2/3} = \frac{4}{3}x^{-2/3}(x+1) = \dfrac{4(x+1)}{3\sqrt[3]{x^2}}$. $C'(x) > 0$ if

$-1 < x < 0$ or $x > 0$ and $C'(x) < 0$ for $x < -1$, so C is increasing on $(-1, \infty)$ and C is decreasing on $(-\infty, -1)$.

(b) $C(-1) = -3$ is a local minimum value.

(d)

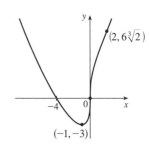

(c) $C''(x) = \frac{4}{9}x^{-2/3} - \frac{8}{9}x^{-5/3} = \frac{4}{9}x^{-5/3}(x-2) = \frac{4(x-2)}{9\sqrt[3]{x^5}}$.

$C''(x) < 0$ for $0 < x < 2$ and $C''(x) > 0$ for $x < 0$ and $x > 2$, so C is

concave downward on $(0, 2)$ and concave upward on $(-\infty, 0)$ and $(2, \infty)$.

There are inflection points at $(0, 0)$ and $\left(2, 6\sqrt[3]{2}\right) \approx (2, 7.56)$.

43. (a) $f(\theta) = 2\cos\theta + \cos^2\theta$, $0 \le \theta \le 2\pi$ $\Rightarrow$ $f'(\theta) = -2\sin\theta + 2\cos\theta(-\sin\theta) = -2\sin\theta(1 + \cos\theta)$.

$f'(\theta) = 0$ $\Leftrightarrow$ $\theta = 0, \pi$, and 2π. $f'(\theta) > 0$ $\Leftrightarrow$ $\pi < \theta < 2\pi$ and $f'(\theta) < 0$ $\Leftrightarrow$ $0 < \theta < \pi$. So f is increasing on $(\pi, 2\pi)$ and f is decreasing on $(0, \pi)$.

(b) $f(\pi) = -1$ is a local minimum value.

(c) $f'(\theta) = -2\sin\theta(1 + \cos\theta)$ $\Rightarrow$

$$f''(\theta) = -2\sin\theta(-\sin\theta) + (1 + \cos\theta)(-2\cos\theta) = 2\sin^2\theta - 2\cos\theta - 2\cos^2\theta$$

$$= 2(1 - \cos^2\theta) - 2\cos\theta - 2\cos^2\theta = -4\cos^2\theta - 2\cos\theta + 2$$

$$= -2(2\cos^2\theta + \cos\theta - 1) = -2(2\cos\theta - 1)(\cos\theta + 1)$$

Since $-2(\cos\theta + 1) < 0$ [for $\theta \ne \pi$], $f''(\theta) > 0$ $\Rightarrow$ $2\cos\theta - 1 < 0$ $\Rightarrow$ $\cos\theta < \frac{1}{2}$ $\Rightarrow$ $\frac{\pi}{3} < \theta < \frac{5\pi}{3}$ and

$f''(\theta) < 0$ $\Rightarrow$ $\cos\theta > \frac{1}{2}$ $\Rightarrow$ $0 < \theta < \frac{\pi}{3}$ or $\frac{5\pi}{3} < \theta < 2\pi$. So f is CU on $\left(\frac{\pi}{3}, \frac{5\pi}{3}\right)$ and f is CD on $\left(0, \frac{\pi}{3}\right)$ and

$\left(\frac{5\pi}{3}, 2\pi\right)$. There are points of inflection at $\left(\frac{\pi}{3}, f\left(\frac{\pi}{3}\right)\right) = \left(\frac{\pi}{3}, \frac{5}{4}\right)$ and $\left(\frac{5\pi}{3}, f\left(\frac{5\pi}{3}\right)\right) = \left(\frac{5\pi}{3}, \frac{5}{4}\right)$.

(d)

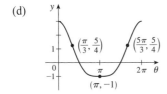

45. $f(x) = \dfrac{x^2}{x^2 - 1} = \dfrac{x^2}{(x+1)(x-1)}$ has domain $(-\infty, -1) \cup (-1, 1) \cup (1, \infty)$.

(a) $\displaystyle\lim_{x \to \pm\infty} f(x) = \lim_{x \to \pm\infty} \frac{x^2/x^2}{(x^2-1)/x^2} = \lim_{x \to \pm\infty} \frac{1}{1 - 1/x^2} = \frac{1}{1 - 0} = 1$, so $y = 1$ is a HA.

$\displaystyle\lim_{x \to -1^-} \frac{x^2}{x^2 - 1} = \infty$ since $x^2 \to 1$ and $(x^2 - 1) \to 0^+$ as $x \to -1^-$, so $x = -1$ is a VA.

$\displaystyle\lim_{x \to 1^+} \frac{x^2}{x^2 - 1} = \infty$ since $x^2 \to 1$ and $(x^2 - 1) \to 0^+$ as $x \to 1^+$, so $x = 1$ is a VA.

(b) $f(x) = \dfrac{x^2}{x^2 - 1}$ $\Rightarrow$ $f'(x) = \dfrac{(x^2 - 1)(2x) - x^2(2x)}{(x^2 - 1)^2} = \dfrac{2x[(x^2 - 1) - x^2]}{(x^2 - 1)^2} = \dfrac{-2x}{(x^2 - 1)^2}$. Since $(x^2 - 1)^2$ is

positive for all x in the domain of f, the sign of the derivative is determined by the sign of $-2x$. Thus, $f'(x) > 0$ if $x < 0$

$(x \ne -1)$ and $f'(x) < 0$ if $x > 0$ $(x \ne 1)$. So f is increasing on $(-\infty, -1)$ and $(-1, 0)$, and f is decreasing on $(0, 1)$

and $(1, \infty)$.

(c) $f'(x) = 0$ $\Rightarrow$ $x = 0$ and $f(0) = 0$ is a local maximum value.

(d) $f''(x) = \dfrac{(x^2-1)^2(-2)-(-2x)\cdot 2(x^2-1)(2x)}{[(x^2-1)^2]^2}$

$= \dfrac{2(x^2-1)[-(x^2-1)+4x^2]}{(x^2-1)^4} = \dfrac{2(3x^2+1)}{(x^2-1)^3}.$

The sign of $f''(x)$ is determined by the denominator; that is, $f''(x) > 0$ if

$|x| > 1$ and $f''(x) < 0$ if $|x| < 1$. Thus, f is CU on $(-\infty, -1)$ and $(1, \infty)$,

and f is CD on $(-1, 1)$. There are no inflection points.

(e)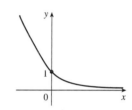

47. (a) $\lim\limits_{x \to -\infty} \left(\sqrt{x^2+1} - x\right) = \infty$ and

$\lim\limits_{x \to \infty} \left(\sqrt{x^2+1} - x\right) = \lim\limits_{x \to \infty} \left(\sqrt{x^2+1} - x\right)\dfrac{\sqrt{x^2+1}+x}{\sqrt{x^2+1}+x} = \lim\limits_{x \to \infty} \dfrac{1}{\sqrt{x^2+1}+x} = 0$, so $y = 0$ is a HA.

(b) $f(x) = \sqrt{x^2+1} - x \;\Rightarrow\; f'(x) = \dfrac{x}{\sqrt{x^2+1}} - 1.$ Since $\dfrac{x}{\sqrt{x^2+1}} < 1$ for all x, $f'(x) < 0$, so f is decreasing on $\mathbb{R}$.

(c) No minimum or maximum

(d) $f''(x) = \dfrac{(x^2+1)^{1/2}(1) - x\cdot\frac{1}{2}(x^2+1)^{-1/2}(2x)}{\left(\sqrt{x^2+1}\right)^2}$

$= \dfrac{(x^2+1)^{1/2} - \dfrac{x^2}{(x^2+1)^{1/2}}}{x^2+1} = \dfrac{(x^2+1)-x^2}{(x^2+1)^{3/2}} = \dfrac{1}{(x^2+1)^{3/2}} > 0,$

so f is CU on $\mathbb{R}$. No IP

(e)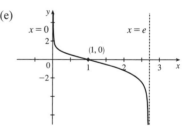

49. $f(x) = \ln(1 - \ln x)$ is defined when $x > 0$ (so that $\ln x$ is defined) and $1 - \ln x > 0$ [so that $\ln(1 - \ln x)$ is defined].

The second condition is equivalent to $1 > \ln x \;\Leftrightarrow\; x < e$, so f has domain $(0, e)$.

(a) As $x \to 0^+$, $\ln x \to -\infty$, so $1 - \ln x \to \infty$ and $f(x) \to \infty$. As $x \to e^-$, $\ln x \to 1^-$, so $1 - \ln x \to 0^+$ and

$f(x) \to -\infty$. Thus, $x = 0$ and $x = e$ are vertical asymptotes. There is no horizontal asymptote.

(b) $f'(x) = \dfrac{1}{1-\ln x}\left(-\dfrac{1}{x}\right) = -\dfrac{1}{x(1-\ln x)} < 0$ on $(0, e)$. Thus, f is decreasing on its domain, $(0, e)$.

(c) $f'(x) \neq 0$ on $(0, e)$, so f has no local maximum or minimum value.

(d) $f''(x) = -\dfrac{-[x(1-\ln x)]'}{[x(1-\ln x)]^2} = \dfrac{x(-1/x)+(1-\ln x)}{x^2(1-\ln x)^2}$

$= -\dfrac{\ln x}{x^2(1-\ln x)^2}$

so $f''(x) > 0 \;\Leftrightarrow\; \ln x < 0 \;\Leftrightarrow\; 0 < x < 1$. Thus, f is CU on $(0, 1)$

and CD on $(1, e)$. There is an inflection point at $(1, 0)$.

(e)

51. (a) $\lim\limits_{x \to \pm\infty} e^{-1/(x+1)} = 1$ since $-1/(x+1) \to 0$, so $y = 1$ is a HA. $\lim\limits_{x \to -1^+} e^{-1/(x+1)} = 0$ since $-1/(x+1) \to -\infty$,

$\lim\limits_{x \to -1^-} e^{-1/(x+1)} = \infty$ since $-1/(x+1) \to \infty$, so $x = -1$ is a VA.

(b) $f(x) = e^{-1/(x+1)} \;\Rightarrow\; f'(x) = e^{-1/(x+1)}\left[-(-1)\dfrac{1}{(x+1)^2}\right]$ [Reciprocal Rule] $= e^{-1/(x+1)}/(x+1)^2 \;\Rightarrow$

$f'(x) > 0$ for all x except -1, so f is increasing on $(-\infty, -1)$ and $(-1, \infty)$.

(c) There is no local maximum or minimum.

(d) $f''(x) = \dfrac{(x+1)^2 e^{-1/(x+1)}\left[1/(x+1)^2\right] - e^{-1/(x+1)}\left[2(x+1)\right]}{\left[(x+1)^2\right]^2}$

(e)

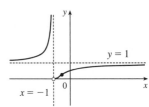

$\qquad = \dfrac{e^{-1/(x+1)}\left[1 - (2x+2)\right]}{(x+1)^4} = -\dfrac{e^{-1/(x+1)}(2x+1)}{(x+1)^4} \quad \Rightarrow$

$f''(x) > 0 \quad \Leftrightarrow \quad 2x + 1 < 0 \quad \Leftrightarrow \quad x < -\frac{1}{2}$, so f is CU on $(-\infty, -1)$

and $\left(-1, -\frac{1}{2}\right)$, and CD on $\left(-\frac{1}{2}, \infty\right)$. f has an IP at $\left(-\frac{1}{2}, e^{-2}\right)$.

53. The nonnegative factors $(x+1)^2$ and $(x-6)^4$ do not affect the sign of $f'(x) = (x+1)^2 (x-3)^5 (x-6)^4$.

So $f'(x) > 0 \quad \Rightarrow \quad (x-3)^5 > 0 \quad \Rightarrow \quad x - 3 > 0 \quad \Rightarrow \quad x > 3$. Thus, f is increasing on the interval $(3, \infty)$.

55. (a)

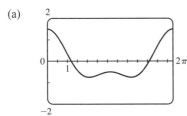

From the graph, we get an estimate of $f(1) \approx 1.41$ as a local maximum

value, and no local minimum value.

$f(x) = \dfrac{x+1}{\sqrt{x^2+1}} \quad \Rightarrow \quad f'(x) = \dfrac{1-x}{(x^2+1)^{3/2}}.$

$f'(x) = 0 \quad \Leftrightarrow \quad x = 1.$ $f(1) = \frac{2}{\sqrt{2}} = \sqrt{2}$ is the exact value.

(b) From the graph in part (a), f increases most rapidly somewhere between $x = -\frac{1}{2}$ and $x = -\frac{1}{4}$. To find the exact value,

we need to find the maximum value of f', which we can do by finding the critical numbers of f'.

$f''(x) = \dfrac{2x^2 - 3x - 1}{(x^2+1)^{5/2}} = 0 \quad \Leftrightarrow \quad x = \dfrac{3 \pm \sqrt{17}}{4}.$ $x = \dfrac{3 + \sqrt{17}}{4}$ corresponds to the *minimum* value of f'.

The maximum value of f' is at $\left(\dfrac{3 - \sqrt{17}}{4}, \sqrt{\dfrac{7}{6} - \dfrac{\sqrt{17}}{6}}\right) \approx (-0.28, 0.69).$

57. $f(x) = \cos x + \frac{1}{2}\cos 2x \quad \Rightarrow \quad f'(x) = -\sin x - \sin 2x \quad \Rightarrow \quad f''(x) = -\cos x - 2\cos 2x$

(a)

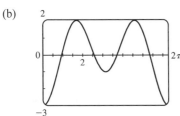

From the graph of f, it seems that f is CD on $(0, 1)$, CU on $(1, 2.5)$, CD on

$(2.5, 3.7)$, CU on $(3.7, 5.3)$, and CD on $(5.3, 2\pi)$. The points of inflection

appear to be at $(1, 0.4)$, $(2.5, -0.6)$, $(3.7, -0.6)$, and $(5.3, 0.4)$.

(b)

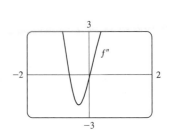

From the graph of f'' (and zooming in near the zeros), it seems that f is CD

on $(0, 0.94)$, CU on $(0.94, 2.57)$, CD on $(2.57, 3.71)$, CU on $(3.71, 5.35)$,

and CD on $(5.35, 2\pi)$. Refined estimates of the inflection points are

$(0.94, 0.44)$, $(2.57, -0.63)$, $(3.71, -0.63)$, and $(5.35, 0.44)$.

59. In Maple, we define f and then use the command

`plot(diff(diff(f,x),x),x=-2..2);`. In Mathematica, we define f

and then use `Plot[Dt[Dt[f,x],x],{x,-2,2}]`. We see that $f'' > 0$ for

$x < -0.6$ and $x > 0.0$ [≈ 0.03] and $f'' < 0$ for $-0.6 < x < 0.0$. So f is CU

on $(-\infty, -0.6)$ and $(0.0, \infty)$ and CD on $(-0.6, 0.0)$.

61. (a) The rate of increase of the population is initially very small, then gets larger until it reaches a maximum at about

$t = 8$ hours, and decreases toward 0 as the population begins to level off.

(b) The rate of increase has its maximum value at $t = 8$ hours.

(c) The population function is concave upward on $(0, 8)$ and concave downward on $(8, 18)$.

(d) At $t = 8$, the population is about 350, so the inflection point is about $(8, 350)$.

63. Most students learn more in the third hour of studying than in the eighth hour, so $K(3) - K(2)$ is larger than $K(8) - K(7)$.

In other words, as you begin studying for a test, the rate of knowledge gain is large and then starts to taper off, so $K'(t)$

decreases and the graph of K is concave downward.

65. $S(t) = At^p e^{-kt}$ with $A = 0.01$, $p = 4$, and $k = 0.07$. We will find the

zeros of f'' for $f(t) = t^p e^{-kt}$.

$f'(t) = t^p(-ke^{-kt}) + e^{-kt}(pt^{p-1}) = e^{-kt}(-kt^p + pt^{p-1})$

$f''(t) = e^{-kt}(-kpt^{p-1} + p(p-1)t^{p-2}) + (-kt^p + pt^{p-1})(-ke^{-kt})$

$\quad = t^{p-2}e^{-kt}[-kpt + p(p-1) + k^2t^2 - kpt]$

$\quad = t^{p-2}e^{-kt}(k^2t^2 - 2kpt + p^2 - p)$

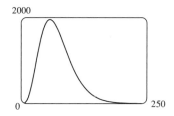

Using the given values of p and k gives us $f''(t) = t^2 e^{-0.07t}(0.0049t^2 - 0.56t + 12)$. So $S''(t) = 0.01f''(t)$ and its zeros

are $t = 0$ and the solutions of $0.0049t^2 - 0.56t + 12 = 0$, which are $t_1 = \frac{200}{7} \approx 28.57$ and $t_2 = \frac{600}{7} \approx 85.71$.

At t_1 minutes, the rate of increase of the level of medication in the bloodstream is at its greatest and at t_2 minutes, the rate of

decrease is the greatest.

67. $f(x) = ax^3 + bx^2 + cx + d \quad \Rightarrow \quad f'(x) = 3ax^2 + 2bx + c$.

We are given that $f(1) = 0$ and $f(-2) = 3$, so $f(1) = a + b + c + d = 0$ and

$f(-2) = -8a + 4b - 2c + d = 3$. Also $f'(1) = 3a + 2b + c = 0$ and

$f'(-2) = 12a - 4b + c = 0$ by Fermat's Theorem. Solving these four equations, we get

$a = \frac{2}{9}$, $b = \frac{1}{3}$, $c = -\frac{4}{3}$, $d = \frac{7}{9}$, so the function is $f(x) = \frac{1}{9}(2x^3 + 3x^2 - 12x + 7)$.

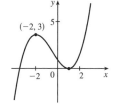

69. $y = \dfrac{1+x}{1+x^2} \quad \Rightarrow \quad y' = \dfrac{(1+x^2)(1) - (1+x)(2x)}{(1+x^2)^2} = \dfrac{1 - 2x - x^2}{(1+x^2)^2} \quad \Rightarrow$

$y'' = \dfrac{(1+x^2)^2(-2-2x) - (1-2x-x^2)\cdot 2(1+x^2)(2x)}{[(1+x^2)^2]^2} = \dfrac{2(1+x^2)[(1+x^2)(-1-x) - (1-2x-x^2)(2x)]}{(1+x^2)^4}$

$\quad = \dfrac{2(-1-x-x^2-x^3-2x+4x^2+2x^3)}{(1+x^2)^3} = \dfrac{2(x^3+3x^2-3x-1)}{(1+x^2)^3} = \dfrac{2(x-1)(x^2+4x+1)}{(1+x^2)^3}$

So $y'' = 0 \quad \Rightarrow \quad x = 1, -2 \pm \sqrt{3}$. Let $a = -2 - \sqrt{3}$, $b = -2 + \sqrt{3}$, and $c = 1$. We can show that $f(a) = \frac{1}{4}(1 - \sqrt{3})$,

$f(b) = \frac{1}{4}\left(1 + \sqrt{3}\right)$, and $f(c) = 1$. To show that these three points of inflection lie on one straight line, we'll show that the slopes m_{ac} and m_{bc} are equal.

$$m_{ac} = \frac{f(c) - f(a)}{c - a} = \frac{1 - \frac{1}{4}\left(1 - \sqrt{3}\right)}{1 - \left(-2 - \sqrt{3}\right)} = \frac{\frac{3}{4} + \frac{1}{4}\sqrt{3}}{3 + \sqrt{3}} = \frac{1}{4}$$

$$m_{bc} = \frac{f(c) - f(b)}{c - b} = \frac{1 - \frac{1}{4}\left(1 + \sqrt{3}\right)}{1 - \left(-2 + \sqrt{3}\right)} = \frac{\frac{3}{4} - \frac{1}{4}\sqrt{3}}{3 - \sqrt{3}} = \frac{1}{4}$$

71. Suppose that f is differentiable on an interval I and $f'(x) > 0$ for all x in I except $x = c$. To show that f is increasing on I, let x_1, x_2 be two numbers in I with $x_1 < x_2$.

> *Case 1* $x_1 < x_2 < c$. Let J be the interval $\{x \in I \mid x < c\}$. By applying the Increasing/Decreasing Test to f on J, we see that f is increasing on J, so $f(x_1) < f(x_2)$.
>
> *Case 2* $c < x_1 < x_2$. Apply the Increasing/Decreasing Test to f on $K = \{x \in I \mid x > c\}$.
>
> *Case 3* $x_1 < x_2 = c$. Apply the proof of the Increasing/Decreasing Test, using the Mean Value Theorem (MVT) on the interval $[x_1, x_2]$ and noting that the MVT does not require f to be differentiable at the endpoints of $[x_1, x_2]$.
>
> *Case 4* $c = x_1 < x_2$. Same proof as in Case 3.
>
> *Case 5* $x_1 < c < x_2$. By Cases 3 and 4, f is increasing on $[x_1, c]$ and on $[c, x_2]$, so $f(x_1) < f(c) < f(x_2)$.

In all cases, we have shown that $f(x_1) < f(x_2)$. Since x_1, x_2 were any numbers in I with $x_1 < x_2$, we have shown that f is increasing on I.

73. (a) Since f and g are positive, increasing, and CU on I with f'' and g'' never equal to 0, we have $f > 0$, $f' \geq 0$, $f'' > 0$, $g > 0$, $g' \geq 0$, $g'' > 0$ on I. Then $(fg)' = f'g + fg'$ $\Rightarrow$ $(fg)'' = f''g + 2f'g' + fg'' \geq f''g + fg'' > 0$ on I $\Rightarrow$ fg is CU on I.

(b) In part (a), if f and g are both decreasing instead of increasing, then $f' \leq 0$ and $g' \leq 0$ on I, so we still have $2f'g' \geq 0$ on I. Thus, $(fg)'' = f''g + 2f'g' + fg'' \geq f''g + fg'' > 0$ on I $\Rightarrow$ fg is CU on I as in part (a).

(c) Suppose f is increasing and g is decreasing [with f and g positive and CU]. Then $f' \geq 0$ and $g' \leq 0$ on I, so $2f'g' \leq 0$ on I and the argument in parts (a) and (b) fails.

> **Example 1.** $I = (0, \infty)$, $f(x) = x^3$, $g(x) = 1/x$. Then $(fg)(x) = x^2$, so $(fg)'(x) = 2x$ and $(fg)''(x) = 2 > 0$ on I. Thus, fg is CU on I.
>
> **Example 2.** $I = (0, \infty)$, $f(x) = 4x\sqrt{x}$, $g(x) = 1/x$. Then $(fg)(x) = 4\sqrt{x}$, so $(fg)'(x) = 2/\sqrt{x}$ and $(fg)''(x) = -1/\sqrt{x^3} < 0$ on I. Thus, fg is CD on I.
>
> **Example 3.** $I = (0, \infty)$, $f(x) = x^2$, $g(x) = 1/x$. Thus, $(fg)(x) = x$, so fg is linear on I.

75. $f(x) = \tan x - x$ $\Rightarrow$ $f'(x) = \sec^2 x - 1 > 0$ for $0 < x < \frac{\pi}{2}$ since $\sec^2 x > 1$ for $0 < x < \frac{\pi}{2}$. So f is increasing on $\left(0, \frac{\pi}{2}\right)$. Thus, $f(x) > f(0) = 0$ for $0 < x < \frac{\pi}{2}$ $\Rightarrow$ $\tan x - x > 0$ $\Rightarrow$ $\tan x > x$ for $0 < x < \frac{\pi}{2}$.

77. Let the cubic function be $f(x) = ax^3 + bx^2 + cx + d$ $\Rightarrow$ $f'(x) = 3ax^2 + 2bx + c$ $\Rightarrow$ $f''(x) = 6ax + 2b$.

So f is CU when $6ax + 2b > 0$ $\Leftrightarrow$ $x > -b/(3a)$, CD when $x < -b/(3a)$, and so the only point of inflection occurs when

$x = -b/(3a)$. If the graph has three x-intercepts x_1, x_2 and x_3, then the expression for $f(x)$ must factor as

$f(x) = a(x - x_1)(x - x_2)(x - x_3)$. Multiplying these factors together gives us

$$f(x) = a[x^3 - (x_1 + x_2 + x_3)x^2 + (x_1x_2 + x_1x_3 + x_2x_3)x - x_1x_2x_3]$$

Equating the coefficients of the x^2-terms for the two forms of f gives us $b = -a(x_1 + x_2 + x_3)$. Hence, the x-coordinate of

the point of inflection is $-\dfrac{b}{3a} = -\dfrac{-a(x_1 + x_2 + x_3)}{3a} = \dfrac{x_1 + x_2 + x_3}{3}$.

79. By hypothesis $g = f'$ is differentiable on an open interval containing c. Since $(c, f(c))$ is a point of inflection, the concavity

changes at $x = c$, so $f''(x)$ changes signs at $x = c$. Hence, by the First Derivative Test, f' has a local extremum at $x = c$.

Thus, by Fermat's Theorem $f''(c) = 0$.

81. Using the fact that $|x| = \sqrt{x^2}$, we have that $g(x) = x\sqrt{x^2}$ $\Rightarrow$ $g'(x) = \sqrt{x^2} + \sqrt{x^2} = 2\sqrt{x^2} = 2|x|$ $\Rightarrow$

$g''(x) = 2x(x^2)^{-1/2} = \dfrac{2x}{|x|} < 0$ for $x < 0$ and $g''(x) > 0$ for $x > 0$, so $(0, 0)$ is an inflection point. But $g''(0)$ does not

exist.

83. (a) $f(x) = x^4 \sin \dfrac{1}{x}$ $\Rightarrow$ $f'(x) = x^4 \cos \dfrac{1}{x} \left(-\dfrac{1}{x^2}\right) + \sin \dfrac{1}{x} (4x^3) = 4x^3 \sin \dfrac{1}{x} - x^2 \cos \dfrac{1}{x}$.

$g(x) = x^4 \left(2 + \sin \dfrac{1}{x}\right) = 2x^4 + f(x)$ $\Rightarrow$ $g'(x) = 8x^3 + f'(x)$.

$h(x) = x^4 \left(-2 + \sin \dfrac{1}{x}\right) = -2x^4 + f(x)$ $\Rightarrow$ $h'(x) = -8x^3 + f'(x)$.

It is given that $f(0) = 0$, so $f'(0) = \lim\limits_{x \to 0} \dfrac{f(x) - f(0)}{x - 0} = \lim\limits_{x \to 0} \dfrac{x^4 \sin \dfrac{1}{x} - 0}{x} = \lim\limits_{x \to 0} x^3 \sin \dfrac{1}{x}$. Since

$-|x^3| \le x^3 \sin \dfrac{1}{x} \le |x^3|$ and $\lim\limits_{x \to 0} |x^3| = 0$, we see that $f'(0) = 0$ by the Squeeze Theorem. Also,

$g'(0) = 8(0)^3 + f'(0) = 0$ and $h'(0) = -8(0)^3 + f'(0) = 0$, so 0 is a critical number of f, g, and h.

For $x_{2n} = \dfrac{1}{2n\pi}$ [n a nonzero integer], $\sin \dfrac{1}{x_{2n}} = \sin 2n\pi = 0$ and $\cos \dfrac{1}{x_{2n}} = \cos 2n\pi = 1$, so $f'(x_{2n}) = -x_{2n}^2 < 0$.

For $x_{2n+1} = \dfrac{1}{(2n+1)\pi}$, $\sin \dfrac{1}{x_{2n+1}} = \sin(2n+1)\pi = 0$ and $\cos \dfrac{1}{x_{2n+1}} = \cos(2n+1)\pi = -1$, so

$f'(x_{2n+1}) = x_{2n+1}^2 > 0$. Thus, f' changes sign infinitely often on both sides of 0.

Next, $g'(x_{2n}) = 8x_{2n}^3 + f'(x_{2n}) = 8x_{2n}^3 - x_{2n}^2 = x_{2n}^2(8x_{2n} - 1) < 0$ for $x_{2n} < \frac{1}{8}$, but

$g'(x_{2n+1}) = 8x_{2n+1}^3 + x_{2n+1}^2 = x_{2n+1}^2(8x_{2n+1} + 1) > 0$ for $x_{2n+1} > -\frac{1}{8}$, so g' changes sign infinitely often on both

sides of 0.

Last, $h'(x_{2n}) = -8x_{2n}^3 + f'(x_{2n}) = -8x_{2n}^3 - x_{2n}^2 = -x_{2n}^2(8x_{2n} + 1) < 0$ for $x_{2n} > -\frac{1}{8}$ and

$h'(x_{2n+1}) = -8x_{2n+1}^3 + x_{2n+1}^2 = x_{2n+1}^2(-8x_{2n+1} + 1) > 0$ for $x_{2n+1} < \frac{1}{8}$, so h' changes sign infinitely often on both

sides of 0.

(b) $f(0) = 0$ and since $\sin \dfrac{1}{x}$ and hence $x^4 \sin \dfrac{1}{x}$ is both positive and negative inifinitely often on both sides of 0, and arbitrarily close to 0, f has neither a local maximum nor a local minimum at 0.

Since $2 + \sin \dfrac{1}{x} \geq 1$, $g(x) = x^4 \left(2 + \sin \dfrac{1}{x}\right) > 0$ for $x \neq 0$, so $g(0) = 0$ is a local minimum.

Since $-2 + \sin \dfrac{1}{x} \leq -1$, $h(x) = x^4 \left(-2 + \sin \dfrac{1}{x}\right) < 0$ for $x \neq 0$, so $h(0) = 0$ is a local maximum.

4.4   Indeterminate Forms and L'Hospital's Rule

Note: The use of l'Hospital's Rule is indicated by an H above the equal sign: $\overset{H}{=}$

1. (a) $\lim\limits_{x \to a} \dfrac{f(x)}{g(x)}$ is an indeterminate form of type $\dfrac{0}{0}$.

(b) $\lim\limits_{x \to a} \dfrac{f(x)}{p(x)} = 0$ because the numerator approaches 0 while the denominator becomes large.

(c) $\lim\limits_{x \to a} \dfrac{h(x)}{p(x)} = 0$ because the numerator approaches a finite number while the denominator becomes large.

(d) If $\lim\limits_{x \to a} p(x) = \infty$ and $f(x) \to 0$ through positive values, then $\lim\limits_{x \to a} \dfrac{p(x)}{f(x)} = \infty$. [For example, take $a = 0$, $p(x) = 1/x^2$, and $f(x) = x^2$.] If $f(x) \to 0$ through negative values, then $\lim\limits_{x \to a} \dfrac{p(x)}{f(x)} = -\infty$. [For example, take $a = 0$, $p(x) = 1/x^2$, and $f(x) = -x^2$.] If $f(x) \to 0$ through both positive and negative values, then the limit might not exist. [For example, take $a = 0$, $p(x) = 1/x^2$, and $f(x) = x$.]

(e) $\lim\limits_{x \to a} \dfrac{p(x)}{q(x)}$ is an indeterminate form of type $\dfrac{\infty}{\infty}$.

3. (a) When x is near a, $f(x)$ is near 0 and $p(x)$ is large, so $f(x) - p(x)$ is large negative. Thus, $\lim\limits_{x \to a} [f(x) - p(x)] = -\infty$.

(b) $\lim\limits_{x \to a} [p(x) - q(x)]$ is an indeterminate form of type $\infty - \infty$.

(c) When x is near a, $p(x)$ and $q(x)$ are both large, so $p(x) + q(x)$ is large. Thus, $\lim\limits_{x \to a} [p(x) + q(x)] = \infty$.

5. This limit has the form $\frac{0}{0}$. We can simply factor and simplify to evaluate the limit.

$$\lim_{x \to 1} \frac{x^2 - 1}{x^2 - x} = \lim_{x \to 1} \frac{(x+1)(x-1)}{x(x-1)} = \lim_{x \to 1} \frac{x+1}{x} = \frac{1+1}{1} = 2$$

7. This limit has the form $\frac{0}{0}$. $\lim\limits_{x \to 1} \dfrac{x^9 - 1}{x^5 - 1} \overset{H}{=} \lim\limits_{x \to 1} \dfrac{9x^8}{5x^4} = \dfrac{9}{5} \lim\limits_{x \to 1} x^4 = \dfrac{9}{5}(1) = \dfrac{9}{5}$

9. This limit has the form $\frac{0}{0}$. $\lim\limits_{x \to (\pi/2)^+} \dfrac{\cos x}{1 - \sin x} \overset{H}{=} \lim\limits_{x \to (\pi/2)^+} \dfrac{-\sin x}{-\cos x} = \lim\limits_{x \to (\pi/2)^+} \tan x = -\infty$.

11. This limit has the form $\frac{0}{0}$. $\lim\limits_{t \to 0} \dfrac{e^t - 1}{t^3} \overset{H}{=} \lim\limits_{t \to 0} \dfrac{e^t}{3t^2} = \infty$ since $e^t \to 1$ and $3t^2 \to 0^+$ as $t \to 0$.

13. This limit has the form $\frac{0}{0}$. $\displaystyle\lim_{x\to 0}\frac{\tan px}{\tan qx} \overset{\text{H}}{=} \lim_{x\to 0}\frac{p\sec^2 px}{q\sec^2 qx} = \frac{p(1)^2}{q(1)^2} = \frac{p}{q}$

15. This limit has the form $\frac{\infty}{\infty}$. $\displaystyle\lim_{x\to\infty}\frac{\ln x}{\sqrt{x}} \overset{\text{H}}{=} \lim_{x\to\infty}\frac{1/x}{\frac{1}{2}x^{-1/2}} = \lim_{x\to\infty}\frac{2}{\sqrt{x}} = 0$

17. $\displaystyle\lim_{x\to 0^+}[(\ln x)/x] = -\infty$ since $\ln x \to -\infty$ as $x\to 0^+$ and dividing by small values of x just increases the magnitude of the quotient $(\ln x)/x$. L'Hospital's Rule does not apply.

19. This limit has the form $\frac{\infty}{\infty}$. $\displaystyle\lim_{x\to\infty}\frac{e^x}{x^3} \overset{\text{H}}{=} \lim_{x\to\infty}\frac{e^x}{3x^2} \overset{\text{H}}{=} \lim_{x\to\infty}\frac{e^x}{6x} \overset{\text{H}}{=} \lim_{x\to\infty}\frac{e^x}{6} = \infty$

21. This limit has the form $\frac{0}{0}$. $\displaystyle\lim_{x\to 0}\frac{e^x - 1 - x}{x^2} \overset{\text{H}}{=} \lim_{x\to 0}\frac{e^x - 1}{2x} \overset{\text{H}}{=} \lim_{x\to 0}\frac{e^x}{2} = \frac{1}{2}$

23. This limit has the form $\frac{0}{0}$. $\displaystyle\lim_{x\to 0}\frac{\tanh x}{\tan x} \overset{\text{H}}{=} \lim_{x\to 0}\frac{\operatorname{sech}^2 x}{\sec^2 x} = \frac{\operatorname{sech}^2 0}{\sec^2 0} = \frac{1}{1} = 1$

25. This limit has the form $\frac{0}{0}$. $\displaystyle\lim_{t\to 0}\frac{5^t - 3^t}{t} \overset{\text{H}}{=} \lim_{t\to 0}\frac{5^t \ln 5 - 3^t \ln 3}{1} = \ln 5 - \ln 3 = \ln\frac{5}{3}$

27. This limit has the form $\frac{0}{0}$. $\displaystyle\lim_{x\to 0}\frac{\sin^{-1} x}{x} \overset{\text{H}}{=} \lim_{x\to 0}\frac{1/\sqrt{1-x^2}}{1} = \lim_{x\to 0}\frac{1}{\sqrt{1-x^2}} = \frac{1}{1} = 1$

29. This limit has the form $\frac{0}{0}$. $\displaystyle\lim_{x\to 0}\frac{1 - \cos x}{x^2} \overset{\text{H}}{=} \lim_{x\to 0}\frac{\sin x}{2x} \overset{\text{H}}{=} \lim_{x\to 0}\frac{\cos x}{2} = \frac{1}{2}$

31. $\displaystyle\lim_{x\to 0}\frac{x + \sin x}{x + \cos x} = \frac{0+0}{0+1} = \frac{0}{1} = 0$. L'Hospital's Rule does not apply.

33. This limit has the form $\frac{0}{0}$. $\displaystyle\lim_{x\to 1}\frac{1 - x + \ln x}{1 + \cos \pi x} \overset{\text{H}}{=} \lim_{x\to 1}\frac{-1 + 1/x}{-\pi \sin \pi x} \overset{\text{H}}{=} \lim_{x\to 1}\frac{-1/x^2}{-\pi^2 \cos \pi x} = \frac{-1}{-\pi^2(-1)} = -\frac{1}{\pi^2}$

35. This limit has the form $\frac{0}{0}$. $\displaystyle\lim_{x\to 1}\frac{x^a - ax + a - 1}{(x-1)^2} \overset{\text{H}}{=} \lim_{x\to 1}\frac{ax^{a-1} - a}{2(x-1)} \overset{\text{H}}{=} \lim_{x\to 1}\frac{a(a-1)x^{a-2}}{2} = \frac{a(a-1)}{2}$

37. This limit has the form $\frac{0}{0}$. $\displaystyle\lim_{x\to 0}\frac{\cos x - 1 + \frac{1}{2}x^2}{x^4} \overset{\text{H}}{=} \lim_{x\to 0}\frac{-\sin x + x}{4x^3} \overset{\text{H}}{=} \lim_{x\to 0}\frac{-\cos x + 1}{12x^2} \overset{\text{H}}{=} \lim_{x\to 0}\frac{\sin x}{24x} \overset{\text{H}}{=} \lim_{x\to 0}\frac{\cos x}{24} = \frac{1}{24}$

39. This limit has the form $\infty \cdot 0$.

$\displaystyle\lim_{x\to\infty} x\sin(\pi/x) = \lim_{x\to\infty}\frac{\sin(\pi/x)}{1/x} \overset{\text{H}}{=} \lim_{x\to\infty}\frac{\cos(\pi/x)(-\pi/x^2)}{-1/x^2} = \pi\lim_{x\to\infty}\cos(\pi/x) = \pi(1) = \pi$

41. This limit has the form $\infty \cdot 0$. We'll change it to the form $\frac{0}{0}$.

$\displaystyle\lim_{x\to 0}\cot 2x \sin 6x = \lim_{x\to 0}\frac{\sin 6x}{\tan 2x} \overset{\text{H}}{=} \lim_{x\to 0}\frac{6\cos 6x}{2\sec^2 2x} = \frac{6(1)}{2(1)^2} = 3$

43. This limit has the form $\infty \cdot 0$. $\displaystyle\lim_{x\to\infty} x^3 e^{-x^2} = \lim_{x\to\infty}\frac{x^3}{e^{x^2}} \overset{\text{H}}{=} \lim_{x\to\infty}\frac{3x^2}{2xe^{x^2}} = \lim_{x\to\infty}\frac{3x}{2e^{x^2}} \overset{\text{H}}{=} \lim_{x\to\infty}\frac{3}{4xe^{x^2}} = 0$

45. This limit has the form $0 \cdot (-\infty)$.

$$\lim_{x \to 1^+} \ln x \, \tan(\pi x/2) = \lim_{x \to 1^+} \frac{\ln x}{\cot(\pi x/2)} \overset{\text{H}}{=} \lim_{x \to 1^+} \frac{1/x}{(-\pi/2)\csc^2(\pi x/2)} = \frac{1}{(-\pi/2)(1)^2} = -\frac{2}{\pi}$$

47. This limit has the form $\infty - \infty$.

$$\lim_{x \to 1} \left(\frac{x}{x-1} - \frac{1}{\ln x} \right) = \lim_{x \to 1} \frac{x \ln x - (x-1)}{(x-1) \ln x} \overset{\text{H}}{=} \lim_{x \to 1} \frac{x(1/x) + \ln x - 1}{(x-1)(1/x) + \ln x} = \lim_{x \to 1} \frac{\ln x}{1 - (1/x) + \ln x}$$

$$\overset{\text{H}}{=} \lim_{x \to 1} \frac{1/x}{1/x^2 + 1/x} \cdot \frac{x^2}{x^2} = \lim_{x \to 1} \frac{x}{1+x} = \frac{1}{1+1} = \frac{1}{2}$$

49. We will multiply and divide by the conjugate of the expression to change the form of the expression.

$$\lim_{x \to \infty} \left(\sqrt{x^2+x} - x \right) = \lim_{x \to \infty} \left(\frac{\sqrt{x^2+x} - x}{1} \cdot \frac{\sqrt{x^2+x}+x}{\sqrt{x^2+x}+x} \right) = \lim_{x \to \infty} \frac{(x^2+x) - x^2}{\sqrt{x^2+x}+x}$$

$$= \lim_{x \to \infty} \frac{x}{\sqrt{x^2+x}+x} = \lim_{x \to \infty} \frac{1}{\sqrt{1+1/x}+1} = \frac{1}{\sqrt{1}+1} = \frac{1}{2}$$

As an alternate solution, write $\sqrt{x^2+x} - x$ as $\sqrt{x^2+x} - \sqrt{x^2}$, factor out $\sqrt{x^2}$, rewrite as $\left(\sqrt{1 + 1/x} - 1 \right)/(1/x)$, and apply l'Hospital's Rule.

51. The limit has the form $\infty - \infty$ and we will change the form to a product by factoring out x.

$$\lim_{x \to \infty} (x - \ln x) = \lim_{x \to \infty} x \left(1 - \frac{\ln x}{x} \right) = \infty \text{ since } \lim_{x \to \infty} \frac{\ln x}{x} \overset{\text{H}}{=} \lim_{x \to \infty} \frac{1/x}{1} = 0.$$

53. $y = x^{x^2} \quad \Rightarrow \quad \ln y = x^2 \ln x$, so $\displaystyle \lim_{x \to 0^+} \ln y = \lim_{x \to 0^+} x^2 \ln x = \lim_{x \to 0^+} \frac{\ln x}{1/x^2} \overset{\text{H}}{=} \lim_{x \to 0^+} \frac{1/x}{-2/x^3} = \lim_{x \to 0^+} \left(-\frac{1}{2} x^2 \right) = 0 \quad \Rightarrow$

$$\lim_{x \to 0^+} x^{x^2} = \lim_{x \to 0^+} e^{\ln y} = e^0 = 1.$$

55. $y = (1-2x)^{1/x} \quad \Rightarrow \quad \ln y = \dfrac{1}{x} \ln(1-2x)$, so $\displaystyle \lim_{x \to 0} \ln y = \lim_{x \to 0} \frac{\ln(1-2x)}{x} \overset{\text{H}}{=} \lim_{x \to 0} \frac{-2/(1-2x)}{1} = -2 \quad \Rightarrow$

$$\lim_{x \to 0} (1-2x)^{1/x} = \lim_{x \to 0} e^{\ln y} = e^{-2}.$$

57. $y = \left(1 + \dfrac{3}{x} + \dfrac{5}{x^2} \right)^x \quad \Rightarrow \quad \ln y = x \ln \left(1 + \dfrac{3}{x} + \dfrac{5}{x^2} \right) \quad \Rightarrow$

$$\lim_{x \to \infty} \ln y = \lim_{x \to \infty} \frac{\ln \left(1 + \dfrac{3}{x} + \dfrac{5}{x^2} \right)}{1/x} \overset{\text{H}}{=} \lim_{x \to \infty} \frac{\left(-\dfrac{3}{x^2} - \dfrac{10}{x^3} \right) \Big/ \left(1 + \dfrac{3}{x} + \dfrac{5}{x^2} \right)}{-1/x^2} = \lim_{x \to \infty} \frac{3 + \dfrac{10}{x}}{1 + \dfrac{3}{x} + \dfrac{5}{x^2}} = 3,$$

so $\displaystyle \lim_{x \to \infty} \left(1 + \frac{3}{x} + \frac{5}{x^2} \right)^x = \lim_{x \to \infty} e^{\ln y} = e^3.$

59. $y = x^{1/x} \quad \Rightarrow \quad \ln y = (1/x) \ln x \quad \Rightarrow \quad \displaystyle \lim_{x \to \infty} \ln y = \lim_{x \to \infty} \frac{\ln x}{x} \overset{\text{H}}{=} \lim_{x \to \infty} \frac{1/x}{1} = 0 \quad \Rightarrow$

$$\lim_{x \to \infty} x^{1/x} = \lim_{x \to \infty} e^{\ln y} = e^0 = 1$$

61. $y = (4x + 1)^{\cot x}$ $\Rightarrow$ $\ln y = \cot x \ln(4x + 1)$, so $\lim\limits_{x \to 0^+} \ln y = \lim\limits_{x \to 0^+} \dfrac{\ln(4x + 1)}{\tan x} \overset{\text{H}}{=} \lim\limits_{x \to 0^+} \dfrac{\frac{4}{4x + 1}}{\sec^2 x} = 4$ $\Rightarrow$

$\lim\limits_{x \to 0^+} (4x + 1)^{\cot x} = \lim\limits_{x \to 0^+} e^{\ln y} = e^4$.

63. $y = (\cos x)^{1/x^2}$ $\Rightarrow$ $\ln y = \dfrac{1}{x^2} \ln \cos x$ $\Rightarrow$ $\lim\limits_{x \to 0^+} \ln y = \lim\limits_{x \to 0^+} \dfrac{\ln \cos x}{x^2} \overset{\text{H}}{=} \lim\limits_{x \to 0^+} \dfrac{-\tan x}{2x} \overset{\text{H}}{=} \lim\limits_{x \to 0^+} \dfrac{-\sec^2 x}{2} = -\dfrac{1}{2}$

$\Rightarrow$ $\lim\limits_{x \to 0^+} (\cos x)^{1/x^2} = \lim\limits_{x \to 0^+} e^{\ln y} = e^{-1/2} = 1/\sqrt{e}$

65.

From the graph, if $x = 500$, $y \approx 7.36$. The limit has the form 1^∞.

Now $y = \left(1 + \dfrac{2}{x}\right)^x$ $\Rightarrow$ $\ln y = x \ln \left(1 + \dfrac{2}{x}\right)$ $\Rightarrow$

$\lim\limits_{x \to \infty} \ln y = \lim\limits_{x \to \infty} \dfrac{\ln(1 + 2/x)}{1/x} \overset{\text{H}}{=} \lim\limits_{x \to \infty} \dfrac{\dfrac{1}{1 + 2/x}\left(-\dfrac{2}{x^2}\right)}{-1/x^2}$

$= 2 \lim\limits_{x \to \infty} \dfrac{1}{1 + 2/x} = 2(1) = 2$ $\Rightarrow$

$\lim\limits_{x \to \infty} \left(1 + \dfrac{2}{x}\right)^x = \lim\limits_{x \to \infty} e^{\ln y} = e^2$ $[\approx 7.39]$

67.

From the graph, it appears that $\lim\limits_{x \to 0} \dfrac{f(x)}{g(x)} = \lim\limits_{x \to 0} \dfrac{f'(x)}{g'(x)} = 0.25$.

We calculate $\lim\limits_{x \to 0} \dfrac{f(x)}{g(x)} = \lim\limits_{x \to 0} \dfrac{e^x - 1}{x^3 + 4x} \overset{\text{H}}{=} \lim\limits_{x \to 0} \dfrac{e^x}{3x^2 + 4} = \dfrac{1}{4}$.

69. $\lim\limits_{x \to \infty} \dfrac{e^x}{x^n} \overset{\text{H}}{=} \lim\limits_{x \to \infty} \dfrac{e^x}{nx^{n-1}} \overset{\text{H}}{=} \lim\limits_{x \to \infty} \dfrac{e^x}{n(n - 1)x^{n-2}} \overset{\text{H}}{=} \cdots \overset{\text{H}}{=} \lim\limits_{x \to \infty} \dfrac{e^x}{n!} = \infty$

71. $\lim\limits_{x \to \infty} \dfrac{x}{\sqrt{x^2 + 1}} \overset{\text{H}}{=} \lim\limits_{x \to \infty} \dfrac{1}{\frac{1}{2}(x^2 + 1)^{-1/2}(2x)} = \lim\limits_{x \to \infty} \dfrac{\sqrt{x^2 + 1}}{x}$. Repeated applications of l'Hospital's Rule result in the

original limit or the limit of the reciprocal of the function. Another method is to try dividing the numerator and denominator

by x: $\lim\limits_{x \to \infty} \dfrac{x}{\sqrt{x^2 + 1}} = \lim\limits_{x \to \infty} \dfrac{x/x}{\sqrt{x^2/x^2 + 1/x^2}} = \lim\limits_{x \to \infty} \dfrac{1}{\sqrt{1 + 1/x^2}} = \dfrac{1}{1} = 1$

73. First we will find $\lim\limits_{n \to \infty} \left(1 + \dfrac{r}{n}\right)^{nt}$, which is of the form 1^∞. $y = \left(1 + \dfrac{r}{n}\right)^{nt}$ $\Rightarrow$ $\ln y = nt \ln\left(1 + \dfrac{r}{n}\right)$, so

$\lim\limits_{n \to \infty} \ln y = \lim\limits_{n \to \infty} nt \ln\left(1 + \dfrac{r}{n}\right) = t \lim\limits_{n \to \infty} \dfrac{\ln(1 + r/n)}{1/n} \overset{\text{H}}{=} t \lim\limits_{n \to \infty} \dfrac{(-r/n^2)}{(1 + r/n)(-1/n^2)} = t \lim\limits_{n \to \infty} \dfrac{r}{1 + i/n} = tr$ $\Rightarrow$

$\lim\limits_{n \to \infty} y = e^{rt}$. Thus, as $n \to \infty$, $A = A_0\left(1 + \dfrac{r}{n}\right)^{nt} \to A_0 e^{rt}$.

75. $\displaystyle\lim_{E\to 0^+} P(E) = \lim_{E\to 0^+}\left(\frac{e^E + e^{-E}}{e^E - e^{-E}} - \frac{1}{E}\right)$

$\displaystyle = \lim_{E\to 0^+}\frac{E(e^E + e^{-E}) - 1(e^E - e^{-E})}{(e^E - e^{-E})\,E} = \lim_{E\to 0^+}\frac{Ee^E + Ee^{-E} - e^E + e^{-E}}{Ee^E - Ee^{-E}}$ [form is $\frac{0}{0}$]

$\displaystyle \overset{H}{=} \lim_{E\to 0^+}\frac{Ee^E + e^E \cdot 1 + E(-e^{-E}) + e^{-E}\cdot 1 - e^E + (-e^{-E})}{Ee^E + e^E\cdot 1 - [E(-e^{-E}) + e^{-E}\cdot 1]}$

$\displaystyle = \lim_{E\to 0^+}\frac{Ee^E - Ee^{-E}}{Ee^E + e^E + Ee^{-E} - e^{-E}} = \lim_{E\to 0^+}\frac{e^E - e^{-E}}{e^E + \dfrac{e^E}{E} + e^{-E} - \dfrac{e^{-E}}{E}}$ [divide by E]

$\displaystyle = \frac{0}{2 + L},\quad\text{where } L = \lim_{E\to 0^+}\frac{e^E - e^{-E}}{E}$ $\left[\text{form is } \frac{0}{0}\right]$ $\overset{H}{=} \displaystyle\lim_{E\to 0^+}\frac{e^E + e^{-E}}{1} = \frac{1+1}{1} = 2$

Thus, $\displaystyle\lim_{E\to 0^+} P(E) = \frac{0}{2+2} = 0$.

77. We see that both numerator and denominator approach 0, so we can use l'Hospital's Rule:

$\displaystyle\lim_{x\to a}\frac{\sqrt{2a^3 x - x^4} - a\sqrt[3]{aax}}{a - \sqrt[4]{ax^3}} \overset{H}{=} \lim_{x\to a}\frac{\frac{1}{2}(2a^3 x - x^4)^{-1/2}(2a^3 - 4x^3) - a\left(\frac{1}{3}\right)(aax)^{-2/3}a^2}{-\frac{1}{4}(ax^3)^{-3/4}(3ax^2)}$

$\displaystyle = \frac{\frac{1}{2}(2a^3 a - a^4)^{-1/2}(2a^3 - 4a^3) - \frac{1}{3}a^3(a^2 a)^{-2/3}}{-\frac{1}{4}(aa^3)^{-3/4}(3aa^2)}$

$\displaystyle = \frac{(a^4)^{-1/2}(-a^3) - \frac{1}{3}a^3(a^3)^{-2/3}}{-\frac{3}{4}a^3(a^4)^{-3/4}} = \frac{-a - \frac{1}{3}a}{-\frac{3}{4}} = \frac{4}{3}\left(\frac{4}{3}a\right) = \frac{16}{9}a$

79. Since $f(2) = 0$, the given limit has the form $\frac{0}{0}$.

$\displaystyle\lim_{x\to 0}\frac{f(2+3x) + f(2+5x)}{x} \overset{H}{=} \lim_{x\to 0}\frac{f'(2+3x)\cdot 3 + f'(2+5x)\cdot 5}{1} = f'(2)\cdot 3 + f'(2)\cdot 5 = 8f'(2) = 8\cdot 7 = 56$

81. Since $\displaystyle\lim_{h\to 0}[f(x+h) - f(x-h)] = f(x) - f(x) = 0$ (f is differentiable and hence continuous) and $\displaystyle\lim_{h\to 0}2h = 0$, we use l'Hospital's Rule:

$\displaystyle\lim_{h\to 0}\frac{f(x+h) - f(x-h)}{2h} \overset{H}{=} \lim_{h\to 0}\frac{f'(x+h)(1) - f'(x-h)(-1)}{2} = \frac{f'(x) + f'(x)}{2} = \frac{2f'(x)}{2} = f'(x)$

$\dfrac{f(x+h) - f(x-h)}{2h}$ is the slope of the secant line between

$(x-h, f(x-h))$ and $(x+h, f(x+h))$. As $h\to 0$, this line gets closer

to the tangent line and its slope approaches $f'(x)$.

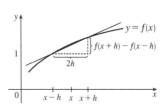

83. (a) We show that $\displaystyle\lim_{x\to 0}\frac{f(x)}{x^n} = 0$ for every integer $n \geq 0$. Let $y = \dfrac{1}{x^2}$. Then

$\displaystyle\lim_{x\to 0}\frac{f(x)}{x^{2n}} = \lim_{x\to 0}\frac{e^{-1/x^2}}{(x^2)^n} = \lim_{y\to\infty}\frac{y^n}{e^y} \overset{H}{=} \lim_{y\to\infty}\frac{ny^{n-1}}{e^y} \overset{H}{=} \cdots \overset{H}{=} \lim_{y\to\infty}\frac{n!}{e^y} = 0 \;\Rightarrow$

$\displaystyle\lim_{x\to 0}\frac{f(x)}{x^n} = \lim_{x\to 0}x^n\frac{f(x)}{x^{2n}} = \lim_{x\to 0}x^n\lim_{x\to 0}\frac{f(x)}{x^{2n}} = 0.$ Thus, $f'(0) = \displaystyle\lim_{x\to 0}\frac{f(x) - f(0)}{x - 0} = \lim_{x\to 0}\frac{f(x)}{x} = 0$.

(b) Using the Chain Rule and the Quotient Rule we see that $f^{(n)}(x)$ exists for $x \neq 0$. In fact, we prove by induction that for

each $n \geq 0$, there is a polynomial p_n and a non-negative integer k_n with $f^{(n)}(x) = p_n(x)f(x)/x^{k_n}$ for $x \neq 0$. This is

true for $n = 0$; suppose it is true for the nth derivative. Then $f'(x) = f(x)(2/x^3)$, so

$$f^{(n+1)}(x) = \left[x^{k_n}\left[p_n'(x)f(x) + p_n(x)f'(x)\right] - k_n x^{k_n-1} p_n(x)f(x)\right]x^{-2k_n}$$
$$= \left[x^{k_n}p_n'(x) + p_n(x)(2/x^3) - k_n x^{k_n-1}p_n(x)\right]f(x)x^{-2k_n}$$
$$= \left[x^{k_n+3}p_n'(x) + 2p_n(x) - k_n x^{k_n+2}p_n(x)\right]f(x)x^{-(2k_n+3)}$$

which has the desired form.

Now we show by induction that $f^{(n)}(0) = 0$ for all n. By part (a), $f'(0) = 0$. Suppose that $f^{(n)}(0) = 0$. Then

$$f^{(n+1)}(0) = \lim_{x \to 0} \frac{f^{(n)}(x) - f^{(n)}(0)}{x - 0} = \lim_{x \to 0} \frac{f^{(n)}(x)}{x} = \lim_{x \to 0} \frac{p_n(x)f(x)/x^{k_n}}{x} = \lim_{x \to 0} \frac{p_n(x)f(x)}{x^{k_n+1}}$$
$$= \lim_{x \to 0} p_n(x) \lim_{x \to 0} \frac{f(x)}{x^{k_n+1}} = p_n(0) \cdot 0 = 0$$

4.5 Summary of Curve Sketching

1. $y = f(x) = x^3 + x = x(x^2 + 1)$ **A.** f is a polynomial, so $D = \mathbb{R}$. **H.**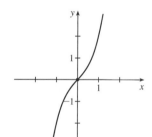

B. x-intercept $= 0$, y-intercept $= f(0) = 0$ **C.** $f(-x) = -f(x)$, so f is

odd; the curve is symmetric about the origin. **D.** f is a polynomial, so there is

no asymptote. **E.** $f'(x) = 3x^2 + 1 > 0$, so f is increasing on $(-\infty, \infty)$.

F. There is no critical number and hence, no local maximum or minimum value.

G. $f''(x) = 6x > 0$ on $(0, \infty)$ and $f''(x) < 0$ on $(-\infty, 0)$, so f is CU on

$(0, \infty)$ and CD on $(-\infty, 0)$. Since the concavity changes at $x = 0$, there is an

inflection point at $(0, 0)$.

3. $y = f(x) = 2 - 15x + 9x^2 - x^3 = -(x-2)(x^2 - 7x + 1)$ **A.** $D = \mathbb{R}$ **B.** y-intercept: $f(0) = 2$; x-intercepts:

$f(x) = 0 \Rightarrow x = 2$ or (by the quadratic formula) $x = \frac{7 \pm \sqrt{45}}{2} \approx 0.15, 6.85$ **C.** No symmetry **D.** No asymptote

E. $f'(x) = -15 + 18x - 3x^2 = -3(x^2 - 6x + 5)$ **H.**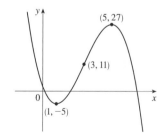
$= -3(x-1)(x-5) > 0 \Leftrightarrow 1 < x < 5$

so f is increasing on $(1, 5)$ and decreasing on $(-\infty, 1)$ and $(5, \infty)$.

F. Local maximum value $f(5) = 27$, local minimum value $f(1) = -5$

G. $f''(x) = 18 - 6x = -6(x-3) > 0 \Leftrightarrow x < 3$, so f is CU on $(-\infty, 3)$

and CD on $(3, \infty)$. IP at $(3, 11)$

5. $y = f(x) = x^4 + 4x^3 = x^3(x+4)$ **A.** $D = \mathbb{R}$ **B.** y-intercept: $f(0) = 0$;   **H.**

x-intercepts: $f(x) = 0 \iff x = -4, 0$ **C.** No symmetry

D. No asymptote   **E.** $f'(x) = 4x^3 + 12x^2 = 4x^2(x+3) > 0 \iff$

$x > -3$, so f is increasing on $(-3, \infty)$ and decreasing on $(-\infty, -3)$.

F. Local minimum value $f(-3) = -27$, no local maximum

G. $f''(x) = 12x^2 + 24x = 12x(x+2) < 0 \iff -2 < x < 0$, so f is CD

on $(-2, 0)$ and CU on $(-\infty, -2)$ and $(0, \infty)$. IP at $(0, 0)$ and $(-2, -16)$

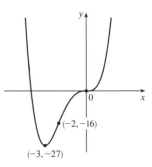

7. $y = f(x) = 2x^5 - 5x^2 + 1$ **A.** $D = \mathbb{R}$ **B.** y-intercept: $f(0) = 1$ **C.** No symmetry   **D.** No asymptote

E. $f'(x) = 10x^4 - 10x = 10x(x^3 - 1) = 10x(x-1)(x^2 + x + 1)$, so $f'(x) < 0 \iff 0 < x < 1$ and $f'(x) > 0 \iff$

$x < 0$ or $x > 1$. Thus, f is increasing on $(-\infty, 0)$ and $(1, \infty)$ and decreasing on $(0, 1)$.   **F.** Local maximum value $f(0) = 1$,

local minimum value $f(1) = -2$ **G.** $f''(x) = 40x^3 - 10 = 10(4x^3 - 1)$ **H.**

so $f''(x) = 0 \iff x = 1/\sqrt[3]{4}$. $f''(x) > 0 \iff x > 1/\sqrt[3]{4}$ and

$f''(x) < 0 \iff x < 1/\sqrt[3]{4}$, so f is CD on $\left(-\infty, 1/\sqrt[3]{4}\right)$ and CU

on $\left(1/\sqrt[3]{4}, \infty\right)$. IP at $\left(\dfrac{1}{\sqrt[3]{4}}, 1 - \dfrac{9}{2\left(\sqrt[3]{4}\right)^2}\right) \approx (0.630, -0.786)$

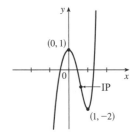

9. $y = f(x) = x/(x-1)$ **A.** $D = \{x \mid x \neq 1\} = (-\infty, 1) \cup (1, \infty)$ **B.** x-intercept $= 0$, y-intercept $= f(0) = 0$

C. No symmetry   **D.** $\lim\limits_{x \to \pm\infty} \dfrac{x}{x-1} = 1$, so $y = 1$ is a HA. $\lim\limits_{x \to 1^-} \dfrac{x}{x-1} = -\infty$, $\lim\limits_{x \to 1^+} \dfrac{x}{x-1} = \infty$, so $x = 1$ is a VA.

E. $f'(x) = \dfrac{(x-1) - x}{(x-1)^2} = \dfrac{-1}{(x-1)^2} < 0$ for $x \neq 1$, so f is   **H.**

decreasing on $(-\infty, 1)$ and $(1, \infty)$.   **F.** No extreme values

G. $f''(x) = \dfrac{2}{(x-1)^3} > 0 \iff x > 1$, so f is CU on $(1, \infty)$ and

CD on $(-\infty, 1)$. No IP

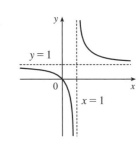

11. $y = f(x) = 1/(x^2 - 9)$ **A.** $D = \{x \mid x \neq \pm 3\} = (-\infty, -3) \cup (-3, 3) \cup (3, \infty)$ **B.** y-intercept $= f(0) = -\frac{1}{9}$, no

x-intercept   **C.** $f(-x) = f(x) \implies f$ is even; the curve is symmetric about the y-axis.   **D.** $\lim\limits_{x \to \pm\infty} \dfrac{1}{x^2 - 9} = 0$, so $y = 0$

is a HA. $\lim\limits_{x \to 3^-} \dfrac{1}{x^2 - 9} = -\infty$, $\lim\limits_{x \to 3^+} \dfrac{1}{x^2 - 9} = \infty$, $\lim\limits_{x \to -3^-} \dfrac{1}{x^2 - 9} = \infty$, $\lim\limits_{x \to -3^+} \dfrac{1}{x^2 - 9} = -\infty$, so $x = 3$ and $x = -3$

are VA.   **E.** $f'(x) = -\dfrac{2x}{(x^2 - 9)^2} > 0 \iff x < 0 \ (x \neq -3)$ so f is increasing on $(-\infty, -3)$ and $(-3, 0)$ and decreasing on $(0, 3)$ and $(3, \infty)$.   **F.** Local maximum value $f(0) = -\frac{1}{9}$.

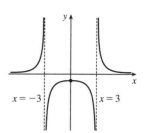

G. $y'' = \dfrac{-2(x^2 - 9)^2 + (2x)2(x^2 - 9)(2x)}{(x^2 - 9)^4} = \dfrac{6(x^2 + 3)}{(x^2 - 9)^3} > 0 \iff$

$x^2 > 9 \iff x > 3$ or $x < -3$, so f is CU on $(-\infty, -3)$ and $(3, \infty)$ and

CD on $(-3, 3)$. No IP

---

13. $y = f(x) = x/(x^2 + 9)$ **A.** $D = \mathbb{R}$ **B.** y-intercept: $f(0) = 0$; x-intercept: $f(x) = 0 \iff x = 0$

C. $f(-x) = -f(x)$, so f is odd and the curve is symmetric about the origin.   **D.** $\lim\limits_{x \to \pm\infty} [x/(x^2 + 9)] = 0$, so $y = 0$ is a

HA; no VA   **E.** $f'(x) = \dfrac{(x^2 + 9)(1) - x(2x)}{(x^2 + 9)^2} = \dfrac{9 - x^2}{(x^2 + 9)^2} = \dfrac{(3 + x)(3 - x)}{(x^2 + 9)^2} > 0 \iff -3 < x < 3$, so f is increasing

on $(-3, 3)$ and decreasing on $(-\infty, -3)$ and $(3, \infty)$.   **F.** Local minimum value $f(-3) = -\frac{1}{6}$, local maximum

value $f(3) = \frac{1}{6}$

$f''(x) = \dfrac{(x^2 + 9)^2(-2x) - (9 - x^2) \cdot 2(x^2 + 9)(2x)}{[(x^2 + 9)^2]^2} = \dfrac{(2x)(x^2 + 9)[-(x^2 + 9) - 2(9 - x^2)]}{(x^2 + 9)^4} = \dfrac{2x(x^2 - 27)}{(x^2 + 9)^3}$

$\qquad = 0 \iff x = 0, \pm\sqrt{27} = \pm 3\sqrt{3}$

G. $f''(x) = \dfrac{(x^2 + 9)^2(-2x) - (9 - x^2) \cdot 2(x^2 + 9)(2x)}{[(x^2 + 9)^2]^2} = \dfrac{(2x)(x^2 + 9)\left[-(x^2 + 9) - 2(9 - x^2)\right]}{(x^2 + 9)^4}$

$\qquad = \dfrac{2x(x^2 - 27)}{(x^2 + 9)^3} = 0 \iff x = 0, \pm\sqrt{27} = \pm 3\sqrt{3}$ **H.**

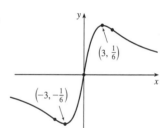

$f''(x) > 0 \iff -3\sqrt{3} < x < 0$ or $x > 3\sqrt{3}$, so f is CU on $\left(-3\sqrt{3}, 0\right)$

and $\left(3\sqrt{3}, \infty\right)$, and CD on $\left(-\infty, -3\sqrt{3}\,\right)$ and $\left(0, 3\sqrt{3}\,\right)$. There are three

inflection points: $(0, 0)$ and $\left(\pm 3\sqrt{3}, \pm\frac{1}{12}\sqrt{3}\,\right)$.

---

15. $y = f(x) = \dfrac{x - 1}{x^2}$ **A.** $D = \{x \mid x \neq 0\} = (-\infty, 0) \cup (0, \infty)$ **B.** No y-intercept; x-intercept: $f(x) = 0 \iff x = 1$

C. No symmetry   **D.** $\lim\limits_{x \to \pm\infty} \dfrac{x - 1}{x^2} = 0$, so $y = 0$ is a HA. $\lim\limits_{x \to 0} \dfrac{x - 1}{x^2} = -\infty$, so $x = 0$ is a VA.

E. $f'(x) = \dfrac{x^2 \cdot 1 - (x - 1) \cdot 2x}{(x^2)^2} = \dfrac{-x^2 + 2x}{x^4} = \dfrac{-(x - 2)}{x^3}$, so $f'(x) > 0 \iff 0 < x < 2$ and $f'(x) < 0 \iff$

$x < 0$ or $x > 2$. Thus, f is increasing on $(0, 2)$ and decreasing on $(-\infty, 0)$ **H.**

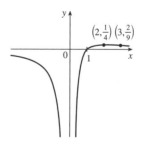

and $(2, \infty)$.   **F.** No local minimum, local maximum value $f(2) = \frac{1}{4}$.

G. $f''(x) = \dfrac{x^3 \cdot (-1) - [-(x - 2)] \cdot 3x^2}{(x^3)^2} = \dfrac{2x^3 - 6x^2}{x^6} = \dfrac{2(x - 3)}{x^4}$.

$f''(x)$ is negative on $(-\infty, 0)$ and $(0, 3)$ and positive on $(3, \infty)$, so f is CD

on $(-\infty, 0)$ and $(0, 3)$ and CU on $(3, \infty)$. IP at $\left(3, \frac{2}{9}\right)$

17. $y = f(x) = \dfrac{x^2}{x^2 + 3} = \dfrac{(x^2 + 3) - 3}{x^2 + 3} = 1 - \dfrac{3}{x^2 + 3}$ **A.** $D = \mathbb{R}$ **B.** y-intercept: $f(0) = 0$;

x-intercepts: $f(x) = 0 \ \Leftrightarrow \ x = 0$ **C.** $f(-x) = f(x)$, so f is even; the graph is symmetric about the y-axis.

D. $\displaystyle\lim_{x \to \pm\infty} \dfrac{x^2}{x^2 + 3} = 1$, so $y = 1$ is a HA. No VA. **E.** Using the Reciprocal Rule, $f'(x) = -3 \cdot \dfrac{-2x}{(x^2 + 3)^2} = \dfrac{6x}{(x^2 + 3)^2}$.

$f'(x) > 0 \ \Leftrightarrow \ x > 0$ and $f'(x) < 0 \ \Leftrightarrow \ x < 0$, so f is decreasing on $(-\infty, 0)$ and increasing on $(0, \infty)$.

F. Local minimum value $f(0) = 0$, no local maximum.

G. $f''(x) = \dfrac{(x^2 + 3)^2 \cdot 6 - 6x \cdot 2(x^2 + 3) \cdot 2x}{[(x^2 + 3)^2]^2}$ **H.**

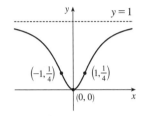

$= \dfrac{6(x^2 + 3)[(x^2 + 3) - 4x^2]}{(x^2 + 3)^4} = \dfrac{6(3 - 3x^2)}{(x^2 + 3)^3} = \dfrac{-18(x + 1)(x - 1)}{(x^2 + 3)^3}$

$f''(x)$ is negative on $(-\infty, -1)$ and $(1, \infty)$ and positive on $(-1, 1)$,

so f is CD on $(-\infty, -1)$ and $(1, \infty)$ and CU on $(-1, 1)$. IP at $\left(\pm 1, \frac{1}{4}\right)$

19. $y = f(x) = x\sqrt{5 - x}$ **A.** The domain is $\{x \mid 5 - x \geq 0\} = (-\infty, 5]$ **B.** y-intercept: $f(0) = 0$;

x-intercepts: $f(x) = 0 \ \Leftrightarrow \ x = 0, 5$ **C.** No symmetry **D.** No asymptote

E. $f'(x) = x \cdot \frac{1}{2}(5 - x)^{-1/2}(-1) + (5 - x)^{1/2} \cdot 1 = \frac{1}{2}(5 - x)^{-1/2}[-x + 2(5 - x)] = \dfrac{10 - 3x}{2\sqrt{5 - x}} > 0 \ \Leftrightarrow$

$x < \frac{10}{3}$, so f is increasing on $\left(-\infty, \frac{10}{3}\right)$ and decreasing on $\left(\frac{10}{3}, 5\right)$.

F. Local maximum value $f\left(\frac{10}{3}\right) = \frac{10}{9}\sqrt{15} \approx 4.3$; no local minimum **H.**

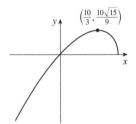

G. $f''(x) = \dfrac{2(5 - x)^{1/2}(-3) - (10 - 3x) \cdot 2\left(\frac{1}{2}\right)(5 - x)^{-1/2}(-1)}{\left(2\sqrt{5 - x}\right)^2}$

$= \dfrac{(5 - x)^{-1/2}[-6(5 - x) + (10 - 3x)]}{4(5 - x)} = \dfrac{3x - 20}{4(5 - x)^{3/2}}$

$f''(x) < 0$ for $x < 5$, so f is CD on $(-\infty, 5)$. No IP

21. $y = f(x) = \sqrt{x^2 + x - 2} = \sqrt{(x + 2)(x - 1)}$ **A.** $D = \{x \mid (x + 2)(x - 1) \geq 0\} = (-\infty, -2] \cup [1, \infty)$

B. y-intercept: none; x-intercepts: -2 and 1 **C.** No symmetry **D.** No asymptote

E. $f'(x) = \frac{1}{2}(x^2 + x - 2)^{-1/2}(2x + 1) = \dfrac{2x + 1}{2\sqrt{x^2 + x - 2}}$, $f'(x) = 0$ if $x = -\frac{1}{2}$, but $-\frac{1}{2}$ is not in the domain.

$f'(x) > 0 \ \Rightarrow \ x > -\frac{1}{2}$ and $f'(x) < 0 \ \Rightarrow \ x < -\frac{1}{2}$, so (considering the domain) f is increasing on $(1, \infty)$ and

f is decreasing on $(-\infty, -2)$. **F.** No local extrema

G. $f''(x) = \dfrac{2(x^2 + x - 2)^{1/2}(2) - (2x + 1) \cdot 2 \cdot \frac{1}{2}(x^2 + x - 2)^{-1/2}(2x + 1)}{\left(2\sqrt{x^2 + x - 2}\right)^2}$ **H.**

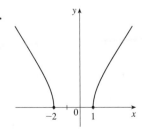

$= \dfrac{(x^2 + x - 2)^{-1/2}\left[4(x^2 + x - 2) - (4x^2 + 4x + 1)\right]}{4(x^2 + x - 2)}$

$= \dfrac{-9}{4(x^2 + x - 2)^{3/2}} < 0$

so f is CD on $(-\infty, -2)$ and $(1, \infty)$. No IP

23. $y = f(x) = x/\sqrt{x^2 + 1}$ **A.** $D = \mathbb{R}$ **B.** y-intercept: $f(0) = 0$; x-intercepts: $f(x) = 0 \Rightarrow x = 0$

C. $f(-x) = -f(x)$, so f is odd; the graph is symmetric about the origin.

D. $\displaystyle\lim_{x \to \infty} f(x) = \lim_{x \to \infty} \frac{x}{\sqrt{x^2 + 1}} = \lim_{x \to \infty} \frac{x/x}{\sqrt{x^2 + 1}/x} = \lim_{x \to \infty} \frac{x/x}{\sqrt{x^2 + 1}/\sqrt{x^2}} = \lim_{x \to \infty} \frac{1}{\sqrt{1 + 1/x^2}} = \frac{1}{\sqrt{1 + 0}} = 1$

and

$\displaystyle\lim_{x \to -\infty} f(x) = \lim_{x \to -\infty} \frac{x}{\sqrt{x^2 + 1}} = \lim_{x \to -\infty} \frac{x/x}{\sqrt{x^2 + 1}/x} = \lim_{x \to -\infty} \frac{x/x}{\sqrt{x^2 + 1}/\left(-\sqrt{x^2}\right)} = \lim_{x \to -\infty} \frac{1}{-\sqrt{1 + 1/x^2}}$

$= \dfrac{1}{-\sqrt{1 + 0}} = -1$ so $y = \pm 1$ are HA.

No VA.

E. $f'(x) = \dfrac{\sqrt{x^2 + 1} - x \cdot \dfrac{2x}{2\sqrt{x^2 + 1}}}{[(x^2 + 1)^{1/2}]^2} = \dfrac{x^2 + 1 - x^2}{(x^2 + 1)^{3/2}} = \dfrac{1}{(x^2 + 1)^{3/2}} > 0$ for all x, so f is increasing on $\mathbb{R}$.

F. No extreme values

H.

G. $f''(x) = -\frac{3}{2}(x^2 + 1)^{-5/2} \cdot 2x = \dfrac{-3x}{(x^2 + 1)^{5/2}}$, so $f''(x) > 0$ for $x < 0$

and $f''(x) < 0$ for $x > 0$. Thus, f is CU on $(-\infty, 0)$ and CD on $(0, \infty)$.

IP at $(0, 0)$

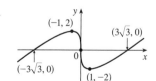

25. $y = f(x) = \sqrt{1 - x^2}/x$ **A.** $D = \{x \mid |x| \le 1,\, x \ne 0\} = [-1, 0) \cup (0, 1]$ **B.** x-intercepts ± 1, no y-intercept

C. $f(-x) = -f(x)$, so the curve is symmetric about $(0, 0)$. **D.** $\displaystyle\lim_{x \to 0^+} \frac{\sqrt{1 - x^2}}{x} = \infty$, $\displaystyle\lim_{x \to 0^-} \frac{\sqrt{1 - x^2}}{x} = -\infty$,

so $x = 0$ is a VA. **E.** $f'(x) = \dfrac{\left(-x^2/\sqrt{1 - x^2}\right) - \sqrt{1 - x^2}}{x^2} = -\dfrac{1}{x^2\sqrt{1 - x^2}} < 0$, so f is decreasing

on $(-1, 0)$ and $(0, 1)$. **F.** No extreme values

H.

G. $f''(x) = \dfrac{2 - 3x^2}{x^3(1 - x^2)^{3/2}} > 0 \Leftrightarrow -1 < x < -\sqrt{\frac{2}{3}}$ or $0 < x < \sqrt{\frac{2}{3}}$, so

f is CU on $\left(-1, -\sqrt{\frac{2}{3}}\right)$ and $\left(0, \sqrt{\frac{2}{3}}\right)$ and CD on $\left(-\sqrt{\frac{2}{3}}, 0\right)$ and $\left(\sqrt{\frac{2}{3}}, 1\right)$.

IP at $\left(\pm\sqrt{\frac{2}{3}}, \pm\frac{1}{\sqrt{2}}\right)$

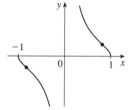

27. $y = f(x) = x - 3x^{1/3}$ **A.** $D = \mathbb{R}$ **B.** y-intercept: $f(0) = 0$; x-intercepts: $f(x) = 0 \Rightarrow x = 3x^{1/3} \Rightarrow$

$x^3 = 27x \Rightarrow x^3 - 27x = 0 \Rightarrow x(x^2 - 27) = 0 \Rightarrow x = 0, \pm 3\sqrt{3}$ **C.** $f(-x) = -f(x)$, so f is odd;

the graph is symmetric about the origin. **D.** No asymptote **E.** $f'(x) = 1 - x^{-2/3} = 1 - \dfrac{1}{x^{2/3}} = \dfrac{x^{2/3} - 1}{x^{2/3}}$.

$f'(x) > 0$ when $|x| > 1$ and $f'(x) < 0$ when $0 < |x| < 1$, so f is increasing on $(-\infty, -1)$ and $(1, \infty)$, and

decreasing on $(-1, 0)$ and $(0, 1)$ [hence decreasing on $(-1, 1)$ since f is

continuous on $(-1, 1)$]. **F.** Local maximum value $f(-1) = 2$, local minimum

value $f(1) = -2$ **G.** $f''(x) = \frac{2}{3}x^{-5/3} < 0$ when $x < 0$ and $f''(x) > 0$

when $x > 0$, so f is CD on $(-\infty, 0)$ and CU on $(0, \infty)$. IP at $(0, 0)$

H.

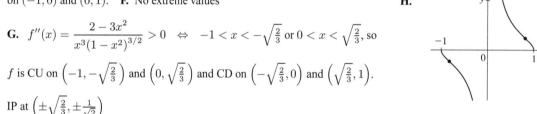

29. $y = f(x) = \sqrt[3]{x^2 - 1}$ **A.** $D = \mathbb{R}$ **B.** y-intercept: $f(0) = -1$; x-intercepts: $f(x) = 0 \iff x^2 - 1 = 0 \iff x = \pm 1$ **C.** $f(-x) = f(x)$, so the curve is symmetric about the y-axis.   **D.** No asymptote

E. $f'(x) = \frac{1}{3}(x^2 - 1)^{-2/3}(2x) = \dfrac{2x}{3\sqrt[3]{(x^2-1)^2}}$. $f'(x) > 0 \iff x > 0$ and $f'(x) < 0 \iff x < 0$, so f is

increasing on $(0, \infty)$ and decreasing on $(-\infty, 0)$.   **F.** Local minimum value $f(0) = -1$

G. $f''(x) = \dfrac{2}{3} \cdot \dfrac{(x^2-1)^{2/3}(1) - x \cdot \frac{2}{3}(x^2-1)^{-1/3}(2x)}{[(x^2-1)^{2/3}]^2}$

$= \dfrac{2}{9} \cdot \dfrac{(x^2-1)^{-1/3}[3(x^2-1) - 4x^2]}{(x^2-1)^{4/3}} = -\dfrac{2(x^2+3)}{9(x^2-1)^{5/3}}$

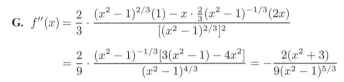

H.

$f''(x) > 0 \iff -1 < x < 1$ and $f''(x) < 0 \iff x < -1$ or $x > 1$, so

f is CU on $(-1, 1)$ and f is CD on $(-\infty, -1)$ and $(1, \infty)$. IP at $(\pm 1, 0)$

31. $y = f(x) = 3\sin x - \sin^3 x$ **A.** $D = \mathbb{R}$ **B.** y-intercept: $f(0) = 0$; x-intercepts: $f(x) = 0 \Rightarrow$
$\sin x\,(3 - \sin^2 x) = 0 \Rightarrow \sin x = 0$ [since $\sin^2 x \le 1 < 3$] $\Rightarrow x = n\pi$, n an integer.
C. $f(-x) = -f(x)$, so f is odd; the graph (shown for $-2\pi \le x \le 2\pi$) is symmetric about the origin and periodic
with period 2π.   **D.** No asymptote   **E.** $f'(x) = 3\cos x - 3\sin^2 x \cos x = 3\cos x\,(1 - \sin^2 x) = 3\cos^3 x$.
$f'(x) > 0 \iff \cos x > 0 \iff x \in \left(2n\pi - \frac{\pi}{2}, 2n\pi + \frac{\pi}{2}\right)$ for each integer n, and $f'(x) < 0 \iff \cos x < 0 \iff$
$x \in \left(2n\pi + \frac{\pi}{2}, 2n\pi + \frac{3\pi}{2}\right)$ for each integer n. Thus, f is increasing on $\left(2n\pi - \frac{\pi}{2}, 2n\pi + \frac{\pi}{2}\right)$ for each integer n,
and f is decreasing on $\left(2n\pi + \frac{\pi}{2}, 2n\pi + \frac{3\pi}{2}\right)$ for each integer n.

F. f has local maximum values $f\left(2n\pi + \frac{\pi}{2}\right) = 2$ and local minimum values $f\left(2n\pi + \frac{3\pi}{2}\right) = -2$.

G. $f''(x) = -9\sin x \cos^2 x = -9\sin x\,(1 - \sin^2 x) = -9\sin x\,(1 - \sin x)(1 + \sin x)$.
$f''(x) < 0 \iff \sin x > 0$ and $\sin x \ne \pm 1 \iff x \in \left(2n\pi, 2n\pi + \frac{\pi}{2}\right) \cup \left(2n\pi + \frac{\pi}{2}, 2n\pi + \pi\right)$ for some integer n.
$f''(x) > 0 \iff \sin x < 0$ and $\sin x \ne \pm 1 \iff x \in \left((2n-1)\pi, (2n-1)\pi + \frac{\pi}{2}\right) \cup \left((2n-1)\pi + \frac{\pi}{2}, 2n\pi\right)$
for some integer n. Thus, f is CD on the intervals $\left(2n\pi, (2n+\frac{1}{2})\pi\right)$ and

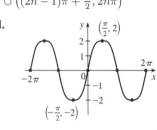

H.

$\left((2n+\frac{1}{2})\pi, (2n+1)\pi\right)$ [hence CD on the intervals $(2n\pi, (2n+1)\pi)$]
for each integer n, and f is CU on the intervals $\left((2n-1)\pi, (2n-\frac{1}{2})\pi\right)$ and
$\left((2n-\frac{1}{2})\pi, 2n\pi\right)$ [hence CU on the intervals $((2n-1)\pi, 2n\pi)$]
for each integer n. f has inflection points at $(n\pi, 0)$ for each integer n.

33. $y = f(x) = x\tan x$, $-\frac{\pi}{2} < x < \frac{\pi}{2}$ **A.** $D = \left(-\frac{\pi}{2}, \frac{\pi}{2}\right)$ **B.** Intercepts are 0   **C.** $f(-x) = f(x)$, so the curve is
symmetric about the y-axis.   **D.** $\displaystyle\lim_{x \to (\pi/2)^-} x\tan x = \infty$ and $\displaystyle\lim_{x \to -(\pi/2)^+} x\tan x = \infty$, so $x = \frac{\pi}{2}$ and $x = -\frac{\pi}{2}$ are VA.

E. $f'(x) = \tan x + x\sec^2 x > 0 \iff 0 < x < \frac{\pi}{2}$, so f increases on $\left(0, \frac{\pi}{2}\right)$

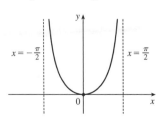

H.

and decreases on $\left(-\frac{\pi}{2}, 0\right)$.   **F.** Absolute and local minimum value $f(0) = 0$.

G. $y'' = 2\sec^2 x + 2x\tan x \sec^2 x > 0$ for $-\frac{\pi}{2} < x < \frac{\pi}{2}$, so f is

CU on $\left(-\frac{\pi}{2}, \frac{\pi}{2}\right)$. No IP

35. $y = f(x) = \frac{1}{2}x - \sin x, 0 < x < 3\pi$ **A.** $D = (0, 3\pi)$ **B.** No y-intercept. The x-intercept, approximately 1.9, can be found using Newton's Method.   **C.** No symmetry   **D.** No asymptote   **E.** $f'(x) = \frac{1}{2} - \cos x > 0$ $\Leftrightarrow$ $\cos x < \frac{1}{2}$ $\Leftrightarrow$ $\frac{\pi}{3} < x < \frac{5\pi}{3}$ or $\frac{7\pi}{3} < x < 3\pi$, so f is increasing on $\left(\frac{\pi}{3}, \frac{5\pi}{3}\right)$ and $\left(\frac{7\pi}{3}, 3\pi\right)$ and decreasing on $\left(0, \frac{\pi}{3}\right)$ and $\left(\frac{5\pi}{3}, \frac{7\pi}{3}\right)$.

F. Local minimum value $f\left(\frac{\pi}{3}\right) = \frac{\pi}{6} - \frac{\sqrt{3}}{2}$, local maximum value
$f\left(\frac{5\pi}{3}\right) = \frac{5\pi}{6} + \frac{\sqrt{3}}{2}$, local minimum value $f\left(\frac{7\pi}{3}\right) = \frac{7\pi}{6} - \frac{\sqrt{3}}{2}$

H.

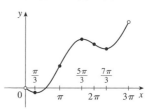

G. $f''(x) = \sin x > 0$ $\Leftrightarrow$ $0 < x < \pi$ or $2\pi < x < 3\pi$, so f is CU on $(0, \pi)$ and $(2\pi, 3\pi)$ and CD on $(\pi, 2\pi)$. IPs at $\left(\pi, \frac{\pi}{2}\right)$ and $(2\pi, \pi)$

37. $y = f(x) = \dfrac{\sin x}{1 + \cos x}$ $\left[\begin{array}{l} \overset{\text{when}}{\underset{\cos x \neq 1}{=}} \dfrac{\sin x}{1 + \cos x} \cdot \dfrac{1 - \cos x}{1 - \cos x} = \dfrac{\sin x \,(1 - \cos x)}{\sin^2 x} = \dfrac{1 - \cos x}{\sin x} = \csc x - \cot x \end{array}\right]$

A. The domain of f is the set of all real numbers except odd integer multiples of π.   **B.** y-intercept: $f(0) = 0$; x-intercepts: $x = n\pi$, n an even integer.   **C.** $f(-x) = -f(x)$, so f is an odd function; the graph is symmetric about the origin and has period 2π.   **D.** When n is an odd integer, $\lim\limits_{x \to (n\pi)^-} f(x) = \infty$ and $\lim\limits_{x \to (n\pi)^+} f(x) = -\infty$, so $x = n\pi$ is a VA for each odd integer n. No HA.   **E.** $f'(x) = \dfrac{(1 + \cos x) \cdot \cos x - \sin x(-\sin x)}{(1 + \cos x)^2} = \dfrac{1 + \cos x}{(1 + \cos x)^2} = \dfrac{1}{1 + \cos x}$. $f'(x) > 0$ for all x except odd multiples of π, so f is increasing on $((2k - 1)\pi, (2k + 1)\pi)$ for each integer k.   **F.** No extreme values

G. $f''(x) = \dfrac{\sin x}{(1 + \cos x)^2} > 0$ $\Rightarrow$ $\sin x > 0$ $\Rightarrow$ $x \in (2k\pi, (2k + 1)\pi)$ and $f''(x) < 0$ on $((2k - 1)\pi, 2k\pi)$ for each integer k. f is CU on $(2k\pi, (2k + 1)\pi)$ and CD on $((2k - 1)\pi, 2k\pi)$ for each integer k. f has IPs at $(2k\pi, 0)$ for each integer k.

H.

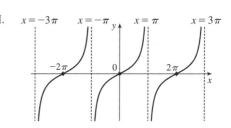

39. $y = f(x) = e^{\sin x}$ **A.** $D = \mathbb{R}$ **B.** y-intercept: $f(0) = e^0 = 1$; x-intercepts: none, since $e^{\sin x} > 0$ **C.** f is periodic with period 2π, so we determine **E–G** for $0 \le x \le 2\pi$.   **D.** No asymptote   **E.** $f'(x) = e^{\sin x} \cos x$. $f'(x) > 0$ $\Leftrightarrow$ $\cos x > 0$ $\Rightarrow$ x is in $\left(0, \frac{\pi}{2}\right)$ or $\left(\frac{3\pi}{2}, 2\pi\right)$ [f is increasing] and $f'(x) < 0$ $\Rightarrow$ x is in $\left(\frac{\pi}{2}, \frac{3\pi}{2}\right)$ [f is decreasing].

F. Local maximum value $f\left(\frac{\pi}{2}\right) = e$ and local minimum value $f\left(\frac{3\pi}{2}\right) = e^{-1}$

G. $f''(x) = e^{\sin x}(-\sin x) + \cos x \,(e^{\sin x} \cos x) = e^{\sin x}(\cos^2 x - \sin x)$. $f''(x) = 0$ $\Leftrightarrow$ $\cos^2 x - \sin x = 0$ $\Leftrightarrow$ $1 - \sin^2 x - \sin x = 0$ $\Leftrightarrow$ $\sin^2 x + \sin x - 1 = 0$ $\Rightarrow$ $\sin x = \frac{-1 \pm \sqrt{5}}{2}$ $\Rightarrow$ $\alpha = \sin^{-1}\left(\frac{-1 + \sqrt{5}}{2}\right) \approx 0.67$ and $\beta = \pi - \alpha \approx 2.48$. $f''(x) < 0$ on (α, β) [f is CD] and $f''(x) > 0$ on $(0, \alpha)$ and $(\beta, 2\pi)$ [f is CU]. The inflection points occur when $x = \alpha, \beta$.

H.

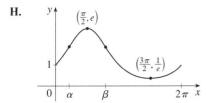

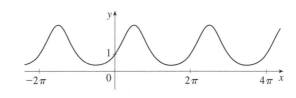

41. $y = 1/(1 + e^{-x})$ **A.** $D = \mathbb{R}$ **B.** No x-intercept; y-intercept $= f(0) = \frac{1}{2}$. **C.** No symmetry

D. $\lim\limits_{x \to \infty} 1/(1 + e^{-x}) = \frac{1}{1+0} = 1$ and $\lim\limits_{x \to -\infty} 1/(1 + e^{-x}) = 0$ since $\lim\limits_{x \to -\infty} e^{-x} = \infty$], so f has horizontal asymptotes

$y = 0$ and $y = 1$. **E.** $f'(x) = -(1 + e^{-x})^{-2}(-e^{-x}) = e^{-x}/(1 + e^{-x})^2$. This is positive for all x, so f is increasing on $\mathbb{R}$.

F. No extreme values **G.** $f''(x) = \dfrac{(1 + e^{-x})^2(-e^{-x}) - e^{-x}(2)(1 + e^{-x})(-e^{-x})}{(1 + e^{-x})^4} = \dfrac{e^{-x}(e^{-x} - 1)}{(1 + e^{-x})^3}$

The second factor in the numerator is negative for $x > 0$ and positive for $x < 0$, **H.**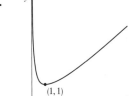

and the other factors are always positive, so f is CU on $(-\infty, 0)$ and CD

on $(0, \infty)$. IP at $\left(0, \frac{1}{2}\right)$

43. $y = f(x) = x - \ln x$ **A.** $D = (0, \infty)$ **B.** y-intercept: none (0 is not in the domain); x-intercept: $f(x) = 0$ $\Leftrightarrow$

$x = \ln x$, which has no solution, so there is no x-intercept. **C.** No symmetry **D.** $\lim\limits_{x \to 0^+} (x - \ln x) = \infty$, so $x = 0$

is a VA. **E.** $f'(x) = 1 - 1/x > 0$ $\Rightarrow$ $1 > 1/x$ $\Rightarrow$ $x > 1$ and **H.**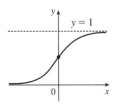

$f'(x) < 0$ $\Rightarrow$ $0 < x < 1$, so f is increasing on $(1, \infty)$ and f is decreasing

on $(0, 1)$. **F.** Local minimum value $f(1) = 1$; no local maximum value

G. $f''(x) = \dfrac{1}{x^2} > 0$ for all x, so f is CU on $(0, \infty)$. No IP

45. $y = f(x) = (1 + e^x)^{-2} = \dfrac{1}{(1 + e^x)^2}$ **A.** $D = \mathbb{R}$ **B.** y-intercept: $f(0) = \frac{1}{4}$. x-intercepts: none [since $f(x) > 0$]

C. No symmetry **D.** $\lim\limits_{x \to \infty} f(x) = 0$ and $\lim\limits_{x \to -\infty} f(x) = 1$, so $y = 0$ and $y = 1$ are HA; no VA

E. $f'(x) = -2(1 + e^x)^{-3}e^x = \dfrac{-2e^x}{(1 + e^x)^3} < 0$, so f is decreasing on $\mathbb{R}$ **F.** No local extrema

G. $f''(x) = (1 + e^x)^{-3}(-2e^x) + (-2e^x)(-3)(1 + e^x)^{-4}e^x$ **H.**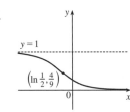

$\qquad = -2e^x(1 + e^x)^{-4}[(1 + e^x) - 3e^x] = \dfrac{-2e^x(1 - 2e^x)}{(1 + e^x)^4}$.

$f''(x) > 0$ $\Leftrightarrow$ $1 - 2e^x < 0$ $\Leftrightarrow$ $e^x > \frac{1}{2}$ $\Leftrightarrow$ $x > \ln \frac{1}{2}$ and

$f''(x) < 0$ $\Leftrightarrow$ $x < \ln \frac{1}{2}$, so f is CU on $\left(\ln \frac{1}{2}, \infty\right)$ and CD on $\left(-\infty, \ln \frac{1}{2}\right)$.

IP at $\left(\ln \frac{1}{2}, \frac{4}{9}\right)$

47. $y = f(x) = \ln(\sin x)$

A. $D = \{x \text{ in } \mathbb{R} \mid \sin x > 0\} = \bigcup\limits_{n=-\infty}^{\infty} (2n\pi, (2n + 1)\pi) = \cdots \cup (-4\pi, -3\pi) \cup (-2\pi, -\pi) \cup (0, \pi) \cup (2\pi, 3\pi) \cup \cdots$

B. No y-intercept; x-intercepts: $f(x) = 0$ $\Leftrightarrow$ $\ln(\sin x) = 0$ $\Leftrightarrow$ $\sin x = e^0 = 1$ $\Leftrightarrow$ $x = 2n\pi + \frac{\pi}{2}$ for each

integer n. **C.** f is periodic with period 2π. **D.** $\lim\limits_{x \to (2n\pi)^+} f(x) = -\infty$ and $\lim\limits_{x \to [(2n+1)\pi]^-} f(x) = -\infty$, so the lines

$x = n\pi$ are VAs for all integers n. **E.** $f'(x) = \dfrac{\cos x}{\sin x} = \cot x$, so $f'(x) > 0$ when $2n\pi < x < 2n\pi + \frac{\pi}{2}$ for each

integer n, and $f'(x) < 0$ when $2n\pi + \frac{\pi}{2} < x < (2n + 1)\pi$. Thus, f is increasing on $\left(2n\pi, 2n\pi + \frac{\pi}{2}\right)$ and decreasing on $\left(2n\pi + \frac{\pi}{2}, (2n+1)\pi\right)$ for each integer n.

F. Local maximum values $f\left(2n\pi + \frac{\pi}{2}\right) = 0$, no local minimum.

G. $f''(x) = -\csc^2 x < 0$, so f is CD on $(2n\pi, (2n+1)\pi)$ for

each integer n. No IP

H.

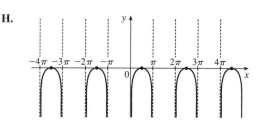

49. $y = f(x) = xe^{-x^2}$ **A.** $D = \mathbb{R}$ **B.** Intercepts are 0 **C.** $f(-x) = -f(x)$, so the curve is symmetric

about the origin. **D.** $\lim\limits_{x \to \pm\infty} xe^{-x^2} = \lim\limits_{x \to \pm\infty} \dfrac{x}{e^{x^2}} \overset{\text{H}}{=} \lim\limits_{x \to \pm\infty} \dfrac{1}{2xe^{x^2}} = 0$, so $y = 0$ is a HA.

E. $f'(x) = e^{-x^2} - 2x^2e^{-x^2} = e^{-x^2}(1 - 2x^2) > 0 \iff x^2 < \frac{1}{2} \iff |x| < \frac{1}{\sqrt{2}}$, so f is increasing on $\left(-\frac{1}{\sqrt{2}}, \frac{1}{\sqrt{2}}\right)$

and decreasing on $\left(-\infty, -\frac{1}{\sqrt{2}}\right)$ and $\left(\frac{1}{\sqrt{2}}, \infty\right)$. **F.** Local maximum value $f\left(\frac{1}{\sqrt{2}}\right) = 1/\sqrt{2e}$, local minimum

value $f\left(-\frac{1}{\sqrt{2}}\right) = -1/\sqrt{2e}$ **G.** $f''(x) = -2xe^{-x^2}(1 - 2x^2) - 4xe^{-x^2} = 2xe^{-x^2}(2x^2 - 3) > 0 \iff$

$x > \sqrt{\frac{3}{2}}$ or $-\sqrt{\frac{3}{2}} < x < 0$, so f is CU on $\left(\sqrt{\frac{3}{2}}, \infty\right)$ and   **H.**

$\left(-\sqrt{\frac{3}{2}}, 0\right)$ and CD on $\left(-\infty, -\sqrt{\frac{3}{2}}\right)$ and $\left(0, \sqrt{\frac{3}{2}}\right)$.

IP are $(0, 0)$ and $\left(\pm\sqrt{\frac{3}{2}}, \pm\sqrt{\frac{3}{2}}\, e^{-3/2}\right)$.

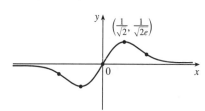

51. $y = f(x) = e^{3x} + e^{-2x}$ **A.** $D = \mathbb{R}$ **B.** y-intercept $= f(0) = 2$; no x-intercept **C.** No symmetry **D.** No asymptote

E. $f'(x) = 3e^{3x} - 2e^{-2x}$, so $f'(x) > 0 \iff 3e^{3x} > 2e^{-2x}$ [multiply by e^{2x}] $\iff$ **H.**

$e^{5x} > \frac{2}{3} \iff 5x > \ln\frac{2}{3} \iff x > \frac{1}{5}\ln\frac{2}{3} \approx -0.081$. Similarly, $f'(x) < 0 \iff$

$x < \frac{1}{5}\ln\frac{2}{3}$. f is decreasing on $\left(-\infty, \frac{1}{5}\ln\frac{2}{3}\right)$ and increasing on $\left(\frac{1}{5}\ln\frac{2}{3}, \infty\right)$.

F. Local minimum value $f\left(\frac{1}{5}\ln\frac{2}{3}\right) = \left(\frac{2}{3}\right)^{3/5} + \left(\frac{2}{3}\right)^{-2/5} \approx 1.96$; no local maximum.

G. $f''(x) = 9e^{3x} + 4e^{-2x}$, so $f''(x) > 0$ for all x, and f is CU on $(-\infty, \infty)$. No IP

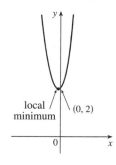

53. $m = f(v) = \dfrac{m_0}{\sqrt{1 - v^2/c^2}}$. The m-intercept is $f(0) = m_0$. There are no v-intercepts. $\lim\limits_{v \to c^-} f(v) = \infty$, so $v = c$ is a VA.

$f'(v) = -\frac{1}{2}m_0(1 - v^2/c^2)^{-3/2}(-2v/c^2) = \dfrac{m_0 v}{c^2(1 - v^2/c^2)^{3/2}} = \dfrac{m_0 v}{\dfrac{c^2(c^2 - v^2)^{3/2}}{c^3}} = \dfrac{m_0 c v}{(c^2 - v^2)^{3/2}} > 0$, so f is

increasing on $(0, c)$. There are no local extreme values.

$f''(v) = \dfrac{(c^2 - v^2)^{3/2}(m_0 c) - m_0 c v \cdot \frac{3}{2}(c^2 - v^2)^{1/2}(-2v)}{[(c^2 - v^2)^{3/2}]^2}$

$= \dfrac{m_0 c(c^2 - v^2)^{1/2}[(c^2 - v^2) + 3v^2]}{(c^2 - v^2)^3} = \dfrac{m_0 c(c^2 + 2v^2)}{(c^2 - v^2)^{5/2}} > 0$,

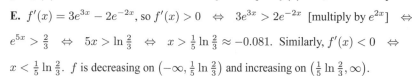

so f is CU on $(0, c)$. There are no inflection points.

55. $y = -\dfrac{W}{24EI}x^4 + \dfrac{WL}{12EI}x^3 - \dfrac{WL^2}{24EI}x^2 = -\dfrac{W}{24EI}x^2\left(x^2 - 2Lx + L^2\right)$

$\quad = \dfrac{-W}{24EI}x^2(x-L)^2 = cx^2(x-L)^2$

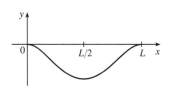

where $c = -\dfrac{W}{24EI}$ is a negative constant and $0 \le x \le L$. We sketch

$f(x) = cx^2(x-L)^2$ for $c = -1$. $f(0) = f(L) = 0$.

$f'(x) = cx^2[2(x-L)] + (x-L)^2(2cx) = 2cx(x-L)[x + (x-L)] = 2cx(x-L)(2x-L)$. So for $0 < x < L$,

$f'(x) > 0 \iff x(x-L)(2x-L) < 0$ [since $c < 0$] $\iff L/2 < x < L$ and $f'(x) < 0 \iff 0 < x < L/2$.

Thus, f is increasing on $(L/2, L)$ and decreasing on $(0, L/2)$, and there is a local and absolute

minimum at the point $(L/2, f(L/2)) = (L/2, cL^4/16)$. $f'(x) = 2c[x(x-L)(2x-L)] \Rightarrow$

$f''(x) = 2c[1(x-L)(2x-L) + x(1)(2x-L) + x(x-L)(2)] = 2c(6x^2 - 6Lx + L^2) = 0 \iff$

$x = \dfrac{6L \pm \sqrt{12L^2}}{12} = \tfrac{1}{2}L \pm \tfrac{\sqrt{3}}{6}L$, and these are the x-coordinates of the two inflection points.

57. $y = \dfrac{x^2 + 1}{x + 1}$.   Long division gives us:

$$
\begin{array}{r}
x - 1 \\
x+1\overline{)\;x^2 \qquad\; +1\;} \\
\underline{x^2 + x\;\;\;} \\
-x+1 \\
\underline{-x-1} \\
2
\end{array}
$$

Thus, $y = f(x) = \dfrac{x^2 + 1}{x+1} = x - 1 + \dfrac{2}{x+1}$ and $f(x) - (x-1) = \dfrac{2}{x+1} = \dfrac{\dfrac{2}{x}}{1 + \dfrac{1}{x}}$ [for $x \ne 0$] $\to 0$ as $x \to \pm\infty$.

So the line $y = x - 1$ is a slant asymptote (SA).

59. $y = \dfrac{4x^3 - 2x^2 + 5}{2x^2 + x - 3}$.   Long division gives us:

$$
\begin{array}{r}
2x - 2 \\
2x^2+x-3\overline{)\;4x^3 - 2x^2 \qquad\; +5\;} \\
\underline{4x^3 + 2x^2 - 6x\;\;\;} \\
-4x^2 + 6x + 5 \\
\underline{-4x^2 - 2x + 6} \\
8x - 1
\end{array}
$$

Thus, $y = f(x) = \dfrac{4x^3 - 2x^2 + 5}{2x^2 + x - 3} = 2x - 2 + \dfrac{8x - 1}{2x^2 + x - 3}$ and $f(x) - (2x - 2) = \dfrac{8x-1}{2x^2+x-3} = \dfrac{\dfrac{8}{x} - \dfrac{1}{x^2}}{2 + \dfrac{1}{x} - \dfrac{3}{x^2}}$

[for $x \ne 0$] $\to 0$ as $x \to \pm\infty$. So the line $y = 2x - 2$ is a SA.

61. $y = f(x) = \dfrac{-2x^2 + 5x - 1}{2x - 1} = -x + 2 + \dfrac{1}{2x - 1}$ **A.** $D = \left\{x \in \mathbb{R} \mid x \ne \tfrac{1}{2}\right\} = \left(-\infty, \tfrac{1}{2}\right) \cup \left(\tfrac{1}{2}, \infty\right)$

B. y-intercept: $f(0) = 1$; x-intercepts: $f(x) = 0 \Rightarrow -2x^2 + 5x - 1 = 0 \Rightarrow x = \dfrac{-5 \pm \sqrt{17}}{-4} \Rightarrow x \approx 0.22, 2.28$.

[continued]

C. No symmetry **D.** $\lim\limits_{x \to (1/2)^-} f(x) = -\infty$ and $\lim\limits_{x \to (1/2)^+} f(x) = \infty$, so $x = \frac{1}{2}$ is a VA.

$\lim\limits_{x \to \pm\infty} [f(x) - (-x + 2)] = \lim\limits_{x \to \pm\infty} \dfrac{1}{2x - 1} = 0$, so the line $y = -x + 2$ is a SA.

E. $f'(x) = -1 - \dfrac{2}{(2x - 1)^2} < 0$ for $x \neq \frac{1}{2}$, so f is decreasing on $\left(-\infty, \frac{1}{2}\right)$

and $\left(\frac{1}{2}, \infty\right)$. **F.** No extreme values **G.** $f'(x) = -1 - 2(2x - 1)^{-2}$ $\Rightarrow$

$f''(x) = -2(-2)(2x - 1)^{-3}(2) = \dfrac{8}{(2x - 1)^3}$, so $f''(x) > 0$ when $x > \frac{1}{2}$ and

$f''(x) < 0$ when $x < \frac{1}{2}$. Thus, f is CU on $\left(\frac{1}{2}, \infty\right)$ and CD on $\left(-\infty, \frac{1}{2}\right)$. No IP

H.

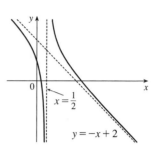

63. $y = f(x) = (x^2 + 4)/x = x + 4/x$ **A.** $D = \{x \mid x \neq 0\} = (-\infty, 0) \cup (0, \infty)$ **B.** No intercept

C. $f(-x) = -f(x)$ $\Rightarrow$ symmetry about the origin **D.** $\lim\limits_{x \to \infty} (x + 4/x) = \infty$ but $f(x) - x = 4/x \to 0$ as $x \to \pm\infty$,

so $y = x$ is a slant asymptote. $\lim\limits_{x \to 0^+} (x + 4/x) = \infty$ and

H.

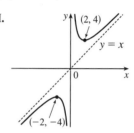

$\lim\limits_{x \to 0^-} (x + 4/x) = -\infty$, so $x = 0$ is a VA. **E.** $f'(x) = 1 - 4/x^2 > 0$ $\Leftrightarrow$

$x^2 > 4$ $\Leftrightarrow$ $x > 2$ or $x < -2$, so f is increasing on $(-\infty, -2)$ and $(2, \infty)$ and

decreasing on $(-2, 0)$ and $(0, 2)$. **F.** Local maximum value $f(-2) = -4$, local

minimum value $f(2) = 4$ **G.** $f''(x) = 8/x^3 > 0$ $\Leftrightarrow$ $x > 0$ so f is CU on

$(0, \infty)$ and CD on $(-\infty, 0)$. No IP

65. $y = f(x) = \dfrac{2x^3 + x^2 + 1}{x^2 + 1} = 2x + 1 + \dfrac{-2x}{x^2 + 1}$ **A.** $D = \mathbb{R}$ **B.** y-intercept: $f(0) = 1$; x-intercept: $f(x) = 0$ $\Rightarrow$

$0 = 2x^3 + x^2 + 1 = (x + 1)(2x^2 - x + 1)$ $\Rightarrow$ $x = -1$ **C.** No symmetry **D.** No VA

$\lim\limits_{x \to \pm\infty} [f(x) - (2x + 1)] = \lim\limits_{x \to \pm\infty} \dfrac{-2x}{x^2 + 1} = \lim\limits_{x \to \pm\infty} \dfrac{-2/x}{1 + 1/x^2} = 0$, so the line $y = 2x + 1$ is a slant asymptote.

E. $f'(x) = 2 + \dfrac{(x^2 + 1)(-2) - (-2x)(2x)}{(x^2 + 1)^2} = \dfrac{2(x^4 + 2x^2 + 1) - 2x^2 - 2 + 4x^2}{(x^2 + 1)^2} = \dfrac{2x^4 + 6x^2}{(x^2 + 1)^2} = \dfrac{2x^2(x^2 + 3)}{(x^2 + 1)^2}$

so $f'(x) > 0$ if $x \neq 0$. Thus, f is increasing on $(-\infty, 0)$ and $(0, \infty)$. Since f is continuous at 0, f is increasing on $\mathbb{R}$.

F. No extreme values

G. $f''(x) = \dfrac{(x^2 + 1)^2 \cdot (8x^3 + 12x) - (2x^4 + 6x^2) \cdot 2(x^2 + 1)(2x)}{[(x^2 + 1)^2]^2}$

$= \dfrac{4x(x^2 + 1)[(x^2 + 1)(2x^2 + 3) - 2x^4 - 6x^2]}{(x^2 + 1)^4} = \dfrac{4x(-x^2 + 3)}{(x^2 + 1)^3}$

so $f''(x) > 0$ for $x < -\sqrt{3}$ and $0 < x < \sqrt{3}$, and $f''(x) < 0$ for

$-\sqrt{3} < x < 0$ and $x > \sqrt{3}$. f is CU on $\left(-\infty, -\sqrt{3}\right)$ and $\left(0, \sqrt{3}\right)$,

and CD on $\left(-\sqrt{3}, 0\right)$ and $\left(\sqrt{3}, \infty\right)$. There are three IPs: $(0, 1)$,

$\left(-\sqrt{3}, -\frac{3}{2}\sqrt{3} + 1\right) \approx (-1.73, -1.60)$, and

$\left(\sqrt{3}, \frac{3}{2}\sqrt{3} + 1\right) \approx (1.73, 3.60)$.

H.

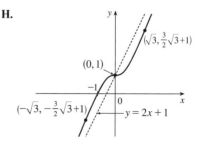

67. $y = f(x) = x - \tan^{-1} x$, $f'(x) = 1 - \dfrac{1}{1+x^2} = \dfrac{1+x^2-1}{1+x^2} = \dfrac{x^2}{1+x^2}$,

$f''(x) = \dfrac{(1+x^2)(2x) - x^2(2x)}{(1+x^2)^2} = \dfrac{2x(1+x^2-x^2)}{(1+x^2)^2} = \dfrac{2x}{(1+x^2)^2}$.

$\lim\limits_{x\to\infty} \left[f(x) - \left(x - \frac{\pi}{2} \right) \right] = \lim\limits_{x\to\infty} \left(\frac{\pi}{2} - \tan^{-1} x \right) = \frac{\pi}{2} - \frac{\pi}{2} = 0$, so $y = x - \frac{\pi}{2}$ is a SA.

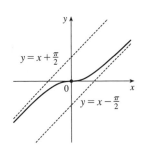

Also, $\lim\limits_{x\to-\infty} \left[f(x) - \left(x + \frac{\pi}{2} \right) \right] = \lim\limits_{x\to-\infty} \left(-\frac{\pi}{2} - \tan^{-1} x \right) = -\frac{\pi}{2} - \left(-\frac{\pi}{2} \right) = 0$,

so $y = x + \frac{\pi}{2}$ is also a SA. $f'(x) \geq 0$ for all x, with equality $\Leftrightarrow$ $x = 0$, so f is

increasing on $\mathbb{R}$. $f''(x)$ has the same sign as x, so f is CD on $(-\infty, 0)$ and CU on

$(0, \infty)$. $f(-x) = -f(x)$, so f is an odd function; its graph is symmetric about the

origin. f has no local extreme values. Its only IP is at $(0, 0)$.

69. $\dfrac{x^2}{a^2} - \dfrac{y^2}{b^2} = 1 \Rightarrow y = \pm \dfrac{b}{a}\sqrt{x^2 - a^2}$. Now

$\lim\limits_{x\to\infty} \left[\dfrac{b}{a}\sqrt{x^2-a^2} - \dfrac{b}{a}x \right] = \dfrac{b}{a} \cdot \lim\limits_{x\to\infty} \left(\sqrt{x^2-a^2} - x \right) \dfrac{\sqrt{x^2-a^2}+x}{\sqrt{x^2-a^2}+x} = \dfrac{b}{a} \cdot \lim\limits_{x\to\infty} \dfrac{-a^2}{\sqrt{x^2-a^2}+x} = 0$,

which shows that $y = \dfrac{b}{a}x$ is a slant asymptote. Similarly,

$\lim\limits_{x\to\infty} \left[-\dfrac{b}{a}\sqrt{x^2-a^2} - \left(-\dfrac{b}{a}x \right) \right] = -\dfrac{b}{a} \cdot \lim\limits_{x\to\infty} \dfrac{-a^2}{\sqrt{x^2-a^2}+x} = 0$, so $y = -\dfrac{b}{a}x$ is a slant asymptote.

71. $\lim\limits_{x\to\pm\infty} \left[f(x) - x^3 \right] = \lim\limits_{x\to\pm\infty} \dfrac{x^4+1}{x} - \dfrac{x^4}{x} = \lim\limits_{x\to\pm\infty} \dfrac{1}{x} = 0$, so the graph of f is asymptotic to that of $y = x^3$.

A. $D = \{x \mid x \neq 0\}$ **B.** No intercept **C.** f is symmetric about the origin. **D.** $\lim\limits_{x\to 0^-} \left(x^3 + \dfrac{1}{x} \right) = -\infty$ and

$\lim\limits_{x\to 0^+} \left(x^3 + \dfrac{1}{x} \right) = \infty$, so $x = 0$ is a vertical asymptote, and as shown above, the graph of f is asymptotic to that of $y = x^3$.

E. $f'(x) = 3x^2 - 1/x^2 > 0 \Leftrightarrow x^4 > \frac{1}{3} \Leftrightarrow |x| > \frac{1}{\sqrt[4]{3}}$, so f is increasing on $\left(-\infty, -\dfrac{1}{\sqrt[4]{3}} \right)$ and $\left(\dfrac{1}{\sqrt[4]{3}}, \infty \right)$ and

decreasing on $\left(-\dfrac{1}{\sqrt[4]{3}}, 0 \right)$ and $\left(0, \dfrac{1}{\sqrt[4]{3}} \right)$. **F.** Local maximum value **H.**

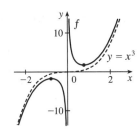

$f\left(-\dfrac{1}{\sqrt[4]{3}} \right) = -4 \cdot 3^{-5/4}$, local minimum value $f\left(\dfrac{1}{\sqrt[4]{3}} \right) = 4 \cdot 3^{-5/4}$

G. $f''(x) = 6x + 2/x^3 > 0 \Leftrightarrow x > 0$, so f is CU on $(0, \infty)$ and CD

on $(-\infty, 0)$. No IP

4.6 Graphing with Calculus and Calculators

1. $f(x) = 4x^4 - 32x^3 + 89x^2 - 95x + 29$ $\Rightarrow$ $f'(x) = 16x^3 - 96x^2 + 178x - 95$ $\Rightarrow$ $f''(x) = 48x^2 - 192x + 178$.

$f(x) = 0$ $\Leftrightarrow$ $x \approx 0.5, 1.60$; $f'(x) = 0$ $\Leftrightarrow$ $x \approx 0.92, 2.5, 2.58$ and $f''(x) = 0$ $\Leftrightarrow$ $x \approx 1.46, 2.54$.

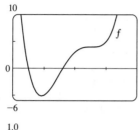
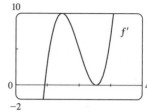

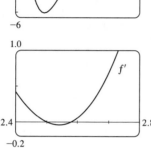
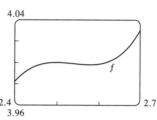

From the graphs of f', we estimate that $f' < 0$ and that f is decreasing on $(-\infty, 0.92)$ and $(2.5, 2.58)$, and that $f' > 0$ and f is increasing on $(0.92, 2.5)$ and $(2.58, \infty)$ with local minimum values $f(0.92) \approx -5.12$ and $f(2.58) \approx 3.998$ and local maximum value $f(2.5) = 4$. The graphs of f' make it clear that f has a maximum and a minimum near $x = 2.5$, shown more clearly in the fourth graph.

From the graph of f'', we estimate that $f'' > 0$ and that f is CU on $(-\infty, 1.46)$ and $(2.54, \infty)$, and that $f'' < 0$ and f is CD on $(1.46, 2.54)$. There are inflection points at about $(1.46, -1.40)$ and $(2.54, 3.999)$.

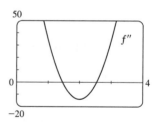

3. $f(x) = x^6 - 10x^5 - 400x^4 + 2500x^3$ $\Rightarrow$ $f'(x) = 6x^5 - 50x^4 - 1600x^3 + 7500x^2$ $\Rightarrow$
$f''(x) = 30x^4 - 200x^3 - 4800x^2 + 1500x$

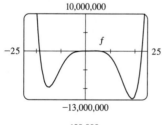
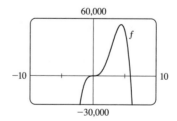

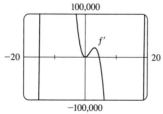
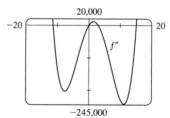

From the graph of f', we estimate that f is decreasing on $(-\infty, -15)$, increasing on $(-15, 4.40)$, decreasing on $(4.40, 18.93)$, and increasing on $(18.93, \infty)$, with local minimum values of $f(-15) \approx -9{,}700{,}000$ and $f(18.93) \approx -12{,}700{,}000$ and local maximum value $f(4.40) \approx 53{,}800$. From the graph of f'', we estimate that f is CU on $(-\infty, -11.34)$, CD on $(-11.34, 0)$, CU on $(0, 2.92)$, CD on $(2.92, 15.08)$, and CU on $(15.08, \infty)$. There is an inflection point at $(0, 0)$ and at about $(-11.34, -6{,}250{,}000)$, $(2.92, 31{,}800)$, and $(15.08, -8{,}150{,}000)$.

5. $f(x) = \dfrac{x}{x^3 - x^2 - 4x + 1} \quad \Rightarrow \quad f'(x) = \dfrac{-2x^3 + x^2 + 1}{(x^3 - x^2 - 4x + 1)^2} \quad \Rightarrow \quad f''(x) = \dfrac{2(3x^5 - 3x^4 + 5x^3 - 6x^2 + 3x + 4)}{(x^3 - x^2 - 4x + 1)^3}$

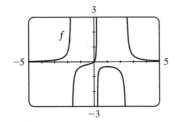
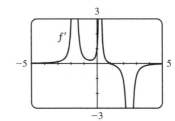
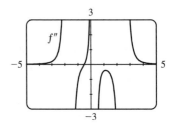

We estimate from the graph of f that $y = 0$ is a horizontal asymptote, and that there are vertical asymptotes at $x = -1.7$, $x = 0.24$, and $x = 2.46$. From the graph of f', we estimate that f is increasing on $(-\infty, -1.7)$, $(-1.7, 0.24)$, and $(0.24, 1)$, and that f is decreasing on $(1, 2.46)$ and $(2.46, \infty)$. There is a local maximum value at $f(1) = -\frac{1}{3}$. From the graph of f'', we estimate that f is CU on $(-\infty, -1.7)$, $(-0.506, 0.24)$, and $(2.46, \infty)$, and that f is CD on $(-1.7, -0.506)$ and $(0.24, 2.46)$. There is an inflection point at $(-0.506, -0.192)$.

7. $f(x) = x^2 - 4x + 7\cos x$, $\ -4 \leq x \leq 4$. $\quad f'(x) = 2x - 4 - 7\sin x \quad \Rightarrow \quad f''(x) = 2 - 7\cos x$.
$f(x) = 0 \ \Leftrightarrow \ x \approx 1.10$; $\ f'(x) = 0 \ \Leftrightarrow \ x \approx -1.49, -1.07$, or 2.89; $\ f''(x) = 0 \ \Leftrightarrow \ x = \pm\cos^{-1}\left(\frac{2}{7}\right) \approx \pm 1.28$.

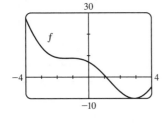
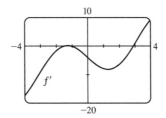
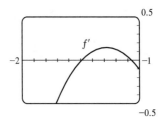

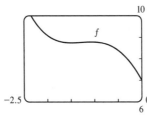
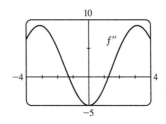

From the graphs of f', we estimate that f is decreasing ($f' < 0$) on $(-4, -1.49)$, increasing on $(-1.49, -1.07)$, decreasing on $(-1.07, 2.89)$, and increasing on $(2.89, 4)$, with local minimum values $f(-1.49) \approx 8.75$ and $f(2.89) \approx -9.99$ and local maximum value $f(-1.07) \approx 8.79$ (notice the second graph of f). From the graph of f'', we estimate that f is CU ($f'' > 0$) on $(-4, -1.28)$, CD on $(-1.28, 1.28)$, and CU on $(1.28, 4)$. There are inflection points at about $(-1.28, 8.77)$ and $(1.28, -1.48)$.

9. $f(x) = 1 + \dfrac{1}{x} + \dfrac{8}{x^2} + \dfrac{1}{x^3}$ $\Rightarrow$ $f'(x) = -\dfrac{1}{x^2} - \dfrac{16}{x^3} - \dfrac{3}{x^4} = -\dfrac{1}{x^4}(x^2 + 16x + 3)$ $\Rightarrow$

$f''(x) = \dfrac{2}{x^3} + \dfrac{48}{x^4} + \dfrac{12}{x^5} = \dfrac{2}{x^5}(x^2 + 24x + 6)$.

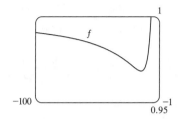

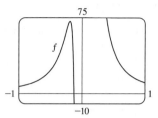

From the graphs, it appears that f increases on $(-15.8, -0.2)$ and decreases on $(-\infty, -15.8)$, $(-0.2, 0)$, and $(0, \infty)$; that f

has a local minimum value of $f(-15.8) \approx 0.97$ and a local maximum value of $f(-0.2) \approx 72$; that f is CD on $(-\infty, -24)$

and $(-0.25, 0)$ and is CU on $(-24, -0.25)$ and $(0, \infty)$; and that f has IPs at $(-24, 0.97)$ and $(-0.25, 60)$.

To find the exact values, note that $f' = 0$ $\Rightarrow$ $x = \dfrac{-16 \pm \sqrt{256 - 12}}{2} = -8 \pm \sqrt{61}$ $[\approx -0.19 \text{ and } -15.81]$.

f' is positive (f is increasing) on $\left(-8 - \sqrt{61}, -8 + \sqrt{61}\right)$ and f' is negative (f is decreasing) on $\left(-\infty, -8 - \sqrt{61}\right)$,

$\left(-8 + \sqrt{61}, 0\right)$, and $(0, \infty)$. $f'' = 0$ $\Rightarrow$ $x = \dfrac{-24 \pm \sqrt{576 - 24}}{2} = -12 \pm \sqrt{138}$ $[\approx -0.25 \text{ and } -23.75]$. f'' is

positive (f is CU) on $\left(-12 - \sqrt{138}, -12 + \sqrt{138}\right)$ and $(0, \infty)$ and f'' is negative (f is CD) on $\left(-\infty, -12 - \sqrt{138}\right)$

and $\left(-12 + \sqrt{138}, 0\right)$.

11. (a) $f(x) = x^2 \ln x$. The domain of f is $(0, \infty)$.

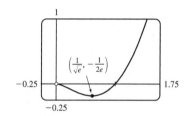

(b) $\displaystyle\lim_{x \to 0^+} x^2 \ln x = \lim_{x \to 0^+} \dfrac{\ln x}{1/x^2} \overset{\text{H}}{=} \lim_{x \to 0^+} \dfrac{1/x}{-2/x^3} = \lim_{x \to 0^+} \left(-\dfrac{x^2}{2}\right) = 0$.

There is a hole at $(0, 0)$.

(c) It appears that there is an IP at about $(0.2, -0.06)$ and a local minimum at $(0.6, -0.18)$. $f(x) = x^2 \ln x$ $\Rightarrow$

$f'(x) = x^2(1/x) + (\ln x)(2x) = x(2\ln x + 1) > 0$ $\Leftrightarrow$ $\ln x > -\dfrac{1}{2}$ $\Leftrightarrow$ $x > e^{-1/2}$, so f is increasing on

$\left(1/\sqrt{e}, \infty\right)$, decreasing on $\left(0, 1/\sqrt{e}\right)$. By the FDT, $f\left(1/\sqrt{e}\right) = -1/(2e)$ is a local minimum value. This point is

approximately $(0.6065, -0.1839)$, which agrees with our estimate.

$f''(x) = x(2/x) + (2\ln x + 1) = 2\ln x + 3 > 0$ $\Leftrightarrow$ $\ln x > -\dfrac{3}{2}$ $\Leftrightarrow$ $x > e^{-3/2}$, so f is CU on $\left(e^{-3/2}, \infty\right)$

and CD on $\left(0, e^{-3/2}\right)$. IP is $\left(e^{-3/2}, -3/(2e^3)\right) \approx (0.2231, -0.0747)$.

13.

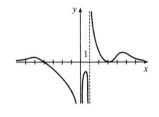

$f(x) = \dfrac{(x+4)(x-3)^2}{x^4(x-1)}$ has VA at $x = 0$ and at $x = 1$ since $\lim\limits_{x \to 0} f(x) = -\infty$,

$\lim\limits_{x \to 1^-} f(x) = -\infty$ and $\lim\limits_{x \to 1^+} f(x) = \infty$.

$f(x) = \dfrac{\dfrac{x+4}{x} \cdot \dfrac{(x-3)^2}{x^2}}{\dfrac{x^4}{x^3} \cdot (x-1)}$ $\begin{bmatrix} \text{dividing numerator} \\ \text{and denominator by } x^3 \end{bmatrix}$ $= \dfrac{(1+4/x)(1-3/x)^2}{x(x-1)} \to 0$

as $x \to \pm\infty$, so f is asymptotic to the x-axis.

Since f is undefined at $x = 0$, it has no y-intercept. $f(x) = 0 \Rightarrow (x+4)(x-3)^2 = 0 \Rightarrow x = -4$ or $x = 3$, so f has x-intercepts -4 and 3. Note, however, that the graph of f is only tangent to the x-axis and does not cross it at $x = 3$, since f is positive as $x \to 3^-$ and as $x \to 3^+$.

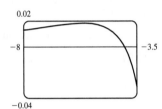

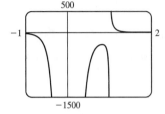

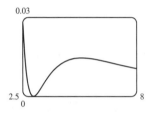

From these graphs, it appears that f has three maximum values and one minimum value. The maximum values are approximately $f(-5.6) = 0.0182$, $f(0.82) = -281.5$ and $f(5.2) = 0.0145$ and we know (since the graph is tangent to the x-axis at $x = 3$) that the minimum value is $f(3) = 0$.

15. $f(x) = \dfrac{x^2(x+1)^3}{(x-2)^2(x-4)^4}$ $\Rightarrow$ $f'(x) = -\dfrac{x(x+1)^2(x^3+18x^2-44x-16)}{(x-2)^3(x-4)^5}$ [from CAS].

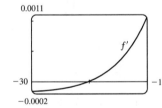

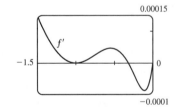

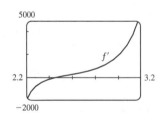

From the graphs of f', it seems that the critical points which indicate extrema occur at $x \approx -20$, -0.3, and 2.5, as estimated in Example 3. (There is another critical point at $x = -1$, but the sign of f' does not change there.) We differentiate again, obtaining $f''(x) = 2\dfrac{(x+1)(x^6+36x^5+6x^4-628x^3+684x^2+672x+64)}{(x-2)^4(x-4)^6}$.

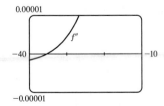

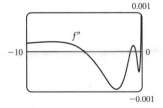

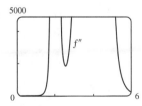

From the graphs of f'', it appears that f is CU on $(-35.3, -5.0)$, $(-1, -0.5)$, $(-0.1, 2)$, $(2, 4)$ and $(4, \infty)$ and CD on $(-\infty, -35.3)$, $(-5.0, -1)$ and $(-0.5, -0.1)$. We check back on the graphs of f to find the y-coordinates of the inflection points, and find that these points are approximately $(-35.3, -0.015)$, $(-5.0, -0.005)$, $(-1, 0)$, $(-0.5, 0.00001)$, and $(-0.1, 0.0000066)$.

17. $y = f(x) = \dfrac{\sqrt{x}}{x^2 + x + 1}$. From a CAS, $y' = -\dfrac{3x^2 + x - 1}{2\sqrt{x}\,(x^2 + x + 1)^2}$ and $y'' = \dfrac{15x^4 + 10x^3 - 15x^2 - 6x - 1}{4x^{3/2}(x^2 + x + 1)^3}$.

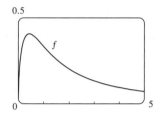 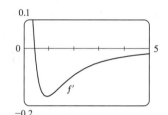 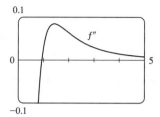

$f'(x) = 0 \iff x \approx 0.43$, so f is increasing on $(0, 0.43)$ and decreasing on $(0.43, \infty)$. There is a local maximum value of $f(0.43) \approx 0.41$. $f''(x) = 0 \iff x \approx 0.94$, so f is CD on $(0, 0.94)$ and CU on $(0.94, \infty)$. There is an inflection point at $(0.94, 0.34)$.

19. $y = f(x) = \sqrt{x + 5\sin x}$, $x \le 20$.

From a CAS, $y' = \dfrac{5\cos x + 1}{2\sqrt{x + 5\sin x}}$ and $y'' = -\dfrac{10\cos x + 25\sin^2 x + 10x\sin x + 26}{4(x + 5\sin x)^{3/2}}$.

We'll start with a graph of $g(x) = x + 5\sin x$. Note that $f(x) = \sqrt{g(x)}$ is only defined if $g(x) \ge 0$. $g(x) = 0 \iff x = 0$ or $x \approx -4.91, -4.10, 4.10$, and 4.91. Thus, the domain of f is $[-4.91, -4.10] \cup [0, 4.10] \cup [4.91, 20]$.

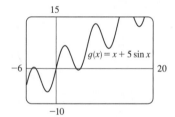 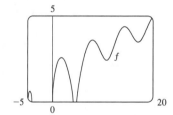 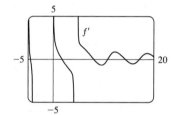

From the expression for y', we see that $y' = 0 \iff 5\cos x + 1 = 0 \Rightarrow x_1 = \cos^{-1}\left(-\frac{1}{5}\right) \approx 1.77$ and $x_2 = 2\pi - x_1 \approx -4.51$ (not in the domain of f). The leftmost zero of f' is $x_1 - 2\pi \approx -4.51$. Moving to the right, the zeros of f' are x_1, $x_1 + 2\pi$, $x_2 + 2\pi$, $x_1 + 4\pi$, and $x_2 + 4\pi$. Thus, f is increasing on $(-4.91, -4.51)$, decreasing on $(-4.51, -4.10)$, increasing on $(0, 1.77)$, decreasing on $(1.77, 4.10)$, increasing on $(4.91, 8.06)$, decreasing on $(8.06, 10.79)$, increasing on $(10.79, 14.34)$, decreasing on $(14.34, 17.08)$, and increasing on $(17.08, 20)$. The local maximum values are $f(-4.51) \approx 0.62$, $f(1.77) \approx 2.58$, $f(8.06) \approx 3.60$, and $f(14.34) \approx 4.39$. The local minimum values are $f(10.79) \approx 2.43$ and $f(17.08) \approx 3.49$.

f is CD on $(-4.91, -4.10)$, $(0, 4.10)$, $(4.91, 9.60)$, CU on $(9.60, 12.25)$, CD on $(12.25, 15.81)$, CU on $(15.81, 18.65)$, and CD on $(18.65, 20)$. There are inflection points at $(9.60, 2.95)$, $(12.25, 3.27)$, $(15.81, 3.91)$, and $(18.65, 4.20)$.

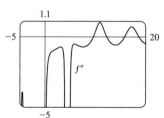

21. $y = f(x) = \dfrac{1 - e^{1/x}}{1 + e^{1/x}}$. From a CAS, $y' = \dfrac{2e^{1/x}}{x^2(1 + e^{1/x})^2}$ and $y'' = \dfrac{-2e^{1/x}(1 - e^{1/x} + 2x + 2xe^{1/x})}{x^4(1 + e^{1/x})^3}$.

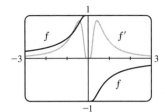

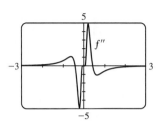

f is an odd function defined on $(-\infty, 0) \cup (0, \infty)$. Its graph has no x- or y-intercepts. Since $\lim\limits_{x \to \pm\infty} f(x) = 0$, the x-axis

is a HA. $f'(x) > 0$ for $x \neq 0$, so f is increasing on $(-\infty, 0)$ and $(0, \infty)$. It has no local extreme values.

$f''(x) = 0$ for $x \approx \pm 0.417$, so f is CU on $(-\infty, -0.417)$, CD on $(-0.417, 0)$, CU on $(0, 0.417)$, and CD on $(0.417, \infty)$.

f has IPs at $(-0.417, 0.834)$ and $(0.417, -0.834)$.

23. (a) $f(x) = x^{1/x}$

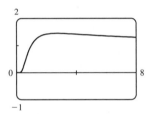

(b) Recall that $a^b = e^{b \ln a}$. $\lim\limits_{x \to 0^+} x^{1/x} = \lim\limits_{x \to 0^+} e^{(1/x) \ln x}$. As $x \to 0^+$, $\dfrac{\ln x}{x} \to -\infty$, so $x^{1/x} = e^{(1/x) \ln x} \to 0$. This

indicates that there is a hole at $(0, 0)$. As $x \to \infty$, we have the indeterminate form ∞^0. $\lim\limits_{x \to \infty} x^{1/x} = \lim\limits_{x \to \infty} e^{(1/x) \ln x}$,

but $\lim\limits_{x \to \infty} \dfrac{\ln x}{x} \overset{\text{H}}{=} \lim\limits_{x \to \infty} \dfrac{1/x}{1} = 0$, so $\lim\limits_{x \to \infty} x^{1/x} = e^0 = 1$. This indicates that $y = 1$ is a HA.

(c) Estimated maximum: $(2.72, 1.45)$. No estimated minimum. We use logarithmic differentiation to find any critical

numbers. $y = x^{1/x} \implies \ln y = \dfrac{1}{x} \ln x \implies \dfrac{y'}{y} = \dfrac{1}{x} \cdot \dfrac{1}{x} + (\ln x)\left(-\dfrac{1}{x^2}\right) \implies y' = x^{1/x}\left(\dfrac{1 - \ln x}{x^2}\right) = 0 \implies$

$\ln x = 1 \implies x = e$. For $0 < x < e$, $y' > 0$ and for $x > e$, $y' < 0$, so $f(e) = e^{1/e}$ is a local maximum value. This

point is approximately $(2.7183, 1.4447)$, which agrees with our estimate.

(d)

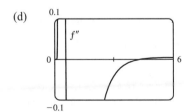

From the graph, we see that $f''(x) = 0$ at $x \approx 0.58$ and $x \approx 4.37$. Since f''

changes sign at these values, they are x-coordinates of inflection points.

25.

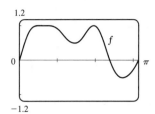 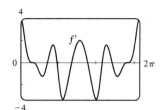 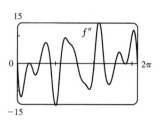

From the graph of $f(x) = \sin(x + \sin 3x)$ in the viewing rectangle $[0, \pi]$ by $[-1.2, 1.2]$, it looks like f has two maxima

and two minima. If we calculate and graph $f'(x) = [\cos(x + \sin 3x)](1 + 3\cos 3x)$ on $[0, 2\pi]$, we see that

the graph of f' appears to be almost tangent to the x-axis at about $x = 0.7$. The graph of

$f'' = -[\sin(x + \sin 3x)](1 + 3\cos 3x)^2 + \cos(x + \sin 3x)(-9\sin 3x)$ is even more interesting near this x-value:

it seems to just touch the x-axis.

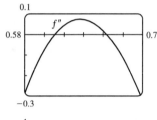 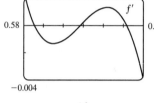

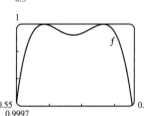 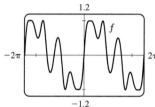

If we zoom in on this place on the graph of f'', we see that f'' actually does cross the axis twice near $x = 0.65$,

indicating a change in concavity for a very short interval. If we look at the graph of f' on the same interval, we see that it

changes sign three times near $x = 0.65$, indicating that what we had thought was a broad extremum at about $x = 0.7$ actually

consists of three extrema (two maxima and a minimum). These maximum values are roughly $f(0.59) = 1$ and $f(0.68) = 1$,

and the minimum value is roughly $f(0.64) = 0.99996$. There are also a maximum value of about $f(1.96) = 1$ and minimum

values of about $f(1.46) = 0.49$ and $f(2.73) = -0.51$. The points of inflection on $(0, \pi)$ are about $(0.61, 0.99998)$,

$(0.66, 0.99998)$, $(1.17, 0.72)$, $(1.75, 0.77)$, and $(2.28, 0.34)$. On $(\pi, 2\pi)$, they are about $(4.01, -0.34)$, $(4.54, -0.77)$,

$(5.11, -0.72)$, $(5.62, -0.99998)$, and $(5.67, -0.99998)$. There are also IP at $(0, 0)$ and $(\pi, 0)$. Note that the function is odd

and periodic with period 2π, and it is also rotationally symmetric about all points of the form $((2n + 1)\pi, 0)$, n an integer.

27. $f(x) = x^4 + cx^2 = x^2(x^2 + c)$. Note that f is an even function. For $c \geq 0$, the only x-intercept is the point $(0, 0)$. We

calculate $f'(x) = 4x^3 + 2cx = 4x(x^2 + \frac{1}{2}c) \Rightarrow f''(x) = 12x^2 + 2c$. If $c \geq 0$, $x = 0$ is the only critical point and there

is no inflection point. As we can see from the examples, there is no change in the basic shape of the graph for $c \geq 0$; it merely

becomes steeper as c increases. For $c = 0$, the graph is the simple curve $y = x^4$. For $c < 0$, there are x-intercepts at 0
and at $\pm\sqrt{-c}$. Also, there is a maximum at $(0,0)$, and there are
minima at $\left(\pm\sqrt{-\frac{1}{2}c}, -\frac{1}{4}c^2\right)$. As $c \to -\infty$, the x-coordinates of
these minima get larger in absolute value, and the minimum points
move downward. There are inflection points at $\left(\pm\sqrt{-\frac{1}{6}c}, -\frac{5}{36}c^2\right)$,
which also move away from the origin as $c \to -\infty$.

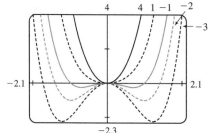

29.

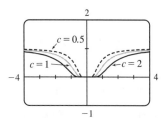 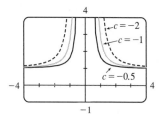

$c = 0$ is a transitional value—we get the graph of $y = 1$. For $c > 0$, we see that there is a HA at $y = 1$, and that the graph

spreads out as c increases. At first glance there appears to be a minimum at $(0,0)$, but $f(0)$ is undefined, so there is no

minimum or maximum. For $c < 0$, we still have the HA at $y = 1$, but the range is $(1, \infty)$ rather than $(0, 1)$. We also have

a VA at $x = 0$. $f(x) = e^{-c/x^2} \Rightarrow f'(x) = e^{-c/x^2}\left(\dfrac{2c}{x^3}\right) \Rightarrow f''(x) = \dfrac{2c(2c - 3x^2)}{x^6 e^{c/x^2}}$.

$f'(x) \neq 0$ and $f'(x)$ exists for all $x \neq 0$ (and 0 is not in the domain of f), so there are no maxima or minima.

$f''(x) = 0 \Rightarrow x = \pm\sqrt{2c/3}$, so if $c > 0$, the inflection points spread out as c increases, and if $c < 0$, there are no IP.

For $c > 0$, there are IP at $\left(\pm\sqrt{2c/3}, e^{-3/2}\right)$. Note that the y-coordinate of the IP is constant.

31. Note that $c = 0$ is a transitional value at which the graph consists of the x-axis. Also, we can see that if we substitute $-c$ for c,

the function $f(x) = \dfrac{cx}{1 + c^2x^2}$ will be reflected in the x-axis, so we investigate only positive values of c (except $c = -1$, as a

demonstration of this reflective property). Also, f is an odd function. $\displaystyle\lim_{x \to \pm\infty} f(x) = 0$, so $y = 0$ is a horizontal asymptote for

all c. We calculate $f'(x) = \dfrac{(1 + c^2x^2)c - cx(2c^2x)}{(1 + c^2x^2)^2} = -\dfrac{c(c^2x^2 - 1)}{(1 + c^2x^2)^2}$. $f'(x) = 0 \Leftrightarrow c^2x^2 - 1 = 0 \Leftrightarrow x = \pm 1/c$.

So there is an absolute maximum value of $f(1/c) = \frac{1}{2}$ and an absolute minimum value of $f(-1/c) = -\frac{1}{2}$. These extrema

have the same value regardless of c, but the maximum points move closer to the y-axis as c increases.

$$f''(x) = \frac{(-2c^3x)(1 + c^2x^2)^2 - (-c^3x^2 + c)[2(1 + c^2x^2)(2c^2x)]}{(1 + c^2x^2)^4}$$

$$= \frac{(-2c^3x)(1 + c^2x^2) + (c^3x^2 - c)(4c^2x)}{(1 + c^2x^2)^3} = \frac{2c^3x(c^2x^2 - 3)}{(1 + c^2x^2)^3}$$

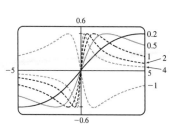

$f''(x) = 0 \Leftrightarrow x = 0$ or $\pm\sqrt{3}/c$, so there are inflection points at $(0,0)$ and

at $\left(\pm\sqrt{3}/c, \pm\sqrt{3}/4\right)$. Again, the y-coordinate of the inflection points does not

depend on c, but as c increases, both inflection points approach the y-axis.

33. $f(x) = cx + \sin x \quad \Rightarrow \quad f'(x) = c + \cos x \quad \Rightarrow \quad f''(x) = -\sin x$

$f(-x) = -f(x)$, so f is an odd function and its graph is symmetric with respect to the origin.

$f(x) = 0 \quad \Leftrightarrow \quad \sin x = -cx$, so 0 is always an x-intercept.

$f'(x) = 0 \quad \Leftrightarrow \quad \cos x = -c$, so there is no critical number when $|c| > 1$. If $|c| \leq 1$, then there are infinitely

many critical numbers. If x_1 is the unique solution of $\cos x = -c$ in the interval $[0, \pi]$, then the critical numbers are $2n\pi \pm x_1$,

where n ranges over the integers. (Special cases: When $c = 1$, $x_1 = 0$; when $c = 0$, $x = \frac{\pi}{2}$; and when $c = -1$, $x_1 = \pi$.)

$f''(x) < 0 \quad \Leftrightarrow \quad \sin x > 0$, so f is CD on intervals of the form $(2n\pi, (2n+1)\pi)$. f is CU on intervals of the form

$((2n-1)\pi, 2n\pi)$. The inflection points of f are the points $(2n\pi, 2n\pi c)$, where n is an integer.

If $c \geq 1$, then $f'(x) \geq 0$ for all x, so f is increasing and has no extremum. If $c \leq -1$, then $f'(x) \leq 0$ for all x, so f is

decreasing and has no extremum. If $|c| < 1$, then $f'(x) > 0 \quad \Leftrightarrow \quad \cos x > -c \quad \Leftrightarrow \quad x$ is in an interval of the form

$(2n\pi - x_1, 2n\pi + x_1)$ for some integer n. These are the intervals on which f is increasing. Similarly, we

find that f is decreasing on the intervals of the form $(2n\pi + x_1, 2(n+1)\pi - x_1)$. Thus, f has local maxima at the points

$2n\pi + x_1$, where f has the values $c(2n\pi + x_1) + \sin x_1 = c(2n\pi + x_1) + \sqrt{1 - c^2}$, and f has local minima at the points

$2n\pi - x_1$, where we have $f(2n\pi - x_1) = c(2n\pi - x_1) - \sin x_1 = c(2n\pi - x_1) - \sqrt{1 - c^2}$.

The transitional values of c are -1 and 1. The inflection points of f move vertically, but not horizontally, when c changes.

When $|c| \geq 1$, there is no extremum. For $|c| < 1$, the maxima are spaced
2π apart horizontally, as are the minima. The horizontal spacing between
maxima and adjacent minima is regular (and equals π) when $c = 0$, but
the horizontal space between a local maximum and the nearest local
minimum shrinks as $|c|$ approaches 1.

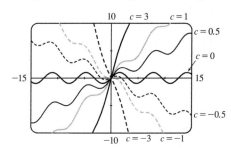

35. If $c < 0$, then $\displaystyle\lim_{x \to -\infty} f(x) = \lim_{x \to -\infty} xe^{-cx} = \lim_{x \to -\infty} \frac{x}{e^{cx}} \overset{\text{H}}{=} \lim_{x \to -\infty} \frac{1}{ce^{cx}} = 0$, and $\displaystyle\lim_{x \to \infty} f(x) = \infty$.

If $c > 0$, then $\displaystyle\lim_{x \to -\infty} f(x) = -\infty$, and $\displaystyle\lim_{x \to \infty} f(x) \overset{\text{H}}{=} \lim_{x \to \infty} \frac{1}{ce^{cx}} = 0$.

If $c = 0$, then $f(x) = x$, so $\displaystyle\lim_{x \to \pm\infty} f(x) = \pm\infty$, respectively.

So we see that $c = 0$ is a transitional value. We now exclude the case $c = 0$, since we know how the function behaves

in that case. To find the maxima and minima of f, we differentiate: $f(x) = xe^{-cx} \quad \Rightarrow$

$f'(x) = x(-ce^{-cx}) + e^{-cx} = (1 - cx)e^{-cx}$. This is 0 when $1 - cx = 0 \quad \Leftrightarrow \quad x = 1/c$. If $c < 0$ then this

represents a minimum value of $f(1/c) = 1/(ce)$, since $f'(x)$ changes from negative to positive at $x = 1/c$;

and if $c > 0$, it represents a maximum value. As $|c|$ increases, the maximum or minimum point gets closer to the origin. To find the inflection points, we differentiate again: $f'(x) = e^{-cx}(1 - cx)$ $\Rightarrow$

$f''(x) = e^{-cx}(-c) + (1 - cx)(-ce^{-cx}) = (cx - 2)ce^{-cx}$. This changes sign when $cx - 2 = 0$ $\Leftrightarrow$ $x = 2/c$. So as $|c|$ increases, the points of inflection get closer to the origin.

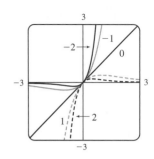

37. (a) $f(x) = cx^4 - 2x^2 + 1$. For $c = 0$, $f(x) = -2x^2 + 1$, a parabola whose vertex, $(0, 1)$, is the absolute maximum. For $c > 0$, $f(x) = cx^4 - 2x^2 + 1$ opens upward with two minimum points. As $c \to 0$, the minimum points spread apart and move downward; they are below the x-axis for $0 < c < 1$ and above for $c > 1$. For $c < 0$, the graph opens downward, and has an absolute maximum at $x = 0$ and no local minimum.

(b) $f'(x) = 4cx^3 - 4x = 4cx(x^2 - 1/c)$ $[c \neq 0]$. If $c \leq 0$, 0 is the only critical number.

$f''(x) = 12cx^2 - 4$, so $f''(0) = -4$ and there is a local maximum at $(0, f(0)) = (0, 1)$, which lies on $y = 1 - x^2$. If $c > 0$, the critical numbers are 0 and $\pm 1/\sqrt{c}$. As before, there is a local maximum at $(0, f(0)) = (0, 1)$, which lies on $y = 1 - x^2$.

$f''\left(\pm 1/\sqrt{c}\right) = 12 - 4 = 8 > 0$, so there is a local minimum at $x = \pm 1/\sqrt{c}$. Here $f\left(\pm 1/\sqrt{c}\right) = c(1/c^2) - 2/c + 1 = -1/c + 1$.

But $\left(\pm 1/\sqrt{c}, -1/c + 1\right)$ lies on $y = 1 - x^2$ since $1 - \left(\pm 1/\sqrt{c}\right)^2 = 1 - 1/c$.

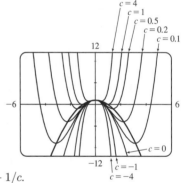

4.7   Optimization Problems

1. (a)

First Number	Second Number	Product
1	22	22
2	21	42
3	20	60
4	19	76
5	18	90
6	17	102
7	16	112
8	15	120
9	14	126
10	13	130
11	12	132

We needn't consider pairs where the first number is larger than the second, since we can just interchange the numbers in such cases. The answer appears to be 11 and 12, but we have considered only integers in the table.

(b) Call the two numbers x and y. Then $x + y = 23$, so $y = 23 - x$. Call the product P. Then

$P = xy = x(23 - x) = 23x - x^2$, so we wish to maximize the function $P(x) = 23x - x^2$. Since $P'(x) = 23 - 2x$,

we see that $P'(x) = 0$ $\Leftrightarrow$ $x = \frac{23}{2} = 11.5$. Thus, the maximum value of P is $P(11.5) = (11.5)^2 = 132.25$ and it

occurs when $x = y = 11.5$.

Or: Note that $P''(x) = -2 < 0$ for all x, so P is everywhere concave downward and the local maximum at $x = 11.5$

must be an absolute maximum.

3. The two numbers are x and $\dfrac{100}{x}$, where $x > 0$. Minimize $f(x) = x + \dfrac{100}{x}$. $f'(x) = 1 - \dfrac{100}{x^2} = \dfrac{x^2 - 100}{x^2}$. The critical

number is $x = 10$. Since $f'(x) < 0$ for $0 < x < 10$ and $f'(x) > 0$ for $x > 10$, there is an absolute minimum at $x = 10$.

The numbers are 10 and 10.

5. If the rectangle has dimensions x and y, then its perimeter is $2x + 2y = 100$ m, so $y = 50 - x$. Thus, the area is

$A = xy = x(50 - x)$. We wish to maximize the function $A(x) = x(50 - x) = 50x - x^2$, where $0 < x < 50$. Since

$A'(x) = 50 - 2x = -2(x - 25)$, $A'(x) > 0$ for $0 < x < 25$ and $A'(x) < 0$ for $25 < x < 50$. Thus, A has an absolute

maximum at $x = 25$, and $A(25) = 25^2 = 625$ m^2. The dimensions of the rectangle that maximize its area are $x = y = 25$ m.

(The rectangle is a square.)

7. We need to maximize Y for $N \geq 0$. $Y(N) = \dfrac{kN}{1 + N^2}$ $\Rightarrow$

$Y'(N) = \dfrac{(1 + N^2)k - kN(2N)}{(1 + N^2)^2} = \dfrac{k(1 - N^2)}{(1 + N^2)^2} = \dfrac{k(1 + N)(1 - N)}{(1 + N^2)^2}$. $Y'(N) > 0$ for $0 < N < 1$ and $Y'(N) < 0$

for $N > 1$. Thus, Y has an absolute maximum of $Y(1) = \frac{1}{2}k$ at $N = 1$.

9. (a)

The areas of the three figures are 12,500, 12,500, and 9000 ft^2. There appears to be a maximum area of at least 12,500 ft^2.

(b) Let x denote the length of each of two sides and three dividers.

Let y denote the length of the other two sides.

(c) Area $A = $ length $\times$ width $= y \cdot x$

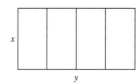

(d) Length of fencing $= 750$ $\Rightarrow$ $5x + 2y = 750$

(e) $5x + 2y = 750$ $\Rightarrow$ $y = 375 - \frac{5}{2}x$ $\Rightarrow$ $A(x) = \left(375 - \frac{5}{2}x\right)x = 375x - \frac{5}{2}x^2$

(f) $A'(x) = 375 - 5x = 0$ $\Rightarrow$ $x = 75$. Since $A''(x) = -5 < 0$ there is an absolute maximum when $x = 75$. Then

$y = \frac{375}{2} = 187.5$. The largest area is $75\left(\frac{375}{2}\right) = 14{,}062.5$ ft^2. These values of x and y are between the values in the first

and second figures in part (a). Our original estimate was low.

11. 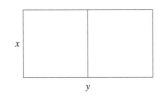 $xy = 1.5 \times 10^6$, so $y = 1.5 \times 10^6/x$. Minimize the amount of fencing, which is

$3x + 2y = 3x + 2(1.5 \times 10^6/x) = 3x + 3 \times 10^6/x = F(x)$.

$F'(x) = 3 - 3 \times 10^6/x^2 = 3(x^2 - 10^6)/x^2$. The critical number is $x = 10^3$ and

$F'(x) < 0$ for $0 < x < 10^3$ and $F'(x) > 0$ if $x > 10^3$, so the absolute minimum

occurs when $x = 10^3$ and $y = 1.5 \times 10^3$.

The field should be 1000 feet by 1500 feet with the middle fence parallel to the short side of the field.

13. Let b be the length of the base of the box and h the height. The surface area is $1200 = b^2 + 4hb$ $\Rightarrow$ $h = (1200 - b^2)/(4b)$.

The volume is $V = b^2h = b^2(1200 - b^2)/4b = 300b - b^3/4$ $\Rightarrow$ $V'(b) = 300 - \frac{3}{4}b^2$.

$V'(b) = 0$ $\Rightarrow$ $300 = \frac{3}{4}b^2$ $\Rightarrow$ $b^2 = 400$ $\Rightarrow$ $b = \sqrt{400} = 20$. Since $V'(b) > 0$ for $0 < b < 20$ and $V'(b) < 0$ for

$b > 20$, there is an absolute maximum when $b = 20$ by the First Derivative Test for Absolute Extreme Values (see page 324).

If $b = 20$, then $h = (1200 - 20^2)/(4 \cdot 20) = 10$, so the largest possible volume is $b^2h = (20)^2(10) = 4000$ cm^3.

15. 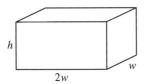 $10 = (2w)(w)h = 2w^2h$, so $h = 5/w^2$. The cost is

$$C(w) = 10(2w^2) + 6[2(2wh) + 2hw] + 6(2w^2)$$
$$= 32w^2 + 36wh = 32w^2 + 180/w$$

$C'(w) = 64w - 180/w^2 = 4(16w^3 - 45)/w^2$ $\Rightarrow$ $w = \sqrt[3]{\frac{45}{16}}$ is the critical number. $C'(w) < 0$ for $0 < w < \sqrt[3]{\frac{45}{16}}$ and

$C'(w) > 0$ for $w > \sqrt[3]{\frac{45}{16}}$. The minimum cost is $C\left(\sqrt[3]{\frac{45}{16}}\right) = 32(2.8125)^{2/3} + 180/\sqrt{2.8125} \approx \$191.28$.

17. The distance from a point (x, y) on the line $y = 4x + 7$ to the origin is $\sqrt{(x-0)^2 + (y-0)^2} = \sqrt{x^2 + y^2}$. However, it is

easier to work with the *square* of the distance; that is, $D(x) = \left(\sqrt{x^2 + y^2}\right)^2 = x^2 + y^2 = x^2 + (4x + 7)^2$. Because the

distance is positive, its minimum value will occur at the same point as the minimum value of D.

$D'(x) = 2x + 2(4x + 7)(4) = 34x + 56$, so $D'(x) = 0$ $\Leftrightarrow$ $x = -\frac{28}{17}$.

$D''(x) = 34 > 0$, so D is concave upward for all x. Thus, D has an absolute minimum at $x = -\frac{28}{17}$. The point closest to the

origin is $(x, y) = \left(-\frac{28}{17}, 4\left(-\frac{28}{17}\right) + 7\right) = \left(-\frac{28}{17}, \frac{7}{17}\right)$.

19. 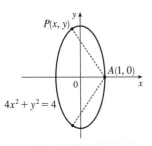 From the figure, we see that there are two points that are farthest away from

$A(1, 0)$. The distance d from A to an arbitrary point $P(x, y)$ on the ellipse is

$d = \sqrt{(x-1)^2 + (y-0)^2}$ and the square of the distance is

$S = d^2 = x^2 - 2x + 1 + y^2 = x^2 - 2x + 1 + (4 - 4x^2) = -3x^2 - 2x + 5$.

$S' = -6x - 2$ and $S' = 0$ $\Rightarrow$ $x = -\frac{1}{3}$. Now $S'' = -6 < 0$, so we know

that S has a maximum at $x = -\frac{1}{3}$. Since $-1 \le x \le 1$, $S(-1) = 4$,

$S\left(-\frac{1}{3}\right) = \frac{16}{3}$, and $S(1) = 0$, we see that the maximum distance is $\sqrt{\frac{16}{3}}$. The corresponding y-values are

$y = \pm\sqrt{4 - 4\left(-\frac{1}{3}\right)^2} = \pm\sqrt{\frac{32}{9}} = \pm\frac{4}{3}\sqrt{2} \approx \pm1.89$. The points are $\left(-\frac{1}{3}, \pm\frac{4}{3}\sqrt{2}\right)$.

21.

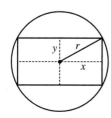

The area of the rectangle is $(2x)(2y) = 4xy$. Also $r^2 = x^2 + y^2$ so $y = \sqrt{r^2 - x^2}$, so the area is $A(x) = 4x\sqrt{r^2 - x^2}$. Now

$$A'(x) = 4\left(\sqrt{r^2 - x^2} - \frac{x^2}{\sqrt{r^2 - x^2}}\right) = 4\frac{r^2 - 2x^2}{\sqrt{r^2 - x^2}}.$$ The critical number is

$x = \frac{1}{\sqrt{2}}r$. Clearly this gives a maximum.

$y = \sqrt{r^2 - \left(\frac{1}{\sqrt{2}}r\right)^2} = \sqrt{\frac{1}{2}r^2} = \frac{1}{\sqrt{2}}r = x$, which tells us that the rectangle is a square. The dimensions are $2x = \sqrt{2}\,r$ and $2y = \sqrt{2}\,r$.

23.

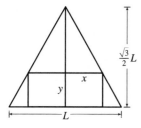

The height h of the equilateral triangle with sides of length L is $\frac{\sqrt{3}}{2}L$,

since $h^2 + (L/2)^2 = L^2 \ \Rightarrow\ h^2 = L^2 - \frac{1}{4}L^2 = \frac{3}{4}L^2 \ \Rightarrow$

$h = \frac{\sqrt{3}}{2}L.$ Using similar triangles, $\dfrac{\frac{\sqrt{3}}{2}L - y}{x} = \dfrac{\frac{\sqrt{3}}{2}L}{L/2} = \sqrt{3} \ \Rightarrow$

$\sqrt{3}\,x = \dfrac{\sqrt{3}}{2}L - y \ \Rightarrow\ y = \dfrac{\sqrt{3}}{2}L - \sqrt{3}\,x \ \Rightarrow\ y = \dfrac{\sqrt{3}}{2}(L - 2x).$

The area of the inscribed rectangle is $A(x) = (2x)y = \sqrt{3}\,x(L - 2x) = \sqrt{3}\,Lx - 2\sqrt{3}\,x^2$, where $0 \le x \le L/2$. Now $0 = A'(x) = \sqrt{3}\,L - 4\sqrt{3}\,x \ \Rightarrow\ x = \sqrt{3}\,L/(4\sqrt{3}) = L/4$. Since $A(0) = A(L/2) = 0$, the maximum occurs when $x = L/4$, and $y = \frac{\sqrt{3}}{2}L - \frac{\sqrt{3}}{4}L = \frac{\sqrt{3}}{4}L$, so the dimensions are $L/2$ and $\frac{\sqrt{3}}{4}L$.

25.

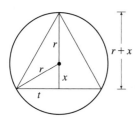

The area of the triangle is

$$A(x) = \tfrac{1}{2}(2t)(r + x) = t(r + x) = \sqrt{r^2 - x^2}(r + x).$$ Then

$$0 = A'(x) = r\frac{-2x}{2\sqrt{r^2 - x^2}} + \sqrt{r^2 - x^2} + x\frac{-2x}{2\sqrt{r^2 - x^2}}$$
$$= -\frac{x^2 + rx}{\sqrt{r^2 - x^2}} + \sqrt{r^2 - x^2} \ \Rightarrow$$

$\dfrac{x^2 + rx}{\sqrt{r^2 - x^2}} = \sqrt{r^2 - x^2} \ \Rightarrow\ x^2 + rx = r^2 - x^2 \ \Rightarrow\ 0 = 2x^2 + rx - r^2 = (2x - r)(x + r) \ \Rightarrow$

$x = \frac{1}{2}r$ or $x = -r$. Now $A(r) = 0 = A(-r) \ \Rightarrow$ the maximum occurs where $x = \frac{1}{2}r$, so the triangle has height

$r + \frac{1}{2}r = \frac{3}{2}r$ and base $2\sqrt{r^2 - \left(\frac{1}{2}r\right)^2} = 2\sqrt{\frac{3}{4}r^2} = \sqrt{3}\,r$.

27.

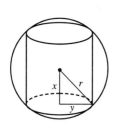

The cylinder has volume $V = \pi y^2(2x)$. Also $x^2 + y^2 = r^2 \ \Rightarrow\ y^2 = r^2 - x^2$, so

$V(x) = \pi(r^2 - x^2)(2x) = 2\pi(r^2 x - x^3)$, where $0 \le x \le r$.

$V'(x) = 2\pi(r^2 - 3x^2) = 0 \ \Rightarrow\ x = r/\sqrt{3}$. Now $V(0) = V(r) = 0$, so there is a

maximum when $x = r/\sqrt{3}$ and $V(r/\sqrt{3}) = \pi(r^2 - r^2/3)(2r/\sqrt{3}) = 4\pi r^3/(3\sqrt{3})$.

29.

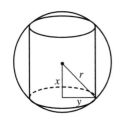

The cylinder has surface area

$$2(\text{area of the base}) + (\text{lateral surface area}) = 2\pi(\text{radius})^2 + 2\pi(\text{radius})(\text{height})$$

$$= 2\pi y^2 + 2\pi y(2x)$$

Now $x^2 + y^2 = r^2 \Rightarrow y^2 = r^2 - x^2 \Rightarrow y = \sqrt{r^2 - x^2}$, so the surface area is

$$S(x) = 2\pi(r^2 - x^2) + 4\pi x\sqrt{r^2 - x^2}, \ 0 \le x \le r$$

$$= 2\pi r^2 - 2\pi x^2 + 4\pi\left(x\sqrt{r^2 - x^2}\right)$$

Thus,

$$S'(x) = 0 - 4\pi x + 4\pi\left[x \cdot \tfrac{1}{2}(r^2 - x^2)^{-1/2}(-2x) + (r^2 - x^2)^{1/2} \cdot 1\right]$$

$$= 4\pi\left[-x - \frac{x^2}{\sqrt{r^2 - x^2}} + \sqrt{r^2 - x^2}\right] = 4\pi \cdot \frac{-x\sqrt{r^2 - x^2} - x^2 + r^2 - x^2}{\sqrt{r^2 - x^2}}$$

$S'(x) = 0 \Rightarrow x\sqrt{r^2 - x^2} = r^2 - 2x^2 \ (\star) \Rightarrow \left(x\sqrt{r^2 - x^2}\right)^2 = (r^2 - 2x^2)^2 \Rightarrow$

$x^2(r^2 - x^2) = r^4 - 4r^2x^2 + 4x^4 \Rightarrow r^2x^2 - x^4 = r^4 - 4r^2x^2 + 4x^4 \Rightarrow 5x^4 - 5r^2x^2 + r^4 = 0.$

This is a quadratic equation in x^2. By the quadratic formula, $x^2 = \frac{5 \pm \sqrt{5}}{10}r^2$, but we reject the root with the $+$ sign since it

doesn't satisfy $(\star)$. [The right side is negative and the left side is positive.] So $x = \sqrt{\frac{5 - \sqrt{5}}{10}}\,r$. Since $S(0) = S(r) = 0$, the

maximum surface area occurs at the critical number and $x^2 = \frac{5 - \sqrt{5}}{10}r^2 \Rightarrow y^2 = r^2 - \frac{5 - \sqrt{5}}{10}r^2 = \frac{5 + \sqrt{5}}{10}r^2 \Rightarrow$

the surface area is

$$2\pi\left(\frac{5 + \sqrt{5}}{10}\right)r^2 + 4\pi\sqrt{\frac{5 - \sqrt{5}}{10}}\sqrt{\frac{5 + \sqrt{5}}{10}}r^2 = \pi r^2\left[2 \cdot \frac{5 + \sqrt{5}}{10} + 4\frac{\sqrt{(5 - \sqrt{5})(5 + \sqrt{5})}}{10}\right] = \pi r^2\left[\frac{5 + \sqrt{5}}{5} + \frac{2\sqrt{20}}{5}\right]$$

$$= \pi r^2\left[\frac{5 + \sqrt{5} + 2 \cdot 2\sqrt{5}}{5}\right] = \pi r^2\left[\frac{5 + 5\sqrt{5}}{5}\right] = \pi r^2\left(1 + \sqrt{5}\right).$$

31.

$xy = 384 \Rightarrow y = 384/x$. Total area is

$A(x) = (8 + x)(12 + 384/x) = 12(40 + x + 256/x)$, so

$A'(x) = 12(1 - 256/x^2) = 0 \Rightarrow x = 16$. There is an absolute minimum

when $x = 16$ since $A'(x) < 0$ for $0 < x < 16$ and $A'(x) > 0$ for $x > 16$.

When $x = 16$, $y = 384/16 = 24$, so the dimensions are 24 cm and 36 cm.

33.

Let x be the length of the wire used for the square. The total area is

$$A(x) = \left(\frac{x}{4}\right)^2 + \frac{1}{2}\left(\frac{10 - x}{3}\right)\frac{\sqrt{3}}{2}\left(\frac{10 - x}{3}\right)$$

$$= \tfrac{1}{16}x^2 + \tfrac{\sqrt{3}}{36}(10 - x)^2, \ 0 \le x \le 10$$

$A'(x) = \frac{1}{8}x - \frac{\sqrt{3}}{18}(10 - x) = 0 \Leftrightarrow \frac{9}{72}x + \frac{4\sqrt{3}}{72}x - \frac{40\sqrt{3}}{72} = 0 \Leftrightarrow x = \frac{40\sqrt{3}}{9 + 4\sqrt{3}}$. Now $A(0) = \left(\frac{\sqrt{3}}{36}\right)100 \approx 4.81$,

$A(10) = \frac{100}{16} = 6.25$ and $A\left(\frac{40\sqrt{3}}{9 + 4\sqrt{3}}\right) \approx 2.72$, so

(a) The maximum area occurs when $x = 10$ m, and all the wire is used for the square.

(b) The minimum area occurs when $x = \frac{40\sqrt{3}}{9 + 4\sqrt{3}} \approx 4.35$ m.

35.

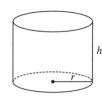

The volume is $V = \pi r^2 h$ and the surface area is

$$S(r) = \pi r^2 + 2\pi rh = \pi r^2 + 2\pi r\left(\frac{V}{\pi r^2}\right) = \pi r^2 + \frac{2V}{r}.$$

$$S'(r) = 2\pi r - \frac{2V}{r^2} = 0 \quad\Rightarrow\quad 2\pi r^3 = 2V \quad\Rightarrow\quad r = \sqrt[3]{\frac{V}{\pi}}\ \text{cm}.$$

This gives an absolute minimum since $S'(r) < 0$ for $0 < r < \sqrt[3]{\dfrac{V}{\pi}}$ and $S'(r) > 0$ for $r > \sqrt[3]{\dfrac{V}{\pi}}$.

When $r = \sqrt[3]{\dfrac{V}{\pi}}$, $h = \dfrac{V}{\pi r^2} = \dfrac{V}{\pi(V/\pi)^{2/3}} = \sqrt[3]{\dfrac{V}{\pi}}\ \text{cm}.$

37.

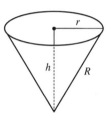

$h^2 + r^2 = R^2 \quad\Rightarrow\quad V = \frac{\pi}{3}r^2 h = \frac{\pi}{3}(R^2 - h^2)h = \frac{\pi}{3}(R^2 h - h^3).$

$V'(h) = \frac{\pi}{3}(R^2 - 3h^2) = 0$ when $h = \frac{1}{\sqrt{3}}R$. This gives an absolute maximum, since

$V'(h) > 0$ for $0 < h < \frac{1}{\sqrt{3}}R$ and $V'(h) < 0$ for $h > \frac{1}{\sqrt{3}}R$. The maximum volume is

$V\left(\frac{1}{\sqrt{3}}R\right) = \frac{\pi}{3}\left(\frac{1}{\sqrt{3}}R^3 - \frac{1}{3\sqrt{3}}R^3\right) = \frac{2}{9\sqrt{3}}\pi R^3.$

39.

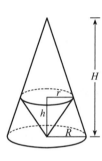

By similar triangles, $\dfrac{H}{R} = \dfrac{H-h}{r}$ **(1)**. The volume of the inner cone is $V = \frac{1}{3}\pi r^2 h$,

so we'll solve **(1)** for h. $\dfrac{Hr}{R} = H - h \quad\Rightarrow$

$h = H - \dfrac{Hr}{R} = \dfrac{HR - Hr}{R} = \dfrac{H}{R}(R - r)$ **(2)**.

Thus, $V(r) = \dfrac{\pi}{3}r^2 \cdot \dfrac{H}{R}(R-r) = \dfrac{\pi H}{3R}(Rr^2 - r^3) \quad\Rightarrow$

$V'(r) = \dfrac{\pi H}{3R}(2Rr - 3r^2) = \dfrac{\pi H}{3R}r(2R - 3r).$

$V'(r) = 0 \quad\Rightarrow\quad r = 0$ or $2R = 3r \quad\Rightarrow\quad r = \frac{2}{3}R$ and from **(2)**, $h = \dfrac{H}{R}\left(R - \frac{2}{3}R\right) = \dfrac{H}{R}\left(\frac{1}{3}R\right) = \frac{1}{3}H.$

$V'(r)$ changes from positive to negative at $r = \frac{2}{3}R$, so the inner cone has a maximum volume of

$V = \frac{1}{3}\pi r^2 h = \frac{1}{3}\pi\left(\frac{2}{3}R\right)^2\left(\frac{1}{3}H\right) = \frac{4}{27}\cdot\frac{1}{3}\pi R^2 H$, which is approximately 15% of the volume of the larger cone.

41. $P(R) = \dfrac{E^2 R}{(R+r)^2} \quad\Rightarrow$

$$P'(R) = \frac{(R+r)^2 \cdot E^2 - E^2 R \cdot 2(R+r)}{[(R+r)^2]^2} = \frac{(R^2 + 2Rr + r^2)E^2 - 2E^2 R^2 - 2E^2 Rr}{(R+r)^4}$$

$$= \frac{E^2 r^2 - E^2 R^2}{(R+r)^4} = \frac{E^2(r^2 - R^2)}{(R+r)^4} = \frac{E^2(r+R)(r-R)}{(R+r)^4} = \frac{E^2(r-R)}{(R+r)^3}$$

$P'(R) = 0 \quad\Rightarrow\quad R = r \quad\Rightarrow\quad P(r) = \dfrac{E^2 r}{(r+r)^2} = \dfrac{E^2 r}{4r^2} = \dfrac{E^2}{4r}.$

The expression for $P'(R)$ shows that $P'(R) > 0$ for $R < r$ and $P'(R) < 0$ for $R > r$. Thus, the maximum value of the

power is $E^2/(4r)$, and this occurs when $R = r$.

43. $S = 6sh - \frac{3}{2}s^2 \cot\theta + 3s^2 \frac{\sqrt{3}}{2}\csc\theta$

(a) $\dfrac{dS}{d\theta} = \frac{3}{2}s^2 \csc^2\theta - 3s^2 \frac{\sqrt{3}}{2}\csc\theta \cot\theta$ or $\frac{3}{2}s^2 \csc\theta \left(\csc\theta - \sqrt{3}\cot\theta\right)$.

(b) $\dfrac{dS}{d\theta} = 0$ when $\csc\theta - \sqrt{3}\cot\theta = 0$ $\Rightarrow$ $\dfrac{1}{\sin\theta} - \sqrt{3}\dfrac{\cos\theta}{\sin\theta} = 0$ $\Rightarrow$ $\cos\theta = \frac{1}{\sqrt{3}}$. The First Derivative Test shows

that the minimum surface area occurs when $\theta = \cos^{-1}\left(\frac{1}{\sqrt{3}}\right) \approx 55°$.

(c)

If $\cos\theta = \frac{1}{\sqrt{3}}$, then $\cot\theta = \frac{1}{\sqrt{2}}$ and $\csc\theta = \frac{\sqrt{3}}{\sqrt{2}}$, so the surface area is

$$S = 6sh - \frac{3}{2}s^2\frac{1}{\sqrt{2}} + 3s^2\frac{\sqrt{3}}{2}\frac{\sqrt{3}}{\sqrt{2}} = 6sh - \frac{3}{2\sqrt{2}}s^2 + \frac{9}{2\sqrt{2}}s^2$$

$$= 6sh + \frac{6}{2\sqrt{2}}s^2 = 6s\left(h + \frac{1}{2\sqrt{2}}s\right)$$

(triangle diagram: hypotenuse $\sqrt{3}$, vertical side $\sqrt{2}$, base 1, angle θ)

45. Here $T(x) = \dfrac{\sqrt{x^2+25}}{6} + \dfrac{5-x}{8}$, $0 \le x \le 5$ $\Rightarrow$ $T'(x) = \dfrac{x}{6\sqrt{x^2+25}} - \dfrac{1}{8} = 0$ $\Leftrightarrow$ $8x = 6\sqrt{x^2+25}$ $\Leftrightarrow$

$16x^2 = 9(x^2+25)$ $\Leftrightarrow$ $x = \frac{15}{\sqrt{7}}$. But $\frac{15}{\sqrt{7}} > 5$, so T has no critical number. Since $T(0) \approx 1.46$ and $T(5) \approx 1.18$, he

should row directly to B.

47. There are $(6-x)$ km over land and $\sqrt{x^2+4}$ km under the river.

We need to minimize the cost C (measured in \$100,000) of the pipeline.

$C(x) = (6-x)(4) + \left(\sqrt{x^2+4}\right)(8)$ $\Rightarrow$

$C'(x) = -4 + 8\cdot\frac{1}{2}(x^2+4)^{-1/2}(2x) = -4 + \dfrac{8x}{\sqrt{x^2+4}}$.

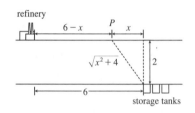

$C'(x) = 0$ $\Rightarrow$ $4 = \dfrac{8x}{\sqrt{x^2+4}}$ $\Rightarrow$ $\sqrt{x^2+4} = 2x$ $\Rightarrow$ $x^2+4 = 4x^2$ $\Rightarrow$ $4 = 3x^2$ $\Rightarrow$ $x^2 = \frac{4}{3}$ $\Rightarrow$

$x = 2/\sqrt{3}$ $[0 \le x \le 6]$. Compare the costs for $x = 0, 2/\sqrt{3}$, and 6. $C(0) = 24 + 16 = 40$,

$C\left(2/\sqrt{3}\right) = 24 - 8/\sqrt{3} + 32/\sqrt{3} = 24 + 24/\sqrt{3} \approx 37.9$, and $C(6) = 0 + 8\sqrt{40} \approx 50.6$. So the minimum cost is about

\$3.79 million when P is $6 - 2/\sqrt{3} \approx 4.85$ km east of the refinery.

49.

(diagram: light source $3k$ at left, k at right, distances x and $10-x$, total 10)

The total illumination is $I(x) = \dfrac{3k}{x^2} + \dfrac{k}{(10-x)^2}$, $0 < x < 10$. Then

$$I'(x) = \dfrac{-6k}{x^3} + \dfrac{2k}{(10-x)^3} = 0 \Rightarrow 6k(10-x)^3 = 2kx^3 \Rightarrow$$

$3(10-x)^3 = x^3$ $\Rightarrow$ $\sqrt[3]{3}(10-x) = x$ $\Rightarrow$ $10\sqrt[3]{3} - \sqrt[3]{3}\,x = x$ $\Rightarrow$ $10\sqrt[3]{3} = x + \sqrt[3]{3}\,x$ $\Rightarrow$

$10\sqrt[3]{3} = \left(1 + \sqrt[3]{3}\right)x$ $\Rightarrow$ $x = \dfrac{10\sqrt[3]{3}}{1 + \sqrt[3]{3}} \approx 5.9$ ft. This gives a minimum since $I''(x) > 0$ for $0 < x < 10$.

51.

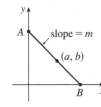

Every line segment in the first quadrant passing through (a, b) with endpoints on the x- and y-axes satisfies an equation of the form $y - b = m(x - a)$, where $m < 0$. By setting $x = 0$ and then $y = 0$, we find its endpoints, $A(0, b - am)$ and $B\left(a - \frac{b}{m}, 0\right)$. The distance d from A to B is given by $d = \sqrt{[(a - \frac{b}{m}) - 0]^2 + [0 - (b - am)]^2}$.

It follows that the square of the length of the line segment, as a function of m, is given by

$$S(m) = \left(a - \frac{b}{m}\right)^2 + (am - b)^2 = a^2 - \frac{2ab}{m} + \frac{b^2}{m^2} + a^2m^2 - 2abm + b^2. \text{ Thus,}$$

$$S'(m) = \frac{2ab}{m^2} - \frac{2b^2}{m^3} + 2a^2m - 2ab = \frac{2}{m^3}(abm - b^2 + a^2m^4 - abm^3)$$

$$= \frac{2}{m^3}[b(am - b) + am^3(am - b)] = \frac{2}{m^3}(am - b)(b + am^3)$$

Thus, $S'(m) = 0 \iff m = b/a$ or $m = -\sqrt[3]{\frac{b}{a}}$. Since $b/a > 0$ and $m < 0$, m must equal $-\sqrt[3]{\frac{b}{a}}$. Since $\frac{2}{m^3} < 0$, we see that $S'(m) < 0$ for $m < -\sqrt[3]{\frac{b}{a}}$ and $S'(m) > 0$ for $m > -\sqrt[3]{\frac{b}{a}}$. Thus, S has its absolute minimum value when $m = -\sqrt[3]{\frac{b}{a}}$.

That value is

$$S\left(-\sqrt[3]{\frac{b}{a}}\right) = \left(a + b\sqrt[3]{\frac{a}{b}}\right)^2 + \left(-a\sqrt[3]{\frac{b}{a}} - b\right)^2 = \left(a + \sqrt[3]{ab^2}\right)^2 + \left(\sqrt[3]{a^2b} + b\right)^2$$

$$= a^2 + 2a^{4/3}b^{2/3} + a^{2/3}b^{4/3} + a^{4/3}b^{2/3} + 2a^{2/3}b^{4/3} + b^2 = a^2 + 3a^{4/3}b^{2/3} + 3a^{2/3}b^{4/3} + b^2$$

The last expression is of the form $x^3 + 3x^2y + 3xy^2 + y^3$ $[= (x + y)^3]$ with $x = a^{2/3}$ and $y = b^{2/3}$, so we can write it as $(a^{2/3} + b^{2/3})^3$ and the shortest such line segment has length $\sqrt{S} = (a^{2/3} + b^{2/3})^{3/2}$.

53. (a) If $c(x) = \dfrac{C(x)}{x}$, then, by Quotient Rule, we have $c'(x) = \dfrac{xC'(x) - C(x)}{x^2}$. Now $c'(x) = 0$ when $xC'(x) - C(x) = 0$

and this gives $C'(x) = \dfrac{C(x)}{x} = c(x)$. Therefore, the marginal cost equals the average cost.

(b) (i) $C(x) = 16,000 + 200x + 4x^{3/2}$, $C(1000) = 16,000 + 200,000 + 40,000\sqrt{10} \approx 216,000 + 126,491$, so

$C(1000) \approx \$342,491$. $c(x) = C(x)/x = \dfrac{16,000}{x} + 200 + 4x^{1/2}$, $c(1000) \approx \$342.49/\text{unit}$. $C'(x) = 200 + 6x^{1/2}$,

$C'(1000) = 200 + 60\sqrt{10} \approx \$389.74/\text{unit}$.

(ii) We must have $C'(x) = c(x) \iff 200 + 6x^{1/2} = \dfrac{16,000}{x} + 200 + 4x^{1/2} \iff 2x^{3/2} = 16,000 \iff$

$x = (8,000)^{2/3} = 400$ units. To check that this is a minimum, we calculate

$c'(x) = \dfrac{-16,000}{x^2} + \dfrac{2}{\sqrt{x}} = \dfrac{2}{x^2}(x^{3/2} - 8000)$. This is negative for $x < (8000)^{2/3} = 400$, zero at $x = 400$,

and positive for $x > 400$, so c is decreasing on $(0, 400)$ and increasing on $(400, \infty)$. Thus, c has an absolute minimum at $x = 400$. [*Note:* $c''(x)$ is *not* positive for all $x > 0$.]

(iii) The minimum average cost is $c(400) = 40 + 200 + 80 = \$320/\text{unit}$.

55. (a) We are given that the demand function p is linear and $p(27{,}000) = 10$, $p(33{,}000) = 8$, so the slope is

$\frac{10-8}{27{,}000-33{,}000} = -\frac{1}{3000}$ and an equation of the line is $y - 10 = \left(-\frac{1}{3000}\right)(x - 27{,}000)$ $\Rightarrow$

$y = p(x) = -\frac{1}{3000}x + 19 = 19 - (x/3000)$.

(b) The revenue is $R(x) = xp(x) = 19x - (x^2/3000)$ $\Rightarrow$ $R'(x) = 19 - (x/1500) = 0$ when $x = 28{,}500$. Since

$R''(x) = -1/1500 < 0$, the maximum revenue occurs when $x = 28{,}500$ $\Rightarrow$ the price is $p(28{,}500) = \$9.50$.

57. (a) As in Example 6, we see that the demand function p is linear. We are given that $p(1000) = 450$ and deduce that

$p(1100) = 440$, since a \$10 reduction in price increases sales by 100 per week. The slope for p is $\frac{440-450}{1100-1000} = -\frac{1}{10}$,

so an equation is $p - 450 = -\frac{1}{10}(x - 1000)$ or $p(x) = -\frac{1}{10}x + 550$.

(b) $R(x) = xp(x) = -\frac{1}{10}x^2 + 550x$. $R'(x) = -\frac{1}{5}x + 550 = 0$ when $x = 5(550) = 2750$.

$p(2750) = 275$, so the rebate should be $450 - 275 = \$175$.

(c) $C(x) = 68{,}000 + 150x$ $\Rightarrow$ $P(x) = R(x) - C(x) = -\frac{1}{10}x^2 + 550x - 68{,}000 - 150x = -\frac{1}{10}x^2 + 400x - 68{,}000$,

$P'(x) = -\frac{1}{5}x + 400 = 0$ when $x = 2000$. $p(2000) = 350$. Therefore, the rebate to maximize profits should be

$450 - 350 = \$100$.

59.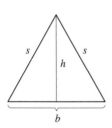
Here $s^2 = h^2 + b^2/4$, so $h^2 = s^2 - b^2/4$. The area is $A = \frac{1}{2}b\sqrt{s^2 - b^2/4}$.

Let the perimeter be p, so $2s + b = p$ or $s = (p - b)/2$ $\Rightarrow$

$A(b) = \frac{1}{2}b\sqrt{(p-b)^2/4 - b^2/4} = b\sqrt{p^2 - 2pb}/4$. Now

$A'(b) = \frac{\sqrt{p^2 - 2pb}}{4} - \frac{bp/4}{\sqrt{p^2 - 2pb}} = \frac{-3pb + p^2}{4\sqrt{p^2 - 2pb}}$.

Therefore, $A'(b) = 0$ $\Rightarrow$ $-3pb + p^2 = 0$ $\Rightarrow$ $b = p/3$. Since $A'(b) > 0$ for $b < p/3$ and $A'(b) < 0$ for $b > p/3$, there

is an absolute maximum when $b = p/3$. But then $2s + p/3 = p$, so $s = p/3$ $\Rightarrow$ $s = b$ $\Rightarrow$ the triangle is equilateral.

61. Note that $|AD| = |AP| + |PD|$ $\Rightarrow$ $5 = x + |PD|$ $\Rightarrow$ $|PD| = 5 - x$.

Using the Pythagorean Theorem for $\triangle PDB$ and $\triangle PDC$ gives us

$L(x) = |AP| + |BP| + |CP| = x + \sqrt{(5-x)^2 + 2^2} + \sqrt{(5-x)^2 + 3^2}$

$\quad = x + \sqrt{x^2 - 10x + 29} + \sqrt{x^2 - 10x + 34}$ $\Rightarrow$

$L'(x) = 1 + \frac{x-5}{\sqrt{x^2 - 10x + 29}} + \frac{x-5}{\sqrt{x^2 - 10x + 34}}$. From the graphs of L

and L', it seems that the minimum value of L is about $L(3.59) = 9.35$ m.

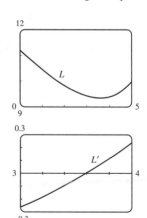

63.

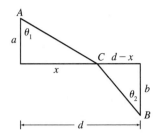

The total time is

$$T(x) = (\text{time from } A \text{ to } C) + (\text{time from } C \text{ to } B)$$

$$= \frac{\sqrt{a^2 + x^2}}{v_1} + \frac{\sqrt{b^2 + (d - x)^2}}{v_2}, \ 0 < x < d$$

$$T'(x) = \frac{x}{v_1\sqrt{a^2 + x^2}} - \frac{d - x}{v_2\sqrt{b^2 + (d - x)^2}} = \frac{\sin\theta_1}{v_1} - \frac{\sin\theta_2}{v_2}$$

The minimum occurs when $T'(x) = 0 \ \Rightarrow \ \dfrac{\sin\theta_1}{v_1} = \dfrac{\sin\theta_2}{v_2}.$

[*Note:* $T''(x) > 0$]

65.

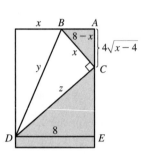

$y^2 = x^2 + z^2$, but triangles CDE and BCA are similar, so

$z/8 = x/\left(4\sqrt{x - 4}\right) \ \Rightarrow \ z = 2x/\sqrt{x - 4}$. Thus, we minimize

$f(x) = y^2 = x^2 + 4x^2/(x - 4) = x^3/(x - 4), \ 4 < x \le 8.$

$$f'(x) = \frac{(x - 4)(3x^2) - x^3}{(x - 4)^2} = \frac{x^2[3(x - 4) - x]}{(x - 4)^2} = \frac{2x^2(x - 6)}{(x - 4)^2} = 0$$

when $x = 6$. $f'(x) < 0$ when $x < 6$, $f'(x) > 0$ when $x > 6$, so the minimum

occurs when $x = 6$ in.

67.

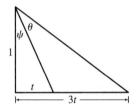

It suffices to maximize $\tan\theta$. Now

$$\frac{3t}{1} = \tan(\psi + \theta) = \frac{\tan\psi + \tan\theta}{1 - \tan\psi\tan\theta} = \frac{t + \tan\theta}{1 - t\tan\theta}. \text{ So}$$

$3t(1 - t\tan\theta) = t + \tan\theta \ \Rightarrow \ 2t = (1 + 3t^2)\tan\theta \ \Rightarrow \ \tan\theta = \dfrac{2t}{1 + 3t^2}.$

Let $f(t) = \tan\theta = \dfrac{2t}{1 + 3t^2} \ \Rightarrow \ f'(t) = \dfrac{2(1 + 3t^2) - 2t(6t)}{(1 + 3t^2)^2} = \dfrac{2(1 - 3t^2)}{(1 + 3t^2)^2} = 0 \ \Leftrightarrow \ 1 - 3t^2 = 0 \ \Leftrightarrow$

$t = \frac{1}{\sqrt{3}}$ since $t \ge 0$. Now $f'(t) > 0$ for $0 \le t < \frac{1}{\sqrt{3}}$ and $f'(t) < 0$ for $t > \frac{1}{\sqrt{3}}$, so f has an absolute maximum when $t = \frac{1}{\sqrt{3}}$

and $\tan\theta = \dfrac{2(1/\sqrt{3})}{1 + 3(1/\sqrt{3})^2} = \dfrac{1}{\sqrt{3}} \ \Rightarrow \ \theta = \frac{\pi}{6}$. Substituting for t and θ in $3t = \tan(\psi + \theta)$ gives us

$\sqrt{3} = \tan\left(\psi + \frac{\pi}{6}\right) \ \Rightarrow \ \psi = \frac{\pi}{6}.$

69.

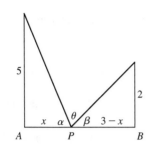

From the figure, $\tan\alpha = \dfrac{5}{x}$ and $\tan\beta = \dfrac{2}{3-x}$. Since

$$\alpha + \beta + \theta = 180° = \pi, \ \theta = \pi - \tan^{-1}\left(\frac{5}{x}\right) - \tan^{-1}\left(\frac{2}{3-x}\right) \ \Rightarrow$$

$$\frac{d\theta}{dx} = -\frac{1}{1 + \left(\dfrac{5}{x}\right)^2}\left(-\frac{5}{x^2}\right) - \frac{1}{1 + \left(\dfrac{2}{3-x}\right)^2}\left[\frac{2}{(3-x)^2}\right]$$

$$= \frac{x^2}{x^2 + 25}\cdot\frac{5}{x^2} - \frac{(3-x)^2}{(3-x)^2 + 4}\cdot\frac{2}{(3-x)^2}.$$

Now $\dfrac{d\theta}{dx} = 0 \ \Rightarrow \ \dfrac{5}{x^2 + 25} = \dfrac{2}{x^2 - 6x + 13} \ \Rightarrow \ 2x^2 + 50 = 5x^2 - 30x + 65 \ \Rightarrow$

$3x^2 - 30x + 15 = 0 \ \Rightarrow \ x^2 - 10x + 5 = 0 \ \Rightarrow \ x = 5 \pm 2\sqrt{5}$. We reject the root with the $+$ sign, since it is larger

than 3. $d\theta/dx > 0$ for $x < 5 - 2\sqrt{5}$ and $d\theta/dx < 0$ for $x > 5 - 2\sqrt{5}$, so θ is maximized when

$|AP| = x = 5 - 2\sqrt{5} \approx 0.53$.

71.

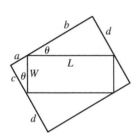

In the small triangle with sides a and c and hypotenuse W, $\sin\theta = \dfrac{a}{W}$ and

$\cos\theta = \dfrac{c}{W}$. In the triangle with sides b and d and hypotenuse L, $\sin\theta = \dfrac{d}{L}$ and

$\cos\theta = \dfrac{b}{L}$. Thus, $a = W\sin\theta$, $c = W\cos\theta$, $d = L\sin\theta$, and $b = L\cos\theta$, so the

area of the circumscribed rectangle is

$$A(\theta) = (a+b)(c+d) = (W\sin\theta + L\cos\theta)(W\cos\theta + L\sin\theta)$$

$$= W^2\sin\theta\cos\theta + WL\sin^2\theta + LW\cos^2\theta + L^2\sin\theta\cos\theta$$

$$= LW\sin^2\theta + LW\cos^2\theta + (L^2 + W^2)\sin\theta\cos\theta$$

$$= LW(\sin^2\theta + \cos^2\theta) + (L^2 + W^2)\cdot\tfrac{1}{2}\cdot 2\sin\theta\cos\theta = LW + \tfrac{1}{2}(L^2 + W^2)\sin 2\theta, \ 0 \le \theta \le \tfrac{\pi}{2}$$

This expression shows, without calculus, that the maximum value of $A(\theta)$ occurs when $\sin 2\theta = 1 \ \Leftrightarrow \ 2\theta = \tfrac{\pi}{2} \ \Rightarrow$

$\theta = \tfrac{\pi}{4}$. So the maximum area is $A\left(\tfrac{\pi}{4}\right) = LW + \tfrac{1}{2}(L^2 + W^2) = \tfrac{1}{2}(L^2 + 2LW + W^2) = \tfrac{1}{2}(L+W)^2$.

73. (a)

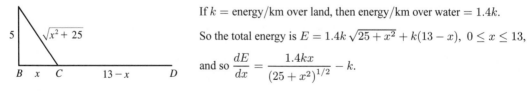

If $k =$ energy/km over land, then energy/km over water $= 1.4k$.

So the total energy is $E = 1.4k\sqrt{25 + x^2} + k(13 - x)$, $0 \le x \le 13$,

and so $\dfrac{dE}{dx} = \dfrac{1.4kx}{(25 + x^2)^{1/2}} - k$.

Set $\dfrac{dE}{dx} = 0$: $1.4kx = k(25 + x^2)^{1/2} \ \Rightarrow \ 1.96x^2 = x^2 + 25 \ \Rightarrow \ 0.96x^2 = 25 \ \Rightarrow \ x = \dfrac{5}{\sqrt{0.96}} \approx 5.1$.

Testing against the value of E at the endpoints: $E(0) = 1.4k(5) + 13k = 20k$, $E(5.1) \approx 17.9k$, $E(13) \approx 19.5k$.

Thus, to minimize energy, the bird should fly to a point about 5.1 km from B.

(b) If W/L is large, the bird would fly to a point C that is closer to B than to D to minimize the energy used flying over water. If W/L is small, the bird would fly to a point C that is closer to D than to B to minimize the distance of the flight.

$$E = W\sqrt{25 + x^2} + L(13 - x) \quad \Rightarrow \quad \frac{dE}{dx} = \frac{Wx}{\sqrt{25 + x^2}} - L = 0 \text{ when } \frac{W}{L} = \frac{\sqrt{25 + x^2}}{x}. \text{ By the same sort of}$$

argument as in part (a), this ratio will give the minimal expenditure of energy if the bird heads for the point x km from B.

(c) For flight direct to D, $x = 13$, so from part (b), $W/L = \frac{\sqrt{25 + 13^2}}{13} \approx 1.07$. There is no value of W/L for which the bird should fly directly to B. But note that $\lim\limits_{x \to 0^+} (W/L) = \infty$, so if the point at which E is a minimum is close to B, then W/L is large.

(d) Assuming that the birds instinctively choose the path that minimizes the energy expenditure, we can use the equation for $dE/dx = 0$ from part (a) with $1.4k = c$, $x = 4$, and $k = 1$: $c(4) = 1 \cdot (25 + 4^2)^{1/2} \quad \Rightarrow \quad c = \sqrt{41}/4 \approx 1.6$.

4.8 Newton's Method

1. (a)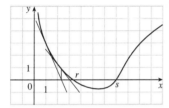

The tangent line at $x = 1$ intersects the x-axis at $x \approx 2.3$, so $x_2 \approx 2.3$. The tangent line at $x = 2.3$ intersects the x-axis at $x \approx 3$, so $x_3 \approx 3.0$.

(b) $x_1 = 5$ would *not* be a better first approximation than $x_1 = 1$ since the tangent line is nearly horizontal. In fact, the second approximation for $x_1 = 5$ appears to be to the left of $x = 1$.

3. Since $x_1 = 3$ and $y = 5x - 4$ is tangent to $y = f(x)$ at $x = 3$, we simply need to find where the tangent line intersects the x-axis. $y = 0 \quad \Rightarrow \quad 5x_2 - 4 = 0 \quad \Rightarrow \quad x_2 = \frac{4}{5}$.

5. $f(x) = x^3 + 2x - 4 \quad \Rightarrow \quad f'(x) = 3x^2 + 2$, so $x_{n+1} = x_n - \dfrac{x_n^3 + 2x_n - 4}{3x_n^2 + 2}$. Now $x_1 = 1 \quad \Rightarrow$

$$x_2 = 1 - \frac{1 + 2 - 4}{3 \cdot 1^2 + 2} = 1 - \frac{-1}{5} = 1.2 \quad \Rightarrow \quad x_3 = 1.2 - \frac{(1.2)^3 + 2(1.2) - 4}{3(1.2)^2 + 2} \approx 1.1797.$$

7. $f(x) = x^5 - x - 1 \quad \Rightarrow \quad f'(x) = 5x^4 - 1$, so $x_{n+1} = x_n - \dfrac{x_n^5 - x_n - 1}{5x_n^4 - 1}$. Now $x_1 = 1 \quad \Rightarrow$

$$x_2 = 1 - \frac{1 - 1 - 1}{5 - 1} = 1 - \left(-\tfrac{1}{4}\right) = 1.25 \quad \Rightarrow \quad x_3 = 1.25 - \frac{(1.25)^5 - 1.25 - 1}{5(1.25)^4 - 1} \approx 1.1785.$$

9. $f(x) = x^3 + x + 3 \quad \Rightarrow \quad f'(x) = 3x^2 + 1$, so $x_{n+1} = x_n - \dfrac{x_n^3 + x_n + 3}{3x_n^2 + 1}$.

Now $x_1 = -1 \quad \Rightarrow$

$$x_2 = -1 - \frac{(-1)^3 + (-1) + 3}{3(-1)^2 + 1} = -1 - \frac{-1 - 1 + 3}{3 + 1} = -1 - \frac{1}{4} = -1.25.$$

Newton's method follows the tangent line at $(-1, 1)$ down to its intersection with the x-axis at $(-1.25, 0)$, giving the second approximation $x_2 = -1.25$.

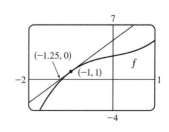

11. To approximate $x = \sqrt[5]{20}$ (so that $x^5 = 20$), we can take $f(x) = x^5 - 20$. So $f'(x) = 5x^4$, and thus,

$x_{n+1} = x_n - \dfrac{x_n^5 - 20}{5x_n^4}$. Since $\sqrt[5]{32} = 2$ and 32 is reasonably close to 20, we'll use $x_1 = 2$. We need to find approximations

until they agree to eight decimal places. $x_1 = 2 \;\Rightarrow\; x_2 = 1.85, x_3 \approx 1.82148614, x_4 \approx 1.82056514,$

$x_5 \approx 1.82056420 \approx x_6$. So $\sqrt[5]{20} \approx 1.82056420$, to eight decimal places.

Here is a quick and easy method for finding the iterations for Newton's method on a programmable calculator.

(The screens shown are from the TI-84 Plus, but the method is similar on other calculators.) Assign $f(x) = x^5 - 20$

to Y_1, and $f'(x) = 5x^4$ to Y_2. Now store $x_1 = 2$ in X and then enter $X - Y_1/Y_2 \to X$ to get $x_2 = 1.85$. By successively

pressing the ENTER key, you get the approximations $x_3, x_4, \ldots$.

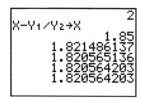
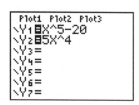

In Derive, load the utility file SOLVE. Enter NEWTON(x^5-20,x,2) and then APPROXIMATE to get

[2, 1.85, 1.82148614, 1.82056514, 1.82056420]. You can request a specific iteration by adding a fourth argument. For

example, NEWTON(x^5-20,x,2,2) gives [2, 1.85, 1.82148614].

In Maple, make the assignments $f := x \to x\hat{\ }5 - 20;$, $g := x \to x - f(x)/D(f)(x);$, and $x := 2.;$. Repeatedly execute

the command $x := g(x);$ to generate successive approximations.

In Mathematica, make the assignments $f[x\_] := x\hat{\ }5 - 20$, $g[x\_] := x - f[x]/f'[x]$, and $x = 2$. Repeatedly execute the

command $x = g[x]$ to generate successive approximations.

13. $f(x) = x^4 - 2x^3 + 5x^2 - 6 \;\Rightarrow\; f'(x) = 4x^3 - 6x^2 + 10x \;\Rightarrow\; x_{n+1} = x_n - \dfrac{x_n^4 - 2x_n^3 + 5x_n^2 - 6}{4x_n^3 - 6x_n^2 + 10x_n}$. We need to find

approximations until they agree to six decimal places. We'll let x_1 equal the midpoint of the given interval, $[1, 2]$.

$x_1 = 1.5 \;\Rightarrow\; x_2 = 1.2625, x_3 \approx 1.218808, x_4 \approx 1.217563, x_5 \approx 1.217562 \approx x_6$. So the root is 1.217562 to six decimal

places.

15. $\sin x = x^2$, so $f(x) = \sin x - x^2 \;\Rightarrow\; f'(x) = \cos x - 2x \;\Rightarrow\;$

$x_{n+1} = x_n - \dfrac{\sin x_n - x_n^2}{\cos x_n - 2x_n}$. From the figure, the positive root of $\sin x = x^2$ is

near 1. $x_1 = 1 \;\Rightarrow\; x_2 \approx 0.891396, x_3 \approx 0.876985, x_4 \approx 0.876726 \approx x_5$. So

the positive root is 0.876726, to six decimal places.

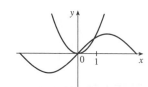

17.

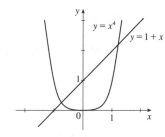

From the graph, we see that there appear to be points of intersection near

$x = -0.7$ and $x = 1.2$. Solving $x^4 = 1 + x$ is the same as solving

$$f(x) = x^4 - x - 1 = 0. \quad f(x) = x^4 - x - 1 \quad \Rightarrow \quad f'(x) = 4x^3 - 1,$$

so $x_{n+1} = x_n - \dfrac{x_n^4 - x_n - 1}{4x_n^3 - 1}.$

$x_1 = -0.7$	$x_1 = 1.2$
$x_2 \approx -0.725253$	$x_2 \approx 1.221380$
$x_3 \approx -0.724493$	$x_3 \approx 1.220745$
$x_4 \approx -0.724492 \approx x_5$	$x_4 \approx 1.220744 \approx x_5$

To six decimal places, the roots of the equation are -0.724492 and 1.220744.

19.

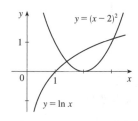

From the graph, we see that there appear to be points of intersection near

$x = 1.5$ and $x = 3$. Solving $(x - 2)^2 = \ln x$ is the same as solving

$$f(x) = (x - 2)^2 - \ln x = 0. \quad f(x) = (x - 2)^2 - \ln x \quad \Rightarrow$$

$f'(x) = 2(x - 2) - 1/x$, so $x_{n+1} = x_n - \dfrac{(x_n - 2)^2 - \ln x_n}{2(x_n - 2) - 1/x_n}.$

$x_1 = 1.5$	$x_1 = 3$
$x_2 \approx 1.406721$	$x_2 \approx 3.059167$
$x_3 \approx 1.412370$	$x_3 \approx 3.057106$
$x_4 \approx 1.412391 \approx x_5$	$x_4 \approx 3.057104 \approx x_5$

To six decimal places, the roots of the equation are 1.412391 and 3.057104.

21. From the graph, there appears to be a point of intersection near $x = 0.6$.

Solving $\cos x = \sqrt{x}$ is the same as solving $f(x) = \cos x - \sqrt{x} = 0$.

$f(x) = \cos x - \sqrt{x} \quad \Rightarrow \quad f'(x) = -\sin x - 1/(2\sqrt{x})$, so

$x_{n+1} = x_n - \dfrac{\cos x_n - \sqrt{x_n}}{-\sin x_n - 1/(2\sqrt{x})}$. Now $x_1 = 0.6 \quad \Rightarrow \quad x_2 \approx 0.641928$,

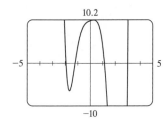

$x_3 \approx 0.641714 \approx x_4$. To six decimal places, the root of the equation is 0.641714.

23.

$$f(x) = x^6 - x^5 - 6x^4 - x^2 + x + 10 \quad \Rightarrow$$

$$f'(x) = 6x^5 - 5x^4 - 24x^3 - 2x + 1 \quad \Rightarrow$$

$$x_{n+1} = x_n - \frac{x_n^6 - x_n^5 - 6x_n^4 - x_n^2 + x_n + 10}{6x_n^5 - 5x_n^4 - 24x_n^3 - 2x_n + 1}.$$

From the graph of f, there appear to be roots near -1.9, -1.2, 1.1, and 3.

$$x_1 = -1.9$$

$$x_2 \approx -1.94278290$$

$$x_3 \approx -1.93828380$$

$$x_4 \approx -1.93822884$$

$$x_5 \approx -1.93822883 \approx x_6$$

$$x_1 = -1.2$$

$$x_2 \approx -1.22006245$$

$$x_3 \approx -1.21997997 \approx x_4$$

$$x_1 = 1.1$$

$$x_2 \approx 1.14111662$$

$$x_3 \approx 1.13929741$$

$$x_4 \approx 1.13929375 \approx x_5$$

$$x_1 = 3$$

$$x_2 \approx 2.99$$

$$x_3 \approx 2.98984106$$

$$x_4 \approx 2.98984102 \approx x_5$$

To eight decimal places, the roots of the equation are -1.93822883, -1.21997997, 1.13929375, and 2.98984102.

25.

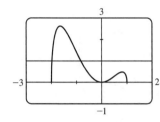

From the graph, $y = x^2\sqrt{2 - x - x^2}$ and $y = 1$ intersect twice, at $x \approx -2$ and

at $x \approx -1$. $f(x) = x^2\sqrt{2 - x - x^2} - 1$ $\Rightarrow$

$$f'(x) = x^2 \cdot \tfrac{1}{2}(2 - x - x^2)^{-1/2}(-1 - 2x) + (2 - x - x^2)^{1/2} \cdot 2x$$

$$= \tfrac{1}{2}x(2 - x - x^2)^{-1/2}[x(-1 - 2x) + 4(2 - x - x^2)]$$

$$= \frac{x(8 - 5x - 6x^2)}{2\sqrt{(2 + x)(1 - x)}},$$

so $x_{n+1} = x_n - \dfrac{x_n^2\sqrt{2 - x_n - x_n^2} - 1}{\dfrac{x_n(8 - 5x_n - 6x_n^2)}{2\sqrt{(2 + x_n)(1 - x_n)}}}$. Trying $x_1 = -2$ won't work because $f'(-2)$ is undefined, so we'll

try $x_1 = -1.95$.

$$x_1 = -1.95$$

$$x_2 \approx -1.98580357$$

$$x_3 \approx -1.97899778$$

$$x_4 \approx -1.97807848$$

$$x_5 \approx -1.97806682$$

$$x_6 \approx -1.97806681 \approx x_7$$

$$x_1 = -0.8$$

$$x_2 \approx -0.82674444$$

$$x_3 \approx -0.82646236$$

$$x_4 \approx -0.82646233 \approx x_5$$

To eight decimal places, the roots of the equation are -1.97806681 and -0.82646233.

27.

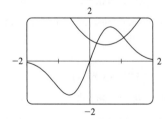

Solving $4e^{-x^2}\sin x = x^2 - x + 1$ is the same as solving

$f(x) = 4e^{-x^2}\sin x - x^2 + x - 1 = 0$.

$f'(x) = 4e^{-x^2}(\cos x - 2x\sin x) - 2x + 1$ $\Rightarrow$

$$x_{n+1} = x_n - \frac{4e^{-x_n^2}\sin x_n - x_n^2 + x_n - 1}{4e^{-x_n^2}(\cos x_n - 2x_n\sin x_n) - 2x_n + 1}.$$

From the figure, we see that the graphs intersect at approximately $x = 0.2$ and $x = 1.1$.

$$x_1 = 0.2$$

$$x_2 \approx 0.21883273$$

$$x_3 \approx 0.21916357$$

$$x_4 \approx 0.21916368 \approx x_5$$

$$x_1 = 1.1$$

$$x_2 \approx 1.08432830$$

$$x_3 \approx 1.08422462 \approx x_4$$

To eight decimal places, the roots of the equation are 0.21916368 and 1.08422462.

29. (a) $f(x) = x^2 - a \implies f'(x) = 2x$, so Newton's method gives

$$x_{n+1} = x_n - \frac{x_n^2 - a}{2x_n} = x_n - \frac{1}{2}x_n + \frac{a}{2x_n} = \frac{1}{2}x_n + \frac{a}{2x_n} = \frac{1}{2}\left(x_n + \frac{a}{x_n}\right).$$

(b) Using (a) with $a = 1000$ and $x_1 = \sqrt{900} = 30$, we get $x_2 \approx 31.666667$, $x_3 \approx 31.622807$, and $x_4 \approx 31.622777 \approx x_5$.
So $\sqrt{1000} \approx 31.622777$.

31. $f(x) = x^3 - 3x + 6 \implies f'(x) = 3x^2 - 3$. If $x_1 = 1$, then $f'(x_1) = 0$ and the tangent line used for approximating x_2 is horizontal. Attempting to find x_2 results in trying to divide by zero.

33. For $f(x) = x^{1/3}$, $f'(x) = \frac{1}{3}x^{-2/3}$ and

$$x_{n+1} = x_n - \frac{f(x_n)}{f'(x_n)} = x_n - \frac{x_n^{1/3}}{\frac{1}{3}x_n^{-2/3}} = x_n - 3x_n = -2x_n.$$

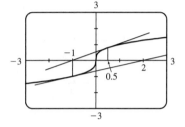

Therefore, each successive approximation becomes twice as large as the previous one in absolute value, so the sequence of approximations fails to converge to the root, which is 0. In the figure, we have $x_1 = 0.5$, $x_2 = -2(0.5) = -1$, and $x_3 = -2(-1) = 2$.

35. (a) $f(x) = x^6 - x^4 + 3x^3 - 2x \implies f'(x) = 6x^5 - 4x^3 + 9x^2 - 2 \implies$
$f''(x) = 30x^4 - 12x^2 + 18x$. To find the critical numbers of f, we'll find the zeros of f'. From the graph of f', it appears there are zeros at approximately $x = -1.3$, -0.4, and 0.5. Try $x_1 = -1.3 \implies$

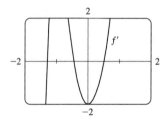

$$x_2 = x_1 - \frac{f'(x_1)}{f''(x_1)} \approx -1.293344 \implies x_3 \approx -1.293227 \approx x_4.$$

Now try $x_1 = -0.4 \implies x_2 \approx -0.443755 \implies x_3 \approx -0.441735 \implies x_4 \approx -0.441731 \approx x_5$. Finally try $x_1 = 0.5 \implies x_2 \approx 0.507937 \implies x_3 \approx 0.507854 \approx x_4$. Therefore, $x = -1.293227$, -0.441731, and 0.507854 are all the critical numbers correct to six decimal places.

(b) There are two critical numbers where f' changes from negative to positive, so f changes from decreasing to increasing.
$f(-1.293227) \approx -2.0212$ and $f(0.507854) \approx -0.6721$, so -2.0212 is the absolute minimum value of f correct to four decimal places.

37.

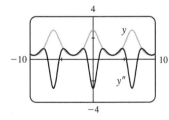

From the figure, we see that $y = f(x) = e^{\cos x}$ is periodic with period 2π. To find the x-coordinates of the IP, we only need to approximate the zeros of y'' on $[0, \pi]$. $f'(x) = -e^{\cos x}\sin x \implies f''(x) = e^{\cos x}(\sin^2 x - \cos x)$. Since $e^{\cos x} \neq 0$, we will use Newton's method with $g(x) = \sin^2 x - \cos x$, $g'(x) = 2\sin x\cos x + \sin x$, and $x_1 = 1$. $x_2 \approx 0.904173$, $x_3 \approx 0.904557 \approx x_4$. Thus, $(0.904557, 1.855277)$ is the IP.

39. We need to minimize the distance from $(0,0)$ to an arbitrary point (x, y) on the

curve $y = (x-1)^2$. $d = \sqrt{x^2 + y^2}$ $\Rightarrow$

$d(x) = \sqrt{x^2 + [(x-1)^2]^2} = \sqrt{x^2 + (x-1)^4}$. When $d' = 0$, d will be

minimized and equivalently, $s = d^2$ will be minimized, so we will use Newton's

method with $f = s'$ and $f' = s''$.

$f(x) = 2x + 4(x-1)^3$ $\Rightarrow$ $f'(x) = 2 + 12(x-1)^2$, so $x_{n+1} = x_n - \dfrac{2x_n + 4(x_n - 1)^3}{2 + 12(x_n - 1)^2}$. Try $x_1 = 0.5$ $\Rightarrow$

$x_2 = 0.4$, $x_3 \approx 0.410127$, $x_4 \approx 0.410245 \approx x_5$. Now $d(0.410245) \approx 0.537841$ is the minimum distance and the point on

the parabola is $(0.410245, 0.347810)$, correct to six decimal places.

41. In this case, $A = 18{,}000$, $R = 375$, and $n = 5(12) = 60$. So the formula $A = \dfrac{R}{i}[1 - (1+i)^{-n}]$ becomes

$18{,}000 = \dfrac{375}{x}[1 - (1+x)^{-60}]$ $\Leftrightarrow$ $48x = 1 - (1+x)^{-60}$ [multiply each term by $(1+x)^{60}$] $\Leftrightarrow$

$48x(1+x)^{60} - (1+x)^{60} + 1 = 0$. Let the LHS be called $f(x)$, so that

$$f'(x) = 48x(60)(1+x)^{59} + 48(1+x)^{60} - 60(1+x)^{59}$$

$$= 12(1+x)^{59}[4x(60) + 4(1+x) - 5] = 12(1+x)^{59}(244x - 1)$$

$x_{n+1} = x_n - \dfrac{48x_n(1+x_n)^{60} - (1+x_n)^{60} + 1}{12(1+x_n)^{59}(244x_n - 1)}$. An interest rate of 1% per month seems like a reasonable estimate for

$x = i$. So let $x_1 = 1\% = 0.01$, and we get $x_2 \approx 0.0082202$, $x_3 \approx 0.0076802$, $x_4 \approx 0.0076291$, $x_5 \approx 0.0076286 \approx x_6$.

Thus, the dealer is charging a monthly interest rate of 0.76286% (or 9.55% per year, compounded monthly).

4.9 Antiderivatives

1. $f(x) = x - 3 = x^1 - 3$ $\Rightarrow$ $F(x) = \dfrac{x^{1+1}}{1+1} - 3x + C = \frac{1}{2}x^2 - 3x + C$

Check: $F'(x) = \frac{1}{2}(2x) - 3 + 0 = x - 3 = f(x)$

3. $f(x) = \frac{1}{2} + \frac{3}{4}x^2 - \frac{4}{5}x^3$ $\Rightarrow$ $F(x) = \frac{1}{2}x + \dfrac{3}{4}\dfrac{x^{2+1}}{2+1} - \dfrac{4}{5}\dfrac{x^{3+1}}{3+1} + C = \frac{1}{2}x + \frac{1}{4}x^3 - \frac{1}{5}x^4 + C$

Check: $F'(x) = \frac{1}{2} + \frac{1}{4}(3x^2) - \frac{1}{5}(4x^3) + 0 = \frac{1}{2} + \frac{3}{4}x^2 - \frac{4}{5}x^3 = f(x)$

5. $f(x) = (x+1)(2x-1) = 2x^2 + x - 1$ $\Rightarrow$ $F(x) = 2\left(\frac{1}{3}x^3\right) + \frac{1}{2}x^2 - x + C = \frac{2}{3}x^3 + \frac{1}{2}x^2 - x + C$

7. $f(x) = 5x^{1/4} - 7x^{3/4}$ $\Rightarrow$ $F(x) = 5\dfrac{x^{1/4+1}}{\frac{1}{4}+1} - 7\dfrac{x^{3/4+1}}{\frac{3}{4}+1} + C = 5\dfrac{x^{5/4}}{5/4} - 7\dfrac{x^{7/4}}{7/4} + C = 4x^{5/4} - 4x^{7/4} + C$

9. $f(x) = 6\sqrt{x} - \sqrt[6]{x} = 6x^{1/2} - x^{1/6}$ $\Rightarrow$

$F(x) = 6\dfrac{x^{1/2+1}}{\frac{1}{2}+1} - \dfrac{x^{1/6+1}}{\frac{1}{6}+1} + C = 6\dfrac{x^{3/2}}{3/2} - \dfrac{x^{7/6}}{7/6} + C = 4x^{3/2} - \frac{6}{7}x^{7/6} + C$

11. $f(x) = \dfrac{10}{x^9} = 10x^{-9}$ has domain $(-\infty, 0) \cup (0, \infty)$, so $F(x) = \begin{cases} \dfrac{10x^{-8}}{-8} + C_1 = -\dfrac{5}{4x^8} + C_1 & \text{if } x < 0 \\ -\dfrac{5}{4x^8} + C_2 & \text{if } x > 0 \end{cases}$

See Example 1(b) for a similar problem.

13. $f(u) = \dfrac{u^4 + 3\sqrt{u}}{u^2} = \dfrac{u^4}{u^2} + \dfrac{3u^{1/2}}{u^2} = u^2 + 3u^{-3/2} \Rightarrow$

$F(u) = \dfrac{u^3}{3} + 3\dfrac{u^{-3/2+1}}{-3/2+1} + C = \dfrac{1}{3}u^3 + 3\dfrac{u^{-1/2}}{-1/2} + C = \dfrac{1}{3}u^3 - \dfrac{6}{\sqrt{u}} + C$

15. $g(\theta) = \cos\theta - 5\sin\theta \Rightarrow G(\theta) = \sin\theta - 5(-\cos\theta) + C = \sin\theta + 5\cos\theta + C$

17. $f(x) = 5e^x - 3\cosh x \Rightarrow F(x) = 5e^x - 3\sinh x + C$

19. $f(x) = \dfrac{x^5 - x^3 + 2x}{x^4} = x - \dfrac{1}{x} + \dfrac{2}{x^3} = x - \dfrac{1}{x} + 2x^{-3} \Rightarrow$

$F(x) = \dfrac{x^2}{2} - \ln|x| + 2\left(\dfrac{x^{-3+1}}{-3+1}\right) + C = \frac{1}{2}x^2 - \ln|x| - \dfrac{1}{x^2} + C$

21. $f(x) = 5x^4 - 2x^5 \Rightarrow F(x) = 5 \cdot \dfrac{x^5}{5} - 2 \cdot \dfrac{x^6}{6} + C = x^5 - \frac{1}{3}x^6 + C.$

$F(0) = 4 \Rightarrow 0^5 - \frac{1}{3} \cdot 0^6 + C = 4 \Rightarrow C = 4$, so $F(x) = x^5 - \frac{1}{3}x^6 + 4.$

The graph confirms our answer since $f(x) = 0$ when F has a local maximum, f is positive when F is increasing, and f is negative when F is decreasing.

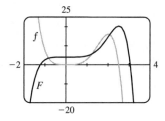

23. $f''(x) = 6x + 12x^2 \Rightarrow f'(x) = 6 \cdot \dfrac{x^2}{2} + 12 \cdot \dfrac{x^3}{3} + C = 3x^2 + 4x^3 + C \Rightarrow$

$f(x) = 3 \cdot \dfrac{x^3}{3} + 4 \cdot \dfrac{x^4}{4} + Cx + D = x^3 + x^4 + Cx + D$ [C and D are just arbitrary constants]

25. $f''(x) = \frac{2}{3}x^{2/3} \Rightarrow f'(x) = \dfrac{2}{3}\left(\dfrac{x^{5/3}}{5/3}\right) + C = \frac{2}{5}x^{5/3} + C \Rightarrow f(x) = \dfrac{2}{5}\left(\dfrac{x^{8/3}}{8/3}\right) + Cx + D = \frac{3}{20}x^{8/3} + Cx + D$

27. $f'''(t) = e^t \Rightarrow f''(t) = e^t + C \Rightarrow f'(t) = e^t + Ct + D \Rightarrow f(t) = e^t + \frac{1}{2}Ct^2 + Dt + E$

29. $f'(x) = 1 - 6x \Rightarrow f(x) = x - 3x^2 + C. \; f(0) = C$ and $f(0) = 8 \Rightarrow C = 8$, so $f(x) = x - 3x^2 + 8.$

31. $f'(x) = \sqrt{x}(6 + 5x) = 6x^{1/2} + 5x^{3/2} \Rightarrow f(x) = 4x^{3/2} + 2x^{5/2} + C.$

$f(1) = 6 + C$ and $f(1) = 10 \Rightarrow C = 4$, so $f(x) = 4x^{3/2} + 2x^{5/2} + 4.$

33. $f'(t) = 2\cos t + \sec^2 t \Rightarrow f(t) = 2\sin t + \tan t + C$ because $-\pi/2 < t < \pi/2.$

$f\left(\frac{\pi}{3}\right) = 2(\sqrt{3}/2) + \sqrt{3} + C = 2\sqrt{3} + C$ and $f\left(\frac{\pi}{3}\right) = 4 \Rightarrow C = 4 - 2\sqrt{3}$, so $f(t) = 2\sin t + \tan t + 4 - 2\sqrt{3}.$

35. $f'(x) = x^{-1/3}$ has domain $(-\infty, 0) \cup (0, \infty)$ $\Rightarrow$ $f(x) = \begin{cases} \frac{3}{2}x^{2/3} + C_1 & \text{if } x > 0 \\ \frac{3}{2}x^{2/3} + C_2 & \text{if } x < 0 \end{cases}$

$f(1) = \frac{3}{2} + C_1$ and $f(1) = 1$ $\Rightarrow$ $C_1 = -\frac{1}{2}$. $f(-1) = \frac{3}{2} + C_2$ and $f(-1) = -1$ $\Rightarrow$ $C_2 = -\frac{5}{2}$.

Thus, $f(x) = \begin{cases} \frac{3}{2}x^{2/3} - \frac{1}{2} & \text{if } x > 0 \\ \frac{3}{2}x^{2/3} - \frac{5}{2} & \text{if } x < 0 \end{cases}$

37. $f''(x) = 24x^2 + 2x + 10$ $\Rightarrow$ $f'(x) = 8x^3 + x^2 + 10x + C$. $f'(1) = 8 + 1 + 10 + C$ and $f'(1) = -3$ $\Rightarrow$

$19 + C = -3$ $\Rightarrow$ $C = -22$, so $f'(x) = 8x^3 + x^2 + 10x - 22$ and hence, $f(x) = 2x^4 + \frac{1}{3}x^3 + 5x^2 - 22x + D$.

$f(1) = 2 + \frac{1}{3} + 5 - 22 + D$ and $f(1) = 5$ $\Rightarrow$ $D = 22 - \frac{7}{3} = \frac{59}{3}$, so $f(x) = 2x^4 + \frac{1}{3}x^3 + 5x^2 - 22x + \frac{59}{3}$.

39. $f''(\theta) = \sin\theta + \cos\theta$ $\Rightarrow$ $f'(\theta) = -\cos\theta + \sin\theta + C$. $f'(0) = -1 + C$ and $f'(0) = 4$ $\Rightarrow$ $C = 5$, so

$f'(\theta) = -\cos\theta + \sin\theta + 5$ and hence, $f(\theta) = -\sin\theta - \cos\theta + 5\theta + D$. $f(0) = -1 + D$ and $f(0) = 3$ $\Rightarrow$ $D = 4$,

so $f(\theta) = -\sin\theta - \cos\theta + 5\theta + 4$.

41. $f''(x) = 2 - 12x$ $\Rightarrow$ $f'(x) = 2x - 6x^2 + C$ $\Rightarrow$ $f(x) = x^2 - 2x^3 + Cx + D$.

$f(0) = D$ and $f(0) = 9$ $\Rightarrow$ $D = 9$. $f(2) = 4 - 16 + 2C + 9 = 2C - 3$ and $f(2) = 15$ $\Rightarrow$ $2C = 18$ $\Rightarrow$

$C = 9$, so $f(x) = x^2 - 2x^3 + 9x + 9$.

43. $f''(x) = 2 + \cos x$ $\Rightarrow$ $f'(x) = 2x + \sin x + C$ $\Rightarrow$ $f(x) = x^2 - \cos x + Cx + D$.

$f(0) = -1 + D$ and $f(0) = -1$ $\Rightarrow$ $D = 0$. $f\left(\frac{\pi}{2}\right) = \pi^2/4 + \left(\frac{\pi}{2}\right)C$ and $f\left(\frac{\pi}{2}\right) = 0$ $\Rightarrow$ $\left(\frac{\pi}{2}\right)C = -\pi^2/4$ $\Rightarrow$

$C = -\frac{\pi}{2}$, so $f(x) = x^2 - \cos x - \left(\frac{\pi}{2}\right)x$.

45. $f''(x) = x^{-2}$, $x > 0$ $\Rightarrow$ $f'(x) = -1/x + C$ $\Rightarrow$ $f(x) = -\ln|x| + Cx + D = -\ln x + Cx + D$ [since $x > 0$].

$f(1) = 0$ $\Rightarrow$ $C + D = 0$ and $f(2) = 0$ $\Rightarrow$ $-\ln 2 + 2C + D = 0$ $\Rightarrow$ $-\ln 2 + 2C - C = 0$ [since $D = -C$] $\Rightarrow$

$-\ln 2 + C = 0$ $\Rightarrow$ $C = \ln 2$ and $D = -\ln 2$. So $f(x) = -\ln x + (\ln 2)x - \ln 2$.

47. Given $f'(x) = 2x + 1$, we have $f(x) = x^2 + x + C$. Since f passes through $(1, 6)$, $f(1) = 6$ $\Rightarrow$ $1^2 + 1 + C = 6$ $\Rightarrow$

$C = 4$. Therefore, $f(x) = x^2 + x + 4$ and $f(2) = 2^2 + 2 + 4 = 10$.

49. b is the antiderivative of f. For small x, f is negative, so the graph of its antiderivative must be decreasing. But both a and c are increasing for small x, so only b can be f's antiderivative. Also, f is positive where b is increasing, which supports our conclusion.

51.

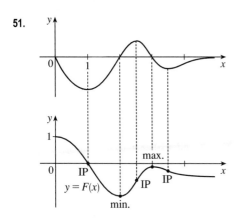

The graph of F must start at $(0, 1)$. Where the given graph, $y = f(x)$, has a local minimum or maximum, the graph of F will have an inflection point.

Where f is negative (positive), F is decreasing (increasing).

Where f changes from negative to positive, F will have a minimum.

Where f changes from positive to negative, F will have a maximum.

Where f is decreasing (increasing), F is concave downward (upward).

53.

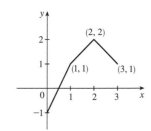

$$f'(x) = \begin{cases} 2 & \text{if } 0 \le x < 1 \\ 1 & \text{if } 1 < x < 2 \\ -1 & \text{if } 2 < x \le 3 \end{cases} \Rightarrow f(x) = \begin{cases} 2x + C & \text{if } 0 \le x < 1 \\ x + D & \text{if } 1 < x < 2 \\ -x + E & \text{if } 2 < x \le 3 \end{cases}$$

$f(0) = -1 \Rightarrow 2(0) + C = -1 \Rightarrow C = -1$. Starting at the point

$(0, -1)$ and moving to the right on a line with slope 2 gets us to the point $(1, 1)$.

The slope for $1 < x < 2$ is 1, so we get to the point $(2, 2)$. Here we have used the fact that f is continuous. We can include the

point $x = 1$ on either the first or the second part of f. The line connecting $(1, 1)$ to $(2, 2)$ is $y = x$, so $D = 0$. The slope for

$2 < x \le 3$ is -1, so we get to $(3, 1)$. $f(3) = 1 \Rightarrow -3 + E = 1 \Rightarrow E = 4$. Thus

$$f(x) = \begin{cases} 2x - 1 & \text{if } 0 \le x \le 1 \\ x & \text{if } 1 < x < 2 \\ -x + 4 & \text{if } 2 \le x \le 3 \end{cases}$$

Note that $f'(x)$ does not exist at $x = 1$ or at $x = 2$.

55. $f(x) = \dfrac{\sin x}{1 + x^2}$, $-2\pi \le x \le 2\pi$

Note that the graph of f is one of an odd function, so the graph of F will

be one of an even function.

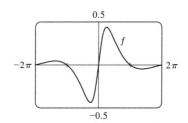

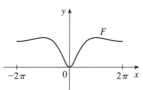

57. $v(t) = s'(t) = \sin t - \cos t \Rightarrow s(t) = -\cos t - \sin t + C$. $s(0) = -1 + C$ and $s(0) = 0 \Rightarrow C = 1$, so

$s(t) = -\cos t - \sin t + 1$.

59. $a(t) = v'(t) = t - 2 \Rightarrow v(t) = \frac{1}{2}t^2 - 2t + C$. $v(0) = C$ and $v(0) = 3 \Rightarrow C = 3$, so $v(t) = \frac{1}{2}t^2 - 2t + 3$ and

$s(t) = \frac{1}{6}t^3 - t^2 + 3t + D$. $s(0) = D$ and $s(0) = 1 \Rightarrow D = 1$, and $s(t) = \frac{1}{6}t^3 - t^2 + 3t + 1$.

61. $a(t) = v'(t) = 10 \sin t + 3 \cos t \Rightarrow v(t) = -10 \cos t + 3 \sin t + C \Rightarrow s(t) = -10 \sin t - 3 \cos t + Ct + D$.

$s(0) = -3 + D = 0$ and $s(2\pi) = -3 + 2\pi C + D = 12 \Rightarrow D = 3$ and $C = \frac{6}{\pi}$. Thus,

$s(t) = -10 \sin t - 3 \cos t + \frac{6}{\pi}t + 3$.

63. (a) We first observe that since the stone is dropped 450 m above the ground, $v(0) = 0$ and $s(0) = 450$.

$v'(t) = a(t) = -9.8 \Rightarrow v(t) = -9.8t + C$. Now $v(0) = 0 \Rightarrow C = 0$, so $v(t) = -9.8t \Rightarrow$

$s(t) = -4.9t^2 + D$. Last, $s(0) = 450 \Rightarrow D = 450 \Rightarrow s(t) = 450 - 4.9t^2$.

(b) The stone reaches the ground when $s(t) = 0$. $450 - 4.9t^2 = 0 \Rightarrow t^2 = 450/4.9 \Rightarrow t_1 = \sqrt{450/4.9} \approx 9.58$ s.

(c) The velocity with which the stone strikes the ground is $v(t_1) = -9.8\sqrt{450/4.9} \approx -93.9$ m/s.

(d) This is just reworking parts (a) and (b) with $v(0) = -5$. Using $v(t) = -9.8t + C$, $v(0) = -5$ $\Rightarrow$ $0 + C = -5$ $\Rightarrow$

$v(t) = -9.8t - 5$. So $s(t) = -4.9t^2 - 5t + D$ and $s(0) = 450$ $\Rightarrow$ $D = 450$ $\Rightarrow$ $s(t) = -4.9t^2 - 5t + 450$.

Solving $s(t) = 0$ by using the quadratic formula gives us $t = \left(5 \pm \sqrt{8845}\right)/(-9.8)$ $\Rightarrow$ $t_1 \approx 9.09$ s.

65. By Exercise 64 with $a = -9.8$, $s(t) = -4.9t^2 + v_0t + s_0$ and $v(t) = s'(t) = -9.8t + v_0$. So

$[v(t)]^2 = (-9.8t + v_0)^2 = (9.8)^2 t^2 - 19.6v_0t + v_0^2 = v_0^2 + 96.04t^2 - 19.6v_0t = v_0^2 - 19.6\left(-4.9t^2 + v_0t\right)$.

But $-4.9t^2 + v_0t$ is just $s(t)$ without the s_0 term; that is, $s(t) - s_0$. Thus, $[v(t)]^2 = v_0^2 - 19.6\left[s(t) - s_0\right]$.

67. Using Exercise 64 with $a = -32$, $v_0 = 0$, and $s_0 = h$ (the height of the cliff), we know that the height at time t is

$s(t) = -16t^2 + h$. $v(t) = s'(t) = -32t$ and $v(t) = -120$ $\Rightarrow$ $-32t = -120$ $\Rightarrow$ $t = 3.75$, so

$0 = s(3.75) = -16(3.75)^2 + h$ $\Rightarrow$ $h = 16(3.75)^2 = 225$ ft.

69. Marginal cost $= 1.92 - 0.002x = C'(x)$ $\Rightarrow$ $C(x) = 1.92x - 0.001x^2 + K$. But $C(1) = 1.92 - 0.001 + K = 562$ $\Rightarrow$

$K = 560.081$. Therefore, $C(x) = 1.92x - 0.001x^2 + 560.081$ $\Rightarrow$ $C(100) = 742.081$, so the cost of producing

100 items is \$742.08.

71. Taking the upward direction to be positive we have that for $0 \leq t \leq 10$ (using the subscript 1 to refer to $0 \leq t \leq 10$),

$a_1(t) = -(9 - 0.9t) = v_1'(t)$ $\Rightarrow$ $v_1(t) = -9t + 0.45t^2 + v_0$, but $v_1(0) = v_0 = -10$ $\Rightarrow$

$v_1(t) = -9t + 0.45t^2 - 10 = s_1'(t)$ $\Rightarrow$ $s_1(t) = -\frac{9}{2}t^2 + 0.15t^3 - 10t + s_0$. But $s_1(0) = 500 = s_0$ $\Rightarrow$

$s_1(t) = -\frac{9}{2}t^2 + 0.15t^3 - 10t + 500$. $s_1(10) = -450 + 150 - 100 + 500 = 100$, so it takes

more than 10 seconds for the raindrop to fall. Now for $t > 10$, $a(t) = 0 = v'(t)$ $\Rightarrow$

$v(t) = \text{constant} = v_1(10) = -9(10) + 0.45(10)^2 - 10 = -55$ $\Rightarrow$ $v(t) = -55$.

At 55 m/s, it will take $100/55 \approx 1.8$ s to fall the last 100 m. Hence, the total time is $10 + \frac{100}{55} = \frac{130}{11} \approx 11.8$ s.

73. $a(t) = k$, the initial velocity is 30 mi/h $= 30 \cdot \frac{5280}{3600} = 44$ ft/s, and the final velocity (after 5 seconds) is

50 mi/h $= 50 \cdot \frac{5280}{3600} = \frac{220}{3}$ ft/s. So $v(t) = kt + C$ and $v(0) = 44$ $\Rightarrow$ $C = 44$. Thus, $v(t) = kt + 44$ $\Rightarrow$

$v(5) = 5k + 44$. But $v(5) = \frac{220}{3}$, so $5k + 44 = \frac{220}{3}$ $\Rightarrow$ $5k = \frac{88}{3}$ $\Rightarrow$ $k = \frac{88}{15} \approx 5.87$ ft/s^2.

75. Let the acceleration be $a(t) = k$ km/h^2. We have $v(0) = 100$ km/h and we can take the initial position $s(0)$ to be 0.

We want the time t_f for which $v(t) = 0$ to satisfy $s(t) < 0.08$ km. In general, $v'(t) = a(t) = k$, so $v(t) = kt + C$, where

$C = v(0) = 100$. Now $s'(t) = v(t) = kt + 100$, so $s(t) = \frac{1}{2}kt^2 + 100t + D$, where $D = s(0) = 0$.

Thus, $s(t) = \frac{1}{2}kt^2 + 100t$. Since $v(t_f) = 0$, we have $kt_f + 100 = 0$ or $t_f = -100/k$, so

$s(t_f) = \frac{1}{2}k\left(-\frac{100}{k}\right)^2 + 100\left(-\frac{100}{k}\right) = 10{,}000\left(\frac{1}{2k} - \frac{1}{k}\right) = -\frac{5{,}000}{k}$. The condition $s(t_f)$ must satisfy is

$-\frac{5{,}000}{k} < 0.08$ $\Rightarrow$ $-\frac{5{,}000}{0.08} > k$ [k is negative] $\Rightarrow$ $k < -62{,}500$ km/h^2, or equivalently,

$k < -\frac{3125}{648} \approx -4.82$ m/s^2.

77. (a) First note that $90 \text{ mi/h} = 90 \times \frac{5280}{3600} \text{ ft/s} = 132 \text{ ft/s}$. Then $a(t) = 4 \text{ ft/s}^2 \Rightarrow v(t) = 4t + C$, but $v(0) = 0 \Rightarrow$ $C = 0$. Now $4t = 132$ when $t = \frac{132}{4} = 33 \text{ s}$, so it takes 33 s to reach 132 ft/s. Therefore, taking $s(0) = 0$, we have $s(t) = 2t^2, 0 \le t \le 33$. So $s(33) = 2178 \text{ ft}$. 15 minutes $= 15(60) = 900 \text{ s}$, so for $33 < t \le 933$ we have $v(t) = 132 \text{ ft/s} \Rightarrow s(933) = 132(900) + 2178 = 120{,}978 \text{ ft} = 22.9125 \text{ mi}$.

(b) As in part (a), the train accelerates for 33 s and travels 2178 ft while doing so. Similarly, it decelerates for 33 s and travels 2178 ft at the end of its trip. During the remaining $900 - 66 = 834 \text{ s}$ it travels at 132 ft/s, so the distance traveled is $132 \cdot 834 = 110{,}088 \text{ ft}$. Thus, the total distance is $2178 + 110{,}088 + 2178 = 114{,}444 \text{ ft} = 21.675 \text{ mi}$.

(c) $45 \text{ mi} = 45(5280) = 237{,}600 \text{ ft}$. Subtract $2(2178)$ to take care of the speeding up and slowing down, and we have $233{,}244 \text{ ft}$ at 132 ft/s for a trip of $233{,}244/132 = 1767 \text{ s}$ at 90 mi/h. The total time is $1767 + 2(33) = 1833 \text{ s} = 30 \text{ min } 33 \text{ s} = 30.55 \text{ min}$.

(d) $37.5(60) = 2250 \text{ s}$. $2250 - 2(33) = 2184 \text{ s}$ at maximum speed. $2184(132) + 2(2178) = 292{,}644$ total feet or $292{,}644/5280 = 55.425 \text{ mi}$.

4 Review

CONCEPT CHECK

1. A function f has an **absolute maximum** at $x = c$ if $f(c)$ is the largest function value on the entire domain of f, whereas f has a **local maximum** at c if $f(c)$ is the largest function value when x is near c. See Figure 4 in Section 4.1.

2. (a) See Theorem 4.1.3.

(b) See the Closed Interval Method before Example 8 in Section 4.1.

3. (a) See Theorem 4.1.4.

(b) See Definition 4.1.6.

4. (a) See Rolle's Theorem at the beginning of Section 4.2.

(b) See the Mean Value Theorem in Section 4.2. Geometric interpretation—there is some point P on the graph of a function f [on the interval (a, b)] where the tangent line is parallel to the secant line that connects $(a, f(a))$ and $(b, f(b))$.

5. (a) See the I/D Test before Example 1 in Section 4.3.

(b) If the graph of f lies above all of its tangents on an interval I, then it is called concave upward on I.

(c) See the Concavity Test before Example 4 in Section 4.3.

(d) An inflection point is a point where a curve changes its direction of concavity. They can be found by determining the points at which the second derivative changes sign.

6. (a) See the First Derivative Test after Example 1 in Section 4.3.

(b) See the Second Derivative Test before Example 6 in Section 4.3.

(c) See the note before Example 7 in Section 4.3.

7. (a) See l'Hospital's Rule and the three notes that follow it in Section 4.4.

(b) Write fg as $\dfrac{f}{1/g}$ or $\dfrac{g}{1/f}$.

(c) Convert the difference into a quotient using a common denominator, rationalizing, factoring, or some other method.

(d) Convert the power to a product by taking the natural logarithm of both sides of $y = f^g$ or by writing f^g as $e^{g \ln f}$.

8. Without calculus you could get misleading graphs that fail to show the most interesting features of a function. See the discussion at the beginning of Section 4.5 and the first paragraph in Section 4.6.

9. (a) See Figure 3 in Section 4.8.

(b) $x_2 = x_1 - \dfrac{f(x_1)}{f'(x_1)}$

(c) $x_{n+1} = x_n - \dfrac{f(x_n)}{f'(x_n)}$

(d) Newton's method is likely to fail or to work very slowly when $f'(x_1)$ is close to 0. It also fails when $f'(x_i)$ is undefined, such as with $f(x) = 1/x - 2$ and $x_1 = 1$.

10. (a) See the definition at the beginning of Section 4.9.

(b) If F_1 and F_2 are both antiderivatives of f on an interval I, then they differ by a constant.

TRUE-FALSE QUIZ

1. False. For example, take $f(x) = x^3$, then $f'(x) = 3x^2$ and $f'(0) = 0$, but $f(0) = 0$ is not a maximum or minimum; $(0, 0)$ is an inflection point.

3. False. For example, $f(x) = x$ is continuous on $(0, 1)$ but attains neither a maximum nor a minimum value on $(0, 1)$. Don't confuse this with f being continuous on the *closed* interval $[a, b]$, which would make the statement true.

5. True. This is an example of part (b) of the I/D Test.

7. False. $f'(x) = g'(x) \;\Rightarrow\; f(x) = g(x) + C$. For example, if $f(x) = x + 2$ and $g(x) = x + 1$, then $f'(x) = g'(x) = 1$, but $f(x) \neq g(x)$.

9. True. The graph of one such function is sketched.

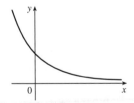

11. True. Let $x_1 < x_2$ where $x_1, x_2 \in I$. Then $f(x_1) < f(x_2)$ and $g(x_1) < g(x_2)$ [since f and g are increasing on I], so $(f + g)(x_1) = f(x_1) + g(x_1) < f(x_2) + g(x_2) = (f + g)(x_2)$.

13. False. Take $f(x) = x$ and $g(x) = x - 1$. Then both f and g are increasing on $(0, 1)$. But $f(x)\,g(x) = x(x - 1)$ is not increasing on $(0, 1)$.

15. True. Let $x_1, x_2 \in I$ and $x_1 < x_2$. Then $f(x_1) < f(x_2)$ [f is increasing] $\Rightarrow$ $\dfrac{1}{f(x_1)} > \dfrac{1}{f(x_2)}$ [f is positive] $\Rightarrow$

$g(x_1) > g(x_2)$ $\Rightarrow$ $g(x) = 1/f(x)$ is decreasing on I.

17. True. If f is periodic, then there is a number p such that $f(x + p) = f(p)$ for all x. Differentiating gives

$f'(x) = f'(x + p) \cdot (x + p)' = f'(x + p) \cdot 1 = f'(x + p)$, so f' is periodic.

19. True. By the Mean Value Theorem, there exists a number c in $(0, 1)$ such that $f(1) - f(0) = f'(c)(1 - 0) = f'(c)$.

Since $f'(c)$ is nonzero, $f(1) - f(0) \neq 0$, so $f(1) \neq f(0)$.

EXERCISES

1. $f(x) = x^3 - 6x^2 + 9x + 1$, $[2, 4]$. $f'(x) = 3x^2 - 12x + 9 = 3(x^2 - 4x + 3) = 3(x - 1)(x - 3)$. $f'(x) = 0$ $\Rightarrow$

$x = 1$ or $x = 3$, but 1 is not in the interval. $f'(x) > 0$ for $3 < x < 4$ and $f'(x) < 0$ for $2 < x < 3$, so $f(3) = 1$ is a local

minimum value. Checking the endpoints, we find $f(2) = 3$ and $f(4) = 5$. Thus, $f(3) = 1$ is the absolute minimum value and

$f(4) = 5$ is the absolute maximum value.

3. $f(x) = \dfrac{3x - 4}{x^2 + 1}$, $[-2, 2]$. $f'(x) = \dfrac{(x^2 + 1)(3) - (3x - 4)(2x)}{(x^2 + 1)^2} = \dfrac{-(3x^2 - 8x - 3)}{(x^2 + 1)^2} = \dfrac{-(3x + 1)(x - 3)}{(x^2 + 1)^2}$.

$f'(x) = 0$ $\Rightarrow$ $x = -\frac{1}{3}$ or $x = 3$, but 3 is not in the interval. $f'(x) > 0$ for $-\frac{1}{3} < x < 2$ and $f'(x) < 0$ for

$-2 < x < -\frac{1}{3}$, so $f\left(-\frac{1}{3}\right) = \frac{-5}{10/9} = -\frac{9}{2}$ is a local minimum value. Checking the endpoints, we find $f(-2) = -2$ and

$f(2) = \frac{2}{5}$. Thus, $f\left(-\frac{1}{3}\right) = -\frac{9}{2}$ is the absolute minimum value and $f(2) = \frac{2}{5}$ is the absolute maximum value.

5. $f(x) = x + \sin 2x$, $[0, \pi]$. $f'(x) = 1 + 2\cos 2x = 0$ $\Leftrightarrow$ $\cos 2x = -\frac{1}{2}$ $\Leftrightarrow$ $2x = \frac{2\pi}{3}$ or $\frac{4\pi}{3}$ $\Leftrightarrow$ $x = \frac{\pi}{3}$ or $\frac{2\pi}{3}$.

$f''(x) = -4\sin 2x$, so $f''\left(\frac{\pi}{3}\right) = -4\sin\frac{2\pi}{3} = -2\sqrt{3} < 0$ and $f''\left(\frac{2\pi}{3}\right) = -4\sin\frac{4\pi}{3} = 2\sqrt{3} > 0$, so

$f\left(\frac{\pi}{3}\right) = \frac{\pi}{3} + \frac{\sqrt{3}}{2} \approx 1.91$ is a local maximum value and $f\left(\frac{2\pi}{3}\right) = \frac{2\pi}{3} - \frac{\sqrt{3}}{2} \approx 1.23$ is a local minimum value. Also $f(0) = 0$

and $f(\pi) = \pi$, so $f(0) = 0$ is the absolute minimum value and $f(\pi) = \pi$ is the absolute maximum value.

7. This limit has the form $\frac{0}{0}$. $\displaystyle\lim_{x \to 0} \frac{\tan \pi x}{\ln(1 + x)} \overset{\text{H}}{=} \lim_{x \to 0} \frac{\pi \sec^2 \pi x}{1/(1 + x)} = \frac{\pi \cdot 1^2}{1/1} = \pi$

9. This limit has the form $\frac{0}{0}$. $\displaystyle\lim_{x \to 0} \frac{e^{4x} - 1 - 4x}{x^2} \overset{\text{H}}{=} \lim_{x \to 0} \frac{4e^{4x} - 4}{2x} \overset{\text{H}}{=} \lim_{x \to 0} \frac{16e^{4x}}{2} = \lim_{x \to 0} 8e^{4x} = 8 \cdot 1 = 8$

11. This limit has the form $\infty \cdot 0$. $\displaystyle\lim_{x \to \infty} x^3 e^{-x} = \lim_{x \to \infty} \frac{x^3}{e^x} \overset{\text{H}}{=} \lim_{x \to \infty} \frac{3x^2}{e^x} \overset{\text{H}}{=} \lim_{x \to \infty} \frac{6x}{e^x} \overset{\text{H}}{=} \lim_{x \to \infty} \frac{6}{e^x} = 0$

13. This limit has the form $\infty - \infty$.

$$\lim_{x \to 1+} \left(\frac{x}{x-1} - \frac{1}{\ln x} \right) = \lim_{x \to 1+} \left(\frac{x \ln x - x + 1}{(x-1)\ln x} \right) \overset{\mathrm{H}}{=} \lim_{x \to 1+} \frac{x \cdot (1/x) + \ln x - 1}{(x-1) \cdot (1/x) + \ln x} = \lim_{x \to 1+} \frac{\ln x}{1 - 1/x + \ln x}$$

$$\overset{\mathrm{H}}{=} \lim_{x \to 1+} \frac{1/x}{1/x^2 + 1/x} = \frac{1}{1+1} = \frac{1}{2}$$

15. $f(0) = 0$, $f'(-2) = f'(1) = f'(9) = 0$, $\lim\limits_{x \to \infty} f(x) = 0$, $\lim\limits_{x \to 6} f(x) = -\infty$,

$f'(x) < 0$ on $(-\infty, -2)$, $(1, 6)$, and $(9, \infty)$, $f'(x) > 0$ on $(-2, 1)$ and $(6, 9)$,

$f''(x) > 0$ on $(-\infty, 0)$ and $(12, \infty)$, $f''(x) < 0$ on $(0, 6)$ and $(6, 12)$

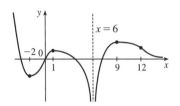

17. f is odd, $f'(x) < 0$ for $0 < x < 2$, $f'(x) > 0$ for $x > 2$,

$f''(x) > 0$ for $0 < x < 3$, $f''(x) < 0$ for $x > 3$, $\lim_{x \to \infty} f(x) = -2$

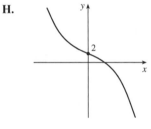

19. $y = f(x) = 2 - 2x - x^3$ **A.** $D = \mathbb{R}$ **B.** y-intercept: $f(0) = 2$.

The x-intercept (approximately 0.770917) can be found using Newton's

Method.   **C.** No symmetry   **D.** No asymptote

E. $f'(x) = -2 - 3x^2 = -(3x^2 + 2) < 0$, so f is decreasing on $\mathbb{R}$.

F. No extreme value   **G.** $f''(x) = -6x < 0$ on $(0, \infty)$ and $f''(x) > 0$ on

$(-\infty, 0)$, so f is CD on $(0, \infty)$ and CU on $(-\infty, 0)$. There is an IP at $(0, 2)$.

H.

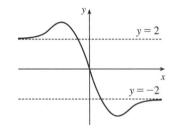

21. $y = f(x) = x^4 - 3x^3 + 3x^2 - x = x(x-1)^3$ **A.** $D = \mathbb{R}$ **B.** y-intercept: $f(0) = 0$; x-intercepts: $f(x) = 0$ $\Leftrightarrow$

$x = 0$ or $x = 1$ **C.** No symmetry   **D.** f is a polynomial function and hence, it has no asymptote.

E. $f'(x) = 4x^3 - 9x^2 + 6x - 1$. Since the sum of the coefficients is 0, 1 is a root of f', so

$f'(x) = (x-1)(4x^2 - 5x + 1) = (x-1)^2(4x-1)$. $f'(x) < 0$ $\Rightarrow$ $x < \frac{1}{4}$, so f is decreasing on $\left(-\infty, \frac{1}{4}\right)$

and f is increasing on $\left(\frac{1}{4}, \infty\right)$.   **F.** $f'(x)$ does not change sign at $x = 1$, so

there is not a local extremum there. $f\left(\frac{1}{4}\right) = -\frac{27}{256}$ is a local minimum value.

G. $f''(x) = 12x^2 - 18x + 6 = 6(2x-1)(x-1)$. $f''(x) = 0$ $\Leftrightarrow$ $x = \frac{1}{2}$

or 1. $f''(x) < 0$ $\Leftrightarrow$ $\frac{1}{2} < x < 1$ $\Rightarrow$ f is CD on $\left(\frac{1}{2}, 1\right)$ and CU on

$\left(-\infty, \frac{1}{2}\right)$ and $(1, \infty)$. There are inflection points at $\left(\frac{1}{2}, -\frac{1}{16}\right)$ and $(1, 0)$.

H.

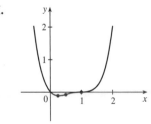

23. $y = f(x) = \dfrac{1}{x(x-3)^2}$ **A.** $D = \{x \mid x \neq 0, 3\} = (-\infty, 0) \cup (0, 3) \cup (3, \infty)$ **B.** No intercepts.   **C.** No symmetry.

D. $\displaystyle\lim_{x \to \pm\infty} \frac{1}{x(x-3)^2} = 0$, so $y = 0$ is a HA. $\displaystyle\lim_{x \to 0^+} \frac{1}{x(x-3)^2} = \infty$, $\displaystyle\lim_{x \to 0^-} \frac{1}{x(x-3)^2} = -\infty$, $\displaystyle\lim_{x \to 3} \frac{1}{x(x-3)^2} = \infty$,

so $x = 0$ and $x = 3$ are VA.   **E.** $f'(x) = -\dfrac{(x-3)^2 + 2x(x-3)}{x^2(x-3)^4} = \dfrac{3(1-x)}{x^2(x-3)^3} \;\Rightarrow\; f'(x) > 0 \;\Leftrightarrow\; 1 < x < 3,$

so f is increasing on $(1, 3)$ and decreasing on $(-\infty, 0)$, $(0, 1)$, and $(3, \infty)$.   **H.**

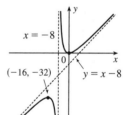

F. Local minimum value $f(1) = \frac{1}{4}$ **G.** $f''(x) = \dfrac{6(2x^2 - 4x + 3)}{x^3(x-3)^4}$.

Note that $2x^2 - 4x + 3 > 0$ for all x since it has negative discriminant.

So $f''(x) > 0 \;\Leftrightarrow\; x > 0 \;\Rightarrow\; f$ is CU on $(0, 3)$ and $(3, \infty)$ and

CD on $(-\infty, 0)$. No IP

---

25. $y = f(x) = \dfrac{x^2}{x+8} = x - 8 + \dfrac{64}{x+8}$ **A.** $D = \{x \mid x \neq -8\}$ **B.** Intercepts are 0 **C.** No symmetry

D. $\displaystyle\lim_{x \to \infty} \frac{x^2}{x+8} = \infty$, but $f(x) - (x-8) = \dfrac{64}{x+8} \to 0$ as $x \to \infty$, so $y = x - 8$ is a slant asymptote.

$\displaystyle\lim_{x \to -8^+} \frac{x^2}{x+8} = \infty$ and $\displaystyle\lim_{x \to -8^-} \frac{x^2}{x+8} = -\infty$, so $x = -8$ is a VA.   **E.** $f'(x) = 1 - \dfrac{64}{(x+8)^2} = \dfrac{x(x+16)}{(x+8)^2} > 0 \;\Leftrightarrow$

$x > 0$ or $x < -16$, so f is increasing on $(-\infty, -16)$ and $(0, \infty)$ and   **H.**

decreasing on $(-16, -8)$ and $(-8, 0)$.

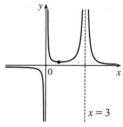

F. Local maximum value $f(-16) = -32$, local minimum value $f(0) = 0$

G. $f''(x) = 128/(x+8)^3 > 0 \;\Leftrightarrow\; x > -8$, so f is CU on $(-8, \infty)$ and

CD on $(-\infty, -8)$. No IP

---

27. $y = f(x) = x\sqrt{2+x}$ **A.** $D = [-2, \infty)$ **B.** y-intercept: $f(0) = 0$; x-intercepts: -2 and 0 **C.** No symmetry

D. No asymptote   **E.** $f'(x) = \dfrac{x}{2\sqrt{2+x}} + \sqrt{2+x} = \dfrac{1}{2\sqrt{2+x}}[x + 2(2+x)] = \dfrac{3x+4}{2\sqrt{2+x}} = 0$ when $x = -\frac{4}{3}$, so f is

decreasing on $\left(-2, -\frac{4}{3}\right)$ and increasing on $\left(-\frac{4}{3}, \infty\right)$.   **F.** Local minimum value $f\left(-\frac{4}{3}\right) = -\frac{4}{3}\sqrt{\frac{2}{3}} = -\frac{4\sqrt{6}}{9} \approx -1.09$,

no local maximum   **H.**

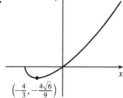

G. $f''(x) = \dfrac{2\sqrt{2+x} \cdot 3 - (3x+4)\dfrac{1}{\sqrt{2+x}}}{4(2+x)} = \dfrac{6(2+x) - (3x+4)}{4(2+x)^{3/2}}$

$= \dfrac{3x+8}{4(2+x)^{3/2}}$

$f''(x) > 0$ for $x > -2$, so f is CU on $(-2, \infty)$. No IP

29. $y = f(x) = \sin^2 x - 2\cos x$ **A.** $D = \mathbb{R}$ **B.** y-intercept: $f(0) = -2$ **C.** $f(-x) = f(x)$, so f is symmetric with respect

to the y-axis. f has period 2π. **D.** No asymptote **E.** $y' = 2\sin x \cos x + 2\sin x = 2\sin x \, (\cos x + 1)$. $y' = 0$ $\Leftrightarrow$

$\sin x = 0$ or $\cos x = -1$ $\Leftrightarrow$ $x = n\pi$ or $x = (2n+1)\pi$. $y' > 0$ when $\sin x > 0$, since $\cos x + 1 \geq 0$ for all x.

Therefore, $y' > 0$ [and so f is increasing] on $(2n\pi, (2n+1)\pi)$; $y' < 0$ [and so f is decreasing] on $((2n-1)\pi, 2n\pi)$.

F. Local maximum values are $f((2n+1)\pi) = 2$; local minimum values are $f(2n\pi) = -2$.

G. $y' = \sin 2x + 2\sin x$ $\Rightarrow$ $y'' = 2\cos 2x + 2\cos x = 2(2\cos^2 x - 1) + 2\cos x = 4\cos^2 x + 2\cos x - 2$

$$= 2(2\cos^2 x + \cos x - 1) = 2(2\cos x - 1)(\cos x + 1)$$

$y'' = 0$ $\Leftrightarrow$ $\cos x = \frac{1}{2}$ or -1 $\Leftrightarrow$ $x = 2n\pi \pm \frac{\pi}{3}$ or $x = (2n+1)\pi$. **H.**

$y'' > 0$ [and so f is CU] on $\left(2n\pi - \frac{\pi}{3}, 2n\pi + \frac{\pi}{3}\right)$; $y'' \leq 0$ [and so f is CD]

on $\left(2n\pi + \frac{\pi}{3}, 2n\pi + \frac{5\pi}{3}\right)$. There are inflection points at $\left(2n\pi \pm \frac{\pi}{3}, -\frac{1}{4}\right)$.

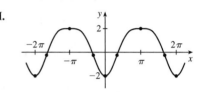

31. $y = f(x) = \sin^{-1}(1/x)$ **A.** $D = \{x \mid -1 \leq 1/x \leq 1\} = (-\infty, -1] \cup [1, \infty)$. **B.** No intercept

C. $f(-x) = -f(x)$, symmetric about the origin **D.** $\displaystyle \lim_{x \to \pm\infty} \sin^{-1}(1/x) = \sin^{-1}(0) = 0$, so $y = 0$ is a HA.

E. $f'(x) = \dfrac{1}{\sqrt{1 - (1/x)^2}}\left(-\dfrac{1}{x^2}\right) = \dfrac{-1}{\sqrt{x^4 - x^2}} < 0$, so f is decreasing on $(-\infty, -1)$ and $(1, \infty)$.

F. No local extreme value, but $f(1) = \frac{\pi}{2}$ is the absolute maximum value **H.**

and $f(-1) = -\frac{\pi}{2}$ is the absolute minimum value.

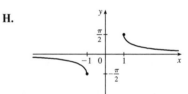

G. $f''(x) = \dfrac{4x^3 - 2x}{2(x^4 - x^2)^{3/2}} = \dfrac{x(2x^2 - 1)}{(x^4 - x^2)^{3/2}} > 0$ for $x > 1$ and $f''(x) < 0$

for $x < -1$, so f is CU on $(1, \infty)$ and CD on $(-\infty, -1)$. No IP

33. $y = f(x) = xe^{-2x}$ **A.** $D = \mathbb{R}$ **B.** y-intercept: $f(0) = 0$; x-intercept: $f(x) = 0$ $\Leftrightarrow$ $x = 0$

C. No symmetry **D.** $\displaystyle \lim_{x \to \infty} xe^{-2x} = \lim_{x \to \infty} \frac{x}{e^{2x}} \overset{\text{H}}{=} \lim_{x \to \infty} \frac{1}{2e^{2x}} = 0$, so $y = 0$ is a HA.

E. $f'(x) = x(-2e^{-2x}) + e^{-2x}(1) = e^{-2x}(-2x + 1) > 0$ $\Leftrightarrow$ $-2x + 1 > 0$ $\Leftrightarrow$ $x < \frac{1}{2}$ and $f'(x) < 0$ $\Leftrightarrow$ $x > \frac{1}{2}$,

so f is increasing on $\left(-\infty, \frac{1}{2}\right)$ and decreasing on $\left(\frac{1}{2}, \infty\right)$. **F.** Local maximum value $f\left(\frac{1}{2}\right) = \frac{1}{2}e^{-1} = 1/(2e)$;

no local minimum value

G. $f''(x) = e^{-2x}(-2) + (-2x + 1)(-2e^{-2x})$ **H.**

$\qquad = 2e^{-2x}[-1 - (-2x + 1)] = 4(x - 1)e^{-2x}$.

$f''(x) > 0$ $\Leftrightarrow$ $x > 1$ and $f''(x) < 0$ $\Leftrightarrow$ $x < 1$, so f is

CU on $(1, \infty)$ and CD on $(-\infty, 1)$. IP at $(1, f(1)) = (1, e^{-2})$

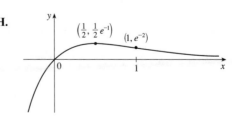

35. $f(x) = \dfrac{x^2 - 1}{x^3} \Rightarrow f'(x) = \dfrac{x^3(2x) - (x^2 - 1)3x^2}{x^6} = \dfrac{3 - x^2}{x^4} \Rightarrow$

$f''(x) = \dfrac{x^4(-2x) - (3 - x^2)4x^3}{x^8} = \dfrac{2x^2 - 12}{x^5}$

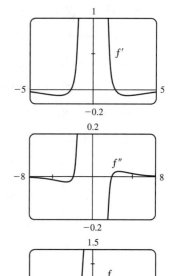

Estimates: From the graphs of f' and f'', it appears that f is increasing on

$(-1.73, 0)$ and $(0, 1.73)$ and decreasing on $(-\infty, -1.73)$ and $(1.73, \infty)$;

f has a local maximum of about $f(1.73) = 0.38$ and a local minimum of about

$f(-1.7) = -0.38$; f is CU on $(-2.45, 0)$ and $(2.45, \infty)$, and CD on

$(-\infty, -2.45)$ and $(0, 2.45)$; and f has inflection points at about

$(-2.45, -0.34)$ and $(2.45, 0.34)$.

Exact: Now $f'(x) = \dfrac{3 - x^2}{x^4}$ is positive for $0 < x^2 < 3$, that is, f is increasing

on $(-\sqrt{3}, 0)$ and $(0, \sqrt{3})$; and $f'(x)$ is negative (and so f is decreasing) on

$(-\infty, -\sqrt{3})$ and $(\sqrt{3}, \infty)$. $f'(x) = 0$ when $x = \pm\sqrt{3}$.

f' goes from positive to negative at $x = \sqrt{3}$, so f has a local maximum of

$f(\sqrt{3}) = \dfrac{(\sqrt{3})^2 - 1}{(\sqrt{3})^3} = \dfrac{2\sqrt{3}}{9}$; and since f is odd, we know that maxima on the

interval $(0, \infty)$ correspond to minima on $(-\infty, 0)$, so f has a local minimum of

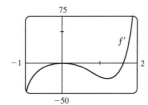

$f(-\sqrt{3}) = -\dfrac{2\sqrt{3}}{9}$. Also, $f''(x) = \dfrac{2x^2 - 12}{x^5}$ is positive (so f is CU) on

$(-\sqrt{6}, 0)$ and $(\sqrt{6}, \infty)$, and negative (so f is CD) on $(-\infty, -\sqrt{6})$ and

$(0, \sqrt{6})$. There are IP at $\left(\sqrt{6}, \dfrac{5\sqrt{6}}{36}\right)$ and $\left(-\sqrt{6}, -\dfrac{5\sqrt{6}}{36}\right)$.

37. $f(x) = 3x^6 - 5x^5 + x^4 - 5x^3 - 2x^2 + 2 \Rightarrow f'(x) = 18x^5 - 25x^4 + 4x^3 - 15x^2 - 4x \Rightarrow$

$f''(x) = 90x^4 - 100x^3 + 12x^2 - 30x - 4$

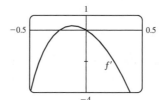
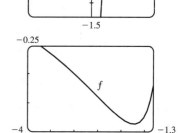
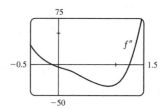

From the graphs of f' and f'', it appears that f is increasing on $(-0.23, 0)$ and $(1.62, \infty)$ and decreasing on $(-\infty, -0.23)$

and $(0, 1.62)$; f has a local maximum of about $f(0) = 2$ and local minima of about $f(-0.23) = 1.96$ and $f(1.62) = -19.2$;

f is CU on $(-\infty, -0.12)$ and $(1.24, \infty)$ and CD on $(-0.12, 1.24)$; and f has inflection points at about $(-0.12, 1.98)$ and $(1.24, -12.1)$.

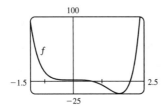

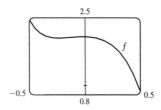

39.

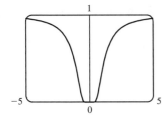

From the graph, we estimate the points of inflection to be about $(\pm 0.82, 0.22)$.

$$f(x) = e^{-1/x^2} \quad \Rightarrow \quad f'(x) = 2x^{-3}e^{-1/x^2} \quad \Rightarrow$$

$$f''(x) = 2[x^{-3}(2x^{-3})e^{-1/x^2} + e^{-1/x^2}(-3x^{-4})] = 2x^{-6}e^{-1/x^2}(2 - 3x^2).$$

This is 0 when $2 - 3x^2 = 0 \Leftrightarrow x = \pm\sqrt{\frac{2}{3}}$, so the inflection points are $\left(\pm\sqrt{\frac{2}{3}}, e^{-3/2}\right)$.

41. $f(x) = \dfrac{\cos^2 x}{\sqrt{x^2 + x + 1}}$, $-\pi \le x \le \pi \quad \Rightarrow \quad f'(x) = -\dfrac{\cos x\,[(2x+1)\cos x + 4(x^2 + x + 1)\sin x]}{2(x^2 + x + 1)^{3/2}} \quad \Rightarrow$

$$f''(x) = -\frac{(8x^4 + 16x^3 + 16x^2 + 8x + 9)\cos^2 x - 8(x^2 + x + 1)(2x + 1)\sin x \cos x - 8(x^2 + x + 1)^2 \sin^2 x}{4(x^2 + x + 1)^{5/2}}$$

$f(x) = 0 \Leftrightarrow x = \pm\frac{\pi}{2}$; $\quad f'(x) = 0 \Leftrightarrow x \approx -2.96, -1.57, -0.18, 1.57, 3.01$;

$f''(x) = 0 \Leftrightarrow x \approx -2.16, -0.75, 0.46$, and 2.21.

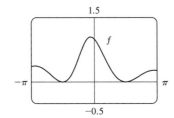

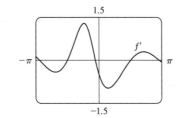

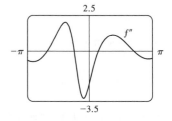

The x-coordinates of the maximum points are the values at which f' changes from positive to negative, that is, -2.96, -0.18, and 3.01. The x-coordinates of the minimum points are the values at which f' changes from negative to positive, that is, -1.57 and 1.57. The x-coordinates of the inflection points are the values at which f'' changes sign, that is, -2.16, -0.75, 0.46, and 2.21.

43. The family of functions $f(x) = \ln(\sin x + C)$ all have the same period and all have maximum values at $x = \frac{\pi}{2} + 2\pi n$. Since the domain of ln is $(0, \infty)$, f has a graph only if $\sin x + C > 0$ somewhere. Since $-1 \le \sin x \le 1$, this happens if $C > -1$, that is, f has no graph if $C \le -1$. Similarly, if $C > 1$, then $\sin x + C > 0$ and f is continuous on $(-\infty, \infty)$. As C increases, the graph of f is shifted vertically upward and flattens out. If $-1 < C \le 1$, f is defined

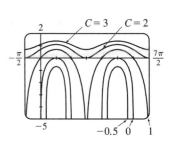

where $\sin x + C > 0 \Leftrightarrow \sin x > -C \Leftrightarrow \sin^{-1}(-C) < x < \pi - \sin^{-1}(-C)$. Since the period is 2π, the domain of f is $\left(2n\pi + \sin^{-1}(-C), (2n+1)\pi - \sin^{-1}(-C)\right)$, n an integer.

45. Let $f(x) = 3x + 2\cos x + 5$. Then $f(0) = 7 > 0$ and $f(-\pi) = -3\pi - 2 + 5 = -3\pi + 3 = -3(\pi - 1) < 0$, and since f is continuous on $\mathbb{R}$ (hence on $[-\pi, 0]$), the Intermediate Value Theorem assures us that there is at least one zero of f in $[-\pi, 0]$. Now $f'(x) = 3 - 2\sin x > 0$ implies that f is increasing on $\mathbb{R}$, so there is exactly one zero of f, and hence, exactly one real root of the equation $3x + 2\cos x + 5 = 0$.

47. Since f is continuous on $[32, 33]$ and differentiable on $(32, 33)$, then by the Mean Value Theorem there exists a number c in $(32, 33)$ such that $f'(c) = \frac{1}{5}c^{-4/5} = \dfrac{\sqrt[5]{33} - \sqrt[5]{32}}{33 - 32} = \sqrt[5]{33} - 2$, but $\frac{1}{5}c^{-4/5} > 0 \Rightarrow \sqrt[5]{33} - 2 > 0 \Rightarrow \sqrt[5]{33} > 2$. Also f' is decreasing, so that $f'(c) < f'(32) = \frac{1}{5}(32)^{-4/5} = 0.0125 \Rightarrow 0.0125 > f'(c) = \sqrt[5]{33} - 2 \Rightarrow \sqrt[5]{33} < 2.0125$. Therefore, $2 < \sqrt[5]{33} < 2.0125$.

49. (a) $g(x) = f(x^2) \Rightarrow g'(x) = 2xf'(x^2)$ by the Chain Rule. Since $f'(x) > 0$ for all $x \neq 0$, we must have $f'(x^2) > 0$ for $x \neq 0$, so $g'(x) = 0 \Leftrightarrow x = 0$. Now $g'(x)$ changes sign (from negative to positive) at $x = 0$, since one of its factors, $f'(x^2)$, is positive for all x, and its other factor, $2x$, changes from negative to positive at this point, so by the First Derivative Test, f has a local and absolute minimum at $x = 0$.

(b) $g'(x) = 2xf'(x^2) \Rightarrow g''(x) = 2[xf''(x^2)(2x) + f'(x^2)] = 4x^2 f''(x^2) + 2f'(x^2)$ by the Product Rule and the Chain Rule. But $x^2 > 0$ for all $x \neq 0$, $f''(x^2) > 0$ [since f is CU for $x > 0$], and $f'(x^2) > 0$ for all $x \neq 0$, so since all of its factors are positive, $g''(x) > 0$ for $x \neq 0$. Whether $g''(0)$ is positive or 0 doesn't matter [since the sign of g'' does not change there]; g is concave upward on $\mathbb{R}$.

51. If $B = 0$, the line is vertical and the distance from $x = -\dfrac{C}{A}$ to (x_1, y_1) is $\left| x_1 + \dfrac{C}{A} \right| = \dfrac{|Ax_1 + By_1 + C|}{\sqrt{A^2 + B^2}}$, so assume $B \neq 0$. The square of the distance from (x_1, y_1) to the line is $f(x) = (x - x_1)^2 + (y - y_1)^2$ where $Ax + By + C = 0$, so we minimize $f(x) = (x - x_1)^2 + \left(-\dfrac{A}{B}x - \dfrac{C}{B} - y_1 \right)^2 \Rightarrow f'(x) = 2(x - x_1) + 2\left(-\dfrac{A}{B}x - \dfrac{C}{B} - y_1 \right)\left(-\dfrac{A}{B} \right)$.

$f'(x) = 0 \Rightarrow x = \dfrac{B^2 x_1 - ABy_1 - AC}{A^2 + B^2}$ and this gives a minimum since $f''(x) = 2\left(1 + \dfrac{A^2}{B^2} \right) > 0$. Substituting this value of x into $f(x)$ and simplifying gives $f(x) = \dfrac{(Ax_1 + By_1 + C)^2}{A^2 + B^2}$, so the minimum distance is $\sqrt{f(x)} = \dfrac{|Ax_1 + By_1 + C|}{\sqrt{A^2 + B^2}}$.

53.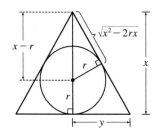

By similar triangles, $\dfrac{y}{x} = \dfrac{r}{\sqrt{x^2 - 2rx}}$, so the area of the triangle is

$$A(x) = \tfrac{1}{2}(2y)x = xy = \frac{rx^2}{\sqrt{x^2 - 2rx}} \quad \Rightarrow$$

$$A'(x) = \frac{2rx\sqrt{x^2 - 2rx} - rx^2(x-r)/\sqrt{x^2 - 2rx}}{x^2 - 2rx} = \frac{rx^2(x - 3r)}{(x^2 - 2rx)^{3/2}} = 0$$

when $x = 3r$.

$A'(x) < 0$ when $2r < x < 3r$, $A'(x) > 0$ when $x > 3r$. So $x = 3r$ gives a minimum and $A(3r) = \dfrac{r(9r^2)}{\sqrt{3}\,r} = 3\sqrt{3}\,r^2$.

55.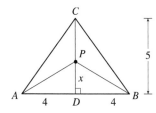

We minimize $L(x) = |PA| + |PB| + |PC| = 2\sqrt{x^2 + 16} + (5 - x)$,

$0 \le x \le 5$. $L'(x) = 2x/\sqrt{x^2 + 16} - 1 = 0 \Leftrightarrow 2x = \sqrt{x^2 + 16} \Leftrightarrow$

$4x^2 = x^2 + 16 \Leftrightarrow x = \frac{4}{\sqrt{3}}$. $L(0) = 13$, $L\left(\frac{4}{\sqrt{3}}\right) \approx 11.9$, $L(5) \approx 12.8$, so the

minimum occurs when $x = \frac{4}{\sqrt{3}} \approx 2.3$.

57. $v = K\sqrt{\dfrac{L}{C} + \dfrac{C}{L}} \Rightarrow \dfrac{dv}{dL} = \dfrac{K}{2\sqrt{(L/C) + (C/L)}}\left(\dfrac{1}{C} - \dfrac{C}{L^2}\right) = 0 \Leftrightarrow \dfrac{1}{C} = \dfrac{C}{L^2} \Leftrightarrow L^2 = C^2 \Leftrightarrow L = C.$

This gives the minimum velocity since $v' < 0$ for $0 < L < C$ and $v' > 0$ for $L > C$.

59. Let x denote the number of $1 decreases in ticket price. Then the ticket price is $\$12 - \$1(x)$, and the average attendance is

$11{,}000 + 1000(x)$. Now the revenue per game is

$$R(x) = (\text{price per person}) \times (\text{number of people per game})$$
$$= (12 - x)(11{,}000 + 1000x) = -1000x^2 + 1000x + 132{,}000$$

for $0 \le x \le 4$ [since the seating capacity is 15,000] $\Rightarrow R'(x) = -2000x + 1000 = 0 \Leftrightarrow x = 0.5$. This is a

maximum since $R''(x) = -2000 < 0$ for all x. Now we must check the value of $R(x) = (12 - x)(11{,}000 + 1000x)$ at

$x = 0.5$ and at the endpoints of the domain to see which value of x gives the maximum value of R.

$R(0) = (12)(11{,}000) = 132{,}000$, $R(0.5) = (11.5)(11{,}500) = 132{,}250$, and $R(4) = (8)(15{,}000) = 120{,}000$. Thus, the

maximum revenue of $132,250 per game occurs when the average attendance is 11,500 and the ticket price is $11.50.

61. $f(x) = x^5 - x^4 + 3x^2 - 3x - 2 \Rightarrow f'(x) = 5x^4 - 4x^3 + 6x - 3$, so $x_{n+1} = x_n - \dfrac{x_n^5 - x_n^4 + 3x_n^2 - 3x_n - 2}{5x_n^4 - 4x_n^3 + 6x_n - 3}$.

Now $x_1 = 1 \Rightarrow x_2 = 1.5 \Rightarrow x_3 \approx 1.343860 \Rightarrow x_4 \approx 1.300320 \Rightarrow x_5 \approx 1.297396 \Rightarrow$

$x_6 \approx 1.297383 \approx x_7$, so the root in $[1, 2]$ is 1.297383, to six decimal places.

63. $f(t) = \cos t + t - t^2 \Rightarrow f'(t) = -\sin t + 1 - 2t$. $f'(t)$ exists for all

t, so to find the maximum of f, we can examine the zeros of f'.

From the graph of f', we see that a good choice for t_1 is $t_1 = 0.3$.

Use $g(t) = -\sin t + 1 - 2t$ and $g'(t) = -\cos t - 2$ to obtain

$t_2 \approx 0.33535293$, $t_3 \approx 0.33541803 \approx t_4$. Since $f''(t) = -\cos t - 2 < 0$

for all t, $f(0.33541803) \approx 1.16718557$ is the absolute maximum.

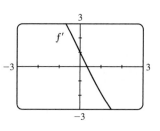

65. $f'(x) = \cos x - (1 - x^2)^{-1/2} = \cos x - \dfrac{1}{\sqrt{1 - x^2}} \Rightarrow f(x) = \sin x - \sin^{-1} x + C$

67. $f'(x) = \sqrt{x^3} + \sqrt[3]{x^2} = x^{3/2} + x^{2/3} \Rightarrow f(x) = \dfrac{x^{5/2}}{5/2} + \dfrac{x^{5/3}}{5/3} + C = \frac{2}{5}x^{5/2} + \frac{3}{5}x^{5/3} + C$

69. $f'(t) = 2t - 3\sin t \Rightarrow f(t) = t^2 + 3\cos t + C$.

$f(0) = 3 + C$ and $f(0) = 5 \Rightarrow C = 2$, so $f(t) = t^2 + 3\cos t + 2$.

71. $f''(x) = 1 - 6x + 48x^2 \Rightarrow f'(x) = x - 3x^2 + 16x^3 + C$. $f'(0) = C$ and $f'(0) = 2 \Rightarrow C = 2$, so

$f'(x) = x - 3x^2 + 16x^3 + 2$ and hence, $f(x) = \frac{1}{2}x^2 - x^3 + 4x^4 + 2x + D$.

$f(0) = D$ and $f(0) = 1 \Rightarrow D = 1$, so $f(x) = \frac{1}{2}x^2 - x^3 + 4x^4 + 2x + 1$.

73. $v(t) = s'(t) = 2t - \dfrac{1}{1 + t^2} \Rightarrow s(t) = t^2 - \tan^{-1} t + C$.

$s(0) = 0 - 0 + C = C$ and $s(0) = 1 \Rightarrow C = 1$, so $s(t) = t^2 - \tan^{-1} t + 1$.

75. (a) Since f is 0 just to the left of the y-axis, we must have a minimum of F at the same place since we are increasing through

$(0, 0)$ on F. There must be a local maximum to the left of $x = -3$, since f changes from positive to negative there.

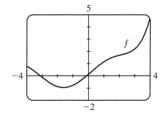

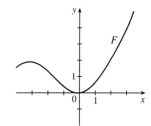

(b) $f(x) = 0.1e^x + \sin x \Rightarrow$

$F(x) = 0.1e^x - \cos x + C$. $F(0) = 0 \Rightarrow$

$0.1 - 1 + C = 0 \Rightarrow C = 0.9$, so

$F(x) = 0.1e^x - \cos x + 0.9$.

(c)

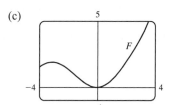

77. Choosing the positive direction to be upward, we have $a(t) = -9.8 \Rightarrow v(t) = -9.8t + v_0$, but $v(0) = 0 = v_0 \Rightarrow$

$v(t) = -9.8t = s'(t) \Rightarrow s(t) = -4.9t^2 + s_0$, but $s(0) = s_0 = 500 \Rightarrow s(t) = -4.9t^2 + 500$. When $s = 0$,

$-4.9t^2 + 500 = 0 \Rightarrow t_1 = \sqrt{\frac{500}{4.9}} \approx 10.1 \Rightarrow v(t_1) = -9.8\sqrt{\frac{500}{4.9}} \approx -98.995$ m/s. Since the canister has been

designed to withstand an impact velocity of 100 m/s, the canister will *not burst*.

79. (a)

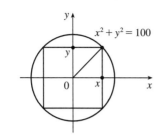

The cross-sectional area of the rectangular beam is

$$A = 2x \cdot 2y = 4xy = 4x\sqrt{100 - x^2},\ 0 \le x \le 10,\ \text{so}$$

$$\frac{dA}{dx} = 4x\left(\tfrac{1}{2}\right)(100 - x^2)^{-1/2}(-2x) + (100 - x^2)^{1/2} \cdot 4$$

$$= \frac{-4x^2}{(100 - x^2)^{1/2}} + 4(100 - x^2)^{1/2} = \frac{4\left[-x^2 + (100 - x^2)\right]}{(100 - x^2)^{1/2}}.$$

$\dfrac{dA}{dx} = 0$ when $-x^2 + (100 - x^2) = 0 \ \Rightarrow\ x^2 = 50 \ \Rightarrow\ x = \sqrt{50} \approx 7.07 \ \Rightarrow\ y = \sqrt{100 - \left(\sqrt{50}\right)^2} = \sqrt{50}$.

Since $A(0) = A(10) = 0$, the rectangle of maximum area is a square.

(b)

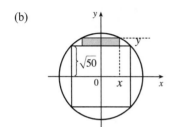

The cross-sectional area of each rectangular plank (shaded in the figure) is

$$A = 2x\left(y - \sqrt{50}\right) = 2x\left[\sqrt{100 - x^2} - \sqrt{50}\right],\ 0 \le x \le \sqrt{50},\ \text{so}$$

$$\frac{dA}{dx} = 2\left(\sqrt{100 - x^2} - \sqrt{50}\right) + 2x\left(\tfrac{1}{2}\right)(100 - x^2)^{-1/2}(-2x)$$

$$= 2(100 - x^2)^{1/2} - 2\sqrt{50} - \frac{2x^2}{(100 - x^2)^{1/2}}$$

Set $\dfrac{dA}{dx} = 0$: $(100 - x^2) - \sqrt{50}\,(100 - x^2)^{1/2} - x^2 = 0 \ \Rightarrow\ 100 - 2x^2 = \sqrt{50}\,(100 - x^2)^{1/2} \ \Rightarrow$

$10{,}000 - 400x^2 + 4x^4 = 50(100 - x^2) \ \Rightarrow\ 4x^4 - 350x^2 + 5000 = 0 \ \Rightarrow\ 2x^4 - 175x^2 + 2500 = 0 \ \Rightarrow$

$x^2 = \dfrac{175 \pm \sqrt{10{,}625}}{4} \approx 69.52$ or $17.98 \ \Rightarrow\ x \approx 8.34$ or 4.24. But $8.34 > \sqrt{50}$, so $x_1 \approx 4.24 \ \Rightarrow$

$y - \sqrt{50} = \sqrt{100 - x_1^2} - \sqrt{50} \approx 1.99$. Each plank should have dimensions about $8\frac{1}{2}$ inches by 2 inches.

(c) From the figure in part (a), the width is $2x$ and the depth is $2y$, so the strength is

$$S = k(2x)(2y)^2 = 8kxy^2 = 8kx(100 - x^2) = 800kx - 8kx^3,\ 0 \le x \le 10.\ dS/dx = 800k - 24kx^2 = 0 \text{ when}$$

$24kx^2 = 800k \ \Rightarrow\ x^2 = \dfrac{100}{3} \ \Rightarrow\ x = \dfrac{10}{\sqrt{3}} \ \Rightarrow\ y = \sqrt{\dfrac{200}{3}} = \dfrac{10\sqrt{2}}{\sqrt{3}} = \sqrt{2}\,x$. Since $S(0) = S(10) = 0$, the

maximum strength occurs when $x = \dfrac{10}{\sqrt{3}}$. The dimensions should be $\dfrac{20}{\sqrt{3}} \approx 11.55$ inches by $\dfrac{20\sqrt{2}}{\sqrt{3}} \approx 16.33$ inches.

81. We first show that $\dfrac{x}{1 + x^2} < \tan^{-1} x$ for $x > 0$. Let $f(x) = \tan^{-1} x - \dfrac{x}{1 + x^2}$. Then

$$f'(x) = \frac{1}{1 + x^2} - \frac{1(1 + x^2) - x(2x)}{(1 + x^2)^2} = \frac{(1 + x^2) - (1 - x^2)}{(1 + x^2)^2} = \frac{2x^2}{(1 + x^2)^2} > 0 \text{ for } x > 0.\text{ So } f(x) \text{ is increasing}$$

on $(0, \infty)$. Hence, $0 < x \ \Rightarrow\ 0 = f(0) < f(x) = \tan^{-1} x - \dfrac{x}{1 + x^2}$. So $\dfrac{x}{1 + x^2} < \tan^{-1} x$ for $0 < x$. We next show

that $\tan^{-1} x < x$ for $x > 0$. Let $h(x) = x - \tan^{-1} x$. Then $h'(x) = 1 - \dfrac{1}{1 + x^2} = \dfrac{x^2}{1 + x^2} > 0$. Hence, $h(x)$ is increasing

on $(0, \infty)$. So for $0 < x$, $0 = h(0) < h(x) = x - \tan^{-1} x$. Hence, $\tan^{-1} x < x$ for $x > 0$, and we conclude that

$$\frac{x}{1 + x^2} < \tan^{-1} x < x \text{ for } x > 0.$$

☐ PROBLEMS PLUS

1. Let $y = f(x) = e^{-x^2}$. The area of the rectangle under the curve from $-x$ to x is $A(x) = 2xe^{-x^2}$ where $x \geq 0$. We maximize

$A(x)$: $A'(x) = 2e^{-x^2} - 4x^2 e^{-x^2} = 2e^{-x^2}\left(1 - 2x^2\right) = 0 \Rightarrow x = \frac{1}{\sqrt{2}}$. This gives a maximum since $A'(x) > 0$

for $0 \leq x < \frac{1}{\sqrt{2}}$ and $A'(x) < 0$ for $x > \frac{1}{\sqrt{2}}$. We next determine the points of inflection of $f(x)$. Notice that

$f'(x) = -2xe^{-x^2} = -A(x)$. So $f''(x) = -A'(x)$ and hence, $f''(x) < 0$ for $-\frac{1}{\sqrt{2}} < x < \frac{1}{\sqrt{2}}$ and $f''(x) > 0$ for $x < -\frac{1}{\sqrt{2}}$

and $x > \frac{1}{\sqrt{2}}$. So $f(x)$ changes concavity at $x = \pm \frac{1}{\sqrt{2}}$, and the two vertices of the rectangle of largest area are at the inflection

points.

3. First, we recognize some symmetry in the inequality: $\dfrac{e^{x+y}}{xy} \geq e^2 \Leftrightarrow \dfrac{e^x}{x} \cdot \dfrac{e^y}{y} \geq e \cdot e$. This suggests that we need to show

that $\dfrac{e^x}{x} \geq e$ for $x > 0$. If we can do this, then the inequality $\dfrac{e^y}{y} \geq e$ is true, and the given inequality follows. $f(x) = \dfrac{e^x}{x} \Rightarrow$

$f'(x) = \dfrac{xe^x - e^x}{x^2} = \dfrac{e^x(x-1)}{x^2} = 0 \Rightarrow x = 1$. By the First Derivative Test, we have a minimum of $f(1) = e$, so

$e^x/x \geq e$ for all x.

5. Let $L = \lim\limits_{x \to 0} \dfrac{ax^2 + \sin bx + \sin cx + \sin dx}{3x^2 + 5x^4 + 7x^6}$. Now L has the indeterminate form of type $\frac{0}{0}$, so we can apply l'Hospital's

Rule. $L = \lim\limits_{x \to 0} \dfrac{2ax + b\cos bx + c\cos cx + d\cos dx}{6x + 20x^3 + 42x^5}$. The denominator approaches 0 as $x \to 0$, so the numerator must also

approach 0 (because the limit exists). But the numerator approaches $0 + b + c + d$, so $b + c + d = 0$. Apply l'Hospital's Rule

again. $L = \lim\limits_{x \to 0} \dfrac{2a - b^2\sin bx - c^2\sin cx - d^2\sin dx}{6 + 60x^2 + 210x^4} = \dfrac{2a - 0}{6 + 0} = \dfrac{2a}{6}$, which must equal 8. $\dfrac{2a}{6} = 8 \Rightarrow a = 24$.

Thus, $a + b + c + d = a + (b + c + d) = 24 + 0 = 24$.

7. Differentiating $x^2 + xy + y^2 = 12$ implicitly with respect to x gives $2x + y + x\dfrac{dy}{dx} + 2y\dfrac{dy}{dx} = 0$, so $\dfrac{dy}{dx} = -\dfrac{2x+y}{x+2y}$.

At a highest or lowest point, $\dfrac{dy}{dx} = 0 \Leftrightarrow y = -2x$. Substituting $-2x$ for y in the original equation gives

$x^2 + x(-2x) + (-2x)^2 = 12$, so $3x^2 = 12$ and $x = \pm 2$. If $x = 2$, then $y = -2x = -4$, and if $x = -2$ then $y = 4$.

Thus, the highest and lowest points are $(-2, 4)$ and $(2, -4)$.

9. $y = x^2 \Rightarrow y' = 2x$, so the slope of the tangent line at $P(a, a^2)$ is $2a$ and the slope of the normal line is $-\dfrac{1}{2a}$ for $a \neq 0$.

An equation of the normal line is $y - a^2 = -\dfrac{1}{2a}(x - a)$. Substitute x^2 for y to find the x-coordinates of the two points of

intersection of the parabola and the normal line. $x^2 - a^2 = -\dfrac{x}{2a} + \dfrac{1}{2} \Rightarrow 2ax^2 + x - 2a^3 - a = 0 \Rightarrow$

$$x = \frac{-1 \pm \sqrt{1 - 4(2a)(-2a^3 - a)}}{2(2a)} = \frac{-1 \pm \sqrt{1 + 16a^4 + 8a^2}}{4a} = \frac{-1 \pm \sqrt{(4a^2 + 1)^2}}{4a} = \frac{-1 \pm (4a^2 + 1)}{4a}$$

$$= \frac{4a^2}{4a} \quad \text{or} \quad \frac{-4a^2 - 2}{4a}, \text{ or equivalently, } a \quad \text{or} \quad -a - \frac{1}{2a}$$

So the point Q has coordinates $\left(-a - \dfrac{1}{2a}, \left(-a - \dfrac{1}{2a}\right)^2\right)$. The square S of the distance from P to Q is given by

$$S = \left(-a - \frac{1}{2a} - a\right)^2 + \left[\left(-a - \frac{1}{2a}\right)^2 - a^2\right]^2 = \left(-2a - \frac{1}{2a}\right)^2 + \left[\left(a^2 + 1 + \frac{1}{4a^2}\right) - a^2\right]^2$$

$$= \left(4a^2 + 2 + \frac{1}{4a^2}\right) + \left(1 + \frac{1}{4a^2}\right)^2 = \left(4a^2 + 2 + \frac{1}{4a^2}\right) + 1 + \frac{2}{4a^2} + \frac{1}{16a^4} = 4a^2 + 3 + \frac{3}{4a^2} + \frac{1}{16a^4}$$

$S' = 8a - \dfrac{6}{4a^3} - \dfrac{4}{16a^5} = 8a - \dfrac{3}{2a^3} - \dfrac{1}{4a^5} = \dfrac{32a^6 - 6a^2 - 1}{4a^5}$. The only real positive zero of the equation $S' = 0$ is

$a = \dfrac{1}{\sqrt{2}}$. Since $S'' = 8 + \dfrac{9}{2a^4} + \dfrac{5}{4a^6} > 0$, $a = \dfrac{1}{\sqrt{2}}$ corresponds to the shortest possible length of the line segment PQ.

11. $f(x) = \left(a^2 + a - 6\right)\cos 2x + (a - 2)x + \cos 1 \quad \Rightarrow \quad f'(x) = -\left(a^2 + a - 6\right)\sin 2x\,(2) + (a - 2)$. The derivative exists

for all x, so the only possible critical points will occur where $f'(x) = 0 \quad \Leftrightarrow \quad 2(a - 2)(a + 3)\sin 2x = a - 2 \quad \Leftrightarrow$

either $a = 2$ or $2(a + 3)\sin 2x = 1$, with the latter implying that $\sin 2x = \dfrac{1}{2(a + 3)}$. Since the range of $\sin 2x$ is $[-1, 1]$,

this equation has no solution whenever either $\dfrac{1}{2(a + 3)} < -1$ or $\dfrac{1}{2(a + 3)} > 1$. Solving these inequalities, we get

$-\frac{7}{2} < a < -\frac{5}{2}$.

13. $A = \left(x_1, x_1^2\right)$ and $B = \left(x_2, x_2^2\right)$, where x_1 and x_2 are the solutions of the quadratic equation $x^2 = mx + b$. Let $P = \left(x, x^2\right)$

and set $A_1 = (x_1, 0)$, $B_1 = (x_2, 0)$, and $P_1 = (x, 0)$. Let $f(x)$ denote the area of triangle PAB. Then $f(x)$ can be expressed

in terms of the areas of three trapezoids as follows:

$$f(x) = \text{area}\,(A_1ABB_1) - \text{area}\,(A_1APP_1) - \text{area}\,(B_1BPP_1)$$

$$= \tfrac{1}{2}\left(x_1^2 + x_2^2\right)(x_2 - x_1) - \tfrac{1}{2}\left(x_1^2 + x^2\right)(x - x_1) - \tfrac{1}{2}\left(x^2 + x_2^2\right)(x_2 - x)$$

After expanding and canceling terms, we get

$f(x) = \tfrac{1}{2}\left(x_2 x_1^2 - x_1 x_2^2 - x x_1^2 + x_1 x^2 - x_2 x^2 + x x_2^2\right) = \tfrac{1}{2}\left[x_1^2(x_2 - x) + x_2^2(x - x_1) + x^2(x_1 - x_2)\right]$

$f'(x) = \tfrac{1}{2}\left[-x_1^2 + x_2^2 + 2x(x_1 - x_2)\right]$. $f''(x) = \tfrac{1}{2}[2(x_1 - x_2)] = x_1 - x_2 < 0$ since $x_2 > x_1$.

$f'(x) = 0 \quad \Rightarrow \quad 2x(x_1 - x_2) = x_1^2 - x_2^2 \quad \Rightarrow \quad x_P = \tfrac{1}{2}(x_1 + x_2)$.

$f(x_P) = \tfrac{1}{2}\left(x_1^2\left[\tfrac{1}{2}(x_2 - x_1)\right] + x_2^2\left[\tfrac{1}{2}(x_2 - x_1)\right] + \tfrac{1}{4}(x_1 + x_2)^2(x_1 - x_2)\right)$

$\quad = \tfrac{1}{2}\left[\tfrac{1}{2}(x_2 - x_1)\left(x_1^2 + x_2^2\right) - \tfrac{1}{4}(x_2 - x_1)(x_1 + x_2)^2\right] = \tfrac{1}{8}(x_2 - x_1)\left[2\left(x_1^2 + x_2^2\right) - \left(x_1^2 + 2x_1 x_2 + x_2^2\right)\right]$

$\quad = \tfrac{1}{8}(x_2 - x_1)\left(x_1^2 - 2x_1 x_2 + x_2^2\right) = \tfrac{1}{8}(x_2 - x_1)(x_1 - x_2)^2 = \tfrac{1}{8}(x_2 - x_1)(x_2 - x_1)^2 = \tfrac{1}{8}(x_2 - x_1)^3$

To put this in terms of m and b, we solve the system $y = x_1^2$ and $y = mx_1 + b$, giving us $x_1^2 - mx_1 - b = 0$ ⇒

$x_1 = \frac{1}{2}\left(m - \sqrt{m^2 + 4b}\right)$. Similarly, $x_2 = \frac{1}{2}\left(m + \sqrt{m^2 + 4b}\right)$. The area is then $\frac{1}{8}(x_2 - x_1)^3 = \frac{1}{8}\left(\sqrt{m^2 + 4b}\right)^3$,

and is attained at the point $P\left(x_P, x_P^2\right) = P\left(\frac{1}{2}m, \frac{1}{4}m^2\right)$.

Note: Another way to get an expression for $f(x)$ is to use the formula for an area of a triangle in terms of the coordinates of

the vertices: $f(x) = \frac{1}{2}\left[\left(x_2 x_1^2 - x_1 x_2^2\right) + \left(x_1 x^2 - x x_1^2\right) + \left(x x_2^2 - x_2 x^2\right)\right]$.

15. Suppose that the curve $y = a^x$ intersects the line $y = x$. Then $a^{x_0} = x_0$ for some $x_0 > 0$, and hence $a = x_0^{1/x_0}$. We find the

maximum value of $g(x) = x^{1/x}$, > 0, because if a is larger than the maximum value of this function, then the curve $y = a^x$

does not intersect the line $y = x$. $g'(x) = e^{(1/x)\ln x}\left(-\frac{1}{x^2}\ln x + \frac{1}{x}\cdot\frac{1}{x}\right) = x^{1/x}\left(\frac{1}{x^2}\right)(1 - \ln x)$. This is 0 only where

$x = e$, and for $0 < x < e$, $f'(x) > 0$, while for $x > e$, $f'(x) < 0$, so g has an absolute maximum of $g(e) = e^{1/e}$. So if

$y = a^x$ intersects $y = x$, we must have $0 < a \le e^{1/e}$. Conversely, suppose that $0 < a \le e^{1/e}$. Then $a^e \le e$, so the graph of

$y = a^x$ lies below or touches the graph of $y = x$ at $x = e$. Also $a^0 = 1 > 0$, so the graph of $y = a^x$ lies above that of $y = x$

at $x = 0$. Therefore, by the Intermediate Value Theorem, the graphs of $y = a^x$ and $y = x$ must intersect somewhere between

$x = 0$ and $x = e$.

17. Note that $f(0) = 0$, so for $x \ne 0$, $\left|\dfrac{f(x) - f(0)}{x - 0}\right| = \left|\dfrac{f(x)}{x}\right| = \dfrac{|f(x)|}{|x|} \le \dfrac{|\sin x|}{|x|} = \dfrac{\sin x}{x}$.

Therefore, $|f'(0)| = \left|\lim\limits_{x\to 0}\dfrac{f(x) - f(0)}{x - 0}\right| = \lim\limits_{x\to 0}\left|\dfrac{f(x) - f(0)}{x - 0}\right| \le \lim\limits_{x\to 0}\dfrac{\sin x}{x} = 1$.

But $f(x) = a_1 \sin x + a_2 \sin 2x + \cdots + a_n \sin nx$ ⇒ $f'(x) = a_1 \cos x + 2a_2 \cos 2x + \cdots + na_n \cos nx$, so

$|f'(0)| = |a_1 + 2a_2 + \cdots + na_n| \le 1$.

Another solution: We are given that $\left|\sum_{k=1}^{n} a_k \sin kx\right| \le |\sin x|$. So for x close to 0, and $x \ne 0$, we have

$\left|\sum\limits_{k=1}^{n} a_k \dfrac{\sin kx}{\sin x}\right| \le 1$ ⇒ $\lim\limits_{x\to 0}\left|\sum\limits_{k=1}^{n} a_k \dfrac{\sin kx}{\sin x}\right| \le 1$ ⇒ $\left|\sum\limits_{k=1}^{n} a_k \lim\limits_{x\to 0}\dfrac{\sin kx}{\sin x}\right| \le 1$. But by l'Hospital's Rule,

$\lim\limits_{x\to 0}\dfrac{\sin kx}{\sin x} = \lim\limits_{x\to 0}\dfrac{k\cos kx}{\cos x} = k$, so $\left|\sum\limits_{k=1}^{n} ka_k\right| \le 1$.

19. (a) Distance = rate × time, so time = distance/rate. $T_1 = \dfrac{D}{c_1}$, $T_2 = \dfrac{2|PR|}{c_1} + \dfrac{|RS|}{c_2} = \dfrac{2h\sec\theta}{c_1} + \dfrac{D - 2h\tan\theta}{c_2}$,

$T_3 = \dfrac{2\sqrt{h^2 + D^2/4}}{c_1} = \dfrac{\sqrt{4h^2 + D^2}}{c_1}$.

(b) $\dfrac{dT_2}{d\theta} = \dfrac{2h}{c_1}\cdot\sec\theta\tan\theta - \dfrac{2h}{c_2}\sec^2\theta = 0$ when $2h\sec\theta\left(\dfrac{1}{c_1}\tan\theta - \dfrac{1}{c_2}\sec\theta\right) = 0$ ⇒

$\dfrac{1}{c_1}\dfrac{\sin\theta}{\cos\theta} - \dfrac{1}{c_2}\dfrac{1}{\cos\theta} = 0$ ⇒ $\dfrac{\sin\theta}{c_1\cos\theta} = \dfrac{1}{c_2\cos\theta}$ ⇒ $\sin\theta = \dfrac{c_1}{c_2}$. The First Derivative Test shows that this gives

a minimum.

(c) Using part (a) with $D = 1$ and $T_1 = 0.26$, we have $T_1 = \dfrac{D}{c_1} \Rightarrow c_1 = \frac{1}{0.26} \approx 3.85 \text{ km/s}$. $T_3 = \dfrac{\sqrt{4h^2 + D^2}}{c_1} \Rightarrow$

$4h^2 + D^2 = T_3^2 c_1^2 \Rightarrow h = \frac{1}{2}\sqrt{T_3^2 c_1^2 - D^2} = \frac{1}{2}\sqrt{(0.34)^2(1/0.26)^2 - 1^2} \approx 0.42 \text{ km}$. To find c_2, we use $\sin\theta = \dfrac{c_1}{c_2}$

from part (b) and $T_2 = \dfrac{2h\sec\theta}{c_1} + \dfrac{D - 2h\tan\theta}{c_2}$ from part (a). From the figure,

$\sin\theta = \dfrac{c_1}{c_2} \Rightarrow \sec\theta = \dfrac{c_2}{\sqrt{c_2^2 - c_1^2}}$ and $\tan\theta = \dfrac{c_1}{\sqrt{c_2^2 - c_1^2}}$, so

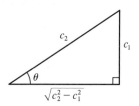

$T_2 = \dfrac{2hc_2}{c_1\sqrt{c_2^2 - c_1^2}} + \dfrac{D\sqrt{c_2^2 - c_1^2} - 2hc_1}{c_2\sqrt{c_2^2 - c_1^2}}$. Using the values for T_2 [given as 0.32],

h, c_1, and D, we can graph $Y_1 = T_2$ and $Y_2 = \dfrac{2hc_2}{c_1\sqrt{c_2^2 - c_1^2}} + \dfrac{D\sqrt{c_2^2 - c_1^2} - 2hc_1}{c_2\sqrt{c_2^2 - c_1^2}}$ and find their intersection points.

Doing so gives us $c_2 \approx 4.10$ and 7.66, but if $c_2 = 4.10$, then $\theta = \arcsin(c_1/c_2) \approx 69.6°$, which implies that point S is to

the left of point R in the diagram. So $c_2 = 7.66 \text{ km/s}$.

21.

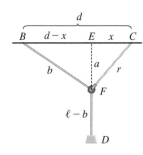

Let $a = |EF|$ and $b = |BF|$ as shown in the figure. Since $\ell = |BF| + |FD|$,

$|FD| = \ell - b$. Now

$|ED| = |EF| + |FD| = a + \ell - b = \sqrt{r^2 - x^2} + \ell - \sqrt{(d-x)^2 + a^2}$

$\qquad = \sqrt{r^2 - x^2} + \ell - \sqrt{(d-x)^2 + \left(\sqrt{r^2 - x^2}\,\right)^2}$

$\qquad = \sqrt{r^2 - x^2} + \ell - \sqrt{d^2 - 2dx + x^2 + r^2 - x^2}$

Let $f(x) = \sqrt{r^2 - x^2} + \ell - \sqrt{d^2 + r^2 - 2dx}$.

$f'(x) = \frac{1}{2}(r^2 - x^2)^{-1/2}(-2x) - \frac{1}{2}(d^2 + r^2 - 2dx)^{-1/2}(-2d) = \dfrac{-x}{\sqrt{r^2 - x^2}} + \dfrac{d}{\sqrt{d^2 + r^2 - 2dx}}$.

$f'(x) = 0 \Rightarrow \dfrac{x}{\sqrt{r^2 - x^2}} = \dfrac{d}{\sqrt{d^2 + r^2 - 2dx}} \Rightarrow \dfrac{x^2}{r^2 - x^2} = \dfrac{d^2}{d^2 + r^2 - 2dx} \Rightarrow$

$d^2 x^2 + r^2 x^2 - 2dx^3 = d^2 r^2 - d^2 x^2 \Rightarrow 0 = 2dx^3 - 2d^2 x^2 - r^2 x^2 + d^2 r^2 \Rightarrow$

$0 = 2dx^2(x - d) - r^2(x^2 - d^2) \Rightarrow 0 = 2dx^2(x - d) - r^2(x + d)(x - d) \Rightarrow 0 = (x - d)[2dx^2 - r^2(x + d)]$

But $d > r > x$, so $x \neq d$. Thus, we solve $2dx^2 - r^2 x - dr^2 = 0$ for x:

$x = \dfrac{-(-r^2) \pm \sqrt{(-r^2)^2 - 4(2d)(-dr^2)}}{2(2d)} = \dfrac{r^2 \pm \sqrt{r^4 + 8d^2 r^2}}{4d}$. Because $\sqrt{r^4 + 8d^2 r^2} > r^2$, the "negative" can be

discarded. Thus, $x = \dfrac{r^2 + \sqrt{r^2}\sqrt{r^2 + 8d^2}}{4d} = \dfrac{r^2 + r\sqrt{r^2 + 8d^2}}{4d} \quad [r > 0] = \dfrac{r}{4d}\left(r + \sqrt{r^2 + 8d^2}\,\right)$. The maximum value

of $|ED|$ occurs at this value of x.

23. $V = \frac{4}{3}\pi r^3 \;\Rightarrow\; \dfrac{dV}{dt} = 4\pi r^2 \dfrac{dr}{dt}$.   But $\dfrac{dV}{dt}$ is proportional to the surface area, so $\dfrac{dV}{dt} = k \cdot 4\pi r^2$ for some constant k.

Therefore, $4\pi r^2 \dfrac{dr}{dt} = k \cdot 4\pi r^2 \;\Leftrightarrow\; \dfrac{dr}{dt} = k =$ constant. An antiderivative of k with respect to t is kt, so $r = kt + C$.

When $t = 0$, the radius r must equal the original radius r_0, so $C = r_0$, and $r = kt + r_0$. To find k we use the fact that when

$t = 3, r = 3k + r_0$ and $V = \frac{1}{2}V_0 \;\Rightarrow\; \frac{4}{3}\pi(3k + r_0)^3 = \frac{1}{2} \cdot \frac{4}{3}\pi r_0^3 \;\Rightarrow\; (3k + r_0)^3 = \frac{1}{2}r_0^3 \;\Rightarrow\; 3k + r_0 = \frac{1}{\sqrt[3]{2}}r_0 \;\Rightarrow$

$k = \frac{1}{3}r_0\left(\dfrac{1}{\sqrt[3]{2}} - 1\right)$. Since $r = kt + r_0$, $r = \frac{1}{3}r_0\left(\dfrac{1}{\sqrt[3]{2}} - 1\right)t + r_0$. When the snowball has melted completely we have

$r = 0 \;\Rightarrow\; \frac{1}{3}r_0\left(\dfrac{1}{\sqrt[3]{2}} - 1\right)t + r_0 = 0$ which gives $t = \dfrac{3\sqrt[3]{2}}{\sqrt[3]{2} - 1}$. Hence, it takes $\dfrac{3\sqrt[3]{2}}{\sqrt[3]{2} - 1} - 3 = \dfrac{3}{\sqrt[3]{2} - 1} \approx 11$ h 33 min

longer.

5 □ INTEGRALS

5.1 Areas and Distances

1. (a) Since f is *increasing*, we can obtain a *lower* estimate by using *left* endpoints. We are instructed to use five rectangles, so $n = 5$.

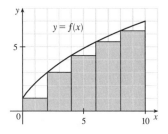

$$L_5 = \sum_{i=1}^{5} f(x_{i-1}) \, \Delta x \qquad [\Delta x = \tfrac{b-a}{n} = \tfrac{10-0}{5} = 2]$$

$$= f(x_0) \cdot 2 + f(x_1) \cdot 2 + f(x_2) \cdot 2 + f(x_3) \cdot 2 + f(x_4) \cdot 2$$

$$= 2\,[f(0) + f(2) + f(4) + f(6) + f(8)]$$

$$\approx 2(1 + 3 + 4.3 + 5.4 + 6.3) = 2(20) = 40$$

Since f is *increasing*, we can obtain an *upper* estimate by using *right* endpoints.

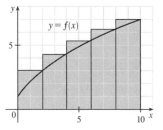

$$R_5 = \sum_{i=1}^{5} f(x_i) \, \Delta x$$

$$= 2\,[f(x_1) + f(x_2) + f(x_3) + f(x_4) + f(x_5)]$$

$$= 2\,[f(2) + f(4) + f(6) + f(8) + f(10)]$$

$$\approx 2(3 + 4.3 + 5.4 + 6.3 + 7) = 2(26) = 52$$

Comparing R_5 to L_5, we see that we have added the area of the rightmost upper rectangle, $f(10) \cdot 2$, to the sum and subtracted the area of the leftmost lower rectangle, $f(0) \cdot 2$, from the sum.

(b) $L_{10} = \displaystyle\sum_{i=1}^{10} f(x_{i-1}) \, \Delta x \qquad [\Delta x = \tfrac{10-0}{10} = 1]$

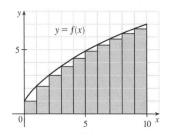

$$= 1\,[f(x_0) + f(x_1) + \cdots + f(x_9)]$$

$$= f(0) + f(1) + \cdots + f(9)$$

$$\approx 1 + 2.1 + 3 + 3.7 + 4.3 + 4.9 + 5.4 + 5.8 + 6.3 + 6.7$$

$$= 43.2$$

$$R_{10} = \sum_{i=1}^{10} f(x_i) \, \Delta x = f(1) + f(2) + \cdots + f(10)$$

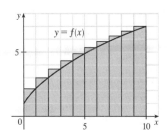

$$= L_{10} + 1 \cdot f(10) - 1 \cdot f(0) \qquad \begin{bmatrix} \text{add rightmost upper rectangle,} \\ \text{subtract leftmost lower rectangle} \end{bmatrix}$$

$$= 43.2 + 7 - 1 = 49.2$$

3. (a) $R_4 = \sum_{i=1}^{4} f(x_i)\,\Delta x \quad \left[\Delta x = \dfrac{\pi/2 - 0}{4} = \dfrac{\pi}{8}\right] \; = \left[\sum_{i=1}^{4} f(x_i)\right]\Delta x$

$= [f(x_1) + f(x_2) + f(x_3) + f(x_4)]\,\Delta x$

$= \left[\cos\frac{\pi}{8} + \cos\frac{2\pi}{8} + \cos\frac{3\pi}{8} + \cos\frac{4\pi}{8}\right]\frac{\pi}{8}$

$\approx (0.9239 + 0.7071 + 0.3827 + 0)\frac{\pi}{8} \approx 0.7908$

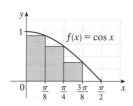
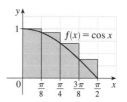

Since f is *decreasing* on $[0, \pi/2]$, an *underestimate* is obtained by using the *right* endpoint approximation, R_4.

(b) $L_4 = \sum_{i=1}^{4} f(x_{i-1})\,\Delta x = \left[\sum_{i=1}^{4} f(x_{i-1})\right]\Delta x$

$= [f(x_0) + f(x_1) + f(x_2) + f(x_3)]\,\Delta x$

$= \left[\cos 0 + \cos\frac{\pi}{8} + \cos\frac{2\pi}{8} + \cos\frac{3\pi}{8}\right]\frac{\pi}{8}$

$\approx (1 + 0.9239 + 0.7071 + 0.3827)\frac{\pi}{8} \approx 1.1835$

L_4 is an overestimate. Alternatively, we could just add the area of the leftmost upper rectangle and subtract the area of the rightmost lower rectangle; that is, $L_4 = R_4 + f(0) \cdot \frac{\pi}{8} - f\left(\frac{\pi}{2}\right) \cdot \frac{\pi}{8}$.

5. (a) $f(x) = 1 + x^2$ and $\Delta x = \dfrac{2-(-1)}{3} = 1 \;\Rightarrow$

$R_3 = 1 \cdot f(0) + 1 \cdot f(1) + 1 \cdot f(2) = 1 \cdot 1 + 1 \cdot 2 + 1 \cdot 5 = 8.$

$\Delta x = \dfrac{2-(-1)}{6} = 0.5 \;\Rightarrow$

$R_6 = 0.5[f(-0.5) + f(0) + f(0.5) + f(1) + f(1.5) + f(2)]$

$= 0.5(1.25 + 1 + 1.25 + 2 + 3.25 + 5)$

$= 0.5(13.75) = 6.875$

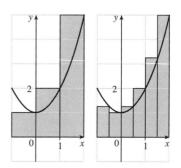

(b) $L_3 = 1 \cdot f(-1) + 1 \cdot f(0) + 1 \cdot f(1) = 1 \cdot 2 + 1 \cdot 1 + 1 \cdot 2 = 5$

$L_6 = 0.5[f(-1) + f(-0.5) + f(0) + f(0.5) + f(1) + f(1.5)]$

$= 0.5(2 + 1.25 + 1 + 1.25 + 2 + 3.25)$

$= 0.5(10.75) = 5.375$

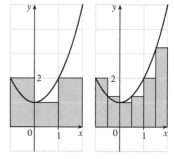

(c) $M_3 = 1 \cdot f(-0.5) + 1 \cdot f(0.5) + 1 \cdot f(1.5)$

$= 1 \cdot 1.25 + 1 \cdot 1.25 + 1 \cdot 3.25 = 5.75$

$M_6 = 0.5[f(-0.75) + f(-0.25) + f(0.25)$
$\qquad + f(0.75) + f(1.25) + f(1.75)]$

$= 0.5(1.5625 + 1.0625 + 1.0625 + 1.5625 + 2.5625 + 4.0625)$

$= 0.5(11.875) = 5.9375$

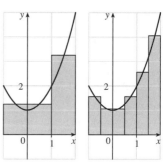

(d) M_6 appears to be the best estimate.

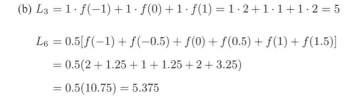

7. Here is one possible algorithm (ordered sequence of operations) for calculating the sums:

    1 Let SUM = 0, X_MIN = 0, X_MAX = 1, N = 10 (depending on which sum we are calculating),

    DELTA_X = (X_MAX - X_MIN)/N, and RIGHT_ENDPOINT = X_MIN + DELTA_X.

    2 Repeat steps 2a, 2b in sequence until RIGHT_ENDPOINT > X_MAX.
      2a Add (RIGHT_ENDPOINT)^4 to SUM.
      2b Add DELTA_X to RIGHT_ENDPOINT.

At the end of this procedure, (DELTA_X)·(SUM) is equal to the answer we are looking for. We find that

$$R_{10} = \frac{1}{10} \sum_{i=1}^{10} \left(\frac{i}{10}\right)^4 \approx 0.2533, \quad R_{30} = \frac{1}{30} \sum_{i=1}^{30} \left(\frac{i}{30}\right)^4 \approx 0.2170, \quad R_{50} = \frac{1}{50} \sum_{i=1}^{50} \left(\frac{i}{50}\right)^4 \approx 0.2101, \text{ and}$$

$$R_{100} = \frac{1}{100} \sum_{i=1}^{100} \left(\frac{i}{100}\right)^4 \approx 0.2050. \text{ It appears that the exact area is 0.2.}$$

The following display shows the program SUMRIGHT and its output from a TI-83 Plus calculator. To generalize the program, we have input (rather than assign) values for Xmin, Xmax, and N. Also, the function, x^4, is assigned to Y_1, enabling us to evaluate any right sum merely by changing Y_1 and running the program.

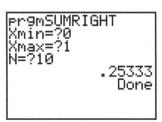

9. In Maple, we have to perform a number of steps before getting a numerical answer. After loading the student package [command: `with(student);`] we use the command

`left_sum:=leftsum(1/(x^2+1),x=0..1,10 [or 30, or 50]);` which gives us the expression in summation notation. To get a numerical approximation to the sum, we use `evalf(left_sum);`. Mathematica does not have a special command for these sums, so we must type them in manually. For example, the first left sum is given by

`(1/10)*Sum[1/(((i-1)/10)^2+1)],{i,1,10}]`, and we use the N command on the resulting output to get a numerical approximation.

In Derive, we use the `LEFT_RIEMANN` command to get the left sums, but must define the right sums ourselves. (We can define a new function using `LEFT_RIEMANN` with k ranging from 1 to n instead of from 0 to $n - 1$.)

(a) With $f(x) = \dfrac{1}{x^2 + 1}, 0 \leq x \leq 1$, the left sums are of the form $L_n = \dfrac{1}{n} \sum_{i=1}^{n} \dfrac{1}{\left(\frac{i-1}{n}\right)^2 + 1}$. Specifically, $L_{10} \approx 0.8100$,

$L_{30} \approx 0.7937$, and $L_{50} \approx 0.7904$. The right sums are of the form $R_n = \dfrac{1}{n} \sum_{i=1}^{n} \dfrac{1}{\left(\frac{i}{n}\right)^2 + 1}$. Specifically, $R_{10} \approx 0.7600$,

$R_{30} \approx 0.7770$, and $R_{50} \approx 0.7804$.

(b) In Maple, we use the `leftbox` (with the same arguments as `left_sum`) and `rightbox` commands to generate the graphs.

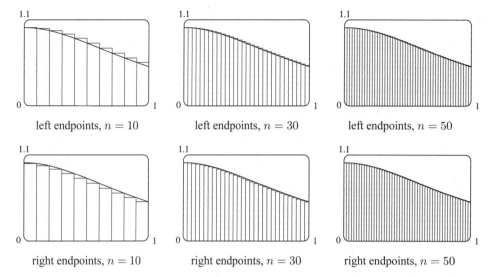

| left endpoints, $n = 10$ | left endpoints, $n = 30$ | left endpoints, $n = 50$ |

| right endpoints, $n = 10$ | right endpoints, $n = 30$ | right endpoints, $n = 50$ |

(c) We know that since $y = 1/(x^2 + 1)$ is a decreasing function on $(0, 1)$, all of the left sums are larger than the actual area, and all of the right sums are smaller than the actual area. Since the left sum with $n = 50$ is about $0.7904 < 0.791$ and the right sum with $n = 50$ is about $0.7804 > 0.780$, we conclude that $0.780 < R_{50} <$ exact area $< L_{50} < 0.791$, so the exact area is between 0.780 and 0.791.

11. Since v is an increasing function, L_6 will give us a lower estimate and R_6 will give us an upper estimate.

$$L_6 = (0 \text{ ft/s})(0.5 \text{ s}) + (6.2)(0.5) + (10.8)(0.5) + (14.9)(0.5) + (18.1)(0.5) + (19.4)(0.5) = 0.5(69.4) = 34.7 \text{ ft}$$

$$R_6 = 0.5(6.2 + 10.8 + 14.9 + 18.1 + 19.4 + 20.2) = 0.5(89.6) = 44.8 \text{ ft}$$

13. Lower estimate for oil leakage: $R_5 = (7.6 + 6.8 + 6.2 + 5.7 + 5.3)(2) = (31.6)(2) = 63.2 \text{ L}$.

Upper estimate for oil leakage: $L_5 = (8.7 + 7.6 + 6.8 + 6.2 + 5.7)(2) = (35)(2) = 70 \text{ L}$.

15. For a decreasing function, using left endpoints gives us an overestimate and using right endpoints results in an underestimate. We will use M_6 to get an estimate. $\Delta t = 1$, so

$$M_6 = 1[v(0.5) + v(1.5) + v(2.5) + v(3.5) + v(4.5) + v(5.5)] \approx 55 + 40 + 28 + 18 + 10 + 4 = 155 \text{ ft}$$

For a very rough check on the above calculation, we can draw a line from $(0, 70)$ to $(6, 0)$ and calculate the area of the triangle: $\frac{1}{2}(70)(6) = 210$. This is clearly an overestimate, so our midpoint estimate of 155 is reasonable.

17. $f(x) = \sqrt[4]{x}$, $1 \le x \le 16$. $\Delta x = (16 - 1)/n = 15/n$ and $x_i = 1 + i\,\Delta x = 1 + 15i/n$.

$$A = \lim_{n \to \infty} R_n = \lim_{n \to \infty} \sum_{i=1}^{n} f(x_i)\,\Delta x = \lim_{n \to \infty} \sum_{i=1}^{n} \sqrt[4]{1 + \frac{15i}{n}} \cdot \frac{15}{n}.$$

19. $f(x) = x \cos x$, $0 \le x \le \frac{\pi}{2}$. $\Delta x = (\frac{\pi}{2} - 0)/n = \frac{\pi}{2}/n$ and $x_i = 0 + i\,\Delta x = \frac{\pi}{2} i/n$.

$$A = \lim_{n \to \infty} R_n = \lim_{n \to \infty} \sum_{i=1}^{n} f(x_i)\,\Delta x = \lim_{n \to \infty} \sum_{i=1}^{n} \frac{i\pi}{2n} \cos\left(\frac{i\pi}{2n}\right) \cdot \frac{\pi}{2n}.$$

21. $\lim\limits_{n\to\infty}\sum\limits_{i=1}^{n}\dfrac{\pi}{4n}\tan\dfrac{i\pi}{4n}$ can be interpreted as the area of the region lying under the graph of $y=\tan x$ on the interval $\left[0,\frac{\pi}{4}\right]$,

since for $y=\tan x$ on $\left[0,\frac{\pi}{4}\right]$ with $\Delta x=\dfrac{\pi/4-0}{n}=\dfrac{\pi}{4n}$, $x_i=0+i\,\Delta x=\dfrac{i\pi}{4n}$, and $x_i^*=x_i$, the expression for the area is

$A=\lim\limits_{n\to\infty}\sum\limits_{i=1}^{n}f\left(x_i^*\right)\Delta x=\lim\limits_{n\to\infty}\sum\limits_{i=1}^{n}\tan\left(\dfrac{i\pi}{4n}\right)\dfrac{\pi}{4n}$. Note that this answer is not unique, since the expression for the area is

the same for the function $y=\tan(x-k\pi)$ on the interval $\left[k\pi,k\pi+\frac{\pi}{4}\right]$, where k is any integer.

23. (a) $y=f(x)=x^5$. $\Delta x=\dfrac{2-0}{n}=\dfrac{2}{n}$ and $x_i=0+i\,\Delta x=\dfrac{2i}{n}$.

$$A=\lim_{n\to\infty}R_n=\lim_{n\to\infty}\sum_{i=1}^{n}f(x_i)\,\Delta x=\lim_{n\to\infty}\sum_{i=1}^{n}\left(\frac{2i}{n}\right)^5\cdot\frac{2}{n}=\lim_{n\to\infty}\sum_{i=1}^{n}\frac{32i^5}{n^5}\cdot\frac{2}{n}=\lim_{n\to\infty}\frac{64}{n^6}\sum_{i=1}^{n}i^5.$$

(b) $\sum\limits_{i=1}^{n}i^5\overset{\text{CAS}}{=}\dfrac{n^2(n+1)^2\left(2n^2+2n-1\right)}{12}$

(c) $\lim\limits_{n\to\infty}\dfrac{64}{n^6}\cdot\dfrac{n^2(n+1)^2\left(2n^2+2n-1\right)}{12}=\dfrac{64}{12}\lim\limits_{n\to\infty}\dfrac{\left(n^2+2n+1\right)\left(2n^2+2n-1\right)}{n^2\cdot n^2}$

$$=\frac{16}{3}\lim_{n\to\infty}\left(1+\frac{2}{n}+\frac{1}{n^2}\right)\left(2+\frac{2}{n}-\frac{1}{n^2}\right)=\tfrac{16}{3}\cdot1\cdot2=\tfrac{32}{3}$$

25. $y=f(x)=\cos x$. $\Delta x=\dfrac{b-0}{n}=\dfrac{b}{n}$ and $x_i=0+i\,\Delta x=\dfrac{bi}{n}$.

$$A=\lim_{n\to\infty}R_n=\lim_{n\to\infty}\sum_{i=1}^{n}f(x_i)\,\Delta x=\lim_{n\to\infty}\sum_{i=1}^{n}\cos\left(\frac{bi}{n}\right)\cdot\frac{b}{n}\overset{\text{CAS}}{=}\lim_{n\to\infty}\left[\frac{b\sin\left(b\left(\dfrac{1}{2n}+1\right)\right)}{2n\sin\left(\dfrac{b}{2n}\right)}-\frac{b}{2n}\right]\overset{\text{CAS}}{=}\sin b$$

If $b=\frac{\pi}{2}$, then $A=\sin\frac{\pi}{2}=1$.

5.2   The Definite Integral

1. $f(x)=3-\frac{1}{2}x$, $2\le x\le14$. $\Delta x=\dfrac{b-a}{n}=\dfrac{14-2}{6}=2$.

Since we are using left endpoints, $x_i^*=x_{i-1}$.

$L_6=\sum\limits_{i=1}^{6}f(x_{i-1})\,\Delta x$

$=(\Delta x)\left[f(x_0)+f(x_1)+f(x_2)+f(x_3)+f(x_4)+f(x_5)\right]$

$=2[f(2)+f(4)+f(6)+f(8)+f(10)+f(12)]$

$=2[2+1+0+(-1)+(-2)+(-3)]=2(-3)=-6$

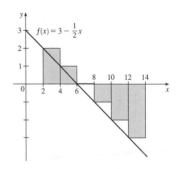

The Riemann sum represents the sum of the areas of the two rectangles above the x-axis minus the sum of the areas of the three rectangles below the x-axis; that is, the *net area* of the rectangles with respect to the x-axis.

3. $f(x) = e^x - 2$, $0 \le x \le 2$. $\Delta x = \dfrac{b-a}{n} = \dfrac{2-0}{4} = \dfrac{1}{2}$.

Since we are using midpoints, $x_i^* = \overline{x}_i = \frac{1}{2}(x_{i-1} + x_i)$.

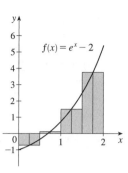

$$M_4 = \sum_{i=1}^{4} f(\overline{x}_i)\,\Delta x = (\Delta x)\left[f(\overline{x}_1) + f(\overline{x}_2) + f(\overline{x}_3) + f(\overline{x}_4)\right]$$

$$= \tfrac{1}{2}\left[f\!\left(\tfrac{1}{4}\right) + f\!\left(\tfrac{3}{4}\right) + f\!\left(\tfrac{5}{4}\right) + f\!\left(\tfrac{7}{4}\right)\right]$$

$$= \tfrac{1}{2}\left[\left(e^{1/4} - 2\right) + \left(e^{3/4} - 2\right) + \left(e^{5/4} - 2\right) + \left(e^{7/4} - 2\right)\right]$$

$$\approx 2.322986$$

The Riemann sum represents the sum of the areas of the three rectangles above the x-axis minus the area of the rectangle below the x-axis; that is, the *net area* of the rectangles with respect to the x-axis.

5. $\Delta x = (b-a)/n = (8-0)/4 = 8/4 = 2$.

(a) Using the right endpoints to approximate $\int_0^8 f(x)\,dx$, we have

$$\sum_{i=1}^{4} f(x_i)\,\Delta x = 2[f(2) + f(4) + f(6) + f(8)] \approx 2[1 + 2 + (-2) + 1] = 4.$$

(b) Using the left endpoints to approximate $\int_0^8 f(x)\,dx$, we have

$$\sum_{i=1}^{4} f(x_{i-1})\,\Delta x = 2[f(0) + f(2) + f(4) + f(6)] \approx 2[2 + 1 + 2 + (-2)] = 6.$$

(c) Using the midpoint of each subinterval to approximate $\int_0^8 f(x)\,dx$, we have

$$\sum_{i=1}^{4} f(\overline{x}_i)\,\Delta x = 2[f(1) + f(3) + f(5) + f(7)] \approx 2[3 + 2 + 1 + (-1)] = 10.$$

7. Since f is increasing, $L_5 \le \int_0^{25} f(x)\,dx \le R_5$.

$$\text{Lower estimate} = L_5 = \sum_{i=1}^{5} f(x_{i-1})\,\Delta x = 5[f(0) + f(5) + f(10) + f(15) + f(20)]$$

$$= 5(-42 - 37 - 25 - 6 + 15) = 5(-95) = -475$$

$$\text{Upper estimate} = R_5 = \sum_{i=1}^{5} f(x_i)\,\Delta x = 5[f(5) + f(10) + f(15) + f(20) + f(25)]$$

$$= 5(-37 - 25 - 6 + 15 + 36) = 5(-17) = -85$$

9. $\Delta x = (10 - 2)/4 = 2$, so the endpoints are 2, 4, 6, 8, and 10, and the midpoints are 3, 5, 7, and 9. The Midpoint Rule

gives $\int_2^{10} \sqrt{x^3 + 1}\,dx \approx \sum_{i=1}^{4} f(\overline{x}_i)\,\Delta x = 2\left(\sqrt{3^3 + 1} + \sqrt{5^3 + 1} + \sqrt{7^3 + 1} + \sqrt{9^3 + 1}\right) \approx 124.1644.$

11. $\Delta x = (1 - 0)/5 = 0.2$, so the endpoints are 0, 0.2, 0.4, 0.6, 0.8, and 1, and the midpoints are 0.1, 0.3, 0.5, 0.7, and 0.9.

The Midpoint Rule gives

$$\int_0^1 \sin(x^2)\,dx \approx \sum_{i=1}^{5} f(\overline{x}_i)\,\Delta x = 0.2\left[\sin(0.1)^2 + \sin(0.3)^2 + \sin(0.5)^2 + \sin(0.7)^2 + \sin(0.9)^2\right] \approx 0.3084.$$

13. In Maple, we use the command `with(student);` to load the sum and box commands, then
`m:=middlesum(sin(x^2),x=0..1,5);` which gives us the sum in summation notation, then `M:=evalf(m);` which
gives $M_5 \approx 0.30843908$, confirming the result of Exercise 11. The command `middlebox(sin(x^2),x=0..1,5)`
generates the graph. Repeating for $n = 10$ and $n = 20$ gives $M_{10} \approx 0.30981629$ and $M_{20} \approx 0.31015563$.

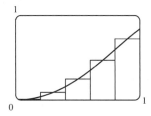 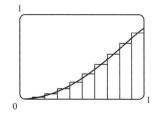 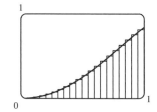

15. We'll create the table of values to approximate $\int_0^\pi \sin x \, dx$ by using the
program in the solution to Exercise 5.1.7 with $Y_1 = \sin x$, $\text{Xmin} = 0$,
$\text{Xmax} = \pi$, and $n = 5, 10, 50$, and 100.

The values of R_n appear to be approaching 2.

n	R_n
5	1.933766
10	1.983524
50	1.999342
100	1.999836

17. On $[2, 6]$, $\lim\limits_{n \to \infty} \sum\limits_{i=1}^{n} x_i \ln(1 + x_i^2) \, \Delta x = \int_2^6 x \ln(1 + x^2) \, dx$.

19. On $[1, 8]$, $\lim\limits_{n \to \infty} \sum\limits_{i=1}^{n} \sqrt{2x_i^* + (x_i^*)^2} \, \Delta x = \int_1^8 \sqrt{2x + x^2} \, dx$.

21. Note that $\Delta x = \dfrac{5 - (-1)}{n} = \dfrac{6}{n}$ and $x_i = -1 + i \, \Delta x = -1 + \dfrac{6i}{n}$.

$$\int_{-1}^{5} (1 + 3x) \, dx = \lim_{n \to \infty} \sum_{i=1}^{n} f(x_i) \, \Delta x = \lim_{n \to \infty} \sum_{i=1}^{n} \left[1 + 3\left(-1 + \frac{6i}{n}\right)\right]\frac{6}{n} = \lim_{n \to \infty} \frac{6}{n} \sum_{i=1}^{n}\left[-2 + \frac{18i}{n}\right]$$

$$= \lim_{n \to \infty} \frac{6}{n}\left[\sum_{i=1}^{n}(-2) + \sum_{i=1}^{n}\frac{18i}{n}\right] = \lim_{n \to \infty} \frac{6}{n}\left[-2n + \frac{18}{n}\sum_{i=1}^{n} i\right]$$

$$= \lim_{n \to \infty} \frac{6}{n}\left[-2n + \frac{18}{n} \cdot \frac{n(n+1)}{2}\right] = \lim_{n \to \infty}\left[-12 + \frac{108}{n^2} \cdot \frac{n(n+1)}{2}\right]$$

$$= \lim_{n \to \infty}\left[-12 + 54\frac{n+1}{n}\right] = \lim_{n \to \infty}\left[-12 + 54\left(1 + \frac{1}{n}\right)\right] = -12 + 54 \cdot 1 = 42$$

23. Note that $\Delta x = \dfrac{2 - 0}{n} = \dfrac{2}{n}$ and $x_i = 0 + i \, \Delta x = \dfrac{2i}{n}$.

$$\int_0^2 (2 - x^2) \, dx = \lim_{n \to \infty} \sum_{i=1}^{n} f(x_i) \, \Delta x = \lim_{n \to \infty} \sum_{i=1}^{n}\left(2 - \frac{4i^2}{n^2}\right)\left(\frac{2}{n}\right) = \lim_{n \to \infty} \frac{2}{n}\left[\sum_{i=1}^{n} 2 - \frac{4}{n^2}\sum_{i=1}^{n} i^2\right]$$

$$= \lim_{n \to \infty} \frac{2}{n}\left(2n - \frac{4}{n^2}\sum_{i=1}^{n} i^2\right) = \lim_{n \to \infty}\left[4 - \frac{8}{n^3} \cdot \frac{n(n+1)(2n+1)}{6}\right]$$

$$= \lim_{n \to \infty}\left(4 - \frac{4}{3} \cdot \frac{n+1}{n} \cdot \frac{2n+1}{n}\right) = \lim_{n \to \infty}\left[4 - \frac{4}{3}\left(1 + \frac{1}{n}\right)\left(2 + \frac{1}{n}\right)\right] = 4 - \frac{4}{3} \cdot 1 \cdot 2 = \frac{4}{3}$$

25. Note that $\Delta x = \dfrac{2-1}{n} = \dfrac{1}{n}$ and $x_i = 1 + i\,\Delta x = 1 + i(1/n) = 1 + i/n$.

$$\int_1^2 x^3\,dx = \lim_{n\to\infty} \sum_{i=1}^n f(x_i)\,\Delta x = \lim_{n\to\infty} \sum_{i=1}^n \left(1+\frac{i}{n}\right)^3 \left(\frac{1}{n}\right) = \lim_{n\to\infty} \frac{1}{n} \sum_{i=1}^n \left(\frac{n+i}{n}\right)^3$$

$$= \lim_{n\to\infty} \frac{1}{n^4} \sum_{i=1}^n \left(n^3 + 3n^2 i + 3ni^2 + i^3\right) = \lim_{n\to\infty} \frac{1}{n^4}\left[\sum_{i=1}^n n^3 + \sum_{i=1}^n 3n^2 i + \sum_{i=1}^n 3ni^2 + \sum_{i=1}^n i^3\right]$$

$$= \lim_{n\to\infty} \frac{1}{n^4}\left[n \cdot n^3 + 3n^2 \sum_{i=1}^n i + 3n \sum_{i=1}^n i^2 + \sum_{i=1}^n i^3\right]$$

$$= \lim_{n\to\infty} \left[1 + \frac{3}{n^2} \cdot \frac{n(n+1)}{2} + \frac{3}{n^3} \cdot \frac{n(n+1)(2n+1)}{6} + \frac{1}{n^4} \cdot \frac{n^2(n+1)^2}{4}\right]$$

$$= \lim_{n\to\infty} \left[1 + \frac{3}{2} \cdot \frac{n+1}{n} + \frac{1}{2} \cdot \frac{n+1}{n} \cdot \frac{2n+1}{n} + \frac{1}{4} \cdot \frac{(n+1)^2}{n^2}\right]$$

$$= \lim_{n\to\infty} \left[1 + \frac{3}{2}\left(1+\frac{1}{n}\right) + \frac{1}{2}\left(1+\frac{1}{n}\right)\left(2+\frac{1}{n}\right) + \frac{1}{4}\left(1+\frac{1}{n}\right)^2\right] = 1 + \frac{3}{2} + \frac{1}{2}\cdot 2 + \frac{1}{4} = 3.75$$

27. $\displaystyle\int_a^b x\,dx = \lim_{n\to\infty} \frac{b-a}{n} \sum_{i=1}^n \left[a + \frac{b-a}{n} i\right] = \lim_{n\to\infty} \left[\frac{a(b-a)}{n} \sum_{i=1}^n 1 + \frac{(b-a)^2}{n^2} \sum_{i=1}^n i\right]$

$$= \lim_{n\to\infty} \left[\frac{a(b-a)}{n} n + \frac{(b-a)^2}{n^2} \cdot \frac{n(n+1)}{2}\right] = a(b-a) + \lim_{n\to\infty} \frac{(b-a)^2}{2}\left(1+\frac{1}{n}\right)$$

$$= a(b-a) + \tfrac{1}{2}(b-a)^2 = (b-a)\left(a + \tfrac{1}{2}b - \tfrac{1}{2}a\right) = (b-a)\tfrac{1}{2}(b+a) = \tfrac{1}{2}\left(b^2 - a^2\right)$$

29. $f(x) = \dfrac{x}{1+x^5}$, $a = 2$, $b = 6$, and $\Delta x = \dfrac{6-2}{n} = \dfrac{4}{n}$. Using Theorem 4, we get $x_i^* = x_i = 2 + i\,\Delta x = 2 + \dfrac{4i}{n}$,

so $\displaystyle\int_2^6 \frac{x}{1+x^5}\,dx = \lim_{n\to\infty} R_n = \lim_{n\to\infty} \sum_{i=1}^n \frac{2+\dfrac{4i}{n}}{1+\left(2+\dfrac{4i}{n}\right)^5} \cdot \frac{4}{n}$.

31. $\Delta x = (\pi - 0)/n = \pi/n$ and $x_i^* = x_i = \pi i/n$.

$$\int_0^\pi \sin 5x\,dx = \lim_{n\to\infty} \sum_{i=1}^n (\sin 5x_i)\left(\frac{\pi}{n}\right) = \lim_{n\to\infty} \sum_{i=1}^n \left(\sin \frac{5\pi i}{n}\right) \frac{\pi}{n} \overset{\text{CAS}}{=} \pi \lim_{n\to\infty} \frac{1}{n} \cot\left(\frac{5\pi}{2n}\right) \overset{\text{CAS}}{=} \pi\left(\frac{2}{5\pi}\right) = \frac{2}{5}$$

33. (a) Think of $\int_0^2 f(x)\,dx$ as the area of a trapezoid with bases 1 and 3 and height 2. The area of a trapezoid is $A = \tfrac{1}{2}(b+B)h$,

so $\int_0^2 f(x)\,dx = \tfrac{1}{2}(1+3)2 = 4$.

(b) $\int_0^5 f(x)\,dx = \int_0^2 f(x)\,dx + \int_2^3 f(x)\,dx + \int_3^5 f(x)\,dx$

                    trapezoid   rectangle   triangle

            $= \tfrac{1}{2}(1+3)2 +$ $3 \cdot 1$ $+$ $\tfrac{1}{2} \cdot 2 \cdot 3$ $= 4+3+3 = 10$

(c) $\int_5^7 f(x)\,dx$ is the negative of the area of the triangle with base 2 and height 3. $\int_5^7 f(x)\,dx = -\tfrac{1}{2} \cdot 2 \cdot 3 = -3$.

(d) $\int_7^9 f(x)\,dx$ is the negative of the area of a trapezoid with bases 3 and 2 and height 2, so it equals

$-\tfrac{1}{2}(B+b)h = -\tfrac{1}{2}(3+2)2 = -5$. Thus,

$\int_0^9 f(x)\,dx = \int_0^5 f(x)\,dx + \int_5^7 f(x)\,dx + \int_7^9 f(x)\,dx = 10 + (-3) + (-5) = 2$.

35. $\int_0^3 \left(\frac{1}{2}x - 1\right) dx$ can be interpreted as the area of the triangle above the x-axis

minus the area of the triangle below the x-axis; that is,

$\frac{1}{2}(1)\left(\frac{1}{2}\right) - \frac{1}{2}(2)(1) = \frac{1}{4} - 1 = -\frac{3}{4}$.

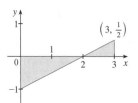

37. $\int_{-3}^0 \left(1 + \sqrt{9 - x^2}\,\right) dx$ can be interpreted as the area under the graph of

$f(x) = 1 + \sqrt{9 - x^2}$ between $x = -3$ and $x = 0$. This is equal to one-quarter

the area of the circle with radius 3, plus the area of the rectangle, so

$\int_{-3}^0 \left(1 + \sqrt{9 - x^2}\,\right) dx = \frac{1}{4}\pi \cdot 3^2 + 1 \cdot 3 = 3 + \frac{9}{4}\pi$.

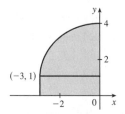

39. $\int_{-1}^2 |x|\, dx$ can be interpreted as the sum of the areas of the two shaded

triangles; that is, $\frac{1}{2}(1)(1) + \frac{1}{2}(2)(2) = \frac{1}{2} + \frac{4}{2} = \frac{5}{2}$.

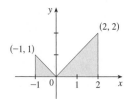

41. $\int_\pi^\pi \sin^2 x \cos^4 x\, dx = 0$ since the limits of intergration are equal.

43. $\int_0^1 (5 - 6x^2)\, dx = \int_0^1 5\, dx - 6 \int_0^1 x^2\, dx = 5(1 - 0) - 6\left(\frac{1}{3}\right) = 5 - 2 = 3$

45. $\int_1^3 e^{x+2}\, dx = \int_1^3 e^x \cdot e^2\, dx = e^2 \int_1^3 e^x\, dx = e^2(e^3 - e) = e^5 - e^3$

47. $\int_{-2}^2 f(x)\, dx + \int_2^5 f(x)\, dx - \int_{-2}^{-1} f(x)\, dx = \int_{-2}^5 f(x)\, dx + \int_{-1}^{-2} f(x)\, dx$ [by Property 5 and reversing limits]

$\qquad\qquad = \int_{-1}^5 f(x)\, dx$ [Property 5]

49. $\int_0^9 [2f(x) + 3g(x)]\, dx = 2\int_0^9 f(x)\, dx + 3\int_0^9 g(x)\, dx = 2(37) + 3(16) = 122$

51. Using Integral Comparison Property 8, $m \le f(x) \le M \;\Rightarrow\; m(2 - 0) \le \int_0^2 f(x)\, dx \le M(2 - 0) \;\Rightarrow$

$2m \le \int_0^2 f(x)\, dx \le 2M$.

53. If $-1 \le x \le 1$, then $0 \le x^2 \le 1$ and $1 \le 1 + x^2 \le 2$, so $1 \le \sqrt{1 + x^2} \le \sqrt{2}$ and

$1[1 - (-1)] \le \int_{-1}^1 \sqrt{1 + x^2}\, dx \le \sqrt{2}\,[1 - (-1)]$ [Property 8]; that is, $2 \le \int_{-1}^1 \sqrt{1 + x^2}\, dx \le 2\sqrt{2}$.

55. If $1 \le x \le 4$, then $1 \le \sqrt{x} \le 2$, so $1(4 - 1) \le \int_1^4 \sqrt{x}\, dx \le 2(4 - 1)$; that is, $3 \le \int_1^4 \sqrt{x}\, dx \le 6$.

57. If $\frac{\pi}{4} \le x \le \frac{\pi}{3}$, then $1 \le \tan x \le \sqrt{3}$, so $1\left(\frac{\pi}{3} - \frac{\pi}{4}\right) \le \int_{\pi/4}^{\pi/3} \tan x\, dx \le \sqrt{3}\left(\frac{\pi}{3} - \frac{\pi}{4}\right)$ or $\frac{\pi}{12} \le \int_{\pi/4}^{\pi/3} \tan x\, dx \le \frac{\pi}{12}\sqrt{3}$.

59. The only critical number of $f(x) = xe^{-x}$ on $[0, 2]$ is $x = 1$. Since $f(0) = 0$, $f(1) = e^{-1} \approx 0.368$, and

$f(2) = 2e^{-2} \approx 0.271$, we know that the absolute minimum value of f on $[0, 2]$ is 0, and the absolute maximum is e^{-1}. By

Property 8, $0 \le xe^{-x} \le e^{-1}$ for $0 \le x \le 2 \;\Rightarrow\; 0(2 - 0) \le \int_0^2 xe^{-x}\, dx \le e^{-1}(2 - 0) \;\Rightarrow\; 0 \le \int_0^2 xe^{-x}\, dx \le 2/e$.

61. $\sqrt{x^4 + 1} \ge \sqrt{x^4} = x^2$, so $\int_1^3 \sqrt{x^4 + 1}\, dx \ge \int_1^3 x^2\, dx = \frac{1}{3}(3^3 - 1^3) = \frac{26}{3}$.

63. Using right endpoints as in the proof of Property 2, we calculate

$$\int_a^b cf(x)\,dx = \lim_{n\to\infty}\sum_{i=1}^n cf(x_i)\,\Delta x = \lim_{n\to\infty} c\sum_{i=1}^n f(x_i)\,\Delta x = c\lim_{n\to\infty}\sum_{i=1}^n f(x_i)\,\Delta x = c\int_a^b f(x)\,dx.$$

65. Since $-|f(x)| \le f(x) \le |f(x)|$, it follows from Property 7 that

$$-\int_a^b |f(x)|\,dx \le \int_a^b f(x)\,dx \le \int_a^b |f(x)|\,dx \quad\Rightarrow\quad \left|\int_a^b f(x)\,dx\right| \le \int_a^b |f(x)|\,dx$$

Note that the definite integral is a real number, and so the following property applies: $-a \le b \le a \quad\Rightarrow\quad |b| \le a$ for all real numbers b and nonnegative numbers a.

67. To show that f is integrable on $[0, 1]$, we must show that $\displaystyle\lim_{n\to\infty}\sum_{i=1}^n f(x_i^*)\,\Delta x$ exists. Let n denote a positive integer and divide

the interval $[0, 1]$ into n equal subintervals $\left[0, \dfrac{1}{n}\right]$, $\left[\dfrac{1}{n}, \dfrac{2}{n}\right]$, $\ldots$, $\left[\dfrac{n-1}{n}, 1\right]$. If we choose x_i^* to be a rational number in the ith

subinterval, then we obtain the Riemann sum $\displaystyle\sum_{i=1}^n f(x_i^*)\cdot\dfrac{1}{n} = 0$, so $\displaystyle\lim_{n\to\infty}\sum_{i=1}^n f(x_i^*)\cdot\dfrac{1}{n} = \lim_{n\to\infty} 0 = 0$. Now suppose we

choose x_i^* to be an irrational number. Then we get $\displaystyle\sum_{i=1}^n f(x_i^*)\cdot\dfrac{1}{n} = \sum_{i=1}^n 1\cdot\dfrac{1}{n} = n\cdot\dfrac{1}{n} = 1$ for each n, so

$\displaystyle\lim_{n\to\infty}\sum_{i=1}^n f(x_i^*)\cdot\dfrac{1}{n} = \lim_{n\to\infty} 1 = 1$. Since the value of $\displaystyle\lim_{n\to\infty}\sum_{i=1}^n f(x_i^*)\,\Delta x$ depends on the choice of the sample points x_i^*, the

limit does not exist, and f is not integrable on $[0, 1]$.

69. $\displaystyle\lim_{n\to\infty}\sum_{i=1}^n \dfrac{i^4}{n^5} = \lim_{n\to\infty}\sum_{i=1}^n \dfrac{i^4}{n^4}\cdot\dfrac{1}{n} = \lim_{n\to\infty}\sum_{i=1}^n \left(\dfrac{i}{n}\right)^4 \dfrac{1}{n}$. At this point, we need to recognize the limit as being of the form

$\displaystyle\lim_{n\to\infty}\sum_{i=1}^n f(x_i)\,\Delta x$, where $\Delta x = (1-0)/n = 1/n$, $x_i = 0 + i\,\Delta x = i/n$, and $f(x) = x^4$. Thus, the definite integral

is $\int_0^1 x^4\,dx$.

71. Choose $x_i = 1 + \dfrac{i}{n}$ and $x_i^* = \sqrt{x_{i-1}x_i} = \sqrt{\left(1 + \dfrac{i-1}{n}\right)\left(1 + \dfrac{i}{n}\right)}$. Then

$$\int_1^2 x^{-2}\,dx = \lim_{n\to\infty}\dfrac{1}{n}\sum_{i=1}^n \dfrac{1}{\left(1 + \frac{i-1}{n}\right)\left(1 + \frac{i}{n}\right)} = \lim_{n\to\infty} n\sum_{i=1}^n \dfrac{1}{(n+i-1)(n+i)}$$

$$= \lim_{n\to\infty} n\sum_{i=1}^n \left(\dfrac{1}{n+i-1} - \dfrac{1}{n+i}\right) \quad \text{[by the hint]} \quad = \lim_{n\to\infty} n\left(\sum_{i=0}^{n-1}\dfrac{1}{n+i} - \sum_{i=1}^n \dfrac{1}{n+i}\right)$$

$$= \lim_{n\to\infty} n\left(\left[\dfrac{1}{n} + \dfrac{1}{n+1} + \cdots + \dfrac{1}{2n-1}\right] - \left[\dfrac{1}{n+1} + \cdots + \dfrac{1}{2n-1} + \dfrac{1}{2n}\right]\right)$$

$$= \lim_{n\to\infty} n\left(\dfrac{1}{n} - \dfrac{1}{2n}\right) = \lim_{n\to\infty} \left(1 - \tfrac{1}{2}\right) = \tfrac{1}{2}$$

5.3   The Fundamental Theorem of Calculus

1. One process undoes what the other one does. The precise version of this statement is given by the Fundamental Theorem of Calculus. See the statement of this theorem and the paragraph that follows it on page 387.

3. (a) $g(x) = \int_0^x f(t)\,dt$.

$g(0) = \int_0^0 f(t)\,dt = 0$

$g(1) = \int_0^1 f(t)\,dt = 1 \cdot 2 = 2$ [rectangle],

$g(2) = \int_0^2 f(t)\,dt = \int_0^1 f(t)\,dt + \int_1^2 f(t)\,dt = g(1) + \int_1^2 f(t)\,dt$

$\quad = 2 + 1 \cdot 2 + \frac{1}{2} \cdot 1 \cdot 2 = 5$ [rectangle plus triangle],

$g(3) = \int_0^3 f(t)\,dt = g(2) + \int_2^3 f(t)\,dt = 5 + \frac{1}{2} \cdot 1 \cdot 4 = 7$,

$g(6) = g(3) + \int_3^6 f(t)\,dt$ [the integral is negative since f lies under the x-axis]

$\quad = 7 + \left[-\left(\frac{1}{2} \cdot 2 \cdot 2 + 1 \cdot 2 \right) \right] = 7 - 4 = 3$

(d)

(b) g is increasing on $(0, 3)$ because as x increases from 0 to 3, we keep adding more area.

(c) g has a maximum value when we start subtracting area; that is, at $x = 3$.

5.
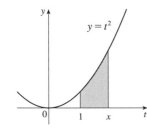

(a) By FTC1 with $f(t) = t^2$ and $a = 1$, $g(x) = \int_1^x t^2\,dt \Rightarrow$

$\quad g'(x) = f(x) = x^2$.

(b) Using FTC2, $g(x) = \int_1^x t^2\,dt = \left[\frac{1}{3}t^3 \right]_1^x = \frac{1}{3}x^3 - \frac{1}{3} \Rightarrow g'(x) = x^2$.

7. $f(t) = \dfrac{1}{t^3 + 1}$ and $g(x) = \displaystyle\int_1^x \dfrac{1}{t^3 + 1}\,dt$, so by FTC1, $g'(x) = f(x) = \dfrac{1}{x^3 + 1}$. Note that the lower limit, 1, could be any

real number greater than -1 and not affect this answer.

9. $f(t) = t^2 \sin t$ and $g(y) = \int_2^y t^2 \sin t\,dt$, so by FTC1, $g'(y) = f(y) = y^2 \sin y$.

11. $F(x) = \displaystyle\int_x^\pi \sqrt{1 + \sec t}\,dt = -\int_\pi^x \sqrt{1 + \sec t}\,dt \Rightarrow F'(x) = -\dfrac{d}{dx}\int_\pi^x \sqrt{1 + \sec t}\,dt = -\sqrt{1 + \sec x}$

13. Let $u = \dfrac{1}{x}$. Then $\dfrac{du}{dx} = -\dfrac{1}{x^2}$. Also, $\dfrac{dh}{dx} = \dfrac{dh}{du}\dfrac{du}{dx}$, so

$h'(x) = \dfrac{d}{dx}\displaystyle\int_2^{1/x} \arctan t\,dt = \dfrac{d}{du}\int_2^u \arctan t\,dt \cdot \dfrac{du}{dx} = \arctan u \dfrac{du}{dx} = -\dfrac{\arctan(1/x)}{x^2}$.

15. Let $u = \tan x$. Then $\dfrac{du}{dx} = \sec^2 x$. Also, $\dfrac{dy}{dx} = \dfrac{dy}{du}\dfrac{du}{dx}$, so

$y' = \dfrac{d}{dx}\displaystyle\int_0^{\tan x} \sqrt{t + \sqrt{t}}\,dt = \dfrac{d}{du}\int_0^u \sqrt{t + \sqrt{t}}\,dt \cdot \dfrac{du}{dx} = \sqrt{u + \sqrt{u}}\,\dfrac{du}{dx} = \sqrt{\tan x + \sqrt{\tan x}}\,\sec^2 x$.

17. Let $w = 1 - 3x$. Then $\dfrac{dw}{dx} = -3$. Also, $\dfrac{dy}{dx} = \dfrac{dy}{dw}\dfrac{dw}{dx}$, so

$y' = \dfrac{d}{dx}\displaystyle\int_{1-3x}^1 \dfrac{u^3}{1 + u^2}\,du = \dfrac{d}{dw}\int_w^1 \dfrac{u^3}{1 + u^2}\,du \cdot \dfrac{dw}{dx} = -\dfrac{d}{dw}\int_1^w \dfrac{u^3}{1 + u^2}\,du \cdot \dfrac{dw}{dx} = -\dfrac{w^3}{1 + w^2}(-3) = \dfrac{3(1 - 3x)^3}{1 + (1 - 3x)^2}$

19. $\displaystyle\int_{-1}^2 (x^3 - 2x)\,dx = \left[\dfrac{x^4}{4} - x^2 \right]_{-1}^2 = \left(\dfrac{2^4}{4} - 2^2 \right) - \left(\dfrac{(-1)^4}{4} - (-1)^2 \right) = (4 - 4) - \left(\dfrac{1}{4} - 1 \right) = 0 - \left(-\dfrac{3}{4} \right) = \dfrac{3}{4}$

21. $\int_1^4 (5 - 2t + 3t^2)\, dt = \left[5t - t^2 + t^3 \right]_1^4 = (20 - 16 + 64) - (5 - 1 + 1) = 68 - 5 = 63$

23. $\int_0^1 x^{4/5}\, dx = \left[\frac{5}{9}x^{9/5} \right]_0^1 = \frac{5}{9} - 0 = \frac{5}{9}$

25. $\displaystyle\int_1^2 \frac{3}{t^4}\, dt = 3\int_1^2 t^{-4}\, dt = 3\left[\frac{t^{-3}}{-3} \right]_1^2 = \frac{3}{-3}\left[\frac{1}{t^3} \right]_1^2 = -1\left(\frac{1}{8} - 1 \right) = \frac{7}{8}$

27. $\int_0^2 x(2 + x^5)\, dx = \int_0^2 (2x + x^6)\, dx = \left[x^2 + \frac{1}{7}x^7 \right]_0^2 = \left(4 + \frac{128}{7} \right) - (0 + 0) = \frac{156}{7}$

29. $\displaystyle\int_1^9 \frac{x - 1}{\sqrt{x}}\, dx = \int_1^9 \left(\frac{x}{\sqrt{x}} - \frac{1}{\sqrt{x}} \right) dx = \int_1^9 (x^{1/2} - x^{-1/2})\, dx = \left[\frac{2}{3}x^{3/2} - 2x^{1/2} \right]_1^9$

$= \left(\frac{2}{3}\cdot 27 - 2\cdot 3 \right) - \left(\frac{2}{3} - 2 \right) = 12 - \left(-\frac{4}{3} \right) = \frac{40}{3}$

31. $\int_0^{\pi/4} \sec^2 t\, dt = \left[\tan t \right]_0^{\pi/4} = \tan \frac{\pi}{4} - \tan 0 = 1 - 0 = 1$

33. $\int_1^2 (1 + 2y)^2\, dy = \int_1^2 (1 + 4y + 4y^2)\, dy = \left[y + 2y^2 + \frac{4}{3}y^3 \right]_1^2 = \left(2 + 8 + \frac{32}{3} \right) - \left(1 + 2 + \frac{4}{3} \right) = \frac{62}{3} - \frac{13}{3} = \frac{49}{3}$

35. $\displaystyle\int_1^9 \frac{1}{2x}\, dx = \frac{1}{2}\int_1^9 \frac{1}{x}\, dx = \frac{1}{2}\left[\ln|x| \right]_1^9 = \frac{1}{2}(\ln 9 - \ln 1) = \frac{1}{2}\ln 9 - 0 = \ln 9^{1/2} = \ln 3$

37. $\displaystyle\int_{1/2}^{\sqrt{3}/2} \frac{6}{\sqrt{1 - t^2}}\, dt = 6\int_{1/2}^{\sqrt{3}/2} \frac{1}{\sqrt{1 - t^2}}\, dt = 6\left[\sin^{-1} t \right]_{1/2}^{\sqrt{3}/2} = 6\left[\sin^{-1}\left(\frac{\sqrt{3}}{2} \right) - \sin^{-1}\left(\frac{1}{2} \right) \right] = 6\left(\frac{\pi}{3} - \frac{\pi}{6} \right) = 6\left(\frac{\pi}{6} \right) = \pi$

39. $\int_{-1}^1 e^{u+1}\, du = \left[e^{u+1} \right]_{-1}^1 = e^2 - e^0 = e^2 - 1$ [or start with $e^{u+1} = e^u e^1$]

41. If $f(x) = \begin{cases} \sin x & \text{if } 0 \le x < \pi/2 \\ \cos x & \text{if } \pi/2 \le x \le \pi \end{cases}$ then

$\int_0^\pi f(x)\, dx = \int_0^{\pi/2} \sin x\, dx + \int_{\pi/2}^\pi \cos x\, dx = \left[-\cos x \right]_0^{\pi/2} + \left[\sin x \right]_{\pi/2}^\pi = -\cos \frac{\pi}{2} + \cos 0 + \sin \pi - \sin \frac{\pi}{2}$

$= -0 + 1 + 0 - 1 = 0$

Note that f is integrable by Theorem 3 in Section 5.2.

43. $f(x) = x^{-4}$ is not continuous on the interval $[-2, 1]$, so FTC2 cannot be applied. In fact, f has an infinite discontinuity at $x = 0$, so $\int_{-2}^1 x^{-4}\, dx$ does not exist.

45. $f(\theta) = \sec\theta\,\tan\theta$ is not continuous on the interval $[\pi/3, \pi]$, so FTC2 cannot be applied. In fact, f has an infinite discontinuity at $x = \pi/2$, so $\int_{\pi/3}^\pi \sec\theta\,\tan\theta\, d\theta$ does not exist.

47. From the graph, it appears that the area is about 60. The actual area is

$\int_0^{27} x^{1/3}\, dx = \left[\frac{3}{4}x^{4/3} \right]_0^{27} = \frac{3}{4}\cdot 81 - 0 = \frac{243}{4} = 60.75$. This is $\frac{3}{4}$ of the

area of the viewing rectangle.

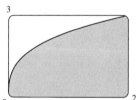

49. It appears that the area under the graph is about $\frac{2}{3}$ of the area of the viewing

rectangle, or about $\frac{2}{3}\pi \approx 2.1$. The actual area is

$\int_0^\pi \sin x \, dx = [-\cos x]_0^\pi = (-\cos \pi) - (-\cos 0) = -(-1) + 1 = 2.$

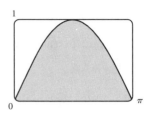

51. $\int_{-1}^2 x^3 \, dx = \left[\frac{1}{4}x^4\right]_{-1}^2 = 4 - \frac{1}{4} = \frac{15}{4} = 3.75$

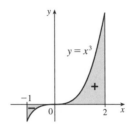

53. $g(x) = \int_{2x}^{3x} \dfrac{u^2 - 1}{u^2 + 1} \, du = \int_{2x}^0 \dfrac{u^2 - 1}{u^2 + 1} \, du + \int_0^{3x} \dfrac{u^2 - 1}{u^2 + 1} \, du = -\int_0^{2x} \dfrac{u^2 - 1}{u^2 + 1} \, du + \int_0^{3x} \dfrac{u^2 - 1}{u^2 + 1} \, du \quad \Rightarrow$

$g'(x) = -\dfrac{(2x)^2 - 1}{(2x)^2 + 1} \cdot \dfrac{d}{dx}(2x) + \dfrac{(3x)^2 - 1}{(3x)^2 + 1} \cdot \dfrac{d}{dx}(3x) = -2 \cdot \dfrac{4x^2 - 1}{4x^2 + 1} + 3 \cdot \dfrac{9x^2 - 1}{9x^2 + 1}$

55. $y = \int_{\sqrt{x}}^{x^3} \sqrt{t} \sin t \, dt = \int_{\sqrt{x}}^1 \sqrt{t} \sin t \, dt + \int_1^{x^3} \sqrt{t} \sin t \, dt = -\int_1^{\sqrt{x}} \sqrt{t} \sin t \, dt + \int_1^{x^3} \sqrt{t} \sin t \, dt \quad \Rightarrow$

$y' = -\sqrt[4]{x}\,(\sin \sqrt{x}\,) \cdot \dfrac{d}{dx}\left(\sqrt{x}\,\right) + x^{3/2} \sin(x^3) \cdot \dfrac{d}{dx}\left(x^3\right) = -\dfrac{\sqrt[4]{x} \sin \sqrt{x}}{2\sqrt{x}} + x^{3/2} \sin(x^3)(3x^2)$

$= 3x^{7/2} \sin(x^3) - \dfrac{\sin \sqrt{x}}{2\sqrt[4]{x}}$

57. $F(x) = \int_1^x f(t) \, dt \quad \Rightarrow \quad F'(x) = f(x) = \int_1^{x^2} \dfrac{\sqrt{1 + u^4}}{u} \, du \quad \left[\text{since } f(t) = \int_1^{t^2} \dfrac{\sqrt{1 + u^4}}{u} \, du\right] \quad \Rightarrow$

$F''(x) = f'(x) = \dfrac{\sqrt{1 + (x^2)^4}}{x^2} \cdot \dfrac{d}{dx}(x^2) = \dfrac{\sqrt{1 + x^8}}{x^2} \cdot 2x = \dfrac{2\sqrt{1 + x^8}}{x}$. So $F''(2) = \sqrt{1 + 2^8} = \sqrt{257}$.

59. By FTC2, $\int_1^4 f'(x) \, dx = f(4) - f(1)$, so $17 = f(4) - 12 \quad \Rightarrow \quad f(4) = 17 + 12 = 29.$

61. (a) The Fresnel function $S(x) = \int_0^x \sin\left(\frac{\pi}{2}t^2\right) dt$ has local maximum values where $0 = S'(x) = \sin\left(\frac{\pi}{2}t^2\right)$ and

S' changes from positive to negative. For $x > 0$, this happens when $\frac{\pi}{2}x^2 = (2n - 1)\pi$ [odd multiples of π] $\Leftrightarrow$

$x^2 = 2(2n - 1) \quad \Leftrightarrow \quad x = \sqrt{4n - 2}$, n any positive integer. For $x < 0$, S' changes from positive to negative where

$\frac{\pi}{2}x^2 = 2n\pi$ [even multiples of π] $\Leftrightarrow \quad x^2 = 4n \quad \Leftrightarrow \quad x = -2\sqrt{n}$. S' does not change sign at $x = 0$.

(b) S is concave upward on those intervals where $S''(x) > 0$. Differentiating our expression for $S'(x)$, we get

$S''(x) = \cos\left(\frac{\pi}{2}x^2\right)\left(2\frac{\pi}{2}x\right) = \pi x \cos\left(\frac{\pi}{2}x^2\right)$. For $x > 0$, $S''(x) > 0$ where $\cos\left(\frac{\pi}{2}x^2\right) > 0 \quad \Leftrightarrow \quad 0 < \frac{\pi}{2}x^2 < \frac{\pi}{2}$ or

$\left(2n - \frac{1}{2}\right)\pi < \frac{\pi}{2}x^2 < \left(2n + \frac{1}{2}\right)\pi$, n any integer $\Leftrightarrow \quad 0 < x < 1$ or $\sqrt{4n - 1} < x < \sqrt{4n + 1}$, n any positive integer.

For $x < 0$, $S''(x) > 0$ where $\cos\left(\frac{\pi}{2}x^2\right) < 0 \quad \Leftrightarrow \quad \left(2n - \frac{3}{2}\right)\pi < \frac{\pi}{2}x^2 < \left(2n - \frac{1}{2}\right)\pi$, n any integer $\Leftrightarrow$

$4n - 3 < x^2 < 4n - 1 \quad \Leftrightarrow \quad \sqrt{4n - 3} < |x| < \sqrt{4n - 1} \quad \Rightarrow \quad \sqrt{4n - 3} < -x < \sqrt{4n - 1} \quad \Rightarrow$

$-\sqrt{4n-3} > x > -\sqrt{4n-1}$, so the intervals of upward concavity for $x < 0$ are $\left(-\sqrt{4n-1}, -\sqrt{4n-3}\right)$, n any positive integer. To summarize: S is concave upward on the intervals $(0, 1)$, $\left(-\sqrt{3}, -1\right)$, $\left(\sqrt{3}, \sqrt{5}\right)$, $\left(-\sqrt{7}, -\sqrt{5}\right)$, $\left(\sqrt{7}, 3\right)$, ...

(c) In Maple, we use `plot({int(sin(Pi*t^2/2),t=0..x),0.2},x=0..2);`. Note that Maple recognizes the Fresnel function, calling it `FresnelS(x)`. In Mathematica, we use `Plot[{Integrate[Sin[Pi*t^2/2],{t,0,x}],0.2},{x,0,2}]`. In Derive, we load the utility file FRESNEL and plot FRESNEL_SIN(x). From the graphs, we see that $\int_0^x \sin\left(\frac{\pi}{2}t^2\right) dt = 0.2$ at $x \approx 0.74$.

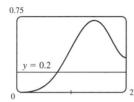
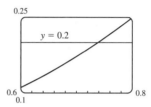

63. (a) By FTC1, $g'(x) = f(x)$. So $g'(x) = f(x) = 0$ at $x = 1, 3, 5, 7$, and 9. g has local maxima at $x = 1$ and 5 (since $f = g'$ changes from positive to negative there) and local minima at $x = 3$ and 7. There is no local maximum or minimum at $x = 9$, since f is not defined for $x > 9$.

(b) We can see from the graph that $\left|\int_0^1 f\, dt\right| < \left|\int_1^3 f\, dt\right| < \left|\int_3^5 f\, dt\right| < \left|\int_5^7 f\, dt\right| < \left|\int_7^9 f\, dt\right|$. So $g(1) = \left|\int_0^1 f\, dt\right|$, $g(5) = \int_0^5 f\, dt = g(1) - \left|\int_1^3 f\, dt\right| + \left|\int_3^5 f\, dt\right|$, and $g(9) = \int_0^9 f\, dt = g(5) - \left|\int_5^7 f\, dt\right| + \left|\int_7^9 f\, dt\right|$. Thus, $g(1) < g(5) < g(9)$, and so the absolute maximum of $g(x)$ occurs at $x = 9$.

(c) g is concave downward on those intervals where $g'' < 0$. But $g'(x) = f(x)$, so $g''(x) = f'(x)$, which is negative on (approximately) $\left(\frac{1}{2}, 2\right)$, $(4, 6)$ and $(8, 9)$. So g is concave downward on these intervals.

(d)

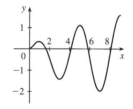

65. $\displaystyle \lim_{n\to\infty} \sum_{i=1}^{n} \frac{i^3}{n^4} = \lim_{n\to\infty} \frac{1-0}{n} \sum_{i=1}^{n} \left(\frac{i}{n}\right)^3 = \int_0^1 x^3\, dx = \left[\frac{x^4}{4}\right]_0^1 = \frac{1}{4}$

67. Suppose $h < 0$. Since f is continuous on $[x+h, x]$, the Extreme Value Theorem says that there are numbers u and v in $[x+h, x]$ such that $f(u) = m$ and $f(v) = M$, where m and M are the absolute minimum and maximum values of f on $[x+h, x]$. By Property 8 of integrals, $m(-h) \le \int_{x+h}^{x} f(t)\, dt \le M(-h)$; that is, $f(u)(-h) \le -\int_x^{x+h} f(t)\, dt \le f(v)(-h)$. Since $-h > 0$, we can divide this inequality by $-h$: $f(u) \le \frac{1}{h} \int_x^{x+h} f(t)\, dt \le f(v)$. By Equation 2,

$\dfrac{g(x+h) - g(x)}{h} = \dfrac{1}{h} \int_x^{x+h} f(t)\, dt$ for $h \ne 0$, and hence $f(u) \le \dfrac{g(x+h) - g(x)}{h} \le f(v)$, which is Equation 3 in the case where $h < 0$.

69. (a) Let $f(x) = \sqrt{x} \Rightarrow f'(x) = 1/(2\sqrt{x}) > 0$ for $x > 0 \Rightarrow f$ is increasing on $(0, \infty)$. If $x \geq 0$, then $x^3 \geq 0$, so

$1 + x^3 \geq 1$ and since f is increasing, this means that $f(1 + x^3) \geq f(1) \Rightarrow \sqrt{1 + x^3} \geq 1$ for $x \geq 0$. Next let

$g(t) = t^2 - t \Rightarrow g'(t) = 2t - 1 \Rightarrow g'(t) > 0$ when $t \geq 1$. Thus, g is increasing on $(1, \infty)$. And since $g(1) = 0$,

$g(t) \geq 0$ when $t \geq 1$. Now let $t = \sqrt{1 + x^3}$, where $x \geq 0$. $\sqrt{1 + x^3} \geq 1$ (from above) $\Rightarrow t \geq 1 \Rightarrow g(t) \geq 0 \Rightarrow$

$(1 + x^3) - \sqrt{1 + x^3} \geq 0$ for $x \geq 0$. Therefore, $1 \leq \sqrt{1 + x^3} \leq 1 + x^3$ for $x \geq 0$.

(b) From part (a) and Property 7: $\int_0^1 1\, dx \leq \int_0^1 \sqrt{1 + x^3}\, dx \leq \int_0^1 (1 + x^3)\, dx \Leftrightarrow$

$\left[x\right]_0^1 \leq \int_0^1 \sqrt{1 + x^3}\, dx \leq \left[x + \frac{1}{4}x^4\right]_0^1 \Leftrightarrow 1 \leq \int_0^1 \sqrt{1 + x^3}\, dx \leq 1 + \frac{1}{4} = 1.25$.

71. $0 < \dfrac{x^2}{x^4 + x^2 + 1} < \dfrac{x^2}{x^4} = \dfrac{1}{x^2}$ on $[5, 10]$, so

$0 \leq \displaystyle\int_5^{10} \dfrac{x^2}{x^4 + x^2 + 1}\, dx < \int_5^{10} \dfrac{1}{x^2}\, dx = \left[-\dfrac{1}{x}\right]_5^{10} = -\dfrac{1}{10} - \left(-\dfrac{1}{5}\right) = \dfrac{1}{10} = 0.1$.

73. Using FTC1, we differentiate both sides of $6 + \displaystyle\int_a^x \dfrac{f(t)}{t^2}\, dt = 2\sqrt{x}$ to get $\dfrac{f(x)}{x^2} = 2\dfrac{1}{2\sqrt{x}} \Rightarrow f(x) = x^{3/2}$.

To find a, we substitute $x = a$ in the original equation to obtain $6 + \displaystyle\int_a^a \dfrac{f(t)}{t^2}\, dt = 2\sqrt{a} \Rightarrow 6 + 0 = 2\sqrt{a} \Rightarrow$

$3 = \sqrt{a} \Rightarrow a = 9$.

75. (a) Let $F(t) = \int_0^t f(s)\, ds$. Then, by FTC1, $F'(t) = f(t) =$ rate of depreciation, so $F(t)$ represents the loss in value over the

interval $[0, t]$.

(b) $C(t) = \dfrac{1}{t}\left[A + \displaystyle\int_0^t f(s)\, ds\right] = \dfrac{A + F(t)}{t}$ represents the average expenditure per unit of t during the interval $[0, t]$,

assuming that there has been only one overhaul during that time period. The company wants to minimize average

expenditure.

(c) $C(t) = \dfrac{1}{t}\left[A + \displaystyle\int_0^t f(s)\, ds\right]$. Using FTC1, we have $C'(t) = -\dfrac{1}{t^2}\left[A + \displaystyle\int_0^t f(s)\, ds\right] + \dfrac{1}{t}f(t)$.

$C'(t) = 0 \Rightarrow t\, f(t) = A + \displaystyle\int_0^t f(s)\, ds \Rightarrow f(t) = \dfrac{1}{t}\left[A + \displaystyle\int_0^t f(s)\, ds\right] = C(t)$.

5.4   Indefinite Integrals and the Net Change Theorem

1. $\dfrac{d}{dx}\left[\sqrt{x^2 + 1} + C\right] = \dfrac{d}{dx}\left[(x^2 + 1)^{1/2} + C\right] = \frac{1}{2}(x^2 + 1)^{-1/2} \cdot 2x + 0 = \dfrac{x}{\sqrt{x^2 + 1}}$

3. $\dfrac{d}{dx}\left[\sin x - \frac{1}{3}\sin^3 x + C\right] = \dfrac{d}{dx}\left[\sin x - \frac{1}{3}(\sin x)^3 + C\right] = \cos x - \frac{1}{3} \cdot 3(\sin x)^2(\cos x) + 0$

$= \cos x(1 - \sin^2 x) = \cos x(\cos^2 x) = \cos^3 x$

5. $\displaystyle\int (x^2 + x^{-2})\, dx = \dfrac{x^3}{3} + \dfrac{x^{-1}}{-1} + C = \frac{1}{3}x^3 - \dfrac{1}{x} + C$

7. $\int (x^4 - \frac{1}{2}x^3 + \frac{1}{4}x - 2)\, dx = \dfrac{x^5}{5} - \dfrac{1}{2}\dfrac{x^4}{4} + \dfrac{1}{4}\dfrac{x^2}{2} - 2x + C = \frac{1}{5}x^5 - \frac{1}{8}x^4 + \frac{1}{8}x^2 - 2x + C$

9. $\int (1-t)(2+t^2)\, dt = \int (2 - 2t + t^2 - t^3)\, dt = 2t - 2\dfrac{t^2}{2} + \dfrac{t^3}{3} - \dfrac{t^4}{4} + C = 2t - t^2 + \frac{1}{3}t^3 - \frac{1}{4}t^4 + C$

11. $\int \dfrac{x^3 - 2\sqrt{x}}{x}\, dx = \int \left(\dfrac{x^3}{x} - \dfrac{2x^{1/2}}{x} \right) dx = \int (x^2 - 2x^{-1/2})\, dx = \dfrac{x^3}{3} - 2\dfrac{x^{1/2}}{1/2} + C = \frac{1}{3}x^3 - 4\sqrt{x} + C$

13. $\int (\sin x + \sinh x)\, dx = -\cos x + \cosh x + C$

15. $\int (\theta - \csc\theta \cot\theta)\, d\theta = \frac{1}{2}\theta^2 + \csc\theta + C$

17. $\int (1 + \tan^2 \alpha)\, d\alpha = \int \sec^2 \alpha\, d\alpha = \tan\alpha + C$

19. $\int \left(\cos x + \frac{1}{2}x \right) dx = \sin x + \frac{1}{4}x^2 + C$. The members of the family in the figure correspond to $C = -5, 0, 5,$ and 10.

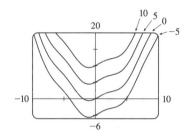

21. $\int_0^2 (6x^2 - 4x + 5)\, dx = \left[6 \cdot \frac{1}{3}x^3 - 4 \cdot \frac{1}{2}x^2 + 5x \right]_0^2 = \left[2x^3 - 2x^2 + 5x \right]_0^2 = (16 - 8 + 10) - 0 = 18$

23. $\int_{-1}^0 (2x - e^x)\, dx = \left[x^2 - e^x \right]_{-1}^0 = (0 - 1) - \left(1 - e^{-1} \right) = -2 + 1/e$

25. $\int_{-2}^2 (3u + 1)^2\, du = \int_{-2}^2 \left(9u^2 + 6u + 1 \right) du = \left[9 \cdot \frac{1}{3}u^3 + 6 \cdot \frac{1}{2}u^2 + u \right]_{-2}^2 = \left[3u^3 + 3u^2 + u \right]_{-2}^2$

$\qquad = (24 + 12 + 2) - (-24 + 12 - 2) = 38 - (-14) = 52$

27. $\int_1^4 \sqrt{t}\,(1+t)\, dt = \int_1^4 (t^{1/2} + t^{3/2})\, dt = \left[\frac{2}{3}t^{3/2} + \frac{2}{5}t^{5/2} \right]_1^4 = \left(\frac{16}{3} + \frac{64}{5} \right) - \left(\frac{2}{3} + \frac{2}{5} \right) = \frac{14}{3} + \frac{62}{5} = \frac{256}{15}$

29. $\int_{-2}^{-1} \left(4y^3 + \dfrac{2}{y^3} \right) dy = \left[4 \cdot \frac{1}{4}y^4 + 2 \cdot \dfrac{1}{-2}y^{-2} \right]_{-2}^{-1} = \left[y^4 - \dfrac{1}{y^2} \right]_{-2}^{-1} = (1 - 1) - \left(16 - \frac{1}{4} \right) = -\frac{63}{4}$

31. $\int_0^1 x\left(\sqrt[3]{x} + \sqrt[4]{x} \right) dx = \int_0^1 (x^{4/3} + x^{5/4})\, dx = \left[\frac{3}{7}x^{7/3} + \frac{4}{9}x^{9/4} \right]_0^1 = \left(\frac{3}{7} + \frac{4}{9} \right) - 0 = \frac{55}{63}$

33. $\int_1^4 \sqrt{5/x}\, dx = \sqrt{5} \int_1^4 x^{-1/2}\, dx = \sqrt{5} \left[2\sqrt{x} \right]_1^4 = \sqrt{5}\,(2 \cdot 2 - 2 \cdot 1) = 2\sqrt{5}$

35. $\int_0^\pi (4\sin\theta - 3\cos\theta)\, d\theta = \left[-4\cos\theta - 3\sin\theta \right]_0^\pi = (4 - 0) - (-4 - 0) = 8$

37. $\displaystyle\int_0^{\pi/4} \dfrac{1 + \cos^2\theta}{\cos^2\theta}\, d\theta = \int_0^{\pi/4} \left(\dfrac{1}{\cos^2\theta} + \dfrac{\cos^2\theta}{\cos^2\theta} \right) d\theta = \int_0^{\pi/4} (\sec^2\theta + 1)\, d\theta$

$\qquad = \left[\tan\theta + \theta \right]_0^{\pi/4} = \left(\tan\frac{\pi}{4} + \frac{\pi}{4} \right) - (0 + 0) = 1 + \frac{\pi}{4}$

39. $\int_1^{64} \frac{1 + \sqrt[3]{x}}{\sqrt{x}}\, dx = \int_1^{64} \left(\frac{1}{x^{1/2}} + \frac{x^{1/3}}{x^{1/2}} \right) dx = \int_1^{64} \left(x^{-1/2} + x^{(1/3)-(1/2)} \right) dx = \int_1^{64} \left(x^{-1/2} + x^{-1/6} \right) dx$

$$= \left[2x^{1/2} + \tfrac{6}{5}x^{5/6} \right]_1^{64} = \left(16 + \tfrac{192}{5} \right) - \left(2 + \tfrac{6}{5} \right) = 14 + \tfrac{186}{5} = \tfrac{256}{5}$$

41. $\int_0^{1/\sqrt{3}} \frac{t^2 - 1}{t^4 - 1}\, dt = \int_0^{1/\sqrt{3}} \frac{t^2 - 1}{(t^2 + 1)(t^2 - 1)}\, dt = \int_0^{1/\sqrt{3}} \frac{1}{t^2 + 1}\, dt = \left[\arctan t \right]_0^{1/\sqrt{3}} = \arctan \left(1/\sqrt{3} \right) - \arctan 0$

$$= \tfrac{\pi}{6} - 0 = \tfrac{\pi}{6}$$

43. $\int_{-1}^{2} (x - 2\,|x|)\, dx = \int_{-1}^{0}[x - 2(-x)]\, dx + \int_0^2 [x - 2(x)]\, dx = \int_{-1}^0 3x\, dx + \int_0^2 (-x)\, dx = 3\left[\tfrac{1}{2}x^2 \right]_{-1}^0 - \left[\tfrac{1}{2}x^2 \right]_0^2$

$$= 3\left(0 - \tfrac{1}{2} \right) - (2 - 0) = -\tfrac{7}{2} = -3.5$$

45. The graph shows that $y = x + x^2 - x^4$ has x-intercepts at $x = 0$ and at $x = a \approx 1.32$. So the area of the region that lies under the curve and above the x-axis is

$\int_0^a \left(x + x^2 - x^4 \right) dx = \left[\tfrac{1}{2}x^2 + \tfrac{1}{3}x^3 - \tfrac{1}{5}x^5 \right]_0^a$

$= \left(\tfrac{1}{2}a^2 + \tfrac{1}{3}a^3 - \tfrac{1}{5}a^5 \right) - 0 \approx 0.84$

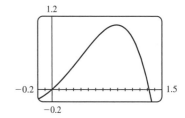

47. $A = \int_0^2 \left(2y - y^2 \right) dy = \left[y^2 - \tfrac{1}{3}y^3 \right]_0^2 = \left(4 - \tfrac{8}{3} \right) - 0 = \tfrac{4}{3}$

49. If $w'(t)$ is the rate of change of weight in pounds per year, then $w(t)$ represents the weight in pounds of the child at age t. We know from the Net Change Theorem that $\int_5^{10} w'(t)\, dt = w(10) - w(5)$, so the integral represents the increase in the child's weight (in pounds) between the ages of 5 and 10.

51. Since $r(t)$ is the rate at which oil leaks, we can write $r(t) = -V'(t)$, where $V(t)$ is the volume of oil at time t. [Note that the minus sign is needed because V is decreasing, so $V'(t)$ is negative, but $r(t)$ is positive.] Thus, by the Net Change Theorem, $\int_0^{120} r(t)\, dt = -\int_0^{120} V'(t)\, dt = -[V(120) - V(0)] = V(0) - V(120)$, which is the number of gallons of oil that leaked from the tank in the first two hours (120 minutes).

53. By the Net Change Theorem, $\int_{1000}^{5000} R'(x)\, dx = R(5000) - R(1000)$, so it represents the increase in revenue when production is increased from 1000 units to 5000 units.

55. In general, the unit of measurement for $\int_a^b f(x)\, dx$ is the product of the unit for $f(x)$ and the unit for x. Since $f(x)$ is measured in newtons and x is measured in meters, the units for $\int_0^{100} f(x)\, dx$ are newton-meters. (A newton-meter is abbreviated N·m.)

57. (a) Displacement $= \int_0^3 (3t - 5)\, dt = \left[\tfrac{3}{2}t^2 - 5t \right]_0^3 = \tfrac{27}{2} - 15 = -\tfrac{3}{2}$ m

(b) Distance traveled $= \int_0^3 |3t - 5|\, dt = \int_0^{5/3} (5 - 3t)\, dt + \int_{5/3}^3 (3t - 5)\, dt$

$$= \left[5t - \tfrac{3}{2}t^2 \right]_0^{5/3} + \left[\tfrac{3}{2}t^2 - 5t \right]_{5/3}^3 = \tfrac{25}{3} - \tfrac{3}{2}\cdot\tfrac{25}{9} + \tfrac{27}{2} - 15 - \left(\tfrac{3}{2}\cdot\tfrac{25}{9} - \tfrac{25}{3} \right) = \tfrac{41}{6} \text{ m}$$

59. (a) $v'(t) = a(t) = t + 4 \;\Rightarrow\; v(t) = \tfrac{1}{2}t^2 + 4t + C \;\Rightarrow\; v(0) = C = 5 \;\Rightarrow\; v(t) = \tfrac{1}{2}t^2 + 4t + 5 \text{ m/s}$

(b) Distance traveled $= \int_0^{10} |v(t)|\, dt = \int_0^{10} \left| \tfrac{1}{2}t^2 + 4t + 5 \right| dt = \int_0^{10} \left(\tfrac{1}{2}t^2 + 4t + 5 \right) dt = \left[\tfrac{1}{6}t^3 + 2t^2 + 5t \right]_0^{10}$

$$= \tfrac{500}{3} + 200 + 50 = 416\tfrac{2}{3} \text{ m}$$

61. Since $m'(x) = \rho(x)$, $m = \int_0^4 \rho(x)\,dx = \int_0^4 \left(9 + 2\sqrt{x}\right)dx = \left[9x + \frac{4}{3}x^{3/2}\right]_0^4 = 36 + \frac{32}{3} - 0 = \frac{140}{3} = 46\frac{2}{3}$ kg.

63. Let s be the position of the car. We know from Equation 2 that $s(100) - s(0) = \int_0^{100} v(t)\,dt$. We use the Midpoint Rule for

$0 \le t \le 100$ with $n = 5$. Note that the length of each of the five time intervals is 20 seconds $= \frac{20}{3600}$ hour $= \frac{1}{180}$ hour.

So the distance traveled is

$\int_0^{100} v(t)\,dt \approx \frac{1}{180}[v(10) + v(30) + v(50) + v(70) + v(90)] = \frac{1}{180}(38 + 58 + 51 + 53 + 47) = \frac{247}{180} \approx 1.4$ miles.

65. From the Net Change Theorem, the increase in cost if the production level is raised

from 2000 yards to 4000 yards is $C(4000) - C(2000) = \int_{2000}^{4000} C'(x)\,dx$.

$\int_{2000}^{4000} C'(x)\,dx = \int_{2000}^{4000} (3 - 0.01x + 0.000006x^2)\,dx = \left[3x - 0.005x^2 + 0.000002x^3\right]_{2000}^{4000} = 60{,}000 - 2{,}000 = \$58{,}000$

67. (a) We can find the area between the Lorenz curve and the line $y = x$ by subtracting the area under $y = L(x)$ from the area

under $y = x$. Thus,

$$\text{coefficient of inequality} = \frac{\text{area between Lorenz curve and line } y = x}{\text{area under line } y = x} = \frac{\int_0^1 [x - L(x)]\,dx}{\int_0^1 x\,dx}$$

$$= \frac{\int_0^1 [x - L(x)]\,dx}{[x^2/2]_0^1} = \frac{\int_0^1 [x - L(x)]\,dx}{1/2} = 2\int_0^1 [x - L(x)]\,dx$$

(b) $L(x) = \frac{5}{12}x^2 + \frac{7}{12}x \;\Rightarrow\; L(50\%) = L\left(\frac{1}{2}\right) = \frac{5}{48} + \frac{7}{24} = \frac{19}{48} = 0.3958\overline{3}$, so the bottom 50% of the households receive

at most about 40% of the income. Using the result in part (a),

$$\text{coefficient of inequality} = 2\int_0^1 [x - L(x)]\,dx = 2\int_0^1 \left(x - \frac{5}{12}x^2 - \frac{7}{12}x\right)dx = 2\int_0^1 \left(\frac{5}{12}x - \frac{5}{12}x^2\right)dx$$

$$= 2\int_0^1 \frac{5}{12}(x - x^2)\,dx = \frac{5}{6}\left[\frac{1}{2}x^2 - \frac{1}{3}x^3\right]_0^1 = \frac{5}{6}\left(\frac{1}{2} - \frac{1}{3}\right) = \frac{5}{6}\left(\frac{1}{6}\right) = \frac{5}{36}$$

5.5   The Substitution Rule

1. Let $u = -x$. Then $du = -dx$, so $dx = -du$. Thus, $\int e^{-x}dx = \int e^u(-du) = -e^u + C = -e^{-x} + C$. Don't forget that it

is often very easy to check an indefinite integration by differentiating your answer. In this case,

$\dfrac{d}{dx}(-e^{-x} + C) = -[e^{-x}(-1)] = e^{-x}$, the desired result.

3. Let $u = x^3 + 1$. Then $du = 3x^2\,dx$ and $x^2\,dx = \frac{1}{3}du$, so

$$\int x^2\sqrt{x^3 + 1}\,dx = \int \sqrt{u}\left(\tfrac{1}{3}\,du\right) = \frac{1}{3}\frac{u^{3/2}}{3/2} + C = \frac{1}{3}\cdot\frac{2}{3}u^{3/2} + C = \tfrac{2}{9}(x^3 + 1)^{3/2} + C.$$

5. Let $u = \cos\theta$. Then $du = -\sin\theta\,d\theta$ and $\sin\theta\,d\theta = -du$, so

$$\int \cos^3\theta\,\sin\theta\,d\theta = \int u^3\,(-du) = -\frac{u^4}{4} + C = -\tfrac{1}{4}\cos^4\theta + C.$$

7. Let $u = x^2$. Then $du = 2x\,dx$ and $x\,dx = \frac{1}{2}du$, so $\int x\sin(x^2)\,dx = \int \sin u\left(\tfrac{1}{2}\,du\right) = -\tfrac{1}{2}\cos u + C = -\tfrac{1}{2}\cos(x^2) + C.$

9. Let $u = 3x - 2$. Then $du = 3\,dx$ and $dx = \frac{1}{3}\,du$, so $\int (3x - 2)^{20}\,dx = \int u^{20}\left(\frac{1}{3}\,du\right) = \frac{1}{3} \cdot \frac{1}{21} u^{21} + C = \frac{1}{63}(3x - 2)^{21} + C$.

11. Let $u = 2x + x^2$. Then $du = (2 + 2x)\,dx = 2(1 + x)\,dx$ and $(x + 1)\,dx = \frac{1}{2}\,du$, so

$$\int (x + 1)\sqrt{2x + x^2}\,dx = \int \sqrt{u}\,\left(\tfrac{1}{2}\,du\right) = \frac{1}{2}\frac{u^{3/2}}{3/2} + C = \tfrac{1}{3}\left(2x + x^2\right)^{3/2} + C.$$

Or: Let $u = \sqrt{2x + x^2}$. Then $u^2 = 2x + x^2 \ \Rightarrow \ 2u\,du = (2 + 2x)\,dx \ \Rightarrow \ u\,du = (1 + x)\,dx$, so

$\int (x + 1)\,\sqrt{2x + x^2}\,dx = \int u \cdot u\,du = \int u^2\,du = \tfrac{1}{3}u^3 + C = \tfrac{1}{3}(2x + x^2)^{3/2} + C.$

13. Let $u = 5 - 3x$. Then $du = -3\,dx$ and $dx = -\frac{1}{3}\,du$, so

$$\int \frac{dx}{5 - 3x} = \int \frac{1}{u}\left(-\tfrac{1}{3}\,du\right) = -\tfrac{1}{3}\ln|u| + C = -\tfrac{1}{3}\ln|5 - 3x| + C.$$

15. Let $u = \pi t$. Then $du = \pi\,dt$ and $dt = \frac{1}{\pi}\,du$, so $\int \sin \pi t\,dt = \int \sin u\,\left(\frac{1}{\pi}\,du\right) = \frac{1}{\pi}(-\cos u) + C = -\frac{1}{\pi}\cos \pi t + C$.

17. Let $u = 3ax + bx^3$. Then $du = (3a + 3bx^2)\,dx = 3(a + bx^2)\,dx$, so

$$\int \frac{a + bx^2}{\sqrt{3ax + bx^3}}\,dx = \int \frac{\frac{1}{3}\,du}{u^{1/2}} = \frac{1}{3}\int u^{-1/2}\,du = \tfrac{1}{3} \cdot 2u^2 + C = \tfrac{2}{3}\sqrt{3ax + bx^3} + C.$$

19. Let $u = \ln x$. Then $du = \dfrac{dx}{x}$, so $\displaystyle\int \frac{(\ln x)^2}{x}\,dx = \int u^2\,du = \tfrac{1}{3}u^3 + C = \tfrac{1}{3}(\ln x)^3 + C$.

21. Let $u = \sqrt{t}$. Then $du = \dfrac{dt}{2\sqrt{t}}$ and $\dfrac{1}{\sqrt{t}}\,dt = 2\,du$, so $\displaystyle\int \frac{\cos \sqrt{t}}{\sqrt{t}}\,dt = \int \cos u\,(2\,du) = 2\sin u + C = 2\sin \sqrt{t} + \dot{C}$.

23. Let $u = \sin \theta$. Then $du = \cos \theta\,d\theta$, so $\int \cos \theta\,\sin^6 \theta\,d\theta = \int u^6\,du = \tfrac{1}{7}u^7 + C = \tfrac{1}{7}\sin^7 \theta + C$.

25. Let $u = 1 + e^x$. Then $du = e^x\,dx$, so $\int e^x\sqrt{1 + e^x}\,dx = \int \sqrt{u}\,du = \tfrac{2}{3}u^{3/2} + C = \tfrac{2}{3}(1 + e^x)^{3/2} + C$.

Or: Let $u = \sqrt{1 + e^x}$. Then $u^2 = 1 + e^x$ and $2u\,du = e^x\,dx$, so

$\int e^x\sqrt{1 + e^x}\,dx = \int u \cdot 2u\,du = \tfrac{2}{3}u^3 + C = \tfrac{2}{3}(1 + e^x)^{3/2} + C.$

27. Let $u = 1 + z^3$. Then $du = 3z^2\,dz$ and $z^2\,dz = \frac{1}{3}\,du$, so

$$\int \frac{z^2}{\sqrt[3]{1 + z^3}}\,dz = \int u^{-1/3}\left(\tfrac{1}{3}\,du\right) = \tfrac{1}{3} \cdot \tfrac{3}{2}u^{2/3} + C = \tfrac{1}{2}(1 + z^3)^{2/3} + C.$$

29. Let $u = \tan x$. Then $du = \sec^2 x\,dx$, so $\int e^{\tan x}\sec^2 x\,dx = \int e^u\,du = e^u + C = e^{\tan x} + C$.

31. Let $u = \sin x$. Then $du = \cos x\,dx$, so $\displaystyle\int \frac{\cos x}{\sin^2 x}\,dx = \int \frac{1}{u^2}\,du = \int u^{-2}\,du = \frac{u^{-1}}{-1} + C = -\frac{1}{u} + C = -\frac{1}{\sin x} + C$

[or $-\csc x + C$].

33. Let $u = \cot x$. Then $du = -\csc^2 x\,dx$ and $\csc^2 x\,dx = -du$, so

$$\int \sqrt{\cot x}\,\csc^2 x\,dx = \int \sqrt{u}\,(-du) = -\frac{u^{3/2}}{3/2} + C = -\tfrac{2}{3}(\cot x)^{3/2} + C.$$

35. $\int \dfrac{\sin 2x}{1 + \cos^2 x}\, dx = 2 \int \dfrac{\sin x \cos x}{1 + \cos^2 x}\, dx = 2I$. Let $u = \cos x$. Then $du = -\sin x\, dx$, so

$$2I = -2 \int \frac{u\, du}{1 + u^2} = -2 \cdot \tfrac{1}{2}\ln(1 + u^2) + C = -\ln(1 + u^2) + C = -\ln(1 + \cos^2 x) + C.$$

Or: Let $u = 1 + \cos^2 x$.

37. $\int \cot x\, dx = \int \dfrac{\cos x}{\sin x}\, dx$. Let $u = \sin x$. Then $du = \cos x\, dx$, so $\int \cot x\, dx = \int \dfrac{1}{u}\, du = \ln|u| + C = \ln|\sin x| + C.$

39. Let $u = \sec x$. Then $du = \sec x \tan x\, dx$, so

$$\int \sec^3 x \tan x\, dx = \int \sec^2 x\, (\sec x \tan x)\, dx = \int u^2\, du = \tfrac{1}{3}u^3 + C = \tfrac{1}{3}\sec^3 x + C.$$

41. Let $u = \sin^{-1} x$. Then $du = \dfrac{1}{\sqrt{1 - x^2}}\, dx$, so $\displaystyle\int \dfrac{dx}{\sqrt{1 - x^2}\,\sin^{-1} x} = \int \dfrac{1}{u}\, du = \ln|u| + C = \ln\left|\sin^{-1} x\right| + C.$

43. Let $u = 1 + x^2$. Then $du = 2x\, dx$, so

$$\int \frac{1 + x}{1 + x^2}\, dx = \int \frac{1}{1 + x^2}\, dx + \int \frac{x}{1 + x^2}\, dx = \tan^{-1} x + \int \frac{\tfrac{1}{2}\, du}{u} = \tan^{-1} x + \tfrac{1}{2}\ln|u| + C$$

$$= \tan^{-1} x + \tfrac{1}{2}\ln\left|1 + x^2\right| + C = \tan^{-1} x + \tfrac{1}{2}\ln\left(1 + x^2\right) + C \quad [\text{since } 1 + x^2 > 0].$$

45. Let $u = x + 2$. Then $du = dx$, so

$$\int \frac{x}{\sqrt[4]{x + 2}}\, dx = \int \frac{u - 2}{\sqrt[4]{u}}\, du = \int \left(u^{3/4} - 2u^{-1/4}\right) du = \tfrac{4}{7}u^{7/4} - 2 \cdot \tfrac{4}{3}u^{3/4} + C$$

$$= \tfrac{4}{7}(x + 2)^{7/4} - \tfrac{8}{3}(x + 2)^{3/4} + C$$

In Exercises 47–50, let $f(x)$ denote the integrand and $F(x)$ its antiderivative (with $C = 0$).

47. $f(x) = x(x^2 - 1)^3$. $\quad u = x^2 - 1 \;\Rightarrow\; du = 2x\, dx$, so

$$\int x(x^2 - 1)^3\, dx = \int u^3 \left(\tfrac{1}{2}\, du\right) = \tfrac{1}{8}u^4 + C = \tfrac{1}{8}(x^2 - 1)^4 + C$$

Where f is positive (negative), F is increasing (decreasing). Where f changes from negative to positive (positive to negative), F has a local minimum (maximum).

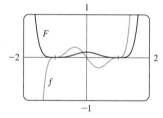

49. $f(x) = \sin^3 x \cos x$. $\quad u = \sin x \;\Rightarrow\; du = \cos x\, dx$, so

$$\int \sin^3 x \cos x\, dx = \int u^3\, du = \tfrac{1}{4}u^4 + C = \tfrac{1}{4}\sin^4 x + C$$

Note that at $x = \frac{\pi}{2}$, f changes from positive to negative and F has a local maximum. Also, both f and F are periodic with period π, so at $x = 0$ and at $x = \pi$, f changes from negative to positive and F has local minima.

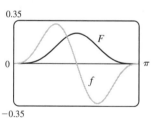

51. Let $u = x - 1$, so $du = dx$. When $x = 0$, $u = -1$; when $x = 2$, $u = 1$. Thus, $\int_0^2 (x - 1)^{25}\, dx = \int_{-1}^1 u^{25}\, du = 0$ by Theorem 7(b), since $f(u) = u^{25}$ is an odd function.

53. Let $u = 1 + 2x^3$, so $du = 6x^2\,dx$. When $x = 0$, $u = 1$; when $x = 1$, $u = 3$. Thus,

$$\int_0^1 x^2 (1 + 2x^3)^5 \, dx = \int_1^3 u^5 \left(\tfrac{1}{6}\,du\right) = \tfrac{1}{6}\left[\tfrac{1}{6} u^6\right]_1^3 = \tfrac{1}{36}(3^6 - 1^6) = \tfrac{1}{36}(729 - 1) = \tfrac{728}{36} = \tfrac{182}{9}.$$

55. Let $u = t/4$, so $du = \tfrac{1}{4}\,dt$. When $t = 0$, $u = 0$; when $t = \pi$, $u = \pi/4$. Thus,

$$\int_0^\pi \sec^2 (t/4) \, dt = \int_0^{\pi/4} \sec^2 u \, (4\,du) = 4\big[\tan u\big]_0^{\pi/4} = 4\big(\tan \tfrac{\pi}{4} - \tan 0\big) = 4(1 - 0) = 4.$$

57. $\int_{-\pi/6}^{\pi/6} \tan^3 \theta \, d\theta = 0$ by Theorem 7(b), since $f(\theta) = \tan^3 \theta$ is an odd function.

59. Let $u = 1/x$, so $du = -1/x^2\,dx$. When $x = 1$, $u = 1$; when $x = 2$, $u = \tfrac{1}{2}$. Thus,

$$\int_1^2 \frac{e^{1/x}}{x^2} \, dx = \int_1^{1/2} e^u \, (-du) = -\big[e^u\big]_1^{1/2} = -(e^{1/2} - e) = e - \sqrt{e}.$$

61. Let $u = 1 + 2x$, so $du = 2\,dx$. When $x = 0$, $u = 1$; when $x = 13$, $u = 27$. Thus,

$$\int_0^{13} \frac{dx}{\sqrt[3]{(1 + 2x)^2}} = \int_1^{27} u^{-2/3} \left(\tfrac{1}{2}\,du\right) = \left[\tfrac{1}{2} \cdot 3 u^{1/3}\right]_1^{27} = \tfrac{3}{2}(3 - 1) = 3.$$

63. Let $u = x^2 + a^2$, so $du = 2x\,dx$ and $x\,dx = \tfrac{1}{2}\,du$. When $x = 0$, $u = a^2$; when $x = a$, $u = 2a^2$. Thus,

$$\int_0^a x\sqrt{x^2 + a^2} \, dx = \int_{a^2}^{2a^2} u^{1/2} \left(\tfrac{1}{2}\,du\right) = \tfrac{1}{2}\left[\tfrac{2}{3} u^{3/2}\right]_{a^2}^{2a^2} = \left[\tfrac{1}{3} u^{3/2}\right]_{a^2}^{2a^2} = \tfrac{1}{3}\left[(2a^2)^{3/2} - (a^2)^{3/2}\right] = \tfrac{1}{3}\left(2\sqrt{2} - 1\right)a^3$$

65. Let $u = x - 1$, so $u + 1 = x$ and $du = dx$. When $x = 1$, $u = 0$; when $x = 2$, $u = 1$. Thus,

$$\int_1^2 x\sqrt{x - 1} \, dx = \int_0^1 (u + 1)\sqrt{u}\,du = \int_0^1 (u^{3/2} + u^{1/2}) \, du = \left[\tfrac{2}{5} u^{5/2} + \tfrac{2}{3} u^{3/2}\right]_0^1 = \tfrac{2}{5} + \tfrac{2}{3} = \tfrac{16}{15}.$$

67. Let $u = \ln x$, so $du = \dfrac{dx}{x}$. When $x = e$, $u = 1$; when $x = e^4$; $u = 4$. Thus,

$$\int_e^{e^4} \frac{dx}{x\sqrt{\ln x}} = \int_1^4 u^{-1/2} \, du = 2\big[u^{1/2}\big]_1^4 = 2(2 - 1) = 2.$$

69. Let $u = e^z + z$, so $du = (e^z + 1)\,dz$. When $z = 0$, $u = 1$; when $z = 1$, $u = e + 1$. Thus,

$$\int_0^1 \frac{e^z + 1}{e^z + z} \, dz = \int_1^{e+1} \frac{1}{u} \, du = \Big[\ln |u|\Big]_1^{e+1} = \ln |e + 1| - \ln |1| = \ln(e + 1).$$

71. From the graph, it appears that the area under the curve is about

$1 + \left(\text{a little more than } \tfrac{1}{2} \cdot 1 \cdot 0.7\right)$, or about 1.4. The exact area is given by

$A = \int_0^1 \sqrt{2x + 1}\,dx$. Let $u = 2x + 1$, so $du = 2\,dx$. The limits change to

$2 \cdot 0 + 1 = 1$ and $2 \cdot 1 + 1 = 3$, and

$A = \int_1^3 \sqrt{u}\left(\tfrac{1}{2}\,du\right) = \tfrac{1}{2}\left[\tfrac{2}{3} u^{3/2}\right]_1^3 = \tfrac{1}{3}\left(3\sqrt{3} - 1\right) = \sqrt{3} - \tfrac{1}{3} \approx 1.399$.

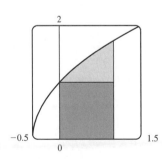

73. First write the integral as a sum of two integrals:

$I = \int_{-2}^2 (x + 3)\sqrt{4 - x^2} \, dx = I_1 + I_2 = \int_{-2}^2 x\sqrt{4 - x^2} \, dx + \int_{-2}^2 3\sqrt{4 - x^2} \, dx$. $I_1 = 0$ by Theorem 7(b), since

$f(x) = x\sqrt{4 - x^2}$ is an odd function and we are integrating from $x = -2$ to $x = 2$. We interpret I_2 as three times the area of

a semicircle with radius 2, so $I = 0 + 3 \cdot \tfrac{1}{2}\left(\pi \cdot 2^2\right) = 6\pi$.

75. First Figure   Let $u = \sqrt{x}$, so $x = u^2$ and $dx = 2u\,du$. When $x = 0$, $u = 0$; when $x = 1$, $u = 1$. Thus,

$$A_1 = \int_0^1 e^{\sqrt{x}}\,dx = \int_0^1 e^u(2u\,du) = 2\int_0^1 ue^u\,du.$$

Second Figure $A_2 = \int_0^1 2xe^x\,dx = 2\int_0^1 ue^u\,du.$

Third Figure   Let $u = \sin x$, so $du = \cos x\,dx$. When $x = 0$, $u = 0$; when $x = \frac{\pi}{2}$, $u = 1$. Thus,

$$A_3 = \int_0^{\pi/2} e^{\sin x}\sin 2x\,dx = \int_0^{\pi/2} e^{\sin x}(2\sin x\,\cos x)\,dx = \int_0^1 e^u(2u\,du) = 2\int_0^1 ue^u\,du.$$

Since $A_1 = A_2 = A_3$, all three areas are equal.

77. The rate is measured in liters per minute. Integrating from $t = 0$ minutes to $t = 60$ minutes will give us the total amount of oil that leaks out (in liters) during the first hour.

$$\int_0^{60} r(t)\,dt = \int_0^{60} 100e^{-0.01t}\,dt \qquad [u = -0.01t, du = -0.01dt]$$

$$= 100\int_0^{-0.6} e^u(-100\,du) = -10{,}000\left[e^u\right]_0^{-0.6} = -10{,}000(e^{-0.6} - 1) \approx 4511.9 \approx 4512 \text{ liters}$$

79. The volume of inhaled air in the lungs at time t is

$$V(t) = \int_0^t f(u)\,du = \int_0^t \tfrac{1}{2}\sin\!\left(\tfrac{2\pi}{5}u\right)du = \int_0^{2\pi t/5} \tfrac{1}{2}\sin v \left(\tfrac{5}{2\pi}\,dv\right) \qquad \left[\text{substitute } v = \tfrac{2\pi}{5}u, dv = \tfrac{2\pi}{5}\,du\right]$$

$$= \tfrac{5}{4\pi}\left[-\cos v\right]_0^{2\pi t/5} = \tfrac{5}{4\pi}\left[-\cos\!\left(\tfrac{2\pi}{5}t\right) + 1\right] = \tfrac{5}{4\pi}\left[1 - \cos\!\left(\tfrac{2\pi}{5}t\right)\right] \text{ liters}$$

81. Let $u = 2x$. Then $du = 2\,dx$, so $\int_0^2 f(2x)\,dx = \int_0^4 f(u)\left(\tfrac{1}{2}\,du\right) = \tfrac{1}{2}\int_0^4 f(u)\,du = \tfrac{1}{2}(10) = 5.$

83. Let $u = -x$. Then $du = -dx$, so

$$\int_a^b f(-x)\,dx = \int_{-a}^{-b} f(u)(-du) = \int_{-b}^{-a} f(u)\,du = \int_{-b}^{-a} f(x)\,dx$$

From the diagram, we see that the equality follows from the fact that we are reflecting the graph of f, and the limits of integration, about the y-axis.

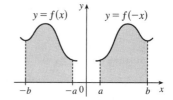

85. Let $u = 1 - x$. Then $x = 1 - u$ and $dx = -du$, so

$$\int_0^1 x^a(1-x)^b\,dx = \int_1^0 (1-u)^a\,u^b(-du) = \int_0^1 u^b(1-u)^a\,du = \int_0^1 x^b(1-x)^a\,dx.$$

87. $\dfrac{x\sin x}{1 + \cos^2 x} = x \cdot \dfrac{\sin x}{2 - \sin^2 x} = x\,f(\sin x)$, where $f(t) = \dfrac{t}{2 - t^2}$. By Exercise 86,

$$\int_0^\pi \frac{x\sin x}{1 + \cos^2 x}\,dx = \int_0^\pi x\,f(\sin x)\,dx = \frac{\pi}{2}\int_0^\pi f(\sin x)\,dx = \frac{\pi}{2}\int_0^\pi \frac{\sin x}{1 + \cos^2 x}\,dx$$

Let $u = \cos x$. Then $du = -\sin x\,dx$. When $x = \pi$, $u = -1$ and when $x = 0$, $u = 1$. So

$$\frac{\pi}{2}\int_0^\pi \frac{\sin x}{1 + \cos^2 x}\,dx = -\frac{\pi}{2}\int_1^{-1} \frac{du}{1 + u^2} = \frac{\pi}{2}\int_{-1}^1 \frac{du}{1 + u^2} = \frac{\pi}{2}\left[\tan^{-1} u\right]_{-1}^1$$

$$= \frac{\pi}{2}[\tan^{-1} 1 - \tan^{-1}(-1)] = \frac{\pi}{2}\left[\frac{\pi}{4} - \left(-\frac{\pi}{4}\right)\right] = \frac{\pi^2}{4}$$

5   Review

1. (a) $\sum_{i=1}^{n} f(x_i^*) \, \Delta x$ is an expression for a Riemann sum of a function f.

   x_i^* is a point in the ith subinterval $[x_{i-1}, x_i]$ and Δx is the length of the subintervals.

   (b) See Figure 1 in Section 5.2.

   (c) In Section 5.2, see Figure 3 and the paragraph beside it.

2. (a) See Definition 5.2.2.

   (b) See Figure 2 in Section 5.2.

   (c) In Section 5.2, see Figure 4 and the paragraph by it (contains "**net area**").

3. See the Fundamental Theorem of Calculus after Example 9 in Section 5.3.

4. (a) See the Net Change Theorem after Example 5 in Section 5.4.

   (b) $\int_{t_1}^{t_2} r(t) \, dt$ represents the change in the amount of water in the reservoir between time t_1 and time t_2.

5. (a) $\int_{60}^{120} v(t) \, dt$ represents the change in position of the particle from $t = 60$ to $t = 120$ seconds.

   (b) $\int_{60}^{120} |v(t)| \, dt$ represents the total distance traveled by the particle from $t = 60$ to 120 seconds.

   (c) $\int_{60}^{120} a(t) \, dt$ represents the change in the velocity of the particle from $t = 60$ to $t = 120$ seconds.

6. (a) $\int f(x) \, dx$ is the family of functions $\{F \mid F' = f\}$. Any two such functions differ by a constant.

   (b) The connection is given by the Net Change Theorem: $\int_a^b f(x) \, dx = \left[\int f(x) \, dx\right]_a^b$ if f is continuous.

7. The precise version of this statement is given by the Fundamental Theorem of Calculus. See the statement of this theorem and the paragraph that follows it at the end of Section 5.3.

8. See the Substitution Rule (5.5.4). This says that it is permissible to operate with the dx after an integral sign as if it were a differential.

1. True by Property 2 of the Integral in Section 5.2.

3. True by Property 3 of the Integral in Section 5.2.

5. False.   For example, let $f(x) = x^2$. Then $\int_0^1 \sqrt{x^2} \, dx = \int_0^1 x \, dx = \frac{1}{2}$, but $\sqrt{\int_0^1 x^2 \, dx} = \sqrt{\frac{1}{3}} = \frac{1}{\sqrt{3}}$.

7. True by Comparison Property 7 of the Integral in Section 5.2.

9. True.   The integrand is an odd function that is continuous on $[-1, 1]$, so the result follows from Theorem 5.5.7(b).

11. False. The function $f(x) = 1/x^4$ is not bounded on the interval $[-2, 1]$. It has an infinite discontinuity at $x = 0$, so it is not integrable on the interval. (If the integral were to exist, a positive value would be expected, by Comparison Property 6 of Integrals.)

13. False. For example, the function $y = |x|$ is continuous on $\mathbb{R}$, but has no derivative at $x = 0$.

15. False. $\int_a^b f(x)\, dx$ is a constant, so $\dfrac{d}{dx}\left(\int_a^b f(x)\, dx\right) = 0$, not $f(x)$ [unless $f(x) = 0$]. Compare the given statement carefully with FTC1, in which the upper limit in the integral is x.

EXERCISES

1. (a)

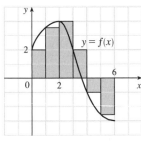

$$L_6 = \sum_{i=1}^{6} f(x_{i-1})\, \Delta x \quad [\Delta x = \tfrac{6-0}{6} = 1]$$

$$= f(x_0)\cdot 1 + f(x_1)\cdot 1 + f(x_2)\cdot 1 + f(x_3)\cdot 1 + f(x_4)\cdot 1 + f(x_5)\cdot 1$$

$$\approx 2 + 3.5 + 4 + 2 + (-1) + (-2.5) = 8$$

The Riemann sum represents the sum of the areas of the four rectangles above the x-axis minus the sum of the areas of the two rectangles below the x-axis.

(b)

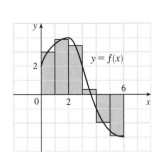

$$M_6 = \sum_{i=1}^{6} f(\overline{x}_i)\, \Delta x \quad [\Delta x = \tfrac{6-0}{6} = 1]$$

$$= f(\overline{x}_1)\cdot 1 + f(\overline{x}_2)\cdot 1 + f(\overline{x}_3)\cdot 1 + f(\overline{x}_4)\cdot 1 + f(\overline{x}_5)\cdot 1 + f(\overline{x}_6)\cdot 1$$

$$= f(0.5) + f(1.5) + f(2.5) + f(3.5) + f(4.5) + f(5.5)$$

$$\approx 3 + 3.9 + 3.4 + 0.3 + (-2) + (-2.9) = 5.7$$

3. $\int_0^1 \left(x + \sqrt{1 - x^2}\right) dx = \int_0^1 x\, dx + \int_0^1 \sqrt{1 - x^2}\, dx = I_1 + I_2$.

I_1 can be interpreted as the area of the triangle shown in the figure and I_2 can be interpreted as the area of the quarter-circle.

Area $= \frac{1}{2}(1)(1) + \frac{1}{4}(\pi)(1)^2 = \frac{1}{2} + \frac{\pi}{4}$.

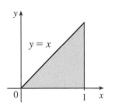
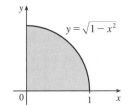

5. $\int_0^6 f(x)\, dx = \int_0^4 f(x)\, dx + \int_4^6 f(x)\, dx \;\Rightarrow\; 10 = 7 + \int_4^6 f(x)\, dx \;\Rightarrow\; \int_4^6 f(x)\, dx = 10 - 7 = 3$

7. First note that either a or b must be the graph of $\int_0^x f(t)\, dt$, since $\int_0^0 f(t)\, dt = 0$, and $c(0) \neq 0$. Now notice that $b > 0$ when c is increasing, and that $c > 0$ when a is increasing. It follows that c is the graph of $f(x)$, b is the graph of $f'(x)$, and a is the graph of $\int_0^x f(t)\, dt$.

9. $\int_1^2 \left(8x^3 + 3x^2\right) dx = \left[8 \cdot \tfrac{1}{4}x^4 + 3 \cdot \tfrac{1}{3}x^3\right]_1^2 = \left[2x^4 + x^3\right]_1^2 = \left(2 \cdot 2^4 + 2^3\right) - (2 + 1) = 40 - 3 = 37$

11. $\int_0^1 \left(1 - x^9\right) dx = \left[x - \tfrac{1}{10}x^{10}\right]_0^1 = \left(1 - \tfrac{1}{10}\right) - 0 = \tfrac{9}{10}$

13. $\int_1^9 \frac{\sqrt{u} - 2u^2}{u}\,du = \int_1^9 (u^{-1/2} - 2u)\,du = \left[2u^{1/2} - u^2\right]_1^9 = (6 - 81) - (2 - 1) = -76$

15. Let $u = y^2 + 1$, so $du = 2y\,dy$ and $y\,dy = \frac{1}{2}\,du$. When $y = 0$, $u = 1$; when $y = 1$, $u = 2$. Thus,

$\int_0^1 y(y^2 + 1)^5\,dy = \int_1^2 u^5\left(\frac{1}{2}\,du\right) = \frac{1}{2}\left[\frac{1}{6}u^6\right]_1^2 = \frac{1}{12}(64 - 1) = \frac{63}{12} = \frac{21}{4}$.

17. $\int_1^5 \frac{dt}{(t - 4)^2}$ does not exist because the function $f(t) = \frac{1}{(t - 4)^2}$ has an infinite discontinuity at $t = 4$;

that is, f is discontinuous on the interval $[1, 5]$.

19. Let $u = v^3$, so $du = 3v^2\,dv$. When $v = 0$, $u = 0$; when $v = 1$, $u = 1$. Thus,

$\int_0^1 v^2 \cos(v^3)\,dv = \int_0^1 \cos u\left(\frac{1}{3}\,du\right) = \frac{1}{3}\left[\sin u\right]_0^1 = \frac{1}{3}(\sin 1 - 0) = \frac{1}{3}\sin 1$.

21. $\int_{-\pi/4}^{\pi/4} \frac{t^4 \tan t}{2 + \cos t}\,dt = 0$ by Theorem 5.5.7(b), since $f(t) = \frac{t^4 \tan t}{2 + \cos t}$ is an odd function.

23. $\int \left(\frac{1 - x}{x}\right)^2 dx = \int \left(\frac{1}{x} - 1\right)^2 dx = \int \left(\frac{1}{x^2} - \frac{2}{x} + 1\right) dx = -\frac{1}{x} - 2\ln|x| + x + C$

25. Let $u = x^2 + 4x$. Then $du = (2x + 4)\,dx = 2(x + 2)\,dx$, so

$\int \frac{x + 2}{\sqrt{x^2 + 4x}}\,dx = \int u^{-1/2}\left(\frac{1}{2}\,du\right) = \frac{1}{2} \cdot 2u^{1/2} + C = \sqrt{u} + C = \sqrt{x^2 + 4x} + C$.

27. Let $u = \sin \pi t$. Then $du = \pi \cos \pi t\,dt$, so $\int \sin \pi t \cos \pi t\,dt = \int u\left(\frac{1}{\pi}\,du\right) = \frac{1}{\pi} \cdot \frac{1}{2}u^2 + C = \frac{1}{2\pi}(\sin \pi t)^2 + C$.

29. Let $u = \sqrt{x}$. Then $du = \frac{dx}{2\sqrt{x}}$, so $\int \frac{e^{\sqrt{x}}}{\sqrt{x}}\,dx = 2\int e^u\,du = 2e^u + C = 2e^{\sqrt{x}} + C$.

31. Let $u = \ln(\cos x)$. Then $du = \frac{-\sin x}{\cos x}\,dx = -\tan x\,dx$, so

$\int \tan x \ln(\cos x)\,dx = -\int u\,du = -\frac{1}{2}u^2 + C = -\frac{1}{2}[\ln(\cos x)]^2 + C$.

33. Let $u = 1 + x^4$. Then $du = 4x^3\,dx$, so $\int \frac{x^3}{1 + x^4}\,dx = \frac{1}{4}\int \frac{1}{u}\,du = \frac{1}{4}\ln|u| + C = \frac{1}{4}\ln\left(1 + x^4\right) + C$.

35. Let $u = 1 + \sec \theta$. Then $du = \sec \theta \tan \theta\,d\theta$, so

$\int \frac{\sec \theta \tan \theta}{1 + \sec \theta}\,d\theta = \int \frac{1}{1 + \sec \theta}(\sec \theta \tan \theta\,d\theta) = \int \frac{1}{u}\,du = \ln|u| + C = \ln|1 + \sec \theta| + C$.

37. Since $x^2 - 4 < 0$ for $0 \le x < 2$ and $x^2 - 4 > 0$ for $2 < x \le 3$, we have $|x^2 - 4| = -(x^2 - 4) = 4 - x^2$ for $0 \le x < 2$ and

$|x^2 - 4| = x^2 - 4$ for $2 < x \le 3$. Thus,

$$\int_0^3 |x^2 - 4|\,dx = \int_0^2 (4 - x^2)\,dx + \int_2^3 (x^2 - 4)\,dx = \left[4x - \frac{x^3}{3}\right]_0^2 + \left[\frac{x^3}{3} - 4x\right]_2^3$$

$$= \left(8 - \frac{8}{3}\right) - 0 + (9 - 12) - \left(\frac{8}{3} - 8\right) = \frac{16}{3} - 3 + \frac{16}{3} = \frac{32}{3} - \frac{9}{3} = \frac{23}{3}$$

In Exercises 39 and 40, let $f(x)$ denote the integrand and $F(x)$ its antiderivative (with $C = 0$).

39. Let $u = 1 + \sin x$. Then $du = \cos x \, dx$, so

$$\int \frac{\cos x \, dx}{\sqrt{1 + \sin x}} = \int u^{-1/2} \, du = 2u^{1/2} + C = 2\sqrt{1 + \sin x} + C.$$

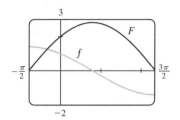

41. From the graph, it appears that the area under the curve $y = x\sqrt{x}$ between $x = 0$ and $x = 4$ is somewhat less than half the area of an 8×4 rectangle, so perhaps about 13 or 14. To find the exact value, we evaluate

$$\int_0^4 x\sqrt{x} \, dx = \int_0^4 x^{3/2} \, dx = \left[\tfrac{2}{5} x^{5/2} \right]_0^4 = \tfrac{2}{5}(4)^{5/2} = \tfrac{64}{5} = 12.8.$$

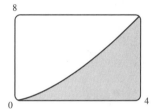

43. $F(x) = \displaystyle\int_0^x \frac{t^2}{1 + t^3} \, dt \quad \Rightarrow \quad F'(x) = \frac{d}{dx} \int_0^x \frac{t^2}{1 + t^3} \, dt = \frac{x^2}{1 + x^3}$

45. Let $u = x^4$. Then $\dfrac{du}{dx} = 4x^3$. Also, $\dfrac{dg}{dx} = \dfrac{dg}{du}\dfrac{du}{dx}$, so

$$g'(x) = \frac{d}{dx} \int_0^{x^4} \cos(t^2) \, dt = \frac{d}{du} \int_0^u \cos(t^2) \, dt \cdot \frac{du}{dx} = \cos(u^2) \frac{du}{dx} = 4x^3 \cos(x^8).$$

47. $y = \displaystyle\int_{\sqrt{x}}^x \frac{e^t}{t} \, dt = \int_{\sqrt{x}}^1 \frac{e^t}{t} \, dt + \int_1^x \frac{e^t}{t} \, dt = -\int_1^{\sqrt{x}} \frac{e^t}{t} \, dt + \int_1^x \frac{e^t}{t} \, dt \quad \Rightarrow$

$$\frac{dy}{dx} = -\frac{d}{dx}\left(\int_1^{\sqrt{x}} \frac{e^t}{t} \, dt \right) + \frac{d}{dx}\left(\int_1^x \frac{e^t}{t} \, dt \right). \text{ Let } u = \sqrt{x}. \text{ Then}$$

$$\frac{d}{dx} \int_1^{\sqrt{x}} \frac{e^t}{t} \, dt = \frac{d}{dx} \int_1^u \frac{e^t}{t} \, dt = \frac{d}{du}\left(\int_1^u \frac{e^t}{t} \, dt \right) \frac{du}{dx} = \frac{e^u}{u} \cdot \frac{1}{2\sqrt{x}} = \frac{e^{\sqrt{x}}}{\sqrt{x}} \cdot \frac{1}{2\sqrt{x}} = \frac{e^{\sqrt{x}}}{2x},$$

so $\dfrac{dy}{dx} = -\dfrac{e^{\sqrt{x}}}{2x} + \dfrac{e^x}{x}.$

49. If $1 \le x \le 3$, then $\sqrt{1^2 + 3} \le \sqrt{x^2 + 3} \le \sqrt{3^2 + 3} \quad \Rightarrow \quad 2 \le \sqrt{x^2 + 3} \le 2\sqrt{3}$, so
$2(3 - 1) \le \int_1^3 \sqrt{x^2 + 3} \, dx \le 2\sqrt{3}(3 - 1)$; that is, $4 \le \int_1^3 \sqrt{x^2 + 3} \, dx \le 4\sqrt{3}$.

51. $0 \le x \le 1 \quad \Rightarrow \quad 0 \le \cos x \le 1 \quad \Rightarrow \quad x^2 \cos x \le x^2 \quad \Rightarrow \quad \int_0^1 x^2 \cos x \, dx \le \int_0^1 x^2 \, dx = \tfrac{1}{3}\left[x^3 \right]_0^1 = \tfrac{1}{3}$ [Property 7].

53. $\cos x \le 1 \quad \Rightarrow \quad e^x \cos x \le e^x \quad \Rightarrow \quad \int_0^1 e^x \cos x \, dx \le \int_0^1 e^x \, dx = \left[e^x \right]_0^1 = e - 1$

55. $\Delta x = (3 - 0)/6 = \tfrac{1}{2}$, so the endpoints are $0, \tfrac{1}{2}, 1, \tfrac{3}{2}, 2, \tfrac{5}{2}$, and 3, and the midpoints are $\tfrac{1}{4}, \tfrac{3}{4}, \tfrac{5}{4}, \tfrac{7}{4}, \tfrac{9}{4}$, and $\tfrac{11}{4}$.

The Midpoint Rule gives

$$\int_0^3 \sin(x^3) \, dx \approx \sum_{i=1}^6 f(\overline{x}_i) \, \Delta x = \tfrac{1}{2}\left[\sin\left(\tfrac{1}{4}\right)^3 + \sin\left(\tfrac{3}{4}\right)^3 + \sin\left(\tfrac{5}{4}\right)^3 + \sin\left(\tfrac{7}{4}\right)^3 + \sin\left(\tfrac{9}{4}\right)^3 + \sin\left(\tfrac{11}{4}\right)^3 \right] \approx 0.280981.$$

57. Note that $r(t) = b'(t)$, where $b(t) =$ the number of barrels of oil consumed up to time t. So, by the Net Change Theorem,

$\int_0^8 r(t) \, dt = b(8) - b(0)$ represents the number of barrels of oil consumed from Jan. 1, 2000, through Jan. 1, 2008.

59. We use the Midpoint Rule with $n = 6$ and $\Delta t = \frac{24-0}{6} = 4$. The increase in the bee population was

$$\int_0^{24} r(t) \, dt \approx M_6 = 4[r(2) + r(6) + r(10) + r(14) + r(18) + r(22)]$$
$$\approx 4[50 + 1000 + 7000 + 8550 + 1350 + 150] = 4(18{,}100) = 72{,}400$$

61. Let $u = 2\sin\theta$. Then $du = 2\cos\theta \, d\theta$ and when $\theta = 0$, $u = 0$; when $\theta = \frac{\pi}{2}$, $u = 2$. Thus,

$\int_0^{\pi/2} f(2\sin\theta)\cos\theta \, d\theta = \int_0^2 f(u)\left(\frac{1}{2}\,du\right) = \frac{1}{2}\int_0^2 f(u)\,du = \frac{1}{2}\int_0^2 f(x)\,dx = \frac{1}{2}(6) = 3.$

63. Area under the curve $y = \sinh cx$ between $x = 0$ and $x = 1$ is equal to 1 $\Rightarrow$

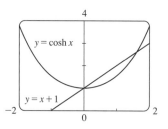

$\int_0^1 \sinh cx \, dx = 1$ $\Rightarrow$ $\frac{1}{c}\left[\cosh cx\right]_0^1 = 1$ $\Rightarrow$ $\frac{1}{c}(\cosh c - 1) = 1$ $\Rightarrow$

$\cosh c - 1 = c$ $\Rightarrow$ $\cosh c = c + 1$. From the graph, we get $c = 0$ and

$c \approx 1.6161$, but $c = 0$ isn't a solution for this problem since the curve

$y = \sinh cx$ becomes $y = 0$ and the area under it is 0. Thus, $c \approx 1.6161$.

65. Using FTC1, we differentiate both sides of the given equation, $\int_0^x f(t)\,dt = xe^{2x} + \int_0^x e^{-t}f(t)\,dt$, and get

$$f(x) = e^{2x} + 2xe^{2x} + e^{-x}f(x) \quad \Rightarrow \quad f(x)\left(1 - e^{-x}\right) = e^{2x} + 2xe^{2x} \quad \Rightarrow \quad f(x) = \frac{e^{2x}(1 + 2x)}{1 - e^{-x}}.$$

67. Let $u = f(x)$ and $du = f'(x)\,dx$. So $2\int_a^b f(x)f'(x)\,dx = 2\int_{f(a)}^{f(b)} u\,du = \left[u^2\right]_{f(a)}^{f(b)} = [f(b)]^2 - [f(a)]^2$.

69. Let $u = 1 - x$. Then $du = -dx$, so $\int_0^1 f(1-x)\,dx = \int_1^0 f(u)(-du) = \int_0^1 f(u)\,du = \int_0^1 f(x)\,dx$.

71. The shaded region has area $\int_0^1 f(x)\,dx = \frac{1}{3}$. The integral $\int_0^1 f^{-1}(y)\,dy$

gives the area of the unshaded region, which we know to be $1 - \frac{1}{3} = \frac{2}{3}$.

So $\int_0^1 f^{-1}(y)\,dy = \frac{2}{3}$.

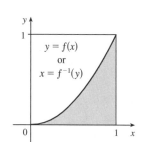

☐ PROBLEMS PLUS

1. Differentiating both sides of the equation $x \sin \pi x = \int_0^{x^2} f(t)\, dt$ (using FTC1 and the Chain Rule for the right side) gives

$\sin \pi x + \pi x \cos \pi x = 2x f(x^2)$. Letting $x = 2$ so that $f(x^2) = f(4)$, we obtain $\sin 2\pi + 2\pi \cos 2\pi = 4f(4)$, so

$f(4) = \frac{1}{4}(0 + 2\pi \cdot 1) = \frac{\pi}{2}$.

3. Differentiating the given equation, $\int_0^x f(t)\, dt = [f(x)]^2$, using FTC1 gives $f(x) = 2f(x)\, f'(x) \Rightarrow$

$f(x)[2f'(x) - 1] = 0$, so $f(x) = 0$ or $f'(x) = \frac{1}{2}$. Since $f(x)$ is never 0, we must have $f'(x) = \frac{1}{2}$ and $f'(x) = \frac{1}{2} \Rightarrow$

$f(x) = \frac{1}{2}x + C$. To find C, we substitute into the given equation to get $\int_0^x \left(\frac{1}{2}t + C\right) dt = \left(\frac{1}{2}x + C\right)^2 \Leftrightarrow$

$\frac{1}{4}x^2 + Cx = \frac{1}{4}x^2 + Cx + C^2$. It follows that $C^2 = 0$, so $C = 0$, and $f(x) = \frac{1}{2}x$.

5. $f(x) = \int_0^{g(x)} \dfrac{1}{\sqrt{1+t^3}}\, dt$, where $g(x) = \int_0^{\cos x} [1 + \sin(t^2)]\, dt$. Using FTC1 and the Chain Rule (twice) we have

$f'(x) = \dfrac{1}{\sqrt{1 + [g(x)]^3}}\, g'(x) = \dfrac{1}{\sqrt{1 + [g(x)]^3}}\, [1 + \sin(\cos^2 x)](-\sin x)$. Now $g\left(\frac{\pi}{2}\right) = \int_0^0 [1 + \sin(t^2)]\, dt = 0$, so

$f'\left(\frac{\pi}{2}\right) = \dfrac{1}{\sqrt{1+0}}\, (1 + \sin 0)(-1) = 1 \cdot 1 \cdot (-1) = -1$.

7. By l'Hospital's Rule and the Fundamental Theorem, using the notation $\exp(y) = e^y$,

$$\lim_{x \to 0} \frac{\int_0^x (1 - \tan 2t)^{1/t}\, dt}{x} \overset{\text{H}}{=} \lim_{x \to 0} \frac{(1 - \tan 2x)^{1/x}}{1} = \exp\left(\lim_{x \to 0} \frac{\ln(1 - \tan 2x)}{x}\right)$$

$$\overset{\text{H}}{=} \exp\left(\lim_{x \to 0} \frac{-2 \sec^2 2x}{1 - \tan 2x}\right) = \exp\left(\frac{-2 \cdot 1^2}{1 - 0}\right) = e^{-2}$$

9. $f(x) = 2 + x - x^2 = (-x + 2)(x + 1) = 0 \Leftrightarrow x = 2$ or $x = -1$. $f(x) \geq 0$ for $x \in [-1, 2]$ and $f(x) < 0$ everywhere

else. The integral $\int_a^b (2 + x - x^2)\, dx$ has a maximum on the interval where the integrand is positive, which is $[-1, 2]$. So

$a = -1, b = 2$. (Any larger interval gives a smaller integral since $f(x) < 0$ outside $[-1, 2]$. Any smaller interval also gives a

smaller integral since $f(x) \geq 0$ in $[-1, 2]$.)

11. (a) We can split the integral $\int_0^n [\![x]\!]\, dx$ into the sum $\displaystyle\sum_{i=1}^n \left[\int_{i-1}^i [\![x]\!]\, dx \right]$. But on each of the intervals $[i-1, i)$ of integration,

$[\![x]\!]$ is a constant function, namely $i - 1$. So the ith integral in the sum is equal to $(i-1)[i - (i-1)] = (i-1)$. So the

original integral is equal to $\displaystyle\sum_{i=1}^n (i-1) = \sum_{i=1}^{n-1} i = \frac{(n-1)n}{2}$.

(b) We can write $\int_a^b [\![x]\!]\, dx = \int_0^b [\![x]\!]\, dx - \int_0^a [\![x]\!]\, dx$.

Now $\int_0^b [\![x]\!]\, dx = \int_0^{[\![b]\!]} [\![x]\!]\, dx + \int_{[\![b]\!]}^b [\![x]\!]\, dx$. The first of these integrals is equal to $\frac{1}{2}([\![b]\!] - 1)\,[\![b]\!]$,

by part (a), and since $[\![x]\!] = [\![b]\!]$ on $[[\![b]\!]\,, b]$, the second integral is just $[\![b]\!]\,(b - [\![b]\!])$. So

$\int_0^b [\![x]\!]\, dx = \frac{1}{2}([\![b]\!] - 1)\,[\![b]\!] + [\![b]\!]\,(b - [\![b]\!]) = \frac{1}{2}\,[\![b]\!]\,(2b - [\![b]\!] - 1)$ and similarly $\int_0^a [\![x]\!]\, dx = \frac{1}{2}\,[\![a]\!]\,(2a - [\![a]\!] - 1)$.

Therefore, $\int_a^b [\![x]\!]\, dx = \frac{1}{2}\,[\![b]\!]\,(2b - [\![b]\!] - 1) - \frac{1}{2}\,[\![a]\!]\,(2a - [\![a]\!] - 1)$.

13. Let $Q(x) = \displaystyle\int_0^x P(t)\, dt = \left[at + \frac{b}{2}t^2 + \frac{c}{3}t^3 + \frac{d}{4}t^4 \right]_0^x = ax + \frac{b}{2}x^2 + \frac{c}{3}x^3 + \frac{d}{4}x^4$. Then $Q(0) = 0$, and $Q(1) = 0$ by the

given condition, $a + \dfrac{b}{2} + \dfrac{c}{3} + \dfrac{d}{4} = 0$. Also, $Q'(x) = P(x) = a + bx + cx^2 + dx^3$ by FTC1. By Rolle's Theorem, applied to

Q on $[0, 1]$, there is a number r in $(0, 1)$ such that $Q'(r) = 0$, that is, such that $P(r) = 0$. Thus, the equation $P(x) = 0$ has a

root between 0 and 1.

More generally, if $P(x) = a_0 + a_1 x + a_2 x^2 + \cdots + a_n x^n$ and if $a_0 + \dfrac{a_1}{2} + \dfrac{a_2}{3} + \cdots + \dfrac{a_n}{n+1} = 0$, then the equation

$P(x) = 0$ has a root between 0 and 1. The proof is the same as before:

Let $Q(x) = \displaystyle\int_0^x P(t)\, dt = a_0 x + \dfrac{a_1}{2}\,x^2 + \dfrac{a_2}{3}\,x^3 + \cdots + \dfrac{a_n}{n+1}\,x^n$. Then $Q(0) = Q(1) = 0$ and $Q'(x) = P(x)$. By

Rolle's Theorem applied to Q on $[0, 1]$, there is a number r in $(0, 1)$ such that $Q'(r) = 0$, that is, such that $P(r) = 0$.

15. Note that $\dfrac{d}{dx}\left(\displaystyle\int_0^x \left[\int_0^u f(t)\, dt \right] du \right) = \displaystyle\int_0^x f(t)\, dt$ by FTC1, while

$$\frac{d}{dx}\left[\int_0^x f(u)(x - u)\, du \right] = \frac{d}{dx}\left[x \int_0^x f(u)\, du \right] - \frac{d}{dx}\left[\int_0^x f(u)u\, du \right]$$

$$= \int_0^x f(u)\, du + x f(x) - f(x)x = \int_0^x f(u)\, du$$

Hence, $\int_0^x f(u)(x - u)\, du = \int_0^x \left[\int_0^u f(t)\, dt \right] du + C$. Setting $x = 0$ gives $C = 0$.

17.
$$\lim_{n \to \infty} \left(\frac{1}{\sqrt{n}\,\sqrt{n+1}} + \frac{1}{\sqrt{n}\,\sqrt{n+2}} + \cdots + \frac{1}{\sqrt{n}\,\sqrt{n+n}} \right)$$

$$= \lim_{n \to \infty} \frac{1}{n}\left(\sqrt{\frac{n}{n+1}} + \sqrt{\frac{n}{n+2}} + \cdots + \sqrt{\frac{n}{n+n}} \right)$$

$$= \lim_{n \to \infty} \frac{1}{n}\left(\frac{1}{\sqrt{1 + 1/n}} + \frac{1}{\sqrt{1 + 2/n}} + \cdots + \frac{1}{\sqrt{1 + 1}} \right)$$

$$= \lim_{n \to \infty} \frac{1}{n}\sum_{i=1}^{n} f\left(\frac{i}{n} \right) \qquad \left[\text{where } f(x) = \frac{1}{\sqrt{1 + x}} \right]$$

$$= \int_0^1 \frac{1}{\sqrt{1 + x}}\, dx = \left[2\sqrt{1 + x} \right]_0^1 = 2\left(\sqrt{2} - 1 \right)$$

6 □ APPLICATIONS OF INTEGRATION

6.1 Areas Between Curves

1. $A = \displaystyle\int_{x=0}^{x=4} (y_T - y_B) \, dx = \int_0^4 \left[(5x - x^2) - x \right] dx = \int_0^4 (4x - x^2) \, dx = \left[2x^2 - \tfrac{1}{3}x^3 \right]_0^4 = \left(32 - \tfrac{64}{3} \right) - (0) = \tfrac{32}{3}$

3. $A = \displaystyle\int_{y=-1}^{y=1} (x_R - x_L) \, dy = \int_{-1}^1 \left[e^y - (y^2 - 2) \right] dy = \int_{-1}^1 \left(e^y - y^2 + 2 \right) dy$

$= \left[e^y - \tfrac{1}{3}y^3 + 2y \right]_{-1}^1 = \left(e^1 - \tfrac{1}{3} + 2 \right) - \left(e^{-1} + \tfrac{1}{3} - 2 \right) = e - \dfrac{1}{e} + \dfrac{10}{3}$

5. $A = \displaystyle\int_{-1}^2 \left[(9 - x^2) - (x + 1) \right] dx$

$= \displaystyle\int_{-1}^2 (8 - x - x^2) \, dx$

$= \left[8x - \dfrac{x^2}{2} - \dfrac{x^3}{3} \right]_{-1}^2$

$= \left(16 - 2 - \tfrac{8}{3} \right) - \left(-8 - \tfrac{1}{2} + \tfrac{1}{3} \right)$

$= 22 - 3 + \tfrac{1}{2} = \tfrac{39}{2}$

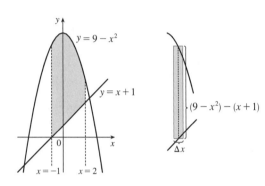

7. The curves intersect when $x = x^2 \iff x^2 - x = 0 \iff$

$x(x - 1) = 0 \iff x = 0$ or 1.

$A = \displaystyle\int_0^1 (x - x^2) \, dx = \left[\tfrac{1}{2}x^2 - \tfrac{1}{3}x^3 \right]_0^1 = \tfrac{1}{2} - \tfrac{1}{3} = \tfrac{1}{6}$

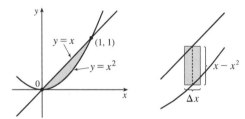

9. $A = \displaystyle\int_1^2 \left(\dfrac{1}{x} - \dfrac{1}{x^2} \right) dx = \left[\ln x + \dfrac{1}{x} \right]_1^2$

$= \left(\ln 2 + \tfrac{1}{2} \right) - (\ln 1 + 1)$

$= \ln 2 - \tfrac{1}{2} \approx 0.19$

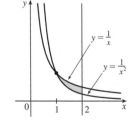
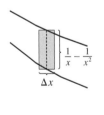

11. $A = \displaystyle\int_0^1 \left(\sqrt{x} - x^2 \right) dx$

$= \left[\tfrac{2}{3}x^{3/2} - \tfrac{1}{3}x^3 \right]_0^1$

$= \tfrac{2}{3} - \tfrac{1}{3} = \tfrac{1}{3}$

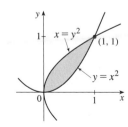
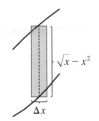

13. $12 - x^2 = x^2 - 6 \Leftrightarrow 2x^2 = 18 \Leftrightarrow$

$x^2 = 9 \Leftrightarrow x = \pm 3$, so

$A = \displaystyle\int_{-3}^{3} \left[(12 - x^2) - (x^2 - 6) \right] dx$

$= 2 \displaystyle\int_{0}^{3} \left(18 - 2x^2 \right) dx$ [by symmetry]

$= 2 \left[18x - \frac{2}{3}x^3 \right]_0^3 = 2 \left[(54 - 18) - 0 \right]$

$= 2(36) = 72$

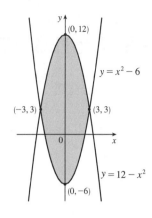
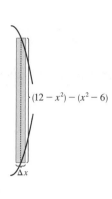

15. The curves intersect when $\tan x = 2 \sin x$ (on $[-\pi/3, \pi/3]$) $\Leftrightarrow$ $\sin x = 2 \sin x \cos x$ $\Leftrightarrow$

$2 \sin x \cos x - \sin x = 0 \Leftrightarrow \sin x \, (2 \cos x - 1) = 0 \Leftrightarrow \sin x = 0$ or $\cos x = \frac{1}{2} \Leftrightarrow x = 0$ or $x = \pm \frac{\pi}{3}$.

$A = \displaystyle\int_{-\pi/3}^{\pi/3} (2 \sin x - \tan x) \, dx$

$= 2 \displaystyle\int_{0}^{\pi/3} (2 \sin x - \tan x) \, dx$ [by symmetry]

$= 2 \left[-2 \cos x - \ln |\sec x| \right]_0^{\pi/3}$

$= 2 \left[(-1 - \ln 2) - (-2 - 0) \right]$

$= 2(1 - \ln 2) = 2 - 2 \ln 2$

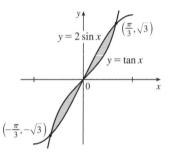
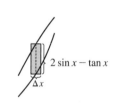

17. $\frac{1}{2} x = \sqrt{x} \Rightarrow \frac{1}{4} x^2 = x \Rightarrow x^2 - 4x = 0 \Rightarrow x(x - 4) = 0 \Rightarrow x = 0$ or 4, so

$A = \displaystyle\int_{0}^{4} \left(\sqrt{x} - \frac{1}{2} x \right) dx + \int_{4}^{9} \left(\frac{1}{2} x - \sqrt{x} \right) dx = \left[\frac{2}{3} x^{3/2} - \frac{1}{4} x^2 \right]_0^4 + \left[\frac{1}{4} x^2 - \frac{2}{3} x^{3/2} \right]_4^9$

$= \left[\left(\frac{16}{3} - 4 \right) - 0 \right] + \left[\left(\frac{81}{4} - 18 \right) - \left(4 - \frac{16}{3} \right) \right] = \frac{81}{4} + \frac{32}{3} - 26 = \frac{59}{12}$

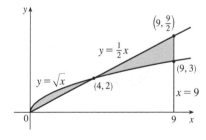

For $4 < x < 9$

19. $2y^2 = 4 + y^2 \iff y^2 = 4 \iff y = \pm 2$, so

$$A = \int_{-2}^{2} \left[(4 + y^2) - 2y^2\right] dy$$

$$= 2 \int_{0}^{2} (4 - y^2) \, dy \qquad \text{[by symmetry]}$$

$$= 2\left[4y - \tfrac{1}{3}y^3\right]_0^2 = 2\left(8 - \tfrac{8}{3}\right) = \tfrac{32}{3}$$

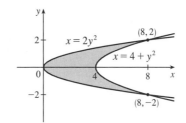
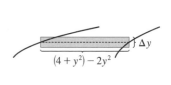

21. The curves intersect when $1 - y^2 = y^2 - 1 \iff 2 = 2y^2 \iff y^2 = 1 \iff y = \pm 1$.

$$A = \int_{-1}^{1} \left[(1 - y^2) - (y^2 - 1)\right] dy$$

$$= \int_{-1}^{1} 2(1 - y^2) \, dy$$

$$= 2 \cdot 2 \int_{0}^{1} (1 - y^2) \, dy$$

$$= 4\left[y - \tfrac{1}{3}y^3\right]_0^1 = 4\left(1 - \tfrac{1}{3}\right) = \tfrac{8}{3}$$

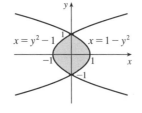
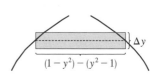

23. Notice that $\cos x = \sin 2x = 2 \sin x \cos x \iff$

$2 \sin x \cos x - \cos x = 0 \iff \cos x (2 \sin x - 1) = 0 \iff$

$2 \sin x = 1$ or $\cos x = 0 \iff x = \tfrac{\pi}{6}$ or $\tfrac{\pi}{2}$.

$$A = \int_{0}^{\pi/6} (\cos x - \sin 2x) \, dx + \int_{\pi/6}^{\pi/2} (\sin 2x - \cos x) \, dx$$

$$= \left[\sin x + \tfrac{1}{2}\cos 2x\right]_0^{\pi/6} + \left[-\tfrac{1}{2}\cos 2x - \sin x\right]_{\pi/6}^{\pi/2}$$

$$= \tfrac{1}{2} + \tfrac{1}{2} \cdot \tfrac{1}{2} - \left(0 + \tfrac{1}{2} \cdot 1\right) + \left(\tfrac{1}{2} - 1\right) - \left(-\tfrac{1}{2} \cdot \tfrac{1}{2} - \tfrac{1}{2}\right) = \tfrac{1}{2}$$

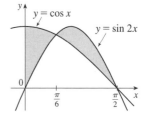

25. The curves intersect when $x^2 = \dfrac{2}{x^2 + 1} \iff$

$x^4 + x^2 = 2 \iff x^4 + x^2 - 2 = 0 \iff$

$(x^2 + 2)(x^2 - 1) = 0 \iff x^2 = 1 \iff x = \pm 1$.

$$A = \int_{-1}^{1} \left(\frac{2}{x^2 + 1} - x^2\right) dx = 2 \int_{0}^{1} \left(\frac{2}{x^2 + 1} - x^2\right) dx$$

$$= 2\left[2 \tan^{-1} x - \tfrac{1}{3}x^3\right]_0^1 = 2\left(2 \cdot \tfrac{\pi}{4} - \tfrac{1}{3}\right) = \pi - \tfrac{2}{3} \approx 2.47$$

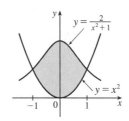
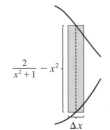

27. $1/x = x \iff 1 = x^2 \iff x = \pm 1$ and $1/x = \frac{1}{4}x \iff$

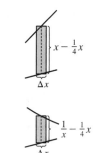
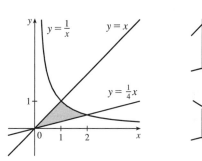

$4 = x^2 \iff x = \pm 2$, so for $x > 0$,

$$A = \int_0^1 \left(x - \frac{1}{4}x\right) dx + \int_1^2 \left(\frac{1}{x} - \frac{1}{4}x\right) dx$$

$$= \int_0^1 \left(\frac{3}{4}x\right) dx + \int_1^2 \left(\frac{1}{x} - \frac{1}{4}x\right) dx$$

$$= \left[\frac{3}{8}x^2\right]_0^1 + \left[\ln|x| - \frac{1}{8}x^2\right]_1^2$$

$$= \frac{3}{8} + \left(\ln 2 - \frac{1}{2}\right) - \left(0 - \frac{1}{8}\right) = \ln 2$$

29. An equation of the line through $(0,0)$ and $(2,1)$ is $y = \frac{1}{2}x$; through $(0,0)$

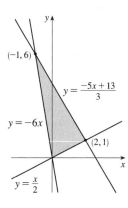

and $(-1,6)$ is $y = -6x$; through $(2,1)$ and $(-1,6)$ is $y = -\frac{5}{3}x + \frac{13}{3}$.

$$A = \int_{-1}^0 \left[\left(-\frac{5}{3}x + \frac{13}{3}\right) - (-6x)\right] dx + \int_0^2 \left[\left(-\frac{5}{3}x + \frac{13}{3}\right) - \frac{1}{2}x\right] dx$$

$$= \int_{-1}^0 \left(\frac{13}{3}x + \frac{13}{3}\right) dx + \int_0^2 \left(-\frac{13}{6}x + \frac{13}{3}\right) dx$$

$$= \frac{13}{3} \int_{-1}^0 (x + 1) dx + \frac{13}{3} \int_0^2 \left(-\frac{1}{2}x + 1\right) dx$$

$$= \frac{13}{3}\left[\frac{1}{2}x^2 + x\right]_{-1}^0 + \frac{13}{3}\left[-\frac{1}{4}x^2 + x\right]_0^2$$

$$= \frac{13}{3}\left[0 - \left(\frac{1}{2} - 1\right)\right] + \frac{13}{3}[(-1 + 2) - 0] = \frac{13}{3} \cdot \frac{1}{2} + \frac{13}{3} \cdot 1 = \frac{13}{2}$$

31. The curves intersect when $\sin x = \cos 2x$ (on $[0, \pi/2]$) $\iff \sin x = 1 - 2\sin^2 x \iff 2\sin^2 x + \sin x - 1 = 0 \iff$

$(2\sin x - 1)(\sin x + 1) = 0 \Rightarrow \sin x = \frac{1}{2} \Rightarrow x = \frac{\pi}{6}$.

$$A = \int_0^{\pi/2} |\sin x - \cos 2x| \, dx$$

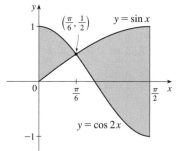

$$= \int_0^{\pi/6} (\cos 2x - \sin x) \, dx + \int_{\pi/6}^{\pi/2} (\sin x - \cos 2x) \, dx$$

$$= \left[\frac{1}{2}\sin 2x + \cos x\right]_0^{\pi/6} + \left[-\cos x - \frac{1}{2}\sin 2x\right]_{\pi/6}^{\pi/2}$$

$$= \left(\frac{1}{4}\sqrt{3} + \frac{1}{2}\sqrt{3}\right) - (0 + 1) + (0 - 0) - \left(-\frac{1}{2}\sqrt{3} - \frac{1}{4}\sqrt{3}\right)$$

$$= \frac{3}{2}\sqrt{3} - 1$$

33. Let $f(x) = \cos^2\left(\frac{\pi x}{4}\right) - \sin^2\left(\frac{\pi x}{4}\right)$ and $\Delta x = \frac{1 - 0}{4}$.

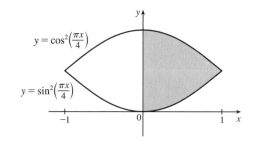

The shaded area is given by

$$A = \int_0^1 f(x) \, dx \approx M_4$$

$$= \frac{1}{4}\left[f\left(\frac{1}{8}\right) + f\left(\frac{3}{8}\right) + f\left(\frac{5}{8}\right) + f\left(\frac{7}{8}\right)\right]$$

$$\approx 0.6407$$

35.

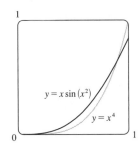

From the graph, we see that the curves intersect at $x = 0$ and $x = a \approx 0.896$, with $x\sin(x^2) > x^4$ on $(0, a)$. So the area A of the region bounded by the curves is

$$A = \int_0^a \left[x\sin(x^2) - x^4 \right] dx = \left[-\tfrac{1}{2}\cos(x^2) - \tfrac{1}{5}x^5 \right]_0^a$$

$$= -\tfrac{1}{2}\cos(a^2) - \tfrac{1}{5}a^5 + \tfrac{1}{2} \approx 0.037$$

37.

From the graph, we see that the curves intersect at

$x = a \approx -1.11$, $x = b \approx 1.25$, and $x = c \approx 2.86$, with

$x^3 - 3x + 4 > 3x^2 - 2x$ on (a, b) and $3x^2 - 2x > x^3 - 3x + 4$

on (b, c). So the area of the region bounded by the curves is

$$A = \int_a^b \left[(x^3 - 3x + 4) - (3x^2 - 2x) \right] dx + \int_b^c \left[(3x^2 - 2x) - (x^3 - 3x + 4) \right] dx$$

$$= \int_a^b (x^3 - 3x^2 - x + 4)\, dx + \int_b^c (-x^3 + 3x^2 + x - 4)\, dx$$

$$= \left[\tfrac{1}{4}x^4 - x^3 - \tfrac{1}{2}x^2 + 4x \right]_a^b + \left[-\tfrac{1}{4}x^4 + x^3 + \tfrac{1}{2}x^2 - 4x \right]_b^c \approx 8.38$$

39. As the figure illustrates, the curves $y = x$ and $y = x^5 - 6x^3 + 4x$

enclose a four-part region symmetric about the origin (since

$x^5 - 6x^3 + 4x$ and x are odd functions of x). The curves intersect

at values of x where $x^5 - 6x^3 + 4x = x$; that is, where

$x(x^4 - 6x^2 + 3) = 0$. That happens at $x = 0$ and where

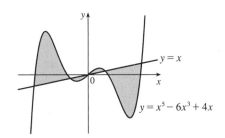

$$x^2 = \frac{6 \pm \sqrt{36 - 12}}{2} = 3 \pm \sqrt{6}; \text{ that is, at } x = -\sqrt{3 + \sqrt{6}}, -\sqrt{3 - \sqrt{6}}, 0, \sqrt{3 - \sqrt{6}}, \text{ and } \sqrt{3 + \sqrt{6}}.$$

The exact area is

$$2\int_0^{\sqrt{3+\sqrt{6}}} \left| (x^5 - 6x^3 + 4x) - x \right| dx = 2\int_0^{\sqrt{3+\sqrt{6}}} \left| x^5 - 6x^3 + 3x \right| dx$$

$$= 2\int_0^{\sqrt{3-\sqrt{6}}} (x^5 - 6x^3 + 3x)\, dx + 2\int_{\sqrt{3-\sqrt{6}}}^{\sqrt{3+\sqrt{6}}} (-x^5 + 6x^3 - 3x)\, dx$$

$$\overset{\text{CAS}}{=} 12\sqrt{6} - 9$$

41. 1 second $= \frac{1}{3600}$ hour, so 10 s $= \frac{1}{360}$ h. With the given data, we can take $n = 5$ to use the Midpoint Rule.

$\Delta t = \frac{1/360 - 0}{5} = \frac{1}{1800}$, so

$$\text{distance }_{\text{Kelly}} - \text{distance }_{\text{Chris}} = \int_0^{1/360} v_K \, dt - \int_0^{1/360} v_C \, dt = \int_0^{1/360} (v_K - v_C) \, dt$$

$$\approx M_5 = \tfrac{1}{1800} \left[(v_K - v_C)(1) + (v_K - v_C)(3) + (v_K - v_C)(5) \right.$$

$$\left. + (v_K - v_C)(7) + (v_K - v_C)(9) \right]$$

$$= \tfrac{1}{1800} \left[(22 - 20) + (52 - 46) + (71 - 62) + (86 - 75) + (98 - 86) \right]$$

$$= \tfrac{1}{1800} (2 + 6 + 9 + 11 + 12) = \tfrac{1}{1800}(40) = \tfrac{1}{45} \text{ mile, or } 117\tfrac{1}{3} \text{ feet}$$

43. Let $h(x)$ denote the height of the wing at x cm from the left end.

$$A \approx M_5 = \frac{200 - 0}{5} \left[h(20) + h(60) + h(100) + h(140) + h(180) \right]$$

$$= 40(20.3 + 29.0 + 27.3 + 20.5 + 8.7) = 40(105.8) = 4232 \text{ cm}^2$$

45. We know that the area under curve A between $t = 0$ and $t = x$ is $\int_0^x v_A(t) \, dt = s_A(x)$, where $v_A(t)$ is the velocity of car A and s_A is its displacement. Similarly, the area under curve B between $t = 0$ and $t = x$ is $\int_0^x v_B(t) \, dt = s_B(x)$.

(a) After one minute, the area under curve A is greater than the area under curve B. So car A is ahead after one minute.

(b) The area of the shaded region has numerical value $s_A(1) - s_B(1)$, which is the distance by which A is ahead of B after 1 minute.

(c) After two minutes, car B is traveling faster than car A and has gained some ground, but the area under curve A from $t = 0$ to $t = 2$ is still greater than the corresponding area for curve B, so car A is still ahead.

(d) From the graph, it appears that the area between curves A and B for $0 \le t \le 1$ (when car A is going faster), which corresponds to the distance by which car A is ahead, seems to be about 3 squares. Therefore, the cars will be side by side at the time x where the area between the curves for $1 \le t \le x$ (when car B is going faster) is the same as the area for $0 \le t \le 1$. From the graph, it appears that this time is $x \approx 2.2$. So the cars are side by side when $t \approx 2.2$ minutes.

47.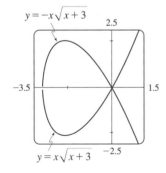

To graph this function, we must first express it as a combination of explicit functions of y; namely, $y = \pm x \sqrt{x + 3}$. We can see from the graph that the loop extends from $x = -3$ to $x = 0$, and that by symmetry, the area we seek is just twice the area under the top half of the curve on this interval, the equation of the top half being $y = -x \sqrt{x + 3}$. So the area is $A = 2 \int_{-3}^0 \left(-x \sqrt{x + 3} \right) dx$. We substitute $u = x + 3$, so $du = dx$ and the limits change to 0 and 3, and we get

$$A = -2 \int_0^3 \left[(u - 3) \sqrt{u} \right] du = -2 \int_0^3 (u^{3/2} - 3u^{1/2}) \, du$$

$$= -2 \left[\tfrac{2}{5} u^{5/2} - 2u^{3/2} \right]_0^3 = -2 \left[\tfrac{2}{5} \left(3^2 \sqrt{3} \right) - 2 \left(3 \sqrt{3} \right) \right] = \tfrac{24}{5} \sqrt{3}$$

49.

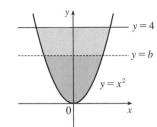

By the symmetry of the problem, we consider only the first quadrant, where

$y = x^2 \ \Rightarrow \ x = \sqrt{y}$. We are looking for a number b such that

$$\int_0^b \sqrt{y}\, dy = \int_b^4 \sqrt{y}\, dy \ \Rightarrow \ \tfrac{2}{3}\Big[y^{3/2}\Big]_0^b = \tfrac{2}{3}\Big[y^{3/2}\Big]_b^4 \ \Rightarrow$$

$$b^{3/2} = 4^{3/2} - b^{3/2} \ \Rightarrow \ 2b^{3/2} = 8 \ \Rightarrow \ b^{3/2} = 4 \ \Rightarrow \ b = 4^{2/3} \approx 2.52.$$

51. We first assume that $c > 0$, since c can be replaced by $-c$ in both equations without changing the graphs, and if $c = 0$ the

curves do not enclose a region. We see from the graph that the enclosed area A lies between $x = -c$ and $x = c$, and by

symmetry, it is equal to four times the area in the first quadrant. The enclosed area is

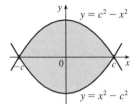

$$A = 4 \int_0^c (c^2 - x^2)\, dx = 4\Big[c^2 x - \tfrac{1}{3}x^3\Big]_0^c = 4\big(c^3 - \tfrac{1}{3}c^3\big) = 4\big(\tfrac{2}{3}c^3\big) = \tfrac{8}{3}c^3$$

So $A = 576 \ \Leftrightarrow \ \tfrac{8}{3}c^3 = 576 \ \Leftrightarrow \ c^3 = 216 \ \Leftrightarrow \ c = \sqrt[3]{216} = 6.$

Note that $c = -6$ is another solution, since the graphs are the same.

53. The curve and the line will determine a region when they intersect at two or

more points. So we solve the equation $x/(x^2 + 1) = mx \ \Rightarrow$

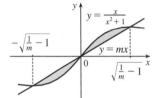

$x = x(mx^2 + m) \ \Rightarrow \ x(mx^2 + m) - x = 0 \ \Rightarrow$

$x(mx^2 + m - 1) = 0 \ \Rightarrow \ x = 0 \ \text{or} \ mx^2 + m - 1 = 0 \ \Rightarrow$

$x = 0 \ \text{or} \ x^2 = \dfrac{1-m}{m} \ \Rightarrow \ x = 0 \ \text{or} \ x = \pm\sqrt{\dfrac{1}{m} - 1}$. Note that if $m = 1$, this has only the solution $x = 0$, and no region

is determined. But if $1/m - 1 > 0 \ \Leftrightarrow \ 1/m > 1 \ \Leftrightarrow \ 0 < m < 1$, then there are two solutions. [Another way of seeing

this is to observe that the slope of the tangent to $y = x/(x^2 + 1)$ at the origin is $y'(0) = 1$ and therefore we must have

$0 < m < 1$.] Note that we cannot just integrate between the positive and negative roots, since the curve and the line cross at

the origin. Since mx and $x/(x^2 + 1)$ are both odd functions, the total area is twice the area between the curves on the interval

$\Big[0, \sqrt{1/m - 1}\,\Big]$. So the total area enclosed is

$$2\int_0^{\sqrt{1/m-1}} \left[\frac{x}{x^2 + 1} - mx\right] dx = 2\Big[\tfrac{1}{2}\ln(x^2 + 1) - \tfrac{1}{2}mx^2\Big]_0^{\sqrt{1/m-1}} = [\ln(1/m - 1 + 1) - m(1/m - 1)] - (\ln 1 - 0)$$

$$= \ln(1/m) - 1 + m = m - \ln m - 1$$

6.2 Volumes

1. A cross-section is a disk with radius $2 - \frac{1}{2}x$, so its area is $A(x) = \pi\left(2 - \frac{1}{2}x\right)^2$.

$$V = \int_1^2 A(x)\, dx = \int_1^2 \pi\left(2 - \tfrac{1}{2}x\right)^2 dx$$

$$= \pi \int_1^2 \left(4 - 2x + \tfrac{1}{4}x^2\right) dx$$

$$= \pi\left[4x - x^2 + \tfrac{1}{12}x^3\right]_1^2$$

$$= \pi\left[\left(8 - 4 + \tfrac{8}{12}\right) - \left(4 - 1 + \tfrac{1}{12}\right)\right]$$

$$= \pi\left(1 + \tfrac{7}{12}\right) = \tfrac{19}{12}\pi$$

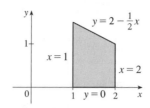
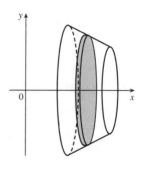

3. A cross-section is a disk with radius $1/x$, so its area is

$A(x) = \pi(1/x)^2$.

$$V = \int_1^2 A(x)\, dx = \int_1^2 \pi\left(\frac{1}{x}\right)^2 dx$$

$$= \pi \int_1^2 \frac{1}{x^2}\, dx = \pi\left[-\frac{1}{x}\right]_1^2$$

$$= \pi\left[-\tfrac{1}{2} - (-1)\right] = \tfrac{\pi}{2}$$

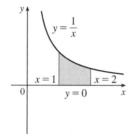
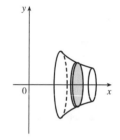

5. A cross-section is a disk with radius $2\sqrt{y}$, so its area is

$$A(y) = \pi\left(2\sqrt{y}\right)^2.$$

$$V = \int_0^9 A(y)\, dy = \int_0^9 \pi\left(2\sqrt{y}\right)^2 dy = 4\pi \int_0^9 y\, dy$$

$$= 4\pi\left[\tfrac{1}{2}y^2\right]_0^9 = 2\pi(81) = 162\pi$$

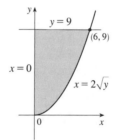
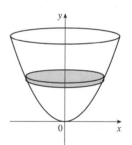

7. A cross-section is a washer (annulus) with inner

radius x^3 and outer radius x, so its area is

$A(x) = \pi(x)^2 - \pi(x^3)^2 = \pi(x^2 - x^6)$.

$$V = \int_0^1 A(x)\, dx = \int_0^1 \pi(x^2 - x^6)\, dx$$

$$= \pi\left[\tfrac{1}{3}x^3 - \tfrac{1}{7}x^7\right]_0^1 = \pi\left(\tfrac{1}{3} - \tfrac{1}{7}\right) = \tfrac{4}{21}\pi$$

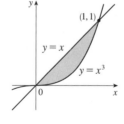
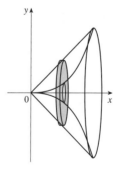

9. A cross-section is a washer with inner radius y^2
and outer radius $2y$, so its area is

$$A(y) = \pi(2y)^2 - \pi(y^2)^2 = \pi(4y^2 - y^4).$$

$$V = \int_0^2 A(y)\,dy = \pi \int_0^2 (4y^2 - y^4)\,dy$$

$$= \pi\left[\tfrac{4}{3}y^3 - \tfrac{1}{5}y^5\right]_0^2 = \pi\left(\tfrac{32}{3} - \tfrac{32}{5}\right) = \tfrac{64}{15}\pi$$

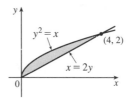

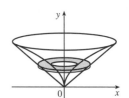

11. A cross-section is a washer with inner radius $1 - \sqrt{x}$ and outer radius $1 - x$, so its area is

$$A(x) = \pi(1-x)^2 - \pi\left(1 - \sqrt{x}\right)^2$$

$$= \pi\left[(1 - 2x + x^2) - \left(1 - 2\sqrt{x} + x\right)\right]$$

$$= \pi\left(-3x + x^2 + 2\sqrt{x}\right).$$

$$V = \int_0^1 A(x)\,dx = \pi \int_0^1 \left(-3x + x^2 + 2\sqrt{x}\right)\,dx$$

$$= \pi\left[-\tfrac{3}{2}x^2 + \tfrac{1}{3}x^3 + \tfrac{4}{3}x^{3/2}\right]_0^1 = \pi\left(-\tfrac{3}{2} + \tfrac{5}{3}\right) = \tfrac{\pi}{6}$$

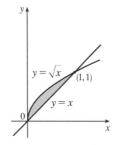

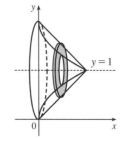

13. A cross-section is a washer with inner radius $(1 + \sec x) - 1 = \sec x$ and outer radius $3 - 1 = 2$, so its area is

$$A(x) = \pi\left[2^2 - (\sec x)^2\right] = \pi(4 - \sec^2 x).$$

$$V = \int_{-\pi/3}^{\pi/3} A(x)\,dx = \int_{-\pi/3}^{\pi/3} \pi(4 - \sec^2 x)\,dx$$

$$= 2\pi \int_0^{\pi/3} (4 - \sec^2 x)\,dx \qquad \text{[by symmetry]}$$

$$= 2\pi\left[4x - \tan x\right]_0^{\pi/3} = 2\pi\left[\left(\tfrac{4\pi}{3} - \sqrt{3}\right) - 0\right]$$

$$= 2\pi\left(\tfrac{4\pi}{3} - \sqrt{3}\right)$$

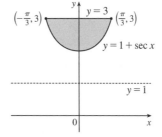

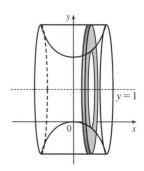

15. $V = \displaystyle\int_{-1}^1 \pi(1 - y^2)^2\,dy = 2\int_0^1 \pi(1 - y^2)^2\,dy$

$$= 2\pi \int_0^1 (1 - 2y^2 + y^4)\,dy$$

$$= 2\pi\left[y - \tfrac{2}{3}y^3 + \tfrac{1}{5}y^5\right]_0^1$$

$$= 2\pi \cdot \tfrac{8}{15} = \tfrac{16}{15}\pi$$

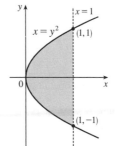

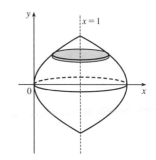

17. $y = x^2 \;\Rightarrow\; x = \sqrt{y}$ for $x \geq 0$. The outer radius is the distance from $x = -1$ to $x = \sqrt{y}$ and the inner radius is the

distance from $x = -1$ to $x = y^2$.

$$V = \int_0^1 \pi\left\{\left[\sqrt{y} - (-1)\right]^2 - \left[y^2 - (-1)\right]^2\right\} dy = \pi \int_0^1 \left[\left(\sqrt{y} + 1\right)^2 - (y^2 + 1)^2\right] dy$$

$$= \pi \int_0^1 \left(y + 2\sqrt{y} + 1 - y^4 - 2y^2 - 1\right) dy = \pi \int_0^1 \left(y + 2\sqrt{y} - y^4 - 2y^2\right) dy$$

$$= \pi \left[\tfrac{1}{2}y^2 + \tfrac{4}{3}y^{3/2} - \tfrac{1}{5}y^5 - \tfrac{2}{3}y^3\right]_0^1 = \pi\left(\tfrac{1}{2} + \tfrac{4}{3} - \tfrac{1}{5} - \tfrac{2}{3}\right) = \tfrac{29}{30}\pi$$

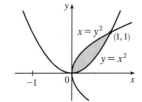 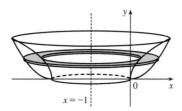

19. $\mathcal{R}_1$ about OA (the line $y = 0$): $V = \int_0^1 A(x)\, dx = \int_0^1 \pi(x^3)^2\, dx = \pi \int_0^1 x^6\, dx = \pi\left[\tfrac{1}{7}x^7\right]_0^1 = \dfrac{\pi}{7}$

21. $\mathcal{R}_1$ about AB (the line $x = 1$):

$$V = \int_0^1 A(y)\, dy = \int_0^1 \pi\left(1 - \sqrt[3]{y}\right)^2 dy = \pi \int_0^1 (1 - 2y^{1/3} + y^{2/3})\, dy = \pi\left[y - \tfrac{3}{2}y^{4/3} + \tfrac{3}{5}y^{5/3}\right]_0^1$$
$$= \pi\left(1 - \tfrac{3}{2} + \tfrac{3}{5}\right) = \dfrac{\pi}{10}$$

23. $\mathcal{R}_2$ about OA (the line $y = 0$):

$$V = \int_0^1 A(x)\, dx = \int_0^1 \left[\pi(1)^2 - \pi\left(\sqrt{x}\right)^2\right] dx = \pi \int_0^1 (1 - x)\, dx = \pi\left[x - \tfrac{1}{2}x^2\right]_0^1 = \pi\left(1 - \tfrac{1}{2}\right) = \dfrac{\pi}{2}$$

25. $\mathcal{R}_2$ about AB (the line $x = 1$):

$$V = \int_0^1 A(y)\, dy = \int_0^1 \left[\pi(1)^2 - \pi(1 - y^2)^2\right] dy = \pi \int_0^1 \left[1 - (1 - 2y^2 + y^4)\right] dy = \pi \int_0^1 (2y^2 - y^4)\, dy$$
$$= \pi\left[\tfrac{2}{3}y^3 - \tfrac{1}{5}y^5\right]_0^1 = \pi\left(\tfrac{2}{3} - \tfrac{1}{5}\right) = \dfrac{7}{15}\pi$$

27. $\mathcal{R}_3$ about OA (the line $y = 0$):

$$V = \int_0^1 A(x)\, dx = \int_0^1 \left[\pi\left(\sqrt{x}\right)^2 - \pi(x^3)^2\right] dx = \pi \int_0^1 (x - x^6)\, dx = \pi\left[\tfrac{1}{2}x^2 - \tfrac{1}{7}x^7\right]_0^1 = \pi\left(\tfrac{1}{2} - \tfrac{1}{7}\right) = \dfrac{5}{14}\pi.$$

Note: Let $\mathcal{R} = \mathcal{R}_1 + \mathcal{R}_2 + \mathcal{R}_3$. If we rotate $\mathcal{R}$ about any of the segments OA, OC, AB, or BC, we obtain a right circular

cylinder of height 1 and radius 1. Its volume is $\pi r^2 h = \pi(1)^2 \cdot 1 = \pi$. As a check for Exercises 19, 23, and 27, we can add the

answers, and that sum must equal π. Thus, $\dfrac{\pi}{7} + \dfrac{\pi}{2} + \dfrac{5\pi}{14} = \left(\dfrac{2 + 7 + 5}{14}\right)\pi = \pi$.

29. $\mathcal{R}_3$ about AB (the line $x = 1$):

$$V = \int_0^1 A(y)\,dy = \int_0^1 \left[\pi(1 - y^2)^2 - \pi\left(1 - \sqrt[3]{y}\right)^2\right] dy = \pi\int_0^1 \left[(1 - 2y^2 + y^4) - (1 - 2y^{1/3} + y^{2/3})\right] dy$$

$$= \pi\int_0^1 (-2y^2 + y^4 + 2y^{1/3} - y^{2/3})\,dy = \pi\left[-\tfrac{2}{3}y^3 + \tfrac{1}{5}y^5 + \tfrac{3}{2}y^{4/3} - \tfrac{3}{5}y^{5/3}\right]_0^1 = \pi\left(-\tfrac{2}{3} + \tfrac{1}{5} + \tfrac{3}{2} - \tfrac{3}{5}\right) = \tfrac{13}{30}\pi$$

Note: See the note in Exercise 27. For Exercises 21, 25, and 29, we have $\tfrac{\pi}{10} + \tfrac{7\pi}{15} + \tfrac{13\pi}{30} = \left(\tfrac{3 + 14 + 13}{30}\right)\pi = \pi$.

31. $V = \pi\displaystyle\int_0^{\pi/4} (1 - \tan^3 x)^2\,dx$

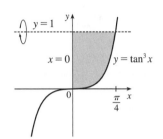

33. $V = \pi\displaystyle\int_0^{\pi}\left[(1 - 0)^2 - (1 - \sin x)^2\right] dx$

$$= \pi\int_0^{\pi}\left[1^2 - (1 - \sin x)^2\right] dx$$

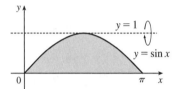

35. $V = \pi\displaystyle\int_{-\sqrt{8}}^{\sqrt{8}}\left\{[3 - (-2)]^2 - \left[\sqrt{y^2 + 1} - (-2)\right]^2\right\} dy$

$$= \pi\int_{-2\sqrt{2}}^{2\sqrt{2}}\left[5^2 - \left(\sqrt{1 + y^2} + 2\right)^2\right] dy$$

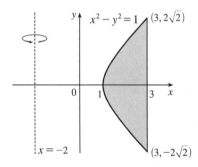

37.

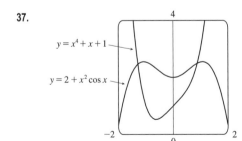

$y = 2 + x^2\cos x$ and $y = x^4 + x + 1$ intersect at

$x = a \approx -1.288$ and $x = b \approx 0.884$.

$$V = \pi\int_a^b \left[(2 + x^2\cos x)^2 - (x^4 + x + 1)^2\right]dx \approx 23.780$$

39. $V = \pi \int_0^{\pi} \left\{ \left[\sin^2 x - (-1)\right]^2 - [0 - (-1)]^2 \right\} dx$

$\overset{CAS}{=} \frac{11}{8}\pi^2$

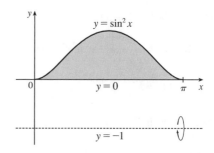

41. $\pi \int_0^{\pi/2} \cos^2 x \, dx$ describes the volume of the solid obtained by rotating the region $\mathcal{R} = \left\{ (x,y) \mid 0 \le x \le \frac{\pi}{2}, 0 \le y \le \cos x \right\}$ of the xy-plane about the x-axis.

43. $\pi \int_0^1 (y^4 - y^8) \, dy = \pi \int_0^1 \left[(y^2)^2 - (y^4)^2 \right] dy$ describes the volume of the solid obtained by rotating the region

$\mathcal{R} = \left\{ (x,y) \mid 0 \le y \le 1, y^4 \le x \le y^2 \right\}$ of the xy-plane about the y-axis.

45. There are 10 subintervals over the 15-cm length, so we'll use $n = 10/2 = 5$ for the Midpoint Rule.

$V = \int_0^{15} A(x) \, dx \approx M_5 = \frac{15-0}{5}[A(1.5) + A(4.5) + A(7.5) + A(10.5) + A(13.5)]$

$\qquad = 3(18 + 79 + 106 + 128 + 39) = 3 \cdot 370 = 1110 \text{ cm}^3$

47. (a) $V = \int_2^{10} \pi \left[f(x) \right]^2 dx \approx \pi \frac{10-2}{4} \left\{ [f(3)]^2 + [f(5)]^2 + [f(7)]^2 + [f(9)]^2 \right\}$

$\qquad \approx 2\pi \left[(1.5)^2 + (2.2)^2 + (3.8)^2 + (3.1)^2 \right] \approx 196 \text{ units}^3$

(b) $V = \int_0^4 \pi \left[(\text{outer radius})^2 - (\text{inner radius})^2 \right] dy$

$\qquad \approx \pi \frac{4-0}{4} \left\{ \left[(9.9)^2 - (2.2)^2 \right] + \left[(9.7)^2 - (3.0)^2 \right] + \left[(9.3)^2 - (5.6)^2 \right] + \left[(8.7)^2 - (6.5)^2 \right] \right\}$

$\qquad \approx 838 \text{ units}^3$

49. We'll form a right circular cone with height h and base radius r by revolving the line $y = \frac{r}{h} x$ about the x-axis.

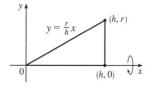

$V = \pi \int_0^h \left(\frac{r}{h} x \right)^2 dx = \pi \int_0^h \frac{r^2}{h^2} x^2 dx = \pi \frac{r^2}{h^2} \left[\frac{1}{3} x^3 \right]_0^h$

$\quad = \pi \frac{r^2}{h^2} \left(\frac{1}{3} h^3 \right) = \frac{1}{3}\pi r^2 h$

Another solution: Revolve $x = -\frac{r}{h} y + r$ about the y-axis.

$V = \pi \int_0^h \left(-\frac{r}{h} y + r \right)^2 dy \overset{*}{=} \pi \int_0^h \left[\frac{r^2}{h^2} y^2 - \frac{2r^2}{h} y + r^2 \right] dy$

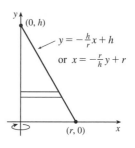

$\quad = \pi \left[\frac{r^2}{3h^2} y^3 - \frac{r^2}{h} y^2 + r^2 y \right]_0^h = \pi \left(\frac{1}{3} r^2 h - r^2 h + r^2 h \right) = \frac{1}{3}\pi r^2 h$

$*$ Or use substitution with $u = r - \frac{r}{h} y$ and $du = -\frac{r}{h} dy$ to get

$\pi \int_r^0 u^2 \left(-\frac{h}{r} du \right) = -\pi \frac{h}{r} \left[\frac{1}{3} u^3 \right]_r^0 = -\pi \frac{h}{r} \left(-\frac{1}{3} r^3 \right) = \frac{1}{3}\pi r^2 h.$

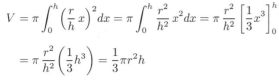

51. $x^2 + y^2 = r^2 \iff x^2 = r^2 - y^2$

$$V = \pi \int_{r-h}^{r} \left(r^2 - y^2\right) dy = \pi \left[r^2 y - \frac{y^3}{3}\right]_{r-h}^{r}$$

$$= \pi \left\{ \left[r^3 - \frac{r^3}{3}\right] - \left[r^2(r - h) - \frac{(r - h)^3}{3}\right] \right\}$$

$$= \pi \left\{ \tfrac{2}{3} r^3 - \tfrac{1}{3}(r - h)\left[3r^2 - (r - h)^2\right] \right\}$$

$$= \tfrac{1}{3}\pi \left\{ 2r^3 - (r - h)\left[3r^2 - (r^2 - 2rh + h^2)\right] \right\}$$

$$= \tfrac{1}{3}\pi \left\{ 2r^3 - (r - h)\left[2r^2 + 2rh - h^2\right] \right\}$$

$$= \tfrac{1}{3}\pi \left(2r^3 - 2r^3 - 2r^2 h + rh^2 + 2r^2 h + 2rh^2 - h^3\right)$$

$$= \tfrac{1}{3}\pi \left(3rh^2 - h^3\right) = \tfrac{1}{3}\pi h^2(3r - h), \text{ or, equivalently, } \pi h^2\left(r - \frac{h}{3}\right)$$

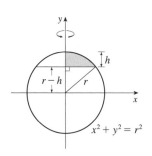

53. For a cross-section at height y, we see from similar triangles that $\dfrac{\alpha/2}{b/2} = \dfrac{h - y}{h}$, so $\alpha = b\left(1 - \dfrac{y}{h}\right)$.

Similarly, for cross-sections having $2b$ as their base and β replacing α, $\beta = 2b\left(1 - \dfrac{y}{h}\right)$. So

$$V = \int_{0}^{h} A(y)\, dy = \int_{0}^{h} \left[b\left(1 - \frac{y}{h}\right)\right]\left[2b\left(1 - \frac{y}{h}\right)\right] dy$$

$$= \int_{0}^{h} 2b^2\left(1 - \frac{y}{h}\right)^2 dy = 2b^2 \int_{0}^{h} \left(1 - \frac{2y}{h} + \frac{y^2}{h^2}\right) dy$$

$$= 2b^2 \left[y - \frac{y^2}{h} + \frac{y^3}{3h^2}\right]_0^{h} = 2b^2\left[h - h + \tfrac{1}{3}h\right]$$

$$= \tfrac{2}{3}b^2 h \quad [\,= \tfrac{1}{3}Bh \text{ where } B \text{ is the area of the base, as with any pyramid.}]$$

55. A cross-section at height z is a triangle similar to the base, so we'll multiply the legs of the base triangle, 3 and 4, by a proportionality factor of $(5 - z)/5$. Thus, the triangle at height z has area

$$A(z) = \frac{1}{2} \cdot 3\left(\frac{5 - z}{5}\right) \cdot 4\left(\frac{5 - z}{5}\right) = 6\left(1 - \frac{z}{5}\right)^2, \text{ so}$$

$$V = \int_0^5 A(z)\, dz = 6 \int_0^5 \left(1 - \frac{z}{5}\right)^2 dz = 6 \int_1^0 u^2(-5\, du) \qquad \begin{bmatrix} u = 1 - z/5, \\ du = -\tfrac{1}{5}\, dz \end{bmatrix}$$

$$= -30\left[\tfrac{1}{3}u^3\right]_1^0 = -30\left(-\tfrac{1}{3}\right) = 10 \text{ cm}^3$$

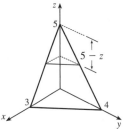

57. If l is a leg of the isosceles right triangle and $2y$ is the hypotenuse,

then $l^2 + l^2 = (2y)^2 \Rightarrow 2l^2 = 4y^2 \Rightarrow l^2 = 2y^2$.

$$V = \int_{-2}^2 A(x)\, dx = 2\int_0^2 A(x)\, dx = 2\int_0^2 \tfrac{1}{2}(l)(l)\, dx = 2\int_0^2 y^2\, dx$$

$$= 2\int_0^2 \tfrac{1}{4}(36 - 9x^2)\, dx = \tfrac{9}{2}\int_0^2 (4 - x^2)\, dx$$

$$= \tfrac{9}{2}\left[4x - \tfrac{1}{3}x^3\right]_0^2 = \tfrac{9}{2}\left(8 - \tfrac{8}{3}\right) = 24$$

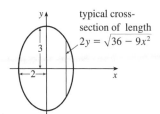

typical cross-section of length $2y = \sqrt{36 - 9x^2}$

59. The cross-section of the base corresponding to the coordinate x has length

$y = 1 - x$. The corresponding square with side s has area

$A(x) = s^2 = (1 - x)^2 = 1 - 2x + x^2$. Therefore,

$$V = \int_0^1 A(x)\, dx = \int_0^1 (1 - 2x + x^2)\, dx$$

$$= \left[x - x^2 + \tfrac{1}{3}x^3\right]_0^1 = \left(1 - 1 + \tfrac{1}{3}\right) - 0 = \tfrac{1}{3}$$

Or: $\int_0^1 (1 - x)^2\, dx = \int_1^0 u^2(-du) \quad [u = 1 - x] = \left[\tfrac{1}{3}u^3\right]_0^1 = \tfrac{1}{3}$

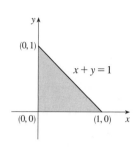

61. The cross-section of the base b corresponding to the coordinate x has length $1 - x^2$. The height h also has length $1 - x^2$,

so the corresponding isosceles triangle has area $A(x) = \tfrac{1}{2}bh = \tfrac{1}{2}(1 - x^2)^2$. Therefore,

$$V = \int_{-1}^1 \tfrac{1}{2}(1 - x^2)^2\, dx$$

$$= 2 \cdot \tfrac{1}{2} \int_0^1 (1 - 2x^2 + x^4)\, dx \qquad \text{[by symmetry]}$$

$$= \left[x - \tfrac{2}{3}x^3 + \tfrac{1}{5}x^5\right]_0^1 = \left(1 - \tfrac{2}{3} + \tfrac{1}{5}\right) - 0 = \tfrac{8}{15}$$

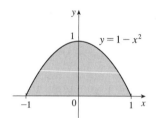

63. (a) The torus is obtained by rotating the circle $(x - R)^2 + y^2 = r^2$ about

the y-axis. Solving for x, we see that the right half of the circle is given by

$x = R + \sqrt{r^2 - y^2} = f(y)$ and the left half by $x = R - \sqrt{r^2 - y^2} = g(y)$. So

$$V = \pi \int_{-r}^r \left\{[f(y)]^2 - [g(y)]^2\right\} dy$$

$$= 2\pi \int_0^r \left[\left(R^2 + 2R\sqrt{r^2 - y^2} + r^2 - y^2\right) - \left(R^2 - 2R\sqrt{r^2 - y^2} + r^2 - y^2\right)\right] dy$$

$$= 2\pi \int_0^r 4R\sqrt{r^2 - y^2}\, dy = 8\pi R \int_0^r \sqrt{r^2 - y^2}\, dy$$

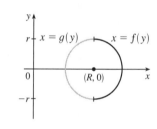

(b) Observe that the integral represents a quarter of the area of a circle with radius r, so

$$8\pi R \int_0^r \sqrt{r^2 - y^2}\, dy = 8\pi R \cdot \tfrac{1}{4}\pi r^2 = 2\pi^2 r^2 R.$$

65. (a) Volume$(S_1) = \int_0^h A(z)\, dz = $ Volume(S_2) since the cross-sectional area $A(z)$ at height z is the same for both solids.

(b) By Cavalieri's Principle, the volume of the cylinder in the figure is the same as that of a right circular cylinder with radius r

and height h, that is, $\pi r^2 h$.

67. The volume is obtained by rotating the area common to two circles of radius r, as shown. The volume of the right half is

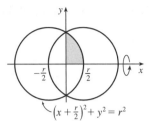

$$V_{\text{right}} = \pi \int_0^{r/2} y^2 \, dx = \pi \int_0^{r/2} \left[r^2 - \left(\tfrac{1}{2}r + x \right)^2 \right] dx$$

$$= \pi \left[r^2 x - \tfrac{1}{3}\left(\tfrac{1}{2}r + x \right)^3 \right]_0^{r/2} = \pi \left[\left(\tfrac{1}{2}r^3 - \tfrac{1}{3}r^3 \right) - \left(0 - \tfrac{1}{24}r^3 \right) \right] = \tfrac{5}{24}\pi r^3$$

So by symmetry, the total volume is twice this, or $\tfrac{5}{12}\pi r^3$.

Another solution: We observe that the volume is the twice the volume of a cap of a sphere, so we can use the formula from Exercise 51 with $h = \tfrac{1}{2}r$: $V = 2 \cdot \tfrac{1}{3}\pi h^2 (3r - h) = \tfrac{2}{3}\pi \left(\tfrac{1}{2}r \right)^2 \left(3r - \tfrac{1}{2}r \right) = \tfrac{5}{12}\pi r^3$.

69. Take the x-axis to be the axis of the cylindrical hole of radius r. A quarter of the cross-section through y, perpendicular to the y-axis, is the rectangle shown. Using the Pythagorean Theorem twice, we see that the dimensions of this rectangle are

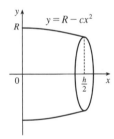

$x = \sqrt{R^2 - y^2}$ and $z = \sqrt{r^2 - y^2}$, so

$\tfrac{1}{4}A(y) = xz = \sqrt{r^2 - y^2}\,\sqrt{R^2 - y^2}$, and

$$V = \int_{-r}^{r} A(y) \, dy = \int_{-r}^{r} 4\sqrt{r^2 - y^2}\,\sqrt{R^2 - y^2}\, dy = 8\int_0^r \sqrt{r^2 - y^2}\,\sqrt{R^2 - y^2}\, dy$$

71. (a) The radius of the barrel is the same at each end by symmetry, since the function $y = R - cx^2$ is even. Since the barrel is obtained by rotating the graph of the function y about the x-axis, this radius is equal to the value of y at $x = \tfrac{1}{2}h$, which is $R - c\left(\tfrac{1}{2}h \right)^2 = R - d = r$.

(b) The barrel is symmetric about the y-axis, so its volume is twice the volume of that part of the barrel for $x > 0$. Also, the barrel is a volume of rotation, so

$$V = 2\int_0^{h/2} \pi y^2 \, dx = 2\pi \int_0^{h/2} \left(R - cx^2 \right)^2 dx = 2\pi \left[R^2 x - \tfrac{2}{3}Rcx^3 + \tfrac{1}{5}c^2 x^5 \right]_0^{h/2}$$

$$= 2\pi \left(\tfrac{1}{2}R^2 h - \tfrac{1}{12}Rch^3 + \tfrac{1}{160}c^2 h^5 \right)$$

Trying to make this look more like the expression we want, we rewrite it as $V = \tfrac{1}{3}\pi h \left[2R^2 + \left(R^2 - \tfrac{1}{2}Rch^2 + \tfrac{3}{80}c^2 h^4 \right) \right]$.

But $R^2 - \tfrac{1}{2}Rch^2 + \tfrac{3}{80}c^2 h^4 = \left(R - \tfrac{1}{4}ch^2 \right)^2 - \tfrac{1}{40}c^2 h^4 = (R - d)^2 - \tfrac{2}{5}\left(\tfrac{1}{4}ch^2 \right)^2 = r^2 - \tfrac{2}{5}d^2$.

Substituting this back into V, we see that $V = \tfrac{1}{3}\pi h \left(2R^2 + r^2 - \tfrac{2}{5}d^2 \right)$, as required.

6.3 Volumes by Cylindrical Shells

1.

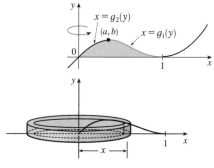

If we were to use the "washer" method, we would first have to locate the local maximum point (a, b) of $y = x(x-1)^2$ using the methods of Chapter 4. Then we would have to solve the equation $y = x(x-1)^2$ for x in terms of y to obtain the functions $x = g_1(y)$ and $x = g_2(y)$ shown in the first figure. This step would be difficult because it involves the cubic formula. Finally we would find the volume using

$$V = \pi \int_0^b \left\{ [g_1(y)]^2 - [g_2(y)]^2 \right\} dy.$$

Using shells, we find that a typical approximating shell has radius x, so its circumference is $2\pi x$. Its height is y, that is, $x(x-1)^2$. So the total volume is

$$V = \int_0^1 2\pi x \left[x(x-1)^2 \right] dx = 2\pi \int_0^1 \left(x^4 - 2x^3 + x^2 \right) dx = 2\pi \left[\frac{x^5}{5} - 2\frac{x^4}{4} + \frac{x^3}{3} \right]_0^1 = \frac{\pi}{15}$$

3. $V = \displaystyle\int_1^2 2\pi x \cdot \frac{1}{x} \, dx = 2\pi \int_1^2 1 \, dx$

$= 2\pi \left[x \right]_1^2 = 2\pi(2-1) = 2\pi$

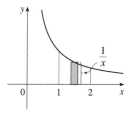
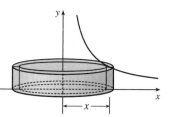

5. $V = \int_0^1 2\pi x e^{-x^2} \, dx$. Let $u = x^2$.

Thus, $du = 2x \, dx$, so

$V = \pi \int_0^1 e^{-u} \, du = \pi \left[-e^{-u} \right]_0^1 = \pi(1 - 1/e)$.

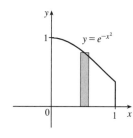

7. The curves intersect when $4(x-2)^2 = x^2 - 4x + 7 \iff 4x^2 - 16x + 16 = x^2 - 4x + 7 \iff$

$3x^2 - 12x + 9 = 0 \iff 3(x^2 - 4x + 3) = 0 \iff 3(x-1)(x-3) = 0$, so $x = 1$ or 3.

$V = 2\pi \int_1^3 \left\{ x \left[(x^2 - 4x + 7) - 4(x-2)^2 \right] \right\} dx = 2\pi \int_1^3 \left[x(x^2 - 4x + 7 - 4x^2 + 16x - 16) \right] dx$

$= 2\pi \int_1^3 \left[x(-3x^2 + 12x - 9) \right] dx = 2\pi(-3) \int_1^3 (x^3 - 4x^2 + 3x) \, dx = -6\pi \left[\frac{1}{4}x^4 - \frac{4}{3}x^3 + \frac{3}{2}x^2 \right]_1^3$

$= -6\pi \left[\left(\frac{81}{4} - 36 + \frac{27}{2} \right) - \left(\frac{1}{4} - \frac{4}{3} + \frac{3}{2} \right) \right] = -6\pi \left(20 - 36 + 12 + \frac{4}{3} \right) = -6\pi \left(-\frac{8}{3} \right) = 16\pi$

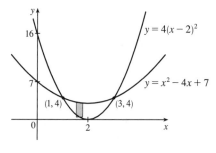
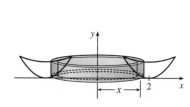

9. $V = \int_1^2 2\pi y(1 + y^2)\,dy = 2\pi \int_1^2 (y + y^3)\,dy = 2\pi \left[\frac{1}{2}y^2 + \frac{1}{4}y^4\right]_1^2$

$= 2\pi\left[(2 + 4) - \left(\frac{1}{2} + \frac{1}{4}\right)\right] = 2\pi\left(\frac{21}{4}\right) = \frac{21}{2}\pi$

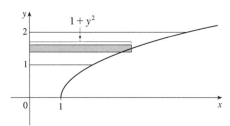

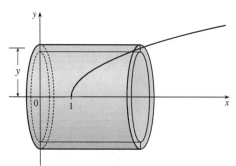

11. $V = 2\pi \int_0^8 \left[y\left(\sqrt[3]{y} - 0\right)\right]dy$

$= 2\pi \int_0^8 y^{4/3}\,dy = 2\pi\left[\frac{3}{7}y^{7/3}\right]_0^8$

$= \frac{6\pi}{7}(8^{7/3}) = \frac{6\pi}{7}(2^7) = \frac{768}{7}\pi$

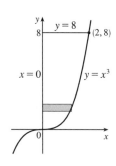

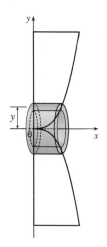

13. The height of the shell is $2 - \left[1 + (y-2)^2\right] = 1 - (y-2)^2 = 1 - \left(y^2 - 4y + 4\right) = -y^2 + 4y - 3.$

$V = 2\pi \int_1^3 y(-y^2 + 4y - 3)\,dy$

$= 2\pi \int_1^3 (-y^3 + 4y^2 - 3y)\,dy$

$= 2\pi\left[-\frac{1}{4}y^4 + \frac{4}{3}y^3 - \frac{3}{2}y^2\right]_1^3$

$= 2\pi\left[\left(-\frac{81}{4} + 36 - \frac{27}{2}\right) - \left(-\frac{1}{4} + \frac{4}{3} - \frac{3}{2}\right)\right]$

$= 2\pi\left(\frac{8}{3}\right) = \frac{16}{3}\pi$

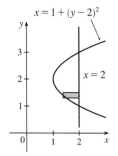

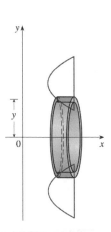

15. The shell has radius $2 - x$, circumference $2\pi(2 - x)$, and height x^4.

$V = \int_0^1 2\pi(2 - x)x^4\,dx$

$= 2\pi \int_0^1 (2x^4 - x^5)\,dx$

$= 2\pi\left[\frac{2}{5}x^5 - \frac{1}{6}x^6\right]_0^1$

$= 2\pi\left[\left(\frac{2}{5} - \frac{1}{6}\right) - 0\right] = 2\pi\left(\frac{7}{30}\right) = \frac{7}{15}\pi$

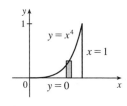

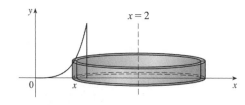

17. The shell has radius $x - 1$, circumference $2\pi(x - 1)$, and height $(4x - x^2) - 3 = -x^2 + 4x - 3$.

$V = \int_1^3 2\pi(x - 1)(-x^2 + 4x - 3)\,dx$

$= 2\pi \int_1^3 (-x^3 + 5x^2 - 7x + 3)\,dx$

$= 2\pi \left[-\frac{1}{4}x^4 + \frac{5}{3}x^3 - \frac{7}{2}x^2 + 3x\right]_1^3$

$= 2\pi \left[\left(-\frac{81}{4} + 45 - \frac{63}{2} + 9\right) - \left(-\frac{1}{4} + \frac{5}{3} - \frac{7}{2} + 3\right)\right]$

$= 2\pi\left(\frac{4}{3}\right) = \frac{8}{3}\pi$

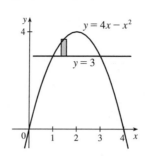

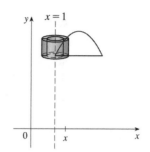

19. The shell has radius $1 - y$, circumference $2\pi(1 - y)$, and height $1 - \sqrt[3]{y}$ $\left[y = x^3 \Leftrightarrow x = \sqrt[3]{y}\right]$.

$V = \int_0^1 2\pi(1 - y)(1 - y^{1/3})\,dy$

$= 2\pi \int_0^1 (1 - y - y^{1/3} + y^{4/3})\,dy$

$= 2\pi \left[y - \frac{1}{2}y^2 - \frac{3}{4}y^{4/3} + \frac{3}{7}y^{7/3}\right]_0^1$

$= 2\pi\left[\left(1 - \frac{1}{2} - \frac{3}{4} + \frac{3}{7}\right) - 0\right]$

$= 2\pi\left(\frac{5}{28}\right) = \frac{5}{14}\pi$

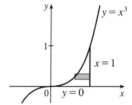

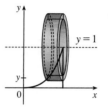

21. $V = \int_1^2 2\pi x \ln x\,dx$

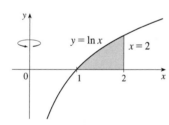

23. $V = \int_0^1 2\pi[x - (-1)]\left(\sin\frac{\pi}{2}x - x^4\right)dx$

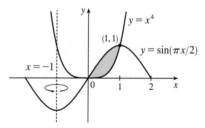

25. $V = \int_0^\pi 2\pi(4 - y)\sqrt{\sin y}\,dy$

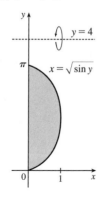

27. $V = \int_0^1 2\pi x \sqrt{1 + x^3}\, dx.$ Let $f(x) = x\sqrt{1 + x^3}.$

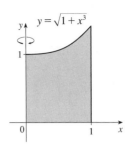

Then the Midpoint Rule with $n = 5$ gives

$\int_0^1 f(x)\, dx \approx \frac{1-0}{5}\left[f(0.1) + f(0.3) + f(0.5) + f(0.7) + f(0.9)\right]$

$\approx 0.2(2.9290)$

Multiplying by 2π gives $V \approx 3.68.$

29. $\int_0^3 2\pi x^5\, dx = 2\pi \int_0^3 x(x^4)\, dx.$ The solid is obtained by rotating the region $0 \le y \le x^4, 0 \le x \le 3$ about the y-axis using cylindrical shells.

31. $\int_0^1 2\pi(3 - y)(1 - y^2)\, dy.$ The solid is obtained by rotating the region bounded by (i) $x = 1 - y^2$, $x = 0$, and $y = 0$ or

(ii) $x = y^2$, $x = 1$, and $y = 0$ about the line $y = 3$ using cylindrical shells.

33.

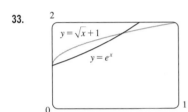

From the graph, the curves intersect at $x = 0$ and $x = a \approx 0.56$,

with $\sqrt{x} + 1 > e^x$ on the interval $(0, a)$. So the volume of the solid

obtained by rotating the region about the y-axis is

$$V = 2\pi \int_0^a x\left[\left(\sqrt{x} + 1\right) - e^x\right]\, dx \approx 0.13.$$

35. $V = 2\pi \displaystyle\int_0^{\pi/2} \left[\left(\frac{\pi}{2} - x\right)\left(\sin^2 x - \sin^4 x\right)\right] dx$

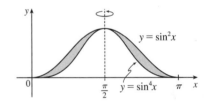

$\overset{CAS}{=} \frac{1}{32}\pi^3$

37. Use shells:

$V = \int_2^4 2\pi x(-x^2 + 6x - 8)\, dx = 2\pi \int_2^4 (-x^3 + 6x^2 - 8x)\, dx$

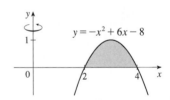

$= 2\pi\left[-\frac{1}{4}x^4 + 2x^3 - 4x^2\right]_2^4$

$= 2\pi[(-64 + 128 - 64) - (-4 + 16 - 16)]$

$= 2\pi(4) = 8\pi$

39. Use shells:

$V = \int_1^4 2\pi[x - (-1)][5 - (x + 4/x)]\, dx$

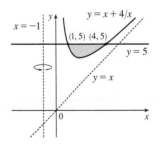

$= 2\pi \int_1^4 (x + 1)(5 - x - 4/x)\, dx$

$= 2\pi \int_1^4 (5x - x^2 - 4 + 5 - x - 4/x)\, dx$

$= 2\pi \int_1^4 (-x^2 + 4x + 1 - 4/x)\, dx = 2\pi\left[-\frac{1}{3}x^3 + 2x^2 + x - 4\ln x\right]_1^4$

$= 2\pi\left[\left(-\frac{64}{3} + 32 + 4 - 4\ln 4\right) - \left(-\frac{1}{3} + 2 + 1 - 0\right)\right]$

$= 2\pi(12 - 4\ln 4) = 8\pi(3 - \ln 4)$

41. Use disks: $x^2 + (y-1)^2 = 1 \quad \Leftrightarrow \quad x = \pm\sqrt{1-(y-1)^2}$

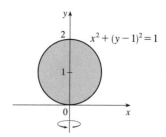

$$V = \pi \int_0^2 \left[\sqrt{1-(y-1)^2}\right]^2 dy = \pi \int_0^2 (2y - y^2)\, dy$$

$$= \pi \left[y^2 - \tfrac{1}{3}y^3\right]_0^2 = \pi\left(4 - \tfrac{8}{3}\right) = \tfrac{4}{3}\pi$$

43. Use shells:

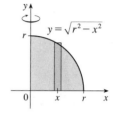

$$V = 2\int_0^r 2\pi x\,\sqrt{r^2 - x^2}\, dx$$

$$= -2\pi \int_0^r (r^2 - x^2)^{1/2}(-2x)\, dx$$

$$= \left[-2\pi \cdot \tfrac{2}{3}(r^2 - x^2)^{3/2}\right]_0^r$$

$$= -\tfrac{4}{3}\pi(0 - r^3) = \tfrac{4}{3}\pi r^3$$

45. $V = 2\pi \int_0^r x\left(-\dfrac{h}{r}x + h\right)dx = 2\pi h \int_0^r \left(-\dfrac{x^2}{r} + x\right)dx$

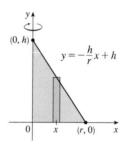

$$= 2\pi h \left[-\dfrac{x^3}{3r} + \dfrac{x^2}{2}\right]_0^r = 2\pi h\,\dfrac{r^2}{6} = \dfrac{\pi r^2 h}{3}$$

6.4 Work

1. $W = Fd = mgd = (40)(9.8)(1.5) = 588$ J

3. $W = \displaystyle\int_a^b f(x)\, dx = \int_0^9 \dfrac{10}{(1+x)^2}\, dx = 10 \int_1^{10} \dfrac{1}{u^2}\, du \quad [u = 1 + x,\ du = dx] = 10\left[-\dfrac{1}{u}\right]_1^{10} = 10\left(-\tfrac{1}{10} + 1\right) = 9$ ft-lb

5. The force function is given by $F(x)$ (in newtons) and the work (in joules) is the area under the curve, given by

$\int_0^8 F(x)\, dx = \int_0^4 F(x)\, dx + \int_4^8 F(x)\, dx = \tfrac{1}{2}(4)(30) + (4)(30) = 180$ J.

7. $10 = f(x) = kx = \tfrac{1}{3}k$ [4 inches $= \tfrac{1}{3}$ foot], so $k = 30$ lb/ft and $f(x) = 30x$. Now 6 inches $= \tfrac{1}{2}$ foot, so

$W = \int_0^{1/2} 30x\, dx = \left[15x^2\right]_0^{1/2} = \dfrac{15}{4}$ ft-lb.

9. (a) If $\int_0^{0.12} kx\, dx = 2$ J, then $2 = \left[\tfrac{1}{2}kx^2\right]_0^{0.12} = \tfrac{1}{2}k(0.0144) = 0.0072k$ and $k = \dfrac{2}{0.0072} = \dfrac{2500}{9} \approx 277.78$ N/m.

Thus, the work needed to stretch the spring from 35 cm to 40 cm is

$\int_{0.05}^{0.10} \dfrac{2500}{9}x\, dx = \left[\dfrac{1250}{9}x^2\right]_{1/20}^{1/10} = \dfrac{1250}{9}\left(\dfrac{1}{100} - \dfrac{1}{400}\right) = \dfrac{25}{24} \approx 1.04$ J.

(b) $f(x) = kx$, so $30 = \dfrac{2500}{9}x$ and $x = \dfrac{270}{2500}$ m $= 10.8$ cm

11. The distance from 20 cm to 30 cm is 0.1 m, so with $f(x) = kx$, we get $W_1 = \int_0^{0.1} kx\,dx = k\left[\frac{1}{2}x^2\right]_0^{0.1} = \frac{1}{200}k$.

Now $W_2 = \int_{0.1}^{0.2} kx\,dx = k\left[\frac{1}{2}x^2\right]_{0.1}^{0.2} = k\left(\frac{4}{200} - \frac{1}{200}\right) = \frac{3}{200}k$. Thus, $W_2 = 3W_1$.

In Exercises 13–20, n is the number of subintervals of length Δx, and x_i^* is a sample point in the ith subinterval $[x_{i-1}, x_i]$.

13. (a) The portion of the rope from x ft to $(x + \Delta x)$ ft below the top of the building weighs $\frac{1}{2}\Delta x$ lb and must be lifted x_i^* ft,

so its contribution to the total work is $\frac{1}{2}x_i^* \Delta x$ ft-lb. The total work is

$$W = \lim_{n \to \infty} \sum_{i=1}^{n} \tfrac{1}{2}x_i^* \Delta x = \int_0^{50} \tfrac{1}{2}x\,dx = \left[\tfrac{1}{4}x^2\right]_0^{50} = \frac{2500}{4} = 625 \text{ ft-lb}$$

Notice that the exact height of the building does not matter (as long as it is more than 50 ft).

(b) When half the rope is pulled to the top of the building, the work to lift the top half of the rope is

$W_1 = \int_0^{25} \frac{1}{2}x\,dx = \left[\frac{1}{4}x^2\right]_0^{25} = \frac{625}{4}$ ft-lb. The bottom half of the rope is lifted 25 ft and the work needed to accomplish

that is $W_2 = \int_{25}^{50} \frac{1}{2} \cdot 25\,dx = \frac{25}{2}\left[x\right]_{25}^{50} = \frac{625}{2}$ ft-lb. The total work done in pulling half the rope to the top of the building

is $W = W_1 + W_2 = \frac{625}{2} + \frac{625}{4} = \frac{3}{4} \cdot 625 = \frac{1875}{4}$ ft-lb.

15. The work needed to lift the cable is $\lim_{n \to \infty} \sum_{i=1}^{n} 2x_i^* \Delta x = \int_0^{500} 2x\,dx = \left[x^2\right]_0^{500} = 250{,}000$ ft-lb. The work needed to lift

the coal is 800 lb · 500 ft = 400,000 ft-lb. Thus, the total work required is $250{,}000 + 400{,}000 = 650{,}000$ ft-lb.

17. At a height of x meters $(0 \le x \le 12)$, the mass of the rope is $(0.8 \text{ kg/m})(12 - x \text{ m}) = (9.6 - 0.8x)$ kg and the mass of the

water is $\left(\frac{36}{12} \text{ kg/m}\right)(12 - x \text{ m}) = (36 - 3x)$ kg. The mass of the bucket is 10 kg, so the total mass is

$(9.6 - 0.8x) + (36 - 3x) + 10 = (55.6 - 3.8x)$ kg, and hence, the total force is $9.8(55.6 - 3.8x)$ N. The work needed to lift

the bucket Δx m through the ith subinterval of $[0, 12]$ is $9.8(55.6 - 3.8x_i^*)\Delta x$, so the total work is

$$W = \lim_{n \to \infty} \sum_{i=1}^{n} 9.8(55.6 - 3.8x_i^*)\Delta x = \int_0^{12}(9.8)(55.6 - 3.8x)\,dx = 9.8\left[55.6x - 1.9x^2\right]_0^{12} = 9.8(393.6) \approx 3857 \text{ J}$$

19. A "slice" of water Δx m thick and lying at a depth of x_i^* m (where $0 \le x_i^* \le \frac{1}{2}$) has volume $(2 \times 1 \times \Delta x)$ m^3, a mass of

$2000\,\Delta x$ kg, weighs about $(9.8)(2000\,\Delta x) = 19{,}600\,\Delta x$ N, and thus requires about $19{,}600x_i^* \Delta x$ J of work for its removal.

So $W = \lim_{n \to \infty} \sum_{i=1}^{n} 19{,}600x_i^* \Delta x = \int_0^{1/2} 19{,}600x\,dx = \left[9800x^2\right]_0^{1/2} = 2450$ J.

21. A rectangular "slice" of water Δx m thick and lying x m above the bottom has width x m and volume $8x\,\Delta x$ m^3. It weighs

about $(9.8 \times 1000)(8x\,\Delta x)$ N, and must be lifted $(5 - x)$ m by the pump, so the work needed is about

$(9.8 \times 10^3)(5 - x)(8x\,\Delta x)$ J. The total work required is

$$W \approx \int_0^3 (9.8 \times 10^3)(5 - x)8x\,dx = (9.8 \times 10^3)\int_0^3(40x - 8x^2)\,dx = (9.8 \times 10^3)\left[20x^2 - \tfrac{8}{3}x^3\right]_0^3$$
$$= (9.8 \times 10^3)(180 - 72) = (9.8 \times 10^3)(108) = 1058.4 \times 10^3 \approx 1.06 \times 10^6 \text{ J}$$

23. Let x measure depth (in feet) below the spout at the top of the tank. A horizontal disk-shaped "slice" of water Δx ft thick and lying at coordinate x has radius $\frac{3}{8}(16-x)$ ft $(\star)$ and volume $\pi r^2 \Delta x = \pi \cdot \frac{9}{64}(16-x)^2 \Delta x$ ft^3. It weighs about $(62.5)\frac{9\pi}{64}(16-x)^2 \Delta x$ lb and must be lifted x ft by the pump, so the work needed to pump it out is about $(62.5)x\frac{9\pi}{64}(16-x)^2 \Delta x$ ft-lb. The total work required is

$$W \approx \int_0^8 (62.5)x\,\frac{9\pi}{64}(16-x)^2\,dx = (62.5)\frac{9\pi}{64}\int_0^8 x(256 - 32x + x^2)\,dx$$

$$= (62.5)\frac{9\pi}{64}\int_0^8 (256x - 32x^2 + x^3)\,dx = (62.5)\frac{9\pi}{64}\left[128x^2 - \frac{32}{3}x^3 + \frac{1}{4}x^4\right]_0^8$$

$$= (62.5)\frac{9\pi}{64}\left(\frac{11,264}{3}\right) = 33,000\pi \approx 1.04 \times 10^5 \text{ ft-lb}$$

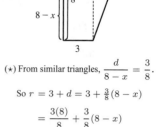

$(\star)$ From similar triangles, $\dfrac{d}{8-x} = \dfrac{3}{8}$.

So $r = 3 + d = 3 + \frac{3}{8}(8-x)$

$\quad = \dfrac{3(8)}{8} + \dfrac{3}{8}(8-x)$

$\quad = \frac{3}{8}(16-x)$

25. If only 4.7×10^5 J of work is done, then only the water above a certain level (call it h) will be pumped out. So we use the same formula as in Exercise 21, except that the work is fixed, and we are trying to find the lower limit of integration:

$$4.7 \times 10^5 \approx \int_h^3 (9.8 \times 10^3)(5-x)8x\,dx = (9.8 \times 10^3)\left[20x^2 - \frac{8}{3}x^3\right]_h^3 \quad\Leftrightarrow$$

$$\frac{4.7}{9.8} \times 10^2 \approx 48 = \left(20 \cdot 3^2 - \frac{8}{3}\cdot 3^3\right) - \left(20h^2 - \frac{8}{3}h^3\right) \quad\Leftrightarrow$$

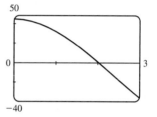

$2h^3 - 15h^2 + 45 = 0$. To find the solution of this equation, we plot $2h^3 - 15h^2 + 45$ between $h = 0$ and $h = 3$. We see that the equation is satisfied for $h \approx 2.0$. So the depth of water remaining in the tank is about 2.0 m.

27. $V = \pi r^2 x$, so V is a function of x and P can also be regarded as a function of x. If $V_1 = \pi r^2 x_1$ and $V_2 = \pi r^2 x_2$, then

$$W = \int_{x_1}^{x_2} F(x)\,dx = \int_{x_1}^{x_2} \pi r^2 P(V(x))\,dx = \int_{x_1}^{x_2} P(V(x))\,dV(x) \qquad [\text{Let } V(x) = \pi r^2 x, \text{ so } dV(x) = \pi r^2\,dx.]$$

$$= \int_{V_1}^{V_2} P(V)\,dV \quad \text{by the Substitution Rule.}$$

29. $W = \displaystyle\int_a^b F(r)\,dr = \int_a^b G\frac{m_1 m_2}{r^2}\,dr = Gm_1 m_2\left[\frac{-1}{r}\right]_a^b = Gm_1 m_2\left(\frac{1}{a} - \frac{1}{b}\right)$

6.5 Average Value of a Function

1. $f_{\text{ave}} = \frac{1}{b-a}\int_a^b f(x)\,dx = \frac{1}{4-0}\int_0^4 (4x - x^2)\,dx = \frac{1}{4}\left[2x^2 - \frac{1}{3}x^3\right]_0^4 = \frac{1}{4}\left[\left(32 - \frac{64}{3}\right) - 0\right] = \frac{1}{4}\left(\frac{32}{3}\right) = \frac{8}{3}$

3. $g_{\text{ave}} = \frac{1}{b-a}\int_a^b g(x)\,dx = \frac{1}{8-1}\int_1^8 \sqrt[3]{x}\,dx = \frac{1}{7}\left[\frac{3}{4}x^{4/3}\right]_1^8 = \frac{3}{28}(16-1) = \frac{45}{28}$

5. $f_{\text{ave}} = \frac{1}{5-0}\int_0^5 te^{-t^2}\,dt = \frac{1}{5}\int_0^{-25} e^u\left(-\frac{1}{2}\,du\right) \qquad [u = -t^2,\ du = -2t\,dt,\ t\,dt = -\frac{1}{2}\,du]$

$$= -\frac{1}{10}\left[e^u\right]_0^{-25} = -\frac{1}{10}(e^{-25} - 1) = \frac{1}{10}(1 - e^{-25})$$

7. $h_{\text{ave}} = \frac{1}{\pi - 0} \int_0^\pi \cos^4 x \sin x\, dx = \frac{1}{\pi} \int_1^{-1} u^4 (-du)$ $[u = \cos x,\ du = -\sin x\, dx]$

$= \frac{1}{\pi} \int_{-1}^1 u^4\, du = \frac{1}{\pi} \cdot 2 \int_0^1 u^4\, du$ [by Theorem 5.5.7] $= \frac{2}{\pi} \left[\frac{1}{5} u^5 \right]_0^1 = \frac{2}{5\pi}$

9. (a) $f_{\text{ave}} = \dfrac{1}{5 - 2} \displaystyle\int_2^5 (x - 3)^2\, dx = \dfrac{1}{3} \left[\dfrac{1}{3}(x - 3)^3 \right]_2^5$

$= \frac{1}{9} \left[2^3 - (-1)^3 \right] = \frac{1}{9}(8 + 1) = 1$

(b) $f(c) = f_{\text{ave}}$ $\Leftrightarrow$ $(c - 3)^2 = 1$ $\Leftrightarrow$

$c - 3 = \pm 1$ $\Leftrightarrow$ $c = 2$ or 4

(c)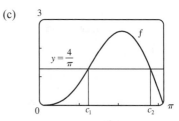

11. (a) $f_{\text{ave}} = \dfrac{1}{\pi - 0} \displaystyle\int_0^\pi (2 \sin x - \sin 2x)\, dx$

$= \frac{1}{\pi} \left[-2 \cos x + \frac{1}{2} \cos 2x \right]_0^\pi$

$= \frac{1}{\pi} \left[\left(2 + \frac{1}{2} \right) - \left(-2 + \frac{1}{2} \right) \right] = \frac{4}{\pi}$

(b) $f(c) = f_{\text{ave}}$ $\Leftrightarrow$ $2 \sin c - \sin 2c = \frac{4}{\pi}$ $\Leftrightarrow$

$c_1 \approx 1.238$ or $c_2 \approx 2.808$

(c)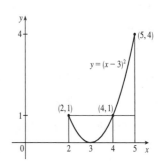

13. f is continuous on $[1, 3]$, so by the Mean Value Theorem for Integrals there exists a number c in $[1, 3]$ such that

$\int_1^3 f(x)\, dx = f(c)(3 - 1)$ $\Rightarrow$ $8 = 2f(c)$; that is, there is a number c such that $f(c) = \frac{8}{2} = 4$.

15. $f_{\text{ave}} = \dfrac{1}{50 - 20} \displaystyle\int_{20}^{50} f(x)\, dx \approx \dfrac{1}{30} M_3 = \dfrac{1}{30} \cdot \dfrac{50 - 20}{3} [f(25) + f(35) + f(45)] = \frac{1}{3}(38 + 29 + 48) = \frac{115}{3} = 38\frac{1}{3}$

17. Let $t = 0$ and $t = 12$ correspond to 9 AM and 9 PM, respectively.

$$T_{\text{ave}} = \frac{1}{12 - 0} \int_0^{12} \left[50 + 14 \sin \frac{1}{12} \pi t \right] dt = \frac{1}{12} \left[50t - 14 \cdot \frac{12}{\pi} \cos \frac{1}{12} \pi t \right]_0^{12}$$

$$= \frac{1}{12} \left[50 \cdot 12 + 14 \cdot \frac{12}{\pi} + 14 \cdot \frac{12}{\pi} \right] = \left(50 + \frac{28}{\pi} \right) {}^\circ \text{F} \approx 59 {}^\circ \text{F}$$

19. $\rho_{\text{ave}} = \dfrac{1}{8} \displaystyle\int_0^8 \dfrac{12}{\sqrt{x + 1}}\, dx = \dfrac{3}{2} \displaystyle\int_0^8 (x + 1)^{-1/2}\, dx = \left[3\sqrt{x + 1} \right]_0^8 = 9 - 3 = 6$ kg/m

21. $V_{\text{ave}} = \frac{1}{5} \int_0^5 V(t)\, dt = \frac{1}{5} \int_0^5 \frac{5}{4\pi} \left[1 - \cos\left(\frac{2}{5} \pi t \right) \right] dt = \frac{1}{4\pi} \int_0^5 \left[1 - \cos\left(\frac{2}{5} \pi t \right) \right] dt$

$= \frac{1}{4\pi} \left[t - \frac{5}{2\pi} \sin\left(\frac{2}{5} \pi t \right) \right]_0^5 = \frac{1}{4\pi} \left[(5 - 0) - 0 \right] = \frac{5}{4\pi} \approx 0.4$ L

23. Let $F(x) = \int_a^x f(t)\, dt$ for x in $[a, b]$. Then F is continuous on $[a, b]$ and differentiable on (a, b), so by the Mean Value

Theorem there is a number c in (a, b) such that $F(b) - F(a) = F'(c)(b - a)$. But $F'(x) = f(x)$ by the Fundamental

Theorem of Calculus. Therefore, $\int_a^b f(t)\, dt - 0 = f(c)(b - a)$.

6 Review

1. (a) See Section 6.1, Figure 2 and Equations 6.1.1 and 6.1.2.

(b) Instead of using "top minus bottom" and integrating from left to right, we use "right minus left" and integrate from bottom to top. See Figures 11 and 12 in Section 6.1.

2. The numerical value of the area represents the number of meters by which Sue is ahead of Kathy after 1 minute.

3. (a) See the discussion in Section 6.2, near Figures 2 and 3, ending in the Definition of Volume.

(b) See the discussion between Examples 5 and 6 in Section 6.2. If the cross-section is a disk, find the radius in terms of x or y and use $A = \pi(\text{radius})^2$. If the cross-section is a washer, find the inner radius r_{in} and outer radius r_{out} and use
$$A = \pi\left(r_{\text{out}}^2\right) - \pi\left(r_{\text{in}}^2\right).$$

4. (a) $V = 2\pi r h \, \Delta r = (\text{circumference})(\text{height})(\text{thickness})$

(b) For a typical shell, find the circumference and height in terms of x or y and calculate
$$V = \int_a^b (\text{circumference})(\text{height})(dx \text{ or } dy), \text{ where } a \text{ and } b \text{ are the limits on } x \text{ or } y.$$

(c) Sometimes slicing produces washers or disks whose radii are difficult (or impossible) to find explicitly. On other occasions, the cylindrical shell method leads to an easier integral than slicing does.

5. $\int_0^6 f(x)\,dx$ represents the amount of work done. Its units are newton-meters, or joules.

6. (a) The average value of a function f on an interval $[a, b]$ is $f_{\text{ave}} = \dfrac{1}{b - a} \displaystyle\int_a^b f(x)\,dx$.

(b) The Mean Value Theorem for Integrals says that there is a number c at which the value of f is exactly equal to the average value of the function, that is, $f(c) = f_{\text{ave}}$. For a geometric interpretation of the Mean Value Theorem for Integrals, see Figure 2 in Section 6.5 and the discussion that accompanies it.

1. The curves intersect when $x^2 = 4x - x^2 \iff 2x^2 - 4x = 0 \iff 2x(x - 2) = 0 \iff x = 0$ or 2.
$$A = \int_0^2 \left[(4x - x^2) - x^2\right] dx = \int_0^2 (4x - 2x^2)\,dx$$
$$= \left[2x^2 - \tfrac{2}{3}x^3\right]_0^2 = \left[\left(8 - \tfrac{16}{3}\right) - 0\right] = \tfrac{8}{3}$$

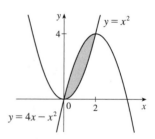

3. If $x \geq 0$, then $|x| = x$, and the graphs intersect when $x = 1 - 2x^2 \iff 2x^2 + x - 1 = 0 \iff (2x - 1)(x + 1) = 0 \iff x = \tfrac{1}{2}$ or -1, but $-1 < 0$. By symmetry, we can double the area from $x = 0$ to $x = \tfrac{1}{2}$.

$$A = 2\int_0^{1/2} \left[(1 - 2x^2) - x\right] dx = 2\int_0^{1/2} (-2x^2 - x + 1)\,dx$$
$$= 2\left[-\tfrac{2}{3}x^3 - \tfrac{1}{2}x^2 + x\right]_0^{1/2} = 2\left[\left(-\tfrac{1}{12} - \tfrac{1}{8} + \tfrac{1}{2}\right) - 0\right]$$
$$= 2\left(\tfrac{7}{24}\right) = \tfrac{7}{12}$$

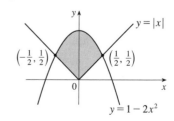

5. $A = \int_0^2 \left[\sin\left(\frac{\pi x}{2}\right) - (x^2 - 2x)\right] dx$

$= \left[-\frac{2}{\pi}\cos\left(\frac{\pi x}{2}\right) - \frac{1}{3}x^3 + x^2\right]_0^2$

$= \left(\frac{2}{\pi} - \frac{8}{3} + 4\right) - \left(-\frac{2}{\pi} - 0 + 0\right) = \frac{4}{3} + \frac{4}{\pi}$

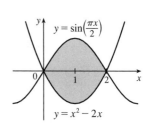

7. Using washers with inner radius x^2 and outer radius $2x$, we have

$V = \pi \int_0^2 \left[(2x)^2 - (x^2)^2\right] dx = \pi \int_0^2 (4x^2 - x^4) \, dx$

$= \pi \left[\frac{4}{3}x^3 - \frac{1}{5}x^5\right]_0^2 = \pi\left(\frac{32}{3} - \frac{32}{5}\right)$

$= 32\pi \cdot \frac{2}{15} = \frac{64}{15}\pi$

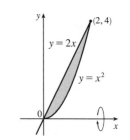

9. $V = \pi \int_{-3}^3 \left\{\left[(9 - y^2) - (-1)\right]^2 - \left[0 - (-1)\right]^2\right\} dy$

$= 2\pi \int_0^3 \left[(10 - y^2)^2 - 1\right] dy = 2\pi \int_0^3 (100 - 20y^2 + y^4 - 1) \, dy$

$= 2\pi \int_0^3 (99 - 20y^2 + y^4) \, dy = 2\pi \left[99y - \frac{20}{3}y^3 + \frac{1}{5}y^5\right]_0^3$

$= 2\pi\left(297 - 180 + \frac{243}{5}\right) = \frac{1656}{5}\pi$

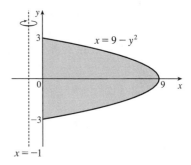

11. The graph of $x^2 - y^2 = a^2$ is a hyperbola with right and left branches.

Solving for y gives us $y^2 = x^2 - a^2 \;\Rightarrow\; y = \pm\sqrt{x^2 - a^2}$.

We'll use shells and the height of each shell is

$\sqrt{x^2 - a^2} - \left(-\sqrt{x^2 - a^2}\right) = 2\sqrt{x^2 - a^2}$.

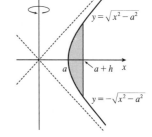

The volume is $V = \int_a^{a+h} 2\pi x \cdot 2\sqrt{x^2 - a^2} \, dx$. To evaluate, let $u = x^2 - a^2$,

so $du = 2x \, dx$ and $x \, dx = \frac{1}{2} du$. When $x = a$, $u = 0$, and when $x = a + h$,

$u = (a + h)^2 - a^2 = a^2 + 2ah + h^2 - a^2 = 2ah + h^2$.

Thus, $V = 4\pi \int_0^{2ah+h^2} \sqrt{u} \left(\frac{1}{2} du\right) = 2\pi \left[\frac{2}{3}u^{3/2}\right]_0^{2ah+h^2} = \frac{4}{3}\pi\left(2ah + h^2\right)^{3/2}$.

13. A shell has radius $\frac{\pi}{2} - x$, circumference $2\pi\left(\frac{\pi}{2} - x\right)$, and height $\cos^2 x - \frac{1}{4}$.

$y = \cos^2 x$ intersects $y = \frac{1}{4}$ when $\cos^2 x = \frac{1}{4} \;\Leftrightarrow\;$

$\cos x = \pm\frac{1}{2} \quad [\,|x| \le \pi/2\,] \;\Leftrightarrow\; x = \pm\frac{\pi}{3}$.

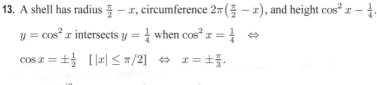

$V = \int_{-\pi/3}^{\pi/3} 2\pi\left(\frac{\pi}{2} - x\right)\left(\cos^2 x - \frac{1}{4}\right) dx$

15. (a) A cross-section is a washer with inner radius x^2 and outer radius x.

$$V = \int_0^1 \pi\left[(x)^2 - (x^2)^2\right] dx = \int_0^1 \pi(x^2 - x^4)\, dx = \pi\left[\tfrac{1}{3}x^3 - \tfrac{1}{5}x^5\right]_0^1 = \pi\left[\tfrac{1}{3} - \tfrac{1}{5}\right] = \tfrac{2}{15}\pi$$

(b) A cross-section is a washer with inner radius y and outer radius $\sqrt{y}$.

$$V = \int_0^1 \pi\left[\left(\sqrt{y}\right)^2 - y^2\right] dy = \int_0^1 \pi(y - y^2)\, dy = \pi\left[\tfrac{1}{2}y^2 - \tfrac{1}{3}y^3\right]_0^1 = \pi\left[\tfrac{1}{2} - \tfrac{1}{3}\right] = \tfrac{\pi}{6}$$

(c) A cross-section is a washer with inner radius $2 - x$ and outer radius $2 - x^2$.

$$V = \int_0^1 \pi\left[(2 - x^2)^2 - (2 - x)^2\right] dx = \int_0^1 \pi(x^4 - 5x^2 + 4x)\, dx = \pi\left[\tfrac{1}{5}x^5 - \tfrac{5}{3}x^3 + 2x^2\right]_0^1 = \pi\left[\tfrac{1}{5} - \tfrac{5}{3} + 2\right] = \tfrac{8}{15}\pi$$

17. (a) Using the Midpoint Rule on $[0, 1]$ with $f(x) = \tan(x^2)$ and $n = 4$, we estimate

$$A = \int_0^1 \tan(x^2)\, dx \approx \tfrac{1}{4}\left[\tan\left(\left(\tfrac{1}{8}\right)^2\right) + \tan\left(\left(\tfrac{3}{8}\right)^2\right) + \tan\left(\left(\tfrac{5}{8}\right)^2\right) + \tan\left(\left(\tfrac{7}{8}\right)^2\right)\right] \approx \tfrac{1}{4}(1.53) \approx 0.38$$

(b) Using the Midpoint Rule on $[0, 1]$ with $f(x) = \pi\tan^2(x^2)$ (for disks) and $n = 4$, we estimate

$$V = \int_0^1 f(x)\, dx \approx \tfrac{1}{4}\pi\left[\tan^2\left(\left(\tfrac{1}{8}\right)^2\right) + \tan^2\left(\left(\tfrac{3}{8}\right)^2\right) + \tan^2\left(\left(\tfrac{5}{8}\right)^2\right) + \tan^2\left(\left(\tfrac{7}{8}\right)^2\right)\right] \approx \tfrac{\pi}{4}(1.114) \approx 0.87$$

19. $\int_0^{\pi/2} 2\pi x \cos x\, dx = \int_0^{\pi/2} (2\pi x) \cos x\, dx$

The solid is obtained by rotating the region $\mathcal{R} = \left\{(x, y) \mid 0 \le x \le \tfrac{\pi}{2}, 0 \le y \le \cos x\right\}$ about the y-axis.

21. $\int_0^\pi \pi(2 - \sin x)^2\, dx$

The solid is obtained by rotating the region $\mathcal{R} = \left\{(x, y) \mid 0 \le x \le \pi, 0 \le y \le 2 - \sin x\right\}$ about the x-axis.

23. Take the base to be the disk $x^2 + y^2 \le 9$. Then $V = \int_{-3}^3 A(x)\, dx$, where $A(x_0)$ is the area of the isosceles right triangle whose hypotenuse lies along the line $x = x_0$ in the xy-plane. The length of the hypotenuse is $2\sqrt{9 - x^2}$ and the length of each leg is $\sqrt{2}\sqrt{9 - x^2}$. $A(x) = \tfrac{1}{2}\left(\sqrt{2}\sqrt{9 - x^2}\right)^2 = 9 - x^2$, so

$$V = 2\int_0^3 A(x)\, dx = 2\int_0^3 (9 - x^2)\, dx = 2\left[9x - \tfrac{1}{3}x^3\right]_0^3 = 2(27 - 9) = 36$$

25. Equilateral triangles with sides measuring $\tfrac{1}{4}x$ meters have height $\tfrac{1}{4}x \sin 60° = \tfrac{\sqrt{3}}{8}x$. Therefore,

$$A(x) = \tfrac{1}{2} \cdot \tfrac{1}{4}x \cdot \tfrac{\sqrt{3}}{8}x = \tfrac{\sqrt{3}}{64}x^2. \quad V = \int_0^{20} A(x)\, dx = \tfrac{\sqrt{3}}{64}\int_0^{20} x^2\, dx = \tfrac{\sqrt{3}}{64}\left[\tfrac{1}{3}x^3\right]_0^{20} = \tfrac{8000\sqrt{3}}{64 \cdot 3} = \tfrac{125\sqrt{3}}{3} \text{ m}^3.$$

27. $f(x) = kx \implies 30 \text{ N} = k(15 - 12) \text{ cm} \implies k = 10 \text{ N/cm} = 1000 \text{ N/m.} \quad 20 \text{ cm} - 12 \text{ cm} = 0.08 \text{ m} \implies$

$$W = \int_0^{0.08} kx\, dx = 1000 \int_0^{0.08} x\, dx = 500\left[x^2\right]_0^{0.08} = 500(0.08)^2 = 3.2 \text{ N·m} = 3.2 \text{ J.}$$

29. (a) The parabola has equation $y = ax^2$ with vertex at the origin and passing through

$(4, 4)$. $4 = a \cdot 4^2 \;\Rightarrow\; a = \frac{1}{4} \;\Rightarrow\; y = \frac{1}{4}x^2 \;\Rightarrow\; x^2 = 4y \;\Rightarrow$

$x = 2\sqrt{y}$. Each circular disk has radius $2\sqrt{y}$ and is moved $4 - y$ ft.

$$W = \int_0^4 \pi \left(2\sqrt{y}\right)^2 62.5(4 - y)\, dy = 250\pi \int_0^4 y(4 - y)\, dy$$

$$= 250\pi \left[2y^2 - \tfrac{1}{3}y^3\right]_0^4 = 250\pi \left(32 - \tfrac{64}{3}\right) = \tfrac{8000\pi}{3} \approx 8378 \text{ ft-lb}$$

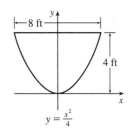

(b) In part (a) we knew the final water level (0) but not the amount of work done. Here

we use the same equation, except with the work fixed, and the lower limit of

integration (that is, the final water level—call it h) unknown: $W = 4000 \;\Leftrightarrow$

$250\pi \left[2y^2 - \tfrac{1}{3}y^3\right]_h^4 = 4000 \;\Leftrightarrow\; \tfrac{16}{\pi} = \left[\left(32 - \tfrac{64}{3}\right) - \left(2h^2 - \tfrac{1}{3}h^3\right)\right] \;\Leftrightarrow$

$h^3 - 6h^2 + 32 - \tfrac{48}{\pi} = 0$. We graph the function $f(h) = h^3 - 6h^2 + 32 - \tfrac{48}{\pi}$

on the interval $[0, 4]$ to see where it is 0. From the graph, $f(h) = 0$ for $h \approx 2.1$.

So the depth of water remaining is about 2.1 ft.

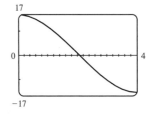

31. $\displaystyle\lim_{h \to 0} f_{\text{ave}} = \lim_{h \to 0} \frac{1}{(x + h) - x} \int_x^{x+h} f(t)\, dt = \lim_{h \to 0} \frac{F(x + h) - F(x)}{h}$, where $F(x) = \int_a^x f(t)\, dt$. But we recognize this

limit as being $F'(x)$ by the definition of a derivative. Therefore, $\displaystyle\lim_{h \to 0} f_{\text{ave}} = F'(x) = f(x)$ by FTC1.

☐ PROBLEMS PLUS

1. (a) The area under the graph of f from 0 to t is equal to $\int_0^t f(x)\,dx$, so the requirement is that $\int_0^t f(x)\,dx = t^3$ for all t. We differentiate both sides of this equation with respect to t (with the help of FTC1) to get $f(t) = 3t^2$. This function is positive and continuous, as required.

(b) The volume generated from $x = 0$ to $x = b$ is $\int_0^b \pi[f(x)]^2\,dx$. Hence, we are given that $b^2 = \int_0^b \pi[f(x)]^2\,dx$ for all $b > 0$. Differentiating both sides of this equation with respect to b using the Fundamental Theorem of Calculus gives

$2b = \pi[f(b)]^2 \quad \Rightarrow \quad f(b) = \sqrt{2b/\pi}$, since f is positive. Therefore, $f(x) = \sqrt{2x/\pi}$.

3. Let a and b be the x-coordinates of the points where the line intersects the curve. From the figure, $R_1 = R_2 \quad \Rightarrow$

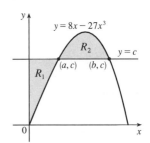

$$\int_0^a \left[c - (8x - 27x^3)\right] dx = \int_a^b \left[(8x - 27x^3) - c\right] dx$$

$$\left[cx - 4x^2 + \tfrac{27}{4}x^4\right]_0^a = \left[4x^2 - \tfrac{27}{4}x^4 - cx\right]_a^b$$

$$ac - 4a^2 + \tfrac{27}{4}a^4 = \left(4b^2 - \tfrac{27}{4}b^4 - bc\right) - \left(4a^2 - \tfrac{27}{4}a^4 - ac\right)$$

$$0 = 4b^2 - \tfrac{27}{4}b^4 - bc = 4b^2 - \tfrac{27}{4}b^4 - b(8b - 27b^3)$$

$$= 4b^2 - \tfrac{27}{4}b^4 - 8b^2 + 27b^4 = \tfrac{81}{4}b^4 - 4b^2$$

$$= b^2\left(\tfrac{81}{4}b^2 - 4\right)$$

So for $b > 0$, $b^2 = \tfrac{16}{81} \quad \Rightarrow \quad b = \tfrac{4}{9}$. Thus, $c = 8b - 27b^3 = 8\left(\tfrac{4}{9}\right) - 27\left(\tfrac{64}{729}\right) = \tfrac{32}{9} - \tfrac{64}{27} = \tfrac{32}{27}$.

5. (a) $V = \pi h^2(r - h/3) = \tfrac{1}{3}\pi h^2(3r - h)$. See the solution to Exercise 6.2.51.

(b) The smaller segment has height $h = 1 - x$ and so by part (a) its volume is

$V = \tfrac{1}{3}\pi(1 - x)^2\left[3(1) - (1 - x)\right] = \tfrac{1}{3}\pi(x - 1)^2(x + 2)$. This volume must be $\tfrac{1}{3}$ of the total volume of the sphere,

which is $\tfrac{4}{3}\pi(1)^3$. So $\tfrac{1}{3}\pi(x - 1)^2(x + 2) = \tfrac{1}{3}\left(\tfrac{4}{3}\pi\right) \quad \Rightarrow \quad (x^2 - 2x + 1)(x + 2) = \tfrac{4}{3} \quad \Rightarrow \quad x^3 - 3x + 2 = \tfrac{4}{3} \quad \Rightarrow$

$3x^3 - 9x + 2 = 0$. Using Newton's method with $f(x) = 3x^3 - 9x + 2$, $f'(x) = 9x^2 - 9$, we get

$$x_{n+1} = x_n - \frac{3x_n^3 - 9x_n + 2}{9x_n^2 - 9}.$$ Taking $x_1 = 0$, we get $x_2 \approx 0.2222$, and $x_3 \approx 0.2261 \approx x_4$, so, correct to four decimal

places, $x \approx 0.2261$.

(c) With $r = 0.5$ and $s = 0.75$, the equation $x^3 - 3rx^2 + 4r^3s = 0$ becomes $x^3 - 3(0.5)x^2 + 4(0.5)^3(0.75) = 0 \quad \Rightarrow$

$x^3 - \tfrac{3}{2}x^2 + 4\left(\tfrac{1}{8}\right)\tfrac{3}{4} = 0 \quad \Rightarrow \quad 8x^3 - 12x^2 + 3 = 0$. We use Newton's method with $f(x) = 8x^3 - 12x^2 + 3$,

$f'(x) = 24x^2 - 24x$, so $x_{n+1} = x_n - \dfrac{8x_n^3 - 12x_n^2 + 3}{24x_n^2 - 24x_n}$. Take $x_1 = 0.5$. Then $x_2 \approx 0.6667$, and $x_3 \approx 0.6736 \approx x_4$.

So to four decimal places the depth is 0.6736 m.

(d) (i) From part (a) with $r = 5$ in., the volume of water in the bowl is

$$V = \tfrac{1}{3}\pi h^2 (3r - h) = \tfrac{1}{3}\pi h^2 (15 - h) = 5\pi h^2 - \tfrac{1}{3}\pi h^3.$$ We are given that $\dfrac{dV}{dt} = 0.2$ in^3/s and we want to find $\dfrac{dh}{dt}$

when $h = 3$. Now $\dfrac{dV}{dt} = 10\pi h\dfrac{dh}{dt} - \pi h^2\dfrac{dh}{dt}$, so $\dfrac{dh}{dt} = \dfrac{0.2}{\pi(10h - h^2)}$. When $h = 3$, we have

$$\dfrac{dh}{dt} = \dfrac{0.2}{\pi(10 \cdot 3 - 3^2)} = \dfrac{1}{105\pi} \approx 0.003 \text{ in/s}.$$

(ii) From part (a), the volume of water required to fill the bowl from the instant that the water is 4 in. deep is

$$V = \tfrac{1}{2} \cdot \tfrac{4}{3}\pi(5)^3 - \tfrac{1}{3}\pi(4)^2(15 - 4) = \tfrac{2}{3} \cdot 125\pi - \tfrac{16}{3} \cdot 11\pi = \tfrac{74}{3}\pi.$$ To find the time required to fill the bowl we divide

this volume by the rate: Time $= \dfrac{74\pi/3}{0.2} = \dfrac{370\pi}{3} \approx 387 \text{ s} \approx 6.5 \text{ min}.$

7. We are given that the rate of change of the volume of water is $\dfrac{dV}{dt} = -kA(x)$, where k is some positive constant and $A(x)$ is

the area of the surface when the water has depth x. Now we are concerned with the rate of change of the depth of the water

with respect to time, that is, $\dfrac{dx}{dt}$. But by the Chain Rule, $\dfrac{dV}{dt} = \dfrac{dV}{dx}\dfrac{dx}{dt}$, so the first equation can be written

$\dfrac{dV}{dx}\dfrac{dx}{dt} = -kA(x)$ $(\star)$. Also, we know that the total volume of water up to a depth x is $V(x) = \int_0^x A(s)\,ds$, where $A(s)$ is

the area of a cross-section of the water at a depth s. Differentiating this equation with respect to x, we get $dV/dx = A(x)$.

Substituting this into equation $\star$, we get $A(x)(dx/dt) = -kA(x) \Rightarrow dx/dt = -k$, a constant.

9. We must find expressions for the areas A and B, and then set them equal and see what this says about the curve C. If

$P = (a, 2a^2)$, then area A is just $\int_0^a (2x^2 - x^2)\,dx = \int_0^a x^2\,dx = \tfrac{1}{3}a^3.$ To find area B, we use y as the variable of

integration. So we find the equation of the middle curve as a function of y: $y = 2x^2 \Leftrightarrow x = \sqrt{y/2}$, since we are

concerned with the first quadrant only. We can express area B as

$$\int_0^{2a^2} \left[\sqrt{y/2} - C(y)\right] dy = \left[\tfrac{4}{3}(y/2)^{3/2}\right]_0^{2a^2} - \int_0^{2a^2} C(y)\,dy = \tfrac{4}{3}a^3 - \int_0^{2a^2} C(y)\,dy$$

where $C(y)$ is the function with graph C. Setting $A = B$, we get $\tfrac{1}{3}a^3 = \tfrac{4}{3}a^3 - \int_0^{2a^2} C(y)\,dy \Leftrightarrow \int_0^{2a^2} C(y)\,dy = a^3.$

Now we differentiate this equation with respect to a using the Chain Rule and the Fundamental Theorem:

$C(2a^2)(4a) = 3a^2 \Rightarrow C(y) = \tfrac{3}{4}\sqrt{y/2}$, where $y = 2a^2$. Now we can solve for y: $x = \tfrac{3}{4}\sqrt{y/2} \Rightarrow$

$x^2 = \tfrac{9}{16}(y/2) \Rightarrow y = \tfrac{32}{9}x^2.$

11. (a) Stacking disks along the y-axis gives us $V = \int_0^h \pi\,[f(y)]^2\,dy.$

(b) Using the Chain Rule, $\dfrac{dV}{dt} = \dfrac{dV}{dh} \cdot \dfrac{dh}{dt} = \pi\,[f(h)]^2\dfrac{dh}{dt}.$

(c) $kA\sqrt{h} = \pi[f(h)]^2\,\dfrac{dh}{dt}$. Set $\dfrac{dh}{dt} = C$: $\pi[f(h)]^2\,C = kA\sqrt{h}$ $\Rightarrow$ $[f(h)]^2 = \dfrac{kA}{\pi C}\sqrt{h}$ $\Rightarrow$ $f(h) = \sqrt{\dfrac{kA}{\pi C}}\,h^{1/4}$; that

is, $f(y) = \sqrt{\dfrac{kA}{\pi C}}\,y^{1/4}$. The advantage of having $\dfrac{dh}{dt} = C$ is that the markings on the container are equally spaced.

13. The cubic polynomial passes through the origin, so let its equation be

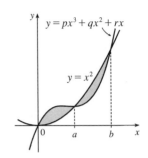

$y = px^3 + qx^2 + rx$. The curves intersect when $px^3 + qx^2 + rx = x^2$ $\Leftrightarrow$

$px^3 + (q-1)x^2 + rx = 0$. Call the left side $f(x)$. Since $f(a) = f(b) = 0$,

another form of f is

$$f(x) = px(x-a)(x-b) = px[x^2 - (a+b)x + ab]$$
$$= p[x^3 - (a+b)x^2 + abx]$$

Since the two areas are equal, we must have $\int_0^a f(x)\,dx = -\int_a^b f(x)\,dx$ $\Rightarrow$

$[F(x)]_0^a = [F(x)]_b^a$ $\Rightarrow$ $F(a) - F(0) = F(a) - F(b)$ $\Rightarrow$ $F(0) = F(b)$, where F is an antiderivative of f.

Now $F(x) = \int f(x)\,dx = \int p[x^3 - (a+b)x^2 + abx]\,dx = p\big[\tfrac{1}{4}x^4 - \tfrac{1}{3}(a+b)x^3 + \tfrac{1}{2}abx^2\big] + C$, so

$F(0) = F(b)$ $\Rightarrow$ $C = p\big[\tfrac{1}{4}b^4 - \tfrac{1}{3}(a+b)b^3 + \tfrac{1}{2}ab^3\big] + C$ $\Rightarrow$ $0 = p\big[\tfrac{1}{4}b^4 - \tfrac{1}{3}(a+b)b^3 + \tfrac{1}{2}ab^3\big]$ $\Rightarrow$

$0 = 3b - 4(a+b) + 6a$ [multiply by $12/(pb^3)$, $b \neq 0$] $\Rightarrow$ $0 = 3b - 4a - 4b + 6a$ $\Rightarrow$ $b = 2a$.

Hence, b is twice the value of a.

15. We assume that P lies in the region of positive x. Since $y = x^3$ is an odd

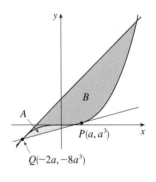

function, this assumption will not affect the result of the calculation. Let

$P = (a, a^3)$. The slope of the tangent to the curve $y = x^3$ at P is $3a^2$, and so

the equation of the tangent is $y - a^3 = 3a^2(x - a)$ $\Leftrightarrow$ $y = 3a^2x - 2a^3$.

We solve this simultaneously with $y = x^3$ to find the other point of intersection:

$x^3 = 3a^2x - 2a^3$ $\Leftrightarrow$ $(x-a)^2(x+2a) = 0$. So $Q = (-2a, -8a^3)$ is

the other point of intersection. The equation of the tangent at Q is

$y - (-8a^3) = 12a^2[x - (-2a)]$ $\Leftrightarrow$ $y = 12a^2x + 16a^3$. By symmetry,

this tangent will intersect the curve again at $x = -2(-2a) = 4a$. The curve lies above the first tangent, and

below the second, so we are looking for a relationship between $A = \int_{-2a}^{a} \big[x^3 - (3a^2x - 2a^3)\big]\,dx$ and

$B = \int_{-2a}^{4a} \big[(12a^2x + 16a^3) - x^3\big]\,dx$. We calculate $A = \big[\tfrac{1}{4}x^4 - \tfrac{3}{2}a^2x^2 + 2a^3x\big]_{-2a}^{a} = \tfrac{3}{4}a^4 - (-6a^4) = \tfrac{27}{4}a^4$, and

$B = \big[6a^2x^2 + 16a^3x - \tfrac{1}{4}x^4\big]_{-2a}^{4a} = 96a^4 - (-12a^4) = 108a^4$. We see that $B = 16A = 2^4A$. This is because our

calculation of area B was essentially the same as that of area A, with a replaced by $-2a$, so if we replace a with $-2a$ in our

expression for A, we get $\tfrac{27}{4}(-2a)^4 = 108a^4 = B$.

7 □ TECHNIQUES OF INTEGRATION

7.1 Integration by Parts

1. Let $u = \ln x$, $dv = x^2\,dx$ $\Rightarrow$ $du = \frac{1}{x}\,dx$, $v = \frac{1}{3}x^3$. Then by Equation 2,

$\int x^2 \ln x\,dx = (\ln x)\left(\frac{1}{3}x^3\right) - \int \left(\frac{1}{3}x^3\right)\left(\frac{1}{x}\right)\,dx = \frac{1}{3}x^3 \ln x - \frac{1}{3}\int x^2\,dx = \frac{1}{3}x^3 \ln x - \frac{1}{3}\left(\frac{1}{3}x^3\right) + C$

$= \frac{1}{3}x^3 \ln x - \frac{1}{9}x^3 + C$ $\left[\text{or } \frac{1}{3}x^3\left(\ln x - \frac{1}{3}\right) + C\right]$

Note: A mnemonic device which is helpful for selecting u when using integration by parts is the LIATE principle of precedence for u:

<u>L</u>ogarithmic

<u>I</u>nverse trigonometric

<u>A</u>lgebraic

<u>T</u>rigonometric

<u>E</u>xponential

If the integrand has several factors, then we try to choose among them a u which appears as high as possible on the list. For example, in $\int xe^{2x}\,dx$ the integrand is xe^{2x}, which is the product of an algebraic function (x) and an exponential function (e^{2x}). Since <u>A</u>lgebraic appears before <u>E</u>xponential, we choose $u = x$. Sometimes the integration turns out to be similar regardless of the selection of u and dv, but it is advisable to refer to LIATE when in doubt.

3. Let $u = x$, $dv = \cos 5x\,dx$ $\Rightarrow$ $du = dx$, $v = \frac{1}{5}\sin 5x$. Then by Equation 2,

$\int x \cos 5x\,dx = \frac{1}{5}x \sin 5x - \int \frac{1}{5}\sin 5x\,dx = \frac{1}{5}x \sin 5x + \frac{1}{25}\cos 5x + C.$

5. Let $u = r$, $dv = e^{r/2}\,dr$ $\Rightarrow$ $du = dr$, $v = 2e^{r/2}$. Then $\int re^{r/2}\,dr = 2re^{r/2} - \int 2e^{r/2}\,dr = 2re^{r/2} - 4e^{r/2} + C.$

7. Let $u = x^2$, $dv = \sin \pi x\,dx$ $\Rightarrow$ $du = 2x\,dx$ and $v = -\frac{1}{\pi}\cos \pi x$. Then

$I = \int x^2 \sin \pi x\,dx = -\frac{1}{\pi}x^2 \cos \pi x + \frac{2}{\pi}\int x \cos \pi x\,dx$ $(\star)$. Next let $U = x$, $dV = \cos \pi x\,dx$ $\Rightarrow$ $dU = dx$,

$V = \frac{1}{\pi}\sin \pi x$, so $\int x \cos \pi x\,dx = \frac{1}{\pi}x \sin \pi x - \frac{1}{\pi}\int \sin \pi x\,dx = \frac{1}{\pi}x \sin \pi x + \frac{1}{\pi^2}\cos \pi x + C_1.$

Substituting for $\int x \cos \pi x\,dx$ in $(\star)$, we get

$I = -\frac{1}{\pi}x^2 \cos \pi x + \frac{2}{\pi}\left(\frac{1}{\pi}x \sin \pi x + \frac{1}{\pi^2}\cos \pi x + C_1\right) = -\frac{1}{\pi}x^2 \cos \pi x + \frac{2}{\pi^2}x \sin \pi x + \frac{2}{\pi^3}\cos \pi x + C$, where $C = \frac{2}{\pi}C_1$.

9. Let $u = \ln(2x + 1)$, $dv = dx$ $\Rightarrow$ $du = \dfrac{2}{2x + 1}\,dx$, $v = x$. Then

$$\int \ln(2x + 1)\,dx = x \ln(2x + 1) - \int \frac{2x}{2x + 1}\,dx = x \ln(2x + 1) - \int \frac{(2x + 1) - 1}{2x + 1}\,dx$$

$$= x \ln(2x + 1) - \int \left(1 - \frac{1}{2x + 1}\right)\,dx = x \ln(2x + 1) - x + \frac{1}{2}\ln(2x + 1) + C$$

$$= \frac{1}{2}(2x + 1)\ln(2x + 1) - x + C$$

11. Let $u = \arctan 4t$, $dv = dt$ $\Rightarrow$ $du = \dfrac{4}{1 + (4t)^2}\,dt = \dfrac{4}{1 + 16t^2}\,dt$, $v = t$. Then

$$\int \arctan 4t\,dt = t \arctan 4t - \int \frac{4t}{1 + 16t^2}\,dt = t \arctan 4t - \frac{1}{8}\int \frac{32t}{1 + 16t^2}\,dt = t \arctan 4t - \frac{1}{8}\ln(1 + 16t^2) + C.$$

13. Let $u = t$, $dv = \sec^2 2t\, dt$ $\Rightarrow$ $du = dt$, $v = \frac{1}{2}\tan 2t$. Then

$\int t\sec^2 2t\, dt = \frac{1}{2}t\tan 2t - \frac{1}{2}\int \tan 2t\, dt = \frac{1}{2}t\tan 2t - \frac{1}{4}\ln|\sec 2t| + C$.

15. First let $u = (\ln x)^2$, $dv = dx$ $\Rightarrow$ $du = 2\ln x \cdot \frac{1}{x}\, dx$, $v = x$. Then by Equation 2,

$I = \int (\ln x)^2\, dx = x(\ln x)^2 - 2\int x\ln x \cdot \frac{1}{x}\, dx = x(\ln x)^2 - 2\int \ln x\, dx$. Next let $U = \ln x$, $dV = dx$ $\Rightarrow$

$dU = 1/x\, dx$, $V = x$ to get $\int \ln x\, dx = x\ln x - \int x\cdot(1/x)\, dx = x\ln x - \int dx = x\ln x - x + C_1$. Thus,

$I = x(\ln x)^2 - 2(x\ln x - x + C_1) = x(\ln x)^2 - 2x\ln x + 2x + C$, where $C = -2C_1$.

17. First let $u = \sin 3\theta$, $dv = e^{2\theta}\, d\theta$ $\Rightarrow$ $du = 3\cos 3\theta\, d\theta$, $v = \frac{1}{2}e^{2\theta}$. Then

$I = \int e^{2\theta}\sin 3\theta\, d\theta = \frac{1}{2}e^{2\theta}\sin 3\theta - \frac{3}{2}\int e^{2\theta}\cos 3\theta\, d\theta$. Next let $U = \cos 3\theta$, $dV = e^{2\theta}\, d\theta$ $\Rightarrow$ $dU = -3\sin 3\theta\, d\theta$,

$V = \frac{1}{2}e^{2\theta}$ to get $\int e^{2\theta}\cos 3\theta\, d\theta = \frac{1}{2}e^{2\theta}\cos 3\theta + \frac{3}{2}\int e^{2\theta}\sin 3\theta\, d\theta$. Substituting in the previous formula gives

$I = \frac{1}{2}e^{2\theta}\sin 3\theta - \frac{3}{4}e^{2\theta}\cos 3\theta - \frac{9}{4}\int e^{2\theta}\sin 3\theta\, d\theta = \frac{1}{2}e^{2\theta}\sin 3\theta - \frac{3}{4}e^{2\theta}\cos 3\theta - \frac{9}{4}I$ $\Rightarrow$

$\frac{13}{4}I = \frac{1}{2}e^{2\theta}\sin 3\theta - \frac{3}{4}e^{2\theta}\cos 3\theta + C_1$. Hence, $I = \frac{1}{13}e^{2\theta}(2\sin 3\theta - 3\cos 3\theta) + C$, where $C = \frac{4}{13}C_1$.

19. Let $u = t$, $dv = \sin 3t\, dt$ $\Rightarrow$ $du = dt$, $v = -\frac{1}{3}\cos 3t$. Then

$\int_0^\pi t\sin 3t\, dt = \left[-\frac{1}{3}t\cos 3t\right]_0^\pi + \frac{1}{3}\int_0^\pi \cos 3t\, dt = \left(\frac{1}{3}\pi - 0\right) + \frac{1}{9}\left[\sin 3t\right]_0^\pi = \frac{\pi}{3}$.

21. Let $u = t$, $dv = \cosh t\, dt$ $\Rightarrow$ $du = dt$, $v = \sinh t$. Then

$\int_0^1 t\cosh t\, dt = \left[t\sinh t\right]_0^1 - \int_0^1 \sinh t\, dt = (\sinh 1 - \sinh 0) - \left[\cosh t\right]_0^1 = \sinh 1 - (\cosh 1 - \cosh 0)$

$\qquad\qquad = \sinh 1 - \cosh 1 + 1$.

We can use the definitions of sinh and cosh to write the answer in terms of e:

$\sinh 1 - \cosh 1 + 1 = \frac{1}{2}(e^1 - e^{-1}) - \frac{1}{2}(e^1 + e^{-1}) + 1 = -e^{-1} + 1 = 1 - 1/e$.

23. Let $u = \ln x$, $dv = x^{-2}\, dx$ $\Rightarrow$ $du = \dfrac{1}{x}\, dx$, $v = -x^{-1}$. By (6),

$\displaystyle\int_1^2 \frac{\ln x}{x^2}\, dx = \left[-\frac{\ln x}{x}\right]_1^2 + \int_1^2 x^{-2}\, dx = -\frac{1}{2}\ln 2 + \ln 1 + \left[-\frac{1}{x}\right]_1^2 = -\frac{1}{2}\ln 2 + 0 - \frac{1}{2} + 1 = \frac{1}{2} - \frac{1}{2}\ln 2$.

25. Let $u = y$, $dv = \dfrac{dy}{e^{2y}} = e^{-2y}dy$ $\Rightarrow$ $du = dy$, $v = -\frac{1}{2}e^{-2y}$. Then

$\displaystyle\int_0^1 \frac{y}{e^{2y}}\, dy = \left[-\frac{1}{2}ye^{-2y}\right]_0^1 + \frac{1}{2}\int_0^1 e^{-2y}dy = \left(-\frac{1}{2}e^{-2} + 0\right) - \frac{1}{4}\left[e^{-2y}\right]_0^1 = -\frac{1}{2}e^{-2} - \frac{1}{4}e^{-2} + \frac{1}{4} = \frac{1}{4} - \frac{3}{4}e^{-2}$.

27. Let $u = \cos^{-1} x$, $dv = dx$ $\Rightarrow$ $du = -\dfrac{dx}{\sqrt{1-x^2}}$, $v = x$. Then

$I = \displaystyle\int_0^{1/2} \cos^{-1} x\, dx = \left[x\cos^{-1} x\right]_0^{1/2} + \int_0^{1/2} \frac{x\, dx}{\sqrt{1-x^2}} = \frac{1}{2}\cdot\frac{\pi}{3} + \int_1^{3/4} t^{-1/2}\left[-\frac{1}{2}dt\right]$, where $t = 1 - x^2$ $\Rightarrow$

$dt = -2x\, dx$. Thus, $I = \frac{\pi}{6} + \frac{1}{2}\int_{3/4}^1 t^{-1/2}\, dt = \frac{\pi}{6} + \left[\sqrt{t}\right]_{3/4}^1 = \frac{\pi}{6} + 1 - \frac{\sqrt{3}}{2} = \frac{1}{6}\left(\pi + 6 - 3\sqrt{3}\right)$.

29. Let $u = \ln(\sin x)$, $dv = \cos x\, dx \;\Rightarrow\; du = \dfrac{\cos x}{\sin x}\, dx$, $v = \sin x$. Then

$I = \int \cos x \ln(\sin x)\, dx = \sin x \ln(\sin x) - \int \cos x\, dx = \sin x \ln(\sin x) - \sin x + C$.

Another method: Substitute $t = \sin x$, so $dt = \cos x\, dx$. Then $I = \int \ln t\, dt = t\ln t - t + C$ (see Example 2) and so $I = \sin x\,(\ln \sin x - 1) + C$.

31. Let $u = (\ln x)^2$, $dv = x^4\, dx \;\Rightarrow\; du = 2\dfrac{\ln x}{x}\, dx$, $v = \dfrac{x^5}{5}$. By (6),

$$\int_1^2 x^4 (\ln x)^2\, dx = \left[\frac{x^5}{5}(\ln x)^2 \right]_1^2 - 2\int_1^2 \frac{x^4}{5}\ln x\, dx = \tfrac{32}{5}(\ln 2)^2 - 0 - 2\int_1^2 \frac{x^4}{5}\ln x\, dx.$$

Let $U = \ln x$, $dV = \dfrac{x^4}{5}\, dx \;\Rightarrow\; dU = \dfrac{1}{x}\, dx$, $V = \dfrac{x^5}{25}$.

Then $\displaystyle\int_1^2 \frac{x^4}{5}\ln x\, dx = \left[\frac{x^5}{25}\ln x \right]_1^2 - \int_1^2 \frac{x^4}{25}\, dx = \tfrac{32}{25}\ln 2 - 0 - \left[\frac{x^5}{125} \right]_1^2 = \tfrac{32}{25}\ln 2 - \left(\tfrac{32}{125} - \tfrac{1}{125} \right)$.

So $\int_1^2 x^4 (\ln x)^2\, dx = \tfrac{32}{5}(\ln 2)^2 - 2\left(\tfrac{32}{25}\ln 2 - \tfrac{31}{125} \right) = \tfrac{32}{5}(\ln 2)^2 - \tfrac{64}{25}\ln 2 + \tfrac{62}{125}$.

33. Let $y = \sqrt{x}$, so that $dy = \tfrac{1}{2}x^{-1/2}\, dx = \dfrac{1}{2\sqrt{x}}\, dx = \dfrac{1}{2y}\, dx$. Thus, $\int \cos\sqrt{x}\, dx = \int \cos y\,(2y\, dy) = 2\int y\cos y\, dy$. Now

use parts with $u = y$, $dv = \cos y\, dy$, $du = dy$, $v = \sin y$ to get $\int y\cos y\, dy = y\sin y - \int \sin y\, dy = y\sin y + \cos y + C_1$,

so $\int \cos\sqrt{x}\, dx = 2y\sin y + 2\cos y + C = 2\sqrt{x}\sin\sqrt{x} + 2\cos\sqrt{x} + C$.

35. Let $x = \theta^2$, so that $dx = 2\theta\, d\theta$. Thus, $\displaystyle\int_{\sqrt{\pi/2}}^{\sqrt{\pi}} \theta^3 \cos(\theta^2)\, d\theta = \int_{\sqrt{\pi/2}}^{\sqrt{\pi}} \theta^2 \cos(\theta^2) \cdot \tfrac{1}{2}(2\theta\, d\theta) = \tfrac{1}{2}\int_{\pi/2}^{\pi} x\cos x\, dx$. Now use

parts with $u = x$, $dv = \cos x\, dx$, $du = dx$, $v = \sin x$ to get

$$\tfrac{1}{2}\int_{\pi/2}^{\pi} x\cos x\, dx = \tfrac{1}{2}\left(\left[x\sin x \right]_{\pi/2}^{\pi} - \int_{\pi/2}^{\pi} \sin x\, dx \right) = \tfrac{1}{2}\left[x\sin x + \cos x \right]_{\pi/2}^{\pi}$$

$$= \tfrac{1}{2}(\pi\sin\pi + \cos\pi) - \tfrac{1}{2}\left(\tfrac{\pi}{2}\sin\tfrac{\pi}{2} + \cos\tfrac{\pi}{2} \right) = \tfrac{1}{2}(\pi\cdot 0 - 1) - \tfrac{1}{2}\left(\tfrac{\pi}{2}\cdot 1 + 0 \right) = -\tfrac{1}{2} - \tfrac{\pi}{4}$$

37. Let $y = 1 + x$, so that $dy = dx$. Thus, $\int x\ln(1+x)\, dx = \int (y-1)\ln y\, dy$. Now use parts with $u = \ln y$, $dv = (y-1)\, dy$,

$du = \tfrac{1}{y}\, dy$, $v = \tfrac{1}{2}y^2 - y$ to get

$$\int (y-1)\ln y\, dy = \left(\tfrac{1}{2}y^2 - y \right)\ln y - \int \left(\tfrac{1}{2}y - 1 \right) dy = \tfrac{1}{2}y(y-2)\ln y - \tfrac{1}{4}y^2 + y + C$$

$$= \tfrac{1}{2}(1+x)(x-1)\ln(1+x) - \tfrac{1}{4}(1+x)^2 + 1 + x + C,$$

which can be written as $\tfrac{1}{2}(x^2 - 1)\ln(1+x) - \tfrac{1}{4}x^2 + \tfrac{1}{2}x + \tfrac{3}{4} + C$.

In Exercises 39 – 42, let $f(x)$ denote the integrand and $F(x)$ its antiderivative (with $C = 0$).

39. Let $u = 2x + 3$, $dv = e^x\, dx \;\Rightarrow\; du = 2\, dx$, $v = e^x$. Then

$\int (2x+3)e^x\, dx = (2x+3)e^x - 2\int e^x\, dx = (2x+3)e^x - 2e^x + C$

$= (2x+1)e^x + C$

We see from the graph that this is reasonable, since F has a minimum where
f changes from negative to positive.

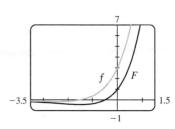

41. Let $u = \frac{1}{2}x^2$, $dv = 2x\sqrt{1+x^2}\,dx$ $\Rightarrow$ $du = x\,dx$, $v = \frac{2}{3}(1+x^2)^{3/2}$.

Then

$$\int x^3\sqrt{1+x^2}\,dx = \frac{1}{2}x^2\left[\frac{2}{3}(1+x^2)^{3/2}\right] - \frac{2}{3}\int x(1+x^2)^{3/2}\,dx$$

$$= \frac{1}{3}x^2(1+x^2)^{3/2} - \frac{2}{3}\cdot\frac{2}{5}\cdot\frac{1}{2}(1+x^2)^{5/2} + C$$

$$= \frac{1}{3}x^2(1+x^2)^{3/2} - \frac{2}{15}(1+x^2)^{5/2} + C$$

Another method: Use substitution with $u = 1+x^2$ to get $\frac{1}{5}(1+x^2)^{5/2} - \frac{1}{3}(1+x^2)^{3/2} + C$.

43. (a) Take $n = 2$ in Example 6 to get $\displaystyle\int \sin^2 x\,dx = -\frac{1}{2}\cos x\sin x + \frac{1}{2}\int 1\,dx = \frac{x}{2} - \frac{\sin 2x}{4} + C.$

(b) $\int \sin^4 x\,dx = -\frac{1}{4}\cos x\sin^3 x + \frac{3}{4}\int \sin^2 x\,dx = -\frac{1}{4}\cos x\sin^3 x + \frac{3}{8}x - \frac{3}{16}\sin 2x + C.$

45. (a) From Example 6, $\displaystyle\int \sin^n x\,dx = -\frac{1}{n}\cos x\,\sin^{n-1}x + \frac{n-1}{n}\int \sin^{n-2}x\,dx.$ Using (6),

$$\int_0^{\pi/2}\sin^n x\,dx = \left[-\frac{\cos x\,\sin^{n-1}x}{n}\right]_0^{\pi/2} + \frac{n-1}{n}\int_0^{\pi/2}\sin^{n-2}x\,dx$$

$$= (0-0) + \frac{n-1}{n}\int_0^{\pi/2}\sin^{n-2}x\,dx = \frac{n-1}{n}\int_0^{\pi/2}\sin^{n-2}x\,dx$$

(b) Using $n = 3$ in part (a), we have $\int_0^{\pi/2}\sin^3 x\,dx = \frac{2}{3}\int_0^{\pi/2}\sin x\,dx = \left[-\frac{2}{3}\cos x\right]_0^{\pi/2} = \frac{2}{3}.$

Using $n = 5$ in part (a), we have $\int_0^{\pi/2}\sin^5 x\,dx = \frac{4}{5}\int_0^{\pi/2}\sin^3 x\,dx = \frac{4}{5}\cdot\frac{2}{3} = \frac{8}{15}.$

(c) The formula holds for $n = 1$ (that is, $2n+1 = 3$) by (b). Assume it holds for some $k \geq 1$. Then

$$\int_0^{\pi/2}\sin^{2k+1}x\,dx = \frac{2\cdot4\cdot6\cdots\cdots(2k)}{3\cdot5\cdot7\cdots\cdots(2k+1)}.$$ By Example 6,

$$\int_0^{\pi/2}\sin^{2k+3}x\,dx = \frac{2k+2}{2k+3}\int_0^{\pi/2}\sin^{2k+1}x\,dx = \frac{2k+2}{2k+3}\cdot\frac{2\cdot4\cdot6\cdots\cdots(2k)}{3\cdot5\cdot7\cdots\cdots(2k+1)}$$

$$= \frac{2\cdot4\cdot6\cdots\cdots(2k)[2(k+1)]}{3\cdot5\cdot7\cdots\cdots(2k+1)[2(k+1)+1]},$$

so the formula holds for $n = k+1$. By induction, the formula holds for all $n \geq 1$.

47. Let $u = (\ln x)^n$, $dv = dx$ $\Rightarrow$ $du = n(\ln x)^{n-1}(dx/x)$, $v = x$. By Equation 2,

$\int(\ln x)^n\,dx = x(\ln x)^n - \int nx(\ln x)^{n-1}(dx/x) = x(\ln x)^n - n\int(\ln x)^{n-1}\,dx.$

49. $\int \tan^n x\,dx = \int \tan^{n-2}x\tan^2 x\,dx = \int \tan^{n-2}x\left(\sec^2 x - 1\right)dx = \int \tan^{n-2}x\sec^2 x\,dx - \int \tan^{n-2}x\,dx$

$$= I - \int \tan^{n-2}x\,dx.$$

Let $u = \tan^{n-2}x$, $dv = \sec^2 x\,dx$ $\Rightarrow$ $du = (n-2)\tan^{n-3}x\sec^2 x\,dx$, $v = \tan x$. Then, by Equation 2,

$$I = \tan^{n-1}x - (n-2)\int \tan^{n-2}x\sec^2 x\,dx$$

$$1I = \tan^{n-1}x - (n-2)I$$

$$(n-1)I = \tan^{n-1}x$$

$$I = \frac{\tan^{n-1}x}{n-1}$$

Returning to the original integral, $\int \tan^n x\,dx = \dfrac{\tan^{n-1}x}{n-1} - \int \tan^{n-2}x\,dx.$

51. By repeated applications of the reduction formula in Exercise 47,

$$\int (\ln x)^3\, dx = x\,(\ln x)^3 - 3\int (\ln x)^2\, dx = x(\ln x)^3 - 3\big[x(\ln x)^2 - 2\int (\ln x)^1\, dx\big]$$

$$= x\,(\ln x)^3 - 3x(\ln x)^2 + 6\big[x(\ln x)^1 - 1\int (\ln x)^0\, dx\big]$$

$$= x\,(\ln x)^3 - 3x(\ln x)^2 + 6x\ln x - 6\int 1\, dx = x\,(\ln x)^3 - 3x(\ln x)^2 + 6x\ln x - 6x + C$$

53. Area $= \int_0^5 xe^{-0.4x}\, dx$. Let $u = x$, $dv = e^{-0.4x}\, dx$ $\Rightarrow$

$du = dx$, $v = -2.5e^{-0.4x}$. Then

area $= \big[-2.5xe^{-0.4x}\big]_0^5 + 2.5\int_0^5 e^{-0.4x}\, dx$

$= -12.5e^{-2} + 0 + 2.5\big[-2.5e^{-0.4x}\big]_0^5$

$= -12.5e^{-2} - 6.25(e^{-2} - 1) = 6.25 - 18.75e^{-2}$ or $\frac{25}{4} - \frac{75}{4}e^{-2}$

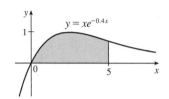

55. The curves $y = x\sin x$ and $y = (x - 2)^2$ intersect at $a \approx 1.04748$ and

$b \approx 2.87307$, so

area $= \int_a^b \big[x\sin x - (x - 2)^2\big]\, dx$

$= \big[-x\cos x + \sin x - \frac{1}{3}(x - 2)^3\big]_a^b$ [by Example 1]

$\approx 2.81358 - 0.63075 = 2.18283$

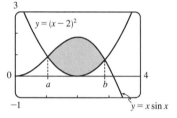

57. $V = \int_0^1 2\pi x\cos(\pi x/2)\, dx$. Let $u = x$, $dv = \cos(\pi x/2)\, dx$ $\Rightarrow$ $du = dx$, $v = \frac{2}{\pi}\sin(\pi x/2)$.

$$V = 2\pi\left[\frac{2}{\pi}x\sin\left(\frac{\pi x}{2}\right)\right]_0^1 - 2\pi \cdot \frac{2}{\pi}\int_0^1 \sin\left(\frac{\pi x}{2}\right)\, dx = 2\pi\left(\frac{2}{\pi} - 0\right) - 4\left[-\frac{2}{\pi}\cos\left(\frac{\pi x}{2}\right)\right]_0^1 = 4 + \frac{8}{\pi}(0 - 1) = 4 - \frac{8}{\pi}.$$

59. Volume $= \int_{-1}^0 2\pi(1 - x)e^{-x}\, dx$. Let $u = 1 - x$, $dv = e^{-x}\, dx$ $\Rightarrow$ $du = -dx$, $v = -e^{-x}$.

$V = 2\pi\big[(1 - x)(-e^{-x})\big]_{-1}^0 - 2\pi\int_{-1}^0 e^{-x}\, dx = 2\pi\big[(x - 1)(e^{-x}) + e^{-x}\big]_{-1}^0 = 2\pi\big[xe^{-x}\big]_{-1}^0 = 2\pi(0 + e) = 2\pi e$

61. The average value of $f(x) = x^2\ln x$ on the interval $[1, 3]$ is $f_{\text{ave}} = \dfrac{1}{3 - 1}\displaystyle\int_1^3 x^2\ln x\, dx = \frac{1}{2}I$.

Let $u = \ln x$, $dv = x^2\, dx$ $\Rightarrow$ $du = (1/x)\, dx$, $v = \frac{1}{3}x^3$.

So $I = \big[\frac{1}{3}x^3\ln x\big]_1^3 - \int_1^3 \frac{1}{3}x^2\, dx = (9\ln 3 - 0) - \big[\frac{1}{9}x^3\big]_1^3 = 9\ln 3 - \big(3 - \frac{1}{9}\big) = 9\ln 3 - \frac{26}{9}$.

Thus, $f_{\text{ave}} = \frac{1}{2}I = \frac{1}{2}\big(9\ln 3 - \frac{26}{9}\big) = \frac{9}{2}\ln 3 - \frac{13}{9}$.

63. Since $v(t) > 0$ for all t, the desired distance is $s(t) = \int_0^t v(w)\, dw = \int_0^t w^2 e^{-w}\, dw$.

First let $u = w^2$, $dv = e^{-w}\, dw$ $\Rightarrow$ $du = 2w\, dw$, $v = -e^{-w}$. Then $s(t) = \big[-w^2 e^{-w}\big]_0^t + 2\int_0^t we^{-w}\, dw$.

Next let $U = w$, $dV = e^{-w}\, dw$ $\Rightarrow$ $dU = dw$, $V = -e^{-w}$. Then

$$s(t) = -t^2 e^{-t} + 2\left(\big[-we^{-w}\big]_0^t + \int_0^t e^{-w}\, dw\right) = -t^2 e^{-t} + 2\left(-te^{-t} + 0 + \big[-e^{-w}\big]_0^t\right)$$

$$= -t^2 e^{-t} + 2(-te^{-t} - e^{-t} + 1) = -t^2 e^{-t} - 2te^{-t} - 2e^{-t} + 2 = 2 - e^{-t}(t^2 + 2t + 2) \text{ meters}$$

65. For $I = \int_1^4 xf''(x)\, dx$, let $u = x$, $dv = f''(x)\, dx$ $\Rightarrow$ $du = dx$, $v = f'(x)$. Then

$I = \big[xf'(x)\big]_1^4 - \int_1^4 f'(x)\, dx = 4f'(4) - 1 \cdot f'(1) - [f(4) - f(1)] = 4 \cdot 3 - 1 \cdot 5 - (7 - 2) = 12 - 5 - 5 = 2.$

We used the fact that f'' is continuous to guarantee that I exists.

67. Using the formula for volumes of rotation and the figure, we see that

$\text{Volume} = \int_0^d \pi b^2 \, dy - \int_0^c \pi a^2 \, dy - \int_c^d \pi[g(y)]^2 \, dy = \pi b^2 d - \pi a^2 c - \int_c^d \pi[g(y)]^2 \, dy$. Let $y = f(x)$,

which gives $dy = f'(x) \, dx$ and $g(y) = x$, so that $V = \pi b^2 d - \pi a^2 c - \pi \int_a^b x^2 f'(x) \, dx$.

Now integrate by parts with $u = x^2$, and $dv = f'(x) \, dx$ $\Rightarrow$ $du = 2x \, dx$, $v = f(x)$, and

$\int_a^b x^2 f'(x) \, dx = \left[x^2 f(x)\right]_a^b - \int_a^b 2x f(x) \, dx = b^2 f(b) - a^2 f(a) - \int_a^b 2x f(x) \, dx$, but $f(a) = c$ and $f(b) = d$ $\Rightarrow$

$V = \pi b^2 d - \pi a^2 c - \pi \left[b^2 d - a^2 c - \int_a^b 2x f(x) \, dx\right] = \int_a^b 2\pi x f(x) \, dx.$

7.2 Trigonometric Integrals

The symbols $\overset{s}{=}$ and $\overset{c}{=}$ indicate the use of the substitutions $\{u = \sin x, du = \cos x \, dx\}$ and $\{u = \cos x, du = -\sin x \, dx\}$, respectively.

1. $\int \sin^3 x \cos^2 x \, dx = \int \sin^2 x \cos^2 x \sin x \, dx = \int (1 - \cos^2 x) \cos^2 x \sin x \, dx \overset{c}{=} \int (1 - u^2) u^2 (-du)$

$\quad = \int (u^2 - 1) u^2 \, du = \int (u^4 - u^2) \, du = \frac{1}{5} u^5 - \frac{1}{3} u^3 + C = \frac{1}{5} \cos^5 x - \frac{1}{3} \cos^3 x + C$

3. $\int_{\pi/2}^{3\pi/4} \sin^5 x \cos^3 x \, dx = \int_{\pi/2}^{3\pi/4} \sin^5 x \cos^2 x \cos x \, dx = \int_{\pi/2}^{3\pi/4} \sin^5 x \, (1 - \sin^2 x) \cos x \, dx \overset{s}{=} \int_1^{\sqrt{2}/2} u^5 (1 - u^2) \, du$

$\quad = \int_1^{\sqrt{2}/2} (u^5 - u^7) \, du = \left[\frac{1}{6} u^6 - \frac{1}{8} u^8\right]_1^{\sqrt{2}/2} = \left(\frac{1/8}{6} - \frac{1/16}{8}\right) - \left(\frac{1}{6} - \frac{1}{8}\right) = -\frac{11}{384}$

5. Let $y = \pi x$, so $dy = \pi \, dx$ and

$\int \sin^2(\pi x) \cos^5(\pi x) \, dx = \frac{1}{\pi} \int \sin^2 y \cos^5 y \, dy = \frac{1}{\pi} \int \sin^2 y \cos^4 y \cos y \, dy$

$\quad = \frac{1}{\pi} \int \sin^2 y \, (1 - \sin^2 y)^2 \cos y \, dy \overset{s}{=} \frac{1}{\pi} \int u^2 (1 - u^2)^2 \, du = \frac{1}{\pi} \int (u^2 - 2u^4 + u^6) \, du$

$\quad = \frac{1}{\pi} \left(\frac{1}{3} u^3 - \frac{2}{5} u^5 + \frac{1}{7} u^7\right) + C = \frac{1}{3\pi} \sin^3 y - \frac{2}{5\pi} \sin^5 y + \frac{1}{7\pi} \sin^7 y + C$

$\quad = \frac{1}{3\pi} \sin^3(\pi x) - \frac{2}{5\pi} \sin^5(\pi x) + \frac{1}{7\pi} \sin^7(\pi x) + C$

7. $\int_0^{\pi/2} \cos^2 \theta \, d\theta = \int_0^{\pi/2} \frac{1}{2}(1 + \cos 2\theta) \, d\theta$ [half-angle identity]

$\quad = \frac{1}{2} \left[\theta + \frac{1}{2} \sin 2\theta\right]_0^{\pi/2} = \frac{1}{2} \left[\left(\frac{\pi}{2} + 0\right) - (0 + 0)\right] = \frac{\pi}{4}$

9. $\int_0^\pi \sin^4(3t) \, dt = \int_0^\pi \left[\sin^2(3t)\right]^2 \, dt = \int_0^\pi \left[\frac{1}{2}(1 - \cos 6t)\right]^2 \, dt = \frac{1}{4} \int_0^\pi (1 - 2\cos 6t + \cos^2 6t) \, dt$

$\quad = \frac{1}{4} \int_0^\pi \left[1 - 2\cos 6t + \frac{1}{2}(1 + \cos 12t)\right] \, dt = \frac{1}{4} \int_0^\pi \left(\frac{3}{2} - 2\cos 6t + \frac{1}{2}\cos 12t\right) \, dt$

$\quad = \frac{1}{4} \left[\frac{3}{2}t - \frac{1}{3}\sin 6t + \frac{1}{24}\sin 12t\right]_0^\pi = \frac{1}{4} \left[\left(\frac{3\pi}{2} - 0 + 0\right) - (0 - 0 + 0)\right] = \frac{3\pi}{8}$

11. $\int (1 + \cos \theta)^2 \, d\theta = \int (1 + 2\cos \theta + \cos^2 \theta) \, d\theta = \theta + 2\sin \theta + \frac{1}{2} \int (1 + \cos 2\theta) \, d\theta$

$\quad = \theta + 2\sin \theta + \frac{1}{2}\theta + \frac{1}{4}\sin 2\theta + C = \frac{3}{2}\theta + 2\sin \theta + \frac{1}{4}\sin 2\theta + C$

13. $\int_0^{\pi/2} \sin^2 x \cos^2 x \, dx = \int_0^{\pi/2} \frac{1}{4}(4\sin^2 x \cos^2 x) \, dx = \int_0^{\pi/2} \frac{1}{4}(2\sin x \cos x)^2 \, dx = \frac{1}{4} \int_0^{\pi/2} \sin^2 2x \, dx$

$\quad = \frac{1}{4} \int_0^{\pi/2} \frac{1}{2}(1 - \cos 4x) \, dx = \frac{1}{8} \int_0^{\pi/2} (1 - \cos 4x) \, dx = \frac{1}{8} \left[x - \frac{1}{4}\sin 4x\right]_0^{\pi/2} = \frac{1}{8}\left(\frac{\pi}{2}\right) = \frac{\pi}{16}$

15. $\displaystyle\int \frac{\cos^5 \alpha}{\sqrt{\sin \alpha}}\, d\alpha = \int \frac{\cos^4 \alpha}{\sqrt{\sin \alpha}} \cos \alpha \, d\alpha = \int \frac{\left(1 - \sin^2 \alpha\right)^2}{\sqrt{\sin \alpha}} \cos \alpha \, d\alpha \overset{s}{=} \int \frac{(1 - u^2)^2}{\sqrt{u}}\, du$

$\displaystyle = \int \frac{1 - 2u^2 + u^4}{u^{1/2}}\, du = \int \left(u^{-1/2} - 2u^{3/2} + u^{7/2}\right) du = 2u^{1/2} - \tfrac{4}{5}u^{5/2} + \tfrac{2}{9}u^{9/2} + C$

$\displaystyle = \tfrac{2}{45}u^{1/2}(45 - 18u^2 + 5u^4) + C = \tfrac{2}{45}\sqrt{\sin \alpha}\,(45 - 18\sin^2 \alpha + 5\sin^4 \alpha) + C$

17. $\displaystyle\int \cos^2 x \tan^3 x \, dx = \int \frac{\sin^3 x}{\cos x}\, dx \overset{c}{=} \int \frac{(1 - u^2)(-du)}{u} = \int \left[\frac{-1}{u} + u\right] du$

$\displaystyle = -\ln|u| + \tfrac{1}{2}u^2 + C = \tfrac{1}{2}\cos^2 x - \ln|\cos x| + C$

19. $\displaystyle\int \frac{\cos x + \sin 2x}{\sin x}\, dx = \int \frac{\cos x + 2\sin x \cos x}{\sin x}\, dx = \int \frac{\cos x}{\sin x}\, dx + \int 2\cos x\, dx \overset{s}{=} \int \frac{1}{u}\, du + 2\sin x$

$\displaystyle = \ln|u| + 2\sin x + C = \ln|\sin x| + 2\sin x + C$

Or: Use the formula $\int \cot x\, dx = \ln|\sin x| + C$.

21. Let $u = \tan x$, $du = \sec^2 x\, dx$. Then $\int \sec^2 x \tan x\, dx = \int u\, du = \tfrac{1}{2}u^2 + C = \tfrac{1}{2}\tan^2 x + C$.

Or: Let $v = \sec x$, $dv = \sec x \tan x\, dx$. Then $\int \sec^2 x \tan x\, dx = \int v\, dv = \tfrac{1}{2}v^2 + C = \tfrac{1}{2}\sec^2 x + C$.

23. $\int \tan^2 x\, dx = \int (\sec^2 x - 1)\, dx = \tan x - x + C$

25. $\int \sec^6 t\, dt = \int \sec^4 t \cdot \sec^2 t\, dt = \int (\tan^2 t + 1)^2 \sec^2 t\, dt = \int (u^2 + 1)^2\, du \qquad [u = \tan t, \, du = \sec^2 t\, dt]$

$= \int (u^4 + 2u^2 + 1)\, du = \tfrac{1}{5}u^5 + \tfrac{2}{3}u^3 + u + C = \tfrac{1}{5}\tan^5 t + \tfrac{2}{3}\tan^3 t + \tan t + C$

27. $\int_0^{\pi/3} \tan^5 x \sec^4 x\, dx = \int_0^{\pi/3} \tan^5 x\,(\tan^2 x + 1)\sec^2 x\, dx = \int_0^{\sqrt{3}} u^5(u^2 + 1)\, du \qquad [u = \tan x, \, du = \sec^2 x\, dx]$

$= \int_0^{\sqrt{3}}(u^7 + u^5)\, du = \left[\tfrac{1}{8}u^8 + \tfrac{1}{6}u^6\right]_0^{\sqrt{3}} = \tfrac{81}{8} + \tfrac{27}{6} = \tfrac{81}{8} + \tfrac{9}{2} = \tfrac{81}{8} + \tfrac{36}{8} = \tfrac{117}{8}$

Alternate solution:

$\int_0^{\pi/3} \tan^5 x \sec^4 x\, dx = \int_0^{\pi/3} \tan^4 x \sec^3 x \sec x \tan x\, dx = \int_0^{\pi/3} (\sec^2 x - 1)^2 \sec^3 x \sec x \tan x\, dx$

$= \int_1^2 (u^2 - 1)^2 u^3\, du \quad [u = \sec x, \, du = \sec x \tan x\, dx] \quad = \int_1^2 (u^4 - 2u^2 + 1)u^3\, du$

$= \int_1^2 (u^7 - 2u^5 + u^3)\, du = \left[\tfrac{1}{8}u^8 - \tfrac{1}{3}u^6 + \tfrac{1}{4}u^4\right]_1^2 = \left(32 - \tfrac{64}{3} + 4\right) - \left(\tfrac{1}{8} - \tfrac{1}{3} + \tfrac{1}{4}\right) = \tfrac{117}{8}$

29. $\int \tan^3 x \sec x\, dx = \int \tan^2 x \sec x \tan x\, dx = \int (\sec^2 x - 1)\sec x \tan x\, dx$

$= \int (u^2 - 1)\, du \quad [u = \sec x, \, du = \sec x \tan x\, dx] \quad = \tfrac{1}{3}u^3 - u + C = \tfrac{1}{3}\sec^3 x - \sec x + C$

31. $\int \tan^5 x\, dx = \int (\sec^2 x - 1)^2 \tan x\, dx = \int \sec^4 x \tan x\, dx - 2\int \sec^2 x \tan x\, dx + \int \tan x\, dx$

$= \int \sec^3 x \sec x \tan x\, dx - 2\int \tan x \sec^2 x\, dx + \int \tan x\, dx$

$= \tfrac{1}{4}\sec^4 x - \tan^2 x + \ln|\sec x| + C \quad [\text{or } \tfrac{1}{4}\sec^4 x - \sec^2 x + \ln|\sec x| + C\,]$

33. $\displaystyle\int \frac{\tan^3 \theta}{\cos^4 \theta}\, d\theta = \int \tan^3 \theta \sec^4 \theta\, d\theta = \int \tan^3 \theta \cdot (\tan^2 \theta + 1) \cdot \sec^2 \theta\, d\theta$

$= \int u^3(u^2 + 1)\, du \qquad [u = \tan \theta, \, du = \sec^2 \theta\, d\theta]$

$= \int (u^5 + u^3)\, du = \tfrac{1}{6}u^6 + \tfrac{1}{4}u^4 + C = \tfrac{1}{6}\tan^6 \theta + \tfrac{1}{4}\tan^4 \theta + C$

35. Let $u = x, dv = \sec x \tan x\, dx \Rightarrow du = dx, v = \sec x$. Then

$\int x \sec x \tan x\, dx = x \sec x - \int \sec x\, dx = x \sec x - \ln|\sec x + \tan x| + C$.

37. $\int_{\pi/6}^{\pi/2} \cot^2 x\, dx = \int_{\pi/6}^{\pi/2}(\csc^2 x - 1)\, dx = \left[-\cot x - x\right]_{\pi/6}^{\pi/2} = \left(0 - \frac{\pi}{2}\right) - \left(-\sqrt{3} - \frac{\pi}{6}\right) = \sqrt{3} - \frac{\pi}{3}$

39. $\int \cot^3 \alpha \csc^3 \alpha\, d\alpha = \int \cot^2 \alpha \csc^2 \alpha \cdot \csc \alpha \cot \alpha\, d\alpha = \int(\csc^2 \alpha - 1)\csc^2 \alpha \cdot \csc \alpha \cot \alpha\, d\alpha$

$\qquad = \int(u^2 - 1)u^2 \cdot (-du) \qquad [u = \csc \alpha, du = -\csc \alpha \cot \alpha\, d\alpha]$

$\qquad = \int(u^2 - u^4)\, du = \frac{1}{3}u^3 - \frac{1}{5}u^5 + C = \frac{1}{3}\csc^3 \alpha - \frac{1}{5}\csc^5 \alpha + C$

41. $I = \int \csc x\, dx = \int \frac{\csc x\,(\csc x - \cot x)}{\csc x - \cot x}\, dx = \int \frac{-\csc x \cot x + \csc^2 x}{\csc x - \cot x}\, dx$. Let $u = \csc x - \cot x \Rightarrow$

$du = (-\csc x \cot x + \csc^2 x)\, dx$. Then $I = \int du/u = \ln|u| = \ln|\csc x - \cot x| + C$.

43. $\int \sin 8x \cos 5x\, dx \overset{2a}{=} \int \frac{1}{2}[\sin(8x - 5x) + \sin(8x + 5x)]\, dx = \frac{1}{2}\int \sin 3x\, dx + \frac{1}{2}\int \sin 13x\, dx$

$\qquad = -\frac{1}{6}\cos 3x - \frac{1}{26}\cos 13x + C$

45. $\int \sin 5\theta \sin \theta\, d\theta \overset{2b}{=} \int \frac{1}{2}[\cos(5\theta - \theta) - \cos(5\theta + \theta)]\, d\theta = \frac{1}{2}\int \cos 4\theta\, d\theta - \frac{1}{2}\int \cos 6\theta\, d\theta = \frac{1}{8}\sin 4\theta - \frac{1}{12}\sin 6\theta + C$

47. $\int \frac{1 - \tan^2 x}{\sec^2 x}\, dx = \int(\cos^2 x - \sin^2 x)\, dx = \int \cos 2x\, dx = \frac{1}{2}\sin 2x + C$

49. Let $u = \tan(t^2) \Rightarrow du = 2t \sec^2(t^2)\, dt$. Then $\int t \sec^2(t^2) \tan^4(t^2)\, dt = \int u^4\left(\frac{1}{2}\, du\right) = \frac{1}{10}u^5 + C = \frac{1}{10}\tan^5(t^2) + C$.

In Exercises 51–54, let $f(x)$ denote the integrand and $F(x)$ its antiderivative (with $C = 0$).

51. Let $u = x^2$, so that $du = 2x\, dx$. Then

$\int x \sin^2(x^2)\, dx = \int \sin^2 u\left(\frac{1}{2}\, du\right) = \frac{1}{2}\int \frac{1}{2}(1 - \cos 2u)\, du$

$\qquad = \frac{1}{4}\left(u - \frac{1}{2}\sin 2u\right) + C = \frac{1}{4}u - \frac{1}{4}\left(\frac{1}{2} \cdot 2\sin u \cos u\right) + C$

$\qquad = \frac{1}{4}x^2 - \frac{1}{4}\sin(x^2)\cos(x^2) + C$

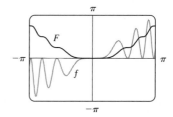

We see from the graph that this is reasonable, since F increases where f is positive and F decreases where f is negative. Note also that f is an odd function and F is an even function.

53. $\int \sin 3x \sin 6x\, dx = \int \frac{1}{2}[\cos(3x - 6x) - \cos(3x + 6x)]\, dx$

$\qquad = \frac{1}{2}\int(\cos 3x - \cos 9x)\, dx$

$\qquad = \frac{1}{6}\sin 3x - \frac{1}{18}\sin 9x + C$

Notice that $f(x) = 0$ whenever F has a horizontal tangent.

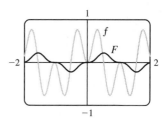

55. $f_{\text{ave}} = \frac{1}{2\pi}\int_{-\pi}^{\pi} \sin^2 x \cos^3 x\, dx = \frac{1}{2\pi}\int_{-\pi}^{\pi} \sin^2 x\,(1 - \sin^2 x)\cos x\, dx$

$\qquad = \frac{1}{2\pi}\int_0^0 u^2(1 - u^2)\, du \qquad [\text{where } u = \sin x]$

$\qquad = 0$

57. $A = \displaystyle\int_{-\pi/4}^{\pi/4} (\cos^2 x - \sin^2 x)\, dx = \int_{-\pi/4}^{\pi/4} \cos 2x\, dx$

$= 2 \displaystyle\int_{0}^{\pi/4} \cos 2x\, dx = 2 \left[\frac{1}{2} \sin 2x\right]_{0}^{\pi/4} = \left[\sin 2x\right]_{0}^{\pi/4}$

$= 1 - 0 = 1$

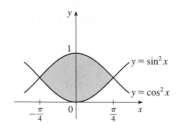

59.

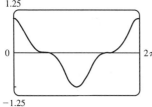

It seems from the graph that $\int_{0}^{2\pi} \cos^3 x\, dx = 0$, since the area below the

x-axis and above the graph looks about equal to the area above the axis and

below the graph. By Example 1, the integral is $\left[\sin x - \frac{1}{3} \sin^3 x\right]_{0}^{2\pi} = 0$.

Note that due to symmetry, the integral of any odd power of $\sin x$ or $\cos x$

between limits which differ by $2n\pi$ (n any integer) is 0.

61. Using disks, $V = \int_{\pi/2}^{\pi} \pi \sin^2 x\, dx = \pi \int_{\pi/2}^{\pi} \frac{1}{2}(1 - \cos 2x)\, dx = \pi \left[\frac{1}{2}x - \frac{1}{4} \sin 2x\right]_{\pi/2}^{\pi} = \pi\left(\frac{\pi}{2} - 0 - \frac{\pi}{4} + 0\right) = \frac{\pi^2}{4}$

63. Using washers,

$V = \int_{0}^{\pi/4} \pi\left[(1 - \sin x)^2 - (1 - \cos x)^2\right] dx$

$\quad = \pi \int_{0}^{\pi/4} \left[(1 - 2\sin x + \sin^2 x) - (1 - 2\cos x + \cos^2 x)\right] dx$

$\quad = \pi \int_{0}^{\pi/4} (2\cos x - 2\sin x + \sin^2 x - \cos^2 x)\, dx$

$\quad = \pi \int_{0}^{\pi/4} (2\cos x - 2\sin x - \cos 2x)\, dx = \pi \left[2\sin x + 2\cos x - \frac{1}{2}\sin 2x\right]_{0}^{\pi/4}$

$\quad = \pi\left[\left(\sqrt{2} + \sqrt{2} - \frac{1}{2}\right) - (0 + 2 - 0)\right] = \pi\left(2\sqrt{2} - \frac{5}{2}\right)$

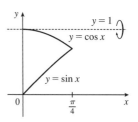

65. $s = f(t) = \int_{0}^{t} \sin \omega u \, \cos^2 \omega u\, du$. Let $y = \cos \omega u \;\Rightarrow\; dy = -\omega \sin \omega u\, du$. Then

$s = -\frac{1}{\omega} \int_{1}^{\cos \omega t} y^2\, dy = -\frac{1}{\omega}\left[\frac{1}{3}y^3\right]_{1}^{\cos \omega t} = \frac{1}{3\omega}(1 - \cos^3 \omega t)$.

67. Just note that the integrand is odd $[f(-x) = -f(x)]$.

Or: If $m \neq n$, calculate

$\displaystyle\int_{-\pi}^{\pi} \sin mx \, \cos nx\, dx = \int_{-\pi}^{\pi} \frac{1}{2}[\sin(m-n)x + \sin(m+n)x]\, dx = \frac{1}{2}\left[-\frac{\cos(m-n)x}{m-n} - \frac{\cos(m+n)x}{m+n}\right]_{-\pi}^{\pi} = 0$

If $m = n$, then the first term in each set of brackets is zero.

69. $\int_{-\pi}^{\pi} \cos mx \, \cos nx\, dx = \int_{-\pi}^{\pi} \frac{1}{2}[\cos(m-n)x + \cos(m+n)x]\, dx$.

If $m \neq n$, this is equal to $\dfrac{1}{2}\left[\dfrac{\sin(m-n)x}{m-n} + \dfrac{\sin(m+n)x}{m+n}\right]_{-\pi}^{\pi} = 0$.

If $m = n$, we get $\int_{-\pi}^{\pi} \frac{1}{2}[1 + \cos(m+n)x]\, dx = \left[\frac{1}{2}x\right]_{-\pi}^{\pi} + \left[\dfrac{\sin(m+n)x}{2(m+n)}\right]_{-\pi}^{\pi} = \pi + 0 = \pi$.

7.3 Trigonometric Substitution

1. Let $x = 3 \sec \theta$, where $0 \le \theta < \frac{\pi}{2}$ or $\pi \le \theta < \frac{3\pi}{2}$. Then

$dx = 3 \sec \theta \tan \theta \, d\theta$ and

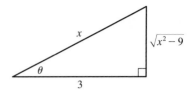

$$\sqrt{x^2 - 9} = \sqrt{9 \sec^2 \theta - 9} = \sqrt{9(\sec^2 \theta - 1)} = \sqrt{9 \tan^2 \theta}$$
$$= 3 \,|\tan \theta| = 3 \tan \theta \text{ for the relevant values of } \theta.$$

$$\int \frac{1}{x^2 \sqrt{x^2 - 9}} \, dx = \int \frac{1}{9 \sec^2 \theta \cdot 3 \tan \theta} \, 3 \sec \theta \tan \theta \, d\theta = \frac{1}{9} \int \cos \theta \, d\theta = \frac{1}{9} \sin \theta + C = \frac{1}{9} \frac{\sqrt{x^2 - 9}}{x} + C$$

Note that $-\sec(\theta + \pi) = \sec \theta$, so the figure is sufficient for the case $\pi \le \theta < \frac{3\pi}{2}$.

3. Let $x = 3 \tan \theta$, where $-\frac{\pi}{2} < \theta < \frac{\pi}{2}$. Then $dx = 3 \sec^2 \theta \, d\theta$ and

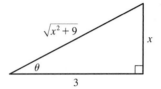

$$\sqrt{x^2 + 9} = \sqrt{9 \tan^2 \theta + 9} = \sqrt{9(\tan^2 \theta + 1)} = \sqrt{9 \sec^2 \theta}$$
$$= 3 \,|\sec \theta| = 3 \sec \theta \text{ for the relevant values of } \theta.$$

$$\int \frac{x^3}{\sqrt{x^2 + 9}} \, dx = \int \frac{3^3 \tan^3 \theta}{3 \sec \theta} \, 3 \sec^2 \theta \, d\theta = 3^3 \int \tan^3 \theta \sec \theta d\theta = 3^3 \int \tan^2 \theta \tan \theta \sec \theta \, d\theta$$

$$= 3^3 \int (\sec^2 \theta - 1) \tan \theta \sec \theta \, d\theta = 3^3 \int (u^2 - 1) \, du \qquad [u = \sec \theta, \, du = \sec \theta \tan \theta \, d\theta]$$

$$= 3^3 \left(\frac{1}{3} u^3 - u \right) + C = 3^3 \left(\frac{1}{3} \sec^3 \theta - \sec \theta \right) + C = 3^3 \left[\frac{1}{3} \frac{(x^2 + 9)^{3/2}}{3^3} - \frac{\sqrt{x^2 + 9}}{3} \right] + C$$

$$= \frac{1}{3} (x^2 + 9)^{3/2} - 9 \sqrt{x^2 + 9} + C \quad \text{or} \quad \frac{1}{3} (x^2 - 18) \sqrt{x^2 + 9} + C$$

5. Let $t = \sec \theta$, so $dt = \sec \theta \tan \theta \, d\theta$, $t = \sqrt{2} \; \Rightarrow \; \theta = \frac{\pi}{4}$, and $t = 2 \; \Rightarrow \; \theta = \frac{\pi}{3}$. Then

$$\int_{\sqrt{2}}^{2} \frac{1}{t^3 \sqrt{t^2 - 1}} \, dt = \int_{\pi/4}^{\pi/3} \frac{1}{\sec^3 \theta \tan \theta} \sec \theta \tan \theta \, d\theta = \int_{\pi/4}^{\pi/3} \frac{1}{\sec^2 \theta} \, d\theta = \int_{\pi/4}^{\pi/3} \cos^2 \theta \, d\theta$$

$$= \int_{\pi/4}^{\pi/3} \frac{1}{2} (1 + \cos 2\theta) \, d\theta = \frac{1}{2} \left[\theta + \frac{1}{2} \sin 2\theta \right]_{\pi/4}^{\pi/3}$$

$$= \frac{1}{2} \left[\left(\frac{\pi}{3} + \frac{1}{2} \frac{\sqrt{3}}{2} \right) - \left(\frac{\pi}{4} + \frac{1}{2} \cdot 1 \right) \right] = \frac{1}{2} \left(\frac{\pi}{12} + \frac{\sqrt{3}}{4} - \frac{1}{2} \right) = \frac{\pi}{24} + \frac{\sqrt{3}}{8} - \frac{1}{4}$$

7. Let $x = 5 \sin \theta$, so $dx = 5 \cos \theta \, d\theta$. Then

$$\int \frac{1}{x^2 \sqrt{25 - x^2}} \, dx = \int \frac{1}{5^2 \sin^2 \theta \cdot 5 \cos \theta} \, 5 \cos \theta \, d\theta = \frac{1}{25} \int \csc^2 \theta \, d\theta$$

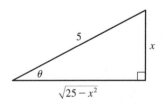

$$= -\frac{1}{25} \cot \theta + C = -\frac{1}{25} \frac{\sqrt{25 - x^2}}{x} + C$$

9. Let $x = 4\tan\theta$, where $-\frac{\pi}{2} < \theta < \frac{\pi}{2}$. Then $dx = 4\sec^2\theta\, d\theta$ and

$$\sqrt{x^2 + 16} = \sqrt{16\tan^2\theta + 16} = \sqrt{16(\tan^2\theta + 1)}$$
$$= \sqrt{16\sec^2\theta} = 4\,|\sec\theta|$$
$$= 4\sec\theta \text{ for the relevant values of } \theta.$$

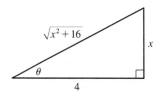

$$\int \frac{dx}{\sqrt{x^2 + 16}} = \int \frac{4\sec^2\theta\, d\theta}{4\sec\theta} = \int \sec\theta\, d\theta = \ln|\sec\theta + \tan\theta| + C_1 = \ln\left|\frac{\sqrt{x^2+16}}{4} + \frac{x}{4}\right| + C_1$$

$$= \ln\left|\sqrt{x^2+16} + x\right| - \ln|4| + C_1 = \ln\left(\sqrt{x^2+16} + x\right) + C, \text{ where } C = C_1 - \ln 4.$$

(Since $\sqrt{x^2+16} + x > 0$, we don't need the absolute value.)

11. Let $2x = \sin\theta$, where $-\frac{\pi}{2} \le \theta \le \frac{\pi}{2}$. Then $x = \frac{1}{2}\sin\theta$, $dx = \frac{1}{2}\cos\theta\, d\theta$,

and $\sqrt{1 - 4x^2} = \sqrt{1 - (2x)^2} = \cos\theta$.

$$\int \sqrt{1 - 4x^2}\, dx = \int \cos\theta\left(\frac{1}{2}\cos\theta\right) d\theta = \frac{1}{4}\int (1 + \cos 2\theta)\, d\theta$$

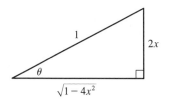

$$= \frac{1}{4}\left(\theta + \frac{1}{2}\sin 2\theta\right) + C = \frac{1}{4}(\theta + \sin\theta\,\cos\theta) + C$$

$$= \frac{1}{4}\left[\sin^{-1}(2x) + 2x\sqrt{1 - 4x^2}\right] + C$$

13. Let $x = 3\sec\theta$, where $0 \le \theta < \frac{\pi}{2}$ or $\pi \le \theta < \frac{3\pi}{2}$. Then

$dx = 3\sec\theta\,\tan\theta\, d\theta$ and $\sqrt{x^2 - 9} = 3\tan\theta$, so

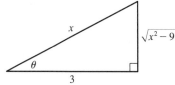

$$\int \frac{\sqrt{x^2 - 9}}{x^3}\, dx = \int \frac{3\tan\theta}{27\sec^3\theta}\, 3\sec\theta\,\tan\theta\, d\theta = \frac{1}{3}\int \frac{\tan^2\theta}{\sec^2\theta}\, d\theta$$

$$= \frac{1}{3}\int \sin^2\theta\, d\theta = \frac{1}{3}\int \frac{1}{2}(1 - \cos 2\theta)\, d\theta = \frac{1}{6}\theta - \frac{1}{12}\sin 2\theta + C = \frac{1}{6}\theta - \frac{1}{6}\sin\theta\,\cos\theta + C$$

$$= \frac{1}{6}\sec^{-1}\left(\frac{x}{3}\right) - \frac{1}{6}\frac{\sqrt{x^2 - 9}}{x}\frac{3}{x} + C = \frac{1}{6}\sec^{-1}\left(\frac{x}{3}\right) - \frac{\sqrt{x^2 - 9}}{2x^2} + C$$

15. Let $x = a\sin\theta$, $dx = a\cos\theta\, d\theta$, $x = 0 \implies \theta = 0$ and $x = a \implies \theta = \frac{\pi}{2}$. Then

$$\int_0^a x^2\sqrt{a^2 - x^2}\, dx = \int_0^{\pi/2} a^2\sin^2\theta\, (a\cos\theta)\, a\cos\theta\, d\theta = a^4\int_0^{\pi/2}\sin^2\theta\,\cos^2\theta\, d\theta = a^4\int_0^{\pi/2}\left[\frac{1}{2}(2\sin\theta\,\cos\theta)\right]^2 d\theta$$

$$= \frac{a^4}{4}\int_0^{\pi/2}\sin^2 2\theta\, d\theta = \frac{a^4}{4}\int_0^{\pi/2}\frac{1}{2}(1 - \cos 4\theta)\, d\theta = \frac{a^4}{8}\left[\theta - \frac{1}{4}\sin 4\theta\right]_0^{\pi/2}$$

$$= \frac{a^4}{8}\left[\left(\frac{\pi}{2} - 0\right) - 0\right] = \frac{\pi}{16}a^4$$

17. Let $u = x^2 - 7$, so $du = 2x\, dx$. Then $\displaystyle\int \frac{x}{\sqrt{x^2 - 7}}\, dx = \frac{1}{2}\int \frac{1}{\sqrt{u}}\, du = \frac{1}{2}\cdot 2\sqrt{u} + C = \sqrt{x^2 - 7} + C.$

19. Let $x = \tan\theta$, where $-\frac{\pi}{2} < \theta < \frac{\pi}{2}$. Then $dx = \sec^2\theta\,d\theta$

and $\sqrt{1 + x^2} = \sec\theta$, so

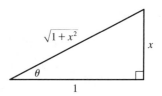

$$\int \frac{\sqrt{1 + x^2}}{x}\,dx = \int \frac{\sec\theta}{\tan\theta}\sec^2\theta\,d\theta = \int \frac{\sec\theta}{\tan\theta}(1 + \tan^2\theta)\,d\theta$$

$$= \int (\csc\theta + \sec\theta\tan\theta)\,d\theta$$

$$= \ln|\csc\theta - \cot\theta| + \sec\theta + C \qquad \text{[by Exercise 7.2.41]}$$

$$= \ln\left|\frac{\sqrt{1 + x^2}}{x} - \frac{1}{x}\right| + \frac{\sqrt{1 + x^2}}{1} + C = \ln\left|\frac{\sqrt{1 + x^2} - 1}{x}\right| + \sqrt{1 + x^2} + C$$

21. Let $x = \frac{3}{5}\sin\theta$, so $dx = \frac{3}{5}\cos\theta\,d\theta$, $x = 0 \Rightarrow \theta = 0$, and $x = 0.6 \Rightarrow \theta = \frac{\pi}{2}$. Then

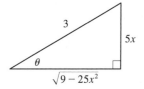

$$\int_0^{0.6} \frac{x^2}{\sqrt{9 - 25x^2}}\,dx = \int_0^{\pi/2} \frac{\left(\frac{3}{5}\right)^2 \sin^2\theta}{3\cos\theta}\left(\frac{3}{5}\cos\theta\,d\theta\right) = \frac{9}{125}\int_0^{\pi/2} \sin^2\theta\,d\theta$$

$$= \frac{9}{125}\int_0^{\pi/2} \frac{1}{2}(1 - \cos 2\theta)\,d\theta = \frac{9}{250}\left[\theta - \frac{1}{2}\sin 2\theta\right]_0^{\pi/2}$$

$$= \frac{9}{250}\left[\left(\frac{\pi}{2} - 0\right) - 0\right] = \frac{9}{500}\pi$$

23. $5 + 4x - x^2 = -(x^2 - 4x + 4) + 9 = -(x - 2)^2 + 9$. Let

$x - 2 = 3\sin\theta$, $-\frac{\pi}{2} \le \theta \le \frac{\pi}{2}$, so $dx = 3\cos\theta\,d\theta$. Then

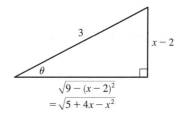

$$\int \sqrt{5 + 4x - x^2}\,dx = \int \sqrt{9 - (x - 2)^2}\,dx = \int \sqrt{9 - 9\sin^2\theta}\,3\cos\theta\,d\theta$$

$$= \int \sqrt{9\cos^2\theta}\,3\cos\theta\,d\theta = \int 9\cos^2\theta\,d\theta$$

$$= \frac{9}{2}\int (1 + \cos 2\theta)\,d\theta = \frac{9}{2}\left(\theta + \frac{1}{2}\sin 2\theta\right) + C$$

$$= \frac{9}{2}\theta + \frac{9}{4}\sin 2\theta + C = \frac{9}{2}\theta + \frac{9}{4}(2\sin\theta\cos\theta) + C$$

$$= \frac{9}{2}\sin^{-1}\left(\frac{x - 2}{3}\right) + \frac{9}{2}\cdot\frac{x - 2}{3}\cdot\frac{\sqrt{5 + 4x - x^2}}{3} + C$$

$$= \frac{9}{2}\sin^{-1}\left(\frac{x - 2}{3}\right) + \frac{1}{2}(x - 2)\sqrt{5 + 4x - x^2} + C$$

25. $x^2 + x + 1 = \left(x^2 + x + \frac{1}{4}\right) + \frac{3}{4} = \left(x + \frac{1}{2}\right)^2 + \left(\frac{\sqrt{3}}{2}\right)^2$. Let

$x + \frac{1}{2} = \frac{\sqrt{3}}{2}\tan\theta$, so $dx = \frac{\sqrt{3}}{2}\sec^2\theta\,d\theta$ and $\sqrt{x^2 + x + 1} = \frac{\sqrt{3}}{2}\sec\theta$.

Then

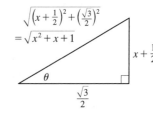

[continued]

$$\int \frac{x}{\sqrt{x^2 + x + 1}}\, dx = \int \frac{\frac{\sqrt{3}}{2}\tan\theta - \frac{1}{2}}{\frac{\sqrt{3}}{2}\sec\theta}\frac{\sqrt{3}}{2}\sec^2\theta\, d\theta$$

$$= \int \left(\tfrac{\sqrt{3}}{2}\tan\theta - \tfrac{1}{2}\right)\sec\theta\, d\theta = \int \tfrac{\sqrt{3}}{2}\tan\theta\sec\theta\, d\theta - \int \tfrac{1}{2}\sec\theta\, d\theta$$

$$= \tfrac{\sqrt{3}}{2}\sec\theta - \tfrac{1}{2}\ln|\sec\theta + \tan\theta| + C_1$$

$$= \sqrt{x^2 + x + 1} - \tfrac{1}{2}\ln\left|\tfrac{2}{\sqrt{3}}\sqrt{x^2 + x + 1} + \tfrac{2}{\sqrt{3}}\left(x + \tfrac{1}{2}\right)\right| + C_1$$

$$= \sqrt{x^2 + x + 1} - \tfrac{1}{2}\ln\left|\tfrac{2}{\sqrt{3}}\left[\sqrt{x^2 + x + 1} + \left(x + \tfrac{1}{2}\right)\right]\right| + C_1$$

$$= \sqrt{x^2 + x + 1} - \tfrac{1}{2}\ln\tfrac{2}{\sqrt{3}} - \tfrac{1}{2}\ln\left(\sqrt{x^2 + x + 1} + x + \tfrac{1}{2}\right) + C_1$$

$$= \sqrt{x^2 + x + 1} - \tfrac{1}{2}\ln\left(\sqrt{x^2 + x + 1} + x + \tfrac{1}{2}\right) + C, \quad \text{where } C = C_1 - \tfrac{1}{2}\ln\tfrac{2}{\sqrt{3}}$$

27. $x^2 + 2x = (x^2 + 2x + 1) - 1 = (x + 1)^2 - 1$. Let $x + 1 = 1\sec\theta$,

so $dx = \sec\theta\tan\theta\, d\theta$ and $\sqrt{x^2 + 2x} = \tan\theta$. Then

$$\int \sqrt{x^2 + 2x}\, dx = \int \tan\theta\,(\sec\theta\tan\theta\, d\theta) = \int \tan^2\theta\sec\theta\, d\theta$$

$$= \int (\sec^2\theta - 1)\sec\theta\, d\theta = \int \sec^3\theta\, d\theta - \int \sec\theta\, d\theta$$

$$= \tfrac{1}{2}\sec\theta\tan\theta + \tfrac{1}{2}\ln|\sec\theta + \tan\theta| - \ln|\sec\theta + \tan\theta| + C$$

$$= \tfrac{1}{2}\sec\theta\tan\theta - \tfrac{1}{2}\ln|\sec\theta + \tan\theta| + C = \tfrac{1}{2}(x + 1)\sqrt{x^2 + 2x} - \tfrac{1}{2}\ln\left|x + 1 + \sqrt{x^2 + 2x}\right| + C$$

(Right-hand diagram: a right triangle with hypotenuse $x+1$, vertical side $\sqrt{(x+1)^2 - 1^2} = \sqrt{x^2 + 2x}$, base 1, angle θ.)

29. Let $u = x^2$, $du = 2x\, dx$. Then

$$\int x\sqrt{1 - x^4}\, dx = \int \sqrt{1 - u^2}\left(\tfrac{1}{2}\, du\right) = \tfrac{1}{2}\int \cos\theta \cdot \cos\theta\, d\theta \qquad \begin{bmatrix} \text{where } u = \sin\theta,\, du = \cos\theta\, d\theta, \\ \text{and } \sqrt{1 - u^2} = \cos\theta \end{bmatrix}$$

$$= \tfrac{1}{2}\int \tfrac{1}{2}(1 + \cos 2\theta)\, d\theta = \tfrac{1}{4}\theta + \tfrac{1}{8}\sin 2\theta + C = \tfrac{1}{4}\theta + \tfrac{1}{4}\sin\theta\,\cos\theta + C$$

$$= \tfrac{1}{4}\sin^{-1} u + \tfrac{1}{4}u\sqrt{1 - u^2} + C = \tfrac{1}{4}\sin^{-1}(x^2) + \tfrac{1}{4}x^2\sqrt{1 - x^4} + C$$

31. (a) Let $x = a\tan\theta$, where $-\frac{\pi}{2} < \theta < \frac{\pi}{2}$. Then $\sqrt{x^2 + a^2} = a\sec\theta$ and

$$\int \frac{dx}{\sqrt{x^2 + a^2}} = \int \frac{a\sec^2\theta\, d\theta}{a\sec\theta} = \int \sec\theta\, d\theta = \ln|\sec\theta + \tan\theta| + C_1 = \ln\left|\frac{\sqrt{x^2 + a^2}}{a} + \frac{x}{a}\right| + C_1$$

$$= \ln\left(x + \sqrt{x^2 + a^2}\right) + C \quad \text{where } C = C_1 - \ln|a|$$

(b) Let $x = a\sinh t$, so that $dx = a\cosh t\, dt$ and $\sqrt{x^2 + a^2} = a\cosh t$. Then

$$\int \frac{dx}{\sqrt{x^2 + a^2}} = \int \frac{a\cosh t\, dt}{a\cosh t} = t + C = \sinh^{-1}\frac{x}{a} + C.$$

33. The average value of $f(x) = \sqrt{x^2 - 1}/x$ on the interval $[1, 7]$ is

$$\frac{1}{7 - 1}\int_1^7 \frac{\sqrt{x^2 - 1}}{x}\, dx = \frac{1}{6}\int_0^\alpha \frac{\tan\theta}{\sec\theta}\cdot\sec\theta\tan\theta\, d\theta \qquad \begin{bmatrix} \text{where } x = \sec\theta,\, dx = \sec\theta\tan\theta\, d\theta, \\ \sqrt{x^2 - 1} = \tan\theta,\, \text{and } \alpha = \sec^{-1} 7 \end{bmatrix}$$

$$= \tfrac{1}{6}\int_0^\alpha \tan^2\theta\, d\theta = \tfrac{1}{6}\int_0^\alpha (\sec^2\theta - 1)\, d\theta = \tfrac{1}{6}\left[\tan\theta - \theta\right]_0^\alpha$$

$$= \tfrac{1}{6}(\tan\alpha - \alpha) = \tfrac{1}{6}\left(\sqrt{48} - \sec^{-1} 7\right)$$

35. Area of $\triangle POQ = \frac{1}{2}(r\cos\theta)(r\sin\theta) = \frac{1}{2}r^2\sin\theta\,\cos\theta$. Area of region $PQR = \int_{r\cos\theta}^r \sqrt{r^2 - x^2}\,dx$.

Let $x = r\cos u \ \Rightarrow\ dx = -r\sin u\,du$ for $\theta \le u \le \frac{\pi}{2}$. Then we obtain

$$\int \sqrt{r^2 - x^2}\,dx = \int r\sin u\,(-r\sin u)\,du = -r^2\int \sin^2 u\,du = -\tfrac{1}{2}r^2(u - \sin u\,\cos u) + C$$
$$= -\tfrac{1}{2}r^2\cos^{-1}(x/r) + \tfrac{1}{2}x\,\sqrt{r^2 - x^2} + C$$

so $\qquad$ area of region $PQR = \frac{1}{2}\Big[-r^2\cos^{-1}(x/r) + x\,\sqrt{r^2 - x^2}\,\Big]_{r\cos\theta}^r$

$$= \tfrac{1}{2}\big[0 - (-r^2\theta + r\cos\theta\,r\sin\theta)\big] = \tfrac{1}{2}r^2\theta - \tfrac{1}{2}r^2\sin\theta\,\cos\theta$$

and thus, (area of sector POR) = (area of $\triangle POQ$) + (area of region PQR) = $\frac{1}{2}r^2\theta$.

37. From the graph, it appears that the curve $y = x^2\sqrt{4 - x^2}$ and the

line $y = 2 - x$ intersect at about $x = a \approx 0.81$ and $x = 2$, with

$x^2\sqrt{4 - x^2} > 2 - x$ on $(a, 2)$. So the area bounded by the curve and the line is

$A \approx \int_a^2 \big[x^2\sqrt{4 - x^2} - (2 - x)\big]\,dx = \int_a^2 x^2\sqrt{4 - x^2}\,dx - \big[2x - \tfrac{1}{2}x^2\big]_a^2$.

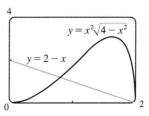

To evaluate the integral, we put $x = 2\sin\theta$, where $-\frac{\pi}{2} \le \theta \le \frac{\pi}{2}$. Then

$dx = 2\cos\theta\,d\theta,\ x = 2 \ \Rightarrow\ \theta = \sin^{-1}1 = \frac{\pi}{2}$, and $x = a \ \Rightarrow\ \theta = \alpha = \sin^{-1}(a/2) \approx 0.416$. So

$$\int_a^2 x^2\sqrt{4 - x^2}\,dx \approx \int_\alpha^{\pi/2} 4\sin^2\theta\,(2\cos\theta)(2\cos\theta\,d\theta) = 4\int_\alpha^{\pi/2}\sin^2 2\theta\,d\theta = 4\int_\alpha^{\pi/2}\tfrac{1}{2}(1 - \cos 4\theta)\,d\theta$$

$$= 2\big[\theta - \tfrac{1}{4}\sin 4\theta\big]_\alpha^{\pi/2} = 2\big[(\tfrac{\pi}{2} - 0) - (\alpha - \tfrac{1}{4}(0.996))\big] \approx 2.81$$

Thus, $A \approx 2.81 - \big[\big(2\cdot 2 - \tfrac{1}{2}\cdot 2^2\big) - \big(2a - \tfrac{1}{2}a^2\big)\big] \approx 2.10$.

39. (a) Let $t = a\sin\theta,\ dt = a\cos\theta\,d\theta,\ t = 0 \ \Rightarrow\ \theta = 0$ and $t = x \ \Rightarrow\ \theta = \sin^{-1}(x/a)$. Then

$$\int_0^x \sqrt{a^2 - t^2}\,dt = \int_0^{\sin^{-1}(x/a)} a\cos\theta\,(a\cos\theta\,d\theta)$$

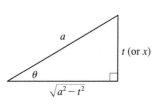

$$= a^2\int_0^{\sin^{-1}(x/a)}\cos^2\theta\,d\theta = \frac{a^2}{2}\int_0^{\sin^{-1}(x/a)}(1 + \cos 2\theta)\,d\theta$$

$$= \frac{a^2}{2}\Big[\theta + \tfrac{1}{2}\sin 2\theta\Big]_0^{\sin^{-1}(x/a)} = \frac{a^2}{2}\Big[\theta + \sin\theta\,\cos\theta\Big]_0^{\sin^{-1}(x/a)}$$

$$= \frac{a^2}{2}\bigg[\Big(\sin^{-1}\Big(\frac{x}{a}\Big) + \frac{x}{a}\cdot\frac{\sqrt{a^2 - x^2}}{a}\Big) - 0\bigg]$$

$$= \tfrac{1}{2}a^2\sin^{-1}(x/a) + \tfrac{1}{2}x\,\sqrt{a^2 - x^2}$$

(b) The integral $\int_0^x \sqrt{a^2 - t^2}\,dt$ represents the area under the curve $y = \sqrt{a^2 - t^2}$ between the vertical lines $t = 0$ and $t = x$.

The figure shows that this area consists of a triangular region and a sector of the circle $t^2 + y^2 = a^2$. The triangular region

has base x and height $\sqrt{a^2 - x^2}$, so its area is $\frac{1}{2}x\,\sqrt{a^2 - x^2}$. The sector has area $\frac{1}{2}a^2\theta = \frac{1}{2}a^2\sin^{-1}(x/a)$.

41. Let the equation of the large circle be $x^2 + y^2 = R^2$. Then the equation of

the small circle is $x^2 + (y - b)^2 = r^2$, where $b = \sqrt{R^2 - r^2}$ is the distance

between the centers of the circles. The desired area is

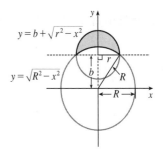

$$A = \int_{-r}^{r} \left[\left(b + \sqrt{r^2 - x^2} \right) - \sqrt{R^2 - x^2} \right] dx$$

$$= 2 \int_{0}^{r} \left(b + \sqrt{r^2 - x^2} - \sqrt{R^2 - x^2} \right) dx$$

$$= 2 \int_{0}^{r} b \, dx + 2 \int_{0}^{r} \sqrt{r^2 - x^2} \, dx - 2 \int_{0}^{r} \sqrt{R^2 - x^2} \, dx$$

The first integral is just $2br = 2r\sqrt{R^2 - r^2}$. The second integral represents the area of a quarter-circle of radius r, so its value

is $\frac{1}{4}\pi r^2$. To evaluate the other integral, note that

$$\int \sqrt{a^2 - x^2} \, dx = \int a^2 \cos^2 \theta \, d\theta \quad [x = a \sin \theta, \, dx = a \cos \theta \, d\theta] \quad = \left(\tfrac{1}{2} a^2 \right) \int (1 + \cos 2\theta) \, d\theta$$

$$= \tfrac{1}{2} a^2 \left(\theta + \tfrac{1}{2} \sin 2\theta \right) + C = \tfrac{1}{2} a^2 (\theta + \sin \theta \, \cos \theta) + C$$

$$= \frac{a^2}{2} \arcsin\left(\frac{x}{a}\right) + \frac{a^2}{2}\left(\frac{x}{a}\right)\frac{\sqrt{a^2 - x^2}}{a} + C = \frac{a^2}{2} \arcsin\left(\frac{x}{a}\right) + \frac{x}{2}\sqrt{a^2 - x^2} + C$$

Thus, the desired area is

$$A = 2r\sqrt{R^2 - r^2} + 2\left(\tfrac{1}{4}\pi r^2\right) - \left[R^2 \arcsin(x/R) + x\sqrt{R^2 - x^2} \right]_{0}^{r}$$

$$= 2r\sqrt{R^2 - r^2} + \tfrac{1}{2}\pi r^2 - \left[R^2 \arcsin(r/R) + r\sqrt{R^2 - r^2} \right] = r\sqrt{R^2 - r^2} + \tfrac{\pi}{2}r^2 - R^2 \arcsin(r/R)$$

43. We use cylindrical shells and assume that $R > r$. $\quad x^2 = r^2 - (y - R)^2 \quad \Rightarrow \quad x = \pm\sqrt{r^2 - (y - R)^2}$,

so $g(y) = 2\sqrt{r^2 - (y - R)^2}$ and

$$V = \int_{R-r}^{R+r} 2\pi y \cdot 2\sqrt{r^2 - (y - R)^2} \, dy = \int_{-r}^{r} 4\pi (u + R)\sqrt{r^2 - u^2} \, du \qquad [\text{where } u = y - R]$$

$$= 4\pi \int_{-r}^{r} u\sqrt{r^2 - u^2} \, du + 4\pi R \int_{-r}^{r} \sqrt{r^2 - u^2} \, du \qquad \begin{bmatrix} \text{where } u = r\sin\theta, \, du = r\cos\theta \, d\theta \\ \text{in the second integral} \end{bmatrix}$$

$$= 4\pi \left[-\tfrac{1}{3}(r^2 - u^2)^{3/2} \right]_{-r}^{r} + 4\pi R \int_{-\pi/2}^{\pi/2} r^2 \cos^2 \theta \, d\theta = -\tfrac{4\pi}{3}(0 - 0) + 4\pi R r^2 \int_{-\pi/2}^{\pi/2} \cos^2 \theta \, d\theta$$

$$= 2\pi R r^2 \int_{-\pi/2}^{\pi/2} (1 + \cos 2\theta) \, d\theta = 2\pi R r^2 \left[\theta + \tfrac{1}{2}\sin 2\theta \right]_{-\pi/2}^{\pi/2} = 2\pi^2 R r^2$$

Another method: Use washers instead of shells, so $V = 8\pi R \int_{0}^{r} \sqrt{r^2 - y^2} \, dy$ as in Exercise 6.2.63(a), but evaluate the

integral using $y = r\sin\theta$.

7.4 Integration of Rational Functions by Partial Fractions

1. (a) $\dfrac{2x}{(x+3)(3x+1)} = \dfrac{A}{x+3} + \dfrac{B}{3x+1}$

(b) $\dfrac{1}{x^3+2x^2+x} = \dfrac{1}{x(x^2+2x+1)} = \dfrac{1}{x(x+1)^2} = \dfrac{A}{x} + \dfrac{B}{x+1} + \dfrac{C}{(x+1)^2}$

3. (a) $\dfrac{x^4+1}{x^5+4x^3} = \dfrac{x^4+1}{x^3\left(x^2+4\right)} = \dfrac{A}{x} + \dfrac{B}{x^2} + \dfrac{C}{x^3} + \dfrac{Dx+E}{x^2+4}$

(b) $\dfrac{1}{(x^2-9)^2} = \dfrac{1}{[(x+3)(x-3)]^2} = \dfrac{1}{(x+3)^2(x-3)^2} = \dfrac{A}{x+3} + \dfrac{B}{(x+3)^2} + \dfrac{C}{x-3} + \dfrac{D}{(x-3)^2}$

5. (a) $\dfrac{x^4}{x^4-1} = \dfrac{(x^4-1)+1}{x^4-1} = 1 + \dfrac{1}{x^4-1}$ [or use long division] $= 1 + \dfrac{1}{(x^2-1)(x^2+1)}$

$= 1 + \dfrac{1}{(x-1)(x+1)(x^2+1)} = 1 + \dfrac{A}{x-1} + \dfrac{B}{x+1} + \dfrac{Cx+D}{x^2+1}$

(b) $\dfrac{t^4+t^2+1}{(t^2+1)(t^2+4)^2} = \dfrac{At+B}{t^2+1} + \dfrac{Ct+D}{t^2+4} + \dfrac{Et+F}{(t^2+4)^2}$

7. $\displaystyle\int \dfrac{x}{x-6}\,dx = \int \dfrac{(x-6)+6}{x-6}\,dx = \int \left(1 + \dfrac{6}{x-6}\right)dx = x + 6\ln|x-6| + C$

9. $\dfrac{x-9}{(x+5)(x-2)} = \dfrac{A}{x+5} + \dfrac{B}{x-2}$. Multiply both sides by $(x+5)(x-2)$ to get $x-9 = A(x-2) + B(x+5)(*)$, or

equivalently, $x-9 = (A+B)x - 2A + 5B$. Equating coefficients of x on each side of the equation gives us $1 = A+B$ **(1)**

and equating constants gives us $-9 = -2A + 5B$ **(2)**. Adding two times **(1)** to **(2)** gives us $-7 = 7B$ ⇔ $B = -1$ and

hence, $A = 2$. [Alternatively, to find the coefficients A and B, we may use substitution as follows: substitute 2 for x in $(*)$ to

get $-7 = 7B$ ⇔ $B = -1$, then substitute -5 for x in $(*)$ to get $-14 = -7A$ ⇔ $A = 2$.] Thus,

$\displaystyle\int \dfrac{x-9}{(x+5)(x-2)}\,dx = \int \left(\dfrac{2}{x+5} + \dfrac{-1}{x-2}\right)dx = 2\ln|x+5| - \ln|x-2| + C.$

11. $\dfrac{1}{x^2-1} = \dfrac{1}{(x+1)(x-1)} = \dfrac{A}{x+1} + \dfrac{B}{x-1}$. Multiply both sides by $(x+1)(x-1)$ to get $1 = A(x-1) + B(x+1)$.

Substituting 1 for x gives $1 = 2B$ ⇔ $B = \frac{1}{2}$. Substituting -1 for x gives $1 = -2A$ ⇔ $A = -\frac{1}{2}$. Thus,

$\displaystyle\int_2^3 \dfrac{1}{x^2-1}\,dx = \int_2^3 \left(\dfrac{-1/2}{x+1} + \dfrac{1/2}{x-1}\right)dx = \left[-\tfrac{1}{2}\ln|x+1| + \tfrac{1}{2}\ln|x-1|\right]_2^3$

$= \left(-\tfrac{1}{2}\ln 4 + \tfrac{1}{2}\ln 2\right) - \left(-\tfrac{1}{2}\ln 3 + \tfrac{1}{2}\ln 1\right) = \tfrac{1}{2}(\ln 2 + \ln 3 - \ln 4) \quad \left[\text{or } \tfrac{1}{2}\ln\tfrac{3}{2}\right]$

13. $\displaystyle\int \dfrac{ax}{x^2-bx}\,dx = \int \dfrac{ax}{x(x-b)}\,dx = \int \dfrac{a}{x-b}\,dx = a\ln|x-b| + C$

15. $\dfrac{x^3 - 2x^2 - 4}{x^3 - 2x^2} = 1 + \dfrac{-4}{x^2(x-2)}$. Write $\dfrac{-4}{x^2(x-2)} = \dfrac{A}{x} + \dfrac{B}{x^2} + \dfrac{C}{x-2}$. Multiplying both sides by $x^2(x-2)$ gives

$-4 = Ax(x-2) + B(x-2) + Cx^2$. Substituting 0 for x gives $-4 = -2B \quad \Leftrightarrow \quad B = 2$. Substituting 2 for x gives

$-4 = 4C \quad \Leftrightarrow \quad C = -1$. Equating coefficients of x^2, we get $0 = A + C$, so $A = 1$. Thus,

$$\int_3^4 \frac{x^3 - 2x^2 - 4}{x^3 - 2x^2}\,dx = \int_3^4 \left(1 + \frac{1}{x} + \frac{2}{x^2} - \frac{1}{x-2}\right)dx = \left[x + \ln|x| - \frac{2}{x} - \ln|x-2|\right]_3^4$$

$$= \left[(4 + \ln 4 - \tfrac{1}{2} - \ln 2) - (3 + \ln 3 - \tfrac{2}{3} - 0)\right] = \tfrac{7}{6} + \ln\tfrac{2}{3}$$

17. $\dfrac{4y^2 - 7y - 12}{y(y+2)(y-3)} = \dfrac{A}{y} + \dfrac{B}{y+2} + \dfrac{C}{y-3} \quad \Rightarrow \quad 4y^2 - 7y - 12 = A(y+2)(y-3) + By(y-3) + Cy(y+2)$. Setting

$y = 0$ gives $-12 = -6A$, so $A = 2$. Setting $y = -2$ gives $18 = 10B$, so $B = \frac{9}{5}$. Setting $y = 3$ gives $3 = 15C$, so $C = \frac{1}{5}$.

Now

$$\int_1^2 \frac{4y^2 - 7y - 12}{y(y+2)(y-3)}\,dy = \int_1^2 \left(\frac{2}{y} + \frac{9/5}{y+2} + \frac{1/5}{y-3}\right)dy = \left[2\ln|y| + \tfrac{9}{5}\ln|y+2| + \tfrac{1}{5}\ln|y-3|\right]_1^2$$

$$= 2\ln 2 + \tfrac{9}{5}\ln 4 + \tfrac{1}{5}\ln 1 - 2\ln 1 - \tfrac{9}{5}\ln 3 - \tfrac{1}{5}\ln 2$$

$$= 2\ln 2 + \tfrac{18}{5}\ln 2 - \tfrac{1}{5}\ln 2 - \tfrac{9}{5}\ln 3 = \tfrac{27}{5}\ln 2 - \tfrac{9}{5}\ln 3 = \tfrac{9}{5}(3\ln 2 - \ln 3) = \tfrac{9}{5}\ln\tfrac{8}{3}$$

19. $\dfrac{1}{(x+5)^2(x-1)} = \dfrac{A}{x+5} + \dfrac{B}{(x+5)^2} + \dfrac{C}{x-1} \quad \Rightarrow \quad 1 = A(x+5)(x-1) + B(x-1) + C(x+5)^2$.

Setting $x = -5$ gives $1 = -6B$, so $B = -\frac{1}{6}$. Setting $x = 1$ gives $1 = 36C$, so $C = \frac{1}{36}$. Setting $x = -2$ gives

$1 = A(3)(-3) + B(-3) + C(3^2) = -9A - 3B + 9C = -9A + \frac{1}{2} + \frac{1}{4} = -9A + \frac{3}{4}$, so $9A = -\frac{1}{4}$ and $A = -\frac{1}{36}$. Now

$$\int \frac{1}{(x+5)^2(x-1)}\,dx = \int \left[\frac{-1/36}{x+5} - \frac{1/6}{(x+5)^2} + \frac{1/36}{x-1}\right]dx = -\frac{1}{36}\ln|x+5| + \frac{1}{6(x+5)} + \frac{1}{36}\ln|x-1| + C.$$

21.

$$
\begin{array}{r}
x \\
x^2 + 4\overline{\big)\,x^3 + 0x^2 + 0x + 4} \\
\underline{x^3 \qquad\quad + 4x} \\
-4x + 4
\end{array}
$$

By long division, $\dfrac{x^3 + 4}{x^2 + 4} = x + \dfrac{-4x + 4}{x^2 + 4}$. Thus,

$$\int \frac{x^3 + 4}{x^2 + 4}\,dx = \int \left(x + \frac{-4x + 4}{x^2 + 4}\right)dx = \int \left(x - \frac{4x}{x^2 + 4} + \frac{4}{x^2 + 2^2}\right)dx$$

$$= \frac{1}{2}x^2 - 4\cdot\frac{1}{2}\ln|x^2 + 4| + 4\cdot\frac{1}{2}\tan^{-1}\!\left(\frac{x}{2}\right) + C = \frac{1}{2}x^2 - 2\ln(x^2 + 4) + 2\tan^{-1}\!\left(\frac{x}{2}\right) + C$$

23. $\dfrac{5x^2 + 3x - 2}{x^3 + 2x^2} = \dfrac{5x^2 + 3x - 2}{x^2(x+2)} = \dfrac{A}{x} + \dfrac{B}{x^2} + \dfrac{C}{x+2}$. Multiply by $x^2(x+2)$ to

get $5x^2 + 3x - 2 = Ax(x+2) + B(x+2) + Cx^2$. Set $x = -2$ to get $C = 3$, and take

$x = 0$ to get $B = -1$. Equating the coefficients of x^2 gives $5 = A + C \quad \Rightarrow \quad A = 2$. So

$$\int \frac{5x^2 + 3x - 2}{x^3 + 2x^2}\,dx = \int \left(\frac{2}{x} - \frac{1}{x^2} + \frac{3}{x+2}\right)dx = 2\ln|x| + \frac{1}{x} + 3\ln|x+2| + C.$$

25. $\dfrac{10}{(x-1)(x^2+9)} = \dfrac{A}{x-1} + \dfrac{Bx+C}{x^2+9}$. Multiply both sides by $(x-1)(x^2+9)$ to get

$10 = A(x^2+9) + (Bx+C)(x-1)$ $(\star)$. Substituting 1 for x gives $10 = 10A \iff A = 1$. Substituting 0 for x gives

$10 = 9A - C \implies C = 9(1) - 10 = -1$. The coefficients of the x^2-terms in $(\star)$ must be equal, so $0 = A + B \implies$

$B = -1$. Thus,

$$\int \frac{10}{(x-1)(x^2+9)}\,dx = \int\left(\frac{1}{x-1} + \frac{-x-1}{x^2+9}\right)dx = \int\left(\frac{1}{x-1} - \frac{x}{x^2+9} - \frac{1}{x^2+9}\right)dx$$

$$= \ln|x-1| - \tfrac{1}{2}\ln(x^2+9) - \tfrac{1}{3}\tan^{-1}\left(\tfrac{x}{3}\right) + C$$

In the second term we used the substitution $u = x^2 + 9$ and in the last term we used Formula 10.

27. $\dfrac{x^3+x^2+2x+1}{(x^2+1)(x^2+2)} = \dfrac{Ax+B}{x^2+1} + \dfrac{Cx+D}{x^2+2}$. Multiply both sides by $(x^2+1)(x^2+2)$ to get

$x^3 + x^2 + 2x + 1 = (Ax+B)(x^2+2) + (Cx+D)(x^2+1) \iff$

$x^3 + x^2 + 2x + 1 = (Ax^3 + Bx^2 + 2Ax + 2B) + (Cx^3 + Dx^2 + Cx + D) \iff$

$x^3 + x^2 + 2x + 1 = (A+C)x^3 + (B+D)x^2 + (2A+C)x + (2B+D)$. Comparing coefficients gives us the following

system of equations:

$$\begin{array}{cccc} A + C = 1 & \textbf{(1)} & \qquad B + D = 1 & \textbf{(2)} \\ 2A + C = 2 & \textbf{(3)} & \qquad 2B + D = 1 & \textbf{(4)} \end{array}$$

Subtracting equation **(1)** from equation **(3)** gives us $A = 1$, so $C = 0$. Subtracting equation **(2)** from equation **(4)** gives us

$B = 0$, so $D = 1$. Thus, $I = \displaystyle\int \frac{x^3+x^2+2x+1}{(x^2+1)(x^2+2)}\,dx = \int\left(\frac{x}{x^2+1} + \frac{1}{x^2+2}\right)dx$. For $\displaystyle\int\frac{x}{x^2+1}\,dx$, let $u = x^2 + 1$

so $du = 2x\,dx$ and then $\displaystyle\int\frac{x}{x^2+1}\,dx = \frac{1}{2}\int\frac{1}{u}\,du = \frac{1}{2}\ln|u| + C = \frac{1}{2}\ln(x^2+1) + C$. For $\displaystyle\int\frac{1}{x^2+2}\,dx$, use

Formula 10 with $a = \sqrt{2}$. So $\displaystyle\int\frac{1}{x^2+2}\,dx = \int\frac{1}{x^2+(\sqrt{2})^2}\,dx = \frac{1}{\sqrt{2}}\tan^{-1}\frac{x}{\sqrt{2}} + C$.

Thus, $I = \dfrac{1}{2}\ln(x^2+1) + \dfrac{1}{\sqrt{2}}\tan^{-1}\dfrac{x}{\sqrt{2}} + C$.

29. $\displaystyle\int\frac{x+4}{x^2+2x+5}\,dx = \int\frac{x+1}{x^2+2x+5}\,dx + \int\frac{3}{x^2+2x+5}\,dx = \frac{1}{2}\int\frac{(2x+2)\,dx}{x^2+2x+5} + \int\frac{3\,dx}{(x+1)^2+4}$

$$= \frac{1}{2}\ln\left|x^2+2x+5\right| + 3\int\frac{2\,du}{4(u^2+1)} \qquad \begin{bmatrix}\text{where } x+1 = 2u, \\ \text{and } dx = 2\,du\end{bmatrix}$$

$$= \frac{1}{2}\ln(x^2+2x+5) + \frac{3}{2}\tan^{-1}u + C = \frac{1}{2}\ln(x^2+2x+5) + \frac{3}{2}\tan^{-1}\left(\frac{x+1}{2}\right) + C$$

31. $\dfrac{1}{x^3-1} = \dfrac{1}{(x-1)(x^2+x+1)} = \dfrac{A}{x-1} + \dfrac{Bx+C}{x^2+x+1} \implies 1 = A(x^2+x+1) + (Bx+C)(x-1)$.

Take $x = 1$ to get $A = \tfrac{1}{3}$. Equating coefficients of x^2 and then comparing the constant terms, we get $0 = \tfrac{1}{3} + B$, $1 = \tfrac{1}{3} - C$, so $B = -\frac{1}{3}$, $C = -\frac{2}{3}$ $\Rightarrow$

$$\int \frac{1}{x^3 - 1}\,dx = \int \frac{\frac{1}{3}}{x - 1}\,dx + \int \frac{-\frac{1}{3}x - \frac{2}{3}}{x^2 + x + 1}\,dx = \frac{1}{3}\ln|x - 1| - \frac{1}{3}\int \frac{x + 2}{x^2 + x + 1}\,dx$$

$$= \frac{1}{3}\ln|x - 1| - \frac{1}{3}\int \frac{x + 1/2}{x^2 + x + 1}\,dx - \frac{1}{3}\int \frac{(3/2)\,dx}{(x + 1/2)^2 + 3/4}$$

$$= \frac{1}{3}\ln|x - 1| - \frac{1}{6}\ln(x^2 + x + 1) - \frac{1}{2}\left(\frac{2}{\sqrt{3}}\right)\tan^{-1}\left(\frac{x + \frac{1}{2}}{\sqrt{3}/2}\right) + K$$

$$= \frac{1}{3}\ln|x - 1| - \frac{1}{6}\ln(x^2 + x + 1) - \frac{1}{\sqrt{3}}\tan^{-1}\left(\frac{1}{\sqrt{3}}(2x + 1)\right) + K$$

33. Let $u = x^4 + 4x^2 + 3$, so that $du = (4x^3 + 8x)\,dx = 4(x^3 + 2x)\,dx$, $x = 0 \Rightarrow u = 3$, and $x = 1 \Rightarrow u = 8$.

Then $\displaystyle\int_0^1 \frac{x^3 + 2x}{x^4 + 4x^2 + 3}\,dx = \int_3^8 \frac{1}{u}\left(\frac{1}{4}\,du\right) = \frac{1}{4}\left[\ln|u|\right]_3^8 = \frac{1}{4}(\ln 8 - \ln 3) = \frac{1}{4}\ln\frac{8}{3}$.

35. $\dfrac{1}{x(x^2 + 4)^2} = \dfrac{A}{x} + \dfrac{Bx + C}{x^2 + 4} + \dfrac{Dx + E}{(x^2 + 4)^2}$ $\Rightarrow$ $1 = A(x^2 + 4)^2 + (Bx + C)x(x^2 + 4) + (Dx + E)x$. Setting $x = 0$

gives $1 = 16A$, so $A = \frac{1}{16}$. Now compare coefficients.

$$1 = \tfrac{1}{16}(x^4 + 8x^2 + 16) + (Bx^2 + Cx)(x^2 + 4) + Dx^2 + Ex$$

$$1 = \tfrac{1}{16}x^4 + \tfrac{1}{2}x^2 + 1 + Bx^4 + Cx^3 + 4Bx^2 + 4Cx + Dx^2 + Ex$$

$$1 = \left(\tfrac{1}{16} + B\right)x^4 + Cx^3 + \left(\tfrac{1}{2} + 4B + D\right)x^2 + (4C + E)x + 1$$

So $B + \frac{1}{16} = 0 \Rightarrow B = -\frac{1}{16}$, $C = 0$, $\frac{1}{2} + 4B + D = 0 \Rightarrow D = -\frac{1}{4}$, and $4C + E = 0 \Rightarrow E = 0$. Thus,

$$\int \frac{dx}{x(x^2 + 4)^2} = \int \left(\frac{\frac{1}{16}}{x} + \frac{-\frac{1}{16}x}{x^2 + 4} + \frac{-\frac{1}{4}x}{(x^2 + 4)^2}\right)dx = \frac{1}{16}\ln|x| - \frac{1}{16}\cdot\frac{1}{2}\ln|x^2 + 4| - \frac{1}{4}\left(-\frac{1}{2}\right)\frac{1}{x^2 + 4} + C$$

$$= \frac{1}{16}\ln|x| - \frac{1}{32}\ln(x^2 + 4) + \frac{1}{8(x^2 + 4)} + C$$

37. $\dfrac{x^2 - 3x + 7}{(x^2 - 4x + 6)^2} = \dfrac{Ax + B}{x^2 - 4x + 6} + \dfrac{Cx + D}{(x^2 - 4x + 6)^2}$ $\Rightarrow$ $x^2 - 3x + 7 = (Ax + B)(x^2 - 4x + 6) + Cx + D$ $\Rightarrow$

$x^2 - 3x + 7 = Ax^3 + (-4A + B)x^2 + (6A - 4B + C)x + (6B + D)$. So $A = 0$, $-4A + B = 1 \Rightarrow B = 1$,

$6A - 4B + C = -3 \Rightarrow C = 1$, $6B + D = 7 \Rightarrow D = 1$. Thus,

$$I = \int \frac{x^2 - 3x + 7}{(x^2 - 4x + 6)^2}\,dx = \int \left(\frac{1}{x^2 - 4x + 6} + \frac{x + 1}{(x^2 - 4x + 6)^2}\right)dx$$

$$= \int \frac{1}{(x - 2)^2 + 2}\,dx + \int \frac{x - 2}{(x^2 - 4x + 6)^2}\,dx + \int \frac{3}{(x^2 - 4x + 6)^2}\,dx$$

$$= I_1 + I_2 + I_3.$$

[continued]

$$I_1 = \int \frac{1}{(x-2)^2 + \left(\sqrt{2}\right)^2}\, dx = \frac{1}{\sqrt{2}}\tan^{-1}\left(\frac{x-2}{\sqrt{2}}\right) + C_1$$

$$I_2 = \frac{1}{2}\int \frac{2x-4}{(x^2-4x+6)^2}\, dx = \frac{1}{2}\int \frac{1}{u^2}\, du = \frac{1}{2}\left(-\frac{1}{u}\right) + C_2 = -\frac{1}{2(x^2-4x+6)} + C_2$$

$$I_3 = 3\int \frac{1}{\left[(x-2)^2 + \left(\sqrt{2}\right)^2\right]^2}\, dx = 3\int \frac{1}{[2(\tan^2\theta + 1)]^2}\sqrt{2}\sec^2\theta\, d\theta \qquad \begin{bmatrix} x-2 = \sqrt{2}\tan\theta, \\ dx = \sqrt{2}\sec^2\theta\, d\theta \end{bmatrix}$$

$$= \frac{3\sqrt{2}}{4}\int \frac{\sec^2\theta}{\sec^4\theta}\, d\theta = \frac{3\sqrt{2}}{4}\int \cos^2\theta\, d\theta = \frac{3\sqrt{2}}{4}\int \frac{1}{2}(1 + \cos 2\theta)\, d\theta$$

$$= \frac{3\sqrt{2}}{8}\left(\theta + \tfrac{1}{2}\sin 2\theta\right) + C_3 = \frac{3\sqrt{2}}{8}\tan^{-1}\left(\frac{x-2}{\sqrt{2}}\right) + \frac{3\sqrt{2}}{8}\left(\tfrac{1}{2}\cdot 2\sin\theta\cos\theta\right) + C_3$$

$$= \frac{3\sqrt{2}}{8}\tan^{-1}\left(\frac{x-2}{\sqrt{2}}\right) + \frac{3\sqrt{2}}{8}\cdot\frac{x-2}{\sqrt{x^2-4x+6}}\cdot\frac{\sqrt{2}}{\sqrt{x^2-4x+6}} + C_3$$

$$= \frac{3\sqrt{2}}{8}\tan^{-1}\left(\frac{x-2}{\sqrt{2}}\right) + \frac{3(x-2)}{4(x^2-4x+6)} + C_3$$

So $I = I_1 + I_2 + I_3 \qquad [C = C_1 + C_2 + C_3]$

$$= \frac{1}{\sqrt{2}}\tan^{-1}\left(\frac{x-2}{\sqrt{2}}\right) + \frac{-1}{2(x^2-4x+6)} + \frac{3\sqrt{2}}{8}\tan^{-1}\left(\frac{x-2}{\sqrt{2}}\right) + \frac{3(x-2)}{4(x^2-4x+6)} + C$$

$$= \left(\frac{4\sqrt{2}}{8} + \frac{3\sqrt{2}}{8}\right)\tan^{-1}\left(\frac{x-2}{\sqrt{2}}\right) + \frac{3(x-2)-2}{4(x^2-4x+6)} + C = \frac{7\sqrt{2}}{8}\tan^{-1}\left(\frac{x-2}{\sqrt{2}}\right) + \frac{3x-8}{4(x^2-4x+6)} + C$$

39. Let $u = \sqrt{x+1}$. Then $x = u^2 - 1$, $dx = 2u\, du \;\Rightarrow$

$$\int \frac{dx}{x\sqrt{x+1}} = \int \frac{2u\, du}{(u^2-1)\, u} = 2\int \frac{du}{u^2-1} = \ln\left|\frac{u-1}{u+1}\right| + C = \ln\left|\frac{\sqrt{x+1}-1}{\sqrt{x+1}+1}\right| + C.$$

41. Let $u = \sqrt{x}$, so $u^2 = x$ and $dx = 2u\, du$. Thus,

$$\int_9^{16} \frac{\sqrt{x}}{x-4}\, dx = \int_3^4 \frac{u}{u^2-4}\, 2u\, du = 2\int_3^4 \frac{u^2}{u^2-4}\, du = 2\int_3^4 \left(1 + \frac{4}{u^2-4}\right) du \qquad \text{[by long division]}$$

$$= 2 + 8\int_3^4 \frac{du}{(u+2)(u-2)} \quad (\star)$$

Multiply $\dfrac{1}{(u+2)(u-2)} = \dfrac{A}{u+2} + \dfrac{B}{u-2}$ by $(u+2)(u-2)$ to get $1 = A(u-2) + B(u+2)$. Equating coefficients we

get $A + B = 0$ and $-2A + 2B = 1$. Solving gives us $B = \tfrac{1}{4}$ and $A = -\tfrac{1}{4}$, so $\dfrac{1}{(u+2)(u-2)} = \dfrac{-1/4}{u+2} + \dfrac{1/4}{u-2}$ and $(\star)$ is

$$2 + 8\int_3^4 \left(\frac{-1/4}{u+2} + \frac{1/4}{u-2}\right) du = 2 + 8\left[-\tfrac{1}{4}\ln|u+2| + \tfrac{1}{4}\ln|u-2|\right]_3^4 = 2 + \left[2\ln|u-2| - 2\ln|u+2|\right]_3^4$$

$$= 2 + 2\left[\ln\left|\frac{u-2}{u+2}\right|\right]_3^4 = 2 + 2\left(\ln\tfrac{2}{6} - \ln\tfrac{1}{5}\right) = 2 + 2\ln\frac{2/6}{1/5}$$

$$= 2 + 2\ln\tfrac{5}{3} \quad \text{or} \quad 2 + \ln\left(\tfrac{5}{3}\right)^2 = 2 + \ln\tfrac{25}{9}$$

43. Let $u = \sqrt[3]{x^2 + 1}$. Then $x^2 = u^3 - 1$, $2x\,dx = 3u^2\,du$ $\Rightarrow$

$$\int \frac{x^3\,dx}{\sqrt[3]{x^2+1}} = \int \frac{(u^3 - 1)\frac{3}{2}u^2\,du}{u} = \frac{3}{2}\int (u^4 - u)\,du$$

$$= \tfrac{3}{10}u^5 - \tfrac{3}{4}u^2 + C = \tfrac{3}{10}(x^2 + 1)^{5/3} - \tfrac{3}{4}(x^2 + 1)^{2/3} + C$$

45. If we were to substitute $u = \sqrt{x}$, then the square root would disappear but a cube root would remain. On the other hand, the substitution $u = \sqrt[3]{x}$ would eliminate the cube root but leave a square root. We can eliminate both roots by means of the substitution $u = \sqrt[6]{x}$. (Note that 6 is the least common multiple of 2 and 3.)

Let $u = \sqrt[6]{x}$. Then $x = u^6$, so $dx = 6u^5\,du$ and $\sqrt{x} = u^3$, $\sqrt[3]{x} = u^2$. Thus,

$$\int \frac{dx}{\sqrt{x} - \sqrt[3]{x}} = \int \frac{6u^5\,du}{u^3 - u^2} = 6\int \frac{u^5}{u^2(u-1)}\,du = 6\int \frac{u^3}{u - 1}\,du$$

$$= 6\int \left(u^2 + u + 1 + \frac{1}{u - 1}\right)du \qquad \text{[by long division]}$$

$$= 6\left(\tfrac{1}{3}u^3 + \tfrac{1}{2}u^2 + u + \ln|u - 1|\right) + C = 2\sqrt{x} + 3\sqrt[3]{x} + 6\sqrt[6]{x} + 6\ln\left|\sqrt[6]{x} - 1\right| + C$$

47. Let $u = e^x$. Then $x = \ln u$, $dx = \dfrac{du}{u}$ $\Rightarrow$

$$\int \frac{e^{2x}\,dx}{e^{2x} + 3e^x + 2} = \int \frac{u^2\,(du/u)}{u^2 + 3u + 2} = \int \frac{u\,du}{(u+1)(u+2)} = \int \left[\frac{-1}{u+1} + \frac{2}{u+2}\right]du$$

$$= 2\ln|u + 2| - \ln|u + 1| + C = \ln\frac{(e^x + 2)^2}{e^x + 1} + C$$

49. Let $u = \tan t$, so that $du = \sec^2 t\,dt$. Then $\displaystyle\int \frac{\sec^2 t}{\tan^2 t + 3\tan t + 2}\,dt = \int \frac{1}{u^2 + 3u + 2}\,du = \int \frac{1}{(u+1)(u+2)}\,du$.

Now $\dfrac{1}{(u+1)(u+2)} = \dfrac{A}{u+1} + \dfrac{B}{u+2}$ $\Rightarrow$ $1 = A(u+2) + B(u+1)$.

Setting $u = -2$ gives $1 = -B$, so $B = -1$. Setting $u = -1$ gives $1 = A$.

Thus, $\displaystyle\int \frac{1}{(u+1)(u+2)}\,du = \int \left(\frac{1}{u+1} - \frac{1}{u+2}\right)du = \ln|u+1| - \ln|u+2| + C = \ln|\tan t + 1| - \ln|\tan t + 2| + C$.

51. Let $u = \ln(x^2 - x + 2)$, $dv = dx$. Then $du = \dfrac{2x - 1}{x^2 - x + 2}\,dx$, $v = x$, and (by integration by parts)

$$\int \ln(x^2 - x + 2)\,dx = x\ln(x^2 - x + 2) - \int \frac{2x^2 - x}{x^2 - x + 2}\,dx = x\ln(x^2 - x + 2) - \int \left(2 + \frac{x - 4}{x^2 - x + 2}\right)dx$$

$$= x\ln(x^2 - x + 2) - 2x - \int \frac{\frac{1}{2}(2x - 1)}{x^2 - x + 2}\,dx + \frac{7}{2}\int \frac{dx}{(x - \frac{1}{2})^2 + \frac{7}{4}}$$

$$= x\ln(x^2 - x + 2) - 2x - \frac{1}{2}\ln(x^2 - x + 2) + \frac{7}{2}\int \frac{\frac{\sqrt{7}}{2}\,du}{\frac{7}{4}(u^2 + 1)} \qquad \begin{bmatrix} \text{where } x - \frac{1}{2} = \frac{\sqrt{7}}{2}u, \\ dx = \frac{\sqrt{7}}{2}\,du, \\ (x - \frac{1}{2})^2 + \frac{7}{4} = \frac{7}{4}(u^2 + 1) \end{bmatrix}$$

$$= (x - \tfrac{1}{2})\ln(x^2 - x + 2) - 2x + \sqrt{7}\tan^{-1} u + C$$

$$= (x - \tfrac{1}{2})\ln(x^2 - x + 2) - 2x + \sqrt{7}\tan^{-1}\frac{2x - 1}{\sqrt{7}} + C$$

53.

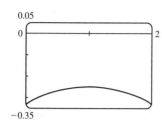

From the graph, we see that the integral will be negative, and we guess that the area is about the same as that of a rectangle with width 2 and height 0.3, so we estimate the integral to be $-(2 \cdot 0.3) = -0.6$. Now

$$\frac{1}{x^2 - 2x - 3} = \frac{1}{(x-3)(x+1)} = \frac{A}{x-3} + \frac{B}{x+1} \quad \Leftrightarrow$$

$1 = (A+B)x + A - 3B$, so $A = -B$ and $A - 3B = 1 \quad \Leftrightarrow \quad A = \frac{1}{4}$

and $B = -\frac{1}{4}$, so the integral becomes

$$\int_0^2 \frac{dx}{x^2 - 2x - 3} = \frac{1}{4}\int_0^2 \frac{dx}{x-3} - \frac{1}{4}\int_0^2 \frac{dx}{x+1} = \frac{1}{4}\Big[\ln|x-3| - \ln|x+1|\Big]_0^2 = \frac{1}{4}\Big[\ln\Big|\frac{x-3}{x+1}\Big|\Big]_0^2$$

$$= \frac{1}{4}\Big(\ln\frac{1}{3} - \ln 3\Big) = -\frac{1}{2}\ln 3 \approx -0.55$$

55. $\displaystyle\int \frac{dx}{x^2 - 2x} = \int \frac{dx}{(x-1)^2 - 1} = \int \frac{du}{u^2 - 1}$ [put $u = x - 1$]

$$= \frac{1}{2}\ln\Big|\frac{u-1}{u+1}\Big| + C \quad \text{[by Equation 6]} \quad = \frac{1}{2}\ln\Big|\frac{x-2}{x}\Big| + C$$

57. (a) If $t = \tan\big(\frac{x}{2}\big)$, then $\frac{x}{2} = \tan^{-1} t$. The figure gives

$$\cos\Big(\frac{x}{2}\Big) = \frac{1}{\sqrt{1+t^2}} \text{ and } \sin\Big(\frac{x}{2}\Big) = \frac{t}{\sqrt{1+t^2}}.$$

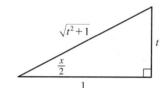

(b) $\cos x = \cos\Big(2 \cdot \frac{x}{2}\Big) = 2\cos^2\Big(\frac{x}{2}\Big) - 1$

$$= 2\Big(\frac{1}{\sqrt{1+t^2}}\Big)^2 - 1 = \frac{2}{1+t^2} - 1 = \frac{1-t^2}{1+t^2}$$

(c) $\frac{x}{2} = \arctan t \quad \Rightarrow \quad x = 2\arctan t \quad \Rightarrow \quad dx = \frac{2}{1+t^2}\,dt$

59. Let $t = \tan(x/2)$. Then, using the expressions in Exercise 57, we have

$$\int \frac{1}{3\sin x - 4\cos x}\,dx = \int \frac{1}{3\Big(\dfrac{2t}{1+t^2}\Big) - 4\Big(\dfrac{1-t^2}{1+t^2}\Big)}\frac{2\,dt}{1+t^2} = 2\int \frac{dt}{3(2t) - 4(1-t^2)} = \int \frac{dt}{2t^2 + 3t - 2}$$

$$= \int \frac{dt}{(2t-1)(t+2)} = \int \Big[\frac{2}{5}\frac{1}{2t-1} - \frac{1}{5}\frac{1}{t+2}\Big]\,dt \quad \text{[using partial fractions]}$$

$$= \frac{1}{5}\Big[\ln|2t-1| - \ln|t+2|\Big] + C = \frac{1}{5}\ln\Big|\frac{2t-1}{t+2}\Big| + C = \frac{1}{5}\ln\Big|\frac{2\tan(x/2) - 1}{\tan(x/2) + 2}\Big| + C$$

61. Let $t = \tan(x/2)$. Then, by Exercise 57,

$$\int_0^{\pi/2} \frac{\sin 2x}{2 + \cos x}\,dx = \int_0^{\pi/2} \frac{2\sin x \cos x}{2 + \cos x}\,dx = \int_0^1 \frac{2 \cdot \dfrac{2t}{1+t^2} \cdot \dfrac{1-t^2}{1+t^2}}{2 + \dfrac{1-t^2}{1+t^2}}\frac{2}{1+t^2}\,dt = \int_0^1 \frac{\dfrac{8t(1-t^2)}{(1+t^2)^2}}{2(1+t^2) + (1-t^2)}\,dt$$

$$= \int_0^1 8t \cdot \frac{1-t^2}{(t^2+3)(t^2+1)^2}\,dt = I$$

If we now let $u = t^2$, then $\dfrac{1 - t^2}{(t^2 + 3)(t^2 + 1)^2} = \dfrac{1 - u}{(u + 3)(u + 1)^2} = \dfrac{A}{u + 3} + \dfrac{B}{u + 1} + \dfrac{C}{(u + 1)^2}$ $\Rightarrow$

$1 - u = A(u + 1)^2 + B(u + 3)(u + 1) + C(u + 3)$. Set $u = -1$ to get $2 = 2C$, so $C = 1$. Set $u = -3$ to get $4 = 4A$, so

$A = 1$. Set $u = 0$ to get $1 = 1 + 3B + 3$, so $B = -1$. So

$$I = \int_0^1 \left[\frac{8t}{t^2 + 3} - \frac{8t}{t^2 + 1} + \frac{8t}{(t^2 + 1)^2} \right] dt = \left[4\ln(t^2 + 3) - 4\ln(t^2 + 1) - \frac{4}{t^2 + 1} \right]_0^1$$

$$= (4\ln 4 - 4\ln 2 - 2) - (4\ln 3 - 0 - 4) = 8\ln 2 - 4\ln 2 - 4\ln 3 + 2 = 4\ln \tfrac{2}{3} + 2$$

63. By long division, $\dfrac{x^2 + 1}{3x - x^2} = -1 + \dfrac{3x + 1}{3x - x^2}$. Now

$\dfrac{3x + 1}{3x - x^2} = \dfrac{3x + 1}{x(3 - x)} = \dfrac{A}{x} + \dfrac{B}{3 - x}$ $\Rightarrow$ $3x + 1 = A(3 - x) + Bx$. Set $x = 3$ to get $10 = 3B$, so $B = \frac{10}{3}$. Set $x = 0$ to

get $1 = 3A$, so $A = \frac{1}{3}$. Thus, the area is

$$\int_1^2 \frac{x^2 + 1}{3x - x^2}\, dx = \int_1^2 \left(-1 + \frac{\frac{1}{3}}{x} + \frac{\frac{10}{3}}{3 - x} \right) dx = \left[-x + \tfrac{1}{3}\ln|x| - \tfrac{10}{3}\ln|3 - x| \right]_1^2$$

$$= \left(-2 + \tfrac{1}{3}\ln 2 - 0 \right) - \left(-1 + 0 - \tfrac{10}{3}\ln 2 \right) = -1 + \tfrac{11}{3}\ln 2$$

65. $\dfrac{P + S}{P\,[(r - 1)P - S]} = \dfrac{A}{P} + \dfrac{B}{(r - 1)P - S}$ $\Rightarrow$ $P + S = A\,[(r - 1)P - S] + BP = [(r - 1)A + B]\,P - AS$ $\Rightarrow$

$(r - 1)A + B = 1, -A = 1$ $\Rightarrow$ $A = -1, B = r$. Now

$$t = \int \frac{P + S}{P\,[(r - 1)P - S]}\, dP = \int \left[\frac{-1}{P} + \frac{r}{(r - 1)P - S} \right] dP = -\int \frac{dP}{P} + \frac{r}{r - 1} \int \frac{r - 1}{(r - 1)P - S}\, dP$$

so $t = -\ln P + \dfrac{r}{r - 1}\ln|(r - 1)P - S| + C$. Here $r = 0.10$ and $S = 900$, so

$t = -\ln P + \dfrac{0.1}{-0.9}\ln|-0.9P - 900| + C = -\ln P - \tfrac{1}{9}\ln(|-1|\,|0.9P + 900|) = -\ln P - \tfrac{1}{9}\ln(0.9P + 900) + C$.

When $t = 0$, $P = 10{,}000$, so $0 = -\ln 10{,}000 - \tfrac{1}{9}\ln(9900) + C$. Thus, $C = \ln 10{,}000 + \tfrac{1}{9}\ln 9900$ [≈ 10.2326], so our

equation becomes

$$t = \ln 10{,}000 - \ln P + \tfrac{1}{9}\ln 9900 - \tfrac{1}{9}\ln(0.9P + 900) = \ln \frac{10{,}000}{P} + \frac{1}{9}\ln \frac{9900}{0.9P + 900}$$

$$= \ln \frac{10{,}000}{P} + \frac{1}{9}\ln \frac{1100}{0.1P + 100} = \ln \frac{10{,}000}{P} + \frac{1}{9}\ln \frac{11{,}000}{P + 1000}$$

67. (a) In Maple, we define $f(x)$, and then use `convert(f,parfrac,x);` to obtain

$$f(x) = \frac{24{,}110/4879}{5x + 2} - \frac{668/323}{2x + 1} - \frac{9438/80{,}155}{3x - 7} + \frac{(22{,}098x + 48{,}935)/260{,}015}{x^2 + x + 5}$$

In Mathematica, we use the command `Apart`, and in Derive, we use `Expand`.

(b) $\int f(x)\,dx = \frac{24{,}110}{4879} \cdot \frac{1}{5}\ln|5x+2| - \frac{668}{323} \cdot \frac{1}{2}\ln|2x+1| - \frac{9438}{80{,}155} \cdot \frac{1}{3}\ln|3x-7|$

$$+ \frac{1}{260{,}015}\int \frac{22{,}098\left(x+\frac{1}{2}\right)+37{,}886}{\left(x+\frac{1}{2}\right)^2 + \frac{19}{4}}\,dx + C$$

$$= \frac{24{,}110}{4879} \cdot \frac{1}{5}\ln|5x+2| - \frac{668}{323} \cdot \frac{1}{2}\ln|2x+1| - \frac{9438}{80{,}155} \cdot \frac{1}{3}\ln|3x-7|$$

$$+ \frac{1}{260{,}015}\left[22{,}098 \cdot \frac{1}{2}\ln\left(x^2+x+5\right) + 37{,}886 \cdot \sqrt{\frac{4}{19}}\tan^{-1}\left(\frac{1}{\sqrt{19/4}}\left(x+\frac{1}{2}\right)\right)\right] + C$$

$$= \frac{4822}{4879}\ln|5x+2| - \frac{334}{323}\ln|2x+1| - \frac{3146}{80{,}155}\ln|3x-7| + \frac{11{,}049}{260{,}015}\ln\left(x^2+x+5\right)$$

$$+ \frac{75{,}772}{260{,}015\sqrt{19}}\tan^{-1}\left[\frac{1}{\sqrt{19}}(2x+1)\right] + C$$

Using a CAS, we get

$$\frac{4822\ln(5x+2)}{4879} - \frac{334\ln(2x+1)}{323} - \frac{3146\ln(3x-7)}{80{,}155}$$

$$+ \frac{11{,}049\ln(x^2+x+5)}{260{,}015} + \frac{3988\sqrt{19}}{260{,}015}\tan^{-1}\left[\frac{\sqrt{19}}{19}(2x+1)\right]$$

The main difference in this answer is that the absolute value signs and the constant of integration have been omitted. Also, the fractions have been reduced and the denominators rationalized.

69. There are only finitely many values of x where $Q(x) = 0$ (assuming that Q is not the zero polynomial). At all other values of x, $F(x)/Q(x) = G(x)/Q(x)$, so $F(x) = G(x)$. In other words, the values of F and G agree at all except perhaps finitely many values of x. By continuity of F and G, the polynomials F and G must agree at those values of x too.

More explicitly: if a is a value of x such that $Q(a) = 0$, then $Q(x) \neq 0$ for all x sufficiently close to a. Thus,

$$F(a) = \lim_{x \to a} F(x) \qquad \text{[by continuity of } F\text{]}$$

$$= \lim_{x \to a} G(x) \qquad \text{[whenever } Q(x) \neq 0\text{]}$$

$$= G(a) \qquad \text{[by continuity of } G\text{]}$$

7.5 Strategy for Integration

1. Let $u = \sin x$, so that $du = \cos x\,dx$. Then $\int \cos x(1 + \sin^2 x)\,dx = \int (1+u^2)\,du = u + \frac{1}{3}u^3 + C = \sin x + \frac{1}{3}\sin^3 x + C$.

3. $\int \frac{\sin x + \sec x}{\tan x}\,dx = \int \left(\frac{\sin x}{\tan x} + \frac{\sec x}{\tan x}\right)dx = \int (\cos x + \csc x)\,dx = \sin x + \ln|\csc x - \cot x| + C$

5. $\int_0^2 \frac{2t}{(t-3)^2}\,dt = \int_{-3}^{-1} \frac{2(u+3)}{u^2}\,du \begin{bmatrix} u = t-3, \\ du = dt \end{bmatrix} = \int_{-3}^{-1}\left(\frac{2}{u} + \frac{6}{u^2}\right)du = \left[2\ln|u| - \frac{6}{u}\right]_{-3}^{-1}$

$= (2\ln 1 + 6) - (2\ln 3 + 2) = 4 - 2\ln 3$ or $4 - \ln 9$

7. Let $u = \arctan y$. Then $du = \dfrac{dy}{1+y^2} \Rightarrow \int_{-1}^1 \dfrac{e^{\arctan y}}{1+y^2}\,dy = \int_{-\pi/4}^{\pi/4} e^u\,du = \left[e^u\right]_{-\pi/4}^{\pi/4} = e^{\pi/4} - e^{-\pi/4}$.

9. $\int_1^3 r^4 \ln r \, dr$ $\begin{bmatrix} u = \ln r, & dv = r^4 \, dr, \\ du = \dfrac{dr}{r} & v = \frac{1}{5} r^5 \end{bmatrix}$ $= \left[\frac{1}{5} r^5 \ln r \right]_1^3 - \int_1^3 \frac{1}{5} r^4 \, dr = \frac{243}{5} \ln 3 - 0 - \left[\frac{1}{25} r^5 \right]_1^3$

$$= \frac{243}{5} \ln 3 - \left(\frac{243}{25} - \frac{1}{25} \right) = \frac{243}{5} \ln 3 - \frac{242}{25}$$

11. $\displaystyle\int \frac{x - 1}{x^2 - 4x + 5} \, dx = \int \frac{(x - 2) + 1}{(x - 2)^2 + 1} \, dx = \int \left(\frac{u}{u^2 + 1} + \frac{1}{u^2 + 1} \right) du \qquad [u = x - 2, \, du = dx]$

$$= \frac{1}{2} \ln(u^2 + 1) + \tan^{-1} u + C = \frac{1}{2} \ln(x^2 - 4x + 5) + \tan^{-1}(x - 2) + C$$

13. $\int \sin^3 \theta \, \cos^5 \theta \, d\theta = \int \cos^5 \theta \sin^2 \theta \sin \theta \, d\theta = -\int \cos^5 \theta \, (1 - \cos^2 \theta)(-\sin \theta) \, d\theta$

$$= -\int u^5 (1 - u^2) \, du \qquad \begin{bmatrix} u = \cos \theta, \\ du = -\sin \theta \, d\theta \end{bmatrix}$$

$$= \int (u^7 - u^5) \, du = \frac{1}{8} u^8 - \frac{1}{6} u^6 + C = \frac{1}{8} \cos^8 \theta - \frac{1}{6} \cos^6 \theta + C$$

Another solution:

$\int \sin^3 \theta \, \cos^5 \theta \, d\theta = \int \sin^3 \theta \, (\cos^2 \theta)^2 \cos \theta \, d\theta = \int \sin^3 \theta \, (1 - \sin^2 \theta)^2 \cos \theta \, d\theta$

$$= \int u^3 (1 - u^2)^2 \, du \qquad \begin{bmatrix} u = \sin \theta, \\ du = \cos \theta \, d\theta \end{bmatrix} \qquad = \int u^3 (1 - 2u^2 + u^4) \, du$$

$$= \int (u^3 - 2u^5 + u^7) \, du = \frac{1}{4} u^4 - \frac{1}{3} u^6 + \frac{1}{8} u^8 + C = \frac{1}{4} \sin^4 \theta - \frac{1}{3} \sin^6 \theta + \frac{1}{8} \sin^8 \theta + C$$

15. Let $x = \sin \theta$, where $-\frac{\pi}{2} \le \theta \le \frac{\pi}{2}$. Then $dx = \cos \theta \, d\theta$ and $(1 - x^2)^{1/2} = \cos \theta$,

so

$$\int \frac{dx}{(1 - x^2)^{3/2}} = \int \frac{\cos \theta \, d\theta}{(\cos \theta)^3} = \int \sec^2 \theta \, d\theta = \tan \theta + C = \frac{x}{\sqrt{1 - x^2}} + C.$$

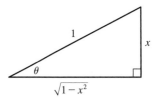

17. $\int x \sin^2 x \, dx$ $\begin{bmatrix} u = x, & dv = \sin^2 x \, dx, \\ du = dx & v = \int \sin^2 x \, dx = \int \frac{1}{2}(1 - \cos 2x) \, dx = \frac{1}{2} x - \frac{1}{2} \sin x \cos x \end{bmatrix}$

$$= \frac{1}{2} x^2 - \frac{1}{2} x \sin x \cos x - \int \left(\frac{1}{2} x - \frac{1}{2} \sin x \cos x \right) dx$$

$$= \frac{1}{2} x^2 - \frac{1}{2} x \sin x \cos x - \frac{1}{4} x^2 + \frac{1}{4} \sin^2 x + C = \frac{1}{4} x^2 - \frac{1}{2} x \sin x \cos x + \frac{1}{4} \sin^2 x + C$$

Note: $\int \sin x \cos x \, dx = \int s \, ds = \frac{1}{2} s^2 + C$ [where $s = \sin x, \, ds = \cos x \, dx$].

A slightly different method is to write $\int x \sin^2 x \, dx = \int x \cdot \frac{1}{2}(1 - \cos 2x) \, dx = \frac{1}{2} \int x \, dx - \frac{1}{2} \int x \cos 2x \, dx$. If we evaluate

the second integral by parts, we arrive at the equivalent answer $\frac{1}{4} x^2 - \frac{1}{4} x \sin 2x - \frac{1}{8} \cos 2x + C$.

19. Let $u = e^x$. Then $\int e^{x + e^x} \, dx = \int e^{e^x} e^x \, dx = \int e^u \, du = e^u + C = e^{e^x} + C$.

21. Let $t = \sqrt{x}$, so that $t^2 = x$ and $2t \, dt = dx$. Then $\int \arctan \sqrt{x} \, dx = \int \arctan t \, (2t \, dt) = I$. Now use parts with

$u = \arctan t, \, dv = 2t \, dt$, $\quad = \dfrac{1}{1 + t^2} \, dt, \, v = t^2$. Thus,

$$I = t^2 \arctan t - \int \frac{t^2}{1 + t^2} \, dt = t^2 \arctan t - \int \left(1 - \frac{1}{1 + t^2} \right) dt = t^2 \arctan t - t + \arctan t + C$$

$$= x \arctan \sqrt{x} - \sqrt{x} + \arctan \sqrt{x} + C \qquad \left[\text{or } (x + 1) \arctan \sqrt{x} - \sqrt{x} + C \right]$$

23. Let $u = 1 + \sqrt{x}$. Then $x = (u-1)^2$, $dx = 2(u-1)\,du$ $\Rightarrow$

$$\int_0^1 \left(1 + \sqrt{x}\right)^8 dx = \int_1^2 u^8 \cdot 2(u-1)\,du = 2\int_1^2 (u^9 - u^8)\,du = \left[\tfrac{1}{5}u^{10} - 2\cdot\tfrac{1}{9}u^9\right]_1^2 = \tfrac{1024}{5} - \tfrac{1024}{9} - \tfrac{1}{5} + \tfrac{2}{9} = \tfrac{4097}{45}.$$

25. $\dfrac{3x^2 - 2}{x^2 - 2x - 8} = 3 + \dfrac{6x + 22}{(x-4)(x+2)} = 3 + \dfrac{A}{x-4} + \dfrac{B}{x+2}$ $\Rightarrow$ $6x + 22 = A(x+2) + B(x-4)$. Setting

$x = 4$ gives $46 = 6A$, so $A = \tfrac{23}{3}$. Setting $x = -2$ gives $10 = -6B$, so $B = -\tfrac{5}{3}$. Now

$$\int \dfrac{3x^2 - 2}{x^2 - 2x - 8}\,dx = \int \left(3 + \dfrac{23/3}{x-4} - \dfrac{5/3}{x+2}\right) dx = 3x + \tfrac{23}{3}\ln|x-4| - \tfrac{5}{3}\ln|x+2| + C.$$

27. Let $u = 1 + e^x$, so that $du = e^x\,dx = (u-1)\,dx$. Then $\displaystyle\int \dfrac{1}{1+e^x}\,dx = \int \dfrac{1}{u}\cdot\dfrac{du}{u-1} = \int \dfrac{1}{u(u-1)}\,du = I$. Now

$\dfrac{1}{u(u-1)} = \dfrac{A}{u} + \dfrac{B}{u-1}$ $\Rightarrow$ $1 = A(u-1) + Bu$. Set $u = 1$ to get $1 = B$. Set $u = 0$ to get $1 = -A$, so $A = -1$.

Thus, $I = \displaystyle\int \left(\dfrac{-1}{u} + \dfrac{1}{u-1}\right) du = -\ln|u| + \ln|u-1| + C = -\ln(1 + e^x) + \ln e^x + C = x - \ln(1 + e^x) + C$.

Another method: Multiply numerator and denominator by e^{-x} and let $u = e^{-x} + 1$. This gives the answer in the

form $-\ln(e^{-x} + 1) + C$.

29. $\displaystyle\int_0^5 \dfrac{3w - 1}{w + 2}\,dw = \int_0^5 \left(3 - \dfrac{7}{w+2}\right) dw = \Big[3w - 7\ln|w+2|\Big]_0^5 = 15 - 7\ln 7 + 7\ln 2$

$\qquad\qquad\qquad = 15 + 7(\ln 2 - \ln 7) = 15 + 7\ln\tfrac{2}{7}$

31. As in Example 5,

$$\int \sqrt{\dfrac{1+x}{1-x}}\,dx = \int \dfrac{\sqrt{1+x}}{\sqrt{1-x}}\cdot\dfrac{\sqrt{1+x}}{\sqrt{1+x}}\,dx = \int \dfrac{1+x}{\sqrt{1-x^2}}\,dx = \int \dfrac{dx}{\sqrt{1-x^2}} + \int \dfrac{x\,dx}{\sqrt{1-x^2}} = \sin^{-1} x - \sqrt{1-x^2} + C.$$

Another method: Substitute $u = \sqrt{(1+x)/(1-x)}$.

33. $3 - 2x - x^2 = -(x^2 + 2x + 1) + 4 = 4 - (x+1)^2$. Let $x + 1 = 2\sin\theta$,

where $-\tfrac{\pi}{2} \le \theta \le \tfrac{\pi}{2}$. Then $dx = 2\cos\theta\,d\theta$ and

$\int \sqrt{3 - 2x - x^2}\,dx = \int \sqrt{4 - (x+1)^2}\,dx = \int \sqrt{4 - 4\sin^2\theta}\,2\cos\theta\,d\theta$

$\qquad = 4\int \cos^2\theta\,d\theta = 2\int(1 + \cos 2\theta)\,d\theta$

$\qquad = 2\theta + \sin 2\theta + C = 2\theta + 2\sin\theta\cos\theta + C$

$\qquad = 2\sin^{-1}\left(\dfrac{x+1}{2}\right) + 2\cdot\dfrac{x+1}{2}\cdot\dfrac{\sqrt{3 - 2x - x^2}}{2} + C$

$\qquad = 2\sin^{-1}\left(\dfrac{x+1}{2}\right) + \dfrac{x+1}{2}\sqrt{3 - 2x - x^2} + C$

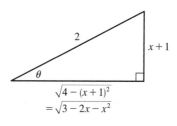

35. Because $f(x) = x^8 \sin x$ is the product of an even function and an odd function, it is odd.

Therefore, $\int_{-1}^1 x^8 \sin x\,dx = 0$ [by (5.5.7)(b)].

37. $\int_0^{\pi/4} \cos^2\theta\,\tan^2\theta\,d\theta = \int_0^{\pi/4} \sin^2\theta\,d\theta = \int_0^{\pi/4} \tfrac{1}{2}(1 - \cos 2\theta)\,d\theta = \left[\tfrac{1}{2}\theta - \tfrac{1}{4}\sin 2\theta\right]_0^{\pi/4} = \left(\tfrac{\pi}{8} - \tfrac{1}{4}\right) - (0 - 0) = \tfrac{\pi}{8} - \tfrac{1}{4}$

39. Let $u = \sec \theta$, so that $du = \sec \theta \tan \theta \, d\theta$. Then $\displaystyle \int \frac{\sec \theta \tan \theta}{\sec^2 \theta - \sec \theta} \, d\theta = \int \frac{1}{u^2 - u} \, du = \int \frac{1}{u(u-1)} \, du = I$. Now

$$\frac{1}{u(u-1)} = \frac{A}{u} + \frac{B}{u-1} \quad \Rightarrow \quad 1 = A(u-1) + Bu. \text{ Set } u = 1 \text{ to get } 1 = B. \text{ Set } u = 0 \text{ to get } 1 = -A, \text{ so } A = -1.$$

Thus, $I = \displaystyle \int \left(\frac{-1}{u} + \frac{1}{u-1} \right) du = -\ln|u| + \ln|u-1| + C = \ln|\sec \theta - 1| - \ln|\sec \theta| + C$ [or $\ln|1 - \cos \theta| + C$].

41. Let $u = \theta$, $dv = \tan^2 \theta \, d\theta = (\sec^2 \theta - 1) \, d\theta \quad \Rightarrow \quad du = d\theta$ and $v = \tan \theta - \theta$. So

$$\int \theta \tan^2 \theta \, d\theta = \theta(\tan \theta - \theta) - \int (\tan \theta - \theta) \, d\theta = \theta \tan \theta - \theta^2 - \ln|\sec \theta| + \tfrac{1}{2}\theta^2 + C$$

$$= \theta \tan \theta - \tfrac{1}{2}\theta^2 - \ln|\sec \theta| + C$$

43. Let $u = 1 + e^x$, so that $du = e^x \, dx$. Then $\int e^x \sqrt{1 + e^x} \, dx = \int u^{1/2} \, du = \tfrac{2}{3}u^{3/2} + C = \tfrac{2}{3}(1 + e^x)^{3/2} + C$.

Or: Let $u = \sqrt{1 + e^x}$, so that $u^2 = 1 + e^x$ and $2u \, du = e^x \, dx$. Then

$\int e^x \sqrt{1 + e^x} \, dx = \int u \cdot 2u \, du = \int 2u^2 \, du = \tfrac{2}{3}u^3 + C = \tfrac{2}{3}(1 + e^x)^{3/2} + C$.

45. Let $t = x^3$. Then $dt = 3x^2 \, dx \quad \Rightarrow \quad I = \int x^5 e^{-x^3} \, dx = \tfrac{1}{3} \int t e^{-t} \, dt$. Now integrate by parts with $u = t$, $dv = e^{-t} \, dt$:

$I = -\tfrac{1}{3}te^{-t} + \tfrac{1}{3} \int e^{-t} \, dt = -\tfrac{1}{3}te^{-t} - \tfrac{1}{3}e^{-t} + C = -\tfrac{1}{3}e^{-x^3}(x^3 + 1) + C$.

47. Let $u = x - 1$, so that $du = dx$. Then

$$\int x^3(x-1)^{-4} \, dx = \int (u+1)^3 u^{-4} \, du = \int (u^3 + 3u^2 + 3u + 1)u^{-4} \, du = \int (u^{-1} + 3u^{-2} + 3u^{-3} + u^{-4}) \, du$$

$$= \ln|u| - 3u^{-1} - \tfrac{3}{2}u^{-2} - \tfrac{1}{3}u^{-3} + C = \ln|x-1| - 3(x-1)^{-1} - \tfrac{3}{2}(x-1)^{-2} - \tfrac{1}{3}(x-1)^{-3} + C$$

49. Let $u = \sqrt{4x+1} \quad \Rightarrow \quad u^2 = 4x + 1 \quad \Rightarrow \quad 2u \, du = 4 \, dx \quad \Rightarrow \quad dx = \tfrac{1}{2}u \, du$. So

$$\int \frac{1}{x\sqrt{4x+1}} \, dx = \int \frac{\tfrac{1}{2}u \, du}{\tfrac{1}{4}(u^2-1) \, u} = 2 \int \frac{du}{u^2-1} = 2\left(\tfrac{1}{2}\right) \ln\left| \frac{u-1}{u+1} \right| + C \qquad \text{[by Formula 19]}$$

$$= \ln\left| \frac{\sqrt{4x+1}-1}{\sqrt{4x+1}+1} \right| + C$$

51. Let $2x = \tan \theta \quad \Rightarrow \quad x = \tfrac{1}{2}\tan \theta$, $dx = \tfrac{1}{2}\sec^2 \theta \, d\theta$, $\sqrt{4x^2 + 1} = \sec \theta$, so

$$\int \frac{dx}{x\sqrt{4x^2+1}} = \int \frac{\tfrac{1}{2}\sec^2 \theta \, d\theta}{\tfrac{1}{2}\tan \theta \sec \theta} = \int \frac{\sec \theta}{\tan \theta} \, d\theta = \int \csc \theta \, d\theta$$

$$= -\ln|\csc \theta + \cot \theta| + C \qquad \text{[or } \ln|\csc \theta - \cot \theta| + C]$$

$$= -\ln\left| \frac{\sqrt{4x^2+1}}{2x} + \frac{1}{2x} \right| + C \qquad \left[\text{or } \ln\left| \frac{\sqrt{4x^2+1}}{2x} - \frac{1}{2x} \right| + C \right]$$

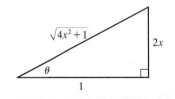

53. $\displaystyle \int x^2 \sinh(mx) \, dx = \frac{1}{m}x^2 \cosh(mx) - \frac{2}{m} \int x \cosh(mx) \, dx \qquad \begin{bmatrix} u = x^2, & dv = \sinh(mx)\,dx, \\ du = 2x\,dx & v = \tfrac{1}{m}\cosh(mx) \end{bmatrix}$

$$= \frac{1}{m}x^2 \cosh(mx) - \frac{2}{m}\left(\tfrac{1}{m}x \sinh(mx) - \tfrac{1}{m} \int \sinh(mx) \, dx \right) \qquad \begin{bmatrix} U = x, & dV = \cosh(mx)\,dx, \\ dU = dx & V = \tfrac{1}{m}\sinh(mx) \end{bmatrix}$$

$$= \frac{1}{m}x^2 \cosh(mx) - \frac{2}{m^2}x \sinh(mx) + \frac{2}{m^3}\cosh(mx) + C$$

55. Let $u = \sqrt{x}$, so that $x = u^2$ and $dx = 2u\, du$. Then $\displaystyle\int \frac{dx}{x + x\sqrt{x}} = \int \frac{2u\, du}{u^2 + u^2 \cdot u} = \int \frac{2}{u(1+u)}\, du = I$.

Now $\dfrac{2}{u(1+u)} = \dfrac{A}{u} + \dfrac{B}{1+u}$ $\Rightarrow$ $2 = A(1+u) + Bu$. Set $u = -1$ to get $2 = -B$, so $B = -2$. Set $u = 0$ to get $2 = A$.

Thus, $I = \displaystyle\int \left(\frac{2}{u} - \frac{2}{1+u}\right) du = 2\ln|u| - 2\ln|1+u| + C = 2\ln\sqrt{x} - 2\ln\left(1 + \sqrt{x}\right) + C$.

57. Let $u = \sqrt[3]{x + c}$. Then $x = u^3 - c$ $\Rightarrow$

$\int x \sqrt[3]{x+c}\, dx = \int (u^3 - c)u \cdot 3u^2\, du = 3\int (u^6 - cu^3)\, du = \frac{3}{7}u^7 - \frac{3}{4}cu^4 + C = \frac{3}{7}(x+c)^{7/3} - \frac{3}{4}c(x+c)^{4/3} + C$

59. Let $u = \sin x$, so that $du = \cos x\, dx$. Then

$\int \cos x \cos^3(\sin x)\, dx = \int \cos^3 u\, du = \int \cos^2 u \cos u\, du = \int (1 - \sin^2 u)\cos u\, du$

$\qquad = \int (\cos u - \sin^2 u \cos u)\, du = \sin u - \frac{1}{3}\sin^3 u + C = \sin(\sin x) - \frac{1}{3}\sin^3(\sin x) + C$

61. Let $y = \sqrt{x}$ so that $dy = \dfrac{1}{2\sqrt{x}}\, dx$ $\Rightarrow$ $dx = 2\sqrt{x}\, dy = 2y\, dy$. Then

$$\int \sqrt{x}\, e^{\sqrt{x}}\, dx = \int y e^y (2y\, dy) = \int 2y^2 e^y\, dy \qquad \begin{bmatrix} u = 2y^2, & dv = e^y\, dy, \\ du = 4y\, dy & v = e^y \end{bmatrix}$$

$$= 2y^2 e^y - \int 4y e^y\, dy \qquad \begin{bmatrix} U = 4y, & dV = e^y\, dy, \\ dU = 4\, dy & V = e^y \end{bmatrix}$$

$$= 2y^2 e^y - \left(4y e^y - \int 4e^y\, dy\right) = 2y^2 e^y - 4y e^y + 4e^y + C$$

$$= 2(y^2 - 2y + 2)e^y + C = 2\left(x - 2\sqrt{x} + 2\right) e^{\sqrt{x}} + C$$

63. Let $u = \cos^2 x$, so that $du = 2\cos x\,(-\sin x)\, dx$. Then

$\displaystyle\int \frac{\sin 2x}{1 + \cos^4 x}\, dx = \int \frac{2\sin x \cos x}{1 + (\cos^2 x)^2}\, dx = \int \frac{1}{1 + u^2}\,(-du) = -\tan^{-1} u + C = -\tan^{-1}(\cos^2 x) + C$.

65. $\displaystyle\int \frac{dx}{\sqrt{x+1} + \sqrt{x}} = \int \left(\frac{1}{\sqrt{x+1} + \sqrt{x}} \cdot \frac{\sqrt{x+1} - \sqrt{x}\sqrt{x}}{\sqrt{x+1} - \sqrt{x}}\right) dx = \int \left(\sqrt{x+1} - \sqrt{x}\right) dx$

$\qquad = \frac{2}{3}\left[(x+1)^{3/2} - x^{3/2}\right] + C$

67. Let $x = \tan\theta$, so that $dx = \sec^2\theta\, d\theta$, $x = \sqrt{3}$ $\Rightarrow$ $\theta = \frac{\pi}{3}$, and $x = 1$ $\Rightarrow$ $\theta = \frac{\pi}{4}$. Then

$$\int_1^{\sqrt{3}} \frac{\sqrt{1 + x^2}}{x^2}\, dx = \int_{\pi/4}^{\pi/3} \frac{\sec\theta}{\tan^2\theta}\sec^2\theta\, d\theta = \int_{\pi/4}^{\pi/3} \frac{\sec\theta\,(\tan^2\theta + 1)}{\tan^2\theta}\, d\theta = \int_{\pi/4}^{\pi/3} \left(\frac{\sec\theta \tan^2\theta}{\tan^2\theta} + \frac{\sec\theta}{\tan^2\theta}\right) d\theta$$

$$= \int_{\pi/4}^{\pi/3} (\sec\theta + \csc\theta \cot\theta)\, d\theta = \left[\ln|\sec\theta + \tan\theta| - \csc\theta\right]_{\pi/4}^{\pi/3}$$

$$= \left(\ln|2 + \sqrt{3}| - \frac{2}{\sqrt{3}}\right) - (\ln|\sqrt{2} + 1| - \sqrt{2}) = \sqrt{2} - \frac{2}{\sqrt{3}} + \ln(2 + \sqrt{3}) - \ln(1 + \sqrt{2})$$

69. Let $u = e^x$. Then $x = \ln u$, $dx = du/u$ $\Rightarrow$

$$\int \frac{e^{2x}}{1+e^x}\,dx = \int \frac{u^2}{1+u}\frac{du}{u} = \int \frac{u}{1+u}\,du = \int\left(1 - \frac{1}{1+u}\right)du = u - \ln|1+u| + C = e^x - \ln(1+e^x) + C.$$

71. Let $\theta = \arcsin x$, so that $d\theta = \dfrac{1}{\sqrt{1-x^2}}\,dx$ and $x = \sin\theta$. Then

$$\int \frac{x + \arcsin x}{\sqrt{1-x^2}}\,dx = \int (\sin\theta + \theta)\,d\theta = -\cos\theta + \tfrac{1}{2}\theta^2 + C$$

$$= -\sqrt{1-x^2} + \tfrac{1}{2}(\arcsin x)^2 + C$$

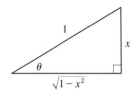

73. $\dfrac{1}{(x-2)(x^2+4)} = \dfrac{A}{x-2} + \dfrac{Bx+C}{x^2+4}$ $\Rightarrow$ $1 = A(x^2+4) + (Bx+C)(x-2) = (A+B)x^2 + (C-2B)x + (4A-2C)$.

So $0 = A+B = C - 2B$, $1 = 4A - 2C$. Setting $x = 2$ gives $A = \tfrac{1}{8}$ $\Rightarrow$ $B = -\tfrac{1}{8}$ and $C = -\tfrac{1}{4}$. So

$$\int \frac{1}{(x-2)(x^2+4)}\,dx = \int\left(\frac{\tfrac{1}{8}}{x-2} + \frac{-\tfrac{1}{8}x - \tfrac{1}{4}}{x^2+4}\right)dx = \frac{1}{8}\int \frac{dx}{x-2} - \frac{1}{16}\int \frac{2x\,dx}{x^2+4} - \frac{1}{4}\int \frac{dx}{x^2+4}$$

$$= \tfrac{1}{8}\ln|x-2| - \tfrac{1}{16}\ln(x^2+4) - \tfrac{1}{8}\tan^{-1}(x/2) + C$$

75. Let $y = \sqrt{1+e^x}$, so that $y^2 = 1 + e^x$, $2y\,dy = e^x\,dx$, $e^x = y^2 - 1$, and $x = \ln(y^2 - 1)$. Then

$$\int \frac{xe^x}{\sqrt{1+e^x}}\,dx = \int \frac{\ln(y^2-1)}{y}(2y\,dy) = 2\int [\ln(y+1) + \ln(y-1)]\,dy$$

$$= 2[(y+1)\ln(y+1) - (y+1) + (y-1)\ln(y-1) - (y-1)] + C \qquad \text{[by Example 7.1.2]}$$

$$= 2[y\ln(y+1) + \ln(y+1) - y - 1 + y\ln(y-1) - \ln(y-1) - y + 1] + C$$

$$= 2[y(\ln(y+1) + \ln(y-1)) + \ln(y+1) - \ln(y-1) - 2y] + C$$

$$= 2\left[y\ln(y^2-1) + \ln\frac{y+1}{y-1} - 2y\right] + C = 2\left[\sqrt{1+e^x}\,\ln(e^x) + \ln\frac{\sqrt{1+e^x}+1}{\sqrt{1+e^x}-1} - 2\sqrt{1+e^x}\right] + C$$

$$= 2x\sqrt{1+e^x} + 2\ln\frac{\sqrt{1+e^x}+1}{\sqrt{1+e^x}-1} - 4\sqrt{1+e^x} + C = 2(x-2)\sqrt{1+e^x} + 2\ln\frac{\sqrt{1+e^x}+1}{\sqrt{1+e^x}-1} + C$$

77. Let $u = x^{3/2}$ so that $u^2 = x^3$ and $du = \tfrac{3}{2}x^{1/2}\,dx$ $\Rightarrow$ $\sqrt{x}\,dx = \tfrac{2}{3}\,du$. Then

$$\int \frac{\sqrt{x}}{1+x^3}\,dx = \int \frac{\tfrac{2}{3}}{1+u^2}\,du = \frac{2}{3}\tan^{-1}u + C = \tfrac{2}{3}\tan^{-1}(x^{3/2}) + C.$$

79. Let $u = x$, $dv = \sin^2 x \cos x\,dx$ $\Rightarrow$ $du = dx$, $v = \tfrac{1}{3}\sin^3 x$. Then

$$\int x\sin^2 x\,\cos x\,dx = \tfrac{1}{3}x\sin^3 x - \int \tfrac{1}{3}\sin^3 x\,dx = \tfrac{1}{3}x\sin^3 x - \tfrac{1}{3}\int(1-\cos^2 x)\sin x\,dx$$

$$= \frac{1}{3}x\sin^3 x + \frac{1}{3}\int(1-y^2)\,dy \qquad \begin{bmatrix} u = \cos x, \\ du = -\sin x\,dx \end{bmatrix}$$

$$= \tfrac{1}{3}x\sin^3 x + \tfrac{1}{3}y - \tfrac{1}{9}y^3 + C = \tfrac{1}{3}x\sin^3 x + \tfrac{1}{3}\cos x - \tfrac{1}{9}\cos^3 x + C$$

81. The function $y = 2xe^{x^2}$ *does* have an elementary antiderivative, so we'll use this fact to help evaluate the integral.

$$\int (2x^2 + 1)e^{x^2}\,dx = \int 2x^2 e^{x^2}\,dx + \int e^{x^2}\,dx = \int x\left(2xe^{x^2}\right)dx + \int e^{x^2}\,dx$$

$$= xe^{x^2} - \int e^{x^2}\,dx + \int e^{x^2}\,dx \quad \begin{bmatrix} u = x, & dv = 2xe^{x^2}\,dx, \\ du = dx & v = e^{x^2} \end{bmatrix} = xe^{x^2} + C$$

7.6 Integration Using Tables and Computer Algebra Systems

Keep in mind that there are several ways to approach many of these exercises, and different methods can lead to different forms of the answer.

1. We could make the substitution $u = \sqrt{2}\,x$ to obtain the radical $\sqrt{7 - u^2}$ and then use Formula 33 with $a = \sqrt{7}$.

Alternatively, we will factor $\sqrt{2}$ out of the radical and use $a = \sqrt{\frac{7}{2}}$.

$$\int \frac{\sqrt{7 - 2x^2}}{x^2}\,dx = \sqrt{2}\int \frac{\sqrt{\frac{7}{2} - x^2}}{x^2}\,dx \overset{33}{=} \sqrt{2}\left[-\frac{1}{x}\sqrt{\frac{7}{2} - x^2} - \sin^{-1}\frac{x}{\sqrt{\frac{7}{2}}}\right] + C$$

$$= -\frac{1}{x}\sqrt{7 - 2x^2} - \sqrt{2}\,\sin^{-1}\left(\sqrt{\tfrac{2}{7}}\,x\right) + C$$

3. Let $u = \pi x \;\Rightarrow\; du = \pi\,dx$, so

$$\int \sec^3(\pi x)\,dx = \frac{1}{\pi}\int \sec^3 u\,du \overset{71}{=} \frac{1}{\pi}\left(\frac{1}{2}\sec u\tan u + \frac{1}{2}\ln|\sec u + \tan u|\right) + C$$

$$= \frac{1}{2\pi}\sec\pi x\tan\pi x + \frac{1}{2\pi}\ln|\sec\pi x + \tan\pi x| + C$$

5. $\displaystyle \int_0^1 2x\cos^{-1}x\,dx \overset{91}{=} 2\left[\frac{2x^2 - 1}{4}\cos^{-1}x - \frac{x\sqrt{1 - x^2}}{4}\right]_0^1 = 2\left[\left(\tfrac{1}{4}\cdot 0 - 0\right) - \left(-\tfrac{1}{4}\cdot\tfrac{\pi}{2} - 0\right)\right] = 2\left(\tfrac{\pi}{8}\right) = \tfrac{\pi}{4}$

7. Let $u = \pi x$, so that $du = \pi\,dx$. Then

$$\int \tan^3(\pi x)\,dx = \int \tan^3 u\left(\tfrac{1}{\pi}\,du\right) = \frac{1}{\pi}\int \tan^3 u\,du \overset{69}{=} \frac{1}{\pi}\left[\tfrac{1}{2}\tan^2 u + \ln|\cos u|\right] + C$$

$$= \frac{1}{2\pi}\tan^2(\pi x) + \frac{1}{\pi}\ln|\cos(\pi x)| + C$$

9. Let $u = 2x$ and $a = 3$. Then $du = 2\,dx$ and

$$\int \frac{dx}{x^2\sqrt{4x^2 + 9}} = \int \frac{\frac{1}{2}\,du}{\frac{u^2}{4}\sqrt{u^2 + a^2}} = 2\int \frac{du}{u^2\sqrt{a^2 + u^2}} \overset{28}{=} -2\frac{\sqrt{a^2 + u^2}}{a^2 u} + C$$

$$= -2\frac{\sqrt{4x^2 + 9}}{9\cdot 2x} + C = -\frac{\sqrt{4x^2 + 9}}{9x} + C$$

11. $\displaystyle \int_{-1}^0 t^2 e^{-t}\,dt \overset{97}{=} \left[\frac{1}{-1}t^2 e^{-t}\right]_{-1}^0 - \frac{2}{-1}\int_{-1}^0 te^{-t}\,dt = e + 2\int_{-1}^0 te^{-t}\,dt \overset{96}{=} e + 2\left[\frac{1}{(-1)^2}(-t - 1)e^{-t}\right]_{-1}^0$

$$= e + 2\left[-e^0 + 0\right] = e - 2$$

13. $\displaystyle \int \frac{\tan^3(1/z)}{z^2}\,dz \quad \begin{bmatrix} u = 1/z, \\ du = -dz/z^2 \end{bmatrix} = -\int \tan^3 u\,du \overset{69}{=} -\frac{1}{2}\tan^2 u - \ln|\cos u| + C$

$$= -\frac{1}{2}\tan^2\left(\tfrac{1}{z}\right) - \ln\left|\cos\left(\tfrac{1}{z}\right)\right| + C$$

15. Let $u = e^x$, so that $du = e^x \, dx$ and $e^{2x} = u^2$. Then

$$\int e^{2x} \arctan(e^x) \, dx = \int u^2 \arctan u \left(\frac{du}{u}\right) = \int u \arctan u \, du$$

$$\stackrel{92}{=} \frac{u^2+1}{2} \arctan u - \frac{u}{2} + C = \frac{1}{2}(e^{2x}+1) \arctan(e^x) - \frac{1}{2}e^x + C$$

17. Let $z = 6 + 4y - 4y^2 = 6 - (4y^2 - 4y + 1) + 1 = 7 - (2y-1)^2$, $u = 2y - 1$, and $a = \sqrt{7}$. Then $z = a^2 - u^2$, $du = 2\, dy$, and

$$\int y \sqrt{6 + 4y - 4y^2} \, dy = \int y \sqrt{z} \, dy = \int \tfrac{1}{2}(u+1) \sqrt{a^2 - u^2} \, \tfrac{1}{2} \, du = \tfrac{1}{4} \int u \sqrt{a^2 - u^2} \, du + \tfrac{1}{4} \int \sqrt{a^2 - u^2} \, du$$

$$= \tfrac{1}{4} \int \sqrt{a^2 - u^2} \, du - \tfrac{1}{8} \int (-2u) \sqrt{a^2 - u^2} \, du$$

$$\stackrel{30}{=} \frac{u}{8} \sqrt{a^2 - u^2} + \frac{a^2}{8} \sin^{-1}\left(\frac{u}{a}\right) - \frac{1}{8} \int \sqrt{w} \, dw \qquad \begin{bmatrix} w = a^2 - u^2, \\ dw = -2u \, du \end{bmatrix}$$

$$= \frac{2y-1}{8} \sqrt{6 + 4y - 4y^2} + \frac{7}{8} \sin^{-1} \frac{2y-1}{\sqrt{7}} - \frac{1}{8} \cdot \frac{2}{3} w^{3/2} + C$$

$$= \frac{2y-1}{8} \sqrt{6 + 4y - 4y^2} + \frac{7}{8} \sin^{-1} \frac{2y-1}{\sqrt{7}} - \frac{1}{12}(6 + 4y - 4y^2)^{3/2} + C$$

This can be rewritten as

$$\sqrt{6 + 4y - 4y^2} \left[\frac{1}{8}(2y-1) - \frac{1}{12}(6 + 4y - 4y^2) \right] + \frac{7}{8} \sin^{-1} \frac{2y-1}{\sqrt{7}} + C$$

$$= \left(\frac{1}{3}y^2 - \frac{1}{12}y - \frac{5}{8} \right) \sqrt{6 + 4y - 4y^2} + \frac{7}{8} \sin^{-1}\left(\frac{2y-1}{\sqrt{7}} \right) + C$$

$$= \frac{1}{24}(8y^2 - 2y - 15) \sqrt{6 + 4y - 4y^2} + \frac{7}{8} \sin^{-1}\left(\frac{2y-1}{\sqrt{7}} \right) + C$$

19. Let $u = \sin x$. Then $du = \cos x \, dx$, so

$$\int \sin^2 x \cos x \ln(\sin x) \, dx = \int u^2 \ln u \, du \stackrel{101}{=} \frac{u^{2+1}}{(2+1)^2} \left[(2+1) \ln u - 1 \right] + C = \tfrac{1}{9} u^3 (3 \ln u - 1) + C$$

$$= \tfrac{1}{9} \sin^3 x \left[3 \ln(\sin x) - 1 \right] + C$$

21. Let $u = e^x$ and $a = \sqrt{3}$. Then $du = e^x \, dx$ and

$$\int \frac{e^x}{3 - e^{2x}} \, dx = \int \frac{du}{a^2 - u^2} \stackrel{19}{=} \frac{1}{2a} \ln \left| \frac{u+a}{u-a} \right| + C = \frac{1}{2\sqrt{3}} \ln \left| \frac{e^x + \sqrt{3}}{e^x - \sqrt{3}} \right| + C.$$

23. $\int \sec^5 x \, dx \stackrel{77}{=} \tfrac{1}{4} \tan x \sec^3 x + \tfrac{3}{4} \int \sec^3 x \, dx \stackrel{77}{=} \tfrac{1}{4} \tan x \sec^3 x + \tfrac{3}{4} \left(\tfrac{1}{2} \tan x \sec x + \tfrac{1}{2} \int \sec x \, dx \right)$

$$\stackrel{14}{=} \tfrac{1}{4} \tan x \sec^3 x + \tfrac{3}{8} \tan x \sec x + \tfrac{3}{8} \ln|\sec x + \tan x| + C$$

25. Let $u = \ln x$ and $a = 2$. Then $du = dx/x$ and

$$\int \frac{\sqrt{4 + (\ln x)^2}}{x} \, dx = \int \sqrt{a^2 + u^2} \, du \stackrel{21}{=} \frac{u}{2} \sqrt{a^2 + u^2} + \frac{a^2}{2} \ln\left(u + \sqrt{a^2 + u^2} \right) + C$$

$$= \tfrac{1}{2}(\ln x) \sqrt{4 + (\ln x)^2} + 2 \ln\left[\ln x + \sqrt{4 + (\ln x)^2} \right] + C$$

27. Let $u = e^x$. Then $x = \ln u$, $dx = du/u$, so

$$\int \sqrt{e^{2x} - 1}\, dx = \int \frac{\sqrt{u^2 - 1}}{u}\, du \overset{41}{=} \sqrt{u^2 - 1} - \cos^{-1}(1/u) + C = \sqrt{e^{2x} - 1} - \cos^{-1}(e^{-x}) + C.$$

29. $\displaystyle \int \frac{x^4\, dx}{\sqrt{x^{10} - 2}} = \int \frac{x^4\, dx}{\sqrt{(x^5)^2 - 2}} = \frac{1}{5} \int \frac{du}{\sqrt{u^2 - 2}}$ $\quad \begin{bmatrix} u = x^5, \\ du = 5x^4\, dx \end{bmatrix}$

$$\overset{43}{=} \tfrac{1}{5} \ln\left| u + \sqrt{u^2 - 2} \right| + C = \tfrac{1}{5} \ln\left| x^5 + \sqrt{x^{10} - 2} \right| + C$$

31. Using cylindrical shells, we get

$$V = 2\pi \int_0^2 x \cdot x\, \sqrt{4 - x^2}\, dx = 2\pi \int_0^2 x^2\, \sqrt{4 - x^2}\, dx \overset{31}{=} 2\pi \left[\frac{x}{8} (2x^2 - 4)\, \sqrt{4 - x^2} + \frac{16}{8} \sin^{-1} \frac{x}{2} \right]_0^2$$

$$= 2\pi[(0 + 2\sin^{-1} 1) - (0 + 2\sin^{-1} 0)] = 2\pi\left(2 \cdot \frac{\pi}{2} \right) = 2\pi^2$$

33. (a) $\displaystyle \frac{d}{du} \left[\frac{1}{b^3} \left(a + bu - \frac{a^2}{a + bu} - 2a \ln|a + bu| \right) + C \right] = \frac{1}{b^3} \left[b + \frac{ba^2}{(a + bu)^2} - \frac{2ab}{(a + bu)} \right]$

$$= \frac{1}{b^3} \left[\frac{b(a + bu)^2 + ba^2 - (a + bu)2ab}{(a + bu)^2} \right] = \frac{1}{b^3} \left[\frac{b^3 u^2}{(a + bu)^2} \right] = \frac{u^2}{(a + bu)^2}$$

(b) Let $t = a + bu \;\Rightarrow\; dt = b\, du$. Note that $u = \dfrac{t - a}{b}$ and $du = \dfrac{1}{b}\, dt$.

$$\int \frac{u^2\, du}{(a + bu)^2} = \frac{1}{b^3} \int \frac{(t - a)^2}{t^2}\, dt = \frac{1}{b^3} \int \frac{t^2 - 2at + a^2}{t^2}\, dt = \frac{1}{b^3} \int \left(1 - \frac{2a}{t} + \frac{a^2}{t^2} \right) dt$$

$$= \frac{1}{b^3} \left(t - 2a \ln|t| - \frac{a^2}{t} \right) + C = \frac{1}{b^3} \left(a + bu - \frac{a^2}{a + bu} - 2a \ln|a + bu| \right) + C$$

35. Maple and Mathematica both give $\int \sec^4 x\, dx = \frac{2}{3} \tan x + \frac{1}{3} \tan x \sec^2 x$, while Derive gives the second

term as $\dfrac{\sin x}{3 \cos^3 x} = \dfrac{1}{3} \dfrac{\sin x}{\cos x} \dfrac{1}{\cos^2 x} = \dfrac{1}{3} \tan x \sec^2 x$. Using Formula 77, we get

$\int \sec^4 x\, dx = \frac{1}{3} \tan x \sec^2 x + \frac{2}{3} \int \sec^2 x\, dx = \frac{1}{3} \tan x \sec^2 x + \frac{2}{3} \tan x + C.$

37. Derive gives $\int x^2\, \sqrt{x^2 + 4}\, dx = \frac{1}{4} x(x^2 + 2)\, \sqrt{x^2 + 4} - 2 \ln\left(\sqrt{x^2 + 4} + x \right)$. Maple gives

$\frac{1}{4} x(x^2 + 4)^{3/2} - \frac{1}{2} x\, \sqrt{x^2 + 4} - 2 \operatorname{arcsinh}\left(\frac{1}{2} x \right)$. Applying the command `convert(%,ln);` yields

$\frac{1}{4} x(x^2 + 4)^{3/2} - \frac{1}{2} x\, \sqrt{x^2 + 4} - 2 \ln\left(\frac{1}{2} x + \frac{1}{2}\, \sqrt{x^2 + 4} \right) = \frac{1}{4} x(x^2 + 4)^{1/2} \left[(x^2 + 4) - 2 \right] - 2 \ln\left[\left(x + \sqrt{x^2 + 4} \right)/2 \right]$

$$= \tfrac{1}{4} x(x^2 + 2)\, \sqrt{x^2 + 4} - 2 \ln\left(\sqrt{x^2 + 4} + x \right) + 2 \ln 2$$

Mathematica gives $\frac{1}{4} x(2 + x^2)\, \sqrt{3 + x^2} - 2 \operatorname{arcsinh}(x/2)$. Applying the `TrigToExp` and `Simplify` commands gives

$\frac{1}{4} \left[x(2 + x^2)\, \sqrt{4 + x^2} - 8 \log\left(\frac{1}{2} \left(x + \sqrt{4 + x^2} \right) \right) \right] = \frac{1}{4} x(x^2 + 2)\, \sqrt{x^2 + 4} - 2 \ln\left(x + \sqrt{4 + x^2} \right) + 2 \ln 2$, so all are

equivalent (without constant).

Now use Formula 22 to get

$$\int x^2 \sqrt{2^2 + x^2}\, dx = \frac{x}{8}(2^2 + 2x^2)\sqrt{2^2 + x^2} - \frac{2^4}{8}\ln\left(x + \sqrt{2^2 + x^2}\right) + C$$

$$= \frac{x}{8}(2)(2 + x^2)\sqrt{4 + x^2} - 2\ln\left(x + \sqrt{4 + x^2}\right) + C$$

$$= \tfrac{1}{4}x(x^2 + 2)\sqrt{x^2 + 4} - 2\ln\left(\sqrt{x^2 + 4} + x\right) + C$$

39. Maple gives $\int x\sqrt{1 + 2x}\, dx = \frac{1}{10}(1 + 2x)^{5/2} - \frac{1}{6}(1 + 2x)^{3/2}$, Mathematica gives $\sqrt{1 + 2x}\left(\frac{2}{5}x^2 + \frac{1}{15}x - \frac{1}{15}\right)$, and Derive

gives $\frac{1}{15}(1 + 2x)^{3/2}(3x - 1)$. The first two expressions can be simplified to Derive's result. If we use Formula 54, we get

$$\int x\sqrt{1 + 2x}\, dx = \frac{2}{15(2)^2}(3 \cdot 2x - 2 \cdot 1)(1 + 2x)^{3/2} + C = \frac{1}{30}(6x - 2)(1 + 2x)^{3/2} + C = \frac{1}{15}(3x - 1)(1 + 2x)^{3/2}.$$

41. Maple gives $\int \tan^5 x\, dx = \frac{1}{4}\tan^4 x - \frac{1}{2}\tan^2 x + \frac{1}{2}\ln(1 + \tan^2 x)$, Mathematica gives

$\int \tan^5 x\, dx = \frac{1}{4}[-1 - 2\cos(2x)]\sec^4 x - \ln(\cos x)$, and Derive gives $\int \tan^5 x\, dx = \frac{1}{4}\tan^4 x - \frac{1}{2}\tan^2 x - \ln(\cos x)$.

These expressions are equivalent, and none includes absolute value bars or a constant of integration. Note that Mathematica's

and Derive's expressions suggest that the integral is undefined where $\cos x < 0$, which is not the case. Using Formula 75,

$\int \tan^5 x\, dx = \frac{1}{5-1}\tan^{5-1} x - \int \tan^{5-2} x\, dx = \frac{1}{4}\tan^4 x - \int \tan^3 x\, dx$. Using Formula 69,

$\int \tan^3 x\, dx = \frac{1}{2}\tan^2 x + \ln|\cos x| + C$, so $\int \tan^5 x\, dx = \frac{1}{4}\tan^4 x - \frac{1}{2}\tan^2 x - \ln|\cos x| + C$.

43. (a) $F(x) = \int f(x)\, dx = \int \dfrac{1}{x\sqrt{1 - x^2}}\, dx \overset{35}{=} -\frac{1}{1}\ln\left|\dfrac{1 + \sqrt{1 - x^2}}{x}\right| + C = -\ln\left|\dfrac{1 + \sqrt{1 - x^2}}{x}\right| + C.$

f has domain $\{x \mid x \neq 0, 1 - x^2 > 0\} = \{x \mid x \neq 0, |x| < 1\} = (-1, 0) \cup (0, 1)$. F has the same domain.

(b) Derive gives $F(x) = \ln\left(\sqrt{1 - x^2} - 1\right) - \ln x$ and Mathematica gives $F(x) = \ln x - \ln\left(1 + \sqrt{1 - x^2}\right)$.

Both are correct if you take absolute values of the logarithm arguments, and both would then have the

same domain. Maple gives $F(x) = -\operatorname{arctanh}\left(1/\sqrt{1 - x^2}\right)$. This function has domain

$\left\{x \mid |x| < 1, -1 < 1/\sqrt{1 - x^2} < 1\right\} = \left\{x \mid |x| < 1, 1/\sqrt{1 - x^2} < 1\right\} = \left\{x \mid |x| < 1, \sqrt{1 - x^2} > 1\right\} = \emptyset,$

the empty set! If we apply the command `convert(%,ln);` to Maple's answer, we get

$-\dfrac{1}{2}\ln\left(\dfrac{1}{\sqrt{1 - x^2}} + 1\right) + \dfrac{1}{2}\ln\left(1 - \dfrac{1}{\sqrt{1 - x^2}}\right)$, which has the same domain, $\emptyset$.

45. Maple gives the antiderivative

$$F(x) = \int \frac{x^2 - 1}{x^4 + x^2 + 1}\, dx = -\frac{1}{2}\ln(x^2 + x + 1) + \frac{1}{2}\ln(x^2 - x + 1).$$

We can see that at 0, this antiderivative is 0. From the graphs, it appears that F has

a maximum at $x = -1$ and a minimum at $x = 1$ [since $F'(x) = f(x)$ changes

sign at these x-values], and that F has inflection points at $x \approx -1.7$, $x = 0$, and

$x \approx 1.7$ [since $f(x)$ has extrema at these x-values].

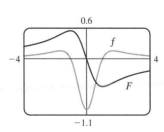

47. Since $f(x) = \sin^4 x \cos^6 x$ is everywhere positive, we know that its antiderivative F is increasing. Maple gives

$$\int f(x)\,dx = -\tfrac{1}{10}\sin^3 x \cos^7 x - \tfrac{3}{80}\sin x \cos^7 x + \tfrac{1}{160}\cos^5 x \sin x + \tfrac{1}{128}\cos^3 x \sin x + \tfrac{3}{256}\cos x \sin x + \tfrac{3}{256}x$$

and this expression is 0 at $x = 0$.

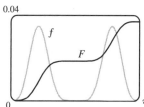

F has a minimum at $x = 0$ and a maximum at $x = \pi$.

F has inflection points where f' changes sign, that is, at $x \approx 0.7$, $x = \pi/2$,

and $x \approx 2.5$.

7.7 Approximate Integration

1. (a) $\Delta x = (b-a)/n = (4-0)/2 = 2$

$$L_2 = \sum_{i=1}^{2} f(x_{i-1})\,\Delta x = f(x_0)\cdot 2 + f(x_1)\cdot 2 = 2\,[f(0) + f(2)] = 2(0.5 + 2.5) = 6$$

$$R_2 = \sum_{i=1}^{2} f(x_i)\,\Delta x = f(x_1)\cdot 2 + f(x_2)\cdot 2 = 2\,[f(2) + f(4)] = 2(2.5 + 3.5) = 12$$

$$M_2 = \sum_{i=1}^{2} f(\overline{x}_i)\Delta x = f(\overline{x}_1)\cdot 2 + f(\overline{x}_2)\cdot 2 = 2\,[f(1) + f(3)] \approx 2(1.6 + 3.2) = 9.6$$

(b)

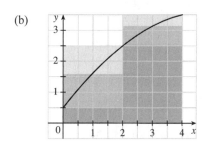

L_2 is an underestimate, since the area under the small rectangles is less than the area under the curve, and R_2 is an overestimate, since the area under the large rectangles is greater than the area under the curve. It appears that M_2 is an overestimate, though it is fairly close to I. See the solution to Exercise 45 for a proof of the fact that if f is concave down on $[a, b]$, then the Midpoint Rule is an overestimate of $\int_a^b f(x)\,dx$.

(c) $T_2 = \left(\tfrac{1}{2}\,\Delta x\right)[f(x_0) + 2f(x_1) + f(x_2)] = \tfrac{2}{2}[f(0) + 2f(2) + f(4)] = 0.5 + 2(2.5) + 3.5 = 9.$

This approximation is an underestimate, since the graph is concave down. Thus, $T_2 = 9 < I$. See the solution to Exercise 45 for a general proof of this conclusion.

(d) For any n, we will have $L_n < T_n < I < M_n < R_n$.

3. $f(x) = \cos(x^2)$, $\Delta x = \tfrac{1-0}{4} = \tfrac{1}{4}$

(a) $T_4 = \tfrac{1}{4\cdot 2}\left[f(0) + 2f\!\left(\tfrac{1}{4}\right) + 2f\!\left(\tfrac{2}{4}\right) + 2f\!\left(\tfrac{3}{4}\right) + f(1)\right] \approx 0.895759$

(b) $M_4 = \tfrac{1}{4}\left[f\!\left(\tfrac{1}{8}\right) + f\!\left(\tfrac{3}{8}\right) + f\!\left(\tfrac{5}{8}\right) + f\!\left(\tfrac{7}{8}\right)\right] \approx 0.908907$

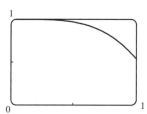

The graph shows that f is concave down on $[0, 1]$. So T_4 is an

underestimate and M_4 is an overestimate. We can conclude that

$0.895759 < \int_0^1 \cos(x^2)\,dx < 0.908907$.

5. $f(x) = x^2 \sin x, \Delta x = \dfrac{b-a}{n} = \dfrac{\pi - 0}{8} = \dfrac{\pi}{8}$

(a) $M_8 = \dfrac{\pi}{8}\left[f\left(\dfrac{\pi}{16}\right) + f\left(\dfrac{3\pi}{16}\right) + f\left(\dfrac{5\pi}{16}\right) + \cdots + f\left(\dfrac{15\pi}{16}\right)\right] \approx 5.932957$

(b) $S_8 = \dfrac{\pi}{8\cdot 3}\left[f(0) + 4f\left(\dfrac{\pi}{8}\right) + 2f\left(\dfrac{2\pi}{8}\right) + 4f\left(\dfrac{3\pi}{8}\right) + 2f\left(\dfrac{4\pi}{8}\right) + 4f\left(\dfrac{5\pi}{8}\right) + 2f\left(\dfrac{6\pi}{8}\right) + 4f\left(\dfrac{7\pi}{8}\right) + f(\pi)\right]$

≈ 5.869247

Actual: $\int_0^\pi x^2 \sin x\, dx \overset{84}{=} \left[-x^2 \cos x\right]_0^\pi + 2\int_0^\pi x \cos x\, dx \overset{83}{=} \left[-\pi^2(-1) - 0\right] + 2\left[\cos x + x \sin x\right]_0^\pi$

$= \pi^2 + 2[(-1+0) - (1+0)] = \pi^2 - 4 \approx 5.869604$

Errors: $E_M = \text{actual} - M_8 = \int_0^\pi x^2 \sin x\, dx - M_8 \approx -0.063353$

$E_S = \text{actual} - S_8 = \int_0^\pi x^2 \sin x\, dx - S_8 \approx 0.000357$

7. $f(x) = \sqrt[4]{1 + x^2}, \Delta x = \dfrac{2-0}{8} = \dfrac{1}{4}$

(a) $T_8 = \dfrac{1}{4\cdot 2}\left[f(0) + 2f\left(\dfrac{1}{4}\right) + 2f\left(\dfrac{1}{2}\right) + \cdots + 2f\left(\dfrac{3}{2}\right) + 2f\left(\dfrac{7}{4}\right) + f(2)\right] \approx 2.413790$

(b) $M_8 = \dfrac{1}{4}\left[f\left(\dfrac{1}{8}\right) + f\left(\dfrac{3}{8}\right) + \cdots + f\left(\dfrac{13}{8}\right) + f\left(\dfrac{15}{8}\right)\right] \approx 2.411453$

(c) $S_8 = \dfrac{1}{4\cdot 3}\left[f(0) + 4f\left(\dfrac{1}{4}\right) + 2f\left(\dfrac{1}{2}\right) + 4f\left(\dfrac{3}{4}\right) + 2f(1) + 4f\left(\dfrac{5}{4}\right) + 2f\left(\dfrac{3}{2}\right) + 4f\left(\dfrac{7}{4}\right) + f(2)\right] \approx 2.412232$

9. $f(x) = \dfrac{\ln x}{1+x}, \Delta x = \dfrac{2-1}{10} = \dfrac{1}{10}$

(a) $T_{10} = \dfrac{1}{10\cdot 2}[f(1) + 2f(1.1) + 2f(1.2) + \cdots + 2f(1.8) + 2f(1.9) + f(2)] \approx 0.146879$

(b) $M_{10} = \dfrac{1}{10}[f(1.05) + f(1.15) + \cdots + f(1.85) + f(1.95)] \approx 0.147391$

(c) $S_{10} = \dfrac{1}{10\cdot 3}[f(1) + 4f(1.1) + 2f(1.2) + 4f(1.3) + 2f(1.4) + 4f(1.5) + 2f(1.6) + 4f(1.7)$

$+ 2f(1.8) + 4f(1.9) + f(2)]$

≈ 0.147219

11. $f(t) = \sin(e^{t/2}), \Delta t = \dfrac{\frac{1}{2} - 0}{8} = \dfrac{1}{16}$

(a) $T_8 = \dfrac{1}{16\cdot 2}\left[f(0) + 2f\left(\dfrac{1}{16}\right) + 2f\left(\dfrac{2}{16}\right) + \cdots + 2f\left(\dfrac{7}{16}\right) + f\left(\dfrac{1}{2}\right)\right] \approx 0.451948$

(b) $M_8 = \dfrac{1}{16}\left[f\left(\dfrac{1}{32}\right) + f\left(\dfrac{3}{32}\right) + f\left(\dfrac{5}{32}\right) + \cdots + f\left(\dfrac{13}{32}\right) + f\left(\dfrac{15}{32}\right)\right] \approx 0.451991$

(c) $S_8 = \dfrac{1}{16\cdot 3}\left[f(0) + 4f\left(\dfrac{1}{16}\right) + 2f\left(\dfrac{2}{16}\right) + \cdots + 4f\left(\dfrac{7}{16}\right) + f\left(\dfrac{1}{2}\right)\right] \approx 0.451976$

13. $f(t) = e^{\sqrt{t}} \sin t, \Delta t = \dfrac{4-0}{8} = \dfrac{1}{2}$

(a) $T_8 = \dfrac{1}{2\cdot 2}\left[f(0) + 2f\left(\dfrac{1}{2}\right) + 2f(1) + 2f\left(\dfrac{3}{2}\right) + 2f(2) + 2f\left(\dfrac{5}{2}\right) + 2f(3) + 2f\left(\dfrac{7}{2}\right) + f(4)\right] \approx 4.513618$

(b) $M_8 = \dfrac{1}{2}\left[f\left(\dfrac{1}{4}\right) + f\left(\dfrac{3}{4}\right) + f\left(\dfrac{5}{4}\right) + f\left(\dfrac{7}{4}\right) + f\left(\dfrac{9}{4}\right) + f\left(\dfrac{11}{4}\right) + f\left(\dfrac{13}{4}\right) + f\left(\dfrac{15}{4}\right)\right] \approx 4.748256$

(c) $S_8 = \dfrac{1}{2\cdot 3}\left[f(0) + 4f\left(\dfrac{1}{2}\right) + 2f(1) + 4f\left(\dfrac{3}{2}\right) + 2f(2) + 4f\left(\dfrac{5}{2}\right) + 2f(3) + 4f\left(\dfrac{7}{2}\right) + f(4)\right] \approx 4.675111$

15. $f(x) = \dfrac{\cos x}{x}, \Delta x = \dfrac{5-1}{8} = \dfrac{1}{2}$

(a) $T_8 = \dfrac{1}{2\cdot 2}\left[f(1) + 2f\left(\dfrac{3}{2}\right) + 2f(2) + \cdots + 2f(4) + 2f\left(\dfrac{9}{2}\right) + f(5)\right] \approx -0.495333$

(b) $M_8 = \dfrac{1}{2}\left[f\left(\dfrac{5}{4}\right) + f\left(\dfrac{7}{4}\right) + f\left(\dfrac{9}{4}\right) + f\left(\dfrac{11}{4}\right) + f\left(\dfrac{13}{4}\right) + f\left(\dfrac{15}{4}\right) + f\left(\dfrac{17}{4}\right) + f\left(\dfrac{19}{4}\right)\right] \approx -0.543321$

(c) $S_8 = \dfrac{1}{2\cdot 3}\left[f(1) + 4f\left(\dfrac{3}{2}\right) + 2f(2) + 4f\left(\dfrac{5}{2}\right) + 2f(3) + 4f\left(\dfrac{7}{2}\right) + 2f(4) + 4f\left(\dfrac{9}{2}\right) + f(5)\right] \approx -0.526123$

17. $f(y) = \dfrac{1}{1+y^5}$, $\Delta y = \dfrac{3-0}{6} = \dfrac{1}{2}$

(a) $T_6 = \frac{1}{2\cdot 2}\left[f(0) + 2f\left(\frac{1}{2}\right) + 2f\left(\frac{2}{2}\right) + 2f\left(\frac{3}{2}\right) + 2f\left(\frac{4}{2}\right) + 2f\left(\frac{5}{2}\right) + f(3)\right] \approx 1.064275$

(b) $M_6 = \frac{1}{2}\left[f\left(\frac{1}{4}\right) + f\left(\frac{3}{4}\right) + f\left(\frac{5}{4}\right) + f\left(\frac{7}{4}\right) + f\left(\frac{9}{4}\right) + f\left(\frac{11}{4}\right)\right] \approx 1.067416$

(c) $S_6 = \frac{1}{2\cdot 3}\left[f(0) + 4f\left(\frac{1}{2}\right) + 2f\left(\frac{2}{2}\right) + 4f\left(\frac{3}{2}\right) + 2f\left(\frac{4}{2}\right) + 4f\left(\frac{5}{2}\right) + f(3)\right] \approx 1.074915$

19. $f(x) = \cos(x^2)$, $\Delta x = \frac{1-0}{8} = \frac{1}{8}$

(a) $T_8 = \frac{1}{8\cdot 2}\left\{f(0) + 2\left[f\left(\frac{1}{8}\right) + f\left(\frac{2}{8}\right) + \cdots + f\left(\frac{7}{8}\right)\right] + f(1)\right\} \approx 0.902333$

$M_8 = \frac{1}{8}\left[f\left(\frac{1}{16}\right) + f\left(\frac{3}{16}\right) + f\left(\frac{5}{16}\right) + \cdots + f\left(\frac{15}{16}\right)\right] = 0.905620$

(b) $f(x) = \cos(x^2)$, $f'(x) = -2x\sin(x^2)$, $f''(x) = -2\sin(x^2) - 4x^2\cos(x^2)$. For $0 \le x \le 1$, sin and cos are positive,

so $|f''(x)| = 2\sin(x^2) + 4x^2\cos(x^2) \le 2\cdot 1 + 4\cdot 1\cdot 1 = 6$ since $\sin(x^2) \le 1$ and $\cos(x^2) \le 1$ for all x,

and $x^2 \le 1$ for $0 \le x \le 1$. So for $n = 8$, we take $K = 6$, $a = 0$, and $b = 1$ in Theorem 3, to get

$|E_T| \le 6\cdot 1^3/(12\cdot 8^2) = \frac{1}{128} = 0.0078125$ and $|E_M| \le \frac{1}{256} = 0.00390625$. [A better estimate is obtained by noting

from a graph of f'' that $|f''(x)| \le 4$ for $0 \le x \le 1$.]

(c) Take $K = 6$ [as in part (b)] in Theorem 3. $|E_T| \le \dfrac{K(b-a)^3}{12n^2} \le 0.0001 \;\Leftrightarrow\; \dfrac{6(1-0)^3}{12n^2} \le 10^{-4} \;\Leftrightarrow\;$

$\dfrac{1}{2n^2} \le \dfrac{1}{10^4} \;\Leftrightarrow\; 2n^2 \ge 10^4 \;\Leftrightarrow\; n^2 \ge 5000 \;\Leftrightarrow\; n \ge 71$. Take $n = 71$ for T_n. For E_M, again take $K = 6$ in

Theorem 3 to get $|E_M| \le 10^{-4} \;\Leftrightarrow\; 4n^2 \ge 10^4 \;\Leftrightarrow\; n^2 \ge 2500 \;\Leftrightarrow\; n \ge 50$. Take $n = 50$ for M_n.

21. $f(x) = \sin x$, $\Delta x = \frac{\pi - 0}{10} = \frac{\pi}{10}$

(a) $T_{10} = \frac{\pi}{10\cdot 2}\left[f(0) + 2f\left(\frac{\pi}{10}\right) + 2f\left(\frac{2\pi}{10}\right) + \cdots + 2f\left(\frac{9\pi}{10}\right) + f(\pi)\right] \approx 1.983524$

$M_{10} = \frac{\pi}{10}\left[f\left(\frac{\pi}{20}\right) + f\left(\frac{3\pi}{20}\right) + f\left(\frac{5\pi}{20}\right) + \cdots + f\left(\frac{19\pi}{20}\right)\right] \approx 2.008248$

$S_{10} = \frac{\pi}{10\cdot 3}\left[f(0) + 4f\left(\frac{\pi}{10}\right) + 2f\left(\frac{2\pi}{10}\right) + 4f\left(\frac{3\pi}{10}\right) + \cdots + 4f\left(\frac{9\pi}{10}\right) + f(\pi)\right] \approx 2.000110$

Since $I = \int_0^\pi \sin x\,dx = \left[-\cos x\right]_0^\pi = 1 - (-1) = 2$, $E_T = I - T_{10} \approx 0.016476$, $E_M = I - M_{10} \approx -0.008248$,

and $E_S = I - S_{10} \approx -0.000110$.

(b) $f(x) = \sin x \;\Rightarrow\; \left|f^{(n)}(x)\right| \le 1$, so take $K = 1$ for all error estimates.

$|E_T| \le \dfrac{K(b-a)^3}{12n^2} = \dfrac{1(\pi - 0)^3}{12(10)^2} = \dfrac{\pi^3}{1200} \approx 0.025839.$ $|E_M| \le \dfrac{|E_T|}{2} = \dfrac{\pi^3}{2400} \approx 0.012919.$

$|E_S| \le \dfrac{K(b-a)^5}{180n^4} = \dfrac{1(\pi - 0)^5}{180(10)^4} = \dfrac{\pi^5}{1{,}800{,}000} \approx 0.000170.$

The actual error is about 64% of the error estimate in all three cases.

(c) $|E_T| \le 0.00001 \;\Leftrightarrow\; \dfrac{\pi^3}{12n^2} \le \dfrac{1}{10^5} \;\Leftrightarrow\; n^2 \ge \dfrac{10^5\pi^3}{12} \;\Rightarrow\; n \ge 508.3$. Take $n = 509$ for T_n.

$|E_M| \le 0.00001 \;\Leftrightarrow\; \dfrac{\pi^3}{24n^2} \le \dfrac{1}{10^5} \;\Leftrightarrow\; n^2 \ge \dfrac{10^5\pi^3}{24} \;\Rightarrow\; n \ge 359.4$. Take $n = 360$ for M_n.

$|E_S| \le 0.00001 \;\Leftrightarrow\; \dfrac{\pi^5}{180n^4} \le \dfrac{1}{10^5} \;\Leftrightarrow\; n^4 \ge \dfrac{10^5\pi^5}{180} \;\Rightarrow\; n \ge 20.3$.

Take $n = 22$ for S_n (since n must be even).

23. (a) Using a CAS, we differentiate $f(x) = e^{\cos x}$ twice, and find that

$f''(x) = e^{\cos x}(\sin^2 x - \cos x)$. From the graph, we see that the maximum

value of $|f''(x)|$ occurs at the endpoints of the interval $[0, 2\pi]$.

Since $f''(0) = -e$, we can use $K = e$ or $K = 2.8$.

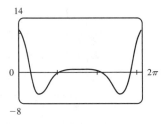

(b) A CAS gives $M_{10} \approx 7.954926518$. (In Maple, use `student[middlesum]`.)

(c) Using Theorem 3 for the Midpoint Rule, with $K = e$, we get $|E_M| \le \dfrac{e(2\pi - 0)^3}{24 \cdot 10^2} \approx 0.280945995$.

With $K = 2.8$, we get $|E_M| \le \dfrac{2.8(2\pi - 0)^3}{24 \cdot 10^2} = 0.289391916$.

(d) A CAS gives $I \approx 7.954926521$.

(e) The actual error is only about 3×10^{-9}, much less than the estimate in part (c).

(f) We use the CAS to differentiate twice more, and then graph

$f^{(4)}(x) = e^{\cos x}(\sin^4 x - 6\sin^2 x \cos x + 3 - 7\sin^2 x + \cos x)$.

From the graph, we see that the maximum value of $\left|f^{(4)}(x)\right|$ occurs at the

endpoints of the interval $[0, 2\pi]$. Since $f^{(4)}(0) = 4e$, we can use $K = 4e$

or $K = 10.9$.

(g) A CAS gives $S_{10} \approx 7.953789422$. (In Maple, use `student[simpson]`.)

(h) Using Theorem 4 with $K = 4e$, we get $|E_S| \le \dfrac{4e(2\pi - 0)^5}{180 \cdot 10^4} \approx 0.059153618$.

With $K = 10.9$, we get $|E_S| \le \dfrac{10.9(2\pi - 0)^5}{180 \cdot 10^4} \approx 0.059299814$.

(i) The actual error is about $7.954926521 - 7.953789422 \approx 0.00114$. This is quite a bit smaller than the estimate in part (h), though the difference is not nearly as great as it was in the case of the Midpoint Rule.

(j) To ensure that $|E_S| \le 0.0001$, we use Theorem 4: $|E_S| \le \dfrac{4e(2\pi)^5}{180 \cdot n^4} \le 0.0001 \quad \Rightarrow \quad \dfrac{4e(2\pi)^5}{180 \cdot 0.0001} \le n^4 \quad \Rightarrow$

$n^4 \ge 5{,}915{,}362 \quad \Leftrightarrow \quad n \ge 49.3$. So we must take $n \ge 50$ to ensure that $|I - S_n| \le 0.0001$.

($K = 10.9$ leads to the same value of n.)

25. $I = \int_0^1 xe^x\,dx = [(x-1)e^x]_0^1$ [parts or Formula 96] $= 0 - (-1) = 1$, $f(x) = xe^x$, $\Delta x = 1/n$

$n = 5$: $\quad L_5 = \frac{1}{5}[f(0) + f(0.2) + f(0.4) + f(0.6) + f(0.8)] \approx 0.742943$

$\qquad\quad R_5 = \frac{1}{5}[f(0.2) + f(0.4) + f(0.6) + f(0.8) + f(1)] \approx 1.286599$

$\qquad\quad T_5 = \frac{1}{5 \cdot 2}[f(0) + 2f(0.2) + 2f(0.4) + 2f(0.6) + 2f(0.8) + f(1)] \approx 1.014771$

$\qquad\quad M_5 = \frac{1}{5}[f(0.1) + f(0.3) + f(0.5) + f(0.7) + f(0.9)] \approx 0.992621$

$\qquad\quad E_L = I - L_5 \approx 1 - 0.742943 = 0.257057$

$\qquad\quad E_R \approx 1 - 1.286599 = -0.286599$

$\qquad\quad E_T \approx 1 - 1.014771 = -0.014771$

$\qquad\quad E_M \approx 1 - 0.992621 = 0.007379$

$n = 10$: $\quad L_{10} = \frac{1}{10}[f(0) + f(0.1) + f(0.2) + \cdots + f(0.9)] \approx 0.867782$

$\qquad R_{10} = \frac{1}{10}[f(0.1) + f(0.2) + \cdots + f(0.9) + f(1)] \approx 1.139610$

$\qquad T_{10} = \frac{1}{10 \cdot 2}\{f(0) + 2[f(0.1) + f(0.2) + \cdots + f(0.9)] + f(1)\} \approx 1.003696$

$\qquad M_{10} = \frac{1}{10}[f(0.05) + f(0.15) + \cdots + f(0.85) + f(0.95)] \approx 0.998152$

$\qquad E_L = I - L_{10} \approx 1 - 0.867782 = 0.132218$

$\qquad E_R \approx 1 - 1.139610 = -0.139610$

$\qquad E_T \approx 1 - 1.003696 = -0.003696$

$\qquad E_M \approx 1 - 0.998152 = 0.001848$

$n = 20$: $\quad L_{20} = \frac{1}{20}[f(0) + f(0.05) + f(0.10) + \cdots + f(0.95)] \approx 0.932967$

$\qquad R_{20} = \frac{1}{20}[f(0.05) + f(0.10) + \cdots + f(0.95) + f(1)] \approx 1.068881$

$\qquad T_{20} = \frac{1}{20 \cdot 2}\{f(0) + 2[f(0.05) + f(0.10) + \cdots + f(0.95)] + f(1)\} \approx 1.000924$

$\qquad M_{20} = \frac{1}{20}[f(0.025) + f(0.075) + f(0.125) + \cdots + f(0.975)] \approx 0.999538$

$\qquad E_L = I - L_{20} \approx 1 - 0.932967 = 0.067033$

$\qquad E_R \approx 1 - 1.068881 = -0.068881$

$\qquad E_T \approx 1 - 1.000924 = -0.000924$

$\qquad E_M \approx 1 - 0.999538 = 0.000462$

n	L_n	R_n	T_n	M_n
5	0.742943	1.286599	1.014771	0.992621
10	0.867782	1.139610	1.003696	0.998152
20	0.932967	1.068881	1.000924	0.999538

n	E_L	E_R	E_T	E_M
5	0.257057	−0.286599	−0.014771	0.007379
10	0.132218	−0.139610	−0.003696	0.001848
20	0.067033	−0.068881	−0.000924	0.000462

Observations:

1. E_L and E_R are always opposite in sign, as are E_T and E_M.

2. As n is doubled, E_L and E_R are decreased by about a factor of 2, and E_T and E_M are decreased by a factor of about 4.

3. The Midpoint approximation is about twice as accurate as the Trapezoidal approximation.

4. All the approximations become more accurate as the value of n increases.

5. The Midpoint and Trapezoidal approximations are much more accurate than the endpoint approximations.

27. $I = \int_0^2 x^4 \, dx = \left[\frac{1}{5}x^5\right]_0^2 = \frac{32}{5} - 0 = 6.4$, $f(x) = x^4$, $\Delta x = \frac{2-0}{n} = \frac{2}{n}$

$n = 6$: $\quad T_6 = \frac{2}{6 \cdot 2}\left\{f(0) + 2\left[f\left(\frac{1}{3}\right) + f\left(\frac{2}{3}\right) + f\left(\frac{3}{3}\right) + f\left(\frac{4}{3}\right) + f\left(\frac{5}{3}\right)\right] + f(2)\right\} \approx 6.695473$

$\qquad M_6 = \frac{2}{6}\left[f\left(\frac{1}{6}\right) + f\left(\frac{3}{6}\right) + f\left(\frac{5}{6}\right) + f\left(\frac{7}{6}\right) + f\left(\frac{9}{6}\right) + f\left(\frac{11}{6}\right)\right] \approx 6.252572$

$\qquad S_6 = \frac{2}{6 \cdot 3}\left[f(0) + 4f\left(\frac{1}{3}\right) + 2f\left(\frac{2}{3}\right) + 4f\left(\frac{3}{3}\right) + 2f\left(\frac{4}{3}\right) + 4f\left(\frac{5}{3}\right) + f(2)\right] \approx 6.403292$

$\qquad E_T = I - T_6 \approx 6.4 - 6.695473 = -0.295473$

$\qquad E_M \approx 6.4 - 6.252572 = 0.147428$

$\qquad E_S \approx 6.4 - 6.403292 = -0.003292$

$n = 12$: $T_{12} = \frac{2}{12 \cdot 2} \left\{ f(0) + 2 \left[f\left(\frac{1}{6}\right) + f\left(\frac{2}{6}\right) + f\left(\frac{3}{6}\right) + \cdots + f\left(\frac{11}{6}\right) \right] + f(2) \right\} \approx 6.474023$

$M_6 = \frac{2}{12} \left[f\left(\frac{1}{12}\right) + f\left(\frac{3}{12}\right) + f\left(\frac{5}{12}\right) + \cdots + f\left(\frac{23}{12}\right) \right] \approx 6.363008$

$S_6 = \frac{2}{12 \cdot 3} \left[f(0) + 4f\left(\frac{1}{6}\right) + 2f\left(\frac{2}{6}\right) + 4f\left(\frac{3}{6}\right) + 2f\left(\frac{4}{6}\right) + \cdots + 4f\left(\frac{11}{6}\right) + f(2) \right] \approx 6.400206$

$E_T = I - T_{12} \approx 6.4 - 6.474023 = -0.074023$

$E_M \approx 6.4 - 6.363008 = 0.036992$

$E_S \approx 6.4 - 6.400206 = -0.000206$

n	T_n	M_n	S_n
6	6.695473	6.252572	6.403292
12	6.474023	6.363008	6.400206

n	E_T	E_M	E_S
6	−0.295473	0.147428	−0.003292
12	−0.074023	0.036992	−0.000206

Observations:

1. E_T and E_M are opposite in sign and decrease by a factor of about 4 as n is doubled.

2. The Simpson's approximation is much more accurate than the Midpoint and Trapezoidal approximations, and E_S seems to decrease by a factor of about 16 as n is doubled.

29. $\Delta x = (b - a)/n = (6 - 0)/6 = 1$

(a) $T_6 = \frac{\Delta x}{2} [f(0) + 2f(1) + 2f(2) + 2f(3) + 2f(4) + 2f(5) + f(6)]$

$\approx \frac{1}{2} [3 + 2(5) + 2(4) + 2(2) + 2(2.8) + 2(4) + 1]$

$= \frac{1}{2} (39.6) = 19.8$

(b) $M_6 = \Delta x [f(0.5) + f(1.5) + f(2.5) + f(3.5) + f(4.5) + f(5.5)]$

$\approx 1[4.5 + 4.7 + 2.6 + 2.2 + 3.4 + 3.2]$

$= 20.6$

(c) $S_6 = \frac{\Delta x}{3} [f(0) + 4f(1) + 2f(2) + 4f(3) + 2f(4) + 4f(5) + f(6)]$

$\approx \frac{1}{3} [3 + 4(5) + 2(4) + 4(2) + 2(2.8) + 4(4) + 1]$

$= \frac{1}{3} (61.6) = 20.5\overline{3}$

31. (a) We are given the function values at the endpoints of 8 intervals of length 0.4, so we'll use the Midpoint Rule with $n = 8/2 = 4$ and $\Delta x = (3.2 - 0)/4 = 0.8$.

$$\int_0^{3.2} f(x)\, dx \approx M_4 = 0.8[f(0.4) + f(1.2) + f(2.0) + f(2.8)] = 0.8[6.5 + 6.4 + 7.6 + 8.8]$$

$$= 0.8(29.3) = 23.44$$

(b) $-4 \leq f''(x) \leq 1 \;\Rightarrow\; |f''(x)| \leq 4$, so use $K = 4$, $a = 0$, $b = 3.2$, and $n = 4$ in Theorem 3.

So $|E_M| \leq \dfrac{4(3.2 - 0)^3}{24(4)^2} = \dfrac{128}{375} = 0.341\overline{3}$.

33. By the Net Change Theorem, the increase in velocity is equal to $\int_0^6 a(t)\,dt$. We use Simpson's Rule with $n = 6$ and $\Delta t = (6-0)/6 = 1$ to estimate this integral:

$$\int_0^6 a(t)\,dt \approx S_6 = \tfrac{1}{3}[a(0) + 4a(1) + 2a(2) + 4a(3) + 2a(4) + 4a(5) + a(6)]$$

$$\approx \tfrac{1}{3}[0 + 4(0.5) + 2(4.1) + 4(9.8) + 2(12.9) + 4(9.5) + 0] = \tfrac{1}{3}(113.2) = 37.7\overline{3}\text{ ft/s}$$

35. By the Net Change Theorem, the energy used is equal to $\int_0^6 P(t)\,dt$. We use Simpson's Rule with $n = 12$ and $\Delta t = \frac{6-0}{12} = \frac{1}{2}$ to estimate this integral:

$$\int_0^6 P(t)\,dt \approx S_{12} = \tfrac{1/2}{3}[P(0) + 4P(0.5) + 2P(1) + 4P(1.5) + 2P(2) + 4P(2.5) + 2P(3)$$

$$+ 4P(3.5) + 2P(4) + 4P(4.5) + 2P(5) + 4P(5.5) + P(6)]$$

$$= \tfrac{1}{6}[1814 + 4(1735) + 2(1686) + 4(1646) + 2(1637) + 4(1609) + 2(1604)$$

$$+ 4(1611) + 2(1621) + 4(1666) + 2(1745) + 4(1886) + 2052]$$

$$= \tfrac{1}{6}(61{,}064) = 10{,}177.\overline{3}\text{ megawatt-hours}$$

37. Let $y = f(x)$ denote the curve. Using cylindrical shells, $V = \int_2^{10} 2\pi x f(x)\,dx = 2\pi \int_2^{10} x f(x)\,dx = 2\pi I_1$. Now use Simpson's Rule to approximate I_1:

$$I_1 \approx S_8 = \tfrac{10-2}{3(8)}[2f(2) + 4 \cdot 3f(3) + 2 \cdot 4f(4) + 4 \cdot 5f(5) + 2 \cdot 6f(6) + 4 \cdot 7f(7) + 2 \cdot 8f(8) + 4 \cdot 9f(9) + 10f(10)]$$

$$\approx \tfrac{1}{3}[2(0) + 12(1.5) + 8(1.9) + 20(2.2) + 12(3.0) + 28(3.8) + 16(4.0) + 36(3.1) + 10(0)]$$

$$= \tfrac{1}{3}(395.2)$$

Thus, $V \approx 2\pi \cdot \tfrac{1}{3}(395.2) \approx 827.7$ or 828 cubic units.

39. Using disks, $V = \int_1^5 \pi(e^{-1/x})^2\,dx = \pi \int_1^5 e^{-2/x}\,dx = \pi I_1$. Now use Simpson's Rule with $f(x) = e^{-2/x}$ to approximate I_1. $I_1 \approx S_8 = \tfrac{5-1}{3(8)}[f(1) + 4f(1.5) + 2f(2) + 4f(2.5) + 2f(3) + 4f(3.5) + 2f(4) + 4f(4.5) + f(5)] \approx \tfrac{1}{6}(11.4566)$

Thus, $V \approx \pi \cdot \tfrac{1}{6}(11.4566) \approx 6.0$ cubic units.

41. $I(\theta) = \dfrac{N^2 \sin^2 k}{k^2}$, where $k = \dfrac{\pi N d \sin \theta}{\lambda}$, $N = 10{,}000$, $d = 10^{-4}$, and $\lambda = 632.8 \times 10^{-9}$. So $I(\theta) = \dfrac{(10^4)^2 \sin^2 k}{k^2}$, where $k = \dfrac{\pi(10^4)(10^{-4})\sin\theta}{632.8 \times 10^{-9}}$. Now $n = 10$ and $\Delta\theta = \dfrac{10^{-6} - (-10^{-6})}{10} = 2 \times 10^{-7}$, so

$$M_{10} = 2 \times 10^{-7}[I(-0.0000009) + I(-0.0000007) + \cdots + I(0.0000009)] \approx 59.4.$$

43. Consider the function f whose graph is shown. The area $\int_0^2 f(x)\,dx$ is close to 2. The Trapezoidal Rule gives

$$T_2 = \tfrac{2-0}{2\cdot 2}[f(0) + 2f(1) + f(2)] = \tfrac{1}{2}[1 + 2\cdot 1 + 1] = 2.$$

The Midpoint Rule gives $M_2 = \tfrac{2-0}{2}[f(0.5) + f(1.5)] = 1[0 + 0] = 0$, so the Trapezoidal Rule is more accurate.

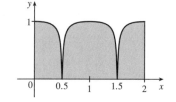

45. Since the Trapezoidal and Midpoint approximations on the interval $[a, b]$ are the sums of the Trapezoidal and Midpoint

approximations on the subintervals $[x_{i-1}, x_i]$, $i = 1, 2, \ldots, n$, we can focus our attention on one such interval. The condition

$f''(x) < 0$ for $a \le x \le b$ means that the graph of f is concave down as in Figure 5. In that figure, T_n is the area of the

trapezoid $AQRD$, $\int_a^b f(x)\, dx$ is the area of the region $AQPRD$, and M_n is the area of the trapezoid $ABCD$, so

$T_n < \int_a^b f(x)\, dx < M_n$. In general, the condition $f'' < 0$ implies that the graph of f on $[a, b]$ lies above the chord joining the

points $(a, f(a))$ and $(b, f(b))$. Thus, $\int_a^b f(x)\, dx > T_n$. Since M_n is the area under a tangent to the graph, and since $f'' < 0$

implies that the tangent lies above the graph, we also have $M_n > \int_a^b f(x)\, dx$. Thus, $T_n < \int_a^b f(x)\, dx < M_n$.

47. $T_n = \frac{1}{2}\Delta x\,[f(x_0) + 2f(x_1) + \cdots + 2f(x_{n-1}) + f(x_n)]$ and

$M_n = \Delta x\,[f(\overline{x}_1) + f(\overline{x}_2) + \cdots + f(\overline{x}_{n-1}) + f(\overline{x}_n)]$, where $\overline{x}_i = \frac{1}{2}(x_{i-1} + x_i)$. Now

$T_{2n} = \frac{1}{2}\left(\frac{1}{2}\Delta x\right)[f(x_0) + 2f(\overline{x}_1) + 2f(x_1) + 2f(\overline{x}_2) + 2f(x_2) + \cdots + 2f(\overline{x}_{n-1}) + 2f(x_{n-1}) + 2f(\overline{x}_n) + f(x_n)]$

so

$\frac{1}{2}(T_n + M_n) = \frac{1}{2}T_n + \frac{1}{2}M_n$

$= \frac{1}{4}\Delta x[f(x_0) + 2f(x_1) + \cdots + 2f(x_{n-1}) + f(x_n)] + \frac{1}{4}\Delta x[2f(\overline{x}_1) + 2f(\overline{x}_2) + \cdots + 2f(\overline{x}_{n-1}) + 2f(\overline{x}_n)]$

$= T_{2n}$

7.8   Improper Integrals

1. (a) Since $\int_1^\infty x^4 e^{-x^4}\, dx$ has an infinite interval of integration, it is an improper integral of Type I.

(b) Since $y = \sec x$ has an infinite discontinuity at $x = \frac{\pi}{2}$, $\int_0^{\pi/2} \sec x\, dx$ is a Type II improper integral.

(c) Since $y = \dfrac{x}{(x-2)(x-3)}$ has an infinite discontinuity at $x = 2$, $\int_0^2 \dfrac{x}{x^2 - 5x + 6}\, dx$ is a Type II improper integral.

(d) Since $\int_{-\infty}^0 \dfrac{1}{x^2 + 5}\, dx$ has an infinite interval of integration, it is an improper integral of Type I.

3. The area under the graph of $y = 1/x^3 = x^{-3}$ between $x = 1$ and $x = t$ is

$A(t) = \int_1^t x^{-3}\, dx = \left[-\frac{1}{2}x^{-2}\right]_1^t = -\frac{1}{2}t^{-2} - \left(-\frac{1}{2}\right) = \frac{1}{2} - 1/(2t^2)$. So the area for $1 \le x \le 10$ is

$A(10) = 0.5 - 0.005 = 0.495$, the area for $1 \le x \le 100$ is $A(100) = 0.5 - 0.00005 = 0.49995$, and the area for

$1 \le x \le 1000$ is $A(1000) = 0.5 - 0.0000005 = 0.4999995$. The total area under the curve for $x \ge 1$ is

$\lim_{t\to\infty} A(t) = \lim_{t\to\infty}\left[\frac{1}{2} - 1/(2t^2)\right] = \frac{1}{2}$.

5. $I = \displaystyle\int_1^\infty \frac{1}{(3x+1)^2}\,dx = \lim_{t\to\infty}\int_1^t \frac{1}{(3x+1)^2}\,dx$. Now

$$\int \frac{1}{(3x+1)^2}\,dx = \frac{1}{3}\int \frac{1}{u^2}\,du \quad [u=3x+1,\,du=3\,dx] \;=\; -\frac{1}{3u}+C = -\frac{1}{3(3x+1)}+C,$$

so $I = \displaystyle\lim_{t\to\infty}\left[-\frac{1}{3(3x+1)}\right]_1^t = \lim_{t\to\infty}\left[-\frac{1}{3(3t+1)}+\frac{1}{12}\right] = 0 + \frac{1}{12} = \frac{1}{12}.$ Convergent

7. $\displaystyle\int_{-\infty}^{-1} \frac{1}{\sqrt{2-w}}\,dw = \lim_{t\to-\infty}\int_t^{-1} \frac{1}{\sqrt{2-w}}\,dw = \lim_{t\to-\infty}\left[-2\sqrt{2-w}\right]_t^{-1} \quad [u=2-w,\,du=-dw]$

$$= \lim_{t\to-\infty}\left[-2\sqrt{3}+2\sqrt{2-t}\right] = \infty. \qquad \text{Divergent}$$

9. $\displaystyle\int_4^\infty e^{-y/2}\,dy = \lim_{t\to\infty}\int_4^t e^{-y/2}\,dy = \lim_{t\to\infty}\left[-2e^{-y/2}\right]_4^t = \lim_{t\to\infty}(-2e^{-t/2}+2e^{-2}) = 0 + 2e^{-2} = 2e^{-2}.$

Convergent

11. $\displaystyle\int_{-\infty}^\infty \frac{x\,dx}{1+x^2} = \int_{-\infty}^0 \frac{x\,dx}{1+x^2} + \int_0^\infty \frac{x\,dx}{1+x^2}$ and

$$\int_{-\infty}^0 \frac{x\,dx}{1+x^2} = \lim_{t\to-\infty}\left[\tfrac{1}{2}\ln(1+x^2)\right]_t^0 = \lim_{t\to-\infty}\left[0 - \tfrac{1}{2}\ln(1+t^2)\right] = -\infty. \qquad \text{Divergent}$$

13. $\displaystyle\int_{-\infty}^\infty xe^{-x^2}\,dx = \int_{-\infty}^0 xe^{-x^2}\,dx + \int_0^\infty xe^{-x^2}\,dx.$

$$\int_{-\infty}^0 xe^{-x^2}\,dx = \lim_{t\to-\infty}\left(-\tfrac{1}{2}\right)\left[e^{-x^2}\right]_t^0 = \lim_{t\to-\infty}\left(-\tfrac{1}{2}\right)\left(1-e^{-t^2}\right) = -\tfrac{1}{2}\cdot 1 = -\tfrac{1}{2},\text{ and}$$

$$\int_0^\infty xe^{-x^2}\,dx = \lim_{t\to\infty}\left(-\tfrac{1}{2}\right)\left[e^{-x^2}\right]_0^t = \lim_{t\to\infty}\left(-\tfrac{1}{2}\right)\left(e^{-t^2}-1\right) = -\tfrac{1}{2}\cdot(-1) = \tfrac{1}{2}.$$

Therefore, $\displaystyle\int_{-\infty}^\infty xe^{-x^2}\,dx = -\tfrac{1}{2}+\tfrac{1}{2} = 0.$ Convergent

15. $\displaystyle\int_{2\pi}^\infty \sin\theta\,d\theta = \lim_{t\to\infty}\int_{2\pi}^t \sin\theta\,d\theta = \lim_{t\to\infty}\left[-\cos\theta\right]_{2\pi}^t = \lim_{t\to\infty}(-\cos t + 1).$ This limit does not exist, so the integral is

divergent.   Divergent

17. $\displaystyle\int_1^\infty \frac{x+1}{x^2+2x}\,dx = \lim_{t\to\infty}\int_1^t \frac{\tfrac{1}{2}(2x+2)}{x^2+2x}\,dx = \tfrac{1}{2}\lim_{t\to\infty}\left[\ln(x^2+2x)\right]_1^t = \tfrac{1}{2}\lim_{t\to\infty}\left[\ln(t^2+2t)-\ln 3\right] = \infty.$

Divergent

19. $\displaystyle\int_0^\infty se^{-5s}\,ds = \lim_{t\to\infty}\int_0^t se^{-5s}\,ds = \lim_{t\to\infty}\left[-\tfrac{1}{5}se^{-5s}-\tfrac{1}{25}e^{-5s}\right] \qquad \begin{bmatrix}\text{by integration by}\\ \text{parts with } u=s\end{bmatrix}$

$$= \lim_{t\to\infty}\left(-\tfrac{1}{5}te^{-5t}-\tfrac{1}{25}e^{-5t}+\tfrac{1}{25}\right) = 0 - 0 + \tfrac{1}{25} \qquad [\text{by l'Hospital's Rule}]$$

$$= \tfrac{1}{25}. \qquad \text{Convergent}$$

21. $\int_1^\infty \frac{\ln x}{x}\,dx = \lim_{t\to\infty}\left[\frac{(\ln x)^2}{2}\right]_1^t \quad \begin{bmatrix}\text{by substitution with}\\ u=\ln x,\, du = dx/x\end{bmatrix} = \lim_{t\to\infty}\frac{(\ln t)^2}{2} = \infty.$ Divergent

23. $\int_{-\infty}^\infty \frac{x^2}{9+x^6}\,dx = \int_{-\infty}^0 \frac{x^2}{9+x^6}\,dx + \int_0^\infty \frac{x^2}{9+x^6}\,dx = 2\int_0^\infty \frac{x^2}{9+x^6}\,dx$ [since the integrand is even].

Now $\int \frac{x^2\,dx}{9+x^6}\ \begin{bmatrix}u=x^3\\ du=3x^2\,dx\end{bmatrix} = \int \frac{\frac{1}{3}\,du}{9+u^2}\ \begin{bmatrix}u=3v\\ du=3\,dv\end{bmatrix} = \int \frac{\frac{1}{3}(3\,dv)}{9+9v^2} = \frac{1}{9}\int \frac{dv}{1+v^2}$

$$= \frac{1}{9}\tan^{-1}v + C = \frac{1}{9}\tan^{-1}\left(\frac{u}{3}\right)+C = \frac{1}{9}\tan^{-1}\left(\frac{x^3}{3}\right)+C,$$

so $2\int_0^\infty \frac{x^2}{9+x^6}\,dx = 2\lim_{t\to\infty}\int_0^t \frac{x^2}{9+x^6}\,dx = 2\lim_{t\to\infty}\left[\frac{1}{9}\tan^{-1}\left(\frac{x^3}{3}\right)\right]_0^t = 2\lim_{t\to\infty}\frac{1}{9}\tan^{-1}\left(\frac{t^3}{3}\right) = \frac{2}{9}\cdot\frac{\pi}{2} = \frac{\pi}{9}.$

Convergent

25. $\int_e^\infty \frac{1}{x(\ln x)^3}\,dx = \lim_{t\to\infty}\int_e^t \frac{1}{x(\ln x)^3}\,dx = \lim_{t\to\infty}\int_1^{\ln t} u^{-3}\,du\ \begin{bmatrix}u=\ln x,\\ du=dx/x\end{bmatrix} = \lim_{t\to\infty}\left[-\frac{1}{2u^2}\right]_1^{\ln t}$

$$= \lim_{t\to\infty}\left[-\frac{1}{2(\ln t)^2}+\frac{1}{2}\right] = 0 + \frac{1}{2} = \frac{1}{2}.$$ Convergent

27. $\int_0^1 \frac{3}{x^5}\,dx = \lim_{t\to 0^+}\int_t^1 3x^{-5}\,dx = \lim_{t\to 0^+}\left[-\frac{3}{4x^4}\right]_t^1 = -\frac{3}{4}\lim_{t\to 0^+}\left(1-\frac{1}{t^4}\right) = \infty.$ Divergent

29. $\int_{-2}^{14}\frac{dx}{\sqrt[4]{x+2}} = \lim_{t\to -2^+}\int_t^{14}(x+2)^{-1/4}\,dx = \lim_{t\to -2^+}\left[\frac{4}{3}(x+2)^{3/4}\right]_t^{14} = \frac{4}{3}\lim_{t\to -2^+}\left[16^{3/4}-(t+2)^{3/4}\right]$

$$= \frac{4}{3}(8-0) = \frac{32}{3}.$$ Convergent

31. $\int_{-2}^3 \frac{dx}{x^4} = \int_{-2}^0 \frac{dx}{x^4}+\int_0^3\frac{dx}{x^4}$, but $\int_{-2}^0\frac{dx}{x^4} = \lim_{t\to 0^-}\left[-\frac{x^{-3}}{3}\right]_{-2}^t = \lim_{t\to 0^-}\left[-\frac{1}{3t^3}-\frac{1}{24}\right] = \infty.$ Divergent

33. There is an infinite discontinuity at $x=1$. $\int_0^{33}(x-1)^{-1/5}\,dx = \int_0^1(x-1)^{-1/5}\,dx + \int_1^{33}(x-1)^{-1/5}\,dx.$ Here

$\int_0^1(x-1)^{-1/5}\,dx = \lim_{t\to 1^-}\int_0^t(x-1)^{-1/5}\,dx = \lim_{t\to 1^-}\left[\frac{5}{4}(x-1)^{4/5}\right]_0^t = \lim_{t\to 1^-}\left[\frac{5}{4}(t-1)^{4/5}-\frac{5}{4}\right] = -\frac{5}{4}$ and

$\int_1^{33}(x-1)^{-1/5}\,dx = \lim_{t\to 1^+}\int_t^{33}(x-1)^{-1/5}\,dx = \lim_{t\to 1^+}\left[\frac{5}{4}(x-1)^{4/5}\right]_t^{33} = \lim_{t\to 1^+}\left[\frac{5}{4}\cdot 16 - \frac{5}{4}(t-1)^{4/5}\right] = 20.$

Thus, $\int_0^{33}(x-1)^{-1/5}\,dx = -\frac{5}{4}+20 = \frac{75}{4}.$ Convergent

35. $I = \int_0^3 \frac{dx}{x^2-6x+5} = \int_0^3\frac{dx}{(x-1)(x-5)} = I_1 + I_2 = \int_0^1\frac{dx}{(x-1)(x-5)}+\int_1^3\frac{dx}{(x-1)(x-5)}.$

Now $\frac{1}{(x-1)(x-5)} = \frac{A}{x-1}+\frac{B}{x-5}\ \Rightarrow\ 1 = A(x-5)+B(x-1).$

Set $x=5$ to get $1=4B$, so $B=\frac{1}{4}$. Set $x=1$ to get $1=-4A$, so $A=-\frac{1}{4}$. Thus

$$I_1 = \lim_{t \to 1^-} \int_0^t \left(\frac{-\frac{1}{4}}{x-1} + \frac{\frac{1}{4}}{x-5} \right) dx = \lim_{t \to 1^-} \left[-\frac{1}{4} \ln|x-1| + \frac{1}{4} \ln|x-5| \right]_0^t$$

$$= \lim_{t \to 1^-} \left[\left(-\tfrac{1}{4} \ln|t-1| + \tfrac{1}{4} \ln|t-5| \right) - \left(-\tfrac{1}{4} \ln|-1| + \tfrac{1}{4} \ln|-5| \right) \right]$$

$$= \infty, \quad \text{since} \ \lim_{t \to 1^-} \left(-\tfrac{1}{4} \ln|t-1| \right) = \infty.$$

Since I_1 is divergent, I is divergent.

37. $\displaystyle\int_{-1}^0 \frac{e^{1/x}}{x^3}\, dx = \lim_{t \to 0^-} \int_{-1}^t \frac{1}{x} e^{1/x} \cdot \frac{1}{x^2}\, dx = \lim_{t \to 0^-} \int_{-1}^{1/t} u e^u \, (-du) \qquad \begin{bmatrix} u = 1/x, \\ du = -dx/x^2 \end{bmatrix}$

$$= \lim_{t \to 0^-} \left[(u-1)e^u \right]_{1/t}^{-1} \quad \begin{bmatrix} \text{use parts} \\ \text{or Formula 96} \end{bmatrix} = \lim_{t \to 0^-} \left[-2e^{-1} - \left(\frac{1}{t} - 1 \right) e^{1/t} \right]$$

$$= -\frac{2}{e} - \lim_{s \to -\infty} (s-1)e^s \quad [s = 1/t] \quad = -\frac{2}{e} - \lim_{s \to -\infty} \frac{s-1}{e^{-s}} \overset{\text{H}}{=} -\frac{2}{e} - \lim_{s \to -\infty} \frac{1}{-e^{-s}}$$

$$= -\frac{2}{e} - 0 = -\frac{2}{e}. \qquad \text{Convergent}$$

39. $I = \int_0^2 z^2 \ln z \, dz = \lim_{t \to 0^+} \int_t^2 z^2 \ln z \, dz = \lim_{t \to 0^+} \left[\frac{z^3}{3^2} (3 \ln z - 1) \right]_t^2 \quad \begin{bmatrix} \text{integrate by parts} \\ \text{or use Formula 101} \end{bmatrix}$

$$= \lim_{t \to 0^+} \left[\tfrac{8}{9}(3 \ln 2 - 1) - \tfrac{1}{9} t^3 (3 \ln t - 1) \right] = \tfrac{8}{3} \ln 2 - \tfrac{8}{9} - \tfrac{1}{9} \lim_{t \to 0^+} \left[t^3 (3 \ln t - 1) \right] = \tfrac{8}{3} \ln 2 - \tfrac{8}{9} - \tfrac{1}{9} L.$$

Now $L = \lim_{t \to 0^+} \left[t^3 (3 \ln t - 1) \right] = \lim_{t \to 0^+} \frac{3 \ln t - 1}{t^{-3}} \overset{\text{H}}{=} \lim_{t \to 0^+} \frac{3/t}{-3/t^4} = \lim_{t \to 0^+} \left(-t^3 \right) = 0.$

Thus, $L = 0$ and $I = \tfrac{8}{3} \ln 2 - \tfrac{8}{9}.$ \quad Convergent

41.

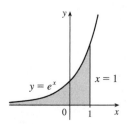

$\text{Area} = \int_{-\infty}^1 e^x \, dx = \lim_{t \to -\infty} \left[e^x \right]_t^1 = e - \lim_{t \to -\infty} e^t = e$

43.

$\text{Area} = \displaystyle\int_{-\infty}^{\infty} \frac{2}{x^2 + 9} \, dx = 2 \cdot 2 \int_0^{\infty} \frac{1}{x^2 + 9} \, dx = 4 \lim_{t \to \infty} \int_0^t \frac{1}{x^2 + 9} \, dx$

$$= 4 \lim_{t \to \infty} \left[\frac{1}{3} \tan^{-1} \frac{x}{3} \right]_0^t = \frac{4}{3} \lim_{t \to \infty} \left[\tan^{-1} \frac{t}{3} - 0 \right] = \frac{4}{3} \cdot \frac{\pi}{2} = \frac{2\pi}{3}$$

45.

$\text{Area} = \int_0^{\pi/2} \sec^2 x \, dx = \lim_{t \to (\pi/2)^-} \int_0^t \sec^2 x \, dx = \lim_{t \to (\pi/2)^-} \left[\tan x \right]_0^t$

$$= \lim_{t \to (\pi/2)^-} (\tan t - 0) = \infty$$

Infinite area

47. (a)

t	$\int_1^t g(x)\,dx$
2	0.447453
5	0.577101
10	0.621306
100	0.668479
1000	0.672957
10,000	0.673407

$$g(x) = \frac{\sin^2 x}{x^2}.$$

It appears that the integral is convergent.

(b) $-1 \le \sin x \le 1 \;\Rightarrow\; 0 \le \sin^2 x \le 1 \;\Rightarrow\; 0 \le \dfrac{\sin^2 x}{x^2} \le \dfrac{1}{x^2}$. Since $\displaystyle\int_1^\infty \frac{1}{x^2}\,dx$ is convergent

[Equation 2 with $p = 2 > 1$], $\displaystyle\int_1^\infty \frac{\sin^2 x}{x^2}\,dx$ is convergent by the Comparison Theorem.

(c)

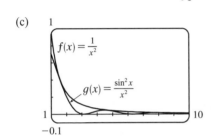

Since $\int_1^\infty f(x)\,dx$ is finite and the area under $g(x)$ is less than the area under $f(x)$ on any interval $[1, t]$, $\int_1^\infty g(x)\,dx$ must be finite; that is, the integral is convergent.

49. For $x > 0$, $\dfrac{x}{x^3 + 1} < \dfrac{x}{x^3} = \dfrac{1}{x^2}$. $\displaystyle\int_1^\infty \frac{1}{x^2}\,dx$ is convergent by Equation 2 with $p = 2 > 1$, so $\displaystyle\int_1^\infty \frac{x}{x^3 + 1}\,dx$ is convergent

by the Comparison Theorem. $\displaystyle\int_0^1 \frac{x}{x^3 + 1}\,dx$ is a constant, so $\displaystyle\int_0^\infty \frac{x}{x^3 + 1}\,dx = \int_0^1 \frac{x}{x^3 + 1}\,dx + \int_1^\infty \frac{x}{x^3 + 1}\,dx$ is also

convergent.

51. For $x > 1$, $f(x) = \dfrac{x + 1}{\sqrt{x^4 - x}} > \dfrac{x + 1}{\sqrt{x^4}} > \dfrac{x}{x^2} = \dfrac{1}{x}$, so $\displaystyle\int_2^\infty f(x)\,dx$ diverges by comparison with $\displaystyle\int_2^\infty \frac{1}{x}\,dx$, which diverges

by Equation 2 with $p = 1 \le 1$. Thus, $\int_1^\infty f(x)\,dx = \int_1^2 f(x)\,dx + \int_2^\infty f(x)\,dx$ also diverges.

53. For $0 < x \le 1$, $\dfrac{\sec^2 x}{x\sqrt{x}} > \dfrac{1}{x^{3/2}}$. Now

$$I = \int_0^1 x^{-3/2}\,dx = \lim_{t \to 0^+} \int_t^1 x^{-3/2}\,dx = \lim_{t \to 0^+} \left[-2x^{-1/2}\right]_t^1 = \lim_{t \to 0^+}\left(-2 + \frac{2}{\sqrt{t}}\right) = \infty, \text{ so } I \text{ is divergent, and by}$$

comparison, $\displaystyle\int_0^1 \frac{\sec^2 x}{x\sqrt{x}}$ is divergent.

55. $\displaystyle\int_0^\infty \frac{dx}{\sqrt{x}\,(1+x)} = \int_0^1 \frac{dx}{\sqrt{x}\,(1+x)} + \int_1^\infty \frac{dx}{\sqrt{x}\,(1+x)} = \lim_{t\to 0^+}\int_t^1 \frac{dx}{\sqrt{x}\,(1+x)} + \lim_{t\to\infty}\int_1^t \frac{dx}{\sqrt{x}\,(1+x)}$. Now

$\displaystyle\int \frac{dx}{\sqrt{x}\,(1+x)} = \int \frac{2u\,du}{u(1+u^2)}\quad \begin{bmatrix} u = \sqrt{x},\, x = u^2, \\ dx = 2u\,du \end{bmatrix} = 2\int \frac{du}{1+u^2} = 2\tan^{-1}u + C = 2\tan^{-1}\sqrt{x} + C$, so

$\displaystyle\int_0^\infty \frac{dx}{\sqrt{x}\,(1+x)} = \lim_{t\to 0^+}\left[2\tan^{-1}\sqrt{x}\right]_t^1 + \lim_{t\to\infty}\left[2\tan^{-1}\sqrt{x}\right]_1^t$

$\displaystyle\qquad\qquad = \lim_{t\to 0^+}\left[2\left(\tfrac{\pi}{4}\right) - 2\tan^{-1}\sqrt{t}\right] + \lim_{t\to\infty}\left[2\tan^{-1}\sqrt{t} - 2\left(\tfrac{\pi}{4}\right)\right] = \tfrac{\pi}{2} - 0 + 2\left(\tfrac{\pi}{2}\right) - \tfrac{\pi}{2} = \pi.$

57. If $p = 1$, then $\displaystyle\int_0^1 \frac{dx}{x^p} = \lim_{t\to 0^+}\int_t^1 \frac{dx}{x} = \lim_{t\to 0^+}\left[\ln x\right]_t^1 = \infty.$ Divergent.

If $p \neq 1$, then $\displaystyle\int_0^1 \frac{dx}{x^p} = \lim_{t\to 0^+}\int_t^1 \frac{dx}{x^p}$ [note that the integral is not improper if $p < 0$]

$\displaystyle\qquad\qquad = \lim_{t\to 0^+}\left[\frac{x^{-p+1}}{-p+1}\right]_t^1 = \lim_{t\to 0^+}\frac{1}{1-p}\left[1 - \frac{1}{t^{p-1}}\right]$

If $p > 1$, then $p - 1 > 0$, so $\dfrac{1}{t^{p-1}} \to \infty$ as $t \to 0^+$, and the integral diverges.

If $p < 1$, then $p - 1 < 0$, so $\dfrac{1}{t^{p-1}} \to 0$ as $t \to 0^+$ and $\displaystyle\int_0^1 \frac{dx}{x^p} = \frac{1}{1-p}\left[\lim_{t\to 0^+}\left(1 - t^{1-p}\right)\right] = \frac{1}{1-p}.$

Thus, the integral converges if and only if $p < 1$, and in that case its value is $\dfrac{1}{1-p}$.

59. First suppose $p = -1$. Then

$\displaystyle\int_0^1 x^p \ln x\,dx = \int_0^1 \frac{\ln x}{x}\,dx = \lim_{t\to 0^+}\int_t^1 \frac{\ln x}{x}\,dx = \lim_{t\to 0^+}\left[\tfrac{1}{2}(\ln x)^2\right]_t^1 = -\tfrac{1}{2}\lim_{t\to 0^+}(\ln t)^2 = -\infty$, so the

integral diverges. Now suppose $p \neq -1$. Then integration by parts gives

$\displaystyle\int x^p \ln x\,dx = \frac{x^{p+1}}{p+1}\ln x - \int \frac{x^p}{p+1}\,dx = \frac{x^{p+1}}{p+1}\ln x - \frac{x^{p+1}}{(p+1)^2} + C.$ If $p < -1$, then $p + 1 < 0$, so

$\displaystyle\int_0^1 x^p \ln x\,dx = \lim_{t\to 0^+}\left[\frac{x^{p+1}}{p+1}\ln x - \frac{x^{p+1}}{(p+1)^2}\right]_t^1 = \frac{-1}{(p+1)^2} - \left(\frac{1}{p+1}\right)\lim_{t\to 0^+}\left[t^{p+1}\left(\ln t - \frac{1}{p+1}\right)\right] = \infty.$

If $p > -1$, then $p + 1 > 0$ and

$\displaystyle\int_0^1 x^p \ln x\,dx = \frac{-1}{(p+1)^2} - \left(\frac{1}{p+1}\right)\lim_{t\to 0^+}\frac{\ln t - 1/(p+1)}{t^{-(p+1)}} \overset{\mathrm{H}}{=} \frac{-1}{(p+1)^2} - \left(\frac{1}{p+1}\right)\lim_{t\to 0^+}\frac{1/t}{-(p+1)t^{-(p+2)}}$

$\displaystyle\qquad\qquad = \frac{-1}{(p+1)^2} + \frac{1}{(p+1)^2}\lim_{t\to 0^+}t^{p+1} = \frac{-1}{(p+1)^2}$

Thus, the integral converges to $-\dfrac{1}{(p+1)^2}$ if $p > -1$ and diverges otherwise.

61. (a) $I = \int_{-\infty}^{\infty} x \, dx = \int_{-\infty}^{0} x \, dx + \int_{0}^{\infty} x \, dx$, and $\int_{0}^{\infty} x \, dx = \lim\limits_{t \to \infty} \int_{0}^{t} x \, dx = \lim\limits_{t \to \infty} \left[\frac{1}{2}x^2\right]_{0}^{t} = \lim\limits_{t \to \infty} \left[\frac{1}{2}t^2 - 0\right] = \infty$,

so I is divergent.

(b) $\int_{-t}^{t} x \, dx = \left[\frac{1}{2}x^2\right]_{-t}^{t} = \frac{1}{2}t^2 - \frac{1}{2}t^2 = 0$, so $\lim\limits_{t \to \infty} \int_{-t}^{t} x \, dx = 0$. Therefore, $\int_{-\infty}^{\infty} x \, dx \neq \lim\limits_{t \to \infty} \int_{-t}^{t} x \, dx$.

63. Volume $= \int_{1}^{\infty} \pi \left(\frac{1}{x}\right)^2 dx = \pi \lim\limits_{t \to \infty} \int_{1}^{t} \frac{dx}{x^2} = \pi \lim\limits_{t \to \infty} \left[-\frac{1}{x}\right]_{1}^{t} = \pi \lim\limits_{t \to \infty} \left(1 - \frac{1}{t}\right) = \pi < \infty$.

65. Work $= \int_{R}^{\infty} F \, dr = \lim\limits_{t \to \infty} \int_{R}^{t} \frac{GmM}{r^2} \, dr = \lim\limits_{t \to \infty} GmM \left(\frac{1}{R} - \frac{1}{t}\right) = \frac{GmM}{R}$. The initial kinetic energy provides the work,

so $\frac{1}{2}mv_0^2 = \frac{GmM}{R} \quad \Rightarrow \quad v_0 = \sqrt{\frac{2GM}{R}}$.

67. We would expect a small percentage of bulbs to burn out in the first few hundred hours, most of the bulbs to burn out after close to 700 hours, and a few overachievers to burn on and on.

(a)
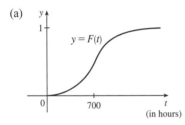

(b) $r(t) = F'(t)$ is the rate at which the fraction $F(t)$ of burnt-out bulbs increases as t increases. This could be interpreted as a fractional burnout rate.

(c) $\int_{0}^{\infty} r(t) \, dt = \lim\limits_{x \to \infty} F(x) = 1$, since all of the bulbs will eventually burn out.

69. $I = \int_{a}^{\infty} \frac{1}{x^2 + 1} \, dx = \lim\limits_{t \to \infty} \int_{a}^{t} \frac{1}{x^2 + 1} \, dx = \lim\limits_{t \to \infty} \left[\tan^{-1} x\right]_{a}^{t} = \lim\limits_{t \to \infty} \left(\tan^{-1} t - \tan^{-1} a\right) = \frac{\pi}{2} - \tan^{-1} a$.

$I < 0.001 \quad \Rightarrow \quad \frac{\pi}{2} - \tan^{-1} a < 0.001 \quad \Rightarrow \quad \tan^{-1} a > \frac{\pi}{2} - 0.001 \quad \Rightarrow \quad a > \tan\left(\frac{\pi}{2} - 0.001\right) \approx 1000$.

71. (a) $F(s) = \int_{0}^{\infty} f(t)e^{-st} \, dt = \int_{0}^{\infty} e^{-st} \, dt = \lim\limits_{n \to \infty} \left[-\frac{e^{-st}}{s}\right]_{0}^{n} = \lim\limits_{n \to \infty} \left(\frac{e^{-sn}}{-s} + \frac{1}{s}\right)$. This converges to $\frac{1}{s}$ only if $s > 0$.

Therefore $F(s) = \frac{1}{s}$ with domain $\{s \mid s > 0\}$.

(b) $F(s) = \int_{0}^{\infty} f(t)e^{-st} \, dt = \int_{0}^{\infty} e^{t}e^{-st} \, dt = \lim\limits_{n \to \infty} \int_{0}^{n} e^{t(1-s)} \, dt = \lim\limits_{n \to \infty} \left[\frac{1}{1-s}e^{t(1-s)}\right]_{0}^{n}$

$= \lim\limits_{n \to \infty} \left(\frac{e^{(1-s)n}}{1-s} - \frac{1}{1-s}\right)$

This converges only if $1 - s < 0 \quad \Rightarrow \quad s > 1$, in which case $F(s) = \frac{1}{s-1}$ with domain $\{s \mid s > 1\}$.

(c) $F(s) = \int_0^\infty f(t)e^{-st}\,dt = \lim\limits_{n\to\infty}\int_0^n te^{-st}\,dt$. Use integration by parts: let $u = t$, $dv = e^{-st}\,dt \Rightarrow du = dt$,

$v = -\dfrac{e^{-st}}{s}$. Then $F(s) = \lim\limits_{n\to\infty}\left[-\dfrac{t}{s}e^{-st} - \dfrac{1}{s^2}e^{-st}\right]_0^n = \lim\limits_{n\to\infty}\left(\dfrac{-n}{se^{sn}} - \dfrac{1}{s^2 e^{sn}} + 0 + \dfrac{1}{s^2}\right) = \dfrac{1}{s^2}$ only if $s > 0$.

Therefore, $F(s) = \dfrac{1}{s^2}$ and the domain of F is $\{s \mid s > 0\}$.

73. $G(s) = \int_0^\infty f'(t)e^{-st}\,dt$. Integrate by parts with $u = e^{-st}$, $dv = f'(t)\,dt \Rightarrow du = -se^{-st}$, $v = f(t)$:

$$G(s) = \lim\limits_{n\to\infty}\left[f(t)e^{-st}\right]_0^n + s\int_0^\infty f(t)e^{-st}\,dt = \lim\limits_{n\to\infty} f(n)e^{-sn} - f(0) + sF(s)$$

But $0 \le f(t) \le Me^{at} \Rightarrow 0 \le f(t)e^{-st} \le Me^{at}e^{-st}$ and $\lim\limits_{t\to\infty} Me^{t(a-s)} = 0$ for $s > a$. So by the Squeeze Theorem,

$\lim\limits_{t\to\infty} f(t)e^{-st} = 0$ for $s > a \Rightarrow G(s) = 0 - f(0) + sF(s) = sF(s) - f(0)$ for $s > a$.

75. We use integration by parts: let $u = x$, $dv = xe^{-x^2}\,dx \Rightarrow du = dx$, $v = -\tfrac{1}{2}e^{-x^2}$. So

$$\int_0^\infty x^2 e^{-x^2}\,dx = \lim\limits_{t\to\infty}\left[-\dfrac{1}{2}xe^{-x^2}\right]_0^t + \dfrac{1}{2}\int_0^\infty e^{-x^2}\,dx = \lim\limits_{t\to\infty}\left[-\dfrac{t}{2e^{t^2}}\right] + \dfrac{1}{2}\int_0^\infty e^{-x^2}\,dx = \dfrac{1}{2}\int_0^\infty e^{-x^2}\,dx$$

(The limit is 0 by l'Hospital's Rule.)

77. For the first part of the integral, let $x = 2\tan\theta \Rightarrow dx = 2\sec^2\theta\,d\theta$.

$$\int \dfrac{1}{\sqrt{x^2+4}}\,dx = \int \dfrac{2\sec^2\theta}{2\sec\theta}\,d\theta = \int \sec\theta\,d\theta = \ln|\sec\theta + \tan\theta|.$$

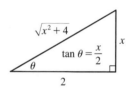

From the figure, $\tan\theta = \dfrac{x}{2}$, and $\sec\theta = \dfrac{\sqrt{x^2+4}}{2}$. So

$$I = \int_0^\infty\left(\dfrac{1}{\sqrt{x^2+4}} - \dfrac{C}{x+2}\right)dx = \lim\limits_{t\to\infty}\left[\ln\left|\dfrac{\sqrt{x^2+4}}{2} + \dfrac{x}{2}\right| - C\ln|x+2|\right]_0^t$$

$$= \lim\limits_{t\to\infty}\left[\ln\dfrac{\sqrt{t^2+4}+t}{2} - C\ln(t+2) - (\ln 1 - C\ln 2)\right]$$

$$= \lim\limits_{t\to\infty}\left[\ln\left(\dfrac{\sqrt{t^2+4}+t}{2\,(t+2)^C}\right) + \ln 2^C\right] = \ln\left(\lim\limits_{t\to\infty}\dfrac{t+\sqrt{t^2+4}}{(t+2)^C}\right) + \ln 2^{C-1}$$

Now $L = \lim\limits_{t\to\infty}\dfrac{t+\sqrt{t^2+4}}{(t+2)^C} \overset{\text{H}}{=} \lim\limits_{t\to\infty}\dfrac{1+t/\sqrt{t^2+4}}{C\,(t+2)^{C-1}} = \dfrac{2}{C\lim\limits_{t\to\infty}(t+2)^{C-1}}$.

If $C < 1$, $L = \infty$ and I diverges.

If $C = 1$, $L = 2$ and I converges to $\ln 2 + \ln 2^0 = \ln 2$.

If $C > 1$, $L = 0$ and I diverges to $-\infty$.

79. No, $I = \int_0^\infty f(x)\,dx$ must be *divergent*. Since $\lim\limits_{x\to\infty} f(x) = 1$, there must exist an N such that if $x \ge N$, then $f(x) \ge \tfrac{1}{2}$.

Thus, $I = I_1 + I_2 = \int_0^N f(x)\,dx + \int_N^\infty f(x)\,dx$, where I_1 is an ordinary definite integral that has a finite value, and I_2 is improper and diverges by comparison with the divergent integral $\int_N^\infty \tfrac{1}{2}\,dx$.

7 Review

1. See Formula 7.1.1 or 7.1.2. We try to choose $u = f(x)$ to be a function that becomes simpler when differentiated (or at least not more complicated) as long as $dv = g'(x)\,dx$ can be readily integrated to give v.

2. See the Strategy for Evaluating $\int \sin^m x \cos^n x\,dx$ on page 462.

3. If $\sqrt{a^2 - x^2}$ occurs, try $x = a \sin \theta$; if $\sqrt{a^2 + x^2}$ occurs, try $x = a \tan \theta$, and if $\sqrt{x^2 - a^2}$ occurs, try $x = a \sec \theta$. See the Table of Trigonometric Substitutions on page 467.

4. See Equation 2 and Expressions 7, 9, and 11 in Section 7.4.

5. See the Midpoint Rule, the Trapezoidal Rule, and Simpson's Rule, as well as their associated error bounds, all in Section 7.7. We would expect the best estimate to be given by Simpson's Rule.

6. See Definitions 1(a), (b), and (c) in Section 7.8.

7. See Definitions 3(b), (a), and (c) in Section 7.8.

8. See the Comparison Theorem after Example 8 in Section 7.8.

1. False. Since the numerator has a higher degree than the denominator, $\dfrac{x(x^2 + 4)}{x^2 - 4} = x + \dfrac{8x}{x^2 - 4} = x + \dfrac{A}{x + 2} + \dfrac{B}{x - 2}$.

3. False. It can be put in the form $\dfrac{A}{x} + \dfrac{B}{x^2} + \dfrac{C}{x - 4}$.

5. False. This is an improper integral, since the denominator vanishes at $x = 1$.

$$\int_0^4 \frac{x}{x^2 - 1}\,dx = \int_0^1 \frac{x}{x^2 - 1}\,dx + \int_1^4 \frac{x}{x^2 - 1}\,dx \text{ and}$$

$$\int_0^1 \frac{x}{x^2 - 1}\,dx = \lim_{t \to 1^-} \int_0^t \frac{x}{x^2 - 1}\,dx = \lim_{t \to 1^-} \left[\tfrac{1}{2} \ln|x^2 - 1| \right]_0^t = \lim_{t \to 1^-} \tfrac{1}{2} \ln|t^2 - 1| = \infty$$

So the integral diverges.

7. False. See Exercise 61 in Section 7.8.

9. (a) True. See the end of Section 7.5.

    (b) False. Examples include the functions $f(x) = e^{x^2}$, $g(x) = \sin(x^2)$, and $h(x) = \dfrac{\sin x}{x}$.

11. False. If $f(x) = 1/x$, then f is continuous and decreasing on $[1, \infty)$ with $\lim\limits_{x \to \infty} f(x) = 0$, but $\int_1^\infty f(x)\,dx$ is divergent.

13. False. Take $f(x) = 1$ for all x and $g(x) = -1$ for all x. Then $\int_a^\infty f(x)\,dx = \infty$ [divergent]

    and $\int_a^\infty g(x)\,dx = -\infty$ [divergent], but $\int_a^\infty [f(x) + g(x)]\,dx = 0$ [convergent].

EXERCISES

1. $\int_0^5 \frac{x}{x+10}\,dx = \int_0^5 \left(1 - \frac{10}{x+10}\right)dx = \left[x - 10\ln(x+10)\right]_0^5 = 5 - 10\ln 15 + 10\ln 10$

$= 5 + 10\ln\frac{10}{15} = 5 + 10\ln\frac{2}{3}$

3. $\int_0^{\pi/2} \frac{\cos\theta}{1+\sin\theta}\,d\theta = \left[\ln(1+\sin\theta)\right]_0^{\pi/2} = \ln 2 - \ln 1 = \ln 2$

5. $\int_0^{\pi/2}\sin^3\theta\cos^2\theta\,d\theta = \int_0^{\pi/2}(1-\cos^2\theta)\cos^2\theta\sin\theta\,d\theta = \int_1^0(1-u^2)u^2\,(-du)$ $\begin{bmatrix} u = \cos\theta, \\ du = -\sin\theta\,d\theta \end{bmatrix}$

$= \int_0^1(u^2 - u^4)\,du = \left[\frac{1}{3}u^3 - \frac{1}{5}u^5\right]_0^1 = \left(\frac{1}{3} - \frac{1}{5}\right) - 0 = \frac{2}{15}$

7. Let $u = \ln t$, $du = dt/t$. Then $\int \frac{\sin(\ln t)}{t}\,dt = \int \sin u\,du = -\cos u + C = -\cos(\ln t) + C$.

9. $\int_1^4 x^{3/2}\ln x\,dx$ $\begin{bmatrix} u = \ln x, & dv = x^{3/2}\,dx, \\ du = dx/x & v = \frac{2}{5}x^{5/2} \end{bmatrix}$ $= \frac{2}{5}\left[x^{5/2}\ln x\right]_1^4 - \frac{2}{5}\int_1^4 x^{3/2}\,dx = \frac{2}{5}(32\ln 4 - \ln 1) - \frac{2}{5}\left[\frac{2}{5}x^{5/2}\right]_1^4$

$= \frac{2}{5}(64\ln 2) - \frac{4}{25}(32-1) = \frac{128}{5}\ln 2 - \frac{124}{25}$ $\left[\text{or } \frac{64}{5}\ln 4 - \frac{124}{25}\right]$

11. Let $x = \sec\theta$. Then

$\int_1^2 \frac{\sqrt{x^2-1}}{x}\,dx = \int_0^{\pi/3}\frac{\tan\theta}{\sec\theta}\sec\theta\tan\theta\,d\theta = \int_0^{\pi/3}\tan^2\theta\,d\theta = \int_0^{\pi/3}(\sec^2\theta - 1)\,d\theta = \left[\tan\theta - \theta\right]_0^{\pi/3} = \sqrt{3} - \frac{\pi}{3}$.

13. Let $t = \sqrt[3]{x}$. Then $t^3 = x$ and $3t^2\,dt = dx$, so $\int e^{\sqrt[3]{x}}\,dx = \int e^t \cdot 3t^2\,dt = 3I$. To evaluate I, let $u = t^2$,

$dv = e^t\,dt \Rightarrow du = 2t\,dt$, $v = e^t$, so $I = \int t^2 e^t\,dt = t^2 e^t - \int 2te^t\,dt$. Now let $U = t$, $dV = e^t\,dt \Rightarrow$

$dU = dt$, $V = e^t$. Thus, $I = t^2 e^t - 2\left[te^t - \int e^t\,dt\right] = t^2 e^t - 2te^t + 2e^t + C_1$, and hence

$3I = 3e^t(t^2 - 2t + 2) + C = 3e^{\sqrt[3]{x}}(x^{2/3} - 2x^{1/3} + 2) + C$.

15. $\frac{x-1}{x^2+2x} = \frac{x-1}{x(x+2)} = \frac{A}{x} + \frac{B}{x+2} \Rightarrow x - 1 = A(x+2) + Bx$. Set $x = -2$ to get $-3 = -2B$, so $B = \frac{3}{2}$. Set $x = 0$

to get $-1 = 2A$, so $A = -\frac{1}{2}$. Thus, $\int \frac{x-1}{x^2+2x}\,dx = \int\left(\frac{-\frac{1}{2}}{x} + \frac{\frac{3}{2}}{x+2}\right)dx = -\frac{1}{2}\ln|x| + \frac{3}{2}\ln|x+2| + C$.

17. Integrate by parts with $u = x$, $dv = \sec x\tan x\,dx \Rightarrow du = dx$, $v = \sec x$:

$\int x\sec x\tan x\,dx = x\sec x - \int \sec x\,dx \overset{14}{=} x\sec x - \ln|\sec x + \tan x| + C$.

19. $\int \frac{x+1}{9x^2+6x+5}\,dx = \int \frac{x+1}{(9x^2+6x+1)+4}\,dx = \int \frac{x+1}{(3x+1)^2+4}\,dx$ $\begin{bmatrix} u = 3x+1, \\ du = 3\,dx \end{bmatrix}$

$= \int \frac{\left[\frac{1}{3}(u-1)\right]+1}{u^2+4}\left(\frac{1}{3}\,du\right) = \frac{1}{3}\cdot\frac{1}{3}\int\frac{(u-1)+3}{u^2+4}\,du$

$= \frac{1}{9}\int\frac{u}{u^2+4}\,du + \frac{1}{9}\int\frac{2}{u^2+2^2}\,du = \frac{1}{9}\cdot\frac{1}{2}\ln(u^2+4) + \frac{2}{9}\cdot\frac{1}{2}\tan^{-1}\left(\frac{1}{2}u\right) + C$

$= \frac{1}{18}\ln(9x^2+6x+5) + \frac{1}{9}\tan^{-1}\left[\frac{1}{2}(3x+1)\right] + C$

21. $\displaystyle\int \frac{dx}{\sqrt{x^2 - 4x}} = \int \frac{dx}{\sqrt{(x^2 - 4x + 4) - 4}} = \int \frac{dx}{\sqrt{(x - 2)^2 - 2^2}}$

$$= \int \frac{2 \sec\theta \tan\theta \, d\theta}{2 \tan\theta} \qquad \begin{bmatrix} x - 2 = 2\sec\theta, \\ dx = 2\sec\theta \tan\theta \, d\theta \end{bmatrix}$$

$$= \int \sec\theta \, d\theta = \ln|\sec\theta + \tan\theta| + C_1$$

$$= \ln\left| \frac{x - 2}{2} + \frac{\sqrt{x^2 - 4x}}{2} \right| + C_1$$

$$= \ln\left| x - 2 + \sqrt{x^2 - 4x} \right| + C, \text{ where } C = C_1 - \ln 2$$

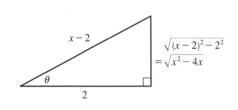

23. Let $x = \tan\theta$, so that $dx = \sec^2\theta \, d\theta$. Then

$$\int \frac{dx}{x\sqrt{x^2 + 1}} = \int \frac{\sec^2\theta \, d\theta}{\tan\theta \sec\theta} = \int \frac{\sec\theta}{\tan\theta} \, d\theta$$

$$= \int \csc\theta \, d\theta = \ln|\csc\theta - \cot\theta| + C$$

$$= \ln\left| \frac{\sqrt{x^2 + 1}}{x} - \frac{1}{x} \right| + C = \ln\left| \frac{\sqrt{x^2 + 1} - 1}{x} \right| + C$$

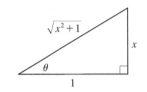

25. $\displaystyle\frac{3x^3 - x^2 + 6x - 4}{(x^2 + 1)(x^2 + 2)} = \frac{Ax + B}{x^2 + 1} + \frac{Cx + D}{x^2 + 2} \;\Rightarrow\; 3x^3 - x^2 + 6x - 4 = (Ax + B)(x^2 + 2) + (Cx + D)(x^2 + 1).$

Equating the coefficients gives $A + C = 3$, $B + D = -1$, $2A + C = 6$, and $2B + D = -4 \;\Rightarrow\;$

$A = 3$, $C = 0$, $B = -3$, and $D = 2$. Now

$$\int \frac{3x^3 - x^2 + 6x - 4}{(x^2 + 1)(x^2 + 2)} \, dx = 3\int \frac{x - 1}{x^2 + 1} \, dx + 2\int \frac{dx}{x^2 + 2} = \frac{3}{2}\ln(x^2 + 1) - 3\tan^{-1}x + \sqrt{2}\tan^{-1}\left(\frac{x}{\sqrt{2}}\right) + C.$$

27. $\displaystyle\int_0^{\pi/2} \cos^3 x \sin 2x \, dx = \int_0^{\pi/2} \cos^3 x \, (2\sin x \cos x) \, dx = \int_0^{\pi/2} 2\cos^4 x \sin x \, dx = \left[-\frac{2}{5}\cos^5 x\right]_0^{\pi/2} = \frac{2}{5}$

29. The product of an odd function and an even function is an odd function, so $f(x) = x^5 \sec x$ is an odd function.

By Theorem 5.5.7(b), $\int_{-1}^{1} x^5 \sec x \, dx = 0$.

31. Let $u = \sqrt{e^x - 1}$. Then $u^2 = e^x - 1$ and $2u \, du = e^x \, dx$. Also, $e^x + 8 = u^2 + 9$. Thus,

$$\int_0^{\ln 10} \frac{e^x \sqrt{e^x - 1}}{e^x + 8} \, dx = \int_0^3 \frac{u \cdot 2u \, du}{u^2 + 9} = 2\int_0^3 \frac{u^2}{u^2 + 9} \, du = 2\int_0^3 \left(1 - \frac{9}{u^2 + 9}\right) du$$

$$= 2\left[u - \frac{9}{3}\tan^{-1}\left(\frac{u}{3}\right)\right]_0^3 = 2\big[(3 - 3\tan^{-1} 1) - 0\big] = 2\left(3 - 3 \cdot \frac{\pi}{4}\right) = 6 - \frac{3\pi}{2}$$

33. Let $x = 2\sin\theta \;\Rightarrow\; (4 - x^2)^{3/2} = (2\cos\theta)^3$, $dx = 2\cos\theta \, d\theta$, so

$$\int \frac{x^2}{(4 - x^2)^{3/2}} \, dx = \int \frac{4\sin^2\theta}{8\cos^3\theta} 2\cos\theta \, d\theta = \int \tan^2\theta \, d\theta = \int (\sec^2\theta - 1) \, d\theta$$

$$= \tan\theta - \theta + C = \frac{x}{\sqrt{4 - x^2}} - \sin^{-1}\left(\frac{x}{2}\right) + C$$

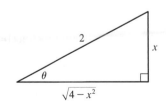

35. $\displaystyle\int \frac{1}{\sqrt{x + x^{3/2}}}\, dx = \int \frac{dx}{\sqrt{x\,(1 + \sqrt{x}\,)}} = \int \frac{dx}{\sqrt{x}\sqrt{1 + \sqrt{x}}}$ $\left[\begin{array}{l} u = 1 + \sqrt{x}, \\ du = \dfrac{dx}{2\sqrt{x}} \end{array}\right] = \int \frac{2\,du}{\sqrt{u}} = \int 2u^{-1/2}\, du$

$$= 4\sqrt{u} + C = 4\sqrt{1 + \sqrt{x}} + C$$

37. $\int (\cos x + \sin x)^2 \cos 2x\, dx = \int \left(\cos^2 x + 2\sin x \cos x + \sin^2 x\right)\cos 2x\, dx = \int (1 + \sin 2x)\cos 2x\, dx$

$$= \int \cos 2x\, dx + \tfrac{1}{2}\int \sin 4x\, dx = \tfrac{1}{2}\sin 2x - \tfrac{1}{8}\cos 4x + C$$

Or: $\int (\cos x + \sin x)^2 \cos 2x\, dx = \int (\cos x + \sin x)^2(\cos^2 x - \sin^2 x)\, dx$

$$= \int (\cos x + \sin x)^3(\cos x - \sin x)\, dx = \tfrac{1}{4}(\cos x + \sin x)^4 + C_1$$

39. We'll integrate $I = \displaystyle\int \frac{xe^{2x}}{(1 + 2x)^2}\, dx$ by parts with $u = xe^{2x}$ and $dv = \dfrac{dx}{(1 + 2x)^2}$. Then $du = (x \cdot 2e^{2x} + e^{2x} \cdot 1)\, dx$

and $v = -\dfrac{1}{2} \cdot \dfrac{1}{1 + 2x}$, so

$$I = -\frac{1}{2} \cdot \frac{xe^{2x}}{1 + 2x} - \int \left[-\frac{1}{2} \cdot \frac{e^{2x}(2x + 1)}{1 + 2x}\right] dx = -\frac{xe^{2x}}{4x + 2} + \frac{1}{2} \cdot \frac{1}{2}e^{2x} + C = e^{2x}\left(\frac{1}{4} - \frac{x}{4x + 2}\right) + C$$

Thus, $\displaystyle\int_0^{1/2} \frac{xe^{2x}}{(1 + 2x)^2}\, dx = \left[e^{2x}\left(\frac{1}{4} - \frac{x}{4x + 2}\right)\right]_0^{1/2} = e\left(\frac{1}{4} - \frac{1}{8}\right) - 1\left(\frac{1}{4} - 0\right) = \frac{1}{8}e - \frac{1}{4}.$

41. $\displaystyle\int_1^\infty \frac{1}{(2x + 1)^3}\, dx = \lim_{t\to\infty} \int_1^t \frac{1}{(2x + 1)^3}\, dx = \lim_{t\to\infty} \int_1^t \tfrac{1}{2}(2x + 1)^{-3}\, 2\, dx = \lim_{t\to\infty}\left[-\frac{1}{4(2x + 1)^2}\right]_1^t$

$$= -\frac{1}{4}\lim_{t\to\infty}\left[\frac{1}{(2t + 1)^2} - \frac{1}{9}\right] = -\frac{1}{4}\left(0 - \frac{1}{9}\right) = \frac{1}{36}$$

43. $\displaystyle\int \frac{dx}{x\ln x}$ $\left[\begin{array}{l} u = \ln x, \\ du = dx/x \end{array}\right] = \int \frac{du}{u} = \ln|u| + C = \ln|\ln x| + C,$ so

$$\int_2^\infty \frac{dx}{x\ln x} = \lim_{t\to\infty} \int_2^t \frac{dx}{x\ln x} = \lim_{t\to\infty}\Big[\ln|\ln x|\Big]_2^t = \lim_{t\to\infty}[\ln(\ln t) - \ln(\ln 2)] = \infty, \text{ so the integral is divergent.}$$

45. $\displaystyle\int_0^4 \frac{\ln x}{\sqrt{x}}\, dx = \lim_{t\to 0^+} \int_t^4 \frac{\ln x}{\sqrt{x}}\, dx \overset{*}{=} \lim_{t\to 0^+}\left[2\sqrt{x}\ln x - 4\sqrt{x}\right]_t^4$

$$= \lim_{t\to 0^+}\left[(2 \cdot 2\ln 4 - 4 \cdot 2) - (2\sqrt{t}\ln t - 4\sqrt{t})\right] \overset{**}{=} (4\ln 4 - 8) - (0 - 0) = 4\ln 4 - 8$$

$(*)$ $\qquad$ Let $u = \ln x$, $dv = \dfrac{1}{\sqrt{x}}\, dx \Rightarrow du = \dfrac{1}{x}\, dx$, $v = 2\sqrt{x}$. Then

$$\int \frac{\ln x}{\sqrt{x}}\, dx = 2\sqrt{x}\ln x - 2\int \frac{dx}{\sqrt{x}} = 2\sqrt{x}\ln x - 4\sqrt{x} + C$$

$(**)$ $\qquad$ $\displaystyle\lim_{t\to 0^+}\left(2\sqrt{t}\ln t\right) = \lim_{t\to 0^+}\frac{2\ln t}{t^{-1/2}} \overset{\text{H}}{=} \lim_{t\to 0^+}\frac{2/t}{-\frac{1}{2}t^{-3/2}} = \lim_{t\to 0^+}\left(-4\sqrt{t}\right) = 0$

47. $\displaystyle\int_0^1 \frac{x - 1}{\sqrt{x}}\, dx = \lim_{t\to 0^+}\int_t^1 \left(\frac{x}{\sqrt{x}} - \frac{1}{\sqrt{x}}\right) dx = \lim_{t\to 0^+}\int_t^1 \left(x^{1/2} - x^{-1/2}\right) dx = \lim_{t\to 0^+}\left[\frac{2}{3}x^{3/2} - 2x^{1/2}\right]_t^1$

$$= \lim_{t\to 0^+}\left[\left(\tfrac{2}{3} - 2\right) - \left(\tfrac{2}{3}t^{3/2} - 2t^{1/2}\right)\right] = -\tfrac{4}{3} - 0 = -\tfrac{4}{3}$$

49. Let $u = 2x + 1$. Then

$$\int_{-\infty}^{\infty} \frac{dx}{4x^2 + 4x + 5} = \int_{-\infty}^{\infty} \frac{\frac{1}{2}\,du}{u^2 + 4} = \frac{1}{2}\int_{-\infty}^{0} \frac{du}{u^2 + 4} + \frac{1}{2}\int_{0}^{\infty} \frac{du}{u^2 + 4}$$

$$= \frac{1}{2}\lim_{t \to -\infty}\left[\frac{1}{2}\tan^{-1}\left(\frac{1}{2}u\right)\right]_t^0 + \frac{1}{2}\lim_{t \to \infty}\left[\frac{1}{2}\tan^{-1}\left(\frac{1}{2}u\right)\right]_0^t = \frac{1}{4}\left[0 - \left(-\frac{\pi}{2}\right)\right] + \frac{1}{4}\left[\frac{\pi}{2} - 0\right] = \frac{\pi}{4}.$$

51. We first make the substitution $t = x + 1$, so $\ln(x^2 + 2x + 2) = \ln\left[(x + 1)^2 + 1\right] = \ln(t^2 + 1)$. Then we use parts with

$u = \ln(t^2 + 1)$, $dv = dt$:

$$\int \ln(t^2 + 1)\,dt = t\,\ln(t^2 + 1) - \int \frac{t(2t)\,dt}{t^2 + 1} = t\,\ln(t^2 + 1) - 2\int \frac{t^2\,dt}{t^2 + 1} = t\,\ln(t^2 + 1) - 2\int\left(1 - \frac{1}{t^2 + 1}\right)dt$$

$$= t\,\ln(t^2 + 1) - 2t + 2\arctan t + C$$

$$= (x + 1)\,\ln(x^2 + 2x + 2) - 2x + 2\arctan(x + 1) + K, \text{ where } K = C - 2$$

[Alternatively, we could have integrated by parts immediately with

$u = \ln(x^2 + 2x + 2)$.] Notice from the graph that $f = 0$ where F has a

horizontal tangent. Also, F is always increasing, and $f \geq 0$.

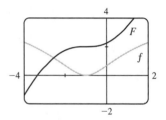

53. From the graph, it seems as though $\int_0^{2\pi} \cos^2 x\,\sin^3 x\,dx$ is equal to 0.

To evaluate the integral, we write the integral as

$I = \int_0^{2\pi} \cos^2 x\,(1 - \cos^2 x)\,\sin x\,dx$ and let $u = \cos x \quad\Rightarrow$

$du = -\sin x\,dx$. Thus, $I = \int_1^1 u^2(1 - u^2)(-du) = 0$.

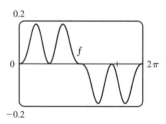

55. $\int \sqrt{4x^2 - 4x - 3}\,dx = \int \sqrt{(2x - 1)^2 - 4}\,dx \quad \begin{bmatrix} u = 2x - 1, \\ du = 2\,dx \end{bmatrix} = \int \sqrt{u^2 - 2^2}\left(\frac{1}{2}\,du\right)$

$\overset{39}{=} \frac{1}{2}\left(\frac{u}{2}\sqrt{u^2 - 2^2} - \frac{2^2}{2}\ln\left|u + \sqrt{u^2 - 2^2}\right|\right) + C = \frac{1}{4}u\sqrt{u^2 - 4} - \ln\left|u + \sqrt{u^2 - 4}\right| + C$

$= \frac{1}{4}(2x - 1)\sqrt{4x^2 - 4x - 3} - \ln\left|2x - 1 + \sqrt{4x^2 - 4x - 3}\right| + C$

57. Let $u = \sin x$, so that $du = \cos x\,dx$. Then

$$\int \cos x\,\sqrt{4 + \sin^2 x}\,dx = \int \sqrt{2^2 + u^2}\,du \overset{21}{=} \frac{u}{2}\sqrt{2^2 + u^2} + \frac{2^2}{2}\ln\left(u + \sqrt{2^2 + u^2}\right) + C$$

$$= \frac{1}{2}\sin x\,\sqrt{4 + \sin^2 x} + 2\ln\left(\sin x + \sqrt{4 + \sin^2 x}\right) + C$$

59. (a) $\dfrac{d}{du}\left[-\dfrac{1}{u}\sqrt{a^2 - u^2} - \sin^{-1}\left(\dfrac{u}{a}\right) + C\right] = \dfrac{1}{u^2}\sqrt{a^2 - u^2} + \dfrac{1}{\sqrt{a^2 - u^2}} - \dfrac{1}{\sqrt{1 - u^2/a^2}} \cdot \dfrac{1}{a}$

$$= (a^2 - u^2)^{-1/2}\left[\dfrac{1}{u^2}(a^2 - u^2) + 1 - 1\right] = \dfrac{\sqrt{a^2 - u^2}}{u^2}$$

(b) Let $u = a\sin\theta \implies du = a\cos\theta\,d\theta,\, a^2 - u^2 = a^2\left(1 - \sin^2\theta\right) = a^2\cos^2\theta.$

$$\int \frac{\sqrt{a^2 - u^2}}{u^2}\,du = \int \frac{a^2\cos^2\theta}{a^2\sin^2\theta}\,d\theta = \int \frac{1 - \sin^2\theta}{\sin^2\theta}\,d\theta = \int \left(\csc^2\theta - 1\right)\,d\theta = -\cot\theta - \theta + C$$

$$= -\frac{\sqrt{a^2 - u^2}}{u} - \sin^{-1}\left(\frac{u}{a}\right) + C$$

61. For $n \geq 0$, $\int_0^\infty x^n\,dx = \lim_{t \to \infty}\left[x^{n+1}/(n+1)\right]_0^t = \infty$. For $n < 0$, $\int_0^\infty x^n\,dx = \int_0^1 x^n\,dx + \int_1^\infty x^n\,dx$. Both integrals are improper. By (7.8.2), the second integral diverges if $-1 \leq n < 0$. By Exercise 7.8.57, the first integral diverges if $n \leq -1$. Thus, $\int_0^\infty x^n\,dx$ is divergent for all values of n.

63. $f(x) = \dfrac{1}{\ln x}$, $\Delta x = \dfrac{b - a}{n} = \dfrac{4 - 2}{10} = \dfrac{1}{5}$

(a) $T_{10} = \frac{1}{5 \cdot 2}\{f(2) + 2[f(2.2) + f(2.4) + \cdots + f(3.8)] + f(4)\} \approx 1.925444$

(b) $M_{10} = \frac{1}{5}[f(2.1) + f(2.3) + f(2.5) + \cdots + f(3.9)] \approx 1.920915$

(c) $S_{10} = \frac{1}{5 \cdot 3}[f(2) + 4f(2.2) + 2f(2.4) + \cdots + 2f(3.6) + 4f(3.8) + f(4)] \approx 1.922470$

65. $f(x) = \dfrac{1}{\ln x} \implies f'(x) = -\dfrac{1}{x(\ln x)^2} \implies f''(x) = \dfrac{2 + \ln x}{x^2(\ln x)^3} = \dfrac{2}{x^2(\ln x)^3} + \dfrac{1}{x^2(\ln x)^2}$. Note that each term of

$f''(x)$ decreases on $[2, 4]$, so we'll take $K = f''(2) \approx 2.022$. $|E_T| \leq \dfrac{K(b - a)^3}{12n^2} \approx \dfrac{2.022(4 - 2)^3}{12(10)^2} = 0.01348$ and

$|E_M| \leq \dfrac{K(b - a)^3}{24n^2} = 0.00674$. $|E_T| \leq 0.00001 \iff \dfrac{2.022(8)}{12n^2} \leq \dfrac{1}{10^5} \iff n^2 \geq \dfrac{10^5(2.022)(8)}{12} \implies n \geq 367.2$.

Take $n = 368$ for T_n. $|E_M| \leq 0.00001 \iff n^2 \geq \dfrac{10^5(2.022)(8)}{24} \implies n \geq 259.6$. Take $n = 260$ for M_n.

67. $\Delta t = \left(\frac{10}{60} - 0\right)/10 = \frac{1}{60}$.

Distance traveled $= \int_0^{10} v\,dt \approx S_{10}$

$$= \frac{1}{60 \cdot 3}[40 + 4(42) + 2(45) + 4(49) + 2(52) + 4(54) + 2(56) + 4(57) + 2(57) + 4(55) + 56]$$

$$= \frac{1}{180}(1544) = 8.5\overline{7}\,\text{mi}$$

69. (a) $f(x) = \sin(\sin x)$. A CAS gives

$$f^{(4)}(x) = \sin(\sin x)[\cos^4 x + 7\cos^2 x - 3]$$
$$+ \cos(\sin x)\left[6\cos^2 x \sin x + \sin x\right]$$

From the graph, we see that $\left|f^{(4)}(x)\right| < 3.8$ for $x \in [0, \pi]$.

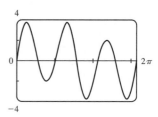

(b) We use Simpson's Rule with $f(x) = \sin(\sin x)$ and $\Delta x = \frac{\pi}{10}$:

$$\int_0^\pi f(x)\,dx \approx \frac{\pi}{10 \cdot 3}\left[f(0) + 4f\left(\frac{\pi}{10}\right) + 2f\left(\frac{2\pi}{10}\right) + \cdots + 4f\left(\frac{9\pi}{10}\right) + f(\pi)\right] \approx 1.786721$$

From part (a), we know that $\left|f^{(4)}(x)\right| < 3.8$ on $[0, \pi]$, so we use Theorem 7.7.4 with $K = 3.8$, and estimate the error

as $|E_S| \leq \dfrac{3.8(\pi - 0)^5}{180(10)^4} \approx 0.000646$.

(c) If we want the error to be less than 0.00001, we must have $|E_S| \leq \dfrac{3.8\pi^5}{180n^4} \leq 0.00001$,

so $n^4 \geq \dfrac{3.8\pi^5}{180(0.00001)} \approx 646{,}041.6 \ \Rightarrow \ n \geq 28.35$. Since n must be even for Simpson's Rule, we must have $n \geq 30$

to ensure the desired accuracy.

71. $\dfrac{x^3}{x^5+2} \leq \dfrac{x^3}{x^5} = \dfrac{1}{x^2}$ for x in $[1,\infty)$. $\displaystyle\int_1^\infty \dfrac{1}{x^2}\,dx$ is convergent by (7.8.2) with $p=2>1$. Therefore, $\displaystyle\int_1^\infty \dfrac{x^3}{x^5+2}\,dx$ is

convergent by the Comparison Theorem.

73. For x in $\left[0, \frac{\pi}{2}\right]$, $0 \leq \cos^2 x \leq \cos x$. For x in $\left[\frac{\pi}{2}, \pi\right]$, $\cos x \leq 0 \leq \cos^2 x$. Thus,

area $= \int_0^{\pi/2}(\cos x - \cos^2 x)\,dx + \int_{\pi/2}^{\pi}(\cos^2 x - \cos x)\,dx$

$= \left[\sin x - \tfrac{1}{2}x - \tfrac{1}{4}\sin 2x\right]_0^{\pi/2} + \left[\tfrac{1}{2}x + \tfrac{1}{4}\sin 2x - \sin x\right]_{\pi/2}^{\pi} = \left[\left(1 - \tfrac{\pi}{4}\right) - 0\right] + \left[\tfrac{\pi}{2} - \left(\tfrac{\pi}{4} - 1\right)\right] = 2$

75. Using the formula for disks, the volume is

$$V = \int_0^{\pi/2} \pi\,[f(x)]^2\,dx = \pi \int_0^{\pi/2}(\cos^2 x)^2\,dx = \pi \int_0^{\pi/2}\left[\tfrac{1}{2}(1+\cos 2x)\right]^2\,dx$$

$$= \tfrac{\pi}{4}\int_0^{\pi/2}(1 + \cos^2 2x + 2\cos 2x)\,dx = \tfrac{\pi}{4}\int_0^{\pi/2}\left[1 + \tfrac{1}{2}(1+\cos 4x) + 2\cos 2x\right]dx$$

$$= \tfrac{\pi}{4}\left[\tfrac{3}{2}x + \tfrac{1}{2}\left(\tfrac{1}{4}\sin 4x\right) + 2\left(\tfrac{1}{2}\sin 2x\right)\right]_0^{\pi/2} = \tfrac{\pi}{4}\left[\left(\tfrac{3\pi}{4} + \tfrac{1}{8}\cdot 0 + 0\right) - 0\right] = \tfrac{3\pi^2}{16}$$

77. By the Fundamental Theorem of Calculus,

$\displaystyle\int_0^\infty f'(x)\,dx = \lim_{t\to\infty}\int_0^t f'(x)\,dx = \lim_{t\to\infty}[f(t) - f(0)] = \lim_{t\to\infty}f(t) - f(0) = 0 - f(0) = -f(0)$.

79. Let $u = 1/x \ \Rightarrow \ x = 1/u \ \Rightarrow \ dx = -(1/u^2)\,du$.

$$\int_0^\infty \frac{\ln x}{1+x^2}\,dx = \int_\infty^0 \frac{\ln(1/u)}{1+1/u^2}\left(-\frac{du}{u^2}\right) = \int_\infty^0 \frac{-\ln u}{u^2+1}\,(-du) = \int_\infty^0 \frac{\ln u}{1+u^2}\,du = -\int_0^\infty \frac{\ln u}{1+u^2}\,du$$

Therefore, $\displaystyle\int_0^\infty \frac{\ln x}{1+x^2}\,dx = -\int_0^\infty \frac{\ln x}{1+x^2}\,dx = 0$.

☐ PROBLEMS PLUS

1.

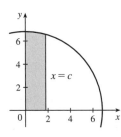

By symmetry, the problem can be reduced to finding the line $x = c$ such that the shaded area is one-third of the area of the

quarter-circle. An equation of the semicircle is $y = \sqrt{49 - x^2}$, so we require that $\int_0^c \sqrt{49 - x^2}\, dx = \frac{1}{3} \cdot \frac{1}{4}\pi(7)^2$ $\Leftrightarrow$

$\left[\frac{1}{2}x\sqrt{49 - x^2} + \frac{49}{2}\sin^{-1}(x/7) \right]_0^c = \frac{49}{12}\pi$ [by Formula 30] $\Leftrightarrow$ $\frac{1}{2}c\sqrt{49 - c^2} + \frac{49}{2}\sin^{-1}(c/7) = \frac{49}{12}\pi$.

This equation would be difficult to solve exactly, so we plot the left-hand side as a function of c, and find that the equation

holds for $c \approx 1.85$. So the cuts should be made at distances of about 1.85 inches from the center of the pizza.

3. The given integral represents the difference of the shaded areas, which appears to

be 0. It can be calculated by integrating with respect to either x or y, so we find x

in terms of y for each curve: $y = \sqrt[3]{1 - x^7}$ $\Rightarrow$ $x = \sqrt[7]{1 - y^3}$ and

$y = \sqrt[7]{1 - x^3}$ $\Rightarrow$ $x = \sqrt[3]{1 - y^7}$, so

$\int_0^1 \left(\sqrt[3]{1 - y^7} - \sqrt[7]{1 - y^3} \right) dy = \int_0^1 \left(\sqrt[7]{1 - x^3} - \sqrt[3]{1 - x^7} \right) dx$. But this

equation is of the form $z = -z$. So $\int_0^1 \left(\sqrt[3]{1 - x^7} - \sqrt[7]{1 - x^3} \right) dx = 0$.

5. The area A of the remaining part of the circle is given by

$A = 4I = 4\int_0^a \left(\sqrt{a^2 - x^2} - \frac{b}{a}\sqrt{a^2 - x^2} \right) dx = 4\left(1 - \frac{b}{a} \right) \int_0^a \sqrt{a^2 - x^2}\, dx$

$\overset{30}{=} \frac{4}{a}(a - b)\left[\frac{x}{2}\sqrt{a^2 - x^2} + \frac{a^2}{2}\sin^{-1}\frac{x}{a} \right]_0^a$

$= \frac{4}{a}(a - b)\left[\left(0 + \frac{a^2}{2}\frac{\pi}{2} \right) - 0 \right] = \frac{4}{a}(a - b)\left(\frac{a^2\pi}{4} \right) = \pi a(a - b)$,

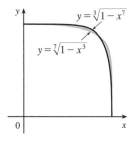

which is the area of an ellipse with semiaxes a and $a - b$.

Alternate solution: Subtracting the area of the ellipse from the area of the circle gives us $\pi a^2 - \pi ab = \pi a\,(a - b)$,

as calculated above. (The formula for the area of an ellipse was derived in Example 2 in Section 7.3.)

7. Recall that $\cos A \cos B = \frac{1}{2}[\cos(A+B) + \cos(A-B)]$. So

$$f(x) = \int_0^\pi \cos t \cos(x-t)\,dt = \frac{1}{2}\int_0^\pi [\cos(t+x-t) + \cos(t-x+t)]\,dt = \frac{1}{2}\int_0^\pi [\cos x + \cos(2t-x)]\,dt$$

$$= \frac{1}{2}\left[t\cos x + \frac{1}{2}\sin(2t-x)\right]_0^\pi = \frac{\pi}{2}\cos x + \frac{1}{4}\sin(2\pi-x) - \frac{1}{4}\sin(-x)$$

$$= \frac{\pi}{2}\cos x + \frac{1}{4}\sin(-x) - \frac{1}{4}\sin(-x) = \frac{\pi}{2}\cos x$$

The minimum of $\cos x$ on this domain is -1, so the minimum value of $f(x)$ is $f(\pi) = -\frac{\pi}{2}$.

9. In accordance with the hint, we let $I_k = \int_0^1 (1-x^2)^k\,dx$, and we find an expression for I_{k+1} in terms of I_k. We integrate I_{k+1} by parts with $u = (1-x^2)^{k+1} \;\Rightarrow\; du = (k+1)(1-x^2)^k(-2x)$, $dv = dx \;\Rightarrow\; v = x$, and then split the remaining integral into identifiable quantities:

$$I_{k+1} = x(1-x^2)^{k+1}\big|_0^1 + 2(k+1)\int_0^1 x^2(1-x^2)^k\,dx = (2k+2)\int_0^1 (1-x^2)^k[1-(1-x^2)]\,dx$$

$$= (2k+2)(I_k - I_{k+1})$$

So $I_{k+1}[1+(2k+2)] = (2k+2)I_k \;\Rightarrow\; I_{k+1} = \dfrac{2k+2}{2k+3}I_k$. Now to complete the proof, we use induction:

$I_0 = 1 = \dfrac{2^0(0!)^2}{1!}$, so the formula holds for $n = 0$. Now suppose it holds for $n = k$. Then

$$I_{k+1} = \frac{2k+2}{2k+3}I_k = \frac{2k+2}{2k+3}\left[\frac{2^{2k}(k!)^2}{(2k+1)!}\right] = \frac{2(k+1)2^{2k}(k!)^2}{(2k+3)(2k+1)!} = \frac{2(k+1)}{2k+2}\cdot\frac{2(k+1)2^{2k}(k!)^2}{(2k+3)(2k+1)!}$$

$$= \frac{[2(k+1)]^2\,2^{2k}(k!)^2}{(2k+3)(2k+2)(2k+1)!} = \frac{2^{2(k+1)}[(k+1)!]^2}{[2(k+1)+1]!}$$

So by induction, the formula holds for all integers $n \geq 0$.

11. $0 < a < b$. Now

$$\int_0^1 [bx + a(1-x)]^t\,dx = \int_a^b \frac{u^t}{(b-a)}\,du \quad [u = bx + a(1-x)] \quad = \left[\frac{u^{t+1}}{(t+1)(b-a)}\right]_a^b = \frac{b^{t+1} - a^{t+1}}{(t+1)(b-a)}.$$

Now let $y = \lim\limits_{t\to 0}\left[\dfrac{b^{t+1}-a^{t+1}}{(t+1)(b-a)}\right]^{1/t}$. Then $\ln y = \lim\limits_{t\to 0}\left[\dfrac{1}{t}\ln\dfrac{b^{t+1}-a^{t+1}}{(t+1)(b-a)}\right]$. This limit is of the form $0/0$,

so we can apply l'Hospital's Rule to get

$$\ln y = \lim_{t\to 0}\left[\frac{b^{t+1}\ln b - a^{t+1}\ln a}{b^{t+1}-a^{t+1}} - \frac{1}{t+1}\right] = \frac{b\ln b - a\ln a}{b-a} - 1 = \frac{b\ln b}{b-a} - \frac{a\ln a}{b-a} - \ln e = \ln\frac{b^{b/(b-a)}}{ea^{a/(b-a)}}.$$

Therefore, $y = e^{-1}\left(\dfrac{b^b}{a^a}\right)^{1/(b-a)}$.

13.

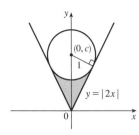

An equation of the circle with center $(0, c)$ and radius 1 is $x^2 + (y - c)^2 = 1^2$, so an equation of the lower semicircle is $y = c - \sqrt{1 - x^2}$. At the points of tangency, the slopes of the line and semicircle must be equal. For $x \geq 0$, we must have

$$y' = 2 \quad \Rightarrow \quad \frac{x}{\sqrt{1 - x^2}} = 2 \quad \Rightarrow \quad x = 2\sqrt{1 - x^2} \quad \Rightarrow \quad x^2 = 4(1 - x^2) \quad \Rightarrow$$

$$5x^2 = 4 \quad \Rightarrow \quad x^2 = \tfrac{4}{5} \quad \Rightarrow \quad x = \tfrac{2}{5}\sqrt{5} \text{ and so } y = 2\left(\tfrac{2}{5}\sqrt{5}\right) = \tfrac{4}{5}\sqrt{5}.$$

The slope of the perpendicular line segment is $-\frac{1}{2}$, so an equation of the line segment is $y - \frac{4}{5}\sqrt{5} = -\frac{1}{2}\left(x - \frac{2}{5}\sqrt{5}\right) \quad \Leftrightarrow$

$y = -\frac{1}{2}x + \frac{1}{5}\sqrt{5} + \frac{4}{5}\sqrt{5} \quad \Leftrightarrow \quad y = -\frac{1}{2}x + \sqrt{5}$, so $c = \sqrt{5}$ and an equation of the lower semicircle is $y = \sqrt{5} - \sqrt{1 - x^2}$.

Thus, the shaded area is

$$2\int_0^{(2/5)\sqrt{5}} \left[\left(\sqrt{5} - \sqrt{1 - x^2}\right) - 2x\right] dx \overset{30}{=} 2\left[\sqrt{5}\,x - \frac{x}{2}\sqrt{1 - x^2} - \frac{1}{2}\sin^{-1}x - x^2\right]_0^{(2/5)\sqrt{5}}$$

$$= 2\left[2 - \frac{\sqrt{5}}{5} \cdot \frac{1}{\sqrt{5}} - \frac{1}{2}\sin^{-1}\left(\frac{2}{\sqrt{5}}\right) - \frac{4}{5}\right] - 2(0)$$

$$= 2\left[1 - \frac{1}{2}\sin^{-1}\left(\frac{2}{\sqrt{5}}\right)\right] = 2 - \sin^{-1}\left(\frac{2}{\sqrt{5}}\right)$$

15. We integrate by parts with $u = \dfrac{1}{\ln(1 + x + t)}$, $dv = \sin t\,dt$, so $du = \dfrac{-1}{(1 + x + t)[\ln(1 + x + t)]^2}$ and $v = -\cos t$. The integral becomes

$$I = \int_0^\infty \frac{\sin t\,dt}{\ln(1 + x + t)} = \lim_{b \to \infty}\left(\left[\frac{-\cos t}{\ln(1 + x + t)}\right]_0^b - \int_0^b \frac{\cos t\,dt}{(1 + x + t)[\ln(1 + x + t)]^2}\right)$$

$$= \lim_{b \to \infty}\frac{-\cos b}{\ln(1 + x + b)} + \frac{1}{\ln(1 + x)} + \int_0^\infty \frac{-\cos t\,dt}{(1 + x + t)[\ln(1 + x + t)]^2} = \frac{1}{\ln(1 + x)} + J$$

where $J = \displaystyle\int_0^\infty \frac{-\cos t\,dt}{(1 + x + t)[\ln(1 + x + t)]^2}$. Now $-1 \leq -\cos t \leq 1$ for all t; in fact, the inequality is strict except

at isolated points. So $-\displaystyle\int_0^\infty \frac{dt}{(1 + x + t)[\ln(1 + x + t)]^2} < J < \int_0^\infty \frac{dt}{(1 + x + t)[\ln(1 + x + t)]^2} \quad \Leftrightarrow$

$-\dfrac{1}{\ln(1 + x)} < J < \dfrac{1}{\ln(1 + x)} \quad \Leftrightarrow \quad 0 < I < \dfrac{2}{\ln(1 + x)}$.

8 ☐ FURTHER APPLICATIONS OF INTEGRATION

8.1 Arc Length

1. $y = 2x - 5 \implies L = \int_{-1}^{3} \sqrt{1 + (dy/dx)^2}\,dx = \int_{-1}^{3} \sqrt{1 + (2)^2}\,dx = \sqrt{5}\,[3 - (-1)] = 4\sqrt{5}$.

The arc length can be calculated using the distance formula, since the curve is a line segment, so

$L = [\text{distance from } (-1, -7) \text{ to } (3, 1)] = \sqrt{[3 - (-1)]^2 + [1 - (-7)]^2} = \sqrt{80} = 4\sqrt{5}$

3. $y = \cos x \implies dy/dx = -\sin x \implies 1 + (dy/dx)^2 = 1 + \sin^2 x$. So $L = \int_0^{2\pi} \sqrt{1 + \sin^2 x}\,dx$.

5. $x = y + y^3 \implies dx/dy = 1 + 3y^2 \implies 1 + (dx/dy)^2 = 1 + (1 + 3y^2)^2 = 9y^4 + 6y^2 + 2$.

So $L = \int_1^4 \sqrt{9y^4 + 6y^2 + 2}\,dy$.

7. $y = 1 + 6x^{3/2} \implies dy/dx = 9x^{1/2} \implies 1 + (dy/dx)^2 = 1 + 81x$. So

$L = \int_0^1 \sqrt{1 + 81x}\,dx = \int_1^{82} u^{1/2} \left(\frac{1}{81}\,du\right) \quad \begin{bmatrix} u = 1 + 81x, \\ du = 81\,dx \end{bmatrix} = \frac{1}{81} \cdot \frac{2}{3} \left[u^{3/2}\right]_1^{82} = \frac{2}{243}\left(82\sqrt{82} - 1\right)$

9. $y = \dfrac{x^5}{6} + \dfrac{1}{10x^3} \implies \dfrac{dy}{dx} = \dfrac{5}{6}x^4 - \dfrac{3}{10}x^{-4} \implies$

$1 + (dy/dx)^2 = 1 + \frac{25}{36}x^8 - \frac{1}{2} + \frac{9}{100}x^{-8} = \frac{25}{36}x^8 + \frac{1}{2} + \frac{9}{100}x^{-8} = \left(\frac{5}{6}x^4 + \frac{3}{10}x^{-4}\right)^2$. So

$L = \int_1^2 \sqrt{\left(\frac{5}{6}x^4 + \frac{3}{10}x^{-4}\right)^2}\,dx = \int_1^2 \left(\frac{5}{6}x^4 + \frac{3}{10}x^{-4}\right)dx = \left[\frac{1}{6}x^5 - \frac{1}{10}x^{-3}\right]_1^2 = \left(\frac{32}{6} - \frac{1}{80}\right) - \left(\frac{1}{6} - \frac{1}{10}\right)$

$= \frac{31}{6} + \frac{7}{80} = \frac{1261}{240}$

11. $x = \frac{1}{3}\sqrt{y}\,(y - 3) = \frac{1}{3}y^{3/2} - y^{1/2} \implies dx/dy = \frac{1}{2}y^{1/2} - \frac{1}{2}y^{-1/2} \implies$

$1 + (dx/dy)^2 = 1 + \frac{1}{4}y - \frac{1}{2} + \frac{1}{4}y^{-1} = \frac{1}{4}y + \frac{1}{2} + \frac{1}{4}y^{-1} = \left(\frac{1}{2}y^{1/2} + \frac{1}{2}y^{-1/2}\right)^2$. So

$L = \int_1^9 \left(\frac{1}{2}y^{1/2} + \frac{1}{2}y^{-1/2}\right)dy = \frac{1}{2}\left[\frac{2}{3}y^{3/2} + 2y^{1/2}\right]_1^9 = \frac{1}{2}\left[\left(\frac{2}{3} \cdot 27 + 2 \cdot 3\right) - \left(\frac{2}{3} \cdot 1 + 2 \cdot 1\right)\right]$

$= \frac{1}{2}\left(24 - \frac{8}{3}\right) = \frac{1}{2}\left(\frac{64}{3}\right) = \frac{32}{3}$.

13. $y = \ln(\sec x) \implies \dfrac{dy}{dx} = \dfrac{\sec x \tan x}{\sec x} = \tan x \implies 1 + \left(\dfrac{dy}{dx}\right)^2 = 1 + \tan^2 x = \sec^2 x$, so

$L = \int_0^{\pi/4} \sqrt{\sec^2 x}\,dx = \int_0^{\pi/4} |\sec x|\,dx = \int_0^{\pi/4} \sec x\,dx = \left[\ln(\sec x + \tan x)\right]_0^{\pi/4}$

$= \ln\left(\sqrt{2} + 1\right) - \ln(1 + 0) = \ln\left(\sqrt{2} + 1\right)$

15. $y = \ln(1 - x^2) \Rightarrow y' = \dfrac{1}{1 - x^2} \cdot (-2x) \Rightarrow$

$1 + \left(\dfrac{dy}{dx}\right)^2 = 1 + \dfrac{4x^2}{(1 - x^2)^2} = \dfrac{1 - 2x^2 + x^4 + 4x^2}{(1 - x^2)^2} = \dfrac{1 + 2x^2 + x^4}{(1 - x^2)^2} = \dfrac{(1 + x^2)^2}{(1 - x^2)^2} \Rightarrow$

$\sqrt{1 + \left(\dfrac{dy}{dx}\right)^2} = \sqrt{\left(\dfrac{1 + x^2}{1 - x^2}\right)^2} = \dfrac{1 + x^2}{1 - x^2} = -1 + \dfrac{2}{1 - x^2}$ [by division] $= -1 + \dfrac{1}{1 + x} + \dfrac{1}{1 - x}$ [partial fractions].

So $L = \displaystyle\int_0^{1/2} \left(-1 + \dfrac{1}{1 + x} + \dfrac{1}{1 - x}\right) dx = \left[-x + \ln|1 + x| - \ln|1 - x|\right]_0^{1/2} = \left(-\tfrac{1}{2} + \ln \tfrac{3}{2} - \ln \tfrac{1}{2}\right) - 0 = \ln 3 - \tfrac{1}{2}$.

17. $y = e^x \Rightarrow y' = e^x \Rightarrow 1 + (y')^2 = 1 + e^{2x}$. So

$$L = \int_0^1 \sqrt{1 + e^{2x}}\, dx = \int_1^e \sqrt{1 + u^2}\, \dfrac{du}{u} \quad \begin{bmatrix} u = e^x,\ \text{so} \\ x = \ln u,\ dx = du/u \end{bmatrix} = \int_1^e \dfrac{\sqrt{1 + u^2}}{u^2}\, u\, du$$

$$= \int_{\sqrt 2}^{\sqrt{1 + e^2}} \dfrac{v}{v^2 - 1}\, v\, dv \quad \begin{bmatrix} v = \sqrt{1 + u^2},\ \text{so} \\ v^2 = 1 + u^2,\ v\, dv = u\, du \end{bmatrix} = \int_{\sqrt 2}^{\sqrt{1 + e^2}} \left(1 + \dfrac{1/2}{v - 1} - \dfrac{1/2}{v + 1}\right) dv$$

$$= \left[v + \tfrac{1}{2} \ln \dfrac{v - 1}{v + 1}\right]_{\sqrt 2}^{\sqrt{1 + e^2}} = \sqrt{1 + e^2} + \tfrac{1}{2} \ln \dfrac{\sqrt{1 + e^2} - 1}{\sqrt{1 + e^2} + 1} - \sqrt 2 - \tfrac{1}{2} \ln \dfrac{\sqrt 2 - 1}{\sqrt 2 + 1}$$

$$= \sqrt{1 + e^2} - \sqrt 2 + \ln(\sqrt{1 + e^2} - 1) - 1 - \ln(\sqrt 2 - 1)$$

Or: Use Formula 23 for $\int (\sqrt{1 + u^2}/u)\, du$, or substitute $u = \tan \theta$.

19. $y = \tfrac{1}{2} x^2 \Rightarrow dy/dx = x \Rightarrow 1 + (dy/dx)^2 = 1 + x^2$. So

$$L = \int_{-1}^1 \sqrt{1 + x^2}\, dx = 2 \int_0^1 \sqrt{1 + x^2}\, dx \quad \text{[by symmetry]} \stackrel{21}{=} 2\left[\tfrac{x}{2} \sqrt{1 + x^2} + \tfrac{1}{2} \ln\left(x + \sqrt{1 + x^2}\right)\right]_0^1 \quad \begin{bmatrix} \text{or substitute} \\ x = \tan \theta \end{bmatrix}$$

$$= 2\left[\left(\tfrac{1}{2} \sqrt 2 + \tfrac{1}{2} \ln\left(1 + \sqrt 2\right)\right) - \left(0 + \tfrac{1}{2} \ln 1\right)\right] = \sqrt 2 + \ln\left(1 + \sqrt 2\right)$$

21.

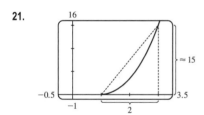

From the figure, the length of the curve is slightly larger than the hypotenuse of the triangle formed by the points $(1, 0)$, $(3, 0)$, and $(3, f(3)) \approx (3, 15)$, where $y = f(x) = \tfrac{2}{3}(x^2 - 1)^{3/2}$. This length is about $\sqrt{15^2 + 2^2} \approx 15$, so we might estimate the length to be 15.5.

$y = \tfrac{2}{3}(x^2 - 1)^{3/2} \Rightarrow y' = (x^2 - 1)^{1/2}(2x) \Rightarrow 1 + (y')^2 = 1 + 4x^2(x^2 - 1) = 4x^4 - 4x^2 + 1 = (2x^2 - 1)^2$,

so, using the fact that $2x^2 - 1 > 0$ for $1 \le x \le 3$,

$L = \int_1^3 \sqrt{(2x^2 - 1)^2}\, dx = \int_1^3 |2x^2 - 1|\, dx = \int_1^3 (2x^2 - 1)\, dx = \left[\tfrac{2}{3} x^3 - x\right]_1^3 = (18 - 3) - \left(\tfrac{2}{3} - 1\right) = \tfrac{46}{3} = 15.\overline{3}$.

23. $y = xe^{-x} \Rightarrow dy/dx = e^{-x} - xe^{-x} = e^{-x}(1 - x) \Rightarrow 1 + (dy/dx)^2 = 1 + e^{-2x}(1 - x)^2$. Let

$f(x) = \sqrt{1 + (dy/dx)^2} = \sqrt{1 + e^{-2x}(1 - x)^2}$. Then $L = \int_0^5 f(x)\, dx$. Since $n = 10$, $\Delta x = \dfrac{5 - 0}{10} = \tfrac{1}{2}$. Now

$L \approx S_{10} = \dfrac{1/2}{3}\left[f(0) + 4f\left(\tfrac{1}{2}\right) + 2f(1) + 4f\left(\tfrac{3}{2}\right) + 2f(2) + 4f\left(\tfrac{5}{2}\right) + 2f(3) + 4f\left(\tfrac{7}{2}\right) + 2f(4) + 4f\left(\tfrac{9}{2}\right) + f(5)\right]$

≈ 5.115840

The value of the integral produced by a calculator is 5.113568 (to six decimal places).

25. $y = \sec x \quad \Rightarrow \quad dy/dx = \sec x \tan x \quad \Rightarrow \quad L = \int_0^{\pi/3} f(x)\,dx$, where $f(x) = \sqrt{1 + \sec^2 x \tan^2 x}$.

Since $n = 10$, $\Delta x = \dfrac{\pi/3 - 0}{10} = \dfrac{\pi}{30}$. Now

$$L \approx S_{10} = \frac{\pi/30}{3}\left[f(0) + 4f\left(\frac{\pi}{30}\right) + 2f\left(\frac{2\pi}{30}\right) + 4f\left(\frac{3\pi}{30}\right) + 2f\left(\frac{4\pi}{30}\right) + 4f\left(\frac{5\pi}{30}\right) \right.$$
$$\left. + 2f\left(\frac{6\pi}{30}\right) + 4f\left(\frac{7\pi}{30}\right) + 2f\left(\frac{8\pi}{30}\right) + 4f\left(\frac{9\pi}{30}\right) + f\left(\frac{\pi}{3}\right) \right] \approx 1.569619.$$

The value of the integral produced by a calculator is 1.569259 (to six decimal places).

27. (a)

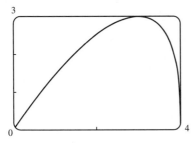

(b)

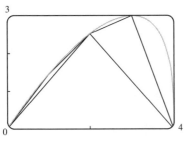

Let $f(x) = y = x\sqrt[3]{4 - x}$. The polygon with one side is just the line segment joining the points $(0, f(0)) = (0, 0)$ and $(4, f(4)) = (4, 0)$, and its length $L_1 = 4$.

The polygon with two sides joins the points $(0, 0)$, $(2, f(2)) = \left(2, 2\sqrt[3]{2}\right)$ and $(4, 0)$. Its length

$$L_2 = \sqrt{(2 - 0)^2 + \left(2\sqrt[3]{2} - 0\right)^2} + \sqrt{(4 - 2)^2 + \left(0 - 2\sqrt[3]{2}\right)^2} = 2\sqrt{4 + 2^{8/3}} \approx 6.43$$

Similarly, the inscribed polygon with four sides joins the points $(0, 0)$, $\left(1, \sqrt[3]{3}\right)$, $\left(2, 2\sqrt[3]{2}\right)$, $(3, 3)$, and $(4, 0)$, so its length

$$L_3 = \sqrt{1 + \left(\sqrt[3]{3}\right)^2} + \sqrt{1 + \left(2\sqrt[3]{2} - \sqrt[3]{3}\right)^2} + \sqrt{1 + \left(3 - 2\sqrt[3]{2}\right)^2} + \sqrt{1 + 9} \approx 7.50$$

(c) Using the arc length formula with $\dfrac{dy}{dx} = x\left[\frac{1}{3}(4 - x)^{-2/3}(-1)\right] + \sqrt[3]{4 - x} = \dfrac{12 - 4x}{3(4 - x)^{2/3}}$, the length of the curve is

$$L = \int_0^4 \sqrt{1 + \left(\frac{dy}{dx}\right)^2}\,dx = \int_0^4 \sqrt{1 + \left[\frac{12 - 4x}{3(4 - x)^{2/3}}\right]^2}\,dx.$$

(d) According to a CAS, the length of the curve is $L \approx 7.7988$. The actual value is larger than any of the approximations in part (b). This is always true, since any approximating straight line between two points on the curve is shorter than the length of the curve between the two points.

29. $y = \ln x \Rightarrow dy/dx = 1/x \Rightarrow 1 + (dy/dx)^2 = 1 + 1/x^2 = (x^2+1)/x^2 \Rightarrow$

$$L = \int_1^2 \sqrt{\frac{x^2+1}{x^2}}\,dx = \int_1^2 \frac{\sqrt{1+x^2}}{x}\,dx \overset{23}{=} \left[\sqrt{1+x^2} - \ln\left|\frac{1+\sqrt{1+x^2}}{x}\right|\right]_1^2$$

$$= \sqrt{5} - \ln\left(\frac{1+\sqrt{5}}{2}\right) - \sqrt{2} + \ln\left(1+\sqrt{2}\right)$$

31. $y^{2/3} = 1 - x^{2/3} \Rightarrow y = (1 - x^{2/3})^{3/2} \Rightarrow$

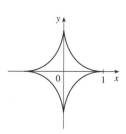

$$\frac{dy}{dx} = \tfrac{3}{2}(1-x^{2/3})^{1/2}\left(-\tfrac{2}{3}x^{-1/3}\right) = -x^{-1/3}(1-x^{2/3})^{1/2} \Rightarrow$$

$$\left(\frac{dy}{dx}\right)^2 = x^{-2/3}(1-x^{2/3}) = x^{-2/3} - 1. \text{ Thus}$$

$$L = 4\int_0^1 \sqrt{1 + (x^{-2/3} - 1)}\,dx = 4\int_0^1 x^{-1/3}\,dx = 4\lim_{t\to 0^+}\left[\tfrac{3}{2}x^{2/3}\right]_t^1 = 6.$$

33. $y = 2x^{3/2} \Rightarrow y' = 3x^{1/2} \Rightarrow 1 + (y')^2 = 1 + 9x$. The arc length function with starting point $P_0(1,2)$ is

$$s(x) = \int_1^x \sqrt{1+9t}\,dt = \left[\tfrac{2}{27}(1+9t)^{3/2}\right]_1^x = \tfrac{2}{27}\left[(1+9x)^{3/2} - 10\sqrt{10}\right].$$

35. $y = \sin^{-1} x + \sqrt{1-x^2} \Rightarrow y' = \dfrac{1}{\sqrt{1-x^2}} - \dfrac{x}{\sqrt{1-x^2}} = \dfrac{1-x}{\sqrt{1-x^2}} \Rightarrow$

$$1 + (y')^2 = 1 + \frac{(1-x)^2}{1-x^2} = \frac{1-x^2+1-2x+x^2}{1-x^2} = \frac{2-2x}{1-x^2} = \frac{2(1-x)}{(1+x)(1-x)} = \frac{2}{1+x} \Rightarrow$$

$\sqrt{1+(y')^2} = \sqrt{\dfrac{2}{1+x}}$. Thus, the arc length function with starting point $(0,1)$ is given by

$$s(x) = \int_0^x \sqrt{1+[f'(t)]^2}\,dt = \int_0^x \sqrt{\frac{2}{1+t}}\,dt = \sqrt{2}\left[2\sqrt{1+t}\right]_0^x = 2\sqrt{2}\left(\sqrt{1+x}-1\right).$$

37. The prey hits the ground when $y = 0 \Leftrightarrow 180 - \tfrac{1}{45}x^2 = 0 \Leftrightarrow x^2 = 45\cdot 180 \Rightarrow x = \sqrt{8100} = 90,$

since x must be positive. $y' = -\tfrac{2}{45}x \Rightarrow 1 + (y')^2 = 1 + \tfrac{4}{45^2}x^2$, so the distance traveled by the prey is

$$L = \int_0^{90} \sqrt{1 + \frac{4}{45^2}x^2}\,dx = \int_0^4 \sqrt{1+u^2}\left(\tfrac{45}{2}\,du\right) \qquad \begin{bmatrix} u = \tfrac{2}{45}x, \\ du = \tfrac{2}{45}\,dx \end{bmatrix}$$

$$\overset{21}{=} \tfrac{45}{2}\left[\tfrac{1}{2}u\sqrt{1+u^2} + \tfrac{1}{2}\ln\left(u + \sqrt{1+u^2}\right)\right]_0^4 = \tfrac{45}{2}\left[2\sqrt{17} + \tfrac{1}{2}\ln\left(4+\sqrt{17}\right)\right] = 45\sqrt{17} + \tfrac{45}{4}\ln\left(4+\sqrt{17}\right) \approx 209.1 \text{ m}$$

39. The sine wave has amplitude 1 and period 14, since it goes through two periods in a distance of 28 in., so its equation is

$y = 1\sin\left(\tfrac{2\pi}{14}x\right) = \sin\left(\tfrac{\pi}{7}x\right)$. The width w of the flat metal sheet needed to make the panel is the arc length of the sine curve

from $x = 0$ to $x = 28$. We set up the integral to evaluate w using the arc length formula with $\tfrac{dy}{dx} = \tfrac{\pi}{7}\cos\left(\tfrac{\pi}{7}x\right)$:

$L = \int_0^{28} \sqrt{1 + \left[\tfrac{\pi}{7}\cos\left(\tfrac{\pi}{7}x\right)\right]^2}\,dx = 2\int_0^{14} \sqrt{1 + \left[\tfrac{\pi}{7}\cos\left(\tfrac{\pi}{7}x\right)\right]^2}\,dx$. This integral would be very difficult to evaluate exactly,

so we use a CAS, and find that $L \approx 29.36$ inches.

41. $y = \int_1^x \sqrt{t^3 - 1}\,dt \Rightarrow dy/dx = \sqrt{x^3 - 1}$ [by FTC1] $\Rightarrow 1 + (dy/dx)^2 = 1 + \left(\sqrt{x^3-1}\right)^2 = x^3 \Rightarrow$

$L = \int_1^4 \sqrt{x^3}\,dx = \int_1^4 x^{3/2}\,dx = \tfrac{2}{5}\left[x^{5/2}\right]_1^4 = \tfrac{2}{5}(32 - 1) = \tfrac{62}{5} = 12.4$

8.2   Area of a Surface of Revolution

1. $y = x^4 \;\Rightarrow\; dy/dx = 4x^3 \;\Rightarrow\; ds = \sqrt{1 + (dy/dx)^2}\, dx = \sqrt{1 + 16x^6}\, dx$

(a) By (7), an integral for the area of the surface obtained by rotating the curve about the x-axis is

$S = \int 2\pi y\, ds = \int_0^1 2\pi x^4 \sqrt{1 + 16x^6}\, dx.$

(b) By (8), an integral for the area of the surface obtained by rotating the curve about the y-axis is

$S = \int 2\pi x\, ds = \int_0^1 2\pi x \sqrt{1 + 16x^6}\, dx.$

3. $y = \tan^{-1} x \;\Rightarrow\; \dfrac{dy}{dx} = \dfrac{1}{1 + x^2} \;\Rightarrow\; ds = \sqrt{1 + \left(\dfrac{dy}{dx}\right)^2}\, dx = \sqrt{1 + \dfrac{1}{(1 + x^2)^2}}\, dx.$

(a) By (7), $S = \int 2\pi y\, ds = \int_0^1 2\pi \tan^{-1} x \sqrt{1 + \dfrac{1}{(1 + x^2)^2}}\, dx.$

(b) By (8), $S = \int 2\pi x\, ds = \int_0^1 2\pi x \sqrt{1 + \dfrac{1}{(1 + x^2)^2}}\, dx.$

5. $y = x^3 \;\Rightarrow\; y' = 3x^2.$ So

$$S = \int_0^2 2\pi y \sqrt{1 + (y')^2}\, dx = 2\pi \int_0^2 x^3 \sqrt{1 + 9x^4}\, dx \qquad [u = 1 + 9x^4,\, du = 36x^3\, dx]$$

$$= \frac{2\pi}{36} \int_1^{145} \sqrt{u}\, du = \frac{\pi}{18} \left[\frac{2}{3} u^{3/2}\right]_1^{145} = \frac{\pi}{27}\left(145\sqrt{145} - 1\right)$$

7. $y = \sqrt{1 + 4x} \;\Rightarrow\; y' = \frac{1}{2}(1 + 4x)^{-1/2}(4) = \dfrac{2}{\sqrt{1 + 4x}} \;\Rightarrow\; \sqrt{1 + (y')^2} = \sqrt{1 + \dfrac{4}{1 + 4x}} = \sqrt{\dfrac{5 + 4x}{1 + 4x}}.$ So

$$S = \int_1^5 2\pi y \sqrt{1 + (y')^2}\, dx = 2\pi \int_1^5 \sqrt{1 + 4x} \sqrt{\dfrac{5 + 4x}{1 + 4x}}\, dx = 2\pi \int_1^5 \sqrt{4x + 5}\, dx$$

$$= 2\pi \int_9^{25} \sqrt{u}\left(\tfrac{1}{4}\, du\right) \quad \begin{bmatrix} u = 4x + 5, \\ du = 4\, dx \end{bmatrix} \quad = \frac{2\pi}{4}\left[\frac{2}{3} u^{3/2}\right]_9^{25} = \frac{\pi}{3}(25^{3/2} - 9^{3/2}) = \frac{\pi}{3}(125 - 27) = \frac{98}{3}\pi$$

9. $y = \sin \pi x \;\Rightarrow\; y' = \pi \cos \pi x \;\Rightarrow\; 1 + (y')^2 = 1 + \pi^2 \cos^2(\pi x).$ So

$$S = \int_0^1 2\pi y \sqrt{1 + (y')^2}\, dx = 2\pi \int_0^1 \sin \pi x \sqrt{1 + \pi^2 \cos^2(\pi x)}\, dx \qquad \begin{bmatrix} u = \pi \cos \pi x, \\ du = -\pi^2 \sin \pi x\, dx \end{bmatrix}$$

$$= 2\pi \int_\pi^{-\pi} \sqrt{1 + u^2}\left(-\frac{1}{\pi^2}\, du\right) = \frac{2}{\pi} \int_{-\pi}^\pi \sqrt{1 + u^2}\, du$$

$$= \frac{4}{\pi} \int_0^\pi \sqrt{1 + u^2}\, du \stackrel{21}{=} \frac{4}{\pi}\left[\frac{u}{2}\sqrt{1 + u^2} + \frac{1}{2}\ln\left(u + \sqrt{1 + u^2}\right)\right]_0^\pi$$

$$= \frac{4}{\pi}\left[\left(\frac{\pi}{2}\sqrt{1 + \pi^2} + \frac{1}{2}\ln\left(\pi + \sqrt{1 + \pi^2}\right)\right) - 0\right] = 2\sqrt{1 + \pi^2} + \frac{2}{\pi}\ln\left(\pi + \sqrt{1 + \pi^2}\right)$$

11. $x = \frac{1}{3}(y^2 + 2)^{3/2} \;\Rightarrow\; dx/dy = \frac{1}{2}(y^2 + 2)^{1/2}(2y) = y\sqrt{y^2 + 2} \;\Rightarrow\; 1 + (dx/dy)^2 = 1 + y^2(y^2 + 2) = (y^2 + 1)^2.$

So $S = 2\pi \int_1^2 y(y^2 + 1)\, dy = 2\pi \left[\frac{1}{4}y^4 + \frac{1}{2}y^2\right]_1^2 = 2\pi\left(4 + 2 - \frac{1}{4} - \frac{1}{2}\right) = \frac{21\pi}{2}.$

13. $y = \sqrt[3]{x} \;\Rightarrow\; x = y^3 \;\Rightarrow\; 1 + (dx/dy)^2 = 1 + 9y^4.$ So

$$S = 2\pi \int_1^2 x \sqrt{1 + (dx/dy)^2}\, dy = 2\pi \int_1^2 y^3 \sqrt{1 + 9y^4}\, dy = \frac{2\pi}{36} \int_1^2 \sqrt{1 + 9y^4}\, 36y^3\, dy = \frac{\pi}{18}\left[\frac{2}{3}(1 + 9y^4)^{3/2}\right]_1^2$$

$$= \frac{\pi}{27}\left(145\sqrt{145} - 10\sqrt{10}\right)$$

15. $x = \sqrt{a^2 - y^2} \Rightarrow dx/dy = \frac{1}{2}(a^2 - y^2)^{-1/2}(-2y) = -y/\sqrt{a^2 - y^2} \Rightarrow$

$$1 + \left(\frac{dx}{dy}\right)^2 = 1 + \frac{y^2}{a^2 - y^2} = \frac{a^2 - y^2}{a^2 - y^2} + \frac{y^2}{a^2 - y^2} = \frac{a^2}{a^2 - y^2} \Rightarrow$$

$$S = \int_0^{a/2} 2\pi \sqrt{a^2 - y^2} \frac{a}{\sqrt{a^2 - y^2}} \, dy = 2\pi \int_0^{a/2} a \, dy = 2\pi a \big[\, y \,\big]_0^{a/2} = 2\pi a\left(\frac{a}{2} - 0\right) = \pi a^2.$$

Note that this is $\frac{1}{4}$ the surface area of a sphere of radius a, and the length of the interval $y = 0$ to $y = a/2$ is $\frac{1}{4}$ the length of the interval $y = -a$ to $y = a$.

17. $y = \ln x \Rightarrow dy/dx = 1/x \Rightarrow 1 + (dy/dx)^2 = 1 + 1/x^2 \Rightarrow S = \int_1^3 2\pi \ln x \sqrt{1 + 1/x^2} \, dx.$

Let $f(x) = \ln x \sqrt{1 + 1/x^2}$. Since $n = 10$, $\Delta x = \frac{3-1}{10} = \frac{1}{5}$. Then

$$S \approx S_{10} = 2\pi \cdot \frac{1/5}{3} \left[f(1) + 4f(1.2) + 2f(1.4) + \cdots + 2f(2.6) + 4f(2.8) + f(3) \right] \approx 9.023754.$$

The value of the integral produced by a calculator is 9.024262 (to six decimal places).

19. $y = \sec x \Rightarrow dy/dx = \sec x \tan x \Rightarrow 1 + (dy/dx)^2 = 1 + \sec^2 x \tan^2 x \Rightarrow$

$$S = \int_0^{\pi/3} 2\pi \sec x \sqrt{1 + \sec^2 x \tan^2 x} \, dx. \text{ Let } f(x) = \sec x \sqrt{1 + \sec^2 x \tan^2 x}. \text{ Since } n = 10, \Delta x = \frac{\pi/3 - 0}{10} = \frac{\pi}{30}.$$

Then $S \approx S_{10} = 2\pi \cdot \frac{\pi/30}{3} \left[f(0) + 4f\left(\frac{\pi}{30}\right) + 2f\left(\frac{2\pi}{30}\right) + \cdots + 2f\left(\frac{8\pi}{30}\right) + 4f\left(\frac{9\pi}{30}\right) + f\left(\frac{\pi}{3}\right) \right] \approx 13.527296.$

The value of the integral produced by a calculator is 13.516987 (to six decimal places).

21. $y = 1/x \Rightarrow ds = \sqrt{1 + (dy/dx)^2} \, dx = \sqrt{1 + (-1/x^2)^2} \, dx = \sqrt{1 + 1/x^4} \, dx \Rightarrow$

$$S = \int_1^2 2\pi \cdot \frac{1}{x} \sqrt{1 + \frac{1}{x^4}} \, dx = 2\pi \int_1^2 \frac{\sqrt{x^4 + 1}}{x^3} \, dx = 2\pi \int_1^4 \frac{\sqrt{u^2 + 1}}{u^2} \left(\tfrac{1}{2} \, du\right) \qquad [u = x^2, \, du = 2x \, dx]$$

$$= \pi \int_1^4 \frac{\sqrt{1 + u^2}}{u^2} \, du \overset{24}{=} \pi \left[-\frac{\sqrt{1 + u^2}}{u} + \ln\left(u + \sqrt{1 + u^2}\right) \right]_1^4$$

$$= \pi \left[-\frac{\sqrt{17}}{4} + \ln\left(4 + \sqrt{17}\right) + \frac{\sqrt{2}}{1} - \ln\left(1 + \sqrt{2}\right) \right] = \frac{\pi}{4} \left[4\ln\left(\sqrt{17} + 4\right) - 4\ln\left(\sqrt{2} + 1\right) - \sqrt{17} + 4\sqrt{2} \right]$$

23. $y = x^3$ and $0 \le y \le 1 \Rightarrow y' = 3x^2$ and $0 \le x \le 1$.

$$S = \int_0^1 2\pi x \sqrt{1 + (3x^2)^2} \, dx = 2\pi \int_0^3 \sqrt{1 + u^2} \, \tfrac{1}{6} \, du \qquad \begin{bmatrix} u = 3x^2, \\ du = 6x \, dx \end{bmatrix} = \frac{\pi}{3} \int_0^3 \sqrt{1 + u^2} \, du$$

$$\overset{21}{=} [\text{or use CAS}] \; \frac{\pi}{3} \left[\tfrac{1}{2} u \sqrt{1 + u^2} + \tfrac{1}{2} \ln\left(u + \sqrt{1 + u^2}\right)\right]_0^3 = \frac{\pi}{3} \left[\tfrac{3}{2}\sqrt{10} + \tfrac{1}{2}\ln\left(3 + \sqrt{10}\right)\right] = \frac{\pi}{6}\left[3\sqrt{10} + \ln\left(3 + \sqrt{10}\right)\right]$$

25. $S = 2\pi \int_1^\infty y \sqrt{1 + \left(\frac{dy}{dx}\right)^2} \, dx = 2\pi \int_1^\infty \frac{1}{x} \sqrt{1 + \frac{1}{x^4}} \, dx = 2\pi \int_1^\infty \frac{\sqrt{x^4 + 1}}{x^3} \, dx.$ Rather than trying to

evaluate this integral, note that $\sqrt{x^4 + 1} > \sqrt{x^4} = x^2$ for $x > 0$. Thus, if the area is finite,

$$S = 2\pi \int_1^\infty \frac{\sqrt{x^4 + 1}}{x^3} \, dx > 2\pi \int_1^\infty \frac{x^2}{x^3} \, dx = 2\pi \int_1^\infty \frac{1}{x} \, dx. \text{ But we know that this integral diverges, so the area } S \text{ is}$$

infinite.

27. Since $a > 0$, the curve $3ay^2 = x(a - x)^2$ only has points with $x \geq 0$.

$[3ay^2 \geq 0 \;\;\Rightarrow\;\; x(a-x)^2 \geq 0 \;\;\Rightarrow\;\; x \geq 0.]$

The curve is symmetric about the x-axis (since the equation is unchanged when y is replaced by $-y$). $y = 0$ when $x = 0$ or a, so the curve's loop extends from $x = 0$ to $x = a$.

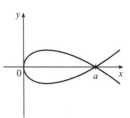

$$\frac{d}{dx}(3ay^2) = \frac{d}{dx}[x(a-x)^2] \;\;\Rightarrow\;\; 6ay\frac{dy}{dx} = x \cdot 2(a-x)(-1) + (a-x)^2 \;\;\Rightarrow\;\; \frac{dy}{dx} = \frac{(a-x)[-2x+a-x]}{6ay} \;\;\Rightarrow$$

$$\left(\frac{dy}{dx}\right)^2 = \frac{(a-x)^2(a-3x)^2}{36a^2y^2} = \frac{(a-x)^2(a-3x)^2}{36a^2} \cdot \frac{3a}{x(a-x)^2} \quad \begin{bmatrix} \text{the last fraction} \\ \text{is } 1/y^2 \end{bmatrix} = \frac{(a-3x)^2}{12ax} \;\;\Rightarrow$$

$$1 + \left(\frac{dy}{dx}\right)^2 = 1 + \frac{a^2 - 6ax + 9x^2}{12ax} = \frac{12ax}{12ax} + \frac{a^2 - 6ax + 9x^2}{12ax} = \frac{a^2 + 6ax + 9x^2}{12ax} = \frac{(a+3x)^2}{12ax} \quad \text{for } x \neq 0.$$

(a) $\displaystyle S = \int_{x=0}^{a} 2\pi y\, ds = 2\pi \int_0^a \frac{\sqrt{x}\,(a-x)}{\sqrt{3a}} \cdot \frac{a+3x}{\sqrt{12ax}}\, dx = 2\pi \int_0^a \frac{(a-x)(a+3x)}{6a}\, dx$

$\displaystyle = \frac{\pi}{3a} \int_0^a (a^2 + 2ax - 3x^2)\, dx = \frac{\pi}{3a}\left[a^2 x + ax^2 - x^3\right]_0^a = \frac{\pi}{3a}(a^3 + a^3 - a^3) = \frac{\pi}{3a} \cdot a^3 = \frac{\pi a^2}{3}.$

Note that we have rotated the top half of the loop about the x-axis. This generates the full surface.

(b) We must rotate the full loop about the y-axis, so we get double the area obtained by rotating the top half of the loop:

$$S = 2 \cdot 2\pi \int_{x=0}^a x\, ds = 4\pi \int_0^a x\frac{a+3x}{\sqrt{12ax}}\, dx = \frac{4\pi}{2\sqrt{3a}} \int_0^a x^{1/2}(a+3x)\, dx = \frac{2\pi}{\sqrt{3a}} \int_0^a (ax^{1/2} + 3x^{3/2})\, dx$$

$$= \frac{2\pi}{\sqrt{3a}}\left[\frac{2}{3}ax^{3/2} + \frac{6}{5}x^{5/2}\right]_0^a = \frac{2\pi\sqrt{3}}{3\sqrt{a}}\left(\frac{2}{3}a^{5/2} + \frac{6}{5}a^{5/2}\right) = \frac{2\pi\sqrt{3}}{3}\left(\frac{2}{3} + \frac{6}{5}\right)a^2 = \frac{2\pi\sqrt{3}}{3}\left(\frac{28}{15}\right)a^2$$

$$= \frac{56\pi\sqrt{3}\,a^2}{45}$$

29. (a) $\dfrac{x^2}{a^2} + \dfrac{y^2}{b^2} = 1 \;\;\Rightarrow\;\; \dfrac{y\,(dy/dx)}{b^2} = -\dfrac{x}{a^2} \;\;\Rightarrow\;\; \dfrac{dy}{dx} = -\dfrac{b^2 x}{a^2 y} \;\;\Rightarrow$

$$1 + \left(\frac{dy}{dx}\right)^2 = 1 + \frac{b^4 x^2}{a^4 y^2} = \frac{b^4 x^2 + a^4 y^2}{a^4 y^2} = \frac{b^4 x^2 + a^4 b^2\left(1 - x^2/a^2\right)}{a^4 b^2\left(1 - x^2/a^2\right)} = \frac{a^4 b^2 + b^4 x^2 - a^2 b^2 x^2}{a^4 b^2 - a^2 b^2 x^2}$$

$$= \frac{a^4 + b^2 x^2 - a^2 x^2}{a^4 - a^2 x^2} = \frac{a^4 - \left(a^2 - b^2\right)x^2}{a^2(a^2 - x^2)}$$

The ellipsoid's surface area is twice the area generated by rotating the first-quadrant portion of the ellipse about the x-axis. Thus,

$$S = 2\int_0^a 2\pi y \sqrt{1 + \left(\frac{dy}{dx}\right)^2}\, dx = 4\pi \int_0^a \frac{b}{a}\sqrt{a^2 - x^2}\,\frac{\sqrt{a^4 - (a^2 - b^2)x^2}}{a\sqrt{a^2 - x^2}}\, dx = \frac{4\pi b}{a^2}\int_0^a \sqrt{a^4 - (a^2 - b^2)x^2}\, dx$$

$$= \frac{4\pi b}{a^2}\int_0^{a\sqrt{a^2 - b^2}} \sqrt{a^4 - u^2}\,\frac{du}{\sqrt{a^2 - b^2}} \quad [u = \sqrt{a^2 - b^2}\,x] \;\;\overset{30}{=}\;\; \frac{4\pi b}{a^2\sqrt{a^2 - b^2}}\left[\frac{u}{2}\sqrt{a^4 - u^2} + \frac{a^4}{2}\sin^{-1}\left(\frac{u}{a^2}\right)\right]_0^{a\sqrt{a^2 - b^2}}$$

$$= \frac{4\pi b}{a^2\sqrt{a^2 - b^2}}\left[\frac{a\sqrt{a^2 - b^2}}{2}\sqrt{a^4 - a^2(a^2 - b^2)} + \frac{a^4}{2}\sin^{-1}\frac{\sqrt{a^2 - b^2}}{a}\right] = 2\pi\left[b^2 + \frac{a^2 b \sin^{-1}\dfrac{\sqrt{a^2 - b^2}}{a}}{\sqrt{a^2 - b^2}}\right]$$

(b) $\dfrac{x^2}{a^2} + \dfrac{y^2}{b^2} = 1 \ \Rightarrow \ \dfrac{x\,(dx/dy)}{a^2} = -\dfrac{y}{b^2} \ \Rightarrow \ \dfrac{dx}{dy} = -\dfrac{a^2 y}{b^2 x} \ \Rightarrow$

$$1 + \left(\dfrac{dx}{dy}\right)^2 = 1 + \dfrac{a^4 y^2}{b^4 x^2} = \dfrac{b^4 x^2 + a^4 y^2}{b^4 x^2} = \dfrac{b^4 a^2(1 - y^2/b^2) + a^4 y^2}{b^4 a^2(1 - y^2/b^2)} = \dfrac{a^2 b^4 - a^2 b^2 y^2 + a^4 y^2}{a^2 b^4 - a^2 b^2 y^2}$$

$$= \dfrac{b^4 - b^2 y^2 + a^2 y^2}{b^4 - b^2 y^2} = \dfrac{b^4 - (b^2 - a^2)y^2}{b^2(b^2 - y^2)}$$

The oblate spheroid's surface area is twice the area generated by rotating the first-quadrant portion of the ellipse about the y-axis. Thus,

$$S = 2\int_0^b 2\pi x\,\sqrt{1 + \left(\dfrac{dx}{dy}\right)^2}\,dy = 4\pi \int_0^b \dfrac{a}{b}\sqrt{b^2 - y^2}\,\dfrac{\sqrt{b^4 - (b^2 - a^2)y^2}}{b\sqrt{b^2 - y^2}}\,dy$$

$$= \dfrac{4\pi a}{b^2}\int_0^b \sqrt{b^4 - (b^2 - a^2)y^2}\,dy = \dfrac{4\pi a}{b^2}\int_0^{b\sqrt{b^2 - a^2}} \sqrt{b^4 - u^2}\,\dfrac{du}{\sqrt{b^2 - a^2}} \qquad \left[u = \sqrt{b^2 - a^2}\,y\right]$$

$$\overset{30}{=} \dfrac{4\pi a}{b^2\sqrt{b^2 - a^2}}\left[\dfrac{u}{2}\sqrt{b^4 - u^2} + \dfrac{b^4}{2}\sin^{-1}\left(\dfrac{u}{b^2}\right)\right]_0^{b\sqrt{b^2 - a^2}}$$

$$= \dfrac{4\pi a}{b^2\sqrt{b^2 - a^2}}\left[\dfrac{b\sqrt{b^2 - a^2}}{2}\sqrt{b^4 - b^2(b^2 - a^2)} + \dfrac{b^4}{2}\sin^{-1}\dfrac{\sqrt{b^2 - a^2}}{b}\right] = 2\pi\left[a^2 + \dfrac{ab^2\sin^{-1}\dfrac{\sqrt{b^2 - a^2}}{b}}{\sqrt{b^2 - a^2}}\right]$$

Notice that this result can be obtained from the answer in part (a) by interchanging a and b.

31. The analogue of $f(x_i^*)$ in the derivation of (4) is now $c - f(x_i^*)$, so

$$S = \lim_{n\to\infty} \sum_{i=1}^{n} 2\pi[c - f(x_i^*)]\,\sqrt{1 + [f'(x_i^*)]^2}\,\Delta x = \int_a^b 2\pi[c - f(x)]\,\sqrt{1 + [f'(x)]^2}\,dx.$$

33. For the upper semicircle, $f(x) = \sqrt{r^2 - x^2}$, $f'(x) = -x/\sqrt{r^2 - x^2}$. The surface area generated is

$$S_1 = \int_{-r}^{r} 2\pi\left(r - \sqrt{r^2 - x^2}\right)\sqrt{1 + \dfrac{x^2}{r^2 - x^2}}\,dx = 4\pi\int_0^r \left(r - \sqrt{r^2 - x^2}\right)\dfrac{r}{\sqrt{r^2 - x^2}}\,dx$$

$$= 4\pi\int_0^r \left(\dfrac{r^2}{\sqrt{r^2 - x^2}} - r\right)dx$$

For the lower semicircle, $f(x) = -\sqrt{r^2 - x^2}$ and $f'(x) = \dfrac{x}{\sqrt{r^2 - x^2}}$, so $S_2 = 4\pi\int_0^r \left(\dfrac{r^2}{\sqrt{r^2 - x^2}} + r\right)dx$.

Thus, the total area is $S = S_1 + S_2 = 8\pi\int_0^r \left(\dfrac{r^2}{\sqrt{r^2 - x^2}}\right)dx = 8\pi\left[r^2\sin^{-1}\left(\dfrac{x}{r}\right)\right]_0^r = 8\pi r^2\left(\dfrac{\pi}{2}\right) = 4\pi^2 r^2$.

35. In the derivation of (4), we computed a typical contribution to the surface area to be $2\pi\dfrac{y_{i-1} + y_i}{2}\,|P_{i-1}P_i|$,

the area of a frustum of a cone. When $f(x)$ is not necessarily positive, the approximations $y_i = f(x_i) \approx f(x_i^*)$ and

$y_{i-1} = f(x_{i-1}) \approx f(x_i^*)$ must be replaced by $y_i = |f(x_i)| \approx |f(x_i^*)|$ and $y_{i-1} = |f(x_{i-1})| \approx |f(x_i^*)|$. Thus,

$2\pi\dfrac{y_{i-1} + y_i}{2}\,|P_{i-1}P_i| \approx 2\pi\,|f(x_i^*)|\,\sqrt{1 + [f'(x_i^*)]^2}\,\Delta x$. Continuing with the rest of the derivation as before,

we obtain $S = \int_a^b 2\pi\,|f(x)|\,\sqrt{1 + [f'(x)]^2}\,dx$.

8.3 Applications to Physics and Engineering

1. The weight density of water is $\delta = 62.5 \ \text{lb/ft}^3$.

(a) $P = \delta d \approx (62.5 \ \text{lb/ft}^3)(3 \ \text{ft}) = 187.5 \ \text{lb/ft}^2$

(b) $F = PA \approx (187.5 \ \text{lb/ft}^2)(5 \ \text{ft})(2 \ \text{ft}) = 1875 \ \text{lb}$.   (*A* is the area of the bottom of the tank.)

(c) As in Example 1, the area of the *i*th strip is $2 \, (\Delta x)$ and the pressure is $\delta d = \delta x_i$. Thus,

$\quad F = \int_0^3 \delta x \cdot 2 \, dx \approx (62.5)(2) \int_0^3 x \, dx = 125 \left[\frac{1}{2}x^2\right]_0^3 = 125 \left(\frac{9}{2}\right) = 562.5 \ \text{lb}$.

In Exercises 3–9, n is the number of subintervals of length Δx and x_i^* is a sample point in the *i*th subinterval $[x_{i-1}, x_i]$.

3. Set up a vertical x-axis as shown, with $x = 0$ at the water's surface and x increasing in the

downward direction. Then the area of the *i*th rectangular strip is $6 \, \Delta x$ and the pressure on

the strip is δx_i^* (where $\delta \approx 62.5 \ \text{lb/ft}^3$). Thus, the hydrostatic force on the strip is

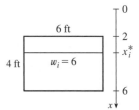

$\delta x_i^* \cdot 6 \, \Delta x$ and the total hydrostatic force $\approx \sum\limits_{i=1}^{n} \delta x_i^* \cdot 6 \, \Delta x$. The total force

$$F = \lim_{n \to \infty} \sum_{i=1}^{n} \delta x_i^* \cdot 6 \, \Delta x = \int_2^6 \delta x \cdot 6 \, dx = 6\delta \int_2^6 x \, dx = 6\delta \left[\frac{1}{2}x^2\right]_2^6 = 6\delta(18 - 2) = 96\delta \approx 6000 \ \text{lb}$$

5. Set up a vertical x-axis as shown. The base of the triangle shown in the figure

has length $\sqrt{3^2 - (x_i^*)^2}$, so $w_i = 2\sqrt{9 - (x_i^*)^2}$, and the area of the *i*th

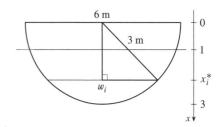

rectangular strip is $2\sqrt{9 - (x_i^*)^2}\,\Delta x$. The *i*th rectangular strip is $(x_i^* - 1)$ m

below the surface level of the water, so the pressure on the strip is $\rho g(x_i^* - 1)$.

The hydrostatic force on the strip is $\rho g(x_i^* - 1) \cdot 2\sqrt{9 - (x_i^*)^2}\,\Delta x$ and the total

force on the plate $\approx \sum\limits_{i=1}^{n} \rho g(x_i^* - 1) \cdot 2\sqrt{9 - (x_i^*)^2}\,\Delta x$. The total force

$$F = \lim \sum_{i=1}^{n} \rho g(x_i^* - 1) \cdot 2\sqrt{9 - (x_i^*)^2}\,\Delta x = 2\rho g \int_1^3 (x - 1)\sqrt{9 - x^2}\,dx$$

$$= 2\rho g \int_1^3 x\sqrt{9 - x^2}\,dx - 2\rho g \int_1^3 \sqrt{9 - x^2}\,dx \overset{30}{=} 2\rho g \left[-\frac{1}{3}(9 - x^2)^{3/2}\right]_1^3 - 2\rho g \left[\frac{x}{2}\sqrt{9 - x^2} + \frac{9}{2}\sin^{-1}\left(\frac{x}{3}\right)\right]_1^3$$

$$= 2\rho g \left[0 + \frac{1}{3}\left(8\sqrt{8}\right)\right] - 2\rho g \left[\left(0 + \frac{9}{2} \cdot \frac{\pi}{2}\right) - \left(\frac{1}{2}\sqrt{8} + \frac{9}{2}\sin^{-1}\left(\frac{1}{3}\right)\right)\right]$$

$$= \frac{32}{3}\sqrt{2}\,\rho g - \frac{9\pi}{2}\rho g + 2\sqrt{2}\,\rho g + 9\left[\sin^{-1}\left(\frac{1}{3}\right)\right]\rho g = \left(\frac{38}{3}\sqrt{2} - \frac{9\pi}{2} + 9\sin^{-1}\left(\frac{1}{3}\right)\right)\rho g$$

$$\approx 6.835 \cdot 1000 \cdot 9.8 \approx 6.7 \times 10^4 \ \text{N}$$

Note: If you set up a typical coordinate system with the water level at $y = -1$, then $F = \int_{-3}^{-1} \rho g(-1 - y)2\sqrt{9 - y^2}\,dy$.

7. Set up a vertical x-axis as shown. Then the area of the ith rectangular strip is

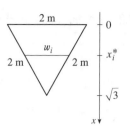

$\left(2 - \dfrac{2}{\sqrt{3}} x_i^*\right) \Delta x.$ $\left[\text{By similar triangles, } \dfrac{w_i}{2} = \dfrac{\sqrt{3} - x_i^*}{\sqrt{3}}, \text{ so } w_i = 2 - \dfrac{2}{\sqrt{3}} x_i^*.\right]$

The pressure on the strip is $\rho g x_i^*$, so the hydrostatic force on the strip is

$\rho g x_i^* \left(2 - \dfrac{2}{\sqrt{3}} x_i^*\right) \Delta x$ and the hydrostatic force on the plate $\approx \sum\limits_{i=1}^{n} \rho g x_i^* \left(2 - \dfrac{2}{\sqrt{3}} x_i^*\right) \Delta x.$

The total force

$$F = \lim_{n \to \infty} \sum_{i=1}^{n} \rho g x_i^* \left(2 - \frac{2}{\sqrt{3}} x_i^*\right) \Delta x = \int_0^{\sqrt{3}} \rho g x \left(2 - \frac{2}{\sqrt{3}} x\right) dx = \rho g \int_0^{\sqrt{3}} \left(2x - \frac{2}{\sqrt{3}} x^2\right) dx$$

$$= \rho g \left[x^2 - \frac{2}{3\sqrt{3}} x^3\right]_0^{\sqrt{3}} = \rho g \left[(3 - 2) - 0\right] = \rho g \approx 1000 \cdot 9.8 = 9.8 \times 10^3 \text{ N}$$

9. Set up coordinate axes as shown in the figure. The length of the ith strip is

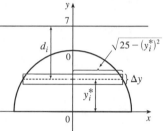

$2\sqrt{25 - (y_i^*)^2}$ and its area is $2\sqrt{25 - (y_i^*)^2} \, \Delta y$. The pressure on this strip is

approximately $\delta d_i = 62.5(7 - y_i^*)$ and so the force on the strip is approximately

$62.5(7 - y_i^*)2\sqrt{25 - (y_i^*)^2} \, \Delta y$. The total force

$$F = \lim_{n \to \infty} \sum_{i=1}^{n} 62.5(7 - y_i^*)2\sqrt{25 - (y_i^*)^2} \, \Delta y = 125 \int_0^5 (7 - y)\sqrt{25 - y^2} \, dy$$

$$= 125 \left\{ \int_0^5 7\sqrt{25 - y^2} \, dy - \int_0^5 y\sqrt{25 - y^2} \, dy \right\} = 125 \left\{ 7 \int_0^5 \sqrt{25 - y^2} \, dy - \left[-\tfrac{1}{3}(25 - y^2)^{3/2}\right]_0^5 \right\}$$

$$= 125 \left\{ 7 \left(\tfrac{1}{4}\pi \cdot 5^2\right) + \tfrac{1}{3}(0 - 125) \right\} = 125 \left(\tfrac{175\pi}{4} - \tfrac{125}{3}\right) \approx 11{,}972 \approx 1.2 \times 10^4 \text{ lb}$$

11. Set up a vertical x-axis as shown. Then the area of the ith rectangular strip is

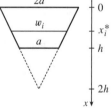

$\dfrac{a}{h}(2h - x_i^*) \Delta x.$ $\left[\text{By similar triangles, } \dfrac{w_i}{2h - x_i^*} = \dfrac{2a}{2h}, \text{ so } w_i = \dfrac{a}{h}(2h - x_i^*).\right]$

The pressure on the strip is δx_i^*, so the hydrostatic force on the plate

$\approx \sum\limits_{i=1}^{n} \delta x_i^* \dfrac{a}{h}(2h - x_i^*) \Delta x.$ The total force

$$F = \lim_{n \to \infty} \sum_{i=1}^{n} \delta x_i^* \frac{a}{h}(2h - x_i^*) \Delta x = \delta \frac{a}{h} \int_0^h x(2h - x) \, dx = \frac{a\delta}{h} \int_0^h \left(2hx - x^2\right) dx$$

$$= \frac{a\delta}{h} \left[hx^2 - \tfrac{1}{3}x^3\right]_0^h = \frac{a\delta}{h}\left(h^3 - \tfrac{1}{3}h^3\right) = \frac{a\delta}{h}\left(\frac{2h^3}{3}\right) = \tfrac{2}{3}\delta a h^2$$

13. By similar triangles, $\dfrac{8}{4\sqrt{3}} = \dfrac{w_i}{x_i^*} \Rightarrow w_i = \dfrac{2x_i^*}{\sqrt{3}}$. The area of the ith

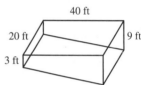

rectangular strip is $\dfrac{2x_i^*}{\sqrt{3}}\,\Delta x$ and the pressure on it is $\rho g\left(4\sqrt{3} - x_i^*\right)$.

$$F = \int_0^{4\sqrt{3}} \rho g\left(4\sqrt{3} - x\right)\frac{2x}{\sqrt{3}}\,dx = 8\rho g\int_0^{4\sqrt{3}} x\,dx - \frac{2\rho g}{\sqrt{3}}\int_0^{4\sqrt{3}} x^2\,dx$$

$$= 4\rho g\left[x^2\right]_0^{4\sqrt{3}} - \frac{2\rho g}{3\sqrt{3}}\left[x^3\right]_0^{4\sqrt{3}} = 192\rho g - \frac{2\rho g}{3\sqrt{3}}\,64\cdot 3\sqrt{3} = 192\rho g - 128\rho g = 64\rho g$$

$$\approx 64(840)(9.8) \approx 5.27\times 10^5\text{ N}$$

15. (a) The top of the cube has depth $d = 1\text{ m} - 20\text{ cm} = 80\text{ cm} = 0.8\text{ m}$.

$$F = \rho g dA \approx (1000)(9.8)(0.8)(0.2)^2 = 313.6 \approx 314\text{ N}$$

(b) The area of a strip is $0.2\,\Delta x$ and the pressure on it is $\rho g x_i^*$.

$$F = \int_{0.8}^1 \rho g x (0.2)\,dx = 0.2\rho g\left[\tfrac{1}{2}x^2\right]_{0.8}^1 = (0.2\rho g)(0.18) = 0.036\rho g = 0.036(1000)(9.8) = 352.8 \approx 353\text{ N}$$

17. (a) The area of a strip is $20\,\Delta x$ and the pressure on it is δx_i.

$$F = \int_0^3 \delta x 20\,dx = 20\delta\left[\tfrac{1}{2}x^2\right]_0^3 = 20\delta\cdot\tfrac{9}{2} = 90\delta$$

$$= 90(62.5) = 5625\text{ lb} \approx 5.63\times 10^3\text{ lb}$$

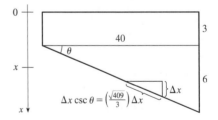

(b) $F = \int_0^9 \delta x 20\,dx = 20\delta\left[\tfrac{1}{2}x^2\right]_0^9 = 20\delta\cdot\tfrac{81}{2} = 810\delta = 810(62.5) = 50{,}625\text{ lb} \approx 5.06\times 10^4\text{ lb.}$

(c) For the first 3 ft, the length of the side is constant at 40 ft. For $3 < x \le 9$, we can use similar triangles to find the length a:

$$\frac{a}{40} = \frac{9-x}{6} \Rightarrow a = 40\cdot\frac{9-x}{6}.$$

$$F = \int_0^3 \delta x 40\,dx + \int_3^9 \delta x(40)\frac{9-x}{6}\,dx = 40\delta\left[\tfrac{1}{2}x^2\right]_0^3 + \frac{20}{3}\delta\int_3^9(9x - x^2)\,dx = 180\delta + \frac{20}{3}\delta\left[\tfrac{9}{2}x^2 - \tfrac{1}{3}x^3\right]_3^9$$

$$= 180\delta + \frac{20}{3}\delta\left[\left(\tfrac{729}{2} - 243\right) - \left(\tfrac{81}{2} - 9\right)\right] = 180\delta + 600\delta = 780\delta = 780(62.5) = 48{,}750\text{ lb} \approx 4.88\times 10^4\text{ lb}$$

(d) For any right triangle with hypotenuse on the bottom,

$$\sin\theta = \frac{\Delta x}{\text{hypotenuse}} \Rightarrow$$

$$\text{hypotenuse} = \Delta x\csc\theta = \Delta x\frac{\sqrt{40^2 + 6^2}}{6} = \frac{\sqrt{409}}{3}\Delta x.$$

$$F = \int_3^9 \delta x 20\frac{\sqrt{409}}{3}\,dx = \tfrac{1}{3}\left(20\sqrt{409}\right)\delta\left[\tfrac{1}{2}x^2\right]_3^9$$

$$= \tfrac{1}{3}\cdot 10\sqrt{409}\,\delta(81 - 9) \approx 303{,}356\text{ lb} \approx 3.03\times 10^5\text{ lb}$$

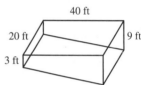

19. $F = \int_2^5 \rho g x\cdot w(x)\,dx$, where $w(x)$ is the width of the plate at depth x. Since $n = 6$, $\Delta x = \frac{5-2}{6} = \frac{1}{2}$, and $F \approx S_6$

$$= \rho g\cdot\frac{1/2}{3}\left[2\cdot w(2) + 4\cdot 2.5\cdot w(2.5) + 2\cdot 3\cdot w(3) + 4\cdot 3.5\cdot w(3.5) + 2\cdot 4\cdot w(4) + 4\cdot 4.5\cdot w(4.5) + 5\cdot w(5)\right]$$

$$= \tfrac{1}{6}\rho g(2\cdot 0 + 10\cdot 0.8 + 6\cdot 1.7 + 14\cdot 2.4 + 8\cdot 2.9 + 18\cdot 3.3 + 5\cdot 3.6)$$

$$= \tfrac{1}{6}(1000)(9.8)(152.4) \approx 2.5\times 10^5\text{ N}$$

21. The moment M of the system about the origin is $M = \sum\limits_{i=1}^{2} m_i x_i = m_1 x_1 + m_2 x_2 = 40 \cdot 2 + 30 \cdot 5 = 230$.

The mass m of the system is $m = \sum\limits_{i=1}^{2} m_i = m_1 + m_2 = 40 + 30 = 70$.

The center of mass of the system is $M/m = \frac{230}{70} = \frac{23}{7}$.

23. $m = \sum\limits_{i=1}^{3} m_i = 6 + 5 + 10 = 21$.

$M_x = \sum\limits_{i=1}^{3} m_i\, y_i = 6(5) + 5(-2) + 10(-1) = 10$; $M_y = \sum\limits_{i=1}^{3} m_i\, x_i = 6(1) + 5(3) + 10(-2) = 1$.

$\overline{x} = \dfrac{M_y}{m} = \dfrac{1}{21}$ and $\overline{y} = \dfrac{M_x}{m} = \dfrac{10}{21}$, so the center of mass of the system is $\left(\frac{1}{21}, \frac{10}{21}\right)$.

25. Since the region in the figure is symmetric about the y-axis, we know

that $\overline{x} = 0$. The region is "bottom-heavy," so we know that $\overline{y} < 2$,

and we might guess that $\overline{y} = 1.5$.

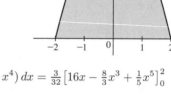

$A = \int_{-2}^{2}(4 - x^2)\,dx = 2\int_{0}^{2}(4 - x^2)\,dx = 2\left[4x - \frac{1}{3}x^3\right]_0^2$

$\qquad = 2\left(8 - \frac{8}{3}\right) = \frac{32}{3}$.

$\overline{x} = \frac{1}{A}\int_{-2}^{2} x(4 - x^2)\,dx = 0$ since $f(x) = x(4 - x^2)$ is an odd

function (or since the region is symmetric about the y-axis).

$\overline{y} = \frac{1}{A}\int_{-2}^{2}\frac{1}{2}(4 - x^2)^2\,dx = \frac{3}{32} \cdot \frac{1}{2} \cdot 2\int_{0}^{2}(16 - 8x^2 + x^4)\,dx = \frac{3}{32}\left[16x - \frac{8}{3}x^3 + \frac{1}{5}x^5\right]_0^2$

$\qquad = \frac{3}{32}\left(32 - \frac{64}{3} + \frac{32}{5}\right) = 3\left(1 - \frac{2}{3} + \frac{1}{5}\right) = 3\left(\frac{8}{15}\right) = \frac{8}{5}$

Thus, the centroid is $(\overline{x}, \overline{y}) = \left(0, \frac{8}{5}\right)$.

27. The region in the figure is "right-heavy" and "bottom-heavy," so we know

$\overline{x} > 0.5$ and $\overline{y} < 1$, and we might guess that $\overline{x} = 0.6$ and $\overline{y} = 0.9$.

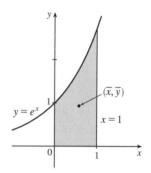

$A = \int_{0}^{1} e^x\,dx = [e^x]_0^1 = e - 1$.

$\overline{x} = \frac{1}{A}\int_{0}^{1} xe^x\,dx = \frac{1}{e-1}[xe^x - e^x]_0^1$ [by parts]

$\qquad = \frac{1}{e-1}[0 - (-1)] = \frac{1}{e-1}$.

$\overline{y} = \frac{1}{A}\int_{0}^{1}\frac{1}{2}(e^x)^2\,dx = \frac{1}{e-1} \cdot \frac{1}{4}[e^{2x}]_0^1 = \frac{1}{4(e-1)}(e^2 - 1) = \frac{e+1}{4}$.

Thus, the centroid is $(\overline{x}, \overline{y}) = \left(\frac{1}{e-1}, \frac{e+1}{4}\right) \approx (0.58, 0.93)$.

29. $A = \int_{0}^{1}(x^{1/2} - x^2)\,dx = \left[\frac{2}{3}x^{3/2} - \frac{1}{3}x^3\right]_0^1 = \left(\frac{2}{3} - \frac{1}{3}\right) - 0 = \frac{1}{3}$.

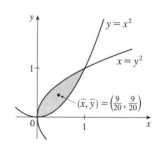

$\overline{x} = \frac{1}{A}\int_{0}^{1} x(x^{1/2} - x^2)\,dx = 3\int_{0}^{1}(x^{3/2} - x^3)\,dx$

$\qquad = 3\left[\frac{2}{5}x^{5/2} - \frac{1}{4}x^4\right]_0^1 = 3\left(\frac{2}{5} - \frac{1}{4}\right) = 3\left(\frac{3}{20}\right) = \frac{9}{20}$.

$\overline{y} = \frac{1}{A}\int_{0}^{1}\frac{1}{2}\left[(x^{1/2})^2 - (x^2)^2\right]\,dx = 3\left(\frac{1}{2}\right)\int_{0}^{1}(x - x^4)\,dx$

$\qquad = \frac{3}{2}\left[\frac{1}{2}x^2 - \frac{1}{5}x^5\right]_0^1 = \frac{3}{2}\left(\frac{1}{2} - \frac{1}{5}\right) = \frac{3}{2}\left(\frac{3}{10}\right) = \frac{9}{20}$.

Thus, the centroid is $(\overline{x}, \overline{y}) = \left(\frac{9}{20}, \frac{9}{20}\right)$.

31. $A = \int_0^{\pi/4} (\cos x - \sin x)\, dx = \big[\sin x + \cos x \big]_0^{\pi/4} = \sqrt{2} - 1.$

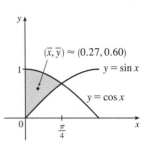

$$\bar{x} = A^{-1} \int_0^{\pi/4} x(\cos x - \sin x)\, dx$$

$$= A^{-1} \big[x(\sin x + \cos x) + \cos x - \sin x \big]_0^{\pi/4} \quad \text{[integration by parts]}$$

$$= A^{-1} \left(\tfrac{\pi}{4}\sqrt{2} - 1 \right) = \frac{\tfrac{1}{4}\pi\sqrt{2} - 1}{\sqrt{2} - 1}.$$

$$\bar{y} = A^{-1} \int_0^{\pi/4} \tfrac{1}{2}(\cos^2 x - \sin^2 x)\, dx = \tfrac{1}{2A} \int_0^{\pi/4} \cos 2x\, dx = \tfrac{1}{4A} \big[\sin 2x \big]_0^{\pi/4} = \frac{1}{4A} = \frac{1}{4\left(\sqrt{2} - 1\right)}.$$

Thus, the centroid is $(\bar{x}, \bar{y}) = \left(\dfrac{\pi\sqrt{2} - 4}{4(\sqrt{2} - 1)}, \dfrac{1}{4(\sqrt{2} - 1)} \right) \approx (0.27, 0.60).$

33. From the figure we see that $\bar{y} = 0$. Now

$$A = \int_0^5 2\sqrt{5 - x}\, dx = 2\left[-\tfrac{2}{3}(5 - x)^{3/2} \right]_0^5 = 2\left(0 + \tfrac{2}{3} \cdot 5^{3/2} \right) = \tfrac{20}{3}\sqrt{5},$$

so

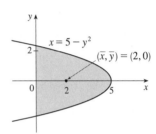

$$\bar{x} = \tfrac{1}{A} \int_0^5 x \big[\sqrt{5 - x} - \left(-\sqrt{5 - x} \right) \big]\, dx = \tfrac{1}{A} \int_0^5 2x\sqrt{5 - x}\, dx$$

$$= \tfrac{1}{A} \int_{\sqrt{5}}^0 2(5 - u^2)u(-2u)\, du \qquad \begin{bmatrix} u = \sqrt{5 - x},\ x = 5 - u^2, \\ u^2 = 5 - x,\ dx = -2u\, du \end{bmatrix}$$

$$= \tfrac{4}{A} \int_0^{\sqrt{5}} u^2(5 - u^2)\, du = \tfrac{4}{A} \left[\tfrac{5}{3}u^3 - \tfrac{1}{5}u^5 \right]_0^{\sqrt{5}} = \tfrac{3}{5\sqrt{5}} \left(\tfrac{25}{3}\sqrt{5} - 5\sqrt{5} \right) = 5 - 3 = 2.$$

Thus, the centroid is $(\bar{x}, \bar{y}) = (2, 0).$

35. The line has equation $y = \tfrac{3}{4}x.$ $A = \tfrac{1}{2}(4)(3) = 6$, so $m = \rho A = 10(6) = 60.$

$$M_x = \rho \int_0^4 \tfrac{1}{2} \left(\tfrac{3}{4}x \right)^2 dx = 10 \int_0^4 \tfrac{9}{32}x^2\, dx = \tfrac{45}{16} \big[\tfrac{1}{3}x^3 \big]_0^4 = \tfrac{45}{16} \left(\tfrac{64}{3} \right) = 60$$

$$M_y = \rho \int_0^4 x \left(\tfrac{3}{4}x \right) dx = \tfrac{15}{2} \int_0^4 x^2\, dx = \tfrac{15}{2} \big[\tfrac{1}{3}x^3 \big]_0^4 = \tfrac{15}{2} \left(\tfrac{64}{3} \right) = 160$$

$\bar{x} = \dfrac{M_y}{m} = \dfrac{160}{60} = \dfrac{8}{3}$ and $\bar{y} = \dfrac{M_x}{m} = \dfrac{60}{60} = 1.$ Thus, the centroid is $(\bar{x}, \bar{y}) = \left(\tfrac{8}{3}, 1 \right).$

37. $A = \displaystyle\int_0^2 (2^x - x^2)\, dx = \left[\dfrac{2^x}{\ln 2} - \dfrac{x^3}{3} \right]_0^2$

$$= \left(\dfrac{4}{\ln 2} - \dfrac{8}{3} \right) - \dfrac{1}{\ln 2} = \dfrac{3}{\ln 2} - \dfrac{8}{3} \approx 1.661418.$$

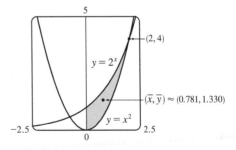

$$\bar{x} = \dfrac{1}{A} \int_0^2 x(2^x - x^2)\, dx = \dfrac{1}{A} \int_0^2 (x2^x - x^3)\, dx$$

$$= \dfrac{1}{A} \left[\dfrac{x2^x}{\ln 2} - \dfrac{2^x}{(\ln 2)^2} - \dfrac{x^4}{4} \right]_0^2 \quad \text{[use parts]}$$

$$= \dfrac{1}{A} \left[\dfrac{8}{\ln 2} - \dfrac{4}{(\ln 2)^2} - 4 + \dfrac{1}{(\ln 2)^2} \right] = \dfrac{1}{A} \left[\dfrac{8}{\ln 2} - \dfrac{3}{(\ln 2)^2} - 4 \right] \approx \dfrac{1}{A}(1.297453) \approx 0.781$$

[continued]

$$\bar{y} = \frac{1}{A}\int_0^2 \tfrac{1}{2}[(2^x)^2 - (x^2)^2]\,dx = \frac{1}{A}\int_0^2 \tfrac{1}{2}(2^{2x} - x^4)\,dx = \frac{1}{A}\cdot\frac{1}{2}\left[\frac{2^{2x}}{2\ln 2} - \frac{x^5}{5}\right]_0^2$$

$$= \frac{1}{A}\cdot\frac{1}{2}\left(\frac{16}{2\ln 2} - \frac{32}{5} - \frac{1}{2\ln 2}\right) = \frac{1}{A}\left(\frac{15}{4\ln 2} - \frac{16}{5}\right) \approx \frac{1}{A}(2.210106) \approx 1.330$$

Thus, the centroid is $(\bar{x}, \bar{y}) \approx (0.781, 1.330)$.

Since the position of a centroid is independent of density when the density is constant, we will assume for convenience that $\rho = 1$ in Exercises 38 and 39.

39. Choose x- and y-axes so that the base (one side of the triangle) lies along the x-axis with the other vertex along the positive y-axis as shown. From geometry, we know the medians intersect at a point $\frac{2}{3}$ of the way from each vertex (along the median) to the opposite side. The median from B goes to the midpoint $\left(\frac{1}{2}(a+c), 0\right)$ of side AC, so the point of intersection of the medians is $\left(\frac{2}{3}\cdot\frac{1}{2}(a+c), \frac{1}{3}b\right) = \left(\frac{1}{3}(a+c), \frac{1}{3}b\right)$.

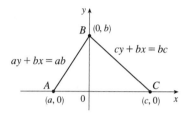

This can also be verified by finding the equations of two medians, and solving them simultaneously to find their point of intersection. Now let us compute the location of the centroid of the triangle. The area is $A = \frac{1}{2}(c-a)b$.

$$\bar{x} = \frac{1}{A}\left[\int_a^0 x\cdot\frac{b}{a}(a-x)\,dx + \int_0^c x\cdot\frac{b}{c}(c-x)\,dx\right] = \frac{1}{A}\left[\frac{b}{a}\int_a^0 (ax-x^2)\,dx + \frac{b}{c}\int_0^c (cx-x^2)\,dx\right]$$

$$= \frac{b}{Aa}\left[\frac{1}{2}ax^2 - \frac{1}{3}x^3\right]_a^0 + \frac{b}{Ac}\left[\frac{1}{2}cx^2 - \frac{1}{3}x^3\right]_0^c = \frac{b}{Aa}\left[-\frac{1}{2}a^3 + \frac{1}{3}a^3\right] + \frac{b}{Ac}\left[\frac{1}{2}c^3 - \frac{1}{3}c^3\right]$$

$$= \frac{2}{a(c-a)}\cdot\frac{-a^3}{6} + \frac{2}{c(c-a)}\cdot\frac{c^3}{6} = \frac{1}{3(c-a)}(c^2 - a^2) = \frac{a+c}{3}$$

and

$$\bar{y} = \frac{1}{A}\left[\int_a^0 \frac{1}{2}\left(\frac{b}{a}(a-x)\right)^2 dx + \int_0^c \frac{1}{2}\left(\frac{b}{c}(c-x)\right)^2 dx\right]$$

$$= \frac{1}{A}\left[\frac{b^2}{2a^2}\int_a^0 (a^2 - 2ax + x^2)\,dx + \frac{b^2}{2c^2}\int_0^c (c^2 - 2cx + x^2)\,dx\right]$$

$$= \frac{1}{A}\left[\frac{b^2}{2a^2}[a^2 x - ax^2 + \tfrac{1}{3}x^3]_a^0 + \frac{b^2}{2c^2}[c^2 x - cx^2 + \tfrac{1}{3}x^3]_0^c\right]$$

$$= \frac{1}{A}\left[\frac{b^2}{2a^2}(-a^3 + a^3 - \tfrac{1}{3}a^3) + \frac{b^2}{2c^2}(c^3 - c^3 + \tfrac{1}{3}c^3)\right] = \frac{1}{A}\left[\frac{b^2}{6}(-a+c)\right] = \frac{2}{(c-a)b}\cdot\frac{(c-a)b^2}{6} = \frac{b}{3}$$

Thus, the centroid is $(\bar{x}, \bar{y}) = \left(\dfrac{a+c}{3}, \dfrac{b}{3}\right)$, as claimed.

Remarks: Actually the computation of $\bar{y}$ is all that is needed. By considering each side of the triangle in turn to be the base, we see that the centroid is $\frac{1}{3}$ of the way from each side to the opposite vertex and must therefore be the intersection of the medians.

The computation of $\overline{y}$ in this problem (and many others) can be simplified by using horizontal rather than vertical approximating rectangles. If the length of a thin rectangle at coordinate y is $\ell(y)$, then its area is $\ell(y)\,\Delta y$, its mass is $\rho\ell(y)\,\Delta y$, and its moment about the x-axis is $\Delta M_x = \rho y\ell(y)\,\Delta y$. Thus,

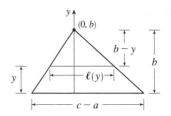

$$M_x = \int \rho y\ell(y)\,dy \qquad \text{and} \qquad \overline{y} = \frac{\int \rho y\ell(y)\,dy}{\rho A} = \frac{1}{A}\int y\ell(y)\,dy$$

In this problem, $\ell(y) = \dfrac{c-a}{b}(b-y)$ by similar triangles, so

$$\overline{y} = \frac{1}{A}\int_0^b \frac{c-a}{b}\,y(b-y)\,dy = \frac{2}{b^2}\int_0^b (by - y^2)\,dy = \frac{2}{b^2}\left[\tfrac{1}{2}by^2 - \tfrac{1}{3}y^3\right]_0^b = \frac{2}{b^2}\cdot\frac{b^3}{6} = \frac{b}{3}$$

Notice that only one integral is needed when this method is used.

41. Divide the lamina into two triangles and one rectangle with respective masses of 2, 2 and 4, so that the total mass is 8. Using the result of Exercise 39, the triangles have centroids $\left(-1, \tfrac{2}{3}\right)$ and $\left(1, \tfrac{2}{3}\right)$. The centroid of the rectangle (its center) is $\left(0, -\tfrac{1}{2}\right)$.

So, using Formulas 5 and 7, we have $\overline{y} = \dfrac{M_x}{m} = \dfrac{1}{m}\displaystyle\sum_{i=1}^{3} m_i\, y_i = \tfrac{1}{8}\left[2\left(\tfrac{2}{3}\right) + 2\left(\tfrac{2}{3}\right) + 4\left(-\tfrac{1}{2}\right)\right] = \tfrac{1}{8}\left(\tfrac{2}{3}\right) = \tfrac{1}{12}$, and $\overline{x} = 0$,

since the lamina is symmetric about the line $x = 0$. Thus, the centroid is $(\overline{x}, \overline{y}) = \left(0, \tfrac{1}{12}\right)$.

43. $\displaystyle\int_a^b (cx + d)\,f(x)\,dx = \int_a^b cx\,f(x)\,dx + \int_a^b d\,f(x)\,dx = c\int_a^b x\,f(x)\,dx + d\int_a^b f(x)\,dx = c\overline{x}A + d\int_a^b f(x)\,dx$ [by (8)]

$$= c\overline{x}\int_a^b f(x)\,dx + d\int_a^b f(x)\,dx = (c\overline{x} + d)\int_a^b f(x)\,dx$$

45. A cone of height h and radius r can be generated by rotating a right triangle about one of its legs as shown. By Exercise 39, $\overline{x} = \tfrac{1}{3}r$, so by the Theorem of Pappus, the volume of the cone is

$$V = Ad = \left(\tfrac{1}{2}\cdot \text{base}\cdot\text{height}\right)\cdot(2\pi\overline{x}) = \tfrac{1}{2}rh\cdot 2\pi\left(\tfrac{1}{3}r\right) = \tfrac{1}{3}\pi r^2 h.$$

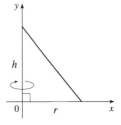

47. Suppose the region lies between two curves $y = f(x)$ and $y = g(x)$ where $f(x) \geq g(x)$, as illustrated in Figure 13. Choose points x_i with $a = x_0 < x_1 < \cdots < x_n = b$ and choose x_i^* to be the midpoint of the ith subinterval; that is, $x_i^* = \overline{x}_i = \tfrac{1}{2}(x_{i-1} + x_i)$. Then the centroid of the ith approximating rectangle R_i is its center $C_i = \left(\overline{x}_i, \tfrac{1}{2}[f(\overline{x}_i) + g(\overline{x}_i)]\right)$. Its area is $[f(\overline{x}_i) - g(\overline{x}_i)]\,\Delta x$, so its mass is

$\rho[f(\overline{x}_i) - g(\overline{x}_i)]\,\Delta x$. Thus, $M_y(R_i) = \rho[f(\overline{x}_i) - g(\overline{x}_i)]\,\Delta x\cdot\overline{x}_i = \rho\overline{x}_i\,[f(\overline{x}_i) - g(\overline{x}_i)]\,\Delta x$ and

$M_x(R_i) = \rho[f(\overline{x}_i) - g(\overline{x}_i)]\,\Delta x\cdot\tfrac{1}{2}[f(\overline{x}_i) + g(\overline{x}_i)] = \rho\cdot\tfrac{1}{2}\left[f(\overline{x}_i)^2 - g(\overline{x}_i)^2\right]\Delta x$. Summing over i and taking the limit

as $n \to \infty$, we get $M_y = \displaystyle\lim_{n\to\infty}\sum_i \rho\overline{x}_i\,[f(\overline{x}_i) - g(\overline{x}_i)]\,\Delta x = \rho\int_a^b x[f(x) - g(x)]\,dx$ and

$M_x = \displaystyle\lim_{n\to\infty}\sum_i \rho\cdot\tfrac{1}{2}\left[f(\overline{x}_i)^2 - g(\overline{x}_i)^2\right]\Delta x = \rho\int_a^b \tfrac{1}{2}\left[f(x)^2 - g(x)^2\right]dx$.

Thus, $\overline{x} = \dfrac{M_y}{m} = \dfrac{M_y}{\rho A} = \dfrac{1}{A}\displaystyle\int_a^b x[f(x) - g(x)]\,dx$ and $\overline{y} = \dfrac{M_x}{m} = \dfrac{M_x}{\rho A} = \dfrac{1}{A}\displaystyle\int_a^b \tfrac{1}{2}\left[f(x)^2 - g(x)^2\right]dx$.

8.4 Applications to Economics and Biology

1. By the Net Change Theorem, $C(2000) - C(0) = \int_0^{2000} C'(x)\,dx$ ⇒

$$C(2000) = 20{,}000 + \int_0^{2000}(5 - 0.008x + 0.000009x^2)\,dx = 20{,}000 + \left[5x - 0.004x^2 + 0.000003x^3\right]_0^{2000}$$

$$= 20{,}000 + 10{,}000 - 0.004(4{,}000{,}000) + 0.000003(8{,}000{,}000{,}000) = 30{,}000 - 16{,}000 + 24{,}000$$

$$= \$38{,}000$$

3. If the production level is raised from 1200 units to 1600 units, then the increase in cost is

$$C(1600) - C(1200) = \int_{1200}^{1600} C'(x)\,dx = \int_{1200}^{1600}(74 + 1.1x - 0.002x^2 + 0.00004x^3)\,dx$$

$$= \left[74x + 0.55x^2 - \tfrac{0.002}{3}x^3 + 0.00001x^4\right]_{1200}^{1600} = 64{,}331{,}733.33 - 20{,}464{,}800 = \$43{,}866{,}933.33$$

5. $p(x) = 10$ ⇒ $\dfrac{450}{x+8} = 10$ ⇒ $x + 8 = 45$ ⇒ $x = 37$.

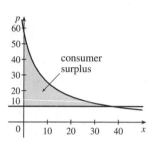

$$\text{Consumer surplus} = \int_0^{37}[p(x) - 10]\,dx = \int_0^{37}\left(\frac{450}{x+8} - 10\right)dx$$

$$= \left[450\ln(x+8) - 10x\right]_0^{37} = (450\ln 45 - 370) - 450\ln 8$$

$$= 450\ln\left(\tfrac{45}{8}\right) - 370 \approx \$407.25$$

7. $P = p_S(x)$ ⇒ $400 = 200 + 0.2x^{3/2}$ ⇒ $200 = 0.2x^{3/2}$ ⇒ $1000 = x^{3/2}$ ⇒ $x = 1000^{2/3} = 100$.

$$\text{Producer surplus} = \int_0^{100}[P - p_S(x)]\,dx = \int_0^{100}[400 - (200 + 0.2x^{3/2})]\,dx = \int_0^{100}\left(200 - \tfrac{1}{5}x^{3/2}\right)dx$$

$$= \left[200x - \tfrac{2}{25}x^{5/2}\right]_0^{100} = 20{,}000 - 8{,}000 = \$12{,}000$$

9. $p(x) = \dfrac{800{,}000e^{-x/5000}}{x + 20{,}000} = 16$ ⇒ $x = x_1 \approx 3727.04$.

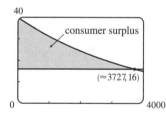

$$\text{Consumer surplus} = \int_0^{x_1}[p(x) - 16]\,dx \approx \$37{,}753$$

11. $f(8) - f(4) = \int_4^8 f'(t)\,dt = \int_4^8 \sqrt{t}\,dt = \left[\tfrac{2}{3}t^{3/2}\right]_4^8 = \tfrac{2}{3}(16\sqrt{2} - 8) \approx \$9.75$ million

13. $N = \displaystyle\int_a^b Ax^{-k}\,dx = A\left[\frac{x^{-k+1}}{-k+1}\right]_a^b = \frac{A}{1-k}\left(b^{1-k} - a^{1-k}\right)$.

Similarly, $\displaystyle\int_a^b Ax^{1-k}\,dx = A\left[\frac{x^{2-k}}{2-k}\right]_a^b = \frac{A}{2-k}\left(b^{2-k} - a^{2-k}\right)$.

Thus, $\overline{x} = \dfrac{1}{N}\displaystyle\int_a^b Ax^{1-k}\,dx = \dfrac{[A/(2-k)](b^{2-k} - a^{2-k})}{[A/(1-k)](b^{1-k} - a^{1-k})} = \dfrac{(1-k)(b^{2-k} - a^{2-k})}{(2-k)(b^{1-k} - a^{1-k})}$.

15. $F = \dfrac{\pi P R^4}{8\eta l} = \dfrac{\pi(4000)(0.008)^4}{8(0.027)(2)} \approx 1.19 \times 10^{-4}$ cm^3/s

17. From (3), $F = \dfrac{A}{\int_0^T c(t)\,dt} = \dfrac{6}{20I}$, where

$$I = \int_0^{10} te^{-0.6t}\,dt = \left[\frac{1}{(-0.6)^2}\,(-0.6t - 1)\,e^{-0.6t}\right]_0^{10} \begin{bmatrix}\text{integrating}\\\text{by parts}\end{bmatrix} = \tfrac{1}{0.36}(-7e^{-6} + 1)$$

Thus, $F = \dfrac{6(0.36)}{20(1 - 7e^{-6})} = \dfrac{0.108}{1 - 7e^{-6}} \approx 0.1099$ L/s or 6.594 L/min.

19. As in Example 2, we will estimate the cardiac output using Simpson's Rule with $\Delta t = (16 - 0)/8 = 2$.

$$\int_0^{16} c(t)\,dt \approx \tfrac{2}{3}[c(0) + 4c(2) + 2c(4) + 4c(6) + 2c(8) + 4c(10) + 2c(12) + 4c(14) + c(16)]$$

$$\approx \tfrac{2}{3}[0 + 4(6.1) + 2(7.4) + 4(6.7) + 2(5.4) + 4(4.1) + 2(3.0) + 4(2.1) + 1.5]$$

$$= \tfrac{2}{3}(109.1) = 72.7\overline{3} \text{ mg} \cdot \text{s/L}$$

Therefore, $F \approx \dfrac{A}{72.7\overline{3}} = \dfrac{7}{72.7\overline{3}} \approx 0.0962$ L/s or 5.77 L/min.

8.5   Probability

1. (a) $\int_{30,000}^{40,000} f(x)\,dx$ is the probability that a randomly chosen tire will have a lifetime between 30,000 and 40,000 miles.

(b) $\int_{25,000}^{\infty} f(x)\,dx$ is the probability that a randomly chosen tire will have a lifetime of at least 25,000 miles.

3. (a) In general, we must satisfy the two conditions that are mentioned before Example 1—namely, (1) $f(x) \geq 0$ for all x, and (2) $\int_{-\infty}^{\infty} f(x)\,dx = 1$. For $0 \leq x \leq 4$, we have $f(x) = \frac{3}{64}x\sqrt{16 - x^2} \geq 0$, so $f(x) \geq 0$ for all x. Also,

$$\int_{-\infty}^{\infty} f(x)\,dx = \int_0^4 \tfrac{3}{64}x\sqrt{16 - x^2}\,dx = -\tfrac{3}{128}\int_0^4 (16 - x^2)^{1/2}(-2x)\,dx = -\tfrac{3}{128}\left[\tfrac{2}{3}(16 - x^2)^{3/2}\right]_0^4$$

$$= -\tfrac{1}{64}\left[(16 - x^2)^{3/2}\right]_0^4 = -\tfrac{1}{64}(0 - 64) = 1.$$

Therefore, f is a probability density function.

(b) $P(X < 2) = \int_{-\infty}^{2} f(x)\,dx = \int_0^2 \tfrac{3}{64}x\sqrt{16 - x^2}\,dx = -\tfrac{3}{128}\int_0^2 (16 - x^2)^{1/2}(-2x)\,dx$

$$= -\tfrac{3}{128}\left[\tfrac{2}{3}(16 - x^2)^{3/2}\right]_0^2 = -\tfrac{1}{64}\left[(16 - x^2)^{3/2}\right]_0^2 = -\tfrac{1}{64}(12^{3/2} - 16^{3/2})$$

$$= \tfrac{1}{64}\left(64 - 12\sqrt{12}\right) = \tfrac{1}{64}\left(64 - 24\sqrt{3}\right) = 1 - \tfrac{3}{8}\sqrt{3} \approx 0.350481$$

5. (a) In general, we must satisfy the two conditions that are mentioned before Example 1—namely, (1) $f(x) \geq 0$ for all x, and (2) $\int_{-\infty}^{\infty} f(x)\,dx = 1$. If $c \geq 0$, then $f(x) \geq 0$, so condition (1) is satisfied. For condition (2), we see that

$$\int_{-\infty}^{\infty} f(x)\,dx = \int_{-\infty}^{\infty} \frac{c}{1 + x^2}\,dx \text{ and }$$

$$\int_0^{\infty} \frac{c}{1 + x^2}\,dx = \lim_{t \to \infty} \int_0^t \frac{c}{1 + x^2}\,dx = c\lim_{t \to \infty}\left[\tan^{-1} x\right]_0^t = c\lim_{t \to \infty}\tan^{-1} t = c\left(\frac{\pi}{2}\right)$$

Similarly, $\displaystyle\int_{-\infty}^{0} \frac{c}{1 + x^2}\,dx = c\left(\frac{\pi}{2}\right)$, so $\displaystyle\int_{-\infty}^{\infty} \frac{c}{1 + x^2}\,dx = 2c\left(\frac{\pi}{2}\right) = c\pi$.

Since $c\pi$ must equal 1, we must have $c = 1/\pi$ so that f is a probability density function.

(b) $P(-1 < X < 1) = \int_{-1}^{1} \frac{1/\pi}{1+x^2}\, dx = \frac{2}{\pi} \int_{0}^{1} \frac{1}{1+x^2}\, dx = \frac{2}{\pi} \left[\tan^{-1} x\right]_0^1 = \frac{2}{\pi}\left(\frac{\pi}{4} - 0\right) = \frac{1}{2}$

7. (a) In general, we must satisfy the two conditions that are mentioned before Example 1—namely, (1) $f(x) \geq 0$ for all x, and (2) $\int_{-\infty}^{\infty} f(x)\, dx = 1$. Since $f(x) = 0$ or $f(x) = 0.1$, condition (1) is satisfied. For condition (2), we see that $\int_{-\infty}^{\infty} f(x)\, dx = \int_0^{10} 0.1\, dx = \left[\frac{1}{10}x\right]_0^{10} = 1$. Thus, $f(x)$ is a probability density function for the spinner's values.

(b) Since all the numbers between 0 and 10 are equally likely to be selected, we expect the mean to be halfway between the endpoints of the interval; that is, $x = 5$.

$$\mu = \int_{-\infty}^{\infty} xf(x)\, dx = \int_0^{10} x(0.1)\, dx = \left[\frac{1}{20}x^2\right]_0^{10} = \frac{100}{20} = 5, \quad \text{as expected.}$$

9. We need to find m so that $\int_m^{\infty} f(t)\, dt = \frac{1}{2}$ $\Rightarrow$ $\lim\limits_{x\to\infty} \int_m^x \frac{1}{5}e^{-t/5}\, dt = \frac{1}{2}$ $\Rightarrow$ $\lim\limits_{x\to\infty} \left[\frac{1}{5}(-5)e^{-t/5}\right]_m^x = \frac{1}{2}$ $\Rightarrow$ $(-1)(0 - e^{-m/5}) = \frac{1}{2}$ $\Rightarrow$ $e^{-m/5} = \frac{1}{2}$ $\Rightarrow$ $-m/5 = \ln\frac{1}{2}$ $\Rightarrow$ $m = -5\ln\frac{1}{2} = 5\ln 2 \approx 3.47$ min.

11. We use an exponential density function with $\mu = 2.5$ min.

(a) $P(X > 4) = \int_4^{\infty} f(t)\, dt = \lim\limits_{x\to\infty} \int_4^x \frac{1}{2.5}e^{-t/2.5}\, dt = \lim\limits_{x\to\infty} \left[-e^{-t/2.5}\right]_4^x = 0 + e^{-4/2.5} \approx 0.202$

(b) $P(0 \leq X \leq 2) = \int_0^2 f(t)\, dt = \left[-e^{-t/2.5}\right]_0^2 = -e^{-2/2.5} + 1 \approx 0.551$

(c) We need to find a value a so that $P(X \geq a) = 0.02$, or, equivalently, $P(0 \leq X \leq a) = 0.98$ $\Leftrightarrow$
$\int_0^a f(t)\, dt = 0.98$ $\Leftrightarrow$ $\left[-e^{-t/2.5}\right]_0^a = 0.98$ $\Leftrightarrow$ $-e^{-a/2.5} + 1 = 0.98$ $\Leftrightarrow$ $e^{-a/2.5} = 0.02$ $\Leftrightarrow$
$-a/2.5 = \ln 0.02$ $\Leftrightarrow$ $a = -2.5\ln\frac{1}{50} = 2.5\ln 50 \approx 9.78$ min ≈ 10 min. The ad should say that if you aren't served within 10 minutes, you get a free hamburger.

13. $P(X \geq 10) = \int_{10}^{\infty} \frac{1}{4.2\sqrt{2\pi}} \exp\left(-\frac{(x - 9.4)^2}{2 \cdot 4.2^2}\right) dx$. To avoid the improper integral we approximate it by the integral from 10 to 100. Thus, $P(X \geq 10) \approx \int_{10}^{100} \frac{1}{4.2\sqrt{2\pi}} \exp\left(-\frac{(x - 9.4)^2}{2 \cdot 4.2^2}\right) dx \approx 0.443$ (using a calculator or computer to estimate the integral), so about 44 percent of the households throw out at least 10 lb of paper a week.

Note: We can't evaluate $1 - P(0 \leq X \leq 10)$ for this problem since a significant amount of area lies to the left of $X = 0$.

15. (a) $P(0 \leq X \leq 100) = \int_0^{100} \frac{1}{8\sqrt{2\pi}} \exp\left(-\frac{(x - 112)^2}{2 \cdot 8^2}\right) dx \approx 0.0668$ (using a calculator or computer to estimate the integral), so there is about a 6.68% chance that a randomly chosen vehicle is traveling at a legal speed.

(b) $P(X \geq 125) = \int_{125}^{\infty} \frac{1}{8\sqrt{2\pi}} \exp\left(-\frac{(x - 112)^2}{2 \cdot 8^2}\right) dx = \int_{125}^{\infty} f(x)\, dx$. In this case, we could use a calculator or computer to estimate either $\int_{125}^{300} f(x)\, dx$ or $1 - \int_0^{125} f(x)\, dx$. Both are approximately 0.0521, so about 5.21% of the motorists are targeted.

17. $P(\mu - 2\sigma \leq X \leq \mu + 2\sigma) = \int_{\mu-2\sigma}^{\mu+2\sigma} \frac{1}{\sigma\sqrt{2\pi}} \exp\left(-\frac{(x-\mu)^2}{2\sigma^2}\right) dx$. Substituting $t = \frac{x-\mu}{\sigma}$ and $dt = \frac{1}{\sigma} dx$ gives us

$$\int_{-2}^{2} \frac{1}{\sigma\sqrt{2\pi}} e^{-t^2/2}(\sigma\,dt) = \frac{1}{\sqrt{2\pi}} \int_{-2}^{2} e^{-t^2/2}\,dt \approx 0.9545.$$

19. (a) First $p(r) = \frac{4}{a_0^3} r^2 e^{-2r/a_0} \geq 0$ for $r \geq 0$. Next,

$$\int_{-\infty}^{\infty} p(r)\,dr = \int_{0}^{\infty} \frac{4}{a_0^3} r^2 e^{-2r/a_0}\,dr = \frac{4}{a_0^3} \lim_{t\to\infty} \int_{0}^{t} r^2 e^{-2r/a_0}\,dr$$

By using parts, tables, or a CAS , we find that $\int x^2 e^{bx}\,dx = (e^{bx}/b^3)(b^2x^2 - 2bx + 2)$. $(\star)$

Next, we use $(\star)$ (with $b = -2/a_0$) and l'Hospital's Rule to get $\frac{4}{a_0^3}\left[\frac{a_0^3}{-8}(-2)\right] = 1$. This satisfies the second condition for

a function to be a probability density function.

(b) Using l'Hospital's Rule, $\frac{4}{a_0^3} \lim_{r\to\infty} \frac{r^2}{e^{2r/a_0}} = \frac{4}{a_0^3} \lim_{r\to\infty} \frac{2r}{(2/a_0)e^{2r/a_0}} = \frac{2}{a_0^2} \lim_{r\to\infty} \frac{2}{(2/a_0)e^{2r/a_0}} = 0.$

To find the maximum of p, we differentiate:

$$p'(r) = \frac{4}{a_0^3}\left[r^2 e^{-2r/a_0}\left(-\frac{2}{a_0}\right) + e^{-2r/a_0}(2r)\right] = \frac{4}{a_0^3} e^{-2r/a_0}(2r)\left(-\frac{r}{a_0} + 1\right)$$

$p'(r) = 0 \iff r = 0$ or $1 = \frac{r}{a_0} \iff r = a_0$ $[a_0 \approx 5.59 \times 10^{-11} \text{ m}].$

$p'(r)$ changes from positive to negative at $r = a_0$, so $p(r)$ has its maximum value at $r = a_0$.

(c) It is fairly difficult to find a viewing rectangle, but knowing the maximum value from part (b) helps.

$$p(a_0) = \frac{4}{a_0^3} a_0^2 e^{-2a_0/a_0} = \frac{4}{a_0} e^{-2} \approx 9{,}684{,}098{,}979$$

With a maximum of nearly 10 billion and a total area under the curve of 1, we know that the "hump" in the graph must be extremely narrow.

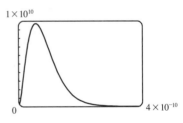

(d) $P(r) = \int_{0}^{r} \frac{4}{a_0^3} s^2 e^{-2s/a_0}\,ds \implies P(4a_0) = \int_{0}^{4a_0} \frac{4}{a_0^3} s^2 e^{-2s/a_0}\,ds$. Using $(\star)$ from part (a) [with $b = -2/a_0$],

$$P(4a_0) = \frac{4}{a_0^3}\left[\frac{e^{-2s/a_0}}{-8/a_0^3}\left(\frac{4}{a_0^2}s^2 + \frac{4}{a_0}s + 2\right)\right]_0^{4a_0} = \frac{4}{a_0^3}\left(\frac{a_0^3}{-8}\right)[e^{-8}(64 + 16 + 2) - 1(2)] = -\tfrac{1}{2}(82e^{-8} - 2)$$

$$= 1 - 41e^{-8} \approx 0.986$$

(e) $\mu = \int_{-\infty}^{\infty} rp(r)\,dr = \frac{4}{a_0^3} \lim_{t\to\infty} \int_{0}^{t} r^3 e^{-2r/a_0}\,dr$. Integrating by parts three times or using a CAS, we find that

$$\int x^3 e^{bx}\,dx = \frac{e^{bx}}{b^4}(b^3x^3 - 3b^2x^2 + 6bx - 6).$$ So with $b = -\frac{2}{a_0}$, we use l'Hospital's Rule, and get

$$\mu = \frac{4}{a_0^3}\left[-\frac{a_0^4}{16}(-6)\right] = \tfrac{3}{2}a_0.$$

8 Review

1. (a) The length of a curve is defined to be the limit of the lengths of the inscribed polygons, as described near Figure 3 in Section 8.1.

   (b) See Equation 8.1.2.

   (c) See Equation 8.1.4.

2. (a) $S = \int_a^b 2\pi f(x) \sqrt{1 + [f'(x)]^2}\, dx$

   (b) If $x = g(y)$, $c \le y \le d$, then $S = \int_c^d 2\pi y \sqrt{1 + [g'(y)]^2}\, dy$.

   (c) $S = \int_a^b 2\pi x \sqrt{1 + [f'(x)]^2}\, dx$ or $S = \int_c^d 2\pi g(y) \sqrt{1 + [g'(y)]^2}\, dy$

3. Let $c(x)$ be the cross-sectional length of the wall (measured parallel to the surface of the fluid) at depth x. Then the hydrostatic force against the wall is given by $F = \int_a^b \delta x c(x)\, dx$, where a and b are the lower and upper limits for x at points of the wall and δ is the weight density of the fluid.

4. (a) The center of mass is the point at which the plate balances horizontally.

   (b) See Equations 8.3.8.

5. If a plane region $\mathcal{R}$ that lies entirely on one side of a line ℓ in its plane is rotated about ℓ, then the volume of the resulting solid is the product of the area of $\mathcal{R}$ and the distance traveled by the centroid of $\mathcal{R}$.

6. See Figure 3 in Section 8.4, and the discussion which precedes it.

7. (a) See the definition in the first paragraph of the subsection *Cardiac Output* in Section 8.4.

   (b) See the discussion in the second paragraph of the subsection *Cardiac Output* in Section 8.4.

8. A probability density function f is a function on the domain of a continuous random variable X such that $\int_a^b f(x)\, dx$ measures the probability that X lies between a and b. Such a function f has nonnegative values and satisfies the relation $\int_D f(x)\, dx = 1$, where D is the domain of the corresponding random variable X. If $D = \mathbb{R}$, or if we define $f(x) = 0$ for real numbers $x \notin D$, then $\int_{-\infty}^{\infty} f(x)\, dx = 1$. (Of course, to work with f in this way, we must assume that the integrals of f exist.)

9. (a) $\int_0^{130} f(x)\, dx$ represents the probability that the weight of a randomly chosen female college student is less than 130 pounds.

   (b) $\mu = \int_{-\infty}^{\infty} x f(x)\, dx = \int_0^{\infty} x f(x)\, dx$

   (c) The median of f is the number m such that $\int_m^{\infty} f(x)\, dx = \frac{1}{2}$.

10. See the discussion near Equation 3 in Section 8.5.

EXERCISES

1. $y = \frac{1}{6}(x^2+4)^{3/2} \Rightarrow dy/dx = \frac{1}{4}(x^2+4)^{1/2}(2x) \Rightarrow$

$1 + (dy/dx)^2 = 1 + \left[\frac{1}{2}x(x^2+4)^{1/2}\right]^2 = 1 + \frac{1}{4}x^2(x^2+4) = \frac{1}{4}x^4 + x^2 + 1 = \left(\frac{1}{2}x^2+1\right)^2$.

Thus, $L = \int_0^3 \sqrt{\left(\frac{1}{2}x^2+1\right)^2}\,dx = \int_0^3 \left(\frac{1}{2}x^2+1\right)dx = \left[\frac{1}{6}x^3 + x\right]_0^3 = \frac{15}{2}$.

3. (a) $y = \dfrac{x^4}{16} + \dfrac{1}{2x^2} = \frac{1}{16}x^4 + \frac{1}{2}x^{-2} \Rightarrow \dfrac{dy}{dx} = \frac{1}{4}x^3 - x^{-3} \Rightarrow$

$1 + (dy/dx)^2 = 1 + \left(\frac{1}{4}x^3 - x^{-3}\right)^2 = 1 + \frac{1}{16}x^6 - \frac{1}{2} + x^{-6} = \frac{1}{16}x^6 + \frac{1}{2} + x^{-6} = \left(\frac{1}{4}x^3 + x^{-3}\right)^2$.

Thus, $L = \int_1^2 \left(\frac{1}{4}x^3 + x^{-3}\right)dx = \left[\frac{1}{16}x^4 - \frac{1}{2}x^{-2}\right]_1^2 = \left(1 - \frac{1}{8}\right) - \left(\frac{1}{16} - \frac{1}{2}\right) = \frac{21}{16}$.

(b) $S = \int_1^2 2\pi x\left(\frac{1}{4}x^3 + x^{-3}\right)dx = 2\pi \int_1^2 \left(\frac{1}{4}x^4 + x^{-2}\right)dx = 2\pi\left[\frac{1}{20}x^5 - \frac{1}{x}\right]_1^2$

$= 2\pi\left[\left(\frac{32}{20} - \frac{1}{2}\right) - \left(\frac{1}{20} - 1\right)\right] = 2\pi\left(\frac{8}{5} - \frac{1}{2} - \frac{1}{20} + 1\right) = 2\pi\left(\frac{41}{20}\right) = \frac{41}{10}\pi$

5. $y = e^{-x^2} \Rightarrow dy/dx = -2xe^{-x^2} \Rightarrow 1 + (dy/dx)^2 = 1 + 4x^2e^{-2x^2}$. Let $f(x) = \sqrt{1 + 4x^2e^{-2x^2}}$. Then

$L = \int_0^3 f(x)\,dx \approx S_6 = \dfrac{(3-0)/6}{3}\left[f(0) + 4f(0.5) + 2f(1) + 4f(1.5) + 2f(2) + 4f(2.5) + f(3)\right] \approx 3.292287$

7. $y = \int_1^x \sqrt{\sqrt{t}-1}\,dt \Rightarrow dy/dx = \sqrt{\sqrt{x}-1} \Rightarrow 1 + (dy/dx)^2 = 1 + \left(\sqrt{x}-1\right) = \sqrt{x}$.

Thus, $L = \int_1^{16}\sqrt{\sqrt{x}}\,dx = \int_1^{16} x^{1/4}\,dx = \frac{4}{5}\left[x^{5/4}\right]_1^{16} = \frac{4}{5}(32-1) = \frac{124}{5}$.

9. As in Example 1 of Section 8.3, $\dfrac{a}{2-x} = \dfrac{1}{2} \Rightarrow 2a = 2 - x$ and $w = 2(1.5+a) = 3 + 2a = 3 + 2 - x = 5 - x$.

Thus, $F = \int_0^2 \rho g x(5-x)\,dx = \rho g\left[\frac{5}{2}x^2 - \frac{1}{3}x^3\right]_0^2 = \rho g\left(10 - \frac{8}{3}\right) = \frac{22}{3}\delta \quad [\rho g = \delta] \approx \frac{22}{3}\cdot 62.5 \approx 458$ lb.

11. $A = \int_0^4 \left(\sqrt{x} - \frac{1}{2}x\right)dx = \left[\frac{2}{3}x^{3/2} - \frac{1}{4}x^2\right]_0^4 = \frac{16}{3} - 4 = \frac{4}{3}$

$\overline{x} = \frac{1}{A}\int_0^4 x\left(\sqrt{x} - \frac{1}{2}x\right)dx = \frac{3}{4}\int_0^4 \left(x^{3/2} - \frac{1}{2}x^2\right)dx$

$= \frac{3}{4}\left[\frac{2}{5}x^{5/2} - \frac{1}{6}x^3\right]_0^4 = \frac{3}{4}\left(\frac{64}{5} - \frac{64}{6}\right) = \frac{3}{4}\left(\frac{64}{30}\right) = \frac{8}{5}$

$\overline{y} = \frac{1}{A}\int_0^4 \frac{1}{2}\left[\left(\sqrt{x}\right)^2 - \left(\frac{1}{2}x\right)^2\right]dx = \frac{3}{4}\int_0^4 \frac{1}{2}\left(x - \frac{1}{4}x^2\right)dx = \frac{3}{8}\left[\frac{1}{2}x^2 - \frac{1}{12}x^3\right]_0^4 = \frac{3}{8}\left(8 - \frac{16}{3}\right) = \frac{3}{8}\left(\frac{8}{3}\right) = 1$

Thus, the centroid is $(\overline{x}, \overline{y}) = \left(\frac{8}{5}, 1\right)$.

13. An equation of the line passing through $(0,0)$ and $(3,2)$ is $y = \frac{2}{3}x$. $A = \frac{1}{2}\cdot 3 \cdot 2 = 3$. Therefore, using Equations 8.3.8,

$\overline{x} = \frac{1}{3}\int_0^3 x\left(\frac{2}{3}x\right)dx = \frac{2}{27}\left[x^3\right]_0^3 = 2$ and $\overline{y} = \frac{1}{3}\int_0^3 \frac{1}{2}\left(\frac{2}{3}x\right)^2 dx = \frac{2}{81}\left[x^3\right]_0^3 = \frac{2}{3}$. Thus, the centroid is $(\overline{x}, \overline{y}) = \left(2, \frac{2}{3}\right)$.

15. The centroid of this circle, $(1, 0)$, travels a distance $2\pi(1)$ when the lamina is rotated about the y-axis. The area of the circle is $\pi(1)^2$. So by the Theorem of Pappus, $V = A(2\pi\bar{x}) = \pi(1)^2 2\pi(1) = 2\pi^2$.

17. $x = 100 \quad \Rightarrow \quad P = 2000 - 0.1(100) - 0.01(100)^2 = 1890$

$$\text{Consumer surplus} = \int_0^{100} [p(x) - P]\, dx = \int_0^{100} \left(2000 - 0.1x - 0.01x^2 - 1890\right) dx$$

$$= \left[110x - 0.05x^2 - \tfrac{0.01}{3}x^3\right]_0^{100} = 11{,}000 - 500 - \tfrac{10{,}000}{3} \approx \$7166.67$$

19. $f(x) = \begin{cases} \frac{\pi}{20}\sin\left(\frac{\pi}{10}x\right) & \text{if } 0 \le x \le 10 \\ 0 & \text{if } x < 0 \text{ or } x > 10 \end{cases}$

(a) $f(x) \ge 0$ for all real numbers x and

$$\int_{-\infty}^{\infty} f(x)\, dx = \int_0^{10} \tfrac{\pi}{20}\sin\left(\tfrac{\pi}{10}x\right) dx = \tfrac{\pi}{20} \cdot \tfrac{10}{\pi}\left[-\cos\left(\tfrac{\pi}{10}x\right)\right]_0^{10} = \tfrac{1}{2}(-\cos\pi + \cos 0) = \tfrac{1}{2}(1+1) = 1$$

Therefore, f is a probability density function.

(b) $P(X < 4) = \int_{-\infty}^{4} f(x)\, dx = \int_0^4 \tfrac{\pi}{20}\sin\left(\tfrac{\pi}{10}x\right) dx = \tfrac{1}{2}\left[-\cos\left(\tfrac{\pi}{10}x\right)\right]_0^4 = \tfrac{1}{2}\left(-\cos\tfrac{2\pi}{5} + \cos 0\right)$

$\approx \tfrac{1}{2}(-0.309017 + 1) \approx 0.3455$

(c) $\mu = \int_{-\infty}^{\infty} x f(x)\, dx = \int_0^{10} \tfrac{\pi}{20} x \sin\left(\tfrac{\pi}{10}x\right) dx$

$= \int_0^{\pi} \tfrac{\pi}{20} \cdot \tfrac{10}{\pi} u(\sin u)\left(\tfrac{10}{\pi}\right) du \qquad [u = \tfrac{\pi}{10}x,\, du = \tfrac{\pi}{10}\, dx]$

$= \tfrac{5}{\pi}\int_0^{\pi} u \sin u\, du \overset{82}{=} \tfrac{5}{\pi}\left[\sin u - u\cos u\right]_0^{\pi} = \tfrac{5}{\pi}[0 - \pi(-1)] = 5$

This answer is expected because the graph of f is symmetric about the line $x = 5$.

$y = \frac{\pi}{20}\sin\left(\frac{\pi x}{10}\right)$

21. (a) The probability density function is $f(t) = \begin{cases} 0 & \text{if } t < 0 \\ \frac{1}{8}e^{-t/8} & \text{if } t \ge 0 \end{cases}$

$$P(0 \le X \le 3) = \int_0^3 \tfrac{1}{8}e^{-t/8}\, dt = \left[-e^{-t/8}\right]_0^3 = -e^{-3/8} + 1 \approx 0.3127$$

(b) $P(X > 10) = \int_{10}^{\infty} \tfrac{1}{8}e^{-t/8}\, dt = \lim_{x\to\infty}\left[-e^{-t/8}\right]_{10}^{x} = \lim_{x\to\infty}\left(-e^{-x/8} + e^{-10/8}\right) = 0 + e^{-5/4} \approx 0.2865$

(c) We need to find m such that $P(X \ge m) = \tfrac{1}{2} \quad \Rightarrow \quad \int_m^{\infty} \tfrac{1}{8}e^{-t/8}\, dt = \tfrac{1}{2} \quad \Rightarrow \quad \lim_{x\to\infty}\left[-e^{-t/8}\right]_m^{x} = \tfrac{1}{2} \quad \Rightarrow$

$\lim_{x\to\infty}\left(-e^{-x/8} + e^{-m/8}\right) = \tfrac{1}{2} \quad \Rightarrow \quad e^{-m/8} = \tfrac{1}{2} \quad \Rightarrow \quad -m/8 = \ln\tfrac{1}{2} \quad \Rightarrow \quad m = -8\ln\tfrac{1}{2} = 8\ln 2 \approx 5.55 \text{ minutes.}$

□ PROBLEMS PLUS

1. $x^2 + y^2 \leq 4y \iff x^2 + (y-2)^2 \leq 4$, so S is part of a circle, as shown
in the diagram. The area of S is

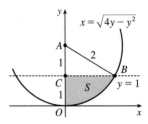

$$\int_0^1 \sqrt{4y - y^2}\, dy \overset{113}{=} \left[\frac{y-2}{2}\sqrt{4y - y^2} + 2\cos^{-1}\left(\frac{2-y}{2}\right) \right]_0^1 \quad [a = 2]$$

$$= -\tfrac{1}{2}\sqrt{3} + 2\cos^{-1}\left(\tfrac{1}{2}\right) - 2\cos^{-1}1$$

$$= -\frac{\sqrt{3}}{2} + 2\left(\frac{\pi}{3}\right) - 2(0) = \frac{2\pi}{3} - \frac{\sqrt{3}}{2}$$

Another method (without calculus): Note that $\theta = \angle CAB = \frac{\pi}{3}$, so the area is

$$\text{(area of sector } OAB) - \text{(area of } \triangle ABC) = \tfrac{1}{2}(2^2)\frac{\pi}{3} - \tfrac{1}{2}(1)\sqrt{3} = \frac{2\pi}{3} - \frac{\sqrt{3}}{2}$$

3. (a) The two spherical zones, whose surface areas we will call S_1 and S_2, are
generated by rotation about the y-axis of circular arcs, as indicated in the figure.
The arcs are the upper and lower portions of the circle $x^2 + y^2 = r^2$ that are
obtained when the circle is cut with the line $y = d$. The portion of the upper arc
in the first quadrant is sufficient to generate the upper spherical zone. That
portion of the arc can be described by the relation $x = \sqrt{r^2 - y^2}$ for
$d \leq y \leq r$. Thus, $dx/dy = -y/\sqrt{r^2 - y^2}$ and

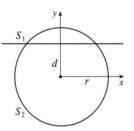

$$ds = \sqrt{1 + \left(\frac{dx}{dy}\right)^2}\, dy = \sqrt{1 + \frac{y^2}{r^2 - y^2}}\, dy = \sqrt{\frac{r^2}{r^2 - y^2}}\, dy = \frac{r\, dy}{\sqrt{r^2 - y^2}}$$

From Formula 8.2.8 we have

$$S_1 = \int_d^r 2\pi x \sqrt{1 + \left(\frac{dx}{dy}\right)^2}\, dy = \int_d^r 2\pi \sqrt{r^2 - y^2}\,\frac{r\, dy}{\sqrt{r^2 - y^2}} = \int_d^r 2\pi r\, dy = 2\pi r(r - d)$$

Similarly, we can compute $S_2 = \int_{-r}^d 2\pi x \sqrt{1 + (dx/dy)^2}\, dy = \int_{-r}^d 2\pi r\, dy = 2\pi r(r + d)$. Note that $S_1 + S_2 = 4\pi r^2$,
the surface area of the entire sphere.

(b) $r = 3960$ mi and $d = r\,(\sin 75°) \approx 3825$ mi,
so the surface area of the Arctic Ocean is about

$$2\pi r(r - d) \approx 2\pi(3960)(135) \approx 3.36 \times 10^6 \text{ mi}^2.$$

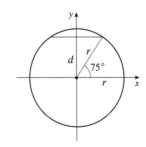

(c) The area on the sphere lies between planes $y = y_1$ and $y = y_2$, where $y_2 - y_1 = h$. Thus, we compute the surface area on

the sphere to be $S = \int_{y_1}^{y_2} 2\pi x \sqrt{1 + \left(\dfrac{dx}{dy}\right)^2}\, dy = \int_{y_1}^{y_2} 2\pi r\, dy = 2\pi r(y_2 - y_1) = 2\pi rh.$

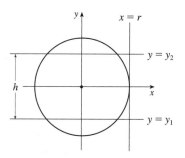

This equals the lateral area of a cylinder of radius r and height h, since such

a cylinder is obtained by rotating the line $x = r$ about the y-axis, so the

surface area of the cylinder between the planes $y = y_1$ and $y = y_2$ is

$$A = \int_{y_1}^{y_2} 2\pi x \sqrt{1 + \left(\frac{dx}{dy}\right)^2}\, dy = \int_{y_1}^{y_2} 2\pi r \sqrt{1 + 0^2}\, dy$$

$$= 2\pi ry \Big|_{y=y_1}^{y_2} = 2\pi r(y_2 - y_1) = 2\pi rh$$

(d) $h = 2r \sin 23.45° \approx 3152$ mi, so the surface area of the

Torrid Zone is $2\pi rh \approx 2\pi(3960)(3152) \approx 7.84 \times 10^7$ mi^2.

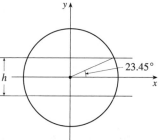

5. (a) Choose a vertical x-axis pointing downward with its origin at the surface. In order to calculate the pressure at depth z,

consider n subintervals of the interval $[0, z]$ by points x_i and choose a point $x_i^* \in [x_{i-1}, x_i]$ for each i. The thin layer of

water lying between depth x_{i-1} and depth x_i has a density of approximately $\rho(x_i^*)$, so the weight of a piece of that layer

with unit cross-sectional area is $\rho(x_i^*)g\,\Delta x$. The total weight of a column of water extending from the surface to depth z

(with unit cross-sectional area) would be approximately $\sum\limits_{i=1}^{n} \rho(x_i^*)g\,\Delta x$. The estimate becomes exact if we take the limit

as $n \to \infty$; weight (or force) per unit area at depth z is $W = \lim\limits_{n\to\infty} \sum\limits_{i=1}^{n} \rho(x_i^*)g\,\Delta x$. In other words, $P(z) = \int_0^z \rho(x)g\,dx.$

More generally, if we make no assumptions about the location of the origin, then $P(z) = P_0 + \int_0^z \rho(x)g\,dx$, where P_0 is

the pressure at $x = 0$. Differentiating, we get $dP/dz = \rho(z)g.$

(b)

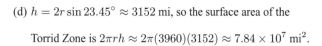

$$F = \int_{-r}^{r} P(L + x) \cdot 2\sqrt{r^2 - x^2}\, dx$$

$$= \int_{-r}^{r} \left(P_0 + \int_0^{L+x} \rho_0 e^{z/H} g\, dz\right) \cdot 2\sqrt{r^2 - x^2}\, dx$$

$$= P_0 \int_{-r}^{r} 2\sqrt{r^2 - x^2}\, dx + \rho_0 gH \int_{-r}^{r} \left(e^{(L+x)/H} - 1\right) \cdot 2\sqrt{r^2 - x^2}\, dx$$

$$= (P_0 - \rho_0 gH)\int_{-r}^{r} 2\sqrt{r^2 - x^2}\, dx + \rho_0 gH \int_{-r}^{r} e^{(L+x)/H} \cdot 2\sqrt{r^2 - x^2}\, dx$$

$$= (P_0 - \rho_0 gH)(\pi r^2) + \rho_0 gH e^{L/H} \int_{-r}^{r} e^{x/H} \cdot 2\sqrt{r^2 - x^2}\, dx$$

7. To find the height of the pyramid, we use similar triangles. The first figure shows a cross-section of the pyramid passing through the top and through two opposite corners of the square base. Now $|BD| = b$, since it is a radius of the sphere, which has diameter $2b$ since it is tangent to the opposite sides of the square base. Also, $|AD| = b$ since $\triangle ADB$ is isosceles. So the height is $|AB| = \sqrt{b^2 + b^2} = \sqrt{2}\, b$.

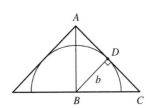

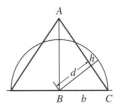

We first observe that the shared volume is equal to half the volume of the sphere, minus the sum of the four equal volumes (caps of the sphere) cut off by the triangular faces of the pyramid. See Exercise 6.2.51 for a derivation of the formula for the volume of a cap of a sphere. To use the formula, we need to find the perpendicular distance h of each triangular face from the surface of the sphere. We first find the distance d from the center of the sphere to one of the triangular faces. The third figure shows a cross-section of the pyramid through the top and through the midpoints of opposite sides of the square base. From similar triangles we find that

$$\frac{d}{b} = \frac{|AB|}{|AC|} = \frac{\sqrt{2}\, b}{\sqrt{b^2 + (\sqrt{2}\, b)^2}} \quad \Rightarrow \quad d = \frac{\sqrt{2}\, b^2}{\sqrt{3b^2}} = \frac{\sqrt{6}}{3}\, b$$

So $h = b - d = b - \frac{\sqrt{6}}{3}b = \frac{3 - \sqrt{6}}{3}b$. So, using the formula $V = \pi h^2(r - h/3)$ from Exercise 6.2.51 with $r = b$, we find that the volume of each of the caps is $\pi\left(\frac{3 - \sqrt{6}}{3}b\right)^2\left(b - \frac{3 - \sqrt{6}}{3 \cdot 3}b\right) = \frac{15 - 6\sqrt{6}}{9} \cdot \frac{6 + \sqrt{6}}{9}\pi b^3 = \left(\frac{2}{3} - \frac{7}{27}\sqrt{6}\right)\pi b^3$. So, using our first observation, the shared volume is $V = \frac{1}{2}\left(\frac{4}{3}\pi b^3\right) - 4\left(\frac{2}{3} - \frac{7}{27}\sqrt{6}\right)\pi b^3 = \left(\frac{28}{27}\sqrt{6} - 2\right)\pi b^3$.

9. We can assume that the cut is made along a vertical line $x = b > 0$, that the disk's boundary is the circle $x^2 + y^2 = 1$, and that the center of mass of the smaller piece (to the right of $x = b$) is $\left(\frac{1}{2}, 0\right)$. We wish to find b to two decimal places. We have

$$\frac{1}{2} = \bar{x} = \frac{\int_b^1 x \cdot 2\sqrt{1 - x^2}\, dx}{\int_b^1 2\sqrt{1 - x^2}\, dx}.$$ Evaluating the

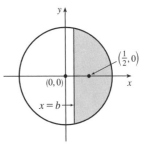

numerator gives us $-\int_b^1 (1 - x^2)^{1/2}(-2x)\, dx = -\frac{2}{3}\left[(1 - x^2)^{3/2}\right]_b^1 = -\frac{2}{3}\left[0 - (1 - b^2)^{3/2}\right] = \frac{2}{3}(1 - b^2)^{3/2}$.

Using Formula 30 in the table of integrals, we find that the denominator is

$\left[x\sqrt{1 - x^2} + \sin^{-1} x\right]_b^1 = \left(0 + \frac{\pi}{2}\right) - \left(b\sqrt{1 - b^2} + \sin^{-1} b\right)$. Thus, we have $\dfrac{1}{2} = \bar{x} = \dfrac{\frac{2}{3}(1 - b^2)^{3/2}}{\frac{\pi}{2} - b\sqrt{1 - b^2} - \sin^{-1} b}$, or,

equivalently, $\frac{2}{3}(1 - b^2)^{3/2} = \frac{\pi}{4} - \frac{1}{2}b\sqrt{1 - b^2} - \frac{1}{2}\sin^{-1} b$. Solving this equation numerically with a calculator or CAS, we obtain $b \approx 0.138173$, or $b = 0.14$ m to two decimal places.

11. If $h = L$, then $P = \dfrac{\text{area under } y = L\sin\theta}{\text{area of rectangle}} = \dfrac{\int_0^\pi L\sin\theta\, d\theta}{\pi L} = \dfrac{[-\cos\theta]_0^\pi}{\pi} = \dfrac{-(-1)+1}{\pi} = \dfrac{2}{\pi}.$

If $h = L/2$, then $P = \dfrac{\text{area under } y = \frac{1}{2}L\sin\theta}{\text{area of rectangle}} = \dfrac{\int_0^\pi \frac{1}{2}L\sin\theta\, d\theta}{\pi L} = \dfrac{[-\cos\theta]_0^\pi}{2\pi} = \dfrac{2}{2\pi} = \dfrac{1}{\pi}.$

9 ☐ DIFFERENTIAL EQUATIONS

9.1 Modeling with Differential Equations

1. $y = x - x^{-1}$ $\Rightarrow$ $y' = 1 + x^{-2}$. To show that y is a solution of the differential equation, we will substitute the expressions for y and y' in the left-hand side of the equation and show that the left-hand side is equal to the right-hand side.

$$\text{LHS} = xy' + y = x(1 + x^{-2}) + (x - x^{-1}) = x + x^{-1} + x - x^{-1} = 2x = \text{RHS}$$

3. (a) $y = e^{rx}$ $\Rightarrow$ $y' = re^{rx}$ $\Rightarrow$ $y'' = r^2 e^{rx}$. Substituting these expressions into the differential equation

$2y'' + y' - y = 0$, we get $2r^2 e^{rx} + re^{rx} - e^{rx} = 0$ $\Rightarrow$ $(2r^2 + r - 1)e^{rx} = 0$ $\Rightarrow$

$(2r - 1)(r + 1) = 0$ [since e^{rx} is never zero] $\Rightarrow$ $r = \frac{1}{2}$ or -1.

(b) Let $r_1 = \frac{1}{2}$ and $r_2 = -1$, so we need to show that every member of the family of functions $y = ae^{x/2} + be^{-x}$ is a solution of the differential equation $2y'' + y' - y = 0$.

$y = ae^{x/2} + be^{-x}$ $\Rightarrow$ $y' = \frac{1}{2}ae^{x/2} - be^{-x}$ $\Rightarrow$ $y'' = \frac{1}{4}ae^{x/2} + be^{-x}$.

$$\begin{aligned}
\text{LHS} = 2y'' + y' - y &= 2\left(\tfrac{1}{4}ae^{x/2} + be^{-x}\right) + \left(\tfrac{1}{2}ae^{x/2} - be^{-x}\right) - \left(ae^{x/2} + be^{-x}\right) \\
&= \tfrac{1}{2}ae^{x/2} + 2be^{-x} + \tfrac{1}{2}ae^{x/2} - be^{-x} - ae^{x/2} - be^{-x} \\
&= \left(\tfrac{1}{2}a + \tfrac{1}{2}a - a\right)e^{x/2} + (2b - b - b)e^{-x} \\
&= 0 = \text{RHS}
\end{aligned}$$

5. (a) $y = \sin x$ $\Rightarrow$ $y' = \cos x$ $\Rightarrow$ $y'' = -\sin x$.

$\text{LHS} = y'' + y = -\sin x + \sin x = 0 \neq \sin x$, so $y = \sin x$ **is not** a solution of the differential equation.

(b) $y = \cos x$ $\Rightarrow$ $y' = -\sin x$ $\Rightarrow$ $y'' = -\cos x$.

$\text{LHS} = y'' + y = -\cos x + \cos x = 0 \neq \sin x$, so $y = \cos x$ **is not** a solution of the differential equation.

(c) $y = \frac{1}{2}x \sin x$ $\Rightarrow$ $y' = \frac{1}{2}(x\cos x + \sin x)$ $\Rightarrow$ $y'' = \frac{1}{2}(-x\sin x + \cos x + \cos x)$.

$\text{LHS} = y'' + y = \frac{1}{2}(-x\sin x + 2\cos x) + \frac{1}{2}x\sin x = \cos x \neq \sin x$, so $y = \frac{1}{2}x\sin x$ **is not** a solution of the differential equation.

(d) $y = -\frac{1}{2}x\cos x$ $\Rightarrow$ $y' = -\frac{1}{2}(-x\sin x + \cos x)$ $\Rightarrow$ $y'' = -\frac{1}{2}(-x\cos x - \sin x - \sin x)$.

$\text{LHS} = y'' + y = -\frac{1}{2}(-x\cos x - 2\sin x) + \left(-\frac{1}{2}x\cos x\right) = \sin x = \text{RHS}$, so $y = -\frac{1}{2}x\cos x$ **is** a solution of the differential equation.

7. (a) Since the derivative $y' = -y^2$ is always negative (or 0 if $y = 0$), the function y must be decreasing (or equal to 0) on any interval on which it is defined.

(b) $y = \dfrac{1}{x + C}$ $\Rightarrow$ $y' = -\dfrac{1}{(x + C)^2}$. $\text{LHS} = y' = -\dfrac{1}{(x + C)^2} = -\left(\dfrac{1}{x + C}\right)^2 = -y^2 = \text{RHS}$

(c) $y = 0$ is a solution of $y' = -y^2$ that is not a member of the family in part (b).

(d) If $y(x) = \dfrac{1}{x+C}$, then $y(0) = \dfrac{1}{0+C} = \dfrac{1}{C}$. Since $y(0) = 0.5$, $\dfrac{1}{C} = \dfrac{1}{2}$ $\Rightarrow$ $C = 2$, so $y = \dfrac{1}{x+2}$.

9. (a) $\dfrac{dP}{dt} = 1.2P\left(1 - \dfrac{P}{4200}\right)$. Now $\dfrac{dP}{dt} > 0$ $\Rightarrow$ $1 - \dfrac{P}{4200} > 0$ [assuming that $P > 0$] $\Rightarrow$ $\dfrac{P}{4200} < 1$ $\Rightarrow$

$P < 4200$ $\Rightarrow$ the population is increasing for $0 < P < 4200$.

(b) $\dfrac{dP}{dt} < 0$ $\Rightarrow$ $P > 4200$

(c) $\dfrac{dP}{dt} = 0$ $\Rightarrow$ $P = 4200$ or $P = 0$

11. (a) This function is increasing *and* also decreasing. But $dy/dt = e^t(y-1)^2 \geq 0$ for all t, implying that the graph of the solution of the differential equation cannot be decreasing on any interval.

(b) When $y = 1$, $dy/dt = 0$, but the graph does not have a horizontal tangent line.

13. (a) P increases most rapidly at the beginning, since there are usually many simple, easily-learned sub-skills associated with learning a skill. As t increases, we would expect dP/dt to remain positive, but decrease. This is because as time progresses, the only points left to learn are the more difficult ones.

(b) $\dfrac{dP}{dt} = k(M-P)$ is always positive, so the level of performance P is increasing. As P gets close to M, dP/dt gets close to 0; that is, the performance levels off, as explained in part (a).

(c)

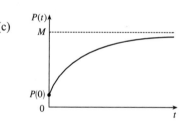

9.2 Direction Fields and Euler's Method

1. (a)

(i)
(ii)
(iii)
(iv)

(b) It appears that the constant functions $y = 0$, $y = -2$, and $y = 2$ are equilibrium solutions. Note that these three values of y satisfy the given differential equation $y' = y\left(1 - \frac{1}{4}y^2\right)$.

3. $y' = 2 - y$. The slopes at each point are independent of x, so the slopes are the same along each line parallel to the x-axis. Thus, III is the direction field for this equation. Note that for $y = 2$, $y' = 0$.

5. $y' = x + y - 1 = 0$ on the line $y = -x + 1$. Direction field IV satisfies this condition. Notice also that on the line $y = -x$ we have $y' = -1$, which is true in IV.

7. (a) $y(0) = 1$

   (b) $y(0) = 2$

   (c) $y(0) = -1$

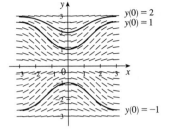

9.

x	y	$y' = 1 + y$
0	0	1
0	1	2
0	2	3
0	-3	-2
0	-2	-1

Note that for $y = -1$, $y' = 0$. The three solution curves sketched go through $(0, 0)$, $(0, -1)$, and $(0, -2)$.

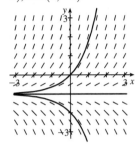

11.

x	y	$y' = y - 2x$
-2	-2	2
-2	2	6
2	2	-2
2	-2	-6

Note that $y' = 0$ for any point on the line $y = 2x$. The slopes are positive to the left of the line and negative to the right of the line. The solution curve in the graph passes through $(1, 0)$.

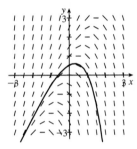

13.

x	y	$y' = y + xy$
0	± 2	± 2
1	± 2	± 4
-3	± 2	∓ 4

Note that $y' = y(x + 1) = 0$ for any point on $y = 0$ or on $x = -1$. The slopes are positive when the factors y and $x + 1$ have the same sign and negative when they have opposite signs. The solution curve in the graph passes through $(0, 1)$.

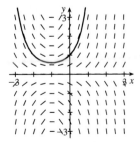

15. In Maple, we can use either `directionfield` (in Maple's share library) or `DEtools[DEplot]` to plot the direction field. To plot the solution, we can either use the initial-value option in `directionfield`, or actually solve the equation.

In Mathematica, we use `PlotVectorField` for the direction field, and the `Plot[Evaluate[...]]` construction to plot the solution, which is

$$y = 2\arctan\left(e^{x^3/3} \cdot \tan\tfrac{1}{2}\right).$$

In Derive, use `Direction_Field` (in utility file ODE_APPR) to plot the direction field. Then use `DSOLVE1(-x^2*SIN(y),1,x,y,0,1)` (in utility file ODE1) to solve the equation. Simplify each result.

17.

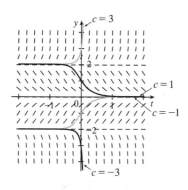

The direction field is for the differential equation $y' = y^3 - 4y$.

$L = \lim\limits_{t \to \infty} y(t)$ exists for $-2 \le c \le 2$;

$L = \pm 2$ for $c = \pm 2$ and $L = 0$ for $-2 < c < 2$.

For other values of c, L does not exist.

19. (a) $y' = F(x, y) = y$ and $y(0) = 1 \Rightarrow x_0 = 0, y_0 = 1$.

  (i) $h = 0.4$ and $y_1 = y_0 + hF(x_0, y_0) \Rightarrow y_1 = 1 + 0.4 \cdot 1 = 1.4$. $x_1 = x_0 + h = 0 + 0.4 = 0.4$,

    so $y_1 = y(0.4) = 1.4$.

  (ii) $h = 0.2 \Rightarrow x_1 = 0.2$ and $x_2 = 0.4$, so we need to find y_2.

    $y_1 = y_0 + hF(x_0, y_0) = 1 + 0.2y_0 = 1 + 0.2 \cdot 1 = 1.2$,

    $y_2 = y_1 + hF(x_1, y_1) = 1.2 + 0.2y_1 = 1.2 + 0.2 \cdot 1.2 = 1.44$.

  (iii) $h = 0.1 \Rightarrow x_4 = 0.4$, so we need to find y_4. $y_1 = y_0 + hF(x_0, y_0) = 1 + 0.1y_0 = 1 + 0.1 \cdot 1 = 1.1$,

    $y_2 = y_1 + hF(x_1, y_1) = 1.1 + 0.1y_1 = 1.1 + 0.1 \cdot 1.1 = 1.21$,

    $y_3 = y_2 + hF(x_2, y_2) = 1.21 + 0.1y_2 = 1.21 + 0.1 \cdot 1.21 = 1.331$,

    $y_4 = y_3 + hF(x_3, y_3) = 1.331 + 0.1y_3 = 1.331 + 0.1 \cdot 1.331 = 1.4641$.

(b)

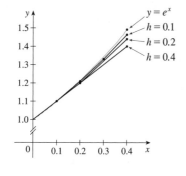

We see that the estimates are underestimates since they are all below the graph of $y = e^x$.

(c) (i) For $h = 0.4$:   (exact value) − (approximate value) $= e^{0.4} - 1.4 \approx 0.0918$

   (ii) For $h = 0.2$:   (exact value) − (approximate value) $= e^{0.4} - 1.44 \approx 0.0518$

   (iii) For $h = 0.1$:   (exact value) − (approximate value) $= e^{0.4} - 1.4641 \approx 0.0277$

Each time the step size is halved, the error estimate also appears to be halved (approximately).

21. $h = 0.5$, $x_0 = 1$, $y_0 = 0$, and $F(x, y) = y - 2x$.

Note that $x_1 = x_0 + h = 1 + 0.5 = 1.5$, $x_2 = 2$, and $x_3 = 2.5$.

$y_1 = y_0 + hF(x_0, y_0) = 0 + 0.5F(1, 0) = 0.5[0 - 2(1)] = -1$.

$y_2 = y_1 + hF(x_1, y_1) = -1 + 0.5F(1.5, -1) = -1 + 0.5[-1 - 2(1.5)] = -3$.

$y_3 = y_2 + hF(x_2, y_2) = -3 + 0.5F(2, -3) = -3 + 0.5[-3 - 2(2)] = -6.5$.

$y_4 = y_3 + hF(x_3, y_3) = -6.5 + 0.5F(2.5, -6.5) = -6.5 + 0.5[-6.5 - 2(2.5)] = -12.25$.

23. $h = 0.1$, $x_0 = 0$, $y_0 = 1$, and $F(x, y) = y + xy$.

Note that $x_1 = x_0 + h = 0 + 0.1 = 0.1$, $x_2 = 0.2$, $x_3 = 0.3$, and $x_4 = 0.4$.

$y_1 = y_0 + hF(x_0, y_0) = 1 + 0.1F(0, 1) = 1 + 0.1[1 + (0)(1)] = 1.1$.

$y_2 = y_1 + hF(x_1, y_1) = 1.1 + 0.1F(0.1, 1.1) = 1.1 + 0.1[1.1 + (0.1)(1.1)] = 1.221$.

$y_3 = y_2 + hF(x_2, y_2) = 1.221 + 0.1F(0.2, 1.221) = 1.221 + 0.1[1.221 + (0.2)(1.221)] = 1.36752$.

$y_4 = y_3 + hF(x_3, y_3) = 1.36752 + 0.1F(0.3, 1.36752) = 1.36752 + 0.1[1.36752 + (0.3)(1.36752)]$

$\qquad = 1.5452976$.

$y_5 = y_4 + hF(x_4, y_4) = 1.5452976 + 0.1F(0.4, 1.5452976)$

$\qquad = 1.5452976 + 0.1[1.5452976 + (0.4)(1.5452976)] = 1.761639264$.

Thus, $y(0.5) \approx 1.7616$.

25. (a) $dy/dx + 3x^2 y = 6x^2$ $\Rightarrow$ $y' = 6x^2 - 3x^2 y$. Store this expression in Y_1 and use the following simple program to

evaluate $y(1)$ for each part, using $H = h = 1$ and $N = 1$ for part (i), $H = 0.1$ and $N = 10$ for part (ii), and so forth.

$\qquad\qquad$ $h \to$ H: $0 \to$ X: $3 \to$ Y:

$\qquad\qquad$ For(I, 1, N): Y + H × $Y_1 \to$ Y: X + H $\to$ X:

$\qquad\qquad$ End(loop):

$\qquad\qquad$ Display Y.   [To see all iterations, include this statement in the loop.]

$\quad$ (i) H = 1, N = 1 $\Rightarrow$ $y(1) = 3$

$\quad$ (ii) H = 0.1, N = 10 $\Rightarrow$ $y(1) \approx 2.3928$

$\quad$ (iii) H = 0.01, N = 100 $\Rightarrow$ $y(1) \approx 2.3701$

$\quad$ (iv) H = 0.001, N = 1000 $\Rightarrow$ $y(1) \approx 2.3681$

(b) $y = 2 + e^{-x^3}$ $\Rightarrow$ $y' = -3x^2 e^{-x^3}$

$\qquad$ LHS $= y' + 3x^2 y = -3x^2 e^{-x^3} + 3x^2 \left(2 + e^{-x^3}\right) = -3x^2 e^{-x^3} + 6x^2 + 3x^2 e^{-x^3} = 6x^2 =$ RHS

$\qquad$ $y(0) = 2 + e^{-0} = 2 + 1 = 3$

(c) The exact value of $y(1)$ is $2 + e^{-1^3} = 2 + e^{-1}$.

    (i) For $h = 1$: (exact value) $-$ (approximate value) $= 2 + e^{-1} - 3 \approx -0.6321$

    (ii) For $h = 0.1$: (exact value) $-$ (approximate value) $= 2 + e^{-1} - 2.3928 \approx -0.0249$

    (iii) For $h = 0.01$: (exact value) $-$ (approximate value) $= 2 + e^{-1} - 2.3701 \approx -0.0022$

    (iv) For $h = 0.001$: (exact value) $-$ (approximate value) $= 2 + e^{-1} - 2.3681 \approx -0.0002$

In (ii)–(iv), it seems that when the step size is divided by 10, the error estimate is also divided by 10 (approximately).

27. (a) $R\dfrac{dQ}{dt} + \dfrac{1}{C}Q = E(t)$ becomes $5Q' + \dfrac{1}{0.05}Q = 60$

or $Q' + 4Q = 12$.

(b) From the graph, it appears that the limiting value of the charge Q is about 3.

(c) If $Q' = 0$, then $4Q = 12 \;\Rightarrow\; Q = 3$ is an equilibrium solution.

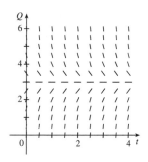

(d)

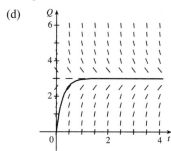

(e) $Q' + 4Q = 12 \;\Rightarrow\; Q' = 12 - 4Q$. Now $Q(0) = 0$, so $t_0 = 0$ and $Q_0 = 0$.

$$Q_1 = Q_0 + hF(t_0, Q_0) = 0 + 0.1(12 - 4 \cdot 0) = 1.2$$

$$Q_2 = Q_1 + hF(t_1, Q_1) = 1.2 + 0.1(12 - 4 \cdot 1.2) = 1.92$$

$$Q_3 = Q_2 + hF(t_2, Q_2) = 1.92 + 0.1(12 - 4 \cdot 1.92) = 2.352$$

$$Q_4 = Q_3 + hF(t_3, Q_3) = 2.352 + 0.1(12 - 4 \cdot 2.352) = 2.6112$$

$$Q_5 = Q_4 + hF(t_4, Q_4) = 2.6112 + 0.1(12 - 4 \cdot 2.6112) = 2.76672$$

Thus, $Q_5 = Q(0.5) \approx 2.77$ C.

9.3 Separable Equations

1. $\dfrac{dy}{dx} = \dfrac{y}{x} \;\Rightarrow\; \dfrac{dy}{y} = \dfrac{dx}{x} \quad [y \neq 0] \;\Rightarrow\; \displaystyle\int \dfrac{dy}{y} = \int \dfrac{dx}{x} \;\Rightarrow\; \ln|y| = \ln|x| + C \;\Rightarrow\;$

$|y| = e^{\ln|x|+C} = e^{\ln|x|}e^C = e^C|x| \;\Rightarrow\; y = Kx$, where $K = \pm e^C$ is a constant. (In our derivation, K was nonzero, but we can restore the excluded case $y = 0$ by allowing K to be zero.)

3. $(x^2 + 1)y' = xy \;\Rightarrow\; \dfrac{dy}{dx} = \dfrac{xy}{x^2 + 1} \;\Rightarrow\; \dfrac{dy}{y} = \dfrac{x\,dx}{x^2 + 1} \quad [y \neq 0] \;\Rightarrow\; \displaystyle\int \dfrac{dy}{y} = \int \dfrac{x\,dx}{x^2 + 1} \;\Rightarrow\;$

$\ln|y| = \tfrac{1}{2}\ln(x^2 + 1) + C \quad [u = x^2 + 1,\, du = 2x\,dx] \quad = \ln(x^2 + 1)^{1/2} + \ln e^C = \ln\!\left(e^C\sqrt{x^2 + 1}\right) \;\Rightarrow\;$

$|y| = e^C\sqrt{x^2 + 1} \;\Rightarrow\; y = K\sqrt{x^2 + 1}$, where $K = \pm e^C$ is a constant. (In our derivation, K was nonzero, but we can restore the excluded case $y = 0$ by allowing K to be zero.)

5. $(1 + \tan y)\, y' = x^2 + 1$ $\Rightarrow$ $(1 + \tan y)\dfrac{dy}{dx} = x^2 + 1$ $\Rightarrow$ $\left(1 + \dfrac{\sin y}{\cos y}\right) dy = (x^2 + 1)\, dx$ $\Rightarrow$

$\displaystyle\int \left(1 - \dfrac{-\sin y}{\cos y}\right) dy = \int (x^2 + 1)\, dx$ $\Rightarrow$ $y - \ln|\cos y| = \frac{1}{3}x^3 + x + C.$

Note: The left side is equivalent to $y + \ln|\sec y|$.

7. $\dfrac{dy}{dt} = \dfrac{te^t}{y\sqrt{1+y^2}}$ $\Rightarrow$ $y\sqrt{1+y^2}\, dy = te^t\, dt$ $\Rightarrow$ $\int y\sqrt{1+y^2}\, dy = \int te^t\, dt$ $\Rightarrow$ $\frac{1}{3}\left(1+y^2\right)^{3/2} = te^t - e^t + C$

[where the first integral is evaluated by substitution and the second by parts] $\Rightarrow$ $1 + y^2 = [3(te^t - e^t + C)]^{2/3}$ $\Rightarrow$

$y = \pm\sqrt{[3(te^t - e^t + C)]^{2/3} - 1}$

9. $\dfrac{du}{dt} = 2 + 2u + t + tu$ $\Rightarrow$ $\dfrac{du}{dt} = (1 + u)(2 + t)$ $\Rightarrow$ $\displaystyle\int \dfrac{du}{1+u} = \int (2 + t)dt$ $[u \neq -1]$ $\Rightarrow$

$\ln|1 + u| = \frac{1}{2}t^2 + 2t + C$ $\Rightarrow$ $|1 + u| = e^{t^2/2 + 2t + C} = Ke^{t^2/2 + 2t}$, where $K = e^C$ $\Rightarrow$ $1 + u = \pm Ke^{t^2/2 + 2t}$ $\Rightarrow$

$u = -1 \pm Ke^{t^2/2 + 2t}$ where $K > 0$. $u = -1$ is also a solution, so $u = -1 + Ae^{t^2/2 + 2t}$, where A is an arbitrary constant.

11. $\dfrac{dy}{dx} = \dfrac{x}{y}$ $\Rightarrow$ $y\, dy = x\, dx$ $\Rightarrow$ $\int y\, dy = \int x\, dx$ $\Rightarrow$ $\frac{1}{2}y^2 = \frac{1}{2}x^2 + C.$ $y(0) = -3$ $\Rightarrow$

$\frac{1}{2}(-3)^2 = \frac{1}{2}(0)^2 + C$ $\Rightarrow$ $C = \frac{9}{2}$, so $\frac{1}{2}y^2 = \frac{1}{2}x^2 + \frac{9}{2}$ $\Rightarrow$ $y^2 = x^2 + 9$ $\Rightarrow$ $y = -\sqrt{x^2 + 9}$ since $y(0) = -3 < 0$.

13. $x\cos x = (2y + e^{3y})\, y'$ $\Rightarrow$ $x\cos x\, dx = (2y + e^{3y})\, dy$ $\Rightarrow$ $\int (2y + e^{3y})\, dy = \int x\cos x\, dx$ $\Rightarrow$

$y^2 + \frac{1}{3}e^{3y} = x\sin x + \cos x + C$ [where the second integral is evaluated using integration by parts].

Now $y(0) = 0$ $\Rightarrow$ $0 + \frac{1}{3} = 0 + 1 + C$ $\Rightarrow$ $C = -\frac{2}{3}$. Thus, a solution is $y^2 + \frac{1}{3}e^{3y} = x\sin x + \cos x - \frac{2}{3}$.

We cannot solve explicitly for y.

15. $\dfrac{du}{dt} = \dfrac{2t + \sec^2 t}{2u}$, $u(0) = -5$. $\int 2u\, du = \int (2t + \sec^2 t)\, dt$ $\Rightarrow$ $u^2 = t^2 + \tan t + C$,

where $[u(0)]^2 = 0^2 + \tan 0 + C$ $\Rightarrow$ $C = (-5)^2 = 25$. Therefore, $u^2 = t^2 + \tan t + 25$, so $u = \pm\sqrt{t^2 + \tan t + 25}$.

Since $u(0) = -5$, we must have $u = -\sqrt{t^2 + \tan t + 25}$.

17. $y'\tan x = a + y$, $0 < x < \pi/2$ $\Rightarrow$ $\dfrac{dy}{dx} = \dfrac{a + y}{\tan x}$ $\Rightarrow$ $\dfrac{dy}{a + y} = \cot x\, dx$ $[a + y \neq 0]$ $\Rightarrow$

$\displaystyle\int \dfrac{dy}{a + y} = \int \dfrac{\cos x}{\sin x}\, dx$ $\Rightarrow$ $\ln|a + y| = \ln|\sin x| + C$ $\Rightarrow$ $|a + y| = e^{\ln|\sin x| + C} = e^{\ln|\sin x|} \cdot e^C = e^C\, |\sin x|$ $\Rightarrow$

$a + y = K\sin x$, where $K = \pm e^C$. (In our derivation, K was nonzero, but we can restore the excluded case

$y = -a$ by allowing K to be zero.) $y(\pi/3) = a$ $\Rightarrow$ $a + a = K\sin\left(\dfrac{\pi}{3}\right)$ $\Rightarrow$ $2a = K\dfrac{\sqrt{3}}{2}$ $\Rightarrow$ $K = \dfrac{4a}{\sqrt{3}}$.

Thus, $a + y = \dfrac{4a}{\sqrt{3}}\sin x$ and so $y = \dfrac{4a}{\sqrt{3}}\sin x - a$.

19. If the slope at the point (x, y) is xy, then we have $\dfrac{dy}{dx} = xy$ $\Rightarrow$ $\dfrac{dy}{y} = x\, dx$ $[y \neq 0]$ $\Rightarrow$ $\displaystyle\int \dfrac{dy}{y} = \int x\, dx$ $\Rightarrow$

$\ln|y| = \frac{1}{2}x^2 + C$. $y(0) = 1$ $\Rightarrow$ $\ln 1 = 0 + C$ $\Rightarrow$ $C = 0$. Thus, $|y| = e^{x^2/2}$ $\Rightarrow$ $y = \pm e^{x^2/2}$, so $y = e^{x^2/2}$

since $y(0) = 1 > 0$. Note that $y = 0$ is not a solution because it doesn't satisfy the initial condition $y(0) = 1$.

21. $u = x + y \;\Rightarrow\; \dfrac{d}{dx}(u) = \dfrac{d}{dx}(x + y) \;\Rightarrow\; \dfrac{du}{dx} = 1 + \dfrac{dy}{dx}$, but $\dfrac{dy}{dx} = x + y = u$, so $\dfrac{du}{dx} = 1 + u \;\Rightarrow$

$\dfrac{du}{1+u} = dx \quad [u \neq -1] \;\Rightarrow\; \displaystyle\int \dfrac{du}{1+u} = \int dx \;\Rightarrow\; \ln|1 + u| = x + C \;\Rightarrow\; |1 + u| = e^{x+C} \;\Rightarrow$

$1 + u = \pm e^C e^x \;\Rightarrow\; u = \pm e^C e^x - 1 \;\Rightarrow\; x + y = \pm e^C e^x - 1 \;\Rightarrow\; y = K e^x - x - 1$, where $K = \pm e^C \neq 0$.

If $u = -1$, then $-1 = x + y \;\Rightarrow\; y = -x - 1$, which is just $y = K e^x - x - 1$ with $K = 0$. Thus, the general solution

is $y = K e^x - x - 1$, where $K \in \mathbb{R}$.

23. (a) $y' = 2x\sqrt{1 - y^2} \;\Rightarrow\; \dfrac{dy}{dx} = 2x\sqrt{1 - y^2} \;\Rightarrow\; \dfrac{dy}{\sqrt{1 - y^2}} = 2x\,dx \;\Rightarrow\; \displaystyle\int \dfrac{dy}{\sqrt{1 - y^2}} = \int 2x\,dx \;\Rightarrow$

$\sin^{-1} y = x^2 + C$ for $-\dfrac{\pi}{2} \leq x^2 + C \leq \dfrac{\pi}{2}$.

(b) $y(0) = 0 \;\Rightarrow\; \sin^{-1} 0 = 0^2 + C \;\Rightarrow\; C = 0$,

so $\sin^{-1} y = x^2$ and $y = \sin(x^2)$ for

$-\sqrt{\pi/2} \leq x \leq \sqrt{\pi/2}$.

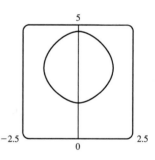

(c) For $\sqrt{1 - y^2}$ to be a real number, we must have $-1 \leq y \leq 1$; that is, $-1 \leq y(0) \leq 1$. Thus, the initial-value problem

$y' = 2x\sqrt{1 - y^2}$, $y(0) = 2$ does *not* have a solution.

25. $\dfrac{dy}{dx} = \dfrac{\sin x}{\sin y}$, $y(0) = \dfrac{\pi}{2}$. So $\int \sin y \, dy = \int \sin x \, dx \;\Leftrightarrow$

$-\cos y = -\cos x + C \;\Leftrightarrow\; \cos y = \cos x - C$. From the initial condition,

we need $\cos \dfrac{\pi}{2} = \cos 0 - C \;\Rightarrow\; 0 = 1 - C \;\Rightarrow\; C = 1$, so the solution is

$\cos y = \cos x - 1$. Note that we cannot take $\cos^{-1}$ of both sides, since that would

unnecessarily restrict the solution to the case where $-1 \leq \cos x - 1 \;\Leftrightarrow\; 0 \leq \cos x$,

as $\cos^{-1}$ is defined only on $[-1, 1]$. Instead we plot the graph using Maple's

`plots[implicitplot]` or Mathematica's `Plot[Evaluate[···]]`.

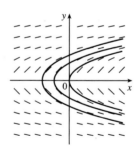

27. (a)

x	y	$y' = 1/y$
0	0.5	2
0	-0.5	-2
0	1	1
0	-1	-1
0	2	0.5

x	y	$y' = 1/y$
0	-2	-0.5
0	4	0.25
0	3	$0.\overline{3}$
0	0.25	4
0	$0.\overline{3}$	3

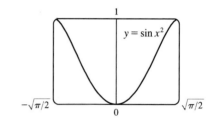

(b) $y' = 1/y \quad \Rightarrow \quad dy/dx = 1/y \quad \Rightarrow$

$\quad y\,dy = dx \quad \Rightarrow \quad \int y\,dy = \int dx \quad \Rightarrow \quad \frac{1}{2}y^2 = x + C \quad \Rightarrow$

$\quad y^2 = 2(x + C) \quad \text{or} \quad y = \pm\sqrt{2(x + C)}.$

(c)

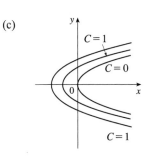

29. The curves $x^2 + 2y^2 = k^2$ form a family of ellipses with major axis on the x-axis. Differentiating gives

$$\frac{d}{dx}(x^2 + 2y^2) = \frac{d}{dx}(k^2) \quad \Rightarrow \quad 2x + 4yy' = 0 \quad \Rightarrow \quad 4yy' = -2x \quad \Rightarrow \quad y' = \frac{-x}{2y}. \text{ Thus, the slope of the tangent line}$$

at any point (x, y) on one of the ellipses is $y' = \dfrac{-x}{2y}$, so the orthogonal trajectories

must satisfy $y' = \dfrac{2y}{x} \quad \Leftrightarrow \quad \dfrac{dy}{dx} = \dfrac{2y}{x} \quad \Leftrightarrow \quad \dfrac{dy}{y} = 2 = \dfrac{dx}{x} \quad \Leftrightarrow$

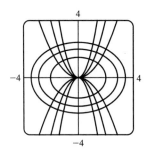

$$\int \frac{dy}{y} = 2\int \frac{dx}{x} \quad \Leftrightarrow \quad \ln|y| = 2\ln|x| + C_1 \quad \Leftrightarrow \quad \ln|y| = \ln|x|^2 + C_1 \quad \Leftrightarrow$$

$|y| = e^{\ln x^2 + C_1} \quad \Leftrightarrow \quad y = \pm x^2 \cdot e^{C_1} = Cx^2.$ This is a family of parabolas.

31. The curves $y = k/x$ form a family of hyperbolas with asymptotes $x = 0$ and $y = 0$. Differentiating gives

$$\frac{d}{dx}(y) = \frac{d}{dx}\left(\frac{k}{x}\right) \quad \Rightarrow \quad y' = -\frac{k}{x^2} \quad \Rightarrow \quad y' = -\frac{xy}{x^2} \quad [\text{since } y = k/x \Rightarrow xy = k] \quad \Rightarrow \quad y' = -\frac{y}{x}. \text{ Thus, the slope}$$

of the tangent line at any point (x, y) on one of the hyperbolas is $y' = -y/x$,

so the orthogonal trajectories must satisfy $y' = x/y \quad \Leftrightarrow \quad \dfrac{dy}{dx} = \dfrac{x}{y} \quad \Leftrightarrow$

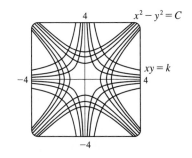

$y\,dy = x\,dx \quad \Leftrightarrow \quad \int y\,dy = \int x\,dx \quad \Leftrightarrow \quad \frac{1}{2}y^2 = \frac{1}{2}x^2 + C_1 \quad \Leftrightarrow$

$y^2 = x^2 + C_2 \quad \Leftrightarrow \quad x^2 - y^2 = C.$ This is a family of hyperbolas with

asymptotes $y = \pm x$.

33. From Exercise 9.2.27, $\dfrac{dQ}{dt} = 12 - 4Q \quad \Leftrightarrow \quad \displaystyle\int \frac{dQ}{12 - 4Q} = \int dt \quad \Leftrightarrow \quad -\frac{1}{4}\ln|12 - 4Q| = t + C \quad \Leftrightarrow$

$\ln|12 - 4Q| = -4t - 4C \quad \Leftrightarrow \quad |12 - 4Q| = e^{-4t - 4C} \quad \Leftrightarrow \quad 12 - 4Q = Ke^{-4t} \quad [K = \pm e^{-4C}] \quad \Leftrightarrow$

$4Q = 12 - Ke^{-4t} \quad \Leftrightarrow \quad Q = 3 - Ae^{-4t} \quad [A = K/4].\quad Q(0) = 0 \quad \Leftrightarrow \quad 0 = 3 - A \quad \Leftrightarrow \quad A = 3 \quad \Leftrightarrow$

$Q(t) = 3 - 3e^{-4t}.$ As $t \to \infty$, $Q(t) \to 3 - 0 = 3$ (the limiting value).

35. $\dfrac{dP}{dt} = k(M - P) \quad \Leftrightarrow \quad \displaystyle\int \frac{dP}{P - M} = \int (-k)\,dt \quad \Leftrightarrow \quad \ln|P - M| = -kt + C \quad \Leftrightarrow \quad |P - M| = e^{-kt + C} \quad \Leftrightarrow$

$P - M = Ae^{-kt} \quad [A = \pm e^C] \quad \Leftrightarrow \quad P = M + Ae^{-kt}.$ If we assume that performance is at level 0 when $t = 0$, then

$P(0) = 0 \quad \Leftrightarrow \quad 0 = M + A \quad \Leftrightarrow \quad A = -M \quad \Leftrightarrow \quad P(t) = M - Me^{-kt}. \quad \displaystyle\lim_{t \to \infty} P(t) = M - M \cdot 0 = M.$

37. (a) If $a = b$, then $\frac{dx}{dt} = k(a - x)(b - x)^{1/2}$ becomes $\frac{dx}{dt} = k(a - x)^{3/2}$ $\Rightarrow$ $(a - x)^{-3/2}\,dx = k\,dt$ $\Rightarrow$

$\int (a - x)^{-3/2}\,dx = \int k\,dt$ $\Rightarrow$ $2(a - x)^{-1/2} = kt + C$ [by substitution] $\Rightarrow$ $\dfrac{2}{kt + C} = \sqrt{a - x}$ $\Rightarrow$

$\left(\dfrac{2}{kt + C}\right)^2 = a - x$ $\Rightarrow$ $x(t) = a - \dfrac{4}{(kt + C)^2}$. The initial concentration of HBr is 0, so $x(0) = 0$ $\Rightarrow$

$0 = a - \dfrac{4}{C^2}$ $\Rightarrow$ $\dfrac{4}{C^2} = a$ $\Rightarrow$ $C^2 = \dfrac{4}{a}$ $\Rightarrow$ $C = 2/\sqrt{a}$ [C is positive since $kt + C = 2(a - x)^{-1/2} > 0$].

Thus, $x(t) = a - \dfrac{4}{(kt + 2/\sqrt{a}\,)^2}$.

(b) $\dfrac{dx}{dt} = k(a - x)(b - x)^{1/2}$ $\Rightarrow$ $\dfrac{dx}{(a - x)\sqrt{b - x}} = k\,dt$ $\Rightarrow$ $\displaystyle\int \dfrac{dx}{(a - x)\sqrt{b - x}} = \int k\,dt$ $(\star)$.

From the hint, $u = \sqrt{b - x}$ $\Rightarrow$ $u^2 = b - x$ $\Rightarrow$ $2u\,du = -dx$, so

$$\int \frac{dx}{(a - x)\sqrt{b - x}} = \int \frac{-2u\,du}{[a - (b - u^2)]u} = -2 \int \frac{du}{a - b + u^2} = -2 \int \frac{du}{\left(\sqrt{a - b}\right)^2 + u^2}$$

$$\overset{17}{=} -2 \left(\frac{1}{\sqrt{a - b}} \tan^{-1} \frac{u}{\sqrt{a - b}} \right)$$

So $(\star)$ becomes $\dfrac{-2}{\sqrt{a - b}} \tan^{-1} \dfrac{\sqrt{b - x}}{\sqrt{a - b}} = kt + C$. Now $x(0) = 0$ $\Rightarrow$ $C = \dfrac{-2}{\sqrt{a - b}} \tan^{-1} \dfrac{\sqrt{b}}{\sqrt{a - b}}$ and we have

$\dfrac{-2}{\sqrt{a - b}} \tan^{-1} \dfrac{\sqrt{b - x}}{\sqrt{a - b}} = kt - \dfrac{2}{\sqrt{a - b}} \tan^{-1} \dfrac{\sqrt{b}}{\sqrt{a - b}}$ $\Rightarrow$ $\dfrac{2}{\sqrt{a - b}} \left(\tan^{-1} \sqrt{\dfrac{b}{a - b}} - \tan^{-1} \sqrt{\dfrac{b - x}{a - b}} \right) = kt$ $\Rightarrow$

$t(x) = \dfrac{2}{k\sqrt{a - b}} \left(\tan^{-1} \sqrt{\dfrac{b}{a - b}} - \tan^{-1} \sqrt{\dfrac{b - x}{a - b}} \right)$.

39. (a) $\dfrac{dC}{dt} = r - kC$ $\Rightarrow$ $\dfrac{dC}{dt} = -(kC - r)$ $\Rightarrow$ $\displaystyle\int \dfrac{dC}{kC - r} = \int -dt$ $\Rightarrow$ $(1/k) \ln|kC - r| = -t + M_1$ $\Rightarrow$

$\ln|kC - r| = -kt + M_2$ $\Rightarrow$ $|kC - r| = e^{-kt + M_2}$ $\Rightarrow$ $kC - r = M_3 e^{-kt}$ $\Rightarrow$ $kC = M_3 e^{-kt} + r$ $\Rightarrow$

$C(t) = M_4 e^{-kt} + r/k$. $C(0) = C_0$ $\Rightarrow$ $C_0 = M_4 + r/k$ $\Rightarrow$ $M_4 = C_0 - r/k$ $\Rightarrow$

$C(t) = (C_0 - r/k)e^{-kt} + r/k$.

(b) If $C_0 < r/k$, then $C_0 - r/k < 0$ and the formula for $C(t)$ shows that $C(t)$ increases and $\lim\limits_{t \to \infty} C(t) = r/k$.

As t increases, the formula for $C(t)$ shows how the role of C_0 steadily diminishes as that of r/k increases.

41. (a) Let $y(t)$ be the amount of salt (in kg) after t minutes. Then $y(0) = 15$. The amount of liquid in the tank is 1000 L at all

times, so the concentration at time t (in minutes) is $y(t)/1000$ kg/L and $\dfrac{dy}{dt} = -\left[\dfrac{y(t)}{1000}\,\dfrac{\text{kg}}{\text{L}}\right]\left(10\,\dfrac{\text{L}}{\text{min}}\right) = -\dfrac{y(t)}{100}\,\dfrac{\text{kg}}{\text{min}}$.

$\displaystyle\int \dfrac{dy}{y} = -\dfrac{1}{100} \int dt$ $\Rightarrow$ $\ln y = -\dfrac{t}{100} + C$, and $y(0) = 15$ $\Rightarrow$ $\ln 15 = C$, so $\ln y = \ln 15 - \dfrac{t}{100}$.

It follows that $\ln\left(\dfrac{y}{15}\right) = -\dfrac{t}{100}$ and $\dfrac{y}{15} = e^{-t/100}$, so $y = 15e^{-t/100}$ kg.

(b) After 20 minutes, $y = 15e^{-20/100} = 15e^{-0.2} \approx 12.3$ kg.

43. Let $y(t)$ be the amount of alcohol in the vat after t minutes. Then $y(0) = 0.04(500) = 20$ gal. The amount of beer in the vat is 500 gallons at all times, so the percentage at time t (in minutes) is $y(t)/500 \times 100$, and the change in the amount of alcohol with respect to time t is $\dfrac{dy}{dt} = \text{rate in} - \text{rate out} = 0.06\left(5\,\dfrac{\text{gal}}{\text{min}}\right) - \dfrac{y(t)}{500}\left(5\,\dfrac{\text{gal}}{\text{min}}\right) = 0.3 - \dfrac{y}{100} = \dfrac{30 - y}{100}\,\dfrac{\text{gal}}{\text{min}}.$

Hence, $\displaystyle\int \dfrac{dy}{30 - y} = \int \dfrac{dt}{100}$ and $-\ln|30 - y| = \frac{1}{100}t + C$. Because $y(0) = 20$, we have $-\ln 10 = C$, so

$-\ln|30 - y| = \frac{1}{100}t - \ln 10 \;\Rightarrow\; \ln|30 - y| = -t/100 + \ln 10 \;\Rightarrow\; \ln|30 - y| = \ln e^{-t/100} + \ln 10 \;\Rightarrow$

$\ln|30 - y| = \ln(10e^{-t/100}) \;\Rightarrow\; |30 - y| = 10e^{-t/100}$. Since y is continuous, $y(0) = 20$, and the right-hand side is never zero, we deduce that $30 - y$ is always positive. Thus, $30 - y = 10e^{-t/100} \;\Rightarrow\; y = 30 - 10e^{-t/100}$. The percentage of alcohol is $p(t) = y(t)/500 \times 100 = y(t)/5 = 6 - 2e^{-t/100}$. The percentage of alcohol after one hour is $p(60) = 6 - 2e^{-60/100} \approx 4.9$.

45. Assume that the raindrop begins at rest, so that $v(0) = 0$. $\quad dm/dt = km$ and $(mv)' = gm \;\Rightarrow\; mv' + vm' = gm \;\Rightarrow$

$mv' + v(km) = gm \;\Rightarrow\; v' + vk = g \;\Rightarrow\; \dfrac{dv}{dt} = g - kv \;\Rightarrow\; \displaystyle\int \dfrac{dv}{g - kv} = \int dt \;\Rightarrow$

$-(1/k)\ln|g - kv| = t + C \;\Rightarrow\; \ln|g - kv| = -kt - kC \;\Rightarrow\; g - kv = Ae^{-kt}.\;\; v(0) = 0 \;\Rightarrow\; A = g.$

So $kv = g - ge^{-kt} \;\Rightarrow\; v = (g/k)(1 - e^{-kt})$. Since $k > 0$, as $t \to \infty$, $e^{-kt} \to 0$ and therefore, $\displaystyle\lim_{t\to\infty} v(t) = g/k$.

47. (a) The rate of growth of the area is jointly proportional to $\sqrt{A(t)}$ and $M - A(t)$; that is, the rate is proportional to the product of those two quantities. So for some constant k, $dA/dt = k\sqrt{A}\,(M - A)$. We are interested in the maximum of the function dA/dt (when the tissue grows the fastest), so we differentiate, using the Chain Rule and then substituting for dA/dt from the differential equation:

$$\frac{d}{dt}\left(\frac{dA}{dt}\right) = k\left[\sqrt{A}\,(-1)\frac{dA}{dt} + (M - A)\cdot\tfrac{1}{2}A^{-1/2}\frac{dA}{dt}\right] = \tfrac{1}{2}kA^{-1/2}\frac{dA}{dt}\left[-2A + (M - A)\right]$$

$$= \tfrac{1}{2}kA^{-1/2}\left[k\sqrt{A}(M - A)\right][M - 3A] = \tfrac{1}{2}k^2(M - A)(M - 3A)$$

This is 0 when $M - A = 0$ [this situation never actually occurs, since the graph of $A(t)$ is asymptotic to the line $y = M$, as in the logistic model] and when $M - 3A = 0 \;\Leftrightarrow\; A(t) = M/3$. This represents a maximum by the First Derivative Test, since $\dfrac{d}{dt}\left(\dfrac{dA}{dt}\right)$ goes from positive to negative when $A(t) = M/3$.

(b) From the CAS, we get $A(t) = M\left(\dfrac{Ce^{\sqrt{M}kt} - 1}{Ce^{\sqrt{M}kt} + 1}\right)^2$. To get C in terms of the initial area A_0 and the maximum area M,

we substitute $t = 0$ and $A = A_0 = A(0)$: $A_0 = M\left(\dfrac{C - 1}{C + 1}\right)^2 \;\Leftrightarrow\; (C + 1)\sqrt{A_0} = (C - 1)\sqrt{M} \;\Leftrightarrow$

$C\sqrt{A_0} + \sqrt{A_0} = C\sqrt{M} - \sqrt{M} \;\Leftrightarrow\; \sqrt{M} + \sqrt{A_0} = C\sqrt{M} - C\sqrt{A_0} \;\Leftrightarrow$

$\sqrt{M} + \sqrt{A_0} = C\left(\sqrt{M} - \sqrt{A_0}\right) \;\Leftrightarrow\; C = \dfrac{\sqrt{M} + \sqrt{A_0}}{\sqrt{M} - \sqrt{A_0}}.$ [Notice that if $A_0 = 0$, then $C = 1$.]

9.4 Models for Population Growth

1. (a) $dP/dt = 0.05P - 0.0005P^2 = 0.05P(1 - 0.01P) = 0.05P(1 - P/100)$. Comparing to Equation 4,

$dP/dt = kP(1 - P/K)$, we see that the carrying capacity is $K = 100$ and the value of k is 0.05.

(b) The slopes close to 0 occur where P is near 0 or 100. The largest slopes appear to be on the line $P = 50$. The solutions

are increasing for $0 < P_0 < 100$ and decreasing for $P_0 > 100$.

(c) 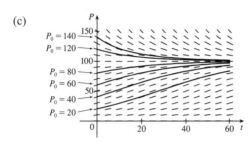 All of the solutions approach $P = 100$ as t increases. As in

part (b), the solutions differ since for $0 < P_0 < 100$ they are

increasing, and for $P_0 > 100$ they are decreasing. Also, some

have an IP and some don't. It appears that the solutions which

have $P_0 = 20$ and $P_0 = 40$ have inflection points at $P = 50$.

(d) The equilibrium solutions are $P = 0$ (trivial solution) and $P = 100$. The increasing solutions move away from $P = 0$ and

all nonzero solutions approach $P = 100$ as $t \to \infty$.

3. (a) $\dfrac{dy}{dt} = ky\left(1 - \dfrac{y}{K}\right) \Rightarrow y(t) = \dfrac{K}{1 + Ae^{-kt}}$ with $A = \dfrac{K - y(0)}{y(0)}$. With $K = 8 \times 10^7$, $k = 0.71$, and

$y(0) = 2 \times 10^7$, we get the model $y(t) = \dfrac{8 \times 10^7}{1 + 3e^{-0.71t}}$, so $y(1) = \dfrac{8 \times 10^7}{1 + 3e^{-0.71}} \approx 3.23 \times 10^7$ kg.

(b) $y(t) = 4 \times 10^7 \Rightarrow \dfrac{8 \times 10^7}{1 + 3e^{-0.71t}} = 4 \times 10^7 \Rightarrow 2 = 1 + 3e^{-0.71t} \Rightarrow e^{-0.71t} = \frac{1}{3} \Rightarrow$

$-0.71t = \ln\frac{1}{3} \Rightarrow t = \dfrac{\ln 3}{0.71} \approx 1.55$ years

5. (a) We will assume that the difference in the birth and death rates is 20 million/year. Let $t = 0$ correspond to the year 1990

and use a unit of 1 billion for all calculations. $k \approx \dfrac{1}{P}\dfrac{dP}{dt} = \dfrac{1}{5.3}(0.02) = \dfrac{1}{265}$, so

$$\frac{dP}{dt} = kP\left(1 - \frac{P}{K}\right) = \frac{1}{265}P\left(1 - \frac{P}{100}\right), \qquad P \text{ in billions}$$

(b) $A = \dfrac{K - P_0}{P_0} = \dfrac{100 - 5.3}{5.3} = \dfrac{947}{53} \approx 17.8679$. $P(t) = \dfrac{K}{1 + Ae^{-kt}} = \dfrac{100}{1 + \frac{947}{53}e^{-(1/265)t}}$, so $P(10) \approx 5.49$ billion.

(c) $P(110) \approx 7.81$, and $P(510) \approx 27.72$. The predictions are 7.81 billion in the year 2100 and 27.72 billion in 2500.

(d) If $K = 50$, then $P(t) = \dfrac{50}{1 + \frac{447}{53}e^{-(1/265)t}}$. So $P(10) \approx 5.48$, $P(110) \approx 7.61$, and $P(510) \approx 22.41$. The predictions

become 5.48 billion in the year 2000, 7.61 billion in 2100, and 22.41 billion in the year 2500.

7. (a) Our assumption is that $\dfrac{dy}{dt} = ky(1 - y)$, where y is the fraction of the population that has heard the rumor.

(b) Using the logistic equation (4), $\dfrac{dP}{dt} = kP\left(1 - \dfrac{P}{K}\right)$, we substitute $y = \dfrac{P}{K}$, $P = Ky$, and $\dfrac{dP}{dt} = K\dfrac{dy}{dt}$,

to obtain $K\dfrac{dy}{dt} = k(Ky)(1 - y)$ $\Leftrightarrow$ $\dfrac{dy}{dt} = ky(1 - y)$, our equation in part (a).

Now the solution to (4) is $P(t) = \dfrac{K}{1 + Ae^{-kt}}$, where $A = \dfrac{K - P_0}{P_0}$.

We use the same substitution to obtain $Ky = \dfrac{K}{1 + \dfrac{K - Ky_0}{Ky_0}e^{-kt}}$ $\Rightarrow$ $y = \dfrac{y_0}{y_0 + (1 - y_0)e^{-kt}}$.

Alternatively, we could use the same steps as outlined in the solution of Equation 4.

(c) Let t be the number of hours since 8 AM. Then $y_0 = y(0) = \dfrac{80}{1000} = 0.08$ and $y(4) = \tfrac{1}{2}$, so

$\dfrac{1}{2} = y(4) = \dfrac{0.08}{0.08 + 0.92e^{-4k}}$. Thus, $0.08 + 0.92e^{-4k} = 0.16$, $e^{-4k} = \dfrac{0.08}{0.92} = \dfrac{2}{23}$, and $e^{-k} = \left(\dfrac{2}{23}\right)^{1/4}$,

so $y = \dfrac{0.08}{0.08 + 0.92(2/23)^{t/4}} = \dfrac{2}{2 + 23(2/23)^{t/4}}$. Solving this equation for t, we get

$2y + 23y\left(\dfrac{2}{23}\right)^{t/4} = 2$ $\Rightarrow$ $\left(\dfrac{2}{23}\right)^{t/4} = \dfrac{2 - 2y}{23y}$ $\Rightarrow$ $\left(\dfrac{2}{23}\right)^{t/4} = \dfrac{2}{23} \cdot \dfrac{1 - y}{y}$ $\Rightarrow$ $\left(\dfrac{2}{23}\right)^{t/4 - 1} = \dfrac{1 - y}{y}$.

It follows that $\dfrac{t}{4} - 1 = \dfrac{\ln[(1 - y)/y]}{\ln\frac{2}{23}}$, so $t = 4\left[1 + \dfrac{\ln((1 - y)/y)}{\ln\frac{2}{23}}\right]$.

When $y = 0.9$, $\dfrac{1 - y}{y} = \dfrac{1}{9}$, so $t = 4\left(1 - \dfrac{\ln 9}{\ln\frac{2}{23}}\right) \approx 7.6$ h or 7 h 36 min. Thus, 90% of the population will have heard

the rumor by 3:36 PM.

9. (a) $\dfrac{dP}{dt} = kP\left(1 - \dfrac{P}{K}\right)$ $\Rightarrow$ $\dfrac{d^2P}{dt^2} = k\left[P\left(-\dfrac{1}{K}\dfrac{dP}{dt}\right) + \left(1 - \dfrac{P}{K}\right)\dfrac{dP}{dt}\right] = k\dfrac{dP}{dt}\left(-\dfrac{P}{K} + 1 - \dfrac{P}{K}\right)$

$= k\left[kP\left(1 - \dfrac{P}{K}\right)\right]\left(1 - \dfrac{2P}{K}\right) = k^2 P\left(1 - \dfrac{P}{K}\right)\left(1 - \dfrac{2P}{K}\right)$

(b) P grows fastest when P' has a maximum, that is, when $P'' = 0$. From part (a), $P'' = 0$ $\Leftrightarrow$ $P = 0, P = K$,

or $P = K/2$. Since $0 < P < K$, we see that $P'' = 0$ $\Leftrightarrow$ $P = K/2$.

11. Following the hint, we choose $t = 0$ to correspond to 1960 and subtract 94,000 from each of the population figures. We then use a calculator to obtain the models and add 94,000 to get the exponential function

$P_E(t) = 1578.3(1.0933)^t + 94{,}000$ and the logistic function

$P_L(t) = \dfrac{32{,}658.5}{1 + 12.75e^{-0.1706t}} + 94{,}000$. P_L is a reasonably accurate

model, while P_E is not, since an exponential model would only be used

for the first few data points.

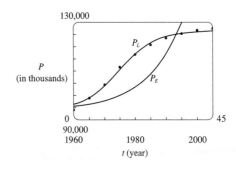

13. (a) $\dfrac{dP}{dt} = kP - m = k\left(P - \dfrac{m}{k}\right)$. Let $y = P - \dfrac{m}{k}$, so $\dfrac{dy}{dt} = \dfrac{dP}{dt}$ and the differential equation becomes $\dfrac{dy}{dt} = ky$.

The solution is $y = y_0 e^{kt}$ $\Rightarrow$ $P - \dfrac{m}{k} = \left(P_0 - \dfrac{m}{k}\right)e^{kt}$ $\Rightarrow$ $P(t) = \dfrac{m}{k} + \left(P_0 - \dfrac{m}{k}\right)e^{kt}$.

(b) Since $k > 0$, there will be an exponential expansion $\Leftrightarrow$ $P_0 - \dfrac{m}{k} > 0$ $\Leftrightarrow$ $m < kP_0$.

(c) The population will be constant if $P_0 - \dfrac{m}{k} = 0$ $\Leftrightarrow$ $m = kP_0$. It will decline if $P_0 - \dfrac{m}{k} < 0$ $\Leftrightarrow$ $m > kP_0$.

(d) $P_0 = 8{,}000{,}000$, $k = \alpha - \beta = 0.016$, $m = 210{,}000$ $\Rightarrow$ $m > kP_0 (= 128{,}000)$, so by part (c), the population was declining.

15. (a) The term -15 represents a harvesting of fish at a constant rate—in this case, 15 fish/week. This is the rate at which fish are caught.

(b)

(c) From the graph in part (b), it appears that $P(t) = 250$ and $P(t) = 750$ are the equilibrium solutions. We confirm this analytically by solving the equation $dP/dt = 0$ as follows: $0.08P(1 - P/1000) - 15 = 0$ $\Rightarrow$

$0.08P - 0.00008P^2 - 15 = 0$ $\Rightarrow$

$-0.00008(P^2 - 1000P + 187{,}500) = 0$ $\Rightarrow$

$(P - 250)(P - 750) = 0$ $\Rightarrow$ $P = 250$ or 750.

(d)

For $0 < P_0 < 250$, $P(t)$ decreases to 0. For $P_0 = 250$, $P(t)$ remains constant. For $250 < P_0 < 750$, $P(t)$ increases and approaches 750. For $P_0 = 750$, $P(t)$ remains constant. For $P_0 > 750$, $P(t)$ decreases and approaches 750.

(e) $\dfrac{dP}{dt} = 0.08P\left(1 - \dfrac{P}{1000}\right) - 15$ $\Leftrightarrow$ $-\dfrac{100{,}000}{8} \cdot \dfrac{dP}{dt} = (0.08P - 0.00008P^2 - 15) \cdot \left(-\dfrac{100{,}000}{8}\right)$ $\Leftrightarrow$

$-12{,}500\dfrac{dP}{dt} = P^2 - 1000P + 187{,}500$ $\Leftrightarrow$ $\dfrac{dP}{(P - 250)(P - 750)} = -\dfrac{1}{12{,}500}\,dt$ $\Leftrightarrow$

$\displaystyle\int\left(\dfrac{-1/500}{P - 250} + \dfrac{1/500}{P - 750}\right)dP = -\dfrac{1}{12{,}500}\,dt$ $\Leftrightarrow$ $\displaystyle\int\left(\dfrac{1}{P - 250} - \dfrac{1}{P - 750}\right)dP = \dfrac{1}{25}\,dt$ $\Leftrightarrow$

$\ln|P - 250| - \ln|P - 750| = \dfrac{1}{25}t + C$ $\Leftrightarrow$ $\ln\left|\dfrac{P - 250}{P - 750}\right| = \dfrac{1}{25}t + C$ $\Leftrightarrow$ $\left|\dfrac{P - 250}{P - 750}\right| = e^{t/25 + C} = ke^{t/25}$ $\Leftrightarrow$

$\dfrac{P - 250}{P - 750} = ke^{t/25}$ $\Leftrightarrow$ $P - 250 = Pke^{t/25} - 750ke^{t/25}$ $\Leftrightarrow$ $P - Pke^{t/25} = 250 - 750ke^{t/25}$ $\Leftrightarrow$

$P(t) = \dfrac{250 - 750ke^{t/25}}{1 - ke^{t/25}}$. If $t = 0$ and $P = 200$, then $200 = \dfrac{250 - 750k}{1 - k}$ $\Leftrightarrow$ $200 - 200k = 250 - 750k$ $\Leftrightarrow$

$550k = 50$ $\Leftrightarrow$ $k = \dfrac{1}{11}$. Similarly, if $t = 0$ and $P = 300$, then

$k = -\dfrac{1}{9}$. Simplifying P with these two values of k gives us

$P(t) = \dfrac{250(3e^{t/25} - 11)}{e^{t/25} - 11}$ and $P(t) = \dfrac{750(e^{t/25} + 3)}{e^{t/25} + 9}$.

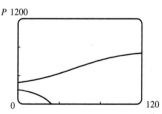

17. (a) $\dfrac{dP}{dt} = (kP)\left(1 - \dfrac{P}{K}\right)\left(1 - \dfrac{m}{P}\right)$. If $m < P < K$, then $dP/dt = (+)(+)(+) = + \;\Rightarrow\; P$ is increasing.

If $0 < P < m$, then $dP/dt = (+)(+)(-) = - \;\Rightarrow\; P$ is decreasing.

(b)

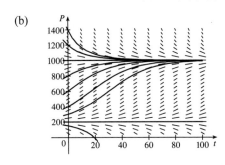

$k = 0.08$, $K = 1000$, and $m = 200 \;\Rightarrow$

$$\frac{dP}{dt} = 0.08P\left(1 - \frac{P}{1000}\right)\left(1 - \frac{200}{P}\right)$$

For $0 < P_0 < 200$, the population dies out. For $P_0 = 200$, the population is steady. For $200 < P_0 < 1000$, the population increases and approaches 1000. For $P_0 > 1000$, the population decreases and approaches 1000.

The equilibrium solutions are $P(t) = 200$ and $P(t) = 1000$.

(c) $\dfrac{dP}{dt} = kP\left(1 - \dfrac{P}{K}\right)\left(1 - \dfrac{m}{P}\right) = kP\left(\dfrac{K-P}{K}\right)\left(\dfrac{P-m}{P}\right) = \dfrac{k}{K}(K-P)(P-m) \;\Leftrightarrow$

$\displaystyle\int \frac{dP}{(K-P)(P-m)} = \int \frac{k}{K}\, dt.$ By partial fractions, $\dfrac{1}{(K-P)(P-m)} = \dfrac{A}{K-P} + \dfrac{B}{P-m}$, so

$A(P-m) + B(K-P) = 1.$

If $P = m$, $B = \dfrac{1}{K-m}$; if $P = K$, $A = \dfrac{1}{K-m}$, so $\dfrac{1}{K-m}\displaystyle\int\left(\dfrac{1}{K-P} + \dfrac{1}{P-m}\right) dP = \int \dfrac{k}{K}\, dt \;\Rightarrow$

$\dfrac{1}{K-m}\left(-\ln|K-P| + \ln|P-m|\right) = \dfrac{k}{K}t + M \;\Rightarrow\; \dfrac{1}{K-m}\ln\left|\dfrac{P-m}{K-P}\right| = \dfrac{k}{K}t + M \;\Rightarrow$

$\ln\left|\dfrac{P-m}{K-P}\right| = (K-m)\dfrac{k}{K}t + M_1 \;\Leftrightarrow\; \dfrac{P-m}{K-P} = De^{(K-m)(k/K)t} \quad [D = \pm e^{M_1}].$

Let $t = 0$: $\dfrac{P_0 - m}{K - P_0} = D$. So $\dfrac{P-m}{K-P} = \dfrac{P_0 - m}{K - P_0}e^{(K-m)(k/K)t}$. Solving for P, we get

$P(t) = \dfrac{m(K-P_0) + K(P_0-m)e^{(K-m)(k/K)t}}{K - P_0 + (P_0-m)e^{(K-m)(k/K)t}}.$

(d) If $P_0 < m$, then $P_0 - m < 0$. Let $N(t)$ be the numerator of the expression for $P(t)$ in part (c). Then

$N(0) = P_0(K-m) > 0$, and $P_0 - m < 0 \;\Leftrightarrow\; \displaystyle\lim_{t\to\infty} K(P_0 - m)e^{(K-m)(k/K)t} = -\infty \;\Rightarrow\; \lim_{t\to\infty} N(t) = -\infty.$

Since N is continuous, there is a number t such that $N(t) = 0$ and thus $P(t) = 0$. So the species will become extinct.

19. (a) $dP/dt = kP\cos(rt - \phi) \;\Rightarrow\; (dP)/P = k\cos(rt - \phi)\, dt \;\Rightarrow\; \int(dP)/P = k\int \cos(rt - \phi)\, dt \;\Rightarrow$

$\ln P = (k/r)\sin(rt - \phi) + C$. (Since this is a growth model, $P > 0$ and we can write $\ln P$ instead of $\ln|P|$.) Since

$P(0) = P_0$, we obtain $\ln P_0 = (k/r)\sin(-\phi) + C = -(k/r)\sin\phi + C \;\Rightarrow\; C = \ln P_0 + (k/r)\sin\phi$. Thus,

$\ln P = (k/r)\sin(rt - \phi) + \ln P_0 + (k/r)\sin\phi$, which we can rewrite as $\ln(P/P_0) = (k/r)[\sin(rt - \phi) + \sin\phi]$ or,

after exponentiation, $P(t) = P_0 e^{(k/r)[\sin(rt - \phi) + \sin\phi]}$.

(b) As k increases, the amplitude increases, but the minimum value stays the same.

As r increases, the amplitude and the period decrease.

A change in ϕ produces slight adjustments in the phase shift and amplitude.

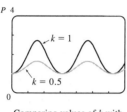

Comparing values of k with $P_0 = 1$, $r = 2$, and $\phi = \pi/2$

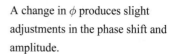

Comparing values of r with $P_0 = 1$, $k = 1$, and $\phi = \pi/2$

Comparing values of ϕ with $P_0 = 1$, $k = 1$, and $r = 2$

$P(t)$ oscillates between $P_0 e^{(k/r)(1+\sin\phi)}$ and $P_0 e^{(k/r)(-1+\sin\phi)}$ (the extreme values are attained when $rt - \phi$ is an odd multiple of $\frac{\pi}{2}$), so $\lim\limits_{t\to\infty} P(t)$ does not exist.

21. By Equation (7), $P(t) = \dfrac{K}{1 + Ae^{-kt}}$. By comparison, if $c = (\ln A)/k$ and $u = \frac{1}{2}k(t - c)$, then

$$1 + \tanh u = 1 + \frac{e^u - e^{-u}}{e^u + e^{-u}} = \frac{e^u + e^{-u}}{e^u + e^{-u}} + \frac{e^u - e^{-u}}{e^u + e^{-u}} = \frac{2e^u}{e^u + e^{-u}} \cdot \frac{e^{-u}}{e^{-u}} = \frac{2}{1 + e^{-2u}}$$

and $e^{-2u} = e^{-k(t-c)} = e^{kc}e^{-kt} = e^{\ln A}e^{-kt} = Ae^{-kt}$, so

$$\frac{1}{2}K\left[1 + \tanh\left(\frac{1}{2}k(t - c)\right)\right] = \frac{K}{2}[1 + \tanh u] = \frac{K}{2} \cdot \frac{2}{1 + e^{-2u}} = \frac{K}{1 + e^{-2u}} = \frac{K}{1 + Ae^{-kt}} = P(t).$$

9.5 Linear Equations

1. $y' + \cos x = y \ \Rightarrow \ y' + (-1)y = -\cos x$ is linear since it can be put into the standard linear form (1), $y' + P(x)\,y = Q(x)$.

3. $yy' + xy = x^2 \ \Rightarrow \ y' + x = x^2/y \ \Rightarrow \ y' - x^2/y = -x$ is not linear since it cannot be put into the standard linear form (1), $y' + P(x)\,y = Q(x)$.

5. Comparing the given equation, $y' + 2y = 2e^x$, with the general form, $y' + P(x)y = Q(x)$, we see that $P(x) = 2$ and the integrating factor is $I(x) = e^{\int P(x)\,dx} = e^{\int 2\,dx} = e^{2x}$. Multiplying the differential equation by $I(x)$ gives

$$e^{2x}y' + 2e^{2x}y = 2e^{3x} \ \Rightarrow \ (e^{2x}y)' = 2e^{3x} \ \Rightarrow \ e^{2x}y = \int 2e^{3x}\,dx \ \Rightarrow \ e^{2x}y = \frac{2}{3}e^{3x} + C \ \Rightarrow \ y = \frac{2}{3}e^x + Ce^{-2x}.$$

7. $xy' - 2y = x^2$ [divide by x] $\ \Rightarrow \ y' + \left(-\dfrac{2}{x}\right)y = x$ $(\star)$.

$I(x) = e^{\int P(x)\,dx} = e^{\int(-2/x)\,dx} = e^{-2\ln|x|} = e^{\ln|x|^{-2}} = e^{\ln(1/x^2)} = 1/x^2$. Multiplying the differential equation $(\star)$ by $I(x)$ gives $\dfrac{1}{x^2}\,y' - \dfrac{2}{x^3}\,y = \dfrac{1}{x} \ \Rightarrow \ \left(\dfrac{1}{x^2}\,y\right)' = \dfrac{1}{x} \ \Rightarrow \ \dfrac{1}{x^2}\,y = \ln|x| + C \ \Rightarrow$

$y = x^2(\ln|x| + C) = x^2\ln|x| + Cx^2$.

9. Since $P(x)$ is the derivative of the coefficient of y' [$P(x) = 1$ and the coefficient is x], we can write the differential equation

$xy' + y = \sqrt{x}$ in the easily integrable form $(xy)' = \sqrt{x} \Rightarrow xy = \frac{2}{3}x^{3/2} + C \Rightarrow y = \frac{2}{3}\sqrt{x} + C/x.$

11. $\sin x \dfrac{dy}{dx} + (\cos x)\, y = \sin(x^2) \Rightarrow [(\sin x)\, y]' = \sin(x^2) \Rightarrow (\sin x)\, y = \int \sin(x^2)\, dx \Rightarrow y = \dfrac{\int \sin(x^2)\, dx + C}{\sin x}.$

13. $(1 + t)\dfrac{du}{dt} + u = 1 + t,\ t > 0$ [divide by $1 + t$] $\Rightarrow \dfrac{du}{dt} + \dfrac{1}{1+t}\, u = 1$ ($\star$), which has the

form $u' + P(t)\, u = Q(t)$. The integrating factor is $I(t) = e^{\int P(t)\, dt} = e^{\int [1/(1+t)]\, dt} = e^{\ln(1+t)} = 1 + t.$

Multiplying ($\star$) by $I(t)$ gives us our original equation back. We rewrite it as $[(1+t)u]' = 1 + t$. Thus,

$(1+t)u = \int (1+t)\, dt = t + \frac{1}{2}t^2 + C \Rightarrow u = \dfrac{t + \frac{1}{2}t^2 + C}{1+t}$ or $u = \dfrac{t^2 + 2t + 2C}{2(t+1)}.$

15. $y' = x + y \Rightarrow y' + (-1)y = x.\ I(x) = e^{\int (-1)\, dx} = e^{-x}$. Multiplying by e^{-x} gives $e^{-x}y' - e^{-x}y = xe^{-x} \Rightarrow$

$(e^{-x}y)' = xe^{-x} \Rightarrow e^{-x}y = \int xe^{-x}\, dx = -xe^{-x} - e^{-x} + C$ [integration by parts with $u = x,\ dv = e^{-x}\, dx$] $\Rightarrow$

$y = -x - 1 + Ce^x.\ y(0) = 2 \Rightarrow -1 + C = 2 \Rightarrow C = 3$, so $y = -x - 1 + 3e^x.$

17. $\dfrac{dv}{dt} - 2tv = 3t^2 e^{t^2},\ v(0) = 5.\ I(t) = e^{\int (-2t)\, dt} = e^{-t^2}$. Multiply the differential equation by $I(t)$ to get

$e^{-t^2}\dfrac{dv}{dt} - 2te^{-t^2}v = 3t^2 \Rightarrow \left(e^{-t^2}v\right)' = 3t^2 \Rightarrow e^{-t^2}v = \int 3t^2\, dt = t^3 + C \Rightarrow v = t^3 e^{t^2} + Ce^{t^2}.$

$5 = v(0) = 0 \cdot 1 + C \cdot 1 = C$, so $v = t^3 e^{t^2} + 5e^{t^2}.$

19. $xy' = y + x^2 \sin x \Rightarrow y' - \dfrac{1}{x}y = x \sin x.\ I(x) = e^{\int (-1/x)\, dx} = e^{-\ln x} = e^{\ln x^{-1}} = \dfrac{1}{x}.$

Multiplying by $\dfrac{1}{x}$ gives $\dfrac{1}{x}y' - \dfrac{1}{x^2}y = \sin x \Rightarrow \left(\dfrac{1}{x}y\right)' = \sin x \Rightarrow \dfrac{1}{x}y = -\cos x + C \Rightarrow y = -x \cos x + Cx.$

$y(\pi) = 0 \Rightarrow -\pi \cdot (-1) + C\pi = 0 \Rightarrow C = -1$, so $y = -x \cos x - x.$

21. $xy' + 2y = e^x \Rightarrow y' + \dfrac{2}{x}y = \dfrac{e^x}{x}.$

$I(x) = e^{\int (2/x)\, dx} = e^{2\ln|x|} = \left(e^{\ln|x|}\right)^2 = |x|^2 = x^2.$

Multiplying by $I(x)$ gives $x^2 y' + 2xy = xe^x \Rightarrow (x^2 y)' = xe^x \Rightarrow$

$x^2 y = \int xe^x\, dx = (x-1)e^x + C$ [by parts] $\Rightarrow$

$y = [(x-1)e^x + C]/x^2$. The graphs for $C = -5, -3, -1, 1, 3, 5,$ and 7 are

shown. $C = 1$ is a transitional value. For $C < 1$, there is an inflection point and

for $C > 1$, there is a local minimum. As $|C|$ gets larger, the "branches" get

further from the origin.

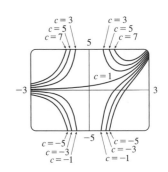

23. Setting $u = y^{1-n}$, $\dfrac{du}{dx} = (1-n)\,y^{-n}\dfrac{dy}{dx}$ or $\dfrac{dy}{dx} = \dfrac{y^n}{1-n}\dfrac{du}{dx} = \dfrac{u^{n/(1-n)}}{1-n}\dfrac{du}{dx}$. Then the Bernoulli differential equation

becomes $\dfrac{u^{n/(1-n)}}{1-n}\dfrac{du}{dx} + P(x)u^{1/(1-n)} = Q(x)u^{n/(1-n)}$ or $\dfrac{du}{dx} + (1-n)P(x)u = Q(x)(1-n)$.

25. Here $y' + \dfrac{2}{x}y = \dfrac{y^3}{x^2}$, so $n = 3$, $P(x) = \dfrac{2}{x}$ and $Q(x) = \dfrac{1}{x^2}$. Setting $u = y^{-2}$, u satisfies $u' - \dfrac{4u}{x} = -\dfrac{2}{x^2}$.

Then $I(x) = e^{\int(-4/x)\,dx} = x^{-4}$ and $u = x^4\left(\int -\dfrac{2}{x^6}\,dx + C\right) = x^4\left(\dfrac{2}{5x^5} + C\right) = Cx^4 + \dfrac{2}{5x}$.

Thus, $y = \pm\left(Cx^4 + \dfrac{2}{5x}\right)^{-1/2}$.

27. (a) $2\dfrac{dI}{dt} + 10I = 40$ or $\dfrac{dI}{dt} + 5I = 20$. Then the integrating factor is $e^{\int 5\,dt} = e^{5t}$. Multiplying the differential equation

by the integrating factor gives $e^{5t}\dfrac{dI}{dt} + 5Ie^{5t} = 20e^{5t}$ $\Rightarrow$ $(e^{5t}I)' = 20e^{5t}$ $\Rightarrow$

$I(t) = e^{-5t}\left[\int 20e^{5t}\,dt + C\right] = 4 + Ce^{-5t}$. But $0 = I(0) = 4 + C$, so $I(t) = 4 - 4e^{-5t}$.

(b) $I(0.1) = 4 - 4e^{-0.5} \approx 1.57$ A

29. $5\dfrac{dQ}{dt} + 20Q = 60$ with $Q(0) = 0$ C. Then the integrating factor is $e^{\int 4\,dt} = e^{4t}$, and multiplying the differential

equation by the integrating factor gives $e^{4t}\dfrac{dQ}{dt} + 4e^{4t}Q = 12e^{4t}$ $\Rightarrow$ $(e^{4t}Q)' = 12e^{4t}$ $\Rightarrow$

$Q(t) = e^{-4t}\left[\int 12e^{4t}\,dt + C\right] = 3 + Ce^{-4t}$. But $0 = Q(0) = 3 + C$ so $Q(t) = 3(1 - e^{-4t})$ is the charge at time t

and $I = dQ/dt = 12e^{-4t}$ is the current at time t.

31. $\dfrac{dP}{dt} + kP = kM$, so $I(t) = e^{\int k\,dt} = e^{kt}$. Multiplying the differential equation

by $I(t)$ gives $e^{kt}\dfrac{dP}{dt} + kPe^{kt} = kMe^{kt}$ $\Rightarrow$ $(e^{kt}P)' = kMe^{kt}$ $\Rightarrow$

$P(t) = e^{-kt}\left(\int kMe^{kt}\,dt + C\right) = M + Ce^{-kt}$, $k > 0$. Furthermore, it is

reasonable to assume that $0 \le P(0) \le M$, so $-M \le C \le 0$.

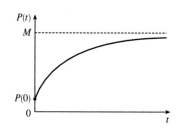

33. $y(0) = 0$ kg. Salt is added at a rate of $\left(0.4\,\dfrac{\text{kg}}{\text{L}}\right)\left(5\,\dfrac{\text{L}}{\text{min}}\right) = 2\,\dfrac{\text{kg}}{\text{min}}$. Since solution is drained from the tank at a rate of

3 L/min, but salt solution is added at a rate of 5 L/min, the tank, which starts out with 100 L of water, contains $(100 + 2t)$ L

of liquid after t min. Thus, the salt concentration at time t is $\dfrac{y(t)}{100 + 2t}\,\dfrac{\text{kg}}{\text{L}}$. Salt therefore leaves the tank at a rate of

$\left(\dfrac{y(t)}{100 + 2t}\,\dfrac{\text{kg}}{\text{L}}\right)\left(3\,\dfrac{\text{L}}{\text{min}}\right) = \dfrac{3y}{100 + 2t}\,\dfrac{\text{kg}}{\text{min}}$. Combining the rates at which salt enters and leaves the tank, we get

$\dfrac{dy}{dt} = 2 - \dfrac{3y}{100 + 2t}$. Rewriting this equation as $\dfrac{dy}{dt} + \left(\dfrac{3}{100 + 2t}\right) y = 2$, we see that it is linear.

$$I(t) = \exp\left(\int \dfrac{3\,dt}{100 + 2t}\right) = \exp\left(\tfrac{3}{2}\ln(100 + 2t)\right) = (100 + 2t)^{3/2}$$

Multiplying the differential equation by $I(t)$ gives $(100 + 2t)^{3/2}\dfrac{dy}{dt} + 3(100 + 2t)^{1/2}y = 2(100 + 2t)^{3/2}$ $\Rightarrow$

$[(100 + 2t)^{3/2}y]' = 2(100 + 2t)^{3/2}$ $\Rightarrow$ $(100 + 2t)^{3/2}y = \tfrac{2}{5}(100 + 2t)^{5/2} + C$ $\Rightarrow$

$y = \tfrac{2}{5}(100 + 2t) + C(100 + 2t)^{-3/2}$. Now $0 = y(0) = \tfrac{2}{5}(100) + C \cdot 100^{-3/2} = 40 + \tfrac{1}{1000}C$ $\Rightarrow$ $C = -40{,}000$, so

$y = \left[\tfrac{2}{5}(100 + 2t) - 40{,}000(100 + 2t)^{-3/2}\right]$ kg. From this solution (no pun intended), we calculate the salt concentration

at time t to be $C(t) = \dfrac{y(t)}{100 + 2t} = \left[\dfrac{-40{,}000}{(100 + 2t)^{5/2}} + \dfrac{2}{5}\right]\dfrac{\text{kg}}{\text{L}}$. In particular, $C(20) = \dfrac{-40{,}000}{140^{5/2}} + \dfrac{2}{5} \approx 0.2275\,\dfrac{\text{kg}}{\text{L}}$

and $y(20) = \tfrac{2}{5}(140) - 40{,}000(140)^{-3/2} \approx 31.85$ kg.

35. (a) $\dfrac{dv}{dt} + \dfrac{c}{m}v = g$ and $I(t) = e^{\int (c/m)\,dt} = e^{(c/m)t}$, and multiplying the differential equation by

$I(t)$ gives $e^{(c/m)t}\dfrac{dv}{dt} + \dfrac{vce^{(c/m)t}}{m} = ge^{(c/m)t}$ $\Rightarrow$ $\left[e^{(c/m)t}v\right]' = ge^{(c/m)t}$. Hence,

$v(t) = e^{-(c/m)t}\left[\int ge^{(c/m)t}\,dt + K\right] = mg/c + Ke^{-(c/m)t}$. But the object is dropped from rest, so $v(0) = 0$ and

$K = -mg/c$. Thus, the velocity at time t is $v(t) = (mg/c)[1 - e^{-(c/m)t}]$.

(b) $\lim\limits_{t \to \infty} v(t) = mg/c$

(c) $s(t) = \int v(t)\,dt = (mg/c)[t + (m/c)e^{-(c/m)t}] + c_1$ where $c_1 = s(0) - m^2g/c^2$.

$s(0)$ is the initial position, so $s(0) = 0$ and $s(t) = (mg/c)[t + (m/c)e^{-(c/m)t}] - m^2g/c^2$.

9.6 Predator-Prey Systems

1. (a) $dx/dt = -0.05x + 0.0001xy$. If $y = 0$, we have $dx/dt = -0.05x$, which indicates that in the absence of y, x declines at a rate proportional to itself. So x represents the predator population and y represents the prey population. The growth of the prey population, $0.1y$ (from $dy/dt = 0.1y - 0.005xy$), is restricted only by encounters with predators (the term $-0.005xy$). The predator population increases only through the term $0.0001xy$; that is, by encounters with the prey and not through additional food sources.

(b) $dy/dt = -0.015y + 0.00008xy$. If $x = 0$, we have $dy/dt = -0.015y$, which indicates that in the absence of x, y would decline at a rate proportional to itself. So y represents the predator population and x represents the prey population. The growth of the prey population, $0.2x$ (from $dx/dt = 0.2x - 0.0002x^2 - 0.006xy = 0.2x(1 - 0.001x) - 0.006xy$), is restricted by a carrying capacity of 1000 [from the term $1 - 0.001x = 1 - x/1000$] and by encounters with predators (the term $-0.006xy$). The predator population increases only through the term $0.00008xy$; that is, by encounters with the prey and not through additional food sources.

3. (a) At $t = 0$, there are about 300 rabbits and 100 foxes. At $t = t_1$, the number of foxes reaches a minimum of about 20 while the number of rabbits is about 1000. At $t = t_2$, the number of rabbits reaches a maximum of about 2400, while the number of foxes rebounds to 100. At $t = t_3$, the number of rabbits decreases to about 1000 and the number of foxes reaches a maximum of about 315. As t increases, the number of foxes decreases greatly to 100, and the number of rabbits decreases to 300 (the initial populations), and the cycle starts again.

(b)

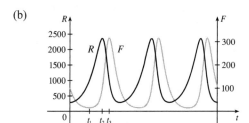

5.

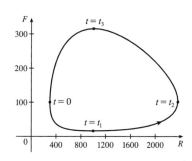

7. $\dfrac{dW}{dR} = \dfrac{-0.02W + 0.00002RW}{0.08R - 0.001RW}$ ⟺ $(0.08 - 0.001W)R\,dW = (-0.02 + 0.00002R)W\,dR$ ⟺

$\dfrac{0.08 - 0.001W}{W}\,dW = \dfrac{-0.02 + 0.00002R}{R}\,dR$ ⟺ $\displaystyle\int\left(\dfrac{0.08}{W} - 0.001\right) dW = \int\left(-\dfrac{0.02}{R} + 0.00002\right) dR$ ⟺

$0.08\ln|W| - 0.001W = -0.02\ln|R| + 0.00002R + K$ ⟺ $0.08\ln W + 0.02\ln R = 0.001W + 0.00002R + K$ ⟺

$\ln\left(W^{0.08}R^{0.02}\right) = 0.00002R + 0.001W + K$ ⟺ $W^{0.08}R^{0.02} = e^{0.00002R + 0.001W + K}$ ⟺

$R^{0.02}W^{0.08} = Ce^{0.00002R}e^{0.001W}$ ⟺ $\dfrac{R^{0.02}W^{0.08}}{e^{0.00002R}e^{0.001W}} = C$. In general, if $\dfrac{dy}{dx} = \dfrac{-ry + bxy}{kx - axy}$, then $C = \dfrac{x^r y^k}{e^{bx}e^{ay}}$.

9. (a) Letting $W = 0$ gives us $dR/dt = 0.08R(1 - 0.0002R)$. $dR/dt = 0$ ⟺ $R = 0$ or 5000. Since $dR/dt > 0$ for $0 < R < 5000$, we would expect the rabbit population to *increase* to 5000 for these values of R. Since $dR/dt < 0$ for $R > 5000$, we would expect the rabbit population to *decrease* to 5000 for these values of R. Hence, in the absence of wolves, we would expect the rabbit population to stabilize at 5000.

(b) R and W are constant ⟹ $R' = 0$ and $W' = 0$ ⟹

$$\begin{cases} 0 = 0.08R(1 - 0.0002R) - 0.001RW \\ 0 = -0.02W + 0.00002RW \end{cases} \Rightarrow \begin{cases} 0 = R[0.08(1 - 0.0002R) - 0.001W] \\ 0 = W(-0.02 + 0.00002R) \end{cases}$$

The second equation is true if $W = 0$ or $R = \dfrac{0.02}{0.00002} = 1000$. If $W = 0$ in the first equation, then either $R = 0$ or

$R = \frac{1}{0.0002} = 5000$ [as in part (a)]. If $R = 1000$, then $0 = 1000[0.08(1 - 0.0002 \cdot 1000) - 0.001W]$ ⟺

$0 = 80(1 - 0.2) - W$ ⟺ $W = 64$.

Case (i): $W = 0$, $R = 0$: both populations are zero

Case (ii): $W = 0$, $R = 5000$: see part (a)

Case (iii): $R = 1000$, $W = 64$: the predator/prey interaction balances and the populations are stable.

(c) The populations of wolves and rabbits fluctuate around 64 and 1000, respectively, and eventually stabilize at those values.

(d)
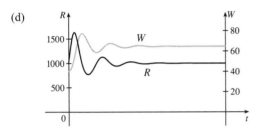

9   Review

CONCEPT CHECK

1. (a) A differential equation is an equation that contains an unknown function and one or more of its derivatives.

(b) The order of a differential equation is the order of the highest derivative that occurs in the equation.

(c) An initial condition is a condition of the form $y(t_0) = y_0$.

2. $y' = x^2 + y^2 \geq 0$ for all x and y. $y' = 0$ only at the origin, so there is a horizontal tangent at $(0,0)$, but nowhere else. The graph of the solution is increasing on every interval.

3. See the paragraph preceding Example 1 in Section 9.2.

4. See the paragraph next to Figure 14 in Section 9.2.

5. A separable equation is a first-order differential equation in which the expression for dy/dx can be factored as a function of x times a function of y, that is, $dy/dx = g(x)f(y)$. We can solve the equation by integrating both sides of the equation $dy/f(y) = g(x)dx$ and solving for y.

6. A first-order linear differential equation is a differential equation that can be put in the form $\dfrac{dy}{dx} + P(x)\,y = Q(x)$, where P and Q are continuous functions on a given interval. To solve such an equation, multiply it by the integrating factor $I(x) = e^{\int P(x)dx}$ to put it in the form $[I(x)\,y]' = I(x)\,Q(x)$ and then integrate both sides to get $I(x)\,y = \int I(x)\,Q(x)\,dx$, that is, $e^{\int P(x)\,dx}y = \int e^{\int P(x)\,dx}Q(x)\,dx$. Solving for y gives us $y = e^{-\int P(x)\,dx}\int e^{\int P(x)\,dx}Q(x)\,dx$.

7. (a) $\dfrac{dy}{dt} = ky$; the relative growth rate, $\dfrac{1}{y}\dfrac{dy}{dt}$, is constant.

(b) The equation in part (a) is an appropriate model for population growth, assuming that there is enough room and nutrition to support the growth.

(c) If $y(0) = y_0$, then the solution is $y(t) = y_0 e^{kt}$.

8. (a) $dP/dt = kP(1 - P/K)$, where K is the carrying capacity.

(b) The equation in part (a) is an appropriate model for population growth, assuming that the population grows at a rate proportional to the size of the population in the beginning, but eventually levels off and approaches its carrying capacity because of limited resources.

9. (a) $dF/dt = kF - aFS$ and $dS/dt = -rS + bFS$.

(b) In the absence of sharks, an ample food supply would support exponential growth of the fish population, that is, $dF/dt = kF$, where k is a positive constant. In the absence of fish, we assume that the shark population would decline at a rate proportional to itself, that is, $dS/dt = -rS$, where r is a positive constant.

TRUE-FALSE QUIZ

1. True. Since $y^4 \geq 0$, $y' = -1 - y^4 < 0$ and the solutions are decreasing functions.

3. False. $x + y$ cannot be written in the form $g(x)f(y)$.

5. True. $e^x y' = y \;\Rightarrow\; y' = e^{-x}y \;\Rightarrow\; y' + (-e^{-x})y = 0$, which is of the form $y' + P(x)y = Q(x)$, so the equation is linear.

7. True. By comparing $\dfrac{dy}{dt} = 2y\left(1 - \dfrac{y}{5}\right)$ with the logistic differential equation (9.4.4), we see that the carrying capacity is 5; that is, $\lim\limits_{t \to \infty} y = 5$.

EXERCISES

1. (a)

(b) $\lim\limits_{t \to \infty} y(t)$ appears to be finite for $0 \leq c \leq 4$. In fact

$\lim\limits_{t \to \infty} y(t) = 4$ for $c = 4$, $\lim\limits_{t \to \infty} y(t) = 2$ for $0 < c < 4$, and

$\lim\limits_{t \to \infty} y(t) = 0$ for $c = 0$. The equilibrium solutions are

$y(t) = 0$, $y(t) = 2$, and $y(t) = 4$.

3. (a)

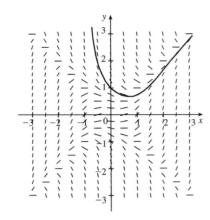

We estimate that when $x = 0.3$, $y = 0.8$, so $y(0.3) \approx 0.8$.

(b) $h = 0.1$, $x_0 = 0$, $y_0 = 1$ and $F(x, y) = x^2 - y^2$. So $y_n = y_{n-1} + 0.1(x_{n-1}^2 - y_{n-1}^2)$. Thus,

$$y_1 = 1 + 0.1(0^2 - 1^2) = 0.9, \; y_2 = 0.9 + 0.1(0.1^2 - 0.9^2) = 0.82, \; y_3 = 0.82 + 0.1(0.2^2 - 0.82^2) = 0.75676.$$

This is close to our graphical estimate of $y(0.3) \approx 0.8$.

(c) The centers of the horizontal line segments of the direction field are located on the lines $y = x$ and $y = -x$.

When a solution curve crosses one of these lines, it has a local maximum or minimum.

5. $y' = xe^{-\sin x} - y\cos x \;\; \Rightarrow \;\; y' + (\cos x)\, y = xe^{-\sin x}$ ($\star$). This is a linear equation and the integrating factor is

$I(x) = e^{\int \cos x \, dx} = e^{\sin x}$. Multiplying ($\star$) by $e^{\sin x}$ gives $e^{\sin x}\, y' + e^{\sin x}(\cos x)\, y = x \;\; \Rightarrow \;\; (e^{\sin x}\, y)' = x \;\; \Rightarrow$

$e^{\sin x}\, y = \frac{1}{2}x^2 + C \;\; \Rightarrow \;\; y = \left(\frac{1}{2}x^2 + C\right)e^{-\sin x}$.

7. $2ye^{y^2}\, y' = 2x + 3\sqrt{x} \;\; \Rightarrow \;\; 2ye^{y^2}\dfrac{dy}{dx} = 2x + 3\sqrt{x} \;\; \Rightarrow \;\; 2ye^{y^2}\, dy = \left(2x + 3\sqrt{x}\right)dx \;\; \Rightarrow$

$\int 2ye^{y^2}\, dy = \int \left(2x + 3\sqrt{x}\right)dx \;\; \Rightarrow \;\; e^{y^2} = x^2 + 2x^{3/2} + C \;\; \Rightarrow \;\; y^2 = \ln(x^2 + 2x^{3/2} + C) \;\; \Rightarrow$

$y = \pm\sqrt{\ln(x^2 + 2x^{3/2} + C)}$

9. $\dfrac{dr}{dt} + 2tr = r \;\; \Rightarrow \;\; \dfrac{dr}{dt} = r - 2tr = r(1 - 2t) \;\; \Rightarrow \;\; \displaystyle\int \dfrac{dr}{r} = \int (1 - 2t)\, dt \;\; \Rightarrow \;\; \ln|r| = t - t^2 + C \;\; \Rightarrow$

$|r| = e^{t - t^2 + C} = ke^{t - t^2}$. Since $r(0) = 5$, $5 = ke^0 = k$. Thus, $r(t) = 5e^{t - t^2}$.

11. $xy' - y = x\ln x \;\; \Rightarrow \;\; y' - \dfrac{1}{x}\, y = \ln x$. $\quad I(x) = e^{\int(-1/x)\, dx} = e^{-\ln|x|} = \left(e^{\ln|x|}\right)^{-1} = |x|^{-1} = 1/x$ since the condition

$y(1) = 2$ implies that we want a solution with $x > 0$. Multiplying the last differential equation by $I(x)$ gives

$\dfrac{1}{x}\, y' - \dfrac{1}{x^2}\, y = \dfrac{1}{x}\ln x \;\; \Rightarrow \;\; \left(\dfrac{1}{x}\, y\right)' = \dfrac{1}{x}\ln x \;\; \Rightarrow \;\; \dfrac{1}{x}\, y = \displaystyle\int \dfrac{\ln x}{x}\, dx \;\; \Rightarrow \;\; \dfrac{1}{x}\, y = \dfrac{1}{2}(\ln x)^2 + C \;\; \Rightarrow$

$y = \frac{1}{2}x(\ln x)^2 + Cx$. Now $y(1) = 2 \;\; \Rightarrow \;\; 2 = 0 + C \;\; \Rightarrow \;\; C = 2$, so $y = \frac{1}{2}x(\ln x)^2 + 2x$.

13. $\dfrac{d}{dx}(y) = \dfrac{d}{dx}(ke^x) \;\; \Rightarrow \;\; y' = ke^x = y$, so the orthogonal trajectories must have $y' = -\dfrac{1}{y} \;\; \Rightarrow \;\; \dfrac{dy}{dx} = -\dfrac{1}{y} \;\; \Rightarrow$

$y\, dy = -dx \;\; \Rightarrow \;\; \int y\, dy = -\int dx \;\; \Rightarrow \;\; \frac{1}{2}y^2 = -x + C \;\; \Rightarrow \;\; x = C - \frac{1}{2}y^2$, which are parabolas with a horizontal axis.

15. (a) Using (4) and (7) in Section 9.4, we see that for $\dfrac{dP}{dt} = 0.1P\left(1 - \dfrac{P}{2000}\right)$ with $P(0) = 100$, we have $k = 0.1$,

$K = 2000$, $P_0 = 100$, and $A = \dfrac{2000 - 100}{100} = 19$. Thus, the solution of the initial-value problem is

$P(t) = \dfrac{2000}{1 + 19e^{-0.1t}}$ and $P(20) = \dfrac{2000}{1 + 19e^{-2}} \approx 560$.

(b) $P = 1200 \iff 1200 = \dfrac{2000}{1 + 19e^{-0.1t}} \iff 1 + 19e^{-0.1t} = \dfrac{2000}{1200} \iff 19e^{-0.1t} = \dfrac{5}{3} - 1 \iff$

$e^{-0.1t} = \left(\dfrac{2}{3}\right)/19 \iff -0.1t = \ln\frac{2}{57} \iff t = -10\ln\frac{2}{57} \approx 33.5$.

17. (a) $\dfrac{dL}{dt} \propto L_\infty - L \implies \dfrac{dL}{dt} = k(L_\infty - L) \implies \displaystyle\int \dfrac{dL}{L_\infty - L} = \int k\,dt \implies -\ln|L_\infty - L| = kt + C \implies$

$\ln|L_\infty - L| = -kt - C \implies |L_\infty - L| = e^{-kt-C} \implies L_\infty - L = Ae^{-kt} \implies L = L_\infty - Ae^{-kt}$.

At $t = 0$, $L = L(0) = L_\infty - A \implies A = L_\infty - L(0) \implies L(t) = L_\infty - [L_\infty - L(0)]e^{-kt}$.

(b) $L_\infty = 53$ cm, $L(0) = 10$ cm, and $k = 0.2 \implies L(t) = 53 - (53 - 10)e^{-0.2t} = 53 - 43e^{-0.2t}$.

19. Let P represent the population and I the number of infected people. The rate of spread dI/dt is jointly proportional to I and

to $P - I$, so for some constant k, $\dfrac{dI}{dt} = kI(P - I) = (kP)I\left(1 - \dfrac{I}{P}\right)$. From Equation 9.4.7 with $K = P$ and k replaced by

kP, we have $I(t) = \dfrac{P}{1 + Ae^{-kPt}} = \dfrac{I_0 P}{I_0 + (P - I_0)e^{-kPt}}$.

Now, measuring t in days, we substitute $t = 7$, $P = 5000$, $I_0 = 160$ and $I(7) = 1200$ to find k:

$1200 = \dfrac{160 \cdot 5000}{160 + (5000 - 160)e^{-5000\cdot 7\cdot k}} \iff 3 = \dfrac{2000}{160 + 4840e^{-35,000k}} \iff 480 + 14{,}520e^{-35,000k} = 2000 \iff$

$e^{-35,000k} = \dfrac{2000 - 480}{14{,}520} \iff -35{,}000k = \ln\dfrac{38}{363} \iff k = \dfrac{-1}{35{,}000}\ln\dfrac{38}{363} \approx 0.00006448$. Next, let

$I = 5000 \times 80\% = 4000$, and solve for t: $4000 = \dfrac{160 \cdot 5000}{160 + (5000 - 160)e^{-k\cdot 5000\cdot t}} \iff 1 = \dfrac{200}{160 + 4840e^{-5000kt}} \iff$

$160 + 4840e^{-5000kt} = 200 \iff e^{-5000kt} = \dfrac{200 - 160}{4840} \iff -5000kt = \ln\dfrac{1}{121} \iff$

$t = \dfrac{-1}{5000k}\ln\dfrac{1}{121} = \dfrac{1}{\frac{1}{7}\ln\frac{38}{363}}\cdot\ln\dfrac{1}{121} = 7\cdot\dfrac{\ln 121}{\ln\frac{363}{38}} \approx 14.875$. So it takes about 15 days for 80% of the population

to be infected.

21. $\dfrac{dh}{dt} = -\dfrac{R}{V}\left(\dfrac{h}{k+h}\right) \implies \displaystyle\int \dfrac{k+h}{h}\,dh = \int\left(-\dfrac{R}{V}\right)dt \implies \int\left(1 + \dfrac{k}{h}\right)dh = -\dfrac{R}{V}\int 1\,dt \implies$

$h + k\ln h = -\dfrac{R}{V}t + C$. This equation gives a relationship between h and t, but it is not possible to isolate h and express it in

terms of t.

23. (a) $dx/dt = 0.4x(1 - 0.000005x) - 0.002xy$, $dy/dt = -0.2y + 0.000008xy$. If $y = 0$, then

$dx/dt = 0.4x(1 - 0.000005x)$, so $dx/dt = 0 \iff x = 0$ or $x = 200{,}000$, which shows that the insect population

increases logistically with a carrying capacity of 200,000. Since $dx/dt > 0$ for $0 < x < 200{,}000$ and $dx/dt < 0$ for

$x > 200{,}000$, we expect the insect population to stabilize at 200,000.

(b) x and y are constant $\Rightarrow$ $x' = 0$ and $y' = 0$ $\Rightarrow$

$$\begin{cases} 0 = 0.4x(1 - 0.000005x) - 0.002xy \\ 0 = -0.2y + 0.000008xy \end{cases} \Rightarrow \begin{cases} 0 = 0.4x[(1 - 0.000005x) - 0.005y] \\ 0 = y(-0.2 + 0.000008x) \end{cases}$$

The second equation is true if $y = 0$ or $x = \frac{0.2}{0.000008} = 25{,}000$. If $y = 0$ in the first equation, then either $x = 0$

or $x = \frac{1}{0.000005} = 200{,}000$. If $x = 25{,}000$, then $0 = 0.4(25{,}000)[(1 - 0.000005 \cdot 25{,}000) - 0.005y]$ $\Rightarrow$

$0 = 10{,}000[(1 - 0.125) - 0.005y]$ $\Rightarrow$ $0 = 8750 - 50y$ $\Rightarrow$ $y = 175$.

Case (i): $y = 0, x = 0$: Zero populations

Case (ii): $y = 0, x = 200{,}000$: In the absence of birds, the insect population is always 200,000.

Case (iii): $x = 25{,}000, y = 175$: The predator/prey interaction balances and the populations are stable.

(c) The populations of the birds and insects fluctuate around 175 and 25,000, respectively, and eventually stabilize at those values.

(d)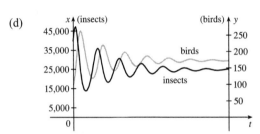

25. (a) $\dfrac{d^2y}{dx^2} = k\sqrt{1 + \left(\dfrac{dy}{dx}\right)^2}$. Setting $z = \dfrac{dy}{dx}$, we get $\dfrac{dz}{dx} = k\sqrt{1 + z^2}$ $\Rightarrow$ $\dfrac{dz}{\sqrt{1 + z^2}} = k\,dx$. Using Formula 25 gives

$\ln\left(z + \sqrt{1 + z^2}\right) = kx + c$ $\Rightarrow$ $z + \sqrt{1 + z^2} = Ce^{kx}$ [where $C = e^c$] $\Rightarrow$ $\sqrt{1 + z^2} = Ce^{kx} - z$ $\Rightarrow$

$1 + z^2 = C^2e^{2kx} - 2Ce^{kx}z + z^2$ $\Rightarrow$ $2Ce^{kx}z = C^2e^{2kx} - 1$ $\Rightarrow$ $z = \dfrac{C}{2}e^{kx} - \dfrac{1}{2C}e^{-kx}$. Now

$\dfrac{dy}{dx} = \dfrac{C}{2}e^{kx} - \dfrac{1}{2C}e^{-kx}$ $\Rightarrow$ $y = \dfrac{C}{2k}e^{kx} + \dfrac{1}{2Ck}e^{-kx} + C'$. From the diagram in the text, we see that $y(0) = a$

and $y(\pm b) = h$. $a = y(0) = \dfrac{C}{2k} + \dfrac{1}{2Ck} + C'$ $\Rightarrow$ $C' = a - \dfrac{C}{2k} - \dfrac{1}{2Ck}$ $\Rightarrow$

$y = \dfrac{C}{2k}(e^{kx} - 1) + \dfrac{1}{2Ck}(e^{-kx} - 1) + a$. From $h = y(\pm b)$, we find $h = \dfrac{C}{2k}(e^{kb} - 1) + \dfrac{1}{2Ck}(e^{-kb} - 1) + a$

and $h = \dfrac{C}{2k}(e^{-kb} - 1) + \dfrac{1}{2Ck}(e^{kb} - 1) + a$. Subtracting the second equation from the first, we get

$0 = \dfrac{C}{k}\dfrac{e^{kb} - e^{-kb}}{2} - \dfrac{1}{Ck}\dfrac{e^{kb} - e^{-kb}}{2} = \dfrac{1}{k}\left(C - \dfrac{1}{C}\right)\sinh kb$.

Now $k > 0$ and $b > 0$, so $\sinh kb > 0$ and $C = \pm 1$. If $C = 1$, then

$y = \dfrac{1}{2k}(e^{kx} - 1) + \dfrac{1}{2k}(e^{-kx} - 1) + a = \dfrac{1}{k}\dfrac{e^{kx} + e^{-kx}}{2} - \dfrac{1}{k} + a = a + \dfrac{1}{k}(\cosh kx - 1)$. If $C = -1$,

then $y = -\dfrac{1}{2k}(e^{kx} - 1) - \dfrac{1}{2k}(e^{-kx} - 1) + a = \dfrac{-1}{k}\dfrac{e^{kx} + e^{-kx}}{2} + \dfrac{1}{k} + a = a - \dfrac{1}{k}(\cosh kx - 1)$.

Since $k > 0$, $\cosh kx \geq 1$, and $y \geq a$, we conclude that $C = 1$ and $y = a + \dfrac{1}{k}(\cosh kx - 1)$, where

$h = y(b) = a + \dfrac{1}{k} \left(\cosh kb - 1\right)$. Since $\cosh(kb) = \cosh(-kb)$, there is no further information to extract from the

condition that $y(b) = y(-b)$. However, we could replace a with the expression $h - \dfrac{1}{k}(\cosh kb - 1)$, obtaining

$y = h + \dfrac{1}{k}\left(\cosh kx - \cosh kb\right)$. It would be better still to keep a in the expression for y, and use the expression for h to

solve for k in terms of a, b, and h. That would enable us to express y in terms of x and the given parameters a, b, and h.

Sadly, it is not possible to solve for k in closed form. That would have to be done by numerical methods when specific

parameter values are given.

(b) The length of the cable is

$$L = \int_{-b}^{b} \sqrt{1 + (dy/dx)^2}\, dx = \int_{-b}^{b} \sqrt{1 + \sinh^2 kx}\, dx = \int_{-b}^{b} \cosh kx\, dx = 2\int_{0}^{b} \cosh kx\, dx$$

$$= 2\left[(1/k)\sinh kx\right]_{0}^{b} = (2/k)\sinh kb$$

☐ PROBLEMS PLUS

1. We use the Fundamental Theorem of Calculus to differentiate the given equation:

$$[f(x)]^2 = 100 + \int_0^x \left\{ [f(t)]^2 + [f'(t)]^2 \right\} dt \quad \Rightarrow \quad 2f(x)f'(x) = [f(x)]^2 + [f'(x)]^2 \quad \Rightarrow$$

$$[f(x)]^2 + [f'(x)]^2 - 2f(x)f'(x) = 0 \quad \Rightarrow \quad [f(x) - f'(x)]^2 = 0 \quad \Leftrightarrow \quad f(x) = f'(x). \text{ We can solve this as a separable}$$

equation, or else use Theorem 9.4.2 with $k = 1$, which says that the solutions are $f(x) = Ce^x$. Now $[f(0)]^2 = 100$, so $f(0) = C = \pm 10$, and hence $f(x) = \pm 10e^x$ are the only functions satisfying the given equation.

3. $f'(x) = \lim\limits_{h \to 0} \dfrac{f(x+h) - f(x)}{h} = \lim\limits_{h \to 0} \dfrac{f(x)\,[f(h) - 1]}{h}$ [since $f(x+h) = f(x)f(h)$]

$= f(x) \lim\limits_{h \to 0} \dfrac{f(h) - 1}{h} = f(x) \lim\limits_{h \to 0} \dfrac{f(h) - f(0)}{h - 0} = f(x)f'(0) = f(x)$

Therefore, $f'(x) = f(x)$ for all x and from Theorem 9.4.2 we get $f(x) = Ae^x$. Now $f(0) = 1 \quad \Rightarrow \quad A = 1 \quad \Rightarrow$ $f(x) = e^x$.

5. "The area under the graph of f from 0 to x is proportional to the $(n+1)$st power of $f(x)$" translates to

$$\int_0^x f(t)\,dt = k[f(x)]^{n+1} \text{ for some constant } k. \text{ By FTC1, } \frac{d}{dx} \int_0^x f(t)\,dt = \frac{d}{dx} \left\{ k[f(x)]^{n+1} \right\} \quad \Rightarrow$$

$$f(x) = k(n+1)[f(x)]^n f'(x) \quad \Rightarrow \quad 1 = k(n+1)[f(x)]^{n-1} f'(x) \quad \Rightarrow \quad 1 = k(n+1)y^{n-1} \frac{dy}{dx} \quad \Rightarrow$$

$$k(n+1)y^{n-1}\,dy = dx \quad \Rightarrow \quad \int k(n+1)y^{n-1}\,dy = \int dx \quad \Rightarrow \quad k(n+1)\frac{1}{n}y^n = x + C.$$

Now $f(0) = 0 \quad \Rightarrow \quad 0 = 0 + C \quad \Rightarrow \quad C = 0$ and then $f(1) = 1 \quad \Rightarrow \quad k(n+1)\dfrac{1}{n} = 1 \quad \Rightarrow \quad k = \dfrac{n}{n+1}$,

so $y^n = x$ and $y = f(x) = x^{1/n}$.

7. Let $y(t)$ denote the temperature of the peach pie t minutes after 5:00 PM and R the temperature of the room. Newton's Law of Cooling gives us $dy/dt = k(y - R)$. Solving for y we get $\dfrac{dy}{y - R} = k\,dt \quad \Rightarrow \quad \ln|y - R| = kt + C \quad \Rightarrow$

$|y - R| = e^{kt+C} \quad \Rightarrow \quad y - R = \pm e^{kt} \cdot e^C \quad \Rightarrow \quad y = Me^{kt} + R$, where M is a nonzero constant. We are given temperatures at three times.

$$
\begin{aligned}
y(0) &= 100 &&\Rightarrow &&100 = M + R &&\Rightarrow &&R = 100 - M \\
y(10) &= 80 &&\Rightarrow &&80 = Me^{10k} + R &&\textbf{(1)} \\
y(20) &= 65 &&\Rightarrow &&65 = Me^{20k} + R &&\textbf{(2)}
\end{aligned}
$$

Substituting $100 - M$ for R in **(1)** and **(2)** gives us

$$-20 = Me^{10k} - M \quad \textbf{(3)} \quad \text{and} \quad -35 = Me^{20k} - M \quad \textbf{(4)}$$

Dividing **(3)** by **(4)** gives us $\dfrac{-20}{-35} = \dfrac{M(e^{10k} - 1)}{M(e^{20k} - 1)} \quad \Rightarrow \quad \dfrac{4}{7} = \dfrac{e^{10k} - 1}{e^{20k} - 1} \quad \Rightarrow \quad 4e^{20k} - 4 = 7e^{10k} - 7 \quad \Rightarrow$

$4e^{20k} - 7e^{10k} + 3 = 0$. This is a quadratic equation in e^{10k}. $\left(4e^{10k} - 3\right)\left(e^{10k} - 1\right) = 0 \;\Rightarrow\; e^{10k} = \frac{3}{4}$ or $1 \;\Rightarrow$

$10k = \ln\frac{3}{4}$ or $\ln 1 \;\Rightarrow\; k = \frac{1}{10}\ln\frac{3}{4}$ since k is a nonzero constant of proportionality. Substituting $\frac{3}{4}$ for e^{10k} in **(3)** gives us

$-20 = M \cdot \frac{3}{4} - M \;\Rightarrow\; -20 = -\frac{1}{4}M \;\Rightarrow\; M = 80$. Now $R = 100 - M$ so $R = 20°\text{C}$.

9. (a) While running from $(L, 0)$ to (x, y), the dog travels a distance

$$s = \int_x^L \sqrt{1 + (dy/dx)^2}\, dx = -\int_L^x \sqrt{1 + (dy/dx)^2}\, dx, \text{ so}$$

$$\frac{ds}{dx} = -\sqrt{1 + (dy/dx)^2}.$$ The dog and rabbit run at the same speed, so the

rabbit's position when the dog has traveled a distance s is $(0, s)$. Since the

dog runs straight for the rabbit, $\dfrac{dy}{dx} = \dfrac{s - y}{0 - x}$ (see the figure).

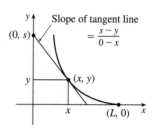

Slope of tangent line $= \dfrac{s-y}{0-x}$

Thus, $s = y - x\dfrac{dy}{dx} \;\Rightarrow\; \dfrac{ds}{dx} = \dfrac{dy}{dx} - \left(x\dfrac{d^2y}{dx^2} + 1\cdot\dfrac{dy}{dx}\right) = -x\dfrac{d^2y}{dx^2}$. Equating the two expressions for $\dfrac{ds}{dx}$

gives us $x\dfrac{d^2y}{dx^2} = \sqrt{1 + \left(\dfrac{dy}{dx}\right)^2}$, as claimed.

(b) Letting $z = \dfrac{dy}{dx}$, we obtain the differential equation $x\dfrac{dz}{dx} = \sqrt{1 + z^2}$, or $\dfrac{dz}{\sqrt{1+z^2}} = \dfrac{dx}{x}$. Integrating:

$\ln x = \displaystyle\int \dfrac{dz}{\sqrt{1+z^2}} \overset{25}{=} \ln\left(z + \sqrt{1+z^2}\right) + C$. When $x = L$, $z = dy/dx = 0$, so $\ln L = \ln 1 + C$. Therefore,

$C = \ln L$, so $\ln x = \ln\left(\sqrt{1+z^2} + z\right) + \ln L = \ln\left[L\left(\sqrt{1+z^2} + z\right)\right] \;\Rightarrow\; x = L\left(\sqrt{1+z^2} + z\right) \;\Rightarrow$

$\sqrt{1+z^2} = \dfrac{x}{L} - z \;\Rightarrow\; 1 + z^2 = \left(\dfrac{x}{L}\right)^2 - \dfrac{2xz}{L} + z^2 \;\Rightarrow\; \left(\dfrac{x}{L}\right)^2 - 2z\left(\dfrac{x}{L}\right) - 1 = 0 \;\Rightarrow$

$z = \dfrac{(x/L)^2 - 1}{2(x/L)} = \dfrac{x^2 - L^2}{2Lx} = \dfrac{x}{2L} - \dfrac{L}{2}\dfrac{1}{x}$ [for $x > 0$]. Since $z = \dfrac{dy}{dx}$, $y = \dfrac{x^2}{4L} - \dfrac{L}{2}\ln x + C_1$.

Since $y = 0$ when $x = L$, $0 = \dfrac{L}{4} - \dfrac{L}{2}\ln L + C_1 \;\Rightarrow\; C_1 = \dfrac{L}{2}\ln L - \dfrac{L}{4}$. Thus,

$$y = \dfrac{x^2}{4L} - \dfrac{L}{2}\ln x + \dfrac{L}{2}\ln L - \dfrac{L}{4} = \dfrac{x^2 - L^2}{4L} - \dfrac{L}{2}\ln\left(\dfrac{x}{L}\right).$$

(c) As $x \to 0^+$, $y \to \infty$, so the dog never catches the rabbit.

11. (a) We are given that $V = \frac{1}{3}\pi r^2 h$, $dV/dt = 60{,}000\pi$ ft^3/h, and $r = 1.5h = \frac{3}{2}h$. So $V = \frac{1}{3}\pi\left(\frac{3}{2}h\right)^2 h = \frac{3}{4}\pi h^3 \;\Rightarrow$

$\dfrac{dV}{dt} = \dfrac{3}{4}\pi \cdot 3h^2 \dfrac{dh}{dt} = \dfrac{9}{4}\pi h^2 \dfrac{dh}{dt}$. Therefore, $\dfrac{dh}{dt} = \dfrac{4(dV/dt)}{9\pi h^2} = \dfrac{240{,}000\pi}{9\pi h^2} = \dfrac{80{,}000}{3h^2}$ $(\star) \;\Rightarrow$

$\int 3h^2\, dh = \int 80{,}000\, dt \;\Rightarrow\; h^3 = 80{,}000t + C$. When $t = 0$, $h = 60$. Thus, $C = 60^3 = 216{,}000$, so

$h^3 = 80{,}000t + 216{,}000$. Let $h = 100$. Then $100^3 = 1{,}000{,}000 = 80{,}000t + 216{,}000 \;\Rightarrow$

$80{,}000t = 784{,}000 \;\Rightarrow\; t = 9.8$, so the time required is 9.8 hours.

(b) The floor area of the silo is $F = \pi \cdot 200^2 = 40{,}000\pi$ ft^2, and the area of the base of the pile is

$A = \pi r^2 = \pi\left(\frac{3}{2}h\right)^2 = \frac{9\pi}{4}h^2$. So the area of the floor which is not covered when $h = 60$ is

$F - A = 40{,}000\pi - 8100\pi = 31{,}900\pi \approx 100{,}217$ ft^2. Now $A = \frac{9\pi}{4}h^2 \Rightarrow dA/dt = \frac{9\pi}{4} \cdot 2h\,(dh/dt)$,

and from $(\star)$ in part (a) we know that when $h = 60$, $dh/dt = \frac{80{,}000}{3(60)^2} = \frac{200}{27}$ ft/h. Therefore,

$dA/dt = \frac{9\pi}{4}(2)(60)\left(\frac{200}{27}\right) = 2000\pi \approx 6283$ ft^2/h.

(c) At $h = 90$ ft, $dV/dt = 60{,}000\pi - 20{,}000\pi = 40{,}000\pi$ ft^3/h. From $(\star)$ in part (a),

$\dfrac{dh}{dt} = \dfrac{4(dV/dt)}{9\pi h^2} = \dfrac{4(40{,}000\pi)}{9\pi h^2} = \dfrac{160{,}000}{9h^2} \Rightarrow \int 9h^2\,dh = \int 160{,}000\,dt \Rightarrow 3h^3 = 160{,}000t + C$. When $t = 0$,

$h = 90$; therefore, $C = 3 \cdot 729{,}000 = 2{,}187{,}000$. So $3h^3 = 160{,}000t + 2{,}187{,}000$. At the top, $h = 100 \Rightarrow$

$3(100)^3 = 160{,}000t + 2{,}187{,}000 \Rightarrow t = \frac{813{,}000}{160{,}000} \approx 5.1$. The pile reaches the top after about 5.1 h.

13. Let $P(a, b)$ be any point on the curve. If m is the slope of the tangent line at P, then $m = y'(a)$, and an equation of the

normal line at P is $y - b = -\dfrac{1}{m}(x - a)$, or equivalently, $y = -\dfrac{1}{m}x + b + \dfrac{a}{m}$. The y-intercept is always 6, so

$b + \dfrac{a}{m} = 6 \Rightarrow \dfrac{a}{m} = 6 - b \Rightarrow m = \dfrac{a}{6 - b}$. We will solve the equivalent differential equation $\dfrac{dy}{dx} = \dfrac{x}{6 - y} \Rightarrow$

$(6 - y)\,dy = x\,dx \Rightarrow \displaystyle\int (6 - y)\,dy = \int x\,dx \Rightarrow 6y - \tfrac{1}{2}y^2 = \tfrac{1}{2}x^2 + C \Rightarrow 12y - y^2 = x^2 + K$.

Since $(3, 2)$ is on the curve, $12(2) - 2^2 = 3^2 + K \Rightarrow K = 11$. So the curve is given by $12y - y^2 = x^2 + 11 \Rightarrow$

$x^2 + y^2 - 12y + 36 = -11 + 36 \Rightarrow x^2 + (y - 6)^2 = 25$, a circle with center $(0, 6)$ and radius 5.

10 □ PARAMETRIC EQUATIONS AND POLAR COORDINATES

10.1 Curves Defined by Parametric Equations

1. $x = 1 + \sqrt{t}, \quad y = t^2 - 4t, \quad 0 \le t \le 5$

t	0	1	2	3	4	5
x	1	2	$1 + \sqrt{2}$	$1 + \sqrt{3}$	3	$1 + \sqrt{5}$
			2.41	2.73		3.24
y	0	-3	-4	-3	0	5

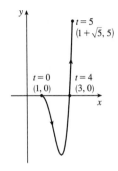

3. $x = 5 \sin t, \quad y = t^2, \quad -\pi \le t \le \pi$

t	$-\pi$	$-\pi/2$	0	$\pi/2$	π
x	0	-5	0	5	0
y	π^2	$\pi^2/4$	0	$\pi^2/4$	π^2
	9.87	2.47		2.47	9.87

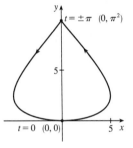

5. $x = 3t - 5, \quad y = 2t + 1$

(a)

t	-2	-1	0	1	2	3	4
x	-11	-8	-5	-2	1	4	7
y	-3	-1	1	3	5	7	9

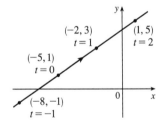

(b) $x = 3t - 5 \implies 3t = x + 5 \implies t = \frac{1}{3}(x + 5) \implies$

$y = 2 \cdot \frac{1}{3}(x + 5) + 1$, so $y = \frac{2}{3}x + \frac{13}{3}$.

7. $x = t^2 - 2, \quad y = 5 - 2t, \quad -3 \le t \le 4$

(a)

t	-3	-2	-1	0	1	2	3	4
x	7	2	-1	-2	-1	2	7	14
y	11	9	7	5	3	1	-1	-3

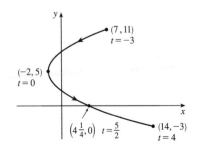

(b) $y = 5 - 2t \implies 2t = 5 - y \implies t = \frac{1}{2}(5 - y) \implies$

$x = \left[\frac{1}{2}(5 - y)\right]^2 - 2$, so $x = \frac{1}{4}(5 - y)^2 - 2, \quad -3 \le y \le 11$.

9. $x = \sqrt{t}$, $y = 1 - t$

(a)

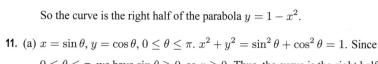

t	0	1	2	3	4
x	0	1	1.414	1.732	2
y	1	0	-1	-2	-3

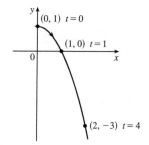

(b) $x = \sqrt{t} \ \Rightarrow \ t = x^2 \ \Rightarrow \ y = 1 - t = 1 - x^2$. Since $t \geq 0$, $x \geq 0$.

So the curve is the right half of the parabola $y = 1 - x^2$.

11. (a) $x = \sin\theta$, $y = \cos\theta$, $0 \leq \theta \leq \pi$. $x^2 + y^2 = \sin^2\theta + \cos^2\theta = 1$. Since

$0 \leq \theta \leq \pi$, we have $\sin\theta \geq 0$, so $x \geq 0$. Thus, the curve is the right half of

the circle $x^2 + y^2 = 1$.

(b)

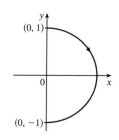

13. (a) $x = \sin t$, $y = \csc t$, $0 < t < \frac{\pi}{2}$. $y = \csc t = \dfrac{1}{\sin t} = \dfrac{1}{x}$. For $0 < t < \frac{\pi}{2}$,

we have $0 < x < 1$ and $y > 1$. Thus, the curve is the portion

of the hyperbola $y = 1/x$ with $y > 1$.

(b)

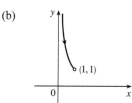

15. (a) $x = e^{2t} \ \Rightarrow \ 2t = \ln x \ \Rightarrow \ t = \frac{1}{2}\ln x$.

$y = t + 1 = \frac{1}{2}\ln x + 1$.

(b)

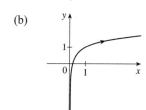

17. (a) $x = \sinh t$, $y = \cosh t \ \Rightarrow \ y^2 - x^2 = \cosh^2 t - \sinh^2 t = 1$. Since

$y = \cosh t \geq 1$, we have the upper branch of the hyperbola $y^2 - x^2 = 1$.

(b)

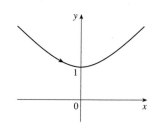

19. $x = 3 + 2\cos t$, $y = 1 + 2\sin t$, $\pi/2 \leq t \leq 3\pi/2$. By Example 4 with $r = 2$, $h = 3$, and $k = 1$, the motion of the particle

takes place on a circle centered at $(3, 1)$ with a radius of 2. As t goes from $\frac{\pi}{2}$ to $\frac{3\pi}{2}$, the particle starts at the point $(3, 3)$ and

moves counterclockwise to $(3, -1)$ [one-half of a circle].

21. $x = 5\sin t$, $y = 2\cos t \ \Rightarrow \ \sin t = \dfrac{x}{5}$, $\cos t = \dfrac{y}{2}$. $\sin^2 t + \cos^2 t = 1 \ \Rightarrow \ \left(\dfrac{x}{5}\right)^2 + \left(\dfrac{y}{2}\right)^2 = 1$. The motion of the

particle takes place on an ellipse centered at $(0, 0)$. As t goes from $-\pi$ to 5π, the particle starts at the point $(0, -2)$ and moves

clockwise around the ellipse 3 times.

23. We must have $1 \le x \le 4$ and $2 \le y \le 3$. So the graph of the curve must be contained in the rectangle $[1, 4]$ by $[2, 3]$.

25. When $t = -1$, $(x, y) = (0, -1)$. As t increases to 0, x decreases to -1 and y

increases to 0. As t increases from 0 to 1, x increases to 0 and y increases to 1.

As t increases beyond 1, both x and y increase. For $t < -1$, x is positive and

decreasing and y is negative and increasing. We could achieve greater accuracy

by estimating x- and y-values for selected values of t from the given graphs and

plotting the corresponding points.

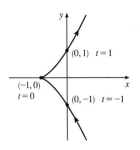

27. When $t = 0$ we see that $x = 0$ and $y = 0$, so the curve starts at the origin. As t

increases from 0 to $\frac{1}{2}$, the graphs show that y increases from 0 to 1 while x

increases from 0 to 1, decreases to 0 and to -1, then increases back to 0, so we

arrive at the point $(0, 1)$. Similarly, as t increases from $\frac{1}{2}$ to 1, y decreases from 1

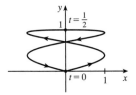

to 0 while x repeats its pattern, and we arrive back at the origin. We could achieve greater accuracy by estimating x- and

y-values for selected values of t from the given graphs and plotting the corresponding points.

29. As in Example 6, we let $y = t$ and $x = t - 3t^3 + t^5$ and use a t-interval of $[-3, 3]$.

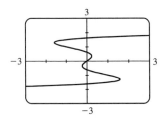

31. (a) $x = x_1 + (x_2 - x_1)t$, $y = y_1 + (y_2 - y_1)t$, $0 \le t \le 1$. Clearly the curve passes through $P_1(x_1, y_1)$ when $t = 0$ and

through $P_2(x_2, y_2)$ when $t = 1$. For $0 < t < 1$, x is strictly between x_1 and x_2 and y is strictly between y_1 and y_2. For

every value of t, x and y satisfy the relation $y - y_1 = \dfrac{y_2 - y_1}{x_2 - x_1}(x - x_1)$, which is the equation of the line through

$P_1(x_1, y_1)$ and $P_2(x_2, y_2)$.

    Finally, any point (x, y) on that line satisfies $\dfrac{y - y_1}{y_2 - y_1} = \dfrac{x - x_1}{x_2 - x_1}$; if we call that common value t, then the given

parametric equations yield the point (x, y); and any (x, y) on the line between $P_1(x_1, y_1)$ and $P_2(x_2, y_2)$ yields a value of

t in $[0, 1]$. So the given parametric equations exactly specify the line segment from $P_1(x_1, y_1)$ to $P_2(x_2, y_2)$.

(b) $x = -2 + [3 - (-2)]t = -2 + 5t$ and $y = 7 + (-1 - 7)t = 7 - 8t$ for $0 \le t \le 1$.

33. The circle $x^2 + (y - 1)^2 = 4$ has center $(0, 1)$ and radius 2, so by Example 4 it can be represented by $x = 2\cos t$,

$y = 1 + 2\sin t$, $0 \le t \le 2\pi$. This representation gives us the circle with a counterclockwise orientation starting at $(2, 1)$.

(a) To get a clockwise orientation, we could change the equations to $x = 2\cos t$, $y = 1 - 2\sin t$, $0 \le t \le 2\pi$.

(b) To get three times around in the counterclockwise direction, we use the original equations $x = 2\cos t$, $y = 1 + 2\sin t$ with

the domain expanded to $0 \le t \le 6\pi$.

(c) To start at $(0, 3)$ using the original equations, we must have $x_1 = 0$; that is, $2\cos t = 0$. Hence, $t = \frac{\pi}{2}$. So we use

$x = 2\cos t$, $y = 1 + 2\sin t$, $\frac{\pi}{2} \le t \le \frac{3\pi}{2}$.

Alternatively, if we want t to start at 0, we could change the equations of the curve. For example, we could use

$x = -2\sin t$, $y = 1 + 2\cos t$, $0 \le t \le \pi$.

35. *Big circle:* It's centered at $(2, 2)$ with a radius of 2, so by Example 4, parametric equations are

$$x = 2 + 2\cos t, \qquad y = 2 + 2\sin t, \qquad 0 \le t \le 2\pi$$

Small circles: They are centered at $(1, 3)$ and $(3, 3)$ with a radius of 0.1. By Example 4, parametric equations are

	(left)	$x = 1 + 0.1\cos t$,	$y = 3 + 0.1\sin t$,	$0 \le t \le 2\pi$
and	*(right)*	$x = 3 + 0.1\cos t$,	$y = 3 + 0.1\sin t$,	$0 \le t \le 2\pi$

Semicircle: It's the lower half of a circle centered at $(2, 2)$ with radius 1. By Example 4, parametric equations are

$$x = 2 + 1\cos t, \qquad y = 2 + 1\sin t, \qquad \pi \le t \le 2\pi$$

To get all four graphs on the same screen with a typical graphing calculator, we need to change the last t-interval to $[0, 2\pi]$ in order to match the others. We can do this by changing t to $0.5t$. This change gives us the upper half. There are several ways to get the lower half—one is to change the "+" to a "−" in the y-assignment, giving us

$$x = 2 + 1\cos(0.5t), \qquad y = 2 - 1\sin(0.5t), \qquad 0 \le t \le 2\pi$$

37. (a) $x = t^3 \;\Rightarrow\; t = x^{1/3}$, so $y = t^2 = x^{2/3}$.

We get the entire curve $y = x^{2/3}$ traversed in a left to right direction.

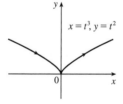

(b) $x = t^6 \;\Rightarrow\; t = x^{1/6}$, so $y = t^4 = x^{4/6} = x^{2/3}$.

Since $x = t^6 \ge 0$, we only get the right half of the curve $y = x^{2/3}$.

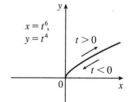

(c) $x = e^{-3t} = (e^{-t})^3$ [so $e^{-t} = x^{1/3}$],

$y = e^{-2t} = (e^{-t})^2 = (x^{1/3})^2 = x^{2/3}$.

If $t < 0$, then x and y are both larger than 1. If $t > 0$, then x and y are between 0 and 1. Since $x > 0$ and $y > 0$, the curve never quite reaches the origin.

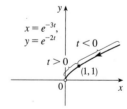

39. The case $\frac{\pi}{2} < \theta < \pi$ is illustrated. C has coordinates $(r\theta, r)$ as in Example 7, and Q has coordinates $(r\theta, r + r\cos(\pi - \theta)) = (r\theta, r(1 - \cos\theta))$ [since $\cos(\pi - \alpha) = \cos\pi\cos\alpha + \sin\pi\sin\alpha = -\cos\alpha$], so P has coordinates $(r\theta - r\sin(\pi - \theta), r(1 - \cos\theta)) = (r(\theta - \sin\theta), r(1 - \cos\theta))$ [since $\sin(\pi - \alpha) = \sin\pi\cos\alpha - \cos\pi\sin\alpha = \sin\alpha$]. Again we have the parametric equations $x = r(\theta - \sin\theta)$, $y = r(1 - \cos\theta)$.

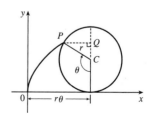

41. It is apparent that $x = |OQ|$ and $y = |QP| = |ST|$. From the diagram,

$x = |OQ| = a\cos\theta$ and $y = |ST| = b\sin\theta$. Thus, the parametric equations are

$x = a\cos\theta$ and $y = b\sin\theta$. To eliminate θ we rearrange: $\sin\theta = y/b \;\Rightarrow$

$\sin^2\theta = (y/b)^2$ and $\cos\theta = x/a \;\Rightarrow\; \cos^2\theta = (x/a)^2$. Adding the two

equations: $\sin^2\theta + \cos^2\theta = 1 = x^2/a^2 + y^2/b^2$. Thus, we have an ellipse.

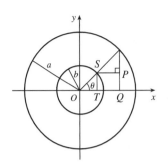

43. $C = (2a\cot\theta, 2a)$, so the x-coordinate of P is $x = 2a\cot\theta$. Let $B = (0, 2a)$.

Then $\angle OAB$ is a right angle and $\angle OBA = \theta$, so $|OA| = 2a\sin\theta$ and

$A = ((2a\sin\theta)\cos\theta, (2a\sin\theta)\sin\theta)$. Thus, the y-coordinate of P

is $y = 2a\sin^2\theta$.

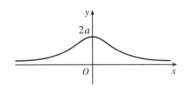

45. (a)

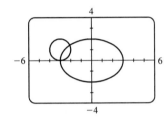

There are 2 points of intersection:

$(-3, 0)$ and approximately $(-2.1, 1.4)$.

(b) A collision point occurs when $x_1 = x_2$ and $y_1 = y_2$ for the same t. So solve the equations:

$$3\sin t = -3 + \cos t \quad \textbf{(1)}$$

$$2\cos t = 1 + \sin t \quad \textbf{(2)}$$

From **(2)**, $\sin t = 2\cos t - 1$. Substituting into **(1)**, we get $3(2\cos t - 1) = -3 + \cos t \;\Rightarrow\; 5\cos t = 0 \; (\star) \;\Rightarrow$

$\cos t = 0 \;\Rightarrow\; t = \frac{\pi}{2}$ or $\frac{3\pi}{2}$. We check that $t = \frac{3\pi}{2}$ satisfies **(1)** and **(2)** but $t = \frac{\pi}{2}$ does not. So the only collision point

occurs when $t = \frac{3\pi}{2}$, and this gives the point $(-3, 0)$. [We could check our work by graphing x_1 and x_2 together as

functions of t and, on another plot, y_1 and y_2 as functions of t. If we do so, we see that the only value of t for which *both*

pairs of graphs intersect is $t = \frac{3\pi}{2}$.]

(c) The circle is centered at $(3, 1)$ instead of $(-3, 1)$. There are still 2 intersection points: $(3, 0)$ and $(2.1, 1.4)$, but there are

no collision points, since $(\star)$ in part (b) becomes $5\cos t = 6 \;\Rightarrow\; \cos t = \frac{6}{5} > 1$.

47. $x = t^2, y = t^3 - ct$. We use a graphing device to produce the graphs for various values of c with $-\pi \le t \le \pi$. Note that all

the members of the family are symmetric about the x-axis. For $c < 0$, the graph does not cross itself, but for $c = 0$ it has a

cusp at $(0, 0)$ and for $c > 0$ the graph crosses itself at $x = c$, so the loop grows larger as c increases.

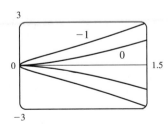
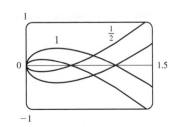

49. Note that all the Lissajous figures are symmetric about the x-axis. The parameters a and b simply stretch the graph in the x- and y-directions respectively. For $a = b = n = 1$ the graph is simply a circle with radius 1. For $n = 2$ the graph crosses itself at the origin and there are loops above and below the x-axis. In general, the figures have $n - 1$ points of intersection, all of which are on the y-axis, and a total of n closed loops.

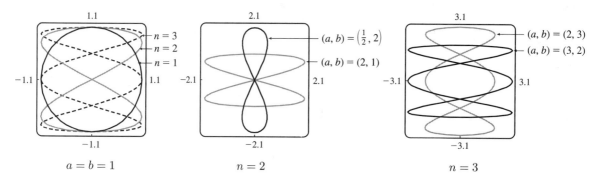

10.2 Calculus with Parametric Curves

1. $x = t \sin t$, $y = t^2 + t$ $\Rightarrow$ $\dfrac{dy}{dt} = 2t + 1$, $\dfrac{dx}{dt} = t \cos t + \sin t$, and $\dfrac{dy}{dx} = \dfrac{dy/dt}{dx/dt} = \dfrac{2t + 1}{t \cos t + \sin t}$.

3. $x = t^4 + 1$, $y = t^3 + t$; $t = -1$. $\dfrac{dy}{dt} = 3t^2 + 1$, $\dfrac{dx}{dt} = 4t^3$, and $\dfrac{dy}{dx} = \dfrac{dy/dt}{dx/dt} = \dfrac{3t^2 + 1}{4t^3}$. When $t = -1$,

$(x, y) = (2, -2)$ and $dy/dx = \dfrac{4}{-4} = -1$, so an equation of the tangent to the curve at the point corresponding to $t = -1$

is $y - (-2) = (-1)(x - 2)$, or $y = -x$.

5. $x = e^{\sqrt{t}}$, $y = t - \ln t^2$; $t = 1$. $\dfrac{dy}{dt} = 1 - \dfrac{2t}{t^2} = 1 - \dfrac{2}{t}$, $\dfrac{dx}{dt} = \dfrac{e^{\sqrt{t}}}{2\sqrt{t}}$, and $\dfrac{dy}{dx} = \dfrac{dy/dt}{dx/dt} = \dfrac{1 - 2/t}{e^{\sqrt{t}}/(2\sqrt{t})} \cdot \dfrac{2t}{2t} = \dfrac{2t - 4}{\sqrt{t}e^{\sqrt{t}}}$.

When $t = 1$, $(x, y) = (e, 1)$ and $\dfrac{dy}{dx} = -\dfrac{2}{e}$, so an equation of the tangent line is $y - 1 = -\dfrac{2}{e}(x - e)$, or $y = -\dfrac{2}{e}x + 3$.

7. (a) $x = 1 + \ln t$, $y = t^2 + 2$; $(1, 3)$. $\dfrac{dy}{dt} = 2t$, $\dfrac{dx}{dt} = \dfrac{1}{t}$, and $\dfrac{dy}{dx} = \dfrac{dy/dt}{dx/dt} = \dfrac{2t}{1/t} = 2t^2$.

At $(1, 3)$, $x = 1 + \ln t = 1$ $\Rightarrow$ $\ln t = 0$ $\Rightarrow$ $t = 1$ and $\dfrac{dy}{dx} = 2$, so an equation of the tangent is $y - 3 = 2(x - 1)$, or $y = 2x + 1$.

(b) $x = 1 + \ln t$ $\Rightarrow$ $x - 1 = \ln t$ $\Rightarrow$ $t = e^{x-1}$, so $y = (e^{x-1})^2 + 2 = e^{2x-2} + 2$ and $\dfrac{dy}{dx} = 2e^{2x-2}$.

When $x = 1$, $\dfrac{dy}{dx} = 2e^0 = 2$, so an equation of the tangent is $y = 2x + 1$, as in part (a).

9. $x = 6 \sin t$, $y = t^2 + t$; $(0, 0)$.

$\dfrac{dy}{dx} = \dfrac{dy/dt}{dx/dt} = \dfrac{2t + 1}{6 \cos t}$. The point $(0, 0)$ corresponds to $t = 0$, so the

slope of the tangent at that point is $\dfrac{1}{6}$. An equation of the tangent is therefore

$y - 0 = \dfrac{1}{6}(x - 0)$, or $y = \dfrac{1}{6}x$.

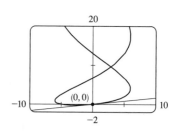

11. $x = 4 + t^2$, $y = t^2 + t^3$ $\Rightarrow$ $\dfrac{dy}{dx} = \dfrac{dy/dt}{dx/dt} = \dfrac{2t + 3t^2}{2t} = 1 + \dfrac{3}{2}t$ $\Rightarrow$

$\dfrac{d^2y}{dx^2} = \dfrac{d}{dx}\left(\dfrac{dy}{dx}\right) = \dfrac{d(dy/dx)/dt}{dx/dt} = \dfrac{(d/dt)\left(1 + \frac{3}{2}t\right)}{2t} = \dfrac{3/2}{2t} = \dfrac{3}{4t}.$

The curve is CU when $\dfrac{d^2y}{dx^2} > 0$, that is, when $t > 0$.

13. $x = t - e^t$, $y = t + e^{-t}$ $\Rightarrow$

$\dfrac{dy}{dx} = \dfrac{dy/dt}{dx/dt} = \dfrac{1 - e^{-t}}{1 - e^t} = \dfrac{1 - \dfrac{1}{e^t}}{1 - e^t} = \dfrac{\dfrac{e^t - 1}{e^t}}{1 - e^t} = \dfrac{e^t - 1}{1 - e^t} = -e^{-t}$ $\Rightarrow$ $\dfrac{d^2y}{dx^2} = \dfrac{\dfrac{d}{dt}\left(\dfrac{dy}{dx}\right)}{dx/dt} = \dfrac{\dfrac{d}{dt}(-e^{-t})}{dx/dt} = \dfrac{e^{-t}}{1 - e^t}.$

The curve is CU when $e^t < 1$ [since $e^{-t} > 0$] $\Rightarrow$ $t < 0$.

15. $x = 2\sin t$, $y = 3\cos t$, $0 < t < 2\pi$.

$\dfrac{dy}{dx} = \dfrac{dy/dt}{dx/dt} = \dfrac{-3\sin t}{2\cos t} = -\dfrac{3}{2}\tan t$, so $\dfrac{d^2y}{dx^2} = \dfrac{\dfrac{d}{dt}\left(\dfrac{dy}{dx}\right)}{dx/dt} = \dfrac{-\frac{3}{2}\sec^2 t}{2\cos t} = -\dfrac{3}{4}\sec^3 t.$

The curve is CU when $\sec^3 t < 0$ $\Rightarrow$ $\sec t < 0$ $\Rightarrow$ $\cos t < 0$ $\Rightarrow$ $\dfrac{\pi}{2} < t < \dfrac{3\pi}{2}$.

17. $x = 10 - t^2$, $y = t^3 - 12t$.

$\dfrac{dy}{dt} = 3t^2 - 12 = 3(t + 2)(t - 2)$, so $\dfrac{dy}{dt} = 0$ $\Leftrightarrow$

$t = \pm 2$ $\Leftrightarrow$ $(x, y) = (6, \mp 16)$.

$\dfrac{dx}{dt} = -2t$, so $\dfrac{dx}{dt} = 0$ $\Leftrightarrow$ $t = 0$ $\Leftrightarrow$ $(x, y) = (10, 0)$.

The curve has horizontal tangents at $(6, \pm 16)$ and a vertical

tangent at $(10, 0)$.

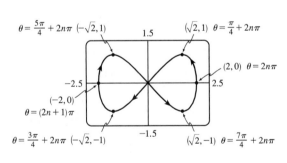

19. $x = 2\cos\theta$, $y = \sin 2\theta$.

$\dfrac{dy}{d\theta} = 2\cos 2\theta$, so $\dfrac{dy}{d\theta} = 0$ $\Leftrightarrow$ $2\theta = \dfrac{\pi}{2} + n\pi$

[n an integer] $\Leftrightarrow$ $\theta = \dfrac{\pi}{4} + \dfrac{\pi}{2}n$ $\Leftrightarrow$

$(x, y) = (\pm\sqrt{2}, \pm 1)$. Also, $\dfrac{dx}{d\theta} = -2\sin\theta$, so

$\dfrac{dx}{d\theta} = 0$ $\Leftrightarrow$ $\theta = n\pi$ $\Leftrightarrow$ $(x, y) = (\pm 2, 0)$.

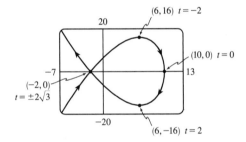

The curve has horizontal tangents at $\left(\pm\sqrt{2}, \pm 1\right)$ (four points), and vertical tangents at $(\pm 2, 0)$.

21. From the graph, it appears that the rightmost point on the curve $x = t - t^6$, $y = e^t$

is about $(0.6, 2)$. To find the exact coordinates, we find the value of t for which the

graph has a vertical tangent, that is, $0 = dx/dt = 1 - 6t^5$ $\Leftrightarrow$ $t = 1/\sqrt[5]{6}$.

Hence, the rightmost point is

$\left(1/\sqrt[5]{6} - 1/\left(6\sqrt[5]{6}\right), e^{1/\sqrt[5]{6}}\right) = \left(5 \cdot 6^{-6/5}, e^{6^{-1/5}}\right) \approx (0.58, 2.01).$

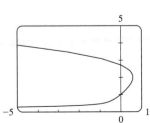

23. We graph the curve $x = t^4 - 2t^3 - 2t^2$, $y = t^3 - t$ in the viewing rectangle $[-2, 1.1]$ by $[-0.5, 0.5]$. This rectangle corresponds approximately to $t \in [-1, 0.8]$.

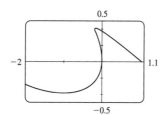

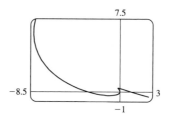

We estimate that the curve has horizontal tangents at about $(-1, -0.4)$ and $(-0.17, 0.39)$ and vertical tangents at

about $(0, 0)$ and $(-0.19, 0.37)$. We calculate $\dfrac{dy}{dx} = \dfrac{dy/dt}{dx/dt} = \dfrac{3t^2 - 1}{4t^3 - 6t^2 - 4t}$. The horizontal tangents occur when

$dy/dt = 3t^2 - 1 = 0 \iff t = \pm\frac{1}{\sqrt{3}}$, so both horizontal tangents are shown in our graph. The vertical tangents occur when

$dx/dt = 2t(2t^2 - 3t - 2) = 0 \iff 2t(2t + 1)(t - 2) = 0 \iff t = 0, -\frac{1}{2}$ or 2. It seems that we have missed one vertical

tangent, and indeed if we plot the curve on the t-interval $[-1.2, 2.2]$ we see that there is another vertical tangent at $(-8, 6)$.

25. $x = \cos t$, $y = \sin t \cos t$. $dx/dt = -\sin t$, $dy/dt = -\sin^2 t + \cos^2 t = \cos 2t$.

$(x, y) = (0, 0) \iff \cos t = 0 \iff t$ is an odd multiple of $\frac{\pi}{2}$. When $t = \frac{\pi}{2}$,

$dx/dt = -1$ and $dy/dt = -1$, so $dy/dx = 1$. When $t = \frac{3\pi}{2}$, $dx/dt = 1$ and

$dy/dt = -1$. So $dy/dx = -1$. Thus, $y = x$ and $y = -x$ are both tangent to the

curve at $(0, 0)$.

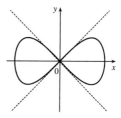

27. $x = r\theta - d\sin\theta$, $y = r - d\cos\theta$.

(a) $\dfrac{dx}{d\theta} = r - d\cos\theta$, $\dfrac{dy}{d\theta} = d\sin\theta$, so $\dfrac{dy}{dx} = \dfrac{d\sin\theta}{r - d\cos\theta}$.

(b) If $0 < d < r$, then $|d\cos\theta| \leq d < r$, so $r - d\cos\theta \geq r - d > 0$. This shows that $dx/d\theta$ never vanishes,

so the trochoid can have no vertical tangent if $d < r$.

29. $x = 2t^3$, $y = 1 + 4t - t^2 \implies \dfrac{dy}{dx} = \dfrac{dy/dt}{dx/dt} = \dfrac{4 - 2t}{6t^2}$. Now solve $\dfrac{dy}{dx} = 1 \iff \dfrac{4 - 2t}{6t^2} = 1 \iff$

$6t^2 + 2t - 4 = 0 \iff 2(3t - 2)(t + 1) = 0 \iff t = \frac{2}{3}$ or $t = -1$. If $t = \frac{2}{3}$, the point is $\left(\frac{16}{27}, \frac{29}{9}\right)$, and if $t = -1$,

the point is $(-2, -4)$.

31. By symmetry of the ellipse about the x- and y-axes,

$$A = 4\int_0^a y\,dx = 4\int_{\pi/2}^0 b\sin\theta\,(-a\sin\theta)\,d\theta = 4ab\int_0^{\pi/2}\sin^2\theta\,d\theta = 4ab\int_0^{\pi/2}\tfrac{1}{2}(1 - \cos 2\theta)\,d\theta$$

$$= 2ab\left[\theta - \tfrac{1}{2}\sin 2\theta\right]_0^{\pi/2} = 2ab\left(\tfrac{\pi}{2}\right) = \pi ab$$

33. The curve $x = 1 + e^t$, $y = t - t^2 = t(1 - t)$ intersects the x-axis when $y = 0$,

that is, when $t = 0$ and $t = 1$. The corresponding values of x are 2 and $1 + e$.

The shaded area is given by

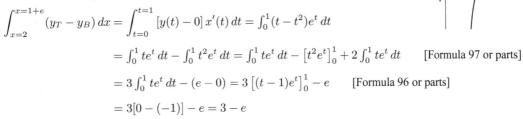

$$\int_{x=2}^{x=1+e} (y_T - y_B)\, dx = \int_{t=0}^{t=1} [y(t) - 0]\, x'(t)\, dt = \int_0^1 (t - t^2)e^t\, dt$$

$$= \int_0^1 te^t\, dt - \int_0^1 t^2 e^t\, dt = \int_0^1 te^t\, dt - \left[t^2 e^t\right]_0^1 + 2\int_0^1 te^t\, dt \qquad \text{[Formula 97 or parts]}$$

$$= 3\int_0^1 te^t\, dt - (e - 0) = 3\left[(t-1)e^t\right]_0^1 - e \qquad \text{[Formula 96 or parts]}$$

$$= 3[0 - (-1)] - e = 3 - e$$

35. $x = r\theta - d\sin\theta,\ y = r - d\cos\theta.$

$$A = \int_0^{2\pi r} y\, dx = \int_0^{2\pi} (r - d\cos\theta)(r - d\cos\theta)\, d\theta = \int_0^{2\pi} (r^2 - 2dr\cos\theta + d^2\cos^2\theta)\, d\theta$$

$$= \left[r^2\theta - 2dr\sin\theta + \tfrac{1}{2}d^2\left(\theta + \tfrac{1}{2}\sin 2\theta\right)\right]_0^{2\pi} = 2\pi r^2 + \pi d^2$$

37. $x = t - t^2,\ y = \frac{4}{3}t^{3/2},\ 1 \le t \le 2.$ $dx/dt = 1 - 2t$ and $dy/dt = 2t^{1/2}$, so

$$(dx/dt)^2 + (dy/dt)^2 = (1 - 2t)^2 + (2t^{1/2})^2 = 1 - 4t + 4t^2 + 4t = 1 + 4t^2.$$

Thus, $L = \int_a^b \sqrt{(dx/dt)^2 + (dy/dt)^2}\, dt = \int_1^2 \sqrt{1 + 4t^2}\, dt \approx 3.1678.$

39. $x = t + \cos t,\ y = t - \sin t,\ 0 \le t \le 2\pi.$ $dx/dt = 1 - \sin t$ and $dy/dt = 1 - \cos t$, so

$$\left(\tfrac{dx}{dt}\right)^2 + \left(\tfrac{dy}{dt}\right)^2 = (1 - \sin t)^2 + (1 - \cos t)^2 = (1 - 2\sin t + \sin^2 t) + (1 - 2\cos t + \cos^2 t) = 3 - 2\sin t - 2\cos t.$$

Thus, $L = \int_a^b \sqrt{(dx/dt)^2 + (dy/dt)^2}\, dt = \int_0^{2\pi} \sqrt{3 - 2\sin t - 2\cos t}\, dt \approx 10.0367.$

41.

$x = 1 + 3t^2,\ y = 4 + 2t^3,\ 0 \le t \le 1.$

$dx/dt = 6t$ and $dy/dt = 6t^2$, so $(dx/dt)^2 + (dy/dt)^2 = 36t^2 + 36t^4$.

Thus, $L = \int_0^1 \sqrt{36t^2 + 36t^4}\, dt = \int_0^1 6t\sqrt{1 + t^2}\, dt$

$$= 6\int_1^2 \sqrt{u}\left(\tfrac{1}{2}du\right) \quad [u = 1 + t^2, du = 2t\, dt]$$

$$= 3\left[\tfrac{2}{3}u^{3/2}\right]_1^2 = 2(2^{3/2} - 1) = 2(2\sqrt{2} - 1)$$

43. $x = \dfrac{t}{1 + t},\ y = \ln(1 + t),\ 0 \le t \le 2.$ $\dfrac{dx}{dt} = \dfrac{(1 + t)\cdot 1 - t\cdot 1}{(1 + t)^2} = \dfrac{1}{(1 + t)^2}$ and $\dfrac{dy}{dt} = \dfrac{1}{1 + t}$,

so $\left(\dfrac{dx}{dt}\right)^2 + \left(\dfrac{dy}{dt}\right)^2 = \dfrac{1}{(1 + t)^4} + \dfrac{1}{(1 + t)^2} = \dfrac{1}{(1 + t)^4}\left[1 + (1 + t)^2\right] = \dfrac{t^2 + 2t + 2}{(1 + t)^4}.$ Thus,

$$L = \int_0^2 \frac{\sqrt{t^2 + 2t + 2}}{(1 + t)^2}\, dt = \int_1^3 \frac{\sqrt{u^2 + 1}}{u^2}\, du \quad \begin{bmatrix} u = t + 1, \\ du = dt \end{bmatrix} \overset{24}{=} \left[-\frac{\sqrt{u^2 + 1}}{u} + \ln\left(u + \sqrt{u^2 + 1}\right)\right]_1^3$$

$$= -\frac{\sqrt{10}}{3} + \ln\left(3 + \sqrt{10}\right) + \sqrt{2} - \ln\left(1 + \sqrt{2}\right)$$

45. $x = e^t \cos t, \ y = e^t \sin t, \ 0 \le t \le \pi$.

$$\left(\tfrac{dx}{dt}\right)^2 + \left(\tfrac{dy}{dt}\right)^2 = [e^t(\cos t - \sin t)]^2 + [e^t(\sin t + \cos t)]^2$$
$$= (e^t)^2(\cos^2 t - 2\cos t \sin t + \sin^2 t)$$
$$+ (e^t)^2(\sin^2 t + 2\sin t \cos t + \cos^2 t)$$
$$= e^{2t}(2\cos^2 t + 2\sin^2 t) = 2e^{2t}$$

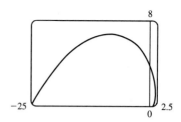

Thus, $L = \int_0^\pi \sqrt{2e^{2t}} \, dt = \int_0^\pi \sqrt{2}\, e^t \, dt = \sqrt{2}\left[e^t\right]_0^\pi = \sqrt{2}\,(e^\pi - 1)$.

47.

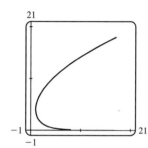

$x = e^t - t, \ y = 4e^{t/2}, \ -8 \le t \le 3$

$$\left(\tfrac{dx}{dt}\right)^2 + \left(\tfrac{dy}{dt}\right)^2 = (e^t - 1)^2 + (2e^{t/2})^2 = e^{2t} - 2e^t + 1 + 4e^t$$
$$= e^{2t} + 2e^t + 1 = (e^t + 1)^2$$

$$L = \int_{-8}^3 \sqrt{(e^t + 1)^2} \, dt = \int_{-8}^3 (e^t + 1) \, dt = \left[e^t + t\right]_{-8}^{3t}$$
$$= (e^3 + 3) - (e^{-8} - 8) = e^3 - e^{-8} + 11$$

49. $x = t - e^t, \ y = t + e^t, \ -6 \le t \le 6$.

$$\left(\tfrac{dx}{dt}\right)^2 + \left(\tfrac{dy}{dt}\right)^2 = (1 - e^t)^2 + (1 + e^t)^2 = (1 - 2e^t + e^{2t}) + (1 + 2e^t + e^{2t}) = 2 + 2e^{2t}, \text{ so } L = \int_{-6}^6 \sqrt{2 + 2e^{2t}} \, dt.$$

Set $f(t) = \sqrt{2 + 2e^{2t}}$. Then by Simpson's Rule with $n = 6$ and $\Delta t = \frac{6 - (-6)}{6} = 2$, we get

$$L \approx \tfrac{2}{3}[f(-6) + 4f(-4) + 2f(-2) + 4f(0) + 2f(2) + 4f(4) + f(6)] \approx 612.3053.$$

51. $x = \sin^2 t, \ y = \cos^2 t, \ 0 \le t \le 3\pi$.

$$(dx/dt)^2 + (dy/dt)^2 = (2\sin t \cos t)^2 + (-2\cos t \sin t)^2 = 8\sin^2 t \cos^2 t = 2\sin^2 2t \quad \Rightarrow$$

Distance $= \int_0^{3\pi} \sqrt{2}\,|\sin 2t| \, dt = 6\sqrt{2} \int_0^{\pi/2} \sin 2t \, dt$ [by symmetry] $= -3\sqrt{2}\left[\cos 2t\right]_0^{\pi/2} = -3\sqrt{2}\,(-1 - 1) = 6\sqrt{2}$.

The full curve is traversed as t goes from 0 to $\frac{\pi}{2}$, because the curve is the segment of $x + y = 1$ that lies in the first quadrant

(since $x, y \ge 0$), and this segment is completely traversed as t goes from 0 to $\frac{\pi}{2}$. Thus, $L = \int_0^{\pi/2} \sin 2t \, dt = \sqrt{2}$, as above.

53. $x = a \sin \theta, \ y = b \cos \theta, \ 0 \le \theta \le 2\pi$.

$$\left(\tfrac{dx}{dt}\right)^2 + \left(\tfrac{dy}{dt}\right)^2 = (a\cos\theta)^2 + (-b\sin\theta)^2 = a^2\cos^2\theta + b^2\sin^2\theta = a^2(1 - \sin^2\theta) + b^2\sin^2\theta$$
$$= a^2 - (a^2 - b^2)\sin^2\theta = a^2 - c^2\sin^2\theta = a^2\left(1 - \frac{c^2}{a^2}\sin^2\theta\right) = a^2(1 - e^2\sin^2\theta)$$

So $L = 4\int_0^{\pi/2} \sqrt{a^2\left(1 - e^2\sin^2\theta\right)}\, d\theta$ [by symmetry] $= 4a\int_0^{\pi/2} \sqrt{1 - e^2\sin^2\theta}\, d\theta$.

55. (a) $x = 11\cos t - 4\cos(11t/2), \ y = 11\sin t - 4\sin(11t/2)$.

Notice that $0 \le t \le 2\pi$ does not give the complete curve because

$x(0) \ne x(2\pi)$. In fact, we must take $t \in [0, 4\pi]$ in order to obtain the

complete curve, since the first term in each of the parametric equations has

period 2π and the second has period $\frac{2\pi}{11/2} = \frac{4\pi}{11}$, and the least common

integer multiple of these two numbers is 4π.

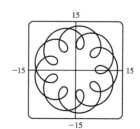

(b) We use the CAS to find the derivatives dx/dt and dy/dt, and then use Formula 1 to find the arc length. Recent versions of Maple express the integral $\int_0^{4\pi} \sqrt{(dx/dt)^2 + (dy/dt)^2}\, dt$ as $88E(2\sqrt{2}\,i)$, where $E(x)$ is the elliptic integral

$$\int_0^1 \frac{\sqrt{1 - x^2 t^2}}{\sqrt{1 - t^2}}\, dt \text{ and } i \text{ is the imaginary number } \sqrt{-1}.$$

Some earlier versions of Maple (as well as Mathematica) cannot do the integral exactly, so we use the command `evalf(Int(sqrt(diff(x,t)^2+diff(y,t)^2),t=0..4*Pi));` to estimate the length, and find that the arc length is approximately 294.03. Derive's `Para_arc_length` function in the utility file `Int_apps` simplifies the integral to $11 \int_0^{4\pi} \sqrt{-4\cos t \, \cos\left(\frac{11t}{2}\right) - 4\sin t \, \sin\left(\frac{11t}{2}\right) + 5}\, dt.$

57. $x = 1 + te^t$, $y = (t^2 + 1)e^t$, $0 \le t \le 1$.

$$\left(\tfrac{dx}{dt}\right)^2 + \left(\tfrac{dy}{dt}\right)^2 = (te^t + e^t)^2 + [(t^2 + 1)e^t + e^t(2t)]^2 = [e^t(t+1)]^2 + [e^t(t^2 + 2t + 1)]^2$$

$$= e^{2t}(t+1)^2 + e^{2t}(t+1)^4 = e^{2t}(t+1)^2[1 + (t+1)^2], \quad \text{so}$$

$$S = \int 2\pi y\, ds = \int_0^1 2\pi(t^2 + 1)e^t \sqrt{e^{2t}(t+1)^2(t^2 + 2t + 2)}\, dt = \int_0^1 2\pi(t^2 + 1)e^{2t}(t+1)\sqrt{t^2 + 2t + 2}\, dt \approx 103.5999$$

59. $x = t^3$, $y = t^2$, $0 \le t \le 1$. $\left(\tfrac{dx}{dt}\right)^2 + \left(\tfrac{dy}{dt}\right)^2 = \left(3t^2\right)^2 + (2t)^2 = 9t^4 + 4t^2.$

$$S = \int_0^1 2\pi y \sqrt{\left(\tfrac{dx}{dt}\right)^2 + \left(\tfrac{dy}{dt}\right)^2}\, dt = \int_0^1 2\pi t^2 \sqrt{9t^4 + 4t^2}\, dt = 2\pi \int_0^1 t^2 \sqrt{t^2(9t^2 + 4)}\, dt$$

$$= 2\pi \int_4^{13} \left(\frac{u - 4}{9}\right) \sqrt{u}\left(\tfrac{1}{18}\, du\right) \quad \begin{bmatrix} u = 9t^2 + 4,\ t^2 = (u - 4)/9, \\ du = 18t\, dt,\ \text{so } t\, dt = \tfrac{1}{18}\, du \end{bmatrix} = \frac{2\pi}{9 \cdot 18} \int_4^{13} (u^{3/2} - 4u^{1/2})\, du$$

$$= \tfrac{\pi}{81}\left[\tfrac{2}{5}u^{5/2} - \tfrac{8}{3}u^{3/2}\right]_4^{13} = \tfrac{\pi}{81} \cdot \tfrac{2}{15}\left[3u^{5/2} - 20u^{3/2}\right]_4^{13}$$

$$= \tfrac{2\pi}{1215}\left[(3 \cdot 13^2 \sqrt{13} - 20 \cdot 13\sqrt{13}) - (3 \cdot 32 - 20 \cdot 8)\right] = \tfrac{2\pi}{1215}\left(247\sqrt{13} + 64\right)$$

61. $x = a\cos^3\theta$, $y = a\sin^3\theta$, $0 \le \theta \le \tfrac{\pi}{2}$. $\left(\tfrac{dx}{d\theta}\right)^2 + \left(\tfrac{dy}{d\theta}\right)^2 = (-3a\cos^2\theta \, \sin\theta)^2 + (3a\sin^2\theta \, \cos\theta)^2 = 9a^2\sin^2\theta \, \cos^2\theta.$

$$S = \int_0^{\pi/2} 2\pi \cdot a\sin^3\theta \cdot 3a\sin\theta \, \cos\theta \, d\theta = 6\pi a^2 \int_0^{\pi/2} \sin^4\theta \, \cos\theta \, d\theta = \tfrac{6}{5}\pi a^2\left[\sin^5\theta\right]_0^{\pi/2} = \tfrac{6}{5}\pi a^2$$

63. $x = t + t^3$, $y = t - \dfrac{1}{t^2}$, $1 \le t \le 2$. $\dfrac{dx}{dt} = 1 + 3t^2$ and $\dfrac{dy}{dt} = 1 + \dfrac{2}{t^3}$, so $\left(\tfrac{dx}{dt}\right)^2 + \left(\tfrac{dy}{dt}\right)^2 = (1 + 3t^2)^2 + \left(1 + \dfrac{2}{t^3}\right)^2$

and $S = \displaystyle\int 2\pi y\, ds = \int_1^2 2\pi\left(t - \dfrac{1}{t^2}\right)\sqrt{(1 + 3t^2)^2 + \left(1 + \dfrac{2}{t^3}\right)^2}\, dt \approx 59.101.$

65. $x = 3t^2$, $y = 2t^3$, $0 \le t \le 5$ $\Rightarrow$ $\left(\tfrac{dx}{dt}\right)^2 + \left(\tfrac{dy}{dt}\right)^2 = (6t)^2 + (6t^2)^2 = 36t^2(1 + t^2)$ $\Rightarrow$

$$S = \int_0^5 2\pi x \sqrt{(dx/dt)^2 + (dy/dt)^2}\, dt = \int_0^5 2\pi(3t^2)6t\sqrt{1 + t^2}\, dt = 18\pi \int_0^5 t^2 \sqrt{1 + t^2}\, 2t\, dt$$

$$= 18\pi \int_1^{26} (u - 1)\sqrt{u}\, du \quad \begin{bmatrix} u = 1 + t^2, \\ du = 2t\, dt \end{bmatrix} = 18\pi \int_1^{26} (u^{3/2} - u^{1/2})\, du = 18\pi\left[\tfrac{2}{5}u^{5/2} - \tfrac{2}{3}u^{3/2}\right]_1^{26}$$

$$= 18\pi\left[\left(\tfrac{2}{5} \cdot 676\sqrt{26} - \tfrac{2}{3} \cdot 26\sqrt{26}\right) - \left(\tfrac{2}{5} - \tfrac{2}{3}\right)\right] = \tfrac{24}{5}\pi\left(949\sqrt{26} + 1\right)$$

67. If f' is continuous and $f'(t) \neq 0$ for $a \leq t \leq b$, then either $f'(t) > 0$ for all t in $[a, b]$ or $f'(t) < 0$ for all t in $[a, b]$. Thus, f is monotonic (in fact, strictly increasing or strictly decreasing) on $[a, b]$. It follows that f has an inverse. Set $F = g \circ f^{-1}$, that is, define F by $F(x) = g(f^{-1}(x))$. Then $x = f(t) \implies f^{-1}(x) = t$, so $y = g(t) = g(f^{-1}(x)) = F(x)$.

69. (a) $\phi = \tan^{-1}\left(\dfrac{dy}{dx}\right) \implies \dfrac{d\phi}{dt} = \dfrac{d}{dt}\tan^{-1}\left(\dfrac{dy}{dx}\right) = \dfrac{1}{1 + (dy/dx)^2}\left[\dfrac{d}{dt}\left(\dfrac{dy}{dx}\right)\right].$ But $\dfrac{dy}{dx} = \dfrac{dy/dt}{dx/dt} = \dfrac{\dot{y}}{\dot{x}} \implies$

$\dfrac{d}{dt}\left(\dfrac{dy}{dx}\right) = \dfrac{d}{dt}\left(\dfrac{\dot{y}}{\dot{x}}\right) = \dfrac{\ddot{y}\dot{x} - \ddot{x}\dot{y}}{\dot{x}^2} \implies \dfrac{d\phi}{dt} = \dfrac{1}{1 + (\dot{y}/\dot{x})^2}\left(\dfrac{\ddot{y}\dot{x} - \ddot{x}\dot{y}}{\dot{x}^2}\right) = \dfrac{\dot{x}\ddot{y} - \ddot{x}\dot{y}}{\dot{x}^2 + \dot{y}^2}.$ Using the Chain Rule, and the

fact that $s = \displaystyle\int_0^t \sqrt{\left(\dfrac{dx}{dt}\right)^2 + \left(\dfrac{dy}{dt}\right)^2}\, dt \implies \dfrac{ds}{dt} = \sqrt{\left(\dfrac{dx}{dt}\right)^2 + \left(\dfrac{dy}{dt}\right)^2} = (\dot{x}^2 + \dot{y}^2)^{1/2},$ we have that

$\dfrac{d\phi}{ds} = \dfrac{d\phi/dt}{ds/dt} = \left(\dfrac{\dot{x}\ddot{y} - \ddot{x}\dot{y}}{\dot{x}^2 + \dot{y}^2}\right)\dfrac{1}{(\dot{x}^2 + \dot{y}^2)^{1/2}} = \dfrac{\dot{x}\ddot{y} - \ddot{x}\dot{y}}{(\dot{x}^2 + \dot{y}^2)^{3/2}}.$ So $\kappa = \left|\dfrac{d\phi}{ds}\right| = \left|\dfrac{\dot{x}\ddot{y} - \ddot{x}\dot{y}}{(\dot{x}^2 + \dot{y}^2)^{3/2}}\right| = \dfrac{|\dot{x}\ddot{y} - \ddot{x}\dot{y}|}{(\dot{x}^2 + \dot{y}^2)^{3/2}}.$

(b) $x = x$ and $y = f(x) \implies \dot{x} = 1, \ddot{x} = 0$ and $\dot{y} = \dfrac{dy}{dx}, \ddot{y} = \dfrac{d^2y}{dx^2}.$

So $\kappa = \dfrac{\left|1 \cdot (d^2y/dx^2) - 0 \cdot (dy/dx)\right|}{[1 + (dy/dx)^2]^{3/2}} = \dfrac{\left|d^2y/dx^2\right|}{[1 + (dy/dx)^2]^{3/2}}.$

71. $x = \theta - \sin\theta \implies \dot{x} = 1 - \cos\theta \implies \ddot{x} = \sin\theta,$ and $y = 1 - \cos\theta \implies \dot{y} = \sin\theta \implies \ddot{y} = \cos\theta.$ Therefore,

$\kappa = \dfrac{\left|\cos\theta - \cos^2\theta - \sin^2\theta\right|}{[(1 - \cos\theta)^2 + \sin^2\theta]^{3/2}} = \dfrac{\left|\cos\theta - (\cos^2\theta + \sin^2\theta)\right|}{(1 - 2\cos\theta + \cos^2\theta + \sin^2\theta)^{3/2}} = \dfrac{\left|\cos\theta - 1\right|}{(2 - 2\cos\theta)^{3/2}}.$ The top of the arch is

characterized by a horizontal tangent, and from Example 2(b) in Section 10.2, the tangent is horizontal when $\theta = (2n - 1)\pi,$

so take $n = 1$ and substitute $\theta = \pi$ into the expression for κ: $\kappa = \dfrac{\left|\cos\pi - 1\right|}{(2 - 2\cos\pi)^{3/2}} = \dfrac{|-1 - 1|}{[2 - 2(-1)]^{3/2}} = \dfrac{1}{4}.$

73. The coordinates of T are $(r\cos\theta, r\sin\theta).$ Since TP was unwound from

arc TA, TP has length $r\theta.$ Also $\angle PTQ = \angle PTR - \angle QTR = \frac{1}{2}\pi - \theta,$

so P has coordinates $x = r\cos\theta + r\theta\cos\left(\frac{1}{2}\pi - \theta\right) = r(\cos\theta + \theta\sin\theta),$

$y = r\sin\theta - r\theta\sin\left(\frac{1}{2}\pi - \theta\right) = r(\sin\theta - \theta\cos\theta).$

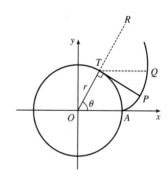

10.3 Polar Coordinates

1. (a) $\left(2, \frac{\pi}{3}\right)$

By adding 2π to $\frac{\pi}{3},$ we obtain the point $\left(2, \frac{7\pi}{3}\right).$ The direction

opposite $\frac{\pi}{3}$ is $\frac{4\pi}{3},$ so $\left(-2, \frac{4\pi}{3}\right)$ is a point that satisfies the $r < 0$

requirement.

(b) $\left(1, -\frac{3\pi}{4}\right)$

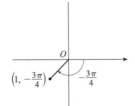

$r > 0$: $\left(1, -\frac{3\pi}{4} + 2\pi\right) = \left(1, \frac{5\pi}{4}\right)$

$r < 0$: $\left(-1, -\frac{3\pi}{4} + \pi\right) = \left(-1, \frac{\pi}{4}\right)$

(c) $\left(-1, \frac{\pi}{2}\right)$

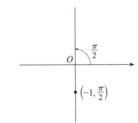

$r > 0$: $\left(-(-1), \frac{\pi}{2} + \pi\right) = \left(1, \frac{3\pi}{2}\right)$

$r < 0$: $\left(-1, \frac{\pi}{2} + 2\pi\right) = \left(-1, \frac{5\pi}{2}\right)$

3. (a)

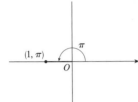

$x = 1\cos\pi = 1(-1) = -1$ and

$y = 1\sin\pi = 1(0) = 0$ give us

the Cartesian coordinates $(-1, 0)$.

(b)

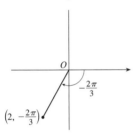

$x = 2\cos\left(-\frac{2\pi}{3}\right) = 2\left(-\frac{1}{2}\right) = -1$ and

$y = 2\sin\left(-\frac{2\pi}{3}\right) = 2\left(-\frac{\sqrt{3}}{2}\right) = -\sqrt{3}$

give us $\left(-1, -\sqrt{3}\right)$.

(c)

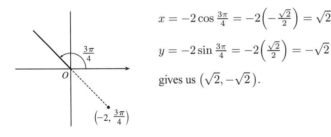

$x = -2\cos\frac{3\pi}{4} = -2\left(-\frac{\sqrt{2}}{2}\right) = \sqrt{2}$ and

$y = -2\sin\frac{3\pi}{4} = -2\left(\frac{\sqrt{2}}{2}\right) = -\sqrt{2}$

gives us $\left(\sqrt{2}, -\sqrt{2}\right)$.

5. (a) $x = 2$ and $y = -2$ $\Rightarrow$ $r = \sqrt{2^2 + (-2)^2} = 2\sqrt{2}$ and $\theta = \tan^{-1}\left(\frac{-2}{2}\right) = -\frac{\pi}{4}$. Since $(2, -2)$ is in the fourth

quadrant, the polar coordinates are (i) $\left(2\sqrt{2}, \frac{7\pi}{4}\right)$ and (ii) $\left(-2\sqrt{2}, \frac{3\pi}{4}\right)$.

(b) $x = -1$ and $y = \sqrt{3}$ $\Rightarrow$ $r = \sqrt{(-1)^2 + \left(\sqrt{3}\right)^2} = 2$ and $\theta = \tan^{-1}\left(\frac{\sqrt{3}}{-1}\right) = \frac{2\pi}{3}$. Since $\left(-1, \sqrt{3}\right)$ is in the second

quadrant, the polar coordinates are (i) $\left(2, \frac{2\pi}{3}\right)$ and (ii) $\left(-2, \frac{5\pi}{3}\right)$.

7. The curves $r = 1$ and $r = 2$ represent circles with center O and radii 1 and 2. The region in the plane satisfying $1 \le r \le 2$ consists of both circles and the shaded region between them in the figure.

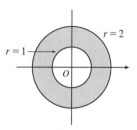

9. The region satisfying $0 \le r < 4$ and $-\pi/2 \le \theta < \pi/6$ does not include the circle $r = 4$ nor the line $\theta = \frac{\pi}{6}$.

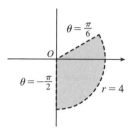

11. $2 < r < 3$, $\frac{5\pi}{3} \le \theta \le \frac{7\pi}{3}$

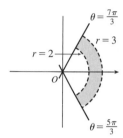

13. Converting the polar coordinates $(2, \pi/3)$ and $(4, 2\pi/3)$ to Cartesian coordinates gives us $\left(2 \cos \frac{\pi}{3}, 2 \sin \frac{\pi}{3}\right) = \left(1, \sqrt{3}\right)$ and $\left(4 \cos \frac{2\pi}{3}, 4 \sin \frac{2\pi}{3}\right) = \left(-2, 2\sqrt{3}\right)$. Now use the distance formula.

$$d = \sqrt{(x_2 - x_1)^2 + (y_2 - y_1)^2} = \sqrt{(-2 - 1)^2 + \left(2\sqrt{3} - \sqrt{3}\right)^2} = \sqrt{9 + 3} = \sqrt{12} = 2\sqrt{3}$$

15. $r = 2 \Leftrightarrow \sqrt{x^2 + y^2} = 2 \Leftrightarrow x^2 + y^2 = 4$, a circle of radius 2 centered at the origin.

17. $r = 3 \sin \theta \Rightarrow r^2 = 3r \sin \theta \Leftrightarrow x^2 + y^2 = 3y \Leftrightarrow x^2 + \left(y - \frac{3}{2}\right)^2 = \left(\frac{3}{2}\right)^2$, a circle of radius $\frac{3}{2}$ centered at $\left(0, \frac{3}{2}\right)$. The first two equations are actually equivalent since $r^2 = 3r \sin \theta \Rightarrow r(r - 3 \sin \theta) = 0 \Rightarrow r = 0$ or $r = 3 \sin \theta$. But $r = 3 \sin \theta$ gives the point $r = 0$ (the pole) when $\theta = 0$. Thus, the single equation $r = 3 \sin \theta$ is equivalent to the compound condition ($r = 0$ or $r = 3 \sin \theta$).

19. $r = \csc \theta \Leftrightarrow r = \dfrac{1}{\sin \theta} \Leftrightarrow r \sin \theta = 1 \Leftrightarrow y = 1$, a horizontal line 1 unit above the x-axis.

21. $x = 3 \Leftrightarrow r \cos \theta = 3 \Leftrightarrow r = 3/\cos \theta \Leftrightarrow r = 3 \sec \theta$.

23. $x = -y^2 \Leftrightarrow r \cos \theta = -r^2 \sin^2 \theta \Leftrightarrow \cos \theta = -r \sin^2 \theta \Leftrightarrow r = -\dfrac{\cos \theta}{\sin^2 \theta} = -\cot \theta \csc \theta$.

25. $x^2 + y^2 = 2cx \Leftrightarrow r^2 = 2cr \cos \theta \Leftrightarrow r^2 - 2cr \cos \theta = 0 \Leftrightarrow r(r - 2c \cos \theta) = 0 \Leftrightarrow r = 0$ or $r = 2c \cos \theta$. $r = 0$ is included in $r = 2c \cos \theta$ when $\theta = \frac{\pi}{2} + n\pi$, so the curve is represented by the single equation $r = 2c \cos \theta$.

27. (a) The description leads immediately to the polar equation $\theta = \frac{\pi}{6}$, and the Cartesian equation $y = \tan\left(\frac{\pi}{6}\right) x = \frac{1}{\sqrt{3}} x$ is slightly more difficult to derive.

(b) The easier description here is the Cartesian equation $x = 3$.

29. $\theta = -\pi/6$

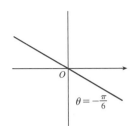

31. $r = \sin\theta \;\Leftrightarrow\; r^2 = r\sin\theta \;\Leftrightarrow\; x^2 + y^2 = y \;\Leftrightarrow$
$x^2 + \left(y - \frac{1}{2}\right)^2 = \left(\frac{1}{2}\right)^2$. The reasoning here is the same
as in Exercise 17. This is a circle of radius $\frac{1}{2}$ centered at $\left(0, \frac{1}{2}\right)$.

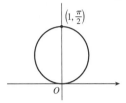

33. $r = 2(1 - \sin\theta)$. This curve is a cardioid.

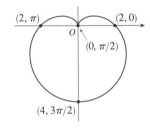

35. $r = \theta, \quad \theta \geq 0$

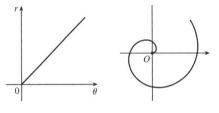

37. $r = 4\sin 3\theta$

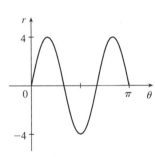
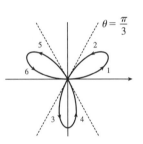

39. $r = 2\cos 4\theta$

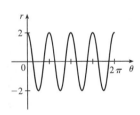
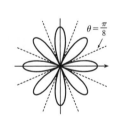

41. $r = 1 - 2\sin\theta$

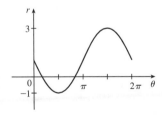
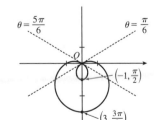

43. $r^2 = 9\sin 2\theta$

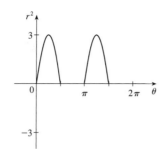
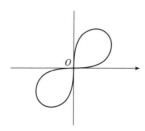

45. $r = 2\cos\left(\frac{3}{2}\theta\right)$

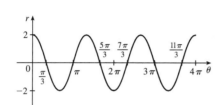
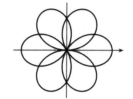

47. $r = 1 + 2\cos 2\theta$

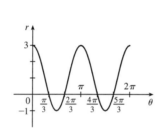
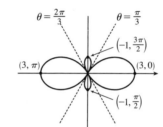

49. For $\theta = 0$, π, and 2π, r has its minimum value of about 0.5. For $\theta = \frac{\pi}{2}$ and $\frac{3\pi}{2}$, r attains its maximum value of 2. We see that the graph has a similar shape for $0 \le \theta \le \pi$ and $\pi \le \theta \le 2\pi$.

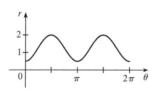
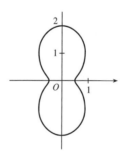

51. $x = (r)\cos\theta = (4 + 2\sec\theta)\cos\theta = 4\cos\theta + 2$. Now, $r \to \infty$ $\Rightarrow$

$(4 + 2\sec\theta) \to \infty$ $\Rightarrow$ $\theta \to \left(\frac{\pi}{2}\right)^-$ or $\theta \to \left(\frac{3\pi}{2}\right)^+$ [since we need only

consider $0 \le \theta < 2\pi$], so $\displaystyle\lim_{r \to \infty} x = \lim_{\theta \to \pi/2^-}(4\cos\theta + 2) = 2$. Also,

$r \to -\infty$ $\Rightarrow$ $(4 + 2\sec\theta) \to -\infty$ $\Rightarrow$ $\theta \to \left(\frac{\pi}{2}\right)^+$ or $\theta \to \left(\frac{3\pi}{2}\right)^-$, so

$\displaystyle\lim_{r \to -\infty} x = \lim_{\theta \to \pi/2^+}(4\cos\theta + 2) = 2$. Therefore, $\displaystyle\lim_{r \to \pm\infty} x = 2$ $\Rightarrow$ $x = 2$ is a vertical asymptote.

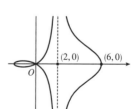

53. To show that $x = 1$ is an asymptote we must prove $\lim\limits_{r \to \pm\infty} x = 1$.

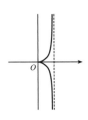

$x = (r)\cos\theta = (\sin\theta\,\tan\theta)\cos\theta = \sin^2\theta$. Now, $r \to \infty \;\;\Rightarrow\;\; \sin\theta\,\tan\theta \to \infty \;\;\Rightarrow$

$\theta \to \left(\frac{\pi}{2}\right)^-$, so $\lim\limits_{r\to\infty} x = \lim\limits_{\theta\to\pi/2^-}\sin^2\theta = 1$. Also, $r \to -\infty \;\;\Rightarrow\;\; \sin\theta\,\tan\theta \to -\infty \;\;\Rightarrow$

$\theta \to \left(\frac{\pi}{2}\right)^+$, so $\lim\limits_{r\to-\infty} x = \lim\limits_{\theta\to\pi/2^+}\sin^2\theta = 1$. Therefore, $\lim\limits_{r\to\pm\infty} x = 1 \;\;\Rightarrow\;\; x = 1$ is

a vertical asymptote. Also notice that $x = \sin^2\theta \ge 0$ for all θ, and $x = \sin^2\theta \le 1$ for all θ. And $x \ne 1$, since the curve is not

defined at odd multiples of $\frac{\pi}{2}$. Therefore, the curve lies entirely within the vertical strip $0 \le x < 1$.

55. (a) We see that the curve $r = 1 + c\sin\theta$ crosses itself at the origin, where $r = 0$ (in fact the inner loop corresponds to

negative r-values,) so we solve the equation of the limaçon for $r = 0 \;\;\Leftrightarrow\;\; c\sin\theta = -1 \;\;\Leftrightarrow\;\; \sin\theta = -1/c$. Now if

$|c| < 1$, then this equation has no solution and hence there is no inner loop. But if $c < -1$, then on the interval $(0, 2\pi)$

the equation has the two solutions $\theta = \sin^{-1}(-1/c)$ and $\theta = \pi - \sin^{-1}(-1/c)$, and if $c > 1$, the solutions are

$\theta = \pi + \sin^{-1}(1/c)$ and $\theta = 2\pi - \sin^{-1}(1/c)$. In each case, $r < 0$ for θ between the two solutions, indicating a loop.

(b) For $0 < c < 1$, the dimple (if it exists) is characterized by the fact that y has a local maximum at $\theta = \frac{3\pi}{2}$. So we determine

for what c-values $\dfrac{d^2y}{d\theta^2}$ is negative at $\theta = \frac{3\pi}{2}$, since by the Second Derivative Test this indicates a maximum:

$y = r\sin\theta = \sin\theta + c\sin^2\theta \;\;\Rightarrow\;\; \dfrac{dy}{d\theta} = \cos\theta + 2c\sin\theta\,\cos\theta = \cos\theta + c\sin 2\theta \;\;\Rightarrow\;\; \dfrac{d^2y}{d\theta^2} = -\sin\theta + 2c\cos 2\theta$.

At $\theta = \frac{3\pi}{2}$, this is equal to $-(-1) + 2c(-1) = 1 - 2c$, which is negative only for $c > \frac{1}{2}$. A similar argument shows that

for $-1 < c < 0$, y only has a local minimum at $\theta = \frac{\pi}{2}$ (indicating a dimple) for $c < -\frac{1}{2}$.

57. $r = 2\sin\theta \;\;\Rightarrow\;\; x = r\cos\theta = 2\sin\theta\,\cos\theta = \sin 2\theta,\; y = r\sin\theta = 2\sin^2\theta \;\;\Rightarrow$

$$\frac{dy}{dx} = \frac{dy/d\theta}{dx/d\theta} = \frac{2\cdot 2\sin\theta\,\cos\theta}{\cos 2\theta \cdot 2} = \frac{\sin 2\theta}{\cos 2\theta} = \tan 2\theta$$

When $\theta = \dfrac{\pi}{6},\; \dfrac{dy}{dx} = \tan\left(2\cdot\dfrac{\pi}{6}\right) = \tan\dfrac{\pi}{3} = \sqrt{3}$. [*Another method:* Use Equation 3.]

59. $r = 1/\theta \;\;\Rightarrow\;\; x = r\cos\theta = (\cos\theta)/\theta,\; y = r\sin\theta = (\sin\theta)/\theta \;\;\Rightarrow$

$$\frac{dy}{dx} = \frac{dy/d\theta}{dx/d\theta} = \frac{\sin\theta(-1/\theta^2) + (1/\theta)\cos\theta}{\cos\theta(-1/\theta^2) - (1/\theta)\sin\theta} \cdot \frac{\theta^2}{\theta^2} = \frac{-\sin\theta + \theta\cos\theta}{-\cos\theta - \theta\sin\theta}$$

When $\theta = \pi,\; \dfrac{dy}{dx} = \dfrac{-0 + \pi(-1)}{-(-1) - \pi(0)} = \dfrac{-\pi}{1} = -\pi$.

61. $r = \cos 2\theta \;\;\Rightarrow\;\; x = r\cos\theta = \cos 2\theta\,\cos\theta,\; y = r\sin\theta = \cos 2\theta\,\sin\theta \;\;\Rightarrow$

$$\frac{dy}{dx} = \frac{dy/d\theta}{dx/d\theta} = \frac{\cos 2\theta\,\cos\theta + \sin\theta\,(-2\sin 2\theta)}{\cos 2\theta\,(-\sin\theta) + \cos\theta\,(-2\sin 2\theta)}$$

When $\theta = \dfrac{\pi}{4},\; \dfrac{dy}{dx} = \dfrac{0(\sqrt{2}/2) + (\sqrt{2}/2)(-2)}{0(-\sqrt{2}/2) + (\sqrt{2}/2)(-2)} = \dfrac{-\sqrt{2}}{-\sqrt{2}} = 1$.

63. $r = 3\cos\theta \;\Rightarrow\; x = r\cos\theta = 3\cos\theta\,\cos\theta,\; y = r\sin\theta = 3\cos\theta\,\sin\theta \;\Rightarrow$

$\frac{dy}{d\theta} = -3\sin^2\theta + 3\cos^2\theta = 3\cos 2\theta = 0 \;\Rightarrow\; 2\theta = \frac{\pi}{2}$ or $\frac{3\pi}{2}$ $\;\Leftrightarrow\;$ $\theta = \frac{\pi}{4}$ or $\frac{3\pi}{4}$.

So the tangent is horizontal at $\left(\frac{3}{\sqrt{2}}, \frac{\pi}{4}\right)$ and $\left(-\frac{3}{\sqrt{2}}, \frac{3\pi}{4}\right)$ $\left[\text{same as } \left(\frac{3}{\sqrt{2}}, -\frac{\pi}{4}\right)\right]$.

$\frac{dx}{d\theta} = -6\sin\theta\,\cos\theta = -3\sin 2\theta = 0 \;\Rightarrow\; 2\theta = 0$ or π $\;\Leftrightarrow\;$ $\theta = 0$ or $\frac{\pi}{2}$. So the tangent is vertical at $(3, 0)$ and $\left(0, \frac{\pi}{2}\right)$.

65. $r = 1 + \cos\theta \;\Rightarrow\; x = r\cos\theta = \cos\theta\,(1 + \cos\theta),\; y = r\sin\theta = \sin\theta\,(1 + \cos\theta) \;\Rightarrow$

$\frac{dy}{d\theta} = (1 + \cos\theta)\cos\theta - \sin^2\theta = 2\cos^2\theta + \cos\theta - 1 = (2\cos\theta - 1)(\cos\theta + 1) = 0 \;\Rightarrow\; \cos\theta = \frac{1}{2}$ or -1 $\;\Rightarrow$

$\theta = \frac{\pi}{3}, \pi$, or $\frac{5\pi}{3}$ $\;\Rightarrow\;$ horizontal tangent at $\left(\frac{3}{2}, \frac{\pi}{3}\right), (0, \pi)$, and $\left(\frac{3}{2}, \frac{5\pi}{3}\right)$.

$\frac{dx}{d\theta} = -(1 + \cos\theta)\sin\theta - \cos\theta\,\sin\theta = -\sin\theta\,(1 + 2\cos\theta) = 0 \;\Rightarrow\; \sin\theta = 0$ or $\cos\theta = -\frac{1}{2}$ $\;\Rightarrow$

$\theta = 0, \pi, \frac{2\pi}{3}$, or $\frac{4\pi}{3}$ $\;\Rightarrow\;$ vertical tangent at $(2, 0), \left(\frac{1}{2}, \frac{2\pi}{3}\right)$, and $\left(\frac{1}{2}, \frac{4\pi}{3}\right)$.

Note that the tangent is horizontal, not vertical when $\theta = \pi$, since $\lim\limits_{\theta \to \pi} \dfrac{dy/d\theta}{dx/d\theta} = 0$.

67. $r = 2 + \sin\theta \;\Rightarrow\; x = r\cos\theta = (2 + \sin\theta)\cos\theta,\; y = r\sin\theta = (2 + \sin\theta)\sin\theta \;\Rightarrow$

$\frac{dy}{d\theta} = (2 + \sin\theta)\cos\theta + \sin\theta\,\cos\theta = \cos\theta \cdot 2(1 + \sin\theta) = 0 \;\Rightarrow\; \cos\theta = 0$ or $\sin\theta = -1$ $\;\Rightarrow$

$\theta = \frac{\pi}{2}$ or $\frac{3\pi}{2}$ $\;\Rightarrow\;$ horizontal tangent at $\left(3, \frac{\pi}{2}\right)$ and $\left(1, \frac{3\pi}{2}\right)$.

$\frac{dx}{d\theta} = (2 + \sin\theta)(-\sin\theta) + \cos\theta\,\cos\theta = -2\sin\theta - \sin^2\theta + 1 - \sin^2\theta = -2\sin^2\theta - 2\sin\theta + 1 \;\Rightarrow$

$\sin\theta = \dfrac{2 \pm \sqrt{4+8}}{-4} = \dfrac{2 \pm 2\sqrt{3}}{-4} = \dfrac{1 - \sqrt{3}}{-2}$ $\left[\dfrac{1 + \sqrt{3}}{-2} < -1\right]$ $\;\Rightarrow$

$\theta_1 = \sin^{-1}\left(-\frac{1}{2} + \frac{1}{2}\sqrt{3}\right)$ and $\theta_2 = \pi - \theta_1$ $\;\Rightarrow\;$ vertical tangent at $\left(\frac{3}{2} + \frac{1}{2}\sqrt{3}, \theta_1\right)$ and $\left(\frac{3}{2} + \frac{1}{2}\sqrt{3}, \theta_2\right)$.

Note that $r(\theta_1) = 2 + \sin\left[\sin^{-1}\left(-\frac{1}{2} + \frac{1}{2}\sqrt{3}\right)\right] = 2 - \frac{1}{2} + \frac{1}{2}\sqrt{3} = \frac{3}{2} + \frac{1}{2}\sqrt{3}$.

69. $r = a\sin\theta + b\cos\theta \;\Rightarrow\; r^2 = ar\sin\theta + br\cos\theta \;\Rightarrow\; x^2 + y^2 = ay + bx \;\Rightarrow$

$x^2 - bx + \left(\frac{1}{2}b\right)^2 + y^2 - ay + \left(\frac{1}{2}a\right)^2 = \left(\frac{1}{2}b\right)^2 + \left(\frac{1}{2}a\right)^2 \;\Rightarrow\; \left(x - \frac{1}{2}b\right)^2 + \left(y - \frac{1}{2}a\right)^2 = \frac{1}{4}(a^2 + b^2)$, and this is a circle

with center $\left(\frac{1}{2}b, \frac{1}{2}a\right)$ and radius $\frac{1}{2}\sqrt{a^2 + b^2}$.

Note for Exercises 71–76: Maple is able to plot polar curves using the `polarplot` command, or using the `coords=polar` option in a regular
`plot` command. In Mathematica, use `PolarPlot`. In Derive, change to `Polar` under `Options State`. If your graphing device cannot
plot polar equations, you must convert to parametric equations. For example, in Exercise 71, $x = r\cos\theta = [1 + 2\sin(\theta/2)]\cos\theta$,
$y = r\sin\theta = [1 + 2\sin(\theta/2)]\sin\theta$.

71. $r = 1 + 2\sin(\theta/2)$. The parameter interval is $[0, 4\pi]$.

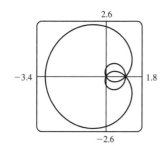

73. $r = e^{\sin\theta} - 2\cos(4\theta)$. The parameter interval is $[0, 2\pi]$.

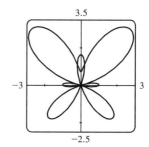

75. $r = 2 - 5\sin(\theta/6)$. The parameter interval is $[-6\pi, 6\pi]$.

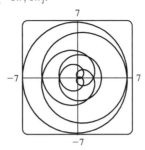

77. It appears that the graph of $r = 1 + \sin\left(\theta - \frac{\pi}{6}\right)$ is the same shape as

the graph of $r = 1 + \sin\theta$, but rotated counterclockwise about the

origin by $\frac{\pi}{6}$. Similarly, the graph of $r = 1 + \sin\left(\theta - \frac{\pi}{3}\right)$ is rotated by

$\frac{\pi}{3}$. In general, the graph of $r = f(\theta - \alpha)$ is the same shape as that of

$r = f(\theta)$, but rotated counterclockwise through α about the origin.

That is, for any point (r_0, θ_0) on the curve $r = f(\theta)$, the point

$(r_0, \theta_0 + \alpha)$ is on the curve $r = f(\theta - \alpha)$, since $r_0 = f(\theta_0) = f((\theta_0 + \alpha) - \alpha)$.

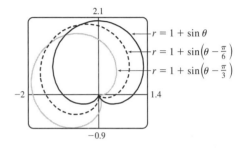

79. (a) $r = \sin n\theta$.

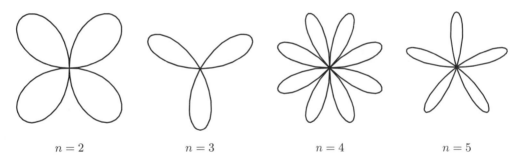

| $n = 2$ | $n = 3$ | $n = 4$ | $n = 5$ |

From the graphs, it seems that when n is even, the number of loops in the curve (called a rose) is $2n$, and when n is odd, the number of loops is simply n. This is because in the case of n odd, every point on the graph is traversed twice, due to the fact that

$$r(\theta + \pi) = \sin[n(\theta + \pi)] = \sin n\theta \cos n\pi + \cos n\theta \sin n\pi = \begin{cases} \sin n\theta & \text{if } n \text{ is even} \\ -\sin n\theta & \text{if } n \text{ is odd} \end{cases}$$

(b) The graph of $r = |\sin n\theta|$ has $2n$ loops whether n is odd or even, since $r(\theta + \pi) = r(\theta)$.

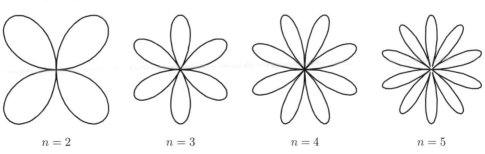

| $n = 2$ | $n = 3$ | $n = 4$ | $n = 5$ |

81. $r = \dfrac{1 - a\cos\theta}{1 + a\cos\theta}$. We start with $a = 0$, since in this case the curve is simply the circle $r = 1$.

As a increases, the graph moves to the left, and its right side becomes flattened. As a increases through about 0.4, the right side seems to grow a dimple, which upon closer investigation (with narrower θ-ranges) seems to appear at $a \approx 0.42$ [the actual value is $\sqrt{2} - 1$]. As $a \to 1$, this dimple becomes more pronounced, and the curve begins to stretch out horizontally, until at $a = 1$ the denominator vanishes at $\theta = \pi$, and the dimple becomes an actual cusp. For $a > 1$ we must choose our parameter interval carefully, since $r \to \infty$ as $1 + a\cos\theta \to 0 \iff \theta \to \pm\cos^{-1}(-1/a)$. As a increases from 1, the curve splits into two parts. The left part has a loop, which grows larger as a increases, and the right part grows broader vertically, and its left tip develops a dimple when $a \approx 2.42$ [actually, $\sqrt{2} + 1$]. As a increases, the dimple grows more and more pronounced. If $a < 0$, we get the same graph as we do for the corresponding positive a-value, but with a rotation through π about the pole, as happened when c was replaced with $-c$ in Exercise 80.

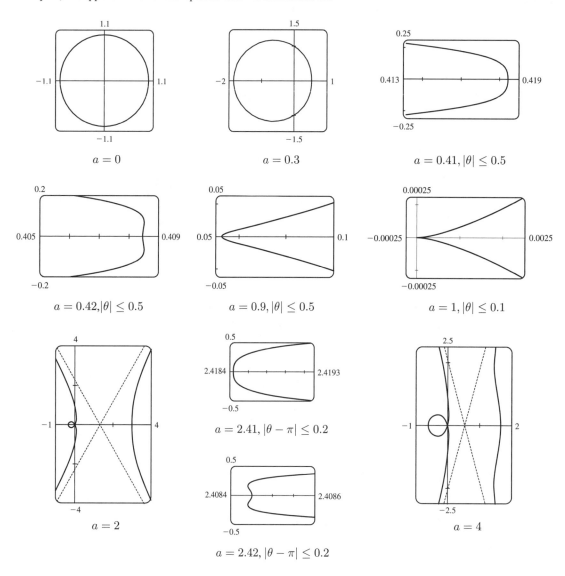

$a = 0$

$a = 0.3$

$a = 0.41, |\theta| \leq 0.5$

$a = 0.42, |\theta| \leq 0.5$

$a = 0.9, |\theta| \leq 0.5$

$a = 1, |\theta| \leq 0.1$

$a = 2$

$a = 2.41, |\theta - \pi| \leq 0.2$

$a = 4$

$a = 2.42, |\theta - \pi| \leq 0.2$

83. $\tan \psi = \tan(\phi - \theta) = \dfrac{\tan \phi - \tan \theta}{1 + \tan \phi \tan \theta} = \dfrac{\dfrac{dy}{dx} - \tan \theta}{1 + \dfrac{dy}{dx} \tan \theta} = \dfrac{\dfrac{dy/d\theta}{dx/d\theta} - \tan \theta}{1 + \dfrac{dy/d\theta}{dx/d\theta} \tan \theta}$

$= \dfrac{\dfrac{dy}{d\theta} - \dfrac{dx}{d\theta} \tan \theta}{\dfrac{dx}{d\theta} + \dfrac{dy}{d\theta} \tan \theta} = \dfrac{\left(\dfrac{dr}{d\theta} \sin \theta + r \cos \theta\right) - \tan \theta \left(\dfrac{dr}{d\theta} \cos \theta - r \sin \theta\right)}{\left(\dfrac{dr}{d\theta} \cos \theta - r \sin \theta\right) + \tan \theta \left(\dfrac{dr}{d\theta} \sin \theta + r \cos \theta\right)} = \dfrac{r \cos \theta + r \cdot \dfrac{\sin^2 \theta}{\cos \theta}}{\dfrac{dr}{d\theta} \cos \theta + \dfrac{dr}{d\theta} \cdot \dfrac{\sin^2 \theta}{\cos \theta}}$

$= \dfrac{r \cos^2 \theta + r \sin^2 \theta}{\dfrac{dr}{d\theta} \cos^2 \theta + \dfrac{dr}{d\theta} \sin^2 \theta} = \dfrac{r}{dr/d\theta}$

10.4 Areas and Lengths in Polar Coordinates

1. $r = \theta^2$, $0 \le \theta \le \frac{\pi}{4}$. $A = \displaystyle\int_0^{\pi/4} \frac{1}{2} r^2 \, d\theta = \int_0^{\pi/4} \frac{1}{2}(\theta^2)^2 \, d\theta = \int_0^{\pi/4} \frac{1}{2} \theta^4 \, d\theta = \left[\frac{1}{10}\theta^5\right]_0^{\pi/4} = \frac{1}{10}\left(\frac{\pi}{4}\right)^5 = \frac{1}{10,240} \pi^5$

3. $r = \sin \theta$, $\frac{\pi}{3} \le \theta \le \frac{2\pi}{3}$.

$A = \displaystyle\int_{\pi/3}^{2\pi/3} \frac{1}{2} \sin^2 \theta \, d\theta = \frac{1}{4} \int_{\pi/3}^{2\pi/3} (1 - \cos 2\theta) \, d\theta = \frac{1}{4}\left[\theta - \frac{1}{2}\sin 2\theta\right]_{\pi/3}^{2\pi/3} = \frac{1}{4}\left[\frac{2\pi}{3} - \frac{1}{2}\sin\frac{4\pi}{3} - \frac{\pi}{3} + \frac{1}{2}\sin\frac{2\pi}{3}\right]$

$= \frac{1}{4}\left[\frac{2\pi}{3} - \frac{1}{2}\left(-\frac{\sqrt{3}}{2}\right) - \frac{\pi}{3} + \frac{1}{2}\left(\frac{\sqrt{3}}{2}\right)\right] = \frac{1}{4}\left(\frac{\pi}{3} + \frac{\sqrt{3}}{2}\right) = \frac{\pi}{12} + \frac{\sqrt{3}}{8}$

5. $r = \sqrt{\theta}$, $0 \le \theta \le 2\pi$. $A = \displaystyle\int_0^{2\pi} \frac{1}{2} r^2 \, d\theta = \int_0^{2\pi} \frac{1}{2}\left(\sqrt{\theta}\right)^2 d\theta = \int_0^{2\pi} \frac{1}{2}\theta \, d\theta = \left[\frac{1}{4}\theta^2\right]_0^{2\pi} = \pi^2$

7. $r = 4 + 3\sin \theta$, $-\frac{\pi}{2} \le \theta \le \frac{\pi}{2}$.

$A = \displaystyle\int_{-\pi/2}^{\pi/2} \frac{1}{2}((4 + 3\sin \theta)^2 \, d\theta = \frac{1}{2}\int_{-\pi/2}^{\pi/2} (16 + 24\sin \theta + 9\sin^2 \theta) \, d\theta = \frac{1}{2}\int_{-\pi/2}^{\pi/2} (16 + 9\sin^2 \theta) \, d\theta$ [by Theorem 5.5.7(b)]

$= \frac{1}{2} \cdot 2 \displaystyle\int_0^{\pi/2} \left[16 + 9 \cdot \frac{1}{2}(1 - \cos 2\theta)\right] d\theta$ [by Theorem 5.5.7(a)]

$= \displaystyle\int_0^{\pi/2} \left(\frac{41}{2} - \frac{9}{2}\cos 2\theta\right) d\theta = \left[\frac{41}{2}\theta - \frac{9}{4}\sin 2\theta\right]_0^{\pi/2} = \left(\frac{41\pi}{4} - 0\right) - (0 - 0) = \frac{41\pi}{4}$

9. The area above the polar axis is bounded by $r = 3\cos \theta$ for $\theta = 0$

to $\theta = \pi/2$ [*not* π]. By symmetry,

$A = 2\int_0^{\pi/2} \frac{1}{2}r^2 \, d\theta = \int_0^{\pi/2}(3\cos \theta)^2 \, d\theta = 3^2 \int_0^{\pi/2} \cos^2 \theta \, d\theta$

$= 9\int_0^{\pi/2} \frac{1}{2}(1 + \cos 2\theta) \, d\theta = \frac{9}{2}\left[\theta + \frac{1}{2}\sin 2\theta\right]_0^{\pi/2} = \frac{9}{2}\left[\left(\frac{\pi}{2} + 0\right) - (0 + 0)\right] = \frac{9\pi}{4}$

Also, note that this is a circle with radius $\frac{3}{2}$, so its area is $\pi\left(\frac{3}{2}\right)^2 = \frac{9\pi}{4}$.

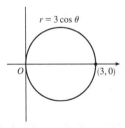

11. The curve goes through the pole when $\theta = \pi/4$, so we'll find the area for

$0 \le \theta \le \pi/4$ and multiply it by 4.

$A = 4\int_0^{\pi/4} \frac{1}{2}r^2 \, d\theta = 2\int_0^{\pi/4}(4\cos 2\theta) \, d\theta$

$= 8\int_0^{\pi/4} \cos 2\theta \, d\theta = 4\left[\sin 2\theta\right]_0^{\pi/4} = 4$

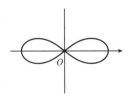

13. One-sixth of the area lies above the polar axis and is bounded by the curve
$r = 2\cos 3\theta$ for $\theta = 0$ to $\theta = \pi/6$.

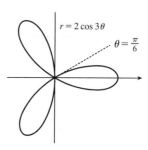

$r = 2\cos 3\theta$

$\theta = \frac{\pi}{6}$

$$A = 6 \int_0^{\pi/6} \tfrac{1}{2}(2\cos 3\theta)^2 \, d\theta = 12 \int_0^{\pi/6} \cos^2 3\theta \, d\theta$$

$$= \tfrac{12}{2} \int_0^{\pi/6}(1 + \cos 6\theta) \, d\theta$$

$$= 6\big[\theta + \tfrac{1}{6}\sin 6\theta\big]_0^{\pi/6} = 6\big(\tfrac{\pi}{6}\big) = \pi$$

15. $A = \int_0^{2\pi} \tfrac{1}{2}(1 + 2\sin 6\theta)^2 \, d\theta = \tfrac{1}{2} \int_0^{2\pi}(1 + 4\sin 6\theta + 4\sin^2 6\theta) \, d\theta$

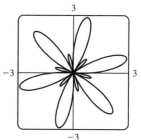

$$= \tfrac{1}{2} \int_0^{2\pi}\big[1 + 4\sin 6\theta + 4 \cdot \tfrac{1}{2}(1 - \cos 12\theta)\big] d\theta$$

$$= \tfrac{1}{2} \int_0^{2\pi}(3 + 4\sin 6\theta - 2\cos 12\theta) \, d\theta$$

$$= \tfrac{1}{2}\big[3\theta - \tfrac{2}{3}\cos 6\theta - \tfrac{1}{6}\sin 12\theta\big]_0^{2\pi}$$

$$= \tfrac{1}{2}\big[(6\pi - \tfrac{2}{3} - 0) - (0 - \tfrac{2}{3} - 0)\big] = 3\pi$$

17. The shaded loop is traced out from $\theta = 0$ to $\theta = \pi/2$.

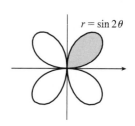

$r = \sin 2\theta$

$$A = \int_0^{\pi/2} \tfrac{1}{2}r^2 \, d\theta = \tfrac{1}{2} \int_0^{\pi/2} \sin^2 2\theta \, d\theta$$

$$= \tfrac{1}{2} \int_0^{\pi/2} \tfrac{1}{2}(1 - \cos 4\theta) \, d\theta = \tfrac{1}{4}\big[\theta - \tfrac{1}{4}\sin 4\theta\big]_0^{\pi/2}$$

$$= \tfrac{1}{4}\big(\tfrac{\pi}{2}\big) = \tfrac{\pi}{8}$$

19. $r = 0 \;\Rightarrow\; 3\cos 5\theta = 0 \;\Rightarrow\; 5\theta = \tfrac{\pi}{2} \;\Rightarrow\; \theta = \tfrac{\pi}{10}.$

$$A = \int_{-\pi/10}^{\pi/10} \tfrac{1}{2}(3\cos 5\theta)^2 \, d\theta = \int_0^{\pi/10} 9\cos^2 5\theta \, d\theta = \tfrac{9}{2} \int_0^{\pi/10}(1 + \cos 10\theta) \, d\theta = \tfrac{9}{2}\big[\theta + \tfrac{1}{10}\sin 10\theta\big]_0^{\pi/10} = \tfrac{9\pi}{20}$$

21.

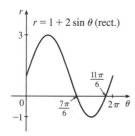

$r = 1 + 2\sin\theta$ (rect.)

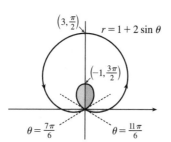

$\big(3, \tfrac{\pi}{2}\big)$

$r = 1 + 2\sin\theta$

$\big(-1, \tfrac{3\pi}{2}\big)$

$\theta = \tfrac{7\pi}{6}$ $\theta = \tfrac{11\pi}{6}$

This is a limaçon, with inner loop traced out between $\theta = \tfrac{7\pi}{6}$ and $\tfrac{11\pi}{6}$ [found by solving $r = 0$].

$$A = 2 \int_{7\pi/6}^{3\pi/2} \tfrac{1}{2}(1 + 2\sin\theta)^2 \, d\theta = \int_{7\pi/6}^{3\pi/2}\big(1 + 4\sin\theta + 4\sin^2\theta\big) \, d\theta = \int_{7\pi/6}^{3\pi/2}\big[1 + 4\sin\theta + 4 \cdot \tfrac{1}{2}(1 - \cos 2\theta)\big] d\theta$$

$$= \big[\theta - 4\cos\theta + 2\theta - \sin 2\theta\big]_{7\pi/6}^{3\pi/2} = \big(\tfrac{9\pi}{2}\big) - \big(\tfrac{7\pi}{2} + 2\sqrt{3} - \tfrac{\sqrt{3}}{2}\big) = \pi - \tfrac{3\sqrt{3}}{2}$$

23. $2\cos\theta = 1 \;\Rightarrow\; \cos\theta = \tfrac{1}{2} \;\Rightarrow\; \theta = \tfrac{\pi}{3} \text{ or } \tfrac{5\pi}{3}.$

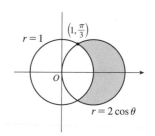

$r = 1$

$\big(1, \tfrac{\pi}{3}\big)$

O

$r = 2\cos\theta$

$$A = 2 \int_0^{\pi/3} \tfrac{1}{2}\big[(2\cos\theta)^2 - 1^2\big] \, d\theta = \int_0^{\pi/3}(4\cos^2\theta - 1) \, d\theta$$

$$= \int_0^{\pi/3}\big\{4\big[\tfrac{1}{2}(1 + \cos 2\theta)\big] - 1\big\} \, d\theta = \int_0^{\pi/3}(1 + 2\cos 2\theta) \, d\theta$$

$$= \big[\theta + \sin 2\theta\big]_0^{\pi/3} = \tfrac{\pi}{3} + \tfrac{\sqrt{3}}{2}$$

25. To find the area inside the leminiscate $r^2 = 8\cos 2\theta$ and outside the circle $r = 2$, we first note that the two curves intersect when $r^2 = 8\cos 2\theta$ and $r = 2$, that is, when $\cos 2\theta = \frac{1}{2}$. For $-\pi < \theta \le \pi$, $\cos 2\theta = \frac{1}{2}$ $\Leftrightarrow$ $2\theta = \pm\pi/3$ or $\pm 5\pi/3$ $\Leftrightarrow$ $\theta = \pm\pi/6$ or $\pm 5\pi/6$. The figure shows that the desired area is 4 times the area between the curves from 0 to $\pi/6$. Thus,

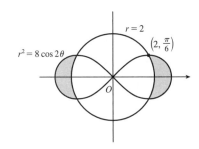

$$A = 4\int_0^{\pi/6} \left[\tfrac{1}{2}(8\cos 2\theta) - \tfrac{1}{2}(2)^2\right] d\theta = 8\int_0^{\pi/6}(2\cos 2\theta - 1)\, d\theta$$

$$= 8\left[\sin 2\theta - \theta\right]_0^{\pi/6} = 8\left(\sqrt{3}/2 - \pi/6\right) = 4\sqrt{3} - 4\pi/3$$

27. $3\cos\theta = 1 + \cos\theta$ $\Leftrightarrow$ $\cos\theta = \frac{1}{2}$ $\Rightarrow$ $\theta = \frac{\pi}{3}$ or $-\frac{\pi}{3}$.

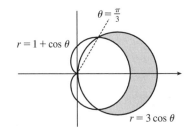

$$A = 2\int_0^{\pi/3} \tfrac{1}{2}\left[(3\cos\theta)^2 - (1+\cos\theta)^2\right] d\theta$$

$$= \int_0^{\pi/3}(8\cos^2\theta - 2\cos\theta - 1)\, d\theta = \int_0^{\pi/3}\left[4(1+\cos 2\theta) - 2\cos\theta - 1\right] d\theta$$

$$= \int_0^{\pi/3}(3 + 4\cos 2\theta - 2\cos\theta)\, d\theta = \left[3\theta + 2\sin 2\theta - 2\sin\theta\right]_0^{\pi/3}$$

$$= \pi + \sqrt{3} - \sqrt{3} = \pi$$

29. $\sqrt{3}\cos\theta = \sin\theta$ $\Rightarrow$ $\sqrt{3} = \dfrac{\sin\theta}{\cos\theta}$ $\Rightarrow$ $\tan\theta = \sqrt{3}$ $\Rightarrow$ $\theta = \frac{\pi}{3}$.

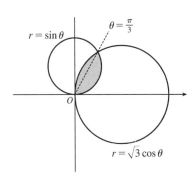

$$A = \int_0^{\pi/3} \tfrac{1}{2}(\sin\theta)^2\, d\theta + \int_{\pi/3}^{\pi/2} \tfrac{1}{2}\left(\sqrt{3}\cos\theta\right)^2 d\theta$$

$$= \int_0^{\pi/3} \tfrac{1}{2}\cdot\tfrac{1}{2}(1 - \cos 2\theta)\, d\theta + \int_{\pi/3}^{\pi/2} \tfrac{1}{2}\cdot 3\cdot\tfrac{1}{2}(1 + \cos 2\theta)\, d\theta$$

$$= \tfrac{1}{4}\left[\theta - \tfrac{1}{2}\sin 2\theta\right]_0^{\pi/3} + \tfrac{3}{4}\left[\theta + \tfrac{1}{2}\sin 2\theta\right]_{\pi/3}^{\pi/2}$$

$$= \tfrac{1}{4}\left[\left(\tfrac{\pi}{3} - \tfrac{\sqrt{3}}{4}\right) - 0\right] + \tfrac{3}{4}\left[\left(\tfrac{\pi}{2} + 0\right) - \left(\tfrac{\pi}{3} + \tfrac{\sqrt{3}}{4}\right)\right]$$

$$= \tfrac{\pi}{12} - \tfrac{\sqrt{3}}{16} + \tfrac{\pi}{8} - \tfrac{3\sqrt{3}}{16} = \tfrac{5\pi}{24} - \tfrac{\sqrt{3}}{4}$$

31. $\sin 2\theta = \cos 2\theta$ $\Rightarrow$ $\dfrac{\sin 2\theta}{\cos 2\theta} = 1$ $\Rightarrow$ $\tan 2\theta = 1$ $\Rightarrow$ $2\theta = \frac{\pi}{4}$ $\Rightarrow$ $\theta = \frac{\pi}{8}$ $\Rightarrow$

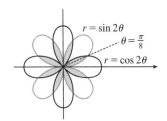

$$A = 8\cdot 2\int_0^{\pi/8} \tfrac{1}{2}\sin^2 2\theta\, d\theta = 8\int_0^{\pi/8} \tfrac{1}{2}(1 - \cos 4\theta)\, d\theta$$

$$= 4\left[\theta - \tfrac{1}{4}\sin 4\theta\right]_0^{\pi/8} = 4\left(\tfrac{\pi}{8} - \tfrac{1}{4}\cdot 1\right) = \tfrac{\pi}{2} - 1$$

33. $\sin 2\theta = \cos 2\theta$ $\Rightarrow$ $\tan 2\theta = 1$ $\Rightarrow$ $2\theta = \frac{\pi}{4}$ $\Rightarrow$ $\theta = \frac{\pi}{8}$

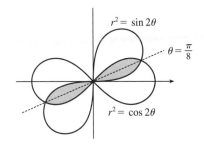

$$A = 4\int_0^{\pi/8} \tfrac{1}{2}\sin 2\theta\, d\theta \quad \left[\text{since } r^2 = \sin 2\theta\right]$$

$$= \int_0^{\pi/8} 2\sin 2\theta\, d\theta = \left[-\cos 2\theta\right]_0^{\pi/8}$$

$$= -\tfrac{1}{2}\sqrt{2} - (-1) = 1 - \tfrac{1}{2}\sqrt{2}$$

35. The darker shaded region (from $\theta = 0$ to $\theta = 2\pi/3$) represents $\frac{1}{2}$ of the desired area plus $\frac{1}{2}$ of the area of the inner loop. From this area, we'll subtract $\frac{1}{2}$ of the area of the inner loop (the lighter shaded region from $\theta = 2\pi/3$ to $\theta = \pi$), and then double that difference to obtain the desired area.

$$A = 2\left[\int_0^{2\pi/3} \frac{1}{2}\left(\frac{1}{2} + \cos\theta\right)^2 d\theta - \int_{2\pi/3}^{\pi} \frac{1}{2}\left(\frac{1}{2} + \cos\theta\right)^2 d\theta\right]$$

$$= \int_0^{2\pi/3} \left(\frac{1}{4} + \cos\theta + \cos^2\theta\right) d\theta - \int_{2\pi/3}^{\pi} \left(\frac{1}{4} + \cos\theta + \cos^2\theta\right) d\theta$$

$$= \int_0^{2\pi/3} \left[\frac{1}{4} + \cos\theta + \frac{1}{2}(1 + \cos 2\theta)\right] d\theta$$

$$\qquad - \int_{2\pi/3}^{\pi} \left[\frac{1}{4} + \cos\theta + \frac{1}{2}(1 + \cos 2\theta)\right] d\theta$$

$$= \left[\frac{\theta}{4} + \sin\theta + \frac{\theta}{2} + \frac{\sin 2\theta}{4}\right]_0^{2\pi/3} - \left[\frac{\theta}{4} + \sin\theta + \frac{\theta}{2} + \frac{\sin 2\theta}{4}\right]_{2\pi/3}^{\pi}$$

$$= \left(\frac{\pi}{6} + \frac{\sqrt{3}}{2} + \frac{\pi}{3} - \frac{\sqrt{3}}{8}\right) - \left(\frac{\pi}{4} + \frac{\pi}{2}\right) + \left(\frac{\pi}{6} + \frac{\sqrt{3}}{2} + \frac{\pi}{3} - \frac{\sqrt{3}}{8}\right)$$

$$= \frac{\pi}{4} + \frac{3}{4}\sqrt{3} = \frac{1}{4}\left(\pi + 3\sqrt{3}\right)$$

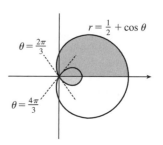

37. The pole is a point of intersection.

$$1 + \sin\theta = 3\sin\theta \quad\Rightarrow\quad 1 = 2\sin\theta \quad\Rightarrow\quad \sin\theta = \frac{1}{2} \quad\Rightarrow$$

$$\theta = \frac{\pi}{6} \text{ or } \frac{5\pi}{6}.$$

The other two points of intersection are $\left(\frac{3}{2}, \frac{\pi}{6}\right)$ and $\left(\frac{3}{2}, \frac{5\pi}{6}\right)$.

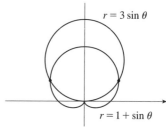

39. $2\sin 2\theta = 1 \quad\Rightarrow\quad \sin 2\theta = \frac{1}{2} \quad\Rightarrow\quad 2\theta = \frac{\pi}{6}, \frac{5\pi}{6}, \frac{13\pi}{6}, \text{ or } \frac{17\pi}{6}.$

By symmetry, the eight points of intersection are given by

$(1, \theta)$, where $\theta = \frac{\pi}{12}, \frac{5\pi}{12}, \frac{13\pi}{12}$, and $\frac{17\pi}{12}$, and

$(-1, \theta)$, where $\theta = \frac{7\pi}{12}, \frac{11\pi}{12}, \frac{19\pi}{12}$, and $\frac{23\pi}{12}$.

[There are many ways to describe these points.]

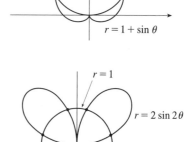

41. The pole is a point of intersection. $\sin\theta = \sin 2\theta = 2\sin\theta\cos\theta \quad\Leftrightarrow$

$\sin\theta\,(1 - 2\cos\theta) = 0 \quad\Leftrightarrow\quad \sin\theta = 0 \text{ or } \cos\theta = \frac{1}{2} \quad\Rightarrow$

$\theta = 0, \pi, \frac{\pi}{3}, \text{ or } -\frac{\pi}{3} \quad\Rightarrow\quad$ the other intersection points are $\left(\frac{\sqrt{3}}{2}, \frac{\pi}{3}\right)$

and $\left(\frac{\sqrt{3}}{2}, \frac{2\pi}{3}\right)$ [by symmetry].

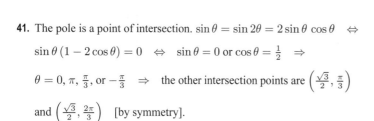
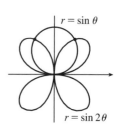

43.

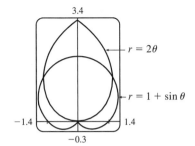

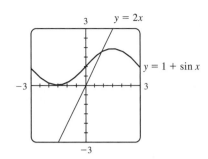

From the first graph, we see that the pole is one point of intersection. By zooming in or using the cursor, we find the θ-values of the intersection points to be $\alpha \approx 0.88786 \approx 0.89$ and $\pi - \alpha \approx 2.25$. (The first of these values may be more easily estimated by plotting $y = 1 + \sin x$ and $y = 2x$ in rectangular coordinates; see the second graph.) By symmetry, the total area contained is twice the area contained in the first quadrant, that is,

$$A = 2\int_0^\alpha \tfrac{1}{2}(2\theta)^2 \, d\theta + 2\int_\alpha^{\pi/2} \tfrac{1}{2}(1+\sin\theta)^2 \, d\theta = \int_0^\alpha 4\theta^2 \, d\theta + \int_\alpha^{\pi/2} \left[1 + 2\sin\theta + \tfrac{1}{2}(1-\cos 2\theta)\right] d\theta$$

$$= \left[\tfrac{4}{3}\theta^3\right]_0^\alpha + \left[\theta - 2\cos\theta + \left(\tfrac{1}{2}\theta - \tfrac{1}{4}\sin 2\theta\right)\right]_\alpha^{\pi/2} = \tfrac{4}{3}\alpha^3 + \left[\left(\tfrac{\pi}{2} + \tfrac{\pi}{4}\right) - \left(\alpha - 2\cos\alpha + \tfrac{1}{2}\alpha - \tfrac{1}{4}\sin 2\alpha\right)\right] \approx 3.4645$$

45. $L = \displaystyle\int_a^b \sqrt{r^2 + (dr/d\theta)^2} \, d\theta = \int_0^{\pi/3} \sqrt{(3\sin\theta)^2 + (3\cos\theta)^2} \, d\theta = \int_0^{\pi/3} \sqrt{9(\sin^2\theta + \cos^2\theta)} \, d\theta$

$\qquad = 3\displaystyle\int_0^{\pi/3} d\theta = 3\big[\theta\big]_0^{\pi/3} = 3\left(\tfrac{\pi}{3}\right) = \pi.$

As a check, note that the circumference of a circle with radius $\tfrac{3}{2}$ is $2\pi\left(\tfrac{3}{2}\right) = 3\pi$, and since $\theta = 0$ to $\pi = \tfrac{\pi}{3}$ traces out $\tfrac{1}{3}$ of the circle (from $\theta = 0$ to $\theta = \pi$), $\tfrac{1}{3}(3\pi) = \pi.$

47. $L = \displaystyle\int_a^b \sqrt{r^2 + (dr/d\theta)^2} \, d\theta = \int_0^{2\pi} \sqrt{(\theta^2)^2 + (2\theta)^2} \, d\theta = \int_0^{2\pi} \sqrt{\theta^4 + 4\theta^2} \, d\theta$

$\qquad = \displaystyle\int_0^{2\pi} \sqrt{\theta^2(\theta^2 + 4)} \, d\theta = \int_0^{2\pi} \theta\sqrt{\theta^2 + 4} \, d\theta$

Now let $u = \theta^2 + 4$, so that $du = 2\theta \, d\theta \quad \left[\theta \, d\theta = \tfrac{1}{2}\, du\right]$ and

$$\int_0^{2\pi} \theta\sqrt{\theta^2 + 4}\, d\theta = \int_4^{4\pi^2 + 4} \tfrac{1}{2}\sqrt{u}\, du = \tfrac{1}{2}\cdot\tfrac{2}{3}\left[u^{3/2}\right]_4^{4(\pi^2 + 1)} = \tfrac{1}{3}[4^{3/2}(\pi^2+1)^{3/2} - 4^{3/2}] = \tfrac{8}{3}[(\pi^2+1)^{3/2} - 1]$$

49. The curve $r = 3\sin 2\theta$ is completely traced with $0 \le \theta \le 2\pi$. $r^2 + \left(\tfrac{dr}{d\theta}\right)^2 = (3\sin 2\theta)^2 + (6\cos 2\theta)^2 \quad \Rightarrow$

$\qquad L = \int_0^{2\pi} \sqrt{9\sin^2 2\theta + 36\cos^2 2\theta}\, d\theta \approx 29.0653$

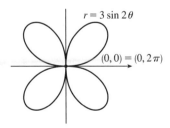

$r = 3\sin 2\theta$

$(0, 0) = (0, 2\pi)$

51. The curve $r = \sin\left(\frac{\theta}{2}\right)$ is completely traced with $0 \le \theta \le 4\pi$. $r^2 + \left(\frac{dr}{d\theta}\right)^2 = \sin^2\left(\frac{\theta}{2}\right) + \left[\frac{1}{2}\cos\left(\frac{\theta}{2}\right)\right]^2 \Rightarrow$

$L = \displaystyle\int_0^{4\pi} \sqrt{\sin^2\left(\frac{\theta}{2}\right) + \frac{1}{4}\cos^2\left(\frac{\theta}{2}\right)} \, d\theta \approx 9.6884$

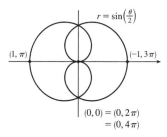

53. The curve $r = \cos^4(\theta/4)$ is completely traced with $0 \le \theta \le 4\pi$.

$$r^2 + (dr/d\theta)^2 = [\cos^4(\theta/4)]^2 + \left[4\cos^3(\theta/4)\cdot(-\sin(\theta/4))\cdot\frac{1}{4}\right]^2$$

$$= \cos^8(\theta/4) + \cos^6(\theta/4)\sin^2(\theta/4)$$

$$= \cos^6(\theta/4)[\cos^2(\theta/4) + \sin^2(\theta/4)] = \cos^6(\theta/4)$$

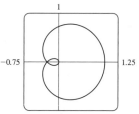

$L = \int_0^{4\pi} \sqrt{\cos^6(\theta/4)} \, d\theta = \int_0^{4\pi} \left|\cos^3(\theta/4)\right| \, d\theta$

$= 2\int_0^{2\pi} \cos^3(\theta/4) \, d\theta$ [since $\cos^3(\theta/4) \ge 0$ for $0 \le \theta \le 2\pi$] $= 8\int_0^{\pi/2} \cos^3 u \, du$ $\left[u = \frac{1}{4}\theta\right]$

$\overset{68}{=} 8\left[\frac{1}{3}(2 + \cos^2 u)\sin u\right]_0^{\pi/2} = \frac{8}{3}[(2\cdot 1) - (3\cdot 0)] = \frac{16}{3}$

55. (a) From (10.2.7),

$$S = \int_a^b 2\pi y \sqrt{(dx/d\theta)^2 + (dy/d\theta)^2} \, d\theta$$

$$= \int_a^b 2\pi y \sqrt{r^2 + (dr/d\theta)^2} \, d\theta \qquad \text{[from the derivation of Equation 10.4.5]}$$

$$= \int_a^b 2\pi r \sin\theta \sqrt{r^2 + (dr/d\theta)^2} \, d\theta$$

(b) The curve $r^2 = \cos 2\theta$ goes through the pole when $\cos 2\theta = 0 \Rightarrow$

$2\theta = \frac{\pi}{2} \Rightarrow \theta = \frac{\pi}{4}$. We'll rotate the curve from $\theta = 0$ to $\theta = \frac{\pi}{4}$ and double

this value to obtain the total surface area generated.

$r^2 = \cos 2\theta \Rightarrow 2r\dfrac{dr}{d\theta} = -2\sin 2\theta \Rightarrow \left(\dfrac{dr}{d\theta}\right)^2 = \dfrac{\sin^2 2\theta}{r^2} = \dfrac{\sin^2 2\theta}{\cos 2\theta}$.

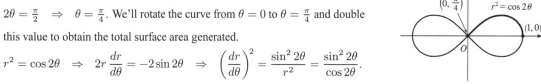

$S = 2\displaystyle\int_0^{\pi/4} 2\pi \sqrt{\cos 2\theta} \sin\theta \sqrt{\cos 2\theta + (\sin^2 2\theta)/\cos 2\theta} \, d\theta = 4\pi\int_0^{\pi/4} \sqrt{\cos 2\theta}\sin\theta \sqrt{\dfrac{\cos^2 2\theta + \sin^2 2\theta}{\cos 2\theta}} \, d\theta$

$= 4\pi\displaystyle\int_0^{\pi/4} \sqrt{\cos 2\theta}\sin\theta \dfrac{1}{\sqrt{\cos 2\theta}} \, d\theta = 4\pi\int_0^{\pi/4} \sin\theta \, d\theta = 4\pi\left[-\cos\theta\right]_0^{\pi/4} = -4\pi\left(\dfrac{\sqrt{2}}{2} - 1\right) = 2\pi\left(2 - \sqrt{2}\right)$

10.5   Conic Sections

1. $x = 2y^2 \Rightarrow y^2 = \frac{1}{2}x.$ $4p = \frac{1}{2},$ so $p = \frac{1}{8}.$ The vertex is $(0, 0),$ the focus is $\left(\frac{1}{8}, 0\right),$ and the directrix is $x = -\frac{1}{8}.$

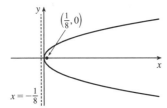

3. $4x^2 = -y \Rightarrow x^2 = -\frac{1}{4}y.$ $4p = -\frac{1}{4},$ so $p = -\frac{1}{16}.$ The vertex is $(0, 0),$ the focus is $\left(0, -\frac{1}{16}\right),$ and the directrix is $y = \frac{1}{16}.$

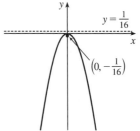

5. $(x+2)^2 = 8(y-3).$ $4p = 8,$ so $p = 2.$ The vertex is $(-2, 3),$ the focus is $(-2, 5),$ and the directrix is $y = 1.$

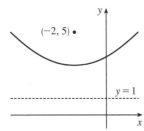

7. $y^2 + 2y + 12x + 25 = 0 \Rightarrow$
$y^2 + 2y + 1 = -12x - 24 \Rightarrow$
$(y+1)^2 = -12(x+2).$ $4p = -12,$ so $p = -3.$
The vertex is $(-2, -1),$ the focus is $(-5, -1),$ and the directrix is $x = 1.$

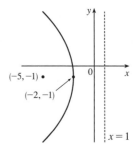

9. The equation has the form $y^2 = 4px,$ where $p < 0.$ Since the parabola passes through $(-1, 1),$ we have $1^2 = 4p(-1),$ so $4p = -1$ and an equation is $y^2 = -x$ or $x = -y^2.$ $4p = -1,$ so $p = -\frac{1}{4}$ and the focus is $\left(-\frac{1}{4}, 0\right)$ while the directrix is $x = \frac{1}{4}.$

11. $\dfrac{x^2}{9} + \dfrac{y^2}{5} = 1 \Rightarrow a = \sqrt{9} = 3, b = \sqrt{5},$
$c = \sqrt{a^2 - b^2} = \sqrt{9 - 5} = 2.$ The ellipse is centered at $(0, 0),$ with vertices at $(\pm 3, 0).$ The foci are $(\pm 2, 0).$

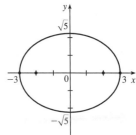

13. $4x^2 + y^2 = 16 \Rightarrow \dfrac{x^2}{4} + \dfrac{y^2}{16} = 1 \Rightarrow$

$a = \sqrt{16} = 4, b = \sqrt{4} = 2,$

$c = \sqrt{a^2 - b^2} = \sqrt{16 - 4} = 2\sqrt{3}.$ The ellipse is

centered at $(0,0)$, with vertices at $(0, \pm 4)$. The foci

are $\left(0, \pm 2\sqrt{3}\right)$.

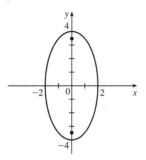

15. $9x^2 - 18x + 4y^2 = 27 \Leftrightarrow$

$9(x^2 - 2x + 1) + 4y^2 = 27 + 9 \Leftrightarrow$

$9(x-1)^2 + 4y^2 = 36 \Leftrightarrow \dfrac{(x-1)^2}{4} + \dfrac{y^2}{9} = 1 \Rightarrow$

$a = 3, b = 2, c = \sqrt{5} \Rightarrow$ center $(1,0)$,

vertices $(1, \pm 3)$, foci $\left(1, \pm\sqrt{5}\right)$

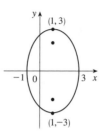

17. The center is $(0,0)$, $a = 3$, and $b = 2$, so an equation is $\dfrac{x^2}{4} + \dfrac{y^2}{9} = 1$. $c = \sqrt{a^2 - b^2} = \sqrt{5}$, so the foci are $\left(0, \pm\sqrt{5}\right)$.

19. $\dfrac{x^2}{144} - \dfrac{y^2}{25} = 1 \Rightarrow a = 12, b = 5, c = \sqrt{144 + 25} = 13 \Rightarrow$

center $(0,0)$, vertices $(\pm 12, 0)$, foci $(\pm 13, 0)$, asymptotes $y = \pm\frac{5}{12}x$.

Note: It is helpful to draw a *2a*-by-*2b* rectangle whose center is the center

of the hyperbola. The asymptotes are the extended diagonals of the

rectangle.

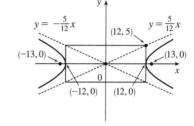

21. $y^2 - x^2 = 4 \Leftrightarrow \dfrac{y^2}{4} - \dfrac{x^2}{4} = 1 \Rightarrow a = \sqrt{4} = 2 = b,$

$c = \sqrt{4 + 4} = 2\sqrt{2} \Rightarrow$ center $(0,0)$, vertices $(0, \pm 2)$,

foci $\left(0, \pm 2\sqrt{2}\right)$, asymptotes $y = \pm x$

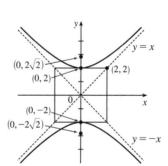

23. $4x^2 - y^2 - 24x - 4y + 28 = 0 \Leftrightarrow$

$4(x^2 - 6x + 9) - (y^2 + 4y + 4) = -28 + 36 - 4 \Leftrightarrow$

$4(x-3)^2 - (y+2)^2 = 4 \Leftrightarrow \dfrac{(x-3)^2}{1} - \dfrac{(y+2)^2}{4} = 1 \Rightarrow$

$a = \sqrt{1} = 1, b = \sqrt{4} = 2, c = \sqrt{1 + 4} = \sqrt{5} \Rightarrow$

center $(3, -2)$, vertices $(4, -2)$ and $(2, -2)$, foci $\left(3 \pm \sqrt{5}, -2\right)$,

asymptotes $y + 2 = \pm 2(x - 3)$.

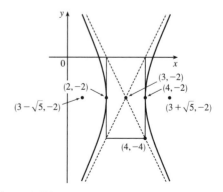

25. $x^2 = y + 1 \iff x^2 = 1(y + 1)$. This is an equation of a *parabola* with $4p = 1$, so $p = \frac{1}{4}$. The vertex is $(0, -1)$ and the

focus is $\left(0, -\frac{3}{4}\right)$.

27. $x^2 = 4y - 2y^2 \iff x^2 + 2y^2 - 4y = 0 \iff x^2 + 2(y^2 - 2y + 1) = 2 \iff x^2 + 2(y - 1)^2 = 2 \iff$

$\dfrac{x^2}{2} + \dfrac{(y-1)^2}{1} = 1$. This is an equation of an *ellipse* with vertices at $(\pm\sqrt{2}, 1)$. The foci are at $(\pm\sqrt{2-1}, 1) = (\pm 1, 1)$.

29. $y^2 + 2y = 4x^2 + 3 \iff y^2 + 2y + 1 = 4x^2 + 4 \iff (y+1)^2 - 4x^2 = 4 \iff \dfrac{(y+1)^2}{4} - x^2 = 1$. This is an equation

of a *hyperbola* with vertices $(0, -1 \pm 2) = (0, 1)$ and $(0, -3)$. The foci are at $\left(0, -1 \pm \sqrt{4+1}\right) = \left(0, -1 \pm \sqrt{5}\right)$.

31. The parabola with vertex $(0, 0)$ and focus $(0, -2)$ opens downward and has $p = -2$, so its equation is $x^2 = 4py = -8y$.

33. The distance from the focus $(-4, 0)$ to the directrix $x = 2$ is $2 - (-4) = 6$, so the distance from the focus to the vertex is

$\frac{1}{2}(6) = 3$ and the vertex is $(-1, 0)$. Since the focus is to the left of the vertex, $p = -3$. An equation is $y^2 = 4p(x + 1) \implies$

$y^2 = -12(x + 1)$.

35. A parabola with vertical axis and vertex $(2, 3)$ has equation $y - 3 = a(x - 2)^2$. Since it passes through $(1, 5)$, we have

$5 - 3 = a(1 - 2)^2 \implies a = 2$, so an equation is $y - 3 = 2(x - 2)^2$.

37. The ellipse with foci $(\pm 2, 0)$ and vertices $(\pm 5, 0)$ has center $(0, 0)$ and a horizontal major axis, with $a = 5$ and $c = 2$,

so $b^2 = a^2 - c^2 = 25 - 4 = 21$. An equation is $\dfrac{x^2}{25} + \dfrac{y^2}{21} = 1$.

39. Since the vertices are $(0, 0)$ and $(0, 8)$, the ellipse has center $(0, 4)$ with a vertical axis and $a = 4$. The foci at $(0, 2)$ and $(0, 6)$

are 2 units from the center, so $c = 2$ and $b = \sqrt{a^2 - c^2} = \sqrt{4^2 - 2^2} = \sqrt{12}$. An equation is $\dfrac{(x-0)^2}{b^2} + \dfrac{(y-4)^2}{a^2} = 1 \implies$

$\dfrac{x^2}{12} + \dfrac{(y-4)^2}{16} = 1$.

41. An equation of an ellipse with center $(-1, 4)$ and vertex $(-1, 0)$ is $\dfrac{(x+1)^2}{b^2} + \dfrac{(y-4)^2}{4^2} = 1$. The focus $(-1, 6)$ is 2 units

from the center, so $c = 2$. Thus, $b^2 + 2^2 = 4^2 \implies b^2 = 12$, and the equation is $\dfrac{(x+1)^2}{12} + \dfrac{(y-4)^2}{16} = 1$.

43. An equation of a hyperbola with vertices $(\pm 3, 0)$ is $\dfrac{x^2}{3^2} - \dfrac{y^2}{b^2} = 1$. Foci $(\pm 5, 0) \implies c = 5$ and $3^2 + b^2 = 5^2 \implies$

$b^2 = 25 - 9 = 16$, so the equation is $\dfrac{x^2}{9} - \dfrac{y^2}{16} = 1$.

45. The center of a hyperbola with vertices $(-3, -4)$ and $(-3, 6)$ is $(-3, 1)$, so $a = 5$ and an equation is

$\dfrac{(y-1)^2}{5^2} - \dfrac{(x+3)^2}{b^2} = 1$. Foci $(-3, -7)$ and $(-3, 9) \implies c = 8$, so $5^2 + b^2 = 8^2 \implies b^2 = 64 - 25 = 39$ and the

equation is $\dfrac{(y-1)^2}{25} - \dfrac{(x+3)^2}{39} = 1$.

47. The center of a hyperbola with vertices $(\pm 3, 0)$ is $(0,0)$, so $a = 3$ and an equation is $\dfrac{x^2}{3^2} - \dfrac{y^2}{b^2} = 1$.

Asymptotes $y = \pm 2x \;\Rightarrow\; \dfrac{b}{a} = 2 \;\Rightarrow\; b = 2(3) = 6$ and the equation is $\dfrac{x^2}{9} - \dfrac{y^2}{36} = 1$.

49. In Figure 8, we see that the point on the ellipse closest to a focus is the closer vertex (which is a distance

$a - c$ from it) while the farthest point is the other vertex (at a distance of $a + c$). So for this lunar orbit,

$(a - c) + (a + c) = 2a = (1728 + 110) + (1728 + 314)$, or $a = 1940$; and $(a + c) - (a - c) = 2c = 314 - 110$,

or $c = 102$. Thus, $b^2 = a^2 - c^2 = 3{,}753{,}196$, and the equation is $\dfrac{x^2}{3{,}763{,}600} + \dfrac{y^2}{3{,}753{,}196} = 1$.

51. (a) Set up the coordinate system so that A is $(-200, 0)$ and B is $(200, 0)$.

$|PA| - |PB| = (1200)(980) = 1{,}176{,}000 \text{ ft} = \frac{2450}{11} \text{ mi} = 2a \;\Rightarrow\; a = \frac{1225}{11}$, and $c = 200$ so

$b^2 = c^2 - a^2 = \dfrac{3{,}339{,}375}{121} \;\Rightarrow\; \dfrac{121x^2}{1{,}500{,}625} - \dfrac{121y^2}{3{,}339{,}375} = 1.$

(b) Due north of $B \;\Rightarrow\; x = 200 \;\Rightarrow\; \dfrac{(121)(200)^2}{1{,}500{,}625} - \dfrac{121y^2}{3{,}339{,}375} = 1 \;\Rightarrow\; y = \dfrac{133{,}575}{539} \approx 248 \text{ mi}$

53. The function whose graph is the upper branch of this hyperbola is concave upward. The function is

$y = f(x) = a\sqrt{1 + \dfrac{x^2}{b^2}} = \dfrac{a}{b}\sqrt{b^2 + x^2}$, so $y' = \dfrac{a}{b}x(b^2 + x^2)^{-1/2}$ and

$y'' = \dfrac{a}{b}\left[(b^2 + x^2)^{-1/2} - x^2(b^2 + x^2)^{-3/2}\right] = ab(b^2 + x^2)^{-3/2} > 0$ for all x, and so f is concave upward.

55. (a) If $k > 16$, then $k - 16 > 0$, and $\dfrac{x^2}{k} + \dfrac{y^2}{k - 16} = 1$ is an *ellipse* since it is the sum of two squares on the left side.

(b) If $0 < k < 16$, then $k - 16 < 0$, and $\dfrac{x^2}{k} + \dfrac{y^2}{k - 16} = 1$ is a *hyperbola* since it is the difference of two squares on the

left side.

(c) If $k < 0$, then $k - 16 < 0$, and there is *no curve* since the left side is the sum of two negative terms, which cannot equal 1.

(d) In case (a), $a^2 = k$, $b^2 = k - 16$, and $c^2 = a^2 - b^2 = 16$, so the foci are at $(\pm 4, 0)$. In case (b), $k - 16 < 0$, so $a^2 = k$,

$b^2 = 16 - k$, and $c^2 = a^2 + b^2 = 16$, and so again the foci are at $(\pm 4, 0)$.

57. $x^2 = 4py \;\Rightarrow\; 2x = 4py' \;\Rightarrow\; y' = \dfrac{x}{2p}$, so the tangent line at (x_0, y_0) is

$y - \dfrac{x_0^2}{4p} = \dfrac{x_0}{2p}(x - x_0)$. This line passes through the point $(a, -p)$ on the

directrix, so $-p - \dfrac{x_0^2}{4p} = \dfrac{x_0}{2p}(a - x_0) \;\Rightarrow\; -4p^2 - x_0^2 = 2ax_0 - 2x_0^2 \;\Leftrightarrow\;$

$x_0^2 - 2ax_0 - 4p^2 = 0 \;\Leftrightarrow\; x_0^2 - 2ax_0 + a^2 = a^2 + 4p^2 \;\Leftrightarrow\;$

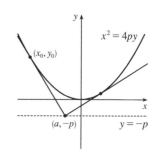

$(x_0 - a)^2 = a^2 + 4p^2 \iff x_0 = a \pm \sqrt{a^2 + 4p^2}$. The slopes of the tangent lines at $x = a \pm \sqrt{a^2 + 4p^2}$

are $\dfrac{a \pm \sqrt{a^2 + 4p^2}}{2p}$, so the product of the two slopes is

$$\frac{a + \sqrt{a^2 + 4p^2}}{2p} \cdot \frac{a - \sqrt{a^2 + 4p^2}}{2p} = \frac{a^2 - (a^2 + 4p^2)}{4p^2} = \frac{-4p^2}{4p^2} = -1,$$

showing that the tangent lines are perpendicular.

59. For $x^2 + 4y^2 = 4$, or $x^2/4 + y^2 = 1$, use the parametrization $x = 2\cos t$, $y = \sin t$, $0 \le t \le 2\pi$ to get

$$L = 4 \int_0^{\pi/2} \sqrt{(dx/dt)^2 + (dy/dt)^2}\, dt = 4 \int_0^{\pi/2} \sqrt{4\sin^2 t + \cos^2 t}\, dt = 4 \int_0^{\pi/2} \sqrt{3\sin^2 t + 1}\, dt$$

Using Simpson's Rule with $n = 10$, $\Delta t = \frac{\pi/2 - 0}{10} = \frac{\pi}{20}$, and $f(t) = \sqrt{3\sin^2 t + 1}$, we get

$$L \approx \frac{4}{3}\left(\frac{\pi}{20}\right)\left[f(0) + 4f\left(\frac{\pi}{20}\right) + 2f\left(\frac{2\pi}{20}\right) + \cdots + 2f\left(\frac{8\pi}{20}\right) + 4f\left(\frac{9\pi}{20}\right) + f\left(\frac{\pi}{2}\right)\right] \approx 9.69$$

61. $\dfrac{x^2}{a^2} - \dfrac{y^2}{b^2} = 1 \;\Rightarrow\; \dfrac{y^2}{b^2} = \dfrac{x^2 - a^2}{a^2} \;\Rightarrow\; y = \pm\dfrac{b}{a}\sqrt{x^2 - a^2}$.

$$A = 2\int_a^c \frac{b}{a}\sqrt{x^2 - a^2}\, dx \overset{39}{=} \frac{2b}{a}\left[\frac{x}{2}\sqrt{x^2 - a^2} - \frac{a^2}{2}\ln\left|x + \sqrt{x^2 - a^2}\right|\,\right]_a^c$$

$$= \frac{b}{a}\left[c\sqrt{c^2 - a^2} - a^2\ln\left|c + \sqrt{c^2 - a^2}\,\right| + a^2 \ln|a|\,\right]$$

Since $a^2 + b^2 = c^2$, $c^2 - a^2 = b^2$, and $\sqrt{c^2 - a^2} = b$.

$$= \frac{b}{a}\left[cb - a^2\ln(c + b) + a^2\ln a\right] = \frac{b}{a}\left[cb + a^2(\ln a - \ln(b + c))\right]$$

$$= b^2 c/a + ab\ln[a/(b + c)], \quad \text{where } c^2 = a^2 + b^2.$$

63. Differentiating implicitly, $\dfrac{x^2}{a^2} + \dfrac{y^2}{b^2} = 1 \;\Rightarrow\; \dfrac{2x}{a^2} + \dfrac{2yy'}{b^2} = 0 \;\Rightarrow\; y' = -\dfrac{b^2 x}{a^2 y}$ $[y \ne 0]$. Thus, the slope of the tangent

line at P is $-\dfrac{b^2 x_1}{a^2 y_1}$. The slope of $F_1 P$ is $\dfrac{y_1}{x_1 + c}$ and of $F_2 P$ is $\dfrac{y_1}{x_1 - c}$. By the formula in Problem 17 on text page 268,

we have

$$\tan\alpha = \frac{\dfrac{y_1}{x_1 + c} + \dfrac{b^2 x_1}{a^2 y_1}}{1 - \dfrac{b^2 x_1 y_1}{a^2 y_1(x_1 + c)}} = \frac{a^2 y_1^2 + b^2 x_1(x_1 + c)}{a^2 y_1(x_1 + c) - b^2 x_1 y_1} = \frac{a^2 b^2 + b^2 c x_1}{c^2 x_1 y_1 + a^2 c y_1} \qquad \left[\begin{array}{l}\text{using } b^2 x_1^2 + a^2 y_1^2 = a^2 b^2, \\ \text{and } a^2 - b^2 = c^2\end{array}\right]$$

$$= \frac{b^2\left(cx_1 + a^2\right)}{cy_1(cx_1 + a^2)} = \frac{b^2}{cy_1}$$

and $\quad \tan\beta = \dfrac{-\dfrac{b^2 x_1}{a^2 y_1} - \dfrac{y_1}{x_1 - c}}{1 - \dfrac{b^2 x_1 y_1}{a^2 y_1(x_1 - c)}} = \dfrac{-a^2 y_1^2 - b^2 x_1(x_1 - c)}{a^2 y_1(x_1 - c) - b^2 x_1 y_1} = \dfrac{-a^2 b^2 + b^2 c x_1}{c^2 x_1 y_1 - a^2 c y_1} = \dfrac{b^2\left(cx_1 - a^2\right)}{cy_1(cx_1 - a^2)} = \dfrac{b^2}{cy_1}$

Thus, $\alpha = \beta$.

10.6 Conic Sections in Polar Coordinates

1. The directrix $y = 6$ is above the focus at the origin, so we use the form with "$+ e \sin \theta$" in the denominator. [See Theorem 6

and Figure 2(c).] $r = \dfrac{ed}{1 + e \sin \theta} = \dfrac{\frac{7}{4} \cdot 6}{1 + \frac{7}{4} \sin \theta} = \dfrac{42}{4 + 7 \sin \theta}$

3. The directrix $x = -5$ is to the left of the focus at the origin, so we use the form with "$- e \cos \theta$" in the denominator.

$r = \dfrac{ed}{1 - e \cos \theta} = \dfrac{\frac{3}{4} \cdot 5}{1 - \frac{3}{4} \cos \theta} = \dfrac{15}{4 - 3 \cos \theta}$

5. The vertex $(4, 3\pi/2)$ is 4 units below the focus at the origin, so the directrix is 8 units below the focus ($d = 8$), and we use the

form with "$-e \sin \theta$" in the denominator. $e = 1$ for a parabola, so an equation is $r = \dfrac{ed}{1 - e \sin \theta} = \dfrac{1(8)}{1 - 1 \sin \theta} = \dfrac{8}{1 - \sin \theta}$.

7. The directrix $r = 4 \sec \theta$ (equivalent to $r \cos \theta = 4$ or $x = 4$) is to the right of the focus at the origin, so we will use the form

with "$+e \cos \theta$" in the denominator. The distance from the focus to the directrix is $d = 4$, so an equation is

$r = \dfrac{ed}{1 + e \cos \theta} = \dfrac{\frac{1}{2}(4)}{1 + \frac{1}{2} \cos \theta} \cdot \dfrac{2}{2} = \dfrac{4}{2 + \cos \theta}$.

9. $r = \dfrac{1}{1 + \sin \theta} = \dfrac{ed}{1 + e \sin \theta}$, where $d = e = 1$.

(a) Eccentricity $= e = 1$

(b) Since $e = 1$, the conic is a parabola.

(c) Since "$+e \sin \theta$" appears in the denominator, the directrix is above the

focus at the origin. $d = |Fl| = 1$, so an equation of the directrix is $y = 1$.

(d) The vertex is at $\left(\frac{1}{2}, \frac{\pi}{2}\right)$, midway between the focus and the directrix.

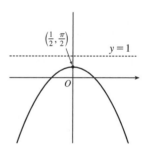

11. $r = \dfrac{12}{4 - \sin \theta} \cdot \dfrac{1/4}{1/4} = \dfrac{3}{1 - \frac{1}{4} \sin \theta}$, where $e = \frac{1}{4}$ and $ed = 3 \;\Rightarrow\; d = 12$.

(a) Eccentricity $= e = \frac{1}{4}$

(b) Since $e = \frac{1}{4} < 1$, the conic is an ellipse.

(c) Since "$-e \sin \theta$" appears in the denominator, the directrix is below the focus

at the origin. $d = |Fl| = 12$, so an equation of the directrix is $y = -12$.

(d) The vertices are $\left(4, \frac{\pi}{2}\right)$ and $\left(\frac{12}{5}, \frac{3\pi}{2}\right)$, so the center is midway between them,

that is, $\left(\frac{4}{5}, \frac{\pi}{2}\right)$.

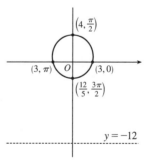

13. $r = \dfrac{9}{6 + 2\cos\theta} \cdot \dfrac{1/6}{1/6} = \dfrac{3/2}{1 + \frac{1}{3}\cos\theta}$, where $e = \frac{1}{3}$ and $ed = \frac{3}{2}$ $\Rightarrow$ $d = \frac{9}{2}$.

(a) Eccentricity $= e = \frac{1}{3}$

(b) Since $e = \frac{1}{3} < 1$, the conic is an ellipse.

(c) Since "$+e\cos\theta$" appears in the denominator, the directrix is to the right of

the focus at the origin. $d = |Fl| = \frac{9}{2}$, so an equation of the directrix is

$x = \frac{9}{2}$.

(d) The vertices are $\left(\frac{9}{8}, 0\right)$ and $\left(\frac{9}{4}, \pi\right)$, so the center is midway between them,

that is, $\left(\frac{9}{16}, \pi\right)$.

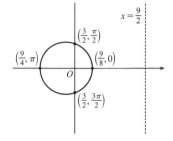

15. $r = \dfrac{3}{4 - 8\cos\theta} \cdot \dfrac{1/4}{1/4} = \dfrac{3/4}{1 - 2\cos\theta}$, where $e = 2$ and $ed = \frac{3}{4}$ $\Rightarrow$ $d = \frac{3}{8}$.

(a) Eccentricity $= e = 2$

(b) Since $e = 2 > 1$, the conic is a hyperbola.

(c) Since "$-e\cos\theta$" appears in the denominator, the directrix is to the left of

the focus at the origin. $d = |Fl| = \frac{3}{8}$, so an equation of the directrix is

$x = -\frac{3}{8}$.

(d) The vertices are $\left(-\frac{3}{4}, 0\right)$ and $\left(\frac{1}{4}, \pi\right)$, so the center is midway between them,

that is, $\left(\frac{1}{2}, \pi\right)$.

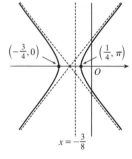

17. (a) $r = \dfrac{1}{1 - 2\sin\theta}$, where $e = 2$ and $ed = 1$ $\Rightarrow$ $d = \frac{1}{2}$. The eccentricity

$e = 2 > 1$, so the conic is a hyperbola. Since "$-e\sin\theta$" appears in the

denominator, the directrix is below the focus at the origin. $d = |Fl| = \frac{1}{2}$,

so an equation of the directrix is $y = -\frac{1}{2}$. The vertices are $\left(-1, \frac{\pi}{2}\right)$ and

$\left(\frac{1}{3}, \frac{3\pi}{2}\right)$, so the center is midway between them, that is, $\left(\frac{2}{3}, \frac{3\pi}{2}\right)$.

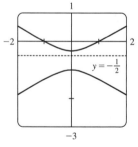

(b) By the discussion that precedes Example 4, the equation

is $r = \dfrac{1}{1 - 2\sin\left(\theta - \frac{3\pi}{4}\right)}$.

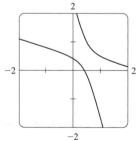

19. For $e < 1$ the curve is an ellipse. It is nearly circular when e is close to 0. As e increases, the graph is stretched out to the right, and grows larger (that is, its right-hand focus moves to the right while its left-hand focus remains at the origin.) At $e = 1$, the curve becomes a parabola with focus at the origin.

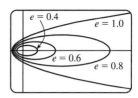

21. $|PF| = e\,|Pl| \quad\Rightarrow\quad r = e[d - r\cos(\pi - \theta)] = e(d + r\cos\theta) \quad\Rightarrow$

$$r(1 - e\cos\theta) = ed \quad\Rightarrow\quad r = \frac{ed}{1 - e\cos\theta}$$

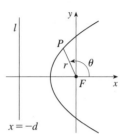

23. $|PF| = e\,|Pl| \quad\Rightarrow\quad r = e[d - r\sin(\theta - \pi)] = e(d + r\sin\theta) \quad\Rightarrow$

$$r(1 - e\sin\theta) = ed \quad\Rightarrow\quad r = \frac{ed}{1 - e\sin\theta}$$

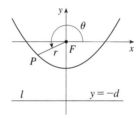

25. We are given $e = 0.093$ and $a = 2.28 \times 10^8$. By (7), we have

$$r = \frac{a(1 - e^2)}{1 + e\cos\theta} = \frac{2.28 \times 10^8[1 - (0.093)^2]}{1 + 0.093\cos\theta} \approx \frac{2.26 \times 10^8}{1 + 0.093\cos\theta}$$

27. Here $2a = $ length of major axis $= 36.18$ AU $\Rightarrow a = 18.09$ AU and $e = 0.97$. By (7), the equation of the orbit is

$$r = \frac{18.09[1 - (0.97)^2]}{1 - 0.97\cos\theta} \approx \frac{1.07}{1 - 0.97\cos\theta}.$$ By (8), the maximum distance from the comet to the sun is

$18.09(1 + 0.97) \approx 35.64$ AU or about 3.314 billion miles.

29. The minimum distance is at perihelion, where $4.6 \times 10^7 = r = a(1 - e) = a(1 - 0.206) = a(0.794) \quad\Rightarrow$

$a = 4.6 \times 10^7/0.794$. So the maximum distance, which is at aphelion, is

$r = a(1 + e) = (4.6 \times 10^7/0.794)(1.206) \approx 7.0 \times 10^7$ km.

31. From Exercise 29, we have $e = 0.206$ and $a(1 - e) = 4.6 \times 10^7$ km. Thus, $a = 4.6 \times 10^7/0.794$. From (7), we can write the

equation of Mercury's orbit as $r = a\dfrac{1 - e^2}{1 - e\cos\theta}$. So since

$$\frac{dr}{d\theta} = \frac{-a(1 - e^2)e\sin\theta}{(1 - e\cos\theta)^2} \quad\Rightarrow$$

$$r^2 + \left(\frac{dr}{d\theta}\right)^2 = \frac{a^2(1 - e^2)^2}{(1 - e\cos\theta)^2} + \frac{a^2(1 - e^2)^2\,e^2\sin^2\theta}{(1 - e\cos\theta)^4} = \frac{a^2(1 - e^2)^2}{(1 - e\cos\theta)^4}\left(1 - 2e\cos\theta + e^2\right)$$

the length of the orbit is

$$L = \int_0^{2\pi} \sqrt{r^2 + (dr/d\theta)^2}\, d\theta = a(1 - e^2) \int_0^{2\pi} \frac{\sqrt{1 + e^2 - 2e\cos\theta}}{(1 - e\cos\theta)^2}\, d\theta \approx 3.6 \times 10^8 \text{ km}$$

This seems reasonable, since Mercury's orbit is nearly circular, and the circumference of a circle of radius a

is $2\pi a \approx 3.6 \times 10^8$ km.

10 Review

CONCEPT CHECK

1. (a) A parametric curve is a set of points of the form $(x, y) = (f(t), g(t))$, where f and g are continuous functions of a variable t.

   (b) Sketching a parametric curve, like sketching the graph of a function, is difficult to do in general. We can plot points on the curve by finding $f(t)$ and $g(t)$ for various values of t, either by hand or with a calculator or computer. Sometimes, when f and g are given by formulas, we can eliminate t from the equations $x = f(t)$ and $y = g(t)$ to get a Cartesian equation relating x and y. It may be easier to graph that equation than to work with the original formulas for x and y in terms of t.

2. (a) You can find $\frac{dy}{dx}$ as a function of t by calculating $\frac{dy}{dx} = \frac{dy/dt}{dx/dt}$ [if $dx/dt \neq 0$].

   (b) Calculate the area as $\int_a^b y\, dx = \int_\alpha^\beta g(t)\, f'(t)dt$ [or $\int_\beta^\alpha g(t)\, f'(t)dt$ if the leftmost point is $(f(\beta), g(\beta))$ rather than $(f(\alpha), g(\alpha))$].

3. (a) $L = \int_\alpha^\beta \sqrt{(dx/dt)^2 + (dy/dt)^2}\, dt = \int_\alpha^\beta \sqrt{[f'(t)]^2 + [g'(t)]^2}\, dt$

   (b) $S = \int_\alpha^\beta 2\pi y \sqrt{(dx/dt)^2 + (dy/dt)^2}\, dt = \int_\alpha^\beta 2\pi g(t) \sqrt{[f'(t)]^2 + [g'(t)]^2}\, dt$

4. (a) See Figure 5 in Section 10.3.

   (b) $x = r\cos\theta,\ y = r\sin\theta$

   (c) To find a polar representation (r, θ) with $r \geq 0$ and $0 \leq \theta < 2\pi$, first calculate $r = \sqrt{x^2 + y^2}$. Then θ is specified by $\cos\theta = x/r$ and $\sin\theta = y/r$.

5. (a) Calculate $\dfrac{dy}{dx} = \dfrac{\dfrac{dy}{d\theta}}{\dfrac{dx}{d\theta}} = \dfrac{\dfrac{d}{d\theta}(y)}{\dfrac{d}{d\theta}(x)} = \dfrac{\dfrac{d}{d\theta}(r\sin\theta)}{\dfrac{d}{d\theta}(r\cos\theta)} = \dfrac{\left(\dfrac{dr}{d\theta}\right)\sin\theta + r\cos\theta}{\left(\dfrac{dr}{d\theta}\right)\cos\theta - r\sin\theta}$, where $r = f(\theta)$.

   (b) Calculate $A = \int_a^b \frac{1}{2}r^2\, d\theta = \int_a^b \frac{1}{2}[f(\theta)]^2\, d\theta$

   (c) $L = \int_a^b \sqrt{(dx/d\theta)^2 + (dy/d\theta)^2}\, d\theta = \int_a^b \sqrt{r^2 + (dr/d\theta)^2}\, d\theta = \int_a^b \sqrt{[f(\theta)]^2 + [f'(\theta)]^2}\, d\theta$

6. (a) A parabola is a set of points in a plane whose distances from a fixed point F (the focus) and a fixed line l (the directrix) are equal.

   (b) $x^2 = 4py;\ y^2 = 4px$

7. (a) An ellipse is a set of points in a plane the sum of whose distances from two fixed points (the foci) is a constant.

(b) $\dfrac{x^2}{a^2} + \dfrac{y^2}{a^2 - c^2} = 1$.

8. (a) A hyperbola is a set of points in a plane the difference of whose distances from two fixed points (the foci) is a constant. This difference should be interpreted as the larger distance minus the smaller distance.

(b) $\dfrac{x^2}{a^2} - \dfrac{y^2}{c^2 - a^2} = 1$

(c) $y = \pm \dfrac{\sqrt{c^2 - a^2}}{a} x$

9. (a) If a conic section has focus F and corresponding directrix l, then the eccentricity e is the fixed ratio $|PF| \,/\, |Pl|$ for points P of the conic section.

(b) $e < 1$ for an ellipse; $e > 1$ for a hyperbola; $e = 1$ for a parabola.

(c) $x = d$: $r = \dfrac{ed}{1 + e \cos \theta}$. $x = -d$: $r = \dfrac{ed}{1 - e \cos \theta}$. $y = d$: $r = \dfrac{ed}{1 + e \sin \theta}$. $y = -d$: $r = \dfrac{ed}{1 - e \sin \theta}$.

TRUE-FALSE QUIZ

1. False. Consider the curve defined by $x = f(t) = (t - 1)^3$ and $y = g(t) = (t - 1)^2$. Then $g'(t) = 2(t - 1)$, so $g'(1) = 0$, but its graph has a *vertical* tangent when $t = 1$. *Note:* The statement is true if $f'(1) \neq 0$ when $g'(1) = 0$.

3. False. For example, if $f(t) = \cos t$ and $g(t) = \sin t$ for $0 \le t \le 4\pi$, then the curve is a circle of radius 1, hence its length is 2π, but $\int_0^{4\pi} \sqrt{[f'(t)]^2 + [g'(t)]^2}\, dt = \int_0^{4\pi} \sqrt{(-\sin t)^2 + (\cos t)^2}\, dt = \int_0^{4\pi} 1\, dt = 4\pi$, since as t increases from 0 to 4π, the circle is traversed twice.

5. True. The curve $r = 1 - \sin 2\theta$ is unchanged if we rotate it through $180°$ about O because
$1 - \sin 2(\theta + \pi) = 1 - \sin(2\theta + 2\pi) = 1 - \sin 2\theta$. So it's unchanged if we replace r by $-r$. (See the discussion after Example 8 in Section 10.3.) In other words, it's the same curve as $r = -(1 - \sin 2\theta) = \sin 2\theta - 1$.

7. False. The first pair of equations gives the portion of the parabola $y = x^2$ with $x \ge 0$, whereas the second pair of equations traces out the whole parabola $y = x^2$.

9. True. By rotating and translating the parabola, we can assume it has an equation of the form $y = cx^2$, where $c > 0$. The tangent at the point (a, ca^2) is the line $y - ca^2 = 2ca(x - a)$; i.e., $y = 2cax - ca^2$. This tangent meets the parabola at the points (x, cx^2) where $cx^2 = 2cax - ca^2$. This equation is equivalent to $x^2 = 2ax - a^2$ [since $c > 0$]. But $x^2 = 2ax - a^2 \Leftrightarrow x^2 - 2ax + a^2 = 0 \Leftrightarrow (x - a)^2 = 0 \Leftrightarrow x = a \Leftrightarrow (x, cx^2) = (a, ca^2)$. This shows that each tangent meets the parabola at exactly one point.

EXERCISES

1. $x = t^2 + 4t$, $y = 2 - t$, $-4 \leq t \leq 1$. $t = 2 - y$, so

$x = (2 - y)^2 + 4(2 - y) = 4 - 4y + y^2 + 8 - 4y = y^2 - 8y + 12$ ⇔

$x + 4 = y^2 - 8y + 16 = (y - 4)^2$. This is part of a parabola with vertex

$(-4, 4)$, opening to the right.

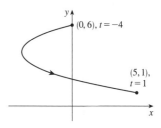

3. $y = \sec\theta = \dfrac{1}{\cos\theta} = \dfrac{1}{x}$. Since $0 \leq \theta \leq \pi/2$, $0 < x \leq 1$ and $y \geq 1$.

This is part of the hyperbola $y = 1/x$.

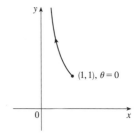

5. Three different sets of parametric equations for the curve $y = \sqrt{x}$ are

(i) $x = t$, $\quad y = \sqrt{t}$

(ii) $x = t^4$, $\quad y = t^2$

(iii) $x = \tan^2 t$, $\quad y = \tan t$, $\quad 0 \leq t < \pi/2$

There are many other sets of equations that also give this curve.

7. (a)

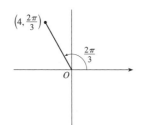

The Cartesian coordinates are $x = 4\cos\frac{2\pi}{3} = 4\left(-\frac{1}{2}\right) = -2$ and

$y = 4\sin\frac{2\pi}{3} = 4\left(\frac{\sqrt{3}}{2}\right) = 2\sqrt{3}$, that is, the point $\left(-2, 2\sqrt{3}\right)$.

(b) Given $x = -3$ and $y = 3$, we have $r = \sqrt{(-3)^2 + 3^2} = \sqrt{18} = 3\sqrt{2}$. Also, $\tan\theta = \dfrac{y}{x}$ ⇒ $\tan\theta = \dfrac{3}{-3}$, and since

$(-3, 3)$ is in the second quadrant, $\theta = \frac{3\pi}{4}$. Thus, one set of polar coordinates for $(-3, 3)$ is $\left(3\sqrt{2}, \frac{3\pi}{4}\right)$, and two others are

$\left(3\sqrt{2}, \frac{11\pi}{4}\right)$ and $\left(-3\sqrt{2}, \frac{7\pi}{4}\right)$.

9. $r = 1 - \cos\theta$. This cardioid is
symmetric about the polar axis.

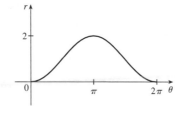

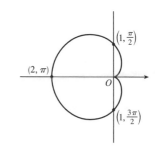

11. $r = \cos 3\theta$. This is a three-leaved rose. The curve is traced twice.

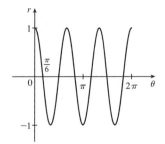

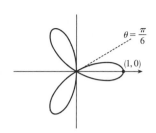

13. $r = 1 + \cos 2\theta$. The curve is symmetric about the pole and both the horizontal and vertical axes.

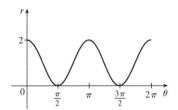

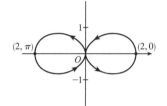

15. $r = \dfrac{3}{1 + 2\sin\theta}$ $\Rightarrow$ $e = 2 > 1$, so the conic is a hyperbola. $de = 3$ $\Rightarrow$ $d = \frac{3}{2}$ and the form "$+2\sin\theta$" imply that the directrix is above the focus at the origin and has equation $y = \frac{3}{2}$. The vertices are $\left(1, \frac{\pi}{2}\right)$ and $\left(-3, \frac{3\pi}{2}\right)$.

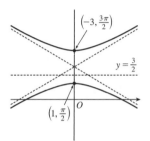

17. $x + y = 2$ $\Leftrightarrow$ $r\cos\theta + r\sin\theta = 2$ $\Leftrightarrow$ $r(\cos\theta + \sin\theta) = 2$ $\Leftrightarrow$ $r = \dfrac{2}{\cos\theta + \sin\theta}$

19. $r = (\sin\theta)/\theta$. As $\theta \to \pm\infty$, $r \to 0$. As $\theta \to 0$, $r \to 1$. In the first figure, there are an infinite number of x-intercepts at $x = \pi n$, n a nonzero integer. These correspond to pole points in the second figure.

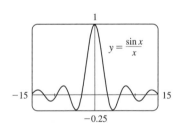

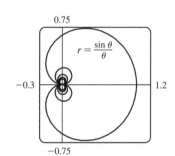

21. $x = \ln t$, $y = 1 + t^2$; $t = 1$. $\dfrac{dy}{dt} = 2t$ and $\dfrac{dx}{dt} = \dfrac{1}{t}$, so $\dfrac{dy}{dx} = \dfrac{dy/dt}{dx/dt} = \dfrac{2t}{1/t} = 2t^2$.

When $t = 1$, $(x, y) = (0, 2)$ and $dy/dx = 2$.

23. $r = e^{-\theta}$ $\Rightarrow$ $y = r\sin\theta = e^{-\theta}\sin\theta$ and $x = r\cos\theta = e^{-\theta}\cos\theta$ $\Rightarrow$

$\dfrac{dy}{dx} = \dfrac{dy/d\theta}{dx/d\theta} = \dfrac{\frac{dr}{d\theta}\sin\theta + r\cos\theta}{\frac{dr}{d\theta}\cos\theta - r\sin\theta} = \dfrac{-e^{-\theta}\sin\theta + e^{-\theta}\cos\theta}{-e^{-\theta}\cos\theta - e^{-\theta}\sin\theta} \cdot \dfrac{-e^{\theta}}{-e^{\theta}} = \dfrac{\sin\theta - \cos\theta}{\cos\theta + \sin\theta}$.

When $\theta = \pi$, $\dfrac{dy}{dx} = \dfrac{0 - (-1)}{-1 + 0} = \dfrac{1}{-1} = -1$.

25. $x = t + \sin t$, $y = t - \cos t$ $\Rightarrow$ $\dfrac{dy}{dx} = \dfrac{dy/dt}{dx/dt} = \dfrac{1 + \sin t}{1 + \cos t}$ $\Rightarrow$

$$\dfrac{d^2 y}{dx^2} = \dfrac{\dfrac{d}{dt}\left(\dfrac{dy}{dx}\right)}{dx/dt} = \dfrac{\dfrac{(1 + \cos t)\cos t - (1 + \sin t)(-\sin t)}{(1 + \cos t)^2}}{1 + \cos t} = \dfrac{\cos t + \cos^2 t + \sin t + \sin^2 t}{(1 + \cos t)^3} = \dfrac{1 + \cos t + \sin t}{(1 + \cos t)^3}$$

27. We graph the curve $x = t^3 - 3t$, $y = t^2 + t + 1$ for $-2.2 \le t \le 1.2$.

By zooming in or using a cursor, we find that the lowest point is about

$(1.4, 0.75)$. To find the exact values, we find the t-value at which

$dy/dt = 2t + 1 = 0$ $\Leftrightarrow$ $t = -\frac{1}{2}$ $\Leftrightarrow$ $(x, y) = \left(\frac{11}{8}, \frac{3}{4}\right)$.

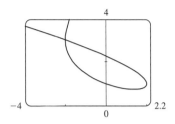

29. $x = 2a\cos t - a\cos 2t$ $\Rightarrow$ $\dfrac{dx}{dt} = -2a\sin t + 2a\sin 2t = 2a\sin t(2\cos t - 1) = 0$ $\Leftrightarrow$

$\sin t = 0$ or $\cos t = \frac{1}{2}$ $\Rightarrow$ $t = 0, \frac{\pi}{3}, \pi$, or $\frac{5\pi}{3}$.

$y = 2a\sin t - a\sin 2t$ $\Rightarrow$ $\dfrac{dy}{dt} = 2a\cos t - 2a\cos 2t = 2a\left(1 + \cos t - 2\cos^2 t\right) = 2a(1 - \cos t)(1 + 2\cos t) = 0$ $\Rightarrow$

$t = 0, \frac{2\pi}{3}$, or $\frac{4\pi}{3}$.

Thus the graph has vertical tangents where

$t = \frac{\pi}{3}, \pi$ and $\frac{5\pi}{3}$, and horizontal tangents where

$t = \frac{2\pi}{3}$ and $\frac{4\pi}{3}$. To determine what the slope is

where $t = 0$, we use l'Hospital's Rule to evaluate

$\lim\limits_{t \to 0} \dfrac{dy/dt}{dx/dt} = 0$, so there is a horizontal tangent

there.

t	x	y
0	a	0
$\frac{\pi}{3}$	$\frac{3}{2}a$	$\frac{\sqrt{3}}{2}a$
$\frac{2\pi}{3}$	$-\frac{1}{2}a$	$\frac{3\sqrt{3}}{2}a$
π	$-3a$	0
$\frac{4\pi}{3}$	$-\frac{1}{2}a$	$-\frac{3\sqrt{3}}{2}a$
$\frac{5\pi}{3}$	$\frac{3}{2}a$	$-\frac{\sqrt{3}}{2}a$

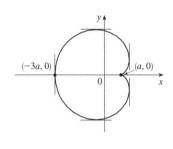

31. The curve $r^2 = 9\cos 5\theta$ has 10 "petals." For instance, for $-\frac{\pi}{10} \le \theta \le \frac{\pi}{10}$, there are two petals, one with $r > 0$ and one

with $r < 0$.

$$A = 10 \int_{-\pi/10}^{\pi/10} \tfrac{1}{2} r^2 \, d\theta = 5 \int_{-\pi/10}^{\pi/10} 9\cos 5\theta \, d\theta = 5 \cdot 9 \cdot 2 \int_{0}^{\pi/10} \cos 5\theta \, d\theta = 18\big[\sin 5\theta\big]_0^{\pi/10} = 18$$

33. The curves intersect when $4\cos\theta = 2$ $\Rightarrow$ $\cos\theta = \frac{1}{2}$ $\Rightarrow$ $\theta = \pm\frac{\pi}{3}$

for $-\pi \le \theta \le \pi$. The points of intersection are $\left(2, \frac{\pi}{3}\right)$ and $\left(2, -\frac{\pi}{3}\right)$.

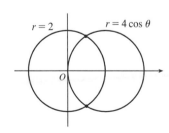

35. The curves intersect where $2\sin\theta = \sin\theta + \cos\theta \implies$

$\sin\theta = \cos\theta \implies \theta = \frac{\pi}{4}$, and also at the origin (at which $\theta = \frac{3\pi}{4}$

on the second curve).

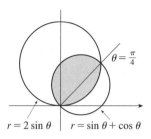

$$A = \int_0^{\pi/4} \tfrac{1}{2}(2\sin\theta)^2\,d\theta + \int_{\pi/4}^{3\pi/4} \tfrac{1}{2}(\sin\theta+\cos\theta)^2\,d\theta$$

$$= \int_0^{\pi/4}(1-\cos 2\theta)\,d\theta + \tfrac{1}{2}\int_{\pi/4}^{3\pi/4}(1+\sin 2\theta)\,d\theta$$

$$= \left[\theta - \tfrac{1}{2}\sin 2\theta\right]_0^{\pi/4} + \left[\tfrac{1}{2}\theta - \tfrac{1}{4}\cos 2\theta\right]_{\pi/4}^{3\pi/4} = \tfrac{1}{2}(\pi-1)$$

$r = 2\sin\theta \qquad r = \sin\theta + \cos\theta$

37. $x = 3t^2$, $y = 2t^3$.

$$L = \int_0^2 \sqrt{(dx/dt)^2+(dy/dt)^2}\,dt = \int_0^2 \sqrt{(6t)^2+(6t^2)^2}\,dt = \int_0^2 \sqrt{36t^2+36t^4}\,dt = \int_0^2 \sqrt{36t^2}\sqrt{1+t^2}\,dt$$

$$= \int_0^2 6\,|t|\,\sqrt{1+t^2}\,dt = 6\int_0^2 t\,\sqrt{1+t^2}\,dt = 6\int_1^5 u^{1/2}\left(\tfrac{1}{2}du\right) \qquad \left[u = 1+t^2,\, du = 2t\,dt\right]$$

$$= 6\cdot\tfrac{1}{2}\cdot\tfrac{2}{3}\left[u^{3/2}\right]_1^5 = 2(5^{3/2}-1) = 2(5\sqrt{5}-1)$$

39. $L = \int_\pi^{2\pi}\sqrt{r^2+(dr/d\theta)^2}\,d\theta = \int_\pi^{2\pi}\sqrt{(1/\theta)^2+(-1/\theta^2)^2}\,d\theta = \displaystyle\int_\pi^{2\pi}\frac{\sqrt{\theta^2+1}}{\theta^2}\,d\theta$

$$\overset{24}{=} \left[-\frac{\sqrt{\theta^2+1}}{\theta}+\ln\left(\theta+\sqrt{\theta^2+1}\right)\right]_\pi^{2\pi} = \frac{\sqrt{\pi^2+1}}{\pi} - \frac{\sqrt{4\pi^2+1}}{2\pi} + \ln\left(\frac{2\pi+\sqrt{4\pi^2+1}}{\pi+\sqrt{\pi^2+1}}\right)$$

$$= \frac{2\sqrt{\pi^2+1}-\sqrt{4\pi^2+1}}{2\pi} + \ln\left(\frac{2\pi+\sqrt{4\pi^2+1}}{\pi+\sqrt{\pi^2+1}}\right)$$

41. $x = 4\sqrt{t}$, $y = \dfrac{t^3}{3} + \dfrac{1}{2t^2}$, $1 \le t \le 4 \implies$

$$S = \int_1^4 2\pi y\sqrt{(dx/dt)^2+(dy/dt)^2}\,dt = \int_1^4 2\pi\left(\tfrac{1}{3}t^3+\tfrac{1}{2}t^{-2}\right)\sqrt{\left(2/\sqrt{t}\right)^2+(t^2-t^{-3})^2}\,dt$$

$$= 2\pi\int_1^4 \left(\tfrac{1}{3}t^3+\tfrac{1}{2}t^{-2}\right)\sqrt{(t^2+t^{-3})^2}\,dt = 2\pi\int_1^4 \left(\tfrac{1}{3}t^5+\tfrac{5}{6}+\tfrac{1}{2}t^{-5}\right)dt = 2\pi\left[\tfrac{1}{18}t^6+\tfrac{5}{6}t-\tfrac{1}{8}t^{-4}\right]_1^4 = \frac{471{,}295}{1024}\pi$$

43. For all c except -1, the curve is asymptotic to the line $x = 1$. For

$c < -1$, the curve bulges to the right near $y = 0$. As c increases, the

bulge becomes smaller, until at $c = -1$ the curve is the straight line $x = 1$.

As c continues to increase, the curve bulges to the left, until at $c = 0$ there

is a cusp at the origin. For $c > 0$, there is a loop to the left of the origin,

whose size and roundness increase as c increases. Note that the x-intercept

of the curve is always $-c$.

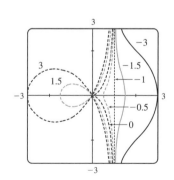

45. $\dfrac{x^2}{9} + \dfrac{y^2}{8} = 1$ is an ellipse with center $(0,0)$.

$a = 3, b = 2\sqrt{2}, c = 1 \;\Rightarrow$

foci $(\pm 1, 0)$, vertices $(\pm 3, 0)$.

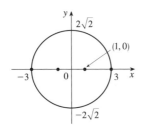

47. $6y^2 + x - 36y + 55 = 0 \;\Leftrightarrow$

$6(y^2 - 6y + 9) = -(x+1) \;\Leftrightarrow$

$(y-3)^2 = -\frac{1}{6}(x+1)$, a parabola with vertex $(-1, 3)$,

opening to the left, $p = -\frac{1}{24} \;\Rightarrow\;$ focus $\left(-\frac{25}{24}, 3\right)$ and

directrix $x = -\frac{23}{24}$.

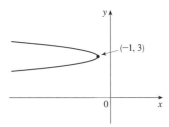

49. The ellipse with foci $(\pm 4, 0)$ and vertices $(\pm 5, 0)$ has center $(0, 0)$ and a horizontal major axis, with $a = 5$ and $c = 4$,

so $b^2 = a^2 - c^2 = 5^2 - 4^2 = 9$. An equation is $\dfrac{x^2}{25} + \dfrac{y^2}{9} = 1$.

51. The center of a hyperbola with foci $(0, \pm 4)$ is $(0, 0)$, so $c = 4$ and an equation is $\dfrac{y^2}{a^2} - \dfrac{x^2}{b^2} = 1$.

The asymptote $y = 3x$ has slope 3, so $\dfrac{a}{b} = \dfrac{3}{1} \;\Rightarrow\; a = 3b$ and $a^2 + b^2 = c^2 \;\Rightarrow\; (3b)^2 + b^2 = 4^2 \;\Rightarrow$

$10b^2 = 16 \;\Rightarrow\; b^2 = \frac{8}{5}$ and so $a^2 = 16 - \frac{8}{5} = \frac{72}{5}$. Thus, an equation is $\dfrac{y^2}{72/5} - \dfrac{x^2}{8/5} = 1$, or $\dfrac{5y^2}{72} - \dfrac{5x^2}{8} = 1$.

53. $x^2 = -(y - 100)$ has its vertex at $(0, 100)$, so one of the vertices of the ellipse is $(0, 100)$. Another form of the equation of a

parabola is $x^2 = 4p(y - 100)$ so $4p(y - 100) = -(y - 100) \;\Rightarrow\; 4p = -1 \;\Rightarrow\; p = -\frac{1}{4}$. Therefore the shared focus is

found at $\left(0, \frac{399}{4}\right)$ so $2c = \frac{399}{4} - 0 \;\Rightarrow\; c = \frac{399}{8}$ and the center of the ellipse is $\left(0, \frac{399}{8}\right)$. So $a = 100 - \frac{399}{8} = \frac{401}{8}$ and

$b^2 = a^2 - c^2 = \dfrac{401^2 - 399^2}{8^2} = 25$. So the equation of the ellipse is $\dfrac{x^2}{b^2} + \dfrac{\left(y - \frac{399}{8}\right)^2}{a^2} = 1 \;\Rightarrow\; \dfrac{x^2}{25} + \dfrac{\left(y - \frac{399}{8}\right)^2}{\left(\frac{401}{8}\right)^2} = 1$,

or $\dfrac{x^2}{25} + \dfrac{(8y - 399)^2}{160{,}801} = 1$.

55. Directrix $x = 4 \;\Rightarrow\; d = 4$, so $e = \frac{1}{3} \;\Rightarrow\; r = \dfrac{ed}{1 + e\cos\theta} = \dfrac{4}{3 + \cos\theta}$.

57. In polar coordinates, an equation for the circle is $r = 2a\sin\theta$. Thus, the coordinates of Q are $x = r\cos\theta = 2a\sin\theta\cos\theta$

and $y = r\sin\theta = 2a\sin^2\theta$. The coordinates of R are $x = 2a\cot\theta$ and $y = 2a$. Since P is the midpoint of QR, we use the

midpoint formula to get $x = a(\sin\theta\cos\theta + \cot\theta)$ and $y = a(1 + \sin^2\theta)$.

1. $x = \int_1^t \dfrac{\cos u}{u}\, du$, $y = \int_1^t \dfrac{\sin u}{u}\, du$, so by FTC1, we have $\dfrac{dx}{dt} = \dfrac{\cos t}{t}$ and $\dfrac{dy}{dt} = \dfrac{\sin t}{t}$. Vertical tangent lines occur when

$\dfrac{dx}{dt} = 0 \iff \cos t = 0$. The parameter value corresponding to $(x, y) = (0, 0)$ is $t = 1$, so the nearest vertical tangent

occurs when $t = \frac{\pi}{2}$. Therefore, the arc length between these points is

$$L = \int_1^{\pi/2} \sqrt{\left(\dfrac{dx}{dt}\right)^2 + \left(\dfrac{dy}{dt}\right)^2}\, dt = \int_1^{\pi/2} \sqrt{\dfrac{\cos^2 t}{t^2} + \dfrac{\sin^2 t}{t^2}}\, dt = \int_1^{\pi/2} \dfrac{dt}{t} = \big[\ln t\big]_1^{\pi/2} = \ln\tfrac{\pi}{2}$$

3. In terms of x and y, we have $x = r\cos\theta = (1 + c\sin\theta)\cos\theta = \cos\theta + c\sin\theta\cos\theta = \cos\theta + \frac{1}{2}c\sin 2\theta$ and

$y = r\sin\theta = (1 + c\sin\theta)\sin\theta = \sin\theta + c\sin^2\theta$. Now $-1 \le \sin\theta \le 1 \Rightarrow -1 \le \sin\theta + c\sin^2\theta \le 1 + c \le 2$, so

$-1 \le y \le 2$. Furthermore, $y = 2$ when $c = 1$ and $\theta = \frac{\pi}{2}$, while $y = -1$ for $c = 0$ and $\theta = \frac{3\pi}{2}$. Therefore, we need a viewing

rectangle with $-1 \le y \le 2$.

To find the x-values, look at the equation $x = \cos\theta + \frac{1}{2}c\sin 2\theta$ and use the fact that $\sin 2\theta \ge 0$ for $0 \le \theta \le \frac{\pi}{2}$ and

$\sin 2\theta \le 0$ for $-\frac{\pi}{2} \le \theta \le 0$. [Because $r = 1 + c\sin\theta$ is symmetric about the y-axis, we only need to consider

$-\frac{\pi}{2} \le \theta \le \frac{\pi}{2}$.] So for $-\frac{\pi}{2} \le \theta \le 0$, x has a maximum value when $c = 0$ and then $x = \cos\theta$ has a maximum value

of 1 at $\theta = 0$. Thus, the maximum value of x must occur on $\left[0, \frac{\pi}{2}\right]$ with $c = 1$. Then $x = \cos\theta + \frac{1}{2}\sin 2\theta \Rightarrow$

$\frac{dx}{d\theta} = -\sin\theta + \cos 2\theta = -\sin\theta + 1 - 2\sin^2\theta \Rightarrow \frac{dx}{d\theta} = -(2\sin\theta - 1)(\sin\theta + 1) = 0$ when $\sin\theta = -1$ or $\frac{1}{2}$

[but $\sin\theta \ne -1$ for $0 \le \theta \le \frac{\pi}{2}$]. If $\sin\theta = \frac{1}{2}$, then $\theta = \frac{\pi}{6}$ and

$x = \cos\frac{\pi}{6} + \frac{1}{2}\sin\frac{\pi}{3} = \frac{3}{4}\sqrt{3}$. Thus, the maximum value of x is $\frac{3}{4}\sqrt{3}$, and,

by symmetry, the minimum value is $-\frac{3}{4}\sqrt{3}$. Therefore, the smallest

viewing rectangle that contains every member of the family of polar curves

$r = 1 + c\sin\theta$, where $0 \le c \le 1$, is $\left[-\frac{3}{4}\sqrt{3}, \frac{3}{4}\sqrt{3}\right] \times [-1, 2]$.

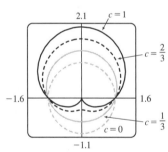

5. (a) If (a, b) lies on the curve, then there is some parameter value t_1 such that $\dfrac{3t_1}{1 + t_1^3} = a$ and $\dfrac{3t_1^2}{1 + t_1^3} = b$. If $t_1 = 0$,

the point is $(0, 0)$, which lies on the line $y = x$. If $t_1 \ne 0$, then the point corresponding to $t = \dfrac{1}{t_1}$ is given by

$x = \dfrac{3(1/t_1)}{1 + (1/t_1)^3} = \dfrac{3t_1^2}{t_1^3 + 1} = b$, $y = \dfrac{3(1/t_1)^2}{1 + (1/t_1)^3} = \dfrac{3t_1}{t_1^3 + 1} = a$. So (b, a) also lies on the curve. [Another way to see

this is to do part (e) first; the result is immediate.] The curve intersects the line $y = x$ when $\dfrac{3t}{1 + t^3} = \dfrac{3t^2}{1 + t^3} \Rightarrow$

$t = t^2 \Rightarrow t = 0$ or 1, so the points are $(0, 0)$ and $\left(\frac{3}{2}, \frac{3}{2}\right)$.

(b) $\dfrac{dy}{dt} = \dfrac{(1+t^3)(6t) - 3t^2(3t^2)}{(1+t^3)^2} = \dfrac{6t - 3t^4}{(1+t^3)^2} = 0$ when $6t - 3t^4 = 3t(2 - t^3) = 0$ $\Rightarrow$ $t = 0$ or $t = \sqrt[3]{2}$, so there are

horizontal tangents at $(0,0)$ and $\left(\sqrt[3]{2}, \sqrt[3]{4}\right)$. Using the symmetry from part (a), we see that there are vertical tangents at

$(0,0)$ and $\left(\sqrt[3]{4}, \sqrt[3]{2}\right)$.

(c) Notice that as $t \to -1^+$, we have $x \to -\infty$ and $y \to \infty$. As $t \to -1^-$, we have $x \to \infty$ and $y \to -\infty$. Also

$$y - (-x - 1) = y + x + 1 = \frac{3t + 3t^2 + (1 + t^3)}{1 + t^3} = \frac{(t+1)^3}{1 + t^3} = \frac{(t+1)^2}{t^2 - t + 1} \to 0 \text{ as } t \to -1. \text{ So } y = -x - 1 \text{ is a}$$

slant asymptote.

(d) $\dfrac{dx}{dt} = \dfrac{(1+t^3)(3) - 3t(3t^2)}{(1+t^3)^2} = \dfrac{3 - 6t^3}{(1+t^3)^2}$ and from part (b) we have $\dfrac{dy}{dt} = \dfrac{6t - 3t^4}{(1+t^3)^2}$. So $\dfrac{dy}{dx} = \dfrac{dy/dt}{dx/dt} = \dfrac{t(2 - t^3)}{1 - 2t^3}$.

Also $\dfrac{d^2y}{dx^2} = \dfrac{\dfrac{d}{dt}\left(\dfrac{dy}{dx}\right)}{dx/dt} = \dfrac{2(1+t^3)^4}{3(1 - 2t^3)^3} > 0 \Leftrightarrow t < \dfrac{1}{\sqrt[3]{2}}$.

So the curve is concave upward there and has a minimum point at $(0,0)$

and a maximum point at $\left(\sqrt[3]{2}, \sqrt[3]{4}\right)$. Using this together with the

information from parts (a), (b), and (c), we sketch the curve.

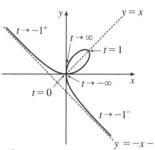

(e) $x^3 + y^3 = \left(\dfrac{3t}{1+t^3}\right)^3 + \left(\dfrac{3t^2}{1+t^3}\right)^3 = \dfrac{27t^3 + 27t^6}{(1+t^3)^3} = \dfrac{27t^3(1+t^3)}{(1+t^3)^3} = \dfrac{27t^3}{(1+t^3)^2}$ and

$3xy = 3\left(\dfrac{3t}{1+t^3}\right)\left(\dfrac{3t^2}{1+t^3}\right) = \dfrac{27t^3}{(1+t^3)^2}$, so $x^3 + y^3 = 3xy$.

(f) We start with the equation from part (e) and substitute $x = r\cos\theta$, $y = r\sin\theta$. Then $x^3 + y^3 = 3xy$ $\Rightarrow$

$r^3\cos^3\theta + r^3\sin^3\theta = 3r^2\cos\theta\sin\theta$. For $r \neq 0$, this gives $r = \dfrac{3\cos\theta\sin\theta}{\cos^3\theta + \sin^3\theta}$. Dividing numerator and denominator

by $\cos^3\theta$, we obtain $r = \dfrac{3\left(\dfrac{1}{\cos\theta}\right)\dfrac{\sin\theta}{\cos\theta}}{1 + \dfrac{\sin^3\theta}{\cos^3\theta}} = \dfrac{3\sec\theta\tan\theta}{1 + \tan^3\theta}$.

(g) The loop corresponds to $\theta \in \left(0, \dfrac{\pi}{2}\right)$, so its area is

$$A = \int_0^{\pi/2} \frac{r^2}{2}\,d\theta = \frac{1}{2}\int_0^{\pi/2}\left(\frac{3\sec\theta\tan\theta}{1 + \tan^3\theta}\right)^2 d\theta = \frac{9}{2}\int_0^{\pi/2}\frac{\sec^2\theta\tan^2\theta}{(1+\tan^3\theta)^2}\,d\theta = \frac{9}{2}\int_0^{\infty}\frac{u^2\,du}{(1+u^3)^2} \quad [\text{let } u = \tan\theta]$$

$$= \lim_{b\to\infty}\frac{9}{2}\left[-\frac{1}{3}(1+u^3)^{-1}\right]_0^b = \frac{3}{2}$$

(h) By symmetry, the area between the folium and the line $y = -x - 1$ is equal to the enclosed area in the third quadrant,

plus twice the enclosed area in the fourth quadrant. The area in the third quadrant is $\frac{1}{2}$, and since $y = -x - 1$ $\Rightarrow$

$r\sin\theta = -r\cos\theta - 1$ $\Rightarrow$ $r = -\dfrac{1}{\sin\theta + \cos\theta}$, the area in the fourth quadrant is

$$\frac{1}{2}\int_{-\pi/2}^{-\pi/4}\left[\left(-\frac{1}{\sin\theta + \cos\theta}\right)^2 - \left(\frac{3\sec\theta\tan\theta}{1 + \tan^3\theta}\right)^2\right]d\theta \overset{\text{CAS}}{=} \frac{1}{2}. \text{ Therefore, the total area is } \frac{1}{2} + 2\left(\frac{1}{2}\right) = \frac{3}{2}.$$

11 ☐ INFINITE SEQUENCES AND SERIES

11.1 Sequences

1. (a) A sequence is an ordered list of numbers. It can also be defined as a function whose domain is the set of positive integers.

   (b) The terms a_n approach 8 as n becomes large. In fact, we can make a_n as close to 8 as we like by taking n sufficiently large.

   (c) The terms a_n become large as n becomes large. In fact, we can make a_n as large as we like by taking n sufficiently large.

3. $a_n = 1 - (0.2)^n$, so the sequence is $\{0.8, 0.96, 0.992, 0.9984, 0.99968, \ldots\}$.

5. $a_n = \dfrac{3(-1)^n}{n!}$, so the sequence is $\left\{\dfrac{-3}{1}, \dfrac{3}{2}, \dfrac{-3}{6}, \dfrac{3}{24}, \dfrac{-3}{120}, \ldots\right\} = \left\{-3, \dfrac{3}{2}, -\dfrac{1}{2}, \dfrac{1}{8}, -\dfrac{1}{40}, \ldots\right\}$.

7. $a_1 = 3$, $a_{n+1} = 2a_n - 1$.   Each term is defined in terms of the preceding term.

   $a_2 = 2a_1 - 1 = 2(3) - 1 = 5$. $a_3 = 2a_2 - 1 = 2(5) - 1 = 9$. $a_4 = 2a_3 - 1 = 2(9) - 1 = 17$.

   $a_5 = 2a_4 - 1 = 2(17) - 1 = 33$. The sequence is $\{3, 5, 9, 17, 33, \ldots\}$.

9. $\left\{1, \frac{1}{3}, \frac{1}{5}, \frac{1}{7}, \frac{1}{9}, \ldots\right\}$.   The denominator of the nth term is the nth positive odd integer, so $a_n = \dfrac{1}{2n - 1}$.

11. $\{2, 7, 12, 17, \ldots\}$.   Each term is larger than the preceding one by 5, so $a_n = a_1 + d(n - 1) = 2 + 5(n - 1) = 5n - 3$.

13. $\left\{1, -\frac{2}{3}, \frac{4}{9}, -\frac{8}{27}, \ldots\right\}$.   Each term is $-\frac{2}{3}$ times the preceding one, so $a_n = \left(-\frac{2}{3}\right)^{n-1}$.

15. The first six terms of $a_n = \dfrac{n}{2n + 1}$ are $\dfrac{1}{3}, \dfrac{2}{5}, \dfrac{3}{7}, \dfrac{4}{9}, \dfrac{5}{11}, \dfrac{6}{13}$. It appears that the sequence is approaching $\dfrac{1}{2}$.

   $$\lim_{n \to \infty} \frac{n}{2n + 1} = \lim_{n \to \infty} \frac{1}{2 + 1/n} = \frac{1}{2}$$

17. $a_n = 1 - (0.2)^n$, so $\lim\limits_{n \to \infty} a_n = 1 - 0 = 1$ by (9).   Converges

19. $a_n = \dfrac{3 + 5n^2}{n + n^2} = \dfrac{(3 + 5n^2)/n^2}{(n + n^2)/n^2} = \dfrac{5 + 3/n^2}{1 + 1/n}$, so $a_n \to \dfrac{5 + 0}{1 + 0} = 5$ as $n \to \infty$.   Converges

21. Because the natural exponential function is continuous at 0, Theorem 7 enables us to write

   $$\lim_{n \to \infty} a_n = \lim_{n \to \infty} e^{1/n} = e^{\lim_{n \to \infty} (1/n)} = e^0 = 1. \quad \text{Converges}$$

23. If $b_n = \dfrac{2n\pi}{1 + 8n}$, then $\lim\limits_{n \to \infty} b_n = \lim\limits_{n \to \infty} \dfrac{(2n\pi)/n}{(1 + 8n)/n} = \lim\limits_{n \to \infty} \dfrac{2\pi}{1/n + 8} = \dfrac{2\pi}{8} = \dfrac{\pi}{4}$. Since $\tan$ is continuous at $\frac{\pi}{4}$, by

   Theorem 7, $\lim\limits_{n \to \infty} \tan\left(\dfrac{2n\pi}{1 + 8n}\right) = \tan\left(\lim\limits_{n \to \infty} \dfrac{2n\pi}{1 + 8n}\right) = \tan\dfrac{\pi}{4} = 1$.   Converges

25. $a_n = \dfrac{(-1)^{n-1} n}{n^2 + 1} = \dfrac{(-1)^{n-1}}{n + 1/n}$, so $0 \le |a_n| = \dfrac{1}{n + 1/n} \le \dfrac{1}{n} \to 0$ as $n \to \infty$, so $a_n \to 0$ by the Squeeze Theorem and

Theorem 6.   Converges

27. $a_n = \cos(n/2)$.   This sequence diverges since the terms don't approach any particular real number as $n \to \infty$.
The terms take on values between -1 and 1.

29. $a_n = \dfrac{(2n-1)!}{(2n+1)!} = \dfrac{(2n-1)!}{(2n+1)(2n)(2n-1)!} = \dfrac{1}{(2n+1)(2n)} \to 0$ as $n \to \infty$.   Converges

31. $a_n = \dfrac{e^n + e^{-n}}{e^{2n} - 1} \cdot \dfrac{e^{-n}}{e^{-n}} = \dfrac{1 + e^{-2n}}{e^n - e^{-n}} \to 0$ as $n \to \infty$ because $1 + e^{-2n} \to 1$ and $e^n - e^{-n} \to \infty$.   Converges

33. $a_n = n^2 e^{-n} = \dfrac{n^2}{e^n}$. Since $\lim\limits_{x \to \infty} \dfrac{x^2}{e^x} \overset{\text{H}}{=} \lim\limits_{x \to \infty} \dfrac{2x}{e^x} \overset{\text{H}}{=} \lim\limits_{x \to \infty} \dfrac{2}{e^x} = 0$, it follows from Theorem 3 that $\lim\limits_{n \to \infty} a_n = 0$.   Converges

35. $0 \le \dfrac{\cos^2 n}{2^n} \le \dfrac{1}{2^n}$ [since $0 \le \cos^2 n \le 1$],   so since $\lim\limits_{n \to \infty} \dfrac{1}{2^n} = 0$, $\left\{ \dfrac{\cos^2 n}{2^n} \right\}$ converges to 0 by the Squeeze Theorem.

37. $a_n = n \sin(1/n) = \dfrac{\sin(1/n)}{1/n}$.   Since $\lim\limits_{x \to \infty} \dfrac{\sin(1/x)}{1/x} = \lim\limits_{t \to 0^+} \dfrac{\sin t}{t}$ [where $t = 1/x$] $= 1$, it follows from Theorem 3

that $\{a_n\}$ converges to 1.

39. $y = \left(1 + \dfrac{2}{x}\right)^x$ $\Rightarrow$ $\ln y = x \ln\left(1 + \dfrac{2}{x}\right)$, so

$$\lim\limits_{x \to \infty} \ln y = \lim\limits_{x \to \infty} \dfrac{\ln(1 + 2/x)}{1/x} \overset{\text{H}}{=} \lim\limits_{x \to \infty} \dfrac{\left(\dfrac{1}{1 + 2/x}\right)\left(-\dfrac{2}{x^2}\right)}{-1/x^2} = \lim\limits_{x \to \infty} \dfrac{2}{1 + 2/x} = 2 \Rightarrow$$

$$\lim\limits_{x \to \infty} \left(1 + \dfrac{2}{x}\right)^x = \lim\limits_{x \to \infty} e^{\ln y} = e^2, \text{ so by Theorem 3, } \lim\limits_{n \to \infty} \left(1 + \dfrac{2}{n}\right)^n = e^2. \quad \text{Convergent}$$

41. $a_n = \ln(2n^2 + 1) - \ln(n^2 + 1) = \ln\left(\dfrac{2n^2 + 1}{n^2 + 1}\right) = \ln\left(\dfrac{2 + 1/n^2}{1 + 1/n^2}\right) \to \ln 2$ as $n \to \infty$.   Convergent

43. $\{0, 1, 0, 0, 1, 0, 0, 0, 1, \dots\}$ diverges since the sequence takes on only two values, 0 and 1, and never stays arbitrarily close to either one (or any other value) for n sufficiently large.

45. $a_n = \dfrac{n!}{2^n} = \dfrac{1}{2} \cdot \dfrac{2}{2} \cdot \dfrac{3}{2} \cdot \dots \cdot \dfrac{(n-1)}{2} \cdot \dfrac{n}{2} \ge \dfrac{1}{2} \cdot \dfrac{n}{2}$ [for $n > 1$] $= \dfrac{n}{4} \to \infty$ as $n \to \infty$, so $\{a_n\}$ diverges.

47.

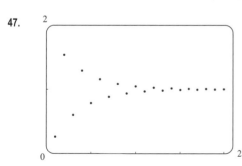

From the graph, it appears that the sequence converges to 1.

$\{(-2/e)^n\}$ converges to 0 by (9), and hence $\{1 + (-2/e)^n\}$

converges to $1 + 0 = 1$.

49.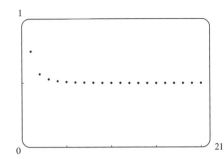

From the graph, it appears that the sequence converges to $\frac{1}{2}$.

As $n \to \infty$,

$$a_n = \sqrt{\frac{3 + 2n^2}{8n^2 + n}} = \sqrt{\frac{3/n^2 + 2}{8 + 1/n}} \quad \Rightarrow \quad \sqrt{\frac{0 + 2}{8 + 0}} = \sqrt{\frac{1}{4}} = \frac{1}{2},$$

so $\lim\limits_{n\to\infty} a_n = \frac{1}{2}$.

51.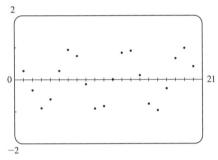

From the graph, it appears that the sequence $\{a_n\} = \left\{\dfrac{n^2 \cos n}{1 + n^2}\right\}$ is

divergent, since it oscillates between 1 and -1 (approximately). To

prove this, suppose that $\{a_n\}$ converges to L. If $b_n = \dfrac{n^2}{1 + n^2}$, then

$\{b_n\}$ converges to 1, and $\lim\limits_{n\to\infty} \dfrac{a_n}{b_n} = \dfrac{L}{1} = L$. But $\dfrac{a_n}{b_n} = \cos n$, so

$\lim\limits_{n\to\infty} \dfrac{a_n}{b_n}$ does not exist. This contradiction shows that $\{a_n\}$ diverges.

53.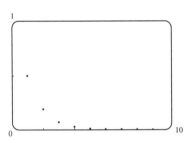

From the graph, it appears that the sequence approaches 0.

$$0 < a_n = \frac{1 \cdot 3 \cdot 5 \cdot \cdots \cdot (2n - 1)}{(2n)^n} = \frac{1}{2n} \cdot \frac{3}{2n} \cdot \frac{5}{2n} \cdot \cdots \cdot \frac{2n - 1}{2n}$$

$$\leq \frac{1}{2n} \cdot (1) \cdot (1) \cdot \cdots \cdot (1) = \frac{1}{2n} \to 0 \text{ as } n \to \infty$$

So by the Squeeze Theorem, $\left\{\dfrac{1 \cdot 3 \cdot 5 \cdot \cdots \cdot (2n - 1)}{(2n)^n}\right\}$ converges to 0.

55. (a) $a_n = 1000(1.06)^n \quad \Rightarrow \quad a_1 = 1060,\ a_2 = 1123.60,\ a_3 = 1191.02,\ a_4 = 1262.48,$ and $a_5 = 1338.23$.

(b) $\lim\limits_{n\to\infty} a_n = 1000 \lim\limits_{n\to\infty} (1.06)^n$, so the sequence diverges by (9) with $r = 1.06 > 1$.

57. If $|r| \geq 1$, then $\{r^n\}$ diverges by (9), so $\{nr^n\}$ diverges also, since $|nr^n| = n\,|r^n| \geq |r^n|$. If $|r| < 1$ then

$$\lim\limits_{x\to\infty} xr^x = \lim\limits_{x\to\infty} \frac{x}{r^{-x}} \overset{\text{H}}{=} \lim\limits_{x\to\infty} \frac{1}{(-\ln r)\,r^{-x}} = \lim\limits_{x\to\infty} \frac{r^x}{-\ln r} = 0, \text{ so } \lim\limits_{n\to\infty} nr^n = 0, \text{ and hence } \{nr^n\} \text{ converges}$$

whenever $|r| < 1$.

59. Since $\{a_n\}$ is a decreasing sequence, $a_n > a_{n+1}$ for all $n \geq 1$. Because all of its terms lie between 5 and 8, $\{a_n\}$ is a bounded sequence. By the Monotonic Sequence Theorem, $\{a_n\}$ is convergent; that is, $\{a_n\}$ has a limit L. L must be less than 8 since $\{a_n\}$ is decreasing, so $5 \leq L < 8$.

61. $a_n = \dfrac{1}{2n + 3}$ is decreasing since $a_{n+1} = \dfrac{1}{2(n + 1) + 3} = \dfrac{1}{2n + 5} < \dfrac{1}{2n + 3} = a_n$ for each $n \geq 1$. The sequence is

bounded since $0 < a_n \leq \frac{1}{5}$ for all $n \geq 1$. Note that $a_1 = \frac{1}{5}$.

63. The terms of $a_n = n(-1)^n$ alternate in sign, so the sequence is not monotonic. The first five terms are $-1, 2, -3, 4$, and -5. Since $\lim\limits_{n\to\infty} |a_n| = \lim\limits_{n\to\infty} n = \infty$, the sequence is not bounded.

65. $a_n = \dfrac{n}{n^2+1}$ defines a decreasing sequence since for $f(x) = \dfrac{x}{x^2+1}$, $f'(x) = \dfrac{(x^2+1)(1) - x(2x)}{(x^2+1)^2} = \dfrac{1-x^2}{(x^2+1)^2} \le 0$ for $x \ge 1$. The sequence is bounded since $0 < a_n \le \frac{1}{2}$ for all $n \ge 1$.

67. For $\left\{ \sqrt{2}, \sqrt{2\sqrt{2}}, \sqrt{2\sqrt{2\sqrt{2}}}, \ldots \right\}$, $a_1 = 2^{1/2}$, $a_2 = 2^{3/4}$, $a_3 = 2^{7/8}, \ldots$, so $a_n = 2^{(2^n-1)/2^n} = 2^{1-(1/2^n)}$.

$\lim\limits_{n\to\infty} a_n = \lim\limits_{n\to\infty} 2^{1-(1/2^n)} = 2^1 = 2$.

Alternate solution: Let $L = \lim\limits_{n\to\infty} a_n$. (We could show the limit exists by showing that $\{a_n\}$ is bounded and increasing.)

Then L must satisfy $L = \sqrt{2 \cdot L} \;\Rightarrow\; L^2 = 2L \;\Rightarrow\; L(L-2) = 0$. $L \ne 0$ since the sequence increases, so $L = 2$.

69. $a_1 = 1$, $a_{n+1} = 3 - \dfrac{1}{a_n}$. We show by induction that $\{a_n\}$ is increasing and bounded above by 3. Let P_n be the proposition that $a_{n+1} > a_n$ and $0 < a_n < 3$. Clearly P_1 is true. Assume that P_n is true. Then $a_{n+1} > a_n \;\Rightarrow\; \dfrac{1}{a_{n+1}} < \dfrac{1}{a_n} \;\Rightarrow\;$

$-\dfrac{1}{a_{n+1}} > -\dfrac{1}{a_n}$. Now $a_{n+2} = 3 - \dfrac{1}{a_{n+1}} > 3 - \dfrac{1}{a_n} = a_{n+1} \;\Leftrightarrow\; P_{n+1}$. This proves that $\{a_n\}$ is increasing and bounded above by 3, so $1 = a_1 < a_n < 3$, that is, $\{a_n\}$ is bounded, and hence convergent by the Monotonic Sequence Theorem.

If $L = \lim\limits_{n\to\infty} a_n$, then $\lim\limits_{n\to\infty} a_{n+1} = L$ also, so L must satisfy $L = 3 - 1/L \;\Rightarrow\; L^2 - 3L + 1 = 0 \;\Rightarrow\; L = \frac{3\pm\sqrt{5}}{2}$. But $L > 1$, so $L = \frac{3+\sqrt{5}}{2}$.

71. (a) Let a_n be the number of rabbit pairs in the nth month. Clearly $a_1 = 1 = a_2$. In the nth month, each pair that is 2 or more months old (that is, a_{n-2} pairs) will produce a new pair to add to the a_{n-1} pairs already present. Thus, $a_n = a_{n-1} + a_{n-2}$, so that $\{a_n\} = \{f_n\}$, the Fibonacci sequence.

(b) $a_n = \dfrac{f_{n+1}}{f_n} \;\Rightarrow\; a_{n-1} = \dfrac{f_n}{f_{n-1}} = \dfrac{f_{n-1} + f_{n-2}}{f_{n-1}} = 1 + \dfrac{f_{n-2}}{f_{n-1}} = 1 + \dfrac{1}{f_{n-1}/f_{n-2}} = 1 + \dfrac{1}{a_{n-2}}$. If $L = \lim\limits_{n\to\infty} a_n$,

then $L = \lim\limits_{n\to\infty} a_{n-1}$ and $L = \lim\limits_{n\to\infty} a_{n-2}$, so L must satisfy $L = 1 + \dfrac{1}{L} \;\Rightarrow\; L^2 - L - 1 = 0 \;\Rightarrow\; L = \frac{1+\sqrt{5}}{2}$

[since L must be positive].

73. (a)

From the graph, it appears that the sequence $\left\{ \dfrac{n^5}{n!} \right\}$ converges to 0, that is, $\lim\limits_{n\to\infty} \dfrac{n^5}{n!} = 0$.

(b)

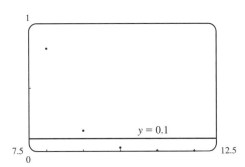

From the first graph, it seems that the smallest possible value of N corresponding to $\varepsilon = 0.1$ is 9, since $n^5/n! < 0.1$ whenever $n \geq 10$, but $9^5/9! > 0.1$. From the second graph, it seems that for $\varepsilon = 0.001$, the smallest possible value for N is 11 since $n^5/n! < 0.001$ whenever $n \geq 12$.

75. Theorem 6: If $\lim\limits_{n\to\infty} |a_n| = 0$ then $\lim\limits_{n\to\infty} -|a_n| = 0$, and since $-|a_n| \leq a_n \leq |a_n|$, we have that $\lim\limits_{n\to\infty} a_n = 0$ by the Squeeze Theorem.

77. To Prove: If $\lim\limits_{n\to\infty} a_n = 0$ and $\{b_n\}$ is bounded, then $\lim\limits_{n\to\infty} (a_n b_n) = 0$.

Proof: Since $\{b_n\}$ is bounded, there is a positive number M such that $|b_n| \leq M$ and hence, $|a_n| \, |b_n| \leq |a_n| \, M$ for all $n \geq 1$. Let $\varepsilon > 0$ be given. Since $\lim\limits_{n\to\infty} a_n = 0$, there is an integer N such that $|a_n - 0| < \dfrac{\varepsilon}{M}$ if $n > N$. Then

$$|a_n b_n - 0| = |a_n b_n| = |a_n| \, |b_n| \leq |a_n| \, M = |a_n - 0| \, M < \frac{\varepsilon}{M} \cdot M = \varepsilon \text{ for all } n > N. \text{ Since } \varepsilon \text{ was arbitrary,}$$

$\lim\limits_{n\to\infty} (a_n b_n) = 0$.

79. (a) First we show that $a > a_1 > b_1 > b$.

$$a_1 - b_1 = \tfrac{a+b}{2} - \sqrt{ab} = \tfrac{1}{2}\left(a - 2\sqrt{ab} + b\right) = \tfrac{1}{2}\left(\sqrt{a} - \sqrt{b}\right)^2 > 0 \quad [\text{since } a > b] \quad \Rightarrow \quad a_1 > b_1. \text{ Also}$$

$a - a_1 = a - \tfrac{1}{2}(a + b) = \tfrac{1}{2}(a - b) > 0$ and $b - b_1 = b - \sqrt{ab} = \sqrt{b}\left(\sqrt{b} - \sqrt{a}\right) < 0$, so $a > a_1 > b_1 > b$. In the same way we can show that $a_1 > a_2 > b_2 > b_1$ and so the given assertion is true for $n = 1$. Suppose it is true for $n = k$, that is, $a_k > a_{k+1} > b_{k+1} > b_k$. Then

$$a_{k+2} - b_{k+2} = \tfrac{1}{2}(a_{k+1} + b_{k+1}) - \sqrt{a_{k+1}b_{k+1}} = \tfrac{1}{2}\left(a_{k+1} - 2\sqrt{a_{k+1}b_{k+1}} + b_{k+1}\right) = \tfrac{1}{2}\left(\sqrt{a_{k+1}} - \sqrt{b_{k+1}}\right)^2 > 0,$$

$$a_{k+1} - a_{k+2} = a_{k+1} - \tfrac{1}{2}(a_{k+1} + b_{k+1}) = \tfrac{1}{2}(a_{k+1} - b_{k+1}) > 0, \text{ and}$$

$$b_{k+1} - b_{k+2} = b_{k+1} - \sqrt{a_{k+1}b_{k+1}} = \sqrt{b_{k+1}}\left(\sqrt{b_{k+1}} - \sqrt{a_{k+1}}\right) < 0 \quad \Rightarrow \quad a_{k+1} > a_{k+2} > b_{k+2} > b_{k+1},$$

so the assertion is true for $n = k + 1$. Thus, it is true for all n by mathematical induction.

(b) From part (a) we have $a > a_n > a_{n+1} > b_{n+1} > b_n > b$, which shows that both sequences, $\{a_n\}$ and $\{b_n\}$, are monotonic and bounded. So they are both convergent by the Monotonic Sequence Theorem.

(c) Let $\lim\limits_{n\to\infty} a_n = \alpha$ and $\lim\limits_{n\to\infty} b_n = \beta$. Then $\lim\limits_{n\to\infty} a_{n+1} = \lim\limits_{n\to\infty} \dfrac{a_n + b_n}{2} \quad \Rightarrow \quad \alpha = \dfrac{\alpha + \beta}{2} \quad \Rightarrow$

$2\alpha = \alpha + \beta \quad \Rightarrow \quad \alpha = \beta$.

81. (a) Suppose $\{p_n\}$ converges to p. Then $p_{n+1} = \dfrac{bp_n}{a + p_n}$ $\Rightarrow$ $\lim\limits_{n \to \infty} p_{n+1} = \dfrac{b \lim\limits_{n \to \infty} p_n}{a + \lim\limits_{n \to \infty} p_n}$ $\Rightarrow$ $p = \dfrac{bp}{a + p}$ $\Rightarrow$

$p^2 + ap = bp$ $\Rightarrow$ $p(p + a - b) = 0$ $\Rightarrow$ $p = 0$ or $p = b - a$.

(b) $p_{n+1} = \dfrac{bp_n}{a + p_n} = \dfrac{\left(\dfrac{b}{a}\right) p_n}{1 + \dfrac{p_n}{a}} < \left(\dfrac{b}{a}\right) p_n$ since $1 + \dfrac{p_n}{a} > 1$.

(c) By part (b), $p_1 < \left(\dfrac{b}{a}\right) p_0$, $p_2 < \left(\dfrac{b}{a}\right) p_1 < \left(\dfrac{b}{a}\right)^2 p_0$, $p_3 < \left(\dfrac{b}{a}\right) p_2 < \left(\dfrac{b}{a}\right)^3 p_0$, etc. In general, $p_n < \left(\dfrac{b}{a}\right)^n p_0$,

so $\lim\limits_{n \to \infty} p_n \leq \lim\limits_{n \to \infty} \left(\dfrac{b}{a}\right)^n \cdot p_0 = 0$ since $b < a$. $\left[\text{By result 9, } \lim\limits_{n \to \infty} r^n = 0 \text{ if } -1 < r < 1. \text{ Here } r = \dfrac{b}{a} \in (0, 1).\right]$

(d) Let $a < b$. We first show, by induction, that if $p_0 < b - a$, then $p_n < b - a$ and $p_{n+1} > p_n$.

For $n = 0$, we have $p_1 - p_0 = \dfrac{bp_0}{a + p_0} - p_0 = \dfrac{p_0(b - a - p_0)}{a + p_0} > 0$ since $p_0 < b - a$. So $p_1 > p_0$.

Now we suppose the assertion is true for $n = k$, that is, $p_k < b - a$ and $p_{k+1} > p_k$. Then

$b - a - p_{k+1} = b - a - \dfrac{bp_k}{a + p_k} = \dfrac{a(b - a) + bp_k - ap_k - bp_k}{a + p_k} = \dfrac{a(b - a - p_k)}{a + p_k} > 0$ because $p_k < b - a$. So

$p_{k+1} < b - a$. And $p_{k+2} - p_{k+1} = \dfrac{bp_{k+1}}{a + p_{k+1}} - p_{k+1} = \dfrac{p_{k+1}(b - a - p_{k+1})}{a + p_{k+1}} > 0$ since $p_{k+1} < b - a$. Therefore,

$p_{k+2} > p_{k+1}$. Thus, the assertion is true for $n = k + 1$. It is therefore true for all n by mathematical induction.

A similar proof by induction shows that if $p_0 > b - a$, then $p_n > b - a$ and $\{p_n\}$ is decreasing.

In either case the sequence $\{p_n\}$ is bounded and monotonic, so it is convergent by the Monotonic Sequence Theorem.
It then follows from part (a) that $\lim\limits_{n \to \infty} p_n = b - a$.

11.2 Series

1. (a) A sequence is an ordered list of numbers whereas a series is the *sum* of a list of numbers.

(b) A series is convergent if the sequence of partial sums is a convergent sequence. A series is divergent if it is not convergent.

3.

n	s_n
1	-2.40000
2	-1.92000
3	-2.01600
4	-1.99680
5	-2.00064
6	-1.99987
7	-2.00003
8	-1.99999
9	-2.00000
10	-2.00000

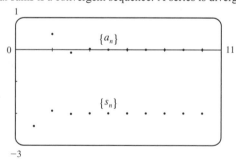

From the graph and the table, it seems that the series converges to -2. In fact, it is a geometric

series with $a = -2.4$ and $r = -\frac{1}{5}$, so its sum is $\sum\limits_{n=1}^{\infty} \dfrac{12}{(-5)^n} = \dfrac{-2.4}{1 - \left(-\frac{1}{5}\right)} = \dfrac{-2.4}{1.2} = -2$.

Note that the dot corresponding to $n = 1$ is part of both $\{a_n\}$ and $\{s_n\}$.

TI-86 Note: To graph $\{a_n\}$ and $\{s_n\}$, set your calculator to Param mode and DrawDot mode. (DrawDot is under

GRAPH, MORE, FORMT (F3).) Now under E(t) = make the assignments: `xt1=t, yt1=12/(-5)^t, xt2=t,`

`yt2=sum seq(yt1,t,1,t,1)`. (sum and seq are under LIST, OPS (F5), MORE.) Under WIND use

`1,10,1,0,10,1,-3,1,1` to obtain a graph similar to the one above. Then use TRACE (F4) to see the values.

5.

n	s_n
1	1.55741
2	−0.62763
3	−0.77018
4	0.38764
5	−2.99287
6	−3.28388
7	−2.41243
8	−9.21214
9	−9.66446
10	−9.01610

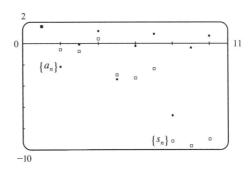

The series $\displaystyle\sum_{n=1}^{\infty} \tan n$ diverges, since its terms do not approach 0.

7.

n	s_n
1	0.29289
2	0.42265
3	0.50000
4	0.55279
5	0.59175
6	0.62204
7	0.64645
8	0.66667
9	0.68377
10	0.69849

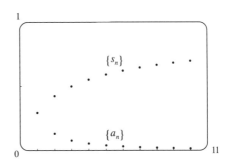

From the graph and the table, it seems that the series converges.

$$\sum_{n=1}^{k} \left(\frac{1}{\sqrt{n}} - \frac{1}{\sqrt{n+1}} \right) = \left(\frac{1}{\sqrt{1}} - \frac{1}{\sqrt{2}} \right) + \left(\frac{1}{\sqrt{2}} - \frac{1}{\sqrt{3}} \right) + \cdots + \left(\frac{1}{\sqrt{k}} - \frac{1}{\sqrt{k+1}} \right)$$

$$= 1 - \frac{1}{\sqrt{k+1}},$$

so $\displaystyle\sum_{n=1}^{\infty} \left(\frac{1}{\sqrt{n}} - \frac{1}{\sqrt{n+1}} \right) = \lim_{k \to \infty} \left(1 - \frac{1}{\sqrt{k+1}} \right) = 1.$

9. (a) $\displaystyle\lim_{n \to \infty} a_n = \lim_{n \to \infty} \frac{2n}{3n+1} = \frac{2}{3}$, so the *sequence* $\{a_n\}$ is convergent by (11.1.1).

(b) Since $\displaystyle\lim_{n \to \infty} a_n = \frac{2}{3} \neq 0$, the *series* $\displaystyle\sum_{n=1}^{\infty} a_n$ is divergent by the Test for Divergence.

11. $3 + 2 + \frac{4}{3} + \frac{8}{9} + \cdots$ is a geometric series with first term $a = 3$ and common ratio $r = \frac{2}{3}$. Since $|r| = \frac{2}{3} < 1$, the series

converges to $\frac{a}{1-r} = \frac{3}{1-2/3} = \frac{3}{1/3} = 9.$

13. $3 - 4 + \frac{16}{3} - \frac{64}{9} + \cdots$ is a geometric series with ratio $r = -\frac{4}{3}$. Since $|r| = \frac{4}{3} > 1$, the series diverges.

15. $\sum\limits_{n=1}^{\infty} 6(0.9)^{n-1}$ is a geometric series with first term $a = 6$ and ratio $r = 0.9$. Since $|r| = 0.9 < 1$, the series converges to

$$\frac{a}{1-r} = \frac{6}{1-0.9} = \frac{6}{0.1} = 60.$$

17. $\sum\limits_{n=1}^{\infty} \frac{(-3)^{n-1}}{4^n} = \frac{1}{4}\sum\limits_{n=1}^{\infty}\left(-\frac{3}{4}\right)^{n-1}$. The latter series is geometric with $a = 1$ and ratio $r = -\frac{3}{4}$. Since $|r| = \frac{3}{4} < 1$, it

converges to $\dfrac{1}{1-(-3/4)} = \frac{4}{7}$. Thus, the given series converges to $\left(\frac{1}{4}\right)\left(\frac{4}{7}\right) = \frac{1}{7}$.

19. $\sum\limits_{n=0}^{\infty} \frac{\pi^n}{3^{n+1}} = \frac{1}{3}\sum\limits_{n=0}^{\infty}\left(\frac{\pi}{3}\right)^n$ is a geometric series with ratio $r = \frac{\pi}{3}$. Since $|r| > 1$, the series diverges.

21. $\sum\limits_{n=1}^{\infty} \frac{1}{2n} = \frac{1}{2}\sum\limits_{n=1}^{\infty}\frac{1}{n}$ diverges since each of its partial sums is $\frac{1}{2}$ times the corresponding partial sum of the harmonic series

$\sum\limits_{n=1}^{\infty}\frac{1}{n}$, which diverges. $\left[\text{If } \sum\limits_{n=1}^{\infty}\frac{1}{2n} \text{ were to converge, then } \sum\limits_{n=1}^{\infty}\frac{1}{n} \text{ would also have to converge by Theorem 8(i).}\right]$

In general, constant multiples of divergent series are divergent.

23. $\sum\limits_{k=2}^{\infty} \frac{k^2}{k^2-1}$ diverges by the Test for Divergence since $\lim\limits_{k\to\infty} a_k = \lim\limits_{k\to\infty} \frac{k^2}{k^2-1} = 1 \neq 0$.

25. Converges.

$$\sum_{n=1}^{\infty} \frac{1+2^n}{3^n} = \sum_{n=1}^{\infty}\left(\frac{1}{3^n} + \frac{2^n}{3^n}\right) = \sum_{n=1}^{\infty}\left[\left(\frac{1}{3}\right)^n + \left(\frac{2}{3}\right)^n\right] \qquad \text{[sum of two convergent geometric series]}$$

$$= \frac{1/3}{1-1/3} + \frac{2/3}{1-2/3} = \frac{1}{2} + 2 = \frac{5}{2}$$

27. $\sum\limits_{n=1}^{\infty} \sqrt[n]{2} = 2 + \sqrt{2} + \sqrt[3]{2} + \sqrt[4]{2} + \cdots$ diverges by the Test for Divergence since

$$\lim_{n\to\infty} a_n = \lim_{n\to\infty} \sqrt[n]{2} = \lim_{n\to\infty} 2^{1/n} = 2^0 = 1 \neq 0.$$

29. $\sum\limits_{n=1}^{\infty} \ln\left(\frac{n^2+1}{2n^2+1}\right)$ diverges by the Test for Divergence since

$$\lim_{n\to\infty} a_n = \lim_{n\to\infty} \ln\left(\frac{n^2+1}{2n^2+1}\right) = \ln\left(\lim_{n\to\infty}\frac{n^2+1}{2n^2+1}\right) = \ln\frac{1}{2} \neq 0.$$

31. $\sum\limits_{n=1}^{\infty} \arctan n$ diverges by the Test for Divergence since $\lim\limits_{n\to\infty} a_n = \lim\limits_{n\to\infty} \arctan n = \frac{\pi}{2} \neq 0$.

33. $\sum\limits_{n=1}^{\infty} \frac{1}{e^n} = \sum\limits_{n=1}^{\infty}\left(\frac{1}{e}\right)^n$ is a geometric series with first term $a = \frac{1}{e}$ and ratio $r = \frac{1}{e}$. Since $|r| = \frac{1}{e} < 1$, the series converges

to $\dfrac{1/e}{1-1/e} = \dfrac{1/e}{1-1/e}\cdot\dfrac{e}{e} = \dfrac{1}{e-1}$. By Example 6, $\sum\limits_{n=1}^{\infty}\dfrac{1}{n(n+1)} = 1$. Thus, by Theorem 8(ii),

$$\sum_{n=1}^{\infty}\left(\frac{1}{e^n} + \frac{1}{n(n+1)}\right) = \sum_{n=1}^{\infty}\frac{1}{e^n} + \sum_{n=1}^{\infty}\frac{1}{n(n+1)} = \frac{1}{e-1} + 1 = \frac{1}{e-1} + \frac{e-1}{e-1} = \frac{e}{e-1}.$$

35. Using partial fractions, the partial sums of the series $\sum\limits_{n=2}^{\infty} \dfrac{2}{n^2 - 1}$ are

$$s_n = \sum_{i=2}^{n} \frac{2}{(i-1)(i+1)} = \sum_{i=2}^{n} \left(\frac{1}{i-1} - \frac{1}{i+1} \right)$$

$$= \left(1 - \frac{1}{3} \right) + \left(\frac{1}{2} - \frac{1}{4} \right) + \left(\frac{1}{3} - \frac{1}{5} \right) + \cdots + \left(\frac{1}{n-3} - \frac{1}{n-1} \right) + \left(\frac{1}{n-2} - \frac{1}{n} \right)$$

This sum is a telescoping series and $s_n = 1 + \dfrac{1}{2} - \dfrac{1}{n-1} - \dfrac{1}{n}$.

Thus, $\sum\limits_{n=2}^{\infty} \dfrac{2}{n^2 - 1} = \lim\limits_{n \to \infty} s_n = \lim\limits_{n \to \infty} \left(1 + \dfrac{1}{2} - \dfrac{1}{n-1} - \dfrac{1}{n} \right) = \dfrac{3}{2}$.

37. For the series $\sum\limits_{n=1}^{\infty} \dfrac{3}{n(n+3)}$, $s_n = \sum\limits_{i=1}^{n} \dfrac{3}{i(i+3)} = \sum\limits_{i=1}^{n} \left(\dfrac{1}{i} - \dfrac{1}{i+3} \right)$ [using partial fractions]. The latter sum is

$$\left(1 - \tfrac{1}{4} \right) + \left(\tfrac{1}{2} - \tfrac{1}{5} \right) + \left(\tfrac{1}{3} - \tfrac{1}{6} \right) + \left(\tfrac{1}{4} - \tfrac{1}{7} \right) + \cdots + \left(\tfrac{1}{n-3} - \tfrac{1}{n} \right) + \left(\tfrac{1}{n-2} - \tfrac{1}{n+1} \right) + \left(\tfrac{1}{n-1} - \tfrac{1}{n+2} \right) + \left(\tfrac{1}{n} - \tfrac{1}{n+3} \right)$$

$$= 1 + \tfrac{1}{2} + \tfrac{1}{3} - \tfrac{1}{n+1} - \tfrac{1}{n+2} - \tfrac{1}{n+3} \quad \text{[telescoping series]}$$

Thus, $\sum\limits_{n=1}^{\infty} \dfrac{3}{n(n+3)} = \lim\limits_{n \to \infty} s_n = \lim\limits_{n \to \infty} \left(1 + \tfrac{1}{2} + \tfrac{1}{3} - \tfrac{1}{n+1} - \tfrac{1}{n+2} - \tfrac{1}{n+3} \right) = 1 + \tfrac{1}{2} + \tfrac{1}{3} = \tfrac{11}{6}$.   Converges

39. For the series $\sum\limits_{n=1}^{\infty} \left(e^{1/n} - e^{1/(n+1)} \right)$,

$$s_n = \sum_{i=1}^{n} \left(e^{1/i} - e^{1/(i+1)} \right) = (e^1 - e^{1/2}) + (e^{1/2} - e^{1/3}) + \cdots + \left(e^{1/n} - e^{1/(n+1)} \right) = e - e^{1/(n+1)}$$

$$\text{[telescoping series]}$$

Thus, $\sum\limits_{n=1}^{\infty} \left(e^{1/n} - e^{1/(n+1)} \right) = \lim\limits_{n \to \infty} s_n = \lim\limits_{n \to \infty} \left(e - e^{1/(n+1)} \right) = e - e^0 = e - 1$.   Converges

41. $0.\overline{2} = \dfrac{2}{10} + \dfrac{2}{10^2} + \cdots$ is a geometric series with $a = \dfrac{2}{10}$ and $r = \dfrac{1}{10}$. It converges to $\dfrac{a}{1-r} = \dfrac{2/10}{1-1/10} = \dfrac{2}{9}$.

43. $3.\overline{417} = 3 + \dfrac{417}{10^3} + \dfrac{417}{10^6} + \cdots$. Now $\dfrac{417}{10^3} + \dfrac{417}{10^6} + \cdots$ is a geometric series with $a = \dfrac{417}{10^3}$ and $r = \dfrac{1}{10^3}$.

It converges to $\dfrac{a}{1-r} = \dfrac{417/10^3}{1-1/10^3} = \dfrac{417/10^3}{999/10^3} = \dfrac{417}{999}$. Thus, $3.\overline{417} = 3 + \dfrac{417}{999} = \dfrac{3414}{999} = \dfrac{1138}{333}$.

45. $1.5\overline{342} = 1.53 + \dfrac{42}{10^4} + \dfrac{42}{10^6} + \cdots$. Now $\dfrac{42}{10^4} + \dfrac{42}{10^6} + \cdots$ is a geometric series with $a = \dfrac{42}{10^4}$ and $r = \dfrac{1}{10^2}$.

It converges to $\dfrac{a}{1-r} = \dfrac{42/10^4}{1-1/10^2} = \dfrac{42/10^4}{99/10^2} = \dfrac{42}{9900}$.

Thus, $1.5\overline{342} = 1.53 + \dfrac{42}{9900} = \dfrac{153}{100} + \dfrac{42}{9900} = \dfrac{15{,}147}{9900} + \dfrac{42}{9900} = \dfrac{15{,}189}{9900}$ or $\dfrac{5063}{3300}$.

47. $\sum\limits_{n=1}^{\infty} \dfrac{x^n}{3^n} = \sum\limits_{n=1}^{\infty} \left(\dfrac{x}{3} \right)^n$ is a geometric series with $r = \dfrac{x}{3}$, so the series converges $\Leftrightarrow |r| < 1 \Leftrightarrow \dfrac{|x|}{3} < 1 \Leftrightarrow |x| < 3$;

that is, $-3 < x < 3$. In that case, the sum of the series is $\dfrac{a}{1-r} = \dfrac{x/3}{1-x/3} = \dfrac{x/3}{1-x/3} \cdot \dfrac{3}{3} = \dfrac{x}{3-x}$.

49. $\sum\limits_{n=0}^{\infty} 4^n x^n = \sum\limits_{n=0}^{\infty} (4x)^n$ is a geometric series with $r = 4x$, so the series converges $\Leftrightarrow |r| < 1 \Leftrightarrow 4|x| < 1 \Leftrightarrow$

$|x| < \frac{1}{4}$. In that case, the sum of the series is $\dfrac{1}{1-4x}$.

51. $\sum\limits_{n=0}^{\infty} \dfrac{\cos^n x}{2^n}$ is a geometric series with first term 1 and ratio $r = \dfrac{\cos x}{2}$, so it converges $\Leftrightarrow |r| < 1$. But $|r| = \dfrac{|\cos x|}{2} \le \dfrac{1}{2}$

for all x. Thus, the series converges for all real values of x and the sum of the series is $\dfrac{1}{1-(\cos x)/2} = \dfrac{2}{2-\cos x}$.

53. After defining f, We use `convert(f,parfrac);` in Maple, `Apart` in Mathematica, or `Expand Rational` and

`Simplify` in Derive to find that the general term is $\dfrac{3n^2 + 3n + 1}{(n^2 + n)^3} = \dfrac{1}{n^3} - \dfrac{1}{(n+1)^3}$. So the nth partial sum is

$$s_n = \sum_{k=1}^{n} \left(\frac{1}{k^3} - \frac{1}{(k+1)^3} \right) = \left(1 - \frac{1}{2^3} \right) + \left(\frac{1}{2^3} - \frac{1}{3^3} \right) + \cdots + \left(\frac{1}{n^3} - \frac{1}{(n+1)^3} \right) = 1 - \frac{1}{(n+1)^3}$$

The series converges to $\lim\limits_{n\to\infty} s_n = 1$. This can be confirmed by directly computing the sum using `sum(f,1..infinity);`

(in Maple), `Sum[f,{n,1,Infinity}]` (in Mathematica), or `Calculus Sum` (from 1 to ∞) and `Simplify` (in Derive).

55. For $n = 1$, $a_1 = 0$ since $s_1 = 0$. For $n > 1$,

$$a_n = s_n - s_{n-1} = \frac{n-1}{n+1} - \frac{(n-1)-1}{(n-1)+1} = \frac{(n-1)n - (n+1)(n-2)}{(n+1)n} = \frac{2}{n(n+1)}$$

Also, $\sum\limits_{n=1}^{\infty} a_n = \lim\limits_{n\to\infty} s_n = \lim\limits_{n\to\infty} \dfrac{1 - 1/n}{1 + 1/n} = 1$.

57. (a) The first step in the chain occurs when the local government spends D dollars. The people who receive it spend a fraction c

of those D dollars, that is, Dc dollars. Those who receive the Dc dollars spend a fraction c of it, that is, Dc^2 dollars.

Continuing in this way, we see that the total spending after n transactions is

$$S_n = D + Dc + Dc^2 + \cdots + Dc^{n-1} = \frac{D(1-c^n)}{1-c} \text{ by (3)}.$$

(b) $\lim\limits_{n\to\infty} S_n = \lim\limits_{n\to\infty} \dfrac{D(1-c^n)}{1-c} = \dfrac{D}{1-c} \lim\limits_{n\to\infty} (1-c^n) = \dfrac{D}{1-c} \quad \left[\text{since } 0 < c < 1 \;\Rightarrow\; \lim\limits_{n\to\infty} c^n = 0 \right]$

$= \dfrac{D}{s} \quad [\text{since } c + s = 1] = kD \quad [\text{since } k = 1/s]$

If $c = 0.8$, then $s = 1 - c = 0.2$ and the multiplier is $k = 1/s = 5$.

59. $\sum\limits_{n=2}^{\infty} (1+c)^{-n}$ is a geometric series with $a = (1+c)^{-2}$ and $r = (1+c)^{-1}$, so the series converges when

$\left|(1+c)^{-1}\right| < 1 \Leftrightarrow |1+c| > 1 \Leftrightarrow 1+c > 1$ or $1+c < -1 \Leftrightarrow c > 0$ or $c < -2$. We calculate the sum of the

series and set it equal to 2: $\dfrac{(1+c)^{-2}}{1-(1+c)^{-1}} = 2 \Leftrightarrow \left(\dfrac{1}{1+c} \right)^2 = 2 - 2\left(\dfrac{1}{1+c} \right) \Leftrightarrow 1 = 2(1+c)^2 - 2(1+c) \Leftrightarrow$

$2c^2 + 2c - 1 = 0 \Leftrightarrow c = \frac{-2 \pm \sqrt{12}}{4} = \frac{\pm\sqrt{3}-1}{2}$. However, the negative root is inadmissible because $-2 < \frac{-\sqrt{3}-1}{2} < 0$.

So $c = \frac{\sqrt{3}-1}{2}$.

61. $e^{s_n} = e^{1 + \frac{1}{2} + \frac{1}{3} + \cdots + \frac{1}{n}} = e^1 e^{1/2} e^{1/3} \cdots e^{1/n} > (1 + 1)\left(1 + \frac{1}{2}\right)\left(1 + \frac{1}{3}\right) \cdots \left(1 + \frac{1}{n}\right)$ $\quad [e^x > 1 + x]$

$= \dfrac{2}{1} \dfrac{3}{2} \dfrac{4}{3} \cdots \dfrac{n+1}{n} = n + 1$

Thus, $e^{s_n} > n + 1$ and $\lim\limits_{n \to \infty} e^{s_n} = \infty$. Since $\{s_n\}$ is increasing, $\lim\limits_{n \to \infty} s_n = \infty$, implying that the harmonic series is

divergent.

63. Let d_n be the diameter of C_n. We draw lines from the centers of the C_i to

the center of D (or C), and using the Pythagorean Theorem, we can write

$1^2 + \left(1 - \frac{1}{2}d_1\right)^2 = \left(1 + \frac{1}{2}d_1\right)^2 \quad \Leftrightarrow$

$1 = \left(1 + \frac{1}{2}d_1\right)^2 - \left(1 - \frac{1}{2}d_1\right)^2 = 2d_1 \text{ [difference of squares]} \quad \Rightarrow \quad d_1 = \frac{1}{2}.$

Similarly,

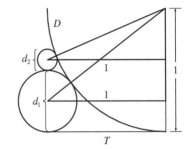

$1 = \left(1 + \frac{1}{2}d_2\right)^2 - \left(1 - d_1 - \frac{1}{2}d_2\right)^2 = 2d_2 + 2d_1 - d_1^2 - d_1 d_2$

$= (2 - d_1)(d_1 + d_2) \quad \Leftrightarrow$

$d_2 = \dfrac{1}{2 - d_1} - d_1 = \dfrac{(1 - d_1)^2}{2 - d_1}, \, 1 = \left(1 + \frac{1}{2}d_3\right)^2 - \left(1 - d_1 - d_2 - \frac{1}{2}d_3\right)^2 \quad \Leftrightarrow \quad d_3 = \dfrac{[1 - (d_1 + d_2)]^2}{2 - (d_1 + d_2)}$, and in general,

$d_{n+1} = \dfrac{\left(1 - \sum_{i=1}^{n} d_i\right)^2}{2 - \sum_{i=1}^{n} d_i}$. If we actually calculate d_2 and d_3 from the formulas above, we find that they are $\dfrac{1}{6} = \dfrac{1}{2 \cdot 3}$ and

$\dfrac{1}{12} = \dfrac{1}{3 \cdot 4}$ respectively, so we suspect that in general, $d_n = \dfrac{1}{n(n+1)}$. To prove this, we use induction: Assume that for all

$k \le n, \, d_k = \dfrac{1}{k(k+1)} = \dfrac{1}{k} - \dfrac{1}{k+1}$. Then $\sum\limits_{i=1}^{n} d_i = 1 - \dfrac{1}{n+1} = \dfrac{n}{n+1}$ [telescoping sum]. Substituting this into our

formula for d_{n+1}, we get $d_{n+1} = \dfrac{\left[1 - \dfrac{n}{n+1}\right]^2}{2 - \left(\dfrac{n}{n+1}\right)} = \dfrac{\dfrac{1}{(n+1)^2}}{\dfrac{n+2}{n+1}} = \dfrac{1}{(n+1)(n+2)}$, and the induction is complete.

Now, we observe that the partial sums $\sum_{i=1}^{n} d_i$ of the diameters of the circles approach 1 as $n \to \infty$; that is,

$\sum\limits_{n=1}^{\infty} a_n = \sum\limits_{n=1}^{\infty} \dfrac{1}{n(n+1)} = 1$, which is what we wanted to prove.

65. The series $1 - 1 + 1 - 1 + 1 - 1 + \cdots$ diverges (geometric series with $r = -1$) so we cannot say that

$0 = 1 - 1 + 1 - 1 + 1 - 1 + \cdots$.

67. $\sum\limits_{n=1}^{\infty} ca_n = \lim\limits_{n \to \infty} \sum\limits_{i=1}^{n} ca_i = \lim\limits_{n \to \infty} c \sum\limits_{i=1}^{n} a_i = c \lim\limits_{n \to \infty} \sum\limits_{i=1}^{n} a_i = c \sum\limits_{n=1}^{\infty} a_n$, which exists by hypothesis.

69. Suppose on the contrary that $\sum (a_n + b_n)$ converges. Then $\sum (a_n + b_n)$ and $\sum a_n$ are convergent series. So by Theorem 8,

$\sum [(a_n + b_n) - a_n]$ would also be convergent. But $\sum [(a_n + b_n) - a_n] = \sum b_n$, a contradiction, since $\sum b_n$ is given to be

divergent.

71. The partial sums $\{s_n\}$ form an increasing sequence, since $s_n - s_{n-1} = a_n > 0$ for all n. Also, the sequence $\{s_n\}$ is bounded

since $s_n \le 1000$ for all n. So by the Monotonic Sequence Theorem, the sequence of partial sums converges, that is, the series

$\sum a_n$ is convergent.

73. (a) At the first step, only the interval $\left(\frac{1}{3}, \frac{2}{3}\right)$ (length $\frac{1}{3}$) is removed. At the second step, we remove the intervals $\left(\frac{1}{9}, \frac{2}{9}\right)$ and $\left(\frac{7}{9}, \frac{8}{9}\right)$, which have a total length of $2 \cdot \left(\frac{1}{3}\right)^2$. At the third step, we remove 2^2 intervals, each of length $\left(\frac{1}{3}\right)^3$. In general, at the nth step we remove 2^{n-1} intervals, each of length $\left(\frac{1}{3}\right)^n$, for a length of $2^{n-1} \cdot \left(\frac{1}{3}\right)^n = \frac{1}{3}\left(\frac{2}{3}\right)^{n-1}$. Thus, the total length of all removed intervals is $\sum_{n=1}^{\infty} \frac{1}{3}\left(\frac{2}{3}\right)^{n-1} = \frac{1/3}{1 - 2/3} = 1$ [geometric series with $a = \frac{1}{3}$ and $r = \frac{2}{3}$]. Notice that at the nth step, the leftmost interval that is removed is $\left(\left(\frac{1}{3}\right)^n, \left(\frac{2}{3}\right)^n\right)$, so we never remove 0, and 0 is in the Cantor set. Also, the rightmost interval removed is $\left(1 - \left(\frac{2}{3}\right)^n, 1 - \left(\frac{1}{3}\right)^n\right)$, so 1 is never removed. Some other numbers in the Cantor set are $\frac{1}{3}, \frac{2}{3}, \frac{1}{9}, \frac{2}{9}, \frac{7}{9}$, and $\frac{8}{9}$.

(b) The area removed at the first step is $\frac{1}{9}$; at the second step, $8 \cdot \left(\frac{1}{9}\right)^2$; at the third step, $(8)^2 \cdot \left(\frac{1}{9}\right)^3$. In general, the area removed at the nth step is $(8)^{n-1}\left(\frac{1}{9}\right)^n = \frac{1}{9}\left(\frac{8}{9}\right)^{n-1}$, so the total area of all removed squares is

$$\sum_{n=1}^{\infty} \frac{1}{9}\left(\frac{8}{9}\right)^{n-1} = \frac{1/9}{1 - 8/9} = 1.$$

75. (a) For $\sum_{n=1}^{\infty} \frac{n}{(n+1)!}$, $s_1 = \frac{1}{1 \cdot 2} = \frac{1}{2}$, $s_2 = \frac{1}{2} + \frac{2}{1 \cdot 2 \cdot 3} = \frac{5}{6}$, $s_3 = \frac{5}{6} + \frac{3}{1 \cdot 2 \cdot 3 \cdot 4} = \frac{23}{24}$,

$s_4 = \frac{23}{24} + \frac{4}{1 \cdot 2 \cdot 3 \cdot 4 \cdot 5} = \frac{119}{120}$. The denominators are $(n+1)!$, so a guess would be $s_n = \frac{(n+1)! - 1}{(n+1)!}$.

(b) For $n = 1$, $s_1 = \frac{1}{2} = \frac{2! - 1}{2!}$, so the formula holds for $n = 1$. Assume $s_k = \frac{(k+1)! - 1}{(k+1)!}$. Then

$$s_{k+1} = \frac{(k+1)! - 1}{(k+1)!} + \frac{k+1}{(k+2)!} = \frac{(k+1)! - 1}{(k+1)!} + \frac{k+1}{(k+1)!(k+2)} = \frac{(k+2)! - (k+2) + k + 1}{(k+2)!}$$

$$= \frac{(k+2)! - 1}{(k+2)!}$$

Thus, the formula is true for $n = k + 1$. So by induction, the guess is correct.

(c) $\lim_{n \to \infty} s_n = \lim_{n \to \infty} \frac{(n+1)! - 1}{(n+1)!} = \lim_{n \to \infty} \left[1 - \frac{1}{(n+1)!}\right] = 1$ and so $\sum_{n=1}^{\infty} \frac{n}{(n+1)!} = 1$.

11.3 The Integral Test and Estimates of Sums

1. The picture shows that $a_2 = \frac{1}{2^{1.3}} < \int_1^2 \frac{1}{x^{1.3}} \, dx$,

$a_3 = \frac{1}{3^{1.3}} < \int_2^3 \frac{1}{x^{1.3}} \, dx$, and so on, so $\sum_{n=2}^{\infty} \frac{1}{n^{1.3}} < \int_1^{\infty} \frac{1}{x^{1.3}} \, dx$. The integral converges by (7.8.2) with $p = 1.3 > 1$, so the series converges.

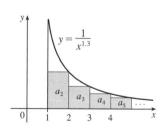

3. The function $f(x) = 1/\sqrt[5]{x} = x^{-1/5}$ is continuous, positive, and decreasing on $[1, \infty)$, so the Integral Test applies.

$\int_1^{\infty} x^{-1/5} \, dx = \lim_{t \to \infty} \int_1^t x^{-1/5} \, dx = \lim_{t \to \infty} \left[\frac{5}{4} x^{4/5}\right]_1^t = \lim_{t \to \infty} \left(\frac{5}{4} t^{4/5} - \frac{5}{4}\right) = \infty$, so $\sum_{n=1}^{\infty} 1/\sqrt[5]{n}$ diverges.

5. The function $f(x) = \dfrac{1}{(2x+1)^3}$ is continuous, positive, and decreasing on $[1, \infty)$, so the Integral Test applies.

$$\int_1^\infty \frac{1}{(2x+1)^3}\,dx = \lim_{t\to\infty} \int_1^t \frac{1}{(2x+1)^3}\,dx = \lim_{t\to\infty}\left[-\frac{1}{4}\frac{1}{(2x+1)^2}\right]_1^t = \lim_{t\to\infty}\left(-\frac{1}{4(2t+1)^2} + \frac{1}{36}\right) = \frac{1}{36}.$$

Since this improper integral is convergent, the series $\displaystyle\sum_{n=1}^\infty \frac{1}{(2n+1)^3}$ is also convergent by the Integral Test.

7. $f(x) = xe^{-x}$ is continuous and positive on $[1, \infty)$. $f'(x) = -xe^{-x} + e^{-x} = e^{-x}(1-x) < 0$ for $x > 1$, so f is decreasing on $[1, \infty)$. Thus, the Integral Test applies.

$$\int_1^\infty xe^{-x}\,dx = \lim_{b\to\infty}\int_1^b xe^{-x}\,dx = \lim_{b\to\infty}\left[-xe^{-x} - e^{-x}\right]_1^b \text{ [by parts] } = \lim_{b\to\infty}\left[-be^{-b} - e^{-b} + e^{-1} + e^{-1}\right] = 2/e$$

since $\displaystyle\lim_{b\to\infty} be^{-b} = \lim_{b\to\infty}(b/e^b) \overset{\text{H}}{=} \lim_{b\to\infty}(1/e^b) = 0$ and $\displaystyle\lim_{b\to\infty} e^{-b} = 0$. Thus, $\sum_{n=1}^\infty ne^{-n}$ converges.

9. The series $\displaystyle\sum_{n=1}^\infty \frac{1}{n^{0.85}}$ is a p-series with $p = 0.85 \le 1$, so it diverges by (1). Therefore, the series $\displaystyle\sum_{n=1}^\infty \frac{2}{n^{0.85}}$ must also diverge,

for if it converged, then $\displaystyle\sum_{n=1}^\infty \frac{1}{n^{0.85}}$ would have to converge [by Theorem 8(i) in Section 11.2].

11. $1 + \dfrac{1}{8} + \dfrac{1}{27} + \dfrac{1}{64} + \dfrac{1}{125} + \cdots = \displaystyle\sum_{n=1}^\infty \frac{1}{n^3}$. This is a p-series with $p = 3 > 1$, so it converges by (1).

13. $1 + \dfrac{1}{3} + \dfrac{1}{5} + \dfrac{1}{7} + \dfrac{1}{9} + \cdots = \displaystyle\sum_{n=1}^\infty \frac{1}{2n-1}$. The function $f(x) = \dfrac{1}{2x-1}$ is

continuous, positive, and decreasing on $[1, \infty)$, so the Integral Test applies.

$$\int_1^\infty \frac{1}{2x-1}\,dx = \lim_{t\to\infty}\int_1^t \frac{1}{2x-1}\,dx = \lim_{t\to\infty}\left[\tfrac{1}{2}\ln|2x-1|\right]_1^t = \tfrac{1}{2}\lim_{t\to\infty}\left(\ln(2t-1) - 0\right) = \infty, \text{ so the series } \sum_{n=1}^\infty \frac{1}{2n-1}$$

diverges.

15. $\displaystyle\sum_{n=1}^\infty \frac{5 - 2\sqrt{n}}{n^3} = 5\sum_{n=1}^\infty \frac{1}{n^3} - 2\sum_{n=1}^\infty \frac{1}{n^{5/2}}$ by Theorem 11.2.8, since $\displaystyle\sum_{n=1}^\infty \frac{1}{n^3}$ and $\displaystyle\sum_{n=1}^\infty \frac{1}{n^{5/2}}$ both converge by (1)

$\left[\text{with } p = 3 > 1 \text{ and } p = \tfrac{5}{2} > 1\right]$. Thus, $\displaystyle\sum_{n=1}^\infty \frac{5 - 2\sqrt{n}}{n^3}$ converges.

17. The function $f(x) = \dfrac{1}{x^2 + 4}$ is continuous, positive, and decreasing on $[1, \infty)$, so we can apply the Integral Test.

$$\int_1^\infty \frac{1}{x^2+4}\,dx = \lim_{t\to\infty}\int_1^t \frac{1}{x^2+4}\,dx = \lim_{t\to\infty}\left[\frac{1}{2}\tan^{-1}\frac{x}{2}\right]_1^t = \frac{1}{2}\lim_{t\to\infty}\left[\tan^{-1}\left(\frac{t}{2}\right) - \tan^{-1}\left(\frac{1}{2}\right)\right]$$

$$= \frac{1}{2}\left[\frac{\pi}{2} - \tan^{-1}\left(\frac{1}{2}\right)\right]$$

Therefore, the series $\displaystyle\sum_{n=1}^\infty \frac{1}{n^2 + 4}$ converges.

19. $\displaystyle\sum_{n=1}^\infty \frac{\ln n}{n^3} = \sum_{n=2}^\infty \frac{\ln n}{n^3}$ since $\dfrac{\ln 1}{1} = 0$. The function $f(x) = \dfrac{\ln x}{x^3}$ is continuous and positive on $[2, \infty)$.

$$f'(x) = \frac{x^3(1/x) - (\ln x)(3x^2)}{(x^3)^2} = \frac{x^2 - 3x^2\ln x}{x^6} = \frac{1 - 3\ln x}{x^4} < 0 \iff 1 - 3\ln x < 0 \iff \ln x > \tfrac{1}{3} \iff$$

$x > e^{1/3} \approx 1.4$, so f is decreasing on $[2, \infty)$, and the Integral Test applies.

$$\int_2^\infty \frac{\ln x}{x^3}\, dx = \lim_{t\to\infty} \int_2^t \frac{\ln x}{x^3}\, dx \overset{(\star)}{=} \lim_{t\to\infty} \left[-\frac{\ln x}{2x^2} - \frac{1}{4x^2} \right]_1^t = \lim_{t\to\infty} \left[-\frac{1}{4t^2}(2\ln t + 1) + \frac{1}{4} \right] \overset{(\star\star)}{=} \frac{1}{4}, \text{ so the series } \sum_{n=2}^\infty \frac{\ln n}{n^3}$$

converges.

$(\star)$: $u = \ln x$, $dv = x^{-3}\, dx$ $\Rightarrow$ $du = (1/x)\, dx$, $v = -\frac{1}{2}x^{-2}$, so

$$\int \frac{\ln x}{x^3}\, dx = -\frac{1}{2}x^{-2}\ln x - \int -\frac{1}{2}x^{-2}(1/x)\, dx = -\frac{1}{2}x^{-2}\ln x + \frac{1}{2}\int x^{-3}\, dx = -\frac{1}{2}x^{-2}\ln x - \frac{1}{4}x^{-2} + C.$$

$(\star\star)$: $\displaystyle \lim_{t\to\infty} \left(-\frac{2\ln t + 1}{4t^2} \right) \overset{H}{=} -\lim_{t\to\infty} \frac{2/t}{8t} = -\frac{1}{4}\lim_{t\to\infty} \frac{1}{t^2} = 0.$

21. $f(x) = \dfrac{1}{x\ln x}$ is continuous and positive on $[2, \infty)$, and also decreasing since $f'(x) = -\dfrac{1+\ln x}{x^2(\ln x)^2} < 0$ for $x > 2$, so we can

use the Integral Test. $\displaystyle \int_2^\infty \frac{1}{x\ln x}\, dx = \lim_{t\to\infty} \left[\ln(\ln x) \right]_2^t = \lim_{t\to\infty} \left[\ln(\ln t) - \ln(\ln 2) \right] = \infty$, so the series $\displaystyle \sum_{n=2}^\infty \frac{1}{n\ln n}$ diverges.

23. The function $f(x) = e^{1/x}/x^2$ is continuous, positive, and decreasing on $[1, \infty)$, so the Integral Test applies.

$[g(x) = e^{1/x}$ is decreasing and dividing by x^2 doesn't change that fact.]

$$\int_1^\infty f(x)\, dx = \lim_{t\to\infty} \int_1^t \frac{e^{1/x}}{x^2}\, dx = \lim_{t\to\infty} \left[-e^{1/x} \right]_1^t = -\lim_{t\to\infty} (e^{1/t} - e) = -(1 - e) = e - 1, \text{ so the series } \sum_{n=1}^\infty \frac{e^{1/n}}{n^2}$$

converges.

25. The function $f(x) = \dfrac{1}{x^3 + x}$ is continuous, positive, and decreasing on $[1, \infty)$, so the Integral Test applies. We use partial

fractions to evaluate the integral:

$$\int_1^\infty \frac{1}{x^3 + x}\, dx = \lim_{t\to\infty} \int_1^t \left[\frac{1}{x} - \frac{x}{1 + x^2} \right] dx = \lim_{t\to\infty} \left[\ln x - \frac{1}{2}\ln(1 + x^2) \right]_1^t = \lim_{t\to\infty} \left[\ln \frac{x}{\sqrt{1 + x^2}} \right]_1^t$$

$$= \lim_{t\to\infty} \left(\ln \frac{t}{\sqrt{1 + t^2}} - \ln \frac{1}{\sqrt{2}} \right) = \lim_{t\to\infty} \left(\ln \frac{1}{\sqrt{1 + 1/t^2}} + \frac{1}{2}\ln 2 \right) = \frac{1}{2}\ln 2$$

so the series $\displaystyle \sum_{n=1}^\infty \frac{1}{n^3 + n}$ converges.

27. We have already shown (in Exercise 21) that when $p = 1$ the series $\displaystyle \sum_{n=2}^\infty \frac{1}{n(\ln n)^p}$ diverges, so assume that $p \neq 1$.

$f(x) = \dfrac{1}{x(\ln x)^p}$ is continuous and positive on $[2, \infty)$, and $f'(x) = -\dfrac{p + \ln x}{x^2(\ln x)^{p+1}} < 0$ if $x > e^{-p}$, so that f is eventually

decreasing and we can use the Integral Test.

$$\int_2^\infty \frac{1}{x(\ln x)^p}\, dx = \lim_{t\to\infty} \left[\frac{(\ln x)^{1-p}}{1 - p} \right]_2^t \quad [\text{for } p \neq 1] = \lim_{t\to\infty} \left[\frac{(\ln t)^{1-p}}{1 - p} - \frac{(\ln 2)^{1-p}}{1 - p} \right]$$

This limit exists whenever $1 - p < 0$ $\Leftrightarrow$ $p > 1$, so the series converges for $p > 1$.

29. Clearly the series cannot converge if $p \geq -\frac{1}{2}$, because then $\lim_{n\to\infty} n(1+n^2)^p \neq 0$. So assume $p < -\frac{1}{2}$. Then

$f(x) = x(1+x^2)^p$ is continuous, positive, and eventually decreasing on $[1, \infty)$, and we can use the Integral Test.

$$\int_1^\infty x(1+x^2)^p dx = \lim_{t\to\infty} \left[\frac{1}{2} \cdot \frac{(1+x^2)^{p+1}}{p+1}\right]_1^t = \frac{1}{2(p+1)} \lim_{t\to\infty} [(1+t^2)^{p+1} - 2^{p+1}].$$

This limit exists and is finite $\Leftrightarrow p+1 < 0 \Leftrightarrow p < -1$, so the series converges whenever $p < -1$.

31. Since this is a p-series with $p = x$, $\zeta(x)$ is defined when $x > 1$. Unless specified otherwise, the domain of a function f is the set of real numbers x such that the expression for $f(x)$ makes sense and defines a real number. So, in the case of a series, it's the set of real numbers x such that the series is convergent.

33. (a) $f(x) = \frac{1}{x^2}$ is positive and continuous and $f'(x) = -\frac{2}{x^3}$ is negative for $x > 0$, and so the Integral Test applies.

$$\sum_{n=1}^\infty \frac{1}{n^2} \approx s_{10} = \frac{1}{1^2} + \frac{1}{2^2} + \frac{1}{3^2} + \cdots + \frac{1}{10^2} \approx 1.549768.$$

$$R_{10} \leq \int_{10}^\infty \frac{1}{x^2} dx = \lim_{t\to\infty}\left[\frac{-1}{x}\right]_{10}^t = \lim_{t\to\infty}\left(-\frac{1}{t}+\frac{1}{10}\right) = \frac{1}{10}, \text{ so the error is at most } 0.1.$$

(b) $s_{10} + \int_{11}^\infty \frac{1}{x^2} dx \leq s \leq s_{10} + \int_{10}^\infty \frac{1}{x^2} dx \Rightarrow s_{10} + \frac{1}{11} \leq s \leq s_{10} + \frac{1}{10} \Rightarrow$

$1.549768 + 0.090909 = 1.640677 \leq s \leq 1.549768 + 0.1 = 1.649768$, so we get $s \approx 1.64522$ (the average of 1.640677 and 1.649768) with error ≤ 0.005 (the maximum of $1.649768 - 1.64522$ and $1.64522 - 1.640677$, rounded up).

(c) $R_n \leq \int_n^\infty \frac{1}{x^2} dx = \frac{1}{n}$. So $R_n < 0.001$ if $\frac{1}{n} < \frac{1}{1000} \Leftrightarrow n > 1000$.

35. $f(x) = 1/(2x+1)^6$ is continuous, positive, and decreasing on $[1, \infty)$, so the Integral Test applies. Using (2),

$$R_n \leq \int_n^\infty (2x+1)^{-6} dx = \lim_{t\to\infty}\left[\frac{-1}{10(2x+1)^5}\right]_n^t = \frac{1}{10(2n+1)^5}. \text{ To be correct to five decimal places, we want}$$

$$\frac{1}{10(2n+1)^5} \leq \frac{5}{10^6} \Leftrightarrow (2n+1)^5 \geq 20,000 \Leftrightarrow n \geq \frac{1}{2}(\sqrt[5]{20,000} - 1) \approx 3.12, \text{ so use } n = 4.$$

$$s_4 = \sum_{n=1}^4 \frac{1}{(2n+1)^6} = \frac{1}{3^6} + \frac{1}{5^6} + \frac{1}{7^6} + \frac{1}{9^6} \approx 0.001\,446 \approx 0.00145.$$

37. $\sum_{n=1}^\infty n^{-1.001} = \sum_{n=1}^\infty \frac{1}{n^{1.001}}$ is a convergent p-series with $p = 1.001 > 1$. Using (2), we get

$$R_n \leq \int_n^\infty x^{-1.001} dx = \lim_{t\to\infty}\left[\frac{x^{-0.001}}{-0.001}\right]_n^t = -1000 \lim_{t\to\infty}\left[\frac{1}{x^{0.001}}\right]_n^t = -1000\left(-\frac{1}{n^{0.001}}\right) = \frac{1000}{n^{0.001}}.$$

We want $R_n < 0.000\,000\,005 \Leftrightarrow \frac{1000}{n^{0.001}} < 5 \times 10^{-9} \Leftrightarrow n^{0.001} > \frac{1000}{5 \times 10^{-9}} \Leftrightarrow$

$n > (2 \times 10^{11})^{1000} = 2^{1000} \times 10^{11,000} \approx 1.07 \times 10^{301} \times 10^{11,000} = 1.07 \times 10^{11,301}.$

39. (a) From the figure, $a_2 + a_3 + \cdots + a_n \leq \int_1^n f(x)\, dx$, so with

$$f(x) = \frac{1}{x}, \frac{1}{2} + \frac{1}{3} + \frac{1}{4} + \cdots + \frac{1}{n} \leq \int_1^n \frac{1}{x}\, dx = \ln n.$$

Thus, $s_n = 1 + \frac{1}{2} + \frac{1}{3} + \frac{1}{4} + \cdots + \frac{1}{n} \leq 1 + \ln n.$

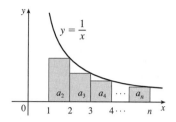

(b) By part (a), $s_{10^6} \leq 1 + \ln 10^6 \approx 14.82 < 15$ and

$s_{10^9} \leq 1 + \ln 10^9 \approx 21.72 < 22.$

41. $b^{\ln n} = \left(e^{\ln b}\right)^{\ln n} = \left(e^{\ln n}\right)^{\ln b} = n^{\ln b} = \dfrac{1}{n^{-\ln b}}$. This is a p-series, which converges for all b such that $-\ln b > 1$ ⟺

$\ln b < -1$ ⟺ $b < e^{-1}$ ⟺ $b < 1/e$ [with $b > 0$].

11.4 The Comparison Tests

1. (a) We cannot say anything about $\sum a_n$. If $a_n > b_n$ for all n and $\sum b_n$ is convergent, then $\sum a_n$ could be convergent or divergent. (See the note after Example 2.)

(b) If $a_n < b_n$ for all n, then $\sum a_n$ is convergent. [This is part (i) of the Comparison Test.]

3. $\dfrac{n}{2n^3 + 1} < \dfrac{n}{2n^3} = \dfrac{1}{2n^2} < \dfrac{1}{n^2}$ for all $n \geq 1$, so $\displaystyle\sum_{n=1}^{\infty} \dfrac{n}{2n^3 + 1}$ converges by comparison with $\displaystyle\sum_{n=1}^{\infty} \dfrac{1}{n^2}$, which converges

because it is a p-series with $p = 2 > 1$.

5. $\dfrac{n+1}{n\sqrt{n}} > \dfrac{n}{n\sqrt{n}} = \dfrac{1}{\sqrt{n}}$ for all $n \geq 1$, so $\displaystyle\sum_{n=1}^{\infty} \dfrac{n+1}{n\sqrt{n}}$ diverges by comparison with $\displaystyle\sum_{n=1}^{\infty} \dfrac{1}{\sqrt{n}}$, which diverges because it is a

p-series with $p = \frac{1}{2} \leq 1$.

7. $\dfrac{9^n}{3 + 10^n} < \dfrac{9^n}{10^n} = \left(\dfrac{9}{10}\right)^n$ for all $n \geq 1$. $\displaystyle\sum_{n=1}^{\infty} \left(\frac{9}{10}\right)^n$ is a convergent geometric series $\left(|r| = \frac{9}{10} < 1\right)$, so $\displaystyle\sum_{n=1}^{\infty} \dfrac{9^n}{3 + 10^n}$

converges by the Comparison Test.

9. $\dfrac{\cos^2 n}{n^2 + 1} \leq \dfrac{1}{n^2 + 1} < \dfrac{1}{n^2}$, so the series $\displaystyle\sum_{n=1}^{\infty} \dfrac{\cos^2 n}{n^2 + 1}$ converges by comparison with the p-series $\displaystyle\sum_{n=1}^{\infty} \dfrac{1}{n^2}$ $[p = 2 > 1]$.

11. $\dfrac{n-1}{n\,4^n}$ is positive for $n > 1$ and $\dfrac{n-1}{n\,4^n} < \dfrac{n}{n\,4^n} = \dfrac{1}{4^n} = \left(\dfrac{1}{4}\right)^n$, so $\displaystyle\sum_{n=1}^{\infty} \dfrac{n-1}{n\,4^n}$ converges by comparison with the convergent

geometric series $\displaystyle\sum_{n=1}^{\infty} \left(\dfrac{1}{4}\right)^n$.

13. $\dfrac{\arctan n}{n^{1.2}} < \dfrac{\pi/2}{n^{1.2}}$ for all $n \geq 1$, so $\displaystyle\sum_{n=1}^{\infty} \dfrac{\arctan n}{n^{1.2}}$ converges by comparison with $\dfrac{\pi}{2}\displaystyle\sum_{n=1}^{\infty} \dfrac{1}{n^{1.2}}$, which converges because it is a

constant times a p-series with $p = 1.2 > 1$.

15. $\dfrac{2 + (-1)^n}{n\sqrt{n}} \leq \dfrac{3}{n\sqrt{n}}$, and $\displaystyle\sum_{n=1}^{\infty} \dfrac{3}{n\sqrt{n}}$ converges because it is a constant multiple of the convergent p-series $\displaystyle\sum_{n=1}^{\infty} \dfrac{1}{n\sqrt{n}}$

$\left[p = \frac{3}{2} > 1\right]$, so the given series converges by the Comparison Test.

17. Use the Limit Comparison Test with $a_n = \dfrac{1}{\sqrt{n^2+1}}$ and $b_n = \dfrac{1}{n}$:

$\lim\limits_{n\to\infty} \dfrac{a_n}{b_n} = \lim\limits_{n\to\infty} \dfrac{n}{\sqrt{n^2+1}} = \lim\limits_{n\to\infty} \dfrac{1}{\sqrt{1+(1/n^2)}} = 1 > 0.$ Since the harmonic series $\sum\limits_{n=1}^{\infty} \dfrac{1}{n}$ diverges, so does

$\sum\limits_{n=1}^{\infty} \dfrac{1}{\sqrt{n^2+1}}.$

19. Use the Limit Comparison Test with $a_n = \dfrac{1+4^n}{1+3^n}$ and $b_n = \dfrac{4^n}{3^n}$:

$\lim\limits_{n\to\infty} \dfrac{a_n}{b_n} = \lim\limits_{n\to\infty} \dfrac{\frac{1+4^n}{1+3^n}}{\frac{4^n}{3^n}} = \lim\limits_{n\to\infty} \dfrac{1+4^n}{1+3^n} \cdot \dfrac{3^n}{4^n} = \lim\limits_{n\to\infty} \dfrac{1+4^n}{4^n} \cdot \dfrac{3^n}{1+3^n} = \lim\limits_{n\to\infty} \left(\dfrac{1}{4^n}+1\right) \cdot \dfrac{1}{\frac{1}{3^n}+1} = 1 > 0$

Since the geometric series $\sum b_n = \sum \left(\frac{4}{3}\right)^n$ diverges, so does $\sum\limits_{n=1}^{\infty} \dfrac{1+4^n}{1+3^n}$. Alternatively, use the Comparison Test with

$\dfrac{1+4^n}{1+3^n} > \dfrac{1+4^n}{3^n+3^n} > \dfrac{4^n}{2(3^n)} = \dfrac{1}{2}\left(\dfrac{4}{3}\right)^n$ or use the Test for Divergence.

21. Use the Limit Comparison Test with $a_n = \dfrac{\sqrt{n+2}}{2n^2+n+1}$ and $b_n = \dfrac{1}{n^{3/2}}$:

$\lim\limits_{n\to\infty} \dfrac{a_n}{b_n} = \lim\limits_{n\to\infty} \dfrac{n^{3/2}\sqrt{n+2}}{2n^2+n+1} = \lim\limits_{n\to\infty} \dfrac{(n^{3/2}\sqrt{n+2})/(n^{3/2}\sqrt{n})}{(2n^2+n+1)/n^2} = \lim\limits_{n\to\infty} \dfrac{\sqrt{1+2/n}}{2+1/n+1/n^2} = \dfrac{\sqrt{1}}{2} = \dfrac{1}{2} > 0.$

Since $\sum\limits_{n=1}^{\infty} \dfrac{1}{n^{3/2}}$ is a convergent p-series $\left[p = \frac{3}{2} > 1\right]$, the series $\sum\limits_{n=1}^{\infty} \dfrac{\sqrt{n+2}}{2n^2+n+1}$ also converges.

23. Use the Limit Comparison Test with $a_n = \dfrac{5+2n}{(1+n^2)^2}$ and $b_n = \dfrac{1}{n^3}$:

$\lim\limits_{n\to\infty} \dfrac{a_n}{b_n} = \lim\limits_{n\to\infty} \dfrac{n^3(5+2n)}{(1+n^2)^2} = \lim\limits_{n\to\infty} \dfrac{5n^3+2n^4}{(1+n^2)^2} \cdot \dfrac{1/n^4}{1/(n^2)^2} = \lim\limits_{n\to\infty} \dfrac{\frac{5}{n}+2}{\left(\frac{1}{n^2}+1\right)^2} = 2 > 0.$ Since $\sum\limits_{n=1}^{\infty} \dfrac{1}{n^3}$ is a convergent

p-series $[p = 3 > 1]$, the series $\sum\limits_{n=1}^{\infty} \dfrac{5+2n}{(1+n^2)^2}$ also converges.

25. If $a_n = \dfrac{1+n+n^2}{\sqrt{1+n^2+n^6}}$ and $b_n = \dfrac{1}{n}$, then $\lim\limits_{n\to\infty} \dfrac{a_n}{b_n} = \lim\limits_{n\to\infty} \dfrac{n+n^2+n^3}{\sqrt{1+n^2+n^6}} = \lim\limits_{n\to\infty} \dfrac{1/n^2+1/n+1}{\sqrt{1/n^6+1/n^4+1}} = 1 > 0,$

so $\sum\limits_{n=1}^{\infty} \dfrac{1+n+n^2}{\sqrt{1+n^2+n^6}}$ diverges by the Limit Comparison Test with the divergent harmonic series $\sum\limits_{n=1}^{\infty} \dfrac{1}{n}$.

27. Use the Limit Comparison Test with $a_n = \left(1+\dfrac{1}{n}\right)^2 e^{-n}$ and $b_n = e^{-n}$: $\lim\limits_{n\to\infty} \dfrac{a_n}{b_n} = \lim\limits_{n\to\infty} \left(1+\dfrac{1}{n}\right)^2 = 1 > 0.$ Since

$\sum\limits_{n=1}^{\infty} e^{-n} = \sum\limits_{n=1}^{\infty} \dfrac{1}{e^n}$ is a convergent geometric series $\left[|r| = \frac{1}{e} < 1\right]$, the series $\sum\limits_{n=1}^{\infty} \left(1+\dfrac{1}{n}\right)^2 e^{-n}$ also converges.

29. Clearly $n! = n(n-1)(n-2)\cdots(3)(2) \geq 2\cdot2\cdot2\cdots\cdots2\cdot2 = 2^{n-1}$, so $\dfrac{1}{n!} \leq \dfrac{1}{2^{n-1}}$. $\sum\limits_{n=1}^{\infty} \dfrac{1}{2^{n-1}}$ is a convergent geometric

series $\left[|r| = \frac{1}{2} < 1\right]$, so $\sum\limits_{n=1}^{\infty} \dfrac{1}{n!}$ converges by the Comparison Test.

31. Use the Limit Comparison Test with $a_n = \sin\left(\dfrac{1}{n}\right)$ and $b_n = \dfrac{1}{n}$. Then $\sum a_n$ and $\sum b_n$ are series with positive terms and

$$\lim_{n\to\infty} \frac{a_n}{b_n} = \lim_{n\to\infty} \frac{\sin(1/n)}{1/n} = \lim_{\theta\to 0} \frac{\sin\theta}{\theta} = 1 > 0. \text{ Since } \sum_{n=1}^{\infty} b_n \text{ is the divergent harmonic series,}$$

$\sum_{n=1}^{\infty} \sin\left(1/n\right)$ also diverges. [Note that we could also use l'Hospital's Rule to evaluate the limit:

$$\lim_{x\to\infty} \frac{\sin(1/x)}{1/x} \overset{H}{=} \lim_{x\to\infty} \frac{\cos(1/x)\cdot\left(-1/x^2\right)}{-1/x^2} = \lim_{x\to\infty} \cos\frac{1}{x} = \cos 0 = 1.]$$

33. $\sum_{n=1}^{10} \dfrac{1}{\sqrt{n^4+1}} = \dfrac{1}{\sqrt{2}} + \dfrac{1}{\sqrt{17}} + \dfrac{1}{\sqrt{82}} + \cdots + \dfrac{1}{\sqrt{10{,}001}} \approx 1.24856.$ Now $\dfrac{1}{\sqrt{n^4+1}} < \dfrac{1}{\sqrt{n^4}} = \dfrac{1}{n^2}$, so the error is

$$R_{10} \leq T_{10} \leq \int_{10}^{\infty} \frac{1}{x^2}\,dx = \lim_{t\to\infty}\left[-\frac{1}{x}\right]_{10}^{t} = \lim_{t\to\infty}\left(-\frac{1}{t} + \frac{1}{10}\right) = \frac{1}{10} = 0.1.$$

35. $\sum_{n=1}^{10} \dfrac{1}{1+2^n} = \dfrac{1}{3} + \dfrac{1}{5} + \dfrac{1}{9} + \cdots + \dfrac{1}{1025} \approx 0.76352.$ Now $\dfrac{1}{1+2^n} < \dfrac{1}{2^n}$, so the error is

$$R_{10} \leq T_{10} = \sum_{n=11}^{\infty} \frac{1}{2^n} = \frac{1/2^{11}}{1 - 1/2} \quad \text{[geometric series]} \approx 0.00098.$$

37. Since $\dfrac{d_n}{10^n} \leq \dfrac{9}{10^n}$ for each n, and since $\sum_{n=1}^{\infty} \dfrac{9}{10^n}$ is a convergent geometric series $\left(|r| = \tfrac{1}{10} < 1\right)$, $0.d_1 d_2 d_3 \ldots = \sum_{n=1}^{\infty} \dfrac{d_n}{10^n}$

will always converge by the Comparison Test.

39. Since $\sum a_n$ converges, $\lim_{n\to\infty} a_n = 0$, so there exists N such that $|a_n - 0| < 1$ for all $n > N \;\Rightarrow\; 0 \leq a_n < 1$ for

all $n > N \;\Rightarrow\; 0 \leq a_n^2 \leq a_n$. Since $\sum a_n$ converges, so does $\sum a_n^2$ by the Comparison Test.

41. (a) Since $\lim_{n\to\infty} \dfrac{a_n}{b_n} = \infty$, there is an integer N such that $\dfrac{a_n}{b_n} > 1$ whenever $n > N$. (Take $M = 1$ in Definition 11.1.5.)

Then $a_n > b_n$ whenever $n > N$ and since $\sum b_n$ is divergent, $\sum a_n$ is also divergent by the Comparison Test.

(b) (i) If $a_n = \dfrac{1}{\ln n}$ and $b_n = \dfrac{1}{n}$ for $n \geq 2$, then $\lim_{n\to\infty} \dfrac{a_n}{b_n} = \lim_{n\to\infty} \dfrac{n}{\ln n} = \lim_{x\to\infty} \dfrac{x}{\ln x} \overset{H}{=} \lim_{x\to\infty} \dfrac{1}{1/x} = \lim_{x\to\infty} x = \infty$,

so by part (a), $\sum_{n=2}^{\infty} \dfrac{1}{\ln n}$ is divergent.

(ii) If $a_n = \dfrac{\ln n}{n}$ and $b_n = \dfrac{1}{n}$, then $\sum_{n=1}^{\infty} b_n$ is the divergent harmonic series and $\lim_{n\to\infty} \dfrac{a_n}{b_n} = \lim_{n\to\infty} \ln n = \lim_{x\to\infty} \ln x = \infty$,

so $\sum_{n=1}^{\infty} a_n$ diverges by part (a).

43. $\lim_{n\to\infty} na_n = \lim_{n\to\infty} \dfrac{a_n}{1/n}$, so we apply the Limit Comparison Test with $b_n = \dfrac{1}{n}$. Since $\lim_{n\to\infty} na_n > 0$ we know that either both

series converge or both series diverge, and we also know that $\sum_{n=1}^{\infty} \dfrac{1}{n}$ diverges [p-series with $p = 1$]. Therefore, $\sum a_n$ must be

divergent.

45. Yes. Since $\sum a_n$ is a convergent series with positive terms, $\lim\limits_{n \to \infty} a_n = 0$ by Theorem 11.2.6, and $\sum b_n = \sum \sin(a_n)$ is a

series with positive terms (for large enough n). We have $\lim\limits_{n \to \infty} \dfrac{b_n}{a_n} = \lim\limits_{n \to \infty} \dfrac{\sin(a_n)}{a_n} = 1 > 0$ by Theorem 3.3.2. Thus, $\sum b_n$

is also convergent by the Limit Comparison Test.

11.5 Alternating Series

1. (a) An alternating series is a series whose terms are alternately positive and negative.

(b) An alternating series $\sum\limits_{n=1}^{\infty} (-1)^{n-1} b_n$ converges if $0 < b_{n+1} \leq b_n$ for all n and $\lim\limits_{n \to \infty} b_n = 0$. (This is the Alternating

Series Test.)

(c) The error involved in using the partial sum s_n as an approximation to the total sum s is the remainder $R_n = s - s_n$ and the

size of the error is smaller than b_{n+1}; that is, $|R_n| \leq b_{n+1}$. (This is the Alternating Series Estimation Theorem.)

3. $\dfrac{4}{7} - \dfrac{4}{8} + \dfrac{4}{9} - \dfrac{4}{10} + \dfrac{4}{11} - \cdots = \sum\limits_{n=1}^{\infty} (-1)^{n-1} \dfrac{4}{n+6}$. Now $b_n = \dfrac{4}{n+6} > 0$, $\{b_n\}$ is decreasing, and $\lim\limits_{n \to \infty} b_n = 0$, so the

series converges by the Alternating Series Test.

5. $\sum\limits_{n=1}^{\infty} a_n = \sum\limits_{n=1}^{\infty} (-1)^{n-1} \dfrac{1}{2n+1} = \sum\limits_{n=1}^{\infty} (-1)^{n-1} b_n$. Now $b_n = \dfrac{1}{2n+1} > 0$, $\{b_n\}$ is decreasing, and $\lim\limits_{n \to \infty} b_n = 0$, so the

series converges by the Alternating Series Test.

7. $\sum\limits_{n=1}^{\infty} a_n = \sum\limits_{n=1}^{\infty} (-1)^n \dfrac{3n-1}{2n+1} = \sum\limits_{n=1}^{\infty} (-1)^n b_n$. Now $\lim\limits_{n \to \infty} b_n = \lim\limits_{n \to \infty} \dfrac{3 - 1/n}{2 + 1/n} = \dfrac{3}{2} \neq 0$. Since $\lim\limits_{n \to \infty} a_n \neq 0$

(in fact the limit does not exist), the series diverges by the Test for Divergence.

9. $b_n = \dfrac{n}{10^n} > 0$ for $n \geq 1$. $\{b_n\}$ is decreasing for $n \geq 1$ since

$\left(\dfrac{x}{10^x}\right)' = \dfrac{10^x(1) - x \cdot 10^x \ln 10}{(10^x)^2} = \dfrac{10^x(1 - x \ln 10)}{(10^x)^2} = \dfrac{1 - x \ln 10}{10^x} < 0$ for $1 - x \ln 10 < 0$ $\Rightarrow$ $x \ln 10 > 1$ $\Rightarrow$

$x > \dfrac{1}{\ln 10} \approx 0.4$. Also, $\lim\limits_{n \to \infty} b_n = \lim\limits_{n \to \infty} \dfrac{n}{10^n} = \lim\limits_{x \to \infty} \dfrac{x}{10^x} \overset{\text{H}}{=} \lim\limits_{x \to \infty} \dfrac{x}{10^x \ln 10} = 0$. Thus, the series $\sum\limits_{n=1}^{\infty} (-1)^n \dfrac{n}{10^n}$

converges by the Alternating Series Test.

11. $b_n = \dfrac{n^2}{n^3 + 4} > 0$ for $n \geq 1$. $\{b_n\}$ is decreasing for $n \geq 2$ since

$\left(\dfrac{x^2}{x^3 + 4}\right)' = \dfrac{(x^3 + 4)(2x) - x^2(3x^2)}{(x^3 + 4)^2} = \dfrac{x(2x^3 + 8 - 3x^3)}{(x^3 + 4)^2} = \dfrac{x(8 - x^3)}{(x^3 + 4)^2} < 0$ for $x > 2$. Also,

$\lim\limits_{n \to \infty} b_n = \lim\limits_{n \to \infty} \dfrac{1/n}{1 + 4/n^3} = 0$. Thus, the series $\sum\limits_{n=1}^{\infty} (-1)^{n+1} \dfrac{n^2}{n^3 + 4}$ converges by the Alternating Series Test.

13. $\sum\limits_{n=2}^{\infty} (-1)^n \dfrac{n}{\ln n}$. $\lim\limits_{n \to \infty} \dfrac{n}{\ln n} = \lim\limits_{x \to \infty} \dfrac{x}{\ln x} \overset{\text{H}}{=} \lim\limits_{x \to \infty} \dfrac{1}{1/x} = \infty$, so the series diverges by the Test for Divergence.

15. $\displaystyle\sum_{n=1}^{\infty} \frac{\cos n\pi}{n^{3/4}} = \sum_{n=1}^{\infty} \frac{(-1)^n}{n^{3/4}}$. $b_n = \dfrac{1}{n^{3/4}}$ is decreasing and positive and $\displaystyle\lim_{n\to\infty} \frac{1}{n^{3/4}} = 0$, so the series converges by the

Alternating Series Test.

17. $\displaystyle\sum_{n=1}^{\infty} (-1)^n \sin\left(\frac{\pi}{n}\right)$. $b_n = \sin\left(\dfrac{\pi}{n}\right) > 0$ for $n \geq 2$ and $\sin\left(\dfrac{\pi}{n}\right) \geq \sin\left(\dfrac{\pi}{n+1}\right)$, and $\displaystyle\lim_{n\to\infty} \sin\left(\frac{\pi}{n}\right) = \sin 0 = 0$, so the series

converges by the Alternating Series Test.

19. $\dfrac{n^n}{n!} = \dfrac{n \cdot n \cdots n}{1 \cdot 2 \cdots n} \geq n \ \Rightarrow \ \displaystyle\lim_{n\to\infty} \frac{n^n}{n!} = \infty \ \Rightarrow \ \displaystyle\lim_{n\to\infty} \frac{(-1)^n n^n}{n!}$ does not exist. So the series diverges by the Test for

Divergence.

21.

n	a_n	s_n
1	1	1
2	-0.35355	0.64645
3	0.19245	0.83890
4	-0.125	0.71390
5	0.08944	0.80334
6	-0.06804	0.73530
7	0.05399	0.78929
8	-0.04419	0.74510
9	0.03704	0.78214
10	-0.03162	0.75051

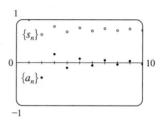

By the Alternating Series Estimation Theorem, the error in the approximation

$\displaystyle\sum_{n=1}^{\infty} \frac{(-1)^{n-1}}{n^{3/2}} \approx 0.75051$ is $|s - s_{10}| \leq b_{11} = 1/(11)^{3/2} \approx 0.0275$ (to four

decimal places, rounded up).

23. The series $\displaystyle\sum_{n=1}^{\infty} \frac{(-1)^{n+1}}{n^6}$ satisfies (i) of the Alternating Series Test because $\dfrac{1}{(n+1)^6} < \dfrac{1}{n^6}$ and (ii) $\displaystyle\lim_{n\to\infty} \frac{1}{n^6} = 0$, so the

series is convergent. Now $b_5 = \dfrac{1}{5^6} = 0.000064 > 0.00005$ and $b_6 = \dfrac{1}{6^6} \approx 0.00002 < 0.00005$, so by the Alternating Series

Estimation Theorem, $n = 5$. (That is, since the 6th term is less than the desired error, we need to add the first 5 terms to get the

sum to the desired accuracy.)

25. The series $\displaystyle\sum_{n=0}^{\infty} \frac{(-1)^n}{10^n \, n!}$ satisfies (i) of the Alternating Series Test because $\dfrac{1}{10^{n+1}(n+1)!} < \dfrac{1}{10^n \, n!}$ and (ii) $\displaystyle\lim_{n\to\infty} \frac{1}{10^n \, n!} = 0$,

so the series is convergent. Now $b_3 = \dfrac{1}{10^3 \, 3!} \approx 0.000\,167 > 0.000\,005$ and $b_4 = \dfrac{1}{10^4 \, 4!} = 0.000\,004 < 0.000\,005$, so by

the Alternating Series Estimation Theorem, $n = 4$ (since the series starts with $n = 0$, not $n = 1$). (That is, since the 5th term

is less than the desired error, we need to add the first 4 terms to get the sum to the desired accuracy.)

27. $b_7 = \dfrac{1}{7^5} = \dfrac{1}{16{,}807} \approx 0.000\,059\,5$, so

$\displaystyle\sum_{n=1}^{\infty} \frac{(-1)^{n+1}}{n^5} \approx s_6 = \sum_{n=1}^{6} \frac{(-1)^{n+1}}{n^5} = 1 - \frac{1}{32} + \frac{1}{243} - \frac{1}{1024} + \frac{1}{3125} - \frac{1}{7776} \approx 0.972\,080$. Adding b_7 to s_6 does not change

the fourth decimal place of s_6, so the sum of the series, correct to four decimal places, is 0.9721.

29. $b_7 = \dfrac{7^2}{10^7} = 0.000\,004\,9$, so

$$\sum_{n=1}^{\infty} \frac{(-1)^{n-1}n^2}{10^n} \approx s_6 = \sum_{n=1}^{6} \frac{(-1)^{n-1}n^2}{10^n} = \frac{1}{10} - \frac{4}{100} + \frac{9}{1000} - \frac{16}{10{,}000} + \frac{25}{100{,}000} - \frac{36}{1{,}000{,}000} = 0.067\,614.$$ Adding b_7 to s_6

does not change the fourth decimal place of s_6, so the sum of the series, correct to four decimal places, is 0.0676.

31. $\displaystyle\sum_{n=1}^{\infty} \frac{(-1)^{n-1}}{n} = 1 - \frac{1}{2} + \frac{1}{3} - \frac{1}{4} + \cdots + \frac{1}{49} - \frac{1}{50} + \frac{1}{51} - \frac{1}{52} + \cdots$. The 50th partial sum of this series is an

underestimate, since $\displaystyle\sum_{n=1}^{\infty} \frac{(-1)^{n-1}}{n} = s_{50} + \left(\frac{1}{51} - \frac{1}{52}\right) + \left(\frac{1}{53} - \frac{1}{54}\right) + \cdots$, and the terms in parentheses are all positive.

The result can be seen geometrically in Figure 1.

33. Clearly $b_n = \dfrac{1}{n+p}$ is decreasing and eventually positive and $\displaystyle\lim_{n\to\infty} b_n = 0$ for any p. So the series converges (by the

Alternating Series Test) for any p for which every b_n is defined, that is, $n + p \neq 0$ for $n \geq 1$, or p is not a negative integer.

35. $\sum b_{2n} = \sum 1/(2n)^2$ clearly converges (by comparison with the p-series for $p = 2$). So suppose that $\sum (-1)^{n-1} b_n$

converges. Then by Theorem 11.2.8(ii), so does $\sum \left[(-1)^{n-1}b_n + b_n\right] = 2\left(1 + \frac{1}{3} + \frac{1}{5} + \cdots\right) = 2\sum \dfrac{1}{2n-1}$. But this

diverges by comparison with the harmonic series, a contradiction. Therefore, $\sum (-1)^{n-1} b_n$ must diverge. The Alternating

Series Test does not apply since $\{b_n\}$ is not decreasing.

11.6   Absolute Convergence and the Ratio and Root Tests

1. (a) Since $\displaystyle\lim_{n\to\infty} \left|\frac{a_{n+1}}{a_n}\right| = 8 > 1$, part (b) of the Ratio Test tells us that the series $\sum a_n$ is divergent.

(b) Since $\displaystyle\lim_{n\to\infty} \left|\frac{a_{n+1}}{a_n}\right| = 0.8 < 1$, part (a) of the Ratio Test tells us that the series $\sum a_n$ is absolutely convergent (and

therefore convergent).

(c) Since $\displaystyle\lim_{n\to\infty} \left|\frac{a_{n+1}}{a_n}\right| = 1$, the Ratio Test fails and the series $\sum a_n$ might converge or it might diverge.

3. $\displaystyle\sum_{n=0}^{\infty} \frac{(-10)^n}{n!}$. Using the Ratio Test, $\displaystyle\lim_{n\to\infty} \left|\frac{a_{n+1}}{a_n}\right| = \lim_{n\to\infty} \left|\frac{(-10)^{n+1}}{(n+1)!} \cdot \frac{n!}{(-10)^n}\right| = \lim_{n\to\infty} \left|\frac{-10}{n+1}\right| = 0 < 1$, so the series is

absolutely convergent.

5. $\displaystyle\sum_{n=1}^{\infty} \frac{(-1)^{n+1}}{\sqrt[4]{n}}$ converges by the Alternating Series Test, but $\displaystyle\sum_{n=1}^{\infty} \frac{1}{\sqrt[4]{n}}$ is a divergent p-series $\left(p = \frac{1}{4} \leq 1\right)$, so the given series

is conditionally convergent.

7. $\lim\limits_{k \to \infty} \left| \dfrac{a_{k+1}}{a_k} \right| = \lim\limits_{k \to \infty} \left[\dfrac{(k+1)\left(\frac{2}{3}\right)^{k+1}}{k\left(\frac{2}{3}\right)^k} \right] = \lim\limits_{k \to \infty} \dfrac{k+1}{k} \left(\dfrac{2}{3} \right)^1 = \dfrac{2}{3} \lim\limits_{k \to \infty} \left(1 + \dfrac{1}{k} \right) = \dfrac{2}{3}(1) = \dfrac{2}{3} < 1$, so the series

$\sum\limits_{n=1}^{\infty} k\left(\frac{2}{3}\right)^k$ is absolutely convergent by the Ratio Test. Since the terms of this series are positive, absolute convergence is the

same as convergence.

9. $\lim\limits_{n \to \infty} \left| \dfrac{a_{n+1}}{a_n} \right| = \lim\limits_{n \to \infty} \left[\dfrac{(1.1)^{n+1}}{(n+1)^4} \cdot \dfrac{n^4}{(1.1)^n} \right] = \lim\limits_{n \to \infty} \dfrac{(1.1)n^4}{(n+1)^4} = (1.1) \lim\limits_{n \to \infty} \dfrac{1}{\dfrac{(n+1)^4}{n^4}} = (1.1) \lim\limits_{n \to \infty} \dfrac{1}{(1+1/n)^4}$

$= (1.1)(1) = 1.1 > 1,$

so the series $\sum\limits_{n=1}^{\infty} (-1)^n \dfrac{(1.1)^n}{n^4}$ diverges by the Ratio Test.

11. Since $0 \le \dfrac{e^{1/n}}{n^3} \le \dfrac{e}{n^3} = e\left(\dfrac{1}{n^3} \right)$ and $\sum\limits_{n=1}^{\infty} \dfrac{1}{n^3}$ is a convergent p-series $[p = 3 > 1]$, $\sum\limits_{n=1}^{\infty} \dfrac{e^{1/n}}{n^3}$ converges, and so

$\sum\limits_{n=1}^{\infty} \dfrac{(-1)^n e^{1/n}}{n^3}$ is absolutely convergent.

13. $\lim\limits_{n \to \infty} \left| \dfrac{a_{n+1}}{a_n} \right| = \lim\limits_{n \to \infty} \left[\dfrac{10^{n+1}}{(n+2)\,4^{2n+3}} \cdot \dfrac{(n+1)\,4^{2n+1}}{10^n} \right] = \lim\limits_{n \to \infty} \left(\dfrac{10}{4^2} \cdot \dfrac{n+1}{n+2} \right) = \dfrac{5}{8} < 1$, so the series $\sum\limits_{n=1}^{\infty} \dfrac{10^n}{(n+1)4^{2n+1}}$

is absolutely convergent by the Ratio Test. Since the terms of this series are positive, absolute convergence is the same as

convergence.

15. $\left| \dfrac{(-1)^n \arctan n}{n^2} \right| < \dfrac{\pi/2}{n^2}$, so since $\sum\limits_{n=1}^{\infty} \dfrac{\pi/2}{n^2} = \dfrac{\pi}{2} \sum\limits_{n=1}^{\infty} \dfrac{1}{n^2}$ converges $(p = 2 > 1)$, the given series $\sum\limits_{n=1}^{\infty} \dfrac{(-1)^n \arctan n}{n^2}$

converges absolutely by the Comparison Test.

17. $\sum\limits_{n=2}^{\infty} \dfrac{(-1)^n}{\ln n}$ converges by the Alternating Series Test since $\lim\limits_{n \to \infty} \dfrac{1}{\ln n} = 0$ and $\left\{ \dfrac{1}{\ln n} \right\}$ is decreasing. Now $\ln n < n$, so

$\dfrac{1}{\ln n} > \dfrac{1}{n}$, and since $\sum\limits_{n=2}^{\infty} \dfrac{1}{n}$ is the divergent (partial) harmonic series, $\sum\limits_{n=2}^{\infty} \dfrac{1}{\ln n}$ diverges by the Comparison Test. Thus,

$\sum\limits_{n=2}^{\infty} \dfrac{(-1)^n}{\ln n}$ is conditionally convergent.

19. $\dfrac{|\cos(n\pi/3)|}{n!} \le \dfrac{1}{n!}$ and $\sum\limits_{n=1}^{\infty} \dfrac{1}{n!}$ converges (use the Ratio Test), so the series $\sum\limits_{n=1}^{\infty} \dfrac{\cos(n\pi/3)}{n!}$ converges absolutely by the

Comparison Test.

21. $\lim\limits_{n \to \infty} \sqrt[n]{|a_n|} = \lim\limits_{n \to \infty} \dfrac{n^2 + 1}{2n^2 + 1} = \lim\limits_{n \to \infty} \dfrac{1 + 1/n^2}{2 + 1/n^2} = \dfrac{1}{2} < 1$, so the series $\sum\limits_{n=1}^{\infty} \left(\dfrac{n^2 + 1}{2n^2 + 1} \right)^n$ is absolutely convergent by the

Root Test.

23. $\lim\limits_{n \to \infty} \sqrt[n]{|a_n|} = \lim\limits_{n \to \infty} \sqrt[n]{\left(1 + \dfrac{1}{n} \right)^{n^2}} = \lim\limits_{n \to \infty} \left(1 + \dfrac{1}{n} \right)^n = e > 1$ (by Equation 3.6.6), so the series $\sum\limits_{n=1}^{\infty} \left(1 + \dfrac{1}{n} \right)^{n^2}$

diverges by the Root Test.

25. Use the Ratio Test with the series

$$1 - \frac{1 \cdot 3}{3!} + \frac{1 \cdot 3 \cdot 5}{5!} - \frac{1 \cdot 3 \cdot 5 \cdot 7}{7!} + \cdots + (-1)^{n-1} \frac{1 \cdot 3 \cdot 5 \cdot \cdots \cdot (2n-1)}{(2n-1)!} + \cdots = \sum_{n=1}^{\infty} (-1)^{n-1} \frac{1 \cdot 3 \cdot 5 \cdot \cdots \cdot (2n-1)}{(2n-1)!}.$$

$$\lim_{n \to \infty} \left| \frac{a_{n+1}}{a_n} \right| = \lim_{n \to \infty} \left| \frac{(-1)^n \cdot 1 \cdot 3 \cdot 5 \cdot \cdots \cdot (2n-1)[2(n+1)-1]}{[2(n+1)-1]!} \cdot \frac{(2n-1)!}{(-1)^{n-1} \cdot 1 \cdot 3 \cdot 5 \cdot \cdots \cdot (2n-1)} \right|$$

$$= \lim_{n \to \infty} \left| \frac{(-1)(2n+1)(2n-1)!}{(2n+1)(2n)(2n-1)!} \right| = \lim_{n \to \infty} \frac{1}{2n} = 0 < 1,$$

so the given series is absolutely convergent and therefore convergent.

27. $\displaystyle \sum_{n=1}^{\infty} \frac{2 \cdot 4 \cdot 6 \cdot \cdots \cdot (2n)}{n!} = \sum_{n=1}^{\infty} \frac{(2 \cdot 1) \cdot (2 \cdot 2) \cdot (2 \cdot 3) \cdot \cdots \cdot (2 \cdot n)}{n!} = \sum_{n=1}^{\infty} \frac{2^n n!}{n!} = \sum_{n=1}^{\infty} 2^n$, which diverges by the Test for

Divergence since $\displaystyle \lim_{n \to \infty} 2^n = \infty$.

29. By the recursive definition, $\displaystyle \lim_{n \to \infty} \left| \frac{a_{n+1}}{a_n} \right| = \lim_{n \to \infty} \left| \frac{5n+1}{4n+3} \right| = \frac{5}{4} > 1$, so the series diverges by the Ratio Test.

31. (a) $\displaystyle \lim_{n \to \infty} \left| \frac{1/(n+1)^3}{1/n^3} \right| = \lim_{n \to \infty} \frac{n^3}{(n+1)^3} = \lim_{n \to \infty} \frac{1}{(1+1/n)^3} = 1.$ Inconclusive

(b) $\displaystyle \lim_{n \to \infty} \left| \frac{(n+1)}{2^{n+1}} \cdot \frac{2^n}{n} \right| = \lim_{n \to \infty} \frac{n+1}{2n} = \lim_{n \to \infty} \left(\frac{1}{2} + \frac{1}{2n} \right) = \frac{1}{2}.$ Conclusive (convergent)

(c) $\displaystyle \lim_{n \to \infty} \left| \frac{(-3)^n}{\sqrt{n+1}} \cdot \frac{\sqrt{n}}{(-3)^{n-1}} \right| = 3 \lim_{n \to \infty} \sqrt{\frac{n}{n+1}} = 3 \lim_{n \to \infty} \sqrt{\frac{1}{1+1/n}} = 3.$ Conclusive (divergent)

(d) $\displaystyle \lim_{n \to \infty} \left| \frac{\sqrt{n+1}}{1+(n+1)^2} \cdot \frac{1+n^2}{\sqrt{n}} \right| = \lim_{n \to \infty} \left[\sqrt{1 + \frac{1}{n}} \cdot \frac{1/n^2 + 1}{1/n^2 + (1+1/n)^2} \right] = 1.$ Inconclusive

33. (a) $\displaystyle \lim_{n \to \infty} \left| \frac{a_{n+1}}{a_n} \right| = \lim_{n \to \infty} \left| \frac{x^{n+1}}{(n+1)!} \cdot \frac{n!}{x^n} \right| = \lim_{n \to \infty} \left| \frac{x}{n+1} \right| = |x| \lim_{n \to \infty} \frac{1}{n+1} = |x| \cdot 0 = 0 < 1$, so by the Ratio Test the

series $\displaystyle \sum_{n=0}^{\infty} \frac{x^n}{n!}$ converges for all x.

(b) Since the series of part (a) always converges, we must have $\displaystyle \lim_{n \to \infty} \frac{x^n}{n!} = 0$ by Theorem 11.2.6.

35. (a) $\displaystyle s_5 = \sum_{n=1}^{5} \frac{1}{n2^n} = \frac{1}{2} + \frac{1}{8} + \frac{1}{24} + \frac{1}{64} + \frac{1}{160} = \frac{661}{960} \approx 0.68854.$ Now the ratios

$$r_n = \frac{a_{n+1}}{a_n} = \frac{n2^n}{(n+1)2^{n+1}} = \frac{n}{2(n+1)} \quad \text{form an increasing sequence, since}$$

$$r_{n+1} - r_n = \frac{n+1}{2(n+2)} - \frac{n}{2(n+1)} = \frac{(n+1)^2 - n(n+2)}{2(n+1)(n+2)} = \frac{1}{2(n+1)(n+2)} > 0. \quad \text{So by Exercise 34(b), the error}$$

in using s_5 is $\displaystyle R_5 \le \frac{a_6}{1 - \lim_{n \to \infty} r_n} = \frac{1/(6 \cdot 2^6)}{1 - 1/2} = \frac{1}{192} \approx 0.00521.$

(b) The error in using s_n as an approximation to the sum is $\displaystyle R_n = \frac{a_{n+1}}{1 - \frac{1}{2}} = \frac{2}{(n+1)2^{n+1}}.$ We want $R_n < 0.00005 \quad \Leftrightarrow$

$\displaystyle \frac{1}{(n+1)2^n} < 0.00005 \quad \Leftrightarrow \quad (n+1)2^n > 20{,}000.$ To find such an n we can use trial and error or a graph. We calculate

$(11+1)2^{11} = 24{,}576$, so $\displaystyle s_{11} = \sum_{n=1}^{11} \frac{1}{n2^n} \approx 0.693109$ is within 0.00005 of the actual sum.

37. (i) Following the hint, we get that $|a_n| < r^n$ for $n \geq N$, and so since the geometric series $\sum_{n=1}^{\infty} r^n$ converges $[0 < r < 1]$, the series $\sum_{n=N}^{\infty} |a_n|$ converges as well by the Comparison Test, and hence so does $\sum_{n=1}^{\infty} |a_n|$, so $\sum_{n=1}^{\infty} a_n$ is absolutely convergent.

(ii) If $\lim_{n\to\infty} \sqrt[n]{|a_n|} = L > 1$, then there is an integer N such that $\sqrt[n]{|a_n|} > 1$ for all $n \geq N$, so $|a_n| > 1$ for $n \geq N$. Thus, $\lim_{n\to\infty} a_n \neq 0$, so $\sum_{n=1}^{\infty} a_n$ diverges by the Test for Divergence.

(iii) Consider $\sum_{n=1}^{\infty} \dfrac{1}{n}$ [diverges] and $\sum_{n=1}^{\infty} \dfrac{1}{n^2}$ [converges]. For each sum, $\lim_{n\to\infty} \sqrt[n]{|a_n|} = 1$, so the Root Test is inconclusive.

39. (a) Since $\sum a_n$ is absolutely convergent, and since $|a_n^+| \leq |a_n|$ and $|a_n^-| \leq |a_n|$ (because a_n^+ and a_n^- each equal either a_n or 0), we conclude by the Comparison Test that both $\sum a_n^+$ and $\sum a_n^-$ must be absolutely convergent.

Or: Use Theorem 11.2.8.

(b) We will show by contradiction that both $\sum a_n^+$ and $\sum a_n^-$ must diverge. For suppose that $\sum a_n^+$ converged. Then so would $\sum \left(a_n^+ - \tfrac{1}{2}a_n\right)$ by Theorem 11.2.8. But $\sum \left(a_n^+ - \tfrac{1}{2}a_n\right) = \sum \left[\tfrac{1}{2}(a_n + |a_n|) - \tfrac{1}{2}a_n\right] = \tfrac{1}{2}\sum |a_n|$, which diverges because $\sum a_n$ is only conditionally convergent. Hence, $\sum a_n^+$ can't converge. Similarly, neither can $\sum a_n^-$.

11.7 Strategy for Testing Series

1. $\dfrac{1}{n + 3^n} < \dfrac{1}{3^n} = \left(\dfrac{1}{3}\right)^n$ for all $n \geq 1$. $\sum_{n=1}^{\infty} \left(\dfrac{1}{3}\right)^n$ is a convergent geometric series $\left[|r| = \tfrac{1}{3} < 1\right]$, so $\sum_{n=1}^{\infty} \dfrac{1}{n + 3^n}$ converges by the Comparison Test.

3. $\lim_{n\to\infty} |a_n| = \lim_{n\to\infty} \dfrac{n}{n+2} = 1$, so $\lim_{n\to\infty} a_n = \lim_{n\to\infty} (-1)^n \dfrac{n}{n+2}$ does not exist. Thus, the series $\sum_{n=1}^{\infty} (-1)^n \dfrac{n}{n+2}$ diverges by the Test for Divergence.

5. $\lim_{n\to\infty} \left|\dfrac{a_{n+1}}{a_n}\right| = \lim_{n\to\infty} \left|\dfrac{(n+1)^2 \, 2^n}{(-5)^{n+1}} \cdot \dfrac{(-5)^n}{n^2 \, 2^{n-1}}\right| = \lim_{n\to\infty} \dfrac{2(n+1)^2}{5n^2} = \dfrac{2}{5} \lim_{n\to\infty} \left(1 + \dfrac{1}{n}\right)^2 = \dfrac{2}{5}(1) = \dfrac{2}{5} < 1$, so the series $\sum_{n=1}^{\infty} \dfrac{n^2 \, 2^{n-1}}{(-5)^n}$ converges by the Ratio Test.

7. Let $f(x) = \dfrac{1}{x \sqrt{\ln x}}$. Then f is positive, continuous, and decreasing on $[2, \infty)$, so we can apply the Integral Test.

Since $\displaystyle\int \dfrac{1}{x\sqrt{\ln x}}\, dx \quad \begin{bmatrix} u = \ln x, \\ du = dx/x \end{bmatrix} = \int u^{-1/2}\, du = 2u^{1/2} + C = 2\sqrt{\ln x} + C$, we find

$\displaystyle\int_2^{\infty} \dfrac{dx}{x\sqrt{\ln x}} = \lim_{t\to\infty} \int_2^t \dfrac{dx}{x\sqrt{\ln x}} = \lim_{t\to\infty} \left[2\sqrt{\ln x}\right]_2^t = \lim_{t\to\infty} \left(2\sqrt{\ln t} - 2\sqrt{\ln 2}\right) = \infty$. Since the integral diverges, the

given series $\sum_{n=2}^{\infty} \dfrac{1}{n\sqrt{\ln n}}$ diverges.

9. $\sum\limits_{k=1}^{\infty} k^2 e^{-k} = \sum\limits_{k=1}^{\infty} \dfrac{k^2}{e^k}$. Using the Ratio Test, we get

$$\lim_{k\to\infty}\left|\frac{a_{k+1}}{a_k}\right| = \lim_{k\to\infty}\left|\frac{(k+1)^2}{e^{k+1}}\cdot\frac{e^k}{k^2}\right| = \lim_{k\to\infty}\left[\left(\frac{k+1}{k}\right)^2\cdot\frac{1}{e}\right] = 1^2\cdot\frac{1}{e} = \frac{1}{e} < 1, \text{ so the series converges.}$$

11. $b_n = \dfrac{1}{n\ln n} > 0$ for $n \geq 2$, $\{b_n\}$ is decreasing, and $\lim\limits_{n\to\infty} b_n = 0$, so the given series $\sum\limits_{n=2}^{\infty} \dfrac{(-1)^{n+1}}{n\ln n}$ converges by the

Alternating Series Test.

13. $\lim\limits_{n\to\infty}\left|\dfrac{a_{n+1}}{a_n}\right| = \lim\limits_{n\to\infty}\left|\dfrac{3^{n+1}(n+1)^2}{(n+1)!}\cdot\dfrac{n!}{3^n n^2}\right| = \lim\limits_{n\to\infty}\dfrac{3(n+1)^2}{(n+1)n^2} = 3\lim\limits_{n\to\infty}\dfrac{n+1}{n^2} = 0 < 1$, so the series $\sum\limits_{n=1}^{\infty}\dfrac{3^n n^2}{n!}$

converges by the Ratio Test.

15. $\lim\limits_{n\to\infty}\left|\dfrac{a_{n+1}}{a_n}\right| = \lim\limits_{n\to\infty}\left|\dfrac{(n+1)!}{2\cdot5\cdot8\cdots(3n+2)[3(n+1)+2]}\cdot\dfrac{2\cdot5\cdot8\cdots(3n+2)}{n!}\right| = \lim\limits_{n\to\infty}\dfrac{n+1}{3n+5} = \dfrac{1}{3} < 1,$

so the series $\sum\limits_{n=0}^{\infty}\dfrac{n!}{2\cdot5\cdot8\cdots(3n+2)}$ converges by the Ratio Test.

17. $\lim\limits_{n\to\infty} 2^{1/n} = 2^0 = 1$, so $\lim\limits_{n\to\infty}(-1)^n\,2^{1/n}$ does not exist and the series $\sum\limits_{n=1}^{\infty}(-1)^n 2^{1/n}$ diverges by the Test for Divergence.

19. Let $f(x) = \dfrac{\ln x}{\sqrt{x}}$. Then $f'(x) = \dfrac{2-\ln x}{2x^{3/2}} < 0$ when $\ln x > 2$ or $x > e^2$, so $\dfrac{\ln n}{\sqrt{n}}$ is decreasing for $n > e^2$.

By l'Hospital's Rule, $\lim\limits_{n\to\infty}\dfrac{\ln n}{\sqrt{n}} = \lim\limits_{n\to\infty}\dfrac{1/n}{1/\left(2\sqrt{n}\right)} = \lim\limits_{n\to\infty}\dfrac{2}{\sqrt{n}} = 0$, so the series $\sum\limits_{n=1}^{\infty}(-1)^n\dfrac{\ln n}{\sqrt{n}}$ converges by the

Alternating Series Test.

21. $\sum\limits_{n=1}^{\infty}\dfrac{(-2)^{2n}}{n^n} = \sum\limits_{n=1}^{\infty}\left(\dfrac{4}{n}\right)^n$. $\lim\limits_{n\to\infty}\sqrt[n]{|a_n|} = \lim\limits_{n\to\infty}\dfrac{4}{n} = 0 < 1$, so the given series is absolutely convergent by the Root Test.

23. Using the Limit Comparison Test with $a_n = \tan\left(\dfrac{1}{n}\right)$ and $b_n = \dfrac{1}{n}$, we have

$$\lim_{n\to\infty}\frac{a_n}{b_n} = \lim_{n\to\infty}\frac{\tan(1/n)}{1/n} = \lim_{x\to\infty}\frac{\tan(1/x)}{1/x} \overset{\text{H}}{=} \lim_{x\to\infty}\frac{\sec^2(1/x)\cdot(-1/x^2)}{-1/x^2} = \lim_{x\to\infty}\sec^2(1/x) = 1^2 = 1 > 0. \text{ Since}$$

$\sum\limits_{n=1}^{\infty} b_n$ is the divergent harmonic series, $\sum\limits_{n=1}^{\infty} a_n$ is also divergent.

25. Use the Ratio Test. $\lim\limits_{n\to\infty}\left|\dfrac{a_{n+1}}{a_n}\right| = \lim\limits_{n\to\infty}\left|\dfrac{(n+1)!}{e^{(n+1)^2}}\cdot\dfrac{e^{n^2}}{n!}\right| = \lim\limits_{n\to\infty}\dfrac{(n+1)n!\cdot e^{n^2}}{e^{n^2+2n+1}n!} = \lim\limits_{n\to\infty}\dfrac{n+1}{e^{2n+1}} = 0 < 1$, so $\sum\limits_{n=1}^{\infty}\dfrac{n!}{e^{n^2}}$

converges.

27. $\displaystyle\int_2^{\infty}\dfrac{\ln x}{x^2}\,dx = \lim_{t\to\infty}\left[-\dfrac{\ln x}{x}-\dfrac{1}{x}\right]_1^t$ [using integration by parts] $\overset{\text{H}}{=} 1$. So $\sum\limits_{n=1}^{\infty}\dfrac{\ln n}{n^2}$ converges by the Integral Test, and since

$\dfrac{k\ln k}{(k+1)^3} < \dfrac{k\ln k}{k^3} = \dfrac{\ln k}{k^2}$, the given series $\sum\limits_{k=1}^{\infty}\dfrac{k\ln k}{(k+1)^3}$ converges by the Comparison Test.

29. $\sum_{n=1}^{\infty} a_n = \sum_{n=1}^{\infty} (-1)^n \dfrac{1}{\cosh n} = \sum_{n=1}^{\infty} (-1)^n b_n$. Now $b_n = \dfrac{1}{\cosh n} > 0$, $\{b_n\}$ is decreasing, and $\lim_{n \to \infty} b_n = 0$, so the series

converges by the Alternating Series Test.

Or: Write $\dfrac{1}{\cosh n} = \dfrac{2}{e^n + e^{-n}} < \dfrac{2}{e^n}$ and $\sum_{n=1}^{\infty} \dfrac{1}{e^n}$ is a convergent geometric series, so $\sum_{n=1}^{\infty} \dfrac{1}{\cosh n}$ is convergent by the

Comparison Test. So $\sum_{n=1}^{\infty} (-1)^n \dfrac{1}{\cosh n}$ is absolutely convergent and therefore convergent.

31. $\lim_{k \to \infty} a_k = \lim_{k \to \infty} \dfrac{5^k}{3^k + 4^k} = $ [divide by 4^k] $\lim_{k \to \infty} \dfrac{(5/4)^k}{(3/4)^k + 1} = \infty$ since $\lim_{k \to \infty} \left(\dfrac{3}{4}\right)^k = 0$ and $\lim_{k \to \infty} \left(\dfrac{5}{4}\right)^k = \infty$.

Thus, $\sum_{k=1}^{\infty} \dfrac{5^k}{3^k + 4^k}$ diverges by the Test for Divergence.

33. Let $a_n = \dfrac{\sin(1/n)}{\sqrt{n}}$ and $b_n = \dfrac{1}{n\sqrt{n}}$. Then $\lim_{n \to \infty} \dfrac{a_n}{b_n} = \lim_{n \to \infty} \dfrac{\sin(1/n)}{1/n} = 1 > 0$, so $\sum_{n=1}^{\infty} \dfrac{\sin(1/n)}{\sqrt{n}}$ converges by limit

comparison with the convergent p-series $\sum_{n=1}^{\infty} \dfrac{1}{n^{3/2}}$ $[p = 3/2 > 1]$.

35. $\lim_{n \to \infty} \sqrt[n]{|a_n|} = \lim_{n \to \infty} \left(\dfrac{n}{n+1}\right)^{n^2/n} = \lim_{n \to \infty} \dfrac{1}{[(n+1)/n]^n} = \dfrac{1}{\lim_{n \to \infty} (1 + 1/n)^n} = \dfrac{1}{e} < 1$, so the series $\sum_{n=1}^{\infty} \left(\dfrac{n}{n+1}\right)^{n^2}$

converges by the Root Test.

37. $\lim_{n \to \infty} \sqrt[n]{|a_n|} = \lim_{n \to \infty} (2^{1/n} - 1) = 1 - 1 = 0 < 1$, so the series $\sum_{n=1}^{\infty} \left(\sqrt[n]{2} - 1\right)^n$ converges by the Root Test.

11.8 Power Series

1. A power series is a series of the form $\sum_{n=0}^{\infty} c_n x^n = c_0 + c_1 x + c_2 x^2 + c_3 x^3 + \cdots$, where x is a variable and the c_n's are

constants called the coefficients of the series.

More generally, a series of the form $\sum_{n=0}^{\infty} c_n (x - a)^n = c_0 + c_1 (x - a) + c_2 (x - a)^2 + \cdots$ is called a power series in

$(x - a)$ or a power series centered at a or a power series about a, where a is a constant.

3. If $a_n = \dfrac{x^n}{\sqrt{n}}$, then $\lim_{n \to \infty} \left|\dfrac{a_{n+1}}{a_n}\right| = \lim_{n \to \infty} \left|\dfrac{x^{n+1}}{\sqrt{n+1}} \cdot \dfrac{\sqrt{n}}{x^n}\right| = \lim_{n \to \infty} \left|\dfrac{x}{\sqrt{n+1}/\sqrt{n}}\right| = \lim_{n \to \infty} \dfrac{|x|}{\sqrt{1 + 1/n}} = |x|$.

By the Ratio Test, the series $\sum_{n=1}^{\infty} \dfrac{x^n}{\sqrt{n}}$ converges when $|x| < 1$, so the radius of convergence $R = 1$. Now we'll check the

endpoints, that is, $x = \pm 1$. When $x = 1$, the series $\sum_{n=1}^{\infty} \dfrac{1}{\sqrt{n}}$ diverges because it is a p-series with $p = \frac{1}{2} \leq 1$. When $x = -1$,

the series $\sum_{n=1}^{\infty} \dfrac{(-1)^n}{\sqrt{n}}$ converges by the Alternating Series Test. Thus, the interval of convergence is $I = [-1, 1)$.

5. If $a_n = \dfrac{(-1)^{n-1} x^n}{n^3}$, then

$$\lim_{n \to \infty} \left|\dfrac{a_{n+1}}{a_n}\right| = \lim_{n \to \infty} \left|\dfrac{(-1)^n x^{n+1}}{(n+1)^3} \cdot \dfrac{n^3}{(-1)^{n-1} x^n}\right| = \lim_{n \to \infty} \left|\dfrac{(-1)x n^3}{(n+1)^3}\right| = \lim_{n \to \infty} \left[\left(\dfrac{n}{n+1}\right)^3 |x|\right] = 1^3 \cdot |x| = |x|. \text{ By the}$$

Ratio Test, the series $\sum\limits_{n=1}^{\infty} \dfrac{(-1)^{n-1}x^n}{n^3}$ converges when $|x| < 1$, so the radius of convergence $R = 1$. Now we'll check the

endpoints, that is, $x = \pm 1$. When $x = 1$, the series $\sum\limits_{n=1}^{\infty} \dfrac{(-1)^{n-1}}{n^3}$ converges by the Alternating Series Test. When $x = -1$,

the series $\sum\limits_{n=1}^{\infty} \dfrac{(-1)^{n-1}(-1)^n}{n^3} = -\sum\limits_{n=1}^{\infty} \dfrac{1}{n^3}$ converges because it is a constant multiple of a convergent p-series $[p = 3 > 1]$.

Thus, the interval of convergence is $I = [-1, 1]$.

7. If $a_n = \dfrac{x^n}{n!}$, then $\lim\limits_{n \to \infty} \left| \dfrac{a_{n+1}}{a_n} \right| = \lim\limits_{n \to \infty} \left| \dfrac{x^{n+1}}{(n+1)!} \cdot \dfrac{n!}{x^n} \right| = \lim\limits_{n \to \infty} \left| \dfrac{x}{n+1} \right| = |x| \lim\limits_{n \to \infty} \dfrac{1}{n+1} = |x| \cdot 0 = 0 < 1$ for *all* real x.

So, by the Ratio Test, $R = \infty$ and $I = (-\infty, \infty)$.

9. If $a_n = (-1)^n \dfrac{n^2 x^n}{2^n}$, then

$$\lim\limits_{n \to \infty} \left| \dfrac{a_{n+1}}{a_n} \right| = \lim\limits_{n \to \infty} \left| \dfrac{(n+1)^2 x^{n+1}}{2^{n+1}} \cdot \dfrac{2^n}{n^2 x^n} \right| = \lim\limits_{n \to \infty} \left| \dfrac{x(n+1)^2}{2n^2} \right| = \lim\limits_{n \to \infty} \left[\dfrac{|x|}{2} \left(1 + \dfrac{1}{n} \right)^2 \right] = \dfrac{|x|}{2}(1)^2 = \tfrac{1}{2}|x|.$$ By the

Ratio Test, the series $\sum\limits_{n=1}^{\infty} (-1)^n \dfrac{n^2 x^n}{2^n}$ converges when $\tfrac{1}{2}|x| < 1 \iff |x| < 2$, so the radius of convergence is $R = 2$.

When $x = \pm 2$, both series $\sum\limits_{n=1}^{\infty} (-1)^n \dfrac{n^2(\pm 2)^n}{2^n} = \sum\limits_{n=1}^{\infty} (\mp 1)^n n^2$ diverge by the Test for Divergence since

$\lim\limits_{n \to \infty} \left| (\mp 1)^n n^2 \right| = \infty$. Thus, the interval of convergence is $I = (-2, 2)$.

11. $a_n = \dfrac{(-2)^n x^n}{\sqrt[4]{n}}$, so $\lim\limits_{n \to \infty} \left| \dfrac{a_{n+1}}{a_n} \right| = \lim\limits_{n \to \infty} \dfrac{2^{n+1}|x|^{n+1}}{\sqrt[4]{n+1}} \cdot \dfrac{\sqrt[4]{n}}{2^n |x|^n} = \lim\limits_{n \to \infty} 2|x| \sqrt[4]{\dfrac{n}{n+1}} = 2|x|$, so by the Ratio Test, the

series converges when $2|x| < 1 \iff |x| < \tfrac{1}{2}$, so $R = \tfrac{1}{2}$. When $x = -\tfrac{1}{2}$, we get the divergent p-series $\sum\limits_{n=1}^{\infty} \dfrac{1}{\sqrt[4]{n}}$

$\left[p = \tfrac{1}{4} \leq 1 \right]$. When $x = \tfrac{1}{2}$, we get the series $\sum\limits_{n=1}^{\infty} \dfrac{(-1)^n}{\sqrt[4]{n}}$, which converges by the Alternating Series Test.

Thus, $I = \left(-\tfrac{1}{2}, \tfrac{1}{2} \right]$.

13. If $a_n = (-1)^n \dfrac{x^n}{4^n \ln n}$, then $\lim\limits_{n \to \infty} \left| \dfrac{a_{n+1}}{a_n} \right| = \lim\limits_{n \to \infty} \left| \dfrac{x^{n+1}}{4^{n+1} \ln(n+1)} \cdot \dfrac{4^n \ln n}{x^n} \right| = \dfrac{|x|}{4} \lim\limits_{n \to \infty} \dfrac{\ln n}{\ln(n+1)} = \dfrac{|x|}{4} \cdot 1$

[by l'Hospital's Rule] $= \dfrac{|x|}{4}$. By the Ratio Test, the series converges when $\dfrac{|x|}{4} < 1 \iff |x| < 4$, so $R = 4$. When

$x = -4$, $\sum\limits_{n=2}^{\infty} (-1)^n \dfrac{x^n}{4^n \ln n} = \sum\limits_{n=2}^{\infty} \dfrac{[(-1)(-4)]^n}{4^n \ln n} = \sum\limits_{n=2}^{\infty} \dfrac{1}{\ln n}$. Since $\ln n < n$ for $n \geq 2$, $\dfrac{1}{\ln n} > \dfrac{1}{n}$ and $\sum\limits_{n=2}^{\infty} \dfrac{1}{n}$ is the

divergent harmonic series (without the $n = 1$ term), $\sum\limits_{n=2}^{\infty} \dfrac{1}{\ln n}$ is divergent by the Comparison Test. When $x = 4$,

$\sum\limits_{n=2}^{\infty} (-1)^n \dfrac{x^n}{4^n \ln n} = \sum\limits_{n=2}^{\infty} (-1)^n \dfrac{1}{\ln n}$, which converges by the Alternating Series Test. Thus, $I = (-4, 4]$.

15. If $a_n = \dfrac{(x-2)^n}{n^2+1}$, then $\displaystyle\lim_{n\to\infty}\left|\dfrac{a_{n+1}}{a_n}\right| = \lim_{n\to\infty}\left|\dfrac{(x-2)^{n+1}}{(n+1)^2+1}\cdot\dfrac{n^2+1}{(x-2)^n}\right| = |x-2|\lim_{n\to\infty}\dfrac{n^2+1}{(n+1)^2+1} = |x-2|$. By the

Ratio Test, the series $\displaystyle\sum_{n=0}^{\infty}\dfrac{(x-2)^n}{n^2+1}$ converges when $|x-2| < 1$ $[R=1]$ $\Leftrightarrow$ $-1 < x-2 < 1$ $\Leftrightarrow$ $1 < x < 3$. When

$x=1$, the series $\displaystyle\sum_{n=0}^{\infty}(-1)^n\dfrac{1}{n^2+1}$ converges by the Alternating Series Test; when $x=3$, the series $\displaystyle\sum_{n=0}^{\infty}\dfrac{1}{n^2+1}$ converges by

comparison with the p-series $\displaystyle\sum_{n=1}^{\infty}\dfrac{1}{n^2}$ $[p=2>1]$. Thus, the interval of convergence is $I=[1,3]$.

17. If $a_n = \dfrac{3^n(x+4)^n}{\sqrt{n}}$, then $\displaystyle\lim_{n\to\infty}\left|\dfrac{a_{n+1}}{a_n}\right| = \lim_{n\to\infty}\left|\dfrac{3^{n+1}(x+4)^{n+1}}{\sqrt{n+1}}\cdot\dfrac{\sqrt{n}}{3^n(x+4)^n}\right| = 3|x+4|\lim_{n\to\infty}\dfrac{\sqrt{n}}{\sqrt{n+1}} = 3|x+4|$.

By the Ratio Test, the series $\displaystyle\sum_{n=1}^{\infty}\dfrac{3^n(x+4)^n}{\sqrt{n}}$ converges when $3|x+4| < 1$ $\Leftrightarrow$ $|x+4| < \frac{1}{3}$ $\left[R=\frac{1}{3}\right]$ $\Leftrightarrow$

$-\frac{1}{3} < x+4 < \frac{1}{3}$ $\Leftrightarrow$ $-\frac{13}{3} < x < -\frac{11}{3}$. When $x = -\frac{13}{3}$, the series $\displaystyle\sum_{n=1}^{\infty}(-1)^n\dfrac{1}{\sqrt{n}}$ converges by the Alternating Series

Test; when $x = -\frac{11}{3}$, the series $\displaystyle\sum_{n=1}^{\infty}\dfrac{1}{\sqrt{n}}$ diverges $\left[p=\frac{1}{2}\le 1\right]$. Thus, the interval of convergence is $I = \left[-\frac{13}{3}, -\frac{11}{3}\right)$.

19. If $a_n = \dfrac{(x-2)^n}{n^n}$, then $\displaystyle\lim_{n\to\infty}\sqrt[n]{|a_n|} = \lim_{n\to\infty}\dfrac{|x-2|}{n} = 0$, so the series converges for all x (by the Root Test).

$R = \infty$ and $I = (-\infty, \infty)$.

21. $a_n = \dfrac{n}{b^n}(x-a)^n$, where $b > 0$.

$$\lim_{n\to\infty}\left|\dfrac{a_{n+1}}{a_n}\right| = \lim_{n\to\infty}\dfrac{(n+1)|x-a|^{n+1}}{b^{n+1}}\cdot\dfrac{b^n}{n|x-a|^n} = \lim_{n\to\infty}\left(1+\dfrac{1}{n}\right)\dfrac{|x-a|}{b} = \dfrac{|x-a|}{b}.$$

By the Ratio Test, the series converges when $\dfrac{|x-a|}{b} < 1$ $\Leftrightarrow$ $|x-a| < b$ $[\text{so } R = b]$ $\Leftrightarrow$ $-b < x-a < b$ $\Leftrightarrow$

$a-b < x < a+b$. When $|x-a| = b$, $\displaystyle\lim_{n\to\infty}|a_n| = \lim_{n\to\infty}n = \infty$, so the series diverges. Thus, $I = (a-b, a+b)$.

23. If $a_n = n!\,(2x-1)^n$, then $\displaystyle\lim_{n\to\infty}\left|\dfrac{a_{n+1}}{a_n}\right| = \lim_{n\to\infty}\left|\dfrac{(n+1)!\,(2x-1)^{n+1}}{n!(2x-1)^n}\right| = \lim_{n\to\infty}(n+1)|2x-1| \to \infty$ as $n\to\infty$

for all $x \ne \frac{1}{2}$. Since the series diverges for all $x \ne \frac{1}{2}$, $R = 0$ and $I = \left\{\frac{1}{2}\right\}$.

25. $\displaystyle\lim_{n\to\infty}\left|\dfrac{a_{n+1}}{a_n}\right| = \lim_{n\to\infty}\left[\dfrac{|4x+1|^{n+1}}{(n+1)^2}\cdot\dfrac{n^2}{|4x+1|^n}\right] = \lim_{n\to\infty}\dfrac{|4x+1|}{(1+1/n)^2} = |4x+1|$, so by the Ratio Test, the series

converges when $|4x+1| < 1$ $\Leftrightarrow$ $-1 < 4x+1 < 1$ $\Leftrightarrow$ $-2 < 4x < 0$ $\Leftrightarrow$ $-\frac{1}{2} < x < 0$, so $R = \frac{1}{4}$. When $x = -\frac{1}{2}$,

the series becomes $\displaystyle\sum_{n=1}^{\infty}\dfrac{(-1)^n}{n^2}$, which converges by the Alternating Series Test. When $x = 0$, the series becomes $\displaystyle\sum_{n=1}^{\infty}\dfrac{1}{n^2}$,

a convergent p-series $[p=2>1]$. $\quad I = \left[-\frac{1}{2}, 0\right]$.

27. If $a_n = \dfrac{x^n}{1 \cdot 3 \cdot 5 \cdot \cdots \cdot (2n-1)}$, then

$$\lim_{n\to\infty} \left| \frac{a_{n+1}}{a_n} \right| = \lim_{n\to\infty} \left| \frac{x^{n+1}}{1 \cdot 3 \cdot 5 \cdot \cdots \cdot (2n-1)(2n+1)} \cdot \frac{1 \cdot 3 \cdot 5 \cdot \cdots \cdot (2n-1)}{x^n} \right| = \lim_{n\to\infty} \frac{|x|}{2n+1} = 0 < 1. \text{ Thus, by the}$$

Ratio Test, the series $\displaystyle\sum_{n=1}^{\infty} \dfrac{x^n}{1 \cdot 3 \cdot 5 \cdot \cdots \cdot (2n-1)}$ converges for *all* real x and we have $R = \infty$ and $I = (-\infty, \infty)$.

29. (a) We are given that the power series $\sum_{n=0}^{\infty} c_n x^n$ is convergent for $x = 4$. So by Theorem 3, it must converge for at least

$-4 < x \leq 4$. In particular, it converges when $x = -2$; that is, $\sum_{n=0}^{\infty} c_n(-2)^n$ is convergent.

(b) It does not follow that $\sum_{n=0}^{\infty} c_n(-4)^n$ is necessarily convergent. [See the comments after Theorem 3 about convergence at

the endpoint of an interval. An example is $c_n = (-1)^n/(n4^n)$.]

31. If $a_n = \dfrac{(n!)^k}{(kn)!} x^n$, then

$$\lim_{n\to\infty} \left| \frac{a_{n+1}}{a_n} \right| = \lim_{n\to\infty} \frac{[(n+1)!]^k \, (kn)!}{(n!)^k \, [k(n+1)]!} |x| = \lim_{n\to\infty} \frac{(n+1)^k}{(kn+k)(kn+k-1)\cdots(kn+2)(kn+1)} |x|$$

$$= \lim_{n\to\infty} \left[\frac{(n+1)}{(kn+1)} \frac{(n+1)}{(kn+2)} \cdots \frac{(n+1)}{(kn+k)} \right] |x|$$

$$= \lim_{n\to\infty} \left[\frac{n+1}{kn+1} \right] \lim_{n\to\infty} \left[\frac{n+1}{kn+2} \right] \cdots \lim_{n\to\infty} \left[\frac{n+1}{kn+k} \right] |x|$$

$$= \left(\frac{1}{k} \right)^k |x| < 1 \quad \Leftrightarrow \quad |x| < k^k \text{ for convergence, and the radius of convergence is } R = k^k.$$

33. No. If a power series is centered at a, its interval of convergence is symmetric about a. If a power series has an infinite radius

of convergence, then its interval of convergence must be $(-\infty, \infty)$, not $[0, \infty)$.

35. (a) If $a_n = \dfrac{(-1)^n \, x^{2n+1}}{n!(n+1)! \, 2^{2n+1}}$, then

$$\lim_{n\to\infty} \left| \frac{a_{n+1}}{a_n} \right| = \lim_{n\to\infty} \left| \frac{x^{2n+3}}{(n+1)!(n+2)! \, 2^{2n+3}} \cdot \frac{n!(n+1)! \, 2^{2n+1}}{x^{2n+1}} \right| = \left(\frac{x}{2} \right)^2 \lim_{n\to\infty} \frac{1}{(n+1)(n+2)} = 0 \text{ for all } x.$$

So $J_1(x)$ converges for all x and its domain is $(-\infty, \infty)$.

(b), (c) The initial terms of $J_1(x)$ up to $n = 5$ are $a_0 = \dfrac{x}{2}$,

$a_1 = -\dfrac{x^3}{16}$, $a_2 = \dfrac{x^5}{384}$, $a_3 = -\dfrac{x^7}{18,432}$, $a_4 = \dfrac{x^9}{1,474,560}$,

and $a_5 = -\dfrac{x^{11}}{176,947,200}$. The partial sums seem to

approximate $J_1(x)$ well near the origin, but as $|x|$ increases,

we need to take a large number of terms to get a good

approximation.

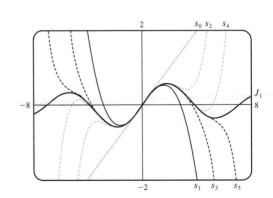

37. $s_{2n-1} = 1 + 2x + x^2 + 2x^3 + x^4 + 2x^5 + \cdots + x^{2n-2} + 2x^{2n-1}$

$$= 1(1 + 2x) + x^2(1 + 2x) + x^4(1 + 2x) + \cdots + x^{2n-2}(1 + 2x) = (1 + 2x)(1 + x^2 + x^4 + \cdots + x^{2n-2})$$

$$= (1 + 2x)\frac{1 - x^{2n}}{1 - x^2} \text{ [by (11.2.3)] with } r = x^2] \to \frac{1 + 2x}{1 - x^2} \text{ as } n \to \infty \text{ [by (11.2.4)], when } |x| < 1.$$

Also $s_{2n} = s_{2n-1} + x^{2n} \to \dfrac{1 + 2x}{1 - x^2}$ since $x^{2n} \to 0$ for $|x| < 1$. Therefore, $s_n \to \dfrac{1 + 2x}{1 - x^2}$ since s_{2n} and s_{2n-1} both

approach $\dfrac{1 + 2x}{1 - x^2}$ as $n \to \infty$. Thus, the interval of convergence is $(-1, 1)$ and $f(x) = \dfrac{1 + 2x}{1 - x^2}$.

39. We use the Root Test on the series $\sum c_n x^n$. We need $\lim\limits_{n \to \infty} \sqrt[n]{|c_n x^n|} = |x| \lim\limits_{n \to \infty} \sqrt[n]{|c_n|} = c\,|x| < 1$ for convergence, or

$|x| < 1/c$, so $R = 1/c$.

41. For $2 < x < 3$, $\sum c_n x^n$ diverges and $\sum d_n x^n$ converges. By Exercise 11.2.69, $\sum (c_n + d_n)\,x^n$ diverges. Since both series

converge for $|x| < 2$, the radius of convergence of $\sum (c_n + d_n)\,x^n$ is 2.

11.9 Representations of Functions as Power Series

1. If $f(x) = \sum\limits_{n=0}^{\infty} c_n x^n$ has radius of convergence 10, then $f'(x) = \sum\limits_{n=1}^{\infty} n c_n x^{n-1}$ also has radius of convergence 10 by

Theorem 2.

3. Our goal is to write the function in the form $\dfrac{1}{1 - r}$, and then use Equation (1) to represent the function as a sum of a power

series. $f(x) = \dfrac{1}{1 + x} = \dfrac{1}{1 - (-x)} = \sum\limits_{n=0}^{\infty} (-x)^n = \sum\limits_{n=0}^{\infty} (-1)^n x^n$ with $|-x| < 1 \iff |x| < 1$, so $R = 1$ and $I = (-1, 1)$.

5. $f(x) = \dfrac{2}{3 - x} = \dfrac{2}{3}\left(\dfrac{1}{1 - x/3}\right) = \dfrac{2}{3} \sum\limits_{n=0}^{\infty} \left(\dfrac{x}{3}\right)^n$ or, equivalently, $2 \sum\limits_{n=0}^{\infty} \dfrac{1}{3^{n+1}}\, x^n$. The series converges when $\left|\dfrac{x}{3}\right| < 1$,

that is, when $|x| < 3$, so $R = 3$ and $I = (-3, 3)$.

7. $f(x) = \dfrac{x}{9 + x^2} = \dfrac{x}{9}\left[\dfrac{1}{1 + (x/3)^2}\right] = \dfrac{x}{9}\left[\dfrac{1}{1 - \{-(x/3)^2\}}\right] = \dfrac{x}{9} \sum\limits_{n=0}^{\infty} \left[-\left(\dfrac{x}{3}\right)^2\right]^n = \dfrac{x}{9} \sum\limits_{n=0}^{\infty} (-1)^n \dfrac{x^{2n}}{9^n} = \sum\limits_{n=0}^{\infty} (-1)^n \dfrac{x^{2n+1}}{9^{n+1}}$

The geometric series $\sum\limits_{n=0}^{\infty} \left[-\left(\dfrac{x}{3}\right)^2\right]^n$ converges when $\left|-\left(\dfrac{x}{3}\right)^2\right| < 1 \iff \dfrac{|x^2|}{9} < 1 \iff |x|^2 < 9 \iff |x| < 3$, so

$R = 3$ and $I = (-3, 3)$.

9. $f(x) = \dfrac{1 + x}{1 - x} = (1 + x)\left(\dfrac{1}{1 - x}\right) = (1 + x) \sum\limits_{n=0}^{\infty} x^n = \sum\limits_{n=0}^{\infty} x^n + \sum\limits_{n=0}^{\infty} x^{n+1} = 1 + \sum\limits_{n=1}^{\infty} x^n + \sum\limits_{n=1}^{\infty} x^n = 1 + 2 \sum\limits_{n=1}^{\infty} x^n$.

The series converges when $|x| < 1$, so $R = 1$ and $I = (-1, 1)$.

A second approach: $f(x) = \dfrac{1 + x}{1 - x} = \dfrac{-(1 - x) + 2}{1 - x} = -1 + 2\left(\dfrac{1}{1 - x}\right) = -1 + 2 \sum\limits_{n=0}^{\infty} x^n = 1 + 2 \sum\limits_{n=1}^{\infty} x^n$.

A third approach:

$$f(x) = \dfrac{1 + x}{1 - x} = (1 + x)\left(\dfrac{1}{1 - x}\right) = (1 + x)(1 + x + x^2 + x^3 + \cdots)$$

$$= (1 + x + x^2 + x^3 + \cdots) + (x + x^2 + x^3 + x^4 + \cdots) = 1 + 2x + 2x^2 + 2x^3 + \cdots = 1 + 2 \sum\limits_{n=1}^{\infty} x^n.$$

11. $f(x) = \dfrac{3}{x^2 - x - 2} = \dfrac{3}{(x-2)(x+1)} = \dfrac{A}{x-2} + \dfrac{B}{x+1}$ $\Rightarrow$ $3 = A(x+1) + B(x-2)$. Let $x = 2$ to get $A = 1$ and

$x = -1$ to get $B = -1$. Thus

$$\dfrac{3}{x^2 - x - 2} = \dfrac{1}{x-2} - \dfrac{1}{x+1} = \dfrac{1}{-2}\left(\dfrac{1}{1-(x/2)}\right) - \dfrac{1}{1-(-x)} = -\dfrac{1}{2}\sum_{n=0}^{\infty}\left(\dfrac{x}{2}\right)^n - \sum_{n=0}^{\infty}(-x)^n$$

$$= \sum_{n=0}^{\infty}\left[-\dfrac{1}{2}\left(\dfrac{1}{2}\right)^n - 1(-1)^n\right]x^n = \sum_{n=0}^{\infty}\left[(-1)^{n+1} - \dfrac{1}{2^{n+1}}\right]x^n$$

We represented f as the sum of two geometric series; the first converges for $x \in (-2, 2)$ and the second converges for $(-1, 1)$.

Thus, the sum converges for $x \in (-1, 1) = I$.

13. (a) $f(x) = \dfrac{1}{(1+x)^2} = \dfrac{d}{dx}\left(\dfrac{-1}{1+x}\right) = -\dfrac{d}{dx}\left[\sum_{n=0}^{\infty}(-1)^n x^n\right]$ [from Exercise 3]

$$= \sum_{n=1}^{\infty}(-1)^{n+1}nx^{n-1} \quad \text{[from Theorem 2(i)]} = \sum_{n=0}^{\infty}(-1)^n(n+1)x^n \text{ with } R = 1.$$

In the last step, note that we *decreased* the initial value of the summation variable n by 1, and then *increased* each

occurrence of n in the term by 1 [also note that $(-1)^{n+2} = (-1)^n$].

(b) $f(x) = \dfrac{1}{(1+x)^3} = -\dfrac{1}{2}\dfrac{d}{dx}\left[\dfrac{1}{(1+x)^2}\right] = -\dfrac{1}{2}\dfrac{d}{dx}\left[\sum_{n=0}^{\infty}(-1)^n(n+1)x^n\right]$ [from part (a)]

$$= -\dfrac{1}{2}\sum_{n=1}^{\infty}(-1)^n(n+1)nx^{n-1} = \dfrac{1}{2}\sum_{n=0}^{\infty}(-1)^n(n+2)(n+1)x^n \text{ with } R = 1.$$

(c) $f(x) = \dfrac{x^2}{(1+x)^3} = x^2 \cdot \dfrac{1}{(1+x)^3} = x^2 \cdot \dfrac{1}{2}\sum_{n=0}^{\infty}(-1)^n(n+2)(n+1)x^n$ [from part (b)]

$$= \dfrac{1}{2}\sum_{n=0}^{\infty}(-1)^n(n+2)(n+1)x^{n+2}$$

To write the power series with x^n rather than x^{n+2}, we will *decrease* each occurrence of n in the term by 2 and *increase*

the initial value of the summation variable by 2. This gives us $\dfrac{1}{2}\sum_{n=2}^{\infty}(-1)^n(n)(n-1)x^n$ with $R = 1$.

15. $f(x) = \ln(5 - x) = -\displaystyle\int\dfrac{dx}{5-x} = -\dfrac{1}{5}\displaystyle\int\dfrac{dx}{1-x/5} = -\dfrac{1}{5}\displaystyle\int\left[\sum_{n=0}^{\infty}\left(\dfrac{x}{5}\right)^n\right]dx = C - \dfrac{1}{5}\sum_{n=0}^{\infty}\dfrac{x^{n+1}}{5^n(n+1)} = C - \sum_{n=1}^{\infty}\dfrac{x^n}{n\,5^n}$

Putting $x = 0$, we get $C = \ln 5$. The series converges for $|x/5| < 1 \iff |x| < 5$, so $R = 5$.

17. $\dfrac{1}{2-x} = \dfrac{1}{2(1-x/2)} = \dfrac{1}{2}\sum_{n=0}^{\infty}\left(\dfrac{x}{2}\right)^n = \sum_{n=0}^{\infty}\dfrac{1}{2^{n+1}}x^n$ for $\left|\dfrac{x}{2}\right| < 1 \iff |x| < 2$. Now

$$\dfrac{1}{(x-2)^2} = \dfrac{d}{dx}\left(\dfrac{1}{2-x}\right) = \dfrac{d}{dx}\left(\sum_{n=0}^{\infty}\dfrac{1}{2^{n+1}}x^n\right) = \sum_{n=1}^{\infty}\dfrac{n}{2^{n+1}}x^{n-1} = \sum_{n=0}^{\infty}\dfrac{n+1}{2^{n+2}}x^n. \text{ So}$$

$$f(x) = \dfrac{x^3}{(x-2)^2} = x^3\sum_{n=0}^{\infty}\dfrac{n+1}{2^{n+2}}x^n = \sum_{n=0}^{\infty}\dfrac{n+1}{2^{n+2}}x^{n+3} \text{ or } \sum_{n=3}^{\infty}\dfrac{n-2}{2^{n-1}}x^n \text{ for } |x| < 2. \text{ Thus, } R = 2 \text{ and } I = (-2, 2).$$

19. $f(x) = \dfrac{x}{x^2 + 16} = \dfrac{x}{16}\left(\dfrac{1}{1 - (-x^2/16)}\right) = \dfrac{x}{16}\displaystyle\sum_{n=0}^{\infty}\left(-\dfrac{x^2}{16}\right)^n = \dfrac{x}{16}\displaystyle\sum_{n=0}^{\infty}(-1)^n \dfrac{1}{16^n}x^{2n} = \displaystyle\sum_{n=0}^{\infty}(-1)^n \dfrac{1}{16^{n+1}}x^{2n+1}.$

The series converges when $\left|-x^2/16\right| < 1 \iff x^2 < 16 \iff |x| < 4$, so $R = 4$. The partial sums are $s_1 = \dfrac{x}{16}$,

$s_2 = s_1 - \dfrac{x^3}{16^2}$, $s_3 = s_2 + \dfrac{x^5}{16^3}$, $s_4 = s_3 - \dfrac{x^7}{16^4}$, $s_5 = s_4 + \dfrac{x^9}{16^5}$, ... Note that s_1 corresponds to the first term of the infinite

sum, regardless of the value of the summation variable and the value of the exponent.

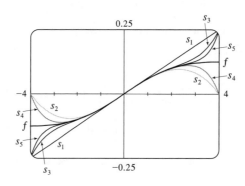

As n increases, $s_n(x)$ approximates f better on the interval of convergence, which is $(-4, 4)$.

21. $f(x) = \ln\left(\dfrac{1+x}{1-x}\right) = \ln(1+x) - \ln(1-x) = \displaystyle\int \dfrac{dx}{1+x} + \int \dfrac{dx}{1-x} = \int \dfrac{dx}{1-(-x)} + \int \dfrac{dx}{1-x}$

$= \displaystyle\int\left[\sum_{n=0}^{\infty}(-1)^n x^n + \sum_{n=0}^{\infty} x^n\right]dx = \int\left[(1 - x + x^2 - x^3 + x^4 - \cdots) + (1 + x + x^2 + x^3 + x^4 + \cdots)\right]dx$

$= \displaystyle\int (2 + 2x^2 + 2x^4 + \cdots)\, dx = \int \sum_{n=0}^{\infty} 2x^{2n}\, dx = C + \sum_{n=0}^{\infty}\dfrac{2x^{2n+1}}{2n+1}$

But $f(0) = \ln\frac{1}{1} = 0$, so $C = 0$ and we have $f(x) = \displaystyle\sum_{n=0}^{\infty}\dfrac{2x^{2n+1}}{2n+1}$ with $R = 1$. If $x = \pm 1$, then $f(x) = \pm 2\displaystyle\sum_{n=0}^{\infty}\dfrac{1}{2n+1}$,

which both diverge by the Limit Comparison Test with $b_n = \dfrac{1}{n}$. The partial sums are $s_1 = \dfrac{2x}{1}$, $s_2 = s_1 + \dfrac{2x^3}{3}$,

$s_3 = s_2 + \dfrac{2x^5}{5}$, ...

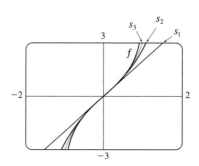

As n increases, $s_n(x)$ approximates f better on the interval of convergence, which is $(-1, 1)$.

23. $\dfrac{t}{1 - t^8} = t \cdot \dfrac{1}{1 - t^8} = t \sum\limits_{n=0}^{\infty} (t^8)^n = \sum\limits_{n=0}^{\infty} t^{8n+1}$ $\Rightarrow$ $\displaystyle\int \dfrac{t}{1 - t^8}\, dt = C + \sum\limits_{n=0}^{\infty} \dfrac{t^{8n+2}}{8n + 2}$. The series for $\dfrac{1}{1 - t^8}$ converges

when $\left|t^8\right| < 1$ $\Leftrightarrow$ $|t| < 1$, so $R = 1$ for that series and also the series for $t/(1 - t^8)$. By Theorem 2, the series for

$\displaystyle\int \dfrac{t}{1 - t^8}\, dt$ also has $R = 1$.

25. By Example 7, $\tan^{-1} x = \sum\limits_{n=0}^{\infty} (-1)^n \dfrac{x^{2n+1}}{2n + 1}$ with $R = 1$, so

$x - \tan^{-1} x = x - \left(x - \dfrac{x^3}{3} + \dfrac{x^5}{5} - \dfrac{x^7}{7} + \cdots \right) = \dfrac{x^3}{3} - \dfrac{x^5}{5} + \dfrac{x^7}{7} - \cdots = \sum\limits_{n=1}^{\infty} (-1)^{n+1} \dfrac{x^{2n+1}}{2n + 1}$ and

$\dfrac{x - \tan^{-1} x}{x^3} = \sum\limits_{n=1}^{\infty} (-1)^{n+1} \dfrac{x^{2n-2}}{2n + 1}$, so

$\displaystyle\int \dfrac{x - \tan^{-1} x}{x^3}\, dx = C + \sum\limits_{n=1}^{\infty} (-1)^{n+1} \dfrac{x^{2n-1}}{(2n + 1)(2n - 1)} = C + \sum\limits_{n=1}^{\infty} (-1)^{n+1} \dfrac{x^{2n-1}}{4n^2 - 1}$. By Theorem 2, $R = 1$.

27. $\dfrac{1}{1 + x^5} = \dfrac{1}{1 - (-x^5)} = \sum\limits_{n=0}^{\infty} (-x^5)^n = \sum\limits_{n=0}^{\infty} (-1)^n x^{5n}$ $\Rightarrow$

$\displaystyle\int \dfrac{1}{1 + x^5}\, dx = \int \sum\limits_{n=0}^{\infty} (-1)^n x^{5n}\, dx = C + \sum\limits_{n=0}^{\infty} (-1)^n \dfrac{x^{5n+1}}{5n + 1}$. Thus,

$I = \displaystyle\int_0^{0.2} \dfrac{1}{1 + x^5}\, dx = \left[x - \dfrac{x^6}{6} + \dfrac{x^{11}}{11} - \cdots \right]_0^{0.2} = 0.2 - \dfrac{(0.2)^6}{6} + \dfrac{(0.2)^{11}}{11} - \cdots$. The series is alternating, so if we use

the first two terms, the error is at most $(0.2)^{11}/11 \approx 1.9 \times 10^{-9}$. So $I \approx 0.2 - (0.2)^6/6 \approx 0.199989$ to six decimal places.

29. We substitute $3x$ for x in Example 7, and find that

$\displaystyle\int x \arctan(3x)\, dx = \int x \sum\limits_{n=0}^{\infty} (-1)^n \dfrac{(3x)^{2n+1}}{2n + 1}\, dx = \int \sum\limits_{n=0}^{\infty} (-1)^n \dfrac{3^{2n+1} x^{2n+2}}{2n + 1}\, dx = C + \sum\limits_{n=0}^{\infty} (-1)^n \dfrac{3^{2n+1} x^{2n+3}}{(2n + 1)(2n + 3)}$

So $\displaystyle\int_0^{0.1} x \arctan(3x)\, dx = \left[\dfrac{3x^3}{1 \cdot 3} - \dfrac{3^3 x^5}{3 \cdot 5} + \dfrac{3^5 x^7}{5 \cdot 7} - \dfrac{3^7 x^9}{7 \cdot 9} + \cdots \right]_0^{0.1}$

$= \dfrac{1}{10^3} - \dfrac{9}{5 \times 10^5} + \dfrac{243}{35 \times 10^7} - \dfrac{2187}{63 \times 10^9} + \cdots$.

The series is alternating, so if we use three terms, the error is at most $\dfrac{2187}{63 \times 10^9} \approx 3.5 \times 10^{-8}$. So

$\displaystyle\int_0^{0.1} x \arctan(3x)\, dx \approx \dfrac{1}{10^3} - \dfrac{9}{5 \times 10^5} + \dfrac{243}{35 \times 10^7} \approx 0.000\,983$ to six decimal places.

31. Using the result of Example 6, $\ln(1 - x) = -\sum\limits_{n=1}^{\infty} \dfrac{x^n}{n}$, with $x = -0.1$, we have

$\ln 1.1 = \ln[1 - (-0.1)] = 0.1 - \dfrac{0.01}{2} + \dfrac{0.001}{3} - \dfrac{0.0001}{4} + \dfrac{0.00001}{5} - \cdots$. The series is alternating, so if we use only

the first four terms, the error is at most $\dfrac{0.00001}{5} = 0.000002$. So $\ln 1.1 \approx 0.1 - \dfrac{0.01}{2} + \dfrac{0.001}{3} - \dfrac{0.0001}{4} \approx 0.09531$.

33. (a) $J_0(x) = \sum\limits_{n=0}^{\infty} \dfrac{(-1)^n x^{2n}}{2^{2n}(n!)^2}$, $J_0'(x) = \sum\limits_{n=1}^{\infty} \dfrac{(-1)^n 2nx^{2n-1}}{2^{2n}(n!)^2}$, and $J_0''(x) = \sum\limits_{n=1}^{\infty} \dfrac{(-1)^n 2n(2n-1)x^{2n-2}}{2^{2n}(n!)^2}$, so

$$x^2 J_0''(x) + xJ_0'(x) + x^2 J_0(x) = \sum_{n=1}^{\infty} \frac{(-1)^n 2n(2n-1)x^{2n}}{2^{2n}(n!)^2} + \sum_{n=1}^{\infty} \frac{(-1)^n 2nx^{2n}}{2^{2n}(n!)^2} + \sum_{n=0}^{\infty} \frac{(-1)^n x^{2n+2}}{2^{2n}(n!)^2}$$

$$= \sum_{n=1}^{\infty} \frac{(-1)^n 2n(2n-1)x^{2n}}{2^{2n}(n!)^2} + \sum_{n=1}^{\infty} \frac{(-1)^n 2nx^{2n}}{2^{2n}(n!)^2} + \sum_{n=1}^{\infty} \frac{(-1)^{n-1} x^{2n}}{2^{2n-2}\left[(n-1)!\right]^2}$$

$$= \sum_{n=1}^{\infty} \frac{(-1)^n 2n(2n-1)x^{2n}}{2^{2n}(n!)^2} + \sum_{n=1}^{\infty} \frac{(-1)^n 2nx^{2n}}{2^{2n}(n!)^2} + \sum_{n=1}^{\infty} \frac{(-1)^n(-1)^{-1}2^2 n^2 x^{2n}}{2^{2n}(n!)^2}$$

$$= \sum_{n=1}^{\infty} (-1)^n \left[\frac{2n(2n-1) + 2n - 2^2 n^2}{2^{2n}(n!)^2}\right] x^{2n}$$

$$= \sum_{n=1}^{\infty} (-1)^n \left[\frac{4n^2 - 2n + 2n - 4n^2}{2^{2n}(n!)^2}\right] x^{2n} = 0$$

(b) $\displaystyle\int_0^1 J_0(x)\,dx = \int_0^1 \left[\sum_{n=0}^{\infty} \frac{(-1)^n x^{2n}}{2^{2n}(n!)^2}\right] dx = \int_0^1 \left(1 - \frac{x^2}{4} + \frac{x^4}{64} - \frac{x^6}{2304} + \cdots\right) dx$

$$= \left[x - \frac{x^3}{3\cdot 4} + \frac{x^5}{5\cdot 64} - \frac{x^7}{7\cdot 2304} + \cdots\right]_0^1 = 1 - \frac{1}{12} + \frac{1}{320} - \frac{1}{16{,}128} + \cdots$$

Since $\frac{1}{16{,}128} \approx 0.000062$, it follows from The Alternating Series Estimation Theorem that, correct to three decimal places,

$\int_0^1 J_0(x)\,dx \approx 1 - \frac{1}{12} + \frac{1}{320} \approx 0.920$.

35. (a) $f(x) = \sum\limits_{n=0}^{\infty} \dfrac{x^n}{n!} \;\Rightarrow\; f'(x) = \sum\limits_{n=1}^{\infty} \dfrac{nx^{n-1}}{n!} = \sum\limits_{n=1}^{\infty} \dfrac{x^{n-1}}{(n-1)!} = \sum\limits_{n=0}^{\infty} \dfrac{x^n}{n!} = f(x)$

(b) By Theorem 9.4.2, the only solution to the differential equation $df(x)/dx = f(x)$ is $f(x) = Ke^x$, but $f(0) = 1$, so $K = 1$ and $f(x) = e^x$.

Or: We could solve the equation $df(x)/dx = f(x)$ as a separable differential equation.

37. If $a_n = \dfrac{x^n}{n^2}$, then by the Ratio Test, $\lim\limits_{n\to\infty} \left|\dfrac{a_{n+1}}{a_n}\right| = \lim\limits_{n\to\infty} \left|\dfrac{x^{n+1}}{(n+1)^2} \cdot \dfrac{n^2}{x^n}\right| = |x| \lim\limits_{n\to\infty} \left(\dfrac{n}{n+1}\right)^2 = |x| < 1$ for

convergence, so $R = 1$. When $x = \pm 1$, $\sum\limits_{n=1}^{\infty} \left|\dfrac{x^n}{n^2}\right| = \sum\limits_{n=1}^{\infty} \dfrac{1}{n^2}$ which is a convergent p-series ($p = 2 > 1$), so the interval of

convergence for f is $[-1, 1]$. By Theorem 2, the radii of convergence of f' and f'' are both 1, so we need only check the

endpoints. $f(x) = \sum\limits_{n=1}^{\infty} \dfrac{x^n}{n^2} \;\Rightarrow\; f'(x) = \sum\limits_{n=1}^{\infty} \dfrac{nx^{n-1}}{n^2} = \sum\limits_{n=0}^{\infty} \dfrac{x^n}{n+1}$, and this series diverges for $x = 1$ (harmonic series)

and converges for $x = -1$ (Alternating Series Test), so the interval of convergence is $[-1, 1)$. $f''(x) = \sum\limits_{n=1}^{\infty} \dfrac{nx^{n-1}}{n+1}$ diverges

at both 1 and -1 (Test for Divergence) since $\lim\limits_{n\to\infty} \dfrac{n}{n+1} = 1 \neq 0$, so its interval of convergence is $(-1, 1)$.

39. By Example 7, $\tan^{-1} x = \sum\limits_{n=0}^{\infty} (-1)^n \dfrac{x^{2n+1}}{2n+1}$ for $|x| < 1$. In particular, for $x = \dfrac{1}{\sqrt{3}}$, we

have $\dfrac{\pi}{6} = \tan^{-1}\left(\dfrac{1}{\sqrt{3}}\right) = \sum\limits_{n=0}^{\infty} (-1)^n \dfrac{(1/\sqrt{3})^{2n+1}}{2n+1} = \sum\limits_{n=0}^{\infty} (-1)^n \left(\dfrac{1}{3}\right)^n \dfrac{1}{\sqrt{3}} \dfrac{1}{2n+1}$, so

$\pi = \dfrac{6}{\sqrt{3}} \sum\limits_{n=0}^{\infty} \dfrac{(-1)^n}{(2n+1)3^n} = 2\sqrt{3} \sum\limits_{n=0}^{\infty} \dfrac{(-1)^n}{(2n+1)3^n}$.

11.10   Taylor and Maclaurin Series

1. Using Theorem 5 with $\sum\limits_{n=0}^{\infty} b_n (x-5)^n$, $b_n = \dfrac{f^{(n)}(a)}{n!}$, so $b_8 = \dfrac{f^{(8)}(5)}{8!}$.

3. Since $f^{(n)}(0) = (n+1)!$, Equation 7 gives the Maclaurin series

$\sum\limits_{n=0}^{\infty} \dfrac{f^{(n)}(0)}{n!} x^n = \sum\limits_{n=0}^{\infty} \dfrac{(n+1)!}{n!} x^n = \sum\limits_{n=0}^{\infty} (n+1)x^n$. Applying the Ratio Test with $a_n = (n+1)x^n$ gives us

$\lim\limits_{n\to\infty} \left| \dfrac{a_{n+1}}{a_n} \right| = \lim\limits_{n\to\infty} \left| \dfrac{(n+2)x^{n+1}}{(n+1)x^n} \right| = |x| \lim\limits_{n\to\infty} \dfrac{n+2}{n+1} = |x| \cdot 1 = |x|$. For convergence, we must have $|x| < 1$, so the

radius of convergence $R = 1$.

5.

n	$f^{(n)}(x)$	$f^{(n)}(0)$
0	$(1-x)^{-2}$	1
1	$2(1-x)^{-3}$	2
2	$6(1-x)^{-4}$	6
3	$24(1-x)^{-5}$	24
4	$120(1-x)^{-6}$	120
⋮	⋮	⋮

$(1-x)^{-2} = f(0) + f'(0)x + \dfrac{f''(0)}{2!}x^2 + \dfrac{f'''(0)}{3!}x^3 + \dfrac{f^{(4)}(0)}{4!}x^4 + \cdots$

$= 1 + 2x + \dfrac{6}{2}x^2 + \dfrac{24}{6}x^3 + \dfrac{120}{24}x^4 + \cdots$

$= 1 + 2x + 3x^2 + 4x^3 + 5x^4 + \cdots = \sum\limits_{n=0}^{\infty} (n+1)x^n$

$\lim\limits_{n\to\infty} \left| \dfrac{a_{n+1}}{a_n} \right| = \lim\limits_{n\to\infty} \left| \dfrac{(n+2)x^{n+1}}{(n+1)x^n} \right| = |x| \lim\limits_{n\to\infty} \dfrac{n+2}{n+1} = |x|(1) = |x| < 1$

for convergence, so $R = 1$.

7.

n	$f^{(n)}(x)$	$f^{(n)}(0)$
0	$\sin \pi x$	0
1	$\pi \cos \pi x$	π
2	$-\pi^2 \sin \pi x$	0
3	$-\pi^3 \cos \pi x$	$-\pi^3$
4	$\pi^4 \sin \pi x$	0
5	$\pi^5 \cos \pi x$	π^5
⋮	⋮	⋮

$\sin \pi x = f(0) + f'(0)x + \dfrac{f''(0)}{2!}x^2 + \dfrac{f'''(0)}{3!}x^3$

$\qquad + \dfrac{f^{(4)}(0)}{4!}x^4 + \dfrac{f^{(5)}(0)}{5!}x^5 + \cdots$

$= 0 + \pi x + 0 - \dfrac{\pi^3}{3!}x^3 + 0 + \dfrac{\pi^5}{5!}x^5 + \cdots$

$= \pi x - \dfrac{\pi^3}{3!}x^3 + \dfrac{\pi^5}{5!}x^5 - \dfrac{\pi^7}{7!}x^7 + \cdots$

$= \sum\limits_{n=0}^{\infty} (-1)^n \dfrac{\pi^{2n+1}}{(2n+1)!} x^{2n+1}$

$\lim\limits_{n\to\infty} \left| \dfrac{a_{n+1}}{a_n} \right| = \lim\limits_{n\to\infty} \left| \dfrac{\pi^{2n+3} x^{2n+3}}{(2n+3)!} \cdot \dfrac{(2n+1)!}{\pi^{2n+1} x^{2n+1}} \right| = \lim\limits_{n\to\infty} \dfrac{\pi^2 x^2}{(2n+3)(2n+2)}$

$= 0 < 1$ for all x, so $R = \infty$.

9.

n	$f^{(n)}(x)$	$f^{(n)}(0)$
0	e^{5x}	1
1	$5e^{5x}$	5
2	$5^2 e^{5x}$	25
3	$5^3 e^{5x}$	125
4	$5^4 e^{5x}$	625
$\vdots$	$\vdots$	$\vdots$

$$e^{5x} = \sum_{n=0}^{\infty} \frac{f^{(n)}(0)}{n!} x^n = \sum_{n=0}^{\infty} \frac{5^n}{n!} x^n.$$

$$\lim_{n\to\infty} \left| \frac{a_{n+1}}{a_n} \right| = \lim_{n\to\infty} \left[\frac{5^{n+1}|x|^{n+1}}{(n+1)!} \cdot \frac{n!}{5^n |x|^n} \right] = \lim_{n\to\infty} \frac{5|x|}{n+1} = 0 < 1$$

for all x, so $R = \infty$.

11.

n	$f^{(n)}(x)$	$f^{(n)}(0)$
0	$\sinh x$	0
1	$\cosh x$	1
2	$\sinh x$	0
3	$\cosh x$	1
4	$\sinh x$	0
$\vdots$	$\vdots$	$\vdots$

$$f^{(n)}(0) = \begin{cases} 0 & \text{if } n \text{ is even} \\ 1 & \text{if } n \text{ is odd} \end{cases} \quad \text{so } \sinh x = \sum_{n=0}^{\infty} \frac{x^{2n+1}}{(2n+1)!}.$$

Use the Ratio Test to find R. If $a_n = \dfrac{x^{2n+1}}{(2n+1)!}$, then

$$\lim_{n\to\infty} \left| \frac{a_{n+1}}{a_n} \right| = \lim_{n\to\infty} \left| \frac{x^{2n+3}}{(2n+3)!} \cdot \frac{(2n+1)!}{x^{2n+1}} \right| = x^2 \cdot \lim_{n\to\infty} \frac{1}{(2n+3)(2n+2)}$$

$$= 0 < 1 \quad \text{for all } x, \text{ so } R = \infty.$$

13.

n	$f^{(n)}(x)$	$f^{(n)}(1)$
0	$x^4 - 3x^2 + 1$	-1
1	$4x^3 - 6x$	-2
2	$12x^2 - 6$	6
3	$24x$	24
4	24	24
5	0	0
6	0	0
$\vdots$	$\vdots$	$\vdots$

$f^{(n)}(x) = 0$ for $n \geq 5$, so f has a finite series expansion about $a = 1$.

$$f(x) = x^4 - 3x^2 + 1 = \sum_{n=0}^{4} \frac{f^{(n)}(1)}{n!} (x-1)^n$$

$$= \frac{-1}{0!} (x-1)^0 + \frac{-2}{1!} (x-1)^1 + \frac{6}{2!} (x-1)^2 + \frac{24}{3!} (x-1)^3 + \frac{24}{4!} (x-1)^4$$

$$= -1 - 2(x-1) + 3(x-1)^2 + 4(x-1)^3 + (x-1)^4$$

A finite series converges for all x, so $R = \infty$.

15. $f(x) = e^x \implies f^{(n)}(x) = e^x$, so $f^{(n)}(3) = e^3$ and $e^x = \sum_{n=0}^{\infty} \dfrac{e^3}{n!}(x-3)^n$. If $a_n = \dfrac{e^3}{n!}(x-3)^n$, then

$$\lim_{n\to\infty} \left| \frac{a_{n+1}}{a_n} \right| = \lim_{n\to\infty} \left| \frac{e^3 (x-3)^{n+1}}{(n+1)!} \cdot \frac{n!}{e^3 (x-3)^n} \right| = \lim_{n\to\infty} \frac{|x-3|}{n+1} = 0 < 1 \text{ for all } x, \text{ so } R = \infty.$$

17.

n	$f^{(n)}(x)$	$f^{(n)}(\pi)$
0	$\cos x$	-1
1	$-\sin x$	0
2	$-\cos x$	1
3	$\sin x$	0
4	$\cos x$	-1
$\vdots$	$\vdots$	$\vdots$

$$\cos x = \sum_{k=0}^{\infty} \frac{f^{(k)}(\pi)}{k!}(x-\pi)^k = -1 + \frac{(x-\pi)^2}{2!} - \frac{(x-\pi)^4}{4!} + \frac{(x-\pi)^6}{6!} - \cdots$$

$$= \sum_{n=0}^{\infty} (-1)^{n+1} \frac{(x-\pi)^{2n}}{(2n)!}$$

$$\lim_{n\to\infty} \left| \frac{a_{n+1}}{a_n} \right| = \lim_{n\to\infty} \left[\frac{|x-\pi|^{2n+2}}{(2n+2)!} \cdot \frac{(2n)!}{|x-\pi|^{2n}} \right] = \lim_{n\to\infty} \frac{|x-\pi|^2}{(2n+2)(2n+1)} = 0 < 1$$

for all x, so $R = \infty$.

19.

n	$f^{(n)}(x)$	$f^{(n)}(9)$
0	$x^{-1/2}$	$\frac{1}{3}$
1	$-\frac{1}{2}x^{-3/2}$	$-\frac{1}{2}\cdot\frac{1}{3^3}$
2	$\frac{3}{4}x^{-5/2}$	$-\frac{1}{2}\cdot\left(-\frac{3}{2}\right)\cdot\frac{1}{3^5}$
3	$-\frac{15}{8}x^{-7/2}$	$-\frac{1}{2}\cdot\left(-\frac{3}{2}\right)\cdot\left(-\frac{5}{2}\right)\cdot\frac{1}{3^7}$
$\vdots$	$\vdots$	$\vdots$

$$\frac{1}{\sqrt{x}} = \frac{1}{3} - \frac{1}{2\cdot 3^3}(x-9) + \frac{3}{2^2\cdot 3^5}\frac{(x-9)^2}{2!}$$

$$- \frac{3\cdot 5}{2^3\cdot 3^7}\frac{(x-9)^3}{3!} + \cdots$$

$$= \frac{1}{3} + \sum_{n=1}^{\infty}(-1)^n\frac{1\cdot 3\cdot 5\cdot\,\cdots\,\cdot(2n-1)}{2^n\cdot 3^{2n+1}\cdot n!}(x-9)^n.$$

$$\lim_{n\to\infty}\left|\frac{a_{n+1}}{a_n}\right| = \lim_{n\to\infty}\left[\frac{1\cdot 3\cdot 5\cdot\,\cdots\,\cdot(2n-1)[2(n+1)-1]\,|x-9|^{n+1}}{2^{n+1}\cdot 3^{[2(n+1)+1]}\cdot(n+1)!}\cdot\frac{2^n\cdot 3^{2n+1}\cdot n!}{1\cdot 3\cdot 5\cdot\,\cdots\,\cdot(2n-1)\,|x-9|^n}\right]$$

$$= \lim_{n\to\infty}\left[\frac{(2n+1)\,|x-9|}{2\cdot 3^2(n+1)}\right] = \frac{1}{9}\,|x-9| < 1$$

for convergence, so $|x-9| < 9$ and $R = 9$.

21. If $f(x) = \sin\pi x$, then $f^{(n+1)}(x) = \pm\pi^{n+1}\sin\pi x$ or $\pm\pi^{n+1}\cos\pi x$. In each case, $\left|f^{(n+1)}(x)\right| \le \pi^{n+1}$, so by Formula 9

with $a = 0$ and $M = \pi^{n+1}$, $|R_n(x)| \le \dfrac{\pi^{n+1}}{(n+1)!}\,|x|^{n+1} = \dfrac{|\pi x|^{n+1}}{(n+1)!}$. Thus, $|R_n(x)| \to 0$ as $n \to \infty$ by Equation 10.

So $\lim\limits_{n\to\infty} R_n(x) = 0$ and, by Theorem 8, the series in Exercise 7 represents $\sin\pi x$ for all x.

23. If $f(x) = \sinh x$, then for all n, $f^{(n+1)}(x) = \cosh x$ or $\sinh x$. Since $|\sinh x| < |\cosh x| = \cosh x$ for all x, we have

$\left|f^{(n+1)}(x)\right| \le \cosh x$ for all n. If d is any positive number and $|x| \le d$, then $\left|f^{(n+1)}(x)\right| \le \cosh x \le \cosh d$, so by

Formula 9 with $a = 0$ and $M = \cosh d$, we have $|R_n(x)| \le \dfrac{\cosh d}{(n+1)!}\,|x|^{n+1}$. It follows that $|R_n(x)| \to 0$ as $n \to \infty$ for

$|x| \le d$ (by Equation 10). But d was an arbitrary positive number. So by Theorem 8, the series represents $\sinh x$ for all x.

25. The general binomial series in (17) is

$$(1+x)^k = \sum_{n=0}^{\infty}\binom{k}{n}x^n = 1 + kx + \frac{k(k-1)}{2!}x^2 + \frac{k(k-1)(k-2)}{3!}x^3 + \cdots.$$

$$(1+x)^{1/2} = \sum_{n=0}^{\infty}\binom{\frac{1}{2}}{n}x^n = 1 + \left(\tfrac{1}{2}\right)x + \frac{\left(\frac{1}{2}\right)\left(-\frac{1}{2}\right)}{2!}x^2 + \frac{\left(\frac{1}{2}\right)\left(-\frac{1}{2}\right)\left(-\frac{3}{2}\right)}{3!}x^3 + \cdots$$

$$= 1 + \frac{x}{2} - \frac{x^2}{2^2\cdot 2!} + \frac{1\cdot 3\cdot x^3}{2^3\cdot 3!} - \frac{1\cdot 3\cdot 5\cdot x^4}{2^4\cdot 4!} + \cdots$$

$$= 1 + \frac{x}{2} + \sum_{n=2}^{\infty}\frac{(-1)^{n-1}\,1\cdot 3\cdot 5\cdot\,\cdots\,\cdot(2n-3)x^n}{2^n\cdot n!}\quad\text{for }|x| < 1,\quad\text{so }R = 1.$$

27. $\dfrac{1}{(2+x)^3} = \dfrac{1}{[2(1+x/2)]^3} = \dfrac{1}{8}\left(1+\dfrac{x}{2}\right)^{-3} = \dfrac{1}{8}\sum\limits_{n=0}^{\infty}\binom{-3}{n}\left(\dfrac{x}{2}\right)^n$. The binomial coefficient is

$$\binom{-3}{n} = \dfrac{(-3)(-4)(-5)\cdot\cdots\cdot(-3-n+1)}{n!} = \dfrac{(-3)(-4)(-5)\cdot\cdots\cdot[-(n+2)]}{n!}$$

$$= \dfrac{(-1)^n\cdot 2\cdot 3\cdot 4\cdot 5\cdot\cdots\cdot(n+1)(n+2)}{2\cdot n!} = \dfrac{(-1)^n(n+1)(n+2)}{2}$$

Thus, $\dfrac{1}{(2+x)^3} = \dfrac{1}{8}\sum\limits_{n=0}^{\infty}\dfrac{(-1)^n(n+1)(n+2)}{2}\dfrac{x^n}{2^n} = \sum\limits_{n=0}^{\infty}\dfrac{(-1)^n(n+1)(n+2)x^n}{2^{n+4}}$ for $\left|\dfrac{x}{2}\right| < 1 \iff |x| < 2$, so $R = 2$.

29. $\sin x = \sum\limits_{n=0}^{\infty}(-1)^n\dfrac{x^{2n+1}}{(2n+1)!} \;\Rightarrow\; f(x) = \sin(\pi x) = \sum\limits_{n=0}^{\infty}(-1)^n\dfrac{(\pi x)^{2n+1}}{(2n+1)!} = \sum\limits_{n=0}^{\infty}(-1)^n\dfrac{\pi^{2n+1}}{(2n+1)!}x^{2n+1}$, $R = \infty$.

31. $e^x = \sum\limits_{n=0}^{\infty}\dfrac{x^n}{n!} \;\Rightarrow\; e^{2x} = \sum\limits_{n=0}^{\infty}\dfrac{(2x)^n}{n!} = \sum\limits_{n=0}^{\infty}\dfrac{2^n\,x^n}{n!}$, so $f(x) = e^x + e^{2x} = \sum\limits_{n=0}^{\infty}\dfrac{1}{n!}x^n + \sum\limits_{n=0}^{\infty}\dfrac{2^n}{n!}x^n = \sum\limits_{n=0}^{\infty}\dfrac{2^n+1}{n!}x^n$,
$R = \infty$.

33. $\cos x = \sum\limits_{n=0}^{\infty}(-1)^n\dfrac{x^{2n}}{(2n)!} \;\Rightarrow\; \cos\left(\tfrac{1}{2}x^2\right) = \sum\limits_{n=0}^{\infty}(-1)^n\dfrac{\left(\tfrac{1}{2}x^2\right)^{2n}}{(2n)!} = \sum\limits_{n=0}^{\infty}(-1)^n\dfrac{x^{4n}}{2^{2n}\,(2n)!}$, so

$f(x) = x\cos\left(\tfrac{1}{2}x^2\right) = \sum\limits_{n=0}^{\infty}(-1)^n\dfrac{1}{2^{2n}(2n)!}x^{4n+1}$, $R = \infty$.

35. We must write the binomial in the form $(1+$ expression$)$, so we'll factor out a 4.

$$\dfrac{x}{\sqrt{4+x^2}} = \dfrac{x}{\sqrt{4(1+x^2/4)}} = \dfrac{x}{2\sqrt{1+x^2/4}} = \dfrac{x}{2}\left(1+\dfrac{x^2}{4}\right)^{-1/2} = \dfrac{x}{2}\sum\limits_{n=0}^{\infty}\binom{-\tfrac{1}{2}}{n}\left(\dfrac{x^2}{4}\right)^n$$

$$= \dfrac{x}{2}\left[1 + \left(-\tfrac{1}{2}\right)\dfrac{x^2}{4} + \dfrac{\left(-\tfrac{1}{2}\right)\left(-\tfrac{3}{2}\right)}{2!}\left(\dfrac{x^2}{4}\right)^2 + \dfrac{\left(-\tfrac{1}{2}\right)\left(-\tfrac{3}{2}\right)\left(-\tfrac{5}{2}\right)}{3!}\left(\dfrac{x^2}{4}\right)^3 + \cdots\right]$$

$$= \dfrac{x}{2} + \dfrac{x}{2}\sum\limits_{n=1}^{\infty}(-1)^n\dfrac{1\cdot 3\cdot 5\cdot\cdots\cdot(2n-1)}{2^n\cdot 4^n\cdot n!}x^{2n}$$

$$= \dfrac{x}{2} + \sum\limits_{n=1}^{\infty}(-1)^n\dfrac{1\cdot 3\cdot 5\cdot\cdots\cdot(2n-1)}{n!\,2^{3n+1}}x^{2n+1} \text{ and } \dfrac{x^2}{4} < 1 \iff \dfrac{|x|}{2} < 1 \iff |x| < 2, \text{ so } R = 2.$$

37. $\sin^2 x = \dfrac{1}{2}(1-\cos 2x) = \dfrac{1}{2}\left[1 - \sum\limits_{n=0}^{\infty}\dfrac{(-1)^n(2x)^{2n}}{(2n)!}\right] = \dfrac{1}{2}\left[1 - 1 - \sum\limits_{n=1}^{\infty}\dfrac{(-1)^n(2x)^{2n}}{(2n)!}\right] = \sum\limits_{n=1}^{\infty}\dfrac{(-1)^{n+1}2^{2n-1}x^{2n}}{(2n)!}$,
$R = \infty$

39. $\cos x = \sum\limits_{n=0}^{\infty}(-1)^n\dfrac{x^{2n}}{(2n)!} \;\Rightarrow\; f(x) = \cos\left(x^2\right) = \sum\limits_{n=0}^{\infty}\dfrac{(-1)^n\left(x^2\right)^{2n}}{(2n)!} = \sum\limits_{n=0}^{\infty}\dfrac{(-1)^n\,x^{4n}}{(2n)!}$, $R = \infty$

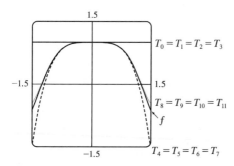

Notice that, as n increases, $T_n(x)$

becomes a better approximation to $f(x)$.

41. $e^x \overset{(11)}{=} \sum\limits_{n=0}^{\infty} \dfrac{x^n}{n!}$, so $e^{-x} = \sum\limits_{n=0}^{\infty} \dfrac{(-x)^n}{n!} = \sum\limits_{n=0}^{\infty} (-1)^n \dfrac{x^n}{n!}$, so

$$f(x) = xe^{-x} = \sum_{n=0}^{\infty} (-1)^n \frac{1}{n!} x^{n+1}$$

$$= x - x^2 + \tfrac{1}{2}x^3 - \tfrac{1}{6}x^4 + \tfrac{1}{24}x^5 - \tfrac{1}{120}x^6 + \cdots$$

$$= \sum_{n=1}^{\infty} (-1)^{n-1} \frac{x^n}{(n-1)!}$$

The series for e^x converges for all x, so the same is true of the series
for $f(x)$; that is, $R = \infty$. From the graphs of f and the first few Taylor
polynomials, we see that $T_n(x)$ provides a closer fit to $f(x)$ near 0 as n increases.

43. $e^x = \sum\limits_{n=0}^{\infty} \dfrac{x^n}{n!}$, so $e^{-0.2} = \sum\limits_{n=0}^{\infty} \dfrac{(-0.2)^n}{n!} = 1 - 0.2 + \dfrac{1}{2!}(0.2)^2 - \dfrac{1}{3!}(0.2)^3 + \dfrac{1}{4!}(0.2)^4 - \dfrac{1}{5!}(0.2)^5 + \dfrac{1}{6!}(0.2)^6 - \cdots$.

But $\dfrac{1}{6!}(0.2)^6 = 8.\overline{8} \times 10^{-8}$, so by the Alternating Series Estimation Theorem, $e^{-0.2} \approx \sum\limits_{n=0}^{5} \dfrac{(-0.2)^n}{n!} \approx 0.81873$, correct to

five decimal places.

45. (a) $1/\sqrt{1-x^2} = \left[1 + \left(-x^2\right)\right]^{-1/2} = 1 + \left(-\tfrac{1}{2}\right)\left(-x^2\right) + \dfrac{\left(-\frac{1}{2}\right)\left(-\frac{3}{2}\right)}{2!}\left(-x^2\right)^2 + \dfrac{\left(-\frac{1}{2}\right)\left(-\frac{3}{2}\right)\left(-\frac{5}{2}\right)}{3!}\left(-x^2\right)^3 + \cdots$

$$= 1 + \sum_{n=1}^{\infty} \frac{1 \cdot 3 \cdot 5 \cdots \cdot (2n-1)}{2^n \cdot n!} x^{2n}$$

(b) $\sin^{-1} x = \displaystyle\int \dfrac{1}{\sqrt{1-x^2}}\, dx = C + x + \sum_{n=1}^{\infty} \dfrac{1 \cdot 3 \cdot 5 \cdots \cdot (2n-1)}{(2n+1)2^n \cdot n!} x^{2n+1}$

$$= x + \sum_{n=1}^{\infty} \frac{1 \cdot 3 \cdot 5 \cdots \cdot (2n-1)}{(2n+1)2^n \cdot n!} x^{2n+1} \quad \text{since } 0 = \sin^{-1} 0 = C.$$

47. $\cos x \overset{(16)}{=} \sum\limits_{n=0}^{\infty} (-1)^n \dfrac{x^{2n}}{(2n)!} \quad \Rightarrow \quad \cos(x^3) = \sum\limits_{n=0}^{\infty} (-1)^n \dfrac{(x^3)^{2n}}{(2n)!} = \sum\limits_{n=0}^{\infty} (-1)^n \dfrac{x^{6n}}{(2n)!} \quad \Rightarrow$

$x\cos(x^3) = \sum\limits_{n=0}^{\infty} (-1)^n \dfrac{x^{6n+1}}{(2n)!} \quad \Rightarrow \quad \displaystyle\int x\cos(x^3)\, dx = C + \sum\limits_{n=0}^{\infty} (-1)^n \dfrac{x^{6n+2}}{(6n+2)(2n)!}$, with $R = \infty$.

49. $\cos x \overset{(16)}{=} \sum\limits_{n=0}^{\infty} (-1)^n \dfrac{x^{2n}}{(2n)!} \quad \Rightarrow \quad \cos x - 1 = \sum\limits_{n=1}^{\infty} (-1)^n \dfrac{x^{2n}}{(2n)!} \quad \Rightarrow \quad \dfrac{\cos x - 1}{x} = \sum\limits_{n=1}^{\infty} (-1)^n \dfrac{x^{2n-1}}{(2n)!} \quad \Rightarrow$

$\displaystyle\int \dfrac{\cos x - 1}{x}\, dx = C + \sum\limits_{n=1}^{\infty} (-1)^n \dfrac{x^{2n}}{2n \cdot (2n)!}$, with $R = \infty$.

51. By Exercise 47, $\displaystyle\int x\cos(x^3)\, dx = C + \sum\limits_{n=0}^{\infty} (-1)^n \dfrac{x^{6n+2}}{(6n+2)(2n)!}$, so

$$\int_0^1 x\cos(x^3)\, dx = \left[\sum_{n=0}^{\infty} (-1)^n \frac{x^{6n+2}}{(6n+2)(2n)!} \right]_0^1 = \sum_{n=0}^{\infty} \frac{(-1)^n}{(6n+2)(2n)!} = \frac{1}{2} - \frac{1}{8 \cdot 2!} + \frac{1}{14 \cdot 4!} - \frac{1}{20 \cdot 6!} + \cdots, \text{ but}$$

$\dfrac{1}{20 \cdot 6!} = \dfrac{1}{14{,}400} \approx 0.000\,069$, so $\displaystyle\int_0^1 x\cos(x^3)\, dx \approx \dfrac{1}{2} - \dfrac{1}{16} + \dfrac{1}{336} \approx 0.440$ (correct to three decimal places) by the

Alternating Series Estimation Theorem.

53. $\sqrt{1+x^4} = (1+x^4)^{1/2} = \sum\limits_{n=0}^{\infty} \binom{1/2}{n}(x^4)^n$, so $\int \sqrt{1+x^4}\,dx = C + \sum\limits_{n=0}^{\infty} \binom{1/2}{n}\dfrac{x^{4n+1}}{4n+1}$ and hence, since $0.4 < 1$,

we have

$$I = \int_0^{0.4} \sqrt{1+x^4}\,dx = \sum_{n=0}^{\infty} \binom{1/2}{n}\frac{(0.4)^{4n+1}}{4n+1}$$

$$= (1)\frac{(0.4)^1}{0!} + \frac{\frac{1}{2}}{1!}\frac{(0.4)^5}{5} + \frac{\frac{1}{2}\left(-\frac{1}{2}\right)}{2!}\frac{(0.4)^9}{9} + \frac{\frac{1}{2}\left(-\frac{1}{2}\right)\left(-\frac{3}{2}\right)}{3!}\frac{(0.4)^{13}}{13} + \frac{\frac{1}{2}\left(-\frac{1}{2}\right)\left(-\frac{3}{2}\right)\left(-\frac{5}{2}\right)}{4!}\frac{(0.4)^{17}}{17} + \cdots$$

$$= 0.4 + \frac{(0.4)^5}{10} - \frac{(0.4)^9}{72} + \frac{(0.4)^{13}}{208} - \frac{5(0.4)^{17}}{2176} + \cdots$$

Now $\dfrac{(0.4)^9}{72} \approx 3.6 \times 10^{-6} < 5 \times 10^{-6}$, so by the Alternating Series Estimation Theorem, $I \approx 0.4 + \dfrac{(0.4)^5}{10} \approx 0.40102$

(correct to five decimal places).

55. $\lim\limits_{x\to 0}\dfrac{x - \tan^{-1} x}{x^3} = \lim\limits_{x\to 0}\dfrac{x - \left(x - \frac{1}{3}x^3 + \frac{1}{5}x^5 - \frac{1}{7}x^7 + \cdots\right)}{x^3} = \lim\limits_{x\to 0}\dfrac{\frac{1}{3}x^3 - \frac{1}{5}x^5 + \frac{1}{7}x^7 - \cdots}{x^3}$

$$= \lim_{x\to 0}\left(\frac{1}{3} - \frac{1}{5}x^2 + \frac{1}{7}x^4 - \cdots\right) = \frac{1}{3}$$

since power series are continuous functions.

57. $\lim\limits_{x\to 0}\dfrac{\sin x - x + \frac{1}{6}x^3}{x^5} = \lim\limits_{x\to 0}\dfrac{\left(x - \frac{1}{3!}x^3 + \frac{1}{5!}x^5 - \frac{1}{7!}x^7 + \cdots\right) - x + \frac{1}{6}x^3}{x^5}$

$$= \lim_{x\to 0}\frac{\frac{1}{5!}x^5 - \frac{1}{7!}x^7 + \cdots}{x^5} = \lim_{x\to 0}\left(\frac{1}{5!} - \frac{x^2}{7!} + \frac{x^4}{9!} - \cdots\right) = \frac{1}{5!} = \frac{1}{120}$$

since power series are continuous functions.

59. From Equation 11, we have $e^{-x^2} = 1 - \dfrac{x^2}{1!} + \dfrac{x^4}{2!} - \dfrac{x^6}{3!} + \cdots$ and we know that $\cos x = 1 - \dfrac{x^2}{2!} + \dfrac{x^4}{4!} - \cdots$ from

Equation 16. Therefore, $e^{-x^2}\cos x = \left(1 - x^2 + \frac{1}{2}x^4 - \cdots\right)\left(1 - \frac{1}{2}x^2 + \frac{1}{24}x^4 - \cdots\right)$. Writing only the terms with

degree ≤ 4, we get $e^{-x^2}\cos x = 1 - \frac{1}{2}x^2 + \frac{1}{24}x^4 - x^2 + \frac{1}{2}x^4 + \frac{1}{2}x^4 + \cdots = 1 - \frac{3}{2}x^2 + \frac{25}{24}x^4 + \cdots$.

61. $\dfrac{x}{\sin x} \overset{(15)}{=} \dfrac{x}{x - \frac{1}{6}x^3 + \frac{1}{120}x^5 - \cdots}$.

$$
\begin{array}{r}
1 + \frac{1}{6}x^2 + \frac{7}{360}x^4 + \cdots \\
x - \frac{1}{6}x^3 + \frac{1}{120}x^5 - \cdots \overline{\smash{\big)}\, x} \\
\underline{x - \frac{1}{6}x^3 + \frac{1}{120}x^5 - \cdots} \\
\frac{1}{6}x^3 - \frac{1}{120}x^5 + \cdots \\
\underline{\frac{1}{6}x^3 - \frac{1}{36}x^5 + \cdots} \\
\frac{7}{360}x^5 + \cdots \\
\underline{\frac{7}{360}x^5 + \cdots} \\
\cdots
\end{array}
$$

From the long division above, $\dfrac{x}{\sin x} = 1 + \frac{1}{6}x^2 + \frac{7}{360}x^4 + \cdots$.

63. $\sum\limits_{n=0}^{\infty}(-1)^n\dfrac{x^{4n}}{n!} = \sum\limits_{n=0}^{\infty}\dfrac{\left(-x^4\right)^n}{n!} = e^{-x^4}$, by (11).

65. $\displaystyle\sum_{n=0}^{\infty} \frac{(-1)^n \pi^{2n+1}}{4^{2n+1}(2n+1)!} = \sum_{n=0}^{\infty} \frac{(-1)^n \left(\frac{\pi}{4}\right)^{2n+1}}{(2n+1)!} = \sin\frac{\pi}{4} = \frac{1}{\sqrt{2}}$, by (15).

67. $3 + \dfrac{9}{2!} + \dfrac{27}{3!} + \dfrac{81}{4!} + \cdots = \dfrac{3^1}{1!} + \dfrac{3^2}{2!} + \dfrac{3^3}{3!} + \dfrac{3^4}{4!} + \cdots = \displaystyle\sum_{n=1}^{\infty} \frac{3^n}{n!} = \sum_{n=0}^{\infty} \frac{3^n}{n!} - 1 = e^3 - 1$, by (11).

69. Assume that $|f'''(x)| \le M$, so $f'''(x) \le M$ for $a \le x \le a + d$. Now $\int_a^x f'''(t)\,dt \le \int_a^x M\,dt \ \Rightarrow$

$f''(x) - f''(a) \le M(x-a) \ \Rightarrow \ f''(x) \le f''(a) + M(x-a)$. Thus, $\int_a^x f''(t)\,dt \le \int_a^x [f''(a) + M(t-a)]\,dt \ \Rightarrow$

$f'(x) - f'(a) \le f''(a)(x-a) + \frac{1}{2}M(x-a)^2 \ \Rightarrow \ f'(x) \le f'(a) + f''(a)(x-a) + \frac{1}{2}M(x-a)^2 \ \Rightarrow$

$\int_a^x f'(t)\,dt \le \int_a^x \left[f'(a) + f''(a)(t-a) + \frac{1}{2}M(t-a)^2\right] dt \ \Rightarrow$

$f(x) - f(a) \le f'(a)(x-a) + \frac{1}{2}f''(a)(x-a)^2 + \frac{1}{6}M(x-a)^3$. So

$f(x) - f(a) - f'(a)(x-a) - \frac{1}{2}f''(a)(x-a)^2 \le \frac{1}{6}M(x-a)^3$. But

$R_2(x) = f(x) - T_2(x) = f(x) - f(a) - f'(a)(x-a) - \frac{1}{2}f''(a)(x-a)^2$, so $R_2(x) \le \frac{1}{6}M(x-a)^3$.

A similar argument using $f'''(x) \ge -M$ shows that $R_2(x) \ge -\frac{1}{6}M(x-a)^3$. So $|R_2(x_2)| \le \frac{1}{6}M\,|x-a|^3$.

Although we have assumed that $x > a$, a similar calculation shows that this inequality is also true if $x < a$.

71. (a) $g(x) = \displaystyle\sum_{n=0}^{\infty} \binom{k}{n} x^n \ \Rightarrow \ g'(x) = \sum_{n=1}^{\infty} \binom{k}{n} n x^{n-1}$, so

$(1+x)g'(x) = (1+x)\displaystyle\sum_{n=1}^{\infty} \binom{k}{n} n x^{n-1} = \sum_{n=1}^{\infty} \binom{k}{n} n x^{n-1} + \sum_{n=1}^{\infty} \binom{k}{n} n x^n$

$= \displaystyle\sum_{n=0}^{\infty} \binom{k}{n+1}(n+1)x^n + \sum_{n=0}^{\infty} \binom{k}{n} n x^n \qquad \begin{bmatrix}\text{Replace } n \text{ with } n+1 \\ \text{in the first series}\end{bmatrix}$

$= \displaystyle\sum_{n=0}^{\infty} (n+1)\frac{k(k-1)(k-2)\cdots(k-n+1)(k-n)}{(n+1)!}x^n + \sum_{n=0}^{\infty}\left[(n)\frac{k(k-1)(k-2)\cdots(k-n+1)}{n!}\right]x^n$

$= \displaystyle\sum_{n=0}^{\infty} \frac{(n+1)k(k-1)(k-2)\cdots(k-n+1)}{(n+1)!}\left[(k-n)+n\right]x^n$

$= k\displaystyle\sum_{n=0}^{\infty} \frac{k(k-1)(k-2)\cdots(k-n+1)}{n!}x^n = k\sum_{n=0}^{\infty} \binom{k}{n} x^n = kg(x)$

Thus, $g'(x) = \dfrac{kg(x)}{1+x}$.

(b) $h(x) = (1+x)^{-k} g(x) \ \Rightarrow$

$\qquad h'(x) = -k(1+x)^{-k-1}g(x) + (1+x)^{-k}g'(x) \qquad$ [Product Rule]

$\qquad\qquad = -k(1+x)^{-k-1}g(x) + (1+x)^{-k}\dfrac{kg(x)}{1+x} \qquad$ [from part (a)]

$\qquad\qquad = -k(1+x)^{-k-1}g(x) + k(1+x)^{-k-1}g(x) = 0$

(c) From part (b) we see that $h(x)$ must be constant for $x \in (-1, 1)$, so $h(x) = h(0) = 1$ for $x \in (-1, 1)$.

Thus, $h(x) = 1 = (1+x)^{-k} g(x) \ \Leftrightarrow \ g(x) = (1+x)^k$ for $x \in (-1, 1)$.

11.11 Applications of Taylor Polynomials

1. (a)

n	$f^{(n)}(x)$	$f^{(n)}(0)$	$T_n(x)$
0	$\cos x$	1	1
1	$-\sin x$	0	1
2	$-\cos x$	-1	$1 - \frac{1}{2}x^2$
3	$\sin x$	0	$1 - \frac{1}{2}x^2$
4	$\cos x$	1	$1 - \frac{1}{2}x^2 + \frac{1}{24}x^4$
5	$-\sin x$	0	$1 - \frac{1}{2}x^2 + \frac{1}{24}x^4$
6	$-\cos x$	-1	$1 - \frac{1}{2}x^2 + \frac{1}{24}x^4 - \frac{1}{720}x^6$

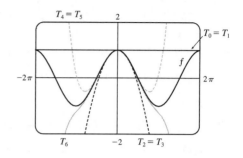

(b)

x	f	$T_0 = T_1$	$T_2 = T_3$	$T_4 = T_5$	T_6
$\frac{\pi}{4}$	0.7071	1	0.6916	0.7074	0.7071
$\frac{\pi}{2}$	0	1	-0.2337	0.0200	-0.0009
π	-1	1	-3.9348	0.1239	-1.2114

(c) As n increases, $T_n(x)$ is a good approximation to $f(x)$ on a larger and larger interval.

3.

n	$f^{(n)}(x)$	$f^{(n)}(2)$
0	$1/x$	$\frac{1}{2}$
1	$-1/x^2$	$-\frac{1}{4}$
2	$2/x^3$	$\frac{1}{4}$
3	$-6/x^4$	$-\frac{3}{8}$

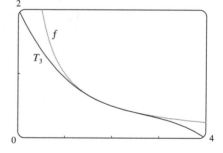

$$T_3(x) = \sum_{n=0}^{3} \frac{f^{(n)}(2)}{n!}(x-2)^n$$

$$= \frac{\frac{1}{2}}{0!} - \frac{\frac{1}{4}}{1!}(x-2) + \frac{\frac{1}{4}}{2!}(x-2)^2 - \frac{\frac{3}{8}}{3!}(x-2)^3$$

$$= \frac{1}{2} - \frac{1}{4}(x-2) + \frac{1}{8}(x-2)^2 - \frac{1}{16}(x-2)^3$$

5.

n	$f^{(n)}(x)$	$f^{(n)}(\pi/2)$
0	$\cos x$	0
1	$-\sin x$	-1
2	$-\cos x$	0
3	$\sin x$	1

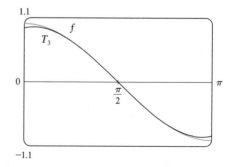

$$T_3(x) = \sum_{n=0}^{3} \frac{f^{(n)}(\pi/2)}{n!}\left(x-\frac{\pi}{2}\right)^n$$

$$= -\left(x-\frac{\pi}{2}\right) + \frac{1}{6}\left(x-\frac{\pi}{2}\right)^3$$

7.

n	$f^{(n)}(x)$	$f^{(n)}(0)$
0	$\arcsin x$	0
1	$\dfrac{1}{\sqrt{1-x^2}}$	1
2	$\dfrac{x}{(1-x^2)^{3/2}}$	0
3	$\dfrac{2x^2+1}{(1-x^2)^{5/2}}$	1

$$T_3(x) = \sum_{n=0}^{3} \frac{f^{(n)}(0)}{n!}\,x^n = x + \frac{x^3}{6}$$

9.

n	$f^{(n)}(x)$	$f^{(n)}(0)$
0	xe^{-2x}	0
1	$(1-2x)e^{-2x}$	1
2	$4(x-1)e^{-2x}$	-4
3	$4(3-2x)e^{-2x}$	12

$$T_3(x) = \sum_{n=0}^{3} \frac{f^{(n)}(0)}{n!}x^n = \frac{0}{1}\cdot 1 + \frac{1}{1}x^1 + \frac{-4}{2}x^2 + \frac{12}{6}x^3 = x - 2x^2 + 2x^3$$

11. You may be able to simply find the Taylor polynomials for

$f(x) = \cot x$ using your CAS. We will list the values of $f^{(n)}(\pi/4)$

for $n = 0$ to $n = 5$.

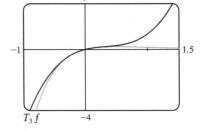

n	0	1	2	3	4	5
$f^{(n)}(\pi/4)$	1	-2	4	-16	80	-512

$$T_5(x) = \sum_{n=0}^{5} \frac{f^{(n)}(\pi/4)}{n!}\left(x - \tfrac{\pi}{4}\right)^n$$

$$= 1 - 2\left(x - \tfrac{\pi}{4}\right) + 2\left(x - \tfrac{\pi}{4}\right)^2 - \tfrac{8}{3}\left(x - \tfrac{\pi}{4}\right)^3 + \tfrac{10}{3}\left(x - \tfrac{\pi}{4}\right)^4 - \tfrac{64}{15}\left(x - \tfrac{\pi}{4}\right)^5$$

For $n = 2$ to $n = 5$, $T_n(x)$ is the polynomial consisting of all the terms up to and including the $\left(x - \tfrac{\pi}{4}\right)^n$ term.

13.

n	$f^{(n)}(x)$	$f^{(n)}(4)$
0	$\sqrt{x}$	2
1	$\frac{1}{2}x^{-1/2}$	$\frac{1}{4}$
2	$-\frac{1}{4}x^{-3/2}$	$-\frac{1}{32}$
3	$\frac{3}{8}x^{-5/2}$	

(a) $f(x) = \sqrt{x} \approx T_2(x) = 2 + \dfrac{1}{4}(x-4) - \dfrac{1/32}{2!}(x-4)^2$

$\qquad\qquad\qquad\qquad = 2 + \dfrac{1}{4}(x-4) - \dfrac{1}{64}(x-4)^2$

(b) $|R_2(x)| \le \dfrac{M}{3!}\,|x-4|^3$, where $|f'''(x)| \le M$. Now $4 \le x \le 4.2 \;\Rightarrow$

$|x-4| \le 0.2 \;\Rightarrow\; |x-4|^3 \le 0.008$. Since $f'''(x)$ is decreasing

on $[4, 4.2]$, we can take $M = |f'''(4)| = \frac{3}{8}4^{-5/2} = \frac{3}{256}$, so

$|R_2(x)| \le \dfrac{3/256}{6}(0.008) = \dfrac{0.008}{512} = 0.000\,015\,625.$

(c)

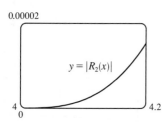

From the graph of $|R_2(x)| = |\sqrt{x} - T_2(x)|$, it seems that the error is less than 1.52×10^{-5} on $[4, 4.2]$.

15.

n	$f^{(n)}(x)$	$f^{(n)}(1)$
0	$x^{2/3}$	1
1	$\frac{2}{3}x^{-1/3}$	$\frac{2}{3}$
2	$-\frac{2}{9}x^{-4/3}$	$-\frac{2}{9}$
3	$\frac{8}{27}x^{-7/3}$	$\frac{8}{27}$
4	$-\frac{56}{81}x^{-10/3}$	

(a) $f(x) = x^{2/3} \approx T_3(x) = 1 + \frac{2}{3}(x-1) - \dfrac{2/9}{2!}(x-1)^2 + \dfrac{8/27}{3!}(x-1)^3$

$$= 1 + \tfrac{2}{3}(x-1) - \tfrac{1}{9}(x-1)^2 + \tfrac{4}{81}(x-1)^3$$

(b) $|R_3(x)| \le \dfrac{M}{4!}|x-1|^4$, where $\left|f^{(4)}(x)\right| \le M$. Now $0.8 \le x \le 1.2 \Rightarrow$

$|x-1| \le 0.2 \Rightarrow |x-1|^4 \le 0.0016$. Since $\left|f^{(4)}(x)\right|$ is decreasing

on $[0.8, 1.2]$, we can take $M = \left|f^{(4)}(0.8)\right| = \frac{56}{81}(0.8)^{-10/3}$, so

$$|R_3(x)| \le \frac{\frac{56}{81}(0.8)^{-10/3}}{24}(0.0016) \approx 0.000\,096\,97.$$

(c)

$$y = |R_3(x)|$$

From the graph of $|R_3(x)| = \left|x^{2/3} - T_3(x)\right|$, it seems that the error is less than $0.000\,053\,3$ on $[0.8, 1.2]$.

17.

n	$f^{(n)}(x)$	$f^{(n)}(0)$
0	$\sec x$	1
1	$\sec x \tan x$	0
2	$\sec x\,(2\sec^2 x - 1)$	1
3	$\sec x \tan x\,(6\sec^2 x - 1)$	

(a) $f(x) = \sec x \approx T_2(x) = 1 + \frac{1}{2}x^2$

(b) $|R_2(x)| \le \dfrac{M}{3!}|x|^3$, where $\left|f^{(3)}(x)\right| \le M$. Now $-0.2 \le x \le 0.2 \Rightarrow |x| \le 0.2 \Rightarrow |x|^3 \le (0.2)^3$.

$f^{(3)}(x)$ is an odd function and it is increasing on $[0, 0.2]$ since $\sec x$ and $\tan x$ are increasing on $[0, 0.2]$,

so $\left|f^{(3)}(x)\right| \le f^{(3)}(0.2) \approx 1.085\,158\,892$. Thus, $|R_2(x)| \le \dfrac{f^{(3)}(0.2)}{3!}(0.2)^3 \approx 0.001\,447$.

(c)

$$y = |R_2(x)|$$

From the graph of $|R_2(x)| = |\sec x - T_2(x)|$, it seems that the error is less than $0.000\,339$ on $[-0.2, 0.2]$.

19.

n	$f^{(n)}(x)$	$f^{(n)}(0)$
0	e^{x^2}	1
1	$e^{x^2}(2x)$	0
2	$e^{x^2}(2+4x^2)$	2
3	$e^{x^2}(12x+8x^3)$	0
4	$e^{x^2}(12+48x^2+16x^4)$	

(a) $f(x) = e^{x^2} \approx T_3(x) = 1 + \dfrac{2}{2!}x^2 = 1 + x^2$

(b) $|R_3(x)| \le \dfrac{M}{4!}|x|^4$, where $\left|f^{(4)}(x)\right| \le M$. Now $0 \le x \le 0.1 \Rightarrow$

$x^4 \le (0.1)^4$, and letting $x = 0.1$ gives

$|R_3(x)| \le \dfrac{e^{0.01}(12 + 0.48 + 0.0016)}{24}(0.1)^4 \approx 0.00006.$

(c)

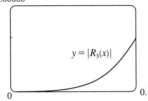

0.00008

$y = |R_3(x)|$

0   0.1

From the graph of $|R_3(x)| = \left|e^{x^2} - T_3(x)\right|$, it appears that the

error is less than $0.000\,051$ on $[0, 0.1]$.

21.

n	$f^{(n)}(x)$	$f^{(n)}(0)$
0	$x \sin x$	0
1	$\sin x + x \cos x$	0
2	$2\cos x - x \sin x$	2
3	$-3\sin x - x \cos x$	0
4	$-4\cos x + x \sin x$	-4
5	$5\sin x + x \cos x$	

(a) $f(x) = x \sin x \approx T_4(x) = \dfrac{2}{2!}(x-0)^2 + \dfrac{-4}{4!}(x-0)^4 = x^2 - \dfrac{1}{6}x^4$

(b) $|R_4(x)| \le \dfrac{M}{5!}|x|^5$, where $\left|f^{(5)}(x)\right| \le M$. Now $-1 \le x \le 1 \Rightarrow$

$|x| \le 1$, and a graph of $f^{(5)}(x)$ shows that $\left|f^{(5)}(x)\right| \le 5$ for $-1 \le x \le 1$.

Thus, we can take $M = 5$ and get $|R_4(x)| \le \dfrac{5}{5!} \cdot 1^5 = \dfrac{1}{24} = 0.041\overline{6}.$

(c)

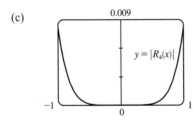

0.009

$y = |R_4(x)|$

-1 1
            0

From the graph of $|R_4(x)| = |x \sin x - T_4(x)|$, it seems that the

error is less than 0.0082 on $[-1, 1]$.

23. From Exercise 5, $\cos x = -\left(x - \frac{\pi}{2}\right) + \frac{1}{6}\left(x - \frac{\pi}{2}\right)^3 + R_3(x)$, where $|R_3(x)| \le \dfrac{M}{4!}\left|x - \frac{\pi}{2}\right|^4$ with

$\left|f^{(4)}(x)\right| = |\cos x| \le M = 1$. Now $x = 80° = (90° - 10°) = \left(\frac{\pi}{2} - \frac{\pi}{18}\right) = \frac{4\pi}{9}$ radians, so the error is

$\left|R_3\left(\frac{4\pi}{9}\right)\right| \le \frac{1}{24}\left(\frac{\pi}{18}\right)^4 \approx 0.000\,039$, which means our estimate would *not* be accurate to five decimal places. However,

$T_3 = T_4$, so we can use $\left|R_4\left(\frac{4\pi}{9}\right)\right| \le \frac{1}{120}\left(\frac{\pi}{18}\right)^5 \approx 0.000\,001$. Therefore, to five decimal places,

$\cos 80° \approx -\left(-\frac{\pi}{18}\right) + \frac{1}{6}\left(-\frac{\pi}{18}\right)^3 \approx 0.17365.$

25. All derivatives of e^x are e^x, so $|R_n(x)| \le \dfrac{e^x}{(n+1)!}|x|^{n+1}$, where $0 < x < 0.1$. Letting $x = 0.1$,

$R_n(0.1) \le \dfrac{e^{0.1}}{(n+1)!}(0.1)^{n+1} < 0.00001$, and by trial and error we find that $n = 3$ satisfies this inequality since

$R_3(0.1) < 0.0000046$. Thus, by adding the four terms of the Maclaurin series for e^x corresponding to $n = 0, 1, 2$, and 3, we can estimate $e^{0.1}$ to within 0.00001. (In fact, this sum is $1.10516\overline{6}$ and $e^{0.1} \approx 1.10517$.)

27. $\sin x = x - \dfrac{1}{3!}x^3 + \dfrac{1}{5!}x^5 - \cdots$. By the Alternating Series

Estimation Theorem, the error in the approximation

$\sin x = x - \dfrac{1}{3!}x^3$ is less than $\left|\dfrac{1}{5!}x^5\right| < 0.01 \quad \Leftrightarrow$

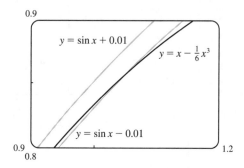

$|x^5| < 120(0.01) \quad \Leftrightarrow \quad |x| < (1.2)^{1/5} \approx 1.037$. The curves

$y = x - \frac{1}{6}x^3$ and $y = \sin x - 0.01$ intersect at $x \approx 1.043$, so

the graph confirms our estimate. Since both the sine function

and the given approximation are odd functions, we need to check the estimate only for $x > 0$. Thus, the desired range of

values for x is $-1.037 < x < 1.037$.

29. $\arctan x = x - \dfrac{x^3}{3} + \dfrac{x^5}{5} - \dfrac{x^7}{7} + \cdots$. By the Alternating Series

Estimation Theorem, the error is less than $\left|-\frac{1}{7}x^7\right| < 0.05 \quad \Leftrightarrow$

$|x^7| < 0.35 \quad \Leftrightarrow \quad |x| < (0.35)^{1/7} \approx 0.8607$. The curves

$y = x - \frac{1}{3}x^3 + \frac{1}{5}x^5$ and $y = \arctan x + 0.05$ intersect at

$x \approx 0.9245$, so the graph confirms our estimate. Since both the

arctangent function and the given approximation are odd functions,

we need to check the estimate only for $x > 0$. Thus, the desired

range of values for x is $-0.86 < x < 0.86$.

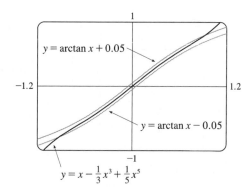

31. Let $s(t)$ be the position function of the car, and for convenience set $s(0) = 0$. The velocity of the car is $v(t) = s'(t)$ and the

acceleration is $a(t) = s''(t)$, so the second degree Taylor polynomial is $T_2(t) = s(0) + v(0)t + \dfrac{a(0)}{2}t^2 = 20t + t^2$. We

estimate the distance traveled during the next second to be $s(1) \approx T_2(1) = 20 + 1 = 21$ m. The function $T_2(t)$ would not be

accurate over a full minute, since the car could not possibly maintain an acceleration of 2 m/s^2 for that long (if it did, its final

speed would be $140 \text{ m/s} \approx 313 \text{ mi/h}!$).

33. $E = \dfrac{q}{D^2} - \dfrac{q}{(D+d)^2} = \dfrac{q}{D^2} - \dfrac{q}{D^2(1+d/D)^2} = \dfrac{q}{D^2}\left[1 - \left(1 + \dfrac{d}{D}\right)^{-2}\right]$.

We use the Binomial Series to expand $(1 + d/D)^{-2}$:

$E = \dfrac{q}{D^2}\left[1 - \left(1 - 2\left(\dfrac{d}{D}\right) + \dfrac{2 \cdot 3}{2!}\left(\dfrac{d}{D}\right)^2 - \dfrac{2 \cdot 3 \cdot 4}{3!}\left(\dfrac{d}{D}\right)^3 + \cdots\right)\right] = \dfrac{q}{D^2}\left[2\left(\dfrac{d}{D}\right) - 3\left(\dfrac{d}{D}\right)^2 + 4\left(\dfrac{d}{D}\right)^3 - \cdots\right]$

$\approx \dfrac{q}{D^2} \cdot 2\left(\dfrac{d}{D}\right) = 2qd \cdot \dfrac{1}{D^3}$

when D is much larger than d; that is, when P is far away from the dipole.

35. (a) If the water is deep, then $2\pi d/L$ is large, and we know that $\tanh x \to 1$ as $x \to \infty$. So we can approximate

$\tanh(2\pi d/L) \approx 1$, and so $v^2 \approx gL/(2\pi) \quad \Leftrightarrow \quad v \approx \sqrt{gL/(2\pi)}$.

(b) From the table, the first term in the Maclaurin series of

$\tanh x$ is x, so if the water is shallow, we can approximate

$\tanh \dfrac{2\pi d}{L} \approx \dfrac{2\pi d}{L}$, and so $v^2 \approx \dfrac{gL}{2\pi} \cdot \dfrac{2\pi d}{L} \quad \Leftrightarrow \quad v \approx \sqrt{gd}$.

n	$f^{(n)}(x)$	$f^{(n)}(0)$
0	$\tanh x$	0
1	$\operatorname{sech}^2 x$	1
2	$-2\operatorname{sech}^2 x \tanh x$	0
3	$2\operatorname{sech}^2 x\,(3\tanh^2 x - 1)$	-2

(c) Since $\tanh x$ is an odd function, its Maclaurin series is alternating, so the error in the approximation

$\tanh \dfrac{2\pi d}{L} \approx \dfrac{2\pi d}{L}$ is less than the first neglected term, which is $\dfrac{|f'''(0)|}{3!}\left(\dfrac{2\pi d}{L}\right)^3 = \dfrac{1}{3}\left(\dfrac{2\pi d}{L}\right)^3$.

If $L > 10d$, then $\dfrac{1}{3}\left(\dfrac{2\pi d}{L}\right)^3 < \dfrac{1}{3}\left(2\pi \cdot \dfrac{1}{10}\right)^3 = \dfrac{\pi^3}{375}$, so the error in the approximation $v^2 = gd$ is less

than $\dfrac{gL}{2\pi} \cdot \dfrac{\pi^3}{375} \approx 0.0132 gL$.

37. (a) L is the length of the arc subtended by the angle θ, so $L = R\theta \quad \Rightarrow$

$\theta = L/R$. Now $\sec\theta = (R+C)/R \quad \Rightarrow \quad R\sec\theta = R+C \quad \Rightarrow$

$C = R\sec\theta - R = R\sec(L/R) - R$.

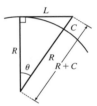

(b) Extending the result in Exercise 17, we have $f^{(4)}(x) = \sec x\,(18\sec^2 x \tan^2 x + 6\sec^4 x - \sec^2 x - \tan^2 x)$,

so $f^{(4)}(0) = 5$, and $\sec x \approx T_4(x) = 1 + \frac{1}{2}x^2 + \frac{5}{24}x^4$. By part (a),

$C \approx R\left[1 + \dfrac{1}{2}\left(\dfrac{L}{R}\right)^2 + \dfrac{5}{24}\left(\dfrac{L}{R}\right)^4\right] - R = R + \dfrac{1}{2}R \cdot \dfrac{L^2}{R^2} + \dfrac{5}{24}R \cdot \dfrac{L^4}{R^4} - R = \dfrac{L^2}{2R} + \dfrac{5L^4}{24R^3}$.

(c) Taking $L = 100$ km and $R = 6370$ km, the formula in part (a) says that

$C = R\sec(L/R) - R = 6370\sec(100/6370) - 6370 \approx 0.785\,009\,965\,44$ km.

The formula in part (b) says that $C \approx \dfrac{L^2}{2R} + \dfrac{5L^4}{24R^3} = \dfrac{100^2}{2 \cdot 6370} + \dfrac{5 \cdot 100^4}{24 \cdot 6370^3} \approx 0.785\,009\,957\,36$ km.

The difference between these two results is only $0.000\,000\,008\,08$ km, or $0.000\,008\,08$ m!

39. Using $f(x) = T_n(x) + R_n(x)$ with $n = 1$ and $x = r$, we have $f(r) = T_1(r) + R_1(r)$, where T_1 is the first-degree Taylor

polynomial of f at a. Because $a = x_n$, $f(r) = f(x_n) + f'(x_n)(r - x_n) + R_1(r)$. But r is a root of f, so $f(r) = 0$

and we have $0 = f(x_n) + f'(x_n)(r - x_n) + R_1(r)$. Taking the first two terms to the left side gives us

$f'(x_n)(x_n - r) - f(x_n) = R_1(r)$. Dividing by $f'(x_n)$, we get $x_n - r - \dfrac{f(x_n)}{f'(x_n)} = \dfrac{R_1(r)}{f'(x_n)}$. By the formula for Newton's

method, the left side of the preceding equation is $x_{n+1} - r$, so $|x_{n+1} - r| = \left|\dfrac{R_1(r)}{f'(x_n)}\right|$. Taylor's Inequality gives us

$|R_1(r)| \le \dfrac{|f''(r)|}{2!}|r - x_n|^2$. Combining this inequality with the facts $|f''(x)| \le M$ and $|f'(x)| \ge K$ gives us

$|x_{n+1} - r| \le \dfrac{M}{2K}|x_n - r|^2$.

11 Review

CONCEPT CHECK

1. (a) See Definition 11.1.1.

(b) See Definition 11.2.2.

(c) The terms of the sequence $\{a_n\}$ approach 3 as n becomes large.

(d) By adding sufficiently many terms of the series, we can make the partial sums as close to 3 as we like.

2. (a) See Definition 11.1.11.

(b) A sequence is monotonic if it is either increasing or decreasing.

(c) By Theorem 11.1.12, every bounded, monotonic sequence is convergent.

3. (a) See (4) in Section 11.2.

(b) The p-series $\sum_{n=1}^{\infty} \dfrac{1}{n^p}$ is convergent if $p > 1$.

4. If $\sum a_n = 3$, then $\lim\limits_{n \to \infty} a_n = 0$ and $\lim\limits_{n \to \infty} s_n = 3$.

5. (a) *Test for Divergence:* If $\lim\limits_{n \to \infty} a_n$ does not exist or if $\lim\limits_{n \to \infty} a_n \neq 0$, then the series $\sum_{n=1}^{\infty} a_n$ is divergent.

(b) *Integral Test:* Suppose f is a continuous, positive, decreasing function on $[1, \infty)$ and let $a_n = f(n)$. Then the series

$\sum_{n=1}^{\infty} a_n$ is convergent if and only if the improper integral $\int_1^{\infty} f(x)\, dx$ is convergent. In other words:

(i) If $\int_1^{\infty} f(x)\, dx$ is convergent, then $\sum_{n=1}^{\infty} a_n$ is convergent.

(ii) If $\int_1^{\infty} f(x)\, dx$ is divergent, then $\sum_{n=1}^{\infty} a_n$ is divergent.

(c) *Comparison Test:* Suppose that $\sum a_n$ and $\sum b_n$ are series with positive terms.

(i) If $\sum b_n$ is convergent and $a_n \leq b_n$ for all n, then $\sum a_n$ is also convergent.

(ii) If $\sum b_n$ is divergent and $a_n \geq b_n$ for all n, then $\sum a_n$ is also divergent.

(d) *Limit Comparison Test:* Suppose that $\sum a_n$ and $\sum b_n$ are series with positive terms. If $\lim\limits_{n \to \infty} (a_n/b_n) = c$, where c is a

finite number and $c > 0$, then either both series converge or both diverge.

(e) *Alternating Series Test:* If the alternating series $\sum_{n=1}^{\infty} (-1)^{n-1} b_n = b_1 - b_2 + b_3 - b_4 + b_5 - b_6 + \cdots$ $[b_n > 0]$

satisfies (i) $b_{n+1} \leq b_n$ for all n and (ii) $\lim\limits_{n \to \infty} b_n = 0$, then the series is convergent.

(f) *Ratio Test:*

(i) If $\lim\limits_{n \to \infty} \left| \dfrac{a_{n+1}}{a_n} \right| = L < 1$, then the series $\sum_{n=1}^{\infty} a_n$ is absolutely convergent (and therefore convergent).

(ii) If $\lim\limits_{n \to \infty} \left| \dfrac{a_{n+1}}{a_n} \right| = L > 1$ or $\lim\limits_{n \to \infty} \left| \dfrac{a_{n+1}}{a_n} \right| = \infty$, then the series $\sum_{n=1}^{\infty} a_n$ is divergent.

(iii) If $\lim\limits_{n \to \infty} \left| \dfrac{a_{n+1}}{a_n} \right| = 1$, the Ratio Test is inconclusive; that is, no conclusion can be drawn about the convergence or

divergence of $\sum a_n$.

(g) *Root Test:*

    (i) If $\lim\limits_{n \to \infty} \sqrt[n]{|a_n|} = L < 1$, then the series $\sum_{n=1}^{\infty} a_n$ is absolutely convergent (and therefore convergent).

    (ii) If $\lim\limits_{n \to \infty} \sqrt[n]{|a_n|} = L > 1$ or $\lim\limits_{n \to \infty} \sqrt[n]{|a_n|} = \infty$, then the series $\sum_{n=1}^{\infty} a_n$ is divergent.

    (iii) If $\lim\limits_{n \to \infty} \sqrt[n]{|a_n|} = 1$, the Root Test is inconclusive..

6. (a) A series $\sum a_n$ is called *absolutely convergent* if the series of absolute values $\sum |a_n|$ is convergent.

   (b) If a series $\sum a_n$ is absolutely convergent, then it is convergent.

   (c) A series $\sum a_n$ is called *conditionally convergent* if it is convergent but not absolutely convergent.

7. (a) Use (3) in Section 11.3.

   (b) See Example 5 in Section 11.4.

   (c) By adding terms until you reach the desired accuracy given by the Alternating Series Estimation Theorem on page 712.

8. (a) $\sum\limits_{n=0}^{\infty} c_n(x - a)^n$

   (b) Given the power series $\sum\limits_{n=0}^{\infty} c_n(x - a)^n$, the radius of convergence is:

    (i) 0 if the series converges only when $x = a$

    (ii) ∞ if the series converges for all x, or

    (iii) a positive number R such that the series converges if $|x - a| < R$ and diverges if $|x - a| > R$.

   (c) The interval of convergence of a power series is the interval that consists of all values of x for which the series converges. Corresponding to the cases in part (b), the interval of convergence is: (i) the single point $\{a\}$, (ii) all real numbers, that is, the real number line $(-\infty, \infty)$, or (iii) an interval with endpoints $a - R$ and $a + R$ which can contain neither, either, or both of the endpoints. In this case, we must test the series for convergence at each endpoint to determine the interval of convergence.

9. (a), (b) See Theorem 11.9.2.

10. (a) $T_n(x) = \sum\limits_{i=0}^{n} \dfrac{f^{(i)}(a)}{i!} (x - a)^i$

   (b) $\sum\limits_{n=0}^{\infty} \dfrac{f^{(n)}(a)}{n!} (x - a)^n$

   (c) $\sum\limits_{n=0}^{\infty} \dfrac{f^{(n)}(0)}{n!} x^n$ [$a = 0$ in part (b)]

   (d) See Theorem 11.10.8.

   (e) See Taylor's Inequality (11.10.9).

11. (a)–(e) See Table 1 on page 743.

12. See the binomial series (11.10.17) for the expansion. The radius of convergence for the binomial series is 1.

TRUE-FALSE QUIZ

1. False. See Note 2 after Theorem 11.2.6.

3. True. If $\lim\limits_{n\to\infty} a_n = L$, then given any $\varepsilon > 0$, we can find a positive integer N such that $|a_n - L| < \varepsilon$ whenever $n > N$.

If $n > N$, then $2n + 1 > N$ and $|a_{2n+1} - L| < \varepsilon$. Thus, $\lim\limits_{n\to\infty} a_{2n+1} = L$.

5. False. For example, take $c_n = (-1)^n/(n6^n)$.

7. False, since $\lim\limits_{n\to\infty}\left|\dfrac{a_{n+1}}{a_n}\right| = \lim\limits_{n\to\infty}\left|\dfrac{1}{(n+1)^3}\cdot\dfrac{n^3}{1}\right| = \lim\limits_{n\to\infty}\left|\dfrac{n^3}{(n+1)^3}\cdot\dfrac{1/n^3}{1/n^3}\right| = \lim\limits_{n\to\infty}\dfrac{1}{(1+1/n)^3} = 1.$

9. False. See the note after Example 2 in Section 11.4.

11. True. See (9) in Section 11.1.

13. True. By Theorem 11.10.5 the coefficient of x^3 is $\dfrac{f'''(0)}{3!} = \dfrac{1}{3} \ \Rightarrow \ f'''(0) = 2.$

Or: Use Theorem 11.9.2 to differentiate f three times.

15. False. For example, let $a_n = b_n = (-1)^n$. Then $\{a_n\}$ and $\{b_n\}$ are divergent, but $a_n b_n = 1$, so $\{a_n b_n\}$ is convergent.

17. True by Theorem 11.6.3. $\left[\sum (-1)^n a_n \text{ is absolutely convergent and hence convergent.}\right]$

19. True. $0.99999\ldots = 0.9 + 0.9(0.1)^1 + 0.9(0.1)^2 + 0.9(0.1)^3 + \cdots = \sum\limits_{n=1}^{\infty}(0.9)(0.1)^{n-1} = \dfrac{0.9}{1-0.1} = 1$ by the formula

for the sum of a geometric series $[S = a_1/(1-r)]$ with ratio r satisfying $|r| < 1$.

EXERCISES

1. $\left\{\dfrac{2+n^3}{1+2n^3}\right\}$ converges since $\lim\limits_{n\to\infty}\dfrac{2+n^3}{1+2n^3} = \lim\limits_{n\to\infty}\dfrac{2/n^3+1}{1/n^3+2} = \dfrac{1}{2}.$

3. $\lim\limits_{n\to\infty} a_n = \lim\limits_{n\to\infty}\dfrac{n^3}{1+n^2} = \lim\limits_{n\to\infty}\dfrac{n}{1/n^2+1} = \infty$, so the sequence diverges.

5. $|a_n| = \left|\dfrac{n\sin n}{n^2+1}\right| \le \dfrac{n}{n^2+1} < \dfrac{1}{n}$, so $|a_n| \to 0$ as $n \to \infty$. Thus, $\lim\limits_{n\to\infty} a_n = 0$. The sequence $\{a_n\}$ is convergent.

7. $\left\{\left(1+\dfrac{3}{n}\right)^{4n}\right\}$ is convergent. Let $y = \left(1+\dfrac{3}{x}\right)^{4x}$. Then

$$\lim_{x\to\infty}\ln y = \lim_{x\to\infty} 4x\ln(1+3/x) = \lim_{x\to\infty}\dfrac{\ln(1+3/x)}{1/(4x)} \overset{\text{H}}{=} \lim_{x\to\infty}\dfrac{\dfrac{1}{1+3/x}\left(-\dfrac{3}{x^2}\right)}{-1/(4x^2)} = \lim_{x\to\infty}\dfrac{12}{1+3/x} = 12, \text{ so}$$

$$\lim_{x\to\infty} y = \lim_{n\to\infty}\left(1+\dfrac{3}{n}\right)^{4n} = e^{12}.$$

9. We use induction, hypothesizing that $a_{n-1} < a_n < 2$. Note first that $1 < a_2 = \frac{1}{3}(1+4) = \frac{5}{3} < 2$, so the hypothesis holds for $n = 2$. Now assume that $a_{k-1} < a_k < 2$. Then $a_k = \frac{1}{3}(a_{k-1}+4) < \frac{1}{3}(a_k+4) < \frac{1}{3}(2+4) = 2$. So $a_k < a_{k+1} < 2$, and the induction is complete. To find the limit of the sequence, we note that $L = \lim\limits_{n\to\infty} a_n = \lim\limits_{n\to\infty} a_{n+1} \ \Rightarrow$

$L = \frac{1}{3}(L+4) \ \Rightarrow \ L = 2.$

11. $\dfrac{n}{n^3 + 1} < \dfrac{n}{n^3} = \dfrac{1}{n^2}$, so $\displaystyle\sum_{n=1}^{\infty} \dfrac{n}{n^3 + 1}$ converges by the Comparison Test with the convergent p-series $\displaystyle\sum_{n=1}^{\infty} \dfrac{1}{n^2}$ $[p = 2 > 1]$.

13. $\displaystyle\lim_{n\to\infty} \left| \dfrac{a_{n+1}}{a_n} \right| = \lim_{n\to\infty} \left[\dfrac{(n+1)^3}{5^{n+1}} \cdot \dfrac{5^n}{n^3} \right] = \lim_{n\to\infty} \left(1 + \dfrac{1}{n} \right)^3 \cdot \dfrac{1}{5} = \dfrac{1}{5} < 1$, so $\displaystyle\sum_{n=1}^{\infty} \dfrac{n^3}{5^n}$ converges by the Ratio Test.

15. Let $f(x) = \dfrac{1}{x \sqrt{\ln x}}$. Then f is continuous, positive, and decreasing on $[2, \infty)$, so the Integral Test applies.

$$\int_2^{\infty} f(x)\, dx = \lim_{t\to\infty} \int_2^t \dfrac{1}{x \sqrt{\ln x}}\, dx \quad \left[u = \ln x,\, du = \dfrac{1}{x}\, dx \right] = \lim_{t\to\infty} \int_{\ln 2}^{\ln t} u^{-1/2}\, du = \lim_{t\to\infty} \left[2\sqrt{u} \right]_{\ln 2}^{\ln t}$$

$$= \lim_{t\to\infty} \left(2\sqrt{\ln t} - 2\sqrt{\ln 2} \right) = \infty,$$

so the series $\displaystyle\sum_{n=2}^{\infty} \dfrac{1}{n \sqrt{\ln n}}$ diverges.

17. $|a_n| = \left| \dfrac{\cos 3n}{1 + (1.2)^n} \right| \le \dfrac{1}{1 + (1.2)^n} < \dfrac{1}{(1.2)^n} = \left(\dfrac{5}{6} \right)^n$, so $\displaystyle\sum_{n=1}^{\infty} |a_n|$ converges by comparison with the convergent geometric

series $\displaystyle\sum_{n=1}^{\infty} \left(\dfrac{5}{6} \right)^n$ $\left[r = \dfrac{5}{6} < 1 \right]$. It follows that $\displaystyle\sum_{n=1}^{\infty} a_n$ converges (by Theorem 3 in Section 11.6).

19. $\displaystyle\lim_{n\to\infty} \left| \dfrac{a_{n+1}}{a_n} \right| = \lim_{n\to\infty} \dfrac{1 \cdot 3 \cdot 5 \cdots (2n-1)(2n+1)}{5^{n+1}\,(n+1)!} \cdot \dfrac{5^n\, n!}{1 \cdot 3 \cdot 5 \cdots (2n-1)} = \lim_{n\to\infty} \dfrac{2n+1}{5(n+1)} = \dfrac{2}{5} < 1$, so the series

converges by the Ratio Test.

21. $b_n = \dfrac{\sqrt{n}}{n+1} > 0$, $\{b_n\}$ is decreasing, and $\displaystyle\lim_{n\to\infty} b_n = 0$, so the series $\displaystyle\sum_{n=1}^{\infty} (-1)^{n-1} \dfrac{\sqrt{n}}{n+1}$ converges by the Alternating

Series Test.

23. Consider the series of absolute values: $\displaystyle\sum_{n=1}^{\infty} n^{-1/3}$ is a p-series with $p = \dfrac{1}{3} \le 1$ and is therefore divergent. But if we apply the

Alternating Series Test, we see that $b_n = \dfrac{1}{\sqrt[3]{n}} > 0$, $\{b_n\}$ is decreasing, and $\displaystyle\lim_{n\to\infty} b_n = 0$, so the series $\displaystyle\sum_{n=1}^{\infty} (-1)^{n-1}\, n^{-1/3}$

converges. Thus, $\displaystyle\sum_{n=1}^{\infty} (-1)^{n-1}\, n^{-1/3}$ is conditionally convergent.

25. $\left| \dfrac{a_{n+1}}{a_n} \right| = \left| \dfrac{(-1)^{n+1}(n+2)3^{n+1}}{2^{2n+3}} \cdot \dfrac{2^{2n+1}}{(-1)^n(n+1)3^n} \right| = \dfrac{n+2}{n+1} \cdot \dfrac{3}{4} = \dfrac{1 + (2/n)}{1 + (1/n)} \cdot \dfrac{3}{4} \to \dfrac{3}{4} < 1$ as $n \to \infty$, so by the Ratio

Test, $\displaystyle\sum_{n=1}^{\infty} \dfrac{(-1)^n(n+1)3^n}{2^{2n+1}}$ is absolutely convergent.

27. $\displaystyle\sum_{n=1}^{\infty} \dfrac{(-3)^{n-1}}{2^{3n}} = \sum_{n=1}^{\infty} \dfrac{(-3)^{n-1}}{(2^3)^n} = \sum_{n=1}^{\infty} \dfrac{(-3)^{n-1}}{8^n} = \dfrac{1}{8} \sum_{n=1}^{\infty} \dfrac{(-3)^{n-1}}{8^{n-1}} = \dfrac{1}{8} \sum_{n=1}^{\infty} \left(-\dfrac{3}{8} \right)^{n-1} = \dfrac{1}{8} \left(\dfrac{1}{1 - (-3/8)} \right)$

$$= \dfrac{1}{8} \cdot \dfrac{8}{11} = \dfrac{1}{11}$$

29. $\displaystyle\sum_{n=1}^{\infty} [\tan^{-1}(n+1) - \tan^{-1} n] = \lim_{n\to\infty} s_n$

$$= \lim_{n\to\infty} [(\tan^{-1} 2 - \tan^{-1} 1) + (\tan^{-1} 3 - \tan^{-1} 2) + \cdots + (\tan^{-1}(n+1) - \tan^{-1} n)]$$

$$= \lim_{n\to\infty} [\tan^{-1}(n+1) - \tan^{-1} 1] = \tfrac{\pi}{2} - \tfrac{\pi}{4} = \tfrac{\pi}{4}$$

31. $1 - e + \dfrac{e^2}{2!} - \dfrac{e^3}{3!} + \dfrac{e^4}{4!} - \cdots = \displaystyle\sum_{n=0}^{\infty} (-1)^n \dfrac{e^n}{n!} = \sum_{n=0}^{\infty} \dfrac{(-e)^n}{n!} = e^{-e}$ since $e^x = \displaystyle\sum_{n=0}^{\infty} \dfrac{x^n}{n!}$ for all x.

33. $\cosh x = \dfrac{1}{2}(e^x + e^{-x}) = \dfrac{1}{2}\left(\displaystyle\sum_{n=0}^{\infty} \dfrac{x^n}{n!} + \sum_{n=0}^{\infty} \dfrac{(-x)^n}{n!} \right)$

$$= \dfrac{1}{2}\left[\left(1 + x + \dfrac{x^2}{2!} + \dfrac{x^3}{3!} + \dfrac{x^4}{4!} + \cdots\right) + \left(1 - x + \dfrac{x^2}{2!} - \dfrac{x^3}{3!} + \dfrac{x^4}{4!} - \cdots\right) \right]$$

$$= \dfrac{1}{2}\left(2 + 2 \cdot \dfrac{x^2}{2!} + 2 \cdot \dfrac{x^4}{4!} + \cdots \right) = 1 + \dfrac{1}{2}x^2 + \sum_{n=2}^{\infty} \dfrac{x^{2n}}{(2n)!} \geq 1 + \dfrac{1}{2}x^2 \quad \text{for all } x$$

35. $\displaystyle\sum_{n=1}^{\infty} \dfrac{(-1)^{n+1}}{n^5} = 1 - \dfrac{1}{32} + \dfrac{1}{243} - \dfrac{1}{1024} + \dfrac{1}{3125} - \dfrac{1}{7776} + \dfrac{1}{16,807} - \dfrac{1}{32,768} + \cdots.$

Since $b_8 = \dfrac{1}{8^5} = \dfrac{1}{32,768} < 0.000031$, $\displaystyle\sum_{n=1}^{\infty} \dfrac{(-1)^{n+1}}{n^5} \approx \sum_{n=1}^{7} \dfrac{(-1)^{n+1}}{n^5} \approx 0.9721.$

37. $\displaystyle\sum_{n=1}^{\infty} \dfrac{1}{2+5^n} \approx \sum_{n=1}^{8} \dfrac{1}{2+5^n} \approx 0.18976224.$ To estimate the error, note that $\dfrac{1}{2+5^n} < \dfrac{1}{5^n}$, so the remainder term is

$$R_8 = \sum_{n=9}^{\infty} \dfrac{1}{2+5^n} < \sum_{n=9}^{\infty} \dfrac{1}{5^n} = \dfrac{1/5^9}{1-1/5} = 6.4 \times 10^{-7} \; \left[\text{geometric series with } a = \tfrac{1}{5^9} \text{ and } r = \tfrac{1}{5}\right].$$

39. Use the Limit Comparison Test. $\displaystyle\lim_{n\to\infty} \left| \dfrac{\left(\frac{n+1}{n}\right)a_n}{a_n} \right| = \lim_{n\to\infty} \dfrac{n+1}{n} = \lim_{n\to\infty} \left(1 + \dfrac{1}{n}\right) = 1 > 0.$

Since $\sum |a_n|$ is convergent, so is $\displaystyle\sum \left| \left(\dfrac{n+1}{n}\right) a_n \right|$, by the Limit Comparison Test.

41. $\displaystyle\lim_{n\to\infty} \left| \dfrac{a_{n+1}}{a_n} \right| = \lim_{n\to\infty} \left[\dfrac{|x+2|^{n+1}}{(n+1)4^{n+1}} \cdot \dfrac{n\,4^n}{|x+2|^n} \right] = \lim_{n\to\infty} \left[\dfrac{n}{n+1} \dfrac{|x+2|}{4} \right] = \dfrac{|x+2|}{4} < 1 \; \Leftrightarrow \; |x+2| < 4$, so $R = 4$.

$|x+2| < 4 \; \Leftrightarrow \; -4 < x+2 < 4 \; \Leftrightarrow \; -6 < x < 2.$ If $x = -6$, then the series $\displaystyle\sum_{n=1}^{\infty} \dfrac{(x+2)^n}{n\,4^n}$ becomes

$\displaystyle\sum_{n=1}^{\infty} \dfrac{(-4)^n}{n4^n} = \sum_{n=1}^{\infty} \dfrac{(-1)^n}{n}$, the alternating harmonic series, which converges by the Alternating Series Test. When $x = 2$, the

series becomes the harmonic series $\displaystyle\sum_{n=1}^{\infty} \dfrac{1}{n}$, which diverges. Thus, $I = [-6, 2)$.

43. $\displaystyle\lim_{n\to\infty} \left| \dfrac{a_{n+1}}{a_n} \right| = \lim_{n\to\infty} \left| \dfrac{2^{n+1}(x-3)^{n+1}}{\sqrt{n+4}} \cdot \dfrac{\sqrt{n+3}}{2^n(x-3)^n} \right| = 2|x-3| \lim_{n\to\infty} \sqrt{\dfrac{n+3}{n+4}} = 2|x-3| < 1 \; \Leftrightarrow \; |x-3| < \tfrac{1}{2}$,

so $R = \tfrac{1}{2}$. $|x-3| < \tfrac{1}{2} \; \Leftrightarrow \; -\tfrac{1}{2} < x-3 < \tfrac{1}{2} \; \Leftrightarrow \; \tfrac{5}{2} < x < \tfrac{7}{2}.$ For $x = \tfrac{7}{2}$, the series $\displaystyle\sum_{n=1}^{\infty} \dfrac{2^n(x-3)^n}{\sqrt{n+3}}$ becomes

$\displaystyle\sum_{n=0}^{\infty} \dfrac{1}{\sqrt{n+3}} = \sum_{n=3}^{\infty} \dfrac{1}{n^{1/2}}$, which diverges $\left[p = \tfrac{1}{2} \leq 1\right]$, but for $x = \tfrac{5}{2}$, we get $\displaystyle\sum_{n=0}^{\infty} \dfrac{(-1)^n}{\sqrt{n+3}}$, which is a convergent

alternating series, so $I = \left[\tfrac{5}{2}, \tfrac{7}{2}\right).$

45.

n	$f^{(n)}(x)$	$f^{(n)}\left(\frac{\pi}{6}\right)$
0	$\sin x$	$\frac{1}{2}$
1	$\cos x$	$\frac{\sqrt{3}}{2}$
2	$-\sin x$	$-\frac{1}{2}$
3	$-\cos x$	$-\frac{\sqrt{3}}{2}$
4	$\sin x$	$\frac{1}{2}$
$\vdots$	$\vdots$	$\vdots$

$$\sin x = f\left(\frac{\pi}{6}\right) + f'\left(\frac{\pi}{6}\right)\left(x - \frac{\pi}{6}\right) + \frac{f''\left(\frac{\pi}{6}\right)}{2!}\left(x - \frac{\pi}{6}\right)^2 + \frac{f^{(3)}\left(\frac{\pi}{6}\right)}{3!}\left(x - \frac{\pi}{6}\right)^3 + \frac{f^{(4)}\left(\frac{\pi}{6}\right)}{4!}\left(x - \frac{\pi}{6}\right)^4 + \cdots$$

$$= \frac{1}{2}\left[1 - \frac{1}{2!}\left(x - \frac{\pi}{6}\right)^2 + \frac{1}{4!}\left(x - \frac{\pi}{6}\right)^4 - \cdots\right] + \frac{\sqrt{3}}{2}\left[\left(x - \frac{\pi}{6}\right) - \frac{1}{3!}\left(x - \frac{\pi}{6}\right)^3 + \cdots\right]$$

$$= \frac{1}{2}\sum_{n=0}^{\infty}(-1)^n \frac{1}{(2n)!}\left(x - \frac{\pi}{6}\right)^{2n} + \frac{\sqrt{3}}{2}\sum_{n=0}^{\infty}(-1)^n \frac{1}{(2n+1)!}\left(x - \frac{\pi}{6}\right)^{2n+1}$$

47. $\dfrac{1}{1+x} = \dfrac{1}{1-(-x)} = \displaystyle\sum_{n=0}^{\infty}(-x)^n = \sum_{n=0}^{\infty}(-1)^n x^n$ for $|x| < 1$ $\Rightarrow$ $\dfrac{x^2}{1+x} = \displaystyle\sum_{n=0}^{\infty}(-1)^n x^{n+2}$ with $R = 1$.

49. $\dfrac{1}{1-x} = \displaystyle\sum_{n=0}^{\infty} x^n$ for $|x| < 1$ $\Rightarrow$ $\ln(1-x) = -\displaystyle\int \frac{dx}{1-x} = -\int \sum_{n=0}^{\infty} x^n\,dx = C - \sum_{n=0}^{\infty}\frac{x^{n+1}}{n+1}$.

$\ln(1-0) = C - 0$ $\Rightarrow$ $C = 0$ $\Rightarrow$ $\ln(1-x) = -\displaystyle\sum_{n=0}^{\infty}\frac{x^{n+1}}{n+1} = \sum_{n=1}^{\infty}\frac{-x^n}{n}$ with $R = 1$.

51. $\sin x = \displaystyle\sum_{n=0}^{\infty}\frac{(-1)^n x^{2n+1}}{(2n+1)!}$ $\Rightarrow$ $\sin(x^4) = \displaystyle\sum_{n=0}^{\infty}\frac{(-1)^n (x^4)^{2n+1}}{(2n+1)!} = \sum_{n=0}^{\infty}\frac{(-1)^n x^{8n+4}}{(2n+1)!}$ for all x, so the radius of

convergence is ∞.

53. $f(x) = \dfrac{1}{\sqrt[4]{16-x}} = \dfrac{1}{\sqrt[4]{16(1-x/16)}} = \dfrac{1}{\sqrt[4]{16}\left(1-\frac{1}{16}x\right)^{1/4}} = \frac{1}{2}\left(1 - \frac{1}{16}x\right)^{-1/4}$

$$= \frac{1}{2}\left[1 + \left(-\frac{1}{4}\right)\left(-\frac{x}{16}\right) + \frac{\left(-\frac{1}{4}\right)\left(-\frac{5}{4}\right)}{2!}\left(-\frac{x}{16}\right)^2 + \frac{\left(-\frac{1}{4}\right)\left(-\frac{5}{4}\right)\left(-\frac{9}{4}\right)}{3!}\left(-\frac{x}{16}\right)^3 + \cdots\right]$$

$$= \frac{1}{2} + \sum_{n=1}^{\infty}\frac{1\cdot 5\cdot 9\cdot\,\cdots\,\cdot(4n-3)}{2\cdot 4^n\cdot n!\cdot 16^n}x^n = \frac{1}{2} + \sum_{n=1}^{\infty}\frac{1\cdot 5\cdot 9\cdot\,\cdots\,\cdot(4n-3)}{2^{6n+1}\,n!}x^n$$

for $\left|-\dfrac{x}{16}\right| < 1$ $\Leftrightarrow$ $|x| < 16$, so $R = 16$.

55. $e^x = \displaystyle\sum_{n=0}^{\infty}\frac{x^n}{n!}$, so $\dfrac{e^x}{x} = \frac{1}{x}\displaystyle\sum_{n=0}^{\infty}\frac{x^n}{n!} = \sum_{n=0}^{\infty}\frac{x^{n-1}}{n!} = x^{-1} + \sum_{n=1}^{\infty}\frac{x^{n-1}}{n!} = \frac{1}{x} + \sum_{n=1}^{\infty}\frac{x^{n-1}}{n!}$ and

$$\int \frac{e^x}{x}\,dx = C + \ln|x| + \sum_{n=1}^{\infty}\frac{x^n}{n\cdot n!}.$$

57. (a)

n	$f^{(n)}(x)$	$f^{(n)}(1)$
0	$x^{1/2}$	1
1	$\frac{1}{2}x^{-1/2}$	$\frac{1}{2}$
2	$-\frac{1}{4}x^{-3/2}$	$-\frac{1}{4}$
3	$\frac{3}{8}x^{-5/2}$	$\frac{3}{8}$
4	$-\frac{15}{16}x^{-7/2}$	$-\frac{15}{16}$
⋮	⋮	⋮

$$\sqrt{x} \approx T_3(x) = 1 + \frac{1/2}{1!}(x-1) - \frac{1/4}{2!}(x-1)^2 + \frac{3/8}{3!}(x-1)^3$$

$$= 1 + \tfrac{1}{2}(x-1) - \tfrac{1}{8}(x-1)^2 + \tfrac{1}{16}(x-1)^3$$

(b)

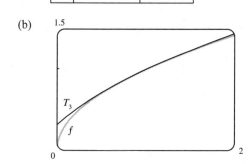

(c) $|R_3(x)| \le \dfrac{M}{4!}|x-1|^4$, where $\left|f^{(4)}(x)\right| \le M$ with

$f^{(4)}(x) = -\frac{15}{16}x^{-7/2}$. Now $0.9 \le x \le 1.1 \;\Rightarrow$

$-0.1 \le x - 1 \le 0.1 \;\Rightarrow\; (x-1)^4 \le (0.1)^4$,

and letting $x = 0.9$ gives $M = \dfrac{15}{16(0.9)^{7/2}}$, so

$$|R_3(x)| \le \frac{15}{16(0.9)^{7/2}\,4!}(0.1)^4 \approx 0.000\,005\,648$$

$$\approx 0.000\,006 = 6 \times 10^{-6}$$

(d)

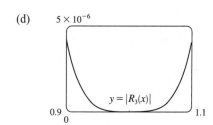

From the graph of $|R_3(x)| = |\sqrt{x} - T_3(x)|$, it appears

that the error is less than 5×10^{-6} on $[0.9, 1.1]$.

59. $\sin x = \displaystyle\sum_{n=0}^{\infty} (-1)^n \frac{x^{2n+1}}{(2n+1)!} = x - \frac{x^3}{3!} + \frac{x^5}{5!} - \frac{x^7}{7!} + \cdots$, so $\sin x - x = -\dfrac{x^3}{3!} + \dfrac{x^5}{5!} - \dfrac{x^7}{7!} + \cdots$ and

$\dfrac{\sin x - x}{x^3} = -\dfrac{1}{3!} + \dfrac{x^2}{5!} - \dfrac{x^4}{7!} + \cdots$. Thus, $\displaystyle\lim_{x \to 0} \frac{\sin x - x}{x^3} = \lim_{x \to 0}\left(-\frac{1}{6} + \frac{x^2}{120} - \frac{x^4}{5040} + \cdots\right) = -\frac{1}{6}$.

61. $f(x) = \displaystyle\sum_{n=0}^{\infty} c_n x^n \;\Rightarrow\; f(-x) = \sum_{n=0}^{\infty} c_n(-x)^n = \sum_{n=0}^{\infty} (-1)^n c_n x^n$

(a) If f is an odd function, then $f(-x) = -f(x) \;\Rightarrow\; \displaystyle\sum_{n=0}^{\infty} (-1)^n c_n x^n = \sum_{n=0}^{\infty} -c_n x^n$. The coefficients of any power series

are uniquely determined (by Theorem 11.10.5), so $(-1)^n c_n = -c_n$.

If n is even, then $(-1)^n = 1$, so $c_n = -c_n \;\Rightarrow\; 2c_n = 0 \;\Rightarrow\; c_n = 0$. Thus, all even coefficients are 0, that is,
$c_0 = c_2 = c_4 = \cdots = 0$.

(b) If f is even, then $f(-x) = f(x) \;\Rightarrow\; \displaystyle\sum_{n=0}^{\infty} (-1)^n c_n x^n = \sum_{n=0}^{\infty} c_n x^n \;\Rightarrow\; (-1)^n c_n = c_n$.

If n is odd, then $(-1)^n = -1$, so $-c_n = c_n \;\Rightarrow\; 2c_n = 0 \;\Rightarrow\; c_n = 0$. Thus, all odd coefficients are 0,

that is, $c_1 = c_3 = c_5 = \cdots = 0$.

1. It would be far too much work to compute 15 derivatives of f. The key idea is to remember that $f^{(n)}(0)$ occurs in the

coefficient of x^n in the Maclaurin series of f. We start with the Maclaurin series for sin: $\sin x = x - \dfrac{x^3}{3!} + \dfrac{x^5}{5!} - \cdots$.

Then $\sin(x^3) = x^3 - \dfrac{x^9}{3!} + \dfrac{x^{15}}{5!} - \cdots$, and so the coefficient of x^{15} is $\dfrac{f^{(15)}(0)}{15!} = \dfrac{1}{5!}$. Therefore,

$f^{(15)}(0) = \dfrac{15!}{5!} = 6 \cdot 7 \cdot 8 \cdot 9 \cdot 10 \cdot 11 \cdot 12 \cdot 13 \cdot 14 \cdot 15 = 10{,}897{,}286{,}400.$

3. (a) From Formula 14a in Appendix D, with $x = y = \theta$, we get $\tan 2\theta = \dfrac{2\tan\theta}{1 - \tan^2\theta}$, so $\cot 2\theta = \dfrac{1 - \tan^2\theta}{2\tan\theta}$ $\Rightarrow$

$2\cot 2\theta = \dfrac{1 - \tan^2\theta}{\tan\theta} = \cot\theta - \tan\theta.$ Replacing θ by $\tfrac{1}{2}x$, we get $2\cot x = \cot\tfrac{1}{2}x - \tan\tfrac{1}{2}x$, or

$\tan\tfrac{1}{2}x = \cot\tfrac{1}{2}x - 2\cot x.$

(b) From part (a) with $\dfrac{x}{2^{n-1}}$ in place of x, $\tan\dfrac{x}{2^n} = \cot\dfrac{x}{2^n} - 2\cot\dfrac{x}{2^{n-1}}$, so the nth partial sum of $\displaystyle\sum_{n=1}^{\infty} \dfrac{1}{2^n}\tan\dfrac{x}{2^n}$ is

$$
\begin{aligned}
s_n &= \frac{\tan(x/2)}{2} + \frac{\tan(x/4)}{4} + \frac{\tan(x/8)}{8} + \cdots + \frac{\tan(x/2^n)}{2^n} \\
&= \left[\frac{\cot(x/2)}{2} - \cot x\right] + \left[\frac{\cot(x/4)}{4} - \frac{\cot(x/2)}{2}\right] + \left[\frac{\cot(x/8)}{8} - \frac{\cot(x/4)}{4}\right] + \cdots \\
&\quad + \left[\frac{\cot(x/2^n)}{2^n} - \frac{\cot(x/2^{n-1})}{2^{n-1}}\right] = -\cot x + \frac{\cot(x/2^n)}{2^n} \quad \text{[telescoping sum]}
\end{aligned}
$$

Now $\dfrac{\cot(x/2^n)}{2^n} = \dfrac{\cos(x/2^n)}{2^n\sin(x/2^n)} = \dfrac{\cos(x/2^n)}{x} \cdot \dfrac{x/2^n}{\sin(x/2^n)} \to \dfrac{1}{x} \cdot 1 = \dfrac{1}{x}$ as $n \to \infty$ since $x/2^n \to 0$

for $x \neq 0$. Therefore, if $x \neq 0$ and $x \neq k\pi$ where k is any integer, then

$$
\sum_{n=1}^{\infty} \frac{1}{2^n}\tan\frac{x}{2^n} = \lim_{n\to\infty} s_n = \lim_{n\to\infty}\left(-\cot x + \frac{1}{2^n}\cot\frac{x}{2^n}\right) = -\cot x + \frac{1}{x}
$$

If $x = 0$, then all terms in the series are 0, so the sum is 0.

5. (a) At each stage, each side is replaced by four shorter sides, each of length
$\tfrac{1}{3}$ of the side length at the preceding stage. Writing s_0 and ℓ_0 for the
number of sides and the length of the side of the initial triangle, we
generate the table at right. In general, we have $s_n = 3 \cdot 4^n$ and
$\ell_n = \left(\tfrac{1}{3}\right)^n$, so the length of the perimeter at the nth stage of construction

is $p_n = s_n\ell_n = 3 \cdot 4^n \cdot \left(\tfrac{1}{3}\right)^n = 3 \cdot \left(\tfrac{4}{3}\right)^n.$

$s_0 = 3$	$\ell_0 = 1$
$s_1 = 3 \cdot 4$	$\ell_1 = 1/3$
$s_2 = 3 \cdot 4^2$	$\ell_2 = 1/3^2$
$s_3 = 3 \cdot 4^3$	$\ell_3 = 1/3^3$
$\vdots$	$\vdots$

(b) $p_n = \dfrac{4^n}{3^{n-1}} = 4\left(\dfrac{4}{3}\right)^{n-1}$. Since $\tfrac{4}{3} > 1$, $p_n \to \infty$ as $n \to \infty$.

(c) The area of each of the small triangles added at a given stage is one-ninth of the area of the triangle added at the preceding stage. Let a be the area of the original triangle. Then the area a_n of each of the small triangles added at stage n is

$a_n = a \cdot \dfrac{1}{9^n} = \dfrac{a}{9^n}$. Since a small triangle is added to each side at every stage, it follows that the total area A_n added to the

figure at the nth stage is $A_n = s_{n-1} \cdot a_n = 3 \cdot 4^{n-1} \cdot \dfrac{a}{9^n} = a \cdot \dfrac{4^{n-1}}{3^{2n-1}}$. Then the total area enclosed by the snowflake

curve is $A = a + A_1 + A_2 + A_3 + \cdots = a + a \cdot \dfrac{1}{3} + a \cdot \dfrac{4}{3^3} + a \cdot \dfrac{4^2}{3^5} + a \cdot \dfrac{4^3}{3^7} + \cdots$. After the first term, this is a

geometric series with common ratio $\dfrac{4}{9}$, so $A = a + \dfrac{a/3}{1 - \frac{4}{9}} = a + \dfrac{a}{3} \cdot \dfrac{9}{5} = \dfrac{8a}{5}$. But the area of the original equilateral

triangle with side 1 is $a = \dfrac{1}{2} \cdot 1 \cdot \sin \dfrac{\pi}{3} = \dfrac{\sqrt{3}}{4}$. So the area enclosed by the snowflake curve is $\dfrac{8}{5} \cdot \dfrac{\sqrt{3}}{4} = \dfrac{2\sqrt{3}}{5}$.

7. (a) Let $a = \arctan x$ and $b = \arctan y$. Then, from Formula 14b in Appendix D,

$$\tan(a - b) = \frac{\tan a - \tan b}{1 + \tan a \tan b} = \frac{\tan(\arctan x) - \tan(\arctan y)}{1 + \tan(\arctan x) \tan(\arctan y)} = \frac{x - y}{1 + xy}$$

Now $\arctan x - \arctan y = a - b = \arctan(\tan(a - b)) = \arctan \dfrac{x - y}{1 + xy}$ since $-\dfrac{\pi}{2} < a - b < \dfrac{\pi}{2}$.

(b) From part (a) we have

$$\arctan \tfrac{120}{119} - \arctan \tfrac{1}{239} = \arctan \frac{\frac{120}{119} - \frac{1}{239}}{1 + \frac{120}{119} \cdot \frac{1}{239}} = \arctan \frac{\frac{28,561}{28,441}}{\frac{28,561}{28,441}} = \arctan 1 = \tfrac{\pi}{4}$$

(c) Replacing y by $-y$ in the formula of part (a), we get $\arctan x + \arctan y = \arctan \dfrac{x + y}{1 - xy}$. So

$$4 \arctan \tfrac{1}{5} = 2\left(\arctan \tfrac{1}{5} + \arctan \tfrac{1}{5}\right) = 2 \arctan \frac{\frac{1}{5} + \frac{1}{5}}{1 - \frac{1}{5} \cdot \frac{1}{5}} = 2 \arctan \tfrac{5}{12} = \arctan \tfrac{5}{12} + \arctan \tfrac{5}{12}$$

$$= \arctan \frac{\frac{5}{12} + \frac{5}{12}}{1 - \frac{5}{12} \cdot \frac{5}{12}} = \arctan \tfrac{120}{119}$$

Thus, from part (b), we have $4 \arctan \tfrac{1}{5} - \arctan \tfrac{1}{239} = \arctan \tfrac{120}{119} - \arctan \tfrac{1}{239} = \tfrac{\pi}{4}$.

(d) From Example 7 in Section 11.9 we have $\arctan x = x - \dfrac{x^3}{3} + \dfrac{x^5}{5} - \dfrac{x^7}{7} + \dfrac{x^9}{9} - \dfrac{x^{11}}{11} + \cdots$, so

$$\arctan \tfrac{1}{5} = \frac{1}{5} - \frac{1}{3 \cdot 5^3} + \frac{1}{5 \cdot 5^5} - \frac{1}{7 \cdot 5^7} + \frac{1}{9 \cdot 5^9} - \frac{1}{11 \cdot 5^{11}} + \cdots$$

This is an alternating series and the size of the terms decreases to 0, so by the Alternating Series Estimation Theorem, the sum lies between s_5 and s_6, that is, $0.197395560 < \arctan \tfrac{1}{5} < 0.197395562$.

(e) From the series in part (d) we get $\arctan \dfrac{1}{239} = \dfrac{1}{239} - \dfrac{1}{3 \cdot 239^3} + \dfrac{1}{5 \cdot 239^5} - \cdots$. The third term is less than

2.6×10^{-13}, so by the Alternating Series Estimation Theorem, we have, to nine decimal places,

$\arctan \tfrac{1}{239} \approx s_2 \approx 0.004184076$. Thus, $0.004184075 < \arctan \tfrac{1}{239} < 0.004184077$.

(f) From part (c) we have $\pi = 16 \arctan \frac{1}{5} - 4 \arctan \frac{1}{239}$, so from parts (d) and (e) we have

$$16(0.197395560) - 4(0.004184077) < \pi < 16(0.197395562) - 4(0.004184075) \quad \Rightarrow$$

$3.141592652 < \pi < 3.141592692$. So, to 7 decimal places, $\pi \approx 3.1415927$.

9. We start with the geometric series $\sum\limits_{n=0}^{\infty} x^n = \dfrac{1}{1-x}$, $|x| < 1$, and differentiate:

$$\sum_{n=1}^{\infty} nx^{n-1} = \frac{d}{dx}\left(\sum_{n=0}^{\infty} x^n\right) = \frac{d}{dx}\left(\frac{1}{1-x}\right) = \frac{1}{(1-x)^2} \text{ for } |x| < 1 \quad \Rightarrow \quad \sum_{n=1}^{\infty} nx^n = x \sum_{n=1}^{\infty} nx^{n-1} = \frac{x}{(1-x)^2}$$

for $|x| < 1$. Differentiate again:

$$\sum_{n=1}^{\infty} n^2 x^{n-1} = \frac{d}{dx}\frac{x}{(1-x)^2} = \frac{(1-x)^2 - x \cdot 2(1-x)(-1)}{(1-x)^4} = \frac{x+1}{(1-x)^3} \quad \Rightarrow \quad \sum_{n=1}^{\infty} n^2 x^n = \frac{x^2 + x}{(1-x)^3} \quad \Rightarrow$$

$$\sum_{n=1}^{\infty} n^3 x^{n-1} = \frac{d}{dx}\frac{x^2 + x}{(1-x)^3} = \frac{(1-x)^3(2x+1) - (x^2+x)3(1-x)^2(-1)}{(1-x)^6} = \frac{x^2 + 4x + 1}{(1-x)^4} \quad \Rightarrow$$

$$\sum_{n=1}^{\infty} n^3 x^n = \frac{x^3 + 4x^2 + x}{(1-x)^4}, |x| < 1.$$ The radius of convergence is 1 because that is the radius of convergence for the

geometric series we started with. If $x = \pm 1$, the series is $\sum n^3(\pm 1)^n$, which diverges by the Test For Divergence, so the

interval of convergence is $(-1, 1)$.

11. $\ln\left(1 - \dfrac{1}{n^2}\right) = \ln\left(\dfrac{n^2 - 1}{n^2}\right) = \ln\dfrac{(n+1)(n-1)}{n^2} = \ln[(n+1)(n-1)] - \ln n^2$

$$= \ln(n+1) + \ln(n-1) - 2\ln n = \ln(n-1) - \ln n - \ln n + \ln(n+1)$$

$$= \ln\frac{n-1}{n} - [\ln n - \ln(n+1)] = \ln\frac{n-1}{n} - \ln\frac{n}{n+1}.$$

Let $s_k = \sum\limits_{n=2}^{k} \ln\left(1 - \dfrac{1}{n^2}\right) = \sum\limits_{n=2}^{k}\left(\ln\dfrac{n-1}{n} - \ln\dfrac{n}{n+1}\right)$ for $k \geq 2$. Then

$$s_k = \left(\ln\frac{1}{2} - \ln\frac{2}{3}\right) + \left(\ln\frac{2}{3} - \ln\frac{3}{4}\right) + \cdots + \left(\ln\frac{k-1}{k} - \ln\frac{k}{k+1}\right) = \ln\frac{1}{2} - \ln\frac{k}{k+1}, \text{ so}$$

$$\sum_{n=2}^{\infty} \ln\left(1 - \frac{1}{n^2}\right) = \lim_{k\to\infty} s_k = \lim_{k\to\infty}\left(\ln\frac{1}{2} - \ln\frac{k}{k+1}\right) = \ln\frac{1}{2} - \ln 1 = \ln 1 - \ln 2 - \ln 1 = -\ln 2.$$

13. (a)

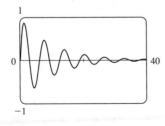

The x-intercepts of the curve occur where $\sin x = 0 \Leftrightarrow x = n\pi$,

n an integer. So using the formula for disks (and either a CAS or

$\sin^2 x = \frac{1}{2}(1 - \cos 2x)$ and Formula 99 to evaluate the integral),

the volume of the nth bead is

$$V_n = \pi \int_{(n-1)\pi}^{n\pi} (e^{-x/10}\sin x)^2\, dx = \pi \int_{(n-1)\pi}^{n\pi} e^{-x/5}\sin^2 x\, dx$$

$$= \frac{250\pi}{101}\left(e^{-(n-1)\pi/5} - e^{-n\pi/5}\right)$$

(b) The total volume is

$$\pi \int_0^\infty e^{-x/5} \sin^2 x \, dx = \sum_{n=1}^\infty V_n = \frac{250\pi}{101} \sum_{n=1}^\infty \left[e^{-(n-1)\pi/5} - e^{-n\pi/5} \right] = \frac{250\pi}{101} \quad \text{[telescoping sum]}.$$

Another method: If the volume in part (a) has been written as $V_n = \frac{250\pi}{101} e^{-n\pi/5}(e^{\pi/5} - 1)$, then we recognize $\sum_{n=1}^\infty V_n$

as a geometric series with $a = \frac{250\pi}{101}(1 - e^{-\pi/5})$ and $r = e^{-\pi/5}$.

15. If L is the length of a side of the equilateral triangle, then the area is $A = \frac{1}{2}L \cdot \frac{\sqrt{3}}{2}L = \frac{\sqrt{3}}{4}L^2$ and so $L^2 = \frac{4}{\sqrt{3}}A$.

Let r be the radius of one of the circles. When there are n rows of circles, the figure shows that

$$L = \sqrt{3}\,r + r + (n-2)(2r) + r + \sqrt{3}\,r = r\big(2n - 2 + 2\sqrt{3}\big), \text{ so } r = \frac{L}{2\big(n + \sqrt{3} - 1\big)}.$$

The number of circles is $1 + 2 + \cdots + n = \dfrac{n(n+1)}{2}$, and so the total area of the circles is

$$A_n = \frac{n(n+1)}{2}\pi r^2 = \frac{n(n+1)}{2}\pi \frac{L^2}{4\big(n + \sqrt{3} - 1\big)^2}$$

$$= \frac{n(n+1)}{2}\pi \frac{4A/\sqrt{3}}{4\big(n + \sqrt{3} - 1\big)^2} = \frac{n(n+1)}{\big(n + \sqrt{3} - 1\big)^2}\frac{\pi A}{2\sqrt{3}} \quad \Rightarrow$$

$$\frac{A_n}{A} = \frac{n(n+1)}{\big(n + \sqrt{3} - 1\big)^2}\frac{\pi}{2\sqrt{3}}$$

$$= \frac{1 + 1/n}{\big[1 + (\sqrt{3} - 1)/n\big]^2}\frac{\pi}{2\sqrt{3}} \to \frac{\pi}{2\sqrt{3}} \quad \text{as } n \to \infty$$

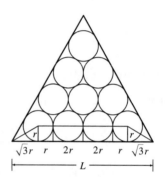

$\sqrt{3}r \quad r \quad 2r \qquad 2r \quad r \quad \sqrt{3}r$

L

17. As in Section 11.9 we have to integrate the function x^x by integrating series. Writing $x^x = (e^{\ln x})^x = e^{x \ln x}$ and using the

Maclaurin series for e^x, we have $x^x = (e^{\ln x})^x = e^{x \ln x} = \sum_{n=0}^\infty \frac{(x \ln x)^n}{n!} = \sum_{n=0}^\infty \frac{x^n (\ln x)^n}{n!}$. As with power series, we can

integrate this series term-by-term: $\displaystyle\int_0^1 x^x \, dx = \sum_{n=0}^\infty \int_0^1 \frac{x^n (\ln x)^n}{n!} \, dx = \sum_{n=0}^\infty \frac{1}{n!}\int_0^1 x^n (\ln x)^n \, dx$. We integrate by parts

with $u = (\ln x)^n$, $dv = x^n \, dx$, so $du = \dfrac{n(\ln x)^{n-1}}{x}\, dx$ and $v = \dfrac{x^{n+1}}{n+1}$:

$$\int_0^1 x^n (\ln x)^n \, dx = \lim_{t \to 0^+} \int_t^1 x^n (\ln x)^n \, dx = \lim_{t \to 0^+} \left[\frac{x^{n+1}}{n+1}(\ln x)^n \right]_t^1 - \lim_{t \to 0^+} \int_t^1 \frac{n}{n+1}x^n (\ln x)^{n-1} \, dx$$

$$= 0 - \frac{n}{n+1}\int_0^1 x^n (\ln x)^{n-1} \, dx$$

(where l'Hospital's Rule was used to help evaluate the first limit). Further integration by parts gives

$$\int_0^1 x^n (\ln x)^k \, dx = -\frac{k}{n+1}\int_0^1 x^n (\ln x)^{k-1} \, dx \text{ and, combining these steps, we get}$$

$$\int_0^1 x^n (\ln x)^n \, dx = \frac{(-1)^n n!}{(n+1)^n}\int_0^1 x^n \, dx = \frac{(-1)^n n!}{(n+1)^{n+1}} \quad \Rightarrow$$

$$\int_0^1 x^x \, dx = \sum_{n=0}^\infty \frac{1}{n!}\int_0^1 x^n (\ln x)^n \, dx = \sum_{n=0}^\infty \frac{1}{n!}\frac{(-1)^n n!}{(n+1)^{n+1}} = \sum_{n=0}^\infty \frac{(-1)^n}{(n+1)^{n+1}} = \sum_{n=1}^\infty \frac{(-1)^{n-1}}{n^n}.$$

19. Let $f(x) = \sum\limits_{m=0}^{\infty} c_m x^m$ and $g(x) = e^{f(x)} = \sum\limits_{n=0}^{\infty} d_n x^n$. Then $g'(x) = \sum\limits_{n=0}^{\infty} n d_n x^{n-1}$, so $n d_n$ occurs as the coefficient

of x^{n-1}. But also

$$g'(x) = e^{f(x)} f'(x) = \left(\sum_{n=0}^{\infty} d_n x^n \right) \left(\sum_{m=1}^{\infty} m c_m x^{m-1} \right)$$

$$= \left(d_0 + d_1 x + d_2 x^2 + \cdots + d_{n-1} x^{n-1} + \cdots \right) \left(c_1 + 2c_2 x + 3c_3 x^2 + \cdots + n c_n x^{n-1} + \cdots \right)$$

so the coefficient of x^{n-1} is $c_1 d_{n-1} + 2c_2 d_{n-2} + 3c_3 d_{n-3} + \cdots + n c_n d_0 = \sum\limits_{i=1}^{n} i c_i d_{n-i}$. Therefore, $n d_n = \sum\limits_{i=1}^{n} i c_i d_{n-i}$.

21. Call the series S. We group the terms according to the number of digits in their denominators:

$$S = \underbrace{\left(\tfrac{1}{1} + \tfrac{1}{2} + \cdots + \tfrac{1}{8} + \tfrac{1}{9} \right)}_{g_1} + \underbrace{\left(\tfrac{1}{11} + \cdots + \tfrac{1}{99} \right)}_{g_2} + \underbrace{\left(\tfrac{1}{111} + \cdots + \tfrac{1}{999} \right)}_{g_3} + \cdots$$

Now in the group g_n, since we have 9 choices for each of the n digits in the denominator, there are 9^n terms.

Furthermore, each term in g_n is less than $\frac{1}{10^{n-1}}$ [except for the first term in g_1]. So $g_n < 9^n \cdot \frac{1}{10^{n-1}} = 9 \left(\frac{9}{10} \right)^{n-1}$.

Now $\sum\limits_{n=1}^{\infty} 9 \left(\frac{9}{10} \right)^{n-1}$ is a geometric series with $a = 9$ and $r = \frac{9}{10} < 1$. Therefore, by the Comparison Test,

$$S = \sum_{n=1}^{\infty} g_n < \sum_{n=1}^{\infty} 9 \left(\tfrac{9}{10} \right)^{n-1} = \frac{9}{1 - 9/10} = 90.$$

23. $u = 1 + \dfrac{x^3}{3!} + \dfrac{x^6}{6!} + \dfrac{x^9}{9!} + \cdots, \ v = x + \dfrac{x^4}{4!} + \dfrac{x^7}{7!} + \dfrac{x^{10}}{10!} + \cdots, \ w = \dfrac{x^2}{2!} + \dfrac{x^5}{5!} + \dfrac{x^8}{8!} + \cdots.$

Use the Ratio Test to show that the series for u, v, and w have positive radii of convergence (∞ in each case), so

Theorem 11.9.2 applies, and hence, we may differentiate each of these series:

$$\frac{du}{dx} = \frac{3x^2}{3!} + \frac{6x^5}{6!} + \frac{9x^8}{9!} + \cdots = \frac{x^2}{2!} + \frac{x^5}{5!} + \frac{x^8}{8!} + \cdots = w$$

Similarly, $\dfrac{dv}{dx} = 1 + \dfrac{x^3}{3!} + \dfrac{x^6}{6!} + \dfrac{x^9}{9!} + \cdots = u$, and $\dfrac{dw}{dx} = x + \dfrac{x^4}{4!} + \dfrac{x^7}{7!} + \dfrac{x^{10}}{10!} + \cdots = v$.

So $u' = w$, $v' = u$, and $w' = v$. Now differentiate the left hand side of the desired equation:

$$\frac{d}{dx}(u^3 + v^3 + w^3 - 3uvw) = 3u^2 u' + 3v^2 v' + 3w^2 w' - 3(u'vw + uv'w + uvw')$$

$$= 3u^2 w + 3v^2 u + 3w^2 v - 3(vw^2 + u^2 w + uv^2) = 0 \quad \Rightarrow$$

$u^3 + v^3 + w^3 - 3uvw = C$. To find the value of the constant C, we put $x = 0$ in the last equation and get

$1^3 + 0^3 + 0^3 - 3(1 \cdot 0 \cdot 0) = C \quad \Rightarrow \quad C = 1$, so $u^3 + v^3 + w^3 - 3uvw = 1$.

☐ APPENDIXES

A   Numbers, Inequalities, and Absolute Values

1. $|5 - 23| = |-18| = 18$

3. $|-\pi| = \pi$ because $\pi > 0$.

5. $\left|\sqrt{5} - 5\right| = -\left(\sqrt{5} - 5\right) = 5 - \sqrt{5}$ because $\sqrt{5} - 5 < 0$.

7. If $x < 2$, $x - 2 < 0$, so $|x - 2| = -(x - 2) = 2 - x$.

9. $|x + 1| = \begin{cases} x + 1 & \text{if } x + 1 \geq 0 \\ -(x + 1) & \text{if } x + 1 < 0 \end{cases} = \{-x - 1 \quad \text{if } x < -1$

11. $\left|x^2 + 1\right| = x^2 + 1$ [since $x^2 + 1 \geq 0$ for all x].

13. $2x + 7 > 3 \iff 2x > -4 \iff x > -2$, so $x \in (-2, \infty)$.

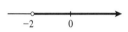

15. $1 - x \leq 2 \iff -x \leq 1 \iff x \geq -1$, so $x \in [-1, \infty)$.

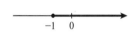

17. $2x + 1 < 5x - 8 \iff 9 < 3x \iff 3 < x$, so $x \in (3, \infty)$.

19. $-1 < 2x - 5 < 7 \iff 4 < 2x < 12 \iff 2 < x < 6$, so $x \in (2, 6)$.

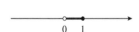

21. $0 \leq 1 - x < 1 \iff -1 \leq -x < 0 \iff 1 \geq x > 0$, so $x \in (0, 1]$.

23. $4x < 2x + 1 \leq 3x + 2$. So $4x < 2x + 1 \iff 2x < 1 \iff x < \frac{1}{2}$, and

$2x + 1 \leq 3x + 2 \iff -1 \leq x$. Thus, $x \in \left[-1, \frac{1}{2}\right)$.

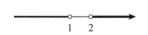

25. $(x - 1)(x - 2) > 0$.

Case 1: (both factors are positive, so their product is positive) $x - 1 > 0 \iff x > 1$,

    and $x - 2 > 0 \iff x > 2$, so $x \in (2, \infty)$.

Case 2: (both factors are negative, so their product is positive) $x - 1 < 0 \iff x < 1$,

    and $x - 2 < 0 \iff x < 2$, so $x \in (-\infty, 1)$.

Thus, the solution set is $(-\infty, 1) \cup (2, \infty)$.

27. $2x^2 + x \leq 1 \iff 2x^2 + x - 1 \leq 0 \iff (2x - 1)(x + 1) \leq 0$.

Case 1: $2x - 1 \geq 0 \iff x \geq \frac{1}{2}$, and $x + 1 \leq 0 \iff x \leq -1$,

which is an impossible combination.

Case 2: $2x - 1 \leq 0 \iff x \leq \frac{1}{2}$, and $x + 1 \geq 0 \iff x \geq -1$, so $x \in \left[-1, \frac{1}{2}\right]$.

Thus, the solution set is $\left[-1, \frac{1}{2}\right]$.

29. $x^2 + x + 1 > 0$ ⇔ $x^2 + x + \frac{1}{4} + \frac{3}{4} > 0$ ⇔ $\left(x + \frac{1}{2}\right)^2 + \frac{3}{4} > 0$. But since

$\left(x + \frac{1}{2}\right)^2 \geq 0$ for every real x, the original inequality will be true for all real x as well.

Thus, the solution set is $(-\infty, \infty)$.

31. $x^2 < 3$ ⇔ $x^2 - 3 < 0$ ⇔ $\left(x - \sqrt{3}\right)\left(x + \sqrt{3}\right) < 0$.

Case 1: $x > \sqrt{3}$ and $x < -\sqrt{3}$, which is impossible.

Case 2: $x < \sqrt{3}$ and $x > -\sqrt{3}$.

Thus, the solution set is $\left(-\sqrt{3}, \sqrt{3}\right)$.

Another method: $x^2 < 3$ ⇔ $|x| < \sqrt{3}$ ⇔ $-\sqrt{3} < x < \sqrt{3}$.

33. $x^3 - x^2 \leq 0$ ⇔ $x^2(x - 1) \leq 0$. Since $x^2 \geq 0$ for all x, the inequality is satisfied when $x - 1 \leq 0$ ⇔ $x \leq 1$.
Thus, the solution set is $(-\infty, 1]$.

35. $x^3 > x$ ⇔ $x^3 - x > 0$ ⇔ $x(x^2 - 1) > 0$ ⇔ $x(x - 1)(x + 1) > 0$. Construct a chart:

Interval	x	$x - 1$	$x + 1$	$x(x - 1)(x + 1)$
$x < -1$	−	−	−	−
$-1 < x < 0$	−	−	+	+
$0 < x < 1$	+	−	+	−
$x > 1$	+	+	+	+

Since $x^3 > x$ when the last column is positive, the solution set is $(-1, 0) \cup (1, \infty)$.

37. $1/x < 4$. This is clearly true for $x < 0$. So suppose $x > 0$. then $1/x < 4$ ⇔
$1 < 4x$ ⇔ $\frac{1}{4} < x$. Thus, the solution set is $(-\infty, 0) \cup \left(\frac{1}{4}, \infty\right)$.

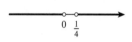

39. $C = \frac{5}{9}(F - 32)$ ⇒ $F = \frac{9}{5}C + 32$. So $50 \leq F \leq 95$ ⇒ $50 \leq \frac{9}{5}C + 32 \leq 95$ ⇒ $18 \leq \frac{9}{5}C \leq 63$ ⇒
$10 \leq C \leq 35$. So the interval is $[10, 35]$.

41. (a) Let T represent the temperature in degrees Celsius and h the height in km. $T = 20$ when $h = 0$ and T decreases by $10°C$
for every km ($1°C$ for each 100-m rise). Thus, $T = 20 - 10h$ when $0 \leq h \leq 12$.

(b) From part (a), $T = 20 - 10h$ ⇒ $10h = 20 - T$ ⇒ $h = 2 - T/10$. So $0 \leq h \leq 5$ ⇒ $0 \leq 2 - T/10 \leq 5$ ⇒
$-2 \leq -T/10 \leq 3$ ⇒ $-20 \leq -T \leq 30$ ⇒ $20 \geq T \geq -30$ ⇒ $-30 \leq T \leq 20$. Thus, the range of
temperatures (in $°C$) to be expected is $[-30, 20]$.

43. $|2x| = 3$ ⇔ either $2x = 3$ or $2x = -3$ ⇔ $x = \frac{3}{2}$ or $x = -\frac{3}{2}$.

45. $|x + 3| = |2x + 1|$ ⇔ either $x + 3 = 2x + 1$ or $x + 3 = -(2x + 1)$. In the first case, $x = 2$, and in the second case,
$x + 3 = -2x - 1$ ⇔ $3x = -4$ ⇔ $x = -\frac{4}{3}$. So the solutions are $-\frac{4}{3}$ and 2.

47. By Property 5 of absolute values, $|x| < 3$ ⇔ $-3 < x < 3$, so $x \in (-3, 3)$.

49. $|x - 4| < 1 \quad \Leftrightarrow \quad -1 < x - 4 < 1 \quad \Leftrightarrow \quad 3 < x < 5$, so $x \in (3, 5)$.

51. $|x + 5| \geq 2 \quad \Leftrightarrow \quad x + 5 \geq 2$ or $x + 5 \leq -2 \quad \Leftrightarrow \quad x \geq -3$ or $x \leq -7$, so $x \in (-\infty, -7] \cup [-3, \infty)$.

53. $|2x - 3| \leq 0.4 \quad \Leftrightarrow \quad -0.4 \leq 2x - 3 \leq 0.4 \quad \Leftrightarrow \quad 2.6 \leq 2x \leq 3.4 \quad \Leftrightarrow \quad 1.3 \leq x \leq 1.7$, so $x \in [1.3, 1.7]$.

55. $1 \leq |x| \leq 4$. So either $1 \leq x \leq 4$ or $1 \leq -x \leq 4 \quad \Leftrightarrow \quad -1 \geq x \geq -4$. Thus, $x \in [-4, -1] \cup [1, 4]$.

57. $a(bx - c) \geq bc \quad \Leftrightarrow \quad bx - c \geq \dfrac{bc}{a} \quad \Leftrightarrow \quad bx \geq \dfrac{bc}{a} + c = \dfrac{bc + ac}{a} \quad \Leftrightarrow \quad x \geq \dfrac{bc + ac}{ab}$

59. $ax + b < c \quad \Leftrightarrow \quad ax < c - b \quad \Leftrightarrow \quad x > \dfrac{c - b}{a}$ [since $a < 0$]

61. $|(x + y) - 5| = |(x - 2) + (y - 3)| \leq |x - 2| + |y - 3| < 0.01 + 0.04 = 0.05$

63. If $a < b$ then $a + a < a + b$ and $a + b < b + b$. So $2a < a + b < 2b$. Dividing by 2, we get $a < \frac{1}{2}(a + b) < b$.

65. $|ab| = \sqrt{(ab)^2} = \sqrt{a^2 b^2} = \sqrt{a^2} \sqrt{b^2} = |a| \, |b|$

67. If $0 < a < b$, then $a \cdot a < a \cdot b$ and $a \cdot b < b \cdot b$ [using Rule 3 of Inequalities]. So $a^2 < ab < b^2$ and hence $a^2 < b^2$.

69. Observe that the sum, difference and product of two integers is always an integer. Let the rational numbers be represented by $r = m/n$ and $s = p/q$ (where m, n, p and q are integers with $n \neq 0$, $q \neq 0$). Now $r + s = \dfrac{m}{n} + \dfrac{p}{q} = \dfrac{mq + pn}{nq}$,

but $mq + pn$ and nq are both integers, so $\dfrac{mq + pn}{nq} = r + s$ is a rational number by definition. Similarly,

$r - s = \dfrac{m}{n} - \dfrac{p}{q} = \dfrac{mq - pn}{nq}$ is a rational number. Finally, $r \cdot s = \dfrac{m}{n} \cdot \dfrac{p}{q} = \dfrac{mp}{nq}$ but mp and nq are both integers, so

$\dfrac{mp}{nq} = r \cdot s$ is a rational number by definition.

B   Coordinate Geometry and Lines

1. Use the distance formula with $P_1(x_1, y_1) = (1, 1)$ and $P_2(x_2, y_2) = (4, 5)$ to get

$$|P_1 P_2| = \sqrt{(4 - 1)^2 + (5 - 1)^2} = \sqrt{3^2 + 4^2} = \sqrt{25} = 5$$

3. The distance from $(6, -2)$ to $(-1, 3)$ is $\sqrt{-1 - 6)^2 + [3 - (-2)]^2} = \sqrt{(-7)^2 + 5^2} = \sqrt{74}$.

5. The distance from $(2, 5)$ to $(4, -7)$ is $\sqrt{(4 - 2)^2 + (-7 - 5)^2} = \sqrt{2^2 + (-12)^2} = \sqrt{148} = 2\sqrt{37}$.

7. The slope m of the line through $P(1, 5)$ and $Q(4, 11)$ is $m = \dfrac{11 - 5}{4 - 1} = \dfrac{6}{3} = 2$.

9. The slope m of the line through $P(-3, 3)$ and $Q(-1, -6)$ is $m = \dfrac{-6 - 3}{-1 - (-3)} = -\dfrac{9}{2}$.

11. Using $A(0, 2)$, $B(-3, -1)$, and $C(-4, 3)$, we have $|AC| = \sqrt{(-4 - 0)^2 + (3 - 2)^2} = \sqrt{(-4)^2 + 1^2} = \sqrt{17}$ and

$|BC| = \sqrt{[-4 - (-3)]^2 + [3 - (-1)]^2} = \sqrt{(-1)^2 + 4^2} = \sqrt{17}$, so the triangle has two sides of equal length, and is isosceles.

13. Using $A(-2,9)$, $B(4,6)$, $C(1,0)$, and $D(-5,3)$, we have

$$|AB| = \sqrt{[4-(-2)]^2 + (6-9)^2} = \sqrt{6^2 + (-3)^2} = \sqrt{45} = \sqrt{9}\sqrt{5} = 3\sqrt{5},$$

$$|BC| = \sqrt{(1-4)^2 + (0-6)^2} = \sqrt{(-3)^2 + (-6)^2} = \sqrt{45} = \sqrt{9}\sqrt{5} = 3\sqrt{5},$$

$$|CD| = \sqrt{(-5-1)^2 + (3-0)^2} = \sqrt{(-6)^2 + 3^2} = \sqrt{45} = \sqrt{9}\sqrt{5} = 3\sqrt{5}, \text{ and}$$

$$|DA| = \sqrt{[-2-(-5)]^2 + (9-3)^2} = \sqrt{3^2 + 6^2} = \sqrt{45} = \sqrt{9}\sqrt{5} = 3\sqrt{5}. \text{ So all sides are of equal length and we have a}$$

rhombus. Moreover, $m_{AB} = \dfrac{6-9}{4-(-2)} = -\dfrac{1}{2}$, $m_{BC} = \dfrac{0-6}{1-4} = 2$, $m_{CD} = \dfrac{3-0}{-5-1} = -\dfrac{1}{2}$, and

$m_{DA} = \dfrac{9-3}{-2-(-5)} = 2$, so the sides are perpendicular. Thus, A, B, C, and D are vertices of a square.

15. For the vertices $A(1,1)$, $B(7,4)$, $C(5,10)$, and $D(-1,7)$, the slope of the line segment AB is $\dfrac{4-1}{7-1} = \dfrac{1}{2}$, the slope of CD

is $\dfrac{7-10}{-1-5} = \dfrac{1}{2}$, the slope of BC is $\dfrac{10-4}{5-7} = -3$, and the slope of DA is $\dfrac{1-7}{1-(-1)} = -3$. So AB is parallel to CD and

BC is parallel to DA. Hence $ABCD$ is a parallelogram.

17. The graph of the equation $x = 3$ is a vertical line with x-intercept 3. The line does not have a slope.

19. $xy = 0 \Leftrightarrow x = 0$ or $y = 0$. The graph consists of the coordinate axes.

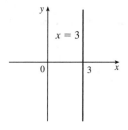

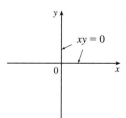

21. By the point-slope form of the equation of a line, an equation of the line through $(2,-3)$ with slope 6 is

$y - (-3) = 6(x-2)$ or $y = 6x - 15$.

23. $y - 7 = \frac{2}{3}(x-1)$ or $y = \frac{2}{3}x + \frac{19}{3}$

25. The slope of the line through $(2,1)$ and $(1,6)$ is $m = \dfrac{6-1}{1-2} = -5$, so an equation of the line is

$y - 1 = -5(x-2)$ or $y = -5x + 11$.

27. By the slope-intercept form of the equation of a line, an equation of the line is $y = 3x - 2$.

29. Since the line passes through $(1,0)$ and $(0,-3)$, its slope is $m = \dfrac{-3-0}{0-1} = 3$, so an equation is $y = 3x - 3$.

Another method: From Exercise 61, $\dfrac{x}{1} + \dfrac{y}{-3} = 1 \Rightarrow -3x + y = -3 \Rightarrow y = 3x - 3$.

31. The line is parallel to the x-axis, so it is horizontal and must have the form $y = k$. Since it goes through the point

$(x,y) = (4,5)$, the equation is $y = 5$.

33. Putting the line $x + 2y = 6$ into its slope-intercept form gives us $y = -\frac{1}{2}x + 3$, so we see that this line has slope $-\frac{1}{2}$. Thus, we want the line of slope $-\frac{1}{2}$ that passes through the point $(1, -6)$: $y - (-6) = -\frac{1}{2}(x - 1)$ $\Leftrightarrow$ $y = -\frac{1}{2}x - \frac{11}{2}$.

35. $2x + 5y + 8 = 0$ $\Leftrightarrow$ $y = -\frac{2}{5}x - \frac{8}{5}$. Since this line has slope $-\frac{2}{5}$, a line perpendicular to it would have slope $\frac{5}{2}$, so the required line is $y - (-2) = \frac{5}{2}[x - (-1)]$ $\Leftrightarrow$ $y = \frac{5}{2}x + \frac{1}{2}$.

37. $x + 3y = 0$ $\Leftrightarrow$ $y = -\frac{1}{3}x$, so the slope is $-\frac{1}{3}$ and the y-intercept is 0.

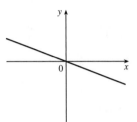

39. $y = -2$ is a horizontal line with slope 0 and y-intercept -2.

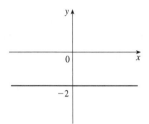

41. $3x - 4y = 12$ $\Leftrightarrow$ $y = \frac{3}{4}x - 3$, so the slope is $\frac{3}{4}$ and the y-intercept is -3.

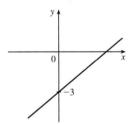

43. $\{(x, y) \mid x < 0\}$

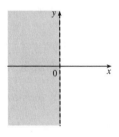

45. $\{(x, y) \mid xy < 0\} = \{(x, y) \mid x < 0 \text{ and } y > 0\}$
$\cup \{(x, y) \mid x > 0 \text{ and } y < 0\}$

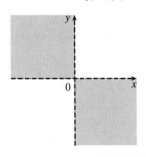

47. $\left\{(x, y) \,\middle|\, |x| \le 2\right\} = \{(x, y) \mid -2 \le x \le 2\}$

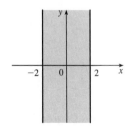

49. $\{(x, y) \mid 0 \le y \le 4, x \le 2\}$

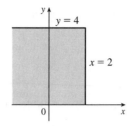

51. $\{(x, y) \mid 1 + x \le y \le 1 - 2x\}$

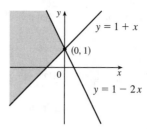

53. Let $P(0, y)$ be a point on the y-axis. The distance from P to $(5, -5)$ is $\sqrt{(5 - 0)^2 + (-5 - y)^2} = \sqrt{5^2 + (y + 5)^2}$. The

distance from P to $(1, 1)$ is $\sqrt{(1 - 0)^2 + (1 - y)^2} = \sqrt{1^2 + (y - 1)^2}$. We want these distances to be

equal: $\sqrt{5^2 + (y + 5)^2} = \sqrt{1^2 + (y - 1)^2}$ $\Leftrightarrow$ $5^2 + (y + 5)^2 = 1^2 + (y - 1)^2$ $\Leftrightarrow$

$25 + (y^2 + 10y + 25) = 1 + (y^2 - 2y + 1)$ $\Leftrightarrow$ $12y = -48$ $\Leftrightarrow$ $y = -4$. So the desired point is $(0, -4)$.

55. (a) Using the midpoint formula from Exercise 54 with $(1, 3)$ and $(7, 15)$, we get $\left(\frac{1+7}{2}, \frac{3+15}{2}\right) = (4, 9)$.

(b) Using the midpoint formula from Exercise 54 with $(-1, 6)$ and $(8, -12)$, we get $\left(\frac{-1+8}{2}, \frac{6+(-12)}{2}\right) = \left(\frac{7}{2}, -3\right)$.

57. $2x - y = 4$ $\Leftrightarrow$ $y = 2x - 4$ $\Rightarrow$ $m_1 = 2$ and $6x - 2y = 10$ $\Leftrightarrow$ $2y = 6x - 10$ $\Leftrightarrow$ $y = 3x - 5$ $\Rightarrow$ $m_2 = 3$.

Since $m_1 \ne m_2$, the two lines are not parallel. To find the point of intersection: $2x - 4 = 3x - 5$ $\Leftrightarrow$ $x = 1$ $\Rightarrow$

$y = -2$. Thus, the point of intersection is $(1, -2)$.

59. With $A(1, 4)$ and $B(7, -2)$, the slope of segment AB is $\frac{-2-4}{7-1} = -1$, so its perpendicular bisector has slope 1. The midpoint

of AB is $\left(\frac{1+7}{2}, \frac{4+(-2)}{2}\right) = (4, 1)$, so an equation of the perpendicular bisector is $y - 1 = 1(x - 4)$ or $y = x - 3$.

61. (a) Since the x-intercept is a, the point $(a, 0)$ is on the line, and similarly since the y-intercept is b, $(0, b)$ is on the line. Hence,

the slope of the line is $m = \dfrac{b - 0}{0 - a} = -\dfrac{b}{a}$. Substituting into $y = mx + b$ gives $y = -\dfrac{b}{a}x + b$ $\Leftrightarrow$ $\dfrac{b}{a}x + y = b$ $\Leftrightarrow$

$\dfrac{x}{a} + \dfrac{y}{b} = 1$.

(b) Letting $a = 6$ and $b = -8$ gives $\dfrac{x}{6} + \dfrac{y}{-8} = 1$ $\Leftrightarrow$ $-8x + 6y = -48$ [multiply by -48] $\Leftrightarrow$ $6y = 8x - 48$ $\Leftrightarrow$

$3y = 4x - 24$ $\Leftrightarrow$ $y = \frac{4}{3}x - 8$.

C   Graphs of Second-Degree Equations

1. An equation of the circle with center $(3, -1)$ and radius 5 is $(x - 3)^2 + (y + 1)^2 = 5^2 = 25$.

3. The equation has the form $x^2 + y^2 = r^2$. Since $(4, 7)$ lies on the circle, we have $4^2 + 7^2 = r^2$ $\Rightarrow$ $r^2 = 65$. So the

required equation is $x^2 + y^2 = 65$.

5. $x^2 + y^2 - 4x + 10y + 13 = 0 \quad\Leftrightarrow\quad x^2 - 4x + y^2 + 10y = -13 \quad\Leftrightarrow$

$(x^2 - 4x + 4) + (y^2 + 10y + 25) = -13 + 4 + 25 = 16 \quad\Leftrightarrow\quad (x - 2)^2 + (y + 5)^2 = 4^2$. Thus, we have a circle with

center $(2, -5)$ and radius 4.

7. $x^2 + y^2 + x = 0 \quad\Leftrightarrow\quad \left(x^2 + x + \frac{1}{4}\right) + y^2 = \frac{1}{4} \quad\Leftrightarrow\quad \left(x + \frac{1}{2}\right)^2 + y^2 = \left(\frac{1}{2}\right)^2$. Thus, we have a circle with center $\left(-\frac{1}{2}, 0\right)$

and radius $\frac{1}{2}$.

9. $2x^2 + 2y^2 - x + y = 1 \quad\Leftrightarrow\quad 2\left(x^2 - \frac{1}{2}x + \frac{1}{16}\right) + 2\left(y^2 + \frac{1}{2}y + \frac{1}{16}\right) = 1 + \frac{1}{8} + \frac{1}{8} \quad\Leftrightarrow$

$2\left(x - \frac{1}{4}\right)^2 + 2\left(y + \frac{1}{4}\right)^2 = \frac{5}{4} \quad\Leftrightarrow\quad \left(x - \frac{1}{4}\right)^2 + \left(y + \frac{1}{4}\right)^2 = \frac{5}{8}$. Thus, we have a circle with center $\left(\frac{1}{4}, -\frac{1}{4}\right)$ and

radius $\frac{\sqrt{5}}{2\sqrt{2}} = \frac{\sqrt{10}}{4}$.

11. $y = -x^2$. Parabola

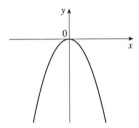

13. $x^2 + 4y^2 = 16 \quad\Leftrightarrow\quad \dfrac{x^2}{16} + \dfrac{y^2}{4} = 1$. Ellipse

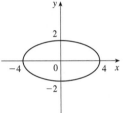

15. $16x^2 - 25y^2 = 400 \quad\Leftrightarrow\quad \dfrac{x^2}{25} - \dfrac{y^2}{16} = 1$. Hyperbola

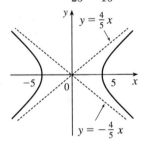

17. $4x^2 + y^2 = 1 \quad\Leftrightarrow\quad \dfrac{x^2}{1/4} + y^2 = 1$. Ellipse

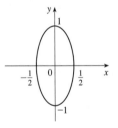

19. $x = y^2 - 1$. Parabola with vertex at $(-1, 0)$

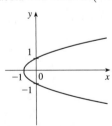

21. $9y^2 - x^2 = 9 \quad\Leftrightarrow\quad y^2 - \dfrac{x^2}{9} = 1$. Hyperbola

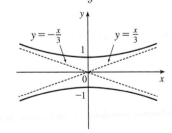

23. $xy = 4$. Hyperbola

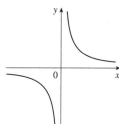

25. $9(x-1)^2 + 4(y-2)^2 = 36 \quad \Leftrightarrow$
$$\frac{(x-1)^2}{4} + \frac{(y-2)^2}{9} = 1. \text{ Ellipse centered at } (1,2)$$

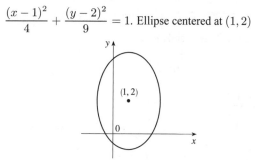

27. $y = x^2 - 6x + 13 = \left(x^2 - 6x + 9\right) + 4 = (x-3)^2 + 4.$
Parabola with vertex at $(3,4)$

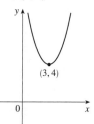

29. $x = 4 - y^2 = -y^2 + 4$. Parabola with vertex at $(4,0)$

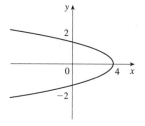

31. $x^2 + 4y^2 - 6x + 5 = 0 \quad \Leftrightarrow$
$(x^2 - 6x + 9) + 4y^2 = -5 + 9 = 4 \quad \Leftrightarrow$
$$\frac{(x-3)^2}{4} + y^2 = 1. \text{ Ellipse centered at } (3,0)$$

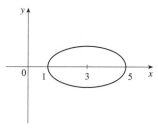

33. $y = 3x$ and $y = x^2$ intersect where $3x = x^2 \quad \Leftrightarrow$
$0 = x^2 - 3x = x(x-3)$, that is, at $(0,0)$ and $(3,9)$.

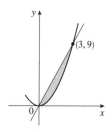

35. The parabola must have an equation of the form $y = a(x-1)^2 - 1$. Substituting $x = 3$ and $y = 3$ into the equation gives
$3 = a(3-1)^2 - 1$, so $a = 1$, and the equation is $y = (x-1)^2 - 1 = x^2 - 2x$. Note that using the other point $(-1,3)$ would
have given the same value for a, and hence the same equation.

37. $\left\{(x,y) \mid x^2 + y^2 \le 1\right\}$

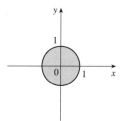

39. $\left\{(x,y) \mid y \ge x^2 - 1\right\}$

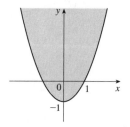

D   Trigonometry

1. $210° = 210° \left(\frac{\pi}{180°}\right) = \frac{7\pi}{6}$ rad

3. $9° = 9° \left(\frac{\pi}{180°}\right) = \frac{\pi}{20}$ rad

5. $900° = 900° \left(\frac{\pi}{180°}\right) = 5\pi$ rad

7. 4π rad $= 4\pi \left(\frac{180°}{\pi}\right) = 720°$

9. $\frac{5\pi}{12}$ rad $= \frac{5\pi}{12} \left(\frac{180°}{\pi}\right) = 75°$

11. $-\frac{3\pi}{8}$ rad $= -\frac{3\pi}{8} \left(\frac{180°}{\pi}\right) = -67.5°$

13. Using Formula 3, $a = r\theta = 36 \cdot \frac{\pi}{12} = 3\pi$ cm.

15. Using Formula 3, $\theta = a/r = \frac{1}{1.5} = \frac{2}{3}$ rad $= \frac{2}{3} \left(\frac{180°}{\pi}\right) = \left(\frac{120}{\pi}\right)° \approx 38.2°$.

17.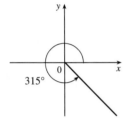

19.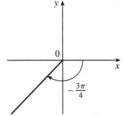

21.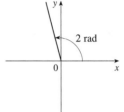

23.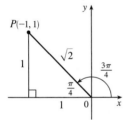

From the diagram we see that a point on the terminal side is $P(-1, 1)$. Therefore, taking $x = -1$, $y = 1$, $r = \sqrt{2}$ in the definitions of the trigonometric ratios, we have $\sin \frac{3\pi}{4} = \frac{1}{\sqrt{2}}$, $\cos \frac{3\pi}{4} = -\frac{1}{\sqrt{2}}$, $\tan \frac{3\pi}{4} = -1$, $\csc \frac{3\pi}{4} = \sqrt{2}$, $\sec \frac{3\pi}{4} = -\sqrt{2}$, and $\cot \frac{3\pi}{4} = -1$.

25.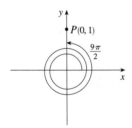

From the diagram we see that a point on the terminal side is $P(0, 1)$. Therefore taking $x = 0$, $y = 1$, $r = 1$ in the definitions of the trigonometric ratios, we have $\sin \frac{9\pi}{2} = 1$, $\cos \frac{9\pi}{2} = 0$, $\tan \frac{9\pi}{2} = y/x$ is undefined since $x = 0$, $\csc \frac{9\pi}{2} = 1$, $\sec \frac{9\pi}{2} = r/x$ is undefined since $x = 0$, and $\cot \frac{9\pi}{2} = 0$.

27.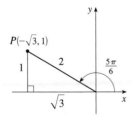

Using Figure 8 we see that a point on the terminal side is $P\left(-\sqrt{3}, 1\right)$. Therefore taking $x = -\sqrt{3}$, $y = 1$, $r = 2$ in the definitions of the trigonometric ratios, we have $\sin \frac{5\pi}{6} = \frac{1}{2}$, $\cos \frac{5\pi}{6} = -\frac{\sqrt{3}}{2}$, $\tan \frac{5\pi}{6} = -\frac{1}{\sqrt{3}}$, $\csc \frac{5\pi}{6} = 2$, $\sec \frac{5\pi}{6} = -\frac{2}{\sqrt{3}}$, and $\cot \frac{5\pi}{6} = -\sqrt{3}$.

29. $\sin \theta = y/r = \frac{3}{5} \Rightarrow y = 3$, $r = 5$, and $x = \sqrt{r^2 - y^2} = 4$ (since $0 < \theta < \frac{\pi}{2}$). Therefore taking $x = 4$, $y = 3$, $r = 5$ in the definitions of the trigonometric ratios, we have $\cos \theta = \frac{4}{5}$, $\tan \theta = \frac{3}{4}$, $\csc \theta = \frac{5}{3}$, $\sec \theta = \frac{5}{4}$, and $\cot \theta = \frac{4}{3}$.

31. $\frac{\pi}{2} < \phi < \pi$ $\Rightarrow$ ϕ is in the second quadrant, where x is negative and y is positive. Therefore

$\sec \phi = r/x = -1.5 = -\frac{3}{2}$ $\Rightarrow$ $r = 3$, $x = -2$, and $y = \sqrt{r^2 - x^2} = \sqrt{5}$. Taking $x = -2$, $y = \sqrt{5}$, and $r = 3$ in the

definitions of the trigonometric ratios, we have $\sin \phi = \frac{\sqrt{5}}{3}$, $\cos \phi = -\frac{2}{3}$, $\tan \phi = -\frac{\sqrt{5}}{2}$, $\csc \phi = \frac{3}{\sqrt{5}}$, and $\cot \theta = -\frac{2}{\sqrt{5}}$.

33. $\pi < \beta < 2\pi$ means that β is in the third or fourth quadrant where y is negative. Also since $\cot \beta = x/y = 3$ which is

positive, x must also be negative. Therefore $\cot \beta = x/y = \frac{3}{1}$ $\Rightarrow$ $x = -3$, $y = -1$, and $r = \sqrt{x^2 + y^2} = \sqrt{10}$. Taking

$x = -3$, $y = -1$ and $r = \sqrt{10}$ in the definitions of the trigonometric ratios, we have $\sin \beta = -\frac{1}{\sqrt{10}}$, $\cos \beta = -\frac{3}{\sqrt{10}}$,

$\tan \beta = \frac{1}{3}$, $\csc \beta = -\sqrt{10}$, and $\sec \beta = -\frac{\sqrt{10}}{3}$.

35. $\sin 35° = \dfrac{x}{10}$ $\Rightarrow$ $x = 10 \sin 35° \approx 5.73576$ cm

37. $\tan \frac{2\pi}{5} = \dfrac{x}{8}$ $\Rightarrow$ $x = 8 \tan \frac{2\pi}{5} \approx 24.62147$ cm

39.

(a) From the diagram we see that $\sin \theta = \dfrac{y}{r} = \dfrac{a}{c}$, and $\sin(-\theta) = \dfrac{-a}{c} = -\dfrac{a}{c} = -\sin \theta$.

(b) Again from the diagram we see that $\cos \theta = \dfrac{x}{r} = \dfrac{b}{c} = \cos(-\theta)$.

41. (a) Using (12a) and (13a), we have

$\frac{1}{2}[\sin(x+y) + \sin(x-y)] = \frac{1}{2}[\sin x \cos y + \cos x \sin y + \sin x \cos y - \cos x \sin y] = \frac{1}{2}(2 \sin x \cos y) = \sin x \cos y.$

(b) This time, using (12b) and (13b), we have

$\frac{1}{2}[\cos(x+y) + \cos(x-y)] = \frac{1}{2}[\cos x \cos y - \sin x \sin y + \cos x \cos y + \sin x \sin y] = \frac{1}{2}(2 \cos x \cos y) = \cos x \cos y.$

(c) Again using (12b) and (13b), we have

$\frac{1}{2}[\cos(x-y) - \cos(x+y)] = \frac{1}{2}[\cos x \cos y + \sin x \sin y - \cos x \cos y + \sin x \sin y]$

$= \frac{1}{2}(2 \sin x \sin y) = \sin x \sin y$

43. Using (12a), we have $\sin\left(\frac{\pi}{2} + x\right) = \sin \frac{\pi}{2} \cos x + \cos \frac{\pi}{2} \sin x = 1 \cdot \cos x + 0 \cdot \sin x = \cos x.$

45. Using (6), we have $\sin \theta \cot \theta = \sin \theta \cdot \dfrac{\cos \theta}{\sin \theta} = \cos \theta.$

47. $\sec y - \cos y = \dfrac{1}{\cos y} - \cos y$ [by (6)] $= \dfrac{1 - \cos^2 y}{\cos y} = \dfrac{\sin^2 y}{\cos y}$ [by (7)] $= \dfrac{\sin y}{\cos y} \sin y = \tan y \sin y$ [by (6)]

49. $\cot^2 \theta + \sec^2 \theta = \dfrac{\cos^2 \theta}{\sin^2 \theta} + \dfrac{1}{\cos^2 \theta}$ [by (6)] $= \dfrac{\cos^2 \theta \cos^2 \theta + \sin^2 \theta}{\sin^2 \theta \cos^2 \theta}$

$= \dfrac{(1 - \sin^2 \theta)(1 - \sin^2 \theta) + \sin^2 \theta}{\sin^2 \theta \cos^2 \theta}$ [by (7)] $= \dfrac{1 - \sin^2 \theta + \sin^4 \theta}{\sin^2 \theta \cos^2 \theta}$

$= \dfrac{\cos^2 \theta + \sin^4 \theta}{\sin^2 \theta \cos^2 \theta}$ [by (7)] $= \dfrac{1}{\sin^2 \theta} + \dfrac{\sin^2 \theta}{\cos^2 \theta} = \csc^2 \theta + \tan^2 \theta$ [by (6)]

51. Using (14a), we have $\tan 2\theta = \tan(\theta + \theta) = \dfrac{\tan\theta + \tan\theta}{1 - \tan\theta\,\tan\theta} = \dfrac{2\tan\theta}{1 - \tan^2\theta}$.

53. Using (15a) and (16a),

$$\sin x \sin 2x + \cos x \cos 2x = \sin x\,(2\sin x \cos x) + \cos x\,(2\cos^2 x - 1) = 2\sin^2 x \cos x + 2\cos^3 x - \cos x$$

$$= 2(1 - \cos^2 x)\,\cos x + 2\cos^3 x - \cos x \quad \text{[by (7)]}$$

$$= 2\cos x - 2\cos^3 x + 2\cos^3 x - \cos x = \cos x$$

Or: $\sin x \sin 2x + \cos x \cos 2x = \cos(2x - x)$ [by 13(b)] $= \cos x$

55. $\dfrac{\sin\phi}{1 - \cos\phi} = \dfrac{\sin\phi}{1 - \cos\phi} \cdot \dfrac{1 + \cos\phi}{1 + \cos\phi} = \dfrac{\sin\phi\,(1 + \cos\phi)}{1 - \cos^2\phi} = \dfrac{\sin\phi\,(1 + \cos\phi)}{\sin^2\phi}$ [by (7)]

$$= \dfrac{1 + \cos\phi}{\sin\phi} = \dfrac{1}{\sin\phi} + \dfrac{\cos\phi}{\sin\phi} = \csc\phi + \cot\phi \quad \text{[by (6)]}$$

57. Using (12a),

$$\sin 3\theta + \sin\theta = \sin(2\theta + \theta) + \sin\theta = \sin 2\theta\,\cos\theta + \cos 2\theta\,\sin\theta + \sin\theta$$

$$= \sin 2\theta\,\cos\theta + (2\cos^2\theta - 1)\,\sin\theta + \sin\theta \quad \text{[by (16a)]}$$

$$= \sin 2\theta\,\cos\theta + 2\cos^2\theta\,\sin\theta - \sin\theta + \sin\theta = \sin 2\theta\,\cos\theta + \sin 2\theta\,\cos\theta \quad \text{[by (15a)]}$$

$$= 2\sin 2\theta\,\cos\theta$$

59. Since $\sin x = \tfrac{1}{3}$ we can label the opposite side as having length 1,
the hypotenuse as having length 3, and use the Pythagorean Theorem
to get that the adjacent side has length $\sqrt{8}$. Then, from the diagram,
$\cos x = \tfrac{\sqrt{8}}{3}$. Similarly we have that $\sin y = \tfrac{3}{5}$. Now use (12a):

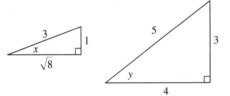

$\sin(x + y) = \sin x \cos y + \cos x \sin y = \tfrac{1}{3} \cdot \tfrac{4}{5} + \tfrac{\sqrt{8}}{3} \cdot \tfrac{3}{5} = \tfrac{4}{15} + \tfrac{3\sqrt{8}}{15} = \tfrac{4 + 6\sqrt{2}}{15}$.

61. Using (13b) and the values for $\cos x$ and $\sin y$ obtained in Exercise 59, we have

$$\cos(x - y) = \cos x \cos y + \sin x \sin y = \tfrac{\sqrt{8}}{3} \cdot \tfrac{4}{5} + \tfrac{1}{3} \cdot \tfrac{3}{5} = \tfrac{8\sqrt{2} + 3}{15}$$

63. Using (15a) and the values for $\sin y$ and $\cos y$ obtained in Exercise 59, we have $\sin 2y = 2\sin y \cos y = 2 \cdot \tfrac{3}{5} \cdot \tfrac{4}{5} = \tfrac{24}{25}$.

65. $2\cos x - 1 = 0 \ \Leftrightarrow \ \cos x = \tfrac{1}{2} \ \Rightarrow \ x = \tfrac{\pi}{3}, \tfrac{5\pi}{3}$ for $x \in [0, 2\pi]$.

67. $2\sin^2 x = 1 \ \Leftrightarrow \ \sin^2 x = \tfrac{1}{2} \ \Leftrightarrow \ \sin x = \pm\tfrac{1}{\sqrt{2}} \ \Rightarrow \ x = \tfrac{\pi}{4}, \tfrac{3\pi}{4}, \tfrac{5\pi}{4}, \tfrac{7\pi}{4}$.

69. Using (15a), we have $\sin 2x = \cos x \ \Leftrightarrow \ 2\sin x \cos x - \cos x = 0 \ \Leftrightarrow \ \cos x(2\sin x - 1) = 0 \ \Leftrightarrow \ \cos x = 0$ or
$2\sin x - 1 = 0 \ \Rightarrow \ x = \tfrac{\pi}{2}, \tfrac{3\pi}{2}$ or $\sin x = \tfrac{1}{2} \ \Rightarrow \ x = \tfrac{\pi}{6}$ or $\tfrac{5\pi}{6}$. Therefore, the solutions are $x = \tfrac{\pi}{6}, \tfrac{\pi}{2}, \tfrac{5\pi}{6}, \tfrac{3\pi}{2}$.

71. $\sin x = \tan x \ \Leftrightarrow \ \sin x - \tan x = 0 \ \Leftrightarrow \ \sin x - \dfrac{\sin x}{\cos x} = 0 \ \Leftrightarrow \ \sin x\left(1 - \dfrac{1}{\cos x}\right) = 0 \ \Leftrightarrow \ \sin x = 0$ or

$1 - \dfrac{1}{\cos x} = 0 \ \Rightarrow \ x = 0, \pi, 2\pi$ or $1 = \dfrac{1}{\cos x} \ \Rightarrow \ \cos x = 1 \ \Rightarrow \ x = 0, 2\pi$. Therefore the solutions
are $x = 0, \pi, 2\pi$.

73. We know that $\sin x = \frac{1}{2}$ when $x = \frac{\pi}{6}$ or $\frac{5\pi}{6}$, and from Figure 13(a), we see that $\sin x \le \frac{1}{2} \;\Rightarrow\; 0 \le x \le \frac{\pi}{6}$ or

$\frac{5\pi}{6} \le x \le 2\pi$ for $x \in [0, 2\pi]$.

75. $\tan x = -1$ when $x = \frac{3\pi}{4}, \frac{7\pi}{4}$, and $\tan x = 1$ when $x = \frac{\pi}{4}$ or $\frac{5\pi}{4}$. From Figure 14(a) we see that $-1 < \tan x < 1 \;\Rightarrow$

$0 \le x < \frac{\pi}{4}, \frac{3\pi}{4} < x < \frac{5\pi}{4}$, and $\frac{7\pi}{4} < x \le 2\pi$.

77. $y = \cos\left(x - \frac{\pi}{3}\right)$. We start with the graph of $y = \cos x$

and shift it $\frac{\pi}{3}$ units to the right.

79. $y = \frac{1}{3} \tan\left(x - \frac{\pi}{2}\right)$. We start with the graph of

$y = \tan x$, shift it $\frac{\pi}{2}$ units to the right and compress it to

$\frac{1}{3}$ of its original vertical size.

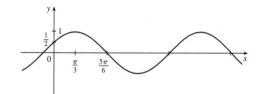

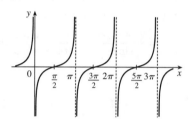

81. $y = |\sin x|$. We start with the graph of $y = \sin x$ and reflect the parts below the x-axis about the x-axis.

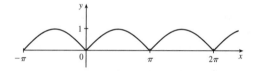

83. From the figure in the text, we see that $x = b\cos\theta$, $y = b\sin\theta$, and from the distance formula we have that the

distance c from (x, y) to $(a, 0)$ is $c = \sqrt{(x-a)^2 + (y-0)^2} \;\Rightarrow$

$$c^2 = (b\cos\theta - a)^2 + (b\sin\theta)^2 = b^2\cos^2\theta - 2ab\cos\theta + a^2 + b^2\sin^2\theta$$
$$= a^2 + b^2(\cos^2\theta + \sin^2\theta) - 2ab\cos\theta = a^2 + b^2 - 2ab\cos\theta \quad [\text{by (7)}]$$

85. Using the Law of Cosines, we have $c^2 = 1^2 + 1^2 - 2(1)(1)\cos(\alpha - \beta) = 2[1 - \cos(\alpha - \beta)]$. Now, using the distance

formula, $c^2 = |AB|^2 = (\cos\alpha - \cos\beta)^2 + (\sin\alpha - \sin\beta)^2$. Equating these two expressions for c^2, we get

$2[1 - \cos(\alpha - \beta)] = \cos^2\alpha + \sin^2\alpha + \cos^2\beta + \sin^2\beta - 2\cos\alpha\cos\beta - 2\sin\alpha\sin\beta \;\Rightarrow$

$1 - \cos(\alpha - \beta) = 1 - \cos\alpha\cos\beta - \sin\alpha\sin\beta \;\Rightarrow\; \cos(\alpha - \beta) = \cos\alpha\cos\beta + \sin\alpha\sin\beta$.

87. In Exercise 86 we used the subtraction formula for cosine to prove the addition formula for cosine. Using that formula with

$x = \frac{\pi}{2} - \alpha$, $y = \beta$, we get $\cos\left[\left(\frac{\pi}{2} - \alpha\right) + \beta\right] = \cos\left(\frac{\pi}{2} - \alpha\right)\cos\beta - \sin\left(\frac{\pi}{2} - \alpha\right)\sin\beta \;\Rightarrow$

$\cos\left[\frac{\pi}{2} - (\alpha - \beta)\right] = \cos\left(\frac{\pi}{2} - \alpha\right)\cos\beta - \sin\left(\frac{\pi}{2} - \alpha\right)\sin\beta$. Now we use the identities given in the problem,

$\cos\left(\frac{\pi}{2} - \theta\right) = \sin\theta$ and $\sin\left(\frac{\pi}{2} - \theta\right) = \cos\theta$, to get $\sin(\alpha - \beta) = \sin\alpha\cos\beta - \cos\alpha\sin\beta$.

89. Using the formula from Exercise 88, the area of the triangle is $\frac{1}{2}(10)(3)\sin 107° \approx 14.34457$ cm^2.

E   Sigma Notation

1. $\sum\limits_{i=1}^{5} \sqrt{i} = \sqrt{1} + \sqrt{2} + \sqrt{3} + \sqrt{4} + \sqrt{5}$

3. $\sum\limits_{i=4}^{6} 3^i = 3^4 + 3^5 + 3^6$

5. $\sum\limits_{k=0}^{4} \dfrac{2k-1}{2k+1} = -1 + \dfrac{1}{3} + \dfrac{3}{5} + \dfrac{5}{7} + \dfrac{7}{9}$

7. $\sum\limits_{i=1}^{n} i^{10} = 1^{10} + 2^{10} + 3^{10} + \cdots + n^{10}$

9. $\sum\limits_{j=0}^{n-1} (-1)^j = 1 - 1 + 1 - 1 + \cdots + (-1)^{n-1}$

11. $1 + 2 + 3 + 4 + \cdots + 10 = \sum\limits_{i=1}^{10} i$

13. $\dfrac{1}{2} + \dfrac{2}{3} + \dfrac{3}{4} + \dfrac{4}{5} + \cdots + \dfrac{19}{20} = \sum\limits_{i=1}^{19} \dfrac{i}{i+1}$

15. $2 + 4 + 6 + 8 + \cdots + 2n = \sum\limits_{i=1}^{n} 2i$

17. $1 + 2 + 4 + 8 + 16 + 32 = \sum\limits_{i=0}^{5} 2^i$

19. $x + x^2 + x^3 + \cdots + x^n = \sum\limits_{i=1}^{n} x^i$

21. $\sum\limits_{i=4}^{8} (3i - 2) = [3(4) - 2] + [3(5) - 2] + [3(6) - 2] + [3(7) - 2] + [3(8) - 2] = 10 + 13 + 16 + 19 + 22 = 80$

23. $\sum\limits_{j=1}^{6} 3^{j+1} = 3^2 + 3^3 + 3^4 + 3^5 + 3^6 + 3^7 = 9 + 27 + 81 + 243 + 729 + 2187 = 3276$

(For a more general method, see Exercise 47.)

25. $\sum\limits_{n=1}^{20} (-1)^n = -1 + 1 - 1 + 1 - 1 + 1 - 1 + 1 - 1 + 1 - 1 + 1 - 1 + 1 - 1 + 1 - 1 + 1 - 1 + 1 = 0$

27. $\sum\limits_{i=0}^{4} (2^i + i^2) = (1 + 0) + (2 + 1) + (4 + 4) + (8 + 9) + (16 + 16) = 61$

29. $\sum\limits_{i=1}^{n} 2i = 2 \sum\limits_{i=1}^{n} i = 2 \cdot \dfrac{n(n+1)}{2}$ [by Theorem 3(c)] $= n(n+1)$

31. $\sum\limits_{i=1}^{n} (i^2 + 3i + 4) = \sum\limits_{i=1}^{n} i^2 + 3 \sum\limits_{i=1}^{n} i + \sum\limits_{i=1}^{n} 4 = \dfrac{n(n+1)(2n+1)}{6} + \dfrac{3n(n+1)}{2} + 4n$

$= \dfrac{1}{6}[(2n^3 + 3n^2 + n) + (9n^2 + 9n) + 24n] = \dfrac{1}{6}(2n^3 + 12n^2 + 34n) = \dfrac{1}{3}n(n^2 + 6n + 17)$

33. $\sum\limits_{i=1}^{n} (i+1)(i+2) = \sum\limits_{i=1}^{n} (i^2 + 3i + 2) = \sum\limits_{i=1}^{n} i^2 + 3 \sum\limits_{i=1}^{n} i + \sum\limits_{i=1}^{n} 2 = \dfrac{n(n+1)(2n+1)}{6} + \dfrac{3n(n+1)}{2} + 2n$

$= \dfrac{n(n+1)}{6} [(2n+1) + 9] + 2n = \dfrac{n(n+1)}{3}(n+5) + 2n$

$= \dfrac{n}{3}[(n+1)(n+5) + 6] = \dfrac{n}{3}(n^2 + 6n + 11)$

35. $\sum\limits_{i=1}^{n} (i^3 - i - 2) = \sum\limits_{i=1}^{n} i^3 - \sum\limits_{i=1}^{n} i - \sum\limits_{i=1}^{n} 2 = \left[\dfrac{n(n+1)}{2}\right]^2 - \dfrac{n(n+1)}{2} - 2n$

$= \dfrac{1}{4}n(n+1)[n(n+1) - 2] - 2n = \dfrac{1}{4}n(n+1)(n+2)(n-1) - 2n$

$= \dfrac{1}{4}n[(n+1)(n-1)(n+2) - 8] = \dfrac{1}{4}n[(n^2 - 1)(n+2) - 8] = \dfrac{1}{4}n(n^3 + 2n^2 - n - 10)$

37. By Theorem 2(a) and Example 3, $\sum\limits_{i=1}^{n} c = c \sum\limits_{i=1}^{n} 1 = cn$.

39. $\sum\limits_{i=1}^{n} [(i+1)^4 - i^4] = (2^4 - 1^4) + (3^4 - 2^4) + (4^4 - 3^4) + \cdots + [(n+1)^4 - n^4]$

$$= (n+1)^4 - 1^4 = n^4 + 4n^3 + 6n^2 + 4n$$

On the other hand,

$$\sum\limits_{i=1}^{n} [(i+1)^4 - i^4] = \sum\limits_{i=1}^{n} (4i^3 + 6i^2 + 4i + 1) = 4\sum\limits_{i=1}^{n} i^3 + 6\sum\limits_{i=1}^{n} i^2 + 4\sum\limits_{i=1}^{n} i + \sum\limits_{i=1}^{n} 1$$

$$= 4S + n(n+1)(2n+1) + 2n(n+1) + n \qquad \left[\text{where } S = \sum\limits_{i=1}^{n} i^3\right]$$

$$= 4S + 2n^3 + 3n^2 + n + 2n^2 + 2n + n = 4S + 2n^3 + 5n^2 + 4n$$

Thus, $n^4 + 4n^3 + 6n^2 + 4n = 4S + 2n^3 + 5n^2 + 4n$, from which it follows that

$$4S = n^4 + 2n^3 + n^2 = n^2(n^2 + 2n + 1) = n^2(n+1)^2 \text{ and } S = \left[\frac{n(n+1)}{2}\right]^2.$$

41. (a) $\sum\limits_{i=1}^{n} \left[i^4 - (i-1)^4\right] = (1^4 - 0^4) + (2^4 - 1^4) + (3^4 - 2^4) + \cdots + [n^4 - (n-1)^4] = n^4 - 0 = n^4$

(b) $\sum\limits_{i=1}^{100} \left(5^i - 5^{i-1}\right) = (5^1 - 5^0) + (5^2 - 5^1) + (5^3 - 5^2) + \cdots + (5^{100} - 5^{99}) = 5^{100} - 5^0 = 5^{100} - 1$

(c) $\sum\limits_{i=3}^{99} \left(\frac{1}{i} - \frac{1}{i+1}\right) = \left(\frac{1}{3} - \frac{1}{4}\right) + \left(\frac{1}{4} - \frac{1}{5}\right) + \left(\frac{1}{5} - \frac{1}{6}\right) + \cdots + \left(\frac{1}{99} - \frac{1}{100}\right) = \frac{1}{3} - \frac{1}{100} = \frac{97}{300}$

(d) $\sum\limits_{i=1}^{n} (a_i - a_{i-1}) = (a_1 - a_0) + (a_2 - a_1) + (a_3 - a_2) + \cdots + (a_n - a_{n-1}) = a_n - a_0$

43. $\lim\limits_{n\to\infty} \sum\limits_{i=1}^{n} \frac{1}{n}\left(\frac{i}{n}\right)^2 = \lim\limits_{n\to\infty} \frac{1}{n^3} \sum\limits_{i=1}^{n} i^2 = \lim\limits_{n\to\infty} \frac{1}{n^3} \frac{n(n+1)(2n+1)}{6} = \lim\limits_{n\to\infty} \frac{1}{6}\left(1 + \frac{1}{n}\right)\left(2 + \frac{1}{n}\right) = \frac{1}{6}(1)(2) = \frac{1}{3}$

45. $\lim\limits_{n\to\infty} \sum\limits_{i=1}^{n} \frac{2}{n}\left[\left(\frac{2i}{n}\right)^3 + 5\left(\frac{2i}{n}\right)\right] = \lim\limits_{n\to\infty} \sum\limits_{i=1}^{n} \left[\frac{16}{n^4}i^3 + \frac{20}{n^2}i\right] = \lim\limits_{n\to\infty} \left[\frac{16}{n^4} \sum\limits_{i=1}^{n} i^3 + \frac{20}{n^2} \sum\limits_{i=1}^{n} i\right]$

$$= \lim\limits_{n\to\infty} \left[\frac{16}{n^4} \frac{n^2(n+1)^2}{4} + \frac{20}{n^2} \frac{n(n+1)}{2}\right] = \lim\limits_{n\to\infty} \left[\frac{4(n+1)^2}{n^2} + \frac{10n(n+1)}{n^2}\right]$$

$$= \lim\limits_{n\to\infty} \left[4\left(1 + \frac{1}{n}\right)^2 + 10\left(1 + \frac{1}{n}\right)\right] = 4 \cdot 1 + 10 \cdot 1 = 14$$

47. Let $S = \sum\limits_{i=1}^{n} ar^{i-1} = a + ar + ar^2 + \cdots + ar^{n-1}$. Multiplying both sides by r gives us

$rS = ar + ar^2 + \cdots + ar^{n-1} + ar^n$. Subtracting the first equation from the second, we find

$(r-1)S = ar^n - a = a(r^n - 1)$, so $S = \dfrac{a(r^n - 1)}{r - 1}$ [since $r \neq 1$].

49. $\sum\limits_{i=1}^{n} (2i + 2^i) = 2\sum\limits_{i=1}^{n} i + \sum\limits_{i=1}^{n} 2 \cdot 2^{i-1} = 2\dfrac{n(n+1)}{2} + \dfrac{2(2^n - 1)}{2 - 1} = 2^{n+1} + n^2 + n - 2.$

For the first sum we have used Theorems 2(a) and 3(c), and for the second, Exercise 47 with $a = r = 2$.

G The Logarithm Defined as an Integral

1. (a) 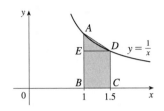 We interpret $\ln 1.5$ as the area under the curve $y = 1/x$ from $x = 1$ to $x = 1.5$. The area of the rectangle $BCDE$ is $\frac{1}{2} \cdot \frac{2}{3} = \frac{1}{3}$. The area of the trapezoid $ABCD$ is $\frac{1}{2} \cdot \frac{1}{2}\left(1 + \frac{2}{3}\right) = \frac{5}{12}$. Thus, by comparing areas, we observe that $\frac{1}{3} < \ln 1.5 < \frac{5}{12}$.

(b) $\ln x = \int_1^x (1/t)\, dt$, so $\ln 1.5 = \int_1^{1.5} (1/t)\, dt$. With $f(t) = 1/t$, $n = 10$, and $\Delta t = \frac{1.5-1}{10} = 0.05$, we have

$$\ln 1.5 = \int_1^{1.5}(1/t)\, dt \approx (0.05)[f(1.025) + f(1.075) + \cdots + f(1.475)] = (0.05)\left[\frac{1}{1.025} + \frac{1}{1.075} + \cdots + \frac{1}{1.475}\right]$$
$$\approx 0.4054$$

3. 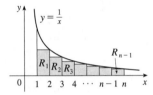 The area of R_i is $\dfrac{1}{i+1}$ and so $\dfrac{1}{2} + \dfrac{1}{3} + \cdots + \dfrac{1}{n} < \displaystyle\int_1^n \frac{1}{t}\, dt = \ln n$.

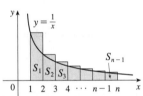 The area of S_i is $\dfrac{1}{i}$ and so $1 + \dfrac{1}{2} + \cdots + \dfrac{1}{n-1} > \displaystyle\int_1^n \frac{1}{t}\, dt = \ln n$.

Thus, $\dfrac{1}{2} + \dfrac{1}{3} + \cdots + \dfrac{1}{n} < \ln n < 1 + \dfrac{1}{2} + \cdots + \dfrac{1}{n-1}$.

5. If $f(x) = \ln(x^r)$, then $f'(x) = (1/x^r)(rx^{r-1}) = r/x$. But if $g(x) = r\ln x$, then $g'(x) = r/x$. So f and g must differ by a constant: $\ln(x^r) = r\ln x + C$. Put $x = 1$: $\ln(1^r) = r\ln 1 + C \Rightarrow C = 0$, so $\ln(x^r) = r\ln x$.

7. Using the third law of logarithms and Equation 10, we have $\ln e^{rx} = rx = r\ln e^x = \ln(e^x)^r$. Since $\ln$ is a one-to-one function, it follows that $e^{rx} = (e^x)^r$.

9. Using Definition 13, the first law of logarithms, and the first law of exponents for e^x, we have

$$(ab)^x = e^{x\ln(ab)} = e^{x(\ln a + \ln b)} = e^{x\ln a + x\ln b} = e^{x\ln a}e^{x\ln b} = a^x b^x.$$

H Complex Numbers

1. $(5 - 6i) + (3 + 2i) = (5 + 3) + (-6 + 2)i = 8 + (-4)i = 8 - 4i$

3. $(2 + 5i)(4 - i) = 2(4) + 2(-i) + (5i)(4) + (5i)(-i) = 8 - 2i + 20i - 5i^2 = 8 + 18i - 5(-1)$
$$= 8 + 18i + 5 = 13 + 18i$$

5. $\overline{12 + 7i} = 12 - 7i$

7. $\dfrac{1+4i}{3+2i} = \dfrac{1+4i}{3+2i} \cdot \dfrac{3-2i}{3-2i} = \dfrac{3-2i+12i-8(-1)}{3^2+2^2} = \dfrac{11+10i}{13} = \dfrac{11}{13} + \dfrac{10}{13}i$

9. $\dfrac{1}{1+i} = \dfrac{1}{1+i} \cdot \dfrac{1-i}{1-i} = \dfrac{1-i}{1-(-1)} = \dfrac{1-i}{2} = \dfrac{1}{2} - \dfrac{1}{2}i$

11. $i^3 = i^2 \cdot i = (-1)i = -i$

13. $\sqrt{-25} = \sqrt{25}\,i = 5i$

15. $\overline{12-5i} = 12 + 15i$ and $|12 - 15i| = \sqrt{12^2 + (-5)^2} = \sqrt{144+25} = \sqrt{169} = 13$

17. $\overline{-4i} = \overline{0-4i} = 0 + 4i = 4i$ and $|-4i| = \sqrt{0^2 + (-4)^2} = \sqrt{16} = 4$

19. $4x^2 + 9 = 0 \iff 4x^2 = -9 \iff x^2 = -\dfrac{9}{4} \iff x = \pm\sqrt{-\dfrac{9}{4}} = \pm\sqrt{\dfrac{9}{4}}\,i = \pm\dfrac{3}{2}i.$

21. By the quadratic formula, $x^2 + 2x + 5 = 0 \iff x = \dfrac{-2 \pm \sqrt{2^2 - 4(1)(5)}}{2(1)} = \dfrac{-2 \pm \sqrt{-16}}{2} = \dfrac{-2 \pm 4i}{2} = -1 \pm 2i.$

23. By the quadratic formula, $z^2 + z + 2 = 0 \iff z = \dfrac{-1 \pm \sqrt{1^2 - 4(1)(2)}}{2(1)} = \dfrac{-1 \pm \sqrt{-7}}{2} = -\dfrac{1}{2} \pm \dfrac{\sqrt{7}}{2}i.$

25. For $z = -3 + 3i$, $r = \sqrt{(-3)^2 + 3^2} = 3\sqrt{2}$ and $\tan\theta = \dfrac{3}{-3} = -1 \implies \theta = \dfrac{3\pi}{4}$ (since z lies in the second quadrant). Therefore, $-3 + 3i = 3\sqrt{2}\left(\cos\dfrac{3\pi}{4} + i\sin\dfrac{3\pi}{4}\right).$

27. For $z = 3 + 4i$, $r = \sqrt{3^2 + 4^2} = 5$ and $\tan\theta = \dfrac{4}{3} \implies \theta = \tan^{-1}\left(\dfrac{4}{3}\right)$ (since z lies in the first quadrant). Therefore, $3 + 4i = 5\left[\cos\left(\tan^{-1}\dfrac{4}{3}\right) + i\sin\left(\tan^{-1}\dfrac{4}{3}\right)\right].$

29. For $z = \sqrt{3} + i$, $r = \sqrt{\left(\sqrt{3}\right)^2 + 1^2} = 2$ and $\tan\theta = \dfrac{1}{\sqrt{3}} \implies \theta = \dfrac{\pi}{6} \implies z = 2\left(\cos\dfrac{\pi}{6} + i\sin\dfrac{\pi}{6}\right).$

For $w = 1 + \sqrt{3}\,i$, $r = 2$ and $\tan\theta = \sqrt{3} \implies \theta = \dfrac{\pi}{3} \implies w = 2\left(\cos\dfrac{\pi}{3} + i\sin\dfrac{\pi}{3}\right).$

Therefore, $zw = 2 \cdot 2\left[\cos\left(\dfrac{\pi}{6} + \dfrac{\pi}{3}\right) + i\sin\left(\dfrac{\pi}{6} + \dfrac{\pi}{3}\right)\right] = 4\left(\cos\dfrac{\pi}{2} + i\sin\dfrac{\pi}{2}\right),$

$z/w = \dfrac{2}{2}\left[\cos\left(\dfrac{\pi}{6} - \dfrac{\pi}{3}\right) + i\sin\left(\dfrac{\pi}{6} - \dfrac{\pi}{3}\right)\right] = \cos\left(-\dfrac{\pi}{6}\right) + i\sin\left(-\dfrac{\pi}{6}\right),$ and $1 = 1 + 0i = 1(\cos 0 + i\sin 0) \implies$

$1/z = \dfrac{1}{2}\left[\cos\left(0 - \dfrac{\pi}{6}\right) + i\sin\left(0 - \dfrac{\pi}{6}\right)\right] = \dfrac{1}{2}\left[\cos\left(-\dfrac{\pi}{6}\right) + i\sin\left(-\dfrac{\pi}{6}\right)\right].$ For $1/z$, we could also use the formula that precedes Example 5 to obtain $1/z = \dfrac{1}{2}\left(\cos\dfrac{\pi}{6} - i\sin\dfrac{\pi}{6}\right).$

31. For $z = 2\sqrt{3} - 2i$, $r = \sqrt{\left(2\sqrt{3}\right)^2 + (-2)^2} = 4$ and $\tan\theta = \dfrac{-2}{2\sqrt{3}} = -\dfrac{1}{\sqrt{3}} \implies \theta = -\dfrac{\pi}{6} \implies$

$z = 4\left[\cos\left(-\dfrac{\pi}{6}\right) + i\sin\left(-\dfrac{\pi}{6}\right)\right].$ For $w = -1 + i$, $r = \sqrt{2}$, $\tan\theta = \dfrac{1}{-1} = -1 \implies \theta = \dfrac{3\pi}{4} \implies$

$w = \sqrt{2}\left(\cos\dfrac{3\pi}{4} + i\sin\dfrac{3\pi}{4}\right).$ Therefore, $zw = 4\sqrt{2}\left[\cos\left(-\dfrac{\pi}{6} + \dfrac{3\pi}{4}\right) + i\sin\left(-\dfrac{\pi}{6} + \dfrac{3\pi}{4}\right)\right] = 4\sqrt{2}\left(\cos\dfrac{7\pi}{12} + i\sin\dfrac{7\pi}{12}\right),$

$z/w = \dfrac{4}{\sqrt{2}}\left[\cos\left(-\dfrac{\pi}{6} - \dfrac{3\pi}{4}\right) + i\sin\left(-\dfrac{\pi}{6} - \dfrac{3\pi}{4}\right)\right] = \dfrac{4}{\sqrt{2}}\left[\cos\left(-\dfrac{11\pi}{12}\right) + i\sin\left(-\dfrac{11\pi}{12}\right)\right] = 2\sqrt{2}\left(\cos\dfrac{13\pi}{12} + i\sin\dfrac{13\pi}{12}\right),$ and

$1/z = \dfrac{1}{4}\left[\cos\left(-\dfrac{\pi}{6}\right) - i\sin\left(-\dfrac{\pi}{6}\right)\right] = \dfrac{1}{4}\left(\cos\dfrac{\pi}{6} + i\sin\dfrac{\pi}{6}\right).$

33. For $z = 1 + i$, $r = \sqrt{2}$ and $\tan\theta = \frac{1}{1} = 1$ $\Rightarrow$ $\theta = \frac{\pi}{4}$ $\Rightarrow$ $z = \sqrt{2}\left(\cos\frac{\pi}{4} + i\sin\frac{\pi}{4}\right)$. So by De Moivre's Theorem,

$$(1+i)^{20} = \left[\sqrt{2}\left(\cos\frac{\pi}{4} + i\sin\frac{\pi}{4}\right)\right]^{20} = (2^{1/2})^{20}\left(\cos\frac{20\cdot\pi}{4} + i\sin\frac{20\cdot\pi}{4}\right) = 2^{10}(\cos 5\pi + i\sin 5\pi)$$

$$= 2^{10}[-1 + i(0)] = -2^{10} = -1024$$

35. For $z = 2\sqrt{3} + 2i$, $r = \sqrt{\left(2\sqrt{3}\right)^2 + 2^2} = \sqrt{16} = 4$ and $\tan\theta = \frac{2}{2\sqrt{3}} = \frac{1}{\sqrt{3}}$ $\Rightarrow$ $\theta = \frac{\pi}{6}$ $\Rightarrow$ $z = 4\left(\cos\frac{\pi}{6} + i\sin\frac{\pi}{6}\right)$.

So by De Moivre's Theorem,

$$\left(2\sqrt{3} + 2i\right)^5 = \left[4\left(\cos\frac{\pi}{6} + i\sin\frac{\pi}{6}\right)\right]^5 = 4^5\left(\cos\frac{5\pi}{6} + i\sin\frac{5\pi}{6}\right) = 1024\left[-\frac{\sqrt{3}}{2} + \frac{1}{2}i\right] = -512\sqrt{3} + 512i.$$

37. $1 = 1 + 0i = 1(\cos 0 + i\sin 0)$. Using Equation 3 with $r = 1$, $n = 8$, and $\theta = 0$, we have

$$w_k = 1^{1/8}\left[\cos\left(\frac{0 + 2k\pi}{8}\right) + i\sin\left(\frac{0 + 2k\pi}{8}\right)\right] = \cos\frac{k\pi}{4} + i\sin\frac{k\pi}{4}, \text{ where } k = 0, 1, 2, \ldots, 7.$$

$w_0 = 1(\cos 0 + i\sin 0) = 1$, $w_1 = 1\left(\cos\frac{\pi}{4} + i\sin\frac{\pi}{4}\right) = \frac{1}{\sqrt{2}} + \frac{1}{\sqrt{2}}i$,

$w_2 = 1\left(\cos\frac{\pi}{2} + i\sin\frac{\pi}{2}\right) = i$, $w_3 = 1\left(\cos\frac{3\pi}{4} + i\sin\frac{3\pi}{4}\right) = -\frac{1}{\sqrt{2}} + \frac{1}{\sqrt{2}}i$,

$w_4 = 1(\cos\pi + i\sin\pi) = -1$, $w_5 = 1\left(\cos\frac{5\pi}{4} + i\sin\frac{5\pi}{4}\right) = -\frac{1}{\sqrt{2}} - \frac{1}{\sqrt{2}}i$,

$w_6 = 1\left(\cos\frac{3\pi}{2} + i\sin\frac{3\pi}{2}\right) = -i$, $w_7 = 1\left(\cos\frac{7\pi}{4} + i\sin\frac{7\pi}{4}\right) = \frac{1}{\sqrt{2}} - \frac{1}{\sqrt{2}}i$

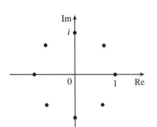

39. $i = 0 + i = 1\left(\cos\frac{\pi}{2} + i\sin\frac{\pi}{2}\right)$. Using Equation 3 with $r = 1$, $n = 3$, and $\theta = \frac{\pi}{2}$, we have

$$w_k = 1^{1/3}\left[\cos\left(\frac{\frac{\pi}{2} + 2k\pi}{3}\right) + i\sin\left(\frac{\frac{\pi}{2} + 2k\pi}{3}\right)\right], \text{ where } k = 0, 1, 2.$$

$w_0 = \left(\cos\frac{\pi}{6} + i\sin\frac{\pi}{6}\right) = \frac{\sqrt{3}}{2} + \frac{1}{2}i$

$w_1 = \left(\cos\frac{5\pi}{6} + i\sin\frac{5\pi}{6}\right) = -\frac{\sqrt{3}}{2} + \frac{1}{2}i$

$w_2 = \left(\cos\frac{9\pi}{6} + i\sin\frac{9\pi}{6}\right) = -i$

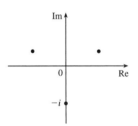

41. Using Euler's formula (6) with $y = \frac{\pi}{2}$, we have $e^{i\pi/2} = \cos\frac{\pi}{2} + i\sin\frac{\pi}{2} = 0 + 1i = i$.

43. Using Euler's formula (6) with $y = \frac{\pi}{3}$, we have $e^{i\pi/3} = \cos\frac{\pi}{3} + i\sin\frac{\pi}{3} = \frac{1}{2} + \frac{\sqrt{3}}{2}i$.

45. Using Equation 7 with $x = 2$ and $y = \pi$, we have $e^{2+i\pi} = e^2 e^{i\pi} = e^2(\cos\pi + i\sin\pi) = e^2(-1 + 0) = -e^2$.

47. Take $r = 1$ and $n = 3$ in De Moivre's Theorem to get

$$[1(\cos\theta + i\sin\theta)]^3 = 1^3(\cos 3\theta + i\sin 3\theta)$$

$$(\cos\theta + i\sin\theta)^3 = \cos 3\theta + i\sin 3\theta$$

$$\cos^3\theta + 3(\cos^2\theta)(i\sin\theta) + 3(\cos\theta)(i\sin\theta)^2 + (i\sin\theta)^3 = \cos 3\theta + i\sin 3\theta$$

$$\cos^3\theta + (3\cos^2\theta\sin\theta)i - 3\cos\theta\sin^2\theta - (\sin^3\theta)i = \cos 3\theta + i\sin 3\theta$$

$$(\cos^3\theta - 3\sin^2\theta\cos\theta) + (3\sin\theta\cos^2\theta - \sin^3\theta)i = \cos 3\theta + i\sin 3\theta$$

Equating real and imaginary parts gives $\cos 3\theta = \cos^3\theta - 3\sin^2\theta\cos\theta$ and $\sin 3\theta = 3\sin\theta\cos^2\theta - \sin^3\theta$.

49. $F(x) = e^{rx} = e^{(a+bi)x} = e^{ax+bxi} = e^{ax}(\cos bx + i \sin bx) = e^{ax} \cos bx + i(e^{ax} \sin bx)$ $\Rightarrow$

$$F'(x) = (e^{ax} \cos bx)' + i(e^{ax} \sin bx)'$$

$$= (ae^{ax} \cos bx - be^{ax} \sin bx) + i(ae^{ax} \sin bx + be^{ax} \cos bx)$$

$$= a[e^{ax}(\cos bx + i \sin bx)] + b[e^{ax}(-\sin bx + i \cos bx)]$$

$$= ae^{rx} + b[e^{ax}(i^2 \sin bx + i \cos bx)]$$

$$= ae^{rx} + bi[e^{ax}(\cos bx + i \sin bx)] = ae^{rx} + bie^{rx} = (a+bi)e^{rx} = re^{rx}$$